FUSED PYRIMIDINES

Part Three
PTERIDINES

D. J. Brown

The Research School of Chemistry
The Australian National University
Canberra

WILEY

AN INTERSCIENCE® PUBLICATION

JOHN WILEY & SONS

NEW YORK · CHICHESTER · BRISBANE · TORONTO · SINGAPORE

An Interscience® Publication

Library of Congress Cataloging-in-Publication Data
(Revised for volume 24, part 3)

Fused pyrimidines.
(The Chemistry of heterocyclic compounds ; v. 24, pt. 3)
"An Interscience publication."
Includes bibliographies.
Contents: pt. 1. Quinazolines, by W. L. F. Armarego—pt. 2. Purines, by J. H. Lister—pt. 3. Pteridines, D. J. Brown.
1. Pyrimidines—Collected works. I. Brown, D. J., ed. II. Series.
QD401.F96 547'.593 68-4274
ISBN 0-471-83041-0 (pt. 3)

Printed in the United States of America

10 9 8 7 6 5 4 3 2 1

Dedicated to
Four Great Living Pteridine Chemists

Adrien Albert

Wolfgang Pfleiderer

Edward C. Taylor

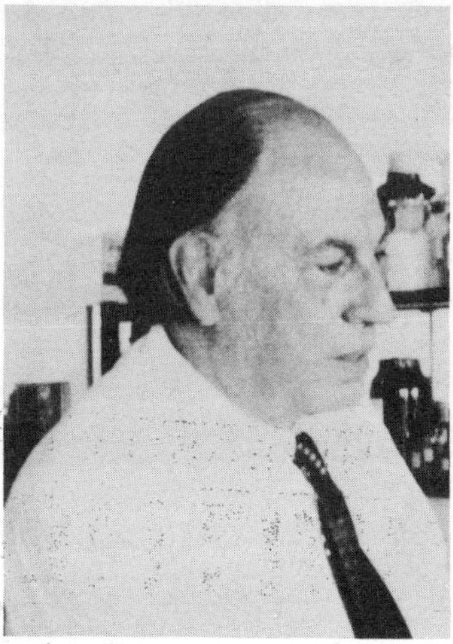

Max Viscontini

Foreword to Fused Pyrimidines

Originally, it was intended to present all the fused pyrimidine systems in one book within this series. However, resurgence of interest in quinazolines and purines, the rapid development of pteridine chemistry, and ever widening exploration of a great many new fused systems embracing the pyrimidine ring, made such a task impossible.

Accordingly, the fused pyrimidine systems will now be covered as several distinct parts of the twenty-fourth volume of the series, Dr. Armarego's *Quinazolines* appeared as Part 1 in 1967; Dr. Lister's *Purines* followed as Part 2 in 1971, with a supplementary work (to cover much subsequent research) in preparation; the current work on *Pteridines* appears as Part 3 after a substantial delay resulting in a change of authorship as late as 1984; and Part 4, covering the more important of the remaining fused pyrimidine systems, is in active preparation.

It has been a privilege to assist the late Dr. Weissberger, Dr. Taylor, and individual authors in organizing this project on *fused pyrimidines* and in maintaining a measure of uniformity and balance within its parts.

DES BROWN

The Australian National University
Canberra, Australia

The Chemistry of Heterocyclic Compounds
Introduction to the Series

The chemistry of heterocyclic compounds is one of the most complex and intriguing branches of organic chemistry. It is of equal interest for its theoretical implications, for the diversity of its synthetic procedures, and for the physiological and industrial significance of heterocyclic compounds.

The Chemistry of Heterocyclic Compounds, published since 1950 under the initial editorship of Arnold Weissberger, and later, until Dr. Weissberger's death in 1984, under our joint editorship, has attempted to make the extraordinarily complex and diverse field of heterocyclic chemistry as organized and readily accessible as possible. Each volume has dealt with syntheses, reactions, properties, structure, physical chemistry and utility of compounds belonging to a specific ring system or class (e.g., pyridines, thiophenes, pyrimidines, three-membered ring systems). This series has become the basic reference collection for information on heterocyclic compounds.

Many broader aspects of heterocyclic chemistry are recognized as disciplines of general significance which impinge on almost all aspects of modern organic and medicinal chemistry, and for this reason we initiated several years ago a parallel series entitled *General Heterocyclic Chemistry* which treated such topics as nuclear magnetic resonance, mass spectra, and photochemistry of heterocyclic compounds, the utility of heterocyclic compounds in organic synthesis, and the synthesis of heterocyclic compounds by means of 1,3-dipolar cycloaddition reactions. These volumes were intended to be of interest to all organic and medicinal chemists, as well as to those whose particular concern is heterocyclic chemistry.

It has become increasingly clear that this arbitrary distinction creates as many problems as it solves, and we have therefore elected to discontinue the more recently initiated series *General Heterocyclic Chemistry*, and to publish all forthcoming volumes in the general area of heterocyclic chemistry in *The Chemistry of Heterocyclic Compounds* series.

Department of Chemistry
Princeton University
Princeton, New Jersey 08544

EDWARD C. TAYLOR

Preface

This is the first attempt to present the detailed chemistry of pteridines as a book, although biochemical, biological, and even a few chemical aspects of pteridines have been well served in *The Biochemistry of Folic Acid and Related Pteridines* (R.L. Blakley: North-Holland, Amsterdam, 1969) and more recently in the several volumes of *Folates and Pterins*, (R.L. Blakley and S.J. Benkovic, Eds. Wiley, New York, 1984–).

The present work is intended as a critical review of pteridine chemistry with emphasis on practical rather than theoretical aspects. The literature from the earliest days until mid-1987 has been used to illustrate syntheses, properties, and reactions but no attempt has been made to include all relevant data, either in the text or in the interspersed tables, which simply serve to diversify the examples succinctly. However, it should be noted that Table XI is intended as a complete catalog of known pK_a values for pteridines; likewise the Appendix Table represents an effort to list alphabetically all simple pteridines (as defined at the head of the table and described up to mid-1987), along with some indication of physical data available and a selection of references to such data and to preparation(s). The enormous gaps in our knowledge of pteridine chemistry will be all too evident to readers of this book: It is hoped that some may be stimulated to remedy the situation.

The widespread appeal of pteridine chemistry is indicated by the following analysis of the origins of more than 17 hundred original publications quoted in this book:

Country	Percentage (%)
United States of America	26.9
Germany (East and West)	20.7
British Commonwealth	19.1
Japan	11.5
Switzerland	8.3
Russia and Eastern Europe	3.5
France	3.1
Ten other countries	6.9

I am greatly indebted to my academic colleagues, Adrien Albert, Wilf Armarego, and Gordon Barlin who have proved ready at all times to discuss matters of fact, interpretation, or presentation; to Margot Anderson and Jenny Rothschild who guided me through the minefield of biological nomenclature; to the library staff, Christine Bloem, Sheila Humphrey, Marie Humphries, and Jennefer Nicholson, for their kindly cooperation and assistance; to Barbara

Cronin, Rosemary Enge, Abira Hassan, Betty Moore, and Akiko Ohnishi, each of whom assisted in some practical way(s) during the preparation of the manuscript; to Lew Mander and succeeding Deans of the Research School of Chemistry, who have provided me with office accommodation and facilities since my retirement from the John Curtin School of Medical Research in 1985; and finally to my wife, Jan, for her understanding, patience, and practical assistance during several years of concentrated writing.

The Australian National University
Canberra, Australia
April 1988

DES BROWN

Contents

List of Tables

FUSED PYRIMIDINES

Part Three

This is a part of the twenty-fourth volume in the series

THE CHEMISTRY OF HETEROCYCLIC COMPOUNDS

CHAPTER I

Introduction to the Pteridines

1. HISTORY

In 1857, Friedrich Wöhler[1509] and H. Hlasiwetz[1510] both reported the formation of new yellow products from prolonged heating of uric acid in water above 100° C in a sealed vessel. It was later suggested that these materials resembled certain naturally occurring (pteridine) pigments[2, 5, 6, 794] but only in 1959, more than 100 years after the original experiments, were such products shown[345, 352] to contain, inter alia, 2,4,6(1*H*,3*H*,5*H*)- and 2,4,7(1*H*,3*H*,8*H*)-pteridinetrione, as well as the 6-methyl and 6-carboxy derivatives of the latter (for more details, see Ch. III, Sect. 2). The first logical synthesis of a pteridine was reported in 1894 by Otto Kühling,[7, 8] who oxidized "tolualloxazin", a *C*-methyl 2,4(1*H*,3*H*)-benzo[*g*]pteridinedione, to 2,4-dioxo-1,2,3,4-tetrahydro-6,7-pteridinedicarboxylic acid and then thermally decarboxylated this product to give "alloxazin", 2,4(1*H*,3*H*)-pteridinedione. In 1901, Siegmund Gabriel and James Colman described the condensation of 4,5-pyrimidinediamine with benzil to give a product that they described simply as "das azin" (6,7-diphenylpteridine),[9] a synthesis unaccountably redescribed by Oskar Isay from the same Berlin laboratory 5 years later[983] and subsequently developed further in 1908 by Franz Sachs and Georg Meyerheim. This synthesis gave several new derivatives of "azin-purin", which was given a purine like numbering system (**1**) that survived for many years in continental Europe.[1380] Meanwhile, in 1907 the same Siegmund Gabriel (this time with Alolf Sonn) reported the first synthesis of a pteridine from a pyrazine by treating 2,3-pyrazinedicarboxamide with two molecules of hypobromite to give an unnamed product,[902] formulated as 2,4(1*H*,3*H*)-pteridinedione but apparently unrecognized as Kuhling's "alloxazin". In 1937, Richard Kuhn and Arthur Cook prepared the same product in Heidelberg by condensing 5,6-diamino-2,4(1*H*,3*H*)-pyrimidinedione with glyoxal,[1012] but recognizing the illogicality of the name "alloxazin" and the difficulties associated with naming derivatives of "azin-purin", they suggested "lumazin" as a definitive name with a more logical numbering system (**2**): Such numbering has now become the standard for all pteridines while derivatives of 2,4(1*H*,3*H*)-pteridinedione are still frequently called lumazines for better or worse!

Parallel with the early development of synthetic pteridine chemistry just outlined, there occurred a totally separate development in our knowledge of

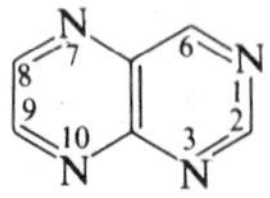

(1)

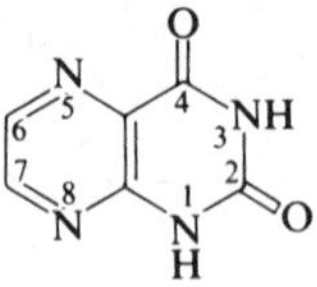

(2)

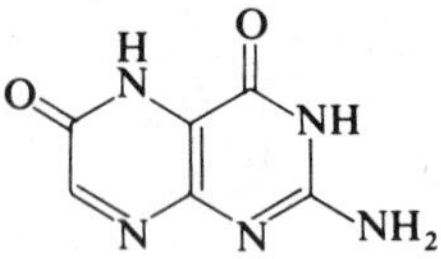

(3)

naturally occurring pteridines. This began in 1889 with Frederick Hopkins' isolation of yellow purine like pigments from the wings of the common English brimstone butterfly.[1,2] He continued the work for the next few years[3–5] and by 1895 it was clear[6] that he had isolated a pure yellow pigment [later characterized as 2-amino-4,6(3*H*,5*H*)-pteridinedione or xanthopterin] and a white pigment which, with hindsight,[794] was almost certainly 2-amino-4,6,7(3*H*,5*H*,8*H*)-pteridinetrione (leucopterin), possibly contaminated with uric acid. The relationship and chemistry of these pigments came under intensive investigation in Freiburg and later München by Clemens Schöpf and Heinrich Wieland from 1924 to 1939. In 1933 during the course of this work they isolated a third pigment called isoxanthopterin [for details and references, see Ch. VI, Sects. 5.A(1), 5.B, and 5.C(1)]. However, because of endemic difficulties with elemental analyses, the constitution of all three compounds remained unknown until 1940 when the München group finally achieved better analytical figures pointing to the pteridine nucleus. Thereupon, Robert Purrmann was able to devise and perform logical syntheses[403–405] of all three compounds (**3**–**5**). At last a highly significant nexus between the synthetic and natural product lines of pteridine research was produced.

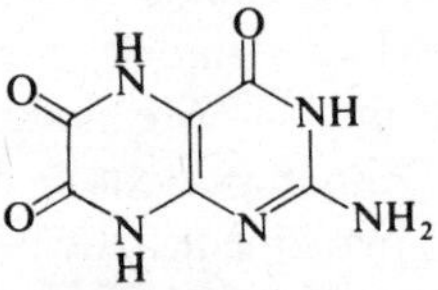

(4)

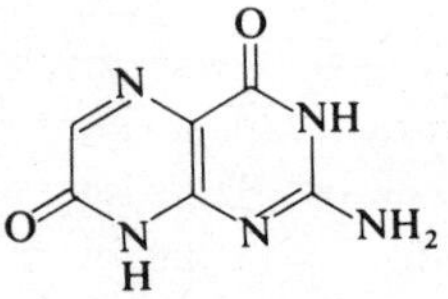

(5)

Thereafter, the development of pteridine chemistry became positively explosive, leading inter alia to (a) the isolation and synthesis of folic acid by the American Cyanamid group (1941–1946); (b) the isolation and characterization of more than 50 other naturally occurring pteridines (see Ch. VI, Sect. 5); (c) the establishment of strong synthetic groups around Michel Polonovski (France: 1944–), E. C. (Ted) Taylor (USA: 1946–), Paul Karrer (Switzerland: 1947–), Rudolf Tschesche (Germany: 1950–), Adrien Albert (Australia: 1950–),

Wolfgang Pfleiderer (Germany: 1954–), and Max Viscontini (Switzerland: 1955–); (d) the preparation of unsubstituted pteridine by William G. M. Jones[250] in 1948; (e) the introduction of methotrexate (**6**) as an anti-leukemia drug by Doris Seeger et al. in 1947 [see Ch. IX, Sect. 2.H(1)]; (f) the discovery of covalent hydration phenomena by Adrien Albert in 1952 (see Ch. VI, Sect. 2.B); and (g) the initiation of periodic international and interdisciplinary pteridine symposia by Michel Polonovski in 1952,[1688] perhaps the greatest stimulant to pteridine research to the present day. The more recent milestones in pteridine research have tended to be biochemically rather than chemically relevant: for example, the isolation of synthetic quinonoid dihydrobiopterin by Sadao Matsuura et al.[1517] in 1986.

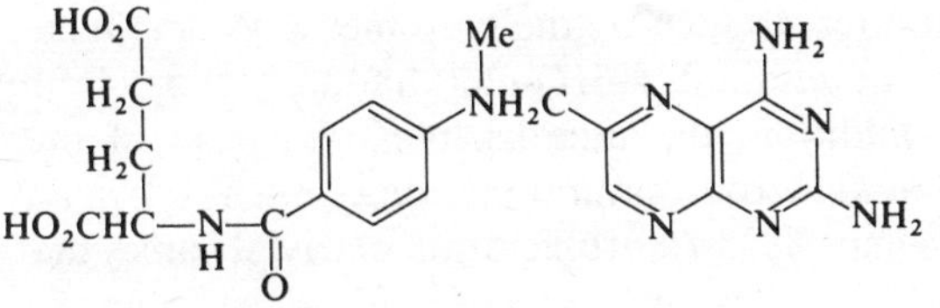

(**6**)

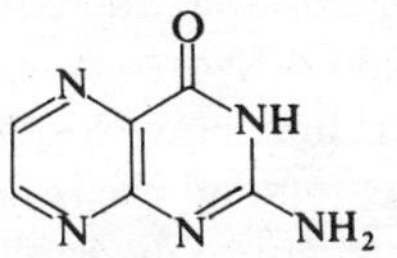

(**7**)

The term *pterin* was introduced by Heinrich Wieland and Clemens Schöpf in 1925 to cover butterfly wing pigments, irrespective of their fundamental structure,[1381] but its meaning was later extended to embrace all insect pigments.[1237] In 1963, Wolfgang Pfleiderer made the excellent suggestion that the word pterin should be restricted to 2-amino-4(3*H*)-pteridinone (**7**) and its derivatives,[334] a convention now generally but not universally accepted. In 1941, Clemens Schöpf in *Überreinstimmung mit den Herren Wieland und Purrmann*, renamed the fundamental system (**1**) as pteridin(e) with the numbering shown.[441] Meanwhile, *Chemical Abstracts*, and hence most publications from the USA, adopted the systematic name pyrimido[4,5-*b*]pyrazine (see the decennial indices covering the period 1927–1946) with the simplified numbering shown for lumazine (**2**). A preferred systematic name, pyrazino[2,3-*d*]pyrimidine, was later substituted,[760] but it has been seldom if ever used because, prior to that change, both the IUPAC and *Chemical Abstracts* had adopted Schöpf's term, pteridine, albeit with the numbering shown for lumazine (**2**).*

Extensive general reviews of pteridine chemistry appeared in 1984,[1536] 1980,[1135] 1967,[824] 1963,[334] 1954,[18] 1952,[26] 1947,[913] 1945,[1036] and 1942[794]; less extensive general reviews in 1982,[870] 1980,[234, 823] 1965,[821] 1962,[793] 1959,[10] 1958,[783] 1954,[119, 289] 1953,[782] 1951,[25, 394] 1950,[803, 1021] 1948,[901, 920] and 1941[900]; reviews of naturally occurring pterins in 1985,[1477, 1478] 1959,[1071] 1957,[807] 1936,[1237] and as noted in the introduction to Chapter VI, Section 5;

---

* A brief 1935 outbreak[415, 1019, 1190] of "pyrimidazine" as a name for the system (with totally illogical numbering) was mercifully ignored, subsequently.

reviews of folic acid chemistry in 1984,[1476] 1977,[822] 1950,[556] and as noted in Chapter VI, Section 5.F(1); essentially historical reviews in 1986,[821, 1688] 1975,[594] 1962,[574] and 1942[1173]; and a fascinating review of ring-separated pteridines in 1982.[898] Other more specialized reviews are mentioned at appropriate points in the following chapters.

2. NOMENCLATURE

As well as the changes in system name just outlined, pteridines continue to suffer the usual vicissitudes of nomenclature common to nitrogenous heterocyclic systems, especially in respect to derivatives with tautomeric substituents. For example, the compound represented by the tautomeric system (**8**) has been called 4-hydroxypteridine, 4-pteridinol, 4(3*H*)-pteridinone, 4-oxo-3,4-dihydropteridine, and so on. In addition, the interdisciplinary nature of pteridine research and the isolation of numerous natural pteridines often well in advance of structural determination, have led to a proliferation of trivial names that have proven virtually impossible to eradicate, especially from the biological and biochemical literature. Most such names are recorded in Table I, each accompanied by a systematic name, a preferred trivial name, or an explanatory cross reference.

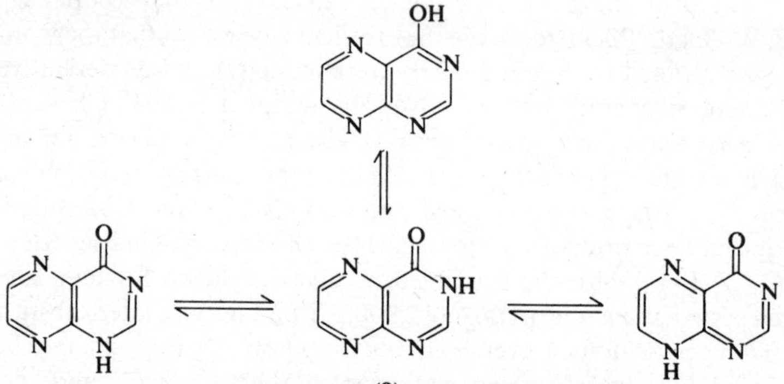

(**8**)

In this book, pteridines are named according to the preferred IUPAC rules, as outlined in the 1979 *blue book* (*Nomenclature of Organic Chemistry, Sections A–F and H*, J. Rigaudy and S. P. Klesney, Eds., Pergamon, Oxford, 1979), with minor modifications. Thus the principal group is cited as a suffix only when it is attached directly to the pteridine nucleus and all other groups are cited as prefixes in alphabetical order. Also cited as prefixes are all substituted-amino, substituted-imino, and ketonic groups. Hydroxy- and mercaptopteridines are always named as the tautomeric pteridinones and pteridinethiones or, if a higher ranking principal group is present, as the oxo- and thioxodihydropteridines, respectively. In contrast, (primary) aminopteridines are always named as pteri-

TABLE I. TRIVIAL NAMES FOR SOME PTERIDINE DERIVATIVES

Trivial Name	Systematic Name or Cross Reference
Allopterorhodin	See Chapter IV, Section 2.B(7)
Alloxazin[a]	See: lumazine
A-Methopterin[a]	See: methotrexate
Aminopterin[a]	4-Amino-4-deoxypteroylglutamic acid
6-Aminopterin	2,6-Diamino-4(3*H*)-pteridinone
Anhydroleucopterin[a]	See: isoxanthopterin
Asperopterin-A	2-Amino-8-methyl-6-β-D-ribofuranosyloxymethyl-4,7(3*H*,8*H*)-pteridinedione
Asperopterin-B	2-Amino-6-hydroxymethyl-8-methyl-4,7(3*H*,8*H*)-pteridinedione
Aurodrosopterin	See: Chapter VI, Section 5.D(11)
Azin-purin[a]	Pteridine
Biolumazine[a]	6-[L-*erythro*]-α,β-Dihydroxypropyl-2,4(1*H*,3*H*)-pteridinedione
Biopterin	2-Amino-6-[L-*erythro*]-α,β-dihydroxypropyl-4(3*H*)-pteridinone
Biopterin glucoside	2-Amino-6-[L-*erythro*]-α,β-dihydroxypropyl-4(3*H*)-pteridinone α-(α-D-glucoside)
Bufochrome	2-Amino-6-[L-*erythro*]-α,β,γ-trihydroxypropyl-4(3*H*)-pteridinone
Chrysopterin	2-Amino-7-methyl-4,6(3*H*,5*H*)-pteridinedione
Ciliapterin	2-Amino-6-[L-*threo*]-α,β-dihydroxypropyl-4(3*H*)-pteridinone
Compound-A[a]	See: compound-V
Compound-C[a]	See: biopterin glucoside
Compound-G[a]	See: dimethylribolumazine
Compound-V	6-Methyl-8-D-ribityl-2,4,7(1*H*,3*H*,8*H*)-pteridinetrione
Cyprino-purple-A1[a]	See: isoxanthopterin
Cyprino-purple-A2[a]	See: ichthyopterin
Cyprino-purple-B	2-Amino-4,7-dioxo-3,4,7,8-tetrahydro-6-pteridinecarboxylic acid
Cyprino-purple-C1	6-β-Acetoxy-α-hydroxypropyl-2-amino-4,7(3*H*,8*H*)-pteridinedione
Cyprino-purple-C2	6-α-Acetoxy-β-hydroxypropyl-2-amino-4,7(3*H*,8*H*)-pteridinedione
Deoxysepiapterin	2-Amino-6-propionyl-7,8-dihydro-4(3*H*)-pteridinone
Dihydroxanthopterin	2-Amino-7,8-dihydro-4,6(3*H*,5*H*)-pteridinedione
Dimethylribolumazine	6,7-Dimethyl-8-D-ribityl-2,4(3*H*,8*H*)-pteridinedione
Drosopterin	See:Chapter VI, Section 5.D(11)
Ekapterin	2-Amino-7-β-carboxy-β-hydroxyethyl-4,6(3*H*,5*H*)-pteridinedione
Erythropterin	2-Amino-7-oxalomethyl-4,6(3*H*,5*H*)-pteridinedione
Euglenapterin	2-Dimethylamino-6-[L-*threo*]-α,β,γ-trihydroxypropyl-4(3*H*)-pteridinone
Folic acid	Pteroylglutamic acid
"Furterene"	6-Fur-2′-yl-2,4,7-pteridinetriamine
6-Hydroxymethyllumazine	6-Hydroxymethyl-2,4(1*H*,3*H*)-pteridinedione
Hynobius blue[a]	See: tetrahydrobiopterin
Ichthyopterin	2-Amino-6-α,β-dihydroxypropyl-4,7(3*H*,8*H*)-pteridinedione
Fluorocyanine[a]	See: ichthyopterin (?)
Isodrosopterin	See: Chapter VI, Section 5.D(11)
Isoleucopterin[a]	4-Amino-2,6,7(1*H*,5*H*,8*H*)-pteridinetrione
Isopterin[a]	4-Amino-2(1*H*)-pteridinone

TABLE I. (*Contd.*)

Trivial Name	Systematic Name or Cross Reference
Isopterorhodin	See: Chapter VI, Section 2.B(7)
Isosepiapterin[a]	See: deoxysepiapterin
Isoxantholumazine[a]	See: violapterin
Isoxanthopterin	2-Amino-4,7(3*H*,8*H*)-pteridinedione
Lepidopterin	2-Amino-7-β-amino-β-carboxyvinyl-4,6(3*H*,5*H*)-pteridinedione
Leucettidine	6-[*S*]-α-Hydroxypropyl-1-methyl-2,4(1*H*,3*H*)-pteridinedione
Leucopterin	2-Amino-4,6,7(3*H*,5*H*,8*H*)-pteridinetrione
Leucopterin-B[a]	See: isoxanthopterin
Luciferesceine	See: Chapter VI, Section 5.F(2)
Luciopterin	8-Methyl-2,4,7(1*H*,3*H*,8*H*)-pteridinetrione
Lumazine	2,4(1*H*,3*H*)-Pteridinedione
6-Lumazinecarboxylic acid	2,4-Dioxo-1,2,3,4-tetrahydro-6-pteridinecarboxylic acid
Mesopterin[a]	See: isoxanthopterin
Methanopterin	See: Chapter VI, Section 5.G(1)
Methotrexate	4-Amino-10-methyl-4-deoxypteroylglutamic acid
6-Methyl-7-hydroxy-ribolumazine[a]	See: compound-V
Monapterin	2-Amino-6-[L-*threo*]-α,β,γ-trihydroxypropyl-4(3*H*)-pteridinone
Neodrosopterin	See: Chapter VI, Section 5.D(11)
Neopterin	2-Amino-6-[D-*erythro*]-α,β,γ-trihydroxypropyl-4(3*H*)-pteridinone
L-Neopterin[a]	See: bufochrome
L-*threo*-Neopterin[a]	See: monapterin
Photolumazine-A	6-L-α,β-Dihydroxyethyl-8-D-ribityl-2,4,7(1*H*,3*H*,8*H*)-pteridinetrione
Photolumazine-B	6-Hydroxymethyl-8-D-ribityl-2,4,7(1*H*,3*H*,8*H*)-pteridinetrione
Photolumazine-C	8-D-Ribityl-2,4,7(1*H*,3*H*,8*H*)-pteridinetrione
Photopterin-A[a]	See: photolumazine-A
Pteridine red(s)	See: Chapter IV, Section 2.B(7)
Pterin	2-Amino-4(3*H*)-pteridinone
6-Pterincarboxylic acid	2-Amino-4-oxo-3,4-dihydro-6-pteridinecarboxylic acid
Pteroic acid	2-Amino-6-*p*-carboxyanilinomethyl-4(3*H*)-pteridinone
Pterorhodin	2-Amino-7-(2′-amino-4′, 6′-dioxo-3′,4′,5′,6′,7′,8′-hexahydro-pteridin-7′-ylidenemethyl)-4,6(3*H*,5*H*)-pteridinedione
Purple substance-II	See: Chapter VI, Section 2.H
Putidolumazine	6-β-Carboxyethyl-8-D-ribityl-2,4,7(1*H*,3*H*,8*H*)-pteridinetrione
Pyrimidazine[a]	Pteridine
Ranachrome-1[a]	See: biopterin
Ranachrome-3	2-Amino-6-hydroxymethyl-4(3*H*)-pteridinone
Ranachrome-4[a]	See: isoxanthopterin
Ranachrome-5[a]	See: 6-pterincarboxylic acid
Rhacophorus-yellow[a]	See: sepiapterin
Rhizopterin	See: Chapter VI, Section 5.G(1)
Rhodopterin[a]	See: pterorhodin
Ribolumazine[a]	See: dimethylribolumazine
Russupteridine-yellow-1	6-Amino-7-formamido-8-D-ribityl-2,4(3*H*,8*H*)-pteridinedione
Russupteridine-yellow-2	See: Chapter VI, Section 5.F(8)
Russupteridine-yellow-4	4-D-Ribityl-1*H*-imidazo[4,5-*g*]pteridine-2,6,8(4*H*,5*H*,7*H*)-trione
Russupteridine-yellow-5	See: Chapter VI, Section 5.F(8)

TABLE I. (*Contd.*)

Trivial Name	Systematic Name or Cross Reference
Sarcinapterin	See: Chapter VI, Section 5.G(1)
Sepialumazine	6-Lactoyl-7,8-dihydro-2,4(1*H*,3*H*)-pteridinedione
Sepiapterin	2-Amino-6-(*S*)-lactoyl-7,8-dihydro-4(3*H*)-pteridinone
Sepiapterin-C	6-Acetyl-2-amino-7,8-dihydro-4(3*H*)-pteridinone
Substance-O	See: Chapter VI, Section 2.H
Surugatoxin	See: Chapter VI, Section 5.G(3)
Tetrahydrobiopterin	2-Amino-6-[L-*erythro*]-α,β-dihydroxypropyl-5,6,7,8-tetrahydro-4(3*H*)-pteridinone
Triamterene	6-Phenyl-2,4,7-pteridinetriamine
Uropterin[a]	See: xanthopterin
Urothione	2-Amino-7-α,β-dihydroxyethyl-6-methylthio-4(3*H*)-thieno[3,2-*g*]pteridinone
Violapterin	2,4,7(1*H*,3*H*,8*H*)-Pteridinetrione
Xanthopterin	2-Amino-4,6(3*H*,5*H*)-pteridinedione
Xanthopterin-Bl[a]	See: sepiapterin
Xanthopterin-B2[a]	See: sepialumazine

[a] Disused trivial name. Not recommended for further use.

dinamines or aminopteridines but never as the tautomeric pteridinimines or iminodihydropteridines. Whenever possible, pteridine *N*-oxides are named as such.

3. THE BASIS OF PTERIDINE CHEMISTRY

The high N/C ratio in pteridine ensures that the system has a greatly depleted π-electron layer and that its aromaticity is therefore significantly reduced, despite the conventional conjugated double bonds in its formula. Moreover, even allowing for a lack of symmetry, each carbon atom capable of bearing a substituent is activated directly by at least one doubly bound ring-nitrogen atom standing in the α- or γ-position to it. This would suggest a semiaromatic system with no great stability towards ring fission, prone to nucleophilic but not electrophilic reactions, and subject to covalent addition reactions. In fact, this is a portrait of pteridine. Naturally, such reactivities are modified considerably by substitution with electron-donating substituents, which increase stability progressively, or by electron-withdrawing substituents, further exaggerating these tendencies.

4. A GENERAL SUMMARY OF PTERIDINE CHEMISTRY

The pteridine ring bearing sundry substituents may be built up in various ways from pyrimidine intermediates (Ch. II); from pyrazine intermediates

(Ch. III, Sect. 1); and from some heterobicyclic or heterotricyclic systems by an initial partial degradation, followed if necessary by appropriate elaboration (Ch. III, Sects. 2 and 3). Since such synthetic processes are not pteridine chemistry, it seems inappropriate to discuss them further at this point but to concentrate on the metathetical and other reactions of pteridines that give the system its unique character. To assist the reader, original references are avoided in favor of extensive cross references to more detailed treatment in later chapters or sections.

A. Electrophilic Reactions

As would be expected, no example of the nitration, nitrosation, sulfonation, or (electrophilic) halogenation of a pteridine nucleus has been reported; even the recorded formation[1238] of a red azo compound from natural xanthopterin and 2,4-dichlorobenzenediazonium chloride could not be repeated[31] on synthetic 2-amino-4,6(3*H*,5*H*)-pteridinedione. Likewise, 6(5*H*)-pteridinone (**9**) gave no colors with several diazotized amines, while efforts to nitrate it simply resulted in an oxidative conversion into 6,7(5*H*,8*H*)-pteridinedione (**10**).[31] Alkylpteridines are susceptible to extranuclear halogenation, sulfonation, and so on. (Ch. IV, Sect. 2.B) but, while phenylpteridines should undergo nitration, and so on, on the benzene ring, such reactions do not appear to have been reported.

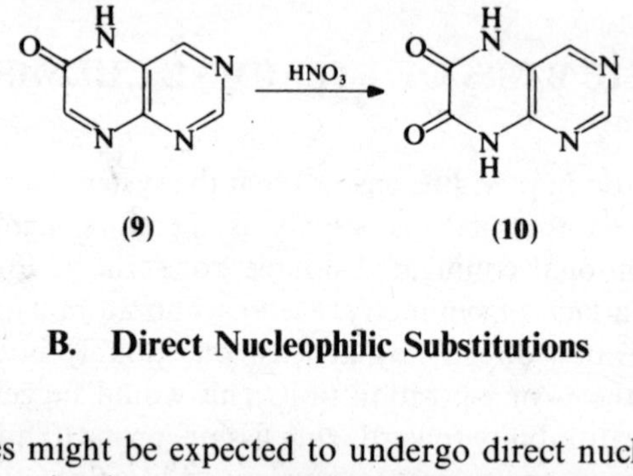

(**9**) (**10**)

B. Direct Nucleophilic Substitutions

While pteridines might be expected to undergo direct nucleophilic substitution, for example, amination, at a vacant 2-, 4-, 6-, or 7-position, such reactions have not been reported, probably on account of the competitive and much more avid (nucleophilic) covalent addition of appropriate reagents to the pteridine nucleus (see Sect. E later). The oxidation of such dihydropteridine adducts will

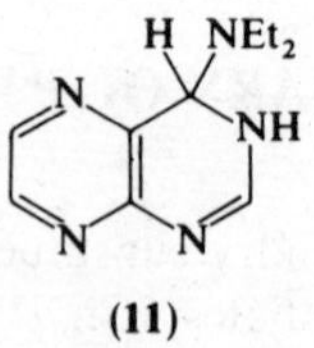

(**11**)

naturally give the very products expected by direct nucleophilic substitution, for example, oxidation of the pteridine 3,4-diethylamine adduct (**11**) gave 4-diethylaminopteridine.

C. Nucleophilic Metatheses

In contrast to the lack of direct nucleophilic substitution reactions for the pteridine system, there are a great many examples in which a halogeno or other leaving group is replaced by a variety of substituents. The initial step in these metatheses is a nucleophilic attack by the reagent at the carbon atom to which the leaving group is attached.

(1) *Replacement of Halogeno Substituents* (Ch. V, Sects. 2 and 4)

The nucleophilic replacement of 2-, 4-, 6-, and 7-chloro substituents takes place very easily except in the presence of powerful electron-donating substituents (e.g., amino groups), which progressively impede such reactions according to the number present. Replacement of extranuclear halogeno substituents differs little from the similar reactions of benzene analogs such as benzyl chloride.

(a) By Amino or Substituted-Amino Groups (Ch. V, Sects. 2.C and 4.B)

All four monochloropteridines undergo aminolysis by ethanolic ammonia or by neat primary or secondary amines at or below room temperature. The more active chlorine in dichloropteridines reacts with similar ease but the resulting chloropteridinamine then reacts less easily to give the pteridinediamine, for example, 6,7-dichloropteridine (**12**) with aqueous ammonia at 20°C gave 6-chloro-7-pteridinamine (**13**) which then required ethanolic ammonia at 150°C to convert it into 6,7-pteridinediamine (**14**). Although no kinetics have been measured for the aminolysis of chloropteridines, it may be cautiously concluded from the conditions required for stepwise preparative aminolyses of di-, tri-, and tetrachloropteridines that the positional order of reactivity for chloro substituents is $7 > 6 > 4 > 2$. A chloropteridine can also react readily with a minimally hindered tertiary amine to give, for example, 2-trimethylammoniopteridine chloride (**15**) (trimethylamine in benzene at 20°C). The aminolysis of extranuclear halogenopteridines has been confined largely to the use of aromatic

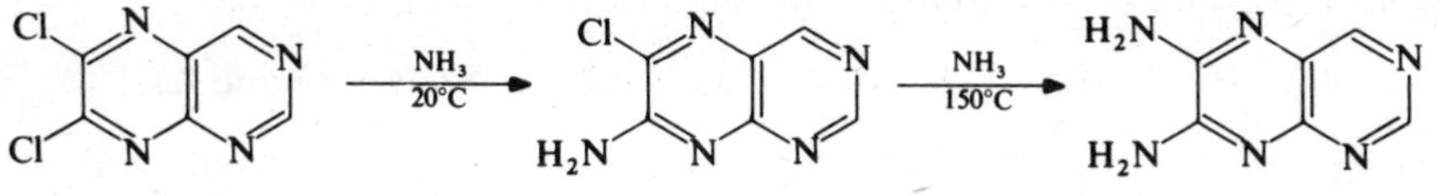

(**12**) (**13**) (**14**)

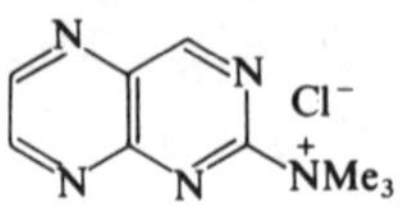

(15)

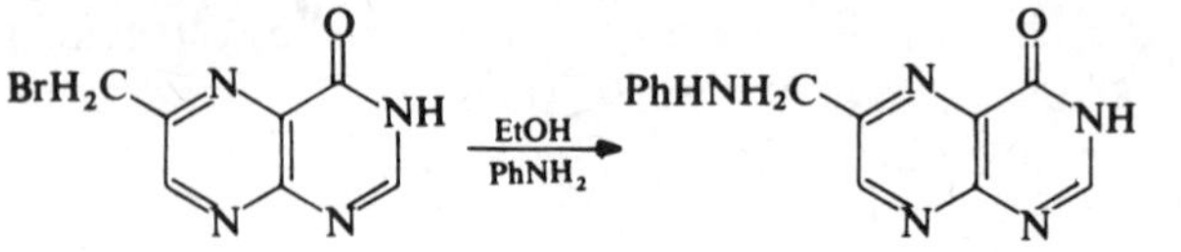

(16) (17)

amines to make pteroylglutamic acid analogs, for example, 6-bromomethyl- (**16**) gave 6-anilinomethyl-4(3*H*)-pteridinone (**17**) by refluxing briefly in ethanolic aniline.

(b) By Hydroxy–Oxo Substituents (Ch. V, Sects. 2.D and 4.C)

Simple chloropteridines undergo alkaline hydrolysis very readily to give the corresponding pteridinones but chloropteridines bearing electron-donating substituents require more severe conditions. For example, 6- or 7-chloropteridine (**18**) gave 6(5*H*)- or 7(8*H*)-pteridinone (**19**) in dilute aqueous sodium carbonate at 20°C for 15 min, whereas 6-chloro-7-pteridinamine (**13**) gave 7-amino-6(5*H*)-pteridinone (**20**) in the same medium only at 100°C for 1 h. Acidic hydrolysis of chloropteridines is also successful but it has been little used. The hydrolysis of extranuclear halogenopteridines to the corresponding alcohols occurs in alkali, as exemplified in the conversion of 2-amino-6-bromomethyl- (**21**, R = Br) into 2-amino-6-hydroxymethyl-8-methyl-4,7(3*H*,8*H*)-pteridinedione (**21**, R = OH) by warming in dilute alkali for 15 min. More interesting is the gentle acidic

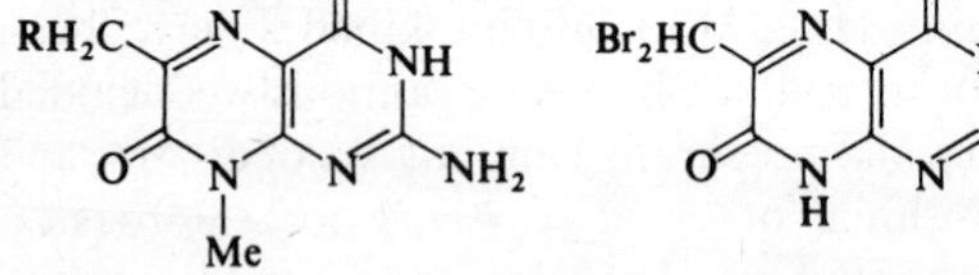

(18) (19)

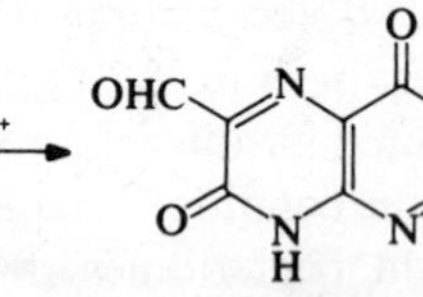

(20)

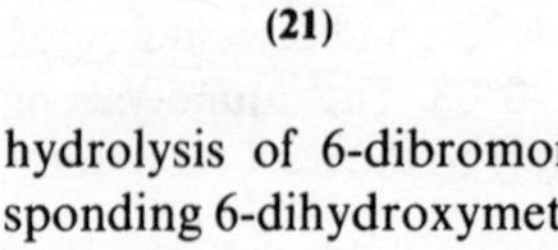

(21)

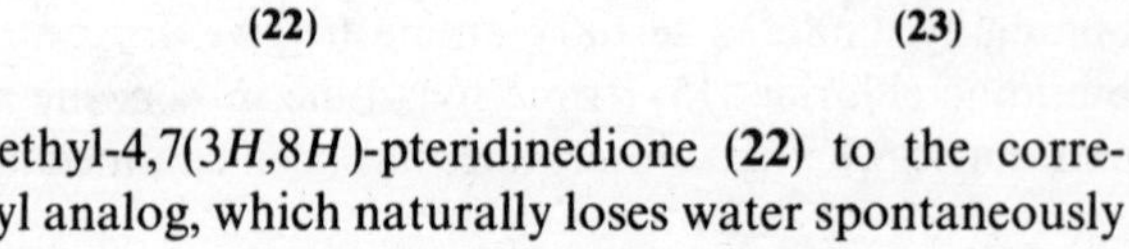

(22) (23)

hydrolysis of 6-dibromomethyl-4,7(3*H*,8*H*)-pteridinedione (**22**) to the corresponding 6-dihydroxymethyl analog, which naturally loses water spontaneously to afford 4,7-dioxo-3,4,7,8-tetrahydro-6-pteridinecarbaldehyde (**23**).

(c) By Alkoxy or Aryloxy Groups (Ch. V, Sects. 2.D and 4.C)

Simple chloropteridines are readily converted into the corresponding alkoxypteridines by alcoholic sodium alkoxide: for example, 4-chloro- gave 4-methoxy-

6,7-dimethylpteridine in methanolic sodium methoxide at 20°C. However, such displacement is slowed significantly by the presence of electron-donating groups, so that 2-amino-6-chloro- (**24**, R = Cl) gave 2-amino-6-methoxy-4(3*H*)-pteridinone (**24**, R = OMe) only by heating in the same medium under reflux for 24 h. Extranuclear aryloxy derivatives have been made slowly from the corresponding halogenomethylpteridines. Thus 6-bromomethyl- (**25**) gave 6-phenoxymethyl-2,4-pteridinediamine (**26**) by stirring with phenol in *N,N*-dimethylacetamide containing potassium *t*-butoxide during 6 days at 20°C.

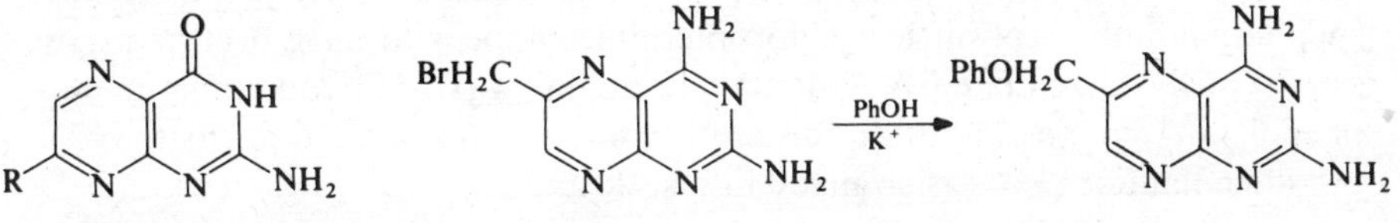

(**24**) (**25**) (**26**)

(d) By Mercapto–Thioxo Substituents (Ch. V, Sect. 2.E)

The thiolysis of simple chloropteridines may be done with aqueous or ethanolic sodium sulfide or sodium hydrogen sulfide; the well known indirect route via a thiouranium salt has not been used in the pteridine series. Conditions are exemplified in the simple conversion of 6-chloropteridine into 6(5*H*)-pteridinethione (**27**) by ethanolic sodium sulfide at 35°C or in the more difficult conversion of 7-chloro-6-phenyl-2,4-pteridinediamine into 2,4-diamino-6-phenyl-7(8*H*)-pteridinethione by ethanolic sodium hydrogen sulfide under reflux for 2 h. An interesting preferential 7-thiolysis of 6,7-dichloro-1,3-dimethyl-2,4(1*H*,3*H*)-pteridinedione (**28**) has been achieved by stirring with a solution of sodium hydrogen sulfide at 25°C to give only 6-chloro-1,3-dimethyl-7-thioxo-7,8-dihydro-2,4(1*H*,3*H*)-pteridinedione (**29**). No thiolyses of extranuclear halogenopteridines appear to have been reported.

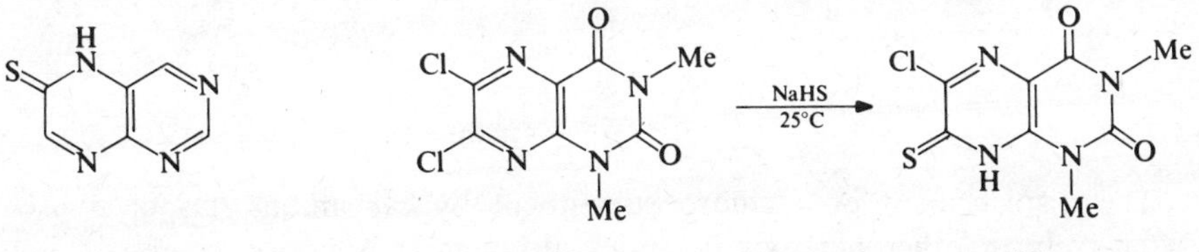

(**27**) (**28**) (**29**)

(e) By Alkylthio or Arylthio Groups (Ch. V, Sects. 2.E and 4.C)

Whereas alkanethiolysis of 6- (**30**, R = H) or 7-chloropteridine (with sodium α-toluenethiolate at 50°C) to give 6- (**31**, R = H) or 7-benzylthiopteridine was complete in 10 min, the conversion of 6-chloro- (**30**, R = NH_2) into 6-benzylthio-2,4-pteridinediamine (**31**, R = NH_2) took 16 h under similar conditions because of deactivation of the leaving group. Although arenethiolysis of the extranuclear

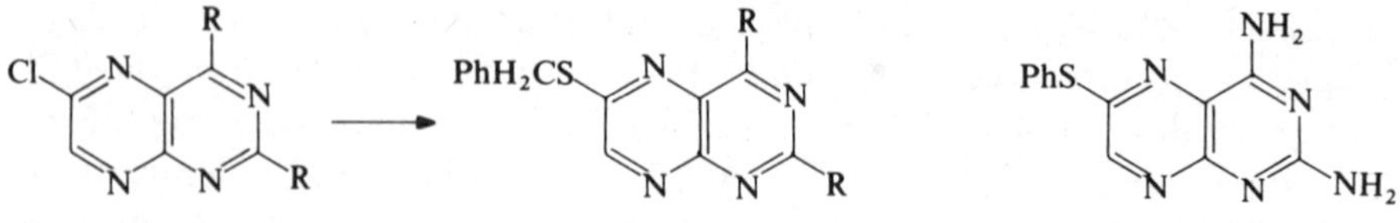

(30) (31) (32)

halogeno compound, 6-bromomethyl-2,4-pteridinediamine, has been done logically with thiophenol (in *N,N*-dimethylacetamide containing potassium carbonate) to give 6-phenylthiomethyl-2,4-pteridinediamine, all existing examples of such arenethiolysis of nuclear chloropteridines appear to have occurred quite satisfactorily in refluxing *N,N*-dimethylformamide (DMF) containing a thiophenol *without* the addition of any base, for example, 6-phenylthio-2,4-pteridinediamine (**32**) was so made in just 20 min.

(f) By Sulfo Groups (Ch. V, Sect. 2.F)

Because direct (electrophilic) sulfonation appears to be impossible, the formation of pteridinesulfonic acids by indirect routes has assumed greater importance. One such method involves the nucleophilic reaction of a chloropteridine with sodium sulfite. For example, both 7-chloro- (**33**, R = H) and 6,7-dichloro-1,3-dimethyl-2,4(1*H*,3*H*)-pteridinedione (**33**, R = Cl) reacted with aqueous sodium sulfite under gentle conditions to give, respectively, sodium 1,3-dimethyl-2,4-dioxo-1,2,3,4-tetrahydro-7-pteridinesulfonate (**34**, R = H) and its 6-chloro derivative (**34**, R = Cl), the latter by a completely preferential displacement of the 7-chloro substituent in its substrate.

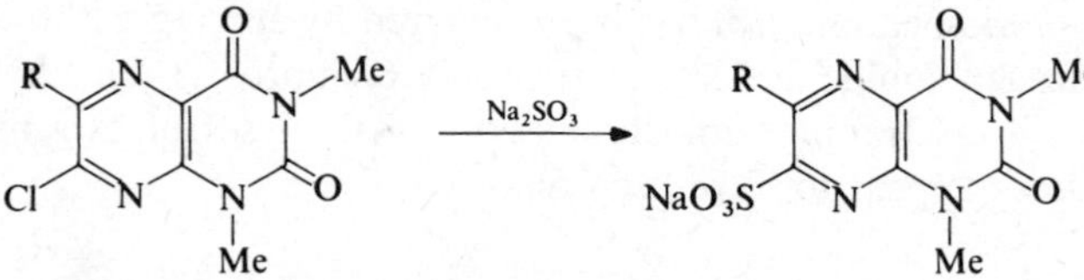

(33) (34)

(g) By Other Groups

The displacement of a chloro substituent by carbanions has been used extensively in other heterocyclic series although it is poorly represented in

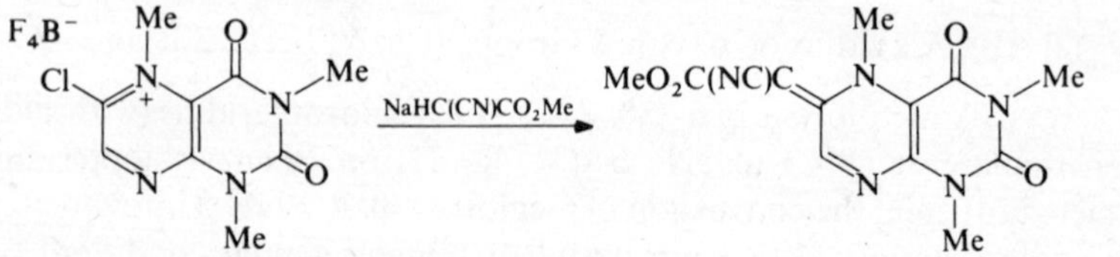

(35) (36)

pteridines. However, the highly activated chlorine in 6-chloro-1,3,5-trimethyl-2,4-dioxo-1,2,3,4-tetrahydro-5-pteridinium tetrafluoroborate (**35**) did react with methanolic methyl sodiocyanoacetate to afford 6-α-cyano-α-methoxycarbonylmethylene-1,3,5-trimethyl-5,6-dihydro-2,4(1*H*,3*H*)-pteridinedione (**36**).[1680]

(2) *Replacement of Alkoxy Groups*

Alkoxypteridines undergo aminolyses and some other nucleophilic replacement reactions less easily than do chloropteridines. Thus 2-amino-6-methoxy-(**37**, R = Me) or 2-amino-6-ethoxy-4(3*H*)-pteridinone (**37**, R = Et) underwent *aminolysis* by neat hydrazine hydrate at 120°C to give 2-amino-6-hydrazino-4(3*H*)-pteridinone (**38**) (Ch. VII, Sect. 2.B); 2-, 4-, or 6-methoxypteridine underwent *hydrolysis* to the corresponding pteridinones by shaking in alkali for 2 h, while 2-methoxy-6,8-dimethyl-7(8*H*)-pteridinone gave 6,8-dimethyl-2,7(1*H*,8*H*)-pteridinedione (**39**) by boiling in dilute hydrochloric acid for 1 h (Ch. VI, Sect. 1.D); 4-methoxypteridine underwent *alcoholysis* (or *transetherification*) by boiling in propanol containing silver oxide (equivalent to silver propoxide) to give 4-propoxypteridine (Ch. VII, Sect. 1.E), but examples of other possible replacement reactions appear to be missing at present in the pteridine series.

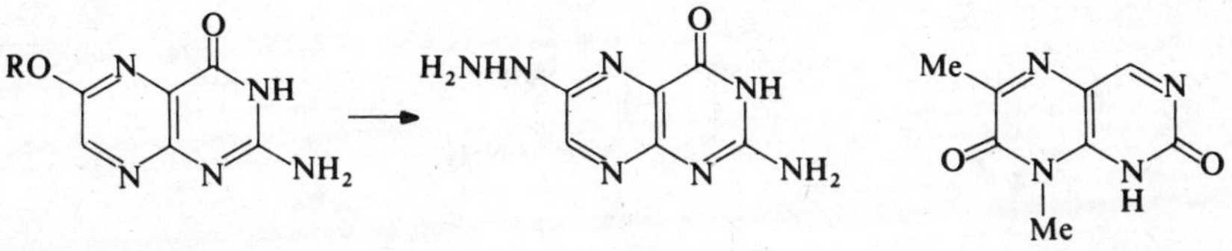

(37) **(38)** **(39)**

(3) *Replacement of Alkylthio Groups*

Alkylthiopteridines seem to undergo nucleophilic replacement reactions about as easily as do alkoxypteridines. Thus *aminolysis* of 4-methylthiopteridine (**41**) with refluxing ethanolic methylamine for 2 h gave 4-methylaminopteridine (**40**), while 7-methylthiopteridine with the less powerful nucleophile (ethanolic ammonia) required heating at 125 °C for 6 h to give 7-pteridinamine (Ch. VIII, Sect. 7.C). The *hydrolysis* of 4-methylthiopteridine (**41**) to 4(3*H*)-pteridinone (**42**) occurred equally easily in dilute aqueous ethanolic alkali at 70 °C or in dilute

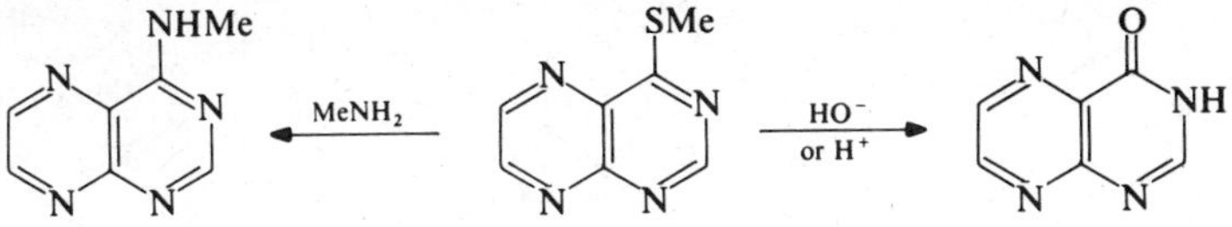

(40) **(41)** **(42)**

acetic acid at 100 °C during 1 h (Ch. VI, Sect. 1.E), and the *thiolysis* of 4-benzylthio-2-pteridinamine to 2-amino-4(3*H*)-pteridinethione took place in refluxing ethanolic sodium hydrogen sulfide during 30 min (Ch. VIII, Sect. 1.E).

(4) *Replacement of Sulfo Groups*

The efficacy of sulfo as a leaving group from heterocycles is often overlooked. It has been minimally so used in the pteridine series. The *aminolysis* of potassium 4-oxo-6,7-diphenyl-3,4-dihydro-2-pteridinesulfonate (**43**, $R=SO_3K$) in boiling morpholine gave 2-morpholino-6,7-diphenyl-4(3*H*)-pteridinone [**43**, $R=N(CH_2CH_2)_2O$] in just 2 min, while the *hydrolysis* of 2-amino-4-oxo-7-sulfo-3,4-dihydro-6-pteridinecarboxylic acid as its dianion (**44**) in boiling alkali gave the decarboxylated product, 2-amino-4,7(3*H*,8*H*)-pteridinedione (**45**) (Ch. VIII, Sect. 5). Other such hydrolyses have been used as part of an integrated route from pteridinethiones to pteridinones by an initial oxidation to a sulfonic acid (or sulfinic acid?) followed by spontaneous alkaline hydrolysis, for example, treatment of 6,7-diphenyl-2,4(1*H*,3*H*)-pteridinedithione with alkaline hydrogen peroxide at 25 °C for 1 day gave 6,7-diphenyl-2,4(1*H*,3*H*)-pteridinedione (Ch. VI, Sect. 1.E).

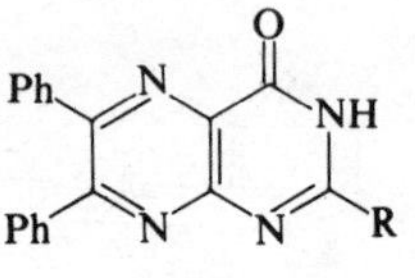

(**43**)

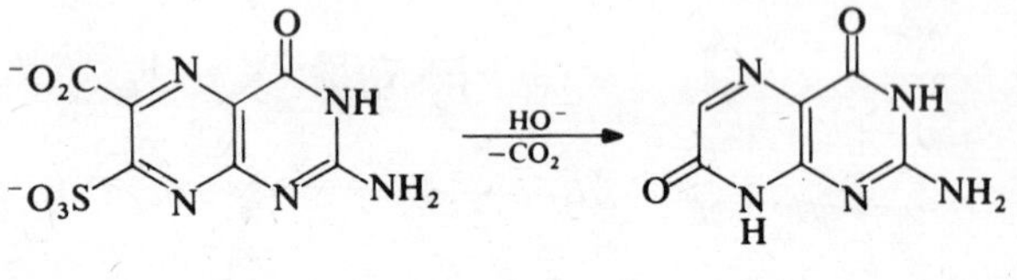

(**44**) (**45**)

(5) *Replacement of Alkylsulfonyl Groups* (Ch. VIII, Sect. 9)

Although alkylsulfonyl and alkylsulfinyl groups are normally even better leaving groups than the chloro substituent, little use has been made of them in the pteridine series, probably because so few pteridine sulfones or sulfoxides have been prepared. However, *hydrolysis* or *alcoholysis* of 2-methylsulfonyl-7-phenyl-4-pteridinamine (**46**, $R=SO_2Me$) occurred in aqueous alkali or methanolic sodium methoxide at room temperature to give 4-amino-7-phenyl-2(1*H*)-

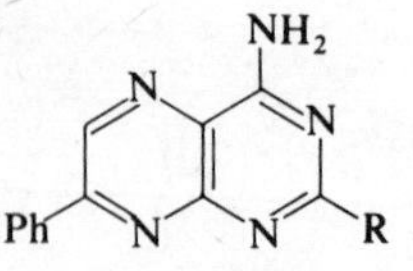

(**46**)

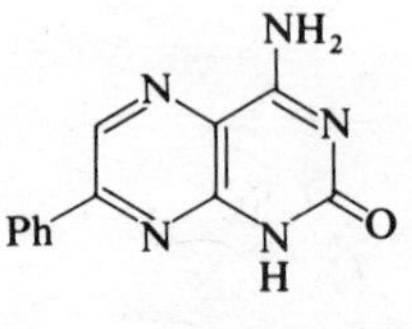

(**47**)

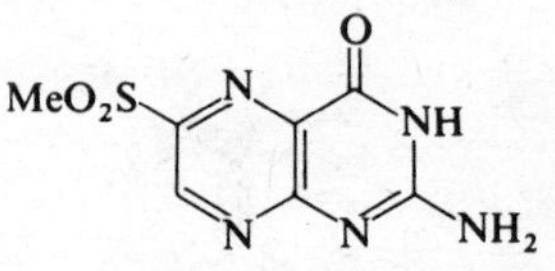

(**48**)

pteridinone (**47**) or 2-methoxy-7-phenyl-4-pteridinamine (**46**, R = OMe), respectively; *alkanethiolysis* or *arenethiolysis* of 2-amino-6-methylsulfonyl-4(3*H*)-pteridinone (**48**) occurred in refluxing DMF containing phenylmethanethiol or thiophenol to give 2-amino-6-benzylthio- or 2-amino-6-phenylthio-4(3*H*)-pteridinone, respectively. The only reported attempt at *aminolysis* failed because the alkylsulfonyl leaving group was so deactivated by the amino groups in the molecule that transamination rather than aminolysis occurred.

D. Other Metatheses

It seems convenient to outline next several other metatheses without trying to classify them into types. Similar metatheses in any extranuclear positions will generally be ignored as being in no way peculiar to the pteridine series.

(1) *Pteridinones to Chloropteridines* (Ch. V, Sects. 1.B and 3.B)

Although a few 2(1*H*)-pteridinones fail to react satisfactorily, most other (tautomeric) pteridinones with one to four oxo substituents react with boiling phosphoryl chloride containing phosphorus pentachloride to afford the corresponding chloropteridines after several hours. The presence of phosphorus pentachloride does appear to be necessary, although few comparative data for chlorinations in neat phosphoryl chloride or in phosphoryl chloride plus an *N*,*N*-dialkylaniline are available. Partial chlorination of pteridinediones or pteridinetriones have been achieved, apparently more by luck than intention. For example, 2-amino-4,6,7(3*H*,5*H*,8*H*)-pteridinetrione (**49**) with phosphoryl chloride–phosphorus pentachloride under mild conditions gave 2-amino-4-chloro-6,7(5*H*,8*H*)-pteridinedione (**50**). Conversely, the phenomenon of an additional direct *C*-chlorination by phosphorus pentachloride (well documented in other series) has also occurred with 2,4,6(1*H*,3*H*,5*H*)-pteridinetrione which gave, not its trichloro analog, but 2,4,6,7-tetrachloropteridine. As in a similar chlorination, 1,3-dimethyl-2,4,7(1*H*,3*H*,8*H*)-pteridinetrione (with only one tautomeric oxo group) gave in phosphoryl bromide a single monobromo product, 7-bromo-1,3-dimethyl-2,4(1*H*,3*H*)-pteridinedione (**51**).

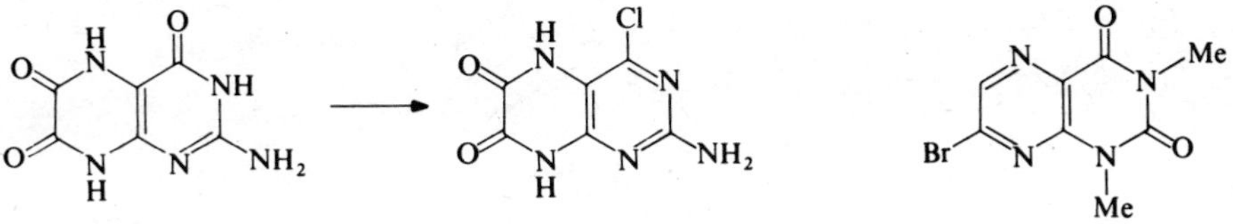

(**49**) (**50**) (**51**)

The formation of (extranuclear) chloroalkyl- from hydroxyalkylpteridines is best done with thionyl chloride or a reagent other than a phosphorus halide, for example, 6-hydroxymethyl- gave 6-chloromethyl-2,4-pteridinediamine in thionyl chloride.

(2) *Pteridinones to Pteridinethiones* (Ch. VI, Sect. 2.G)

Both tautomeric and nontautomeric pteridinones usually undergo direct thiation to the corresponding pteridinethiones by heating with (pure) phosphorus pentasulfide in pyridine, dioxane, or a similar solvent. Although the reaction occurs at all positions, 2-pteridinones are rather less amenable to thiation and may take longer times or give poor yields. For example, brief treatment of 1,3-dimethyl-2,4,7(1*H*,3*H*,8*H*)-pteridinetrione with phosphorus pentasulfide in pyridine gave 1,3-dimethyl-7-thioxo-7,8-dihydro-2,4(1*H*,3*H*)-pteridinedione (**52**), similar but prolonged treatment in dioxane gave 1,3-dimethyl-4,7-dithioxo-3,4,7,8-tetrahydro-2(1*H*)-pteridinone (**53**), and only repeated treatments of the latter product gave the fully thiated 1,3-dimethyl-2,4,7(1*H*,3*H*,8*H*)-pteridinetrithione (**54**).

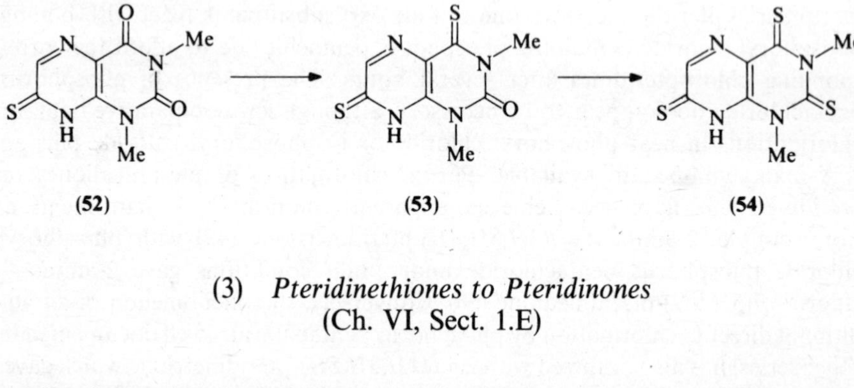

(**52**) (**53**) (**54**)

(3) *Pteridinethiones to Pteridinones* (Ch. VI, Sect. 1.E)

The direct acidic hydrolysis of pteridinethiones to the corresponding pteridinones is possible sometimes, for example, 6,7(5*H*,8*H*)-pteridinedione (**55**, X = Y = O) was made by heating 6-thioxo-5,6-dihydro-7(8*H*)-pteridinone (**55**, X = S, Y = O), its 7-thioxo isomer (**55**, X = O, Y = S), or 6,7(5*H*,8*H*)-pteridinedithione (**55**, X = Y = S) in dilute hydrochloric acid for 1 h. However, it

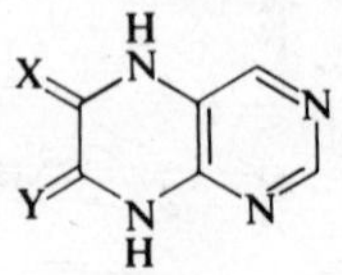

(**55**)

is often better to convert the thione either into an alkylthiopteridine or into a pteridinesulfonic acid prior to the hydrolytic step.

(4) *Pteridinones to Pteridinamines* (Ch. VI, Sect. 2.E)

It is customary to convert pteridinones into pteridinamines either via chloropteridines of via pteridinethiones and the derived alkylthiopteridines. However, when the substrate and product are sufficiently stable, the transformation may be done directly by heating with amines at a high temperature or, better, with substituted phosphoramides. The latter type of reaction is illustrated by the conversion of 2-amino-4(3*H*)-pteridinone (**56**) into 2,4-pteridinediamine (**57**) by heating with neat phenyl phosphorodiamidate and a trace of amine hydrochloride catalyst at 215 °C. Other such amidic reagents may be used to give methylamino-[phenyl *N*,*N*′-dimethylphosphorodiamidate: $PhOP(:O)(NHMe)_2$] or dimethylaminopteridines [phenyl *N*,*N*,*N*′,*N*′-tetramethylphosphorodiamidate: $PhOP(:O)(NMe_2)_2$].

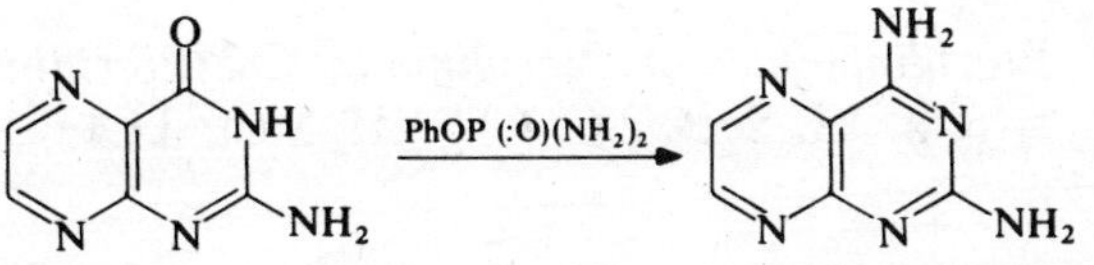

(**56**) (**57**)

(5) *Pteridinethiones to Pteridinamines or Pteridinimines* (Ch. VIII, Sect. 2.E)

The aminolysis of tautomeric pteridinethiones requires quite vigorous conditions that may sometimes induce unwanted side reactions such as transamination. Thus while 6,7-diphenyl-2(1*H*)-pteridinethione (**59**, R = H) reacted with refluxing ethanolic benzylamine during 5 h to give 2-benzylamino-6,7-diphenylpteridine (**58**), the similar but amino-deactivated substrate, 4-amino-6,7-diphenyl-2(1*H*)-pteridinethione (**59**, $R = NH_2$), required prolonged heating

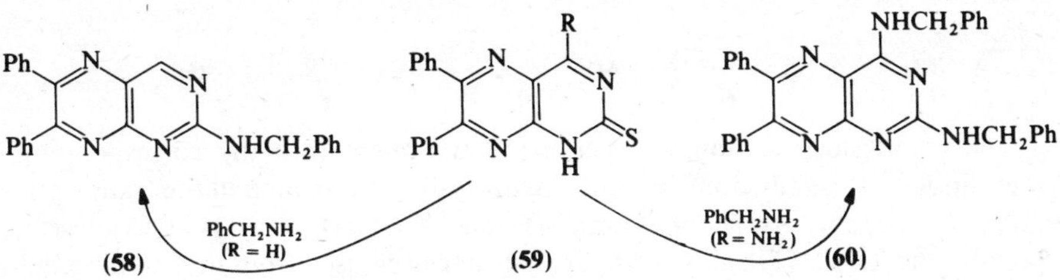

(**58**) (**59**) (**60**)

with neat benzylamine, which induced an additional transamination to give 2,4-bisbenzylamino-6,7-diphenylpteridine (**60**) as the only product. For such reasons, it is sometimes best to *S*-alkylate the thione before attempting a less vigorous aminolysis.

Nontautomeric pteridinethiones can occasionally undergo aminolysis to the corresponding pteridinimine. For example, 1,3-dimethyl-6,7-diphenyl-4-thioxo-3,4-dihydro-2(1*H*)-pteridinone (**61**) with ammonia or butylamine under appropriate conditions gave 4-imino- (**62**, R = H) or 4-butylimino-1,3-dimethyl-6,7-diphenyl-3,4-dihydro-2(1*H*)-pteridinone (**62**, R = Bu). The possibility of such a product undergoing Dimroth rearrangement should not be overlooked.

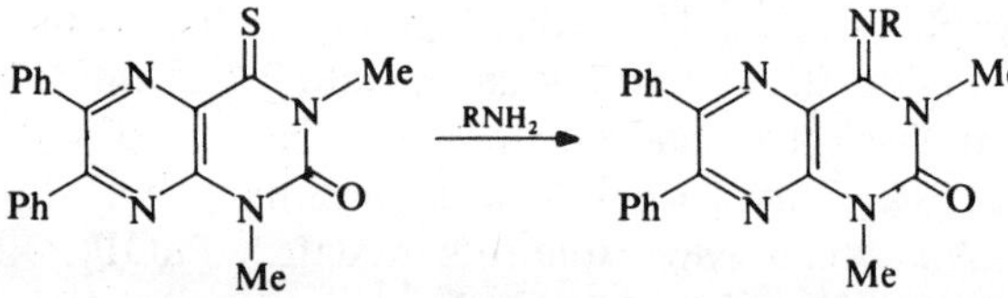

(**61**) (**62**)

(6) *Pteridinamines or Pteridinimines to Pteridinones* (Ch. VI, Sect. 2.C; Ch. VII, Sect. 3.C)

The conversion of tautomeric pteridinamines to pteridinones may be done by alkaline or acidic hydrolysis or by treatment with nitrous acid (primary amines only). Thus 4-pteridinamine in dilute alkali at 100°C for 5 min gave 4(3*H*)-pteridinone; 7-pteridinamine in dilute hydrochloric acid at room temperature gave 7(8*H*)-pteridinone; and 4,6,7-triphenyl-2-pteridinamine with nitrous acid in glacial acetic acid gave 4,6,7-triphenyl-2(1*H*)-pteridinone (**63**). Preferential partial hydrolysis of a pteridinediamine or pteridinetriamine can be achieved sometimes. For example, 2,4-bismethylamino-7-phenylpteridine (**64**) in boiling hydrochloric acid gave 2-methylamino-7-phenyl-4(3*H*)-pteridinone (**65**).

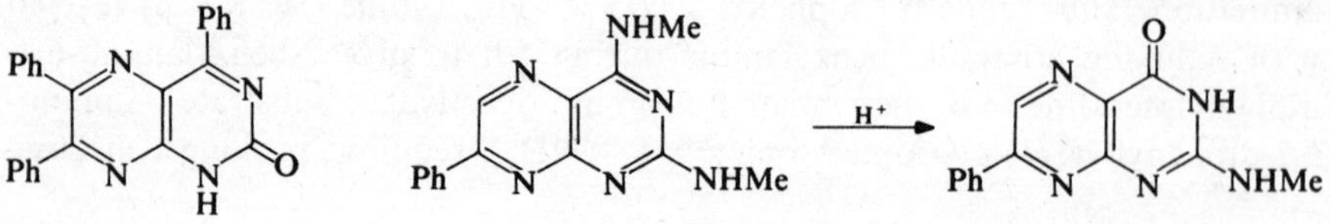

(**63**) (**64**) (**65**)

The conversion of (nontautomeric) pteridinimines into the corresponding pteridinones is usually done by acidic hydrolysis as illustrated in the transformation of 1-methyl-4(1*H*)-pteridinimine (**66**, X = NH) into 1-methyl-4(1*H*)-pteridinone (**66**, X = O); alkaline hydrolysis has been used from time to time but

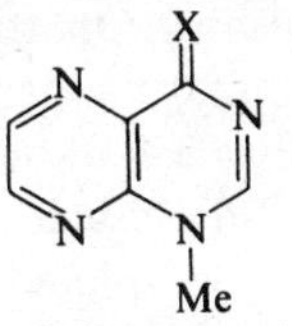

(66)

it is inapplicable when the substrate has its *N*-alkyl group adjacent to the imino group, because of almost certain Dimroth rearrangement instead of hydrolysis.

E. Addition Reactions

(1) *Covalent Hydration and Similar Additions*

The highly polarized bonds of pteridine make it very prone to the nucleophilic addition of water, alcohols, amines, and Michael-like reagents especially at the 3,4-, 7,8-, and (to a less extent) the 5,6-bond. Although not confined to the pteridine series, the phenomenon was discovered and investigated initially therein.

In aqueous solution, pteridine exists mainly as its 3,4-hydrate, 4-hydroxy-3,4-dihydropteridine (**67**). When such a solution is acidified, the 3,4-hydrated cation (**68**) is formed initially but this slowly equilibrates with the 5,6,7,8-dihydrated cation (**69**), although the indicated location of the positive charge at N-3 may well be incorrect. As might be expected, the addition of *C*-methyl groups to pteridine discourages hydration at the point(s) involved but does not inhibit it

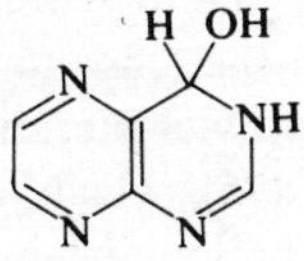

(67)

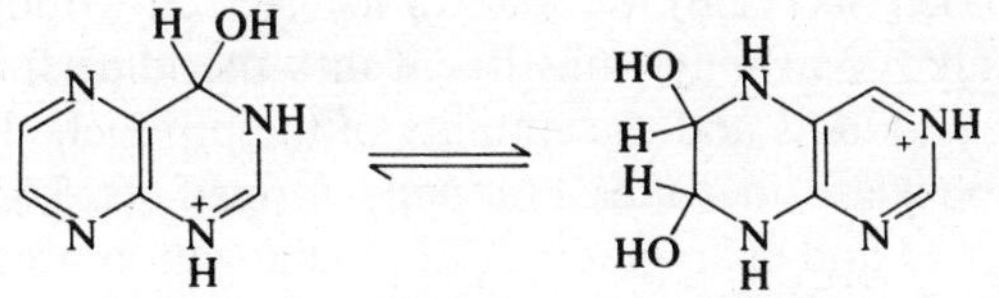

(68) **(69)**

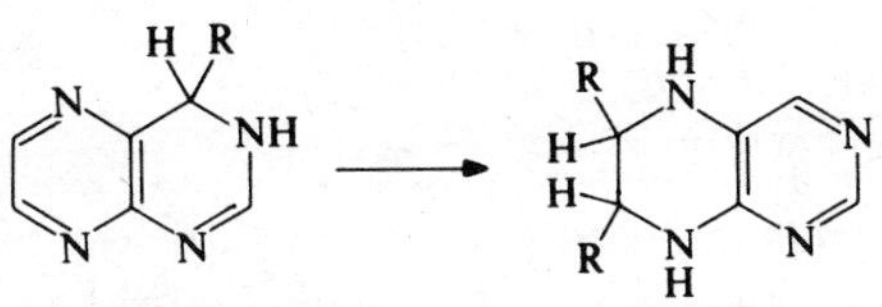

(70) **(71)**

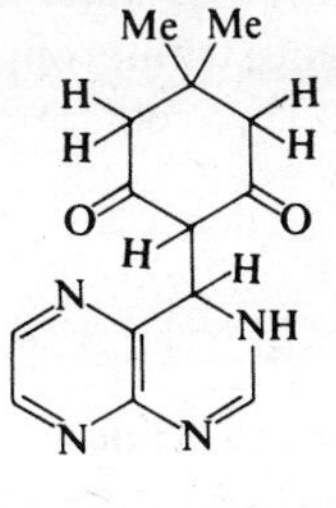

(72)

completely, especially in the cations. On treatment under appropriate conditions with a stronger nucleophile, methanol or methylamine, the neutral molecule of pteridine gives initially a 3,4-adduct (**70**, R = OMe or NHMe), which is eventually replaced by a 5,6,7,8-di-adduct (**71**, R = OMe or NHMe), both of which proved sufficiently stable to be isolated and characterized. Most Michael reagents add to the 3,4-bond of pteridine to give, for example, the dimedone adduct (**72**). Simple alkyl- and arylpteridines behave in a broadly similar way. Gentle oxidation of the adducts (**67**) and (**70**, R = NH_2) gave 4(3*H*)-pteridinone and 4-pteridinamine, respectively [Ch. IV, Sects. 1.C and 2.B(1)].

Both 2(1*H*)- and 6(5*H*)-pteridinone undergo covalent hydration to give their 3,4- and 7,8-hydrate (**73**), respectively. The same substrates and many of their derivatives give similar adducts with alcohols, amines, acetone, sodium bisulfite, acetylacetone, diethyl malonate, and many other such reagents. Most of the adducts have been isolated and characterized. The 5,6-bond may also be involved as in the reaction of 2-amino-7,8-dihydro-4(3*H*)-pteridinone with mercaptoacetic acid to give the 5,6-adduct, 2-amino-6-carboxymethylthio-5,6,7,8-tetrahydro-4(3*H*)-pteridinone (**74**) (Ch. VI, Sect. 2.B).

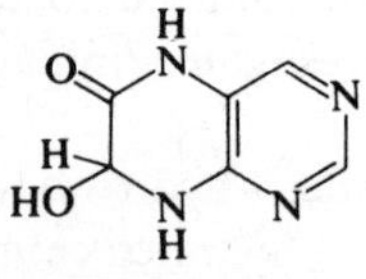

(73)

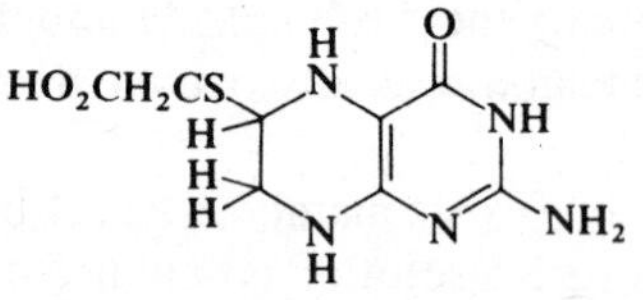

(74)

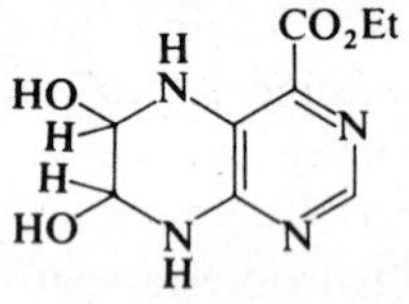

(75)

Although the neutral molecules of 2- and 6-pteridinamine will form adducts only with the more powerful nucleophilic Michael reagents, their cations readily form covalent hydrates and alcoholates, in both cases across the 3,4-bond (Ch. IX, Sect. 2.B). Because of its electron-withdrawing nature, the insertion of an alkoxycarbonyl substituent into pteridine(s) increases both the tendency to form adducts and the stability of the products. For example, ethyl 4-pteridinecarboxylate in water gradually formed its 5,6,7,8-dihydrate (**75**), which was isolated and characterized. Rearomatization occurred only on boiling in *t*-butyl alcohol during 8 h (Ch. X, Sect. 4.E). For the same reason, simple chloropteridines form covalent adducts such as 7-chloro-3,4-dihydro-4-pteridinamine long before the quite facile aminolysis or other nucleophilic displacement reaction begins; trifluoromethylpteridines also undergo adduct formation with great ease (Ch. V, Sect. 2.A and 4.D).

(2) *Addition of Phenyllithium*

The addition of Grignardlike reagents to pteridines has been little explored. However, phenyllithium adds to the 7,8-bond of 2-methylthio-4,6-diphenylpteridine to give, after hydrolysis and oxidation, 2-methylthio-4,6,7-

triphenylpteridine; in contrast, the same reagent with 2-methylthio-4,7-diphenylpteridine gave, not the same product as expected, but 2-methylthio-4,7,7-triphenyl-7,8-dihydropteridine, which naturally could not be oxidized to a formally aromatic structure [Ch. IV, Sect. 2.A(2)].

F. Oxidative Reactions

(1) *Nuclear Oxidation* (Ch. IX, Sects. 1.C, 2.A, and 4.A)

Unlike pyrimidines, which can be raised to an oxidation state above that of the parent heterocycle by the formation of alloxan derivatives, pteridines cannot be so oxidized. However, di- and tetrahydropteridines can generally be dehydrogenated with little difficulty. The *oxidation of dihydropteridines* usually arises when such compounds have been obtained by primary synthesis or by adduct formation from a less-substituted pteridine. For example, 3,4-dihydro-2-pteridinamine (**76**), prepared by primary synthesis from a pyrazine, was oxidized to 2-pteridinamine by permanganate in pyridine. 6-Methyl-7,8-dihydro-2,4(1*H*,3*H*)-pteridinedione (**77**), made by a Boon synthesis, gave 6-methyl-2,4(1*H*,3*H*)-pteridinedione on oxidation with hydrogen peroxide or permanganate. The covalent hydrate (**78**) of 2(1*H*)-pteridinone gave 2,4(1*H*,3*H*)-pteridinedione on prolonged aeration of an aqueous solution or by permanganate oxidation.

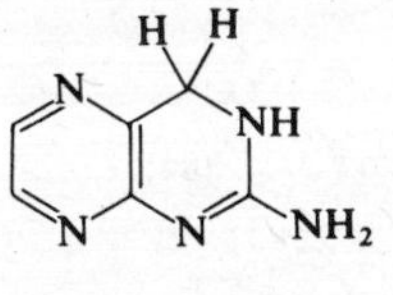

(**76**)

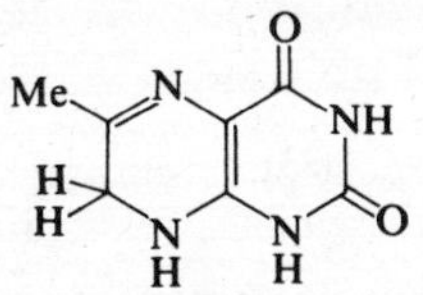

(**77**)

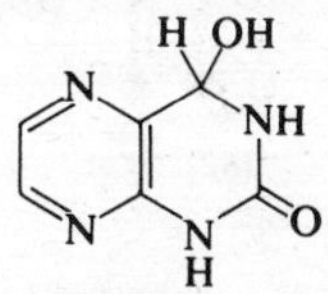

(**78**)

The *oxidation of tetrahydropteridines* is biologically important (Ch. IX, Sect. 7) but chemically less so, because most tetrahydropteridines are made by nuclear reduction. However, oxygenation of 2-amino-6,7-dimethyl-5,6,7,8-tetrahydro-4(3*H*)-pteridinone (**80**) in aqueous diethylamine gave the corresponding 7,8-dihydro compound (**79**) quite effectively, while oxygenation of the same substrate (**80**) in strong aqueous base gave 2-amino-6,7-dimethyl-4(3*H*)-pteridinone

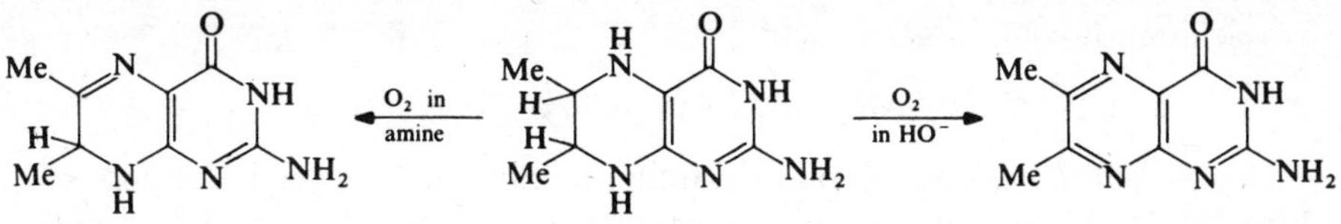

(**79**) (**80**) (**81**)

(**81**). Attempts to aromatize 5,6,7,8-tetrahydropteridine by a variety of methods failed to afford any isolable pteridine, but 5,6,7,8-tetrahydro-2,4-pteridinediamine did give the parent diamine on treatment with fresh manganese dioxide in DMF.

(2) *Formation of N-Oxides* (Ch. VII, Sect. 5.B)

Although pteridine undergoes oxidation by peroxide or peroxycarboxylic acids to give 4(3*H*)-pteridinone (from the 3,4-hydrate of the substrate) rather than a pteridine *N*-oxide, many substituted pteridines do give a 5- or 8-oxide or even a 5,8-dioxide on such treatment. The site of oxidation appears to be controlled by steric factors introduced by adjacent or *peri* groups. Thus treatment of 2,4(1*H*,3*H*)-pteridinedione or its 6,7-dimethyl derivative with hydrogen peroxide in formic acid gave the respective 8-oxide (**82**, R = H or Me); treatment of 1-methyl- or 1,3-dimethyl-2,4(1*H*,3*H*)-pteridinedione with hydrogen peroxide in trifluoroacetic acid gave the respective 5-oxide (**83**, R = H or Me); and similar but prolonged treatment of 2-amino-6,7-dimethyl-4(3*H*)-pteridinone gave the 5,8-dioxide (**84**). Some 1- or 3-oxides have been made by nonoxidative routes.

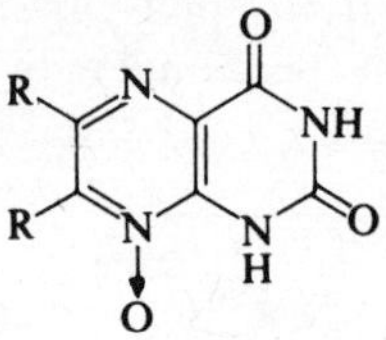

(**82**)

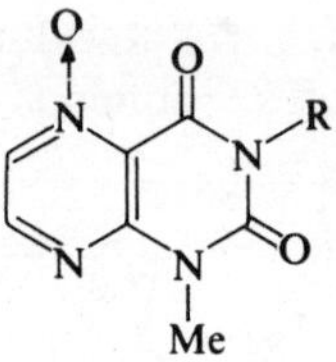

(**83**)

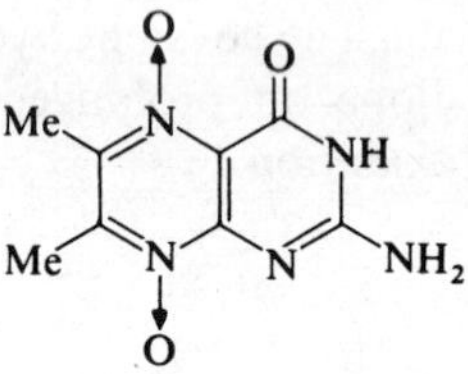

(**84**)

(3) *Oxidative Modification of Groups*

Oxidative changes to substituents on the pteridine nucleus differ little from similar changes in regular aromatic systems. Some such modifications are summarized briefly here.

(a) *C*-Methyl to *C*-Formyl [Ch. IV, Sect. 2.B(6)]

This oxidation is exemplified in the conversion of 2-amino-6-methyl-4(3*H*)-pteridinone (**86**) into 2-amino-4-oxo-3,4-dihydro-6-pteridinecarbaldehyde (**85**) by selenium dioxide in acetic acid.

(b) *C*-Alkyl to *C*-Acyl (Ch. X, Sect. 11.D)

This is seen in the conversion of 2-amino-6-ethyl- into 6-acetyl-2-amino-7,7-dimethyl-7,8-dihydro-4(3*H*)-pteridinone (**87**) by aeration in warm butanolic acetic acid.

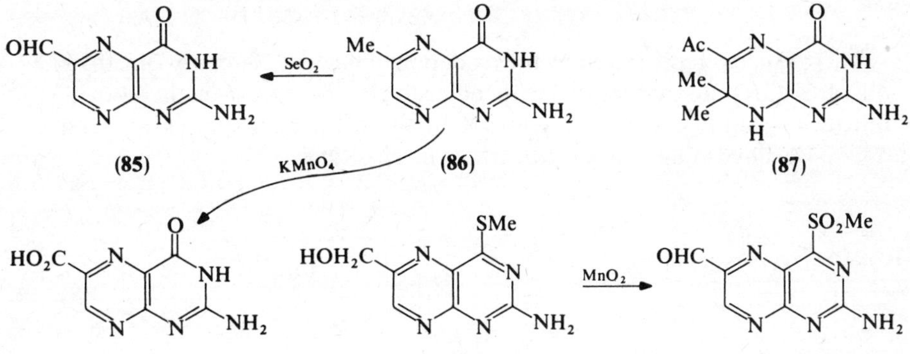

(85) (86) (87)

(88) (89) (90)

(c) Alkyl to Carboxy [Ch. IV, Sect. 2.B(6)]

This very common process is typified in the oxidation of this substrate (**86**) to 2-amino-4-oxo-3,4-dihydro-6-pteridinecarboxylic acid (**88**) by permanganate; ethyl or even quite complicated substituted-alkyl side chains are amenable to such oxidation but phenyl substituents are not.

(d) Hydroxyalkyl to Formyl (Ch. VI, Sect. 4.B)

Such oxidations may be done with manganese dioxide, as in the conversion of 6-hydroxymethyl-4-methylthio-2-pteridinamine (**89**) into 2-amino-4-methylsulfonyl-6-pteridinecarbaldehyde (**90**), or with periodate, as in the transformation of 2-amino-6-α,β,γ-trihydroxypropyl-4(3*H*)-pteridinone into 2-amino-4-oxo-3,4-dihydro-6-pteridinecarbaldehyde.

(e) Hydroxyalkyl to Carboxy (Ch. VI, Sect. 4.B)

This process has been largely used to identify alcohols of ambiguous orientation. Thus doubtful 7-hydroxymethyl-2,4-pteridinediamine (**91**, $R = CH_2OH$) with permanganate gave 2,4-diamino-7-pteridinecarboxylic acid (**91**, $R = CO_2H$), which could be identified as authentic.

(f) *C*-Formyl to Carboxy (Ch. X, Sect. 1.B)

This little-used process is exemplified in the conversion of 2-amino-4-oxo-3,4-dihydro-6-pteridinecarbaldehyde (**92**) into the carboxylic acid (**93**), either with alkaline hydrogen peroxide or with permanganate.

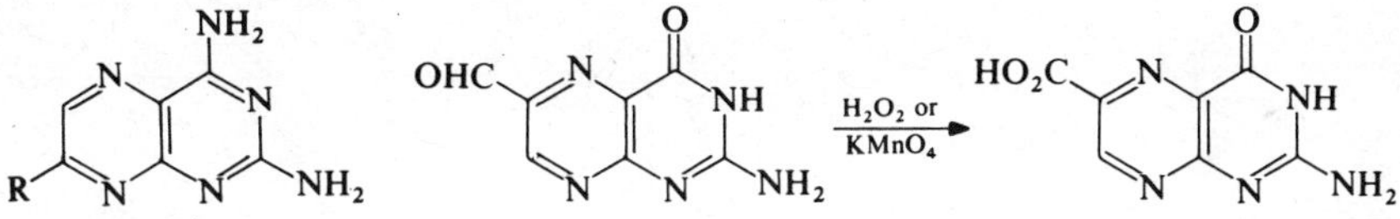

(91) (92) (93)

(g) *C*-Acyl to Carboxy (Ch. X, Sect. 1.B)

This change is represented by the oxidation of 6-acetyl-2-amino-7-methyl-7,8-dihydro-4(3*H*)-pteridinone (**95**) with iodine to give 2-amino-7-methyl-4-oxo-3,4-dihydro-6-pteridinecarboxylic acid (**94**) or with permanganate to give 2-amino-4-oxo-3,4-dihydro-6,7-pteridinedicarboxylic acid (**96**).

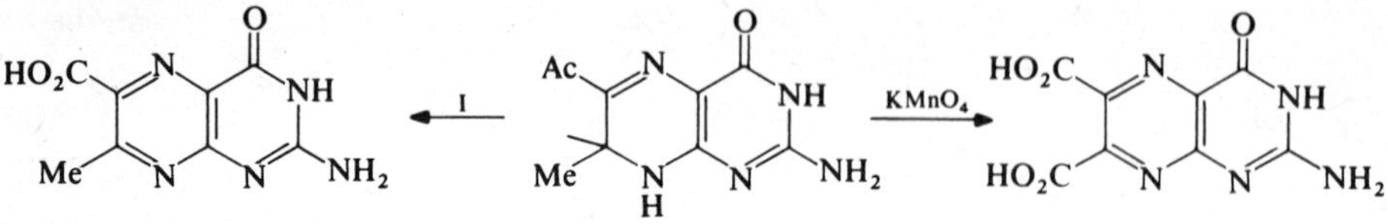

(94) **(95)** **(96)**

(h) Pteridinethiones to Dipteridinyl Disulfides (Ch. VIII, Sect. 2.B)

As in other series, tautomeric pteridinethiones are converted into disulfides by prolonged aeration or by treatment with iodine. For example, 2-amino-4-pentyloxy-7(8*H*)-pteridinethione with iodine in chloroform–aqueous bicarbonate gave bis(2-amino-4-pentyloxy-7-pteridinyl) disulfide (**97**).

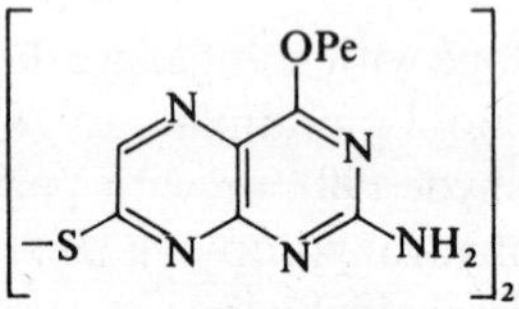

(97)

(i) Thioxo to Sulfeno, Sulfino, or Sulfo (Ch. VIII, Sect. 2.B)

These oxidation steps are exemplified by the conversion of 1,3,6-trimethyl-7-thioxo-7,8-dihydro-2,4(1*H*,3*H*)-pteridinedione into 1,3,6-trimethyl-2,4-dioxo-1,2,3,4-tetrahydro-7-pteridinesulfenic acid (**98**) by treatment with hydrogen peroxide under very mild conditions; of 1,3-dimethyl-6-thioxo-5,6-dihydro-2,4(1*H*,3*H*)-pteridinedione into 1,3-dimethyl-2,4-dioxo-1,2,3,4-tetrahydro-6-pteridinesulfinic acid (**99**) by treatment with manganese dioxide in aqueous sodium bicarbonate; and of the same substrate into 1,3-dimethyl-2,4-dioxo-

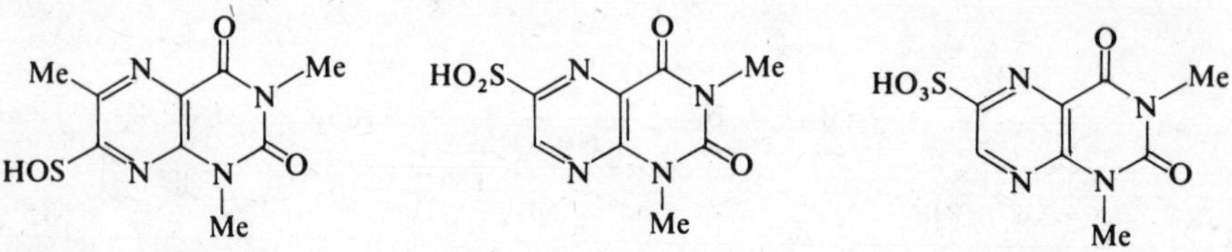

(98) **(99)** **(100)**

1,2,3,4-tetrahydro-6-pteridinesulfonic acid (**100**) by treatment with permanganate. Other reagents may be used but there are few data available.

(j) Alkylthio to Alkylsulfinyl or Alkylsulfonyl (Ch. VIII, Sect. 8)

Most of the few known pteridine sulfoxides were made by controlled oxidation of the corresponding thioethers with *m*-chloroperoxybenzoic acid: for example, 1,3-dimethyl-6-methylthio- so gave 1,3-dimethyl-6-methylsulfinyl-2,4(1*H*,3*H*)-pteridinedione (**101**). Sulfones may be made similarly but more easily by using an excess of the peroxycarboxylic acid or by treatment with permanganate. The latter reagent was used to convert 2-amino-6-methylthio- into 2-amino-6-methylsulfonyl-4(3*H*)-pteridinone (**102**). Extranuclear sulfoxides or sulfones can be made rather similarly.

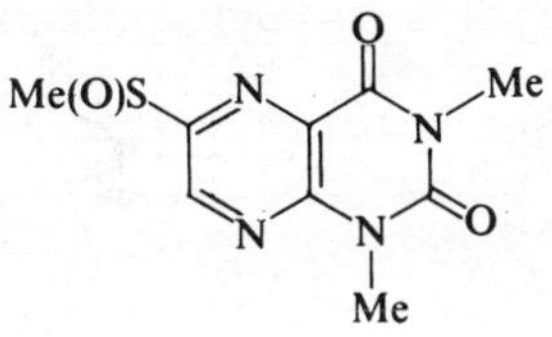

(**101**)

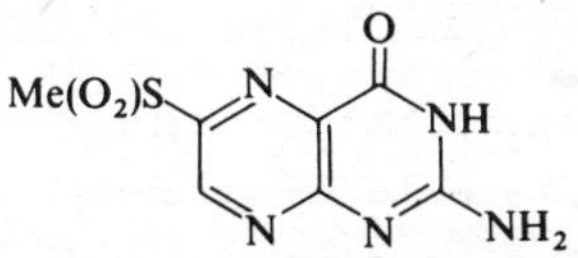

(**102**)

G. Reductive Reactions

(1) *Nuclear Reduction* (Ch. XI, Sects. 1.B, 2.B, and 3.B)

The pteridine nucleus is readily reduced by a variety of reagents such as catalytic hydrogenation, sodium borohydride, sodium dithionite, and so on. Pteridines tend to reduce at the same C–N bonds that might be expected to undergo covalent additions and in which the carbon atom is preferably unsubstituted or carries only an alkyl group. Reduction of appropriately substituted pteridines stops naturally at the dihydro stage and reduction of other pteridines can usually be stopped at the dihydro stage before much tetrahydropteridine has been formed. The commonly encountered systems are 3,4-, 5,6-, and 7,8-dihydropteridines and 5,6,7,8-tetrahydropteridines, although members of other di- and tetrahydro systems have been reported occasionally.

A few typical examples include the reduction of 2(1*H*)-pteridinone to its 3,4-dihydro derivative (**103**) by hydrogenation, borohydride, or dithionite; of 7(8*H*)-pteridinone to its 5,6-dihydro derivative (**104**) by borohydride or potassium amalgam; of 6(5*H*)-pteridinone to its 7,8-dihydro derivative (**105**) by hydrogenation, borohydride, dithionite, sodium amalgam, or electrolytic reduction; of 2-amino-6,7-diphenyl-5,6-dihydro-4(3*H*)-pteridinone (**106**) to its 5,6,7,8-tetrahydro analog (**107**) by borohydride; of 4,6-dimethyl-7,8-dihydro-2-

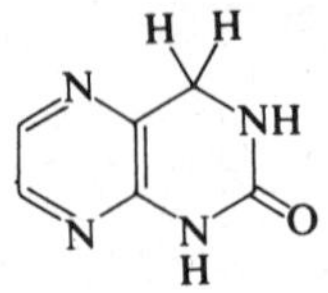

(103)

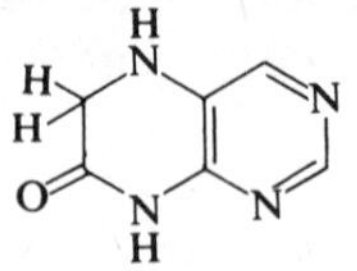

(104)

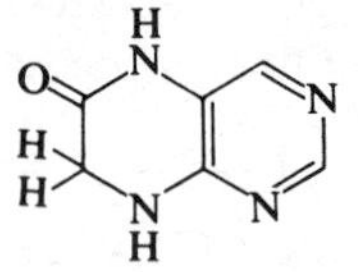

(105)

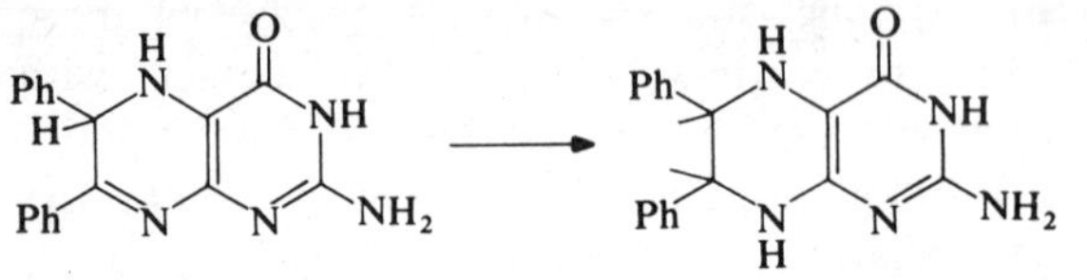

(106) **(107)**

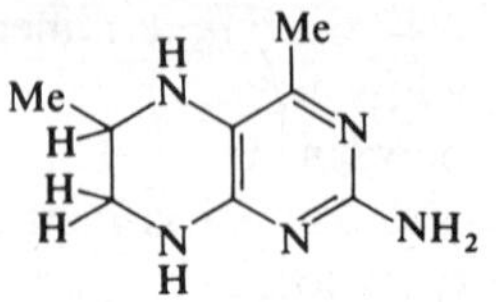

(108)

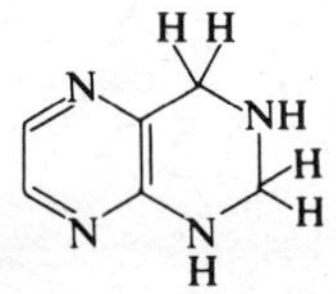

(109)

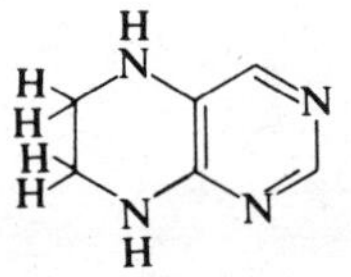

(110)

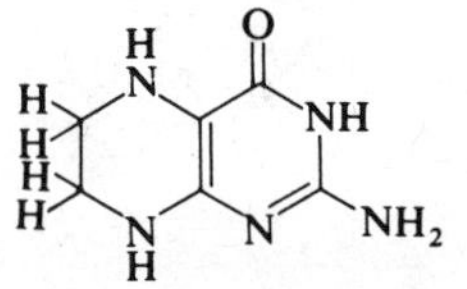

(111)

pteridinamine to its 5,6,7,8-tetrahydro analog (**108**) by hydrogenation; of pteridine to a separable mixture of 1,2,3,4-tetrahydro- (**109**) and 5,6,7,8-tetrahydropteridine (**110**) by borohydride, although use of lithium aluminum hydride gave only the latter product; and of 2-amino-4(3*H*)-pteridinone to its 5,6,7,8-tetrahydro analog (**111**) by hydrogenation.

(2) *Reductive Removal of Groups*

The pteridine nucleus is so easily reduced that the reductive removal of substituents is accompanied frequently by partial reduction of one or the other ring. Thus while *dechlorination* of 2-amino-4-chloro-6,7(5*H*, 8*H*)-pteridinedione (**112**, R = Cl) to 2-amino-6,7(5*H*,8*H*)-pteridinedione (**112**, R = H) can be achieved in warm hydriodic acid, the more usual result is seen in the partial dechlorination and nuclear reduction of 2,4,6,7-tetrachloropteridine (**113**) by lithium aluminum hydride to give 2,4-dichloro-5,6,7,8-tetrahydropteridine (**114**, R = Cl),

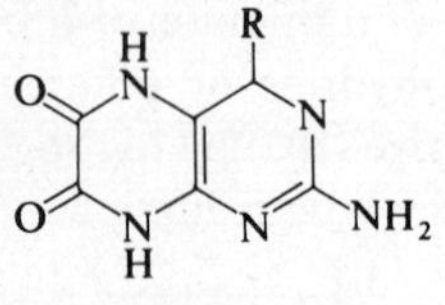

(112)

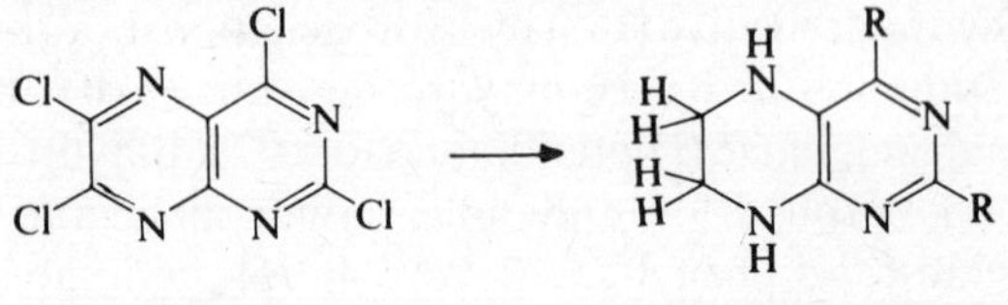

(113) **(114)**

which may be further dechlorinated by hydrogenolysis to give 5,6,7,8-tetrahydropteridine (**114**, R = H) (Ch. V, Sect. 2.A).

Probably because of nuclear reduction, the *desulfurization* of pteridinethiones with Raney-nickel, and so on, has remained virtually unused. However, this common reaction has been applied occasionally and with care to alkylthiopteridines. Thus treatment of 1,3-dimethyl-7-methylthio-6-propionyl-2,4(1*H*,3*H*)-pteridinedione (**116**) with acetone-deactivated Raney-nickel gave 1,3-dimethyl-6-propionyl-2,4(1*H*,3*H*)-pteridinedione (**115**), while the use of regular Raney-nickel in methanol gave the reduced product, 6-α-hydroxypropyl-1,3-dimethyl-2,4(1*H*,3*H*)-pteridinedione (**117**) (Ch. VIII, Sect. 7.A).

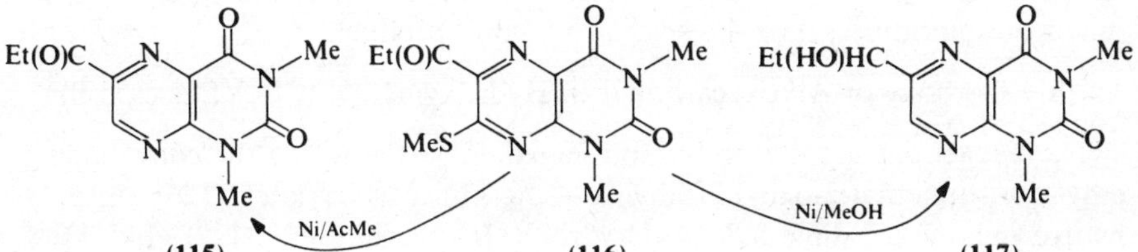

(**115**) (**116**) (**117**)

The *deoxygenation* of pteridinones is confined in practice to the 7-position and it is accompanied by 7,8-reduction. Thus treatment of 2-amino-4,6,7(3*H*,5*H*,8*H*)-pteridinetrione (**118**) with sodium amalgam gave 2-amino-7,8-dihydro-4,6(3*H*,5*H*)-pteridinedione (**119**), which proved easily oxidizable to 2-amino-4,6(3*H*,5*H*)-pteridinedione (**120**). Electrolytic reduction has proven particularly useful for such deoxygenations (Ch. VI, Sect. 2.C). The reductive removal of *N*-oxide functions is also usually accompanied by nuclear reduction. 2-Amino-6-methyl-4(3*H*)-pteridinone 8-oxide (**121**), with dithionite gave 2-amino-6-methyl-7,8-dihydro-4(3*H*)-pteridinone (**122**), which was easily aromatized by potassium permanganate (Ch. VII, Sect. 6.A).

Reductive *debenzylation* can clearly occur in the usual way. Treatment of 8-benzyl-4-chloro-5,6,7,8-tetrahydropteridine (**123**) with sodium in liquid am-

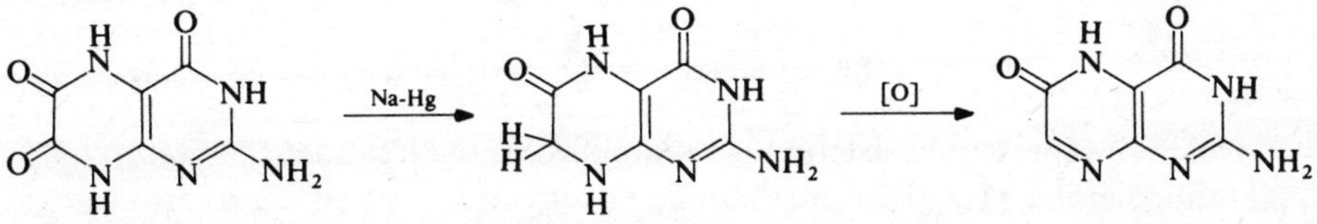

(**118**) (**119**) (**120**)

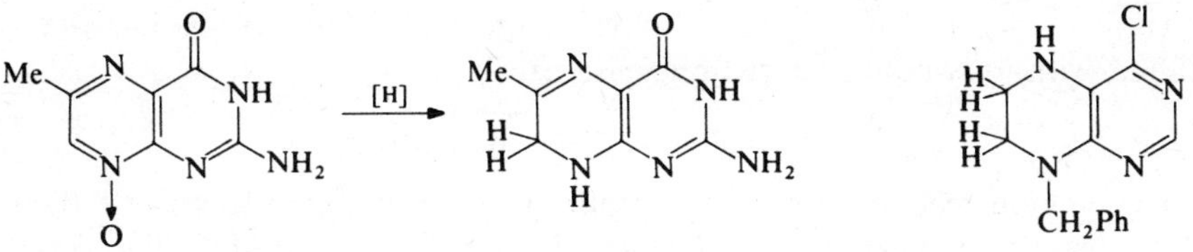

(**121**) (**122**) (**123**)

monia removed both the chloro and benzyl substituent to give 5,6,7,8-tetrahydropteridine.[43]

(3) *Reductive Modification of Groups*

(a) *C*-Formyl to Hydroxymethyl (Ch. VI, Sect. 3.B)

This reaction is illustrated by the treatment of 2-amino-4-oxo-3,4-dihydro-6-pteridinecarbaldehyde (**124**) with sodium borohydride to give 2-amino-6-hydroxymethyl-4(3*H*)-pteridinone (**125**). The Cannizzaro reaction may be used for the same conversion, naturally with the formation of 2-amino-4-oxo-3,4-dihydro-6-pteridinecarboxylic acid as a second product.

(b) Carboxy or Alkoxycarbonyl to Hydroxymethyl (Ch. VI, Sect. 3.B)

These reactions are meagrely represented, for example, by the conversion of ethyl 6,7-dimethyl-2-oxo-1,2-dihydro-4-pteridinecarboxylate (**126**) into 4-hydroxymethyl-6,7-dimethyl-3,4-dihydro-2(1*H*)-pteridinone (**127**) by borohydride, a reaction in which additional nuclear reduction occurred.

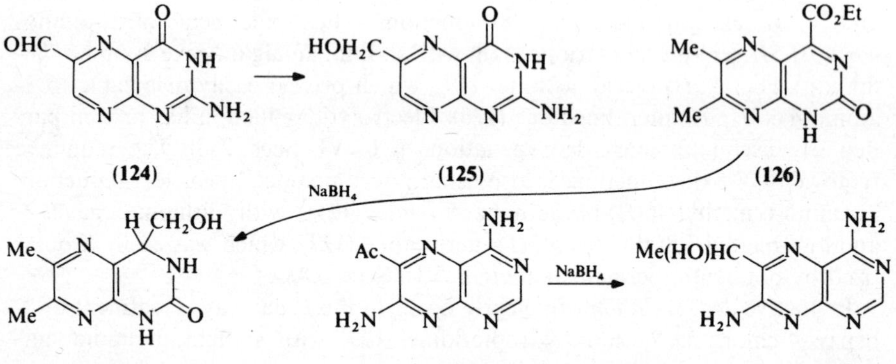

(124) (125) (126)

(127) (128) (129)

(c) *C*-Acyl to Hydroxyalkyl or Alkyl (Ch. VI, Sect. 3.B; Ch. X, Sect. 12.B)

This well-represented reaction is exemplified in the treatment of 6-acetyl-4,7-pteridinediamine (**128**) with methanolic borohydride to give 6-α-hydroxyethyl-4,7-pteridinediamine (**129**). Cathodic reduction may be used but it is prone to afford pinacols, 2,3-bispteridinylbutane-2,3-diols, as primary products. Total reduction of the acyl group to an alkyl group may occur occasionally, for example, with hydriodic acid–red phosphorus.

(d) Cyano to Aminomethyl (Ch. IX, Sect. 3.B)

This common reaction is virtually confined in the pteridine series to tetrahydro substrates that can suffer no further nuclear reduction during the process. Thus 2-amino-6-methyl-4-oxo-3,4,5,6,7,8-hexahydro-6-pteridinecarbonitrile

(**130**), on hydrogenation over platinum in trifluoroacetic acid, gave 2-amino-6-aminomethyl-6-methyl-5,6,7,8-tetrahydro-4(3*H*)-pteridinone (**131**).

(e) Sulfo, Sulfino, or Sulfeno to Thioxo (Ch. VIII, Sect. 1.E)

Most such reductions appear to have escaped attention in the literature to date, as has the reduction of disulfides to pteridinethiones. However, 1,3,6-trimethyl-2,4-dioxo-1,2,3,4-tetrahydro-7-pteridinesulfenic acid (**132**) did give 1,3,6-trimethyl-7-thioxo-7,8-dihydro-2,4(1*H*,3*H*)-pteridinedione (**133**) on gentle treatment with sodium borohydride. The related reductions of alkylsulfonyl-pteridines→alkylsulfinylpteridines→alkylthiopteridines also appear to have been neglected.

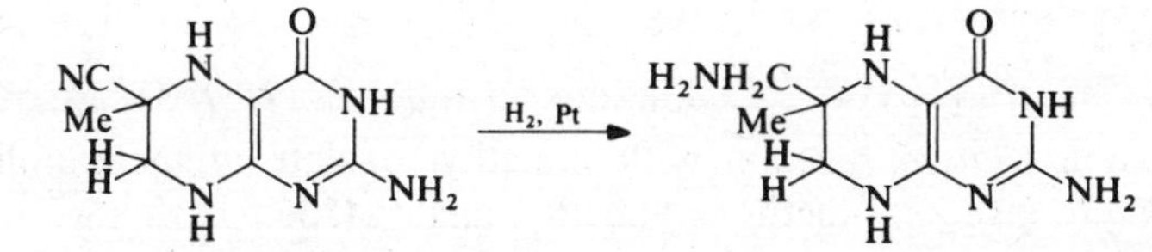

(**130**) (**131**)

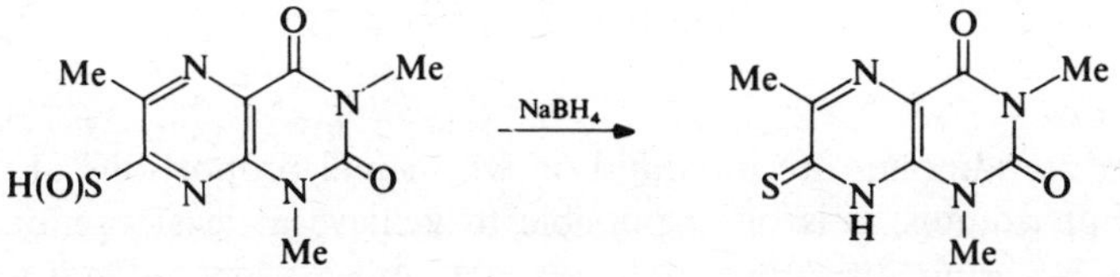

(**132**) (**133**)

H. Other Substituent Modifications

Oxidative and reductive modifications applicable to substituents attached to pteridines have been mentioned earlier; the more important remaining types of substituent modification are summarized here.

(1) *Modification of Alkyl Groups* [Ch. IV, Sects. 2.B(2), (5), and (8)]

Because they are attached to the π-deficient ring carbon atoms of pteridine, methyl or α-methylene substituents are activated considerable. The *deuteration* of methyl groups on pteridine is not easy although 2-amino-6-methyl-4(3*H*)-pteridinone has been converted into 2-amino-6-trideuteromethyl-4(3*H*)-pteridinone (**134**) by prolonged heating with NaOD–D_2O at 100°C. The *styrylation* of methyl groups appears to occur quite normally, as in the conversion of 7-methyl- into 7-styryl-2,4-pteridinediamine (**135**) by using benzaldehyde in

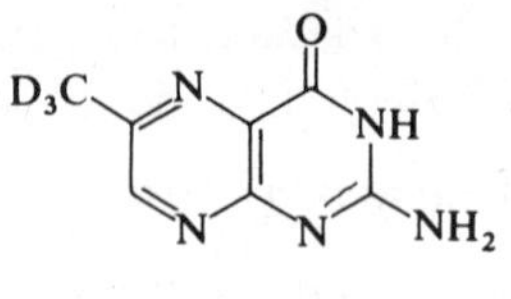

(134)

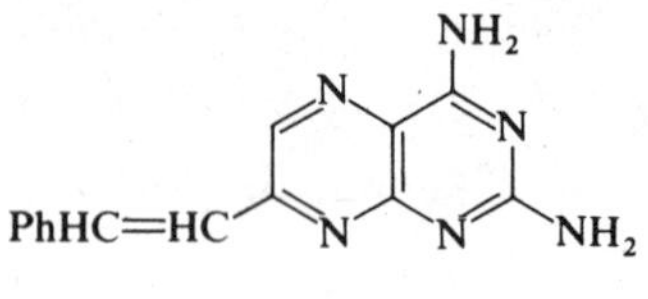

(135)

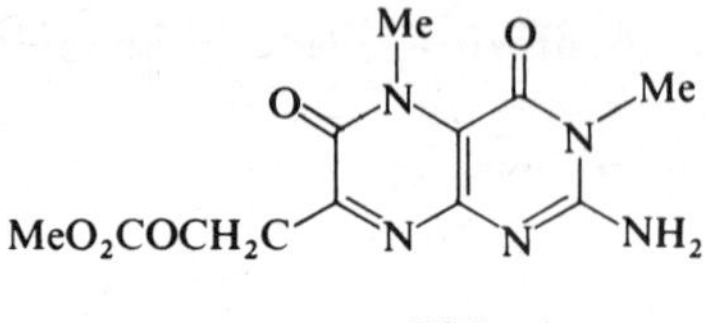

(136)

alcoholic alkali at 100°C. 2-Amino-3,5,7-trimethyl-4,6(3*H*,5*H*)-pteridinedione underwent a *Claisen reaction* with dimethyl oxalate in methanolic sodium methoxide to give its 7-methoxalylmethyl analog (**136**).

(2) *Modification of Oxo Substituents* (Ch. VI, Sect. 2.F)

Becausê pteridinones are in equilibria with small proportions of the corresponding pteridinols, it is often possible to achieve at least predominant *O-alkylation* by using an appropriate reagent under proper conditions. Thus treatment of 2,4-diamino-6(5*H*)-pteridinone with methanolic hydrogen chloride gave 6-methoxy-2,4-pteridinediamine, apparently without any *N*-methylated product. Diazomethane in methanolic ether appears to be less *O*-selective and gives quite variable results. Thus 7(8*H*)-pteridinone gave mainly 8-methyl-7(8*H*)-pteridinone with only a trace of 7-methoxypteridine; in contrast, 1,3-dimethyl-2,4,6(1*H*,3*H*,5*H*)-pteridinetrione (**137**) yielded only the 6-methoxy-1,3-dimethyl-2,4(1*H*,3*H*)-pteridinedione (**138**). The use of methyl iodide or dimethyl sulfate on the preformed sodium salt of the pteridinone seldom favors *O*- at the expense of *N*-methylation but a mixture is usually formed. For example, the same substrate (**137**) with dimethyl sulfate at pH 9 gave a separable mixture of the methoxy derivative (**138**) and 1,3,5-trimethyl-2,4,6(1*H*,3*H*,5*H*)-pteridine-

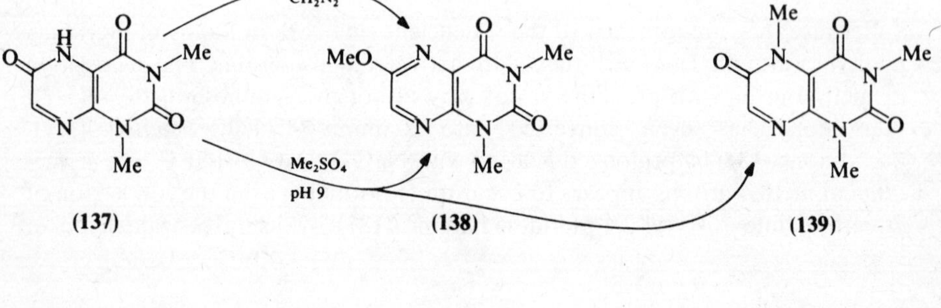

(137) **(138)** **(139)**

trione (**139**). Invariably, the trimethylsilylation of pteridinones with hexamethyldisilazane strongly favors the production of trimethylsiloxypteridines.

Pteridinones seldom undergo *O-acylation* and acetoxypteridines are usually made by treatment of pteridine *N*-oxides with acetic anhydride (Ch. VII, Sect. 6.D). Extranuclear hydroxypteridines do undergo acylation normally as exemplified in the conversion of 2-amino-6-hydroxymethyl-4(3*H*)-pteridinone into 2-acetamido-6-acetoxymethyl-4(3*H*)-pteridinone (Ch. VI, Sect. 4.C).

(3) *Modification of Thioxo Substituents* (Ch. VIII, Sect. 2.C)

In contrast to pteridinones, pteridinethiones undergo *S-alkylation* exclusively and easily. Thus 2(1*H*)-, 4(3*H*)-, or 7(8*H*)-pteridinethione, on shaking in aqueous or ethanolic alkali with methyl iodide, gave 2-, 4-, or 7-methylthiopteridine, respectively; likewise, 3,6,7-trimethyl-2-thioxo-1,2-dihydro-4(3*H*)-pteridinone (**140**) with dibromomethane in DMF containing potassium carbonate gave bis(3,6,7-trimethyl-4-oxo-3,4-dihydro-2-pteridinylthio)methane (**141**). Pteridinethiones undergo *S*-trimethylsilylation readily.

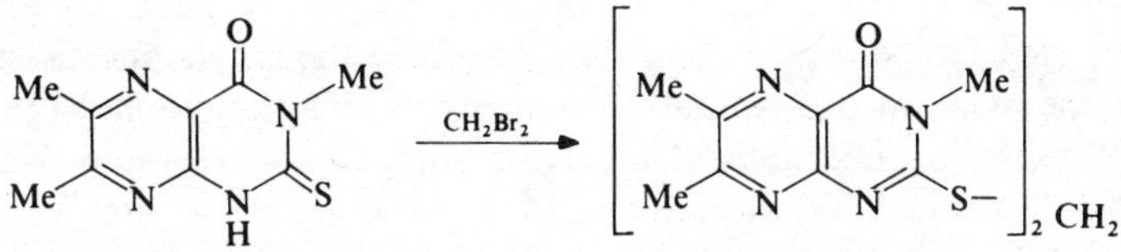

(140) **(141)**

(4) *Modification of Amino Groups*

Amino groups undergo several real modifications and two types of apparent modification.

(a) Amino to Acylamino (Ch, IX, Sect. 2.C)

This process is quite straightforward and has been widely used. For example, 2,4-pteridinediamine with hot acetic anhydride gave 2,4-diacetamidopteridine; secondary amino groups react similarly as indicated in the conversion of 1,3-dimethyl-7-methylamino-2,4,6(1*H*,3*H*,5*H*)-pteridinetrione by acetic anhydride–potassium acetate into 6-acetoxy-1,3-dimethyl-7-*N*-methylacetamido-2,4(1*H*,3*H*)-pteridinedione (**142**). The acylation of ring NH in reduced pteridines is equally easy (Ch. XI, Sects. 2.C and 4.B).

(b) Amino to Alkylideneamino (Ch. IX, Sect. 2.E)

The formation of Schiff bases from primary pteridinamines iś quite normal but has been little used in the pteridine series. However, 2-hydrazino- gave 2-

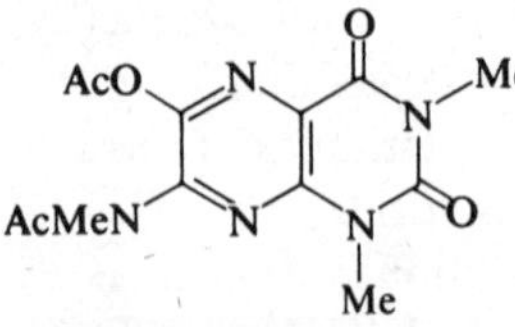

(142)

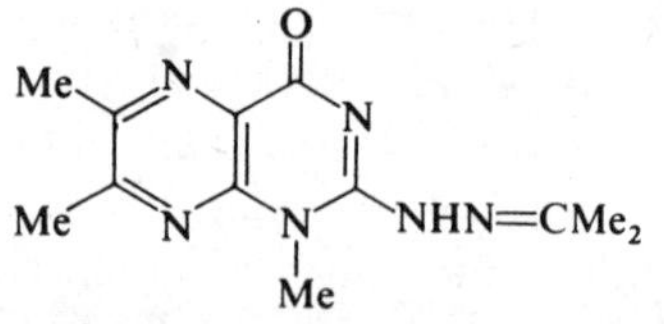

(143)

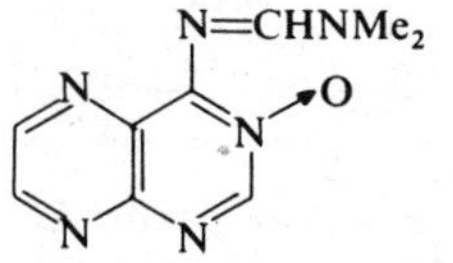

(144)

isopropylidenehydrazino-1,6,7-trimethyl-4(1*H*)-pteridinone (**143**) by dissolution in warm acetone, while 4-pteridinamine 3-oxide gave 4-dimethylaminomethyleneaminopteridine 3-oxide (**144**) in hot DMF dimethyl acetal.

(c) Amino to Alkylamino (Ch. IX, Sects. 2.D, 5, and 6.B)

Although pteridinamines undergo trimethylsilylation on the amino group, direct alkylation thereon appears to be unknown. Thus treatment of 4-pteridinamine with methyl iodide results in the formation of 1-methyl-4(1*H*)-pteridinimine (**145**), representing modification of an amino to an imino group. However, when the *N*-alkyl group is adjacent to the imino or potential imino group, Dimroth rearrangement can occur subsequently under alkaline conditions by ring fission and recyclization to afford the alkylamino isomer. For example, methylation of 2-amino-6,7-diphenyl-4(3*H*)-pteridinone (**146**) gave a separable mixture of 2-amino-1-methyl-6,7-diphenyl-4(1*H*)- and 2-amino-3-methyl-6,7-diphenyl-4(3*H*)-pteridinone (**147**). The latter, as its anion (**148**), underwent Dimroth rearrangement to afford the new anion (**149**) and then on acidification it gave 2-methylamino-6,7-diphenyl-4(3*H*)-pteridinone (**150**); although this appears to represent a modification of the primary amino group in (**146, 147**) to a secondary methylamino group in (**150**), in fact the two complete groups in (**147**) were simply interchanged.

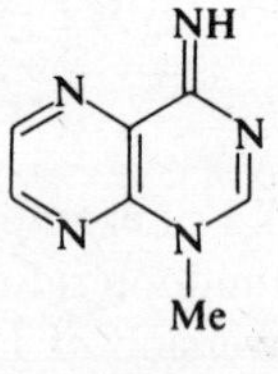

(145)

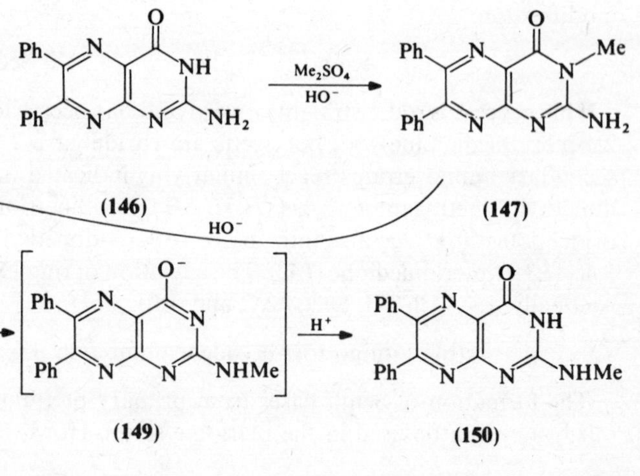

(146) **(147)** **(148)** **(149)** **(150)**

(d) Transamination (Ch. IX, Sect. 2.F)

The direct interchange of an amino or substituted-amino group with another such group by simply heating with an appropriate amine has been widely reported in the pteridine series. For example, 2,4-pteridinediamine in boiling hydrazine hydrate gave 4-hydrazino-2-pteridinamine, an apparent modification of the 4-amino group of the substrate.

(5) *Modification of Carboxy and Related Groups*

Although the modification of carboxylike groups in the pteridine series differs little from that in regular aromatic systems, some of the more important of such reactions are mentioned briefly next.

(a) Carboxy to Related Groups

The conversion of pteridinecarboxylic acids into *esters* is exemplified in the formation of methyl 2,4-diamino-6-pteridinecarboxylate (**151**) (methanolic hydrogen chloride), ethyl 2,4,6-trioxo-1,2,3,4,5,6-hexahydro-7-pteridinecarboxylate (ethanolic sulfuric acid), and methyl 7-methoxy-1,3-dimethyl-2,4-dioxo-1,2,3,4-tetrahydro-6-pteridinecarboxylate (diazomethane on 1,3-dimethyl-2,4,7-trioxo-1,2,3,4,7,8-hexahydro-6-pteridinecarboxylic acid) (Ch. X, Sect. 2.C). The conversion of carboxylic acids into *amides* is best done indirectly, for example, by converting 1,3-dimethyl-2,4-dioxo-1,2,3,4-tetrahydro-6-pteridinecarboxylic acid (**152**, R = OH) into the corresponding carbonyl chloride (**152**, R = Cl) (thionyl chloride) and then into the carboxamide (**152**, R = NH_2); alternatively, an intermediate ester may be used (Ch. X, Sect. 2.D).

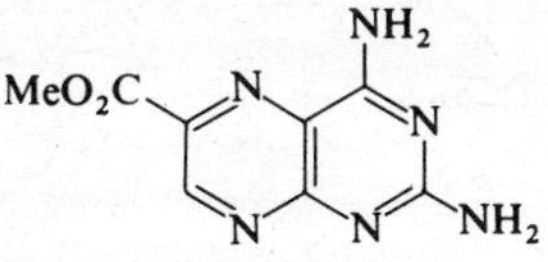

(**151**)

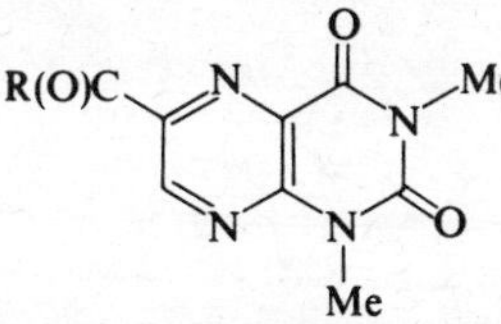

(**152**)

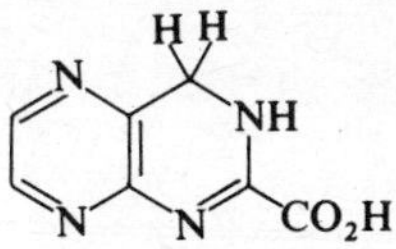

(**153**)

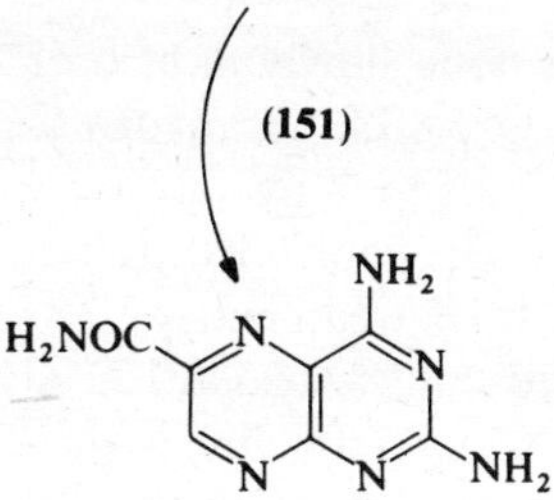

(**154**)

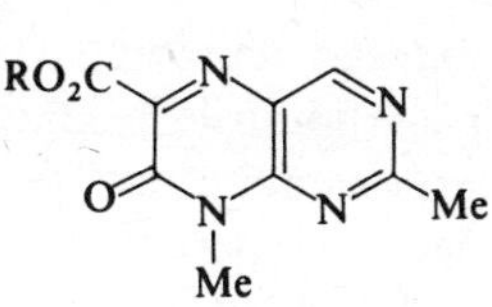

(**155**)

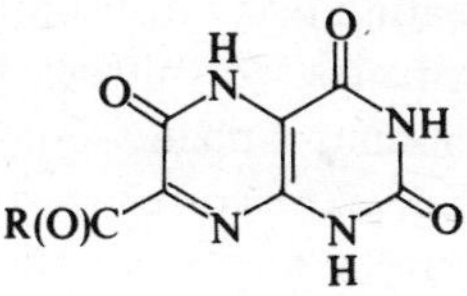

(**156**)

(b) Alkoxycarbonyl to Related Groups

The conversion of esters into *carboxylic acids* may be done by acidic or basic hydrolysis, as illustrated in the formation of 4-pteridinecarboxylic acid (aqueous sodium hydrogen carbonate) or 3,4-dihydro-2-pteridinecarboxylic acid (**153**) (very dilute hydrochloric acid) (Ch. X, Sect. 1.C). The formation of *amides* from esters is usually done with neat or alcoholic amine. For example, methyl 2,4-diamino-6-pteridinecarboxylate (**151**) with ammonia in ethylene glycol at 100°C gave the corresponding carboxamide (**154**) (Ch. X, Sect. 4.E). The conversion of one ester into another by *transesterification* is illustrated in the change from ethyl (**155**, R = Et) to methyl 2,8-dimethyl-7-oxo-7,8-dihydro-6-pteridinecarboxylate (**155**, R = Me) by prolonged boiling in methanol containing a trace of thorium nitrate (Ch. X, Sect. 3.B).

(c) Carbamoyl to Related Groups

The hydrolysis of carboxamides to *carboxylic acids* is less easy than the hydrolysis of the corresponding esters. For example, *N*-methyl-2,4,6-trioxo-1,2,3,4,5,6-hexahydro-7-pteridinecarboxamide (**156**, R = NHMe) required alkali under reflux for 2 h to give the acid (**156**, R = OH) (Ch. X, Sect. 1.C).The dehydration of amides to *nitriles* is represented in the formation (in abysmal yield) of 4-pteridinecarbonitrile (**157**) from the corresponding amide by brief heating in propionic anhydride; thionyl chloride has also been used as a dehydrating agent (Ch. X, Sects. 6 and 7.C). The conversion of thioamides into *carboxamidines* is illustrated by the butylaminolysis that gave 4-amino-*N*-butyl-7-butylamino-*N*′-β-methoxyethyl-2-phenyl-6-pteridinecarboxamidine (**158**) (Ch. X, Sect. 6).

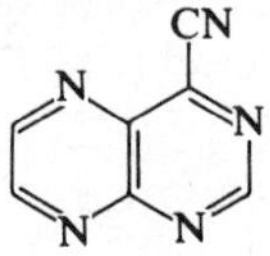

(157)

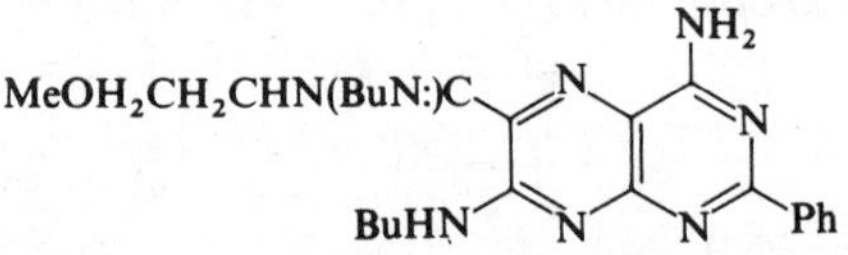

(158)

(d) Cyano to Related Groups

The hydrolysis of nitriles to *carboxylic acids* may be done slowly in acid or alkali: For instance, 2-amino-5,6,7,7-tetramethyl-4-oxo-3,4,5,6,7,8-hexahydro-6-pteridinecarboxylic acid (**159**) was formed from the corresponding nitrile by heating in hydrobromic acid for 17 h (Ch. X, Sect. 1.C). The controlled partial hydrolysis of nitriles to *amides* is represented in the brief treatment of 2,4-diamino-7-phenyl-6-pteridinecarbonitrile in concentrated sulfuric acid at 100 °C to give the corresponding carboxamide (**160**); the Radziszewski alkaline hydrolysis of nitriles in the presence of peroxide and the thiolysis of nitriles to thioamides appear to be unrepresented in the literature (Ch. X, Sect. 5.C). The formation of a *carboximidate* (imino ether) from a nitrile is exemplified in the

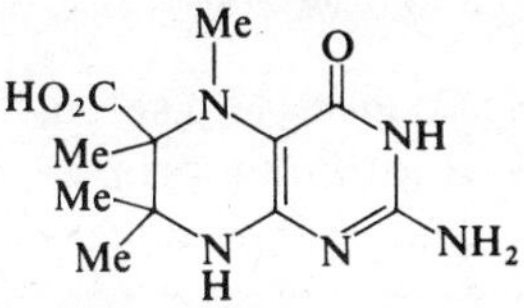

(159)

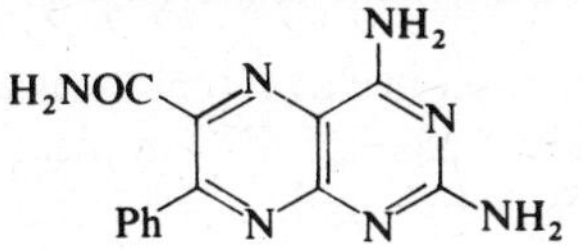

(160)

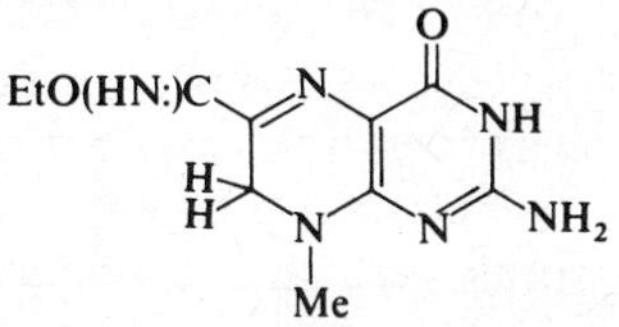

(161)

alcoholysis leading to ethyl 2-amino-8-methyl-4-oxo-3,4,7,8-tetrahydro-6-pteridinecarboximidate (**161**), which underwent a facile hydrolysis to the corresponding carboxylate (Ch. X, Sect. 8).

(e) *C*-Formyl to Related Groups

The condensation of pteridinecarbaldehydes with active methylene groups to give *styryl-type* products is limited by the activity of the methylene reagent. Thus 1,3-dimethyl-2,4,7-trioxo-1,2,3,4,7,8-hexahydro-6-pteridinecarbaldehyde with malonic acid in pyridine gave (after spontaneous monodecarboxylation) 6-β-carboxyvinyl-1,3-dimethyl-2,4,7(1*H*,3*H*,8*H*)-pteridinetrione (**162**), but *p*-chlorotoluene will not so react. However, this difficulty can be overcome by using, for example, the corresponding Wittig reagent, diethyl *p*-chlorobenzylphosphonate, which gave with the appropriate pteridinecarbaldehyde, 2-amino-6-*p*-chlorostyryl-7,7-dimethyl-7,8-dihydro-4(3*H*)-pteridinone (**163**) (Ch. X, Sect. 10.B). The rather similar conversion of pteridinecarbaldehydes by primary amines into *Schiff bases* (or *hydrazones*) is easier. 2-Amino-4-oxo-3,4-dihydro-7-pteridinecarbaldehyde reacted with ethyl *p*-aminobenzoate to give 2-amino-7-*p*-ethoxycarbonylphenyliminomethyl-4(3*H*)-pteridinone (**164**) (Ch. IX, Sect. 3.C; Ch. X, Sect. 10.C). The conversions of pteridinecarbaldehydes into the corresponding *acetals*, *oximes*, or *semicarbazones* appear to occur by the usual procedures, although only a few such derivatives have been prepared (Ch. X, Sect. 10.D).

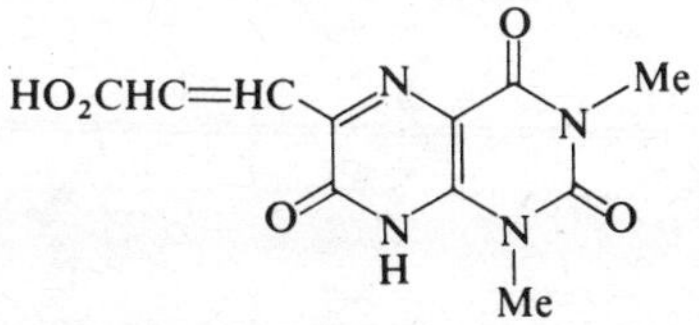

(162)

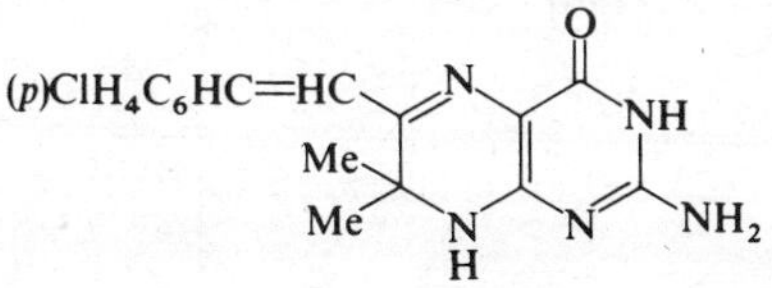

(163)

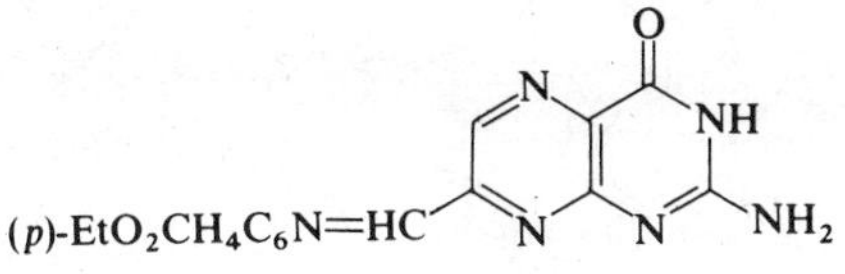

(164)

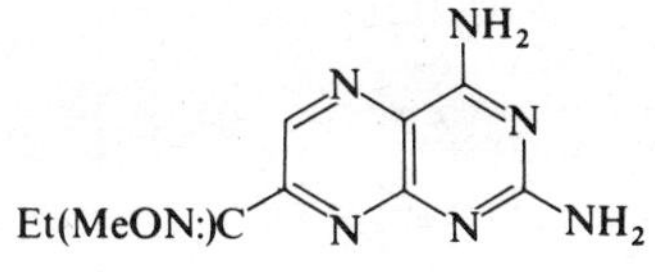

(165)

(f) *C*-Acyl to Related Groups

The conversion of pteridine ketones into their *oximes*, *hydrazones*, and so on, has been little used but the few cases, for example, 7-α-methoxyiminopropyl- (**165**) from 7-propionyl-2,4-pteridinediamine and *O*-methylhydroxylamine, appear to be unexceptional (Ch. X, Sect. 12.C).

I. Decarboxylation (Ch. X, Sect. 2.A)

The conditions required for thermal decarboxylation of pteridinecarboxylic acids vary considerably but do not appear to conform to any pattern. For example, 2,4-diamino-7-oxo-7,8-dihydro-6-pteridinecarboxylic acid (**166**) decarboxylated only at 350°C, whereas 2-amino-6-α-aminocarboxymethyl-4,7(3*H*,8*H*)-pteridinedione (**167**, R = CO_2H) lost carbon dioxide in dilute hydrochloric acid at 50 °C to give 2-amino-6-aminomethyl-4,7(3*H*,8*H*)-pteridinedione (**167**, R = H). Even preferential monodecarboxylation of pteridinedicarboxylic acids may or may not be achieved. Thus 2,4-dioxo-1,2,3,4-tetrahydro-6,7-pteridinedicarboxylic acid in refluxing quinoline underwent preferential 6-decarboxylation to afford only 2,4-dioxo-1,2,3,4-tetrahydro-7-pteridinecarboxylic acid (**168**), whereas 2-amino-4-oxo-3,4-dihydro-6,7-pteridinedicarboxylic acid (**169**) at 250 °C gave a mixture of 2-amino-4-oxo-3,4-dihydro-6-pteridinecarboxylic acid (**170**) and the isomeric 7-pteridinecarboxylic acid (**171**), both of which gave 2-amino-4(3*H*)-pteridinone (**172**) at 280 °C.

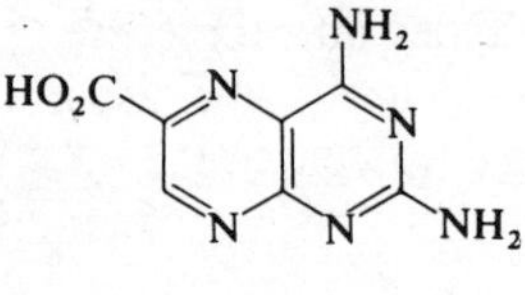

(**166**)

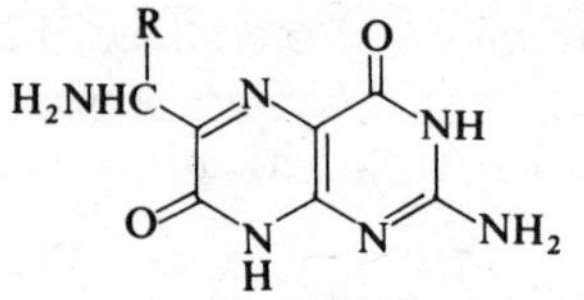

(**167**)

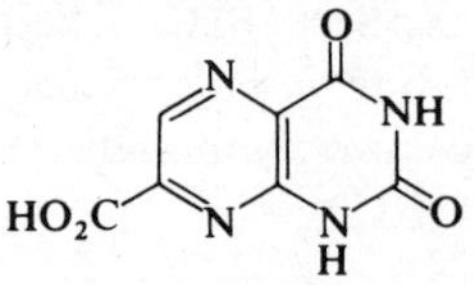

(**168**)

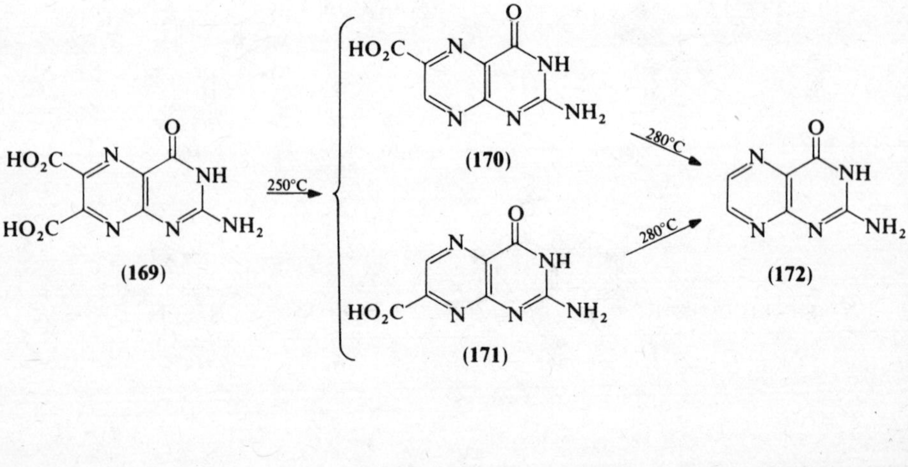

(**169**) (**170**) (**171**) (**172**)

It is possible to induce loss of carbon dioxide from some 6-pteridinecarboxylic acids by reduction, usually with aluminum amalgam in aqueous ammonia or alkali. A typical substrate for this reaction is 2-amino-4,7-dioxo-3,4,7,8-tetrahydro-6-pteridinecarboxylic acid.

J. Homolytic *C*-Acylation (Ch. X, Sect. 11.C)

Pteridines with electron-donating groups in the 2- and 4-position undergo Menisci homolytic *C*-acylation at the 7-position or, if that is occupied, at the 6-position. Thus simultaneous treatment of 2,4(1*H*,3*H*)-pteridinedione (**173**) and propionaldehyde in dilute acetic acid at <20 °C with aqueous ferrous sulfate and aqueous *t*-butyl hydroperoxide gave 7-propionyl-2,4(1*H*,3*H*)-pteridinedione (**174**); appropriate variations gave such products as 6-acetyl-1,3,7-trimethyl-2,4(1*H*,3*H*)-pteridinedione (**175**), 7-butyryl-2,4-pteridinediamine, and 2-methylthio-7-phenethylcarbonyl-4-pteridinamine (**176**). The reaction may have a far wider scope than is presently believed. Homolytic hydroxyalkylation by alcohols with persulfate has been successful (Ch. VI, Sect. 3.D).

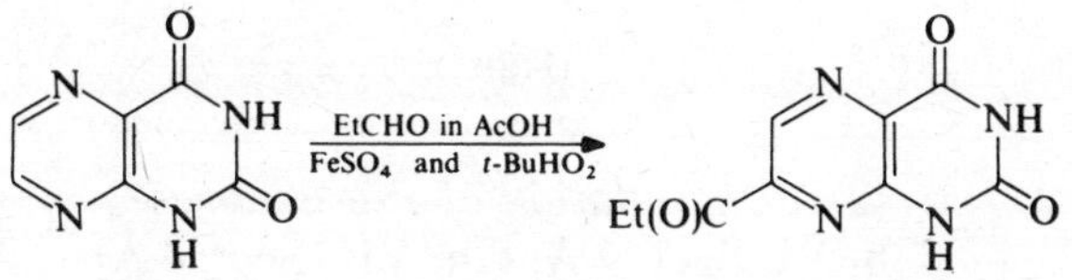

(**173**) (**174**)

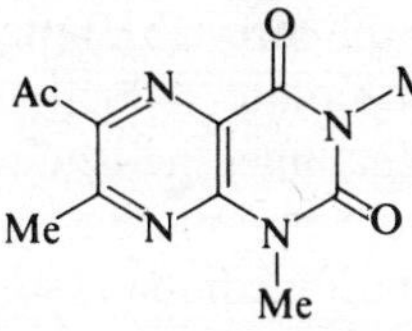

(**175**)

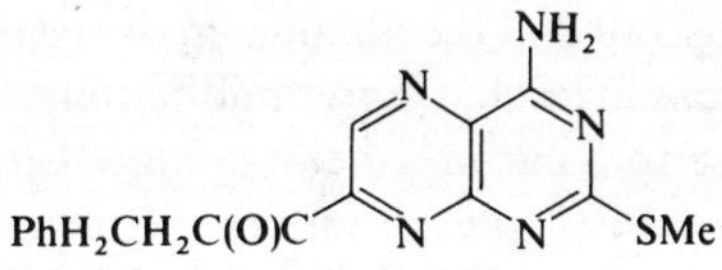

(**176**)

K. Degradation to Pyrazines or Pyrimidines

It is probably true to say that every pteridine should suffer ring cleavage under sufficiently drastic hydrolytic or aminolytic conditions to give (initially at least) a pyrazine or pyrimidine. Under such conditions, most other attached groups will be eliminated or become oxo or amino substituents, so the most important data on such cleavages are those involving pteridine itself, pteridinones, and (to a less extent) pteridinamines. As a rule the functionally substituted ring of a pteridine will undergo cleavage. Thus 2- or 4-substituted pteridines will give pyrazines, whereas 6- or 7-substituted pteridines will give pyrimidines; when there are no functional substituents or when both rings are so substituted, the degradation products are likely to be pyrazines. The pteridine *N*-oxides form an exception to this rule because the ring bearing the *N*-oxide function generally survives irrespective of other substituents. The less common reductive or oxidative degradations appear to follow broadly similar lines.

(1) *Degradation of Pteridine* (Ch. IV, Sect. 1.C)

Pteridine (**177**) undergoes slow fission of its pyrimidine ring in aqueous sodium carbonate to give 3-aminomethyleneamino-2-pyrazinecarbaldehyde (**178**), while in dilute acid it gives 3-amino-2-pyrazinecarbaldehyde (**179**); both were isolated as oximes. Destabilization of the pyrimidine ring by covalent 3,4-hydration under the reaction conditions may or may not contribute to the formation of such pyrazines.

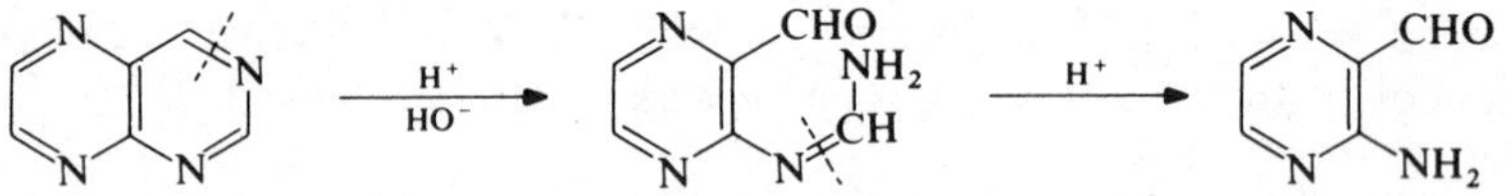

(177) **(178)** **(179)**

(2) *Degradation of Pteridinones* (Ch. VI, Sect. 2.A; Ch. VII, Sect. 4.A)

Fission of both tautomeric and nontautomeric 6- or 7-pteridinones to *pyrimidines* is illustrated in the alkaline hydrolysis of 7,8-dihydro-6(5*H*)-pteridinone to 4-carboxymethylamino-5-pyrimidinamine, the dithionite reduction of 7(8*H*)-pteridinone to 5-carboxymethylamino-4-pyrimidinamine, the acidic hydrolysis of 8-methyl-7(8*H*)-pteridinone to 4-methylamino-5-pyrimidinamine, and the alkaline hydrolysis of 5,8-dimethyl-6,7(5*H*,8*H*)-pteridinedione (**180**) to 4,5-bis-methylaminopyrimidine (**181**); exceptionally, 2-methyl-4(3*H*)-pteridinone (**182**) underwent hydrazinolysis to give 5,6-diamino-2-methyl-4(3*H*)-pyrimidinone.

The more numerous degradations to *pyrazines* are typified by the hydrolysis of 2,4(1*H*,3*H*)-pteridinedione (**184**) to 3-amino-2-pyrazinecarboxylic acid (**183**) (alkali at 170 °C) or to 2-pyrazinamine (**185**) (concentrated sulfuric acid at

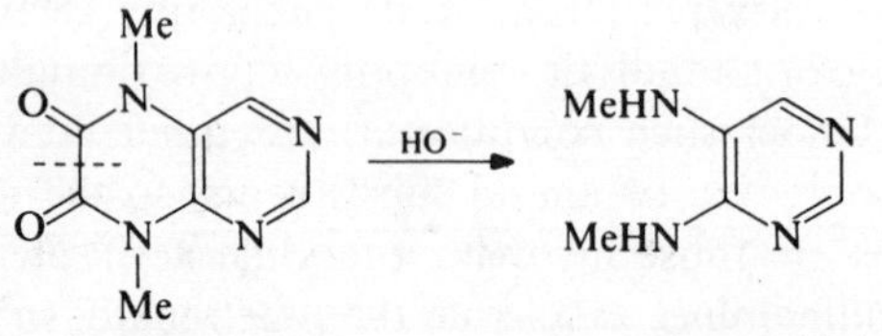

(180) **(181)**

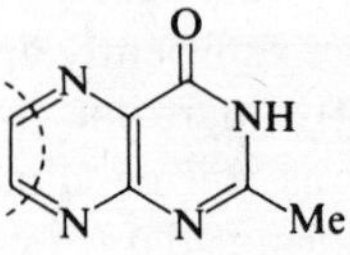

(182)

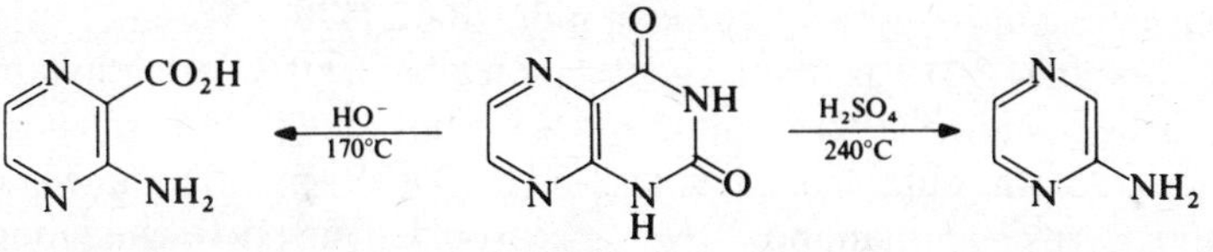

(183) (184) (185)

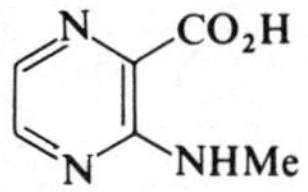

(186)

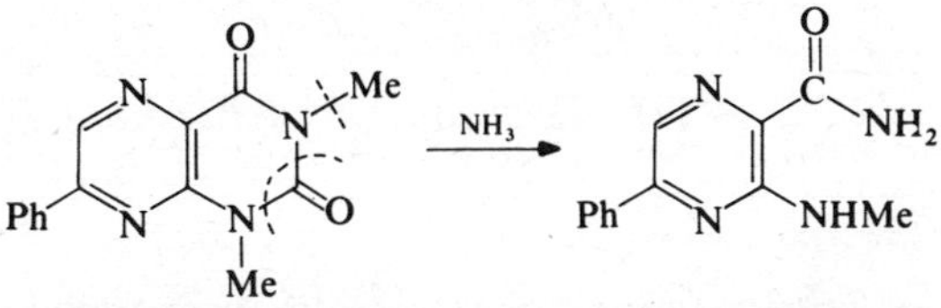

(187) (188)

240 °C), the alkaline hydrolysis of 4-amino-1-methyl-2(1*H*)-pteridinone to 3-methylamino-2-pyrazinecarboxylic acid (**186**), and the several-stage aminolysis of 1,3-dimethyl-7-phenyl-2,4(1*H*,3*H*)-pteridinedione (**187**) to 3-methylamino-5-phenyl-2-pyrazinecarboxamide (**188**).

(3) *Degradation of Pteridinamines and Pteridinimines* (Ch. IX, Sects. 2.A and 6.A)

Under vigorous hydrolytic conditions, many pteridinamines will become pteridinones prior to ring fission. However, under aminolytic conditions pteridinamines will be degraded in their own right. For example, 4-pteridinamine (**189**) reacts with ammonia at 170 °C to give the ammonium salt of 3-amino-2-pyrazinecarboxylic acid (**183**) while with refluxing hydrazine hydrate it gives 3-

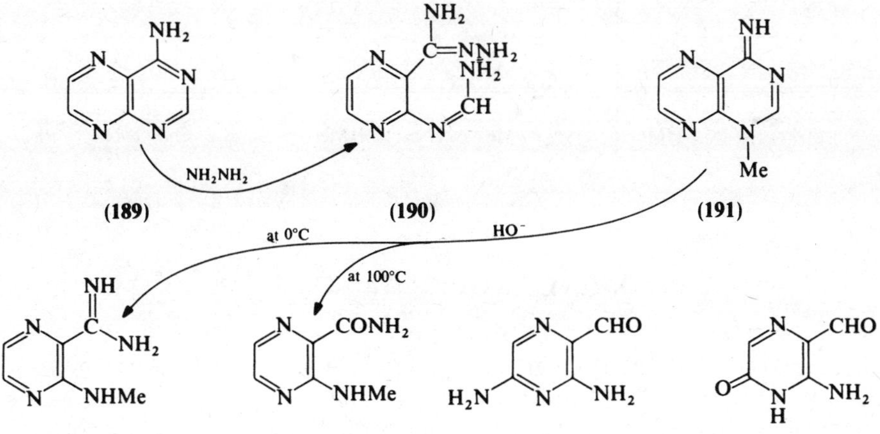

(189) (190) (191)

(192) (193) (194) (195)

aminomethyleneamino-2-pyrazinecarboxamidrazone (**190**). When 1-methyl-4(1*H*)-pteridinimine (**191**) was treated gently with alkali at 0 °C, 3-methylamino-2-pyrazinecarboxamidine (**192**) resulted but with hot alkali the corresponding carboxamide (**193**) was obtained; likewise, 3-methyl-7(3*H*)-pteridinimine in pH 4 buffer at 100 °C gave 3,5-diamino-2-pyrazinecarbaldehyde (**194**) while in alkali at 100 °C, 3-amino-5-oxo-4,5-dihydro-2-pyrazinecarbaldehyde (**195**) was obtained, probably by initial hydrolysis of the imino group.

(4) *Degradation of Pteridine N-Oxides* (Ch. VII, Sect. 6.C)

Degradation of pteridine *N*-oxides is exceptional in that the ring bearing the *N*-oxide is stabilized thereby and hence generally survives, irrespective of the locations of other substituents. Thus 4(1*H*)-pteridinone 3-oxide (**196**) in hot aqueous hydroxylamine gave, not a pyrazine, but 5,6-diamino-4(1*H*)-pyrimidinone 3-oxide (**197**); 1,6,7-trimethyl-2,4(1*H*,3*H*)-pteridinedione 5-oxide (**198**) in hot alkali gave 5,6-dimethyl-3-methylamino-2-pyrazinecarboxylic acid 1-oxide (**200**), via the amide (**199**); but, exceptionally, 2-phenyl-4-pteridinamine 3-oxide gave the oximelike product, 3-(α-hydroxyiminobenzyl)amino-2-pyrazinecarboxylic acid (**201**) on gentle alkaline hydrolysis.

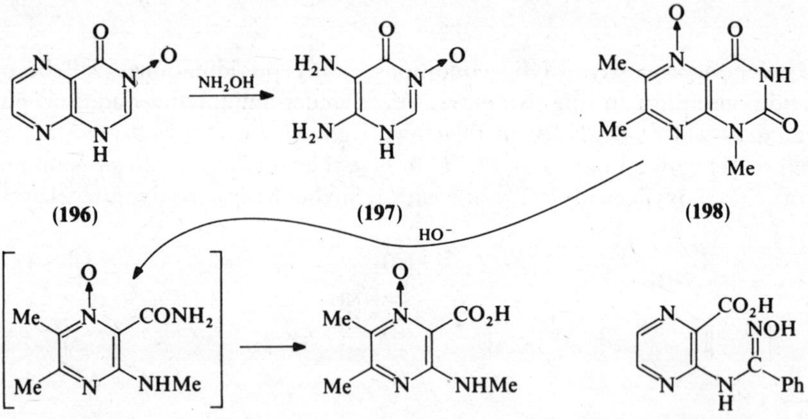

(**196**) (**197**) (**198**)

(**199**) (**200**) (**201**)

5. PHYSICAL PROPERTIES OF PTERIDINES

Several *electron-density* diagrams for pteridine were calculated about 30 years ago.[787, 915, 1170, cf. 26, 824] Although these diagrams differ considerably, the one (**202**) that appears to be most acceptable[26] indicated that the electron deficiencies at peripheral carbon atoms are in the order $4 \gtrsim 7 > 2 > 6$. Perhaps it

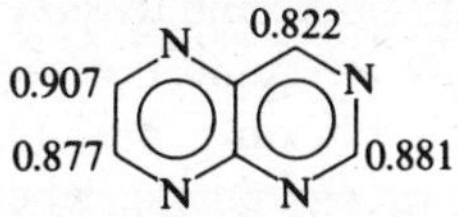

(202)

is not surprising that this does not correspond with the positional order of reactivity for aminolysis of simple chloropteridines (7 > 6 > 4 > 2; see Ch. V, Sect. 2.C) in view of the propensity of some chloropteridines to undergo covalent addition of amines prior to aminolysis; this calculated order is closer to, but not identical with, that for chemical shifts of signals for C-protons in pteridine (4 > 2 > 7 > 6; see Table XIII), which should also broadly reflect electron deficiencies at the carbon atoms.

Existing data on the *ionization constants* of pteridines is remarkably adequate, mainly on account of conscientious measurements in laboratories at Canberra and Konstanz for most of the new pteridines produced therein. Because of great variations in the position and degree of covalent hydration in different pteridines and their ionic species in aqueous solution, the interpretation of measured pK_a values for many pteridines (Table XI) is by no means simple. For example, although the intrinsic pK_a of pteridine is ~ -2.0, measurement in aqueous solution reveals a basic pK_a value at ~ 4.1 and an acidic value at ~ 11.2, probably representing the formation of a hydrated cation (**203**) and/or (**204**) and a hydrated anion (**205**), respectively (Ch. XII, Sect. 1). When the formation of a covalent hydrate is not instantaneous, it is often possible to measure the pK_a value(s) for anhydrous pteridines by employing rapid-reaction techniques. For example, the acidic pK_a for anhydrous 2(1*H*)-pteridinethione is ~ 6.5 but that for the hydrated species (**206**) is ~ 9.7.

Although not all *ultraviolet spectra* (uv) recorded for pteridines represent pure ionic species, the ready availability of pK_a values has stimulated the laudable practice of reporting a spectrum for each ionic species in a buffer of appropriate

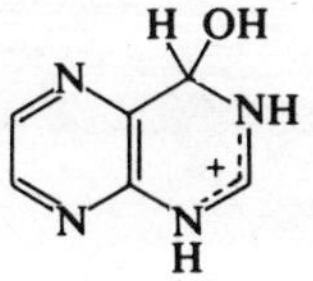

(203)

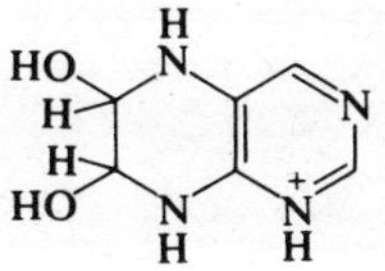

(204)

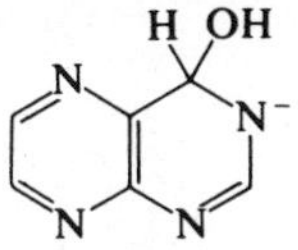

(205)

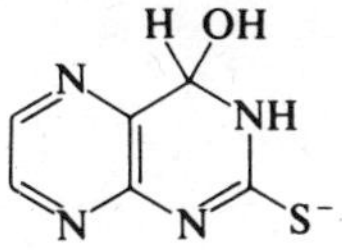

(206)

pH. Naturally, the spectra of pteridine and its simple alkyl derivatives in suitable aprotic solvents, for example, cyclohexane, reveal much more fine structure than in aqueous media. Thus in cyclohexane, pteridine shows at least 10 spectral features (maxima, inflexions, or shoulders), whereas in an aqueous phase it shows only 3 or 4 such features, according to the species present. The fine structure just mentioned includes an interesting band of low intensity at 387 nm, representing an unusual $n \rightarrow \pi$ transition that is evident only in highly nitrogenous heteroaromatics (Ch. XII, Sect. 2). The uv spectral data for a selection of simple pteridine derivatives (Table XII) indicates how important covalent hydration is to the spectra of susceptible compounds. As in other series, comparisons of the uv spectra of (tautomeric) pteridinones with those of appropriate methoxypteridines and *N*-methylpteridinones have been used to determine the predominant tautomer(s) present in aqueous solutions of the parent pteridinones. In every case, a pteridinone form predominates heavily over the hydroxypteridine form; similarly, pteridinethiones exist as such, as do pteridinamines.

Data for *proton nuclear magnetic resonance* (nmr) *spectra* of a range of simple pteridines are presented in Table XIII. Such spectra are particularly useful in detecting the position(s) of covalent addition or hydrogenation. Signals for protons at such sites undergo a very marked upfield shift while other protons undergo only a minor shift in the same direction. Moreover, at sufficiently high field strengths, ^{1}H nmr spectra can indicate the stereoconfiguration and even preferred conformation(s) in reduced pteridines, a facility of immense value in studying the biochemistry of tetrahydrobiopterin and other pteridines involved in enzymatic reactions (Ch. XII, Sect. 3).

Leading references to the *other spectra* and *polarography* of representative pteridines have been collected in Table XIV: Such properties are discussed briefly in Chapter XII, Sections 4–8 and where relevant to individual pteridines in other chapters.

CHAPTER II

Primary Syntheses from Pyrimidines

Pteridines may be made from pyrimidine intermediates by three major and several minor methods. Of these, the most important involves the reaction of 4,5-pyrimidinediamines with α-dicarbonyl reagents, a procedure first used by Siegmund Gabriel and James Colman* who condensed 6-methyl-4,5-pyrimidinediamine (**2**) with benzil (**1**) to afford 4-methyl-6,7-diphenylpteridine (**3**)[9]; unsymmetrical dicarbonyl reagents often give two isomeric pteridines. The second major method consists of the condensation of a 5-nitroso-4-pyrimidinamine with an α-carbonylmethylene compound, as first recorded by Geoffrey Timmis[1179] in the preparation of 7-methyl-6-phenyl-2,4-pteridinediamine (**6**) from 5-nitroso-2,4,6-pyrimidinetriamine (**5**) and phenylacetone (**4**); the reaction is rightly called the *Timmis Synthesis* and its products are normally unambiguous in structure. The third major method comprises several stages, typically the aminolysis of a 4-chloro-5-nitropyrimidine by an α-amino carbonyl compound to give a 4-(substituted amino)-5-nitropyrimidine, the reduction of the latter with subsequent spontaneous cyclization to a 6-substituted-7,8-dihydropteridine, and a final oxidation to the corresponding pteridine. Pioneering work on the formation of dihydropteridines in this way is associated with the names of George Ramage,[873] Michel Polonovski,[396] William Boon,[689] and their colleagues but the first complete synthetic sequence was that of William Boon and Thomas Leigh.[727] In 1951, they described the condensation of aminoacetone with 2-amino-6-chloro-5-phenylazo-4(3*H*)-pyrimidinone to afford 6-acetonylamino-2-amino-5-phenylazo-4(3*H*)-pyrimidinone (**7**) and its subsequent conversion, via (**8**), into 2-amino-6-methyl-7,8-dihydro-4(3*H*)-pteridinone (**9**) and then into 2-amino-6-methyl-4(3*H*)-pteridinone (**10**). The procedure normally leads to products of unambiguous structure and for pragmatic simplicity it is known as the *Boon Synthesis*.

Minor routes from pyrimidines to pteridines are grouped in Section 4 at the end of this chapter.

---

* Some reviewers have overlooked the prior claim of these authors by naming this reaction after Oskar Isay who first used it only 5 years later.[983] It seems logical and just to rename this reaction as the Gabriel and Colman Synthesis.

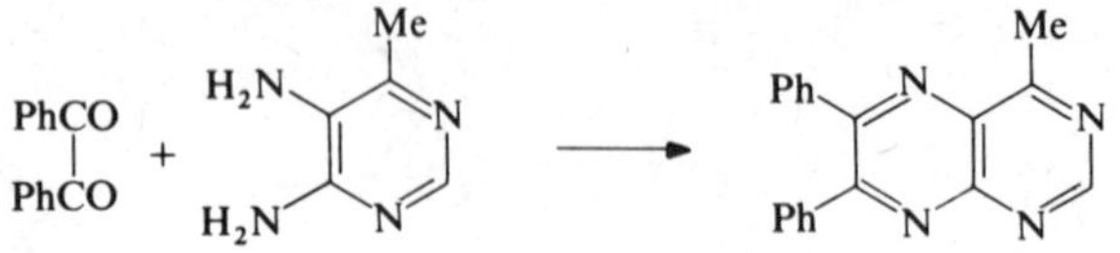

(1) (2) (3)

A Gabriel and Colman Synthesis

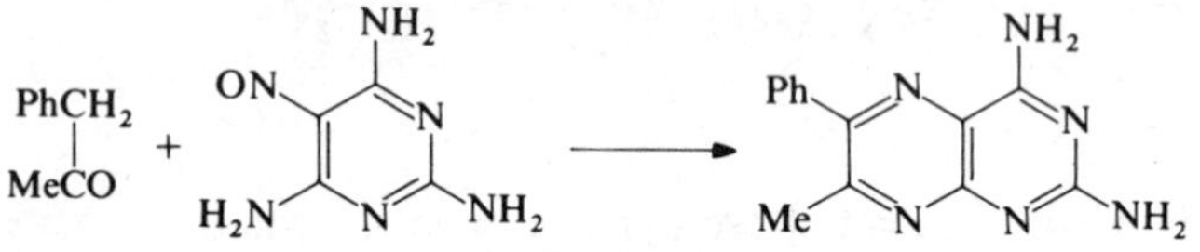

(4) (5) (6)

A Timmis Synthesis

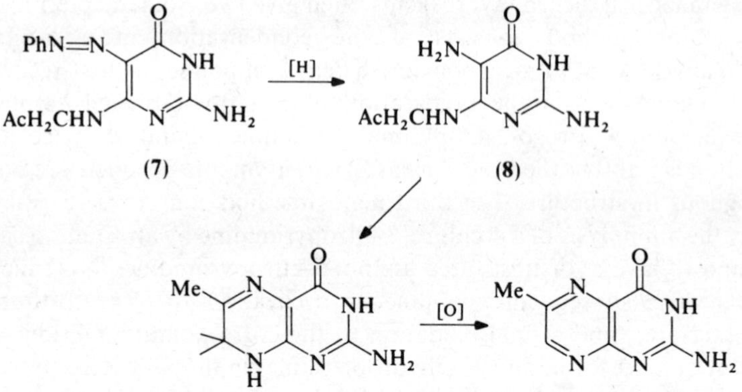

(9) (10)

A Boon Synthesis

1. THE GABRIEL AND COLMAN SYNTHESIS FROM 4,5-PYRIMIDINEDIAMINES AND α-DICARBONYL COMPOUNDS

It will be evident that groups attached to the pyrimidinediamine (**2**) will determine the substitution pattern at the 1- to 5-, and 8-position of the resulting pteridine, whereas the nature of the dicarbonyl compound (**1**) will determine the 6- and 7-substitution (or otherwise) in the pteridine. For practical purposes, treatment of this widely used synthesis is subdivided next into condensations involving the dialdehyde, glyoxal (for 6,7-diunsubstituted pteridines); aldehydoketones [for 6(7)-unsubstituted-7(6)-alkylpteridines]; aldehydoesters (or acids) [for 6(7)-unsubstituted-7(6)-pteridinones]; diketones (for 6,7-dialkylpteridines); ketoesters [for 6(7)-alkyl-7(6)-pteridinones]; and diesters (for 6,7-pteridinediones). In each category, reagents with structures equivalent to the appropriate α-dicarbonyl compounds are included: For example, ethyl dichloroacetate reacts as an aldehydoester.

A. Use of the α-Dialdehyde, Glyoxal

For synthetic purposes, glyoxal is readily available as its bis(sodium hydrogen sulfite) monohydrate, as a hydrated polymeric powder (~80%), and as a crude aqueous solution (30–40%) from which a quite usable solid polymer may deposit on prolonged refrigeration. All such forms have been used for pteridine syntheses but for any particular condensation, all forms are seldom equally successful. Thus the condensation of 4,5-pyrimidinediamine (**11**, R = H) with glyoxal bisulfite in aqueous solution gave pteridine (**12**, R = H) in < 20% yield,[30, 148, 250] whereas the use of alcoholic crude polymer and alcoholic *hydrated polymer* gave 63 and 93% yield, respectively[30, 65]; 2-deuteropteridine (**12**, R = D) was made similarly from 2-deutero-4,5-pyrimidinediamine (**11**, R = D).[294]

In much the same way, 2-methyl-4,5-pyrimidinediamine (**11**, R = Me) gave 2-methylpteridine (**12**, R = Me) (62%)[33, 119]; 2-phenyl-4,5-pyrimidinediamine (**11**, R = Ph) gave 2-phenylpteridine (**12**, R = Ph) (55%)[152]; 2-styryl-4,5-pyrimidinediamine (**11**, R = CH:CHPh) gave 2-styrylpteridine (**12**, R = CH:CHPh)[105]; 2,4,5-pyrimidinetriamine (**11**, R = NH_2) gave 2-pteridinamine (**12**, R = NH_2) (~50%)[30]; 2-methylamino- (**11**, R = NHMe) and 2-dimethylamino-4,5-pyrimidinediamine (**11**, R = NMe_2) gave 2-methylamino- (**12**, R = NHMe)[32] and 2-dimethylaminopteridine (**12**, R = NMe_2),[30] respectively; 2-chloro- (**11**, R = Cl), 2-methylthio- (**11**, R = SMe), and 2-methoxy-4,5-pyrimidinediamine (**11**, R = OMe) gave 2-chloro- (**12**, R = Cl),[30, 124, 1138] 2-methylthio- (**12**, R = SMe),[33] and 2-methoxypteridine (**12**, R = OMe),[32] respectively; 4,5-diamino-2(1*H*)-pyrimidinone (**13**, X = O) gave 2(1*H*)-pteridinone (**14**,

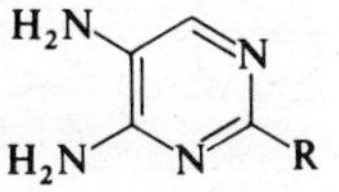

(11)

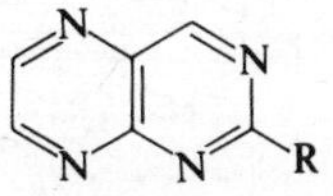

(12)

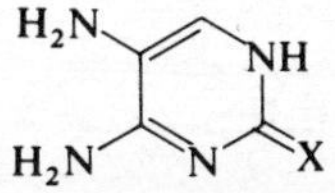

(13)

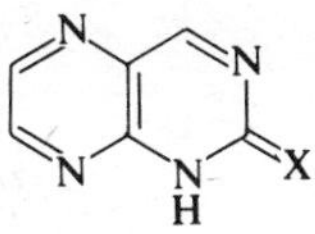

(14)

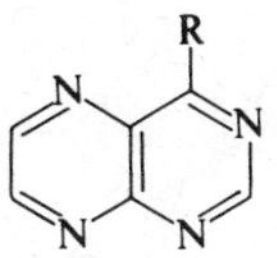

(15)

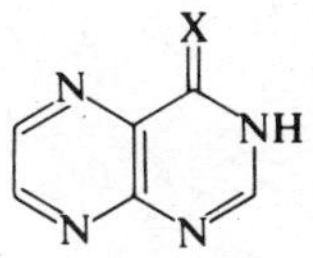

(16)

X = O) (80–93%)[14, 30]; and 4,5-diamino-2(1*H*)-pyrimidinethione (**13**, X = S) gave 2(1*H*)-pteridinethione (**14**, X = S) (52%).[878]

Likewise, appropriate 6-substituted-4,5-pyrimidinediamines with glyoxal afforded the following 4-monosubstituted-pteridines: 4-methylpteridine (**15**, R = Me) (43–88%, presumably according to the quality of glyoxal used),[33, 64, 281] 4-phenylpteridine (**15**, R = Ph) (45%),[151] 4-pteridinamine (**15**, R = NH_2) (best yield 80% at pH 10),[28, 30, 185] 4-dimethylaminopteridine (**15**, R = NMe_2) (70–82%),[32, 280] 4-methoxypteridine (**15**, R = OMe) (methanolic polyglyoxal: 75% yield),[32] 4-chloropteridine (**15**, R = Cl) (buffered aqueous glyoxal under gentle conditions: 20%; abnormal product in boiling methanolic polyglyoxal),[32] 4-methylthiopteridine (**15**, R = SMe) (75%),[33] 4-trifluoromethylpteridine (**15**, R = CF_3) (polyglyoxal in bis-*β*-methoxyethyl ether at 155°C),[156] 4-pyrimidinecarboxamide (**15**, R = $CONH_2$) (similar conditions: ~ 25%),[157] ethyl 4-pteridinecarboxylate (**15**, R = CO_2Et) (similar conditions: 60%; preparation in aqueous glyoxal gave a stable covalent dihydrate, from which the free ester could be recovered only be prolonged boiling in *t*-butyl alcohol),[147–149] 4(3*H*)-pteridinone (**16**, X = O) (aqueous glyoxal or its bisulfite: 80 or 68%),[28, 30, 34, 1508] and 4(3*H*)-pteridinethione (**16**, X = S) (aqueous glyoxal: 83%).[33]

The preparation of simple ring-*N*-alkylated pteridinones or pteridinimines is similarly possible, as exemplified in the condensation of 4,5-diamino-3-methyl-2(3*H*)-pyrimidinone (**17**) with aqueous glyoxal at pH 5 to give 1-methyl-2(1*H*)-pteridinone (**18**),[120] in the analogous preparation of the isomeric 3-methyl-2(3*H*)-pteridinone (**19**) (> 80%),[34] in the conversion of 5-amino-4-benzylamino-2(1*H*)-pyrimidinone (**20**) by alcoholic polyglyoxal into 8-benzyl-2(8*H*)-pteridinone (**21**),[189] and in the similar conversions: 5,6-diamino-l-methyl-4(1*H*)-pyrimidinone into 1-methyl-4(1*H*)-pteridinone,[131] 5,6-diamino-3-methyl-4(3*H*)-pyrimidinone into 3-methyl-4(3*H*)-pteridinone (**22**, X = O),[126] 5-amino-6-methylamino-4(3*H*)-pyrimidinone into 8-methyl-4(8*H*)-pteridinone (**23**),[248] 5,6-diamino-3-methyl-4(3*H*)-pyrimidinimine hydriodide into 3-methyl-4(3*H*)-pteridinimine (**22**, X = NH) (as hydriodide; base underwent Dimroth

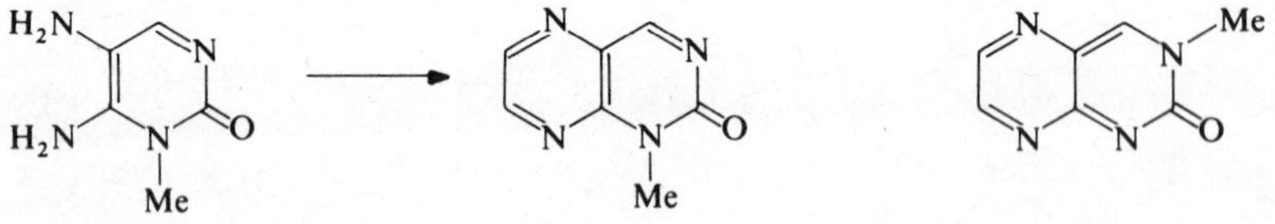

(17) **(18)** **(19)**

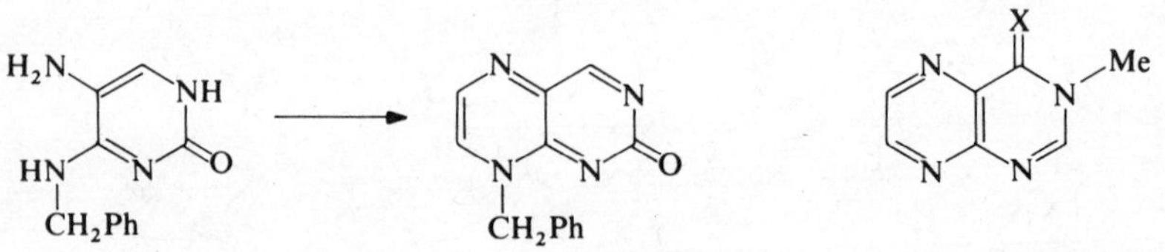

(20) **(21)** **(22)**

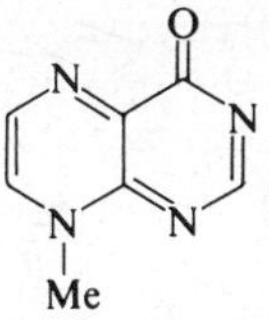

(23)

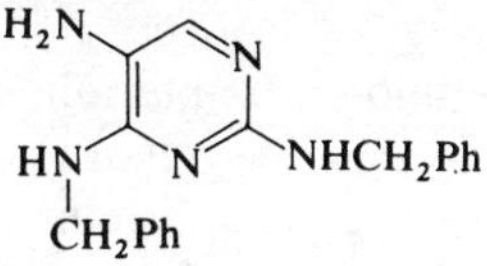

(24)

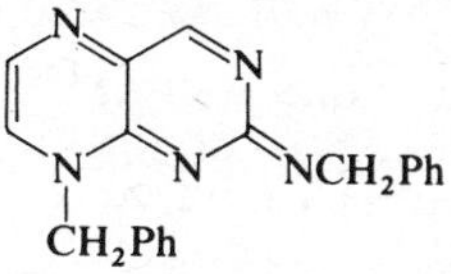

(25)

rearrangement to 4-methylaminopteridine),[132] and 2,4-bisbenzylamino-5-pyrimidinamine (**24**) into 8-benzyl-2-benzylimino-2,8-dihydropteridine (**25**).[189]

The formation of 2,4-disubstituted-pteridines using glyoxal is well represented. Being less soluble on the whole, the products are more easily isolated than their monosubstituted analogs just cited and, therefore, yields tend to be higher. Thus, 2-methyl-6-phenyl-4,5-pyrimidinediamine and aqueous polyglyoxal gave 2-methyl-4-phenylpteridine (**26**, R = Me) in 82% yield[151] and appropriate pyrimidinediamines gave 4-methyl-2-phenylpteridine (85%),[152] 2-methylthio-4-phenylpteridine (**26**, R = SMe) (alcoholic polyglyoxal: 68–81%),[1118, 1436] 4-methyl-2-methylthiopteridine (56%),[46] 2-chloro-4-methylpteridine (80%),[281, 959] 4-methyl-2-pteridinamine (**27**, R = Me) (30–77%),[39, 185, 281] 2-diethylamino-4-methylpteridine,[281] 2-methyl-4-pteridinamine (~ 70%),[185] 4-methyl-2(1*H*)-pteridinone (**28**, X = O) (60%),[38] 2-methyl-4(3*H*)-pteridinone (50%),[38] 2-phenyl-4(3*H*)-pteridinone (**29**, X = O) (85%),[101, 663] 4-methyl-2(1*H*)-pteridinethione (**28**, X = S) (57%),[46] and 2-phenyl-4(3*H*)-pteridinethione (**29**, X = S) (54%).[1436]

Many products with two functional groups were made similarly: 2,4-dichloropteridine (acetonic glyoxal: 8%),[486] 2,4-dimethoxypteridine (monomeric or polyglyoxal in ethanol: ~ 15% yield),[338, 486] 4-methoxy-2-pteridinamine (**27**, R = OMe) (methanolic polyglyoxal),[130, 133, 368] 2-methoxy-4-pteridinamine (~ 15%),[185] 4-isopropoxy-2-pteridinamine (**27**, R = O *i*-Pr) (~ 30%),[370] 4-methylthio-2-pteridinamine (**27**, R = SMe) (30–84%),[305, 1468] the isomeric 2-methylthio-4-pteridinamine (~ 30%),[1467] 4-benzylthio-2-pteridinamine (**27**,

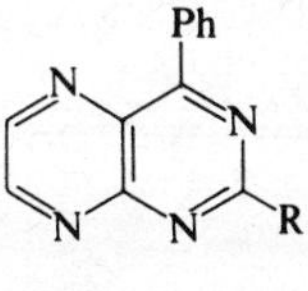

(26)

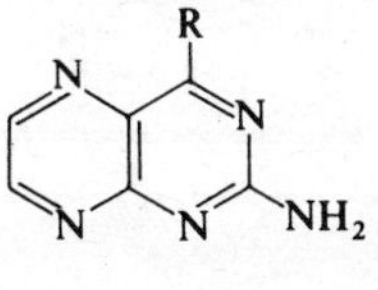

(27)

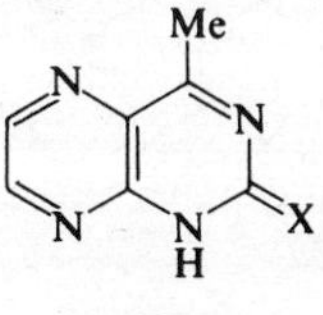

(28)

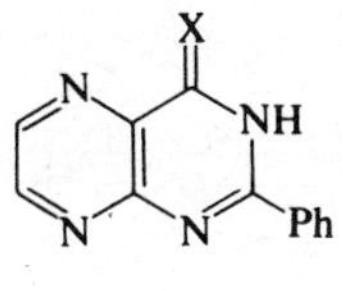

(29)

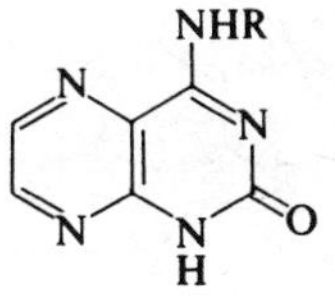

(30)

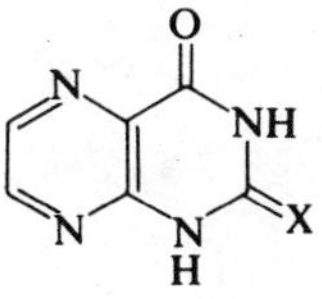

(31)

R = SCH_2Ph) (75%),[305, 1101] 2,4-pteridinediamine (**27**, R = NH_2) (glyoxal bisulfite: 70–97%[267, 284, 475]; polyglyoxal: 46–66%[267]),[730] 2-methylthio-4(3*H*)-pteridinone (80%),[437] 2-ethylthio-4(3*H*)-pteridinone (~65%),[399, 1234] 2-amino-4(3*H*)-pteridinone (pterin; aqueous glyoxal: 60–85%[509, 883, 1084]; [4a-^{14}C][1032]; [8a-^{14}C][911]; [6,7-D] from glyoxal bisulfite: 93%[646]), 2-methylamino-4(3*H*)-pteridinone,[130, 133, 368] 4-amino-2(1*H*)-pteridinone (**30**, R = H) (glyoxal bisulfite: 80%),[362, 471] 4-methylamino-2(1*H*)-pteridinone (**30**, R = Me),[362] 2-amino-4(3*H*)-pteridinethione,[305] 4-amino-2(1*H*)-pteridinethione,[916] 2,4(1*H*,3*H*)-pteridinedione (lumazine: **31**, X = O) (ethanolic polyglyoxal: 76%[367]; aqueous polyglyoxal: 78%[1012]; glyoxal bisulfite in acid: 87%[30, 747]; glyoxal bisulfite in base: 87%[1438]: purification[30]), 2-thioxo-1,2-dihydro-4(3*H*)-pteridinone (**31**, X = S),[399, 437] and 2,4(1*H*,3*H*)-pteridinedithione (38%).[437] In addition, some 2,4-disubstituted pteridines with extra *N*-substitution were made from appropriate pyrimidines with glyoxal: 2-methoxy-3-methyl-4(3*H*)-pteridinone (**32**, R = OMe) (~40%),[338] 1-methyl-2-methylthio-4(1*H*)-pteridinone (**33**, R = SMe) (33%),[437] the 3-methyl isomer (50%),[437] 2-amino-3-methyl-4(3*H*)-pteridinone (**32**, R = NH_2) (methanolic polyglyoxal: 70%[130, 133]; glyoxal bisulfite: < 20%[368, 511]), 2-amino-1-methyl-4(1*H*)-pteridinone (**33**, R = NH_2),[133, 368] 2-amino-8-methyl-4(8*H*)-pteridinone (**34**, R = Me) (as hydrochloride: 50–71%)[130, 133, 359, 422, cf. 907], 4-amino-1-methyl-2(1*H*)-pteridinone,[362] 2-amino-8-phenyl-4(8*H*)-pteridinone (**34**, R = Ph) (48%),[377] 1- and 3-methyl-2,4(1*H*,3*H*)-pteridinedione (40 and 50%, respectively),[338] 8-methyl-2,4(3*H*,8*H*)-pteridinedione (~60%),[1545] 8-β-hydroxyethyl-2,4(3*H*,8*H*)-pteridinedione (~30%) (also other analogs),[422, 1545] 1,3-dimethyl-2,4(1*H*,3*H*)-pteridinedione [aqueous glyoxal: 74%[1096]; glyoxal bisulfite: 30%[1435]; an interesting variant involved Schiff base formation between 5,6-diamino-1,3-dimethyl-2,4(1*H*,3*H*)-pyrimidinedione and glycolic aldehyde, followed by oxidative cyclization by mercuric chloride in dimethyl sulfoxide: 62% yield[1121]], 1-methyl-2-thioxo-1,2-dihydro-4(3*H*)-pteridinone,[437] the 3-methyl isomer,[446] the 1,3-dimethyl homolog,[437] 8-methyl-2-thioxo-2,3-dihydro-4(8*H*)-pteridinone (**35**),[634] 2(3*H*)-

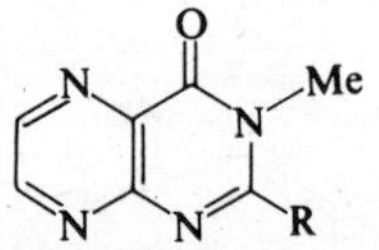

(32)

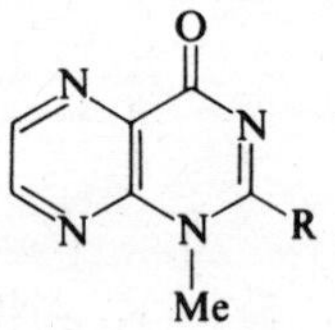

(33)

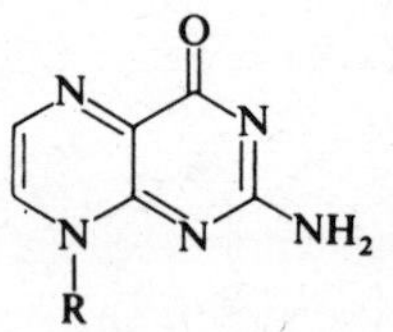

(34)

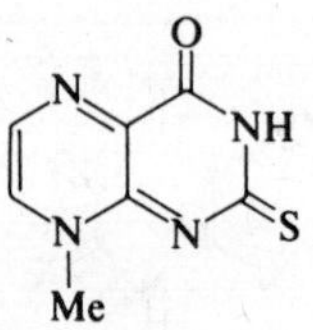

(35)

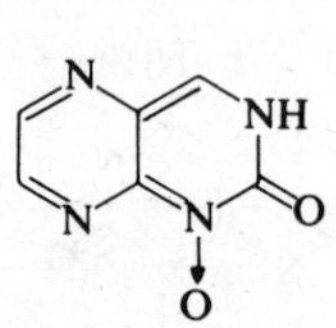

(36)

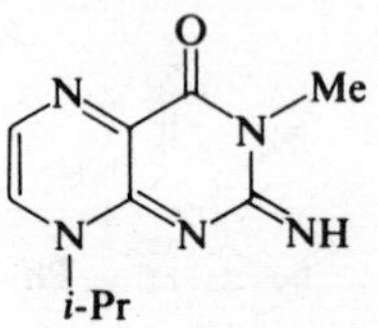

(37)

pteridinone 1-oxide (**36**) (aqueous glyoxal: 70%),[1098] 2,4(3*H*,8*H*)-pteridinedione 1-oxide (50%),[1098] 1-benzyloxy-2,4(1*H*,3*H*)-pteridinedione (18%),[1098] and 2-imino-8-isopropyl-3-methyl-2,3-dihydro-4(8*H*)-pteridinone (**37**) (as hydrochloride: 37%).[377]

Further examples of the use of glyoxal in the Gabriel and Colman synthesis are given in Table II. In addition, polymethylene-bridged pteridines have been made using glyoxal, for example, 8-amino-2,3-dihydro-5,7(1*H*,6*H*)-imidazo[1,2-*c*]pyrimidinedione (**38**) with aqueous glyoxal at 25°C followed by treatment with methanolic hydrogen chloride gave 2,4-dioxo-1,2,3,4-tetrahydro-1,8-ethanopteridin-8-ium chloride (**39**) in 53% yield.[1339]

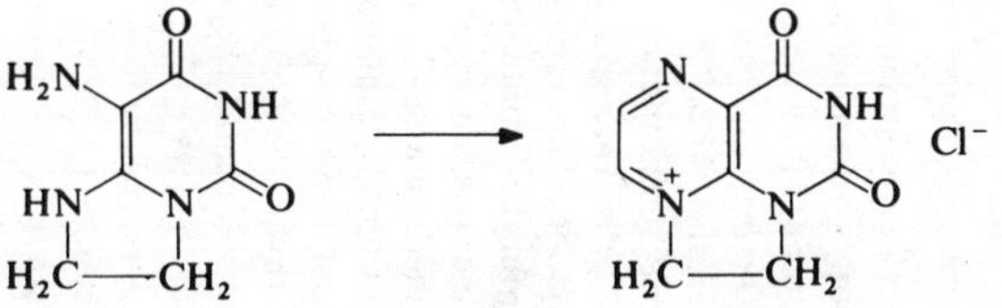

(**38**) (**39**)

B. Use of α-Aldehydoketones

Because α-aldehydoketones are necessarily unsymmetrical, their condensations with 4,5-pyrimidinediamines will lead to pteridines of ambiguous structure bearing a substituent in either the 6- or the 7-position. Fortunately, the reactivities of the carbonyl groups differ as do also those of the two amino groups. Accordingly, either the 6- or the 7-substituted pteridine usually predominates in the mixture of isomeric products and this tendency may be reinforced or reversed by the appropriate use of protective groups or variation in reaction conditions, such as in pH or the type of solvent. Products invariably need homogeneity tests and subsequent structural confirmation.

(1) *Simple Cases*

As might be expected in simple cases, the more reactive aldehydo-carbonyl group tends to react first with the more *aromatic* 5-amino group of the pyrimidine intermediate to give, after cyclization, the 6-unsubstituted-7-substituted pteridine. Thus methylglyoxal (**40**) reacted with 4,5-pyrimidinediamine (**41**) to give a 60% yield of 7-methylpteridine (**42**), even in the presence of sodium bisulfite, which might be expected to complex preferentially with the aldehydo group and thereby reverse the orientation of the product (as it does in some less simple cases). No trace of the 6-methyl isomer could be detected and the structure was proven both by oxidative hydroxylation to 7-methyl-4(3*H*)-pteridinone (**43**) and by subsequent degradation to 3-amino-5-methyl-2-

TABLE II. ADDITIONAL EXAMPLES OF THE GABRIEL AND COLMAN SYNTHESIS WITH THE DIALDEHYDE, GLYOXAL

Pyrimidine Intermediate	Form of Glyoxal (G) and Reaction Conditions	Pteridine (Yield)	References
6-Ethyl-2,4,5-pyrimidinetriamine	Aqueous G-bisulfite; reflux; 80 min	4-Ethyl-2-pteridinamine (4%)	185
6-Propyl-2,4,5-pyrimidinetriamine	Aqueous G-bisulfite; reflux; 80 min	4-Propyl-2-pteridinamine	185
6-Isopropyl-2,4,5-pyrimidinetriamine	Aqueous G-bisulfite; reflux; 80 min	4-Isopropyl-2-pteridinamine	185
6-*β*-Dimethylaminoethylamino-2-phenyl-4,5-pyrimidinediamine	Aqueous alcoholic G; reflux; 1 h	4-*β*-Dimethylaminoethylamino-2-phenylpteridine (61%)	1436
6-Benzylamino-4,5-pyrimidinediamine	Aqueous G; 100°C; 15 min	4-Benzylaminopteridine	280
6-Phenethylamino-4,5-pyrimidinediamine	Aqueous G; 100°C; 15 min	4-Phenethylaminopteridine (65%)	280
6-Cyclohexylamino-4,5-pyrimidinediamine	Aqueous ethanolic C; 100°C; 30 min	4-Cyclohexylaminopteridine (93%)	280
6-Furfurylamino-4,5-pyrimidinediamine	Aqueous ethanolic G; 100°C; 30 min	4-Furfurylaminopteridine (68%)	280
2-Dimethylamino-4,5,6-pyrimidinetriamine	Aqueous poly-G; 70°C; 5 min	2-Dimethylamino-4-pteridinamine (54%)	267
6-Dimethylamine-2,4,5-pyrimidinetriamine	Aqueous poly-G; 70°C; 5 min	4-Dimethylamino-2-pteridinamine (54%)	267
2,6-Bisdimethylamino-4,5-pyrimidinediamine	Aqueous poly-G; 75°C; 5 min	2,4-Bisdimethylaminopteridine (57%)	267
2-Dibutylamino-4,5,6-pyrimidinetriamine	Aqueous G; reflux; 105 min	2-Dibutylamino-4-pteridinamine (~20%)	185
2-Methylamino-4,5,6-pyrimidinetriamine	Aqueous poly-G; reflux; 30 min	2-Methylamino-4-pteridinamine[a] (52%)	1678
6-*p*-Dimethylaminostyryl-2,4,5-pyrimidinetriamine	Ethanolic poly-G; reflux; 20 min	4-*p*-Dimethylaminostyryl-2-pteridinamine[a] (68%)	282
6-*p*-Dimethylaminophenethyl-2,4,5-pyrimidinetriamine	Ethanolic poly-G; reflux	4-*p*-Dimethylaminophenethyl-2-pteridinamine (~20%)	282
2-Diethylamino-6-*p*-dimethylaminophenethyl-4,5-pyrimidinediamine	Ethanolic poly-G; reflux; 30 min	2-Diethylamino-6-*p*-dimethylaminophenethylpteridine	282
2-Chloro-6-trifluoromethyl-4,5-pyrimidinediamine	Poly-G in *t*-BuOH;[b] reflux; 8 h	2-Chloro-4-trifluoromethylpteridine (28%)	159
2-Methoxy-6-trifluoromethyl-4,5-pyrimidinediamine	Poly-G in *t*-BuOH; reflux; 1 h	2-Methoxy-4-trifluoromethylpteridine (51%)	159
2-Ethoxy-6-trifluoromethyl-4,5-pyrimidinediamine	Poly-G in *t*-BuOH; reflux; 75 min	2-Ethoxy-4-trifluoromethylpteridine (60%)	159

2-Methylthio-6-trifluoromethyl-4,5-pyrimidinediamine	Poly-G in *t*-PeOH; reflux; 90 min	2-Methylthio-4-trifluoromethylpteridine (60%)	159
6-Trifluoromethyl-2,4,5-pyrimidinetriamine	Poly-G in *t*-PeOH; reflux; 30 min	4-Trifluoromethyl-2-pteridinamine (45%)	159
2-Dimethylamino-6-trifluoromethyl-4,5-pyrimidinediamine	Ethanolic poly-G; reflux; 30 min	2-Dimethylamino-4-trifluoromethylpteridine (67%)	159
Ethyl 5,6-diamino-2-chloro-4-pyrimidinecarboxylate	Poly-G in toluene[c]; reflux; 16 h	Ethyl 2-chloro-4-pteridinecarboxylate (17%)	758
6-Benzyloxy-2,4,5-pyrimidinetriamine	Methanolic poly-G; reflux; 2 h	4-Benzyloxy-2-pteridinamine (~30%)	370
2-Dimethylamino-6-methoxy-4,5-pyrimidinediamine	Aqueous ethanolic poly-G; reflux; 1 h	2-Dimethylamino-4-methoxypteridine (67%)	416
Ethyl 5,6-diamino-2-ethoxy-4-pyrimidinecarboxylate	Poly-G in *t*-BuOH[c]; reflux; 75 min	Ethyl 2-ethoxy-4-pteridinecarboxylate (28%)	758
2-Methylthio-4,5,6-pyrimidinetriamine	Poly-G in aqueous acetic acid; reflux; 30 min	2-Methylthio-4-pteridinamine[a] (89%)	1678
Ethyl 5,6-diamino-2-methylthio-4-pyrimidinecarboxylate	Poly-G in diglyme[c]; reflux; 15 min	Ethyl 2-methylthio-4-pteridinecarboxylate (38%)	758
6-Carboxymethoxy-2,4,5-pyrimidinetriamine	Aqueous G at pH 6; 100°C; 45 min	4-Carboxymethoxy-2-pteridinamine (*c*. 50%)	883
Ethyl 2,5,6-triamino-4-pyrimidinecarboxylate	Poly-G in toluene; reflux; 48 h (or in water; 20°C; 15 min)	Ethyl 2-amino-4-pteridinecarboxylate (35%)	758
Ethyl 5,6-diamino-2-dimethylamino-4-pyrimidinecarboxylate	Ethanolic poly-G; reflux; 30 min	Ethyl 2-dimethylamino-4-pteridinecarboxylate (49%)	758
5,6-Diamino-2-*p*-tolyl-4(3*H*)-pyrimidinone	Aqueous G; reflux; 60 min	2-*p*-Tolyl-4(3*H*)-pteridinone	664
5,6-Diamino-2-(5′-methylfur-2′-yl)-4(3*H*)-pyrimidinone	G in aqueous AcOH; reflux; 10 min	2-(5′-Methylfur-2′-yl)-4(3*H*)-pteridinone	664
4,5-Diamino-6-benzylamino-2(1*H*)-pyrimidinone	Buffered aqueous G; 100°C; 2 h	4-Benzylamino-2(1*H*)-pteridinone (~30%)	362
4,5-Diamino-6-anilino-2(1*H*)-pyrimidinone	Buffered aqueous G; 100°C; 2 h	4-Anilino-2(1*H*)-pteridinone (~40%)	362
5,6-Diamino-2-dibenzylamino-4(3*H*)-pyrimidinone	Aqueous G; 90°C; 1 h	2-Dibenzylamino-4(3*H*)-pteridinone (52%)	182
5,6-Diamino-2-dimethylamino-4(3*H*)-pyrimidinone	Aqueous G plus H_2SO_3; 35°C; 2 h	2-Dimethylamino-4(3*H*)-pteridinone (68%)	511, cf. 368
5,6-Diamino-2-carboxymethylamino-4(3*H*)-pyrimidinone	Aqueous G-bisulfite; reflux; 1 h	2-Carboxymethylamino-4(3*H*)-pteridinone (~50%)	461
5,6-Diamino-2′-dimethylamino-2,4′-bipyrimidin-4(3*H*)-one	Aqueous ethanolic G; reflux; 30 min	2-(2′-Dimethylaminopyrimidin-4′-yl)-4(3*H*)-pteridinone (76%)	122

TABLE II. (*Contd.*)

Pyrimidine Intermediate	Form of Glyoxal (G) and Reaction Conditions	Pteridine (Yield)	References
5,6-Diamino-2′-methoxy-2,4′-bipyrimidin-4(3*H*)-one	Aqueous G; 100°C; 15 min	2-(2′-Methoxypyrimidin-4′-yl)-4(3*H*)-pteridinone	122
5,6-Diamino-2-fur-2′-yl-4(3*H*)-pyrimidinone	Aqueous G; reflux; 5 min	2-Fur-2′-yl-4(3*H*)-pteridinone	663, cf. 667
2-Acetylsulfanilamido-5,6-diamino-4(3*H*)-pyrimidinone	G in aqueous NH_3; 70°C; 30 min	2-Acetylsulfanilamido-4(3*H*)-pteridinone (74%)	186
2,5-Diamino-6-isopropylamino-4(3*H*)-pyrimidinone	Methanolic poly-G; reflux; 90 min	2-Amino-8-isopropyl-4(8*H*)-pteridinone (46%)	377
2,5-Diamino-6-*β*-hydroxyethylamino-4(3*H*)-pyrimidinone	Methanolic poly-G; reflux; 30 min	2-Amino-8-*β*-hydroxyethyl-4(8*H*)-pteridinone (40%)	422
2,5-Diamino-6-benzylamino-4(3*H*)-pyrimidinone	Methanolic poly-G; reflux; 2 h	2-Amino-8-benzyl-4(8*H*)-pteridinone (42%)	377
2,5-Diamino-6-*p*-toluidino-4(3*H*)-pyrimidinone	Methanolic poly-G; reflux; 2 h	2-Amino-8-*p*-tolyl-4(8*H*)-pteridinone (34%)	377
2,5-Diamino-6-*p*-chlorophenyl-4(3*H*)-pyrimidinone	Methanolic poly-G; reflux; 2 h	2-Amino-8-*p*-chlorophenyl-4(8*H*)-pteridinone (42%)	377
5-Amino-2-dimethylamino-6-methylamino-4(3*H*)-pyrimidinone	Aqueous ethanolic poly-G; reflux; 30 min	2-Dimethylamino-8-methyl-4(8*H*)-pteridinone (~40%)	359
5-Amino-2-dimethylamino-6-*β*-hydroxyethylamino-4(3*H*)-pyrimidinone	Aqueous ethanolic poly-G (?); 1 h	2-Dimethylamino-8-*β*-hydroxyethyl-4(8*H*)-pteridinone (~40%)	359
Ethyl 5,6-diamino-2-oxo-1,2-dihydro-4-pyrimidinecarboxylate	Aqueous G; reflux; 10 min	Ethyl 2-oxo-1,2-dihydro-4-pteridinecarboxylate[d]	758
Ethyl 5,6-diamino-2-thioxo-1,2-dihydro-4-pyrimidinecarboxylate	Aqueous G; reflux; 3 h	Ethyl 2-thioxo-1,2-dihydro-4-pteridinecarboxylate[d]	758
4,5-Diamino-6-trifluoromethyl-2(1*H*)-pyrimidinethione	Aqueous G; reflux; 15 min	4-Trifluoromethyl-2(1*H*)-pteridinethione[d] (~50%)	159
5,6-Diamino-1-phenyl-2,4(1*H*,3*H*)-pyrimidinedione	Acidic aqueous ethanolic G-hydrate; reflux	1-Phenyl-2,4(1*H*,3*H*)-pteridinedione (81%)	108

5,6-Diamino-1-*p*-chlorophenyl-2,4(1*H*,3*H*)-pyrimidinedione	Poly-G (?)	1-*p*-Chlorophenyl-2,4(1*H*,3*H*)-pteridinedione[a]	797
5,6-Diamino-1-β-hydroxyethyl-2,4(1*H*,3*H*)-pyrimidinedione	G in aqueous AcOH; reflux; 30 min	1-β-Hydroxyethyl-2,4(1*H*,3*H*)-pteridinedione (92%)	1339
5-Amino-6-β-hydroxyethylamino-2,4(1*H*,3*H*)-pyrimidinedione	Methanolic poly-G; reflux; 30 min	8-β-Hydroxyethyl-2,4(3*H*,8*H*)-pteridinedione[a] (70%)	422, 1199, cf. 1545
5-Amino-6-β-hydroxyethylamino-2-thioxo-1,2-dihydro-4(3*H*)-pyrimidinone	Poly-G (?)	8-β-Hydroxyethyl-2-thioxo-2,8-dihydro-4(3*H*)-pteridinone	634
5-Amino-6-*s*-butylamino-3-methyl-2,4(1*H*,3*H*)-pyrimidinedione	Poly-G (?)	8-*s*-Butyl-3-methyl-2,4(3*H*,8*H*)-pteridinedione[a]	1199
2,5-Diamino-6-benzylamino-3-methyl-4(3*H*)-pyrimidinone	Aqueous ethanolic poly-G; reflux; 15 min	8-Benzyl-2-imino-3-methyl-2,3-dihydro-4(8*H*)-pteridinone (31%)	377
2,5-Diamino-3-methyl-6-methylamino-4(3*H*)-pyrimidinone	Ethanolic poly-G; reflux; 30 min	2-Imino-3,8-dimethyl-2,8-dihydro-4(3*H*)-pteridinone (~60%)	359

[a] And analogs.
[b] Aqueous glyoxal gave the covalent 3,4-hydrate.
[c] Aqueous glyoxal gave a covalent dihydrate.
[d] As a covalent 3,4-hydrate.

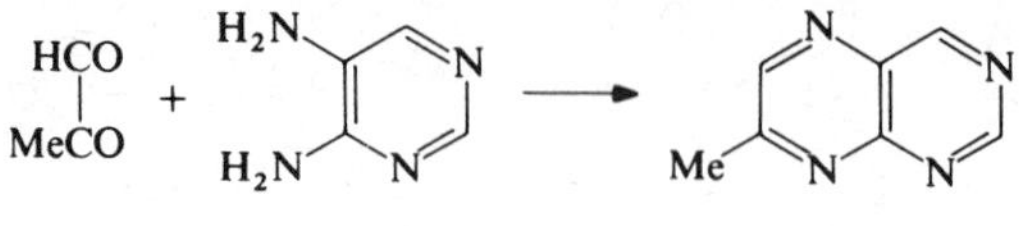

(40) (41) (42)

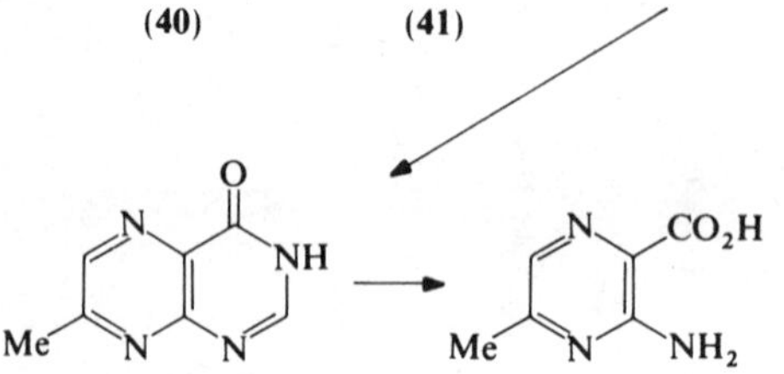

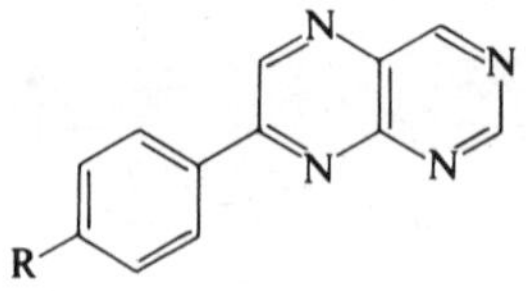

(43) (44) (45)

pyrazinecarboxylic acid (**44**) of known configuration.[33] 2-Deutero-7-methylpteridine was made similarly[294] from 2-deutero-4,5-pyrimidinediamine. Other 7-alkyl- or 7-aryl-pteridines (of nmr-confirmed structures), which have been made from relatively simple pyrimidine intermediates, include 7-methyl-2-phenylpteridine (ethanolic methylglyoxal under reflux: 79%),[152] 7-methyl-4-phenylpteridine (aqueous ethanolic methyl glyoxal under reflux: 95%),[151] 2,7-dimethyl-4-phenylpteridine (likewise: 70%),[151] 4,7-dimethyl-2-phenylpteridine (87%),[152] 7-phenylpteridine (**45**, R = H) (ethanolic phenylglyoxal: 95%),[959, 1287] 7-*p*-(methoxy-, bromo-, nitro-, or cyanophenyl)pteridine (**45**, R = OMe, Br, NO_2, or CN) (appropriately *p*-substituted phenylglyoxal as cited previously),[1287] and 4,7-diphenylpteridine (phenylglyoxal similarly: 92%).[959]

(2) *Cases with One Functional Group*

The presence of even one extra functional or sterically bulky group in the pyrimidinediamine may upset this absolute preference for 7-substituted products with aldehydoketones; moreover, the formation of 6-isomers may be stimulated therein by devices such as modification of the dicarbonyl reagent by bisulfite complexation or hydrazone formation. Thus 6-*t*-butyl-2-chloro-4,5-pyrimidinediamine (**46**, R = *t*-Bu) and phenylglyoxal in refluxing ethanol gave 4-*t*-butyl-2-chloro-7-phenylpteridine (**47**, R = *t*-Bu) contaminated with ~10% of the 6-phenyl isomer[959]; in contrast, similar treatment of 2-chloro-4,5-pyrimidinediamine (**46**, R = H) apparently gave only 2-chloro-7-phenylpteridine (**47**, R = H).[959] 6-Trifluoromethyl-4,5-pyrimidinediamine with

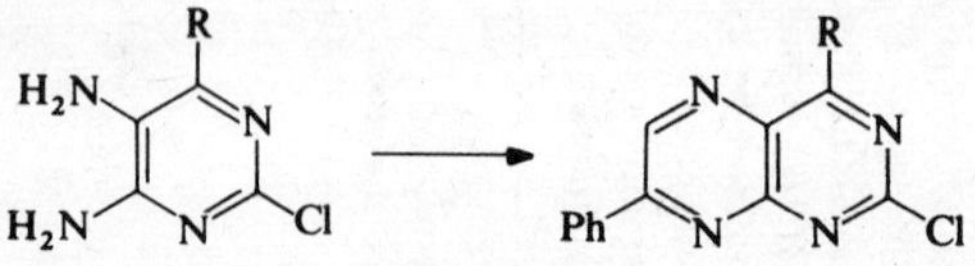

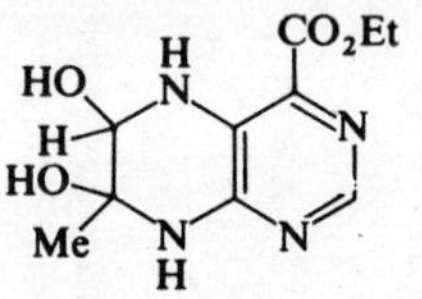

(46) (47) (48)

methylglyoxal in boiling *t*-butyl alcohol apparently gave only one product, 7-methyl-4-trifluoromethylpteridine (~55%).[156] However, ethyl 5,6-diamino-4-pyrimidinecarboxylate and aqueous methylglyoxal (~1 mol) under reflux for 5 min gave a separable mixture of ethyl 7-methyl-4-peteridinecarboxylate and its 6-methyl isomer in ~5:1 ratio[147]; when the proportion of methylglyoxal was increased to >2 mol, none of the 6-methyl isomer could be detected and the yield of the 7-methyl product increased to ~ 60%, a phenomenon not easily explained.[147] Prolongation of this reaction or its performance under strongly acidic conditions gave only one product, the covalent dihydrate (**48**) of the 7-methyl isomer, which underwent dehydration in boiling *t*-butyl alcohol.[148, 149] 2-Methylthio-6-phenyl-4,5-pyrimidinediamine and phenylglyoxal in hot 2-ethoxyethanol gave a separable mixture (85% yield) of 2-methylthio-4,7-diphenylpteridine and its 4,6-diphenyl isomer, in which the former predominated (10:1).[1118] 2,4,5-Pyrimidinetriamine and its 6-methyl derivative condensed with aqueous methylglyoxal at pH 7 to give (authenticated) 7-methyl- and 4,7-dimethyl-2-pteridinamine, respectively, without any trace of an isomeric product in either case[39]; 2-phenyl-4,5,6-pyrimidinetriamine likewise gave only 7-methyl-2-phenyl-4-pteridinamine (~60%),[328] but addition of hydrazine to the methylglyoxal prior to admixture with the pyrimidine, changed the course of the reaction to give only 6-methyl-2-phenyl-4-pteridinamine (29% yield).[1354] 4,5-Diamino-2(1*H*)-pyrimidinone reacted with aqueous methylglyoxal at 100°C to give only 7-methyl-2(1*H*)-pteridinone (87%), confirmed in structure because it differed from a sample of 6-methyl-2(1*H*)-pteridinone prepared by an unambiguous synthesis.[38] Under similar conditions, in the presence of sodium sulfite, 5,6-diamino-4(3*H*)-pyrimidinone (**50**, X = O) gave 7-methyl-4(3*H*)-pteridinone (**49**, X = O) (~60% yield), but after initial complexation of the methylglyoxal with sodium bisulfite, an otherwise similar reaction at 30°C gave 6-methyl-4(3*H*)-pteridinone (**51**) (56%) containing only a trace of the 7-methyl isomer.[32] Even in the presence of sodium bisulfite, the reaction of 5,6-diamino-4(3*H*)-pyrimidinethione (**50**, X = S) with methylglyoxal gave only one product (50%), identified as 7-methyl-4(3*H*)-pteridinethione (**49**, X = S) by hydrolytic degradation to 3-amino-5-methyl-2-pyrazinecarboxylic acid (**44**).[33] 5,6-Diamino-2-phenyl-4(3*H*)-pyrimidinone and aqueous methylglyoxal (without sodium bisulfite) gave only 7-methyl-2-phenyl-4(3*H*)-pteridinone (87%), which was identified by degradation to 3-amino-5-methyl-2-pyrazinecarboxylic acid (**44**)[152]; 5,6-diamino-2-methyl-4(3*H*)-pyrimidinone likewise gave 2,7-dimethyl-4(3*H*)-pteridinone (~40%), confirmed in structure by the unambiguous synthesis of its 2,6-dimethyl isomer.[153]

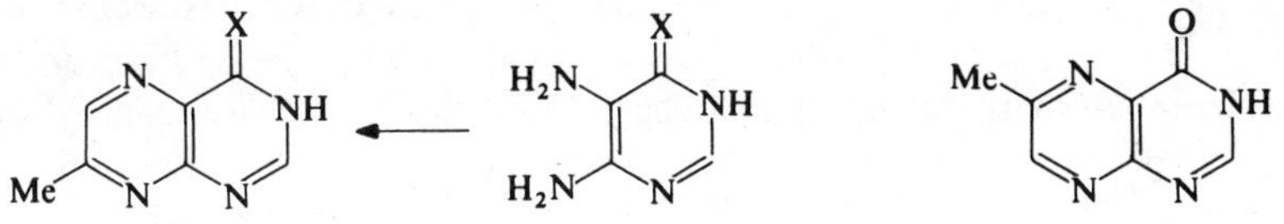

(**49**) (**50**) (**51**)

(3) *Cases with Two Functional Groups*

The condensation of aldehydoketones with 4,5-pyrimidinediamines bearing two extra functional groups has been investigated extensively for many years because it is a key step in the syntheses of folic acid, methotrexate, and related molecules of enormous biological importance, all of which require 6- without 7-substitution. Much of the early work in this area was done in the days before paper or thin layer chromatography and even more of it in pre-nmr days. Moreover, molecules such as 2-amino-7-methyl-4(3*H*)-pteridinone (**52**) and its 6-methyl isomer (**53**) differ quite minimally in their uv spectra and even in their chromatographic behavior in most solvent systems; the infrared (ir) spectra of the pure isomers do differ in some respects but this could be used effectively, neither for estimation of individual isomers in mixtures during separation, nor for detection of perhaps 10% of an unwanted isomer in an otherwise *pure* product. Thus while most of the older work in this area was done with great care and skill using the means at hand, the isomeric purity of many of the products described (and, occasionally, even their identity) must be suspect. Fortunately, nmr techniques now provide an effective answer in the more modern work within the area.

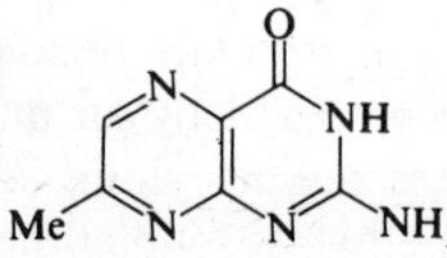

(52)

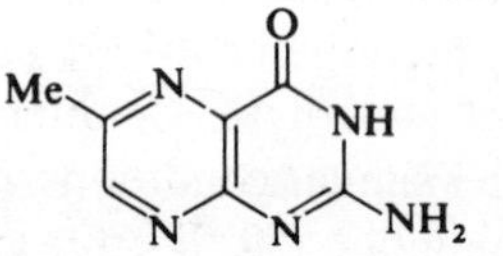

(53)

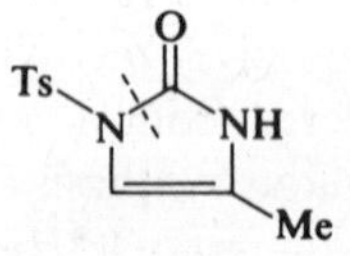

(54)

(a) The 6- or 7-Alkyl-2-amino-4(3*H*)-pteridinones

It appears that methylglyoxal reacts with 2,5,6-triamino-4(3*H*)-pyrimidinone in neutral or mildly acidic aqueous solution to give a mixture of 2-amino-7-methyl-4(3*H*)-pteridinone (**52**) and its 6-methyl isomer (**53**), in which the former predominates so greatly that the minor isomer may sometimes be removed by normal purification.[192, 449, 748, 1212, 1373] A well-described procedure for preparing 2-amino-7-methyl-4(3*H*)-pteridinone is that of Wolfgang Pfleiderer and co-workers.[389] They condensed methylglyoxal with the pyrimidine in aqueous alkali at room temperature to give a crude product (96%), which was purified by acetylation to 2-acetamido-7-methyl-4(3*H*)-pteridinone with subsequent saponification to the required product (**52**). A similar result was obtained by using methylglyoxal diethyl (or dimethyl) acetal in dilute hydrochloric acid followed by purification through the sodium salt.[201, 748, 1084] Other acetallike equivalents of methylglyoxal have been used also: of these, methylglyoxal dithioacetal, 1-acetoxy-1-(phenylthio)acetone, and 1-acetoxy-1-(ethylthio)-acetone were reported to give only the 7-methyl isomer (**52**) (30–40%) under acidic conditions[1377]; 1-chloro-1-(phenylthio)acetone and 1-chloro-1-

(ethylthio)acetone gave a mixture of the 7- (**52**) and 6-isomer (**53**) in neutral aqueous media[1377]; 1-acetamido-1-bromoacetone and its 1-chloro analog appear to have given only the 7-methyl isomer (**52**) (68–88%) under mildly basic conditions,[1475] but in the presence of hydrazine or sulfite, a mixture of the 6- and 7-methyl isomers resulted[545]; 5-methyl-3-tosyl-2(1*H*)-imidazolone (**54**) gave (in aqueous piperidine) the well-authenticated 7-isomer (**52**) (82%) by a mechanism that must involve aerial oxidation at some stage[1231]; and 1,1-dichloro- or 1,1-dibromoacetone has been reported to give a small yield of the 7-methyl isomer above pH 4 but a comparable yield of the 6-methyl isomer about pH 1.5.[1254, 1375] Two variations of the pyrimidine intermediate are seen in the condensation of 2,4-diamino-5-formamido-4(3*H*)-pyrimidinone (**55**) with aqueous methylglyoxal under reflux to give an unseparated mixture of the isomers (**52**) and (**53**) in 1:4 ratio,[940] and in the peculiar reaction of 2,4-diamino-5-hydroxy-4(3*H*)-pyrimidinone (divicine) (**56**, R = OH) with methylglyoxal in concentrated aqueous ammonia to give a similar mixture.[548]

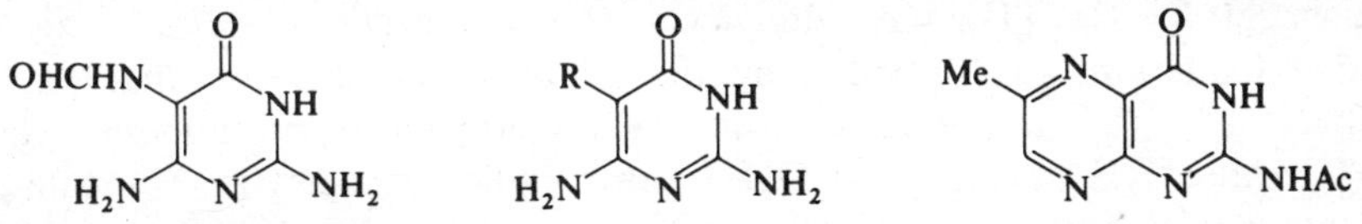

(**55**) (**56**) (**57**)

It is very doubtful if any known procedure for condensing methylglyoxal (or an equivalent reagent) with 2,5,6-triamino-4(3*H*)-pyrimidinone gives 2-amino-6-methyl-4(3*H*)-pteridinone (**53**), initially uncontaminated with an appreciable proportion of its 7-methyl isomer (**52**) and therefore not requiring considerable purification. Mixtures high in the 6-methyl isomer have been achieved (and subsequently purified) in several ways. (i) Pretreatment of methylglyoxal with *hydrazine hydrate* followed by condensation with the pyrimidine in aqueous buffer under nitrogen gave a mixture from which the 6-methyl isomer (**53**) was isolated in rather poor yield via its crystalline sodium salt.[193] Two modified procedures improved the yield of pure product to 37 and 47%, respectively,[389] but the use of purified methylglyoxal dihydrazone proved[1475] to have no particular advantage. (ii) The condensation was also directed towards the 6-methyl product in the presence of *β-mercaptoethanol*, which presumably formed a thioacetal or hemithioacetal with the aldehydo group of methylglyoxal. The crude material (98%) contained ~75% of the 6-methyl isomer that was isolated after purification in 53% overall yield.[449] The procedure was adapted to the formation of [4a-^{13}C]-2-amino-6-methyl-4(3*H*)-pteridinone.[939] (iii) Pretreatment of the methylglyoxal with *sodium bisulfite* produced a similar effect on the condensation. Based on earlier work,[84, 1492] the final procedure of Paul Waring and Wilf Armarego[1443] gave an 80% yield of a mixture containing 90% of the 6-methyl isomer (**53**), from which pure 2-acetamido-6-methyl-4(3*H*)-pteridinone (**57**) was made in 90% yield.

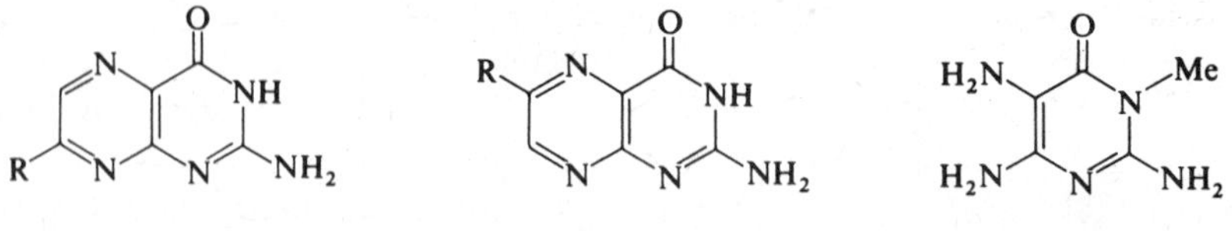

(58) (59) (60)

The formation of other 7- or 6-alkyl (or aryl)-2-amino-4(3H)-pteridinones by condensation of alkylglyoxals or their equivalents with the pyrimidine (**56**, R = NH_2) is exemplified in 2-amino-7-phenyl-4(3H)-pteridinone (**58**, R = Ph) (phenylglyoxal at pH 9: 41%),[449] its 6-phenyl isomer (**59**, R = Ph) (phenylglyoxal at pH 4: 40%[449, cf. 176, 1006]; α,α-dichloroacetophenone: 60%[176, 1005, 1006]); 2-amino-7-ethyl-4(3H)-pteridinone (**58**, R = Et) (1-benzamido-1-chlorobutan-2-one at pH 8 and 100°C: 80%)[1475]; 2-amino-7-hexyl-4(3H)-pteridinone (**58**, R = Hx) (hexylglyoxal methyl hemithioacetal, i.e., 1-hydroxy-1-methylthio-octan-2-one, in acetic acid at 100°C: 74%; for analogs see Table III),[420, 1683] its 6-hexyl isomer (**59**, R = Hx) [1-methylsulfinyloctan-2-one in acetic acid: 15% (the required oxidation was provided by the S-oxide function); for analogs see Table III][420, 1683]; 2-amino-6-α(or β)-carboxyethyl-4(3H)-pteridinone (**59**, R = $CHMeCO_2H$ or $CH_2CH_2CO_2H$) [ethyl α-(diethoxyacetyl)-propionate or diethyl α-(diethoxyacetyl)succinate, respectively, in aqueous buffer: 60–70% crude][906]; 2-amino-6-benzyl-4(3H)-pteridinone (**59**, R = CH_2Ph) (benzylglyoxal dimethyl acetal in acetic acid: 46%)[419]; its 7-benzyl isomer (benzylglyoxal dimethyl acetal in hydrochloric acid: 31% including a trace of the 6-isomer)[419]; and more complicated examples.[881, 1213, 1361] The same pyrimidine (**56**, R = NH_2) with phenethylglyoxal diethyl acetal under mildly or strongly acidic conditions gave a separable mixture of 2-amino-6-phenethyl-4(3H)-pteridinone and its 7-phenethyl isomer[90]; with pyridin-4-ylmethylglyoxal diethyl acetal it gave an inseparable mixture of 2-amino-6(and 7)-pyridin-4′-ylmethyl-4(3H)-pteridinone[1396]; with free hexylglyoxal or its dimethyl acetal it gave a mixture of 2-amino-6(and 7)-hexyl-4(3H)-pteridinone[420]; and with phenylglyoxal diethyl acetal it gave a difficultly separable mixture of 2-amino-6(and 7)-phenyl-4(3H)-pteridinone.[67] Likewise, 2,5,6-triamino-3-methyl-4(3H)-pyrimidinone (**60**) and phenylglyoxal diethyl acetal gave a separable mixture of 2-amino-3-methyl-6(and 7)-phenyl-4(3H)-pteridinone (**61**) and (**62**) but with free phenylglyoxal it gave only the 7-phenyl isomer (**62**) (>95%) via the isolated intermediate, 2,6-diamino-3-methyl-5-phenacylideneamino-4(3H)-pyrimidinone (**63**).[67] 5,6-Diamino-2-dimethylamino-4(3H)-pyrimidinone and phenylglyoxal in acidic

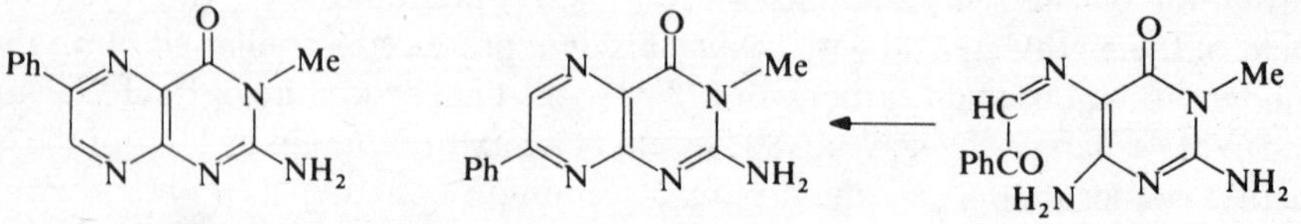

(61) (62) (63)

TABLE III. ADDITIONAL EXAMPLES OF THE GABRIEL AND COLMAN SYNTHESIS WITH ALDEHYDO KETONES

Pyrimidine Intermediate	Reagent and Reaction Conditions	Pteridine (Yield)	References
2-Phenyl-4,5,6-pyrimidinetriamine	Phenylglyoxal; ethanolic buffer; 80 °C; 30 min	2,7-Diphenyl-4-pteridinamine (~90%)	328
6-Methylthio-2,4,5-pyrimidinetriamine	1,3-Dihydroxyacetone	6-Hydroxymethyl-4-methylthio-2-pteridinamine (62%)	815
4,5,6-Pyrimidinetriamine	*p*-Chlorophenylglyoxal; aqueous dioxane; reflux; 1 h	7-*p*-Chlorophenyl-4-pteridinamine[a] (80%)	489
4,5,6-Pyrimidinetriamine	*o*-Tolylglyoxal; aqueous dioxane; reflux; 1 h	7-*o*-Tolyl-4-pteridinamine (50%)	489
2,5,6-Triamino-4(3*H*)-pyramidinone	*t*-Butylglyoxal; aqueous ethanol; 25 °C; 3 days	2-Amino-7-*t*-butyl-4(3*H*)-pteridinone (~35%)	541
2,5,6-Triamino-4(3*H*)-pyrimidinone	1-Methylsulfinylnonan-2-one; AcOH; reflux	2-Amino-6-heptyl-4(3*H*)-pteridinone[b] (17%)	420, 1683
2,5,6-Triamino-4(3*H*)-pyrimidinone	1-Hydroxy-1-methylthio-4-phenylbutan-2-one; AcOH; reflux	2-Amino-7-phenethyl-4(3*H*)-pteridinone[b] (59%)	420, 1683
2,5,6-Triamino-4(3*H*)-pyrimidinone	1-Phthalimidoheptan-2-one; aqueous piperidine; air; 20 °C	2-Amino-7-pentyl-4(3*H*)-pteridinone[b] (83%)	1491
2,5,6-Triamino-4(3*H*)-pyrimidinone	Methyl γ,γ-dimethoxyacetoacetate; AcOH; 100 °C; *or* 4-hydroxybut-2-ynoic acid; NaOAc; neat; 165 °C; air	2-Amino-6-carboxymethyl-4(3*H*)-pteridinone (~30%)	502, 1084
2,5,6-Triamino-4(3*H*)-pyrimidinone	But-2-yn-1,4-diol; NaOAc; neat; 110 °C; air	2-Amino-6-β-hydroxyethyl-4(3*H*)-pteridinone (35%)	502
2,4,5,6-Pyrimidinetetramine	*p*-Acetamidophenylglyoxal; aqueous sodium bisulfite; 25 °C	6-*p*-Acetamidophenyl-2,4-pteridinediamine (~25%)	1257
5,6-Diamino-2,4(1*H*,3*H*)-pyrimidinedione	*p*-Methoxyphenylglyoxal	7-*p*-Methoxyphenyl-2,4(1*H*,3*H*)-pteridinedione[c]	1287
5,6-Diamino-1-*o*-methoxyphenyl-2,4(1*H*,3*H*)-pyrimidinedione	Phenylglyoxal; aqueous alcoholic AcOH; reflux; 30 min	1-*o*-Methoxyphenyl-7-phenyl-2,4(1*H*,3*H*)-pteridinedione[c] (90%)	797
5,6-Diamino-1-*m*-chlorophenyl-2-thioxo-1,2-dihydro-4(3*H*)-pyrimidinone	Phenylglyoxal; aqueous alcoholic AcOH; reflux	1-*m*-Chlorophenyl-2-thioxo-1,2-dihydro-4(3*H*)-pteridinone[c]	797
5,6-Diamino-1-methyl-2,4(1*H*,3*H*)-pyrimidinedione	D-Galactose: *p*-toluidine; hydrazine; aqueous AcOH	1-Methyl-6-[D-*lyxo*]-$\alpha,\beta,\gamma,\delta$-tetrahydroxybutyl-2,4(1*H*,3*H*)-pteridinedione (18%)	962

TABLE III. (*Contd.*)

Pyrimidine Intermediate	Reagent and Reaction Conditions	Pteridine (Yield)	References
5,6-Diamino-2-thioxo-1,2-dihydro-4(3*H*)-pyrimidinone	[L-*lyxo*]-6-Deoxyhexulose; aqueous AcOH and AcONa	2-Thioxo-7-[L-*lyxo*]-α,β,γ-trihydroxybutyl-1,2-dihydro-4(3*H*)-pteridinone[c] (7%)	1247
2,5,6-Triamino-4(3*H*)-pyramidinone	L-Ascorbic acid and aqueous iodine–KI; reflux	2-Amino-4-oxo-6-[L-*threo*]-α,β,γ-trihydroxypropyl-3,4-dihydro-7-pteridinecarboxylic acid[c]	1057
2,5,6-Triamino-4(3*H*)-pyrmidinone	Ribose-5-phosphate; aqueous boric acid and hydrazine; 25 °C; 2 days	2-Amino-6 (and 7)-[D-*erythro*]-α,β,γ-trihydroxypropyl-4(3*H*)-pteridinone 3′-phosphate	505
5,6-Diamino-2-methylthio-4(3*H*)-pyrimidinone	5-Deoxy-L-arabinose phenylhydrazone; ferricyanide and oxygen	6-[L-*erythro*]-α,β-Dihydroxypropyl-2-methylthio-4(3*H*)-pteridinone (~30%)	458
5,6-Diamino-2-dimethylamino-4(3*H*)-pyrimidinone	Aqueous methylglyoxal; reflux; 5 min	2-Dimethylamino-7-methyl-4(3*H*)-pteridinone (45%)	1710

[a] And a substituted-amino analog.
[b] And homologs.
[c] And analog(s).

ethanol has been reported to give only 2-dimethylamino-6-phenyl-4(3*H*)-pteridinone, while at pH 4 only the 7-phenyl isomer was formed.[110]

Condensation of the pyrimidine (**56**, R = NH_2) with hydroxymethylglyoxal in boiling water gave 2-amino-7-hydroxymethyl-4(3*H*)-pteridinone (**58**, R = CH_2OH), its 6-hydroxymethyl isomer (**59**, R = CH_2OH), or a mixture of both[776]; later work suggested that the main product was the first of these (**58**, R = CH_2OH).[1212] When 1,1,3-tribromoacetone or 2,2,3-tribromopropionaldehyde was used in place of hydroxymethylglyoxal, the same product was reported, presumably arising via the bromomethylpteridine (**58**, R = CH_2Br)[1375]; the first of these reagents also gave an isomeric product, the 6-hydroxymethylpteridine (**59**, R = CH_2OH), which could be made in a pure state by using 2,3-dibromopropionaldehyde for the condensation in the presence of iodine as an oxidizing agent.[1375] Other reagents of low oxidation state were also used in place of hydroxymethylglyoxal, presumably relying on aerial oxidation at some stage for success. Thus 1,3-dihydroxyacetone, pretreated with hydrazine or cysteine and then allowed to react with the pyrimidine (**56**, R = NH_2) in water containing sodium acetate and boric acid, gave 40% of pure 2-amino-6-hydroxymethyl-4(3*H*)-pteridinone (**59**, R = CH_2OH)[193, 1212, cf. 96, 774]; [8a-^{14}C]-[911] and [4a-^{14}C]-tagged[1032] products were made similarly. However, in the absence of hydrazine, a similar mixture gave, not the expected 7-hydroxymethyl isomer, but 2-amino-7-methyl-4(3*H*)-pteridinone (**58**, R = Me) in reasonable yield[193, 1212]; a similar result was obtained by replacing dihydroxyacetone by 1,3-diacetoxyacetone,[1375] compare reference 1373, 1,3-bis-(*N*-carboxy-*N*-formylanilino)acetone,[1375] and other such reagents.[1026]

Further examples of the use of such low oxidation state reagents (usually ketoalcohols, aldehydoalcohols, or their equivalents) instead of aldehydoketones, are seen in their condensation with appropriate pyrimidines to afford 2-amino-7-methyl-4(3*H*)-pteridinone (**58**, R = Me) (hydroxyacetone–air: ~15%[193]; bromoacetone–hydrogen peroxide or air: 40%[1092, 1375]; α-bromopropionaldehyde diethyl acetal–air: 28%[1092]; acetamidoacetone plus diethylamine–air: 50%[545, 1154]; *N,N*-diacetylaminoacetone plus diethylamine–air: 67%[942]); 2-amino-6-methyl-4(3*H*)-pteridinone (**59**, R = Me) [chloroacetone plus hydrazine at 25°C/air: ~ 25%[1147]; hydroxyacetone (acetol) plus β-mercaptoethanol–air: 73%[202, 509]; hydroxyacetone plus hydrazine–air: ~30% after removal of some 7-methyl isomer[993, 1212]; acetamidoacetone plus hydrazine–air: 50% (including a little 7-methyl isomer)[545, 1154]; acetoxyacetone plus 2-mercaptoethanol: 79% (including a trace of the 7-methyl isomer)[1340]; *N,N*-diacetylaminoacetone plus hydrazine–air: 61%[942]; α-acetamidopropionaldehyde plus triethylamine–air: 30%[944]]; 2-amino-7-ethyl-4(3*H*)-pteridinone (**58**, R = Et) (1-benzamidobutan-2-one–air: 67%[942]; 1-phthalimidobutan-2-one–air: 71%[1491]); 2-amino-6-ethyl-4(3*H*)-pteridinone (plus some 7-ethyl isomer) (1-bromobutan-2-one–air: 31%[1092]; 1-acetamido- or 1-benzamidoacetone plus hydrazine–air: 30–35%[942]); 2-amino-6-phenyl-4(3*H*)-pteridinone (**59**, R = Ph) (α-bromophenylacetaldehyde–air: 44%)[1092]; 4-methyoxy-7-phenyl-2-pteridinamine (**64**) (α-chloroacetophenone–air: 40%)[1092];

2-amino-4-oxo-3,4-dihydro-6-pteridinecarboxylic acid (**59**, R=CO_2H) (ethyl α-bromo-β,β-diethoxypropionate–iodine: ~10%)[1084]; 2-amino-6-diethoxymethyl-4(3*H*)-pteridinone [**59**, R=$CH(OEt)_2$] (α-bromo-β,β-diethoxypropionaldehyde–hydrogen peroxide: ~65%)[464, 1300]; 2-amino-6-pyridiniomethyl-4(3*H*)-pteridinone iodide (**65**) [*N*-(β-bromo-β-formylethyl)-pyridinium bromide then potassium iodide–air: 15%][1180, 1394]; 2-amino-6-bromomethyl-4(3*H*)-pteridinone (**59**, R=CH_2Br) (α,β-dibromopropionaldehyde–air: 70%)[731]; and others.[649, 715, 1220, 1403]

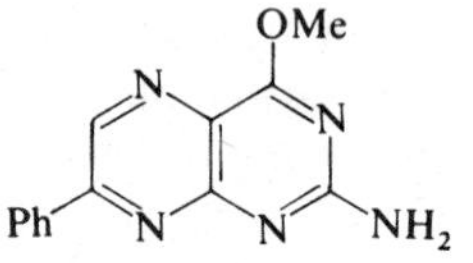

(64)

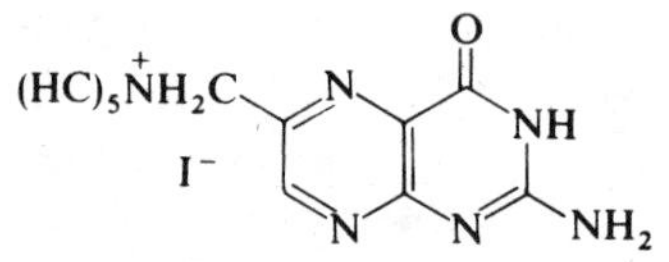

(65)

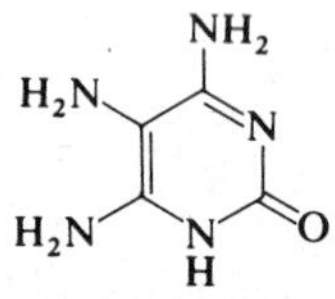

(66)

Three syntheses related to those just cited are included here. 4,5,6-Triamino-2(1*H*)-pyrimidinone (**66**) and methylglyoxal with sodium bisulfite gave a single product, 4-amino-6-methyl-2(1*H*)-pteridinone (**67**, R=H) (83%), which was purified through its crystalline sodium salt and confirmed in orientation by alkaline hydrolysis to 6-methyl-2,4(1*H*,3*H*)-pteridinedione[168]; the same pyrimidine (**66**) with dihydroxyacetone, cysteine, and a stream of air gave 4-amino-6-hydroxymethyl-2(1*H*)-pteridinone (**67**, R=OH).[96] The condensation of 2,5,6-triamino-4(3*H*)-pyrimidinone (**56**, R=NH_2) with hydroxyiminoacetone (simply the aldoxime of methylglyoxal) in cool aqueous sodium sulfite gave, not the expected 2-amino-7(or 6)-methyl-4(3*H*)-pteridinone (**58** or **59**; R=Me), but 2,6-diamino-7-methyl-4(3*H*)-pteridinone (**68**). This was confirmed in structure by nitrous acid deamination to 7-methyl-2,4,6(1*H*,3*H*,5*H*)-pteridinetrione of proven structure.[279] Covalent addition of the liberated hydroxylamine over the 5,6 bond to give (**69**), reduction of the 6-hydroxyamino to an amino group by sulfite ion, and a final dehydrogenation in air, may account for the formation of the anomalous product (**68**).

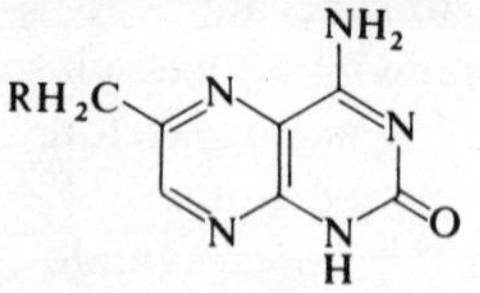

(67)

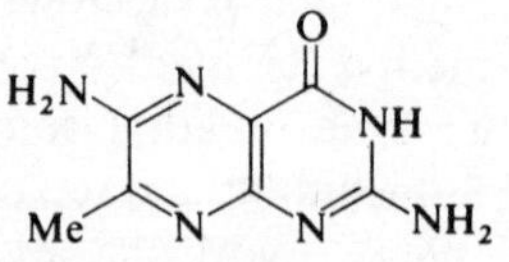

(68)

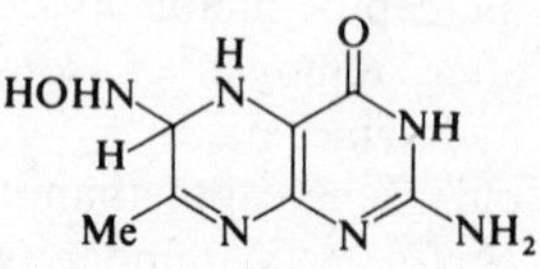

(69)

(b) The 6- or 7-Alkyl-2,4-pteridinediamines

As early as 1947, 2,4,5,6-pyrimidinetetramine was allowed to react with aqueous methylglyoxal to give a 90% yield of a single product,[284] later shown[748] to be 7-methyl-2,4-pteridinediamine (**70**) by degradation to 3-amino-

5-methyl-2-pyrazinecarboxylic acid; modified procedures in dilute acidic media have been reported.[730, 1263] When the same reaction was carried out in aqueous sodium sulfite with methylglyoxal pretreated with sodium bisulfite, the product (70%) was mainly 6-methyl-2,4-pteridinediamine (**71**) from which a little of the 7-methyl isomer (**70**) could be removed easily, albeit with some loss.[1263] Reagents other than methylglyoxal have been tried with the pyrimidinetetramine. Thus benzamidoacetone in aqueous piperidine (with aerial oxidation?) gave 7-methyl-2,4-pteridinediamine (**70**) in 82% yield[944]; *N*-tosylaminoacetone under similar conditions gave a mixture of the isomers (**70**) and (**71**)[944]; 3-*p*-chlorobenzenesulfonyl-5-methyl-2(1*H*)-imidazolone gave a similar mixture,[1231] from which the 6-methyl-2,4-pteridinediamine (**71**) was separated[945] in a pure state by preferential permanganate oxidation of the 7-isomer. Methylglyoxal dihydrazone gave mainly the 6-methyl-2,4-pteridinediamine (**71**) (64%) containing a little of the 7-isomer.[1475] 2,3-Dibromopropionaldehyde at pH 8 gave (mainly?) 6-methyl-2,4-pteridinediamine (**71**),[1254] and dihydroxyacetone in aqueous sodium sulfite with a stream of air gave pure 6-methyl-2,4-pteridinediamine (**71**) but in only 20% yield.[123, cf. 595] Dihydroxyacetone can also give the expected 6-hydroxymethyl derivative. Treatment of the pyrimidinetetramine with dihydroxyacetone in aqueous cysteine (as a complexing agent) through which a stream of oxygen was passed, gave 6-hydroxymethyl-2,4-pteridinediamine (70–87%), free of the 6-methyl analog (**71**).[113, 1465, cf. 96, 595, 1522, 1757] 7-Hydroxymethyl-2,4-pteridinediamine has been made in low yield from the pyrimidinetetramine and hydroxymethylglyoxal in aqueous sodium pyrosulfite solution at 30°C.[123]

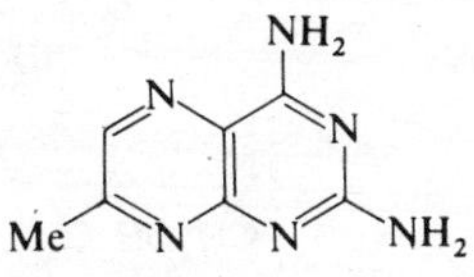

(**70**)

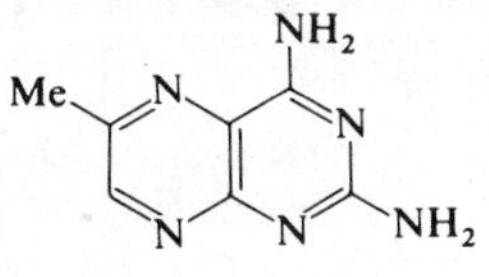

(**71**)

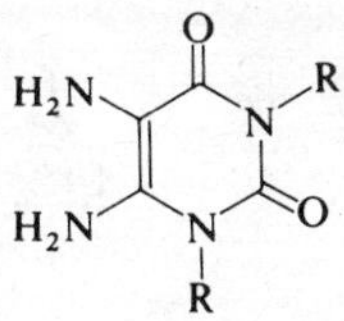

(**72**)

Less simple examples include the formation of 7-*phenyl*-2,4-*pteridinediamine* (phenylglyoxal in refluxing aqueous ethanol: 83%)[541, 1006]; 6-*phenyl*-2,4-*pteridinediamine* (phenylglyoxal plus sodium bisulfite in aqueous sodium sulfite, although no evidence of isomeric purity was presented[1257]; the method of Frederick King and Philip Spensley[1006] using dichloroacetophenone actually gave only the 7-phenyl isomer[327]; under the circumstances, it is wise to make the 6-phenyl isomer unambiguously by the Timmis synthesis[327, 541] or from a pyrazine[484] intermediate); 7-*adamant*-1′-*yl*-2,4-*pteridinediamine* (adamant-1′-ylglyoxal in methanol)[973]; its 6-*adamantyl isomer* (adamantylglyoxal hydrazone in methanol)[973]; 6-*phenethyl*-2,4-*pteridinediamine* (phenethylglyoxal diethyl acetal in aqueous ethanol: 23%; identified by hydrolysis to the known corresponding pterin)[90]; and separable mixtures of 6- and 7-*p*-*chlorophenyl*-2,4-

pteridinediamine (*p*-chlorophenylglyoxal in acid aqueous dioxane: high in 7-isomer; same reagent in aqueous sodium sulfite–sodium bisulfite: high in 6-isomer).[489]

(c) The 6- or 7-Alkyl-2,4(1*H*,3*H*)-pteridinediones

Neither 6- nor 7-methyl-2,4(1*H*,3*H*)-pteridinedione appears to have been made directly by the Gabriel and Colman synthesis, although mixed products (?) have been described from the condensation of 5,6-diamino-2,4(1*H*,3*H*)-pyrimidinedione (**72**, R=H) and methylglyoxal.[1012, 1438] Both are available from hydrolysis of the corresponding pterins[655] and by other means. However, the same pyrimidine (**72**, R=H) and dihydroxyacetone in the presence of cysteine and air did give 6-hydroxymethyl-2,4(1*H*,3*H*)-pteridinedione (**73**, R=CH_2OH) in low yield after chromographic purification.[1031, cf. 1546] More successful preparations from (**72**, R=H) afforded *7-phenyl*-2,4(1*H*,3*H*)-*pteridinedione* (phenylglyoxal in aqueous ethanol: 70%[367]; earlier attempts in acidic or ammoniacal media[1438, 1473] appear to have given a mixture of 7- and 6-isomers[176]); *7-p-chlorophenyl*-2,4(1*H*,3*H*)-*pteridinedione* (*p*-chlorophenylglyoxal in aqueous ethanolic ammonia under reflux: 85%)[412]; and *6-p-chlorophenyl*-2,4(1*H*,3*H*)-*pteridinedione* (**73**, R=*p*-ClC_6H_4) (same reagent in 85% sulfuric acid at 25°C: 27% yield; *p*-chlorophenylglyoxal dimethyl acetal in 85% sulfuric acid at 100°C: 66%).[412] 5,6-Diamino-2-thioxo-1,2-dihydro-4(3*H*)-pyrimidinone and phenylglyoxal gave 6- and/or *7-phenyl-2-thioxo*-1,2-*dihydro*-4(3*H*)-*pteridinone* (50%).[1190]

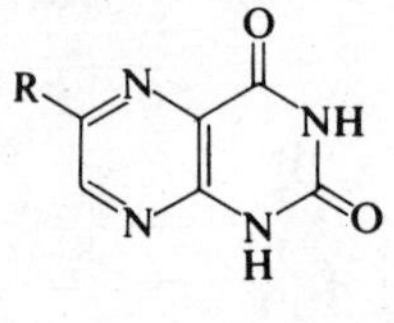

(**73**)

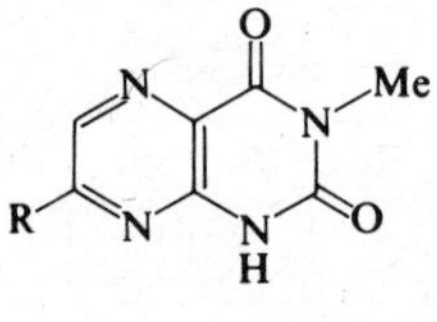

(**74**)

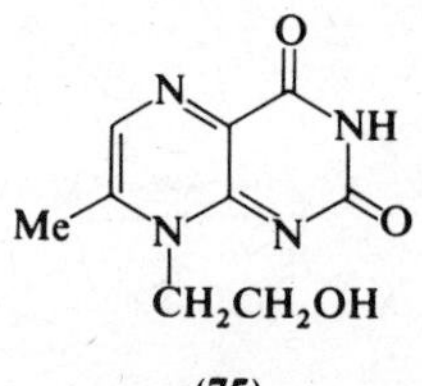

(**75**)

Some 6- or 7-alkyl-2,4-pteridinediones with *N*-substituents have been made similarly. For example, 5,6-diamino-3-methyl-2,4(1*H*,3*H*)-pyrimidinedione and *t*-butyl- or phenylglyoxal gave *7-t-butyl-3-methyl*- or *3-methyl-7-phenyl*-2,4(1*H*,3*H*)-*pteridinedione* (**74**, R=*t*-Bu or Ph) in 50–60% yield, respectively[367]; 5,6-diamino-1,3-dimethyl-2,4(1*H*,3*H*)-pyrimidinedione (**72**, R=Me) with methylglyoxal[108] or with glyceraldehyde plus mercuric chloride as an oxidizing agent[1121] gave 1,3,7-*trimethyl*-2,4(1*H*,3*H*)-*pteridinedione* (55, 62%); the same pyrimidine (**72**, R=Me) with isopropyl- or *t*-butylglyoxal gave low yields of 7-*isopropyl*- or 7-*t-butyl*-1,3-*dimethyl*-2,4(1*H*,3*H*)-*pteridinedione*, respectively[108]; this pyrimidine (**72**, R=Me) with ethanolic 1,3-di-*p*-toluoylethylene also gave some 1,3-*dimethyl-7-p-tolyl*-2,4(1*H*,3*H*)-*pteridinedione* and other separable products[1365]; 5-amino-6-*β*-hydroxyethylamino-2,4(1*H*,3*H*)-pyrimidinedione with methylglyoxal gave 8-*β-hydroxyethyl-7-methyl*-2,4(3*H*,8*H*)-*pteridinedione* (**75**)

(42%)[1144]; 5-amino-6-methylamino-2,4(1*H*,3*H*)-pyrimidinedione with methylglyoxal gave 7,8-*dimethyl*-2,4(3*H*,8*H*)-*pteridinedione*[1278]; 5,6-diamino-1-phenyl-2,4(1*H*,3*H*)-pyrimidinedione with phenylglyoxal gave 1,7-*diphenyl*-2,4(1*H*,3*H*)-*pteridinedione*[797]; and 5,6-diamino-1-phenyl-2-thioxo-1,2-dihydro-4(3*H*)-pyrimidinone with phenylglyoxal gave 1,7-*diphenyl-2-thioxo*-1,2-*dihydro*-4(3*H*)-*pteridinone*.[497]

(d) The Folic Acid-Type Synthesis

The original synthesis of *pteroylglutamic acid* (**76**, $n=2$), the simplest of the folic acids, was outlined first by American Cyanamid workers in 1946.[1174, cf. 1520] It consisted of a one-pot reaction between three starting materials: the pyrimidine (**76a**), a three-carbon fragment (**76b**), and the complex amine side chain (**76c**). Although many variations of this method appeared subsequently, all retained this essential character and are easily recognizable as a special case of the Gabriel and Colman synthesis in which the three-carbon fragment is actually or potentially an α-aldehydoketone. The nature of the fragment (**76b**) is the basis for subclassification to follow.

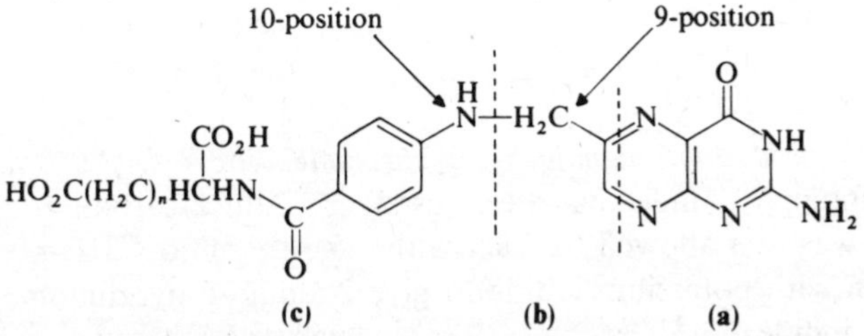

(**76**)

2,3-*Dibromopropionaldehyde* is the original and most commonly used three-carbon fragment. Pteroylglutamic acid (**76**, $n=2$) was best made by treating an aqueous solution of 2,5,6-triamino-4(3*H*)-pyrimidinone and *p*-aminobenzoylglutamic acid at pH 4 with ethanolic 2,3-dibromopropionaldehyde at 25°C. After long stirring, which allowed time for the necessary aerial oxidation, the crude product (~50%) contained some 25% of bioactive material.[1174, 1185, 1235, cf. 529, 836, 1171, 1290] One major impurity was 2-amino-6-methyl-4(3*H*)-pteridinone.[1253] A similar procedure using ethyl *p*-aminobenzoate in place of the aminobenzoylglutamic acid gave a crude ester which, on saponification, gave *pteroic acid*, 2-amino-6-*p*-carboxyanilinomethyl-4(3*H*)-pteridinone (**77**, R=OH).[1235] Other such variations are typified in the formation of pteroylaspartic acid (**76**, $n=1$),[1255] pteroylalanine (**77**, R=$NHCHMeCO_2H$),[1364] pteroylaminomalonic acid (**77**, R=$NHCH(CO_2H)_2$],[1364] pteroyl-α,β-glutamyldiglutamic acid,[1085] and others.[408, 678, 775, 802, 837, 838, 1205, 1261] The simple analogs, 2-amino-6-anilinomethyl-4(3*H*)-pteridinone[1375] and 2-amino-6-*o*(*m* and *p*)-toluidinomethyl-4(3*H*)-pteridinone (in making these three isomers, sodium dichromate

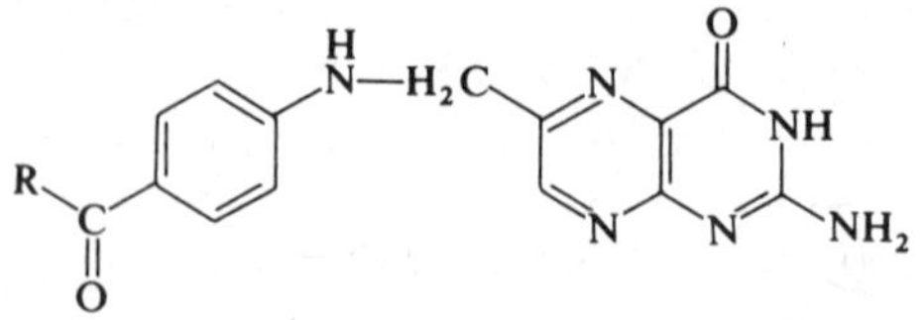

(77)

was added to avoid reliance upon air for the oxidation) were similarly prepared.[832, 1207] Variations were also introduced by using different pyrimidine intermediates, under otherwise similar conditions. Thus 2,4,5,6-pyrimidinetetramine, 2,3-dibromopropionaldehyde, and *p*-aminobenzoylglutamic acid gave the 4-amino analog of pteroylglutamic acid, known as 4-aminopteroylglutamic acid or (commercially) by the very confusing name "Aminopterin", which turned out to be a potent antagonist of its parent (**76**, $n=2$).[1255, 1263, 1265] The same pyrimidine led to 4-aminopteroylaminomalonic acid,[1364] 4-aminopteroylserine,[1364] and others.[408, 1255] Likewise, 6-dimethylamino-2,4,5-pyrimidinetriamine gave 4-dimethylaminopteroylglutamic acid, the 4-dimethylamino analog of (**76**, $n=2$),[1188] and 2-dimethylamino-4,5,6-pyrimidinetriamine gave N^2,N^2-dimethyl-4-aminopteroylglutamic acid, the 4-amino-2-dimethylamino analog of (**76**, $n=2$).[1189]

2-*Bromo*-3-*pyridiniopropionaldehyde bromide*, or *N*-(*β*-bromo-*β*-formylethyl)pyridinium bromide, has been used as a three-carbon fragment. For example, it was first allowed to react with 2,5,6-triamino-4(3*H*)-pteridinone in air and then with potassium iodide to give 2-amino-6-pyridiniomethyl-4(3*H*)-pteridinone iodide (**65**),[1174, 1180, 1236, 1394] which was subsequently heated with *p*-aminobenzoylglutamic acid or *p*-aminobenzoic acid in ethylene glycol containing sodium methoxide to afford a small yield of pteroylglutamic acid (**76**, $n=2$) or pteroic acid (**77**, R=OH), respectively.[1174, 1236]

Reductone (2,3-dihydroxypropenal) may be used as a three-carbon fragment but in a different way. For example, it was first allowed to react with diethyl *p*-aminobenzoylglutamate to give a crude Schiff base (**78**, R=Et), which was subsequently condensed with 2,5,6-triamino-4(3*H*)-pyrimidinone (**79**) in ethylene glycol at 130°C to give (after a gentle acid hydrolysis) pteroylglutamic acid (**76**, $n=2$). The oxidation state of the Schiff base precluded any necessity for an aerial or other oxidative step.[651] Similar procedures with appropriate amino acids were used to furnish pteroic acid (**77**, R=OH), 2-amino-6-*p*-

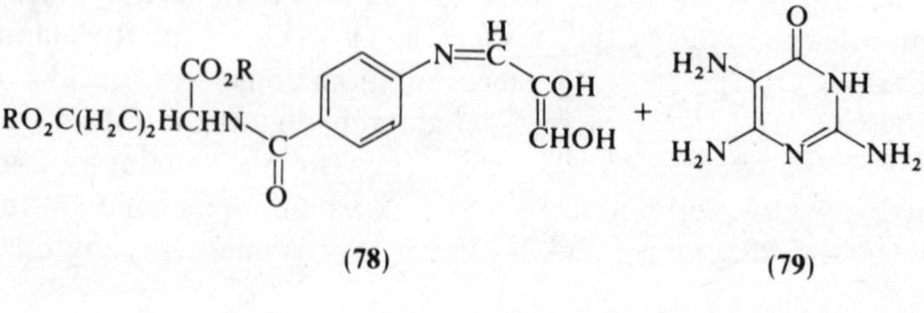

(78) (79)

sulfamoylanilinomethyl-4(3*H*)-pteridinone and analogs.[926, 1177] See also a related procedure.[527]

1,3-*Dibromoacetone* and related compounds have been used to make pteroylglutamic acid (**76**, $n=2$).[1252] An oxidative step is clearly necessary, a by-product consisting of 2-amino-7-methyl-4(3*H*)-pteridinone has been isolated, and the mechanism has been investigated.[1253]

2,2,3-*Tribromopropionaldehyde* and related compounds have been used in the folic acid-type synthesis* without any necessity for an oxidative step. Thus 2,5,6-triamino-4(3*H*)-pyrimidinone, *p*-aminobenzoic acid, and the tribromopropionaldehyde in ethylene glycol at 100°C gave pteroic acid (**77**, R=OH).[1374] Likewise, 2,2,3-trichlorobutyraldehyde, *p*-aminobenzoylglutamic acid, and 2,5,6-triamino-4(3*H*)-pyrimidinone gave pteroylglutamic acid (**76**, $n=2$) bearing an extra *C*-methyl group at the 9-position[964]; a further appropriate variation gave 9,10-dimethylated 4-aminopteroylglutamic acid.[964]

1,1,3-*Tribromoacetone* and related compounds are also exact equivalents of α-aldehydoketones as three-carbon fragments. 1,1,3-Tribromoacetone was first so used in the synthesis of pteroylglutamic acid (**76**, $n=2$) under normal conditions.[1374, 1493] However, it was later shown[1351] that the addition of sodium bisulfite to the reaction mixture resulted in a greatly improved yield of a better product and that 1,1,3-tribromoacetone could be replaced without adverse effect by 1-bromo-3,3-dichloro- or 1,1,3-trichloroacetone. Similar conditions were used to condense *m*-aminobenzoylglutamic acid, 1,1,3-trichloroacetone, and 2,5,6-triamino-4(3*H*)-pyrimidinone to give the meta-isomer of pteroylglutamic acid.[1321] Such trihalogenoacetones were also used to make [2-^{14}C]-[1372] and [7,9-^{14}C]-pteroylglutamic acid[1376]; pteroic acid (**77**, R=OH),[1374] and several analogs of the latter in which the carboxyl group was replaced by an arsonic acid group [As(:O)$(OH)_2$].[652, cf. 805] 1,3-Dichloro-3-(phenylthio)acetone has been used to make pteroylglutamic acid.[1377]

(e) Use of Sugars as Pro-α-Aldehydoketones

Earlier work on the condensation of glucose with *o*-phenylenediamine to give 2-α,β,γ,δ-tetrahyroxybutylquinoxaline,[1494, 1495] appeared to stimulate an investigation by Paul Karrer and colleagues in Zürich on the essentially more complicated condensation of 2,5,6-triamino-4(3*H*)-pyrimidinone with glucose and other sugars.[1210] This paper brought immediate responses from London[1176] and Kalamazoo,[1143] where independent studies had been in progress, and the resulting discussions of structures and so on,[192, 193, 1147, 1211, 1219] soon involved German workers as well.[706, 937, 962, 1387] A brief history of this fascinating episode has been published.[1456] From such beginnings, the condensation of diaminopyrimidines has been studied with all manner of sugars. The following examples are classified according to the length of the polyhydroxyalkyl side chain on the resulting pteridine. In nearly all cases, an aerial oxidative step is necessary.

---

* This version has been known as the Waller reaction.[1290]

The *α,β,γ,δ,ε-pentahydroxypentyl* side chain is poorly represented by a 2-amino-6-α,β,γ,δ,ε-pentahydroxypentyl-4(3*H*)-pteridinone (**80**, $n=4$) of unknown stereo configuration, which was made from 2,5,6-triamino-4(3*H*)-pyrimidinone (**84**) and D-galactoheptose or D-glucoheptose in the presence of hydrazine.[912]

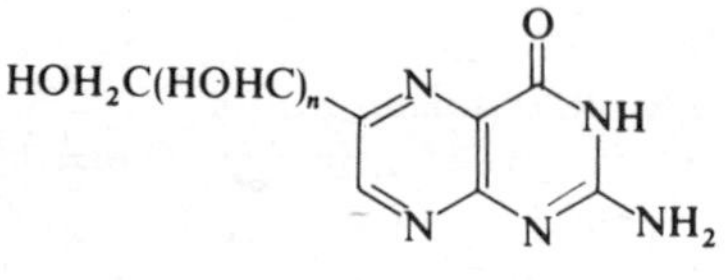

(**80**)

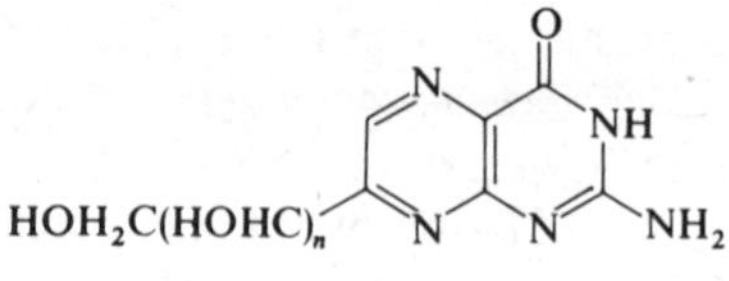

(**81**)

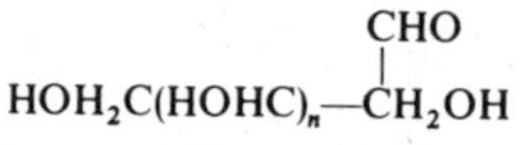

(**82**)

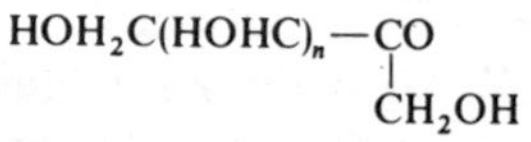

(**83**)

The *α,β,γ,δ-tetrahydroxybutyl* side chain arises from aldo or keto-hexoses. Such condensations with diaminopyrimidines are quite *dirty* and crude products require considerable purification, usually by chromatography; in some of the earlier papers, described products are by no means pure. Condensation of 2,5,6-triamino-4(3*H*)-pyrimidinone (**84**) with glucose (**82**, $n=3$) gave a product high in 2-amino-7-[D-*arabino*]-α,β,γ,δ-tetrahydroxybutyl-4(3*H*)-pteridinone (**81**, $n=3$) and low in the 6-[D-*arabino*]-isomer (**80**, $n=3$).[192, 1143, 1210, 1219, 1378] The same reaction in the presence of hydrazine gave a product in which the 6-[D-*arabino*]-isomer (**80**, $n=3$) predominated,[192, 1143, 1147, 1176, 1211, 1378] and a similar reaction using fructose (**83**, $n=3$) with or without hydrazine, also gave mainly the 6-[D-*arabino*]-isomer (**80**, $n=3$).[937, 1210, 1211] These results are broadly true for other aldoses and ketoses as well.[1456] In the presence of hydrazine, 1-*p*-toluidino-1-deoxyfructose (also called *p*-tolyl-D-isoglucosamine) can replace fructose to give 2-amino-6-[D-*arabino*]-α,β,γ,δ-tetrahydroxybutyl-4(3*H*)-pteridinone (**80**, $n=3$)[193, 576, 1060, 1373, 1378] but without hydrazine, trihydroxybutylpteridines were formed (see next). Other hexoses may be used in such condensations: D-galactose gave mainly 2-amino-7-[D-*lyxo*]-α,β,γ,δ-tetrahydroxybutyl-4(3*H*)-pteridinone (**81**, $n=3$)[1210]; D-galactose in the presence of toluidine and hydrazine gave mainly the 6-[D-*lyxo*]-isomer (**80**, $n=3$)[962]; and L-sorbose with hydrazine gave the 6-[L-*xylo*]-isomer (**80**, $n=3$).[1147]

Condensation of pyrimidines other than (**84**) with hexoses is exemplified in that of 2,4,5,6-pyrimidinetetramine with glucose or 1-*p*-toluidino-1-deoxy-D-fructose in the presence of hydrazine to give mainly 6-[D-*arabino*]-*α,β,γ,δ-tetrahydroxybutyl-2,4-pteridinediamine*[1058, 1257]; of 5,6-diamino-1(or 3)-methyl-2,4(1*H*,3*H*)-pyrimidinedione (**85**, $R^1=H$, $R^2=Me$; or $R^1=Me$, $R^2=H$) with D-glucose in aqueous acetic acid to give 1- and 3-*methyl-7-*[D-*arabino*]-*α,β,γ,δ-tetrahydroxybutyl-2,4(1H,3H)-pteridinedione*, respectively[862, cf. 937, 1243]; of the first of these pyrimidines, viz (**85**, $R^1=H$, $R^2=Me$), with 1-*p*-toluidino-1-deoxy-

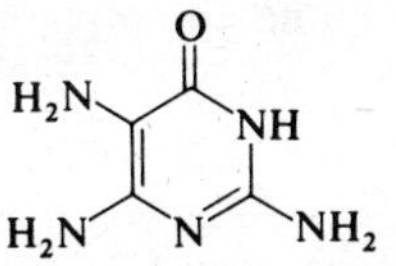
(84)

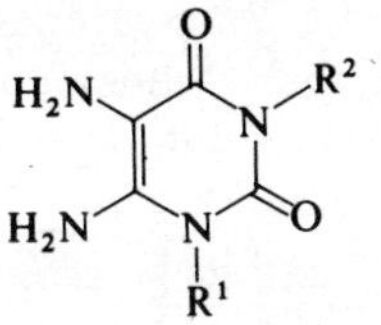
(85)

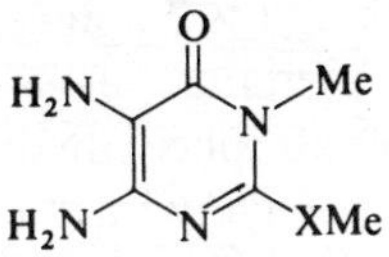
(86)

D-fructose in the presence of hydrazine to give 1-*methyl*-6-[D-*arabino*]-*α,β,γ,δ-tetrahydroxybutyl*-2,4(1*H*,3*H*)*pteridinedione*[962, cf. 937]; of the pyrimidine (**85**, $R^1=R^2=H$) with D-glucose or D-galactose to give 7-[D-*arabino*]-*α,β,γ,δ-tetrahydroxybutyl*-2,4(1*H*,3*H*)-*pteridinedione* or *its* [D-*lyxo*]-*isomer*, respectively[706]; of the pyrimidine (**85**, $R^1=R^2=Me$) with D-glucose or with 1-*p*-toluidino-l-deoxy-D-fructose–hydrazine to give 1,3-*dimethyl*-7(or 6)-[D-*arabino*]-*α,β,γ,δ-tetrahydroxybutyl*-2,4(1*H*,3*H*)-*pteridinedione*, respectively[862, 962] ; and of the pyrimidine (**86**, X=O or S) with D-glucose in aqueous ethanolic acetic acid to give 2-*methoxy*-3-*methyl*- or 3-*methyl*-2-*methylthio*-7-[D-*arabino*]-*α,β,γ,δ-tetrahydroxybutyl*-4(3*H*)-*pteridinone*, respectively.[863]

A *trihydroxybutyl* side chain was obtained unexpectedly when 2,5,6-triamino-4(3*H*)-pyrimidinone (**84**) reacted with 1-*p*-toluidino-1-deoxy-D-fructose in the absence of hydrazine to give an inseparable mixture of 2-amino-6(and 7)-[D-*erythro*]-β,γ,δ-trihydroxybutyl-4(3*H*)-pteridinone by an obscure mechanism.[193, 1373] A different trihydroxybutyl side chain was a more logical result when the pyrimidine (**84**) was condensed with L-rhamnose in the presence of hydrazine to give only 2-amino-6-[L-*arabino*]-α,β,γ-trihydroxybutyl-4(3*H*)-pteridinone.[330] This also occurred in the condensation of 5,6-diamino-2,4(1*H*,3*H*)-pyrimidinedione (**85**, $R^1=R^2=H$) with 6-deoxy-[L-*lyxo*]-hexulose to give 7-[L-*lyxo*]-α,β,γ-trihydroxybutyl-2,4(1*H*,3*H*)-pteridinedione (21%) of confirmed structure.[1247]

The *α,β,γ-trihydroxypropyl* side chain is seen on the following pteridines, made from appropriate pyrimidinediamines and the given sugar: 2-*amino*-6-[D-*erythro*]-*α,β,γ-trihydroxypropyl*-4(3*H*)-*pteridinone* (neopterin: **80**, $n=2$) (D-arabinose phenylhydrazone in methanol containing 2-mercaptoethanol: 37%[1216, 1496]; D-ribose, *p*-toluidine, and hydrazine in dilute acetic acid followed by removal of the 7-isomer: 4% yield[330, 904, 905, 911, 1422, 1563]; or 1-benzylamino-1-deoxy-D-ribulose in methanol containing 2-mercaptoethanol: 6%[509]); the 6-[L-*erythro*]-*isomer* (L-arabinose phenylhydrazone: 42%[1216]; L-arabinose, *p*-toluidine, and hydrazine: 86% crude yield[962]; L-arabinose and hydrazine[330]; or 1-benzylamino-1-deoxy-L-ribulose in methanolic 2-mercaptoethanol: ~5%[1495]); the 6-[D-*threo*]-*isomer* (D-xylose phenylhydrazone: 33%[1216, 1495,1563]; D-xylose and hydrazine[330]; or 1-benzylamino-1-deoxy-Dxylulose in methanolic 2-mercaptoethanol[1495]); and the 6-[*L-threo*]-*isomer* (L-monapterin) (L-xylose phenylhydrazone: 57–67%[1216, 1496, 1563]; 1-benzylamino-1-deoxy-L-xylulose in methanolic 2-mercaptoethanol followed by oxygen[509]). In addition, *2-amino-7-[D-erythro]-α,β,γ-trihydroxypropyl-4(3H)-pteridinone* (**81**, $n=2$) and its [*L-erthro*]-, [*D-threo*]-, and [*L-threo*]-*isomers* have been made from the pyrimidine (**84**) with appropriate pentoses in boiling dilute acetic acid, although their isomeric purities should be accepted with some reservation.[1210, 1556] Analogs lacking the 2-amino-4-oxo configuration include, for example, *2-dimethylamino-6-[D(and L)-erythro]-α,β,γ-trihydroxypropyl-4(3H)-pteridinone* (D- and L-arabinose phenylhydrazone)[109, 1710]; their [*D*(and *L*)-*threo*] *isomers* (D- and L-xylose phenylhydrazone; the L-form proved identical with the natural product, euglenapterin)[109, 1710]; *7-[D-threo]-α,β,γ-trihydroxypropyl-1-2,4(1H,3H)-pteridinedione* [from (**85**, $R^1=R^2=H$) and D-xylose in water][706]; its 7-[*D-lyxo*]-*isomer* (similarly from galactose)[706]; *2-methylthio-6-[D(and L)-erythro]-α,β,γ-trihydroxypropyl-4(3H)-pteridinone* (D- and L-arabinose phenylhydrazone followed by oxidation with potassium ferricyanide and oxygen: 60–70%)[458]; and *2-ε-carboxypentylamino-* or *2-carboxymethylamino-6-[L-erythro]-α,β,γ-trihydroxypropyl-4(3H)-pteridinone* (L-arabinose phenylhydrazone followed by aerial oxidation: 40 and 53%).[461]

An *α,β-dihydroxypropyl* side chain occurs on the important natural product, biopterin, 2-amino-6-[*L-erythro*]-α,β-dihydroxypropyl-4(3*H*)-pteridinone (**87**, $R=H$). Although several synthetic procedures are now available for biopterin, careful and time consuming purification of the product is invariably needed. Thus condensation of 2,5,6-triamino-4(3*H*)-pyrimidinone (i) with 5-deoxy-DL-ribulose in the presence of 2-mercaptoethanol gave, after oxidation and extensive purification from 2-amino-4(3*H*)-pteridinone and other by-products, DL-biopterin in 12% yield[1327]; (ii) with 1-(*N*-benzyl-*N'*-phenylhydrazino)-1,5-dideoxy-L-ribulose gave L-biopterin (10–15%)[1497]; (iii) with 5-deoxy-L-arabinose phenylhydrazone followed by oxidation gave L-biopterin (~40%)[429, 454, 455, 645, 1439]; and (iv) with other sugars and/or by other procedures gave some biopterin.[500, 564, 571, 1043, 1324, 1326, 1456, 1564] Other pteridines with a dihydroxypropyl side chain include *2-amino-6-[L-threo]-α,β-dihydroxypropyl-4(3H)-pteridinone*, its [*D-threo*]-*isomer*, and its [*D-erythro*]-*isomer* (from the phenylhydrazones of 5-deoxy-L-xylose, 5-deoxy-D-xylose, and 5-deoxy-D-arabinose, respectively)[454]; *2-ε-carboxypentylamino-6-[L-erythro]-α,β-dihydroxypropyl-4(3H)-pteridinone* (~10%) (from 5,6-diamino-2-ε-carboxypentylamino-4(3*H*)-pyrimidinone and 5-deoxy-L-arabinose phenylhydrazone, followed by oxidation)[461]; *6-[L-erythro]-α,β-dihydroxypropyl-2-dimethylamino-4(3H)-pteridinone* (**87**, $R=Me$) (from 5,6-diamino-2-dimethylamino-4(3*H*)-

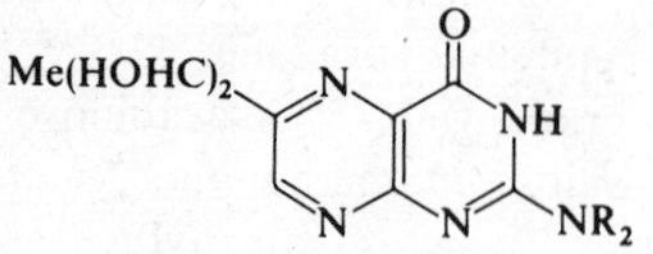

(**87**)

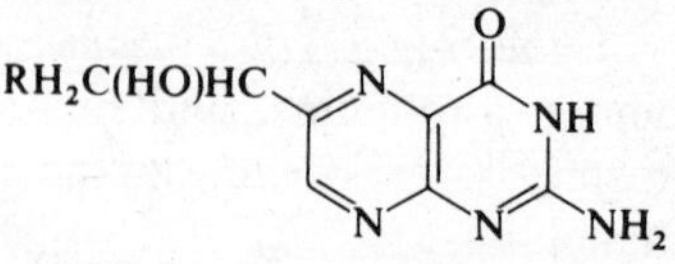

(**88**)

pyrimidinone as just cited: 51% yield)[1439], 6-[*L-erythro*]-*α,β-dihydroxypropyl-2,4-pteridinediamine* (from 2,4,5,6-pyrimidinetetramine with 5-deoxy-L-arabinose, *p*-toluidine, and hydrazine: <10% yield)[1058]; and 6-[*L-erythro*]-*α,β-dihydroxypropyl-2,4(1H,3H)-pteridinedione* (biolumazine) (from 5,6-diamino-2,4(1*H*,3*H*)-pyrimidinedione and 5-deoxy-L-arabinose phenylhydrazone: 42% yield).[1439]

α,β-Dihydroxyethyl and *α-hydroxyethyl* side chains are represented in 2-amino-6-[*αS* (and *αR*)]-α,β-dihydroxyethyl-4(3*H*)-pteridinone (**88**, R=OH) [from the pyrimidine (**84**) and D- or L-threose phenylhydrazone, respectively, followed by oxidation: ~25%] and in 2-amino-6-[*αS* (and *αR*)]-α-hydroxyethyl-4(3*H*)-pteridinone (**88**, R=H) [from the same pyrimidine and 4-deoxy-L(or D)-erythrose phenylhydrazone as just cited: ~12%]. In all cases, much 2-amino-4(3*H*)-pteridinone was formed during oxidation.[455]

C. Use of the α-Aldehydoester, Ethyl Glyoxylate, or an Equivalent

Condensations between pyrimidinediamines and ethyl glyoxylate can give only a pteridin-6-one, a pteridin-7-one, or a mixture of both products. Groups in the 1- to 5- and 8-position of the product may be varied by appropriate presubstitution in the pyrimidinediamine but any possible changes to ethyl glyoxylate, such as the use of its acetal or an homologous alkyl group in place of ethyl, can at best affect only the ratio of 6- to 7-pteridinone or the overall yield. As with aldehydoketones, the ratio of isomeric products may be controlled to some extent by the reaction conditions. The most popular form of ethyl glyoxylate is its ethyl hemiacetal (**89**),[1499] now commercially available and sometimes called ethyl ethoxyglycolate.

(1) *Simple Cases*

The condensation of 4,5-pyrimidinediamine (**90**) with ethyl glyoxylate hemiacetal (**89**) at pH 6 gave a mixture of 7(8*H*)-pteridinone (**91**) and 6(5*H*)-pteridinone (**93**) in which the former predominated. The products could be separated at pH 2 where the 7-isomer was insoluble but the 6-isomer [which is more basic because it exists as its covalent 7,8-hydrate (**92**)] remained in solution as the cation. When the condensation was carried out at pH 0, the 6-isomer (**93**) predominated in the products.[31] For practical purposes, the condensation at pH 10 in 1*M* sodium carbonate under reflux gave almost exclusively the 7-isomer (**91**), initially as its sodium salt from which it could be recovered in 77% yield.[41] In contrast, the condensation at pH 0 in 1*M* sulfuric acid at 37°C gave almost exclusively the 6-isomer (**93**) as its stable hydrate (**92**) in 85% yield.[19] Likewise, condensation of 4-methylamino-5-pyrimidinamine with the hemiacetal (**89**) in water gave (anhydrous) 8-methyl-7(8*H*)-pteridinone (**94**, R=H) (85%; no isomeric product is structurally possible)[34] while condensation of 5-methylamino-

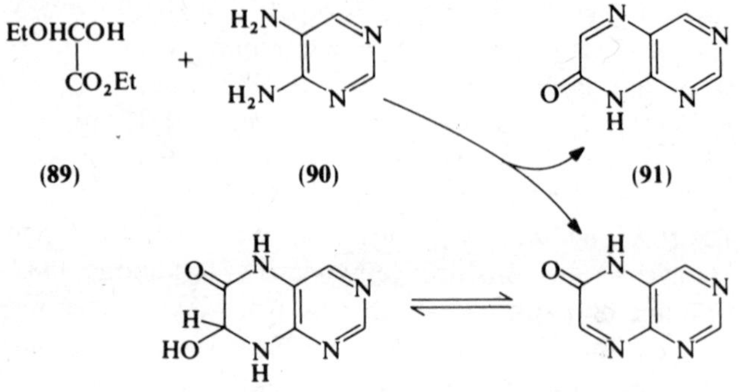

(89) (90) (91)

(92) (93)

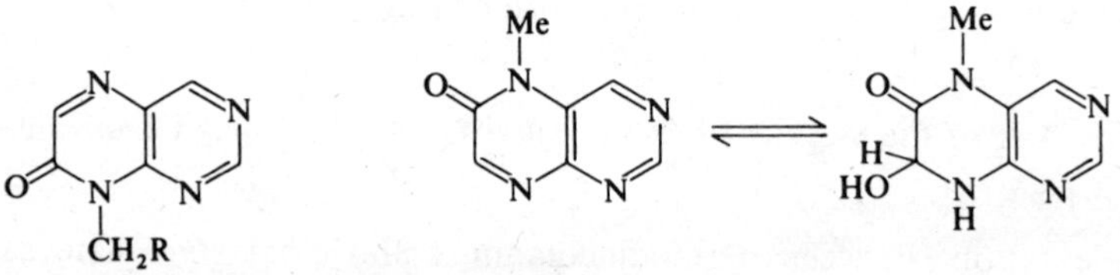

(94) (95) (96)

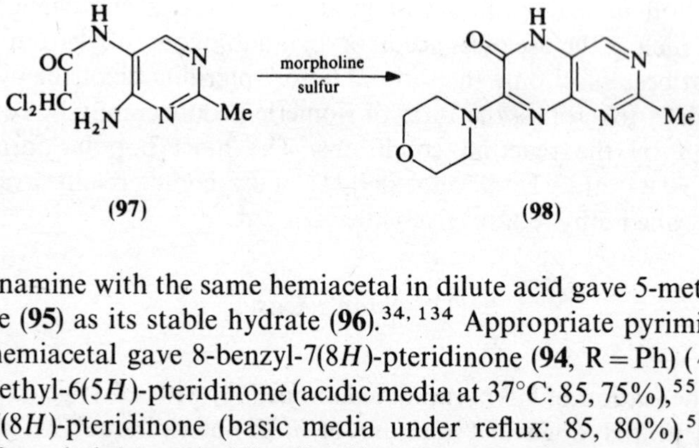

(97) (98)

4-pyrimidinamine with the same hemiacetal in dilute acid gave 5-methyl-6(5*H*)-pteridinone (**95**) as its stable hydrate (**96**).[34, 134] Appropriate pyrimidines with the same hemiacetal gave 8-benzyl-7(8*H*)-pteridinone (**94**, R = Ph) (~80%),[125] 2- and 4-methyl-6(5*H*)-pteridinone (acidic media at 37°C: 85, 75%),[55] and 2- and 4-methyl-7(8*H*)-pteridinone (basic media under reflux: 85, 80%).[55] Condensation of 2-methyl-4,5-pyrimidinediamine with the aldehydoester equivalent, dichloroacetic acid, stopped prior to ring closure to give only 5-dichloroacetamido-2-methyl-4-pyrimidinamine (**97**). On treatment with morpholine containing sulfur, this gave 2-methyl-7-morpholino-6(5*H*)-pteridinone (**98**), which clearly arose through oxidation of the covalent 7,8-adduct of morpholine with 2-methyl-6(5*H*)-pteridinone.[575, 591]

(2) *Cases with a 2- or a 4-Functional Group*

The condensation of ethyl glyoxylate ethyl hemiacetal (**89**) or equivalent with a pyrimidinediamine already bearing one other functional group is illustrated in

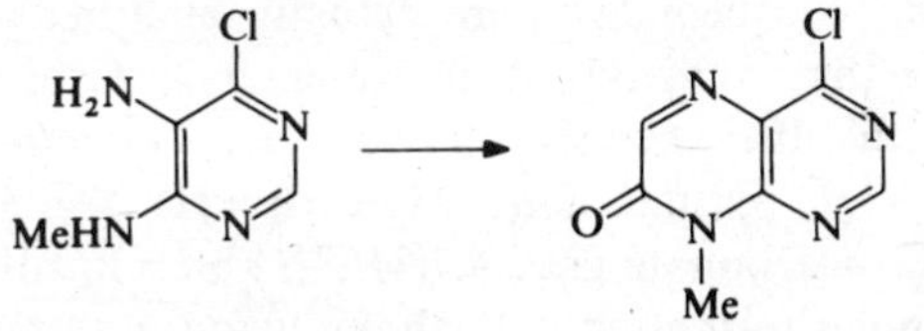

(99) **(100)**

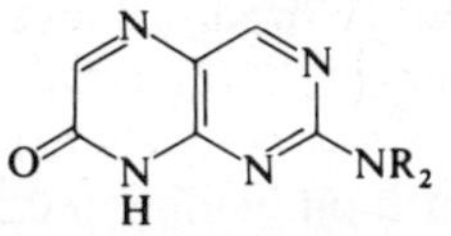

(101)

the following transformations: 4-chloro-6-methylamino-5-pyrimidinamine (**99**) to 4-chloro-8-methyl-7(8*H*)-pteridinone (**100**) (in boiling water for 15 min: >80%)[248]; 2,4,5-pyrimidinetriamine to 2-amino-7(8*H*)-pteridinone (**101**, R = H) (>85%)[383]; 2-dimethylamino-4,5-pyrimidinediamine to 2-dimethylamino-7(8*H*)-pteridinone (**101**, R = Me) (25°C for 2 h, then in boiling aqueous sodium bicarbonate: ~40%)[381]; 2-methylamino-4,5-pyrimidinediamine to 2-methylamino-7(8*H*)-pteridinone (as just cited ~20%)[132]; 2,4-bisethylamino-5-pyrimidinamine to 8-ethyl-2-ethylamino-7(8*H*)-pteridinone (as just cited: ~25%)[387]; 4,5,6-pyrimidinetriamine to 4-amino-7(8*H*)-pteridinone (in ethanol at 25°C for 12 h: 58%[224]; in water followed by bicarbonate and separation of isomers: ~20%[445]) and to 4-amino-6(5*H*)-pteridinone (in 1*M* sulfuric acid for 4 days at 37°C: ~75%[445]); 2-phenyl-4,5,6-pyrimidinetriamine to 4-amino-2-phenyl-7(8*H*)-pteridinone (in methanol for 12 h: 42%)[224]; 6-dimethylamino-4,5-pyrimidinediamine to 4-dimethylamino-7(8*H*)-pteridinone (in 1*M* sodium carbonate followed by separation of isomers at pH 2: ~30%) and 4-dimethylamino-6(5*H*)-pteridinone (in 1*M* sulfuric acid: ~50%)[445]; 6-methylamino-4,5-pyrimidinediamine to a separable mixture of 4-amino-8-methyl-7(8*H*)-pteridinone, 4-methylamino-7(8*H*)-pteridinone, and 4-methylamino-6(5*H*)-pteridinone (alkaline medium: an acidic medium favored the first product)[445]; 6-imino-1-methyl-1,6-dihydro-4,5-pyrimidinediamine (**102**) to 4-amino-1-methyl-7(1*H*)-pteridinone (**103**) (~75%)[132]; 4,5-diamino-2(1*H*)-pyrimidinone to 2,7(1*H*,8*H*)-pteridinedione [pH 7 at 50°C: separable mixture of 2,7-isomer (65%) and 2,6-isomer (25%)] and 2,6(1*H*,5*H*)-pteridinedione (in 1*M* hydrochloric acid: 80% as monohydrate)[41]; 4,5-diamino-6-methyl-2(1*H*)-pyrimidinone to 4-methyl-2,7(1*H*,8*H*)-pteridinedione (~60%)[248]; 4,5-diamino-6-methyl-2(1*H*)-pyrimidinone to 4-methyl-2,6(1*H*,5*H*)-pteridinedione (1*M* sulfuric acid at 25°C: <20%)[40]; 5-amino-1-methyl-4-methylamino-2(1*H*)-pyrimidinone to 3,8-dimethyl-2,7(3*H*,8*H*)-pteridinedione (**104**) (~50%)[134]; and 5,6-diamino-4(3*H*)-pyrimidinone to 4,6(3*H*,5*H*)-pteridinedione (1*M* sulfuric acid under reflux: 52%).[29, cf. 346] Many of these reactions were done in two

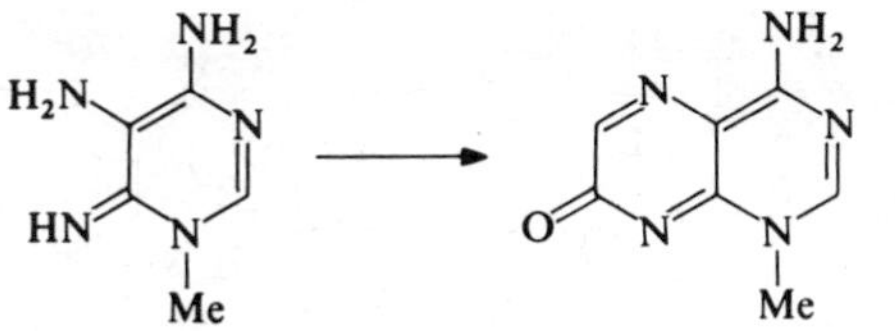

(102) **(103)**

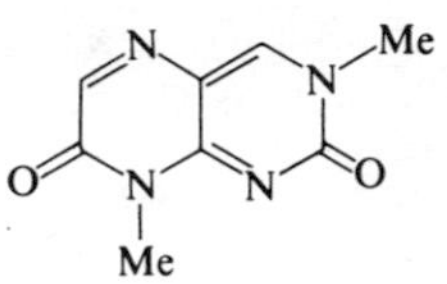

(104)

stages. Such cases, in which the intermediates were actually analyzed and characterized, are exemplified by the condensation in water at 25°C of 5,6-diamino-4(3*H*)-pyrimidinone with the hemiacetal to give 6-amino-5-ethoxycarbonylmethyleneamino-4(3*H*)-pyrimidinone (**105**, R=H) (~75%), which on boiling in 0.5*M* sodium bicarbonate gave 4,7(3*H*,8*H*)-pteridinedione (**106**, R=H) (~70%)[346]; in the similar formation of the homologous intermediate (**105**, R=Me) and then 3-methyl-4,7(3*H*,8*H*)-pteridinedione (**106**, R=Me)[346]; and in the synthesis of the isomeric 8-methyl-4,7(3*H*,8*H*)-pteridinedione.[445]

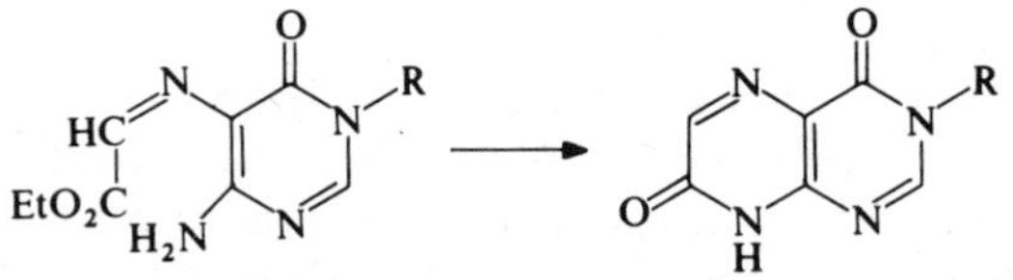

(**105**) (**106**)

(3) *Cases with Both a 2- and a 4- Functional Group*

Condensations of 2,4-disubstituted-4,5-pyrimidinediamines with the hemiacetal (**89**) differ but little from simpler examples already discussed. Accordingly, many routine cases are summarized in Table IV.

Representative examples of 2,4-*diamino*-6(*or* 7)-*pteridinones*, so made from appropriate pyrimidines, include 2,4-diamino-7(8*H*)-pteridinone (pH 6 at 100°C: 57%)[268]; 2,4-diamino-6(5*H*)-pteridinone (sodium glyoxylate in 1*M* sulfuric acid at 80°C: 70%)[268]; 2-amino-4-dimethylamino-7(8*H*)-pteridinone (under reflux: 72%)[357, cf. 268]; and 2-amino-4-dimethylamino-6(5*H*)-pteridinone (in 2*M* hydrochloric acid: 23%).[268]

Some examples of 2-*amino*-4,6(*or* 4,7)-*pteridinediones* and related compounds so made, include the parent natural products, xanthopterin (**107**) and isoxanthopterin (**108**), 2-amino-4,6(3*H*,5*H*)- and 2-amino-4,7(3*H*,8*H*)-pteridinedione, respectively. 2,5,6-Triamino-4(3*H*)-pyrimidinone has been converted into xanthopterin in at least five ways: (a) by heating in 78% sulfuric acid with the glyoxylic acid–barium bisulfite complex at 95°C for 1 h (~30% yield)[1410]; (b) by treatment with neat dichloroacetic acid to give an intermediate, 2,6-diamino-5-dichloroacetamido-4(3*H*)-pyrimidinone (**109**, X=Cl), which cyclized on heating its silver salt in the presence of silver carbonate (overall yield: 6%), or by a

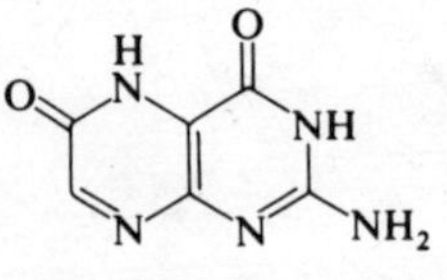

(**107**)

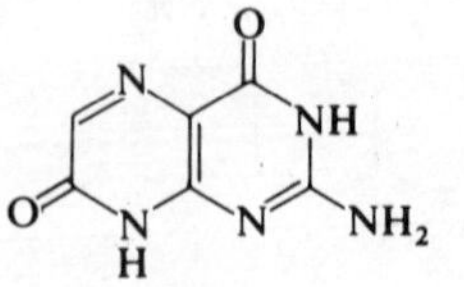

(**108**)

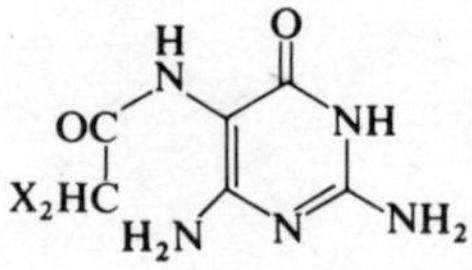

(**109**)

TABLE IV. ADDITIONAL EXAMPLES OF THE GABRIEL AND COLMAN SYNTHESIS WITH THE ALDEHYDO ESTER, ETHYL GLYOXYLATE ETHYL HEMIACETAL

Pyrimidine Intermediate	Reaction Conditions	Pteridine (Yield)	Reference
2-Dimethylamino-4-ethylamino-5-pyrimidinamine	1*M* $NaHCO_3$; reflux; 15 min	2-Dimethylamino-8-ethyl-7(8*H*)-pteridinone (~45%)	387
4-Dimethylamino-6-methylamino-5-pyrimidinamine	Ethanol; reflux; 1 h	4-Dimethylamino-8-methyl-7(8*H*)-pteridinone (~40%)	445
4-Dimethylamino-6-2′,3′,4′,6′-tetra-acetyl-β-D-glucopyranosylamino-5-pyrimidinamine	Ethanol; reflux; 2 h	4-Dimethylamino-8-2′,3′,4′,6′-tetra-acetyl-β-D-glucopyranosyl-7(8*H*)-pteridinone (~10%)	386
5-Amino-4-benzylamino-2(1*H*)-pyrimidinone	Ethanol; reflux	8-Benzyl-2,7(1*H*,8*H*)-pteridinedione	189
5-Amino-6-methylamino-4(3*H*)-pyrimidinone	Water; 25 °C; then $NaHCO_3$; reflux	6-Hydroxy-8-methyl-4(8*H*)-pteridinone[a] (~15%)	248
5,6-Diamino-1-methyl-4(1*H*)-pyrimidinone	Aqueous buffer; 25 °C; then 1*M* NaOH; 100 °C; 30 min	1-Methyl-4,7(3*H*,8*H*)-pteridinedione (~40%)	132
2,6-Bisdimethylamino-4,5-pyrimidinediamine	Aqueous EtOH; NaOAc; 100 °C; 15 min	2,4-Bisdimethylamino-7(8*H*)-pteridinone (43%)	268
2-Dimethylamino-4,5,6-pyrimidinetriamine	pH 6; reflux; 10 min	4-Amino-2-dimethylamino-7(8*H*)-pteridinone (28%)	268
2-Dimethylamino-4,5,6-pyrimidinetriamine	1*M* Sulfuric acid; 25 °C; 2 h	4-Amino-2-dimethylamino-6(5*H*)-pteridinone (38%)	268
4-Dimethylamino-6-methylamino-2,5-pyrimidinediamine	Methanol; reflux; 1 h	2-Amino-4-dimethylamino-8-methyl-7(8*H*)-pteridinone (63%)	357
6-Isopropoxy-2,4,5-pyrimidinetriamine	Methanol; reflux; 4 h	2-Amino-4-isopropoxy-7(8*H*)-pteridinone (~60%)	371
4,5-Diamino-6-dimethylamino-2(1*H*)-pyrimidinone	Aqueous NaOAc; 100 °C; 1 h	4-Dimethylamino-2,7(1*H*,8*H*)-pteridinedione (~40%)	363
6-Methoxy-2,4,5-pyrimidinetriamine	Methanol; 25 °C; 1 h; then $NaHCO_3$; reflux	2-Amino-4-methoxy-7(8*H*)-pteridinone (~50%)	380
6-Benzyloxy-2,4,5-pyrimidinetriamine	Water; 25 °C; 3 h; then $NaHCO_3$; reflux	2-Amino-4-benzyloxy-7(8*H*)-pteridinone (~55%)	380

TABLE IV. (*Contd.*)

Pyrimidine Intermediate	Reaction Conditions	Pteridine (Yield)	Reference
4-Isopropoxy-6-methylamino-2,5-pyrimidinediamine	Methanol; reflux; 90 min	2-Amino-4-isopropoxy-8-methyl-7(8*H*)-pteridinone (~10%)	371
4-Benzylamino-6-benzyloxy-2-dimethylamino5-pyrimidinamine	MeOH; 25 °C; then methanolic NaOMe; reflux; 80 min	8-Benzyl-4-benzyloxy-2-dimethylamino-7(8*H*)-pteridinone[b] (27%)	246
2-Dimethylamino-4-methoxy-6-methylamino-5-pyrimidinamine	Water; 25 °C; then methanolic $NaHCO_3$; reflux	2-Dimethylamino-4-methoxy-8-methyl-7(8*H*)pteridinone (12%)	417
2,5-Diamino-6-β-hydroxyethylamino-4(3*H*)-pyrimidinone	Water; 25 °C; then $NaHCO_3$; reflux	2-Amino-8-β-hydroxyethyl-4,7(3*H*,8*H*)-pteridinedione (72%)	422
2,5-Diamino-6-methylamino-4(3*H*)-pyrimidinone	Water; 25 °C; then $NaHCO_3$; reflux; 45 min	2-Amino-8-methyl-4,7(3*H*,8*H*)-pteridinedione (~25%)	383
5-Amino-2-dimethylamino-6-methylamino-4(3*H*)-pyrimidinone	Ethanol: 25 °C; then ethanolic NaOEt; reflux; 3 h	2-Dimethylamino-8-methyl-4,7(3*H*,8*H*)-pteridinedione (~70%)	360
4,5-Diamino-6-anilino-2(1*H*)-pyrimidinone	Water; reflux; 2 h	4-Amino-8-phenyl-2,7(1*H*,8*H*)-pteridinedione[c] (~50%)	363
2,5-Diamino-6-D-sorbitylamino-4(3*H*)-pyrimidinone	Water; reflux; then $NaHCO_3$; reflux; 15 min	2-Amino-8-D-sorbityl-4,7(3*H*,8*H*)-pteridinedione (44%)	172
2-Dimethylamino-4-methoxy-6-2′,3′,5′-tri-*O*-benzoyl-β-D-ribofuranosylamino-5-pyrimidinamine	Methanol; 60 °C; then *t*-BuOK; methanol; reflux	2-Dimethylamino-4-methoxy-8-β-D-ribofuranosyl-7(8*H*)-pteridinone[d] (23%)	417
4,5,6-Triamino-2(1*H*)-pyrimidinethione	1*M* sodium carbonate; 100 °C; 1 h	4-Amino-2-thioxo-1,2-dihydro-7(8*H*)-pteridinone (~65%)	445
5-Amino-6-D-ribitylamino-2,4(1*H*,3*H*)-pyrimidinedione	Aqueous AcOH; 25 °C; then $NaHCO_3$; reflux; 15 min	8-D-Ribityl-2,4,7(1*H*,3*H*,8*H*)-pteridinetrione (54%)	172
5,6-Diamino-2-dimethylamino-4(3*H*)-pyrimidinone	80% Sulfuric acid; 50 °C; 20 min	2-Dimethylamino-4,6(3*H*,5*H*)-pteridinedione (16%)	1710

[a] Necessarily in the 6-hydroxy form.
[b] And the 4-methoxy analog.
[c] And the 8-benzyl homolog.
[d] And analogs.[417, 592, 1090]

variation using dibromoacetic acid to give the dibromo intermediate (**109**, X = Br) and cyclizing in aqueous sodium bicarbonate at 50°C (65%)[404, 528]; (c) by heating with chloral hydrate in 80% sulfuric acid at 100°C (36%)[188, cf. 405]; (d) by brief heating at 100°C in 80% sulfuric acid to which had been added potassium diacetoxyacetate (yield: 40–45%)[273, 1011]; and (e) by heating at 100°C in 80% sulfuric acid with ethyl glyoxylate ethyl hemiacetal (78%).[94, 269] Isoxanthopterin (**108**) was first made[405] from its 6-carboxylic acid but it is better made directly from 2,5,6-triamino-4(3*H*)-pyrimidinone by treatment with aqueous ethyl glyoxylate ethyl hemiacetal at pH 3 to give the intermediate, 2,6-diamino-5-ethoxycarbonylmethyleneamino-4(3*H*)-pyrimidinone (**110**), followed by cyclization in boiling 1*M* ammonia or sodium bicarbonate (overall yield: 50%).[61, 62, 339, 1267]

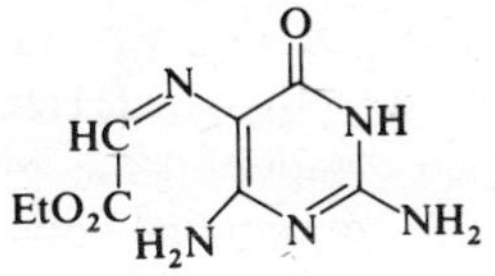

(**110**)

Some related pteridines, made from the hemiacetal (**89**) and appropriate pyrimidines, are exemplified in 2-dimethylamino-4-methoxy-7(8*H*)-pteridinone (two-stage: 50%)[417]; 2-amino-4-pentyloxy-7(8*H*)-pteridinone (two-stage: 49%; one-pot: 45%)[436]; 4-benzyloxy-2-dimethylamino-7(8*H*)-pteridinone (71%)[246]; 2-dimethylamino-4,7(3*H*,8*H*)-pteridinedione (~60%)[360]; 4-amino-2,7(1*H*,8*H*)-pteridinedione (70%)[41, 363]; 2-amino-4-methoxy-8-methyl-7(8*H*)-pteridinone (**111**) (~80%)[371]; 2-amino-8-methyl-4,7(3*H*,8*H*)-pteridinedione (glyoxylic acid hydrate: 45%)[422]; 2-amino-3-methyl-4,6(3*H*,5*H*)-pteridinedione (in 78% sulfuric acid at 90°C for 3 min: ~50%)[369]; 2-amino-3-methyl-4,7(3*H*,8*H*)-pteridinedione (two-stage: ~40%)[383]; 4-amino-1-methyl-2,7(1*H*,8*H*)-pteridinedione (~35%)[363]; 4-amino-8-methyl-2,7(1*H*,8*H*)-pteridinedione (**112**) (~30%)[363]; 2-amino-8-D-ribityl-4,7(3*H*,8*H*)pteridinedione (42%)[172]; 4-amino-2-methylthio-7(8*H*)-pteridinone (**113**, R = Me) (34%)[445, 1126]; its 2-pentylthio analog (**113**, R = Pe)[621]; 4-dimethylamino-2-methylthio-7(8*H*)-pteridinone (29%)[445, 1126]; and related compounds.[1541, 1542]

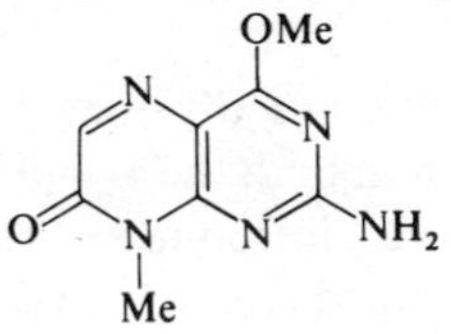

(**111**)

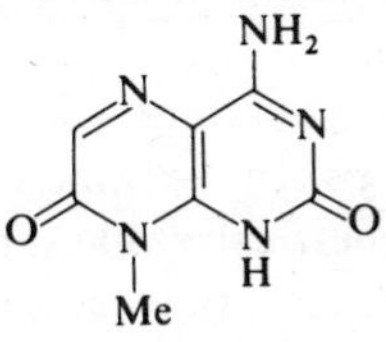

(**112**)

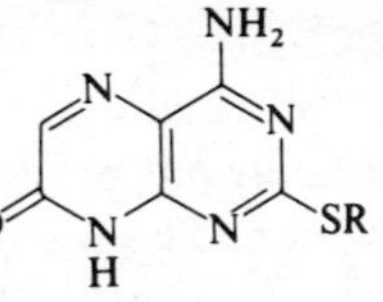

(**113**)

The parent 2,4,6- and 2,4,7-*pteridinetriones*, as well as related compounds, may be made by condensation of glyoxylic acid derivatives with pyrimidines. Thus

2,4,6(1*H*,3*H*,5*H*)-pteridinetrione (**114**, R=H) was made from 5,6-diamino-2,4(1*H*,3*H*)-pyrimidinedione with the glyoxylic acid–sodium bisulfite complex in 78% sulfuric acid at 100°C for 1 h (yield unstated)[1369]; with ethyl glyoxylate ethyl hemiacetal in 78% sulfuric acid at 90°C for 2–3 min: 50% yield[41, 340, 1116]; or with neat dichloroacetic acid, first at 110°C and then at 170°C for 3 h: 37%.[101] The isomeric 2,4,7(1*H*,3*H*,8*H*)-pteridinetrione (**115**, R=H) was best made from the same pyrimidine and the hemiacetal (**89**), initially in water at 25°C and subsequently in 1*M* sodium bicarbonate under reflux to cyclize the intermediate: ~60% yield.[339] The following *N*-alkylated derivatives have been made similarly: 1-methyl-,[339] 3-methyl-,[339] 8-methyl- (**115**, R=Me),[339] 1,3-dimethyl-,[339, 1275, cf. 1380] 1,8-dimethyl-,[716] 1,3,8-trimethyl-,[339, 1000] 1-phenyl-,[343] 3-methyl-1-phenyl-,[344] 8-ethyl- (**115**, R=Et),[317] 8-benzyl- (**115**, R=CH_2Ph),[317] and 8-*β*-hydroxyethyl-2,4,7(1*H*,3*H*,8*H*)-pteridinetrione (**115**, R=CH_2CH_2OH)[317, 1339]; also, by condensation in 78% sulfuric acid, 3-methyl- and 1,3-dimethyl-2,4,6(1*H*,3*H*,5*H*)-pteridinetrione (**114**, R=Me).[340] The last mentioned compound was probably also made from 5,6-diamino-1,3-dimethyl-2,4(1*H*,3*H*)-pyrimidinedione and dichloroacetic acid, although it was named originally as the 2,4,7-isomer.[1380]

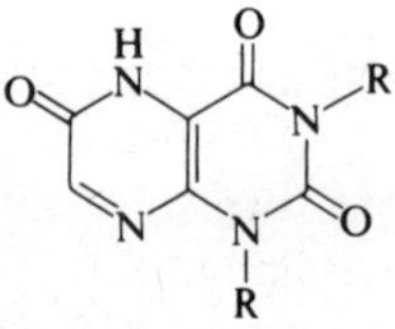

(114)

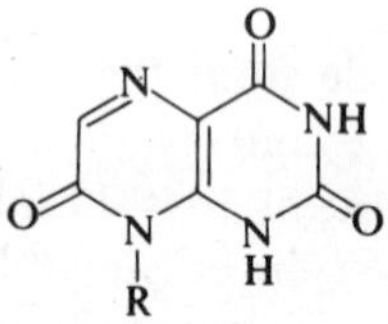

(115)

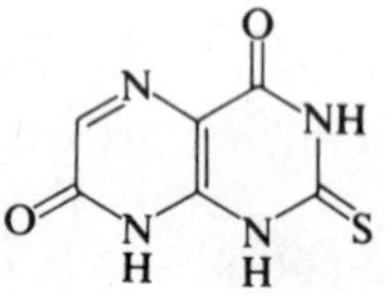

(116)

Similar procedures gave the alkylthio and thioxo analogs: 2-methylthio-4,7(3*H*,8*H*)-pteridinedione (67%)[1000]; its 3-methyl-derivative[344]; 2-thioxo-1,2-dihydro-4,7(3*H*,8*H*)-pteridinedione (**116**) (62%)[1000]; its 1-methyl, 3-methyl, 1,3-dimethyl, and 1,3,8-trimethyl derivatives[1000]; 1,3-dimethyl-4-thioxo-3,4-dihydro-2,7(1*H*,8*H*)-pteridinedione[1000]; and 1,3-dimethyl-2,4-dithioxo-1,2,3,4-tetrahydro-7(8*H*)-pteridinone (68%).[1000]

D. The Use of α-Diketones

The reaction of a 4,5-pyrimidinediamine with a symmetrical α-diketone can give only one product, a pteridine with the same substituent in both the 6- and 7-position. The use of an unsymmetrical α-diketone, however, introduces structural ambiguity and the great majority of workers have avoided this issue. A few brave souls have indeed employed such unsymmetrical diketones and, in some cases, have even deduced the structures of their products, without being able to propose any general rule(s) for forecasting the predominant isomers in such cases.[106, 142, 164, 327, 537, 674, 1112, 1217, 1348, 1349] As in previous sections, the discussion of α-diketone condensations is subdivided in the following section according to the group(s) in the 2- and/or 4-position of the resulting pteridines.

(1) *Simple Cases without a Functional Group*

The condensation of 6-methyl-4,5-pyrimidinediamine with neat benzil at 160°C to give 4-methyl-6,7-diphenylpteridine (**117**, R = Me) in ~60% yield constitutes the original Gabriel and Colman synthesis[9]; 4,5-pyrimidinediamine and benzil likewise gave 6,7-diphenylpteridine (**117**, R = H).[983] A more convenient medium proved to be refluxing methanol containing a little acetic acid, in which the methyldiphenylpteridine (**117**, R = Me) was prepared in 84% yield.[1227] Broadly similar procedures converted appropriate pyrimidinediamines into 6,7-dimethyl- (biacetyl: 90%),[33] 2,6,7-trimethyl- (90%),[33] 4,6,7-trimethyl- (71%),[66] 2,4,6,7-tetramethyl-,[66] 6,7-diethyl- (hexane-3,4-dione: 53%),[32] 6,7-dimethyl-2-phenyl- (80%),[152] 6,7-dimethyl-4-phenyl- (71%),[151] 4,6,7-trimethyl-2-phenyl- (82%),[152] 2,6,7-trimethyl-4-phenyl- (70%),[151] 2,4-dimethyl-6,7-diphenyl- (no solvent, 210°C),[657] 4,6,7-triphenyl- (86%),[959] and 6,7-dimethyl-2-styrylpteridine.[105] The use of a reagent of low oxidation state, viz. benzoin, with 6-methyl-4,5-pyrimidinediamine gave 4-methyl-6,7-diphenyl-7,8(?)-dihydropteridine of then unproven structure.[1227]

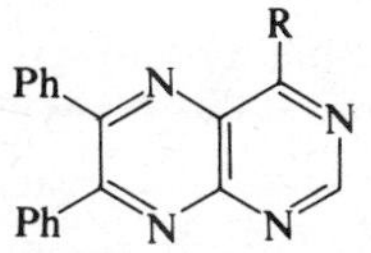

(**117**)

(2) *Cases with a 2-Functional Group Only*

Condensation of 2-chloro-4,5-pyrimidinediamine with biacetyl in boiling benzene–methanol (5:1) gave 2-chloro-6,7-dimethylpteridine (**118**, R = H) in 64% yield[170]; 2-chloro-4,6,7-trimethylpteridine (**118**, R = Me) (~55%) and 2-chloro-6,7-dimethyl-4-phenylpteridine (**118**, R = Ph) (in dilute ethanol: 57%) were made similarly.[124, 1436] Fusion of 2-chloro-6-ethyl-4,5-pyrimidinediamine with benzil at 145°C gave 2-chloro-4-ethyl-6,7-diphenylpteridine (7-chloro-2,3-diphenyl-5-ethylpyrimidazine)[415]; 2-chloro-6-phenyl-4,5-pyrimidinediamine and benzil in 2-ethoxyethanol at 120°C gave 2-chloro-4,6,7-triphenylpteridine (58%).[1118] Condensations giving 2-pteridinamines are exemplified in the formation of 6,7-dimethyl- (96%),[39] 4,6,7-trimethyl- (98%),[39, 418] 6,7-diethyl- (81%),[32] 6,7-diisopropyl- (31%),[400] 6,7-diphenyl- (44%),[400] and 4-methyl-6,7-diphenyl-2-pteridinamine.[744] They are also found in the preparation of 6,7-pentamethylene-2-pteridinamine (36%) obtained from 2,4,5-pyrimidinetriamine

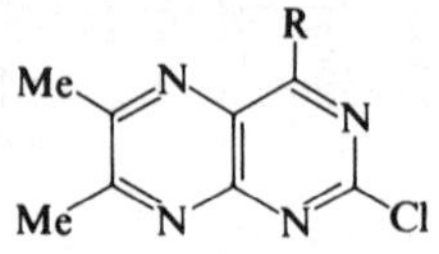

(118)

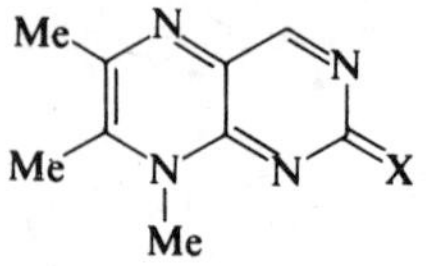

(119)

and cycloheptane-1,2-dione in refluxing methanol,[400] of 6,7,8-trimethyl-2(8*H*)-pteridinimine (**119**, X=NH) (~25%) from 4-methylamino-2,5-pyrimidinediamine and biacetyl,[190] of 8-benzyl-2-benzylimino-6,7-dimethyl-2,8-dihydropteridine from 2,4-bisbenzylamino-5-pyrimidinamine and biacetyl,[189] and of 6,7,8-trimethyl-2-methylimino-2,8-dihydropteridine (**119**, X=NMe) from 2,4-bismethylamino-5-pyrimidinamine and biacetyl.[190, 248]

Similar procedures gave 2-pteridinones and related compounds, exemplified by 6,7-dimethyl-2(1*H*)-pteridinone (93%),[134] 6,7-diethyl-2(1*H*)-pteridinone (hexane-3,4-dione: >80%),[32] 4,6,7-trimethyl-2(1*H*)-pteridinone (>95%),[38, 56] 1,6,7-trimethyl-2(1*H*)-pteridinone [from 4,5-diamino-3-methyl-2(3*H*)-pyrimidinone; isolated as its ethanolate],[120] 3,6,7-trimethyl-2(3*H*)-pteridinone (as hydrate: ~80%),[134] 6,7,8-trimethyl-2(8*H*)-pteridinone (**119**, X=O) (~75%),[134] 2-ethoxy-4-methyl-6,7-diphenylpteridine (56%),[398, 1230] 6,7-dimethyl-2(1*H*)-pteridinethione (68%),[170] the 6,7-diphenyl analog (90%),[400] 6,7,8-trimethyl-2(8*H*)-pteridinethione (**119**, X=S) (~75%),[248] 6,7-dimethyl-2-methylthio-4-phenylpteridine (73%),[1436] 2-methylthio-4,6,7-triphenylpteridine (43%),[1118] and 2-methylthio-6,7-diphenylpteridine (86%).[1138] The use of a reagent of low oxidation state is illustrated by the reaction of 2-ethoxy-6-methyl-4,5-pyrimidinediamine with benzoin to give 2-ethoxy-4-methyl-6,7-diphenyl-7,8-dihydropteridine (**120**), proven in structure by unambiguous synthesis.[398, 1230] It is also seen in the formation of 8-methyl-6,7-diphenyl-7,8-dihydro-2(1*H*)-pteridinone.[189]

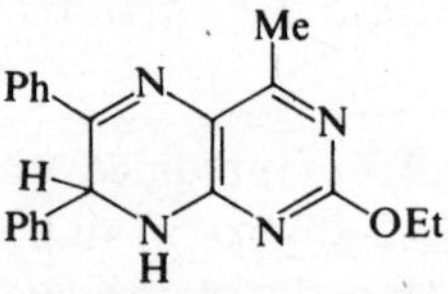

(120)

(3) *Cases with a 4-Functional Group Only*

The reaction of 6-chloro-4,5-pyrimidinediamine with biacetyl in benzene–methanol gave 4-chloro-6,7-dimethylpteridine (**121**, R=H) (73%)[170]; its 2-styryl derivative (**121**, R=CH:CHPh) was made similarly in ethanol (~55% yield).[105] 6-Trifluoromethyl-4,5-pyrimidinediamine and biacetyl in refluxing *t*-butyl alcohol gave 6,7-dimethyl-4-trifluoromethylpteridine (>80%)[156]

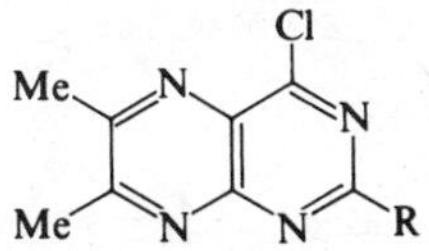

(121)

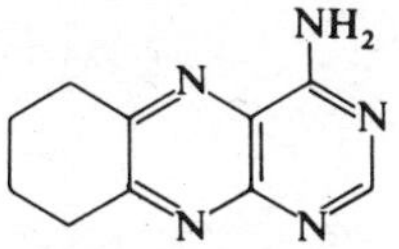

(122)

and similar reactions in ethanol gave ethyl 6,7-dimethyl-4-pteridinecarboxylate (~70%)[147, 149] and 6,7-dimethyl-4-pteridinecarboxamide.[157] When the former was made in water, a covalent dihydrate was isolated. This could be dehydrated by refluxing in *t*-butyl alcohol.[148] Condensations yielding 4-amino- or 4-(substituted-amino)-pteridines are illustrated in the formation of 6,7-dimethyl-4-pteridinamine (64–90%),[129, 170] the 6,7-diethyl homolog (hexane-3,4-dione: 71%),[32] 4-diethylamino-6,7-diphenylpteridine (~90%),[688] 6,7-tetramethylene-4-pteridinamine (**122**) (cyclohexane-1,2-dione: 29%),[400] 3,6,7-trimethyl-4(3*H*)-pteridinimine (from 6-imino-1-methyl-1,6-dihydro-4,5-pyrimidinediamine and biacetyl: ~60%, as hydriodide),[132] and 6,7,8-trimethyl-4-methylimino-4,8-dihydropteridine (from 4,6-bismethylamino-5-pyrimidinamine and biacetyl).[129]

Similarly prepared were the 4-oxo- or alkoxypteridines such as 6,7-dimethyl-4(3*H*)-pteridinone (81%),[30] the 6,7-diethyl homolog (73%),[32] the 6,7-diphenyl analog (67–75%),[32, 1316] 2,6,7-trimethyl-4(3*H*)-pteridinone (62%),[170, 1508] 6,7-dimethyl-2-phenyl-4(3*H*)-pteridinone,[663] 2-methyl-6,7-diphenyl-4(3*H*)-pteridinone (84%),[1228] 2,6,7-triphenyl-4(3*H*)-pteridinone,[663] 6,7,8-trimethyl-4(8*H*)-pteridinone (from 5-amino-6-methylamino-4(3*H*)-pyrimidinone and biacetyl: ~40%),[134] 3,6,7-trimethyl-4(3*H*)-pteridinone,[126] 6,7-dimethyl-2-phenyl-4(3*H*)-pteridinethione (58%),[1436] and 6,7-dimethyl-4-methylthiopteridine (62%).[129] The low oxidation state reagents, benzoin and *p,p'*-dimethoxybenzoin, have been used to make 2-methyl-6,7-diphenyl-5,6(or 7,8)-dihydro-4(3*H*)-pteridinone and its 6,7-bis-*p*-methoxyphenyl analog, respectively, both of unproven configuration.[1228]

(4) *Cases with Two Amino-Type Functional Groups*

Biacetyl may be condensed with 2,4,5,6-pyrimidinetetramine in water,[284, 475, 884] aqueous ethanol,[541] or aqueous acetic acid [749] to give >80% yield of 6,7-dimethyl-2,4-pteridinediamine (**123**, R = H); a similar procedure with fully deuterated biacetyl gave 6,7-bistrideuteromethyl-2,4-pteridinediamine.[1193] The homologous 6,7-diethyl- (aqueous acetic acid: 43%), 6,7-dipropyl- (42%), 6,7-diisopropyl- (aqueous alcohol: 90%), 6,7-dibutyl- (aqueous acetic acid: 40%), 6,7-diisobutyl- (30%), 6,7-di-*s*-butyl- (21%), 6,7-dipentyl- (15%), 6,7-diisopentyl- (14%), and other 6,7-dialkyl-2,4-pteridinediamine molecules were made likewise from the same pyrimidinetetramine and appropriate alkanediones.[142, 749] The analogous condensation of the tetramine with benzil

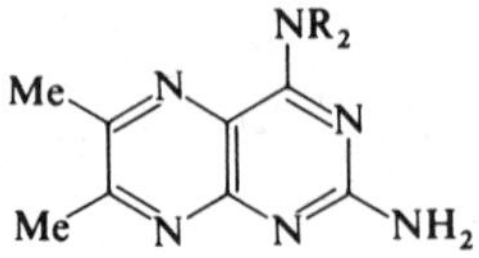

(123)

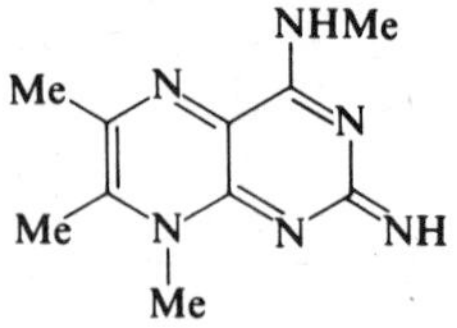

(124)

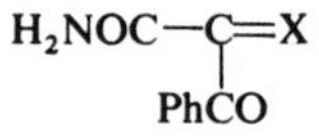

(125)

has been done in aqueous ethanol to give 6,7-diphenyl-2,4-pteridinediamine, in up to 80% yield,[284, 475, 541] and examples of the formation of substituted-phenyl analogs are given in Table V. A remarkably successful reaction was that of 2,4,5,6-pyrimidinetetramine with α,α′-dibromobiacetyl in aqueous ethanolic hydrogen bromide at 25°C to give 6,7-bisbromomethyl-2,4-pteridinediamine as hydrobromide in 84% yield.[850] The use of other pyrimidine intermediates is illustrated in the condensation of an appropriate alkanedione with 4,6-bismethylamino-2,5-pyrimidinediamine to give 6,7,8-trimethyl-4-methylamino-2(8*H*)-pteridinimine (**124**) (hydrochloride: 29%),[1354] with 2-dimethylamino-4,5,6-pyrimidinetriamine to give 2-dimethylamino-6,7-dimethyl-4-pteridinamine,[1189] with 6-dimethylamino-2,4,5-pyrimidinetriamine to give 4-dimethylamino-6,7-dimethyl-2-pteridinamine (**123**, R = Me) or its 6,7-diphenyl analog (~70%),[1188] and with 6-methylamino-2,4,5-pyrimidinetriamine to give 4-methylamino-6,7-diphenyl-2-pteridinamine.[110] Ambiguous condensations are represented by the reaction of 2,4,5,6-pyrimidinetetramine with the unsymmetrical 2,3-dioxo-3-phenylpropionamide (**125**, X = O) to a single product, proven subsequently by an unambiguous Timmis synthesis to be 2,4-diamino-7-phenyl-6-pteridinecarboxamide (**127**, R = NH_2); the same product was obtained by using α-benzoyl-α,α-dibromoacetamide (**126**, R = Br) or α-benzoyl-α-bromoacetamide (**126**, R = H) (a low oxidation state reagent requiring an aerial oxidation step in the condensation) in place of the α-diketone (**125**, X = O).[327] The 2-phenyl analog (**127**, R = Ph) was made similarly by using all three reagents with 2-phenyl-4,5,6-pyrimidinetriamine, and likewise the corresponding nitriles, 2,4-diamino-7-phenyl- and 4-amino-2,7-diphenyl-6-pteridinecarbonitrile, using α-benzoyl-α-bromoacetonitrile in each case; all structures were proven.[327] Use of the oxime, α-hydroxyiminobenzoylacetamide (**125**, X = NOH) in place of the diketone (**125**, X = O) with 2-phenyl-4,5,6-pyrimidinetriamine reversed the orientation of the product to give 4-amino-2,6-diphenyl-7-pteridinecarboxamide (**128**).[327]

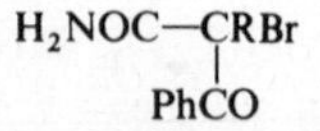

(126)

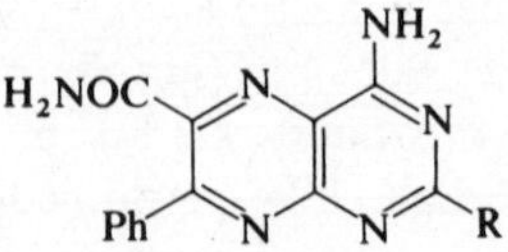

(127)

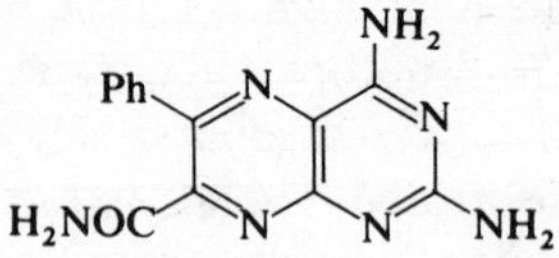

(128)

TABLE V. ADDITIONAL EXAMPLES OF THE GABRIEL AND COLMAN SYNTHESIS WITH α-DIKETONES

Pyrimidine Intermediate	Reagent and Reaction Conditions	Pteridine (Yield)	References
2,4-Dianilino-5-pyrimidinamine	Biacetyl; aqueous ethanol; reflux	6,7-Dimethyl-8-phenyl-2-phenylimino-2,8-dihydropteridine (~30%)	190
2,4-Dianilino-5-pyrimidinamine	Benzil; aqueous ethanol; reflux	6,7,8-Triphenyl-2-phenylimino-2,8-dihydropteridine (~60%)	190
2,4-Bismethylamino-5-pyramidinamine	2,5-Dimethylhexane-3,4-dione; aqueous ethanol; reflux	6,7-Di-isopropyl-8-methyl-2-methylimino-2,8-dihydropteridine (~50%)	248
6-*p*-Dimethylaminostyryl-2,4,5-pyrimidinetriamine	Benzil; ethanol; reflux	4-*p*-Dimethylaminostyryl-6,7-diphenyl-2-pteridinamine[a] (42%)	282
2-γ-Diethylaminopropylamino-4,5-pyrimidinediamine	Benzil; methanol; reflux	2-γ-Diethylaminopropylamino-6,7-diphenylpteridine[a] (71%)	400
5-Amino-4-benzylamino-2(1*H*)-pyrimidinone	Biacetyl; ethanol; reflux	8-Benzyl-6,7-dimethyl-2(8*H*)-pteridinone	189
5-Amino-4-methylamino-2(1*H*)-pyrimidinone	Benzil; aqueous ethanol; reflux; 15 min; *or* DMF; reflux	8-Methyl-6,7-diphenyl-2(8*H*)-pteridinone (79, 30%)	189, 248
5-Amino-4-methylamino-2(1*H*)-pyrimidinone	2,5-Dimethylhexan-3,4-dione; aqueous ethanol; reflux	6,7-Di-isopropyl-8-methyl-2(8*H*)-pteridinone (~80%)	248
5,6-Diamino-*N*-ethyl-4-pyrimidinecarboxamide	Biacetyl; ethanol; reflux; 10 min	*N*-Ethyl-6,7-dimethyl-4-pteridinecarboxamide (~60%)	157
6-β-Dimethylaminoethylamino-2-phenyl-4,5-pyrimidinediamine	Biacetyl; ethanol: reflux; 1 h	4-β-Dimethylaminoethylamino-6,7-dimethyl-2-phenylpteridine (61%)	1436
4,5,6-Pyrimidinetriamine	2,5-Dimethylhexane-3,4-dione; aqueous alcoholic AcOH; reflux; 6 h	6,7-Di-isopropyl-4-pteridinamine (48%)	400
4,5,6-Pyrimidinetriamine	Benzil; aqueous ethanol; reflux; 2 h	6,7-Diphenyl-4-pteridinamine (50%)	400
2-Phenyl-4,5,6-pyrimidinetriamine	Benzil; aqueous ethanol; reflux; 2 h	2,6,7-Triphenyl-4-pteridinamine (55%)	475
5,6-Diamino-2-methyl-4(3*H*)-pyrimidinone	*p,p′*-Dimethoxybenzil; ethanolic AcOH; reflux; 30 min	6,7-Bis-*p*-methoxyphenyl-2-methyl-4(3*H*)-pteridinone (~80%)	1228
5,6-Diamino-2-phenyl-4(3*H*)-pyrimidinone	*p,p′*-Dimethylbenzil; AcOH; reflux; 1 h	2-Phenyl-6,7-di-*p*-tolyl-4(3*H*)-pteridinone[a]	663
5,6-Diamino-2-fur-2′-yl-4(3*H*)-pyrimidinone	Benzil; aqueous AcOH; reflux; 1 h	2-Fur-2′-yl-6,7-diphenyl-4(3*H*)-pteridinone[a]	663
5,6-Diamino-2-*p*-tolyl-4(3*H*)-pyrimidinone	Biacetyl; water; reflux; 1 h	6,7-Dimethyl-2-*p*-tolyl-4(3*H*)-pteridinone[a]	664

TABLE V. (*Contd.*)

Pyrimidine Intermediate	Reagent and Reaction Conditions	Pteridine (Yield)	References
6-*p*-Chloroanilino-2,4,5-pyrimidinetriamine	Biacetyl; water; 30 °C; 4 h	4-*p*-Chloroanilino-6,7-dimethyl-2-pteridinamine[a]	1206
2,4,5,6-Pyrimidinetetramine	Cyclohexane-1,2-dione; aqueous ethanol; reflux; 2 h	6,7-Tetramethylene-2,4-pteridinediamine[a] (85%)	400
2,4,5,6-Pyrimidinetetramine	*p,p'*-Diaminobenzil; water; reflux; 1 h	6,7-Bis-*p*-aminophenyl-2,4-pteridinediamine[a] (>95%)	141, 730
2,4,5,6-Pyrimidinetetramine	*p,p'*-Dimethoxybenzil; aqueous AcOH; reflux; 2 h	6,7-Bis-*p*-methoxyphenyl-2,4-pteridinediamine	142
2,4,5,6-Pyrimidinetetramine	α,α'-Bis-3',4'-dichlorophenylbiacetyl; pH 6; reflux; 5 h	6,7-Bis-3',4'-dichlorobenzyl-2,4-pteridinediamine[a] (43%)	419
2,4,5,6-Pyrimidinetetramine	2,5-Bis-3',4'-dimethylphenyl-2,5-dihydro-3,4-thiophenediol *S,S*-dioxide[b]; pH 6; reflux; 7h	6,7-Bis-3',4'-dimethylbenzyl-2,4-pteridinediamine[a]	419
2,5-Diamino-6-methylamino-4(3*H*)-pyrimidinone	Dimethyl α,α'-dioxosuccinate; aqueous ethanol; reflux; 15 min	Dimethyl 2-amino-8-methyl-4-oxo-4,8-dihydro-6,7-pteridinedicarboxylate (HCl) (71%)	307
2,5-Diamino-3-methyl-6-methylamino-4(3*H*)-pyrimidinone	Dimethyl α,α'-dioxosuccinate; aqueous ethanol; reflux; 1 h	Dimethyl 2-imino-3,8-dimethyl-4-oxo-2,3,4,8-tetrahydro-6,7-pteridinedicarboxylate (HCl) (27%)	307
2-*p*-Acetamidobenzenesulfonamido-5,6-diamino-4(3*H*)-pyrimidinone	Biacetyl; dilute ammonia; 70 °C; 10 min	2-*p*-Acetamidobenzenesulfonamido-6,7-dimethyl-4(3*H*)-pteridinone[a] (96%)	186
5,6-Diamino-2-β-carboxyethylamino-4(3*H*)-pyrimidinone	Biacetyl; pH 3; reflux; 1 h	2-β-Carboxyethylamino-6,7-dimethyl-4(3*H*)-pteridinone[a] (70%)	461
5,6-Diamino-2-dimethylamino-4(3*H*)-pyrimidinone	Biacetyl; water; reflux; 45 min	2-Dimethylamino-6,7-dimethyl-4(3*H*)-pteridinone	1189
5,6-Diamino-2-dimethylamino-4(3*H*)-pyrimidinone	Benzil; DMF; reflux; 40 min	2-Dimethylamino-6,7-diphenyl-4(3*H*)-pteridinone (~35%)	359
2,5-Diamino-6-D-ribitylamino-4(3*H*)-pyrimidinone	Biacetyl; water; 90 °C; 30 min	2-Amino-6,7-dimethyl-8-D-ribityl-4(8*H*)-pteridinone (51%)	172

2,5-Diamino-3-methyl-6-methylamino-4(3*H*)-pyrimidinone	Benzil; ethanolic DMF; reflux; 2 h	2-Imino-3,8-dimethyl-6,7-diphenyl-2,8-dihydro-4(3*H*)-pteridinone[a] (~70%)	359
5,6-Diamino-2-dimethylamino-3-methyl-4(3*H*)-pyrimidinone	Biacetyl; aqueous ethanol; reflux; 1 h	2-Dimethylamino-3,6,7-trimethyl-4(3*H*)-pteridinone (57%)	416
4-Methoxy-6-methylamino-2,5-pyrimidinediamine	Biacetyl; methanol; reflux; 2 h	4-Methoxy-6,7,8-trimethyl-2(8*H*)-pteridinimine[c] (~60%)	359
5-Amino-2-dimethylamino-6-β-hydroxyethylamino-4(3*H*)-pyrimidinone	Biacetyl; aqueous methanol; reflux; 30 min	2-Dimethylamino-8-β-hydroxyethyl-6,7-dimethyl-4(8*H*)-pteridinone[a] (~60%)	359
2,5-Diamino-6-anilino-3-methyl-4(3*H*)-pyrimidinone	Biacetyl; aqueous ethanol; reflux; 15 min	2-Imino-3,6,7-trimethyl-8-phenyl-2,3-dihydro-4(8*H*)-pteridinone[c]	397
4,5-Diamino-6-β-hydroxyethylamino-2(1*H*)-pyrimidinone	Biacetyl; water; 95 °C; 1 h	4-β-Hydroxyethylamino-6,7-dimethyl-2(1*H*)-pteridinone (~45%)	362
2,5-Diamino-6-methylamino-4(3*H*)-pyrimidinone	3,4-Dimethyl-*o*-benzoquinone; dilute sodium hydroxide; 25 °C; 3 days	2-Amino-7,8,10-trimethyl-4(10*H*)-benzo[*g*]pteridinone (~10%)	162
2,5,6-Triamino-4(3*H*)-pyrimidinone	Cyclododecane-1,2-dione; aqueous ethanol; reflux; 2 h	2-Amino-6,7-decamethylene-4(3*H*)-pteridinone (21%)	419
5,6-Diamino-2,4(1*H*,3*H*)-pyrimidinedione	3,4:3′,4′-Bismethylenedioxybenzil; AcOH; 12 h	6,7-Bis(3′,4′-methylenedioxyphenyl)-2,4(1*H*,3*H*)-pteridinedione (low yield)	1473
5,6-Diamino-1-β-hydroxyethyl-2,4(1*H*,3*H*)-pyrimidinedione	Benzil; aqueous alcoholic AcOH; reflux; 1 h	1-β-Hydroxyethyl-6,7-diphenyl-2,4(1*H*,3*H*)-pteridinedione (91%)	1339
5,6-Diamino-1-*p*-tolyl-2,4(1*H*,3*H*)-pyrimidinedione	Biacetyl; ethanolic AcOH; reflux; 12 h	6,7-Dimethyl-1-*p*-tolyl-2,4(1*H*,3*H*)-pteridinedione[a]	797
5,6-Diamino-1-phenyl-2-thioxo-1,2-dihydro-4(3*H*)-pyrimidinone	Benzil; ethanolic AcOH; reflux; 8 h	1,6,7-Triphenyl-2-thioxo-1,2-dihydro-4(3*H*)-pteridinone[a]	797
5,6-Diamino-1,3-dimethyl-2,4(1*H*,3*H*)-pyrimidinedione	*p*,*p*′-Dimethylbenzil; ethanolic AcOH; reflux; 8 h	1,3-Dimethyl-6,7-di-*p*-tolyl-2,4(1*H*,3*H*)-pteridinedione[a] (61%)	1349
5-Amino-6-benzylamino-2,4(1*H*,3*H*)-pyrimidinedione	Biacetyl; water; 80 °C; 30 min	8-Benzyl-6,7-dimethyl-2,4(3*H*,8*H*)-pteridinedione (~65%)	378
5-Amino-6-anilino-3-methyl-2,4(1*H*,3*H*)-pyrimidinedione	Benzil; ethanol; reflux; 3h	3-Methyl-6,7,8-triphenyl-2,4(3*H*,8*H*)-pteridinedione[a] (67%)	409
5-Amino-6-β-hydroxyethylamino-2,4(1*H*,3*H*)-pyrimidinedione	Benzil	8-β-Hydroxyethyl-6,7-diphenyl-2,4(3*H*,8*H*)-pteridinedione[a]	1199

TABLE V. (*Contd.*)

Pyrimidine Intermediate	Reagent and Reaction Conditions	Pteridine (Yield)	References
5-Amino-6-D-erythritylamino-2,4(1*H*,3*H*)-pyrimidinedione	Biacetyl	8-D-Erythrityl-6,7-dimethyl-2,4(3*H*,8*H*)-pteridinedione[a]	713
5-Amino-6-D-mannitylamino-2,4(1*H*,3*H*)-pyrimidinedione	Biacetyl; water; reflux	8-D-Mannityl-6,7-dimethyl-2,4(3*H*,8*H*)-pteridinedione[a] (66%)	1444
5-Amino-6-D-arabinitylamino-2,4(1*H*,3*H*)-pyrimidinedione	Biacetyl; pH 4.5; 80 °C; 40 min	8-D-Arabinityl-6,7-dimethyl-2,4(3*H*,8*H*)-pteridinedione[a] (52%)	537
5-Amino-6-benzylamino-2,4(1*H*,3*H*)-pyrimidinedione	Hexafluoro-2,3-butanedione; DMF; 25 °C; 7 h	8-Benzyl-6,7-bistrifluoromethyl-2,4(3*H*,8*H*)-pteridinedione[a,d] (57%)	1570
5,6-Diamino-1,3-dimethyl-2,4(1*H*,3*H*)-pyrimidinedione	Hexafluoro-2,3-butanedione; DMF; 25 °C; 3 h	1,3-Dimethyl-6,7-bistrifluoromethyl-2,4(1*H*,3*H*)-pteridinedione[a] (67%)	1571
5,6-Diamino-2-thioxo-1,2-dihydro-4(3*H*)-pyrimidinone	Hexafluoro-2,3-butanedione; DMF; 25 °C; 12 h	2-Thioxo-6,7-bistrifluoromethyl-1,2-dihydro-4(3*H*)-pteridinone (85%)	1571

[a] And analog(s).
[b] Equivalent of α,α′-bis-3,4-dimethylphenylbiacetyl.
[c] And homolog(s).
[d] Underwent thermal rearrangement at 210 °C to give 1-benzyl-6,7-bistrifluoromethyl-2,4(1*H*,3*H*)-pteridinedione (36%)

(5) *Cases with an Amino- and an Oxo-Type Functional Group*

The reaction of 2,5,6-triamino-4(3*H*)-pyrimidinone with biacetyl or an equivalent reagent to give 2-amino-6,7-dimethyl-4(3*H*)-pteridinone has been described several times: the pyrimidine hydrochloride with biacetyl in aqueous ethanol gave >95%[541]; the pyrimidine bisulfite with biacetyl in water gave 72%[747]; the pyrimidine sulfate with biacetyl in aqueous ethanolic sodium acetate gave 82%[1038, 1380]; the pyrimidine sulfate with 2-acetamidobutan-2-one in the presence of diethylamine and oxygen gave 61%[942]; the pyrimidine sulfate with 3-acetamido-3-bromobutan-2-one in the presence of 2-mercaptoethanol gave 30%[544]; 5-acetamido-2,6-diamino-4(3*H*)-pyrimidinone and biacetyl in aqueous hydrochloric acid gave ~35%[1373]; and 6,7,7-trimethylpyrimido[4,5-*b*][1,4]oxazine-2,4-diamine (**129**) (a precursor of the appropriate pyrimidine) with biacetyl gave 73%.[1092] In addition, 2-amino-6,7-bistrideuteromethyl-4(3*H*)-pteridinone[1193] and a multi-^{15}N-labeled 2-amino-6,7-dimethyl-4(3*H*)-pteridinone[1469] were so prepared. Similar reactions gave 2-amino-6,7-diethyl-(64%),[32] 2-amino-6,7-bisbromomethyl- (from α,α′-dibromobiacetyl: ~70%),[680] 2-amino-6,7-diphenyl- (directly from benzil: 20–80%[359, 747, 825]; or indirectly, from benzoin via the 5,6- or 7,8-dihydro derivative under neutral or acidic conditions, respectively, with subsequent oxidation: yields unstated[1333]), and several cases of 2-amino-6,7-bis-*p*-substituted-phenyl-4(3*H*)-pteridinones.[891] In addition, 2,5,6-triamino-4(3*H*)-pyrimidinone and dimethyl α,α′-dioxosuccinate gave dimethyl 2-amino-4-oxo-3,4-dihydro-6,7-pteridinedicarboxylate (**130**) (80%)[307, 601]; 6-methoxy-2,4,5-pyrimidinetriamine and benzil gave 4-methoxy-6,7-diphenyl-2-pteridinamine (**131**, R = Me) (51%)[1500]; similar reactions gave the 4-ethoxy (**131**, R = Et), 4-β-hydroxyethoxy (**131**, R = CH_2CH_2OH), and other analogs;[627, 1500] 6-benzyloxy-, 6-isopropoxy-, or 6-methoxy-2,4,5-pyrimidinetriamine and biacetyl gave 4-benzyloxy-, 4-isopropoxy-, or 4-methoxy-6,7-dimethyl-2-pteridinamine, respectively, in >80% yield[370]; 4,5,6-triamino-2(1*H*)-pyrimidinone and an appropriate α-diketone gave 4-amino-6,7-dimethyl- (**132**, R = Me) (>90%),[130, 362] 4-amino-6,7-diethyl- (**132**, R = Et) (~70%),[32] and 4-amino-6,7-diphenyl-2(1*H*)-pteridinone (**132**, R = Ph) (>70%)[362, 471]; 4,5-diamino-6-dimethyl(amino-2(1*H*)-pyrimidinone gave 4-dimethylamino-6,7-dimethyl-2(1*H*)-pteridinone (~50%)[362]; and 2-methoxy-4,5,6-pyrimidinetriamine gave 2-methoxy-6,7-dimethyl- and 2-methoxy-6,7-diphenyl-4-pteridinamine.[362]

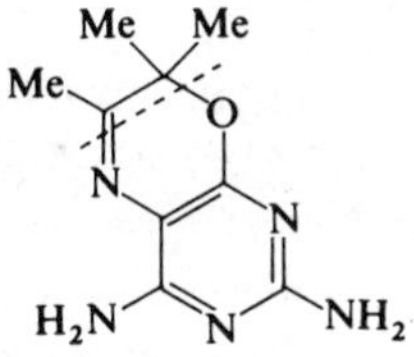

(**129**)

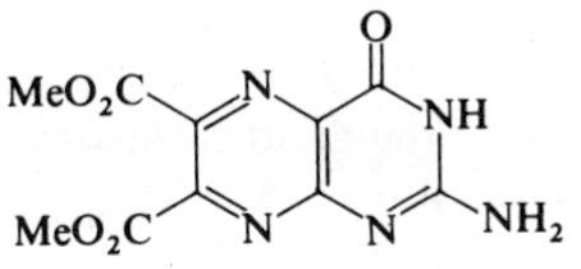

(**130**)

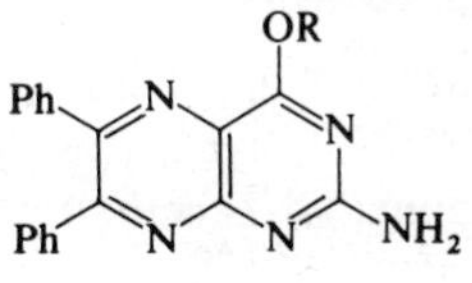

(**131**)

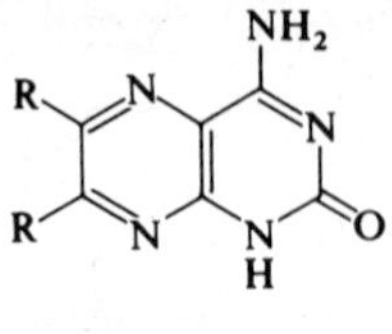

(132)

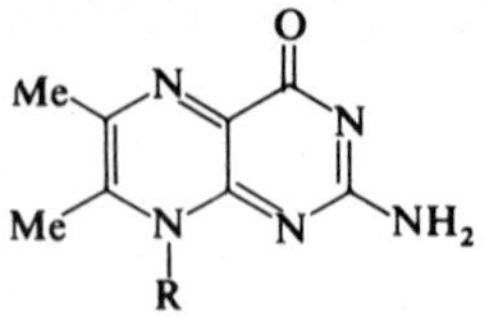

(133)

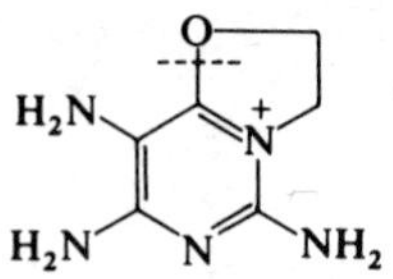

(134)

N-Alkylated derivatives are well represented in the present category. For example, 5,6-diamino-2-methylamino-4(3*H*)-pyrimidinone and biacetyl or benzil at 70°C gave 6,7-dimethyl-2-methylamino- (~15%) or 2-methylamino-6,7-diphenyl-4(3*H*)-pteridinone, respectively[1189]; 2-dimethylamino-6-methoxy-4,5-pyrimidinediamine gave 2-dimethylamino-4-methoxy-6,7-dimethylpteridine (67%)[416]; 4-ethoxy-6-methylamino-2,5-pyrimidinediamine and biacetyl gave 4-ethoxy-6,7,8-trimethyl-2(8*H*)-pteridinimine[190]; 2,5,6-triamino-3-methyl-4(3*H*)-pyrimidinone with biacetyl or benzil at pH 6 gave 2-amino-3,6,7-trimethyl- (52%)[166, 1189] or 2-amino-3-methyl-6,7-diphenyl-4(3*H*)-pteridinone (89%),[68] respectively; 2,5,6-triamino-1-methyl-4(1*H*)-pyrimidinone with benzil gave 2-amino-1-methyl-6,7-diphenyl-4(1*H*)-pteridinone (61–85%)[68, 111]; 2,5-diamino-6-methylamino-4(3*H*)-pyrimidinone and biacetyl gave 2-amino-6,7,8-trimethyl-4(8*H*)-pteridinone (**133**, R = Me) (40–65%)[190, 359, 422]; similar condensations gave the 8-ethyl (**133**, R = Et),[660] 8-benzyl (**133**, R = CH_2Ph) (24%),[377] 8-phenyl (**133**, R = Ph),[377] and other analogs;[377, 422] and 2,5-diamino-6-methylamino-4(3*H*)-pyrimidinone with benzil gave 2-amino-8-methyl-6,7-diphenyl-4(8*H*)-pteridinone (~40%).[190, 359] Treatment of the cation, 5,7,8-triamino-2,3-dihydrooxazolo [3,2-*c*]pyrimidinium (**134**), with sodium acetate caused fission of the five-membered ring and subsequent treatment with benzil gave a separable mixture of 2-amino-3-*β*-hydroxyethyl-6,7-diphenyl-4(3*H*)-pteridinone (33%) and the 3-*β*-acetoxyethyl analog (7%)[627, 1500]; other such reactions were also described.[1500] The formation of further types of *N*-alkylated derivatives is exemplified in the reaction of 4,5,6-triamino-1-methyl-2(1*H*)-pyrimidinone with biacetyl or benzil to afford 4-amino-1,6,7-trimethyl- (~40%) or 4-amino-1-methyl-6,7-diphenyl-2(1*H*)-pteridinone (~35%)[362]; of 4,5-diamino-6-anilino-2(1*H*)-pyrimidinone with biacetyl to give 4-amino-6,7-dimethyl-8-phenyl-2(8*H*)-pteridinone (~40%)[362]; and of 4,5-diamino-6-benzylamino-2(1*H*)-pyrimidinone with biacetyl to give 4-amino-8-benzyl-6,7-dimethyl-2(8*H*)-pteridinone (~20%).[362]

Analogous condensations gave thio analogs such as 6,7-dimethyl-4-methylthio- and 4-methylthio-6,7-diphenyl-2-pteridinamine (from 6-methylthio-2,4,5-pyrimidinetriamine with biacetyl or benzil, respectively: 88 and 82%)[1468]; 4-amino-6,7-dimethyl-2(1*H*)-pteridinethione (40–93%)[472, 916]; the 6,7-diphenyl analog (84%)[472, 699]; the 6,7-diisopropyl analog (50%)[400]; the 6,7-dimethyl-2-methylthio-4-pteridinamine (59%)[400]; the 6,7-diphenyl analog (47%)[472, 699] 2-ethylthio-6,7-diphenyl-4-pteridinamine (70%)[955]; and 2-amino-6,7-dimethyl-4(3*H*)-pteridinethione (~60%).[1508]

Ambiguous condensations using an unsymmetrical α-diketone are represented by the reaction of 2,5,6-triamino-4(3*H*)-pyrimidinone with ethyl 2,3-dioxobutyrate to give a single product (after saponification), which was formulated without real proof as 2-amino-7-methyl-4-oxo-3,4-dihydro-6-pteridinecarboxylic acid.[1363] These condensations are also represented by the reaction of the same pyrimidinone with pentane-2,3,4-trione to give the 6-acetyl-2-amino-7-methyl-4(3*H*)-pteridinone of subsequently proven orientation.[1217] 2,6-Diamino-5-methylamino-4(3*H*)-pyrimidinone with desyl chloride (i.e., α-benzoyl-α-chlorotoluene) appears to give exclusively 2-amino-5-methyl-6,7-diphenyl-5,6-dihydro-4(3*H*)-pteridinone in 28% yield.[509, 1333]

(6) *Cases with Two Oxo-Type Functional Groups*

Condensation of 5,6-diamino-2,4(1*H*,3*H*)-pyrimidinedione with biacetyl in boiling water gave 6,7-dimethyl-2,4(1*H*,3*H*)-pteridinedione (**135**, R = Me) in 70% yield.[747, 1096, 1438] However, use of the pyrimidine hydrochloride in acidified aqueous ethanol gave a 76% yield[367] and the pyrimidine sulfate in boiling water has also been used successfully.[106, 1012] The [8a-^{13}C]-labeled compound has been made similarly.[960] The same pyrimidine with hexane-3,4-dione gave 6,7-diethyl-2,4(1*H*,3*H*)-pteridinedione (**135**, R = Et) (67%),[32] and with benzil, it gave the 6,7-diphenyl analog (**135**, R = Ph) (~ 65%).[399, 747, 1438, 1473] Likewise, 5,6-diamino-2-methoxy-4(3*H*)-pyrimidinone with benzil gave 2-methoxy-6,7-diphenyl-4(3*H*)-pteridinone (31%)[283] and the following thio analogs were made by similar methods: 6,7-diphenyl-2-thioxo-1,2-dihydro-4(3*H*)-pteridinone (fusion of reactants at 200°C for 5 h: ~ 25%[399]; in acidic aqueous ethanol: 40%[437]), 6,7-dimethyl-2,4(1*H*,3*H*)-pteridinedithione (47%),[170] 6,7-diphenyl-2,4(1*H*,3*H*)-pteridinedithione (30%),[437] 6,7-dimethyl-2-thioxo-1,2-dihydro-4(3*H*)-pteridinone (65–76%),[437, 916] 2-methylthio-6,7-diphenyl-4(3*H*)-pteridinone (62%),[437] its 2-ethylthio homolog (> 70%),[399, 1234] 6,7-dimethyl-2-methylthio-4(3*H*)-pteridinone (60–70%),[437, 654] its 2-ethylthio homolog (~ 85%),[1234] and others using either benzils or benzoins.[955] The condensation of dimeric biacetyl, that is, 5-acetyl-2-hydroxy-2,5-dimethyl-2,3,4,5-tetrahydrofuran-3-one (**136**), with 5,6-diamino-2,4(1*H*,3*H*)-pyrimidinedione is said to give 7-(2′-hydroxy-2′-methyl-3′-oxobutyl)-6-methyl-2,4(1*H*,3*H*)-pteridinedione.[106, 164, 960] 4-Amino-6,6,7-trimethyl-7*H*-pyrimido[4,5-

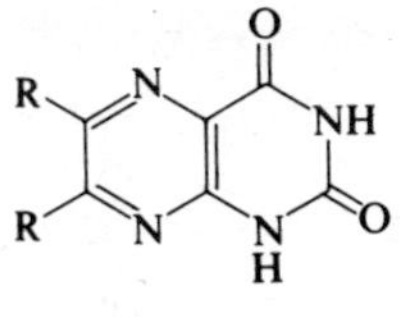

(135)

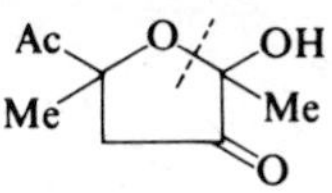

(136)

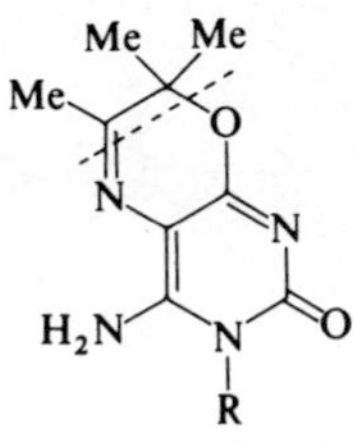

(137)

b][1,4]oxazin-2(3*H*)-one (**137**, R = H) and its 3-methyl derivative (**137**, R = Me) acted as pro-pyrimidinediamines to give with biacetyl, 6,7-dimethyl- (73%) and 1,6,7-trimethyl-2,4(1*H*,3*H*)-pteridinedione (38%), respectively.[1092] By employing appropriate cyclic α-diketones, numerous 2,4(1*H*,3*H*)-pteridinediones with one or more 6,7-fused carbocyclic rings have been made.[106, 1012, 1173, 1234]

A variety of 1- and/or 3-alkylated derivatives may be made by similar condensations with α-diketones. For example, 5,6-diamino-1-methyl-2,4-(1*H*,3*H*)-pyrimidinedione[956] and biacetyl gave 1,6,7-trimethyl-2,4(1*H*,3*H*)-pteridinedione (40–76%)[362, 367, cf. 1380] and similar condensations gave 1-methyl-6,7-diphenyl- (60–76%),[362, 962, 1019] 6,7-diisopropyl-1-methyl- (11%),[367] 6,7-di-*t*-butyl-1-methyl- (0.5%),[367] 1-β-hydroxyethyl-6,7-dimethyl- (71%),[1339] and 6,7-dimethyl-1-phenyl-2,4(1*H*,3*H*)-pteridinedione.[797] Other 1-alkylated derivatives so made include 1,6,7-trimethyl-2-methylthio-4(1*H*)-pteridinone (23%),[437] its 6,7-diphenyl analog (23%),[437] 1-methyl-6,7-diphenyl-2-thioxo-1,2-dihydro-4(3*H*)-pteridinone (55%),[111, 437] its 1,6,7-trimethyl analog (70%),[437] and 6-acetyl-1,7-dimethyl-2,4(1*H*,3*H*)-pteridinedione (of unproven orientation).[1380]

Examples of 3-alkylated derivatives are less numerous but include 3,6,7-trimethyl-2,4(1*H*,3*H*)-pteridinedione (**138**, R = Me) [from 5,6-diamino-3-methyl-2,4(1*H*,3*H*)-pyrimidinedione with biacetyl in aqueous ethanol under reflux: 74%],[367] the 6,7-diphenyl analog (**138**, R = Ph) (~70%),[362] 3,6,7-trimethyl-2-thioxo-1,2-dihydro-4(3*H*)-pteridinone (70%),[446] the 6,7-dipyridin-2′-yl analog (8%),[446] the 6,7-diphenyl analog (50%),[446] 3,6,7-trimethyl-2-methylthio-4(3*H*)-pteridinone (71%),[362, 437] and its 6,7-diphenyl analog (48%).[367]

The formation of 1,3-dialkylated compounds in this category is illustrated by the reaction of 5,6-diamino-1,3-dimethyl-2,4(1*H*,3*H*)-pyrimidinedione with biacetyl in acidic aqueous ethanol to give 1,3,6,7-tetramethyl-2,4(1*H*,3*H*)-pteridinedione (60%[367, 1380, 1435]; improved to 73% by continuous extraction of the crude product with light petroleum[1096]). Other similarly made products are 6,7-diisopropyl-1,3-dimethyl-2,4(1*H*,3*H*)-pteridinedione (18%),[367] the 6,7-diphenyl analog (70–80%),[962, 1096, 1149, 1435] 1,3,6,7-tetramethyl-2-thioxo-1,2-dihydro-4(3*H*)-pteridinone (34%),[437] its 6,7-diphenyl analog (16%),[437] and 6-acetyl-1,3,7-trimethyl-2,4(1*H*,3*H*)-pteridinedione (of unproven configuration).[1380] The condensation of 5,6-diamino-1,3-dimethyl-2,4(1*H*,3*H*)-pyrimidinedione with unsymmetrical α-diketones of the type Ar–CO–CO–R (in which Ar is phenyl or a *p*-substituted phenyl group and R is an aliphatic group

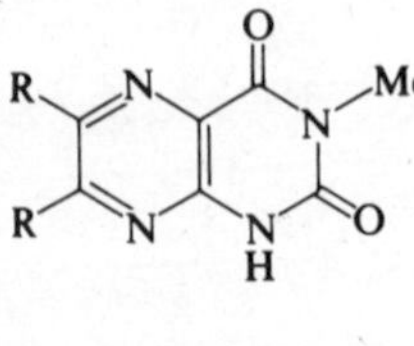

(**138**)

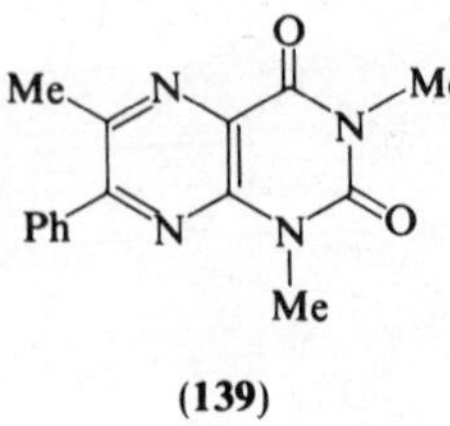

(**139**)

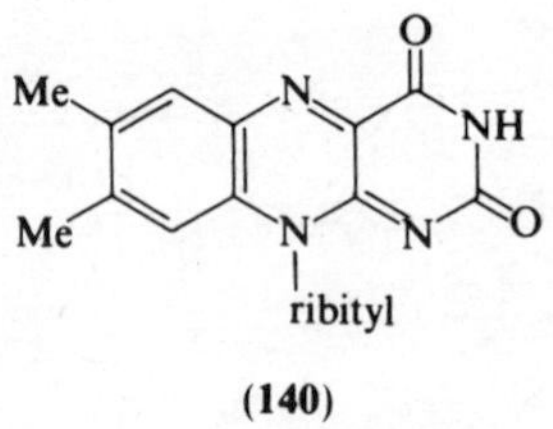

(**140**)

from methyl to butyl) has been studied in respect of overall yield and ratio of isomers under standardized conditions. Both phenomena are affected profoundly by the size of the aliphatic group.[1112, 1348, 1349] Isomeric structures such as 1,3,6-trimethyl-7-phenyl- (**139**) and 1,3,7-trimethyl-6-phenyl-2,4(1*H*,3*H*)-pteridinedione were checked by unambiguous syntheses[1348] and the nmr data were used to determine ratios.

Because of their kinship to riboflavine (**140**), 6,7,8-trimethyl-2,4(3*H*,8*H*)-pteridinedione (**141**, R = Me) and related compounds, especially those with an 8-alditylor other sugarlike group, have been studied widely. Thus 5-amino-6-methylamino-2,4(1*H*,3*H*)-pyrimidinedione and an excess of biacetyl in water at 80°C gave the parent trimethylpteridinedione (**141**, R = Me) (~ 45%).[107, 165, 290, 378, 634, 659] Analogous procedures gave, among other simple examples, 6,7-diisopropyl-8-methyl- (70%),[248, 537, 1545] 8-ethyl-6-7-dimethyl- (**141**, R = Et) (~ 50%)[378] 6,7-dimethyl-8-propyl- (**141**, R = Pr),[1553] 8-isopropyl-6,7-dimethyl- (**141**, R = *i*-Pr) (75%),[375] 8-*β*-hydroxyethyl-6,7-dimethyl- (**141**, R = CH_2CH_2OH) (50%),[378, 409, 422, 634, cf. 165] 8-*β*-hydroxypropyl-6,7-dimethyl- (**141**, R = $CH_2CHOHMe$) (39%),[1686] 6,7-di-isopropyl-3,8-dimethyl- (20%),[375] 8-isopropyl-3,6,7-trimethyl- (40–58%),[375, 409] 8-*β*-hydroxyethyl-6,7-diphenyl-,[1545] and 8-*s*-butyl-6,7-dimethyl-2,4(3*H*,8*H*)-pteridinedione (**141**, R = *s*-Bu) (60%).[409] Compounds with sugar-derived groups in the 8-position include, for example, 6,7-dimethyl-8-D-ribityl- (**141**, R = ribityl) [from 5-amino-6-D-ribitylamino-2,4(1*H*,3*H*)-pyrimidinedione and biacetyl in water at 80°C: 28–70%],[164, 172, 378, 422, 1437, 1543, 1734] 6,7-diethyl-8-D-ribityl- (58%),[172] 6,7-diphenyl-8-D-ribityl- (44%),[674] 6,7-dimethyl-8-D-sorbityl- (**141**, R = sorbityl) (61%),[172, 1444] 8-L-lyxityl-6,7-dimethyl- (36%),[1444] 6,7-dimethyl-8-D-xylityl- (59%),[537] and 8-(2′-deoxy-D-ribityl)-6,7-dimethyl-2,4(3*H*,8*H*)-pteridinedione.[713] The condensations of 5-amino-6-D-ribitylamino-2,4(1*H*,3*H*)-pyrimidinedione with the unsymmetrical diketones, hexane-2,3-dione and 1-phenylbutane-2,3-dione, gave only one isolable product in each case. By careful study of nmr data, they were shown to be 7-methyl-6-propyl- and 6-benzyl-7-methyl-8-D-ribityl-2,4(3*H*,8*H*)-pteridinedione, respectively, in both of which the larger alkyl group occupied the 6-position.[674] Whether this points to a general rule of thumb, remains to be seen. Related syntheses are exemplified in those of riboflavine (**140**) [from 5-amino-6-D-ribitylamino-2,4(1*H*,3*H*)-pyrimidinedione with either 3,4-dimethyl-*o*-benzoquinone (29%)[163] or dimeric biacetyl (**136**) (< 10%)[165]]; the simple analogs, 7,8,10-trimethyl- (lumiflavine) and 10-*β*-hydroxyethyl-7,8-dimethyl-2,4(3*H*,10*H*)-benzo[*g*]pteri-

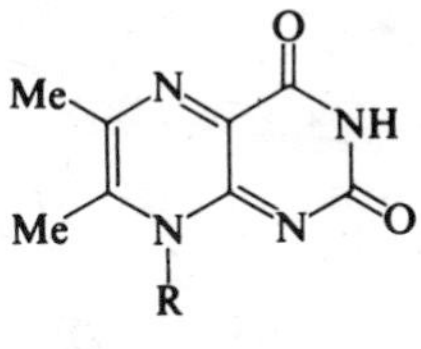

(141)

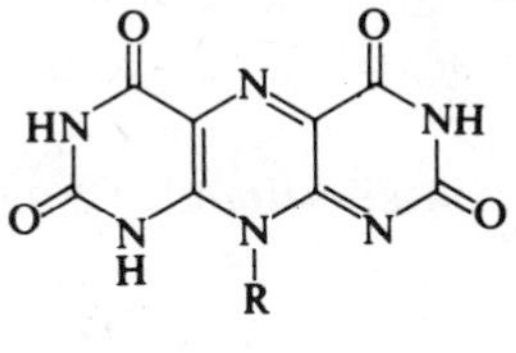

(142)

dinedione (21%)[165]; the 10-D-mannityl and 10-D-sorbityl analogs of riboflavine (**140**)[163]; 10-methyl-2,4,6,8(1*H*,3*H*,7*H*,10*H*)-pyrimido[5,4-*g*]pteridinetetrone (**142**, R = Me), its 10-ethyl analog (**142**, R = Et) and its 10-D-ribityl analog, and so on[379]; and 1,8-ethano-6-methyl-7-methylene-7,8-dihydro-2,4(1*H*,3*H*)-pteridinedione (**145**, *n* = 2) or its 1,8-propano analog (**145**, *n* = 3) {from biacetyl with 8-amino-2,3-dihydro-5,7(1*H*,6*H*)-imidazo[1,2-*c*]pyrimidinedione (**143**) or 9-amino-3,4-dihydro-6,8(1*H*,7*H*)-2*H*-pyrimido[1,2-*c*]pyrimidinedione (**144**), respectively}.[1339]

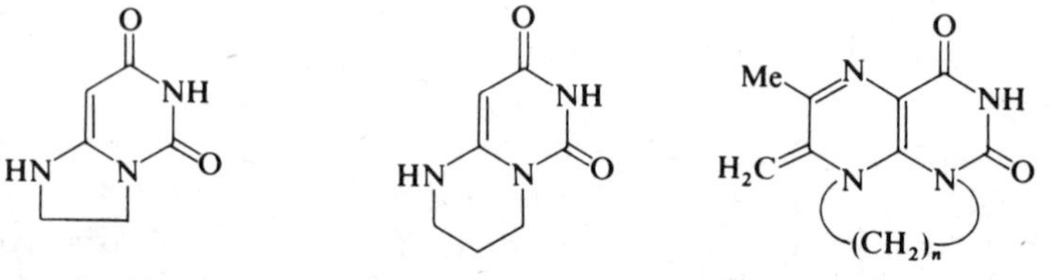

(**143**) (**144**) (**145**)

(7) *Cases with Two Other Functional Groups*

The reaction of 2,6-dichloro-4,5-pyrimidinediamine with biacetyl in refluxing benzene–methanol gave 2,4-dichloro-6,7-dimethylpteridine (53%).[170] Appropriate pyrimidines with biacetyl in refluxing *t*-butyl alcohol also gave ethyl 2-amino-6,7-dimethyl-4-pteridinecarboxylate (61%); its 2-chloro (42%), 2-dimethylamino (98%), 2-ethoxy (77%), and 2-methylthio (88%) analogs; 6,7-dimethyl-4-trifluoromethyl-2-pteridinamine (99%); and 2-ethoxy (and dimethylamino)-6,7-dimethyl-4-trifluoromethylpteridine (72 and 40%).[1440]

E. The Use of α-Ketoesters

The condensation of 4,5-pyrimidinediamines with α-ketoesters is more versatile than that with α-aldehydoesters because the keto portion may be varied to introduce different alkyl, substituted-alkyl, aryl, or even a carboxyl group into the 6- or 7-position of the resulting 7- or 6-pteridinone. This, combined with the variations possible at the 1-, 2-, 3-, 4-, 5-, and 8-positions by presubstitution of the pyrimidine portion, has lead to much use of this category of the Gabriel and Colman synthesis. The only major drawback is an essential ambiguity, so that the orientation of the resulting 6-alkyl-7-pteridinone or 7-alkyl-6-pteridinone must be checked in all but a few rare cases (noted in the next section in passing) when one or other isomer is precluded by valency.

(1) *Simple Cases without a Functional Group in the Pyrimidine Ring*

The condensation of 4,5-pyrimidinediamine with ethyl pyruvate in ~5*M* hydrochloric acid at room temperature gave 7-methyl-6(5*H*)-pteridinone (**146**,

R = Me) (60%). In contrast, condensation in water at room temperature gave the isomeric 6-methyl-7(8*H*)-pteridinone (**147**, R = Me) (50%), while condensation in 15*M* sulfuric acid caused an initial hydrolysis of the pyruvic ester followed by combination of two molecules of pyruvic acid and subsequent reaction with the diamine to afford 7-(2′-carboxyprop-1′-enyl)-6(5*H*)-pteridinone (**146**, R = CH:CMeCO_2H) (27%).[55] Likewise, 4,5-pyrimidinediamine or its 6-methyl derivative with diethyl mesoxalate in dilute potassium carbonate at 25°C gave ethyl 7-oxo-7,8-dihydro-6-pteridinecarboxylate (**147**, R = CO_2Et) (~65%) or its 4-methyl derivative (80%), respectively[57]; the same diamine with ethoxalylacetone (ethyl 2,4-dioxopentanoate) at pH 5 gave 6-acetonyl-7(8*H*)-pteridinone (**147**, R = CH_2Ac) (35%)[55]; the same diamine and 1,3-dimethylalloxan, which reacts as the mono-(*N*-methylamide) of mesoxalic acid, in dilute sulfuric acid gave (unexpectedly) only *N*-methyl-7-oxo-7,8-dihydro-6-pteridinecarboxamide (**147**, R = CONHMe) (65%), of subsequently confirmed structure[57]; and 4,5-pyrimidinediamine with 4-benzoyl-3-hydroxy-2(5*H*)-furanone (**149**) in hot glacial acetic acid gave 7-phenacyl-6(5*H*)-pteridinone (**146**, R = CH_2COPh) (35%).[658]

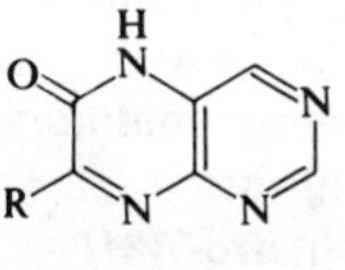

(146)

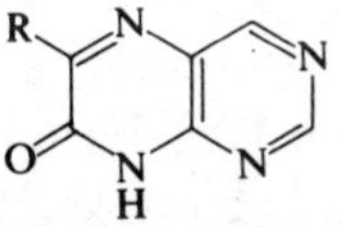

(147)

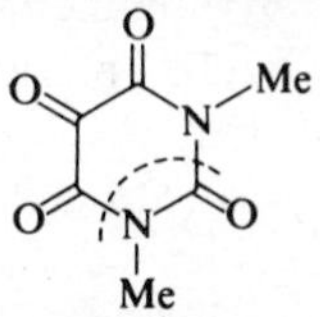

(148)

When 4-methylamino-5-pyrimidinamine reacted with ethyl pyruvate in boiling water, the only structurally possible product, 6,8-dimethyl-7(8*H*)-pteridinone (**150**, R = Me) was formed in 70% yield[34]; appropriate 4-substituted-amino-5-pyrimidinamines and ethyl pyruvate gave (also without ambiguity) 8-*β*-dimethylaminoethyl-6-methyl- (**150**, R = $CH_2CH_2NMe_2$), 8-*γ*-dimethylaminopropyl-6-methyl- (**150**, R = $CH_2CH_2CH_2NMe_2$), 6-methyl-8-*β*-morpholinoethyl-, and 6-methyl-8-*γ*-morpholinopropyl-7(8*H*)-pteridinone (as hydrochlorides)[1490]; 4-methylamino-, 2-methyl-4-methylamino- or 4-methylamino-2-phenyl-5-pyrimidinamine with diethyl mesoxalate in ethanol or aqueous ethanol under reflux gave (unambiguously) ethyl 8-methyl-7-oxo-7,8-dihydro-6-pteridinecarboxylate (**151**, R = H) (93%), its 2,8-dimethyl homolog (**151**, R = Me) (40%), or its 8-methyl-2-phenyl analog (**151**, R = Ph) (58%), respectively.[1311]

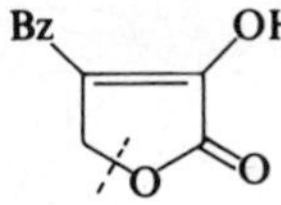

(149)

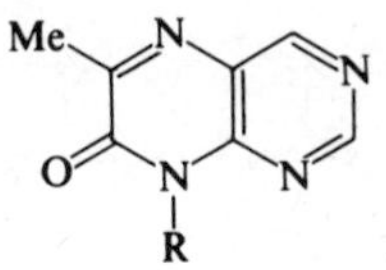

(150)

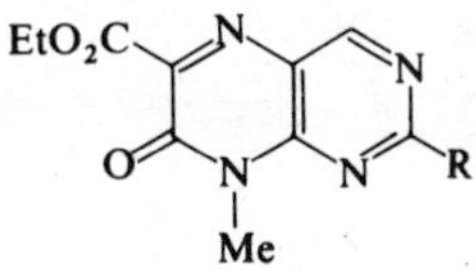

(151)

(2) *Cases with One Functional Group in the Pyrimidine Ring*

2,4,5-Pyrimidinetriamine and ethyl sodiooxaloacetate in acetic acid at 100°C gave (after concomitant decarboxylation) 2-amino-6-methyl-7(8*H*)-pteridinone (of unproven orientation)[279]; its 4,6-dimethyl analog (46%) (also of unproven orientation) was made similarly[279]; 2-ethylamino-4,5-pyrimidinediamine with diethyl or disodium mesoxalate in water under reflux gave good yields of ethyl 2-ethylamino-7-oxo-7,8-dihydro-6-pteridinecarboxylate or the free acid, respectively[387]; and 2-dimethylamino-4,5-pyrimidinediamine with ethyl pyruvate in aqueous ethanol gave 2-dimethylamino-6-methyl-7(8*H*)-pteridinone (43%).[357] The numerous 8-alkylated-2-substituted-7-pteridinones, which may be so made unambiguously, are exemplified by 2-dimethylamino-6,8-dimethyl-7(8*H*)-pteridinone (from 2-dimethylamino-4-methylamino-5-pyrimidinamine and ethyl pyruvate: 60%),[1150] the 8-cyclohexyl analog (~40%),[1150] the 8-pyridin-2′-yl analog (~25%),[139] 6,8-dimethyl-2-methylamino-7(8*H*)-pteridinone (from 2,4-bismethylamino-5-pyrimidinamine and ethyl pyruvate: ~30%),[248] 8-methyl-2-methylamino-6-phenyl-7(8*H*)-pteridinone (from the same pyrimidine and ethyl benzoylformate: ~55%),[248] 2-methoxy-6,8-dimethyl-7(8*H*)-pteridinone (~75%),[248] 6,8-dimethyl-2,7(1*H*,8*H*)-pteridinedione (**152**, R=Me) (from 5-amino-4-methylamino-2(1*H*)-pyrimidinone and ethyl pyruvate: 80%),[248] its 6-phenyl analog (**152**, R=Ph) (using ethyl benzoylformate: ~35%),[248] and 6,8-dimethyl-2-thioxo-1,2-dihydro-7(8*H*)-pteridinone (~70%).[248] They are exemplified also by the following carboxylic esters and acids: ethyl or methyl 2-amino-8-methyl- (from 4-methylamino-2,5-pyrimidinediamine with diethyl or dimethyl mesoxalate: 72 and 54%),[1311] ethyl 8-methyl-2-methylamino-,[1311] ethyl 4,8-dimethyl-2-methylamino- (81%),[1311] ethyl 8-ethyl-2-ethylamino- (~20%),[387] ethyl 2-dimethylamino-8-ethyl- (**153**, R=Et) (~30%),[387] and ethyl 2-methoxy-8-methyl-7-oxo-7,8-dihydro-6-pteridinecarboxylate (70%),[1311] as well as 2-amino-8-methyl- (from 4-methylamino-2,5-pyrimidinediamine and disodium mesoxalate in water: 35%),[1311] 8-ethyl-2-ethylamino- (in water: ~40%; in dilute alkali: ~60%),[387] and 2-dimethylamino-8-ethyl-7-oxo-7,8-dihydro-6-pteridinecarboxylic acid (**153**, R=H) (~60%).[387] In addition, an interesting preparation must be noted of the above-mentioned 8-ethyl-2-ethylamino-7-oxo-7,8-dihydro-6-pteridinecarboxylic acid (69%) by condensation in 1*M* alkali of 2,4-bisethylamino-5-pyrimidinamine with alloxan (**154**), which under such conditions acted as a source of mesoxalic acid.[387, 481]

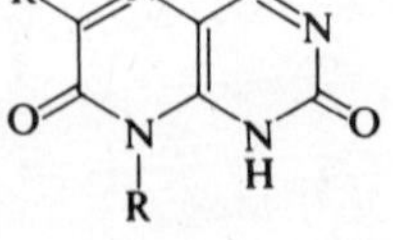

(**152**)

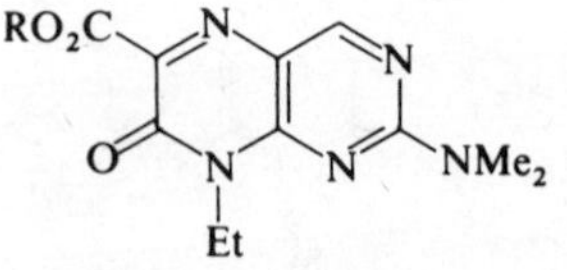

(**153**)

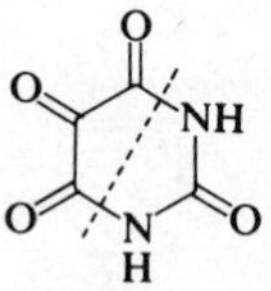

(**154**)

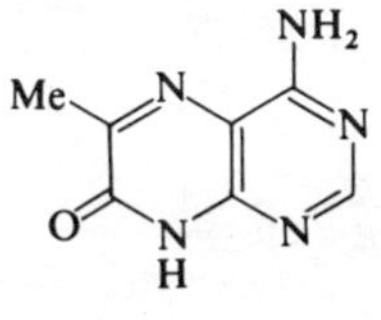

(155)

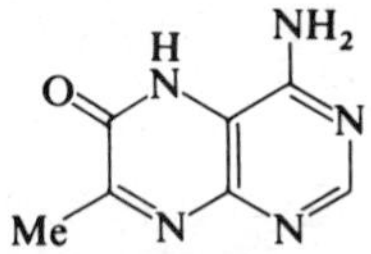

(156)

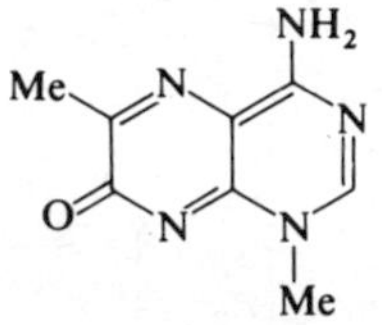

(157)

Examples of such syntheses, which give products with a functional group at the 4-position, are seen in the condensation of 4,5,6-pyrimidinetriamine with ethyl pyruvate in 1*M* sodium carbonate to give a separable mixture of 4-amino-6-methyl-7(8*H*)-pteridinone (**155**) (~ 5%) and the isomeric 4-amino-7-methyl-6(5*H*)-pteridinone (**156**) (~ 20%)[445]; in the condensation of the same pyrimidine with ethyl sodiooxaloacetate in acetic acid, which is said (without proof) to give only the 7-pteridinone (**155**) in 70% yield[279]; in the condensation of 2-phenyl-4,5,6-pyrimidinetriamine with ethanolic ethyl or methyl benzoylformate to give 4-amino-2,7-diphenyl-6(5*H*)-pteridinone (71 and 85%) but with ethanolic benzoylformic acid to give the isomeric 4-amino-2,6-diphenyl-7(8*H*)-pteridinone (92%), both products of proven orientation[526]; in the condensation of 4,5,6-pyrimidinetriamine with cyclohexylcarbonylformic acid in 1*M* sulfuric acid to give 4-amino-7-cyclohexyl-6(5*H*)-pteridinone (~ 60%) but in pH 5 buffer to give the isomeric 4-amino-6-cyclohexyl-7(8*H*)-pteridinone (< 5% yield), with satisfactory spectral evidence for orientations[924]; in the condensation of 6-imino-1-methyl-1,6-dihydro-4,5-pyrimidinediamine with ethyl pyruvate in 1*M* sodium carbonate to give 4-amino-1,6-dimethyl-7(1*H*)-pteridinone (**157**) (~ 35%), with spectral evidence for orientation[132]; in the condensation of 2-phenyl-4,5,6-pyrimidinetriamine with diethyl mesoxalate to give ethyl 4-amino-7-oxo-2-phenyl-7,8-dihydro-6-pteridinecarboxylate (77%)[1357]; in the unambiguous syntheses of 4-chloro-6,8-dimethyl-7(8*H*)-pteridinone (~ 85%),[248] 6,8-dimethyl-4-methylamino-7(8*H*)-pteridinone (~ 65%),[248] and 4-dimethylamino-6-methyl-8-(2′,3′,4′,6′-tetra-*O*-acetyl-*β*-D-glucopyranosyl)-7(8*H*)-pteridinone[386]; in the condensation of 5,6-diamino-4(3*H*)-pyrimidinone with ethyl sodiooxaloacetate to give 6-ethoxycarbonylmethyl-4,7(3*H*,8*H*)-pteridinedione (70%) of proven orientation,[29] or with mesoxalic acid in aqueous acetic acid to give 4,6-dioxo-3,4,5,6-tetrahydro-7-pteridinecarboxylic acid (~ 80%) with orientation proven by decarboxylation[346]; and in the esoteric condensation of ethyl *α*-benzamido-*α*-(2′,3′,4′,5′-tetrachloro-6′-hydroxyphenoxy)-acetate with 4,5,6-pyrimidinetriamine to give 4-amino-6-benzamido-5,6-dihydro-7(8*H*)-pteridinone (91%), which lost benzamide in trifluoroacetic acid to afford 4-amino-7(8*H*)-pteridinone (82% overall yield).[1386] See also examples in Table VI.

(3) *Cases with Two Amino-Type Groups in the Pyrimidine Ring*

2,4,5,6-Pyrimidinetetramine reacted with pyruvic acid in 1*M* sulfuric acid to give 2,4-diamino-7-methyl-6(5*H*)-pteridinone (95%), initially containing ~ 3%

TABLE VI. ADDITIONAL EXAMPLES OF THE GABRIEL AND COLMAN SYNTHESIS WITH α-KETO ESTERS

Pyrimidine Intermediate	Reagent and Reaction Conditions	Pteridine (Yield)	References
4,5-Pyrimidinediamine	Ethyl acetylpyruvate; glacial AcOH	7-Acetonyl-6(5*H*)-pteridinone	580
2-Dimethylamino-4-β-D-glucopyranosyl-amino-5-pyrimidinamine	Ethyl pyruvate; DMF; 100°C; 1 h	2-Dimethylamino-8-β-D-glucopyranosyl-6-methyl-7(8*H*)-pteridinone[a] (~40%)	1150
2,4-Bisethylamino-6-methyl-5-pyrimidin-amine	Disodium mesoxalate; dilute alkali; reflux; 1 h	8-Ethyl-2-ethylamino-4-methyl-7-oxo-7,8-dihydro-6-pteridinecarboxylic acid (~40%)	387
2,4-Bisethylamino-6-methyl-5-pyrimidin-amine	Diethyl mesoxalate; water; reflux; 15 min	Ethyl 8-ethyl-2-ethylamino-4-methyl-7-oxo-7,8-dihydro-6-pteridinecarboxylate (~85%)	387
2-Dimethylamino-4-ethylamino-6-phenyl-5-pyrimidinamine	Diethyl mesoxalate; ethanol; reflux; 2 h	Ethyl 2-dimethylamino-8-ethyl-7-oxo-4-phenyl-7,8-dihydro-6-pteridinecarboxylate[a] (74%)	1311
5-Amino-4-methylamino-2(1*H*)-pyrimidinone	Disodium mesoxalate; water; reflux; 2 h	8-Methyl-2,7-dioxo-1,2,7,8-tetrahydro-6-pteridinecarboxylic acid (65%)	1311
5-Amino-4-methylamino-2(1*H*)-pyrimidinone	Diethyl mesoxalate; water; reflux; 90 min	Ethyl 8-methyl-2,7-dioxo-1,2,7,8-tetrahydro-6-pteridinecarboxylate (67%)	1311
4,5,6-Pyrimidinetriamine	Cyclopentanecarbonylformic acid; ethanol; 100°C; 1 h	4-Amino-7-cyclopentyl-6(5*H*)-pteridinone	924
4-Dimethylamino-6-methylamino-5-pyrimidinamine	Ethyl pyruvate; ethanol; reflux; 1 h	4-Dimethylamino-6,8-dimethyl-7(8*H*)-pteridinone (~75%)	445
5,6-Diamino-3-methyl-4(3*H*)-pyrimidinone	Mesoxalic acid; aqueous AcOH; reflux; 15 min	3-Methyl-4,6-dioxo-3,4,5,6-tetrahydro-7-pteridinecarboxylic acid (~60%)	346
5,6-Diamino-2-methyl-4(3*H*)-pyrimidinone	Ethyl sodiooxaloacetate; AcOH; 100°C; 2 h	2,6-Dimethyl-4,7(3*H*,8*H*)-pteridinedione[b]	279
5-Amino-6-methylamino-4(3*H*)-pyrimidinone	Ethyl acetylpyruvate; pH 1; 100°C; 30 min	6-Acetonyl-8-methyl-4,7(3*H*,8*H*)-pteridine-dione (42%)	712
2,4,5,6-Pyrimidinetetramine	Ethyl sodio-*p*-toluoylpyruvate; dilute AcOH (?); 100°C; 2 h	2,4-Diamino-7-*p*-methylphenacyl-6(5*H*)-pteri-dinone[a] (~80%)	982
2-Dimethylamino-4-methoxy-6-methylamino-5-pyrimidinamine	Ethyl pyruvate; aqueous ethanol; reflux; 15 min	2-Dimethylamino-4-methoxy-6,8-dimethyl-7(8*H*)-pteridinone (42%)	417

2,5,6-Triamino-4(3*H*)-pyrimidinone	Ethyl 5-benzyloxy-2,4-dioxovalerate; aqueous ethanolic morpholine; reflux; 10 min	2-Amino-6-γ-benzyloxyacetonyl-4,7(3*H*,8*H*)-pteridinedione[a] (69%)	499
2,5,6-Triamino-4(3*H*)-pyrimidinone	Ethyl 5,6-dibenzyloxy-2,4-dioxohexanoate; methanolic morpholine; reflux	2-Amino-6-(3′,4′-dibenzyloxy-2′-oxobutyl)-4,7(3*H*,8*H*)-pteridinedione (48%)	1262
2,5,6-Triamino-4(3*H*)-pyrimidinone	α-Benzoyl-α-phenyliminoacetonitrile; 2*M* hydrochloric acid; reflux; 15 min	2-Amino-6-phenyl-4,7(3*H*,8*H*)-pteridinedione[c] (~10%)	1011
2,5,6-Triamino-4(3*H*)-pyrimidinone	Ethyl α-ethoxalyl-α-methoxyacetate; aqueous AcOH; 100°C; 1 h	2-Amino-6-methoxymethyl-4,7(3*H*,8*H*)-pteridinedione[a]	1083
4-Methoxy-6-methylamino-2,5-pyrimidinediamine	Ethyl pyruvate; methanol; reflux; 2 h	2-Amino-4-methoxy-6,8-dimethyl-7(8*H*)-pteridinone (~75%)	371
2,5-Diamino-6-2′,3′,4′,6′-tetra-*O*-acetyl-D-glucosylamino-4(3*H*)-pyrimidinone	Ethyl pyruvate; methanol; reflux; 2 h	2-Amino-8-D-glucosyl-6-methyl-4,7(3*H*,8*H*)-pteridinedione[a] (84%[d])	1090
2,5-Diamino-6-D-ribitylamino-4(3*H*)-pyrimidinone	Pyruvic acid; water; reflux; 1 h	2-Amino-6-methyl-8-D-ribityl-4,7(3*H*,8*H*)-pteridinedione[a] (63%)	172
2,5-Diamino-6-methyl-4(3*H*)-pyrimidinone	Diethyl mesoxalate; water; 100°C; 15 min	Ethyl 2-amino-8-methyl-4,7-dioxo-3,4,7,8-tetrahydro-6-pteridinecarboxylate[a] (~60%)	385
2,5-Diamino-6-β-hydroxyethylamino-4(3*H*)-pyrimidinone	Diethyl mesoxalate; pH 5; reflux; 1 h	Ethyl 2-amino-8-β-hydroxyethyl-4,7-dioxo-3,4,7,8-tetrahydro-6-pteridinecarboxylate (90%)	882, 1090
2,5,6-Triamino-4(3*H*)-pyrimidinone	Ethyl 5-phthalimido-2,4-dioxovalerate; aqueous alcoholic hydrochloric acid; reflux; 40 min	2-Amino-7-γ-phthalimidoacetonyl-4,6(3*H*,5*H*)-pteridinedione[a] (49%)	504
2,5,6-Triamino-4(3*H*)-pyrimidinone	Ethyl *p*-chlorobenzoylpyruvate; ethanol; reflux; 90 min	2-Amino-7-*p*-chlorophenacyl-4,6(3*H*,5*H*)-pteridinedione[a,e] (~75%)	982
2,5,6-Triamino-4(3*H*)-pyrimidinone	Ethyl *p*-chlorobenzoylpyruvate; dilute ammonia; then intermediate cyclized in hot Me_2SO	2-Amino-6-*p*-chlorophenacyl-4,7(3*H*,8*H*)-pteridinedione[a,e]	982
2,5-Diamino-6-β,γ-dihydroxypropylamino-4(3*H*)-pyrimidinone	Alloxan; water; pH ?; reflux; 2 h	2-Amino-8-β,γ-dihydroxypropyl-4,7-dioxo-3,4,7,8-tetrahydro-6-pteridinecarboxylic acid[a] (79%)	480
2,5-Diamino-6-D-ribitylamino-4(3*H*)-pyrimidinone	Alloxan; water; reflux; 2 h	2-Amino-4,7-dioxo-8-D-ribityl-3,4,7,8-tetrahydro-6-pteridinecarboxylic acid[a]	480
4,5,6-Triamino-1-methyl-2(1*H*)-pyrimidinone	Ethyl pyruvate; water; 100°C; 2 h	4-Amino-1,6-dimethyl-2,7(1*H*,8*H*)-pteridinedione[f] (~30%)	363

TABLE VI. (*Contd.*)

Pyrimidine Intermediate	Reagent and Reaction Conditions	Pteridine (Yield)	References
4,5-Diamino-6-anilino-2(1*H*)-pyrimidinone	Ethyl pyruvate; aqueous NaOAc; 100°C; 2 h	4-Amino-6-methyl-8-phenyl-2,7(1*H*,8*H*)-pteridinedione[a] (~50%)	363
5,6-Diamino-2-methylthio-4(3*H*)-pyrimidinone	Phenylpyruvic acid; acidified water; reflux; "some time"	7-Benzyl-2-methylthio-4,6(3*H*,5*H*)-pteridinedione[a] (41%)	92
5,6-Diamino-3-methyl-1-phenyl-2-thioxo-1,2-dihydro-4(3*H*)-pyrimidinone	Phenylpyruvic acid; aqueous ethanolic NaOAc; reflux; 24 h (?)	7-Benzyl-3-methyl-1-phenyl-2-thioxo-1,2-dihydro-4,6(3*H*,8*H*)-pteridinedione	990
5,6-Diamino-1,3-dimethyl-2,4(1*H*,3*H*)-pyrimidinedione	Phenylpyruvic acid; aqueous ethanolic NaOAc; reflux; 10–15 h	7-Benzyl-1,3-dimethyl-2,4,6(1*H*,3*H*,5*H*)-pteridinetrione[a] (85%)	710
5,6-Diamino-3-methyl-2-methylthio-4(3*H*)-pyrimidinone	1,3-Dimethylalloxan; water; reflux; 15 min	3,*N*-Dimethyl-2-methylthio-4,6-dioxo-3,4,5,6-tetrahydro-7-pteridinecarboxamide (~40%)	342
5,6-Diamino-2-thioxo-1,2-dihydro-4(3*H*)-pyrimidinone	Diethyl mesoxalate; 1*M* AcOH; reflux; 30 min	4,7-Dioxo-2-thioxo-1,2,3,4,7,8-hexahydro-6-pteridinecarboxylic acid	399
5-Amino-6-D-ribitylamino-2,4(1*H*,3*H*)-pyrimidinedione	Diethyl mesoxalate; pH 1; 100°C; 30 min	Ethyl 2,4,7-trioxo-8-D-ribityl-1,2,3,4,7,8-hexahydro-6-pteridinecarboxylate (33%)	1218
5-Amino-6-*β*-hydroxyethylamino-2,4(1*H*,3*H*)-pyrimidinedione	*p*-Hydroxybenzoylformic acid; water; 100°C; 1 h	8-*β*-Hydroxyethyl-6-*p*-hydroxyphenyl-2,4,7(1*H*,3*H*,8*H*)-pteridinetrione[a] (29%)	1218
5-Amino-6-D-ribitylamino-2,4(1*H*,3*H*)-pyrimidinedione	2-Oxoglutaric acid; water; 100°C; 30 min	6-*β*-Carboxyethyl-8-D-ribityl-2,4,7(1*H*,3*H*,8*H*)-pteridinetrione (26%)	1218, 1248

[a] And analog(s).
[b] Orientation unproven.
[c] For mechanism, see reference.
[d] Yield before deacetylation.
[e] Predominantly as the phenacylidene tautomer(s).
[f] Orientation proven.

of the isomer, 2,4-diamino-6-methyl-7(8*H*)-pteridinone. Replacement of pyruvic acid by ethyl sodiooxaloacetate gave a less satisfactory mixture of isomers in the ratio 5:2.[874] In contrast, when both condensations were carried out at pH 5, the reaction with pyruvic acid gave an unsatisfactory mixture, whereas the reaction with ethyl sodiooxaloacetate afforded 2,4-diamino-6-methyl-7(8*H*)-pteridinone (67%), uncontaminated by its isomer.[874] The same pyrimidinetetramine with benzoylformic acid in aqueous sodium acetate or with ethyl benzoylformate in dilute acetic acid gave isomer free 2,4-diamino-6-phenyl-7(8*H*)-pteridinone in > 70% yield.[128, 1195] In aqueous ethanolic hydrochloric acid, a separable 1:1 mixture of this 7-pteridinone and the isomeric 2,4-diamino-7-phenyl-6(5*H*)-pteridinone was obtained.[128] Other condensations with 2,4,5,6-pyrimidinetetramine are exemplified in the formation of 7-acetonyl-2,4-diamino-6(5*H*)-pteridinone (**158**) (ethyl acetylpyruvate in 2*M* hydrochloric acid: isomer free),[580] 2,4-diamino-7-phenacyl-6(5*H*)-pteridinone (ethyl benzoylpyruvate: ~55%; the product appeared to exist predominantly as its phenacylidene tautomer according to nmr studies),[982] 2,4-diamino-7-oxo-7,8-dihydro-6-pteridinecarboxylic acid (ethyl mesoxalate in 1*M* sulfuric acid, dilute acetic acid, or at pH 5: 87–90%),[874, 1234] 2,4-diamino-6-ethoxycarbonylmethyl-7(8*H*)-pteridinone (ethyl sodiooxaloacetate* in dilute[1195] or glacial[850] acetic acid: 72 and 85%; diethyl acetylenedicarboxylate in aqueous ethanolic sodium acetate: ~ 35%[247]); its 6-methoxycarbonylmethyl homolog (dimethyl acetylenedicarboxylate: ~ 50%),[247] its 6-carboxymethyl analog (80%),[506] and 2,4-diamino-6-α,β-diethoxycarbonylethyl-7(8*H*)-pteridinone (potassium diethyl oxalosuccinate: 59%).[1195] Similarly, 4,6-bismethylamino-2,5-pyrimidinediamine with diethyl mesoxalate in dilute sulfuric acid gave 2-amino-8-methyl-4-methylamino-7-oxo-7,8-dihydro-6-pteridinecarboxylic acid (**159**, R = Me)[882]; and 6-methylamino-2,4,5-pyrimidinetriamine with diethyl mesoxalate at pH 3 gave a separable mixture of 2-amino-4-methylamino-7-oxo-7,8-dihydro-6-pteridinecarboxylic acid, 2,4-diamino-8-methyl-7-oxo-7,8-dihydro-6-pteridinecarboxylic acid (**159**, R = H), and its ethyl ester.[882]

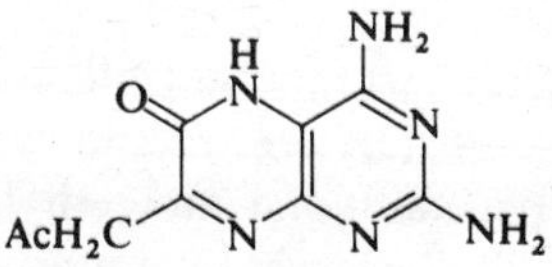

(**158**)

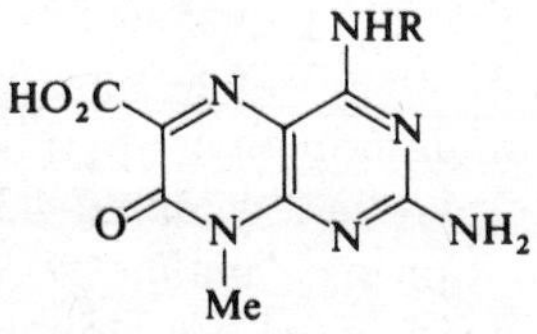

(**159**)

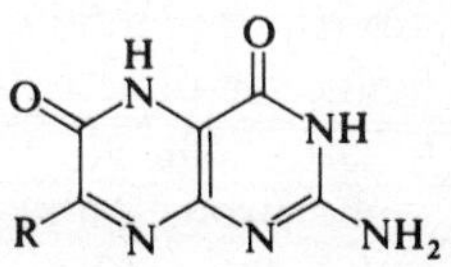

(**160**)

* This common reagent is really the diethyl ester.

(4) *Cases with an Amino- and an Oxo-Type Group in the Pyrimidine Ring*

The simplest possible condensation in this category is that of 2,5,6-triamino-4(3*H*)-pyrimidinone with pyruvic acid in 1*M* sulfuric acid to give 2-amino-7-methyl-4,6(3*H*,5*H*)-pteridinedione (**160**, R = Me) (76%) or in aqueous acetic acid to give a separable mixture of the same material (**160**, R = Me) (42% yield) and its isomer, 2-amino-6-methyl-4,7(3*H*,8*H*)-pteridinedione (**161**, R = Me) (10% yield).[878] The latter compound (**161**, R = Me) was better made with methyl pyruvate in water (~ 50% yield)[339] or indirectly by condensation of the pyrimidine with dimethyl acetylenedicarboxylate or diethyl oxaloacetate, followed by hydrolysis and decarboxylation of the (unisolated) 6-alkoxycarbonylmethyl-2-amino-4,7(3*H*,8*H*)-pteridinedione (up to 69% overall yield).[874, 1371] Other simple examples include the formation of a separable mixture of 2-amino-7-benzyl-4,6(3*H*,5*H*)-pteridinedione (**160**, R = CH_2Ph) (~40% yield) and 2-amino-6-benzyl-4,7(3*H*,8*H*)-pteridinedione (**161**, R = CH_2Ph) (~30% yield) using phenylpyruvic acid[718, cf. 707, 710]; 2-amino-6-phenyl-4,7(3*H*,8*H*)-pteridinedione (40%) using ethyl benzoylformate[1195]; 2-dimethylamino-6-methyl-4,7(3*H*,8*H*)-pteridinedione (ethyl pyruvate in boiling water: ~40%)[360]; 2-dimethylamino-4-methoxy-6-methyl-7(8*H*)-pteridinone (ethyl pyruvate in ethanol: 78%)[417]; 2-amino-4-benzyloxy-6-methyl-7(8*H*)-pteridinone (methanolic ethyl pyruvate: ~80%)[372]; its 4-isopropoxy analog (> 80%)[371]; 7-isobutyl-4-isopropoxy-2-β,β,β-trifluoro-α-hydroxyethylamino-6(5*H*)-pteridinone (**163**, R = $CHOHCF_3$) [from 6-isopropoxy-2,4,5-pyrimidinetriamine and 4-isobutyl-2-trifluoromethyl-2,5-dihydro-1,3-oxazol-5-one (**162**). The mechanism appears to involve hydrolytic splitting of the oxazolone to isopropylpyruvic acid and α-amino-β,β,β-trifluoroethanol, the former being used in the pteridine synthesis and the latter in a subsequent transamination at the 2-position. Several analogs were made similarly[1379]; a separable mixture of 2-amino-3,7-dimethyl-4,6(3*H*,5*H*)- and 2-amino-3,6-dimethyl-4,7(3*H*,8*H*)-pteridinedione (ethyl pyruvate: ~40 and 20% yield, respectively)[383]; 2-amino-6-hydroxymethyl-4,7(3*H*,8*H*)-pteridinedione (bromopyruvic acid in boiling water: ~35%)[499]; 2-amino-6,8-dimethyl- and 2-amino-3,6,8-trimethyl-4,7(3*H*,8*H*)-pteridinedione (each ~75%[383, cf. 712]; 2-amino-6-methyl-8-pyridin-2′-yl-4,7(3*H*,8*H*)-pteridinedione (from 2,5-diamino-6-pyridin-2′-ylamino-4(3*H*)-pyrimidinone and ethyl pyruvate: ~10%)[139]; its 8-phenyl analog (71%)[1090]; 7-acetonyl-2-amino-4,6(3*H*,5*H*)-pteridinedione (with ethyl acetylpyruvate: 63%, plus a separable

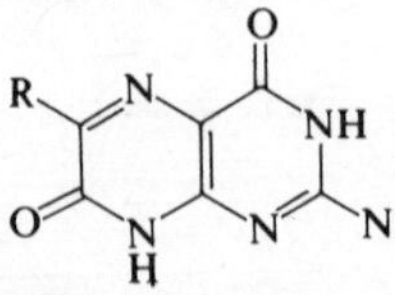

(161)

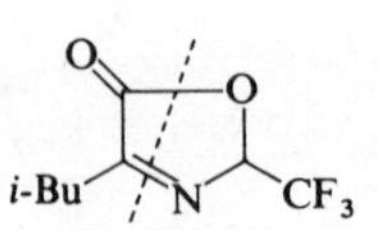

(162)

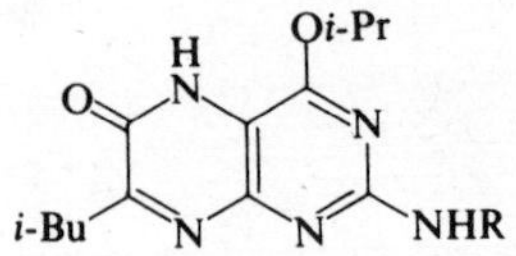

(163)

purine)[580, 712, 1083 1313]; 6-acetonyl-2-amino-4,7(3*H*,8*H*)-pteridinedione (initially alkaline medium then slightly acidic: ~50% isomer free)[1083]; 6-acetonyl-2-amino-8-methyl- and 6-acetonyl-2-amino-8-phenyl-4,7(3*H*,8*H*)-pteridinedione: 50 and 60%; the latter accompanied by a separable purine)[712]; 2-amino-4-benzyloxy-6,8-dimethyl-7(8*H*)-pteridinone (ethyl pyruvate: ~85%)[372]; its 4-isopropoxy analog (~75%)[371]; 2-amino-8-β-hydroxyethyl-4-isopropoxy-6-methyl-7(8*H*)-pteridinone (~60%)[371]; 2-amino-8-β-hydroxyethyl-6-methyl-4,7(3*H*,8*H*)-pteridinedione (60–79%)[422, 1090]; and 2-amino-8-ethyl-6-methyl-4,7(3*H*,8*H*)-pteridinedione (~30%).[660]

Appropriate α-ketoesters afford 6- or 7-pteridinecarboxylic acids (or esters) as illustrated by the condensation of 2,5,6-triamino-4(3*H*)-pyrimidinone in water with diethyl mesoxalate[405, 480, cf. 726] or, better, alloxan (**154**)[480, 481] to give 2-amino-4,7-dioxo-3,4,7,8-tetrahydro-6-pteridinecarboxylic acid (**164**, R = OH) in 85 or 97% yield, respectively. When the condensation with diethyl mesoxalate was done in 1*M* sulfuric acid, the isomeric 2-amino-4,6-dioxo-3,4,5,6-tetrahydro-7-pteridinecarboxylic acid (42%) resulted.[405] Other examples include the formation of 2-amino-*N*-methyl-4,7-dioxo-3,4,7,8-tetrahydro-6-pteridinecarboxamide (**164**, R = NHMe) [using 1,3-dimethylalloxan (**148**): ~45%],[342] ethyl 2-dimethylamino-4,7-dioxo-3,4,7,8-tetrahydro-6-pteridinecarboxylate (diethyl mesoxalate in neutral aqueous or ethanolic solution: 40 and 80%),[360] ethyl 2-amino-4-isopropoxy-7-oxo-7,8-dihydro-6-pteridinecarboxylate (diethyl mesoxalate in aqueous methanol: ~75%),[371] its 4-methylthio analog (diethyl mesoxalate and boron trifluoride in glacial acetic acid: 66%),[1468] 2-amino-3-methyl-4,7-dioxo-3,4,7,8-tetrahydro-6-pteridinecarboxylic acid (diethyl mesoxalate in 2*M* acetic acid: ~55%),[385] its ethyl ester (in neutral aqueous solution: ~30%),[385] ethyl 2-amino-4-isopropoxy (and methoxy)-8-methyl-7-oxo-7,8-dihydro-6-pteridinecarboxylate (70 and 55%),[371] ethyl 2-dimethylamino-8-methyl-4,7-dioxo-3,4,7,8-tetrahydro-6-pteridinecarboxylate (~70%),[360] 2-amino-8-ethyl-4,7-dioxo-3,4,7,8-tetrahydro-6-pteridinecarboxylic acid (using alloxan: 67%),[480, 481] ethyl 3,8-dimethyl-2-methylamino-4,7-dioxo-3, 4, 7, 8-tetrahydro-6-pteridinecarboxylate (~80%),[1565] and the corresponding carboxylic acid (87%) (diethyl mesoxalate with an alkaline work-up).[1214] Some 6/7-acetonylpteridinediones have been mentioned previously and other ketones so made include, for example, 2-amino-7-phenacyl-4,6(3*H*,5*H*)-pteridinedione (**165**, R = Ph) (ethyl benzoylpyruvate in 1*M* acetic acid: ~85%),[982] 2-amino-7-α'-methoxyacetonyl-4,6(3*H*,5*H*)-

(164)

(165)

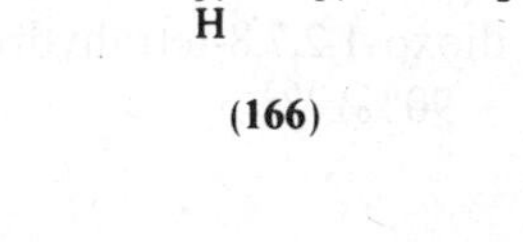

(166)

pteridinedione (**165**, R = CH_2OMe) (ethyl 5-methoxy-2,4-dioxovalerate in aqueous ethanol: 36%),[504] and its 7-α′-(methylthio)acetonyl analog (**165**, R = CH_2SMe) (37%).[1295] The orientation of this phenacyl derivative (**165**, R = Ph) may be reversed by doing the initial condensation in dilute ammonia and subsequently cyclizing the intermediate (**166**) in concentrated sulfuric acid, to give 2-amino-6-phenacyl-4,7(3*H*,8*H*)-pteridinedione (~25% overall).[982] Related compounds bearing carbonyl groups well removed from the nucleus may be made similarly. For example, 2,5,6-triamino-4(3*H*)-pyrimidinone with ethyl sodiooxaloacetate (in hot acetic acid–sodium acetate[1304] or 1*M* acetic acid[1195]) or with diethyl acetylenedicarboxylate in ethanol at 25°C,[247] gave up to 60% of 2-amino-6-ethoxycarbonylmethyl-4,7(3*H*,8*H*)-pteridinedione (**167**, R = Et) or, if an appropriate work-up was used,[1061, 1083] the free carboxymethyl analog (**167**, R = H). The orientation was reversed by using ethyl sodiooxaloacetate in ethanol containing hydrochloric acid to afford 2-amino-7-ethoxycarbonylmethyl-4,6(3*H*,5*H*)-pteridinedione in rather poor yield.[497, 1055] Similar products include 2-amino-6-α,β-diethoxycarbonylethyl-4,7(3*H*,8*H*)-pteridinedione (using potassium diethyl oxalosuccinate: 50%),[1195] the free dicarboxyethyl analog,[1083] 2-amino-6-β-carboxyethyl-4,7(3*H*,8*H*)-pteridinedione (using ethyl 2-oxoglutarate: ~30%),[1083] and the α-carboxyethyl isomer (using ethyl α-ethoxalylpropionate).[1083]

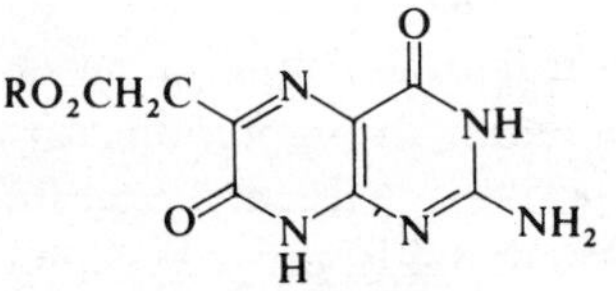

(167)

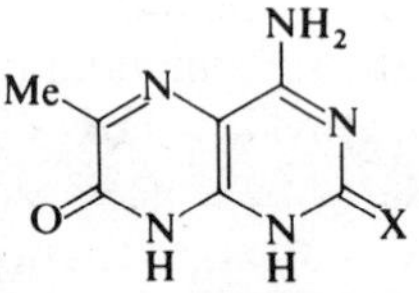

(168)

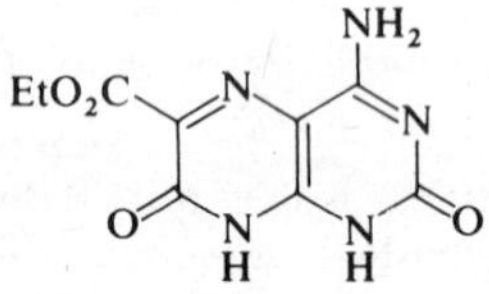

(169)

Finally, some derivatives of 4-amino-2-pteridinone or the corresponding thione have been made by such condensations. For example, 4,5,6-triamino-2(1*H*)-pyrimidinone with ethyl sodiooxaloacetate in acetic acid[279] or with ethyl pyruvate in aqueous buffer[363] gave 4-amino-6-methyl-2,7(1*H*,8*H*)-pteridinedione (**168**, X = O) (~75%). Similar procedures afforded 4-amino-6-methyl-2-thioxo-1,2-dihydro-7(8*H*)-pteridinone (**168**, X = S) (ethyl sodiooxaloacetate or pyruvic acid in aqueous acetic acid: 68%),[916] the isomeric 4-amino-7-methyl-2-thioxo-1,2-dihydro-6(5*H*)-pteridinone (pyruvic acid in 1*M* sulfuric acid: 60%),[916] 4-amino-6-benzyl-2-thioxo-1,2-dihydro-7(8*H*)-pteridinone (phenylpyruvic acid in slightly acidic water: 30%),[92] 4-amino-6-benzyl-2-methylthio-7(8*H*)-pteridinone (54%),[92] 4-dimethylamino-6-methyl-2,7(1*H*,8*H*)-pteridinedione,[363] 4-amino-6-carboxymethyl-2-thioxo-1,2-dihydro-7(8*H*)-pteridinone,[699] 4-amino-7-carboxymethyl-2-methylthio-6(5*H*)-pteridinone,[699] 4-amino-6,8-dimethyl-2,7(1*H*,8*H*)-pteridinedione,[363] and ethyl 4-amino-2,7-dioxo-1,2,7,8-tetrahydro-6-pteridinecarboxylate (**169**) (diethyl mesoxalate: ~90%).[363]

(5) *Cases with Two Oxo-Type Groups in the Pyrimidine Ring*

The reaction of 5,6-diamino-2,4(1*H*,3*H*)-pyrimidinedione with refluxing aqueous methyl pyruvate gave 6-methyl-2,4,7(1*H*,3*H*,8*H*)-pteridinetrione (~70%)[339] The same pyrimidinedione with pyruvic acid in dilute acetic acid containing hydrazine also gave 6-methyl-2,4,7(1*H*,3*H*,8*H*)-pteridinetrione (~20%), whereas the reaction in 1*M* sulfuric acid gave the isomeric 7-methyl-2,4,6(1*H*,3*H*,5*H*)-pteridinetrione (**170**, R = H) in about the same yield.[1109, 1110] The latter compound was made indirectly by condensation of the same pyrimidine with ethyl acetylpyruvate in dilute hydrochloric acid to give 7-acetonyl-2,4,6(1*H*,3*H*,5*H*)-pteridinetrione (**170**, R = Ac), which was then split hydrolytically in aqueous sodium hydroxide. Again, the yield was only 20%[340, 352]; the 1,7- and 3,7-dimethyl analogs were made by similar indirect routes from appropriate pyrimidines.[340] Simple products from direct syntheses are represented by 6-phenyl-2,4,7(1*H*,3*H*,8*H*)-pteridinetrione (methyl benzoylformate: 59%),[413] the 7-benzyl compound (**170**, R = Ph) (phenylpyruvic acid: 70%),[710] 7-benzyl-2-thioxo-1,2-dihydro-4,6(3*H*,5*H*)-pteridinedione (phenylpyruvic acid in aqueous ethanolic sodium acetate),[990] 2,4-dimethoxy-6-methyl-7(8*H*)-pteridinone (ethyl pyruvate: 69%),[1192] 7-methyl-2-thioxo-1,2-dihydro-4,6(3*H*,5*H*)-pteridinedione (pyruvic acid in 1*M* sulfuric acid: 42%),[916] the isomeric 6-methyl-2-thioxo-1,2-dihydro-4,7(3*H*,8*H*)-pteridinedione (pyruvic acid in 2*M* acetic acid: 55%),[916] 6-phenacyl-2,4,7(1*H*,3*H*,8*H*)-pteridinetrione (ethyl benzoylpyruvate in refluxing pyridine: 75%),[1524] the 7-phenacyl isomer (similarly, but in boiling 1*M* hydrochloric acid: 30%),[1524] and others.[708, 714]

Examples with 1- and/or 3-alkyl groups include 1,6- and 3,6-dimethyl-2,4,7(1*H*,3*H*,8*H*)-pteridinetrione (80 and 60%),[339, cf. 1380] 3-methyl- and 1,3-dimethyl-6-phenyl-2,4,7(1*H*,3*H*,8*H*)-pteridinetrione (ethyl benzoylformate: 80 and 62%),[367] 1-methyl-6-phenyl-2,4,7(1*H*,3*H*,8*H*)-pteridinetrione (41%),[413] 7-benzyl-1-methyl-2-thioxo-1,2-dihydro-4,6(3*H*,5*H*)-pteridinedione (ethyl phenylpyruvate: 68%),[990] its 1-phenyl analog (ethyl phenylpyruvate: 81%[990]; 3-phenyl-2-thioxopropionic acid in ethanol: 55%[810]), 3,6-dimethyl-1-phenyl-2,4,7(1*H*,3*H*,8*H*)-pteridinetrione (**171**, R = Ph) (aqueous ethanolic ethyl pyruvate: 87%), a separable mixture of 1,3,6-trimethyl-2,4,7(1*H*,3*H*,8*H*)- (**171**, R = Me) and 1,3,7-trimethyl-2,4,6(1*H*,3*H*,5*H*)-pteridinetrione in which the former predominated,[238, 337, 1380] 1,3,6-trimethyl-2-thioxo-1,2-dihydro-4,7(3*H*,8*H*)-pteridinedione (70%),[1000] and others.[709, 1524]

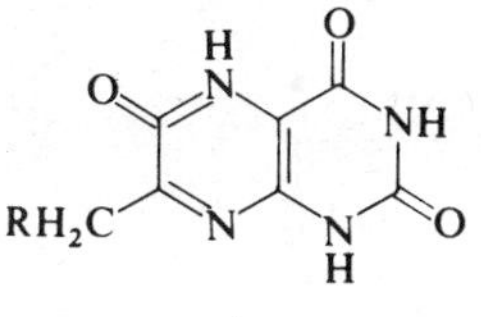

(170)

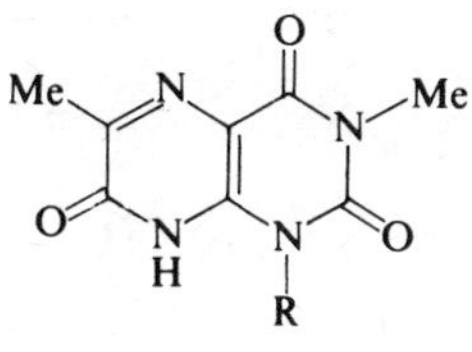

(171)

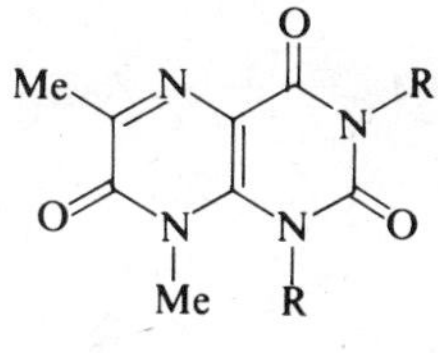

(172)

There is also a good representation of 8-alkylated derivatives. For example, 5-amino-6-methylamino-2,4(1*H*,3*H*)-pyrimidinedione and ethyl pyruvate at pH 3 gave 6,8-dimethyl-2,4,7(1*H*,3*H*,8*H*)-pteridinetrione (**172**, R=H) in ~35% yield[248, 1441]; the use of methyl pyruvate improved the yield to ~50%.[339] Similar reactions gave 8-ethyl-6-methyl- (methyl pyruvate: ~50%),[378] 8-benzyl-6-methyl-(~65%),[378] 8-ethyl-6-phenyl- (11%),[413] 1,6,8-trimethyl- (ethyl pyruvate: 25%),[716] 3,6,8-trimethyl- (~80%),[248] 1,3,6,8-tetramethyl- (**172**, R=Me) (~10%),[339] 8-*β*-hydroxyethyl-6-methyl- (methyl pyruvate: ~55%[378, 1082]; ethyl acetylpyruvate: 4% plus a purine[712]), and 6-benzyl-8-*β*-hydroxyethyl-2,4,7(1*H*,3*H*,8*H*)-pteridinetrione (ethyl phenylpyruvate: 66%).[674] Appropriate pyrimidines also gave, 1,3,6,8-tetramethyl-2-thioxo-1,2-dihydro-4,7(3*H*,8*H*)-pteridinedione (52%),[1000] 6-methyl-8-D-ribityl-2,4,7(1*H*,3*H*,8*H*)-pteridinetrione (up to 59%),[378, 674, 1082, 1544] 6-hydroxymethyl-8-D-ribityl-2,4,7(1*H*,3*H*,8*H*)-pteridinetrione (hydroxypyruvic acid: ~4%),[1430] and related compounds.[1430]

The 6- or 7-alkyl group may be replaced by a carbonyl-containing group as exemplified in the condensation of 5,6-diamino-2,4(1*H*,3*H*)-pyrimidinedione with mesoxalic acid,[1304] diethyl mesoxalate,[341, 1234] alloxan,[480, 481] or alloxanic acid[480] to give 2,4,7-trioxo-1,2,3,4,7,8-hexahydro-6-pteridinecarboxylic acid (**173**) in ~55% (plus a small yield of the isomer), ~90, 55, or 55%, respectively. The mechanisms with alloxan or alloxanic acid are by no means simple.[480] In contrast, the same pyrimidine reacted with 1,3-dimethylalloxan to afford *N*-methyl-2,4,6-trioxo-1,2,3,4,5,6-hexahydro-7-pteridinecarboxamide (**174**, R=H) (~40%),[342, 345] proven in structure by hydrolysis to the corresponding carboxylic acid and subsequent decarboxylation.[342] Similar reactions afforded 1,*N*-dimethyl- (~45%),[342] 1,3,*N*-trimethyl- (**174**, R=Me),[117, 336] or 1,3,5,*N*-tetramethyl-2,4,6-trioxo-1,2,3,4,5,6-hexahydro-7-pteridinecarboxamide (~55%)[347] (the last compound is a rare example of a 5-alkylated-6-pteridinone being made from a 5-alkylamino-4-pyrimidinamine). Further illustrations of the use of a traditional reagent (diethyl mesoxalate), as distinct from alloxan, and so on, are provided in the formation of 1-methyl- (50–90%),[341, 1083] 3-methyl- (55%),[341] 8-methyl- (~40%),[341] 1,3,8-trimethyl- (~85%),[341] 8-ethyl- (~40%),[317] 8-*β*-hydroxyethyl (~50%),[317] 8-benzyl- (~55%),[317] and 8-benzyl-3-methyl-2,4,7-trioxo-1,2,3,4,7,8-hexahydro-6-pteridinecarboxylic acid,[317] all naturally via the esters that were seldom isolated; mesoxalic acid itself was used to make 1,3-dimethyl-2,4,7-trioxo-1,2,3,4,7,8-hexahydro-6-pteridinecarboxylic acid.[1380] Examples of products with carbonyl groups further away from the ring include 7-acetonyl-2,4,6(1*H*,3*H*,5*H*)-pteridinetrione (ethyl acetylpyruvate in 2*M*

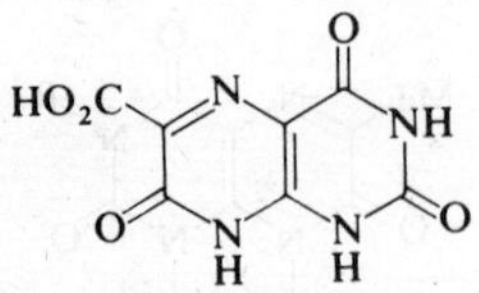

(**173**)

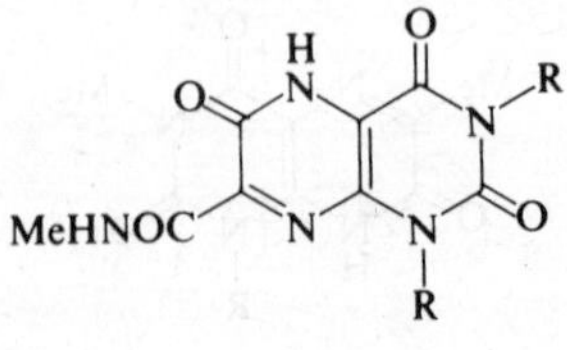

(**174**)

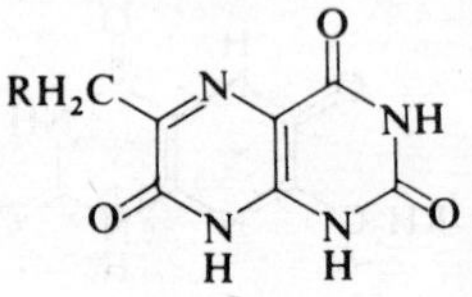

(**175**)

hydrochloric acid or methanol: ~50%)[1083, 1304]; 6-acetonyl-2,4,7(1*H*,3*H*,8*H*)-pteridinetrione (**175**, R=Ac) (same reagent at pH 7)[1083]; 6-carboxymethyl- (**175**, R=CO_2H) (ethyl oxaloacetate in glacial acetic acid: 50%),[1304] 6-α-carboxyethyl- (ethyl α-ethoxalylpropionate: 56%),[1083] and 6-β-carboxyethyl-2,4,7(1*H*,3*H*,8*H*)-pteridinetrione (**175**, R=CH_2CO_2H) (2-oxoglutaric acid: 21%)[1218, 1248]; 6-ethoxycarbonylmethyl-1,3-dimethyl-2,4,7(1*H*,3*H*,8*H*)-pteridinetrione (ethyl sodiooxaloacetate: ~40%)[337]; 6-β-carboxyethyl-1,3,8-trimethyl-2,4,7(1*H*,3*H*,8*H*)-pteridinetrione (2-oxoglutaric acid: 46%)[674]; and 6-β-carboxyethyl-8-β-hydroxyethyl-2,4,7(1*H*,3*H*,8*H*)-pteridinetrione (2-oxoglutaric acid at pH 2: 30–68%).[1218, 1449] The condensation of 5,6-diamino-1,3-dimethyl-2,4(1*H*,3*H*,)-pyrimidinedione with freshly distilled pyruvic acid without solvent under reflux gave three products: 1,3,7-trimethyl-2,4,6(1*H*,3*H*,5*H*)-, 1,3,6-trimethyl-2,4,7(1*H*,3*H*,8*H*)-, and 7-(2′-carboxyprop-1′-enyl)-1,3-dimethyl-2,4,6(1*H*,3*H*,5*H*)-pteridinetrione, the last needing two molecules of pyruvic acid for its formation.[337] Under appropriate conditions, 5-amino-6-β-hydroxyethylamino-2,4(1*H*,3*H*)-pyrimidinedione and 2,6-dioxoheptanedioic acid gave 1,3-bis(8′-β-hydroxyethyl-2′,4′,7′-trioxo-1′,2′,3′,4′,7′,8′-hexahydro-6′pteridinyl)-propane in 25% yield; the 8′-D-ribityl analog was made similarly.[674]

F. Use of the α-Diester, Diethyl Oxalate, or Equivalent

The condensation of 4,5-pyrimidinediamines with oxalic acid or diethyl oxalate involves no structural ambiguity in the products.

(1) *Simple Cases without a Functional Group in the Pyrimidine Ring*

When 4,5-pyrimidinediamine and oxalic acid dihydrate were heated together at 160°C, preferably under a slight vacuum to assist water removal, 6,7(5*H*,8*H*)-pteridinedione (**176**, R=H) resulted in 88–94% yield.[31, 208] Similar procedures gave 2-phenyl- (~55%),[152] 4-methyl- (**176**, R=Me) (83%),[64] and 4-phenyl-6,7(5*H*,8*H*)-pteridinedione (**176**, R=Ph) (~60%).[151] Heating 5-methylamino-4-pyrimidinamine with diethyl oxalate at 180°C for 6 min gave <20% of 5-methyl-6,7(5*H*,8*H*)-pteridinedione[134]. Analogous treatment of 4-methylamino-5-pyrimidinamine with dimethyl oxalate at 150°C for 1 h gave initially only the oxalate salt of the pyrimidine, which on fusion at 200°C afforded a low yield of 8-methyl-6,7(5*H*,8*H*)-pteridinedione.[34] An unusual reaction is also best classified here: 4,5-Pyrimidinediamine (**178**) reacted with ethyl (*C*-ethoxyformimidoyl) formate (**177**) (the half-imino ester of diethyl oxalate) in boiling ethanol to give 6-amino-7(8*H*)-pteridinone (**179**) (80%) of proven orientation; 6-amino-2,4,7(1*H*,3*H*,8*H*)-pteridinetrione and its 1,3-dimethyl derivative were made similarly in 60 and 90% yield, respectively.[1093]

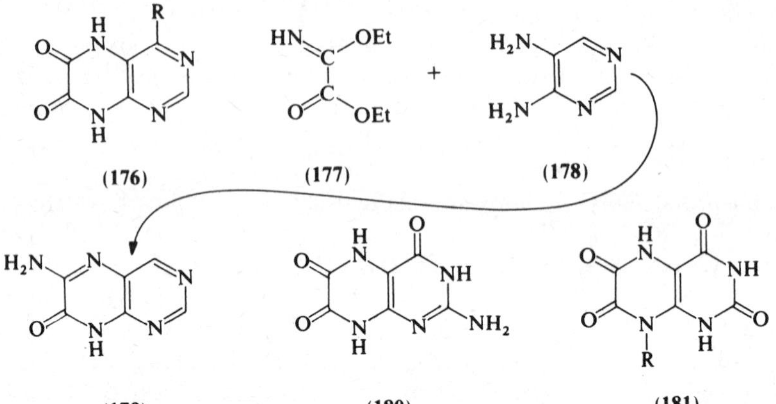

(176) (177) (178) (179) (180) (181)

(2) *Cases with One Functional Group in the Pyrimidine Ring*

A typical reaction is that of 2-chloro-6-phenyl-4,5-pyrimidinediamine and dimethyl oxalate in refluxing water during 20 h to give 2-chloro-4-phenyl-6,7(5*H*,8*H*)-pteridinedione (37%)[522]; diethyl oxalate gave a lower yield and oxalyl chloride gave a different (unidentified) product.[522] Although 6-styryl-2,4,5-pyrimidinetriamine reacted with oxalic acid at 160°C to give 2-amino-4-styryl-6,7(5*H*,8*H*)-pteridinedione,[421] and 2,4,5-pyrimidinetriamine did likewise to give 2-amino-6,7(5*H*,8*H*)-pteridinedione (~70%),[41, cf. 536, 1152] the apparently similar condensation of 4,5-diamino-2(1*H*)-pyrimidinone with oxalic acid or dimethyl oxalate under appropriate conditions gave a product (up to 95%), which showed anomalous chromatographic behavior, possibly due to it existing as an easily disturbed equilibrium mixture of 2,6,7(1*H*,5*H*,8*H*)-pteridinetrione and its covalent 3,4-hydrate.[41] 2-Dimethylamino-8-ethyl-6,7(5*H*,8*H*)-pteridinedione (63%) and 4,8-dimethyl-2,6,7(1*H*,5*H*,8*H*)-pteridinetrione (44%) were made by using ethanolic or aqueous dimethyl oxalate, respectively.[208]

Products with a functional group at the 4-position include 4-amino- (49%),[208] 4-methoxy- (71%),[208] 4-dimethylamino-8-methyl- (46%),[208] 4-amino-2-phenyl- (84%),[522] and 4-amino-2-methyl-6,7(5*H*,8*H*)-pteridinedione (oxalic acid–sodium oxalate at 250°C)[918]; also 4,6,7(3*H*,5*H*,8*H*)-pteridinetrione (oxalic acid at 165°C: 81%[208]; refluxing aqueous bromal: ~35%[101]) and its 2-methyl,[918] 2-phenyl (diethyl oxalate in ethoxyethanol under reflux for 6 h: 76–84%),[101, 522, 663, 1398] 2-fur-2′-yl,[663] and 2-*p*-tolyl[664] derivatives.

(3) *Cases with Two Functional Groups in the Pyrimidine Ring*

The reaction of 2,4,5,6-pyrimidinetetramine with oxalic acid at 260°C gave 2,4-diamino-6,7(5*H*,8*H*)-pteridinedione (49%).[208, 730] Similarly, a mixture of

2,5,6-triamino-4(3*H*)-pyrimidinone and oxalic acid at 260°C was used by Robert Purrmann for his original preparation of 2-amino-4,6,7(3*H*,5*H*,8*H*)-pteridinetrione (leucopterin) (**180**) in 90% yield.[403, 1152] Subsequently, variations in detail were introduced,[71, 800, 983, 1158, 1268] but the two definitive procedures[60, 343] remained essentially that of Purrmann. Analogous preparations include the conversion of appropriate pyrimidinediamines into 2-amino-3-methyl- (oxalic acid at 225°C: ~80%; aqueous oxalic acid under reflux for 8 h: ~65%),[384] 2-amino-8-methyl- (oxalic acid or aqueous oxalic acid: 50–60%),[384] 2-amino-8-ethyl- (oxalic acid at 260°C),[885] 2-amino-8-β-hydroxyethyl- (diethyl oxalate in ethanolic sodium ethoxide: yield quite poor),[183] 2-amino-3,8-dimethyl- (oxalic acid or aqueous oxalic acid: 20 and 75%),[384] 2-amino-8-D-ribityl (and sorbityl)- (aqueous diethyl oxalate: both 58%),[172] 2-dimethylamino- (oxalic acid at 240°C: ~40%),[360] and 2-dimethylamino-8-methyl-4,6,7(3*H*,5*H*,8*H*)-pteridinetrione (aqueous diethyl oxalate: ~80%).[360] They also convert into 2-amino-4-isopropoxy-6,7(5*H*,8*H*)-pteridinedione (diethyl oxalate without solvent at 160–180°C: ~85%),[371] its 8-methyl derivative (likewise: ~90%),[371] 4-amino-2,6,7(1*H*,5*H*,8*H*)-pteridinetrione (diethyl oxalate under reflux: ~35%[363]; oxalic acid–potassium oxalate at 250°C),[1369] its 1-methyl derivative (oxalic acid–potassium oxalate at 250°C: ~40%),[363] 4-amino-2-thioxo-1,2-dihydro-6,7(5*H*,8*H*)-pteridinedione (oxalic acid at 240°C: 90%),[1369] 4-amino-2-methylthio-6,7(5*H*,8*H*)-pteridinedione,[885] and its 8-D-glucosyl derivative (ethanolic diethyl oxalate plus sodium ethoxide).[885] It is interesting that 6-amino-5-formamidopyrimidines may be used in place of 5,6-diaminopyrimidines for such reactions. For example, 2-amino-5-formamido-6-methylamino-4(3*H*)-pyrimidinone and aqueous oxalic acid gave a better yield (~90%) of 2-amino-8-methyl-4,6,7(3*H*,5*H*,8*H*)-pteridinetrione than did 2,5-diamino-6-methylamino-4(3*H*)-pyrimidinone (previously cited)[384]; several other such reactions have been tried but the yield was not always improved by pre-formylation of the 5-amino group.[384]

The fusion of 5,6-diamino-2,4(1*H*,3*H*)-pyrimidinedione (base) with oxalic acid at 260°C gave a 62% yield of 2,4,6,7(1*H*,3*H*,5*H*,8*H*)-pteridinetetrone (**181**, R = H).[404, 1152, 1158] The procedure was improved subsequently by using the more readily obtained pyrimidine sulfate and adding some sodium acetate to the melt. The product (~70%) was much cleaner.[441] The procedure was improved further[706] to give an 85% yield but, despite a reduced yield of ~45%, it has proven better to use the pyrimidine with diethyl oxalate in ethylene glycol under reflux[41] or with bromal in water under reflux.[101] Another method involved treatment of the pyrimidine with trichloroacetic acid at 100°C, isolation of 4-amino-5-trichloroacetamido-2,4(1*H*,3*H*)-pyrimidinedione, and a final thermal cyclization at 280°C. The yield was very low.[101] Analogous procedures also gave 1-methyl- (oxalic acid at 240°C),[343, 404, 1152] 8-D-ribityl- (**181**, R = ribityl) (diethyl oxalate in ethanol at 25°C: ~35%),[1444] 1,3-dimethyl- (oxalic acid at 240°C: 58%),[1435] 3-methyl- (similarly: ~70%),[343] 8-methyl- (**181**, R = Me) (similarly: ~55%),[343] 1,3,8-trimethyl- (~15%),[343] and 1,3,5,8-tetramethyl-2,4,6,7(1*H*,3*H*,5*H*,8*H*)-pteridinetetrone [neat diethyl oxalate with 1,3-dimethyl-

5,6-bismethylamino-2,4(1*H*,3*H*)-pyrimidinedione under reflux: <10%].[343] Heating 5,6-diamino-2-thioxo-1,2-dihydro-4(3*H*)-pyrimidinone with oxalic acid at 260°C gave 2-thioxo-1,2-dihydro-4,6,7(3*H*,5*H*,8*H*)-pteridinetrione (60%).[878]

2. THE TIMMIS SYNTHESIS FROM 5-NITROSO-4-PYRIMIDINAMINES WITH α-CARBONYLMETHYLENE OR RELATED COMPOUNDS

From its modest introduction in 1949,[1179] the Timmis synthesis has been widely used especially for cases in which Gabriel and Colman syntheses will give products of ambiguous structure.[493, 559, 579] Considerable versatility may be achieved by altering the substitution pattern at the 1-, 2-, 3-, and 6-position of the 5-nitroso-4-pyrimidinamine (**186**) and by mono- but not disubstitution of the 4-amino group. In addition, the carbonylmethylene reagent (always provided that its methylene group remains sufficiently activated) may consist of an *α-methylenealdehyde*, which will give a 6-substituted-7-unsubstituted-pteridine: for example, phenylacetaldehyde (**182**) will give a 6-phenyl-pteridine; an *α-methyleneketone*, which will give a 7-substituted-pteridine, with or without a 6-substituent according to its nature: acetophenone (**183**, R = H) will give a 7-phenyl-6-unsubstituted-pteridine but propiophenone (**183**, R = Me) will give a 6-methyl-7-phenylpteridine; an *α-methylene-ester* (methylene-acyl chloride or sometimes a methylene-amide), which will give a 7-pteridinone, with or without a 6-substituent: e.g. ethyl phenylacetate (**184**, R = OEt) or phenylacetyl chloride (**184**, R = Cl) will give a 6-phenyl-7-pteridinone; or an *α-methylenenitrile* (or methyleneamide), which will give a 7-pteridinamine, with or without a 6-substituent: e.g. phenylacetonitrile (**185**, R = Ph) will give a 6-phenyl-7-pteridinamine. It should be noted that α,α′-dimethyleneketones (with a methylene group on each side of the carbonyl group) may be used but unsymmetrical versions could give products of ambiguous structure. Other equivalents and important variations will be discussed under the appropriate general headings next.

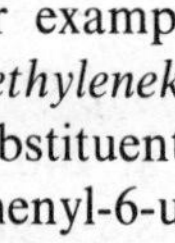
(182)

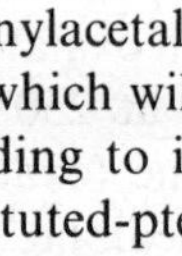
(183)

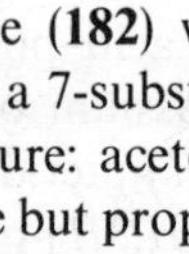
(184)

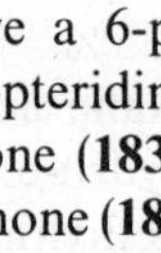
(185)

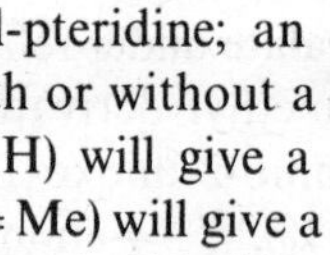
(186)

A. Use of α-Methylenealdehydes

A straightforward example is the condensation of 6-amino-3-methyl-2-methylthio-5-nitroso-4(3*H*)-pyrimidinone (**187**) with phenylacetaldehyde (**182**),

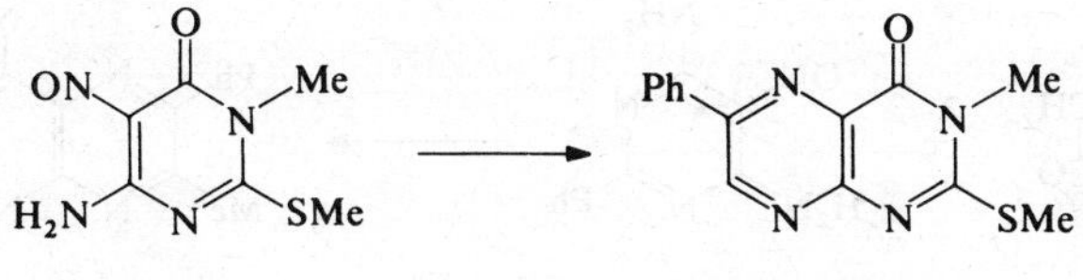

(187) (188)

by boiling without solvent under a water separator (Dean and Stark?) for 30 min, to give 3-methyl-2-methylthio-6-phenyl-4(3*H*)-pteridinone (**188**) in 21% yield[367]. In a similar manner, 6-amino-1,3-dimethyl-5-nitroso-2,4(1*H*,3*H*)-pyrimidinedione with phenylacetaldehyde gave 1,3-dimethyl-6-phenyl-2,4(1*H*,3*H*)-pteridinedione (41%).[176] Rather differently, 5-nitroso-2-phenyl-4,6-pyrimidinediamine and phenylacetaldehyde in boiling 2-ethoxyethanol containing potassium acetate gave 2,6-diphenyl-4-pteridinamine (<20%).[326] Possibly for fear of competitive Schiff-base formation at the unprotected amino group(s), di- or tri-acetylated 5-nitroso-2,4,6-pyrimidinetriamine has been allowed to react with phenylacetaldehyde in boiling ethanol containing potassium acetate to give partly acetylated products, which underwent alkaline hydrolysis to afford 6-phenyl-2,4-pteridinediamine in ~55 or 78% yield, respectively.[327, 541] Less simple examples include the reaction of 6-imino-1,3-dimethyl-5-phenoxyimino-5,6-dihydro-2,4(1*H*,3*H*)-pyrimidinedione (**190**) with *N*-but-1′-enylmorpholine (**189**) (equivalent to butyraldehyde) in boiling toluene to afford 6-ethyl-1,3-dimethyl 2,4(1*H*,3*H*)-pteridinedione (**191**) (36%)[1291]; and of 6-amino-1,3-dimethyl-5-nitroso-2,4(1*H*,3*H*)-pyrimidinedione with β-phenylethylamine hydrochloride (in which the β-methylene is activated by the phenyl group and by the quaternary nitrogen, but in which there is an alcohol rather than an aldehyde equivalent and hence a requirement for oxidation at some stage of the reaction) to give, on fusion at 180°C, 1,3-dimethyl-6-phenyl-2,4(1*H*,3*H*)-pteridinedione (41%) accompanied by a separable purine in smaller amount.[552]

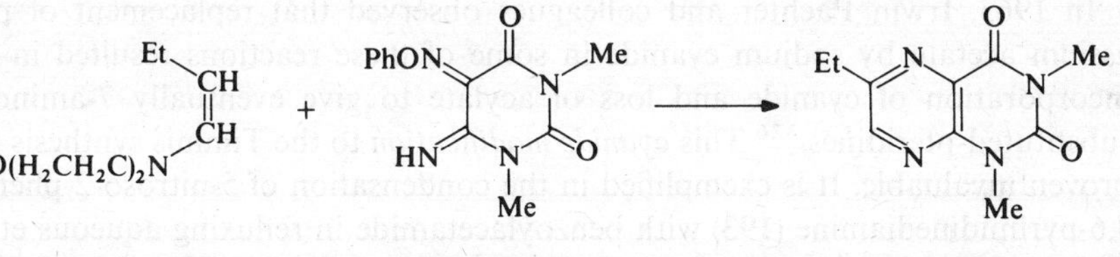

(189) (190) (191)

B. Use of α-Methyleneketones

The use of such a ketone is illustrated by the condensation of 5-nitroso-2-phenyl-4,6-pyrimidinediamine (**193**) with phenylacetone (**192**), in refluxing butanol containing potassium acetate, to give 7-methyl-2,6-diphenyl-4-

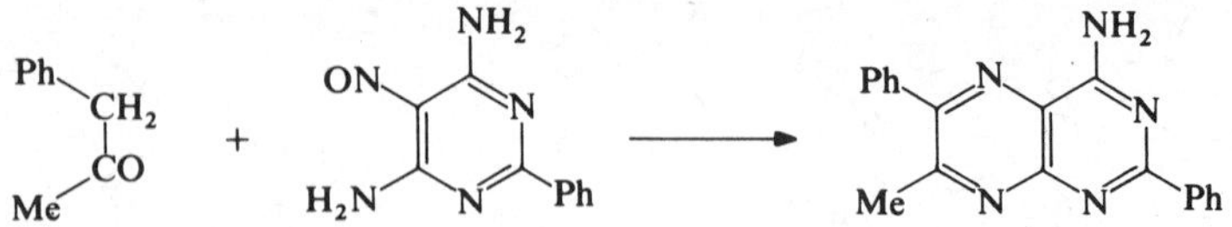

(192) (193) (194)

pteridinamine (**194**) in 42% yield.[326] Naturally, it would be possible for phenylacetone to react differently, so that the carbonyl-activated terminal methyl group was used in the condensation, thereby affording the isomeric 7-benzyl-2-phenyl-4-pteridinamine. However, the much greater activation of the methylene group by *both* of its neighbors, ensures that only one product (**194**) is formed. Other examples, of which some suffer from this theoretical ambiguity, are seen in the formation (from appropriate 5-nitrosopyrimidines) of 1,3-dimethyl-7-phenyl-2,4(1*H*,3*H*)-pteridinedione [acetophenone (**183**, R = H) under reflux with azeotropic removal of water: 35%],[176] 6,7-diphenyl-2,4-pteridinediamine [benzyl phenyl ketone (**183**, R = Ph) in glacial acetic acid],[1179] 7-methyl-6-phenyl-2,4-pteridinediamine [5-nitroso-2,4,6-pyrimidinetriamine with phenylacetone (**192**) in glacial acetic acid[1179]; the triacetylated (but not the diacetylated) pyrimidine with phenylacetone in ethanolic potassium acetate followed by alkaline deacetylation: 64%[525]], 4-amino-2,7-diphenyl-6-pteridinecarboxamide (**195**, R = Ph) (benzoylacetamide in ethanol containing potassium acetate: 70%),[326] 1,3,6-trimethyl-7-phenyl-2,4(1*H*,3*H*)-pteridinedione [neat propiophenone (**183**, R = Me) under reflux: 14%],[1348] 1,3,7-trimethyl-6-phenyl-2,4(1*H*,3*H*)-pteridinedione [neat phenylacetone (**192**) under reflux: 24%],[1348] 2,4-diamino-7-methyl-6-pteridinecarboxylic acid (from 5-nitroso-2,4,6-pyrimidinetriamine with methyl acetoacetate in refluxing β-methoxyethanolic sodium methoxyethoxide: 19%),[307] and 2,4-diamino-7-phenyl-6-pteridinecarboxamide (from diacetylated 5-nitroso-2,4,6-pyrimidinetriamine with benzoylacetamide in ethanolic potassium acetate followed by alkaline hydrolysis: ~85%).[327]

In 1963, Irwin Pachter and colleagues observed that replacement of potassium acetate by sodium cyanide in some of these reactions resulted in the incorporation of cyanide and loss of acylate to give eventually 7-amino-6-substituted-pteridines.[326] This *cyanide modification* to the Timmis synthesis has proven invaluable. It is exemplified in the condensation of 5-nitroso-2-phenyl-4,6-pyrimidinediamine (**193**) with benzoylacetamide in refluxing aqueous ethanolic sodium cyanide to give, not (**195**, R = Ph) as in the potassium acetate-induced reaction previously cited, but 4,7-diamino-2-phenyl-6-pteri-

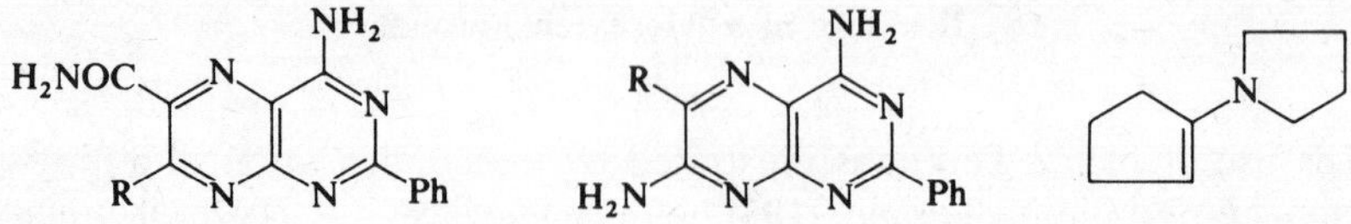

(195) (196) (197)

dinecarboxamide (**195**, R = NH_2),[326] in effect the product obtained[1466] by using cyanoacetamide on the same pyrimidine substrate in a normal Timmis synthesis (see: Sect. 2.D. in this chapter). Other examples of the cyanide modification include the preparation of the following:

2,6-Diphenyl-4,7-pteridinediamine (**196**, R = Ph) [from the pyrimidine (**193**) with phenylacetaldehyde plus sodium cyanide, phenylacetone plus sodium cyanide, or phenylacetylacetonitrile plus sodium cyanide: all equivalent to phenylacetonitrile, which itself gave the same product; yields: up to 91%.[326]

6-Methyl-2-phenyl-4,7-pteridinediamine (**196**, R = Me) [from the pyrimidine (**193**) with α-benzoylpropionitrile plus sodium cyanide or ethyl α-cyanopropionate plus sodium cyanide: both equivalent to propionitrile; here the cyanide is not finally incorporated although it does serve to induce deacylation].[326]

6-Ethyl-2-phenyl-4,7-pteridinediamine (**196**, R = Et) (α-benzoylbutyronitrile plus sodium cyanide: ~ 80%).[326]

6-Cyclohexyl-2-phenyl-4,7-pteridinediamine (α-benzoyl-α-cyclohexylacetonitrile plus sodium cyanide: ~ 70%).[326]

6-Benzyl-2-phenyl-4,7-pteridinediamine (α-benzoyl-α-benzylacetonitrile plus sodium cyanide: ~ 30%).[326]

The reaction of 2-methylthio-5-nitroso-4,6-pyrimidinediamine with α-benzoylpropionitrile in the presence of either potassium acetate or sodium cyanide gave a separable mixture of 6-methyl-2-methylthio-4,7-pteridinediamine and 6-methyl-2-methylthio-7-phenyl-4-pteridinamine.[326] Further examples of the cyanide modification will be found in subsequent sections.

Miscellaneous exceptional reactions that are best classified within this category include that of 5-nitroso-2-phenyl-4,6-pyrimidinediamine with *N*-cyclopent-1′-enylpyrrolidine (**197**), which will be recognized as a reactive equivalent of cyclopentanone, to give 2-phenyl-6,7-trimethylene-4-pteridinamine (**198**)[1354]; of 6-amino-1,3-dimethyl-5-nitroso-2,4(1*H*,3*H*)-pyrimidinedione with *N*-cyclohex-1′-enylmorpholine (likewise) to give 1,3-dimethyl-6,7-tetramethylene-2,4(1*H*,3*H*)-pteridinedione (78%), subsequently aromatized by sulfur at 240°C to give 1,3-dimethyl-2,4(1*H*,3*H*)-benzo[*g*]pteridinedione (1,3-dimethylalloxazine: 58%; analogs similarly)[316]; of 6-amino-1,3-dimethyl-5-nitroso- (**199**, X = O) or 6-amino-1,3-dimethyl-5-phenylazo-2,4(1*H*,3*H*)-pyrimidinedione (**199**, X = NPh) with dimethyl acetylenedicarboxylate in DMF at 180°C to give,

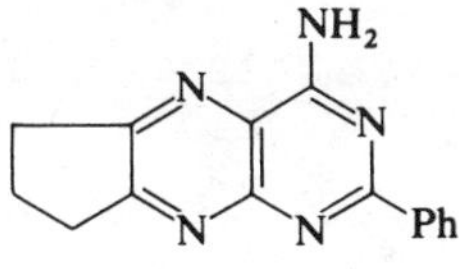

(**198**)

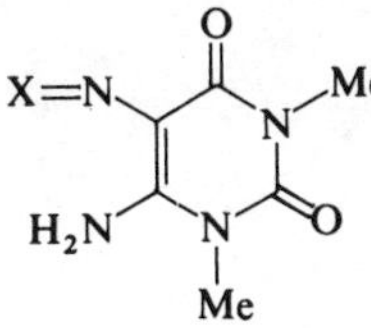

(**199**)

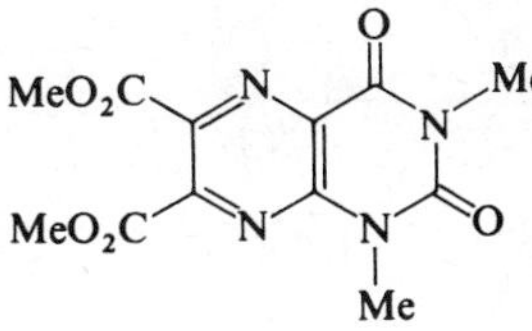

(**200**)

by a complicated mechanism, dimethyl 1,3-dimethyl-2,4-dioxo-1,2,3,4-tetrahydro-6,7-pteridinedicarboxylate (**200**) in 63 or 76% yield, respectively[315, 1388]; of the same nitroso compound (**199**, X = O) with propiolamide to give 1,3-dimethyl-2,4-dioxo-1,2,3,4-tetrahydro-6-pteridinecarboxamide in no less than 95% yield[1388]; of 2,6-diamino-5-nitroso-4(3*H*)-pyrimidinone with 2-aminoethanol or 2-amino-1-methylethanol in ethylene glycol at 200°C (and involving oxidation at some stage) to give 2-amino-4(3*H*)-pteridinone or its 7-methyl derivative, respectively, in 20–30% yield[941]; of 5-nitroso-2-phenyl-4,6-pyrimidinediamine with α-diethoxyphosphinyl-α′-phenylacetone (anion) (**201**) in THF at 25°C to give (with loss of the phosphinyl part) 2,7-diphenyl-4-pteridinamine (75%)[1312]; of 6-amino-1,3-dimethyl-5-nitroso-2,4(1*H*,3*H*)-pyrimidinedione with phenacylidenetriphenylphosphorane (**202**) (made *in situ* from phenacyl bromide and triphenylphosphine) to give 1,3-dimethyl-7-phenyl-2,4(1*H*,3*H*)-pteridinedione (67%)[1389]; of the same substrate with *N*-phenacylpyridinium bromide (**203**) in pyridine–sodium hydroxide to give 1,3-dimethyl-7-phenyl-2,4,6(1*H*,3*H*,5*H*)-pteridinetrione (71%)[1279]; and of 2,6-diamino-5-nitroso-4(3*H*)-pyrimidinone with α,α′-dihydroxyacetone (**204**) in sodium hydroxide to give a product that on oxidation afforded 2-amino-4-oxo-3,4-dihydro-7-pteridinecarboxylic acid[279]: a most peculiar reaction.

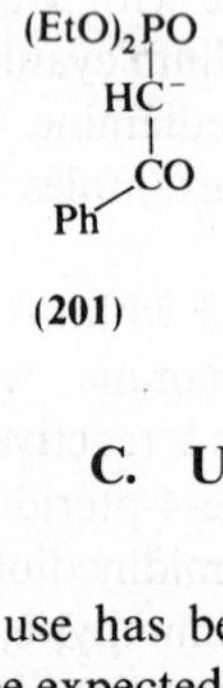

(**201**)

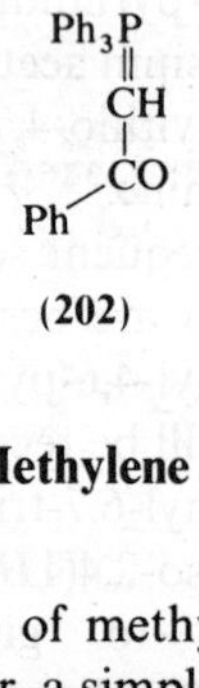

(**202**)

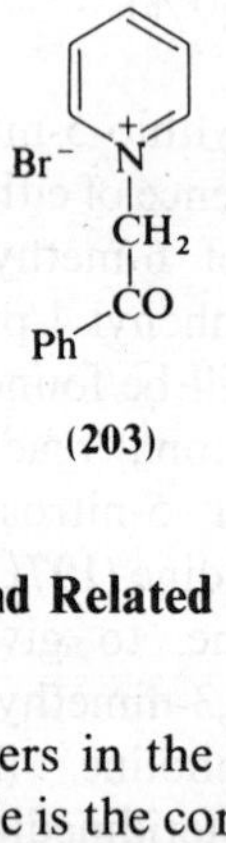

(**203**)

(**204**)

C. Use of α-Methylene esters and Related Compounds

Less use has been made of methylene esters in the Timmis synthesis than might be expected. However, a simple example is the condensation of 5-nitroso-4,6-pyrimidinediamine (**206**) with ethyl phenylacetate (**205**) in ethanolic sodium ethoxide under reflux to give 4-amino-6-phenyl-7(8*H*)-pteridinone (**207**) in 41% yield.[224] Others include the preparation from appropriate 5-nitrosopyrimidines of 2,4-diamino-6-phenyl-7(8*H*)-pteridinone (neat phenylacetyl chloride at

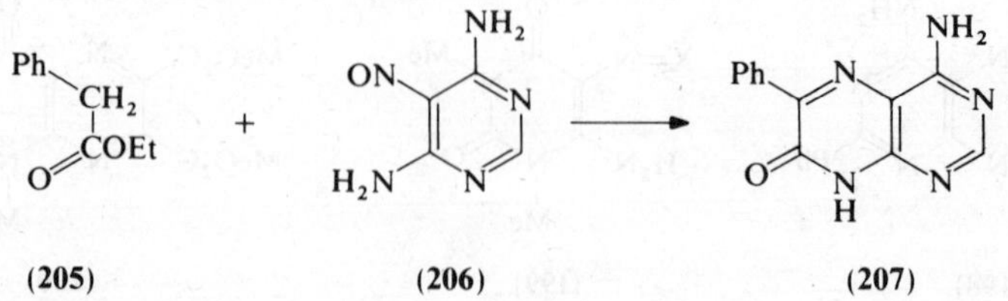

(**205**) (**206**) (**207**)

140°C: ~ 90%),[128] 2,4-diamino-7-oxo-7,8-dihydro-6-pteridinecarboxylic acid (diethyl malonate in 2-ethoxyethanolic ethoxide under reflux, followed by hydrolysis: ~ 65%),[1229] ethyl 4-amino-2-benzylthio(or methylthio)-7-oxo-7,8-dihydro-6-pteridinecarboxylate (diethyl malonate in ethanolic ethoxide: 27, 47%),[1126] the homologous methyl ester of the latter (dimethyl malonate: 68%),[1126] ethyl 4-amino-7-oxo-2-phenyl-7,8-dihydro-6-pteridinecarboxylate or the free carboxylic acid (diethyl malonate in ethanolic ethoxide, according to conditions and work-up),[320] 2-amino-4-isopropoxy-7-oxo-7,8-dihydro-6-pteridinecarbonitrile (from 6-isopropoxy-5-nitroso-2,4-pyrimidinediamine with neat methyl cyanoacetate at >120°C, the ester part reacting in preference to the nitrile part: ~ 60%),[371] and 7-ethoxy-2,6-diphenyl-4-pteridinamine (from 5-nitroso-2-phenyl-4,6-pyrimidinediamine with 2-phenylpropionitrile in anhydrous ethanolic ethoxide: 14%; the displacement of CN is easy to explain under these conditions but loss of Me is not).[526] Ethyl sodiodiethoxyphosphinylacetate (**208**, R = H) has been used with 5-nitroso-2-phenyl-4,6-pyrimidinediamine at <50°C to afford 4-amino-2-phenyl-7(8*H*)-pteridinone (**209a**) (90%)[1312, 1390]; at 150°C, *N*-ethylation ensued to give 4-amino-8-ethyl-2-phenyl-7(8*H*)-pteridinone (**209b**) (74%).[1390] The same or other pyrimidines with the phosphinylacetate (**208**, R = H) or ethyl sodio-α-diethoxyphosphinylpropionate (**208**, R = Me) under appropriate conditions gave 4-amino-2-methylthio-7(8*H*)-pteridinone (**209c**) (88%),[1390] its 8-ethyl derivative (**209d**) (77%),[1390] 4-amino-2-dimethylamino-8-ethyl-7(8*H*)-pteridinone (**209e**) (74%),[1390] its 2-diethylamino analog (**209f**) (71%),[1390] 4-amino-6-methyl-2-phenyl-7(8*H*)-pteridinone (**209g**) (74%),[1312] and its 2-dephenyl analog (**209h**) (78%).[1312] The few known examples of the use of methylene-amides are covered in Section D to follow.

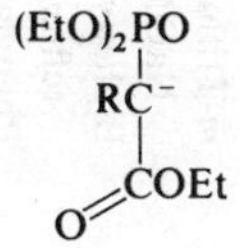

(208)

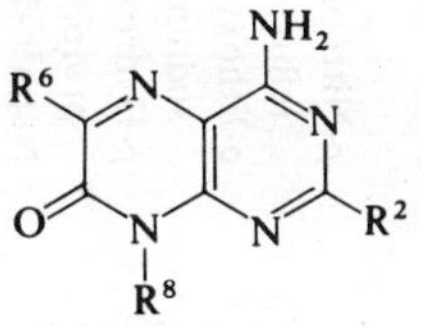

(209)

	R^2	R^6	R^8
a:	Ph	H	H
b:	Ph	H	Et
c:	SMe	H	H
d:	SMe	H	Et
e:	NMe_2	H	Et
f:	NEt_2	H	Et
g:	Ph	Me	H
h:	H	Me	H

D. Use of α-Methylenenitriles

Because of the diuretic activity associated with 6-phenyl-2,4,7-pteridinetriamine (**212**) (triamterene) and related compounds,[577, 579] work in this category of the Timmis synthesis has been brisk. The ordinary use of α-methylenenitriles is illustrated in the condensation of 5-nitroso-2,4,6-

TABLE VII. ADDITIONAL EXAMPLES OF THE TIMMIS SYNTHESIS

Pyrimidine Intermediate	Reagent and Reaction Conditions	Pteridine (Yield)	References
6-Amino-1,3-dimethyl-5-nitroso-2,4(1*H*,3*H*)-pyrimidinedione	*p*-Methoxyphenylacetone; reflux; 90 min	6-*p*-Methoxyphenyl-1,3,7-trimethyl-2,4(1*H*,3*H*)-pteridinedione (18%)	1348
6-Amino-1,3-dimethyl-5-nitroso-2,4(1*H*,3*H*)-pyrimidinedione	*p*-Chlorophenacylidenetriphenylphosphorane; NaOH; aqueous THF; reflux; 30 min	7-*p*-Chlorophenyl-1,3-dimethyl-2,4(1*H*,3*H*)-pteridinedione[a] (55%)	1389
6-Amino-1,3-dimethyl-5-nitroso-2,4(1*H*,3*H*)-pyrimidinedione	*N*-*p*-Bromophenacylpyridinium bromide; NaOH; pyridine	7-*p*-Bromophenyl-1,3-dimethyl-2,4,6(1*H*,3*H*,5*H*)-pteridinetrione[a] (68%)	1279
5-Nitroso-2,4,6-pyrimidinetriamine	*p*-Nitrophenylacetyl chloride; 140°C; 10 min	2,4-Diamino-6-*p*-nitrophenyl-7(8*H*)-pteridinone[a] (~90%)	128
5-Nitroso-2,4,6-pyrimidinetriamine	*p*-Nitrophenylacetonitrile; AcOH; NaOAc; reflux; 20 h	6-*p*-Nitrophenyl-2,4,7-pteridinetriamine (~85%)	1281
5-Nitroso-2,4,6-pyrimidinetriamine	*o*-Ethoxyphenylacetonitrile; 2-ethoxyethanolic sodium ethoxyethoxide; reflux; 1–2 h	6-*o*-Ethoxyphenyl-2,4,7-pteridinetriamine[a]	1281
5-Nitroso-2,4,6-pyrimidinetriamine	*p*-Hydroxyphenylacetonitrile (dihydropyran-protected); final hydrolysis	6-*p*-Hydroxyphenyl-2,4,7-pteridinetriamine	1078
5-Nitroso-2,4,6-pyrimidinetriamine	Naphth-1′-ylacetonitrile; 2-ethoxyethanolic sodium ethoxyethoxide; reflux; 2 h	6-Naphth-1′-yl-2,4,7-pteridinetriamine[a] (94%)	1151, 1281
5-Nitroso-2,4,6-pyrimidinetriamine	*o*-Iodophenylacetonitrile; 2-ethoxyethanolic sodium ethoxyethoxide; reflux; 2 h	6-*o*-Iodophenyl-2,4,7-pteridinetriamine[a] (33%)	1151
5-Nitroso-2,4,6-pyrimidinetriamine	*m*-Trifluoromethylphenylacetonitrile; DMF; NaOMe; reflux; 2 h	6-*m*-Trifluoromethylphenyl-2,4,7-pteridinetriamine[a] (41%)	523
5-Nitroso-2,4,6-pyrimidinetriamine	Biphenyl-4′-ylacetonitrile; DMF; NaOMe; reflux; 2 h	6-Biphenyl-4′-yl-2,4,7-pteridinetriamine (47%)	523
2-Cyanomethyl-5-nitroso-4,6-pyrimidinediamine	Phenylacetonitrile; DMF; NaOMe; reflux; 5 min	2-Cyanomethyl-6-phenyl-4,7-pteridinediamine[a] (71%)	523
6-Methylthio-5-nitroso-2,4-pyrimidinediamine	Phenylacetonitrile; DMF; NaOMe; reflux; 1 h	4-Methylthio-6-phenyl-2,7-pteridinediamine[a] (60%)	523

5-Nitroso-2,4,6-pyrimidinetriamine	2′-Ethylcyclohex-1′-enylacetonitrile; 2-ethoxyethanolic sodium ethoxyethoxide; reflux	6-(2′-Ethylcyclohex-1′-enyl)-2,4,7-pteridinetriamine[a]	682
5-Nitroso-2,4,6-pyrimidinetriamine	3′,5′-Dimethylisoxazol-4′-ylacetonitrile; DMF; NaOMe; reflux; 30 min	6-(3′,5′-Dimethylisoxazol-4′-yl)-2,4,7-pteridinetriamine[a] (68%)	524
5-Nitroso-2-*o*-tolyl-4,6-pyrimidinediamine	Cyanoacetamide; DMF; NaOMe; reflux; <15 min	4,7-Diamino-2-*o*-tolyl-6-pteridinecarboxamide[a] (89%)	1357
2-Anilino-5-nitroso-4,6-pyrimidinediamine	Cyanoacetamide; DMF; NaOMe; reflux; <15 min	4,7-Diamino-2-anilino-6-pteridinecarboxamide[a] (66%)	1357
5-Nitroso-2-phenyl-4,6-pyridinediamine	α-Cyano-*N*,*N*-dimethylacetamide; DMF; NaOMe; reflux; 15 min	4,7-Diamino-*N*,*N*-dimethyl-2-phenyl-6-pteridinecarboxamide[a] (67%)	1357
5-Nitroso-2-phenyl-4,6-pyrimidinediamine	α-Cyano-*N*-isopropylacetamide; DMF; NaOMe; reflux; <15 min	4,7-Diamino-*N*-isopropyl-2-phenyl-6-pteridinecarboxamide (87%)	1357, 1466
6-Dimethylamino-5-nitroso-2,4-pyrimidinediamine	Cyanoacetamide; Me_2SO; KO*i*-Bu; 95 °C; 24 h	2,7-Diamino-4-dimethylamino-6-pteridinecarboxamide[a] (62%)	1357
5-Nitroso-2-propyl-4,6-pyrimidinediamine	Cyanoacetamide; ethanolic NaOEt; reflux; 10 min	4,7-Diamino-2-propyl-6-pteridinecarboxamide[a] (81%)	321
5-Nitroso-2-phenyl-4,6-pyrimidinediamine	α-Cyano-*N*-β-diethylaminoethylacetamide; ethanolic NaOEt; reflux; 10 min	4,7-Diamino-*N*-β-diethylaminoethyl-2-phenyl-6-pteridinecarboxamide[a] (>95%)	321
2-Morpholino-5-nitroso-4,6-pyrimidinediamine	α-Cyano-*N*-γ-dimethylaminopropylacetamide; 2-ethoxyethanolic sodium ethoxyethoxide; reflux	4,7-Diamino-*N*-γ-dimethylaminopropyl-2-morpholino-6-pteridinecarboxamide[a] (88%)	321
5-Nitroso-2-phenyl-4,6-pyrimidinediamine	α-Cyano-*N*,*N*-pentamethyleneacetamide; DMF; NaOMe; reflux; 10 min	4,7-Diamino-*N*,*N*-pentamethylene-2-phenyl-6-pteridinecarboxamide[a] (71%)	321
5-Nitroso-2-thien-2′-yl-4,6-pyrimidinediamine	*N*,*N*′-Bis-β-ethoxyethylmalondiamide; ethanolic NaOEt; reflux; 5 min	4-Amino-*N*-β-ethoxyethyl-7-β-ethoxyethylamino-2-thien-2′-yl-6-pteridinecarboxamide[a,b] (28%)	320

[a] And analog(s).
[b] Plus the 7-oxo-7,8-dihydro analog (18%).

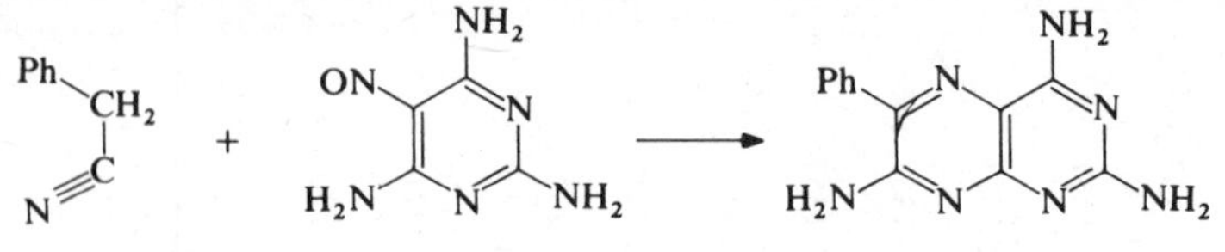

(210) (211) (212)

pyrimidinetriamine (**211**) with phenylacetonitrile (**210**) in refluxing 2-ethoxyethanol containing a little sodium 2-ethoxyethoxide to give this pteridinetriamine (**212**) (55–72%)[1151, 1281] or, from appropriately labeled intermediates, its [7-^{14}C]- and [3′-^{3}H]-analogs.[771] Sixty 6-aryl analogs were made rather similarly[523, 1151, 1281] (see Table VII for examples) as were also (with phenylacetonitrile itself or some indicated variations) 2-methylthio- and 2-ethylthio-6-phenyl-4,7-pteridinediamine (refluxing ethanolic sodium ethoxide)[1281]; 2,7-diamino-6-phenyl-4(3*H*)-pteridinone (refluxing ethylene glycol plus sodium)[1281]; 7-amino-2-methyl-6-phenyl-4(3*H*)-pteridinone[1281]; 6-phenyl-4,7-pteridinediamine[1281]; 7-amino-6-phenyl- and 7-amino-2,6-diphenyl-4(3*H*)-pteridinone[1281]; 4,7-diamino-6-phenyl-2(1*H*)-pteridinone (ethylene glycol–sodium under reflux)[1281]; the corresponding thione (ethanolic sodium ethoxide; with a little 2-ethoxy-6-phenyl-4,7-pteridinediamine as a by-product)[1281]; 4-methoxy-6-phenyl-2,7-pteridinediamine (refluxing methanolic DMF plus sodium methoxide: ~ 10%)[328]: 7-imino-8-methyl-4-methylamino-6-phenyl-7,8-dihydro-2-pteridinamine (**213**) (from 4,6-bismethylamino-5-nitroso-2-pyrimidinamine in DMF–sodium methoxide under reflux briefly: 22%)[523]; 2-benzyl-6-phenyl-4,7-pteridinediamine (DMF plus sodium methoxide under reflux for 5 min: 18%)[523]; 6-phenyl-2-piperidino-4,7-pteridinediamine (similarly: 51%)[523]; 4-dimethylamino-6-phenyl-2,7-pteridinediamine (similarly but for 2 h: 76%)[523]; and 4-butoxy-6-phenyl-2,7-pteridinediamine (dimethyl sulfoxide plus potassium *t*-butoxide under reflux for 2 h: 25%).[523] The use of cycloalkyl- and heteroarylacetonitriles is exemplified in the formation of 6-cyclopent-1′-enyl-2,4,7-pteridinetriamine (from 5-nitroso-2,4,6-pyrimidinetriamine and cyclopent-1′-enylacetonitrile),[682] 6-pyridin-4′-yl-2,4,7-pteridinetriamine (from the same pyrimidine and pyridin-4′-ylacetonitrile in ethanolic sodium ethoxide under reflux for 24 h: 65%[1396]; or in DMF–sodium ethoxide under reflux for 3 min: 48%[524]), 6-(4′-methylthien-2′-yl)-2,4,7-pteridinetriamine (**214**) [from the same pyrimidine and (4′-methylthien-2′-yl)acetonitrile in refluxing DMF containing

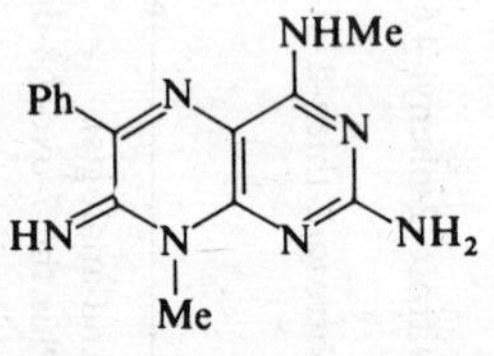

(213)

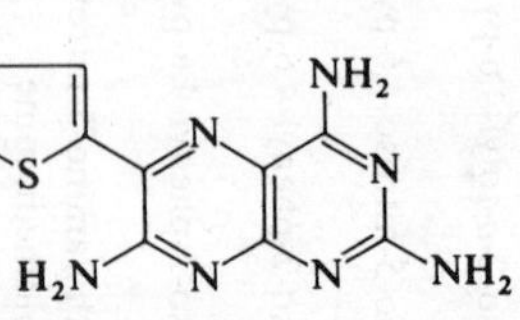

(214)

sodium methoxide for 1 min: 58%],[524] and other such pteridines in Table VII. The reaction of 6-isopropoxy-5-nitroso-2,4-pyrimidinediamine with malononitrile in glycol at 120°C gave 2,7-diamino-4-isopropoxy-6-pteridinecarbonitrile (~45%).[371]

When a 5-nitroso-4-pyrimidinamine reacts with acylacetonitriles (including carboxy-, alkoxycarbonyl-, or carbamoylacetonitriles) in the presence of an alkaline catalyst (preferably sodium cyanide) it is the cyano group that reacts with the amino group of the pyrimidine, thereby leaving the acyl group as a 6-substituent on the resulting pteridine[578]; this should be compared, for example, with the reaction of 6-isopropoxy-5-nitroso-2,4-pyrimidinediamine with neat methyl cyanoacetate, which gave 2-amino-4-isopropoxy-7-oxo-7,8-dihydro-6-pteridinecarbonitrile.[371] Thus 5-nitroso-2-phenyl-4,6-pyrimidinediamine (**217**) with benzoylacetonitrile (**216**), in refluxing aqueous ethanol containing sodium cyanide, gave 6-benzoyl-2-phenyl-4,7-pteridinediamine (**218**) (>90%) and not 4-amino-2,7-diphenyl-6-pteridinecarbonitrile (**215**).[328] Likewise, a separable mixture of both products, with (**218**) predominating, was formed when benzoylacetonitrile was replaced under these conditions by phenacylpyridinium bromide (**203**),[328] usually considered to be the equivalent of benzoylacetonitrile in the

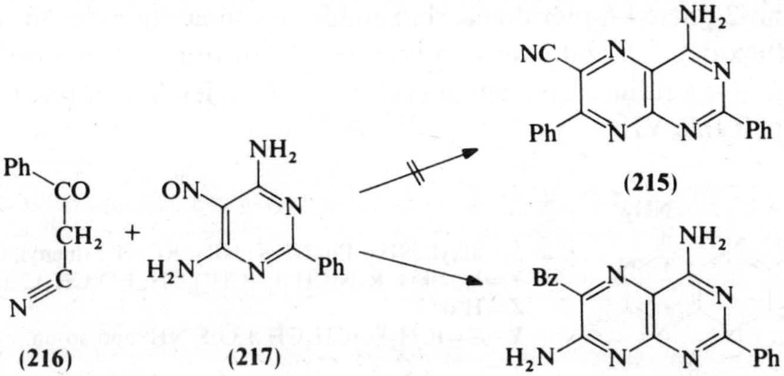

(**216**) (**217**) (**215**) (**218**)

presence of cyanide (see Sect. B just cited). This behavior is confirmed by the formation from suitable pyrimidines of 6-benzoyl-4,7-pteridinediamine (benzoylacetonitrile and sodium cyanide: ~85%), its 2-methyl and 2-methylthio derivatives (phenacylpyridinium bromide and sodium cyanide: 70–80%), 6-acetyl-4,7-pteridinediamine (acetonylpyridinium chloride and sodium cyanide: >80%), its 2-methyl and 2-methylthio derivatives (likewise: ~80%), 4,7-diamino-2-phenyl-6-pteridinecarboxamide (carbamoylmethylpyridinium bromide and sodium cyanide: >80%), 6-benzoyl-2,4,7-pteridinetriamine (from the diacetylated pyrimidine with benzoylacetonitrile and potassium acetate, followed by alkaline deacetylation: ~70%), and the 2-piperidino analog (likewise: ~85%).[328] Some alkyl 4,7-diamino-2-phenyl-6-pteridinecarboxylates may be made from 5-nitroso-2-phenyl-4,6-pyrimidinediamine with the indicated reagents. (a) Ethyl cyanoacetate/ethanolic sodium ethoxide gave the ethyl ester

(40%) plus the carboxylic acid (30%); (b) isopropyl cyanoacetate–sodium isopropoxide in isopropyl alcohol gave the isopropyl ester (49%) plus carboxylic acid (39%); (c) methyl cyanoacetate–methanolic sodium methoxide gave the methyl ester (47%) plus carboxylic acid (~20%); (d) ethyl cyanoacetate–ethanolic sodium cyanide gave the ethyl ester only (75%); (e) diphenyl sulfoacetate–ethanolic sodium acetate gave the ethyl ester only (28%); (f) cyanoacetylurethane–ethanolic sodium cyanide gave the ethyl ester only (54%); and (g) cyanoacetylurea–ethanolic sodium cyanide gave the ethyl ester only (32%).[1357] The following were formed likewise: 4,7-diamino-2-phenyl-6-pteridinecarboxamide (cyanoacetylurea in DMF containing sodium methoxide: 31%),[1357] 4,7-diamino-2-phenyl-6-pteridinecarboxylic acid (just cited; also from cyanoacetic acid in DMF containing sodium methoxide: 56%; or in 2-ethoxyethanolic sodium ethoxyethoxide: 98%),[1357] 4,7-diamino-6-pteridinecarboxylic acid (cyanoacetic acid in 2-ethoxyethanolic sodium ethoxyethoxide: ~50%),[324] 4,7-diamino-6-pteridinecarboxamide (cyanoacetamide likewise: ~90%),[324] 2,4,7-triamino-6-pteridinecarboxamide (likewise: ~55%),[324] 2,4,7-triamino-6-pteridinecarboxylic acid (cyanoacetic acid likewise),[324] 4,7-diamino-2-methylthio-6-pteridinecarboxamide (cyanoacetamide likewise),[324] 4,7-diamino-2-methylthio-6-pteridinecarboxylic acid (cyanoacetic acid likewise),[324] 4,7-diamino-2-phenyl-6-pteridinecarboxamide (cyanoacetamide in ethanolic sodium ethoxide),[1466] and over two hundred 2-substituted-*N*,*N*-disubstituted-4,7-diamino-6-pteridinecarboxamides (**219**),[321, 1466] a few examples of which are included in Table VII.

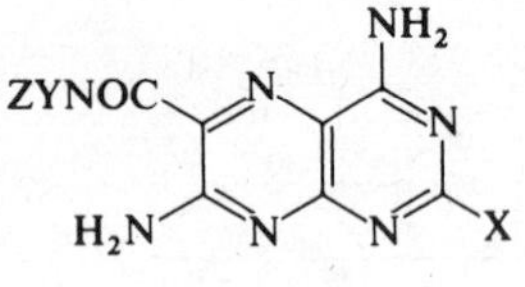

(219)

X = alkyl, NH_2, Ph, NHR, NR_2, RC_6H_4, thienyl, and so on
Y = H, alkyl, $R_2N(CH_2)_n$, $(CH_2)_nCH$, $RO(CH_2)_n$, and so on
Z = H or Y
Y + Z = $(CH_2)_n$, $(CH_2CH_2)_2O/S/NH$, and so on.

Although the carbamoyl group is clearly less reactive than the cyano group in cyanoacetamides, it can take an active part in the Timmis synthesis if there is no cyano group present. This is illustrated by the condensation of 5-nitroso-2-phenyl-4,6-pyrimidinediamine (**221**) with malondiamide (**220**, R = H) in refluxing ethanolic sodium ethoxide for 20 min, to give 4,7-diamino-2-phenyl-6-pteridinecarboxamide (**222**, R = H) (~60%).[320] The use of *N*,*N*′-dimethylmalondiamide (**220**, R = Me) gave the homologous product, 4-amino-*N*-methyl-7-methylamino-2-phenyl-6-pteridinecarboxamide (**222**, R = Me) in low yield, accompanied by 4-amino-*N*-methyl-7-oxo-2-phenyl-7,8-dihydro-6-pteridinecarboxamide in rather larger amount. This could be formed either by hydrolysis or (more likely) by an alternative elimination of methylamine during the condensation.[320] Several analogous amides, for example 4-amino-*N*-butyl-7-butylamino-2-phenyl-6-pteridinecarboxamide (**222**, R = Bu) (4%), were made similarly but usually in poor yield for the same reason.[320]

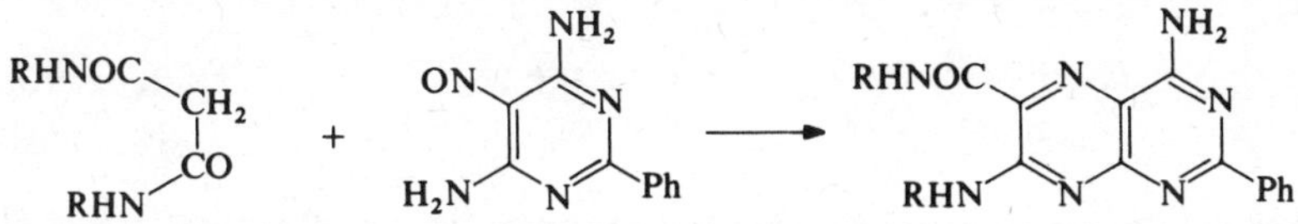

(220) (221) (222)

As in other categories, phosphinyl-activated methylenenitriles have been used. α-Diethoxyphosphinylacetonitrile or its α-phenyl derivative with 5-nitroso-2-phenyl-4,6-pyrimidinediamine gave 2-phenyl- or 2,6-diphenyl-4,7-pteridinediamine, respectively, in ~30% yield.[1312] Another unusual but practical sequence is seen in the condensation of 4,6-diacetamido-5-nitroso-2-pyrimidinamine (**224**) with *N*-cyanomethylpyridinium chloride (**223**) in aqueous alcoholic sodium hydroxide containing thiophenol, to give 6-phenylthio-2,4,7-pteridinetriamine (**225**, R = H) in 71% yield. The stages at which deacetylation and phenylthiolysis occur are not clear.[329] The reaction proved quite versatile, giving inter alia 2-phenyl-6-phenylthio-4,7-pteridinediamine (66%), 4-acetamido-6-phenylthio-2,7-pteridinediamine (**225**, R = Ac) (72%), its hexylthio analog (73%), and 4-acetamido-2-amino-6-phenylthio-7(8*H*)-pteridinone (using *N*-ethoxycarbonylmethylpyridinium chloride: 56%), according to the substrates, reagents, and conditions used.[329] No experimental details appear to have been published for a direct condensation (mentioned[329] in passing) of the pyrimidine (**224**) with α-phenylthioacetonitrile in DMF containing sodium acetate to give (**225**, R = Ac).

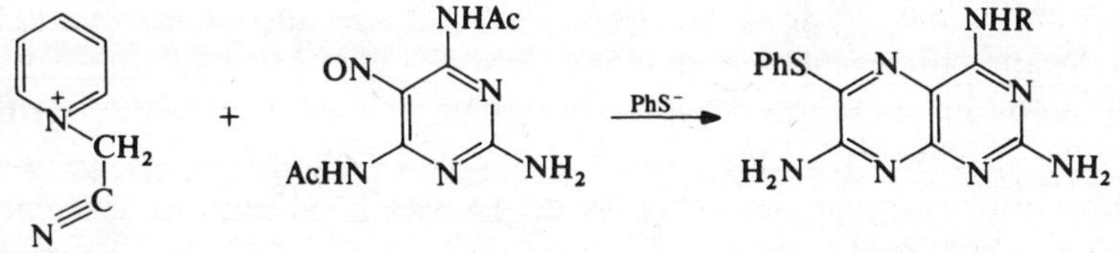

(223) (224) (225)

It will be recalled that 5-nitroso-2-phenyl-4,6-pyrimidinediamine with *N*-phenacylpyridinium bromide in the presence of sodium cyanide gave mainly 6-benzoyl-2-phenyl-4,7-pteridinediamine (**218**). However, when sodium cyanide was replaced by potassium acetate, 2,7-diphenyl-4-pteridinamine 5-oxide (**228**) resulted in poor yield, presumably via the intermediate hydroxylamine (**226**) and the nitrone (**227**).[328, 578] Similar reactions gave 7-methyl-2-phenyl-4-pteridinamine 5-oxide (*N*-acetonylpyridinium chloride: ~85%) and 7-methyl-4-pteridinamine 5-oxide (*N*-acetonylpyridinium chloride: ~70%).[328] In the case of *N*-(α-cyanobenzyl)pyridinium benzenesulfonate (**229**), this type of reaction occurred even in the presence of sodium cyanide (which, in fact, proved to be an excellent *catalyst*). Under these conditions, 5-nitroso-2-phenyl-4,6-, 2-methylthio-5-nitroso-4,6-, or 6-methylthio-5-nitroso-2,4-pyrimidinediamine

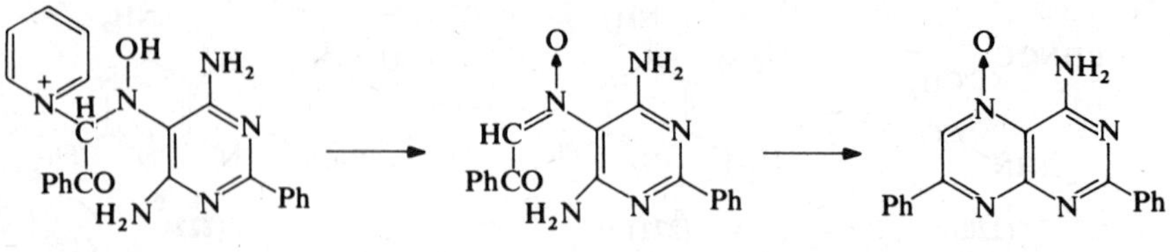

(226) (227) (228)

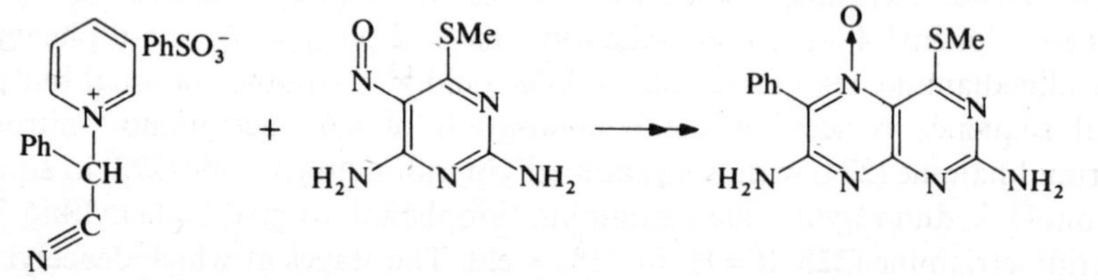

(229) (230) (231)

(230) gave 2,6-diphenyl-4,7-, 2-methylthio-6-phenyl-4,7-, or 4-methylthio-6-phenyl-2,7-pteridinediamine 5-oxide (231), respectively, each in good yield.[328]

3. THE BOON SYNTHESIS FROM 4-CHLORO-5-NITRO (OR ARYLAZO)PYRIMIDINES AND α-AMINOCARBONYL COMPOUNDS VIA 7,8-DIHYDROPTERIDINES

Like the Timmis synthesis, the Boon synthesis was introduced specifically to avoid the ambiguities inherent in many Gabriel and Colman syntheses. Its steps have been outlined in formulas (7)–(10). Although a 4-chloro-5-nitropyrimidine is now the usual substrate for a Boon synthesis, a 4-chloro-5-arylazo- or even a 4-chloro-5-nitrosopyrimidine may be used instead. Indeed, the last two substrates may have some advantage in that the arylazo or nitroso group may be inserted either before or after displacement of the chloro substituent by an α-aminocarbonyl compound (or α-aminonitrile).

It will be evident that a Boon synthesis may cease, either by choice or in some cases by necessity, at the 7,8-dihydropteridine stage. In addition, the general procedure is applicable to the formation of tetrahydropteridines, either as intermediates or as final products, as required.

A. Complete Boon Syntheses

Apart from variations at the 1-, 2-, 3-, 4-, or 8-position, arising from presubstitution of the pyrimidine substrate, the Boon synthesis can give pteridines with a small range of substituents at the 6- or 7-position, as determined by the nature of the α-aminocarbonyl or α-aminonitrile reagent employed.

(1) *From 4-Chloro-5-nitropyrimidines*

The formation of 6-*unsubstituted pteridines* is illustrated in the treatment of 2-amino-6-chloro-5-nitro-4(3*H*)-pyrimidinone (**232**) with aminoacetaldehyde diethyl acetal to give 2-amino-6-*β*,*β*-diethoxyethylamino-5-nitro-4(3*H*)-pyrimidinone (**233**, R = H) and then 2-amino-6-*β*-formylethylamino-5-nitro-4(3*H*)-pyrimidinone (**234**, R = H), which on dithionite reduction afforded 2-amino-7,8-dihydro-4(3*H*)-pteridinone (**235**, R = H) and by subsequent permanganate oxidation, 2-amino-4(3*H*)-pteridinone (**236**).[453, cf. 81] When *N*-methylaminoacetaldehyde diethyl acetal was used, the analogous sequence, (**232**) → (**233**, R = Me) → (**234**, R = Me) → (**235**, R = Me), occurred,[422, 684] followed by spontaneous aerial oxidation to 2-amino-8-methyl-4(8*H*)-pteridinone (**237**).[422]

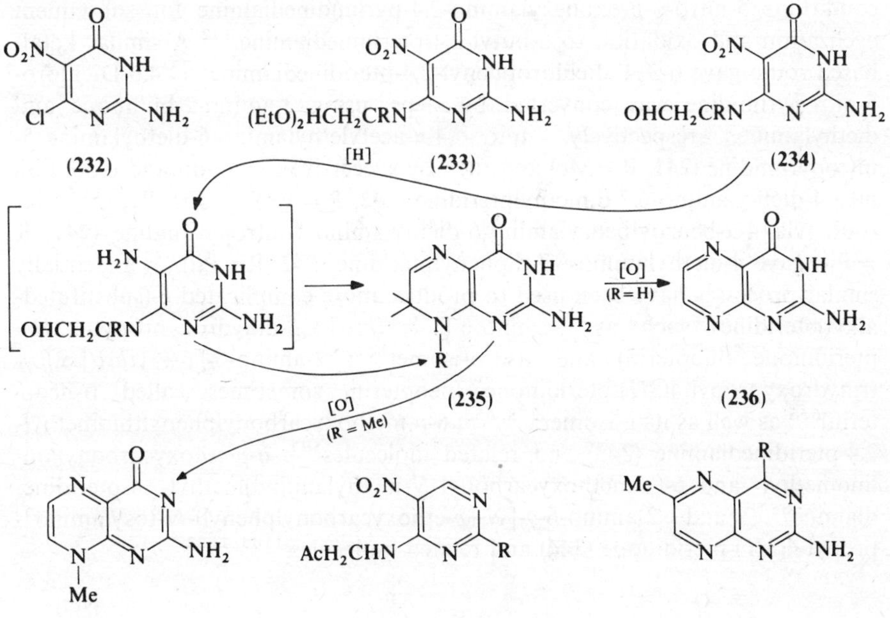

(**232**) (**233**) (**234**)

(**235**) (**236**)

(**237**) (**238**) (**239**)

The formation of 6-*alkylpteridines* requires the use of ketonic reagents such as aminoacetone. For example, 2,4-dichloro-5-nitropyrimidine was converted by aminoacetone into 4-acetonylamino-2-chloro-5-nitropyrimidine (**238**, R = Cl) and then by ammonia into 4-acetonylamino-5-nitro-2-pyrimidinamine (**238**, R = NH_2), which underwent reductive cyclization to 6-methyl-7,8-dihydro-2-pteridinamine (65%) and subsequent permanganate oxidation to 6-methyl-2-pteridinamine (**239**, R = H).[39, 688] A rather similar route from 2,4-dichloro-6-methyl-5-nitropyrimidine gave 4-acetonylamino-6-methyl-5-nitro-2-pyrimidinamine,[1088] 4,6-dimethyl-7,8-dihydro-2-pteridinamine,[1088] and then 4,6-

dimethyl-2-pteridinamine (**239**, R = Me).[39] 2-Diethylamino-4,6-dimethylpteridine was made similarly via its 7,8-dihydro derivative.[1088] 4-Acetonylamino-2-chloro-5-nitropyrimidine (**238**, R = Cl) (just cited) underwent mild hydrolysis to 4-acetonylamino-5-nitro-2(1*H*)-pyrimidinone and subsequent reductive cyclization to the 7,8-dihydropteridinone followed by oxidation to 6-methyl-2(1*H*)-pteridinone.[42] The sequence 6-chloro-5-nitro-2,4(1*H*,3*H*)-pyrimidinedione, 6-acetonylamino-5-nitro-2,4(1*H*,3*H*)-pyrimidinedione, 6-methyl-7,8-dihydro-2,4(1*H*,3*H*)-pteridinedione, and a final permanganate (67% yield) or peroxide (71% yield) oxidation gave 6-methyl-2,4(1*H*,3*H*)-pteridinedione.[1129, 1232] 2-Amino-6-chloro- gave 6-acetonylamino-2-amino-3-methyl-5-nitro-4(3*H*)-pyrimidinone, which on reduction gave a 7,8-dihydropteridine and then, by treatment with hydrogen peroxide 2-amino-3,6-dimethyl-4(3*H*)-pteridinone (**240**) (structure[1752]).[1232] 6-Chloro-5-nitro-2,4-pyrimidinediamine with the ketal, β,β-dimethoxyhexylamine gave under appropriate conditions 5-nitro-6-β-oxohexylamino-2,4-pyrimidinediamine for subsequent cyclization and oxidation to 6-butyl-2,4-pteridinediamine.[742] A similar ketal-based route gave 6-3′,4′-dichlorophenyl-2,4-pteridinediamine.[173] 4,6-Dichloro-5-nitropyrimidine was converted by steps, using 3-amino-2-butanone and diethylamine, respectively, into 4-α-acetylethylamino-6-diethylamino-5-nitropyrimidine (**241**, R = Me) and then by successive reduction and oxidation into 4-diethylamino-6,7-dimethylpteridine (**242**, R = Me).[688] Finally, a similar route via 4-α-benzoylbenzylamino-6-diethylamino-5-nitropyrimidine (**241**, R = Ph) gave 4-diethylamino-6,7-diphenylpteridine (**242**, R = Ph).[688] Essentially similar processes have been used to produce more complicated 6-(substituted-alkyl)pteridines such as 2-amino-6-[L-*erythro*]-α,β-dihydroxypropyl-4(3*H*)-pteridinone (biopterin) and its D-isomer[653]; 2-amino-6-[D-*erythro*]-α,β,γ-trihydroxypropyl-4(3*H*)-pteridinone (neopterin, sometimes called D-neopterin)[653] as well as its L-isomer[278, 653]; 6-*p*-methoxycarbonylphenylthiomethyl-2,4-pteridinediamine (**243**) and related molecules[597]; 6-*p*-ethoxycarbonylanilinomethyl- and 6-(*p*-methoxycarbonyl-*N*-methylanilino)methyl-2,4-pteridinediamine[857]; and 2-amino-6-γ-[*N*-(*p*-ethoxycarbonylphenyl)-*N*-tosylamino]-propyl-4(3*H*)-pteridinone (**244**) and related molecules.[1105, 1582]

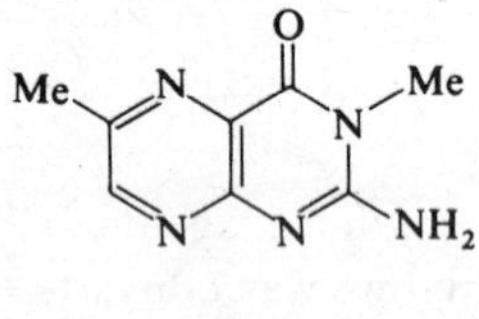

(240)

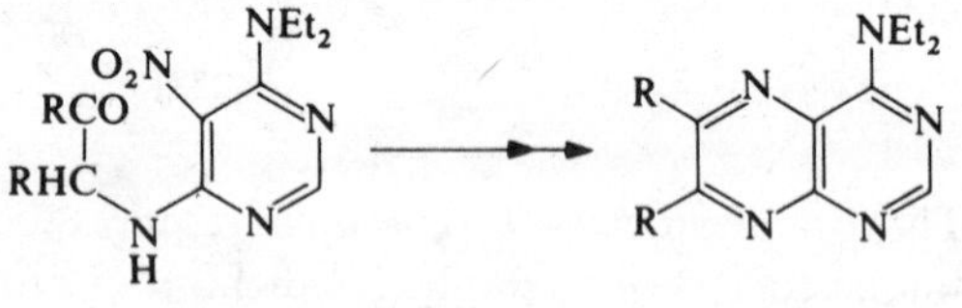

(241) **(242)**

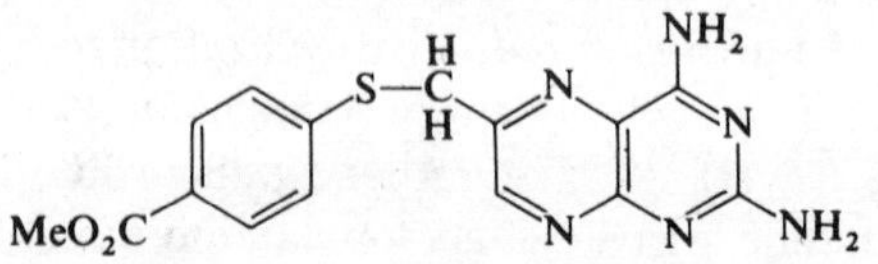

(243)

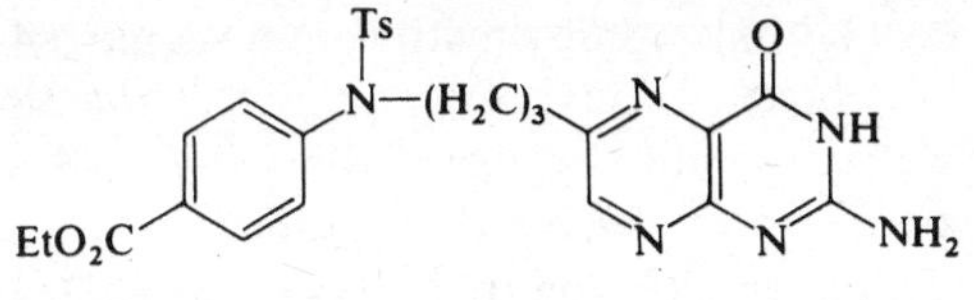

(244)

The preparation of 6-*aminopteridines* needs an aminonitrile such as aminoacetonitrile. Thus 2-amino-6-chloro-5-nitro-4(3*H*)-pyrimidinone reacted with aminoacetonitrile to give the 6-cyanomethylamino analog (**245**), which on dithionite reduction gave 2,6-diamino-7,8-dihydro-4(3*H*)-pteridinone (**246**) and then by permanganate oxidation, 2,6-diamino-4(3*H*)-pteridinone (**247**) in 48% overall yield.[450] In a similar manner, 6-chloro-1,3-dimethyl-5-nitro-2,4(1*H*,3*H*)-pyrimidinedione and aminoacetonitrile gave its 6-cyanomethylamino analog and then 6-amino-1,3-dimethyl-2,4(1*H*,3*H*)-pteridinedione, via the peroxide oxidation of its 7,8-dihydro derivative.[1275] An aminoacetamide has also been used.[1132]

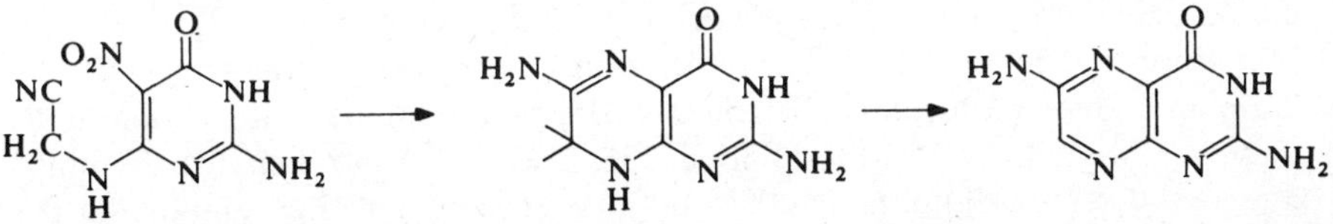

(245) (246) (247)

The formation of 6-*pteridinones* by the Boon synthesis requires an amino ester (or equivalent), such as ethyl aminoacetate. As illustration, 2-amino-6-chloro-5-nitro-4(3*H*)-pyrimidinone reacted with the ethyl ester of glycine to give 2-amino-6-(ethoxycarbonylmethyl)amino-5-nitro-4(3*H*)-pyrimidinone (**248**, R = OEt) (89%), which cyclized on reduction to afford 2-amino-7,8-dihydro-4,6(3*H*,5*H*)-pteridinedione (**249**) (80–90%),[349, 450, 1232] and by subsequent permanganate oxidation, 2-amino-4,6(3*H*,5*H*)-pteridinedione (xanthopterin) (**250**) ($>$ 95%)[451]; the dihydropteridine (**249**) was equally well made by condensation of this chloropyrimidine with aminoacetamide to give 2-amino-6-carbamoylmethylamino-5-nitro-4(3*H*)-pyrimidinone (**248**, R = NH_2) (83%) which, on reduction, lost ammonia to give (**249**) (89%).[450] Likewise, ethyl 2,6-dichloro-5-nitro-4-pyrimidinecarboxylate was converted successively

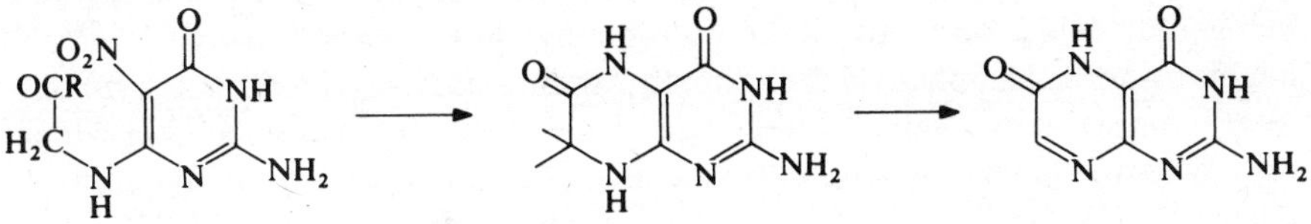

(248) (249) (250)

into ethyl 2-chloro-6-ethoxycarbonylmethylamino-5-nitro-4-pyrimidinecarboxylate,[750] ethyl 6-ethoxycarbonylmethylamino-5-nitro-2-oxo-1,2-dihydro-4-pyrimidinecarboxylate, 2,6-dioxo-1,2,5,6,7,8-hexahydro-4-pteridinecarboxylic acid, and 2,6-dioxo-1,2,5,6-tetrahydro-4-pteridinecarboxylic acid.[150] It is interesting that reduction of 1-(2′-amino-5′-nitro-6′-oxo-1′,6′-dihydropyrimidin-4′-ylamino)-1-deoxy-D-[*erythro*]pentulose and a related compound caused loss of the sugar portions with the formation of 2-amino-7,8-dihydro-4,6(3*H*,5*H*)-pteridinedione.[451, 452] See also examples in Table VIII.

(2) *From 4-Chloro-5-arylazopyrimidines*

The formation of 6-*unsubstituted-pteridines* appears to be represented by only two examples. 6-Chloro-5-*p*-chlorophenylazo-2-dimethylamino-4(3*H*)-pyrimidinone was converted by α-amino-α-phenylacetaldehyde dimethyl acetal into 5-*p*-chlorophenylazo-6-β,β-dimethoxy-α-phenylethylamino-2-dimethylamino-4(3*H*)-pyrimidinone (**251**), which was hydrolyzed to the (unisolated) aldehyde and then submitted to reductive cyclization to afford, after spontaneous oxidation during the work-up, 2-dimethylamino-7-phenyl-4(3*H*)-pteridinone (**252**) (~ 50%).[110] The second example involved the conversion of 2-amino-6-chloro-5-*p*-chlorophenylazo-4(3*H*)-pyrimidinone into 2-amino-8-methyl-7,8-dihydro-4(3*H*)-pteridinone (**235**, R = Me)[684] and then into 2-amino-8-methyl-4(8*H*)-pteridinone (**237**)[422] by a route similar to that employing the corresponding 5-nitropyrimidine intermediate (**232**) [see Sect. (1) in this chapter].

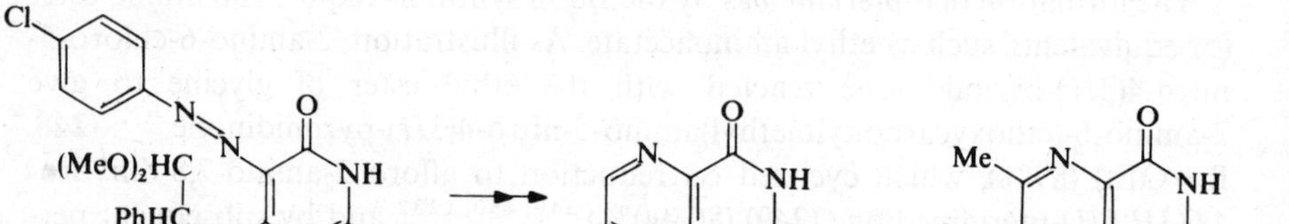

(**251**) (**252**) (**253**)

The production of 6-*alkylpteridines* is better represented. For example, 2-amino-6-chloro-5-phenylazo-4(3*H*)-pyrimidinone was converted by aminoacetone into the 6-acetonylamino analog that was reduced by zinc and acetic acid to give 2-amino-6-methyl-7,8-dihydro-4(3*H*)-pteridinone and then by permanganate oxidation to give 2-amino-6-methyl-4(3*H*)-pteridinone (**253**).[729, 940] 6-Chloro-5-*p*-chlorophenylazo-2-dimethylamino-4(3*H*)-pyrimidinone underwent aminolysis by phenacylamine to give the 6-phenacylamino analog which, on reductive cyclization and subsequent oxidation, gave 2-dimethylamino-6-phenyl-4(3*H*)-pteridinone.[110] 6-*p*-Chlorophenyl-2-dimethylamino-4(3*H*)-pteridinone was made in a similar manner.[110] Analogous routes gave 6-*p*-

TABLE VIII. ADDITIONAL EXAMPLES OF THE BOON SYNTHESIS

Unreduced Pyrimidine Intermediate	Final Product Described (Yield)	References
4-α-Benzoylbenzylamino-5-nitro-2-pyrimidinamine	6,7-Diphenyl-2-pteridinamine (94%)	688
6-α-Benzoylbenzylamino-5-nitro-4-pyrimidinamine	6,7-Diphenyl-4-pteridinamine[a] (90%)	688
6-(δ-*p*-Ethoxycarbonyl-β-oxobutyl)-5-nitro-2,4-pyrimidinediamine	6-β-Ethoxycarbonylethyl-2,4-pteridinediamine[a]	173
6-α′-Phenylacetonylamino-5-nitro-2,4-pyrimidinediamine	6-Benzyl-2,4-pteridinediamine[a] (66%)	742
2-Amino-6-[β-oxo-δ-(*N*-phenylacetamido)butyl]amino-5-phenylazo-4(3*H*)-pyramidinone	2-Amino-6-β-(*N*-phenylacetamido)ethyl-4(3*H*)-pteridinone[a,b] (59%)	743
2-Amino-6-[δ-(*N*-*p*-ethoxycarbonylphenylacetamido)-β-oxobutyl]-amino-5-phenzylazo-4(3*H*)-pyrimidinone	2-Amino-6-β-(*N*-*p*-ethoxycarbonylphenylacetamido)ethyl-4(3*H*)-pteridinone[a] (~70%)	735
2-Amino-6-α,α-dimethylphenacylamino-5-nitro-4(3*H*)-pyrimidinone	2-Amino-7,7-dimethyl-6-phenyl-7,8-dihydro-4(3*H*)-pteridinone (~25%)	388
2-Amino-6-(1′-β-carboxyethyl-1′-methylacetonylamino)-5-nitro-4(3*H*)-pyrimidinone	2-Amino-7-β-carboxyethyl-6,7-dimethyl-7,8-dihydro-4(3*H*)-pteridinone (57%)	1502
2-Amino-6-α,α,α′-trimethylacetonylamino-5-nitro-4(3*H*)-pyrimidinone	2-Amino-6-ethyl-7,7-dimethyl-4(3*H*)-pteridinone[a] (43%)	1501
4-Chloro-6-α-methoxycarbonylmethylamino-2-methyl-5-nitropyrimidine	4-Chloro-2-methyl-7,8-dihydro-6(5*H*)-pteridinone[a]	689
2,4-Bis-α-methoxycarbonylmethylamino-5-nitropyrimidine	2-α-Methoxycarbonylmethylamino-7,8-dihydro-6(5*H*)-pteridinone[a]	689, 873
6-α-Ethoxycarbonyl-β-methylpropylamino-5-nitro-4-pyrimidinamine	4-Amino-7-isopropyl-6(5*H*)-pteridinone[a] (82%)	733
Ethyl 2-amino-6-α-ethoxycarbonylmethylamino-5-nitro-4-pyrimidinecarboxylate	Ethyl 2-amino-6-oxo-5,6,7,8-tetrahydro-4-pteridinecarboxylate[a]	750
2-Dimethylamino-6-(α-ethoxycarbonyl-α-methylethyl)amino-5-nitro-4(3*H*)-pyrimidinone	2-Dimethylamino-7,7-dimethyl-7,8-dihydro-4,6(3*H*,5*H*)-pteridinedione (19%)	349
2-Amino-6-β-hydroxyimino-δ-(*p*-methoxycarbonylphenyl)pentyl-5-nitro-4(3*H*)-pyrimidinone	2-Amino-6-β-(*p*-carboxyphenyl)propyl-4(3*H*)-pteridinone[a]	1690
6-β-Ethylenedioxypropylamino-1,3-dimethyl-5-nitro-2,4(1*H*,3*H*)-pyrimidinedione	1,3,6-trimethyl-2,4(1*H*,3*H*)-pteridinedione (80%)	1704

[a] And analog(s).
[b] Hydrolysis gave 2-amino-6-β-anilinoethyl-4(3*H*)-pteridinone.

chlorophenyl-2-methylamino-4(3*H*)-pteridinone, 6-*p*-chlorophenyl-2-dimethylamino-7-phenyl-4(3*H*)-pteridinone, and the isomeric 7-*p*-chlorophenyl-2-dimethylamino-6-phenyl-4(3*H*)-pteridinone.[110] More complicated cases are illustrated in Table VIII. A warning on care in the oxidation step is sounded by the reductive cyclization of 6-*N*-acetonyl-*N*-methylamino-2-amino-5-phenylazo-4(3*H*)-pyrimidinone (**254**) to give 2-amino-6,8-dimethyl-7,8-dihydro-4(3*H*)-pteridinone (**255**), which on oxidation gave, not the expected 2-amino-6,8-dimethyl-4(8*H*)-pteridinone (**256**), but either 2-amino-6-hydroxymethyl-8-methyl-4(8*H*)-pteridinone (**257**) (permanganate) or 2-amino-6-hydroxymethyl-8-methyl-4,7(3*H*,8*H*)-pteridinedione (**258**) (platinum oxide–oxygen; 24 h).[1370]

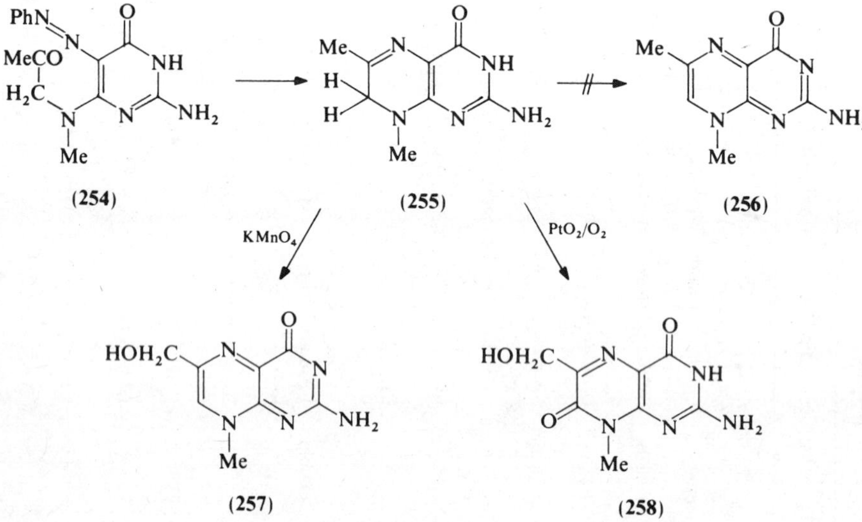

Although there appears to be no example of the formation of a 6-aminopteridine via a 5-phenylazopyrimidine, several 6-*pteridinones* have been so made. Thus 4-chloro-5-*p*-chlorophenylazo-2,6-bismethylaminopyrimidine was converted by glycine ethyl ester into 5-*p*-chlorophenylazo-2,4-bisdimethylamino-6-α-ethoxycarbonylmethylaminopyrimidine and then into 2,4-bisdimethylamino-6(5*H*)-pteridinone (**259**) via its 7,8-dihydro derivative[268]; and 2-amino-6-chloro-5-phenylazo-4(3*H*)-pyrimidinone with glycine ethyl ester gave the 6-α-ethoxycarbonylmethylamino analog, which then gave 2-amino-4,6(3*H*,5*H*)-pteridinedione (xanthopterin) (**250**) via its 7,8-dihydro derivative (**249**).[727] An interesting variation has been introduced into the last-mentioned preparation by doing the aminolysis step prior to introducing the arylazo group. Although the overall process is said to be less satisfactory by the new sequence, viz. 2-amino-6-chloro-4(3*H*)-pyrimidinone → 2-amino-6-α-ethoxycarbonylmethylamino-4(3*H*)-pyrimidinone → 2-amino-5-*p*-chlorophenylazo-6-α-ethoxycarbonylmethylamino-4(3*H*)-pyrimidinone → 2-amino-7,8-dihydro-4,6(3*H*,5*H*)-pteridinedione → xanthopterin, the principle remains of poten-

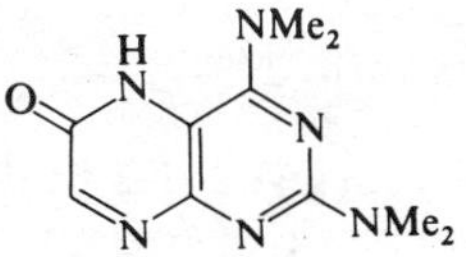

(259)

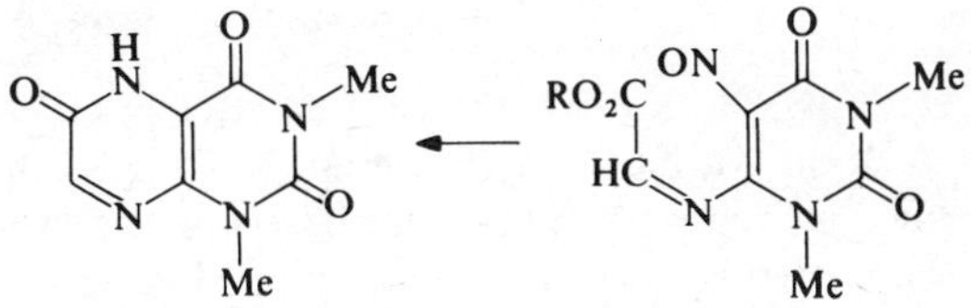

(260) (261)

tial importance.[883] A similar procedure is seen in the sequence of reactions, 6-chloro-1,3-dimethyl-2,4(1*H*,3*H*)-pyrimidinedione → 6-α-ethoxycarbonylmethylamino-1,3-dimethyl-2,4(1*H*,3*H*)-pyrimidinedione → its 5-*p*-chlorophenylazo derivative → 1,3-dimethyl-7,8-dihydro-2,4,6(1*H*,3*H*,5*H*)-pteridinetrione → 1,3-dimethyl-2,4,6(1*H*,3*H*,5*H*)-pteridinetrione (**260**).[340] It is also seen in the similar formation of its 1,3,7-trimethyl analog.[340]

(3) *From Nitrosopyrimidines and Related Compounds*

There is no straightforward example of a Boon synthesis from a chloronitrosopyrimidine but related sequences are exemplified in the treatment of 6-amino-1,3-dimethyl-5-nitroso-2,4(1*H*,3*H*)-pyrimidinedione with diethyl mesoxalate to give (after saponification and monodecarboxylation during the reaction) the 6-carboxymethyleneamino derivative (**261**, R = H), which on dithionite reduction gave 1,3-dimethyl-2,4,6(1*H*,3*H*,5*H*)-pteridinetrione (**260**) *without* passing through any dihydropteridine.[364] The key intermediate (**261**, R = H) was also made by treatment of 6-amino-1,3-dimethyl-2,4(1*H*,3*H*)-pyrimidinedione with methyl dichloroacetate to give the 6-methoxycarbonylmethyleneamino derivative (**261**, R = Me) followed by gentle saponification.[364] Another route of interest and importance involved the lead tetraacetate oxidation of 5-nitroso-2-phenyl-4,6-pyrimidinediamine to give 5-phenyl[1,2,5]oxadiazolo[3,4-*d*]pyrimidin-7-amine (**262**, R = H), which on treatment with aminoacetaldehyde diethyl acetal or aminoacetone dimethyl acetal produced 7-β,β-diethoxyethylamino-[**262**, R = $CH_2CH(OEt)_2$] or 7-β,β-dimethoxypropylamino-5-phenyl[1,2,5]-oxadiazolo[3,4-*d*]pyrimidine [**262**, R = $CH_2CMe(OMe)_2$], respectively. Subsequent reductive fission of the five-membered ring of each, followed by spontaneous recyclization and aerial oxidation, gave 2-phenyl-4-pteridinamine (**263**, R = H) and its 6-methyl derivative (**263**, R = Me).[482]

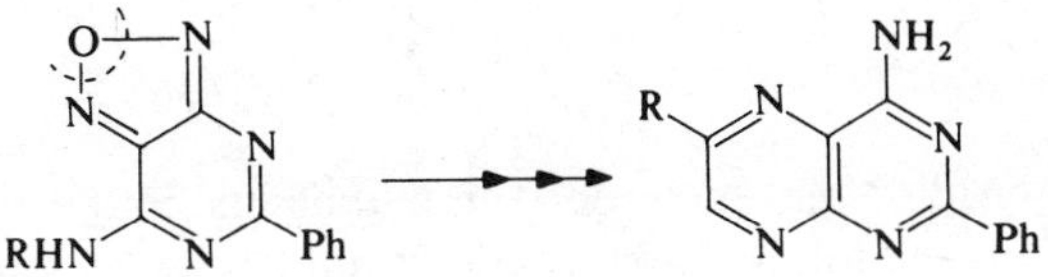

(262) (263)

B. Incomplete Boon Syntheses

A Boon synthesis may terminate at the 7,8-dihydropteridine stage either from choice or necessity. The usual basis for the latter is the presence of two alkyl groups at the 7-position, thereby precluding oxidation to a formally aromatic pteridine. In this section, no distinction will be drawn between 5-nitro, 5-arylazo, or other pyrimidine substrates.

The formation of a *6-unsubstituted-7,8-dihydropteridine* is seen in the aminolysis of 2,4-dichloro-5-nitropyrimidine by aminoacetaldehyde diethyl acetal to give 2-chloro-4-β,β-diethoxyethylamino-5-nitropyrimidine that was immediately hydrolyzed by alkali to give 4-β,β-diethoxyethylamino-5-nitro-2(1*H*)-pyrimidinone (**264**) and then, by acidic hydrolysis, to give 4-α-formylmethylamino-5-nitro-2(1*H*)-pyrimidinone. On reduction of the nitro group, the resulting 5-aminopyrimidinone cyclized spontaneously to 7,8-dihydro-2(1*H*)-pteridinone (**265**) (23% for the last stage) but no subsequent oxidation appears to have been carried out.[42]

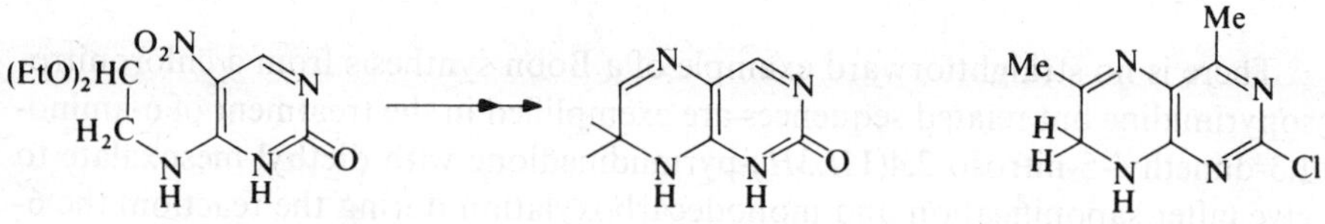

(264) (265) (266)

The preparation of *6-alkyl–aryl-7,8-dihydropteridines* is illustrated by the reaction of 2,4-dichloro-6-methyl-5-nitropyrimidine with aminoacetone to give 4-acetonylamino-2-chloro-6-methyl-5-nitropyrimidine which on hydrogenation gave 2-chloro-4,6-dimethyl-7,8-dihydropteridine (**266**) (77%)[1088]; by the similar formation of 4,6-dimethyl-7,8-dihydro-2(1*H*)-pteridinone[1088]; and by the reaction of 2,4-dichloro-6-methyl-5-nitropyrimidine with α-benzoylbenzylamine to give 4-α-benzoylbenzylamino-2-chloro-6-methyl-5-nitropyrimidine (**267**), which on stannous chloride reduction gave 2-chloro-4-methyl-6,7-diphenyl-7,8-dihydropteridine (**268**).[398, 1230]

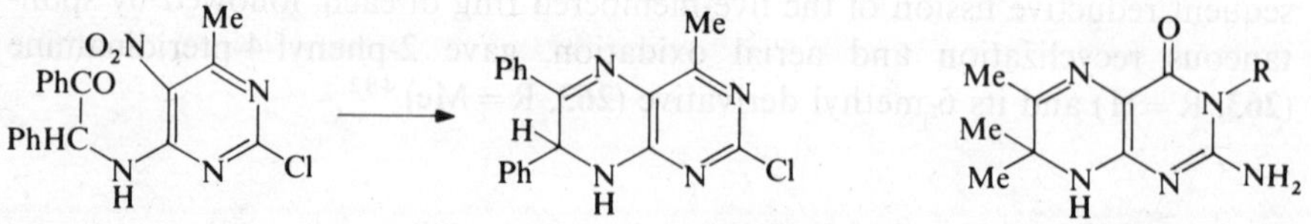

(267) (268) (269)

The dihydropteridines just mentioned should be susceptible to aromatization, but products from the following reactions cannot be so oxidized. 2-Amino-6-chloro-5-nitro-4(3*H*)-pyrimidinone reacted with α-amino-α,α-dimethylacetone to give its 6-α,α-dimethylacetonylamino analog, which on reduction gave 2-amino-6,7,7-trimethyl-7,8-dihydro-4(3*H*)-pteridinone (**269**, R=H) (76%).[388, 1501] A similar route gave the 3,6,7,7-tetramethyl homolog (**269**, R=Me).[388] 6-Chloro-5-nitro-2,4(1*H*,3*H*)-pyrimidinedione with α-amino-α,α-dimethylacetone oxime gave 6-β-hydroxyimino-α,α-dimethylpropylamino-5-nitro-2,4(1*H*,3*H*)-pyrimidinedione, which on dithionite reduction gave 6,7,7-trimethyl-7,8-dihydro-2,4(1*H*,3*H*)-pteridinedione (**270**) (67%).[1144] A similar route using free α-amino-α,α-dimethylacetone gave the same product (**270**) in 81% yield.[1503] 6-Chloro-5-nitro-2,4-pyrimidinediamine and α-amino-α,α-dimethylacetone gave 6-α,α-dimethylacetonylamino-5-nitro-2,4-pyrimidinediamine that underwent reductive cyclization to 6,7,7-trimethyl-7,8-dihydro-2,4-pteridinediamine (95%).[1503] 2-Amino-6-chloro-5-nitro4(3*H*)-pyrimidinone with 3-amino-4-hydroxy-3-methylbutan-2-one gave 2-amino-6-α-hydroxymethyl-α-methylacetonylamino-5-nitro-4(3*H*)-pyrimidinone and then 2-amino-7-hydroxymethyl-6,7-dimethyl-7,8-dihydro-4(3*H*)-pteridinone (**271**) (49%).[1502] 2-Amino-6-chloro-5-nitro-4(3*H*)-pyrimidinone with 3-amino-1-hydroxy-3-methylbutan-2-one oxime gave 2-amino-6-(4′-hydroxy-3′-hydroxyimino-2′-methylbutan-2′-yl)amino-5-nitro-4(3*H*)-pyrimidinone and then on reduction, 2-amino-6-hydroxymethyl-7,7-dimethyl-7,8-dihydro-4(3*H*)-pteridinone.[1501] Finally, 2-amino-6-chloro-5-nitro-4(3*H*)-pyrimidinone with 2-amino-2-methylcyclohexanone oxime gave 2-amino-6-(2′-hydroxyimino-l′-methylcyclohexylamino)-5-nitro-4(3*H*)-pyrimidinone, which then afforded 2-amino-7-methyl-6,7-tetramethylene-4(3*H*)-pteridinone (**272**) (50%).[628, 1501] Other examples are known.[1749]

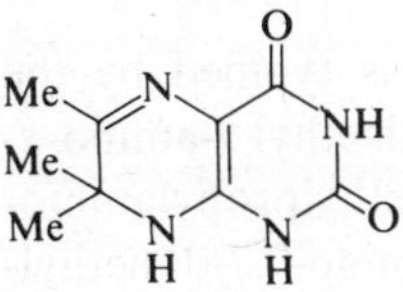
(**270**)

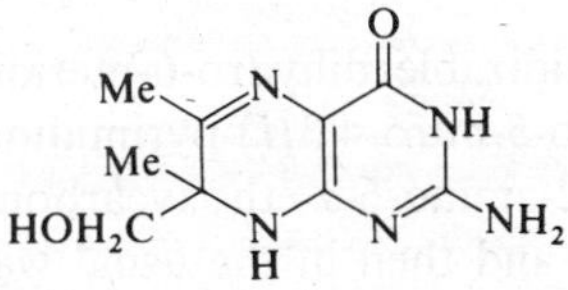
(**271**)

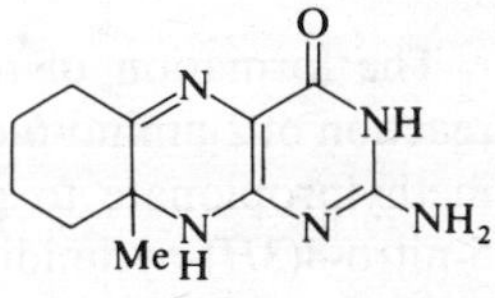
(**272**)

The formation of potentially oxidizable 7,8-*dihydro-6-pteridinones* is well exemplified. 2,4-Dichloro-5-nitropyrimidine and glycine ethyl ester gave 2-chloro-4-α-ethoxycarbonylmethylamino-5-nitropyrimidine which, on catalytic hydrogenation, gave 2-chloro-7,8-dihydro-6(5*H*)-pteridinone (**273**, R=Cl)[689, 873]; when hydriodic acid–red phosphorus was used as the reductant, dechlorination occurred as well, to afford 7,8-dihydro-6(5*H*)-pteridinone (**273**, R=H) (77%).[35, cf. 689] Likewise, 2,4-dichloro-5-nitropyrimidine with alanine methyl ester gave, after hydrogenation of the intermediate, 2-chloro-7-methyl-7,8-dihydro-6(5*H*)-pteridinone (80%)[55]; 2-diethylamino-4-α-methoxycarbonylmethylamino-6-methyl-5-nitropyrimidine gave 2-diethylamino-4-methyl-7,8-dihydro-6(5*H*)-pteridinone[689, 873]; 4-α-carboxymethylamino-6-methyl-5-nitro-2-pyrimidinamine on stannous chloride reduction gave

2-amino-4-methyl-7,8-dihydro-6(5*H*)-pteridinone[396]; 6-α-ethoxycarbonylethylamino-5-nitro-4-pyrimidinamine gave 4-amino-7-methyl-6(5*H*)-pteridinone[733]; ethyl 2-dimethylamino- or ethyl 2-ethoxy-6-α-ethoxycarbonylmethylamino-5-nitro-4-pyrimidinecarboxylate on gentle reduction gave the 5-amino analogs (analyzed), which on refluxing in aqueous ethanol gave ethyl 2-dimethylamino- (**274**, R=NMe_2) and ethyl 2-ethoxy-6-oxo-5,6,7,8-tetrahydro-4-pteridinecarboxylate (**274**, R=OEt), respectively[150,750]; and ethyl 5-amino-6-α-ethoxycarbonylmethylamino-2-thioxo-1,2-dihydro-4-pyrimidinecarboxylate likewise gave ethyl 6-oxo-2-thioxo-1,2,5,6,7,8-hexahydro-4-pteridinecarboxylate (~70%).[750] In addition, the following dihydroxanthopterin derivatives have been made by similar methods (although not described in the given references, the oxidation of some may well have been reported elsewhere): 2-amino-3-methyl- (16, 72%),[349,1232] 2-amino-7-methyl- (23%),[349] 2-amino-8-methyl- (54, 70%),[349,1232] 2-amino-3,8-dimethyl- (73, 84%),[349,1232] 2-dimethylamino- (46%),[349] and 2-dimethylamino-8-methyl-7,8-dihydro-4,6(3*H*,5*H*)-pteridinedione (**275**) (72%).[349]

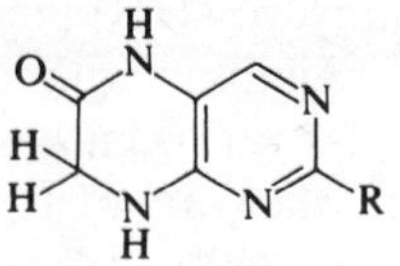

(**273**)

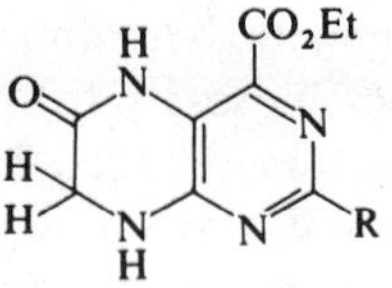

(**274**)

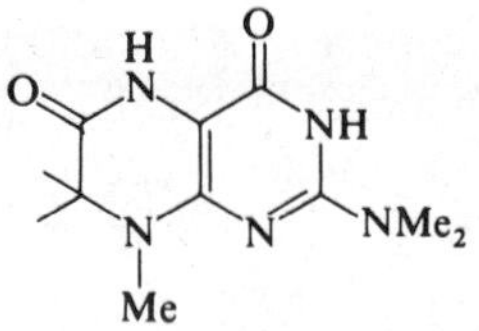

(**275**)

The formation of nonoxidizable dihydro-6-pteridinones is typified by the reaction of 2-amino-6-chloro-5-nitro-4(3*H*)-pyrimidinone with ethyl α-amino-α-methylpropionate to give 2-amino-6-α-ethoxycarbonyl-α-methylpropylamino-5-nitro-4(3*H*)-pyrimidinone and then in the usual way, 2-amino-7,7-dimethyl-7,8-dihydro-4,6(3*H*,5*H*)-pteridinedione (**276**, R=H) (23%)[349,450]; by the similar reactions leading to its 3-methyl derivative (**276**, R=Me)[349,358]; and by the (indirect) conversion of 2-amino-6-D-glucopyranosylamino-3-methyl-5-nitro-4(3*H*)-pyrimidinone (**277**) into 2-amino-3-methyl-7-[D-*arabino*]-tetrahydroxybutyl-7,8-dihydro-4,6(3*H*,5*H*)-pteridinedione (**278**), which resisted oxidation by a stream of oxygen over a long period in the presence of platinum.[358]

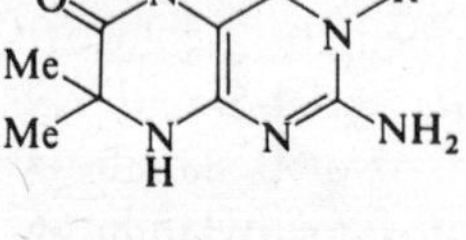

(**276**)

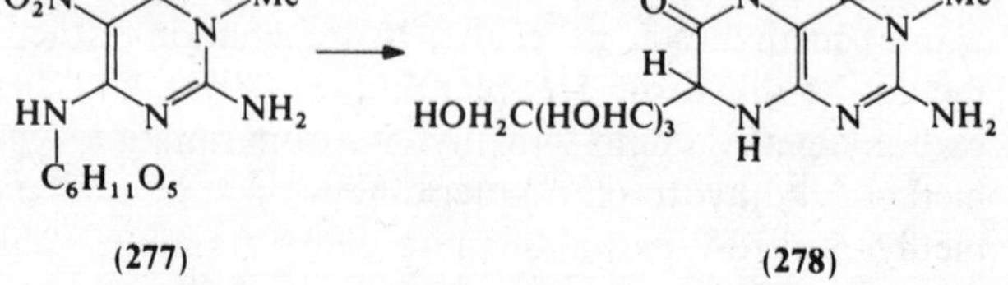

(**277**)

(**278**)

C. The Boon like Synthesis of 5,6,7,8-Tetrahydropteridines

A few 5,6,7,8-tetrahydropteridines have been made from pyrimidines by a process akin to the Boon synthesis. As explored to date, the scope is quite limited. Treatment of 2,4-dichloro-5-nitropyrimidine with *N*-benzyl-β-chloroethylamine gave 4-(*N*-benzyl-β-chloroethylamino)-2-chloro-5-nitropyrimidine (**279**, R=Cl) which, on hydrogenation over Raney-nickel in ethanol, gave 8-benzyl-2-chloro-5,6,7,8-tetrahydropteridine (**281**).[683] A better yield was obtained by initial treatment of the dichloronitropyrimidine with β-benzylaminoethanol to give 4-(*N*-benzyl-β-hydroxyethylamino)-2-chloro-5-nitropyrimidine (**279**, R=OH), hydrogenation to 4-(*N*-benzyl-β-hydroxyethylamino)-2-chloro-5-pyrimidinamine (**280**), and a final cyclization in phosphoryl chloride to give

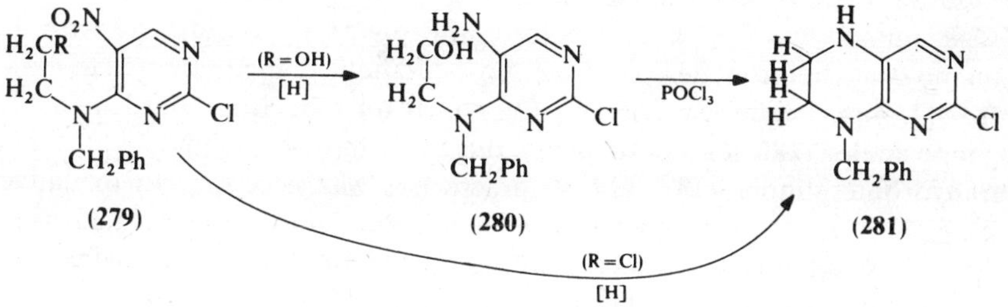

(**279**) (**280**) (**281**)

(**281**).[683] The second of these methods was applied successfully to the preparation of 2-chloro-8-ethyl-[683] and 8-benzyl-4-chloro-5,6,7,8-tetrahydropteridine[43] and also (with some variations) to the formation of 8-benzyl-5,6,7,8-tetrahydro-4(3*H*)-pteridinone.[43] Both methods were used to make 8-benzyl-2-chloro-4-methyl-5,6,7,8-tetrahydropteridine but the second again gave a better yield.[118] These pteridines proved to be handy substrates for reductive debenzylation and/or dechlorination to unsubstituted,[43, 683] 8-benzyl-,[683] 4-methyl-,[118] and 8-benzyl-4-methyl-5,6,7,8-tetrahydropteridine;[118] and also to 5,6,7,8-tetrahydro-4(3*H*)-pteridinone.[43] Attempts to make these and other such simple tetrahydropteridines from simple 4-β-chloroethylamino-5-nitropyrimidines failed because of the preferred cyclization to tetrahydroimidazo-[1,2-*c*]pyrimidines[1505]; however, hydrogenation of 4-bis(β-chloroethyl)amino-6-chloro-5-nitropyrimidine did give a small yield of 4-chloro-8-β-chloroethyl-5,6,7,8-tetrahydropteridine as well as an imidazopyrimidine and an imidazopteridine.[1452]

Two other reaction sequences are appropriate for inclusion here. As mentioned in Section 3.A.(2), the dithionite reduction of 2-amino-5-*p*-chlorophenylazo-6-*N*-(formylmethyl)-*N*-methylamino-4(3*H*)-pyrimidinone (**282**) gave 2-amino-8-methyl-7,8-dihydro-4(3*H*)-pteridinone (**283**) in the usual way. However, when hydrogenation over platinum was used instead of dithionite, the 2-amino-8-methyl-5,6,7,8-tetrahydro-4(3*H*)-pteridinone (**284**) (~45%) was produced, probably by further reduction of the dihydropteridinone (**283**) but

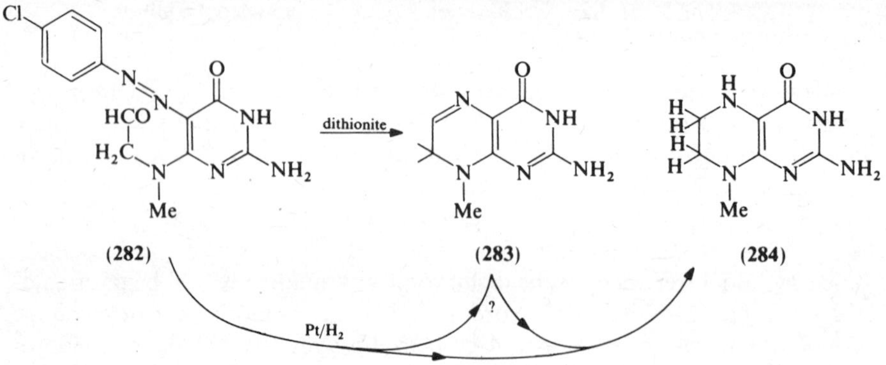

possibly by prereduction of the formyl group in the pyrimidine (**282**) prior to cyclization.[684] Further examples have been reported recently.[1756]

The aminolysis of 2-amino-6-chloro-5-nitro-4(3*H*)-pyrimidinone with 1,3-diamino-2-methylpropane gave 2-amino-6-(β-amino-β-methylpropylamino)-5-nitro-4(3*H*)-pyrimidinone (**285**, R = O), which on reduction afforded the 2,5-diamino analog (**285**, R = H). In theory, this could have lost ammonia to give the tetrahydropteridinone (**287**) but, in practice, it was necessary to oxidatively cyclize (**285**) with bromine to give a dihydropteridinone, formulated as the quinonoidal imine (**286**), and then to reduce the latter to afford 2-amino-6,6-dimethyl-5,6,7,8-tetrahydro-4(3*H*)-pteridinone (**287**).[626, 725]

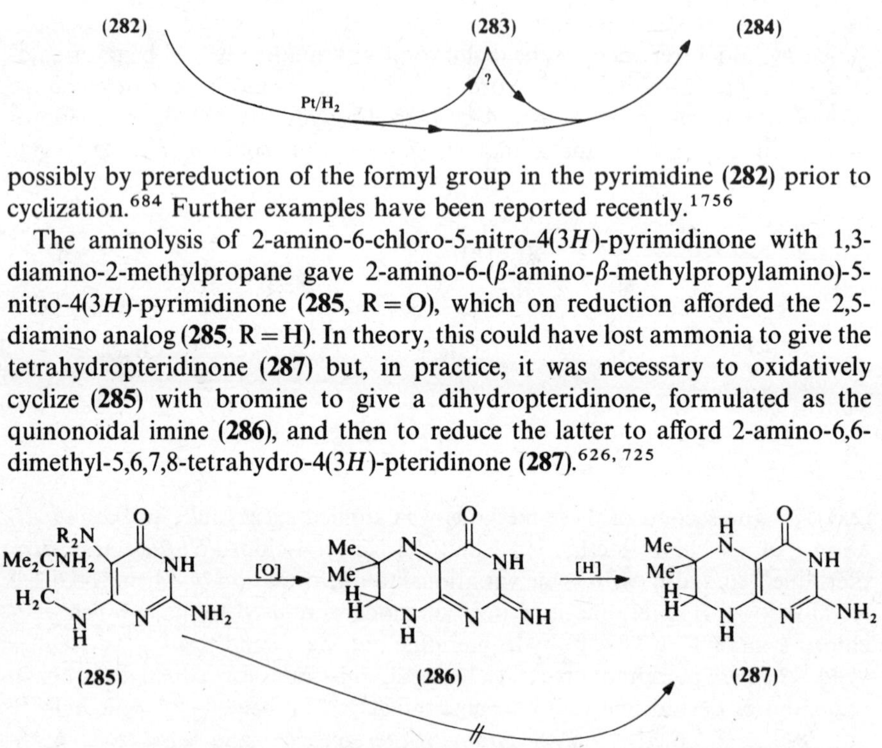

4. MINOR SYNTHETIC ROUTES FROM PYRIMIDINES TO PTERIDINES

There are several minor, but not unimportant, syntheses of pteridines or hydropteridines from pyrimidines. They are best classified according to the pyrimidine substrates used.

A. Minor Syntheses from 4,5-Pyrimidinediamines

(1) *By Initial Acylation or Alkylation of the* 4-*Amino Group*

Treatment of 2,5,6-triamino-4(3*H*)-pyrimidinone with neat chloroacetic acid at 100 °C gave 2,6-diamino-5-α-chloroacetamido-4(3*H*)-pyrimidinone (**288**)

(95%), which underwent cyclization in sodium bicarbonate solution at 95°C to furnish, not β-dihydroxanthopterin (**289**),[965] but 2,4-diamino-7*H*-pyrimido[4,5-*b*][1,4]oxazin-6(5*H*)-one (69%), of proven structure.[883] In hot concentrated sulfuric acid the latter gave 2-amino-4,6(3*H*,5*H*)-pteridinedione.[965] However, it seems that 4-anilino-2,6-dimethyl-5-pyrimidinamine with chloroacetyl chloride gave 4-anilino-5-α-chloroacetamido-2,6-dimethylpyrimidine which, on treatment with aqueous silver acetate followed by silver carbonate, afforded 2,4-dimethyl-8-phenyl-7,8-dihydro-6(5*H*)-pteridinone (~25% overall). No attempt at oxidation was reported.[885] In contrast, sodio-5-formamido-4,6-pyrimidinediamine reacted with ethyl chloroacetate in warm DMF to give 4-amino-5-formyl-5,6-dihydro-7(8*H*)-pteridinone.[736]

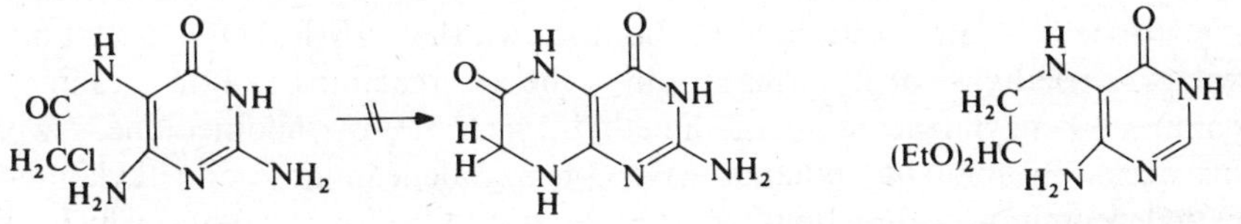

(**288**) (**289**) (**290**)

When a mixture of 5,6-diamino-4(3*H*)-pyrimidinone and diethoxyacetaldehyde was hydrogenated in ethanol over Raney-nickel, reductive alkylation of the 5-amino group occurred to give 6-amino-5-β,β-diethoxyethylamino-4(3*H*)-pyrimidinone (**290**) which, on boiling in dilute hydrochloric acid for a minute, afforded 5,6-dihydro-4(3*H*)-pteridinone (88%).[43] Likewise, hydrogenation of the same diaminopyrimidinone and ethyl glyoxylate hemiacetal gave 6-amino-5-α-ethoxycarbonylmethylamino-4(3*H*)-pyrimidinone and then by acidic treatment gave 5,6-dihydro-4,7(3*H*,8*H*)-pteridinedione (**291**) (80%).[43]

In attempting to make a 7-substituted-purine, 2-methylthio-5-D-ribosylamino-4,6-pyrimidinediamine was treated inter alia with triethyl orthoformate. Instead of supplying the 8-carbon for a purine, this reagent produced intramolecular dehydration to give 2-methylthio-7-α,β,γ-trihydroxypropyl-4-pteridinamine (**292**).[1467]

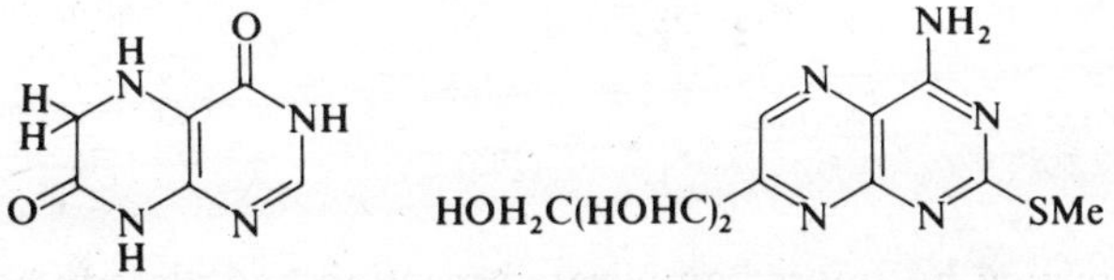

(**291**) (**292**)

(2) *By Initial Conversion of the 5-Amino Group to a Schiff Base*

About 1976, Fumio Yoneda and his colleagues observed that treatment of 5,6-diamino-2,4(1*H*,3*H*)-pyrimidinedione with benzaldehyde in refluxing DMF gave 4-amino-5-benzylideneamino-2,4(1*H*,3*H*)-pyrimidinedione (**293**, Ar = Ph,

Y = H) and then, by refluxing with an excess of triethyl orthoformate in DMF gave 6-phenyl-2,4(1*H*,3*H*)-pteridinedione (**295**, Ar = Ph, R = Y = H) (70%).[543] The mechanism of the cyclization appeared to involve the formation of an (unisolated) ethoxymethylamino intermediate (**294**, Ar = Ph, R = Y = H) because, when triethyl orthoacetate was used in some related reactions, analogous intermediates such as 6-α-ethoxyethylideneamino-5-*p*-methoxybenzylidene-amino-1,3-dimethyl-2,4(1*H*,3*H*)-pyrimidinedione (**294**, Ar = *p*-$MeOC_6H_4$, R = Y = Me) were isolated and subsequently converted thermally into pteridines such as 6-*p*-methoxyphenyl-1,3,7-trimethyl-2,4(1*H*,3*H*)-pteridinedione (**295**, R = *p*-$MeOC_6H_4$, R = Y = Me).[543] As implied in formulas (**293**)–(**295**), the reaction is quite versatile: ~35 pteridines (**295**, Ar = substituted-phenyl, thienyl, pyridinyl, etc.; R = H or Me; Y = H or Me) have been made, invariably in good yield.[543] In addition, it has been shown that DMF diethyl acetal may replace triethyl orthoformate in these reactions. For example, 6-amino-5-benzylideneamino-1,3-dimethyl-2,4(1*H*,3*H*)-pyrimidinedione with this acetal in refluxing ethanol gave 5-benzylideneamino-6-dimethylamino-methyleneamino-1,3-dimethyl-2,4(1*H*,3*H*)-pyrimidinedione (**296**), which in tetramethylene sulfone at 200 °C gave 1,3-dimethyl-6-phenyl-2,4(1*H*,3*H*)-pteridinedione (**295**, Ar = Ph, R = H, Y = Me) in good yield[543, 1387]; substituted-phenyl analogs were made similarly.[543]

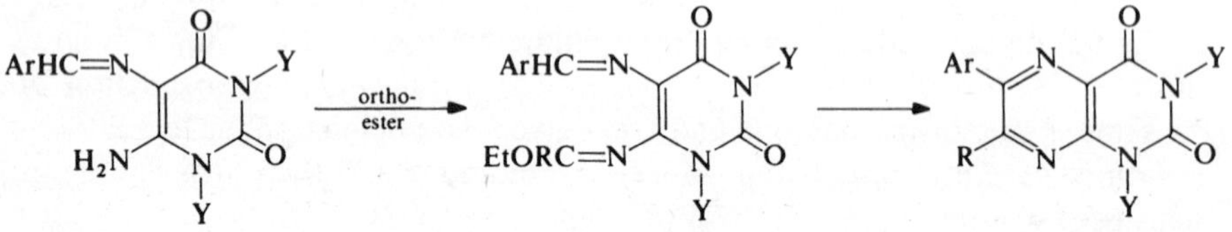

(**293**) (**294**) (**295**)

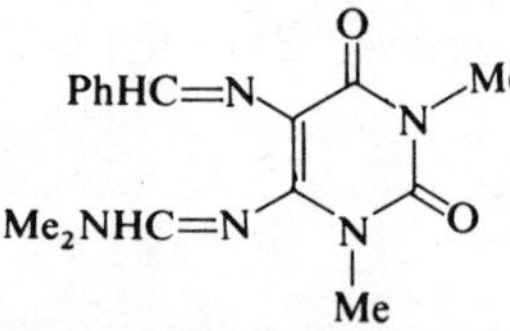

(**296**)

A closely related but more complicated reaction had already been observed by Wolfgang Pfleiderer and Heinz-Ulrich Blank in 1968.[1506] For example, 5,6-diamino-1,3-dimethyl-2,4(1*H*,3*H*)-pyrimidinedione with an excess of benzaldehyde at 180°C afforded 1,3-dimethyl-6,7diphenyl-2,4(1*H*,3*H*)-pteridinedione (**295**, Ar = R = Ph, Y = Me) in 33%yield; when the benzylidene compound (**293**, Ar = Ph, Y = Me) was used as substrate, the same product was obtained in 47% yield.[1506] Several analogs were made similarly: for example, 6,7-diphenyl-2,4-pteridinediamine and 2-amino-6,7-diphenyl-4(3*H*)-pteridinone.[1506] 6,7-Diarylpteridines were also obtained when the benzylidene compounds (**293**) were treated with a limited amount (rather than an excess) of triethyl orthoformate,[543] or when they were heated in formamide.[233] Using the latter technique, a separable purine by-product was produced in each case, sometimes in greater amount than the pteridine; a mechanism has been proposed.[233, 543]

A third synthesis of this type was reported by Frederick Blicke and Henry Godt in 1954[1435, cf. 395] and subsequently developed by Irwin Pachter[1415] and others. The initial example involved treatment of 5,6-diamino-1,3-dimethyl-2,4(1*H*,3*H*)-pyrimidinedione with hydrogen cyanide and formaldehyde to give 6-amino-5-α-cyanomethylamino-1,3-dimethyl-2,4(1*H*,3*H*)-pyrimidinedione (**297**) (presumably by addition of HCN to the 5-methyleneamino derivative) and subsequent alkaline cyclization to 7-amino-1,3-dimethyl-5,6-dihydro-2,4(1*H*,3*H*)-pteridinedione (**298**) followed by peroxide oxidation to 7-amino-1,3-dimethyl-2,4(1*H*,3*H*)-pteridinedione (~27% overall).[1435]

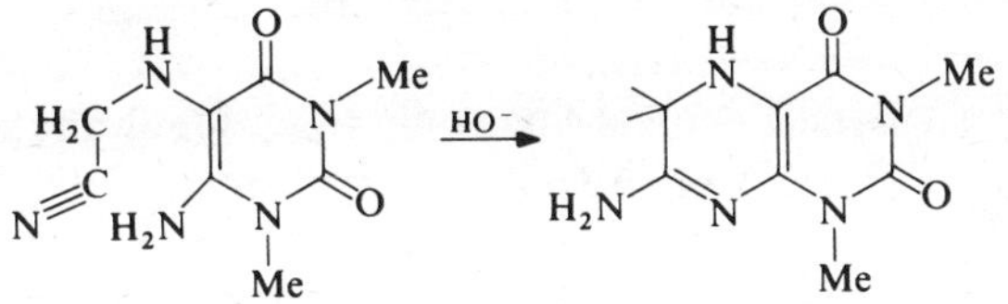

(**297**) (**298**)

Variations in the pyrimidine substrate and in the aldehyde to give a variety of 7-pteridinamines are exemplified in the reaction of 2,4,5,6-pyrimidinetetramine with acetaldehyde and hydrogen cyanide to give 5-α-cyanoethylamino-2,4,6-pyrimidinetriamine (**299**, R = Me) and then 6-methyl-2,4,7-pteridinetriamine (**300**, R = Me) (41%) via oxidation of its 5,6-dihydro derivative[1415]; in the similar formation of 6-isopropyl- (**300**, R = *i*-Pr) (from isobutyraldehyde: 28%), 6-cyclohex-3′-enyl- (from cyclohex-3′-enecarbaldehyde: 33%), 6-benzyl- (**300**, R = CH_2Ph) (from phenylacetaldehyde: 35%), 6-phenethyl- (from β-phenylpropionaldehyde: 60%), and 6-phenyl-2,4,7-pteridinetriamine (**300**, R = Ph) (from benzaldehyde: 33%)[1415]; in the formation from 2-phenyl-4,5,6-pyrimidinetriamine of 6-methyl-2-phenyl- (with acetaldehyde: 36%), 6-benzyl-2-phenyl- (with phenylacetaldehyde: 34%), and 2,6-diphenyl-4,7-pteridinediamine (with benzaldehyde: 21%)[1415]; in the conversion of 5,6-diamino-1-methyl-4(1*H*)-pyrimidinone into the 5-α-cyanomethylamino derivative and then into 7-amino-1-methyl-4(1*H*)-pteridinone via its 5,6-dihydro derivative[131]; in the similar conversion of 5,6-diamino- and 5-amino-6-methylamino-4(3*H*)-pyrimidi-

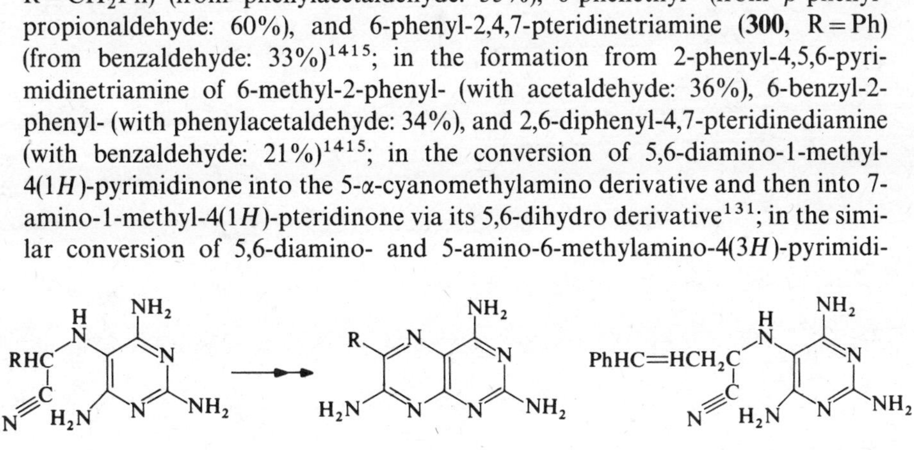

(**299**) (**300**) (**301**)

none into 7-amino-4(3*H*)- and 7-amino-8-methyl-4(8*H*)-pteridinone, respectively[131]; and in the eventual conversion of 2,4,5,6-pyrimidinetetramine by *p*-trifluoromethylbenzaldehyde–hydrogen cyanide into 6-*p*-trifluoromethylphenyl-2,4,7-pteridinetriamine.[523] By using an unsaturated aldehyde, the necessity for an oxidative step was removed and the pteridine was obtained directly on cyclization. For example, 2,4,5,6-pyrimidinetetramine with cinnamaldehyde–hydrogen cyanide gave 5-α-cyanocinnamylamino-2,4,6-pyrimidinetriamine (**301**), which cyclized directly into 6-phenethyl-2,4,7-pteridinetriamine (**300**, R = CH_2CH_2Ph).[1415] Likewise, the same pyrimidine substrate and phenylpropionaldehyde–hydrogen cyanide gave 6-styryl-2,4,7-pteridinetriamine (**300**, R = CH:CHPh).[1415]

Finally, treatment of 5,6-diamino-1,3-dimethyl-2,4(1*H*,3*H*)-pyrimidinedione with hydroxyacetaldehyde gave an uncharacterized Schiff base that on further treatment with mercuric chloride in dimethyl sulfoxide afforded 1,3-dimethyl-2,4(1*H*,3*H*)-pteridinedione (62%).[1264]

B. Pteridine 5-Oxides Directly from 4-(Substituted Amino)-5-nitropyrimidines

The cyclization of appropriate 4-(substituted-amino)-5-nitropyrimidines to give pteridine 5-oxides was first achieved by George Tennant in 1979. Thus he briefly reported[490] the acylation of 2-dimethylamino-5-nitro-4-pyrimidinamine with cyanoacetyl chloride to give 4-α-cyanoacetamido-2-dimethylamino-5-nitropyrimidine (**302**, R = H), which underwent cyclization in alkali at 40 °C to afford 2-dimethylamino-7-oxo-7,8-dihydro-6-pteridinecarbonitrile 5-oxide (**303**, R = H) (55%). This further underwent (a) dithionite treatment with hydrolysis of the cyano group and removal of the *N*-oxide function to give 2-dimethylamino-7(8*H*)-pteridinone (**304**) and (b) hydrolysis in hot alkali to give 2-

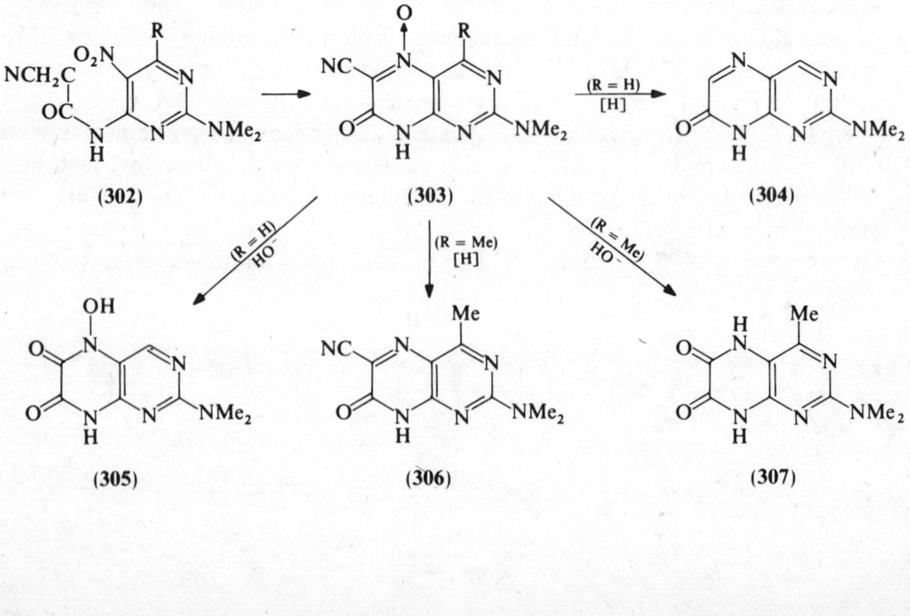

dimethylamino-5-hydroxy-6,7(5*H*,8*H*)-pteridinedione (**305**).[490] The homolog, 2-dimethylamino-4-methyl-7-oxo-7,8-dihydro-6-pteridinecarbonitrile 5-oxide (**303**, R = Me), was made similarly from the pyrimidine (**302**, R = Me) but behaved rather differently: (a) dithionite reduction affected only the *N*-oxide function to give 2-dimethylamino-4-methyl-7-oxo-7,8-dihydro-6-pteridinecarbonitrile (**306**) and (b) hot alkali caused both hydrolysis and *N*-oxide removal to give 2-dimethylamino-4-methyl-6,7(5*H*,8*H*)-pteridinedione (**307**).[490]

Some 7-dimethylamino-6-phenylpteridine 5-oxides were made by an analogous synthesis.[764] For example, *N*,*N*-dimethyl-α-phenylacetamidine and 6-chloro-5-nitro-4-pyrimidinamine gave 6-α-dimethylaminophenethylideneamino-5-nitro-4-pyrimidinamine (**308**, R = NH_2), which in ethanolic sodium ethoxide at 25 °C gave 7-dimethylamino-6-phenyl-4-pteridinamine 5-oxide (**309**, R = NH_2) from which the oxide function was removed by dithionite reduction to give 7-dimethylamino-6-phenyl-4-pteridinamine (**310**, R = NH_2) in 50% yield.[764] Similar procedures on appropriate substrates (**308**) gave the 4-diethylamino, 4-piperidino, 4-morpholino, 4-cyclohexylamino, and 4-ethoxy analogs (**309**) and then the corresponding 4-substituted-7-dimethylamino-6-phenylpteridines [**310**, R = NEt_2, $(CH_2)_5N$, $O(CH_2CH_2)_2N$, $C_6H_{11}NH$, or OEt]; hydrolysis of the ethoxy *N*-oxide (**309**, R = OEt) gave 7-dimethylamino-6-phenyl-4(3*H*)-pteridinone 5-oxide and then the deoxypteridine.[764]

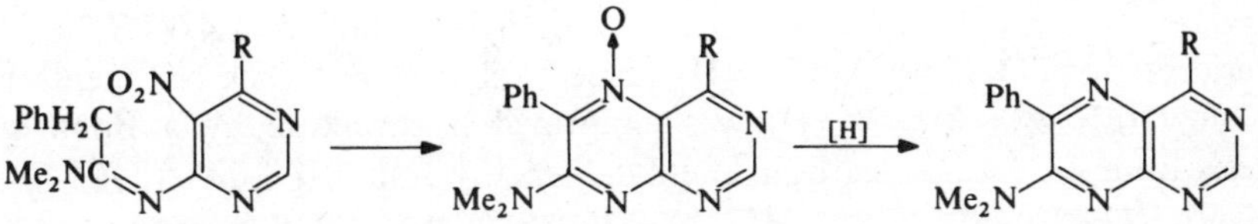

(**308**) (**309**) (**310**)

C. Pteridines by Cyclization of 4-(Substituted Amino)-5-unsubstituted Pyrimidines

There are several distinct procedures within this category. In the first of these, 2,6-diamino-4(3*H*)-pyrimidinone (**311**, R = H) and ethylenediamine were heated in glycol containing cupric acetate at 200 °C for several hours and the cooled reaction mixture was then treated with hydrogen peroxide to give eventually 2-amino-4(3*H*)-pteridinone (**312**) (~15%). Similar treatment of 2,6-diamino-5-nitroso-4(3*H*)-pyrimidinone with β-aminoethanol, but without peroxide treatment, gave the same product (**312**) (36%) but the precise mechanism in each case is obscure.[546, 547]

The second involves irradiation of 6-azido-1,3-dimethyl-2,4(1*H*,3*H*)-pyrimidinedione (**314**) with α-aminoketones (**313**, R = aryl) or α-aminoesters (**313**, R = OEt, Y = H or alkyl) to give 7,8-dihydropteridines. Thus irradiation of the azide (**314**) with phenacylamine (**313**, R = Ph, Y = H) or its *p*-bromo derivative in THF gave 1,3-dimethyl-6-phenyl-7,8-dihydro-2,4(1*H*,3*H*)-pteridinedione

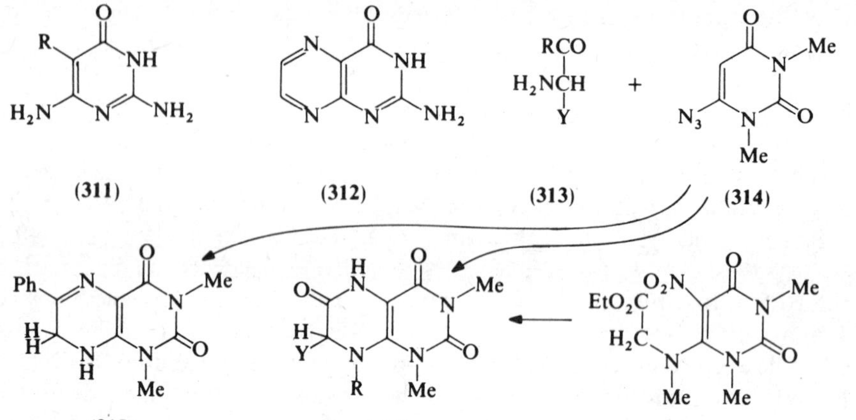

(311) (312) (313) (314)

(315) (316) (317)

(315) (75%) or the 6-*p*-bromophenyl analog (70%).[1241] The same substrate (314) with the ethyl esters of glycine (313, R = OEt, Y = H), alanine (313, R = OEt, Y = Me), phenylalanine (313, R = CH_2Ph, Y = H), methionine (313, R = CH_2CH_2SMe, Y = H), or *N*-methylglycine gave in >60% yield 1,3-dimethyl- (316, R = Y = H), 1,3,7-trimethyl- (316, R = H, Y = Me), 7-benzyl-1,3-dimethyl- (316, R = H, Y = CH_2Ph), 1,3-dimethyl-7-β-methylthioethyl- (316, R = H, Y = CH_2CH_2SMe), or 1,3,8-trimethyl-7,8-dihydro-2,4,6(1*H*,3*H*,5*H*)-pteridinetrione (316, R = Me, Y = H), respectively.[1240, 1241] The last mentioned product (316, R = Me, Y = H) was confirmed in structure by a Boon synthesis from 6-(*N*-ethoxycarbonylmethyl-*N*-methylamino)-1,3-dimethyl-5-nitro-2,4(1*H*,3*H*)-pyrimidinedione (317).[1241] It is evident that this reaction has a considerable potential.

Treatment of 6-allylamino-1,3-dimethyl-2,4(1*H*,3*H*)-pyrimidinedione (318) with nitrous acid in aprotic solvents produced a deep color (of the 5-nitroso derivative) and then a purine; the same treatment in water produced no such color but the oxime, 6-hydroxyiminomethyl-1,3-dimethyl-2,4(1*H*,3*H*)-pteridinedione (319, X = NOH), which was confirmed in structure by synthesis from 1,3-dimethyl-2,4-dioxo-1,2,3,4-tetrahydro-6-pteridinecarbaldehyde (319, X = O) with hydroxylamine.[196] A plausable mechanism involving an initial addition of nitrous acid to the allyl group has been proposed.[196]

The reaction of 1,3-dimethyl-6-(α-methylbenzylidene)hydrazino-2,4(1*H*,3*H*)-pyrimidinedione (320) with aqueous nitrous acid followed by dithionite reduc-

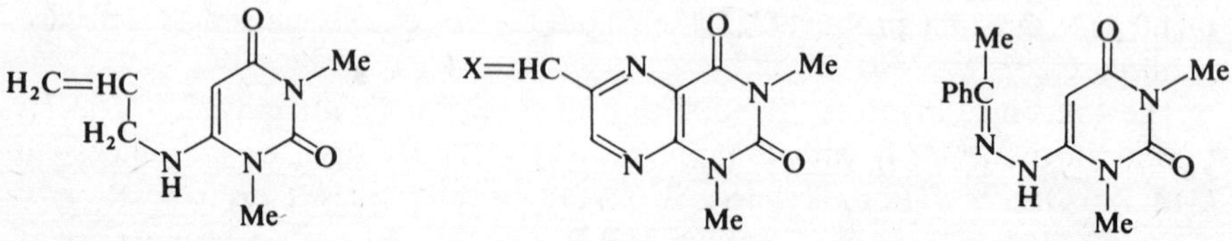

(318) (319) (320)

tion gave 1,3-dimethyl-6-phenyl-2,4(1*H*,3*H*)-pteridinedione (**321**); *para*-substituted intermediates (**320**) produced *p*-chloro-, *p*-bromo-, and *p*-methoxyphenyl, as well as the *p*-tolyl analogs of the product (**321**).[1292] No mechanism has been proposed.

Irradiation of 6-azido-1,3-dimethyl-2,4(1*H*,3*H*)-pyrimidinedione (**314**) in methanol gave 1,3,6,8-tetramethyl-2,4,7,9(1*H*,3*H*,6*H*,8*H*)-pyrimido[4,5-*g*]pteridinetetrone (**322**), which was confirmed in structure by a more logical synthesis.[549]

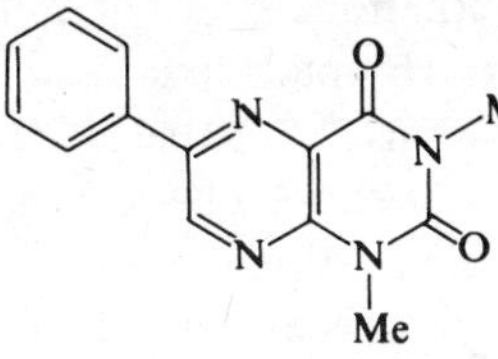

(**321**)

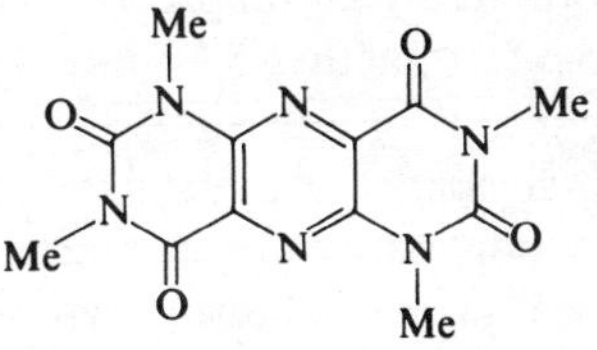

(**322**)

D. Pteridines from Alloxan Derivatives

Although simply the reverse of a Gabriel and Colman synthesis, condensation of the α-dicarbonylpyrimidine, alloxan (**324**), with the diamine, α,α′-diaminomaleonitrile (**323**), in boiling acetic acid to give 2,4-dioxo-1,2,3,4-tetrahydro-6,7-pteridinedicarbonitrile (**325**) (40%),[1198] is one of the very few examples of such a synthesis. A more complicated example is the condensation of *N*-methylalloxan with 2-β-hydroxyethyl-1-naphthylamine or with the isomeric 1-β-hydroxyethyl-2-naphthylamine to give 7-β-hydroxyethyl-10-methyl-9,11(7*H*,10*H*)-naphtho[1,2-*g*]pteridinedione (**326**) or 12-β-hydroxyethyl-9-methyl-8,10(9*H*,12*H*)-naphtho[2,1-*g*]pteridinedione (**327**), respectively.[211]

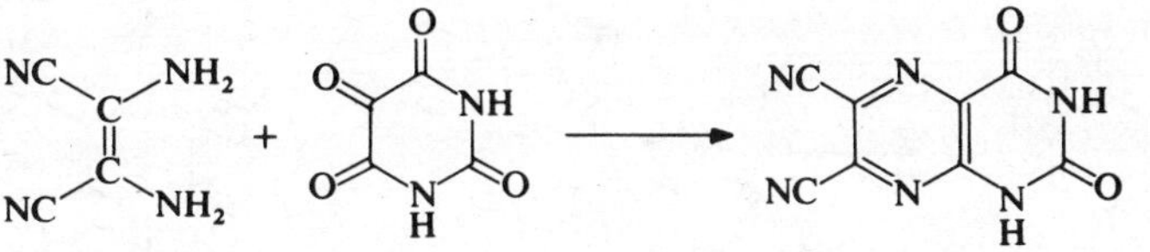

(**323**) (**324**) (**325**)

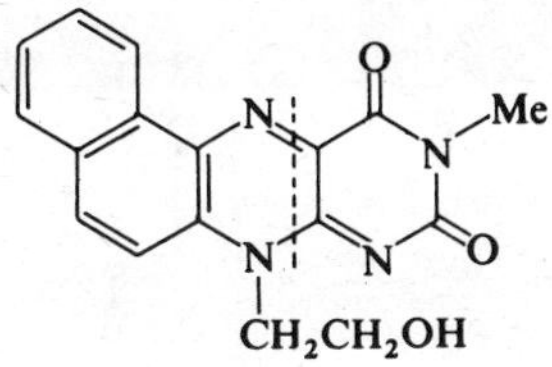

(**326**)

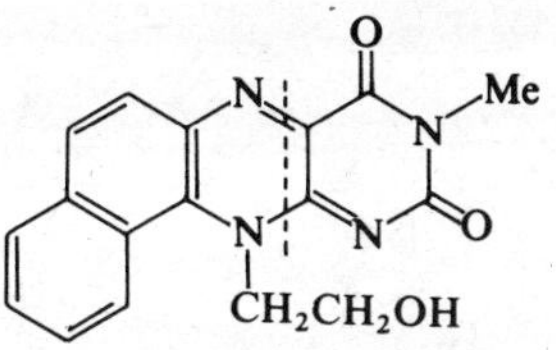

(**327**)

E. Hydropteridines from 5-Halogenopyrimidines and α-Diamines

The use of 5-halogenopyrimidines in hydropteridine syntheses is rare because the 5-halogeno substituent on pyrimidine is normally quite unreactive. However, in the case of 5-halogeno-2,4(1*H*,3*H*)-pyrimidinediones (halogenouracils), and so on, the halogeno substituent undergoes aminolysis and other nucleophilic displacements relatively easily.[1507] Therefore, it proved practical to condense 5-bromo-6-chloro-1,3-dimethyl-2,4(1*H*,3*H*)-pyrimidinedione (**329**) with (2*S*)-2-methylaminopropylamine (**328**) in ethanolic triethylamine to give (6*S*)-1,3,5,6-tetramethyl-5,6,7,8-tetrahydro-2,4(1*H*,3*H*)-pteridinedione (**330**) in good yield and of subsequently confirmed structure.[456] None of the 1,3,7,8-tetramethyl isomer was detected, quite possibly because of the preferentially rapid reaction between the highly active chloro substituent in (**329**) and the unhindered primary amino group in (**328**). Other examples have been reported.[1412]

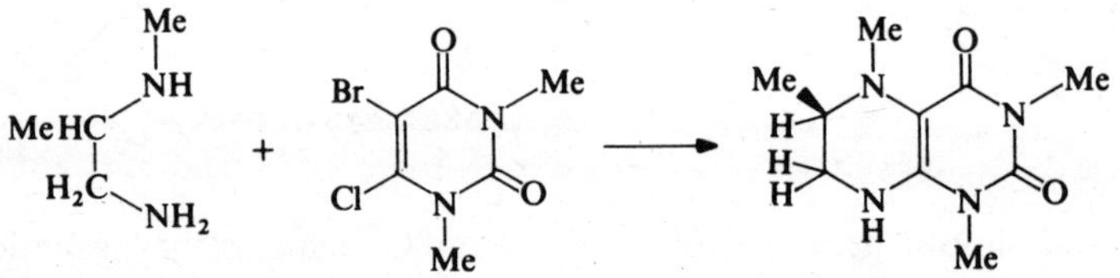

(**328**) (**329**) (**330**)

It is also possible to condense such substituted ethylenediamines with 5-halogeno-4-pyrimidinamines or even 5-halogeno-4-pyrimidinones to give hydropteridines. Understandably, 2,6-diamino-5-bromo-4(3*H*)-pyrimidinone (**332**) reacted at 110 °C with neat *N*,*N*′-dimethylethylenediamine to involve preferentially the halogeno and 6-amino substituents of (**332**), thereby affording only 2-amino-5,8-dimethyl-5,6,7,8-tetrahydro-4(3*H*)-pteridinone (**331**) (50%); 6-

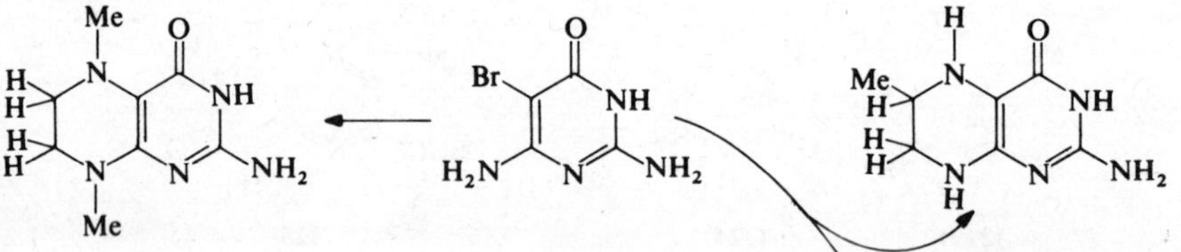

(**331**) (**332**) (**333**)

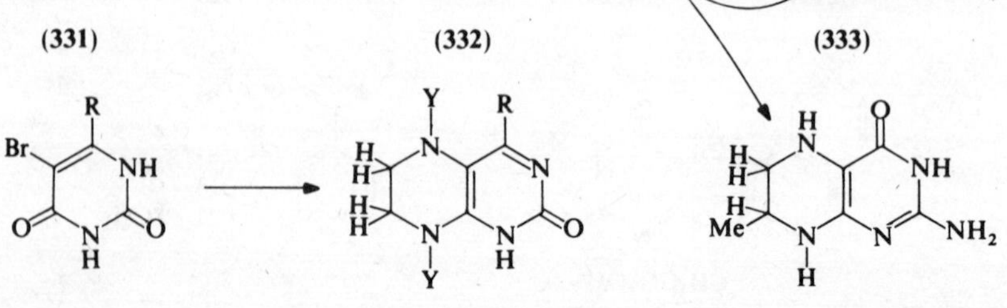

(**334**) (**335**) (**336**)

amino-5-bromo-2,4(1*H*,3*H*)-pyrimidinedione behaved in like manner to give 5,8-dimethyl-5,6,7,8-tetrahydro-2,4(1*H*,3*H*)-pteridinedione (65%).[886] When the 2,4-diamino substrate (**332**) was treated with the unsymmetrical (*S*)-α,β-propanediamine in boiling acetic acid, a separable mixture of (*S*)-2-amino-6-methyl- (**333**) and (*S*)-2-amino-7-methyl-5,6,7,8-tetrahydro-4(3*H*)-pteridinone (**336**) was obtained.[296] In the absence of an amino group on the pyrimidine substrate, an oxo substituent became involved. Thus 5-bromo-2,4(1*H*,3*H*)-pyrimidinedione (**334**, R = H) or its 6-methyl derivative (**334**, R = Me) with aqueous ethanolic ethylenediamine under reflux for a long period gave good yields of 5,6,7,8-tetrahydro-2(1*H*)-pteridinone (**335**, R = Y = H) or its 4-methyl derivative (**335**, R = Me, Y = H), respectively. Similar reactions with neat *N*,*N'*-dimethylethylenediamine at 160 °C gave >80% of 5,8-dimethyl-5,6,7,8-tetrahydro-2(1*H*)-pteridinone (**335**, R = H, Y = Me) or 4,5,8-trimethyl-5,6,7,8-tetrahydro-2(1*H*)-pteridinone (**335**, R = Y = Me), according to the substrate.[586, 886]

CHAPTER III

Primary Syntheses from Pyrazines or Other Heterocycles

In theory there are many ways of building a pyrimidine ring onto a pyrazine intermediate to form a pteridine. For example, the pyrazine intermediate may require the addition of C-2 and the completion of the 1,2- and 2,3-bonds to make a pteridine, as indicated in the skeletal formula (**1**); all the ring atoms may be already attached to the pyrazine, which therefore needs only cyclization by completing the 2,3-bond, as indicated in formula (**5**); and so on. Such possibilities, for which actual examples exist at the present time, are indicated in skeletal formulas (**1**)–(**7**) and these provide a basis for subclassification of the pyrazine→pteridine syntheses to be discussed in this chapter.

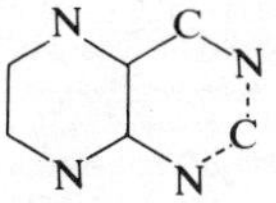

(**1**)

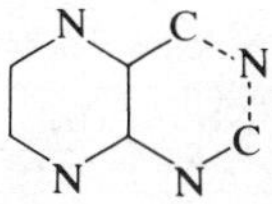

(**2**)

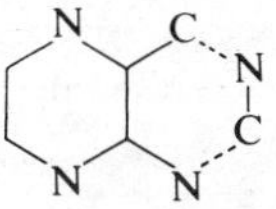

(**3**)

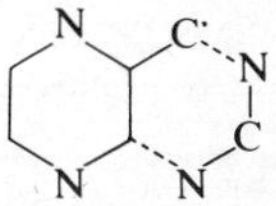

(**4**)

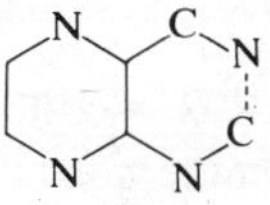

(**5**)

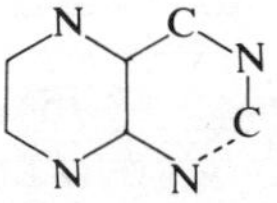

(**6**)

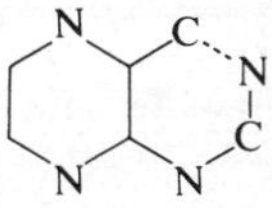

(**7**)

The remaining syntheses of pteridines from purines or other (mainly tricyclic) heterosystems are of little chemical significance, although some from purines may be of biochemical interest as models for the biosynthesis of some pteridines. The formation of pteridines during the thermolysis of amino acid mixtures is now firmly based.[229]

1. SYNTHESES FROM PYRAZINES

For the most part, syntheses of pteridines from pyrazines are unambiguous, and for this reason they are especially valuable. The difficulty of obtaining appropriate pyrazine intermediates has been minimized by the publication of Gordon Barlin's monograph which, inter alia, contains a useful table listing such known syntheses.[1480] In addition, there are several valuable reviews of these syntheses by Ted Taylor (Pteridine Ted),[465, 473, 599, 625] who has personally contributed very significantly to this area of research.

A. By Addition of C-2 as in Skeletal Formula (1)

Most of the pteridines prepared by the addition of C-2 to a pyrazine carry a 4-oxo or other substituent. However, a few quite *simple pteridines* have been made in this way, usually via hydropteridines. Thus 3-aminomethyl-2-pyrazinamine (**8**, R=H) in refluxing neat triethyl orthoformate gave 3,4-dihydropteridine (**9**, R=Y=H) (74%), which underwent oxidation by manganese dioxide to give pteridine (**10**, R=Y=H) (52%).[51] The same substrate (**8**, R=H) with triethyl orthoacetate gave 2-methyl-3,4-dihydropteridine (**9**, R=H, Y=Me) (70%) and then 2-methylpteridine (**10**, R=H, Y=Me) (80%).[51, 54] Similar treatment of 3-aminomethyl-5-methyl-2-pyrazinamine (**8**, R=Me) with triethyl orthoformate gave 6-methyl-3,4-dihydropteridine (**9**, R=Me, Y=H) (69%) and then the long-sought 6-methylpteridine (**10**, R=Me, Y=H) (74%).[51] In addition, 3-aminomethyl-2-pyrazinamine (**8**, R=H) reacted with formaldehyde to give 1,2,3,4-tetrahydropteridine in 43% yield.[51]

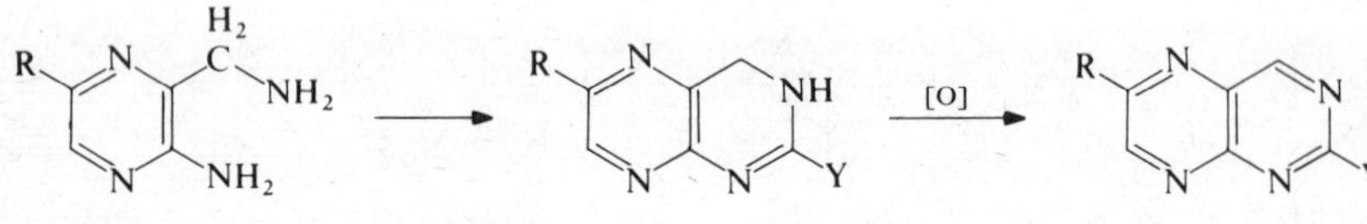

(**8**) (**9**) (**10**)

The formation of 4-*pteridinamines* by the supply of C-2 is confined to a few derivatives. For example, 3-amino-2-pyrazinecarboxamide *O*-acetyloxime (**11**, R=Ac) with triethyl orthoformate under reflux gave 4-acetoxyaminopteridine (**12**, R=Ac) (29%)[1004]; similar treatment of the corresponding *O*-benzoyloxime (**11**, R=Bz) or the *O*-methyloxime (**11**, R=Me) gave 4-benzoyloxyaminopteridine (**12**, R=Bz) (34%) or 4-methoxyaminopteridine (**12**, R=Me) (54%), respectively.[631, 1004] However, when the *O*-unsubstituted substrate, 3-amino-2-pyrazinecarboxamide oxime (**11**, R=H), was treated with neat triethyl orthoformate or with diethoxymethyl acetate in toluene, 4-pteridinamine 3-oxide (**13**) resulted in good yield.[999, 1003] A distantly related reaction is that of 3-imidazol-2′-yl-2-pyrazinamine (**14**, R=H) with triethyl orthoformate, triethyl orthoacetate, or triethyl orthopropionate to give imidazo[1,2-*c*]pteridine (**15**, R=Y=H), its 6-methyl derivative (**15**, R=H, Y=Me), or its 6-ethyl derivative (**15**, R=H, Y=Et), respectively, each in >80% yield.[459] 3-Imidazol-2′-yl-5,6-dimethyl-2-pyrazinamine (**14**, R=Me) similarly gave 2,3-dimethylimidazo[1,2-*c*]pteri-

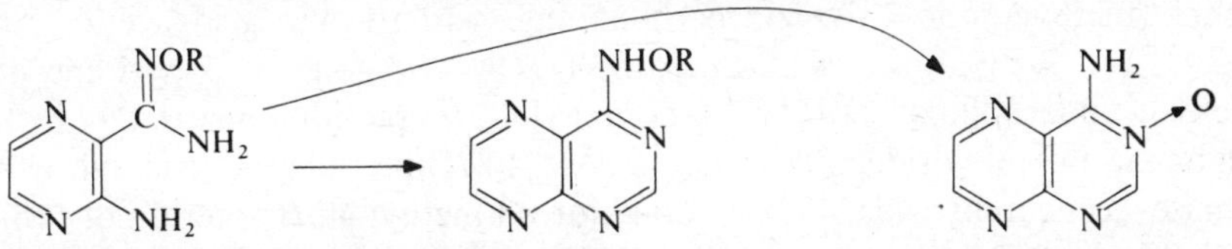

(11) (12) (13)

(14) (15)

dine (**15**, R = Me, Y = H), its 2,3,6-trimethyl homolog (**15**, R = Y = Me), and its 6-ethyl-2,3-dimethyl homolog (**15**, R = Me, Y = Et).[459]

The preparation of 4-*pteridinones*, 4-*pteridinethiones*, and related compounds by this general procedure is very well represented and several useful routes, specifically for obtaining suitable pyrazine substrates, have been reported.[322, 470, 517] Thus, 3-amino-2-pyrazinecarboxamide (**16**, X = O) was converted into 4(3*H*)-pteridinone (**17**, X = O) in several ways: (a) by refluxing in a mixture of triethyl orthoformate and acetic anhydride (yield: 80%)[28, 30]; (b) by boiling in formic acid followed by evaporation and, either heating at 225°C (44%) or boiling in ethanolic sodium ethoxide, to complete cyclization (60%)[30]; (c) by heating in formamide at 195°C (84%)[20]; (d) by refluxing with formamide in butanolic sodium butoxide (75%)[20]; or (e) by heating with butanolic formamidine acetate (64%).[20] When applied to 3-amino-2-pyrazinecarbothioamide (**16**, X = S), the first of these methods afforded 4(3*H*)-pteridinethione (**17**, X = S) in 84% yield.[30] Broadly similar procedures also converted appropriately substituted 3-amino-2-pyrazinecarboxamides into 2-methyl-4(3*H*)-pteridinone (triethyl orthoacetate–acetic anhydride: 62%)[20]; 2-trifluoromethyl-4(3*H*)-pteridinone (trifluoroacetic acid–trifluoroacetic anhydride in a sealed tube at 110°C: 81%)[20]; 2-trifluoromethyl-4(3*H*)-pteridinethione (trifluoroacetamide–ethanolic sodium ethoxide: 64%)[20]; 3-methyl-4(3*H*)-pteridinone (from 3-amino-*N*-methyl-2-pyrazinecarboxamide with triethyl orthoformate–acetic anhydride,[187] or formic acid–acetic anhydride: 47%)[32]; 1-methyl-4(1*H*)-pteridinone

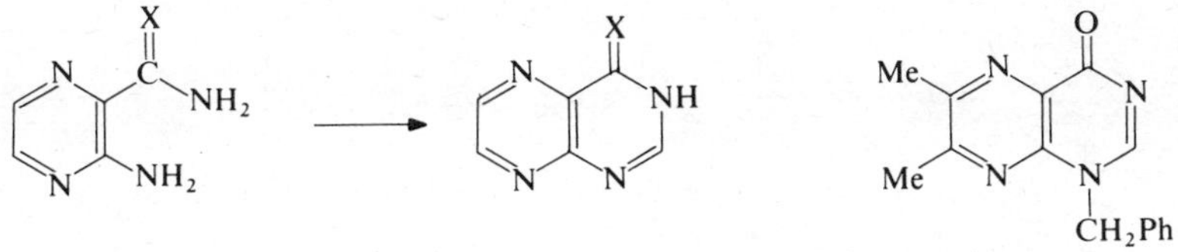

(16) (17) (18)

(from 3-methylamino-2-pyrazinecarboxamide with formic acid–acetic anhydride)[538]; 2-pyridin-3′-yl-1,2-dihydro-4(3*H*)-pteridinone (3-pyridinecarbaldehyde under reflux: 8%)[1140]; 2,6-dimethyl-4(3*H*)-pteridinone (triethyl orthoacetate–acetic anhydride: 40%)[153]; 6-chloro-4(3*H*)-pteridinone (triethyl orthoformate–acetic anhydride: 57%)[974]; 6-chloro-3-methyl-4(3*H*)-pteridinone (likewise: 82%)[974]; 3,7-dimethyl-4(3*H*)-pteridinone (likewise: 62%)[167]; 7-methyl-4(3*H*)-pteridinone (likewise: 71%)[167]; 3-benzyl-6,7-diphenyl-4(3*H*)-pteridinone (formic acid–sodium acetate–acetic anhydride: 41%[468]; triethyl orthoformate–acetic anhydride: 72%[474]); 3-butyl-6,7-diphenyl-4(3*H*)-pteridinone (formic acid–acetic anhydride: 33%; triethyl orthoformate–acetic anhydride: 77%)[474]; 6,7-diphenyl-4(3*H*)-pteridinethione (from 3-amino-5,6-diphenyl-2-pyrazinecarbothioamide with triethyl orthoformate–acetic anhydride: 55%)[474]; 3-butyl-6,7-diphenyl-4(3*H*)-pteridinethione (formic acid–sodium acetate–acetic anhydride: 68%; triethyl orthoformate–acetic anhydride: 61%)[474]; 2,6,7-triphenyl-4(3*H*)-pteridinone (refluxing benzoyl chloride: 60%)[476]; its 3-benzyl derivative (likewise: 72%)[476]; 2,6,7-triphenyl-4(3*H*)-pteridinethione (likewise)[476]; 2-anilino-6,7-diphenyl-4(3*H*)-pteridinethione (from 3-amino-5,6-diphenyl-2-pyrazinecarboxamide with phenyl isothiocyanate in refluxing pyridine: 84%)[476]; 1-benzyl-6,7-dimethyl-4(1*H*)-pteridinone (**18**) (from 3-benzylamino-5,6-dimethyl-2-pyrazinecarboxamide with formic acid in acetic anhydride: 60%)[1008]; its 6,7-diphenyl analog (likewise 59%)[1008]; 1-benzyl- and 1-cyclohexyl-4(1*H*)-pteridinone (DMF diethyl acetal under reflux: 60, 65%)[1008]; 1-cyclohexyl-6,7-diphenyl-4(1*H*)-pteridinone (likewise: 54%)[1008]; a variety[187] of 2-alkyl-3-alkyl(or aryl)-4(3*H*)-pteridinones, exemplified in Table IX; and 3-methyl-6-phenyl-4(3*H*)-pteridinone, its 8-oxide, and homologous compounds.[1758]

Pteridine *N*-oxides or their derivatives may be produced by this reaction. Thus 3-amino-6-methyl-2-pyrazinecarboxamide 4-oxide (**19**, R = Me) with triethyl orthoformate in DMF at 140°C gave 6-methyl-4(3*H*)-pteridinone 8-oxide (**20**, R = Me) (64%)[475, 485]; the 6-phenyl analog (**20**, R = Ph) was made

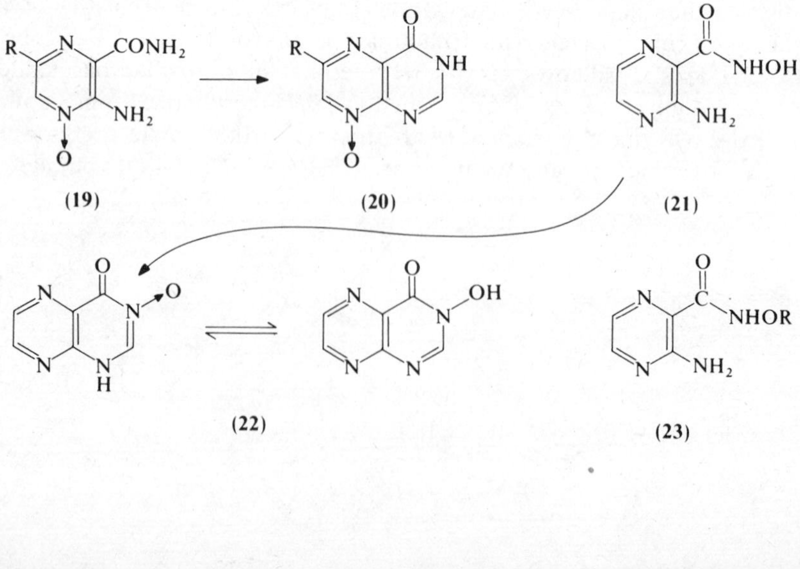

(19) (20) (21) (22) (23)

TABLE IX. ADDITIONAL EXAMPLES OF THE PREPARATION OF PTERIDINES FROM PYRAZINES

Pyrazine Intermediate	Reagent and Conditions	Pteridine (Yield)	References
3-Amino-*N*-isopropyl-2-pyrazinecarboxamide	Triethyl orthoformate; acetic anhydride; reflux; 5 h	3-Isopropyl-4(3*H*)-pteridinone[a]	187
3-Amino-*N*-*p*-tolyl-2-pyrazinecarboxamide	Triethyl orthoacetate; acetic anhydride; reflux; 5 h	2-Methyl-3-*p*-tolyl-4(3*H*)-pteridinone[a]	187
3-Amino-*N*-pyridin-3′-yl-2-pyrazinecarboxamide	Triethyl orthoacetate; acetic anhydride; reflux; 5 h	2-Methyl-3-pyridin-3′-yl-4(3*H*)-pteridinone	187
Methyl 3-amino-5-methyl-2-pyrazinecarbohydroxamate	Formic acid; acetic anhydride; 80 °C; 90 min	3-Methoxy-7-methyl-4(3*H*)-pteridinone (48%)	759
Methyl 3-amino-5,6-dimethyl-2-pyrazinecarbohydroxamate	Acetic acid; acetic anhydride; 100 °C; 2.5 h	3-Methoxy-2,6,7-trimethyl-4(3*H*)-pteridinone (56%)	759
3-Amino-5-methyl-2-pyrazinecarbohydroxamic acid	Triethyl orthoformate; acetic anhydride; reflux; 2 h	7-Methyl-4(1*H*)-pteridinone 3-oxide	154
3-Amino-*N*′-benzylidene-2-pyrazinecarbohydrazide	Triethyl orthoformate; acetic anhydride; reflux; 2 h	3-Benzylideneamino-4(3*H*)-pteridinone (>95%)	487
6-Chloro-3-dimethylaminomethyleneamino-2-pyrazinecarbonitrile	Ammonium acetate; water; reflux; 2 h	6-Chloro-4-pteridinamine (87%)	53
6-Chloro-3-dimethylaminomethyleneamine-2-pyrazinecarbonitrile	Methylamine acetate; water; reflux; 1 h	6-Chloro-4-methylaminopteridine (65%)	53
3-Amino-6-cyclopropyl-2-pyrazinecarbonitrile	Guanidine; methanol; reflux; 30 min	6-Cyclopropyl-2,4-pteridinediamine[b] (85%)	729
3-Amino-6-propyl-2-pyrazinecarbonitrile	Guanidine; methanolic NaOMe; reflux; 12 h	6-Propyl-2,4-pteridinediamine[a] (80%)	484
3-Amino-6-chloro-5-ethylamino-2-pyrazinecarbonitrile	Guanidine; methanol; reflux; 30 min	6-Chloro-7-ethylamino-2,4-pteridinediamine[a] (86%)	729

TABLE IX. (*Contd.*)

Pyrazine Intermediate	Reagent and Conditions	Pteridine (Yield)	References
3-Amino-6-prop-1′-enyl-2-pyrazinecarbonitrile	Guanidine; methanolic NaOMe; reflux; 6 h	6-Prop-1′-enyl-2,4-pteridinediamine[a] (88%)	1297
3-Amino-6-phthalimidomethyl-2-pyrazinecarbonitrile	Guanidine acetate; DMF; 120 °C; 48 h	6-Phthalimidomethyl-2,4-pteridinediamine[a] (~55%)	1302
3-Amino-5-β-methoxypropyl-2-pyrazinecarbonitrile	Guanidine; methanolic NaOMe; reflux; 18 h	7-β-Methoxypropyl-2,4-pteridinediamine[a] (60%)	1305
3-Amino-6-*p*-chlorophenylthiomethyl-2-pyrazinecarbonitrile	Guanidine; methanol; reflux; 2 h	6-*p*-Chlorophenylthiomethyl-2,4-pteridinediamine[a] (90%)	1299, 1353
3-Amino-6-(*N*-3′4′-dichlorophenyl-*N*-propylamino)methyl-2-pyrazinecarbonitrile	Guanidine; methanol; reflux	6-(*N*-3′,4′-Dichlorophenyl-*N*-propylamino)methyl-2,4-pteridinediamine[a] (56%)	1355
3-Amino-6-2′-hydroxycyclohexylmethyl-2-pyrazinecarbonitrile	Guanidine; methanol; reflux; 24 h	6-2′-Hydroxycyclohexylmethyl-2,4-pteridinediamine[a] (52%)	1299
3-Amino-6-phenyl-2-pyrazinecarbonitrile 4-oxide	Guanidine; methanolic NaOMe; reflux; 16 h	6-Phenyl-2,4-pteridinediamine 8-oxide[a] (>95%)	484
Ethyl 3-amino-6-[D-*arabino*]-α,β,γ,δ-tetrahydroxybutyl-2-pyrazinecarboxylate 4-oxide	Guanidine; DMF; 75 °C; 12 h	2-Amino-6-[D-*arabino*]-α,β,γ,δ-tetrahydroxybutyl-4(3*H*)-pteridinone 8-oxide[a]	1318
Ethyl 3-amino-6-phenyl-2-pyrazinecarboxylate 4-oxide	Guanidine; DMF; reflux; 4 h	2-Amino-6-phenyl-4(3*H*)-pteridinone 8-oxide[a] (65%)	478, 485

[a] And analog(s).
[b] Several methotrexate analogs have been made similarly.[1724]

similarly (80%) (using *N*,*N*-dimethylacetamide as solvent) from the pyrazine (**19**, R = Ph)[475, 485]; 3-amino-2-pyrazinecarbohydroxamic acid (**21**) with triethyl orthoformate in acetic anhydride gave 4(1*H*)-pteridinone 3-oxide–3-hydroxy-4(3*H*)-pteridinone (**22**) in 67% yield[540]; methyl 3-amino-2-pyrazinecarbohydroxamate (**23**, R = Me) with formic acid–acetic anhydride at 80°C gave 3-methoxy-4(3*H*)-pteridinone (71%)[759]; methyl 3-amino-5,6-dimethyl-2-pyrazinecarbohydroxamate likewise gave 3-methoxy-6,7-dimethyl-4(3*H*)-pteridinone (60%)[759]; methyl 3-amino-2-pyrazinecarbohydroxamate with acetic acid–acetic anhydride at 100°C gave 3-methoxy-2-methyl-4(3*H*)-pteridinone (44%)[759]; 3-amino-2-pyrazinecarbohydroxamic acid with triethyl orthoformate–acetic anhydride gave 2-methyl-4(1*H*)-pteridinone 3-oxide (~40%)[154]; and 3-amino-2-pyrazinecarbohydrazide, as its isopropylidene derivative (**24**), with ethyl orthoformate–acetic anhydride gave 3-isopropylideneamino-4(3*H*)-pteridinone (**25**) (91%), which on brief acid hydrolysis liberated 3-amino-4(3*H*)-pteridinone (**26**) (83%).[487]

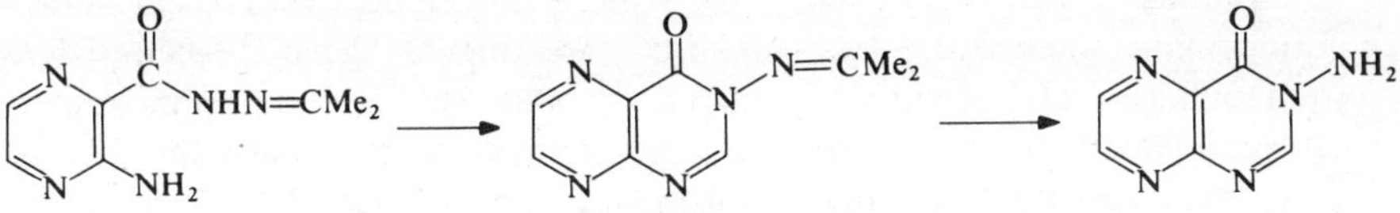

(**24**) (**25**) (**26**)

Other examples of note are the condensation of methyl 3-amino-6-cyano-5-methoxy-2-pyrazinecarboximidate (**27**) with triethyl orthoformate in acetic anhydride to give 4,7-dimethoxy-6-pteridinecarbonitrile (**28**) (96%)[1136] and the reaction of 3-amino-2-quinoxalinecarboxamide with acetic anhydride to give, after hydrolysis of an acetyl derivative, 2-methyl-4(3*H*)-benzo[*g*]pteridinone (**29**).[914]

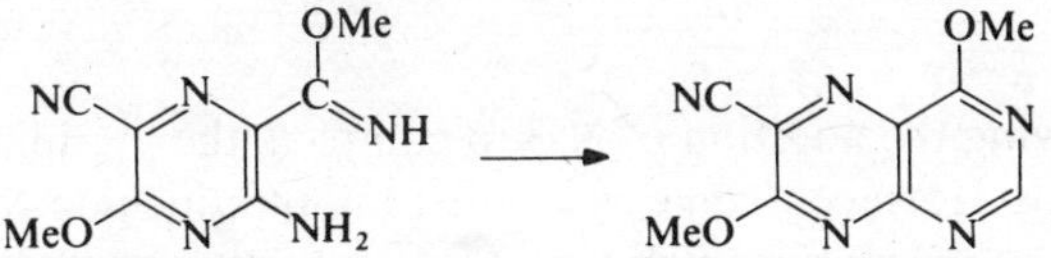

(**27**) (**28**)

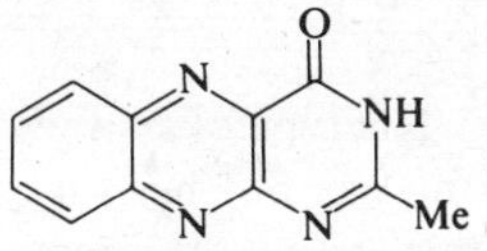

(**29**)

The formation of 2,4-pteridinediones and related compounds in this way is meagrely illustrated in the reaction of 3-benzylamino-2-pyrazinecarboxamide in boiling neat ethyl chloroformate during 30 h to give 1-benzyl-2,4(1*H*,3*H*)-pteridinedione (**30**, R = H) in 55% yield[1008]; in the similar formation of its 6,7-dimethyl (**30**, R = Me) and 6,7-diphenyl (**30**, R = Ph) derivatives[1008]; in the conversion of *N*,*N*′-dimethyl-3-methylamino-5-oxo-4,5-dihydro-2,6-pyrazinedicarboxamide (**31**), by treatment with ethyl chloroformate followed by ethanolic sodium ethoxide, into 1,3,*N*-trimethyl-2,4,7-trioxo-1,2,3,4,7,8-hexahydro-6-pteridinecarboxamide (**32**) (~50%)[336]; in the similar (two stage) conversion of

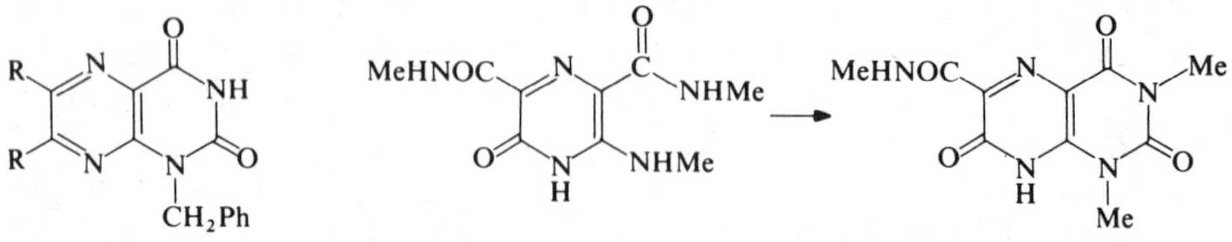

(30) (31) (32)

3-amino-*N*-butyl-5,6-diphenyl-2-pyrazinecarbothioamide into 3-butyl-6,7-diphenyl-4-thioxo-3,4-dihydro-2-(1*H*)-pteridinone (73%)[474]; in the similar conversion of 3-amino-*N*-butyl-2-pyrazinecarboxamide into 3-butyl-2,4(1*H*,3*H*)-pteridinedione (89%)[876]; and in the conversion of 3-amino-6-[L-*threo*]-α,β,γ-trihydroxypropyl-2-pyrazinecarboxamide (**33**), by tetramethylurea diethyl acetal (**34**) in DMF, into 2-dimethylamino-6-[L-*threo*]-α,β,γ-trihydroxypropyl-4(3*H*)-pteridinone (**35**) (euglenapterin: 77%).[985] In addition, 3-benzylamino-5,6-tetramethylene-2-pyrazinecarboxamide, with neat ethyl chloroformate followed by alkaline treatment, gave 1-benzyl-6,7-tetramethylene-4(1*H*)-pteridinone (60%)[1008]; and 3,6-diamino-2-quinoxalinecarboxamide with ethyl orthoformate–acetic anhydride gave 8-acetamido- and then by saponification 8-amino-4(3*H*)benzo[*g*]pteridinone.[323]

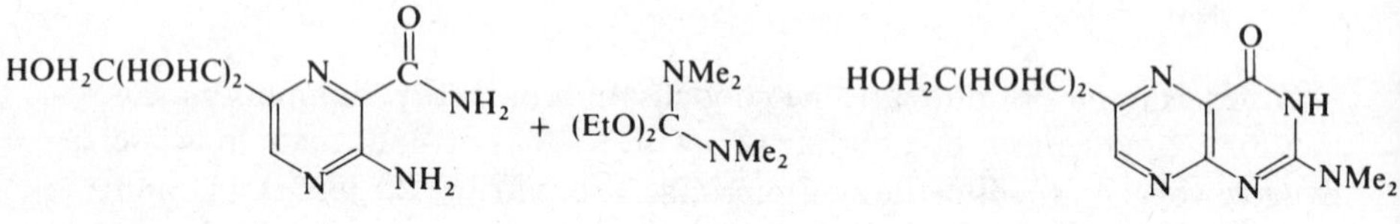

(33) (34) (35)

B. By Addition of N-3 as in Skeletal Formula (2)

The type of synthesis involving the addition of N-3 to complete the pteridine ring has not been used widely. However, some valuable examples include the conversion of 3-acetamido-2-pyrazinecarbaldehyde (**37**, R = Ac) by ethanolic ammonia at 0°C into 2-methylpteridine (**36**) (34%)[52]; of 3-ethoxycarbonylamino-2-pyrazinecarbaldehyde (**37**, R = CO_2Et) by ethanolic ammonia at 25°C into 2(1*H*)-pteridinone (**38**) (55%)[52]; of 3-ethoxymethyleneamino- (**40**, R = OEt) or 3-dimethylaminomethyleneamino-2-pyrazinecarbonitrile (**40**, R = NMe_2), by ammonia under gentle conditions, into 4-pteridinamine (**39**) in 81 and 96%

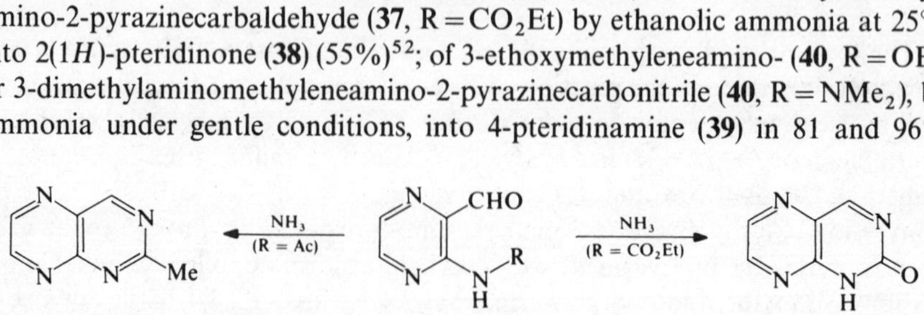

(36) (37) (38)

yield, respectively[53, cf. 1003]; of the ethoxy substrate (**40**, R=OEt) by ethanolic methylamine at 0°C into 3-methyl-4(3*H*)-pteridinimine (**41**) (65%)[53]; of the dimethylamino substrate (**40**, R=NMe_2), by methanolic methylamine acetate under reflux, into 4-methylaminopteridine (**44**) (87%), presumably formed by a Dimroth rearrangement of the imine (**41**) under the conditions used[53]; and of the same substrate (**40**, R=NMe_2), by refluxing methanolic hydroxylamine hydrochloride, into 4-pteridinamine 3-oxide (**43**) (80%),[631, 999, 1003] possibly via the adduct (**42**).

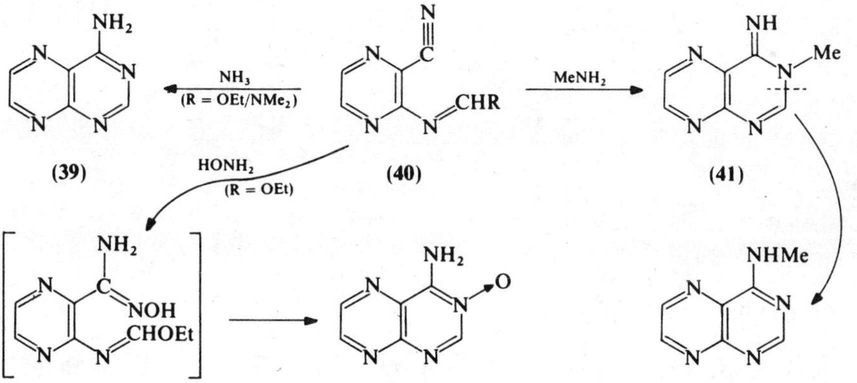

(**39**) (**40**) (**41**)

(**42**) (**43**) (**44**)

C. By Addition of C-2+N-3 as in Skeletal Formula (3)

This procedure for preparing pteridines by the addition of C-2+N-3 to appropriate pyrazines has been widely used to make 2,4-pteridinediamines with potential diuretic activity. It has been employed to make a very limited number of other pteridines. Thus the formation of *simple pteridines* is represented by the condensation of 3-amino-2-pyrazinecarbonitrile (**45**) with formamidine acetate in refluxing pentanol to give 4-pteridinamine (**46**) in 60% yield[53]; also, by the reaction of 3-benzoyl-5,6-diphenyl-2-pyrazinamine with tagged urea at 200°C to give (3-^{15}N)-4,6,7-triphenyl-2(1*H*)-pteridinone (**47**) in 65% yield.[1120] Significant failures have been reported in attempts to condense 3-hydroxymethyl-2-pyrazinamine with urea, urethane, or cyanic acid[42]; 3-amino-2-pyrazine-

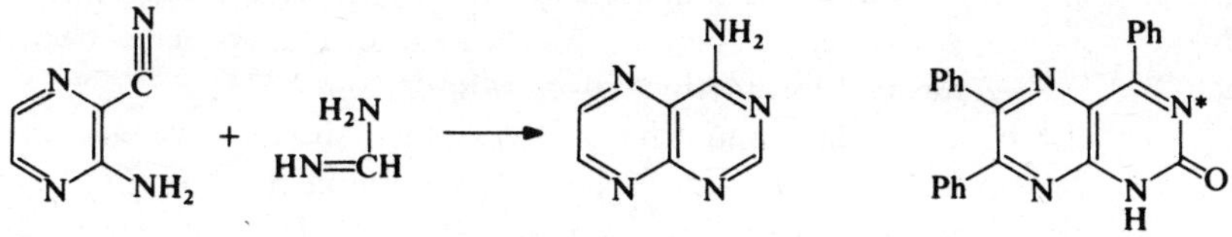

(**45**) (**46**) (**47**)

carboxylic acid with formamide[30]; 3-amino-2-pyrazinecarbonitrile or 3-amino-2-pyrazinecarboxamide with formamide[30]; the same carboxamide with formaldehyde[30]; and others.[30] However, formamide containing formic acid proved successful in converting 3-amino-*N*-benzyl(or butyl)-5,6-diphenyl-2-pyrazinecarboxamide into 6,7-diphenyl-4(3*H*)-pteridinone (60, 52%),[474] and benzamidine converted 3-amino-5,6-tetramethylene-2-pyrazinecarbonitrile into 2-phenyl-6,7-tetramethylene-4-pteridineamine (34%).[1408]

The formation of 2,4-*pteridinediamines* in this way is illustrated by that of 7-methyl- (**49**) [from 3-amino-5-methyl-2-pyrazinecarbonitrile (**48**) with guanidine

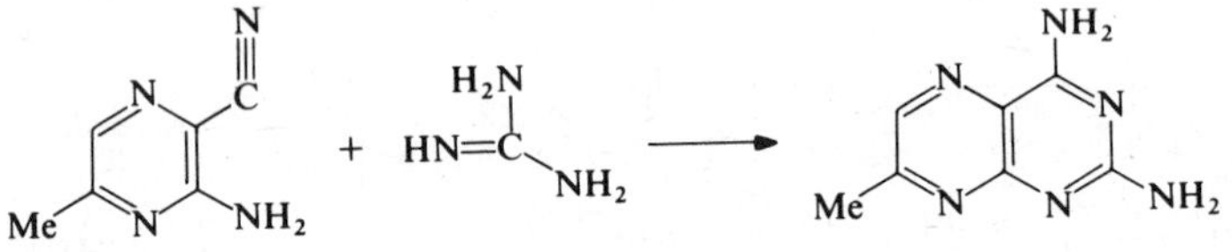

(**48**) (**49**)

* ^{15}N atom.

in methanolic sodium methoxide under reflux: 81%],[1305] 6-styryl- (95%),[484] 7-styryl- (89%),[1305] 6-methyl- (40–80%[484, 729]; partially deuterated: 73% yield[84]), 6-chloro- (from 3-amino-5-chloro-2-pyrazinecarbonitrile with methanolic guanidine under reflux for 30 min: 75%),[729] 6-methoxy- (from the same chloropyrazine with guanidine in an excess of methanolic sodium methoxide under reflux for 5 h: 62%),[1301] 6-bromo- (as its chloro analog just cited: 50%),[729] 7-methoxy- (as its 6-isomer just cited: 74%),[1301] 6-dimethoxymethyl- (84%),[1306] and 7-dimethoxymethyl-2,4-pteridinediamine (71%)[1310]; also in the formation of [2-^{13}C]-6-hydroxymethyl- (from 6-acetoxymethyl-3-amino-2-pyrazinecarbonitrile and guanidine in methanolic sodium methoxide: 85%; also the untagged analog: 84%),[761, 1299] 6-vinyl- (63%),[1297] 6-phenylethynyl- (88%),[1753] 6-hex-1′-enyl- (85%),[1753] 6-methoxymethyl- (85%),[1297, 1472, 1728] 6-[DL-*threo*]-α,β-dihydroxypropyl-,[623] 6-hydroxymethyl-7-methoxy- (from 6-acetoxymethyl-3-amino-5-chloro-2-pyrazinecarbonitrile and guanidine in methanolic sodium methoxide; note two incidental reactions: yield 54%),[1296] 6-γ-hydroxybutyl- (81%),[1302] 6-β-ethoxycarbonylethyl- (52%),[1302] 6-naphth-2′-ylthiomethyl- (54%),[1353] 6-*p*-chlorophenoxymethyl- (40%)[1353]; 6-*p*-ethoxycarbonylphenylthiomethyl- (67%),[261] 6-(*N*-ethyl-*p*-methoxyanilino)methyl- (33%),[1355] 6-benzylthiomethyl- (85%),[1299] and 6,7-tetramethylene-2,4-pteridinediamine (72%)[1408]; also 7-α-methoxyethyl-6-methylthio-2,4-thieno[3,2-*g*]-pteridinediamine (from 3-amino-6-α-methoxyethyl-7-methylthio-2-thieno[2,3-*b*]-pyrazinecarbonitrile and guanidine).[624] Methotrexate analogs have been so made,[1533, 1724] as has 6-chloromethyl-2,4-pteridinediamine.[1757]

Numerous 8-oxides of these and other 2,4-pteridinediamines have been made similarly. They include, for example, 2,4-pteridinediamine 8-oxide (**51**, R = H) [from 3-amino-2-pyrazinecarbonitrile 4-oxide (**50**, R = H), and methanolic guanidine under reflux for 16 h: 90%][477, 484]; its 6-methyl (**51**, R = Me)

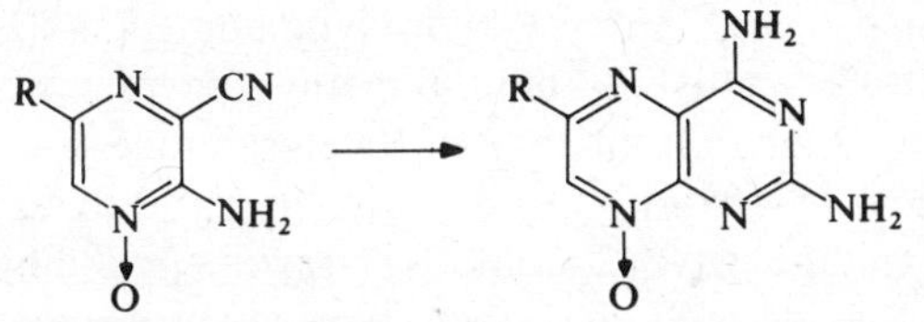

(50) (51)

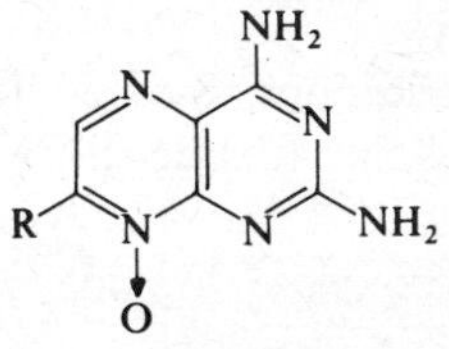

(52)

(81%), 6-ethyl (**51**, R = Et) (70%), 6-propyl (**51**, R = Pr) (86%), and other such derivatives[478, 484]; 6-methoxymethyl-2,4-pteridinediamine 8-oxide (81%)[1297, 1728]; 6-hydroxyiminomethyl-2,4-pteridinediamine 8-oxide (**51**, ·R = CH : NOH)[469]; 7-methyl-2,4-pteridinediamine 8-oxide (**52**, R = Me) (from 3-amino-5-methyl-2-pyrazinecarbonitrile 4-oxide with guanidine: 83%)[1305]; 6-naphth-2′-ylthiomethyl-2,4-pteridinediamine 8-oxide (and some analogs)[1353]; 6-(3′,4′-dichloroanilinomethyl)-2,4-pteridinediamine 8-oxide (39%) (and several analogs)[1355]; and 2,4-diamino-6-methyl-7(3*H*)-pteridinone 8-oxide (**54**) [from 3-amino-4-hydroxy-6-methyl-5-oxo-4,5-dihydro-2-pyrazinecarbonitrile (**53**) in refluxing methanolic guanidine followed by evaporation and heating in DMF to complete cyclization: 23%].[1298] More complicated examples have been reported.[1533, 1724, 1728]

There are a few examples of the formation of 2-*amino*-4-*pteridinones* and 2,4-*pteridinediones* in this way, nearly all of them 8-oxides. Thus 3-amino-5,6-diphenyl-2-pyrazinecarboxamide (**55**, R = H) or its *N*-benzyl derivative (**55**, R = CH_2Ph) with phenyl isocyanate in refluxing pyridine for 3 days gave the same product, 3,6,7-triphenyl-2,4(1*H*,3*H*)-pteridinedione (**56**, X = O) (thereby proving that both C-2 and N-3 were supplied by the reagent rather than the pyrazine)[476]; the 2-thio analog, 3,6,7-triphenyl-2-thioxo-1,2-dihydro-4(3*H*)-pteridinone (**56**, X = S) was made similarly from the pyrazine (**55**, R = H) and phenyl isothiocyanate[476]; ethyl 3-amino-6-hydroxyaminomethyl-2-pyrazinecarb-

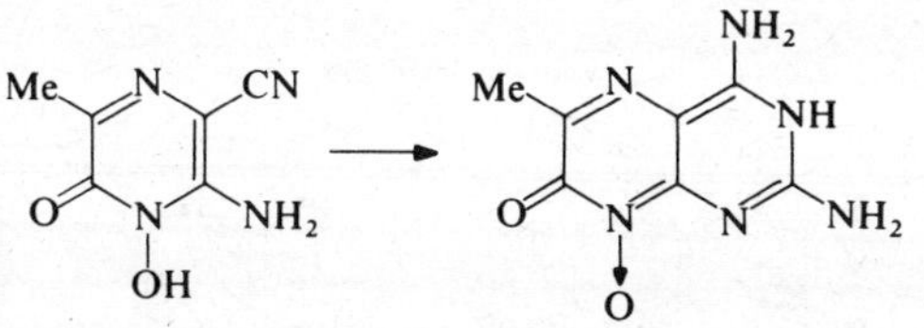

(53) (54)

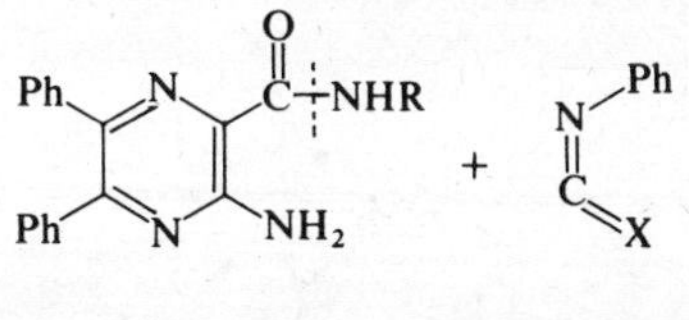

(55)

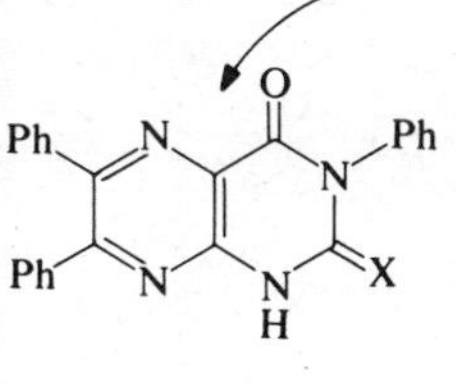

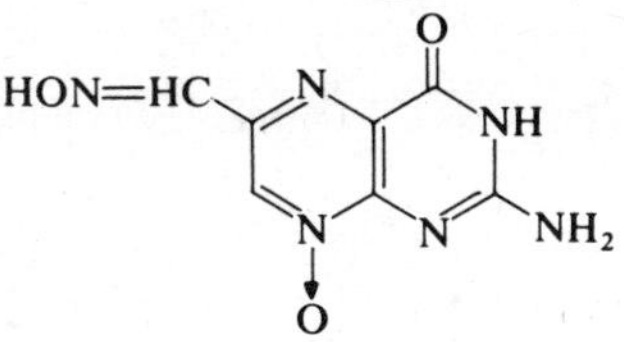

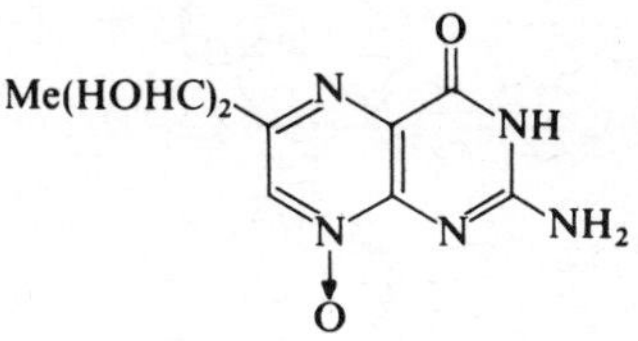

(56) (57) (58)

oxylate 4-oxide with guanidine gave 2-amino-6-hydroxyiminomethyl-4(3*H*)-pteridinone 8-oxide (**57**) (72%)[479]; ethyl 3-amino-6-methyl-2-pyrazinecarboxylate 4-oxide likewise gave 2-amino-6-methyl-4(3*H*)-pteridinone 8-oxide (70%)[478, 485]; benzyl 3-amino-6-[L-*erythro*]-α,β-dihydroxypropyl-2-pyrazinecarboxylate 4-oxide with guanidine gave 2-amino-6-[L-*erythro*]-α,β-dihydroxypropyl-4(3*H*)-pteridinone 8-oxide (**58**) and then by reductive removal of the oxide, [L-*erythro*]-biopterin[1285, 1318]; and the [8a-^{13}C] version of (**58**) was prepared similarly.[772] Treatment of 3-methylamino-2-pyrazinecarbonitrile with sodium hydride followed by methyl isocyanate has been reported to give 4-imino-1,3-dimethyl-3,4-dihydro-2(1*H*)-pteridinone (75% yield of free base).[1733]

D. By Addition of N-1 + C-2 + N-3 as in Skeletal Formula (4)

This synthesis has been little used although intermediate pyrazines are quite easily obtained and yields appear to be good. Thus methyl 3-chloro-2-pyrazinecarboxylate (**59**, R = Y = H) and guanidine carbonate were mixed and heated at 170°C for 30 min to give 2-amino-4(3*H*)-pteridinone (**60**, R = Y = H) in 89% yield[175]; similar treatment of the pyrazines (**59**, R = Me, Y = H), (**59**, R = Ph, Y = H), or (**59**, R = Y = Ph) gave 2-amino-7-methyl- (**60**, R = Me, Y = H) (80%),[175] 2-amino-7-phenyl- (**60**, R = Ph, Y = H) (88%),[176] or 2-amino-6,7-diphenyl-4(3*H*)-pteridinone (**60**, R = Y = Ph) (70%),[175] respectively. When methanolic guanidine was used for the preparation of (**60**, R = Y = H), the yield was <20%.[175] A subtly different approach is illustrated by the fusion of 3-chloro-5,6-diphenyl-2-pyrazinecarbonitrile (**61**) with guanidine carbonate to give 6,7-diphenyl-2,4-pterdinediamine (65%); with urea to give 4-amino-6,7-diphenyl-2(1*H*)-pteridinone (**62**, X = O) (59%); or with thiourea to give 4-amino-6,7-diphenyl-2(1*H*)-pteridinethione (**62**, X = S) (51%).[483]

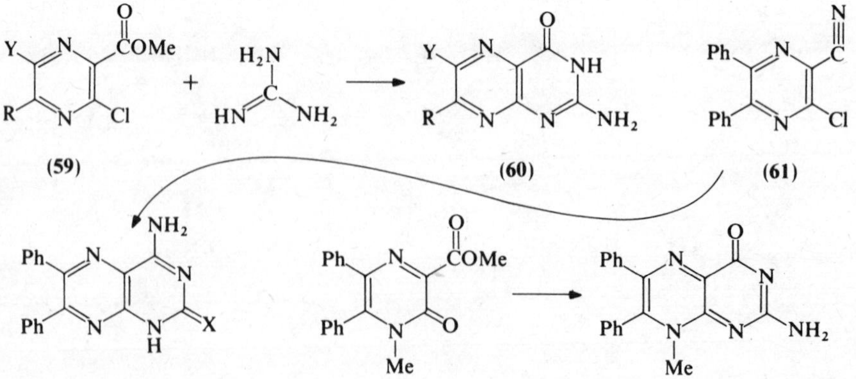

(**59**) (**60**) (**61**)

(**62**) (**63**) (**64**)

Although somewhat unexpected, fusion of methyl 4-methyl-3-oxo-5,6-diphenyl-3,4-dihydro-2-pyrazinecarboxylate (**63**) with guanidine carbonate afforded 2-amino-8-methyl-6,7-diphenyl-4(8*H*)-pteridinone (**64**) in 93% yield; the reaction was unsuccessful in the absence of phenyl substituents.[737]

E. By Completion of the 2,3-Bond as in Skeletal Formula (5)

In this category, all the ring atoms needed for the complete pteridine are already in place on the intermediate pyrazine and only cyclization is required. Thus 3-acetamido-2-pyrazinecarboxamide (**65**, R = Me, Y = H) in 1*M* alkali at 22°C gave, on acidification, 2-methyl-4(3*H*)-pteridinone (**66**, R = Me, Y = H) (81%). Appropriately substituted pyrazines (**65**) likewise gave 2-phenyl- (**66**, R = Ph, Y = H) (91%), 6-bromo-2-methyl- (**66**, R = Me, Y = Br) (83%), and 6-bromo-2-phenyl-4(3*H*)-pteridinone (**66**, R = Ph, Y = Br) (80%).[20] In a rather similar way, 3-formamido-5,*N*-dimethyl-2-pyrazinecarboxamide (**67**) in hot

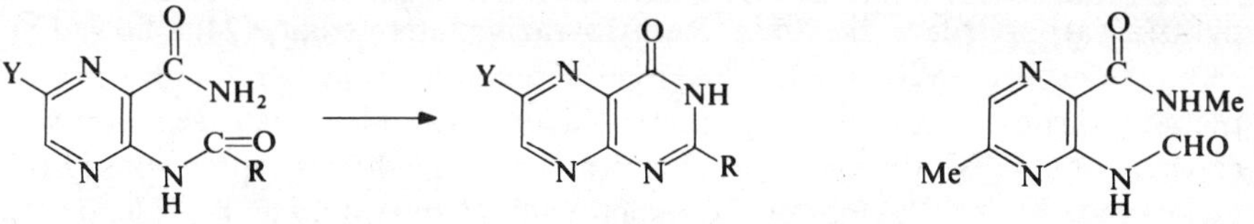

(65) **(66)** **(67)**

aqueous pyridine for 8 h, or in 5% sodium bicarbonate under reflux for 5 min, gave 3,7-dimethyl-4(3*H*)-pteridinone in 44 or 59% yield, respectively[167]; and 3-propionamido-2-pyrazinecarboxamide (**65**, R = Et, Y = H) or the corresponding ester, ethyl 3-propionamido-2-pyrazinecarboxylate, in an excess of ammonia for 24 h gave 2-ethyl-4(3*H*)-pteridinone (**66**, R = Et, Y = H) in 70% yield.[980] Such cyclizations can also be done in ways other than dissolution in alkaline media. For example, 3-acetamido-2-pyrazinecarboxamide (**65**, R = Me, Y = H) underwent thermal cyclization at 230°C during 5 min to give 2-methyl-4(3*H*)-pteridinone (**66**, R = Me, Y = H) in 52% yield[1004]; 3-acetamido- (**68**, R = Me) or 3-butyramido-2-pyrazinecarbohydrazide (**68**, R = Pr) in refluxing propyl alcohol for 2 h gave 3-amino-2-methyl- (**69**, R = Me) (87%) or 3-amino-2-propyl-4(3*H*)-pteridinone (**69**, R = Pr) (91%), respectively[763]; 3-benzamido-2-pyrazinecarbohydrazide (**68**, R = Ph) in refluxing butanolic triethylamine for 10 h gave 3-amino-2-phenyl-4(3*H*)-pteridinone (**69**, R = Ph) (53%)[763]; and 3-acetamide-*N*-

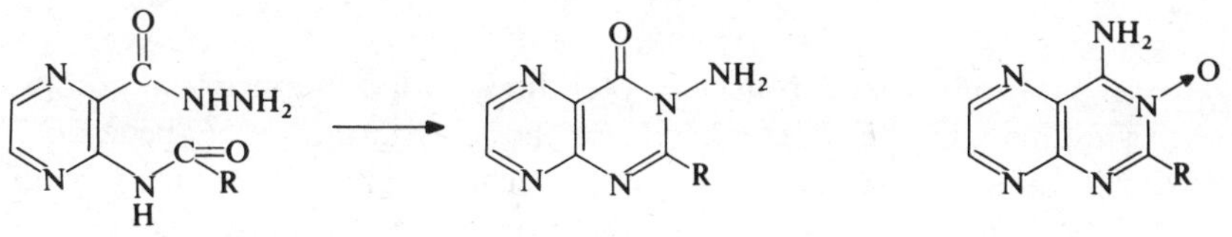

(68) **(69)** **(70)**

benzyl-2-pyrazinecarboxamide in phosphoryl chloride (but not in alkali or by thermal means) gave 3-benzyl-2-methyl-4(3*H*)-pteridinone.[980]

Some related cyclizations are illustrated by the formation of 2-methyl-4-pteridinamine 3-oxide (**70**, R = Me) from 3-acetamido-2-pyrazinecarboxamide oxime in refluxing glacial acetic acid (79% yield) or in polyphosphoric acid at 80°C (40% yield)[999]; of 2-phenyl-4-pteridinamine 3-oxide from 3-benzamido-2-pyrazinecarboxamide oxime in sulfuric acid at 70°C (43% yield) or in polyphosphoric acid at 100°C (36% yield)[631, 1004]; of 4-pteridinamine 3-oxide (**70**, R = H) from 3-formamido-2-pyrazinecarboxamide dioxime (**71**)[1017]; of 4(3*H*)-pteridinethione (25%) from 3-dimethylaminomethyleneamino-2-pyrazinecarbonitrile by treatment with sodium hydrogen sulfide to give the intermediate thioamide (**72**) followed by spontaneous cyclization under the alkaline conditions pertaining[53]; of 3-methoxy-4(3*H*)-pteridinone (74%) from methyl 3-methoxyaminomethyleneamino-2-pyrazinecarbohydroxamate (**73**) in pH 4 buffer at 80°C for 3 h[759]; of its 6-methyl, 7-methyl and 6,7-dimethyl homologs likewise from appropriately 5- and/or 6-methylated pyrazines[759]; of ethyl 3,4-dihydro-2-pteridinecarboxylate (**75**) by hydrogenation of 3-ethoxalylamino-2-pyrazinecarbonitrile to the aminomethylpyrazine intermediate (**74**) followed by spontaneous cyclization (yield 17%) (and, if required, aromatization to ethyl 2-pteridinecarboxylate with manganese dioxide) (*note*: When hydrogenation was carried out at 70°C, the initial product was ethyl 1,2,3,4-tetrahydro-2-pteridinecarboxylate)[51, 54, 585]; and of 2,6-dimethyl-4(3*H*)-pteridinone 8-oxide (80%) and the 5,8-dioxide (50%) from 3-acetamido-6-methyl-2-pyrazinecarbonitrile 4-oxide or 1,4-dioxide, respectively, in aqueous alkali at 100°C for 1 min.[484]

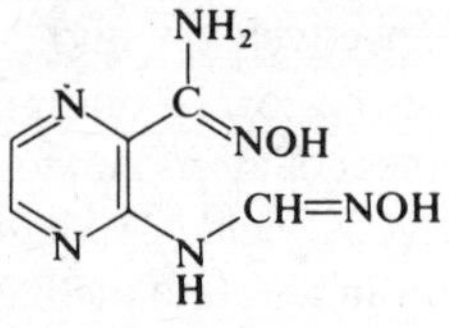

(71)

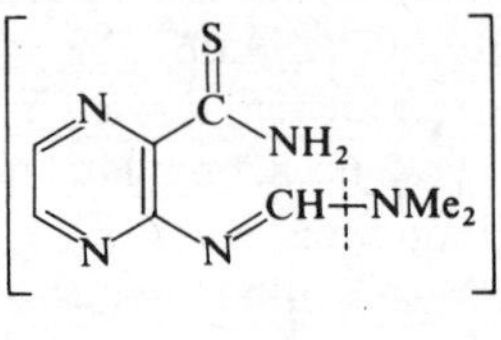

(72)

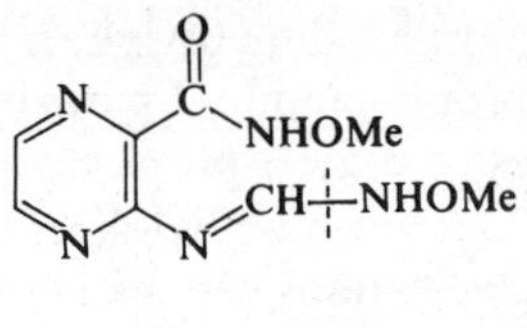

(73)

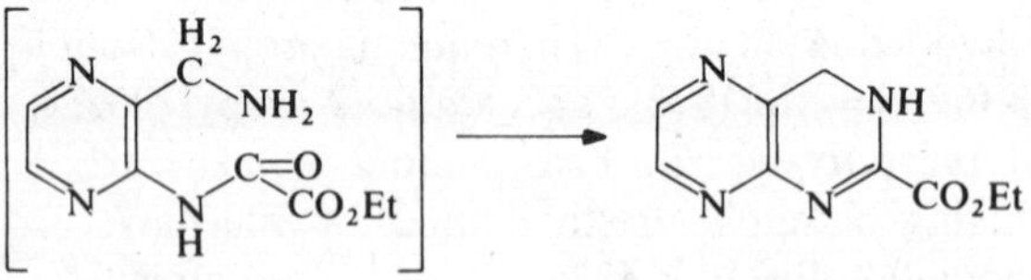

(74) **(75)**

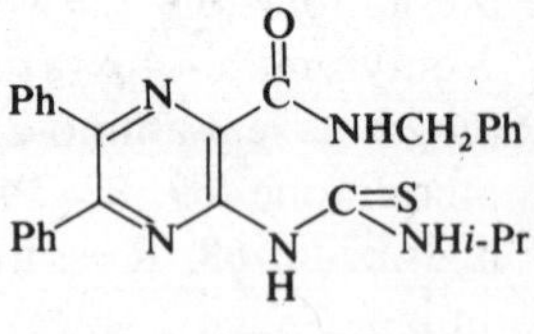

(76)

The preparation by such routes, or pteridines with two functional groups in the pyrimidine ring, is exemplified in the cyclization of *N*-benzyl-3-*N'*-isopropyl(thioureido)-5,6-diphenyl-2-pyrazinecarboxamide (**76**) in ethanolic sodium ethoxide to give 3-benzyl-2-isopropylamino-6,7-diphenyl-4(3*H*)-pteridinone (83%)[476]; in the treatment of 2,3-pyrazinedicarboxamide (**77**) with

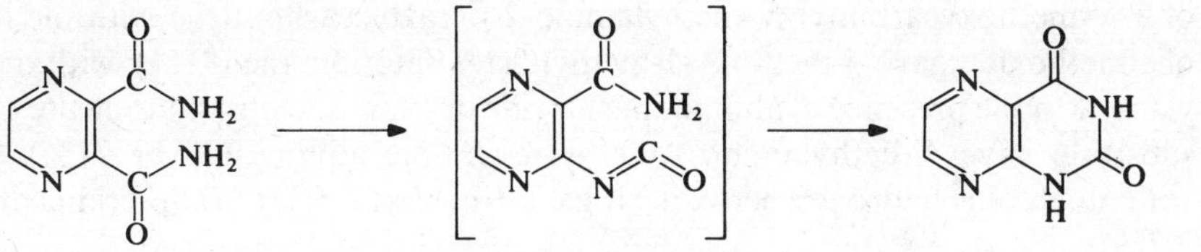

(77) (78) (79)

hypobromite under controlled conditions to give the rearranged intermediate (**78**) and then 2,4(1*H*,3*H*)-pteridinedione (**79**) in 40% yield[902]; in the cyclization of 3-(*N*-ethoxycarbonyl-*N*-methylamino)-*N*-methyl-5-oxo-4,5-dihydro-2-pyrazinecarboxamide (**80**) to give 1,3-dimethyl-2,4,7(1*H*,3*H*,8*H*)-pteridinetrione (~50%)[364]; in the cyclization of 3-ethoxycarbonylamino-*N*-methyl-5,6-diphenyl-2-pyrazinecarboxamide in ethanolic sodium ethoxide to give 3-methyl-6,7-diphenyl-2,4(1*H*,3*H*)-pteridinedione (~55%)[111]; in the treatment of 3-*N*-ethoxycarbonyl-*N*-methylamino-5-phenyl-2-pyrazinecarbonitrile in refluxing ethanolic sodium ethoxide to give 1-methyl-7-phenyl-2,4(1*H*,3*H*)-pteridinedione (34%)[174]; in the cyclization of *N*-benzyl or *N*-butyl-3-ethoxycarbonylamino-5,6-diphenyl-2-pyrazinecarboxamide in refluxing ethanolic sodium ethoxide to give 3-benzyl- (41%) or 3-butyl-6,7-diphenyl-2,4(1*H*,3*H*)-pteridinedione (89%), respectively[474]; in the conversion of 3-ethoxycarbonylamino-*N*,*N*-pentamethylene-5,6-diphenyl-2-pyrazinecarboxamide by ethanolic ammonia at 155°C into 6,7-diphenyl-2,4(1*H*,3*H*)-pteridinedione (90%)[474]; in the cyclization of 3-ethoxycarbonylamino-6-methyl- or 3-ethoxycarbonylamino-6-phenyl-2-pyrazinecarboxamide 4-oxide, by boiling in methanolic sodium methoxide, to give 6-methyl- or 6-phenyl-2,4(1*H*,3*H*)-pteridinedione 8-oxide in 86 and 90% yield, respectively[485]; in the treatment of disodium 2,3-pyrazinedicarbohydroxamate with benzenesulfonyl chloride in THF to give (via an isocyanate intermediate) 3-benzenesulfonyloxy- (**81**, R = SO_2Ph) and then in alkali, 3-hydroxy-2,4(1*H*,3*H*)-pteridinedione (**81**, R = OH) in good yield overall[1445]; in the treatment of 3-ethoxycarbonylamino-2-quinoxalinecarboxamide with ethanolic potassium hydroxide or (better) ethanolic sodium ethoxide to give 2,4(1*H*,3*H*)-benzo[*g*]pteridinedione (alloxazine) in ~60% yield[914]; and in the remarkable conversion of 6,7-diphenyl-2,4(1*H*,3*H*)-pteridinedione by hydrazine into (unisolated) 5,6-diphenyl-3-ureido-2-pyrazinecarbohydrazide (**82**) and then (in part) by loss of ammonia into 3-amino-6,7-diphenyl-2,4(1*H*,3*H*)-pteridinedione.[468] Treatment

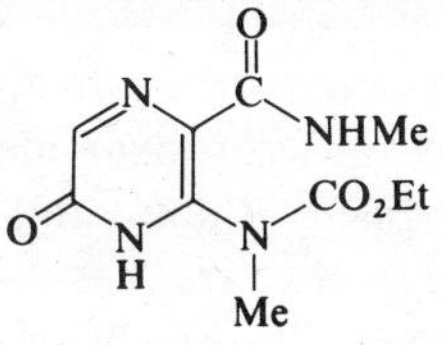

(80)

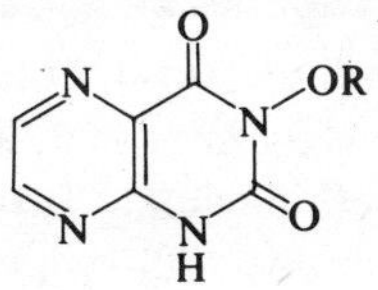

(81)

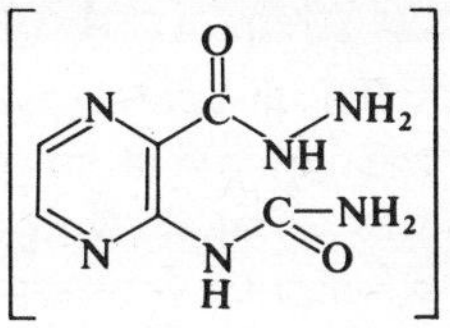

(82)

of 3-*N*-methoxycarbonyl-*N*-methylamino-2-pyrazinecarbonitrile with methanolic methoxide gave 4-methoxy-1-methyl-2(1*H*)-pteridinone (81%); with diethylamine in the presence of aluminum chloride or titanium tetrachloride the same substrate gave 4-diethylamino-1-methyl-2(1*H*)-pteridinone (98 or 77%); and with alkali plus hydrogen peroxide, it gave 1-methyl-2,4(1*H*,3*H*)-pteridinedione (67%).[1733]

F. By Completion of the 1,2-Bond as in Skeletal Formula (6)

This cyclization of pyrazines, to which all the atoms required to produce a pteridine ring are attached, is represented by only two examples. 3-Guanidinomethyl-2-pyrazinamine (**83**) was liberated from its hydrochloride and, after evaporation of solvent, was heated at 100°C to give 3,4-dihydro-2-pteridinamine (**84**), isolated initially as its *p*-toluenesulfonate salt (58%). The base was then liberated and oxidized by manganese dioxide to afford 2-pteridinamine (**85**) in 74% yield.[51, 585] Similarly, 3-ethoxycarbonylaminomethyl-2-pyrazinamine (**86**) in refluxing ethanolic sodium ethoxide gave 3,4-dihydro-2(1*H*)-pteridinone (**87**) (68%), which underwent ferricyanide oxidation to afford 2(1*H*)-pteridinone (**88**), isolated as its covalent hydrate in 78% yields.[51, 54]

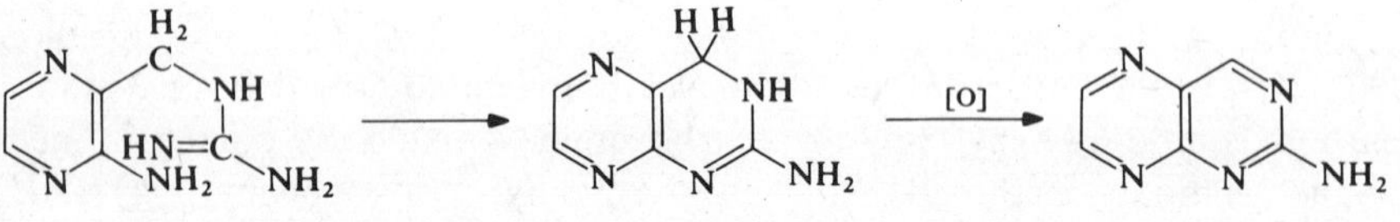

(83) **(84)** **(85)**

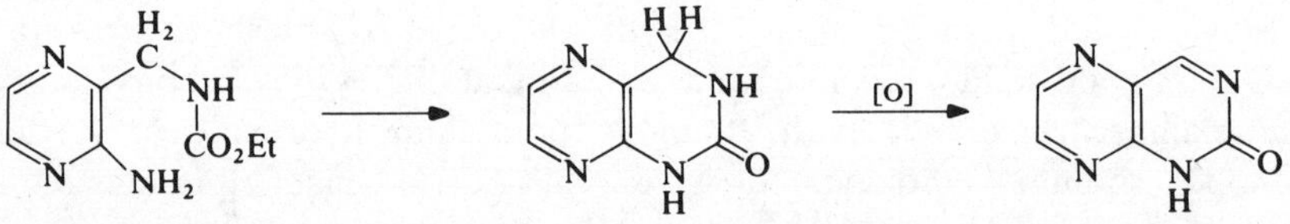

(86) **(87)** **(88)**

G. By Completion of the 3,4-Bond as in Skeletal Formula (7)

The last category to be represented at present also employs a pyrazine with all the required atoms in place. Thus 3-amino-5,6-diphenyl-2-pyrazinecarboxamide was treated with phenyl isothiocyanate to give 3-*N*′-phenyl(thioureido)-5,6-diphenyl-2-pyrazinecarboxamide (**89**), which was characterized and subsequently cyclized in boiling pyridine during 3 days to furnish 3,6,7-triphenyl-2-thioxo-1,2-dihydro-4(3*H*)-pteridinone (**90**).[476] The second example is the

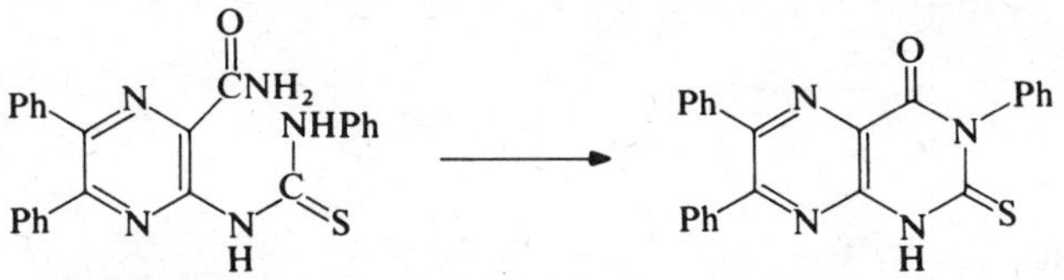

(89) (90)

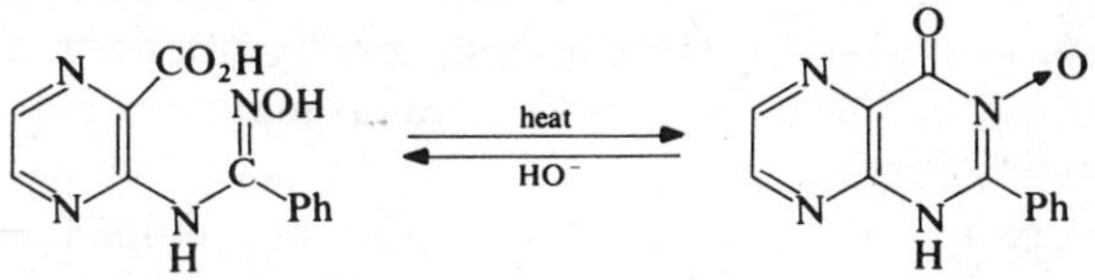

(91) (92)

thermal cyclization at 200°C of 3-α-(hydroxyimino)benzylamino-2-pyrazinecarboxylic acid (**91**) to give 2-phenyl-4(1*H*)-pteridinone 3-oxide (**92**) in 65% yield; treatment with hot alkali caused reversion to the starting material (**91**) (61%).[1004]

2. SYNTHESES FROM PURINES

It seems likely that the first synthesis of a pteridine from a purine was carried out prior to 1857, long before the structure of the pteridine system was known. Thus the great Friedrich Wöhler heated uric acid and water in a sealed tube at 140°C for 2 weeks and reported[1509] the formation of small amounts of a new yellow substance; similar results were published in the same year by H. Hlasiwetz.[1510] Nearly 40 years later Frederick Hopkins repeated much of this work at a higher temperature and concluded that the yellow substance was closely akin to one of the (pteridine) wing pigments of the sulfur-yellow *Pieridae* (butterflies)[2, 5, 6]; after a further 45 years, he returned to this work and suggested (incorrectly) that this product contained xanthopterin,[794] which had meanwhile been obtained elsewhere and characterized by others. The matter was only finally decided by Wolfgang Pfleiderer in 1959. The result was that at least 12 substances arose from heating uric acid in water at 190–220°C.[345] The main products proved to be 2,4,7(1*H*,3*H*,8*H*)-pteridinetrione, its 6-methyl derivative, 2,4,6(1*H*,3*H*,5*H*)-pteridinetrione, 2,4,7-trioxo-1,2,3,4,7,8-hexahydro-6-pteridinecarboxylic acid, and two isomeric pyrimidopteridines.[345, 352]

Following a study of the instability of simple purines,[1511] it occurred to Adrien Albert that such purines might serve as propyrimidinediamines on incubation with α-dicarbonyl compounds in a virtual Gabriel and Colman synthesis of pteridines.[16, 1551] Thus 2(1*H*)-purinone (**93**) and aqueous glyoxal in dilute sulfuric acid at 37°C for 24 h gave 84% (52% isolated) of 2(1*H*)-pteridinone (**95**), presumably via 4-amino-5-formamido- (**94**, R = CHO) and

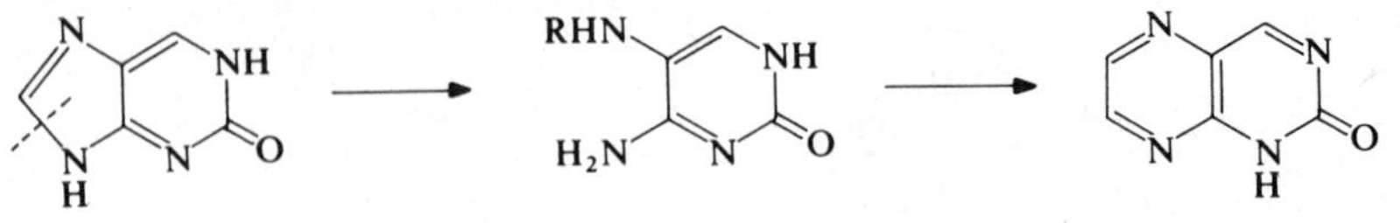

(93) **(94)** **(95)**

4,5-diamino-2(1*H*)-pyrimidinone (**94**, R = H)[14]; the same substrate (**93**) with glyoxylic acid or ethyl glyoxylate hemiacetal in dilute acid gave 2,6(1*H*,5*H*)-pteridinedione (70–85%)[14]; 6(1*H*)-purinone (hypoxanthine) and this hemiacetal in 1*M* hydrochloric acid at 120°C gave a minute yield of 4,6(3*H*,5*H*)-pteridinedione[14]; 2-amino-6(1*H*)-purinone (guanine) on similar treatment gave a separable mixture of 2-amino-4,6(3*H*,5*H*)-pteridinedione (xanthopterin) and 2,4,6(1*H*,3*H*,5*H*)-pteridinetrione in <1% yield[14]; and 9-methyl-2(1*H*)-purinone with biacetyl in dilute acid at 37°C gave 6,7,8-trimethyl-2(8*H*)-pteridinone (~7%).[14]

Quarternization of a purine at the 7–9-position destabilized the imidazole ring, so that the resulting pyrimidine can undergo recyclization, involving an appropriate 7–9 substituent, to afford a pteridine.[361] Thus, 2-amino-9-methyl-6(1*H*)-purinone reacted with ethyl bromoacetate to give 2-amino-7-ethoxycarbonylmethyl-9-methyl-6-oxo-1,6-dihydro-7-purinium bromide (**96**, R = CO_2Et), which underwent ring fission and ester saponification in alkali to give the intermediate, 2-amino-5-*N*-(carboxymethyl)formamido-6-methylamino-4(3*H*)-pyrimidinone (**97**, R = CO_2H) and then, after deformylation and aerial oxidation, 2-amino-8-methyl-4,7(3*H*,8*H*)-pteridinedione (**98**) in 63% yield.[181] Use of the corresponding nitrile (**96**, R = CN) gave the same product (**98**) on account of

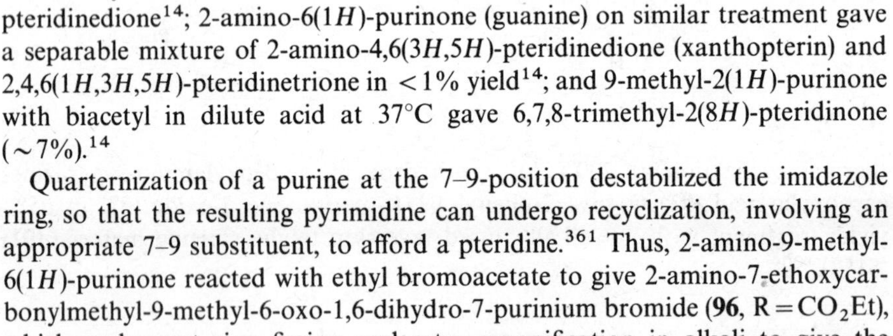

(96) **(97)** **(98)**

hydrolysis, perhaps of the 7-iminopteridine corresponding to (**98**).[181] In a subtly different way, 2-amino-7-ethoxycarbonylmethyl-6-oxo-9-(2′,3′,5′tri-*O*-acetyl-*β*-D-ribofuranosyl)-1,6-dihydro-7-purinium bromide afforded only 2-amino-4,7(3*H*,8*H*)-pteridinedione on similar treatment, presumably because of loss of the sugar during the acidic work-up procedure.[856] Likewise, 6-benzamido-9-benzyl-7-phenacyl-7-purinium bromide in alkali gave first 4-benzamido-6-benzylamino-5-*N*-phenacylformamidopyrimidine and subsequently 4-benzylamino-7-phenylpteridine (41%).[1100]

A patently exciting purine to pteridine transformation has been described.[1385] "When 8-azido-1,3,7-trimethyl-2,6(1*H*,3*H*)-purinedione was heated at 130°C in a covered Petri dish, the compound soon exploded to give a dark brown solid that was leeched with boiling ethanolic chloroform and the soluble material was chromatographed to yield 7-amino-1,3-dimethyl-2,4(1*H*,3*H*)-pteridinedione in 18–23% yield"; a mechanism was proposed.[1385] Irradiation of theophylline [1,3-dimethyl-2,6(1*H*,3*H*)-purinedione] with Rose Bengal in aqueous ethylenediamine gave an intermediate pyrimidine, which on acid treatment gave 1,3-dimethyl-2,4(1*H*,3*H*)pteridinedione by a rather complicated mechanism.[1017]

3. SYNTHESES FROM OTHER BI- AND TRICYCLIC SYSTEMS

The use of *oxadiazolopyrimidines* as substrates is exemplified in the conversion of 7-α-ethoxycarbonylmethylamino-5-[1,2,5]oxadiazolo[3,4-*d*] pyrimidinamine (**99**) by hydrogenation into 2,4-diamino-7,8-dihydro-6(5*H*)-pteridinone (**100**) and then by iodine oxidation to 2,4-diamino-6(5*H*)-pteridinone (~35% overall)[1293]; of 7-methoxy[1,2,5]oxadiazolo[3,4-*d*]pyrimidine 1-oxide (**101**, R = OMe) or 7-[1,2,5]oxadiazolo[3,4-*d*]pyrimidinamine 1-oxide (**101**, R = NH_2), by treatment with ethyl methyl ketone in methanolic ammonia at 25°C, into 6,7-dimethyl-4-pteridinamine 5,8-dioxide in 58% yield[723]; of the same substrates (**101**, R = OMe or NH_2), by α-acetyl-*N*,*N*-dimethylacetamide–methanolic ammonia, into 4-amino-6,*N*,*N*-trimethyl-7-pteridinecarboxamide 5,8-dioxide (30%)[723]; of 5,7-dimethoxy[1,2,5]oxadiazolo[3,4-*d*]pyrimidine 1-oxide or 5-methoxy-7-[1,2,5]oxadiazolo[3,4-*d*]pyrimidinamine 1-oxide, by α-acetyl-*N*,*N*-dimethylacetamide–methanolic ammonia under reflux, into 4-amino-2-methoxy-6,*N*,*N*-trimethyl-7-pteridinecarboxamide 5,8-dioxide (**103**) (81%)[723]; and of several like substrates into pteridines,[723] although the mechanisms remain in some doubt.

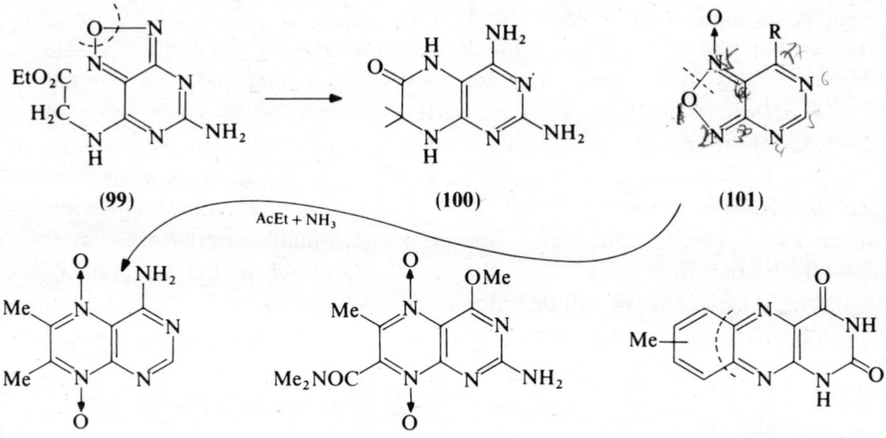

(**99**) (**100**) (**101**)

(**102**) (**103**) (**104**)

The conversion of a *pyrimido*[4,5-*b*] [1,4]*oxazine* into the corresponding dihydropteridine by ammonia has been postulated but appears to remain unproven.[1416]

Oxidative degradation of *tolualloxazin* (**104**) was carried out by Otto Kühling as early as 1894 to afford, first 2,4-dioxo-1,2,3,4-tetrahydro-6,7-pteridinedicarboxylic acid, and then, by decarboxylation at 320°C, 2,4(1*H*,3*H*)-pteridinedione.[7, 8] No other such syntheses appear to have been attempted.

The conversion of *heterotricyclic systems* into pteridines covers a wide area quite sparsely; in some cases the tricyclic substrate is itself made from a pteridine! Thus 2-amino-7-methyl-4,6(3*H*,5*H*)-pteridinedione and pyruvic acid afforded the lactone, 2-amino-8-methyl-7*H*-pyrano[2,3-*g*]pteridine-4(3*H*),7-dione (**105**), which underwent fission by methanolic hydrogen chloride to give 2-amino-7-(2′-methoxycarbonylprop-1′-enyl)-4,6(3*H*,5*H*)-pteridinedione (**106**)[1055]; 2-thioxo-1,2-dihydro-4(3*H*)-pteridinone with α,β-dibromoethane

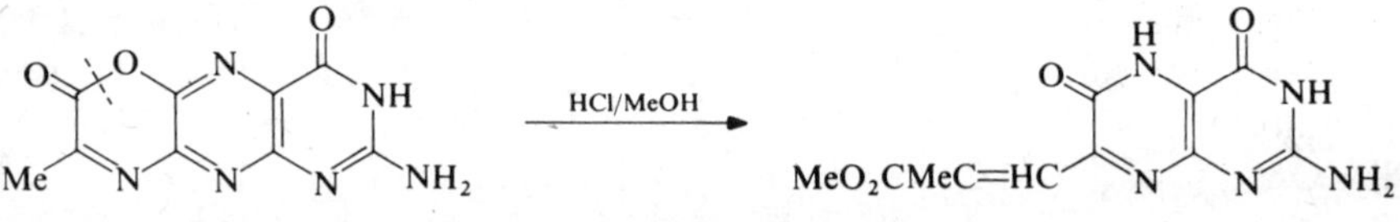

(105) **(106)**

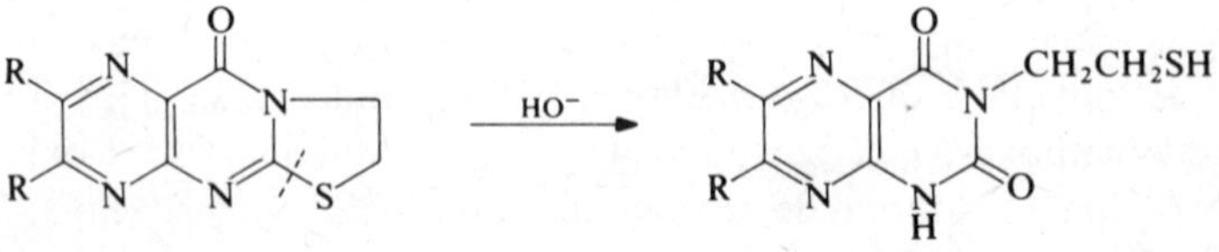

(107) **(108)**

gave 7,8-dihydro-10*H*-thiazolo[2,3-*b*]pteridin-10-one (**107**, R = H), which underwent hydrolytic fission in boiling alkali to afford 51% of 3-β-mercaptoethyl-2,4(1*H*,3*H*)-pteridinedione (**108**, R = H)[876]; similar reactions with obvious variations likewise gave 3-β-mercaptoethyl-6,7-dimethyl-2,4(1*H*,3*H*)-pteridinedione (**108**, R = Me) as well as 3-γ-mercaptopropyl-2,4(1*H*,3*H*)-pteridinedione and its 6,7-dimethyl derivative, all in > 80% yield {in the preparation of intermediate (**107**, R = Me), an isomeric by-product was isolated: It proved to be 2,3-dimethyl-8,9-dihydro-5*H*-thiazolo[3,2-*a*]pteridin-5-one (**109**), which underwent fission to 1-β-mercaptoethyl-6,7-dimethyl-2,4(1*H*,3*H*)-pteridinedione}[876]; 8-acetoxy-2-amino-7-methyl-4(3*H*)-furo[2,3-*g*]pteridinone (**110**) in dilute alkali gave 2-amino-7-α-hydroxypropionyl-4,6(3*H*,5*H*)-pteridinedione (64%)[496]; 3,5,7- trimethyl-2,6,8(3*H*,5*H*,7*H*)-[1,3]ox-

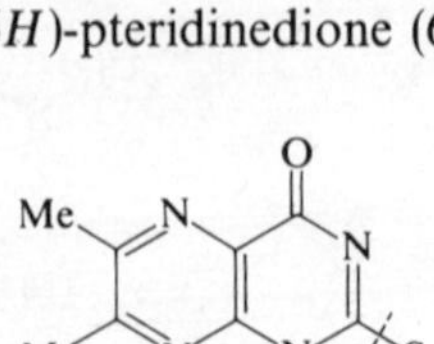

(109)

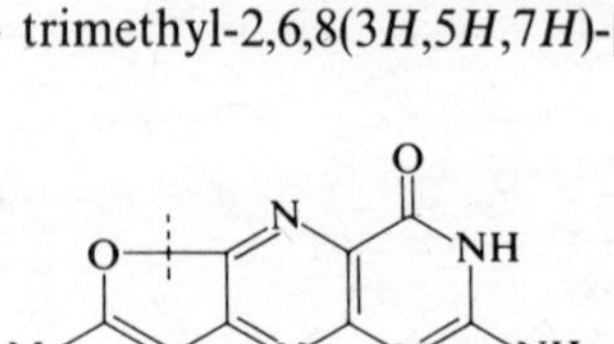

(110)

azolo[5,4-*g*]pteridinetrione (**111**) in refluxing aqueous ethanol gave 1,3-dimethyl-7-methylamino-2,4,6(1*H*,3*H*,5*H*)-pteridinetrione (~50%)[1294]; 7-amino-3-methyl-5(6*H*)-isothiazolo[4,5-*g*]pteridinone (**112**), on dithionite reduction, gave 6-acetyl-2-amino-7-thioxo-7,8-dihydro-4(3*H*)-pteridinone (**113**)

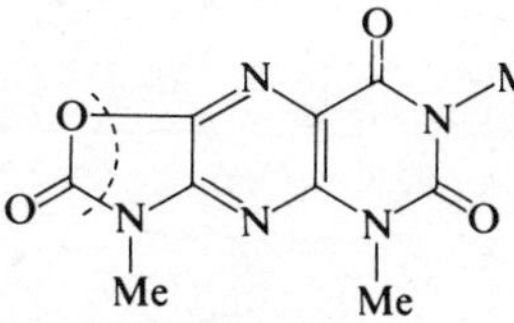

(**111**)

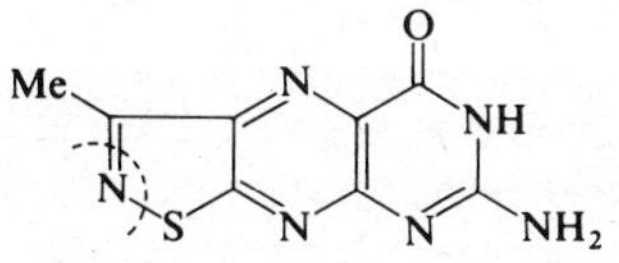

(**112**)

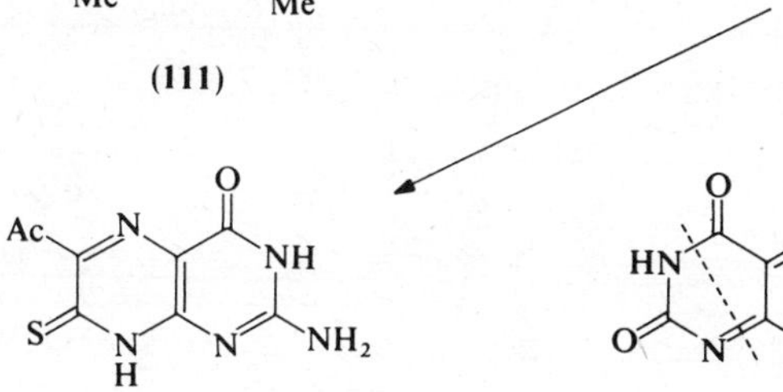

(**113**)

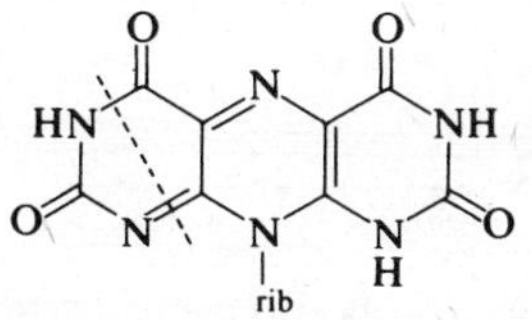

(**114**)

(~80%)[1309]; and 10-D-ribityl-2,4,6,8(1*H*,3*H*,7*H*,10*H*)- pyrimido[5,4-*g*]pteridinetetrone (**114**) underwent alkaline degradation to give 2,4,7-trioxo-8-D-ribityl-1,2,3,4,7,8-hexahydro-6-pteridinecarboxylic acid (60%).[674] A more complicated example involved treatment of 2-amino-6,7-dimethyl-4(3*H*)-pteridinone (**115**) with acrylonitrile in refluxing aqueous pyridine to give 2,3-dimethyl-8,9-dihydro-11*H*-pyrimido[2,1-*b*]pteridine-7(6*H*),11-dione (**116**),

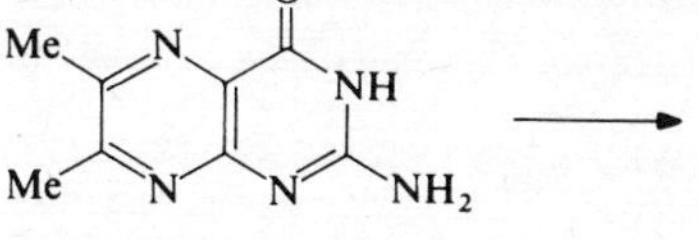

(**115**)

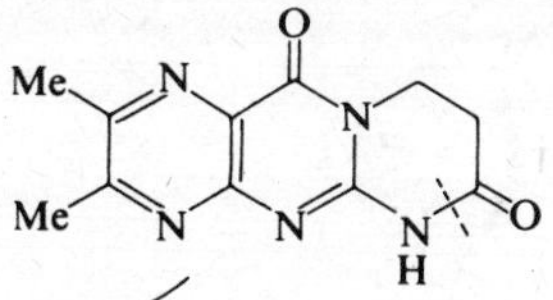

(**116**)

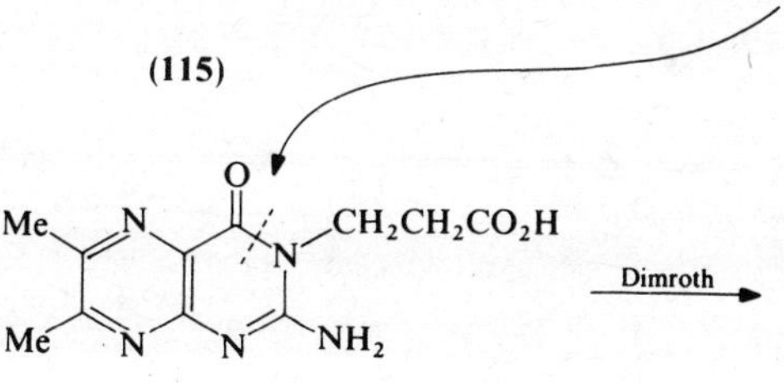

(**117**)

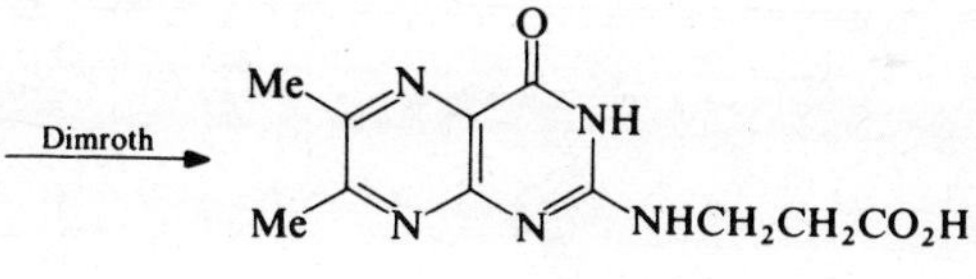

(**118**)

which in 0.1*M* sodium borate gave 2-amino-3-β-carboxyethyl-6,7-dimethyl-4(3*H*)-pteridinone (**117**) and then by Dimroth rearrangement in dilute alkali, 2-β-carboxyethylamino-6,7-dimethyl-4(3*H*)-pteridinone (**118**).[654] Appropriate variations gave several analogs.[654]

CHAPTER IV

Pteridine and Its Alkyl and Aryl Derivatives

Although some naturally occurring and related pteridines have been known for a century, the first monosubstituted pteridine was prepared by G. B. (Trudy) Elion and George Hitchings[878] only in 1947 and pteridine itself by William G. M. Jones[250] the following year.

1. PTERIDINE

Pteridine undergoes few reactions, but it has been studied widely by theochemists because it is an unsymmetrical tetraazanaphthalene and the parent of derivatives with diverse biological roles.

A. Preparation of Pteridine

All the reported procedures for making pteridine (**2**) directly are based on that of Jones[250] who employed the condensation of 4,5-pyrimidinediamine (**1**) with aqueous glyoxal bisulfite and obtained $<15\%$ yield.[30] The use of aqueous polyglyoxal at 25°C for 1 h gave $\sim 25\%$[148]; aqueous alcoholic glyoxal gave 50%[30]; alcoholic polyglyoxal under reflux for 30 min gave 63%[30]; and alcoholic polyglyoxal hydrate under reflux, followed by evaporation in a vacuum at 30°C and subsequent sublimation, gave 93% of virtually pure pteridine.[65] 2-Deutero-4,5-pyrimidinediamine with polyglyoxal gave 2-deuteropteridine.[294]

An indirect route to pteridine involved reduction of 3-amino-2-pyrazinecarbonitrile (**3**) to 3-aminomethyl-2-pyrazinamine (**4**) (81%) followed by cyclization with triethyl orthoformate to give 3,4-dihydropteridine (**5**) (74%) and a final oxidation by manganese dioxide to afford pteridine (**2**) in 52% yield.[51]

5,6,7,8-Tetrahydropteridine (**8**) was first made by Peter Brook and George Ramage[683] who dechlorinated and debenzylated 8-benzyl-2-chloro-5,6,7,8-tetrahydropteridine (**6**) by treatment with sodium–liquid ammonia to give the

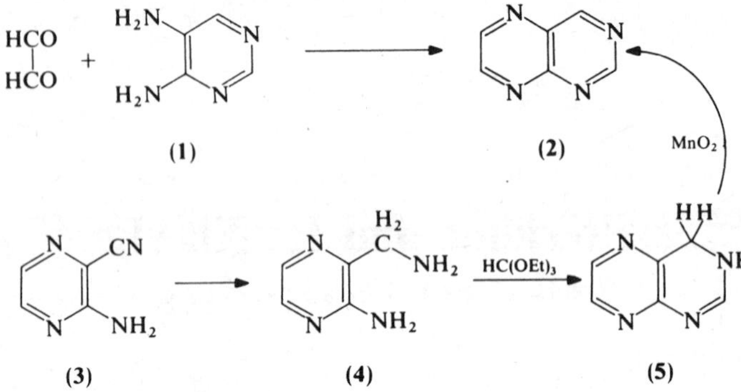

product (**8**) in ~25% yield. When the reduction time was shortened, the intermediate 8-benzyl-5,6,7,8-tetrahydropteridine (**7**) was isolated.[683] A similar procedure reduced 8-benzyl-4-chloro-5,6,7,8-tetrahydropteridine (**9**) to 5,6,7,8-tetrahydropteridine (**8**) in 14% yield.[43] The best way to make the tetrahydropteridine (**8**) is by hydrogenolysis of its 2,4-dichloro derivative (**10**) over palladium (60% yield) or by treatment of pteridine (**2**) with lithium aluminum hydride (58% yield).[486] However, all efforts to oxidize the tetrahydropteridine (**8**) to pteridine have failed[486]; the isomeric 1,2,3,4-tetrahydropteridine also proved remarkably stable.[51]

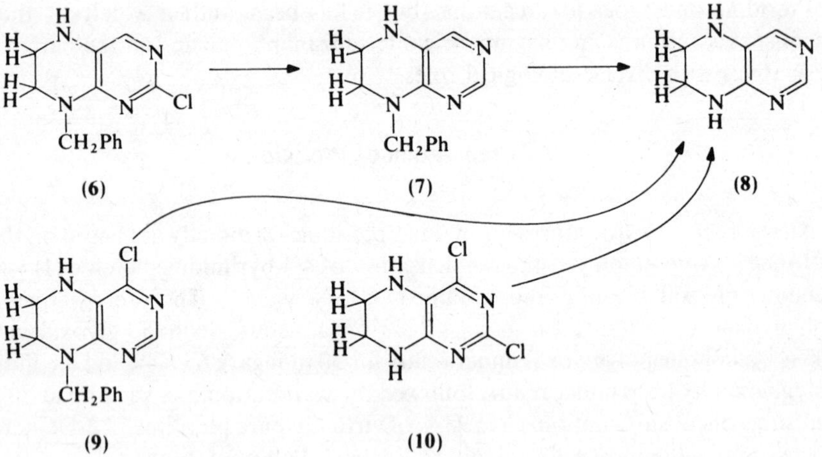

B. Properties and Structure of Pteridine

Pteridine appears to exist in two (interchangeable) pale yellow crystalline forms: that obtained from crystallization from benzene or light petroleum had a mp 138.5°C while that obtained by sublimation melted slightly higher at 140°C.[30, 51, 65, 250] However, the difference is so small, that this phenomenon must be considered unproven. It is soluble in seven parts of water at room temperature and quite soluble in all other solvents.[30] Although it darkens on storage, especially in light, there is little quantitative change and most of the specimen may be recovered in its original purity by sublimation. The solid is sternutatory but odorless.[30] It is steam volatile to the extent of ~35 mg/100 mL of distillate and the partition coefficient between chloroform and water is ~2.0.[30] An aqueous solution shows an apparent cationic pK_a of 4.12 [cf. 6.36 for 3,4-dihydropteridine (**5**)[51] and 6.63 for 5,6,7,8-tetrahydropteridine (**8**)[43]] and an apparent anionic pK_a of 12.2, both due to the covalent hydrate (**11**)[30, 34]: The natural cationic pK_a of anhydrous material was first estimated[1145] to be ~2.6 but later measurements suggested[1513, cf. 1481] a figure of ~ −2.0. It forms a picrate, mp 117–118°C,[250] and a dihydrated monooxalate, mp >128°C,[250] which is, in fact,[1145] the monohydrated oxalate of 3,4-dihydro-4-pteridinol (**11**). Finally, its dipole moment is 2.7 (experimental)[1523] or 2.42 to 2.52 (theoretical calculation).[781, 1523]

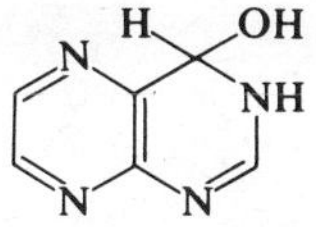

(**11**)

Experimental studies on the spectra of pteridine, usually in conjunction with those of simple derivatives and/or analogous systems, have been recorded.

Infrared. Some 50 well-defined bands are shown by pteridine (as a potassium bromide disc) in the range 1600 to 400 cm^{-1} [286]

Ultraviolet. Pteridine in water[30, 250] is quite simple [λ_{max}(log ε) at pH 6: <220 nm (>3.83), 298 (3.87), 309 (3.83); at pH 2.1: >220 (>3.84), 300 (3.92)] but a cyclohexane solution[286, 724] shows considerable additional fine structure [210 (4.04), 235 (3.46), 292 (3.80), 296 (3.84), 301 (3.87), 308 (3.75), 380 (1.92), 387 (1.92), 395 (1.89)], which has been carefully assigned to an $n \rightarrow \pi$ and three $\pi \rightarrow \pi$ bands.[285, 286, 1089]

Proton nmr. In deuterochloroform the spectrum[51, 293] is quite simple [δ at 33°C: 9.80, s, H-4; 9.67, s, H-2; 9.33, d (J 1.7 Hz), H-7; 9.15, d, H-6] but the assignments of H-6 and H-7 remain unproven despite confirmation[294] of those for H-2 and H-4 by unambiguous 2-deuteration; [cf. 1037] the spectra in aqueous media are radically different, being of covalently hydrated species.[51]

^{13}C nmr. The latest spectrum[959] of pteridine in deuterochloroform (chemical shifts in ppm from internal trimethylsilane: 135.3, C-4a; 148.4, C-6; 153.0, C-7; 154.4, C-8a; 159.5, C-2; 164.1, C-4) differs but slightly from an earlier one,[843] which had corrected erroneous assignments of C-2 and C-4 in an even earlier version[851]; the spectrum in water is very complicated because of the two distinct covalent hydrates present.[843]

Mass spectrum. The initial disintegration of pteridine is by the successive loss of two HCN fragments from the pyrimidine ring. Deuterium-labeling experiments have shown that this occurs in two ways.[207] Thus pteridine (**12**) loses N-3 + C-4 + H-4 to give a fragment (**13**) (75%) and (concomitantly) N-1 + C-2 + H-2 to give a fragment (**14**) or (**15**) (19%); subsequent loss of a second HCN fragment from both intermediates of *m/e* 105 gives a single dehydropyrazine cation (**16**) of *m/e* 78.[207, 1707] Mass spectral data for fragmentation of the bistrimethylsilyl derivative of pteridine (hydrate?) have been reported.[1457]

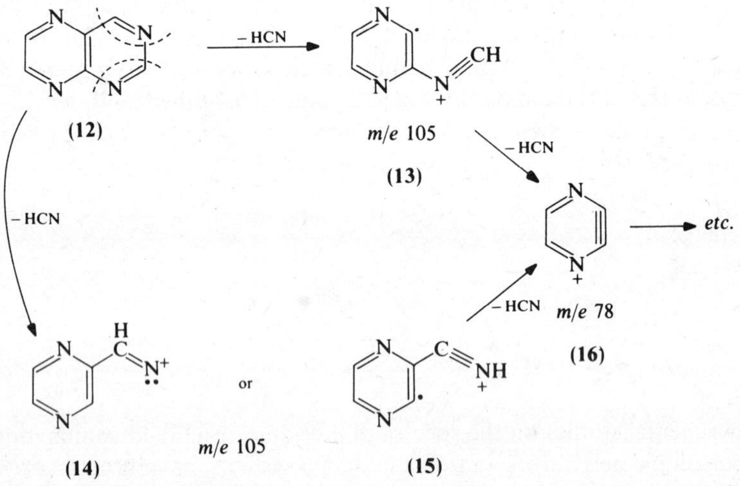

Phosphorescence. The phosphorescence spectrum of pteridine in host naphthalene at 1.5°K has been measured and tentative vibration assignments have been made; the total phosphorescence decay rate has also been reported.[722]

Photoelectron Spectra. The spectrum of pteridine has been measured recently[1513] and with its help a new value for the ionization of pteridine (as distinct from its covalent hydrates) has been determined as pK_a −2, with preferred protonation in the pyrimidine ring.[1513, cf. 312]

The polarographic reduction of pteridine has been studied in DMF[1360] and in aqueous solution over the range pH 1 to 12.[890] Earlier work suggested some opening of the pyrimidine ring during polarography in neutral media.[792, 1417, 1418] The crystal and molecular structure of pteridine was first measured in 1956.[963] It was concluded that pteridine was flat or nearly so and that the molecule crystallized in a noncentrosymmetric space group with four molecules per unit cell. However, some of the bond lengths differed markedly from those expected on theoretical grounds[915] and a redetermination of crystal and molecular structure[1303, 1407] appears to have confirmed inadequacies in the earlier determination. The new bond lengths (**17**) and angles (**18**) are claimed to be more consistent with expected values based on molecular

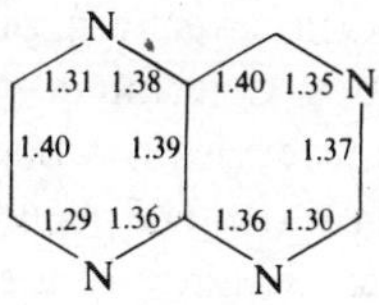

(17)

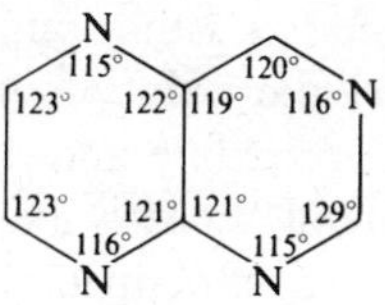

(18)

orbital calculations.[957, 1303] Nevertheless, more recent theoretical calculations[1432, 1483, 1512, 1751] have suggested, for example, that the N-3 to C-4 bond (1.35 Å) in formula (**17**) is anomalously long and Jill Gready[1512] has concluded that "errors in the pteridine x-ray structure determination may be greater than reported." Clearly, a third experimental determination of structure by state-of-the-art techniques is needed.

Rationalization or prediction of the electronic spectra of pteridine and related compounds, with the aid of variously modified molecular orbital calculation, has proven popular over many years.[922, 923, 938, 950, 951, 972, 1016, 1020, 1250, 1521] In contrast, observed proton chemical shifts for pteridine have been used to obtain an idea of π-electron distribution in the system.[702] There are also related papers with less well defined aims.[781, 787, 828, 1153, 1165, 1170]

C. Reactions of Pteridine

As a result of the extreme π-electron deficiency at each carbon atom in pteridine, the parent heterocycle undergoes no electrophilic reactions. Thus its reactivity is confined virtually to degradative ring fissions and to avid addition reactions with water, alcohols, amines, and some Michael-type reagents.

Following an early observation that pteridine (**20**) was highly unstable in hot dilute acid or alkali,[32] it was shown that 1*M* sulfuric acid at ~100°C for 5 min gave 3-amino-2-pyrazinecarbaldehyde (**19**, X=O), which could be isolated in >85% yield as its oxime (**19**, X=NOH)[34]; from buffer of pH 2.5 at 100°C for 10 min, the free aldehyde (**19**, X=O) could itself be isolated in 80% yield[65]; and in 1*M* sodium carbonate at 25°C for 24 h, pteridine (**20**) gave 3-aminomethyleneamino-2-pyrazinecarbaldehyde (**21**, X=O), isolated in the presence of hydroxylamine as the oxime (**21**; X=NOH) (60% yield).[34] Products formed by condensation of two or three molecules of the amino aldehyde (**19**, X=O) were also isolated under appropriate conditions.[65] The ring fission of pteridine and its derivatives to afford pyrazines has been reviewed.[466, 1480b] Treatment of

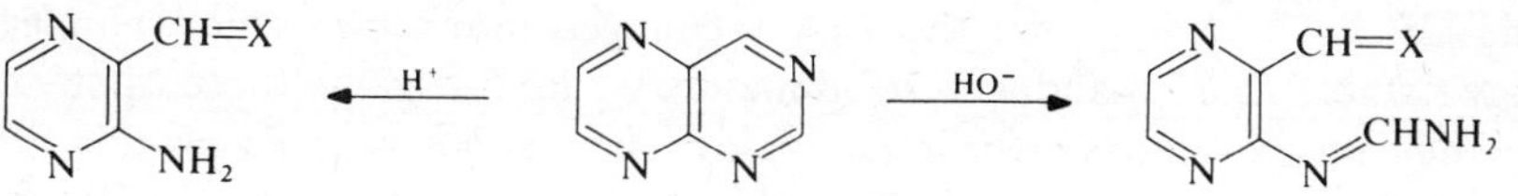

(19) (20) (21)

pteridine with methyl iodide under mild conditions gave a pitchlike product containing hydriodic acid: its nature was not determined.[34] The thermal graphitization of pteridine and other such heterocycles has been examined.[636]

The surprising difference[281] between the uv spectrum of pteridine at pH 7 and 13 was subsequently explained correctly by the existence of a second (anionic) pK_a at ~12.2, representing the formation of a resonance-stabilized anion (**22**) from the covalent hydrate, 3,4-dihydro-4-pteridinol (**11**)[34]; the kinetics of hydration ⇌ dehydration were then investigated thoroughly[242, 1145] and, taking those data into consideration, the true anionic pK_a of the covalent hydrate (**11**) emerged as 11.2.[1145]

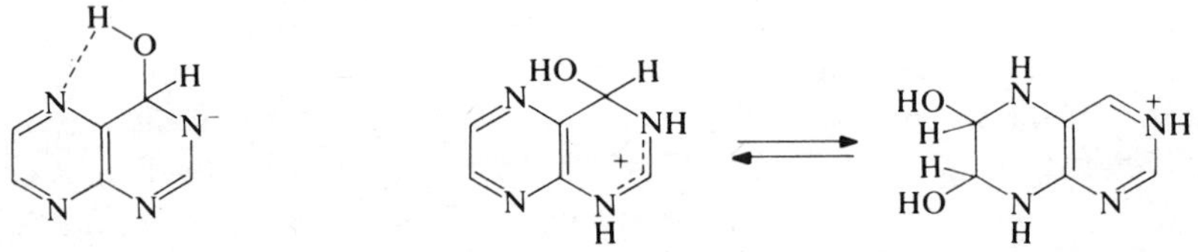

(**22**) (**23**) (**24**)

A slow upward drift in pH values during the titration of pteridine with an equivalent of acid was initially explained[34] as *signifying decomposition*, presumably in terms of the ring-fission reaction (described previously) under much less gentle conditions; subsequently, kinetic parameters for what was considered to be such a ring-opening–ring-closing reaction, involving various species of the anhydrous and hydrated molecules, were measured and discussed.[242, 1145] However, later pmr studies indicated that the 3,4-hydrated cation (**23**) of pteridine in aqueous solution gradually reached equilibrium with a 5,6,7,8-dihydrated cation (**24**) and that the changes previously attributed to reversible ring-fission simply involved a change in hydration site.[27] This finding was broadly confirmed by subsequent ^{13}C nmr studies.[843] The general acid–base catalysis, solvent deuterium isotope effects, and transition state characterizations of the (initial) reversible hydration of pteridine have been investigated in detail.[765, 790, 1146] The covalent hydration of pteridine, its derivatives, and other nitrogenous heterocycles has been reviewed expertly by Adrien Albert (the original investigator of this area) in 1962,[582] 1967,[12] and 1976.[11]

Not surprisingly, even partially hydrated pteridine reacted with peroxyphthalic acid or hydrogen peroxide to afford, not an *N*-oxide (72%) as first reported,[30] but 4(3*H*)-pteridinone.[26, 1145]

When followed by nmr, it was evident that pteridine in methyl alcohol at room temperature steadily became the 3,4-adduct, 4-methoxy-3,4-dihydropteridine (**25**, R = Me), and that this slowly changed into the thermodynamically more stable 5,6,7,8-adduct, 6,7-dimethoxy-5,6,7,8-tetrahydropteridine (**26**, R = Me); after several days the latter could even be isolated in 56% yield.[48] Ethyl alcohol behaved similarly to give the adducts (**25**, R = Et) and (**26**, R = Et), albeit more slowly. After 6 weeks the latter was isolated in 62% yield.[48] The secondary

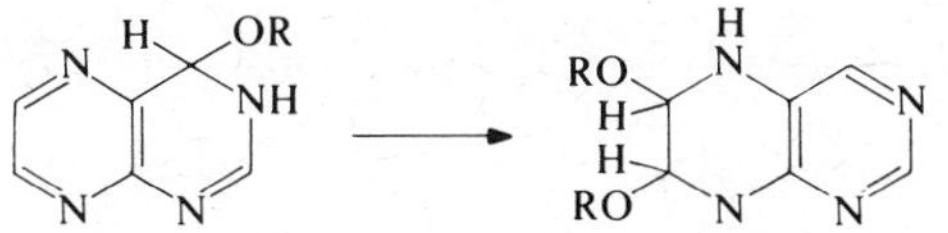

(25) (26)

alcohol, 2-propanol, likewise gave a similar mono-adduct (**25**, R = *i*-Pr) very slowly and only after several months could any evidence of the di-adduct (**26**, R = *i*-Pr) be obtained. However, neutralization of a solution of pteridine cation in isopropyl alcohol did give the stable di-adduct (**26**, R = *i*-Pr).[48] *t*-Butyl alcohol showed no tendency to form either type of adduct.[48] As might be expected, the pteridine cation underwent adduct formation with alcohols much more quickly; the pteridine anion also seemed to form adducts but the picture was less clear.[48]

When pteridine in pH 10 buffer was treated with one equivalent of similarly buffered ammonia, there was good uv and nmr spectral evidence for the formation of a 3,4-adduct (**27**, R = H).[184] However, when ammonia was passed into a solution of pteridine in ethanolic ammonia, a di-adduct spectrally akin to 5,6,7,8-tetrahydropteridine was precipitated. Although purification proved impossible without some decomposition, the similarly made bis(methylamine)-adduct could be purified for analysis and its structure was confirmed as 6,7-bismethylamino-5,6,7,8-tetrahydropteridine (**28**) by excellent nmr data.[52, 1700] Other evidence for the 3,4-adducts (**27**) has recently been provided by dissolution of pteridine in liquid ammonia containing one redox equivalent of potassium permanganate. After 10 min, a 50% yield of 4-pteridineamine was obtained; when ammonia was replaced by neat ethylamine, 4-ethylaminopteridine resulted in comparable yield.[934]

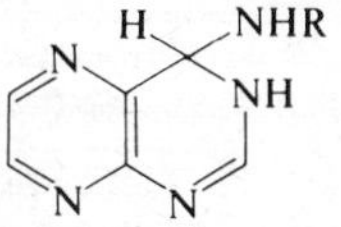

(27)

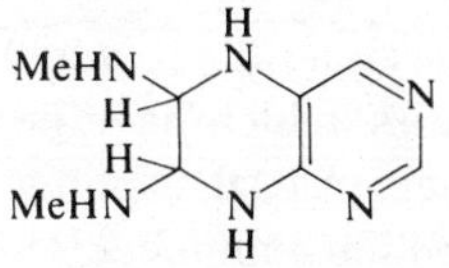

(28)

Michael-type reagents react with pteridine in several ways. Dimedone (5,5-dimethyl-1,3-cyclohexanedione), barbituric acid, 2-thiobarbituric acid, diethyl malonate, and ethyl benzoylacetate all added across the 3,4-bond to give 4-substituted-3,4-dihydropteridines such as 4-(4′,4′-dimethyl-2′,6′-dioxocyclohexyl)-3,4-dihydropteridine (**29**).[49] Ethyl acetoacetate, benzyl acetoacetate, and acetylacetone added (as enols) across both the 5,6- and the 7,8-bond to afford tricyclic adducts such as ethyl 7-methyl-5,5a,8a,9-tetrahydro-8-furo[2,3-*g*]pteridinecarboxylate (**30**).[49] Both malononitrile and cyanoacetamide underwent addition to the 3,4-bond, followed by ring fission and recyclization with

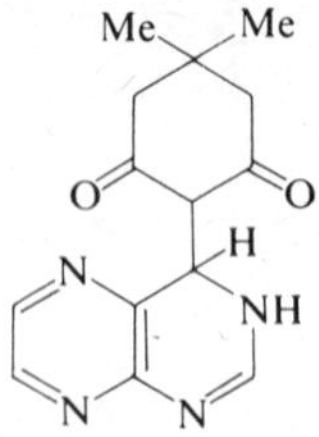

(29)

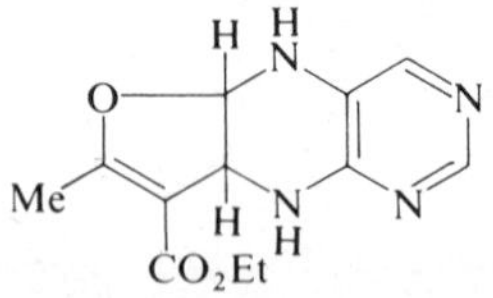

(30)

elimination of nitrogen, to give 6-amino-7-pyrido[2,3-*b*]pyrazinecarbonitrile and 6-amino-7-pyrido[2,3-*b*]pyrazinecarboxamide, respectively.[49]

2. SIMPLE ALKYL- AND ARYLPTERIDINES

This section is mainly concerned with the preparation and reactions of alkyl- or arylpteridines that bear no other type of group. However, in discussing the direct introduction of an alkyl–aryl group, the modification of an existing group to afford an alkyl–aryl group, or the reactions in which an alkyl–aryl group is directly involved, molecules bearing other types of (incidental) group have not been excluded.

A. Preparation of Alkyl- and Arylpteridines

(1) *By Primary Syntheses*

Most alkyl- or arylpteridines, whether simple or otherwise, have been made by primary syntheses, which have been covered systematically in Chapters II and III. For convenience, the following illustrations are included here. Thus 2-methyl-4,5-pyrimidinediamine (**31**) with polyglyoxal in methanol gave 2-methylpteridine (**32**)[33]; 6-methyl-4,5-pyrimidinediamine with polyglyoxal in methanol[33, 281] or (better) with *glyoxal hydrate polymer* in ethanol gave 4-methylpteridine[64]; 4,5-pyrimidinediamine with methylglyoxal in aqueous sodium bisulfite gave 7-methylpteridine[33, cf. 294]; the same substrate with methanolic biacetyl or 3,4-hexanedione gave 6,7-dimethyl-[33] or 6,7-diethylpteridine,[32] respectively; 2-methyl-4,5-pyrimidinediamine with methanolic biacetyl gave 2,6,7-trimethylpteridine[33]; 6-methyl-4,5-pyrimidinediamine with ethanolic biacetyl gave 4,6,7-trimethylpteridine[66]; 2,6-dimethyl-4,5-pyrimidinediamine similarly gave 2,4,6,7-tetramethylpteridine[66]; and 2-styryl-4,5-pyrimidinediamine with polyglyoxal or biacetyl gave 2-styrylpteridine or its 6,7-dimethyl derivative, respectively.[105] Broadly similar condensations afforded 6,7-diphenyl-,[983] 2-phenyl-,[152] 4-phenyl-,[151] 4-methyl-2-phenyl-,[152] 7-methyl-2-

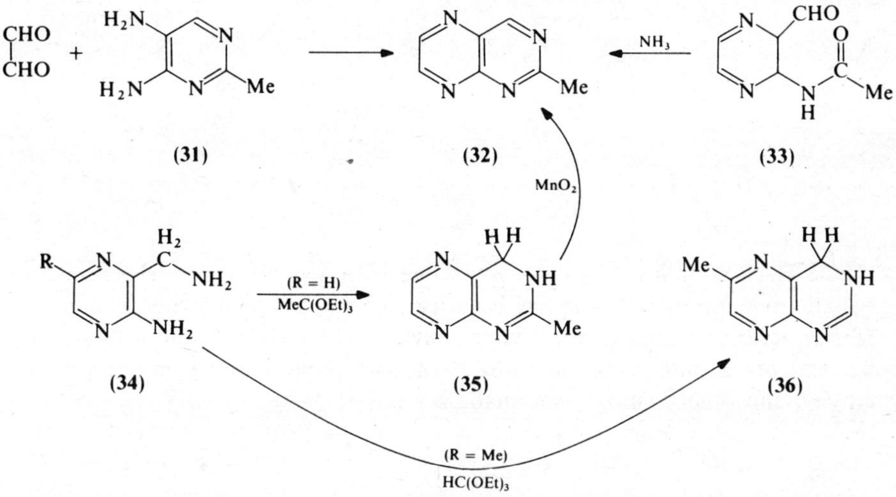

phenyl-,[152] 4,7-dimethyl-2-phenyl-,[152] 6,7-dimethyl-2-phenyl-,[152] 4,6,7-trimethyl-2-phenyl-,[152] 2-methyl-4-phenyl-,[151] 7-methyl-4-phenyl-,[151] 2,7-dimethyl-4-phenyl-,[151] 6,7-dimethyl-4-phenyl-,[151] and 2,6,7-trimethyl-4-phenylpteridine.[151]

Primary syntheses from pyrazines include the cyclization of 3-acetamido-2-pyrazinecarbaldehyde (**33**) in ethanolic ammonia to give 2-methylpteridine (**32**)[52]; the reduction of 3-amino-2-pyrazinecarbonitrile to 3-aminomethyl-2-pyrazineamine (**34**, R = H) and subsequent treatment with triethyl orthoacetate to give 2-methyl-3,4-dihydropteridine (**35**), followed by oxidation with manganese dioxide to afford 2-methylpteridine (**32**)[51]; and the similar conversion of 3-aminomethyl-5-methyl-2-pyrazinamine (**34**, R = Me) by triethyl orthoformate into 6-methyl-3,4-dihydropteridine (**36**) and then by oxidation to 6-methylpteridine.[51, 54]

(2) *By C-Alkylation or Arylation*

This process is but poorly represented by the methylation of 7(8*H*)-pteridine with dimethyl sulfate at pH 8 to give mainly the expected 8-methyl derivative (**37**, R = H) but also up to 5% of 6,8-dimethyl-7(8*H*)-pteridinone (**37**, R = Me) of subsequently proven structure[34]; by the diazomethane methylation of (**37**, R = H) to give (**37**, R = Me) in ~75% yield[31]; by the addition of phenyllithium to 2-methylthio-4,6-diphenylpteridine to give, after hydrolysis and permanganate dehydrogenation, 2-methylthio-4,6,7-triphenylpteridine (**38**) in 75% yield[1120]; and by the unexpected addition of phenyllithium to 2-methylthio-4,7-diphenylpteridine to give, after hydrolysis, 2-methylthio-4,7,7-triphenyl-7,8-dihydropteridine (**39**), which was naturally immune to dehydrogenation.[1120]

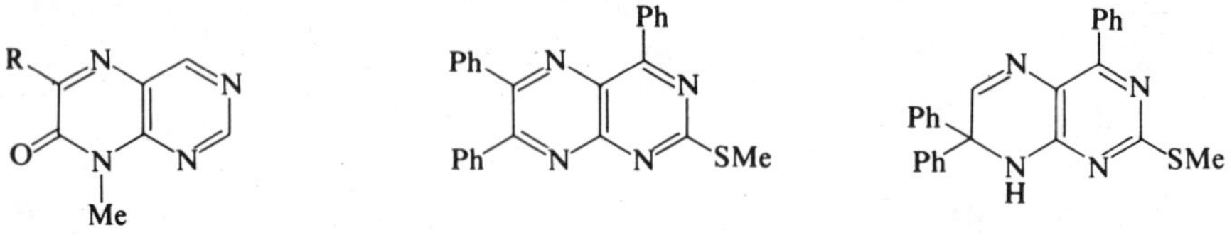

(37) (38) (39)

The Heck coupling of a chloropteridine with acetylenes in the presence of palladium–cuprous catalysts has been used to convert 6-chloro-2-pivalamido-4(3*H*)-pteridinone into 6-hex-1′-ynyl-2-pivalamido-4(3*H*)-pteridinone and then by hydrolysis into 2-amino-6-hex-1′-ynyl-4(3*H*)-pteridinone; the 6-phenylethynyl and other analogs were made similarly.[1753]

(3) *From Derivatives*

2-Amino-6-carboxymethyl-4,7(3*H*,8*H*)-pteridinedione (**40**, R = CO_2H) underwent decarboxylation in hot dilute acid to afford 2-amino-6-methyl-4,7(3*H*,8*H*)-pteridinedione (**40**, R = H)[1195]; 6-ethoxycarbonylmethyl-4,7(3*H*,8*H*)-pteridinedione underwent saponification in alkali and subsequent decarboxylation in hot dilute acid to give 6-methyl-4,7(3*H*,8*H*)-pteridinedione (90%)[29]; and 2,4-diamino-6-ethoxycarbonylmethyl-7(8*H*)-pteridinone (**41**, R = CO_2Et) underwent hydrolysis and decarboxylation on boiling in dilute sulfuric acid to give 2,4-diamino-6-methyl-7(8*H*)-pteridinone (**41**, R = H).[1195] Deacylation of a methyl group is also possible sometimes. Thus, 2-amino-6-*p*-chlorophenacyl-4,7(3*H*,8*H*)-pteridinedione (**40**, R = *p*-ClC_6H_4CO) in hot dilute alkali for 30 min or in hot dilute hydrochloric acid for 28 h gave a separable mixture of 2-amino-6-methyl-4,7(3*H*,8*H*)-pteridinedione (**40**, R = H) and *p*-chlorobenzoic acid; the 6-*p*-hydroxyphenacyl analog (**40**, R = *p*-HOC_6H_4CO) in alkali for 3 h likewise gave the same pteridine (**40**, R = H) and *p*-hydroxybenzoic acid.[982] In addition, the reaction has been applied to 6-phenacyl- (**42**, R = H, Y = Bz) and 1,3-dimethyl-6-phenacyl-2,4,7(1*H*,3*H*,8*H*)-pteridinetrione (**42**, R = Me, Y = Bz), which in hot alkali afforded 6-methyl- (**42**, R = Y = H) and 1,3,6-trimethyl-2,4,7-(1*H*,3*H*,8*H*)-pteridinetrione (**42**, R = Me, Y = H); to 7-phenacyl- and 1,3-dimethyl-7-phenacyl-2,4,6(1*H*,3*H*,5*H*)-pteridinetrione that gave 7-methyl- and 1,3,7-trimethyl-2,4,6(1*H*,3*H*,5*H*)-pteridinetrione: and to several of their *p*-sub-

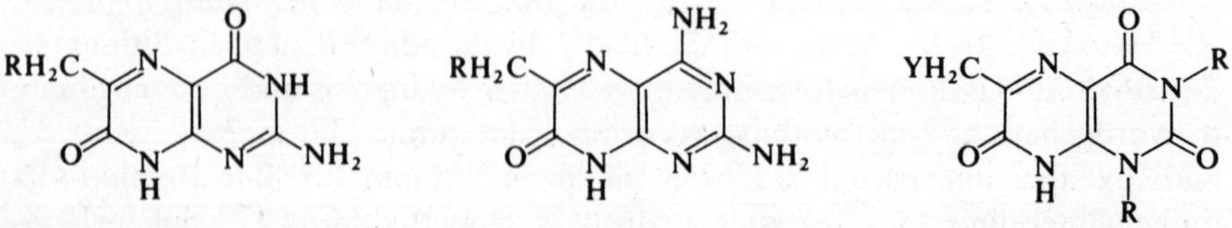

(40) (41) (42)

stituted-phenacyl analogs, which gave the same pteridines as just cited.[1524] Pteroylglutamic acid underwent a virtual *deamination* of the 6-methyl group in an electrolytic reduction to afford (after peroxide reoxidation of the pteridine ring) 2-amino-6-methyl-4(3*H*)-pteridinone.[1482]

Although little used in this series, it should be borne in mind that simple alkylpteridines may be made by removal of unwanted groups at other positions. For example, 4-methyl-5,6,7,8-tetrahydropteridine may be made from its 8-benzyl-2-chloro or its 8-benzyl-2-chloro-5-formyl derivative by treatment with sodium in liquid ammonia,[118] as well as by the lithium aluminum hydride reduction of 4-methylpteridine.[683]

B. Reactions of Alkyl- and Arylpteridines

Alkylpteridines undergo several types of reaction, most of which directly involve the alkyl group. Although it should be possible for pteridine-attached aryl groups to undergo electrophilic and other reactions, no examples appear to have been reported yet. The existence of certain 6-(substituted-isopropyl)pteridines as stable 6-(substituted-isopropylidene)-7,8-dihydropteridine tautomers has been discussed.[1698]

(1) *Covalent Additions*

Like pteridine and many of its other derivatives, simple alkyl- and arylpteridines undergo covalent addition of water in aqueous solution. The extent and position(s) of such hydration are naturally modified by the nature and position(s) of the alkyl–aryl group(s). Thus, 2-, 6-, and 7-methylpteridine have been shown to exist in neutral aqueous solution as their respective 3,4-hydrates, but only to a very limited extent[51,242,1145]; 4-methylpteridine under similar conditions was hydrated to an even less extent, presumably on account of steric hindrance by the methyl group.[242,1145] Although, like pteridine, the cations of 2-, 6-, and 7-methylpteridine initially formed 3,4-hydrates that were slowly converted in part into the 5,6,7,8-dihydrates, the steric hindrance of the 6- or 7-methyl group is visible in the respective ratios of mono- to dihydrate at equilibrium: pteridine 0.25, 2-methylpteridine 3, 6-methylpteridine 4, and 7-methylpteridine 7; for 6,7-dimethyl- and 2,6,7-trimethylpteridine the comparable ratios were ∞, that is, no hydration on the pyrazine ring was detectable.[27,51] In contrast, the equilibrium for 4-methylpteridine (cation) was zero, that is, there was no detectable 3,4-hydrate,[27] although 4,6,7-trimethyl- and 2,4,6,7-tetramethylpteridine, having no unhindered hydration sites, did show evidence of some 3,4-hydration.[66] None of the following phenylpteridines were appreciably hydrated in aqueous solution as neutral molecules but their cations in acidic solution were hydrated as shown: 2-phenylpteridine, initially 3,4-monohydrate changing to 5,6,7,8-dihydrate[152]; 4-phenylpteridine, 5,6,7,8-

dihydrate[151]; 4-methyl- and 4,7-dimethyl-2-phenylpteridine, 5,6,7,8-dihydrates[152]; 2-methyl-4-phenylpteridine, 5,6,7,8-dihydrate[151]; 7-methyl- and 6,7-dimethyl-2-phenylpteridine, mainly 3,4-monohydrates[152]; and 2,7-dimethyl-, 6,7-dimethyl-, and 2,6,7-trimethyl-4-phenylpteridine were unstable as cations.[151] Independent evidence for hydration was furnished by the following facile oxidations by hydrogen peroxide in acidic media: 7-methyl-2-phenylpteridine to 7-methyl-2-phenyl-4(3*H*)-pteridinone (~75%),[152] 6,7-dimethyl-2-phenylpteridine to 6,7-dimethyl-2-phenyl-4(3*H*)-pteridinone (80%),[152] 2-phenylpteridine (**43**) to 2-phenyl-4(3*H*)-pteridinone (**44**) (3 h: ~75%) or 2-phenyl-4,6,7(3*H*,5*H*,8*H*)-pteridinetrione (**45**) (1 week),[152] 4-phenylpteridine to

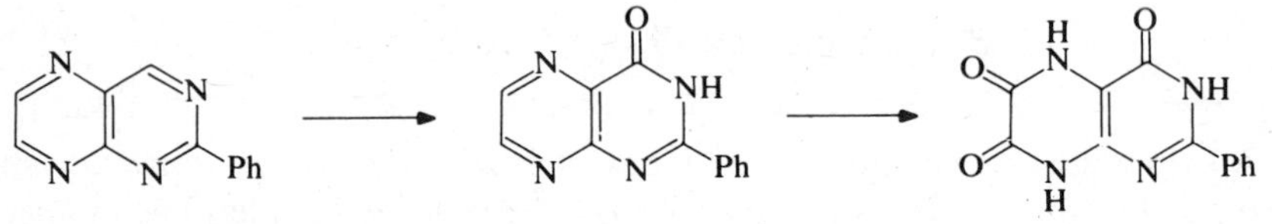

(**43**) (**44**) (**45**)

4-phenyl-6,7(5*H*,8*H*)-pteridinedione (~20%),[151] and 4-methylpteridine to 4-methyl-6,7(5*H*,8*H*)-pteridinedione (70%).[64] In addition, 4-methylpteridine underwent covalent addition by barbituric acid to give 4-methyl-7-(2′,4′,6′-trioxo-1′,2′,3′,4′,5′,6′-hexahydro-5′-pyrimidinyl)-7,8-dihydropteridine (97%) or 4-methyl-6,7-bis-(2′,4′,6′-trioxo-1′,2′,3′,4′,5′,6′-hexahydro-5′-pyrimidinyl)-5,6,7,8-tetrahydropteridine (92%), according to the amount of barbituric acid[64]; also by sodium bisulfite to give 4-methyl-5,6,7,8-tetrahydro-6,7-pteridinedisulfonic acid as its monosodium salt (56%)[64]; by benzylamine to give 6,7-bisbenzylamino-4-methyl-5,6,7,8-tetrahydropteridine (59%),[50] and by ethyl acetoacetate.[50] Both 2-methyl- and 6,7-dimethylpteridine also formed such adducts, for example, 4-methoxy-6,7-dimethyl-3,4-dihydropteridine,[50] while 7-phenyl- and 6,7-diphenylpteridine underwent addition of ethylamine or even *t*-butylamine to give, for example, 4-ethylamino-7-phenyl-3,4-dihydropteridine (nmr evidence).[1167]

(2) *Deuteration of Methyl Groups*

There appears to be no example of the α-deuteration of a simple alkylpteridine but 2-amino-6-methyl-4(3*H*)-pteridinone gave its 6-trideuteromethyl analog (**46**) by heating in a sealed vessel with 2*M* sodium deuteroxide at 100°C for 24 h; no deuteration of the 7-proton occurred under these conditions or even at 120°C.[84] The kinetics for deuteration of the 7-methyl group in 6,7,8-trimethyl-2,4-(1*H*,8*H*)-pteridinedione have been measured at 35°C. The reaction showed general acid and base catalysis but at best was extremely slow.[448, 1273] For practical purposes it has proven advantageous to prepare, for example, 7-deutero-6-trideuteromethyl-2,4-pteridinediamine by primary synthesis from a

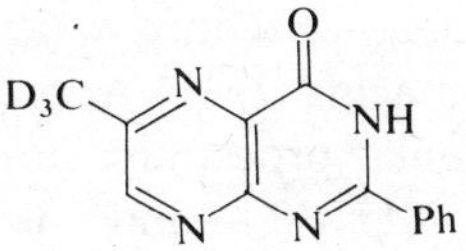

(46)

predeuterated pyrazine intermediate.[84] The deuteration of pteroylglutamic acid on the 6-methylene group and at the 7-position was facilitated by *N*-nitrosation at the adjacent NH group, and subsequent removal of the nitroso group with dithionite–NaOD at room temperature.[258]

(3) *Halogenation of Methyl Groups*

Apparently the halogenation of simple alkylpteridines has not been attempted but bromination of 2-amino-6-methyl-4(3*H*)-pteridinone and related compounds has been used extensively; for example, in the synthesis of pteroylglutamic acid and its analogs. Thus treatment of a preheated solution of 6- or 7-methyl-4(3*H*)-pteridinone in acetic acid with bromine in acetic acid rapidly gave 6- (**47**, R = H) or 7-bromomethyl-4(3*H*)-pteridinone in 86 or 65% yield, respectively.[121] 2-Amino-6-methyl-4(3*H*)-pteridinone reacted with *N*-bromosuccinimide (NBS), in concentrated sulfuric acid at 20°C or in chloroacetic acid containing a little dibenzoyl peroxide at 105°C, to give its 6-bromomethyl analog (**47**, R = NH_2) in >70% yield.[408, 1155] The same reaction has been carried out with neat bromine at 150°C in a sealed tube. It was also carried out with bromine in hot acetic acid (no details) or with bromine in 48% hydrobromic acid at 100°C (80% crude yield), but an attempt to make the corresponding chloromethyl product (using sulfuryl chloride containing dibenzoyl peroxide under reflux) gave poor yields.[569, cf. 112] Dibromination of the same substrate, 2-amino-6-methyl-4(3*H*)-pteridinone, has also been done using bromine in hydrobromic acid at 100°C, to give 2-amino-6-dibromomethyl-4(3*H*)-pteridinone (~70%) as hydrobromide.[112, 1358, cf. 943, 1739] 2-Amino-7-methyl-4(3*H*)-pteridinone was converted by bromine in hydrobromic acid into its 7-bromomethyl or 7-dibromomethyl analog, according to the conditions and proportion of bromine.[112, 1359] Finally, 2-amino-6,7-dimethyl-4(3*H*)-pteridinone (**48**, R = Y = H) likewise gave its 7-bromomethyl analog (**48**, R = H, Y = Br) with little or none of the isomer (**48**, R = Br, Y = H), which had to be made by primary synthesis of the bisbromomethyl analog (**48**,

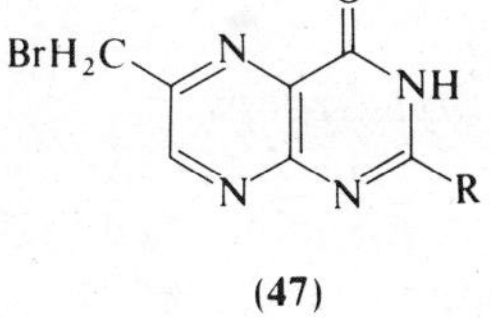

(47)

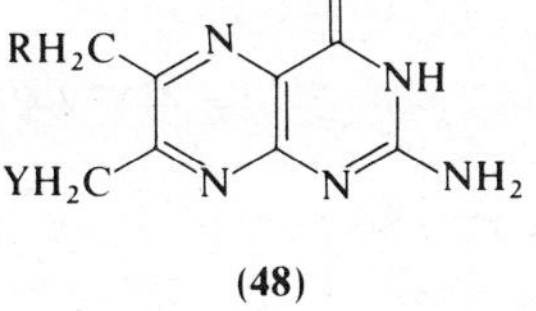

(48)

R = Y = Br) followed by preferential reductive dehalogenation of the 7-bromomethyl group with hydriodic acid.[112, 680] Appropriate methylpteridines were also converted by broadly similar procedures into 2,4-diacetamido-6-dibromomethylpteridine (NBS: 68%),[123] 2-amino-6-bromomethyl-4,7(3*H*,8*H*)-pteridinedione (~90%),[1061, 1371] 2-amino-6-bromomethyl-8-methyl-4,7(3*H*,8*H*)-pteridinedione (not characterized but converted into the 6-hydroxymethyl analog in 22% yield overall),[1214] 6-bromomethyl-4,7(3*H*,8*H*)-pteridinedione (unisolated: aminolysis confirmed its presence),[279] 7-bromomethyl-1-methyl-2,4(1*H*,3*H*)-pteridinedione (90%),[232] 6,7-bisbromomethyl-1-methyl-2,4(1*H*,3*H*)-pteridinedione (65%),[232] 6-dibromomethyl-4,7(3*H*,8*H*)-pteridinedione (bromine in boiling acetic acid: 90% as hydrobromide),[29] 7-dibromomethyl-1-methyl-2,4(1*H*,3*H*)-pteridinedione (80%),[232] 6,7-bisdibromomethyl-1-methyl-2,4(1*H*,3*H*)-pteridinedione (bromine in boiling acetic acid for 5 h: 65%),[232] and 2,4-diamino-6-chloromethyl-7(3*H*)-pteridinone 8-oxide (**49**) (chlorine in glacial acetic acid at 100°C in a sealed tube: 67% as hydrochloride).[1296]

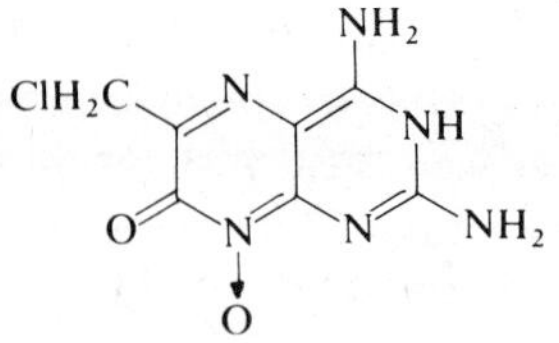

(**49**)

(4) *Sulfonation of an Alkyl Group*

2-Amino-7-ethyl-4,6(3*H*,5*H*)-pteridinedione in concentrated sulfuric acid at 110°C has been reported to give 2-amino-7-α-sulfoethyl-4,6(3*H*,5*H*)pteridinedione (**50**, R = Me), in 18% yield[504]; similar treatment of 7-acetonyl-2-amino-4,6(3*H*,5*H*)-pteridinedione apparently gave the 7-sulfomethyl analog (**50**, R = H) (36%) with loss of acetyl.[504]

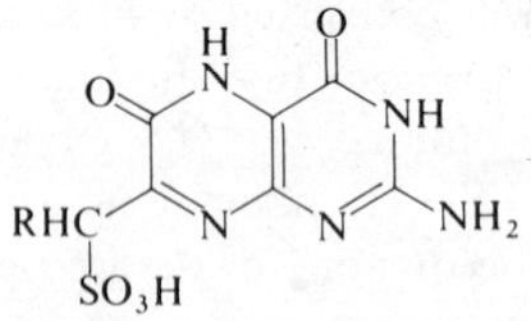

(**50**)

(5) *Styryl from Methyl Groups*

Early attempts to prepare styryl from methylpteridines "led to destruction of the pteridine nucleus" and the required products were made by styrylation of methylpyrimidine intermediates and subsequent cyclization with glyoxal, and so on,[282, 421] a tendency that has continued.[105, 484, 1305, 1415] However, it is possible to convert some methyl- into styrylpteridines as illustrated in the formation of 7-styryl-2,4-(1*H*,3*H*)-pteridinedione (**51**) (benzaldehyde–alcoholic sodium hydroxide under reflux: 53%), 7-styryl-2,4-pteridinediamine (similarly: 80%), 7-furfurylidenemethyl-2,4-pteridinediamine (similarly: 59%), and others.[252, 1010]

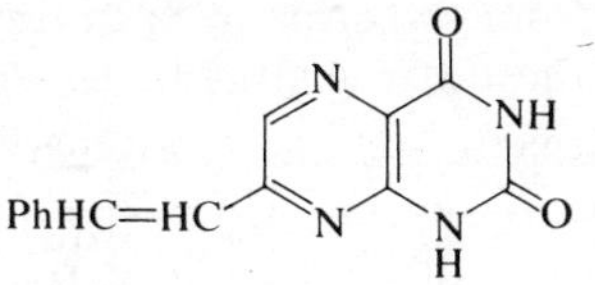

(**51**)

(6) *Oxidative Reactions of Alkyl Groups*

The peculiar conversion of 6,7,8-trimethyl-2,4(3*H*,8*H*)-pteridinedione, by permanganate at pH > 3.5, into 6,8-dimethyl-2,4,7(1*H*,3*H*,8*H*)-pteridinetrione (**52**) has been examined kinetically and a mechanism has been proposed.[1131, 1273] Analogous reactions are the conversions of 8-β-hydroxyethyl-7-methyl- or 8-β-hydroxyethyl-6,7-dimethyl-2,4(3*H*,8*H*)-pteridinetrione by aqueous permanganate into 8-β-hydroxyethyl-2,4,7(1*H*,3*H*,8*H*)-pteridinetrione (26%)[1144] or its 6-methyl derivative (40%),[422] respectively, and the conversion of 8-β-hydroxy ethyl-2,4(1*H*,3*H*)-pteridinedione by prolonged refluxing in dilute alkali into 8-β-hydroxyethyl-2,4,7(1*H*,3*H*,8*H*)-pteridinetrione (24%)[422]: Such reactions appear to be peculiar to the 7-position. Equally unexpected was hydroxylation of the *C*-methyl group in 2-amino-6,8-dimethyl-7,8-dihydro-4(3*H*)-pteridinone (**53**) on

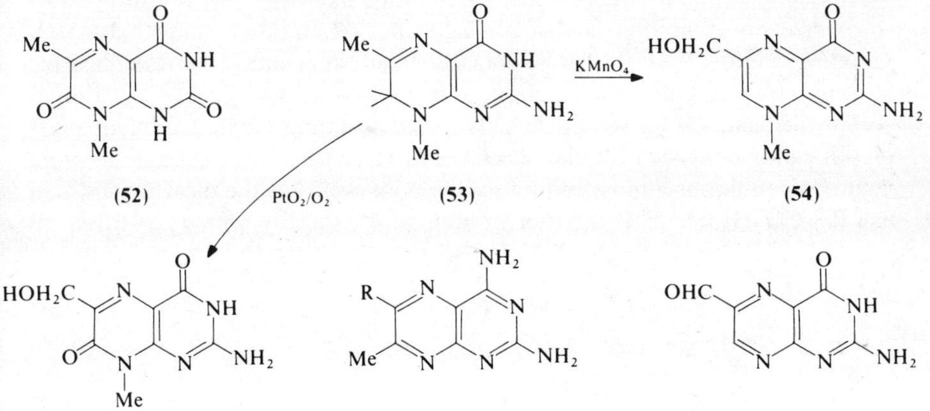

oxidation. Alkaline permanganate at 25°C gave 2-amino-6-hydroxymethyl-8-methyl-4(8*H*)-pteridinone (**54**) in 55% yield, while a stream of oxygen in the presence of platinum oxide went one step further to afford 2-amino-6-hydroxymethyl-8-methyl-4,7(3*H*,8*H*)-pteridinedione (**55**) in 30% yield.[1370] 6,7-Dimethyl-2,4-pteridinediamine (**56**, R = Me) underwent in vivo oxidative metabolism in rats to give 6-hydroxymethyl-7-methyl-2,4-pteridinediamine (**56**, R = CH_2OH) and 2,4-diamino-7-methyl-6-pteridinecarboxylic acid (**56**, R = CO_2H), isolated from the urine in 3:2 ratio.[847]

The conventional oxidation of a *C*-alkyl to a *C*-formyl group is represented in pteridines. Thus treatment of 2-amino-6-methyl-4(3*H*)-pteridinone with selenium dioxide in refluxing acetic acid containing a little nickel chloride gave 2-amino-4-oxo-3,4-dihydro-6-pteridinecarbaldehyde (**57**) in 63% yield.[940, cf. 1739] 2-Amino-6-methyl-4,7(3*H*,8*H*)-pteridinedione behaved similarly with selenium dioxide in acetic acid containing some sulfuric acid to give 2-amino-4,7-dioxo-3,4,7,8-tetrahydro-6-pteridinecarbaldehyde (55%).[1061] Larger alkyl groups can be similarly oxidized to acyl groups (see Ch. X, Sect. 11.D).

The oxidation of alkylpteridines to the corresponding pteridinecarboxylic acids has been used extensively, mainly to distinguish 6- from 7-methyl groups. The oxidation of 2-amino-6-methyl-4(3*H*)-pteridinone (**58**, R = Me) is probably best done as per Wolfgang Pfleiderer et al.,[389] who slowly added 5% potassium permanganate solution to a boiling solution of the substrate in dilute alkali and finally isolated a 44% yield of pure 2-amino-4-oxo-3,4-dihydro-6-pteridinecarboxylic acid (**58**, R = CO_2H); less detailed descriptions have appeared[1253, 1254, 1492] and the same product (**58**, R = CO_2H) has been isolated in comparable yield from analogous oxidation of the 6-hexyl, 6-nonyl or 6-(4′-methylcyclohexyl) derivative of 2-amino-4(3*H*)-pteridinone.[420] A rather analogous procedure,[389] based on earlier descriptions,[192, 193, 1084, 1254] was used to convert 2-amino-7-methyl-4(3*H*)-pteridinone into 2-amino-4-oxo-3,4-dihydro-7-pteridinecarboxylic acid (77%); the same product was obtained satisfactorily by oxidizing 2-amino-7-hexyl-4(3*H*)-pteridinone likewise.[420] It is also possible to oxidize preferentially the 7-methyl group in 2-amino-6,7-dimethyl-4(3*H*)-pteridinone (**59**, R = Me). Alkaline potassium permanganate (2 equivalents) gave a separable mixture of 2-amino-6-methyl-4-oxo-3,4-dihydro-7-pteridinecarboxylic acid (**59**, R = CO_2H) (70%) and 2-amino-4-oxo-3,4-dihydro-6,7-pteridinedicarboxylic acid (**60**, R = CO_2H) (13%).[307, 601] The use of more vigorous conditions almost certainly gave poor yields of the dicarboxylic acid (**60**, R = CO_2H).[601, 748] It is better to prepare it either by primary synthesis of

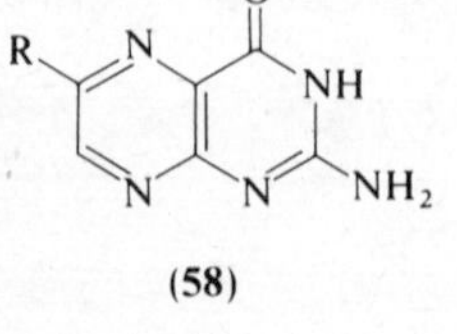

(**58**)

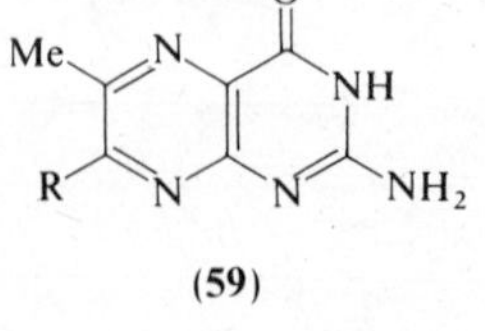

(**59**)

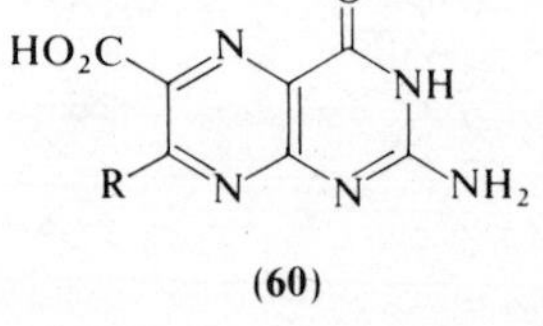

(**60**)

the corresponding ester followed by saponification,[307] or by primary synthesis of 2-amino-7-methyl-4-oxo-3,4-dihydro-6-pteridinecarboxylic acid (**60**, R = Me) followed by permanganate oxidation of the methyl group.[1363]

The oxidation of 7-methyl-2,4-(1*H*,3*H*)-pteridinedione (**61**, R = Me) by permanganate gave 2,4-dioxo-1,2,3,4-tetrahydro-7-pteridinecarboxylic acid (**61**, R = CO_2H) (~60%)[748, 1110] but the isomer (**62**, R = H, Y = CO_2H) has been made invariable by a route other than oxidation, despite the fact that 6-ethyl-1,3-dimethyl-2,4-(1*H*,3*H*)-pteridinedione (**62**, R = Me, Y = Et) has been oxidized satisfactorily to 1,3-dimethyl-2,4-dioxo-1,2,3,4-tetrahydro-6-pteridinecarboxylic acid (**62**, R = Me, Y = CO_2H) (45%)[1291]; 1,3,7-trimethyl-2,4(1*H*,3*H*)-pteridinedione likewise afforded 1,3-dimethyl-2,4-dioxo-1,2,3,4-tetrahydro-7-pteridinecarboxylic acid (26%).[108] 6,7-Dimethyl-2,4-(1*H*,3*H*)-pteridinedione was readily oxidized to 2,4-dioxo-1,2,3,4-tetrahydro-6,7-pteridinedicarboxylic acid (83%) by 4 molar equivalents of potassium permanganate.[748]

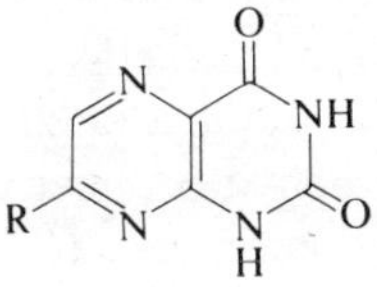
(**61**)

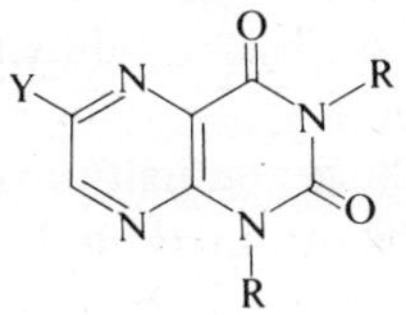
(**62**)

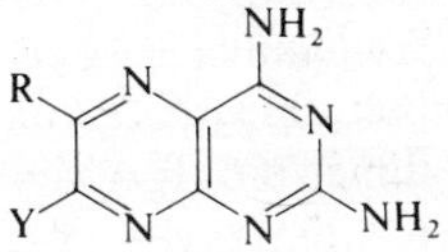
(**63**)

Both 6- (**63**, R = Me, Y = H) and 7-methyl-2,4-pteridinediamine (**63**, R = H, Y = Me) have been oxidized by permanganate in hot water to give 2,4-diamino-6-pteridinecarboxylic (**63**, R = CO_2H, Y = H) (40–55%)[123, 1354] and the isomeric 7-carboxylic acid (**63**, R = H, Y = CO_2H) (75%),[748] respectively; 6,7-dimethyl-2,4-pteridinediamine (**63**, R = Y = Me) likewise gave 2,4-diamino-6,7-pteridinedicarboxylic acid (**63**, R = Y = CO_2H) in unstated yield[748]; and 6-methyl-2-phenyl-4-pteridineamine and 6-methyl-2-phenyl-4(3*H*)-pteridinone by a similar procedure gave 4-amino-2-phenyl- and 4-oxo-2-phenyl-3,4-dihydro-6-pteridinecarboxylic acid, in 8 and 51% yield, respectively.[1354] When more alkaline conditions were used during the oxidation of 6- and 7-methyl-2,4-pteridinediamine (cf. just cited), the respective products were 2-amino-4-oxo-3,4-dihydro-6-pteridinecarboxylic acid (**58**, R = CO_2H) (~35%) and 2-amino-4-oxo-3,4-dihydro-7-pteridinecarboxylic acid (61%).[1263]

(7) *Dimeric Methylpteridines and Related Compounds*

There is a long history of dimers or near dimers of methylpteridines, often containing oxo, amino, or other groups. The simplest example is a hydrated dimer, formed in 47% yield by warming 4-methylpteridine in dilute sulfuric acid.[64] Based on good spectral and chemical evidence it has been formulated as 7-(6′,7′-dihydroxy-5′,6′,7′,8′-tetrahydropteridin-4′-ylmethyl)-4-methyl-5,6,7,8-

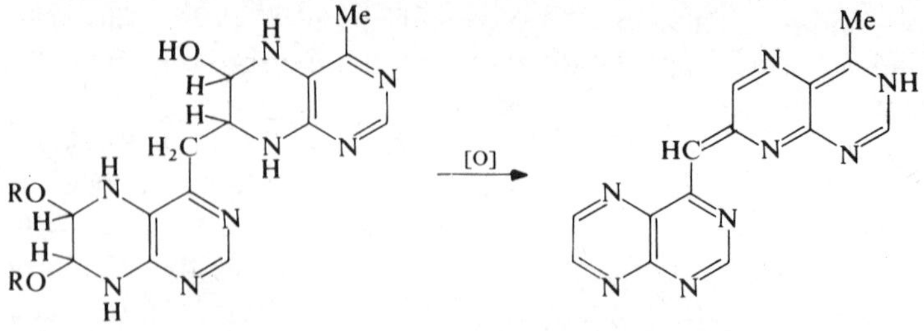

(64) (65)

tetrahydro-6-pteridinol (**64**, R=H) formed by the covalent addition of one methylpteridine to the 7,8-bond of another; prolonged boiling of the dimer trihydrate in ethanol containing a little *p*-toluenesulfonic acid gave the diethanolate (**64**, R=Et).[64] Oxidation of the hydrate (**64**, R=H) afforded a deep red compound, formulated on analytical and spectral evidence as the through-conjugated molecule, 4-(4′-methyl-3′,7′-dihydropteridin-7′-ylidenemethyl)-pteridine (**65**).[64] Similarly, 6,7-dimethylpteridine in alkali gave a monohydrated dimer (**66**, R=Me, Y=OH), which gradually oxidized to the dehydro dimer, 6,7-dimethyl-4-(6′-methyl-3′,7′-dihydropteridin-7′-ylidenemethyl)pteridine, (**67**, R=Me); the oxidation was accelerated by using methylamine in THF as the medium.[50] Under the latter conditions, 7-methylpteridine quickly gave the oxidized molecule, 4-(3′,7′-dihydropteridin-7′-ylidenemethyl)-7-methyl-pteridine (**67**, R=H), presumably via the dimer–methylamine adduct (**66**, R=H, Y=NHMe).[50] Among other such dimers and dehydro dimers,[57] the

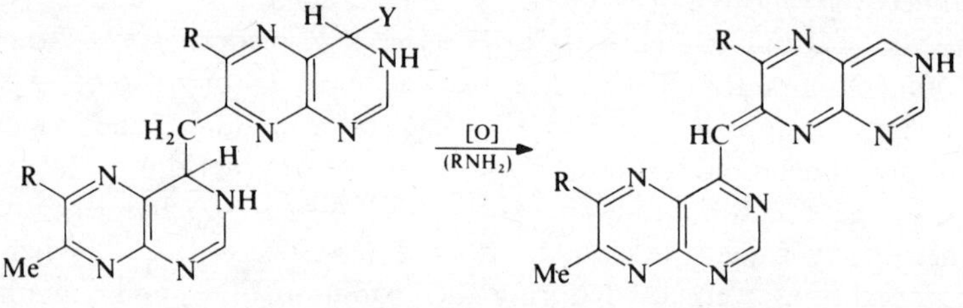

(66) (67)

mixed dimer from 6-methyl-7(8*H*)-pteridinone and 7-methyl-6(5*H*)-pteridinone in a slightly acidic medium, viz. 6-(7′-methyl-6′-oxo-5′,6′,7′,8′-tetra-hydropteridin-7′-ylmethyl)-7(8*H*)-pteridinone (**68**), proved interesting because it was completely resolved into its enantiomers by passage through a cellulose column.[57, 58] Perhaps the most important such dehydro dimer is *pterorohodin* (or rhodopterin), which was a deep red substance, initially observed as an artifact during the isolation of butterfly wing pigments in the early days[6, 407, 794] but subsequently isolated as a primary wing pigment from certain butterflies[355] and as an eye pigment.[1525] Its constitution was narrowed to one of two isomers in 1944[407]; 5 years later it was proven by synthesis to be 2-amino-7-(2′-amino-

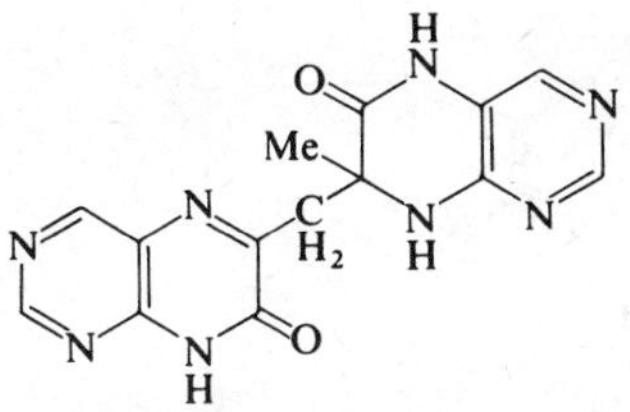

(68)

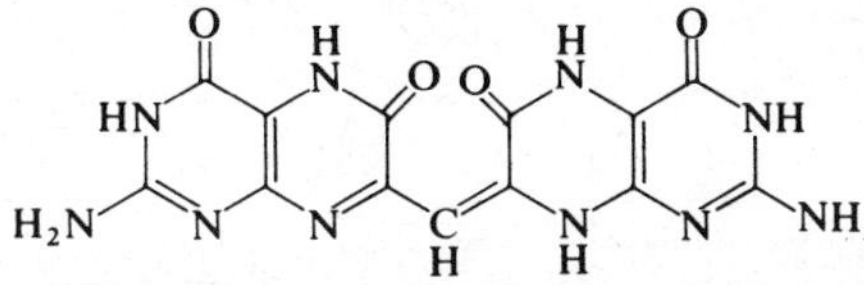

(69)

4′,6′-dioxo-3′,4′,5′,6′,7′,8′-hexahydropteridin-7′-ylidenemethyl)-4,6(3*H*,5*H*)-pteridinedione (**69**), tautomeric with the symmetrical molecule, bis(2-amino-4,6-dioxo-3,4,5,6-tetrahydropteridin-7-yl)methane,[425] although the method of its formation as an artifact continued to receive attention.[501] Similar structures have been assigned,[425] to isopterorhodin and allopterorhodin (which do not appear to occur naturally or even as artifacts) and to the various *pteridine reds* studied by Paul Karrer and colleagues.[257, 991–994, 1209, 1212] Other pterorhodin analogs have been reported.[345, 352] It has also been claimed, on real but slender mass spectrographic evidence, that 6,7,8-trimethyl-2,4(3*H*,8*H*)-pteridinedione, in which all addition sites are effectively blocked, underwent gentle oxidation by potassium *trans*-1,2-diaminocyclohexanetetracetatomanganate under nitrogen to give 1,2-bis(6′,8′-dimethyl-2′,4′-dioxo-2′,3′,4′,8′-tetrahydropteridin-7′-yl)ethane or the corresponding bispteridinyl ethylene, according to the amount of oxidant.[1131] Drosopterin and related compounds are discussed in Chapter VI, Section 5.D(11). A review of heterocyclic oligomers, including those formed by pteridines, has been presented expertly by Adrien Albert and Hiroshi Yamamoto.[63]

(8) *Other Reactions*

Although all *C*-methyl groups on pteridine should be prone to acylation by the *Claisen reaction*, the only example on hand is the conversion of 2-amino-3,5,7-trimethyl-4,6(3*H*,5*H*)pteridinedione (**70**, R = H) into the 7-methoxalylmethyl analog (**70**, R = $COCO_2Me$) ($\sim$80%) by treatment with dimethyl oxalate in methanolic potassium methoxide at room temperature for 24 h.[348]

2-Amino-6-phenethyl-4(3*H*)-pteridinone (**71**) undergoes an interesting *cyclization reaction* on treatement with fluorosulfonic acid in trifluoroacetic acid

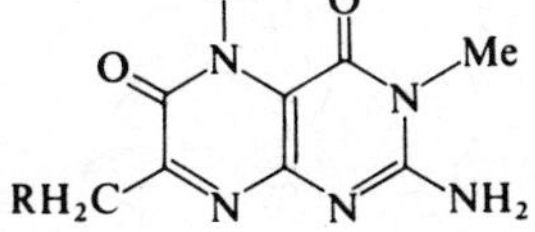

(70)

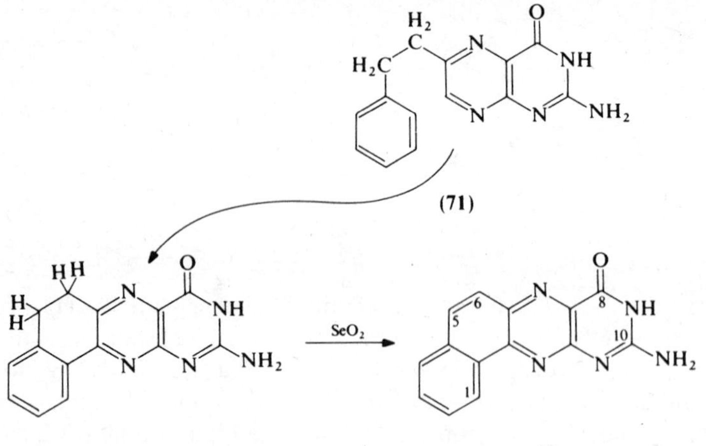

(71)

(72) (73)

at 25°C to give 10-amino-5,6-dihydro-8(9*H*)-naphtho[2,1-*g*]pteridinone (**72**) (74%) that was aromatized, by stirring with selenium dioxide in refluxing acetic acid, to afford 10-amino-8(9*H*)-naphtho[2,1-*g*]pteridinone (**73**) (77%); a mechanism was proposed.[1194]

3. THE *N*-ALKYLPTERIDINIUM SALTS

Most *N*-alkylated pteridines are *N*-alkylpteridinones, *N*-alkylpteridinethiones, or *N*-alkylpteridinimines in which the alkyl group has replaced a tautomeric hydrogen atom associated with the other substituent. Such compounds, for example, (**74**, X = O, S, or NH), are discussed in Chapters VII, IX, or X respectively.

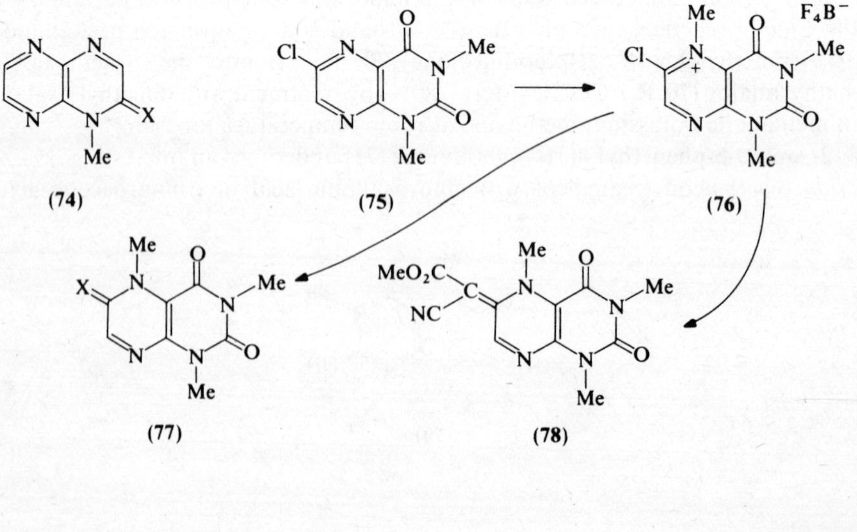

(74) (75) (76)

(77) (78)

However, there is at least one truly quaternary *N*-alkylated pteridine known. Thus treatment of 6-chloro-1,3-dimethyl-2,4(1*H*,3*H*)-pteridinedione with trimethyloxonium tetrafluoroborate (Me_3OBF_4) in methylene chloride gave a precipitate of 6-chloro-1,3,5-trimethyl-2,4-dioxo-1,2,3,4-tetrahydro-6-pteridinium tetrafluoroborate (**76**). As might be expected, the chloro substituent in this salt proved extremely reactive towards nucleophiles, so that with water or aqueous sodium hydrogen sulfide at room temperature for a few minutes it gave 1,3,5-trimethyl-2,4,6(1*H*,3*H*,5*H*)-pteridinetrione (**77**, X=O) (73% overall yield) or 1,3,5-trimethyl-6-thioxo-5,6-dihydro-2,4(1*H*,3*H*)-pteridinedione (**77**, X=S) (78%), respectively[1680]; the same salt (**76**) with aniline in refluxing methylene chloride for 1 h gave 1,3,5-trimethyl-6-phenylimino-5,6-dihydro-2,4(1*H*,3*H*)-pteridinedione (**77**, X=NPh) (38%), while with methanolic methyl sodiocyanoacetate at 25°C it gave 6-α-cyano-α-methoxycarbonylmethylene-1,3,5-trimethyl-5,6-dihydro-2,4(1*H*,3*H*)-pteridinedione (**78**) in 30% yield.[1680]

CHAPTER V

Halogenopteridines

Because they are easily made and then undergo all manner of nucleophilic displacement reactions, the 2-, 4-, 6-, and 7-chloropteridines have proven to be invaluable intermediates in pteridine syntheses. The same may be said of most extranuclear chloro- or bromopteridines with the exception of halogenophenyl derivatives which, of their nature, are unreactive.

1. PREPARATION OF NUCLEAR HALOGENOPTERIDINES

For obvious pragmatic reasons, most of the known nuclear halogenopteridines are in fact chloropteridines and most have been made by primary synthesis or by the action of phosphoryl chloride and/or phosphorus pentachloride on pteridinones.

A. By Primary Syntheses

The primary synthesis of bromo- or chloropteridines from pyrimidine or pyrazine intermediates has been included in Chapters II and III, respectively. However, a few typical simple examples are given here. Thus, by using the Gabriel and Colman syntheses, 2-chloro-4,5-pyrimidinediamine (**2**, R = H) with glyoxal (**1**, Y = H) gave 2-chloropteridine (**3**, R = Y = H),[1138] 2-chloro-6-methyl-4,5-pyrimidinediamine (**2**, R = Me) with biacetyl (**1**, Y = Me) gave 2-chloro-4,6,7-trimethylpteridine (**3**, R = Y = Me),[124] 2-chloro-6-trifluoromethyl-4,5-pyrimidinediamine (**2**, R = CF_3) with glyoxal (**1**, Y = H) gave 2-chloro-4-trifluoromethylpteridine (**3**, R = CF_3, Y = H),[159] and 2-chloro-6-phenyl-4,5-pyrimidinediamine (**2**, R = Ph) with dimethyl oxalate (**1**, Y = OMe) gave 2-chloro-4-phenyl-6,7(5*H*, 8*H*)-pteridinedione, the preferred tautomer of (**3**,

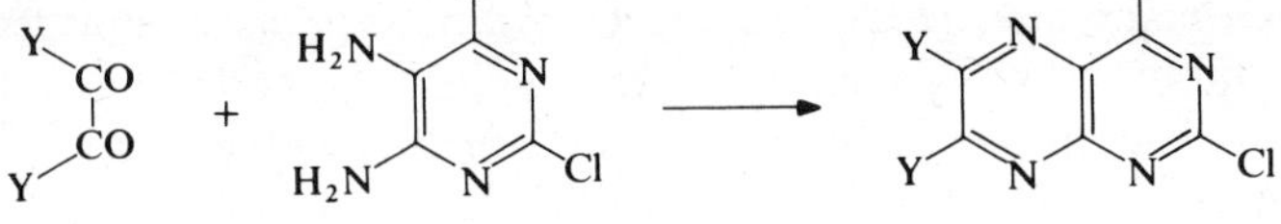

(**1**) (**2**) (**3**)

R = Ph, Y = OH).[522] Likewise, condensation of 6-chloro-4,5-pyrimidinediamine with glyoxal afforded 4-chloropteridine,[32] the same pyrimidine with biacetyl gave 4-chloro-6,7-dimethylpteridine,[170] and 4-chloro-6-methylamino-5-pyrimidinamine with ethyl glyoxylate hemiacetal afforded 4-chloro-8-methyl-7(8*H*)-pteridinone (**4**).[248] The condensation of 2,6-dichloro-4,5-pyrimidinediamine (**2**, R = Cl) with glyoxal (**1**, Y = H) gave 2,4-dichloropteridine (**3**, R = Cl, Y = H).[486] A Boon synthesis, by aminolysis of 2,4-dichloro-6-methyl-5-nitropyrimidine to 4-acetonylamino-2-chloro-6-methyl-5-nitropyrimidine (**5**) followed by hydrogenation and spontaneous cyclization, gave 2-chloro-4,6-dimethyl-7,8-dihydropteridine (**6**)[1088]; 4-α-benzoylbenzylamino-2-chloro-6-methyl-5-nitropyrimidine on reduction gave 2-chloro-4-methyl-6,7-diphenyl-7,8-dihydropteridine[398]; and cyclization of 4-(*N*-benzyl-β-hydroxyethylamino)-6-chloro-5-pyrimidinamine with phosphorus trichloride gave 8-benzyl-4-chloro-5,6,7,8-tetrahydropteridine.[43]

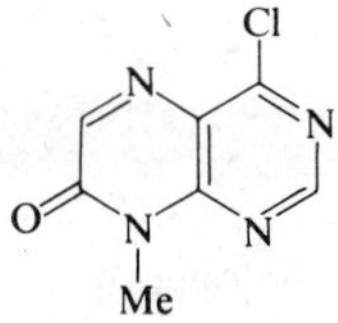

(4)

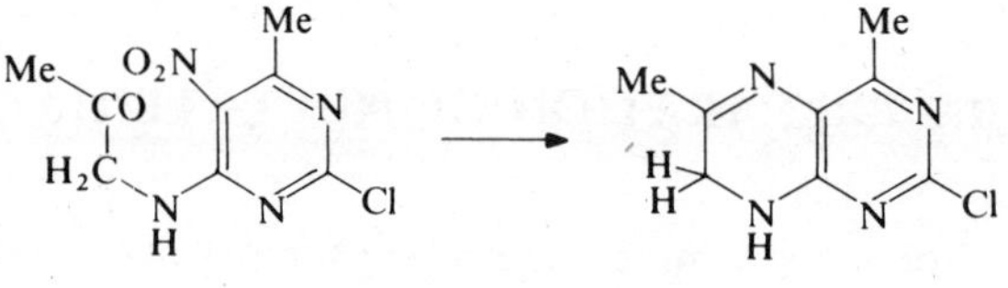

(5) (6)

The use of a pyrazine substrate is exemplified in the conversion of 6-chloro-3-dimethylaminomethyleneamino-2-pyrazinecarbonitrile (**7**) by methylamine acetate into 6-chloro-4-methlyaminopteridine (**9**), via Dimroth rearrangement of the initial imine (**8**).[53] Another example is the gentle alkaline treatment of 3-benzamido-6-bromo-2-pyrazinecarboxamide to afford 6-bromo-2-phenyl-4(3*H*)-pteridinone (80%).[20]

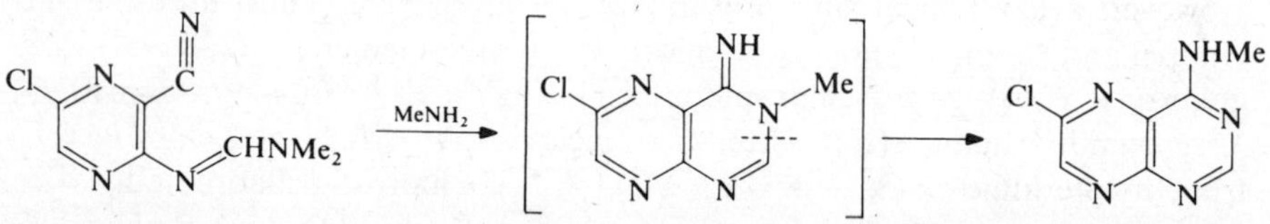

(7) (8) (9)

B. From Pteridinones

Although many *monopteridinones* react with phosphorus chlorides successfully to afford the corresponding chloropyrimidines, 2-pteridinones appear to be exceptional. The only example is the transformation of 4,6,7-triphenyl-2(1*H*)-pteridinone, by a mixture of phosphoryl chloride and phosphorus pentachloride at 100°C for 1 h, into 2-chloro-4,6,7-triphenylpteridine (**10**) (35%),[1120] which is better made by primary synthesis in any case.[1118] The use of 4-pteridinones as substrates is illustrated by the conversion of 2-amino-4(3*H*)-pteridinone into 4-chloro-2-pteridinamine (**11**, R=Y=H) (phosphoryl chloride: no details or yield),[668] of 2-methylamino-6,7-diphenyl-4(3*H*)-pteridinone into 4-chloro-2-methylamino-6,7-diphenylpteridine (**11**, R=Me, Y=Ph) (phosphorus pentachloride in refluxing phosphoryl chloride for 2 h: >90% of crude hydrochloride?),[1189] of 2-amino-6,7-diphenyl-4(3*H*)-pteridinone into 4-chloro-6,7-diphenyl-2-pteridinamine (**11**, R=H, Y=Ph) (similarly: 71–81%),[141, 891] and of 2-(2′-methoxypyrimidin-4′-yl)-4(3*H*)-pteridinone into 4-chloro-2-(2′-methoxypyrimidin-4′-yl)pteridine (phosphorus pentachloride in refluxing phosphoryl chloride for 4 h: 72%).[122] Hydrated 6(5*H*)-pteridinone was used as the substrate to prepare 6-chloropteridine (**12**, R=Y=H) [phosphorus pentachloride in

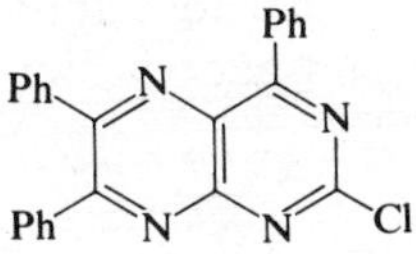
(10)

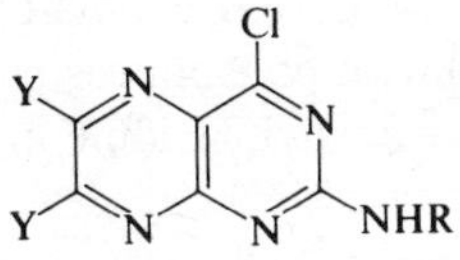
(11)

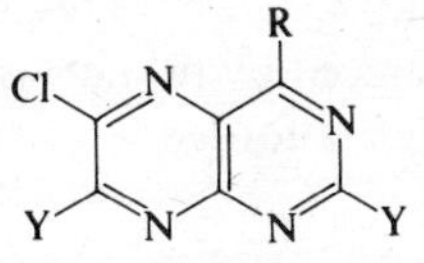
(12)

refluxing phosphoryl chloride for 2 h: 65%; the illogical use of 7,8-dihydro-6(5*H*)-pteridinone gave a 35% yield][31, 35]; 4-amino-2,7-diphenyl-6(5*H*)-pteridinone under similar conditions gave 6-chloro-2,7-diphenyl-4-pteridinamine (**12**, R=NH_2, Y=Ph) (66%)[526]; and appropriate 7-pteridinones afforded 7-chloropteridine (phosphorus pentachloride in refluxing pentachloroethane for 10 min: ~85%; or phosphorus pentachloride in refluxing phosphorus trichloride for 7 h: 30–50%),[33, 35] 7-chloro-6-methylpteridine (phosphorus pentachloride in refluxing phosphorus trichloride for 6 h: yield 50% on a small scale but much lower as the scale was increased),[55] 7-chloro-4-pteridinamine (phosphorus pentachloride in refluxing phosphoryl chloride for 90 min: ~30%),[445] 7-chloro-4-dimethylaminopteridine (likewise: ~40%),[381] 7-chloro-2,4-pteridinediamine (phosphoryl chloride),[95,] 7-chloro-4-pentyloxy-2-pteridinamine (phosphoryl chloride),[95, 1681] 7-chloro-6-phenyl-2,4-pteridinamine (phosphorus pentachloride in refluxing phosphoryl chloride for 3 h: 37%),[525] and 7-chloro-2,6-diphenyl-4-pteridinamine (likewise: 75%).[526] The conversion of 4-methyl-7(8*H*)-pteridinone into 7-chloro-4-dichloromethylpteridine by phosphorus pentachloride in refluxing pentachloroethane represents a warning to be heeded.[64]

The conversion of *pteridinediones* into mono- or dichloropteridines is illustrated in the treatment of 4,6(3*H*,5*H*)-pteridinedione with phosphorus pentachloride in phosphoryl chloride under reflux for 4 h to give (unstable) 4,6-dichloropteridine (**13**, R=H) (~50%),[41] of 6,7-diphenyl-2,4(1*H*,3*H*)-pteridinedione as just cited to give 2,4-dichloro-6,7-diphenylpteridine (92%),[1315] of 4,7(3*H*,8*H*)-pteridinedione with phosphorus pentachloride in refluxing pentachloroethane for 30 min to give 4,7-dichloropteridine (~60% yield crude but only 15% yield of pure material),[131] of 6,7(5*H*,8*H*)-pteridinedione with phosphorus pentachloride in benzoyl chloride at 140°C to give 6,7-dichloropteridine (65%),[35] and of 2-amino-4,7(3*H*,8*H*)-pteridinedione with phosphorus pentachloride–phosphoryl chloride for 1 h to give the monochlorinated product, 2-amino-7-chloro-4(3*H*)-pteridinone (**14**) (~70%),[380] just possibly by partial hydrolysis of 4,7-dichloro-2-pteridinamine during the work-up? The same explanation may, or may not, account for the formation of 2-amino-7-chloro-4-oxo-3,4-dihydro- from 2-amino-4,7-dioxo-3,4,7,8-tetrahydro-6-pteridinecarboxylic acid under rather similar conditions (~80%).[1084] Since the substrate has but one tautomeric oxo–hydroxy substituent available for reaction, it is not surprising that 2-amino-3-methyl-4,7(3*H*,8*H*)-pteridinedione (**15**) with phosphorus pentachloride–phosphoryl chloride gave only the monochloro derivative, 2-amino-7-chloro-3-methyl-4(3*H*)-pteridinone (**16**) (~35%).[383] Treatment of 2-chloro-4-phenyl-6,7(5*H*,8*H*)-pteridinedione (produced by primary synthesis) with phosphorus pentachloride–phosphoryl chloride afforded 2,6,7-trichloro-4-phenylpteridine in up to 37% yield.[522]

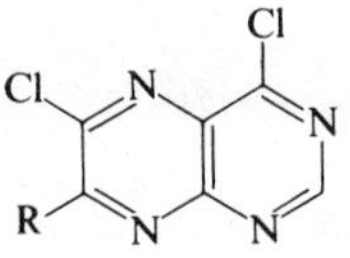

(13)

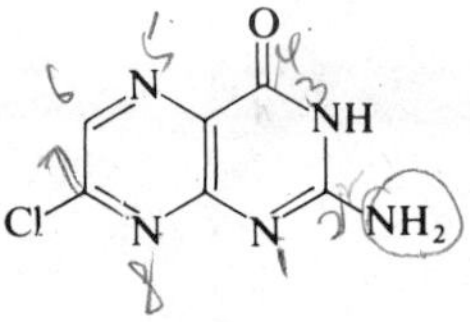

(14)

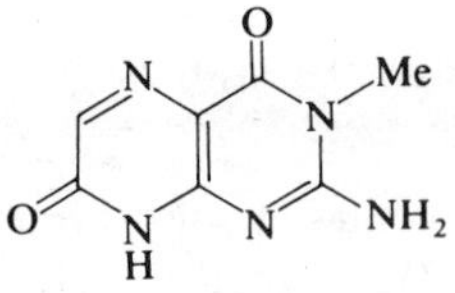

(15)

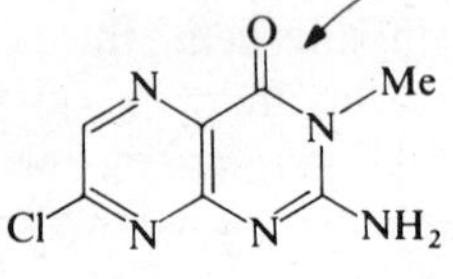

(16)

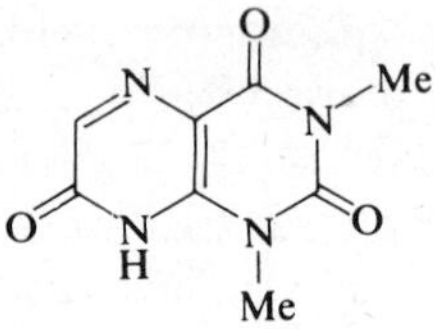

(17)

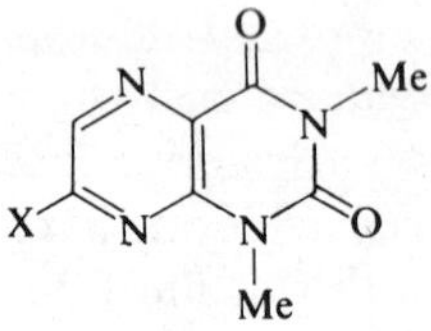

(18)

The complete or partial halogenation of *pteridinetriones* is exemplified in the regular formation of 2,4,7-trichloropteridine (phosphorus pentachloride in refluxing phosphoryl chloride for 4 h: 77%),[41] 4,6,7-trichloropteridine (likewise: 23%),[41] 4,6,7-trichloro-2-phenylpteridine (similarly: 78% crude but only 20% yield of pure sublimed material),[522, 1398] and 2,4,7-trichloro-6-phenylpteridine (same reagents for 8 h: ~80%).[1251] This is also shown in the conversion of 1,3-dimethyl-2,4,6(1*H*,3*H*,5*H*)- and 1,3-dimethyl-2,4,7(1*H*,3*H*,8*H*)-pteridinetrione (**17**) (by prolonged refluxing in phosphoryl chloride containing, in the latter case, some potassium chloride) into 6- (50%) and 7-chloro-1,3-dimethyl-2,4(1*H*,3*H*)-pteridinedione (**18**, X = Cl) (90%), respectively[1275]; in the similar conversion of the same triones by phosphoryl bromide at 120°C into 6- (80%) and 7-bromo-1,3-dimethyl-2,4(1*H*, 3*H*)-pteridinedione (**18**, X = Br) (81%), respectively[1275]; in the conversion of 1,3-dimethyl-7-phenyl-2,4,6(1*H*,3*H*,5*H*)-

pteridinetrione, by phosphoryl chloride–*N*,*N*-dimethylaniline, into 6-chloro-1,3-dimethyl-7-phenyl-2,4(1*H*,3*H*)-pteridinedione (94%)[1279]; and in the selective conversion of 1- and 3-methyl-2,4,7(1*H*,3*H*,8*H*)-pteridinetrione, by phosphoryl chloride containing potassium chloride at 80°C, into 7-chloro-1-methyl- (57%) and 7-chloro-3-methyl-2,4(1*H*,3*H*)-pteridinedione (85%), respectively.[354, 1709] 4-Amino-2,6,7(1*H*,5*H*,8*H*)-pteridinetrione with phosphorus pentachloride in phosphoryl chloride under mild conditions also gave a monochloro derivative, hopefully formulated as 4-amino-2-chloro-6,7(5*H*,8*H*)-pteridinedione, in unstated yield.[1369] The isomeric substrate, 2-amino-4,6,7(3*H*,5*H*,8*H*)-pteridinetrione gave the initially unidentified 2-amino-4-chloro-6,7(5*H*,8*H*)-pteridinedione,[531] which was later dechlorinated by hydriodic acid to give 2-amino-6,7(5*H*,8*H*)-pteridinedione, identical with a product synthesized unambiguously.[536] In contrast, 2,4,6(1*H*,3*H*,5*H*)-pteridinetrione with phosphorus pentachloride in phosphoryl chloride gave, not 2,4,6-trichloropteridine, but 2,4,6,7-tetrachloropteridine (**19**),[1251] a *C*-chlorination phenomenon reported occasionally in the pyrimidine[1527] and other heterocyclic series.

Chlorination of *pteridinetetrones* is seen in the conversion of the parent substrate, 2,4,6,7(1*H*, 3*H*,5*H*,8*H*)-pteridinetetrone, by phosphorus pentachloride in refluxing phosphoryl chloride for 8 h, into 2,4,6,7-tetrachloropteridine (**19**) in 33% yield.[441, 486] It is also shown in the conversion of 1,3-dimethyl-2,4,6,7(1*H*,3*H*,5*H*,8*H*)-pteridinetetrone by the use of similar reagents at 90°C for 48 h to give a 46% yield of 6,7-dichloro-1,3-dimethyl-2,4(1*H*,3*H*)-pteridinedione (**20**).[633, 1681]

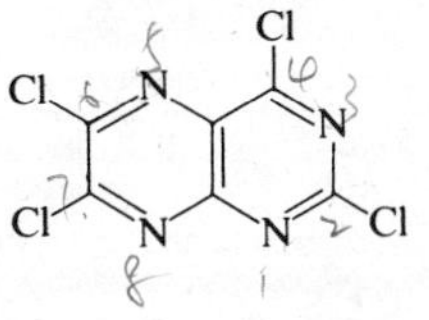

(**19**)

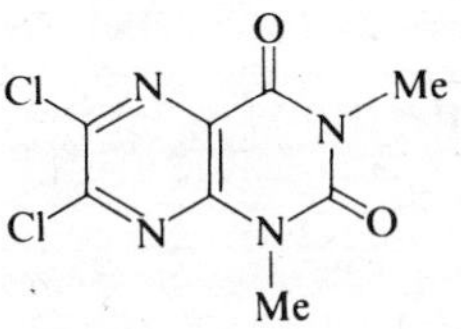

(**20**)

C. From Pteridine *N*-Oxides

In view of the number of pteridine *N*-oxides reported, their conversion into *C*-chloropteridines is relatively rare. However, the process is typified by treatment of 1,3-dimethyl-2,4(1*H*,3*H*)-pteridinedione 5-oxide (**21**) with acetyl chloride at −65°C followed by trifluoroacetic acid at −30°C, to give 6-chloro-1,3-dimethyl-2,4(1*H*,3*H*)-pteridinedione (**22**) in 89% yield[1275]; in the similarly induced conversions of 3-methyl-7-phenyl-2,4(1*H*,3*H*)-pteridinedione 5-oxide and the corresponding 5,8-dioxide into 6-chloro-3-methyl-7-phenyl-2,4(1*H*,3*H*)-pteridinedione (74%) and its 8-oxide (65%), respectively[331]; in the more complicated conversion of 6-hydroxyiminomethyl-2,4-pteridinediamine 8-oxide (**23**) by phosphoryl chloride–DMF into 7-chloro-2,4-bisdimethylaminomethyleneamino-6-pteridinecarbonitrile (**24**) in 93% yield[469]; and in the un-

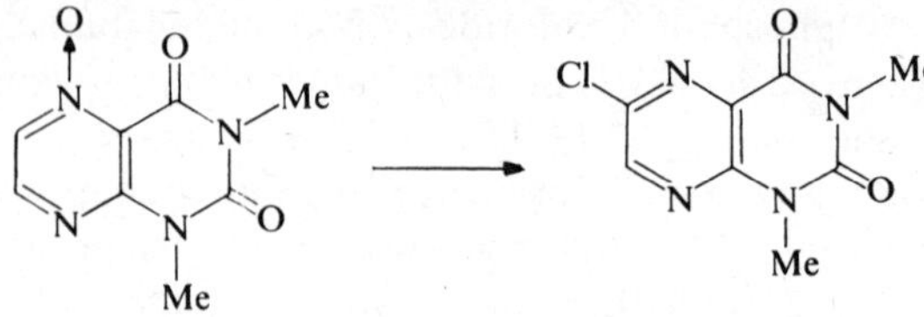

(21) (22)

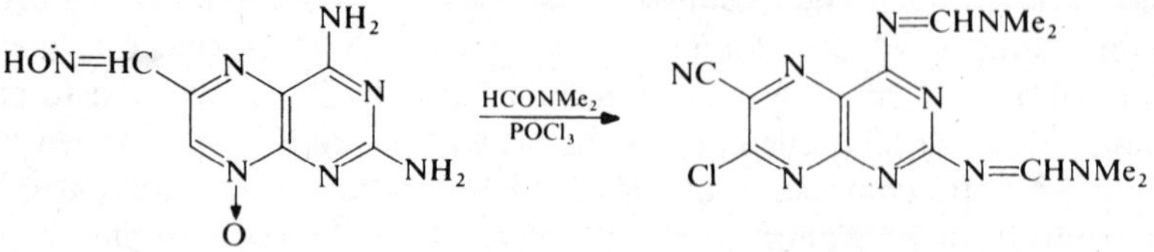

(23) (24)

usual β-rearrangement of an *N*-oxide involved during transformation of 2-amino-4(3*H*)-pteridinone 8-oxide (**25**) by acetyl chloride–trifluoroacetic acid into 2-amino-6-chloro-4(3*H*)-pteridinone (**26**) (>95%) or into the same product using phosphoryl chloride–trifluoroacetic acid (94% yield) or diphenylimidoyl chloride (73% yield).[1307] An analogous reaction was employed to convert 2,4-pteridinediamine 8-oxide into 6-chloro-2,4-pteridinediamine (**27**) (91%).[1307]

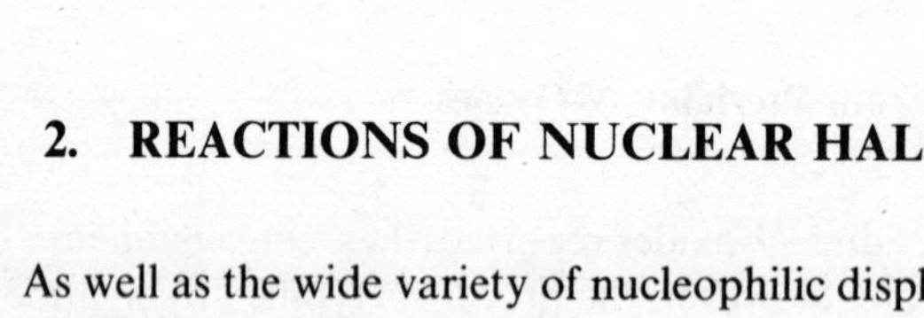

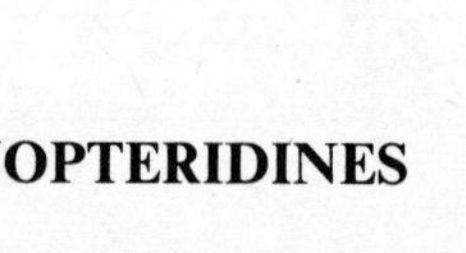

(25) (26) (27)

2. REACTIONS OF NUCLEAR HALOGENOPTERIDINES

As well as the wide variety of nucleophilic displacement reactions expected of them, chloropteridines can undergo even more facile covalent adduct formation in appropriate cases; in addition, some may be degraded to chloropyrazines under quite gentle conditions.

A. Formation of Covalent Adducts and Degradative Reactions

The addition of any halogeno substituent to the pteridine nucleus further increases the π-electron depletion at the carbon atoms, thereby facilitating both covalent adduct-formation and ring-fission reactions. Thus 6- and 7-chloropteridine reacted with ammonia in benzene at 5°C to give 6-chloro- (**28a**) and 7-chloro-3,4-dihydro-4-pteridinamine (**29a**), respectively. Although characterized, each adduct decomposed slowly in air or instantaneously in aqueous or alcoholic solution, to give the original chloropteridine in >95% yield.[146] 6-Chloro-4-cyclohexylamino- (**28b**), 4-benzylamino-6-chloro- (**28c**), 4-benzylthio-6-chloro- (**28d**), 4-benzylamino-7-chloro- (**29c**), and 4-benzylthio-7-chloro-3,4-dihydropteridine (**29d**) were made similarly and all were characterized.[146] The use of ammonia, benzylamine, or the benzyl mercaptan anion in a protic solvent at room temperature gave the regular 6- or 7-substituted pteridine, for example, 6- or 7-pteridinamine.[146] Treatment of 2-chloropteridine with diethylamine containing potassium permanganate at −75°C for 20 min gave 2-chloro-4-diethylaminopteridine (58%), clearly via the 3,4-adduct; on lengthening the reaction time to 12 h, 2,4-bisdiethylaminopteridine (56%) resulted.[934, cf. 1700] In dilute acid, 6,7-dichloropteridine formed a hydrate that was stable as its salt, 6,7-dichloro-4-hydroxy-3,4-dihydropteridine hemisulfate.[35]

a R = NH_2
b R = $C_6H_{11}NH$
c R = $PhCH_2NH$
d R = $PhCH_2S$

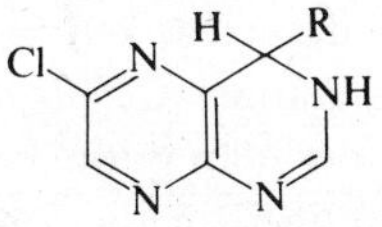

(28)

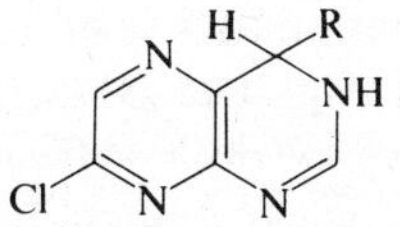

(29)

Degradation of 6-chloropteridine (**30**, R = H) at 75°C and pH 4 for 30 min gave 6-chloro-3-formamido-2-pyrazinecarbaldehyde (**31**, R = H) but at pH 3 for 4 h, the product was 3-amino-6-chloro-2-pyrazinecarbaldehyde (**32**, R = H).[35] Degradation of 6,7-dichloropteridine (**30**, R = Cl) in 0.1*M* hydrochloric acid at 90°C for 5 min gave 5,6-dichloro-3-formamido-2-pyrazinecarbaldehyde (**31**, R = Cl) but in 0.05*M* sulfuric acid at 90°C for 30 min the product was 3-amino-5,6-dichloro-2-pyrazinecarbaldehyde (**32**, R = Cl).[35] Finally, the acidic degradation of 7-chloropteridine gave 5-chloro-3-formamido- or 3-amino-5-chloro-2-pyrazinecarbaldehyde according to the severity of conditions.[35]

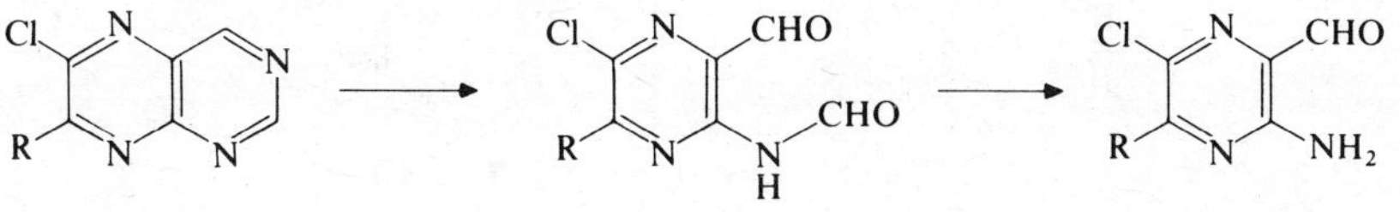

(30) **(31)** **(32)**

B. Reductive Dehalogenation

The pteridine ring is so easily reduced that reductive dechlorination usually results in nuclear reduction as well; exceptions are provided by the formation of 1,3-dimethyl-7-phenyl-2,4(1*H*,3*H*)-pteridinedione (**33**, R = H) (47%) by careful hydrogenation of its 6-chloro derivative (**33**, R = Cl) over palladium in ethanol,[1279] by the similar formation of 4,6-dimethyl-7,8-dihydropteridine from its 2-chloro derivative,[43] by the formation of 2- or 4-amino-6,7(5*H*,8*H*)-pteridinedione on treatment of their respective 4- or 2-chloro derivatives in warm concentrated hydriodic acid for 10 min,[536, 1369] and by the similar formation of 2-amino-4-oxo-3,4-dihydro-6-pteridinecarboxylic acid from its 7-chloro derivative.[1084] The more usual situation is exemplified in the partial dechlorination and nuclear reduction of 2,4,6,7-tetrachloropteridine by four molecular equivalents of lithium aluminum hydride to give 2,4-dichloro-5,6,7,8-tetrahydropteridine (**34**) in 96% yield[486]; in the partial dechlorination, partial hydrolysis, and nuclear reduction of the same tetrachloropteridine by hydrogenation in benzene over palladium or platinum to give 2,4-dichloro-7,8-dihydro-6(5*H*)-pteridinone (**35**, R = Cl), which underwent complete dechlorination on a subsequent hydrogenation over palladium in methanol, to afford 7,8-dihydro-6(5*H*)-pteridinone (**35**, R = H) in low yield[486]; and in the reductive dechlorination of 2-amino-6-chloro-4(3*H*)-pteridinone by dithionite, borohydride, or hydrogenation to give an unidentified nuclear-reduced 2-amino-4(3*H*)-pteridinone, which was treated with permanganate to afford 2-amino-4(3*H*)-pteridinone in ~50% yield.[1307] Dechlorination at an initial oxidation state that precludes further nuclear reduction, is exemplified in the formation of 5,6,7,8-tetrahydropteridine (60%) from its 2,4-dichloro derivative (**34**) by hydrogenation over palladium[486]; of 4,6-dimethyl-5,6,7,8-tetrahydropteridine (**36**, R = H) (~80%) from its 2-chloro derivative (**36**, R = Cl) similarly[1088]; and of 5,6,7,8-tetrahydropteridine from its 8-benzyl-4-chloro derivative (**37**) (14%) by treat-

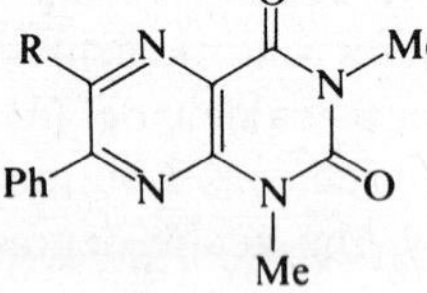

(33)

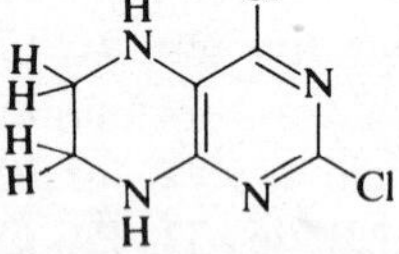

(34)

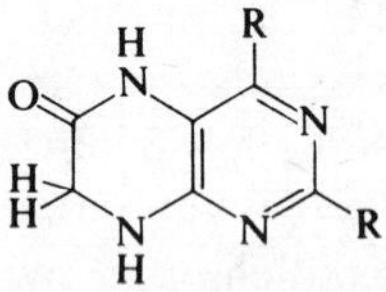

(35)

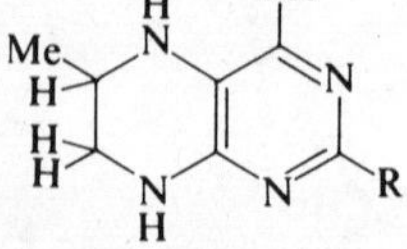

(36)

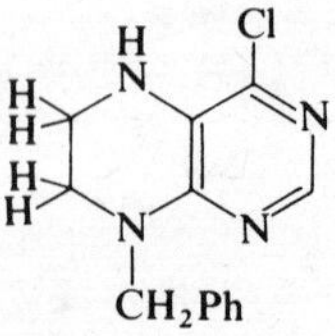

(37)

ment with sodium in liquid ammonia.[43] Treatment of 2-chloro-4,6,7-triphenylpteridine with potassium amide in liquid ammonia gave mainly 4,6,7-triphenyl-2-pteridinamine but also some 4,6,7-triphenylpteridine.[1122]

C. Aminolysis

No kinetic work has been done on the relative reactivity of a chloro substituent at the 2-, 4-, 6-, or 7-position of pteridine or on the effects of additional substituents on such reactivity. However, some qualitative pointers emerge during consideration of the preparative aminolyses described next.

The aminolysis of *simple 2-chloropteridines* is illustrated by the conversion of 2-chloropteridine into 2-hydrazinopteridine (**38**, R = NH_2) (methanolic hydrazine hydrate at 20°C for 10 min: 72%),[136] 2-ethylaminopteridine (**38**, R = Et)

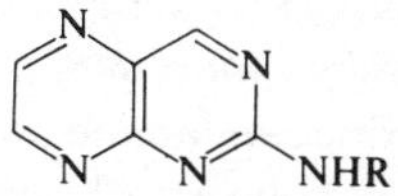

(**38**)

(neat ethylamine at 20°C for 3 h: unstated yield),[934] 2-propylaminopteridine (neat propylamine in excess, initially at room temperature for 5 min: 26%),[124] 2-isopropylaminopteridine (similarly: 20%),[124] 2-butylaminopteridine (similarly: 44%),[124] its *s*- and *t*-butylamino isomers (30°C for 45 min: 40%, 21%),[124] 2-diethylaminopteridine (neat amine at 20°C for 5 min: 25%),[124] and 2-dipropylaminopteridine (similarly: 20%)[124]; by the conversion of 2-chloro-4,6,7-trimethylpteridine into 2-*s*-butylamino- (48%), 2-isobutylamino- (51%), and 2-allylamino-4,6,7-trimethylpteridine (64%), using neat amine at room temperature in each case[124]; by the conversion of 2-chloropteridine, by trimethylamine in anhydrous benzene at 20°C for 3 h, into 2-trimethylammoniopteridine chloride[124]; by the conversion of 2-chloro-4,6,7-triphenylpteridine into 4,6,7-triphenyl-2-pteridinamine (ethanolic ammonia at 100°C for 30 min in a sealed tube: >95%[1118]; or potassium amide in liquid ammonia[1122]: 70%); and in the conversion of 2-chloro- into 2-benzylamino-4-trifluoromethylpteridine (benzylamine in refluxing benzene for 2 h: ~25%).[159]

Aminolysis of *simple 4-chloropteridines* is represented meagrely in the transformation of 4-chloro-6,7-dimethylpteridine (**39**, R = Cl) into 6,7-dimethyl-4-pteridinamine (**39**, R = NH_2) by an excess of ethanolic ammonia at 25°C for 15 min (77%)[170] or into 4-furfurylamino-6,7-dimethylpteridine by furfurylamine in hot benzene for 1 h.[662]

Aminolysis of *simple 6-chloropteridines* is illustrated by the conversion of 6-chloropteridine (**40**, R = Cl) into 6-pteridinamine (**40**, R = NH_2) by ammonia in benzene at 20°C for 1 h (53% yield),[31] or better, by aqueous ammonia at 5°C for

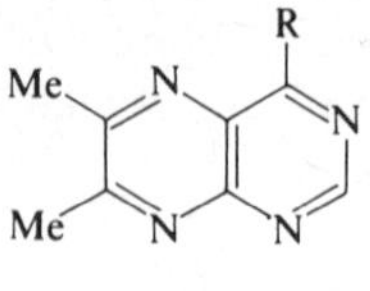

(39)

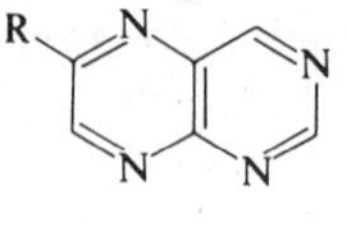

(40)

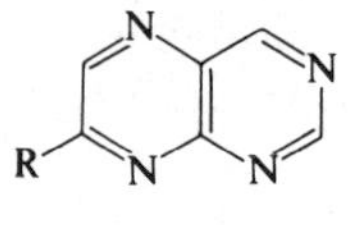

(41)

24 h (65%)[146]; also of the same substrate (**40**, R = Cl) into 6-dimethylaminopteridine (**40**, R = NMe_2) by methanolic dimethylamine at 20°C for 4 h (78%), into 6-cyclohexylaminopteridine (ethanolic cyclohexylamine at 20°C for 2 h: 51%), or into 6-benzylaminopteridine (aqueous benzylamine with stirring for 1 h: ~85%).[146]

Aminolysis of *simple 7-chloropteridines* is exemplified in the formation, from the parent 7-chloropteridine (**41**, R = Cl), of 7-pteridinamine (**41**, R = NH_2) (ammonia in benzene[33]; aqueous ammonia at 5°C for 24 h: 70%[146]; or boiling ethanolic ammonia: *good yield*),[146] 7-methylaminopteridine (**41**, R = NHMe) (ethanolic methylamine under reflux for 5 min: 82%),[131] 7-dimethylaminopteridine (**41**, R = NMe_2) (ethanolic dimethylamine under reflux for 15 min: 94%),[146] 7-cyclohexylaminopteridine (ethanolic amine similarly: 64%),[146] and 7-benzylaminopteridine (similarly: 83%).[146]

The *chloropteridinamines* should undergo aminolysis rather less easily than simple chloropteridines because of the π-electron donation from the amino group(s) in the former. However, practical chemists have such a strong tendency towards overkill when devising conditions for nucleophilic reactions that any discernible difference between conditions for the two categories may or may not be meaningful. Typical reactions are represented in the formation of 2,4-bisethylaminopteridine (**42**, R = NHEt) (51%) from its 2-chloro analog (**42**, R = Cl) by stirring in neat ethylamine at 20°C for 3 h[934]; of 4-β-dimethylaminoethylamino-2-(2′-methoxypyrimidin-4′-yl)pteridine (34%) from its 4-chloro analog by refluxing in neat β-dimethylaminoethylamine for 1 h[122]; of 7-methylamino- (refluxing butanolic methylamine for 6 h: 43%), 7-pentylamino- (refluxing neat amine for 2 h: 34%), 7-hexylamino- (refluxing neat amine for 5 h: 43%), 7-benzylamino- (refluxing neat amine for 2.5 h: 47%), 7-dimethylamino- (dimethylamine in DMF at 100°C for 6 h: 52%), 7-β-hydroxyethylamino- (ethanolic β-aminoethanol under reflux for 1 h: 92%), 7-hydrazino- (95% hydrazine under reflux for 5 min: 69% yield), and other such 7-substituted-6-phenyl-2,4-pteridinediamines (**43**, R = NHMe, etc.) from 7-chloro-6-phenyl-2,4-pteridinediamine (**43**, R = Cl)[525]; of 6,7-diphenyl-2,4-pteridinediamine (**43**,

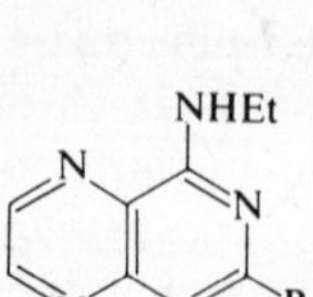

(42)

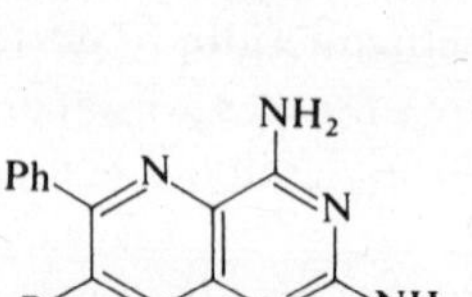

(43)

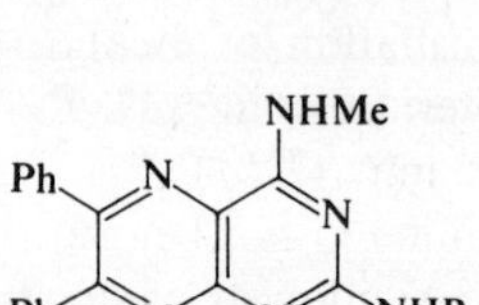

(44)

R = Ph) from 4-chloro-2-methylamino-6,7-diphenylpteridine by methanolic ammonia at 155°C for 16 h: Note the concomitant transamination at the 2-position[1189]; of 4-methylamino-6,7-diphenyl-2-pteridinamine (**44**, R = H) or 2,4-bismethylamino-6,7-diphenylpteridine (**44**, R = Me) from 4-chloro-6,7-diphenyl-2-pteridinamine with ethanolic methylamine at 110–120°C or 155°C, respectively (with additional transamination under the latter conditions)[110, cf. 141]; of 4-dimethylamino-, 4-diethylamino-, or 4-hydrazino-6,7-diphenyl-2-pteridinamine from the same substrate (46–70% yield),[891] and of other amines likewise.[848]

The *chloropteridinones* should likewise undergo aminolysis at a reduced rate on account of some electron donation to the π system from the tautomeric oxo–hydroxy substituent(s). For practical purposes, this appears to be unimportant and the process is exemplified in the conversion of 6-chloro-1,3-dimethyl- (**45**, X = Cl, Y = H) into 6-hydrazino-1,3-dimethyl- (**45**, X = $NHNH_2$, Y = H) (refluxing 80% hydrazine hydrate for 15 min: 70% yield), 1,3-dimethyl-6-methylamino- (**45**, X = NHMe, Y = H) (methylamine hydrochloride in refluxing *N*-methylformamide for 20 h: 33%), or 6-dimethylamino-1,3-dimethyl-2,4(1*H*,3*H*)-pteridinedione (**45**, X = NMe_2, Y = H) (ethanolic dimethylamine at 100°C in a sealed tube for 90 min: 70%).[1275] Aminolysis also converts 7-chloro-1,3-dimethyl- (**45**, X = H, Y = Cl) into 7-amino-1,3-dimethyl- (**45**, X = H, Y = NH_2) (ammonia in aqueous dioxane at 20°C for 14 h: 90%), 1,3-dimethyl-7-methylamino- (**45**, X = H, Y = NHMe) (alcoholic methylamine at 25°C for 30 min: 89%), or 7-dimethylamino-1,3-dimethyl-2,4(1*H*,3*H*)-pteridinedione (**45**, X = H, Y = NMe_2) (ethanolic dimethylamine at 25°C for 30 min: 90%).[1275] From these data it may be concluded cautiously that the 7- is considerably more activated than the 6-chloro substituent towards aminolysis. Other examples include the transformation of 6-chloro-3-methyl-4(3*H*)-pteridinone to 6-dimethylamino- (72%), 6-benzylamino- (60%), and other 6-(substituted-amino)-3-methyl-4(3*H*)-pteridinones (amine in β-methoxyethanol at 100°C for 2.5 h)[974]; 6-chloro-7(8*H*)-pteridinone to 6-amino-7(8*H*)-pteridinone (refluxing ethanolic ammonia for 90 min: ~50%)[36]; and this substrate (**45**, X = H, Y = Cl) to 7-ethoxycarbonylmethylamino-1,3-dimethyl-2,4(1*H*,3*H*)-pteridinedione (**45**, X = H, Y = $NHCH_2CO_2Et$) (glycine ethyl ester: *good yield*).[629] A less simple aminolysis is illustrated by the treatment of 6-chloro-1,3-dimethyl-7-propionyl-2,4(1*H*,3*H*)-pteridinedione (**46**, X = O) with hydrazine or methylhydrazine to give the intermediate hydrazone (**46**, X = :NNH_2 or :NNHMe), which underwent cyclization to afford 3-ethyl-5,7-dimethyl- (**47**, R = H) or 3-ethyl-1,5,7-

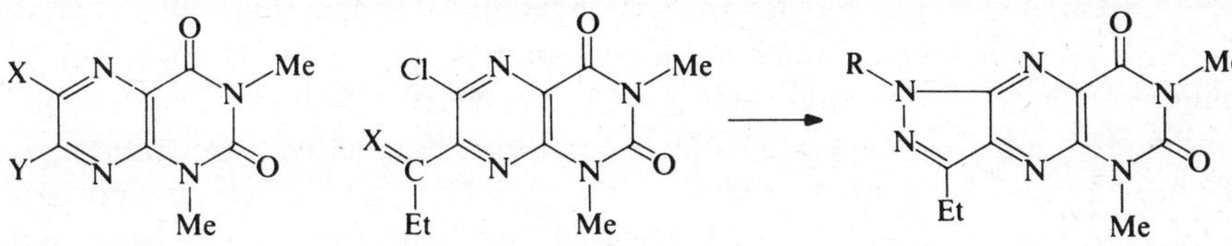

(45) **(46)** **(47)**

trimethyl-6,8(5*H*,7*H*)-1*H*-pyrazolo[3,4-*g*]pteridinedione (**47**, R = Me), respectively.[629] 6-Chloro-1,3,5-trimethyl-2,4-dioxo-1,2,3,4-tetrahydro-5-pteridinium tetrafluoroborate reacted readily with aniline in dichloromethane to give 1,3,5-trimethyl-6-phenylimino-5,6-dihydro-2,4(1*H*,3*H*)-pteridinedione (38%).[1680, 1759]

The aminolysis of *dichloropteridines* can sometimes be done in two stages. Thus, bearing in mind the previously mentioned greater activity of the 7-chloro substituent in (**45**, X = H, Y = Cl) compared with the 6-chloro substituent in (**45**, X = Cl, Y = H), it is not surprising that 6,7-dichloropteridine (**49**) reacted with (aqueous) ammonia at room temperature to give only 6-chloro-7-pteridinamine

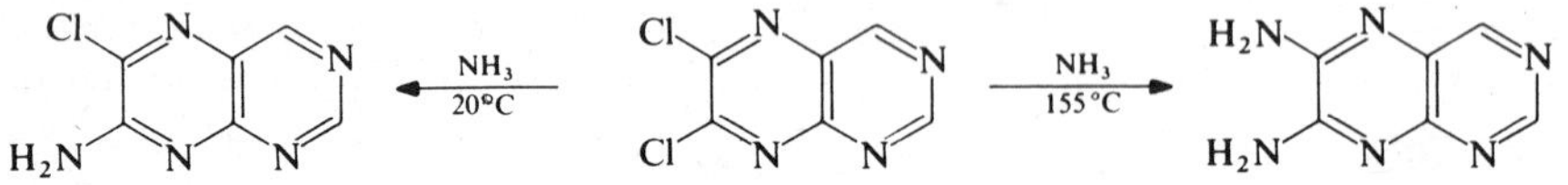

(**48**) (**49**) (**50**)

(**48**) (71% yield) but with (alcoholic) ammonia at 155°C to give 6,7-pteridinediamine (**50**) (46% yield).[36] Likewise, 4,7-dichloropteridine and ethanolic ammonia at <5°C gave 4-chloro-7-pteridinamine (**51**, R = Cl) (~65%), which underwent a second aminolysis by ethanolic methylamine at 80°C to give 4-methylamino-7-pteridinamine (**51**, R = NHMe), thus suggesting that the 7-chloro substituent is considerably more active than the 4-chloro substituent.[131] Along the same lines, 2,4-dichloro-6,7-dimethylpteridine (**52**, R = Cl) reacted with ethanolic ammonia (2 mol) at room temperature for 1 h to give (entirely?) 2-chloro-6,7-dimethyl-4-pteridinamine (**52**, R = NH_2) (55% yield) (confirmed in structure by primary synthesis from 2-chloro-4,5,6-pyrimidinetriamine and biacetyl), thereby suggesting that a 4-chloro- is more activated than a 2-chloro substituent.[170]

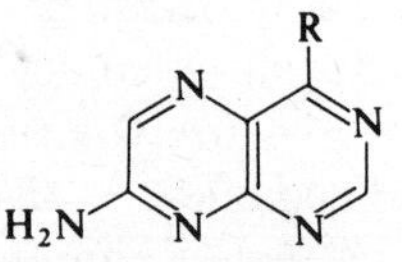

(**51**)

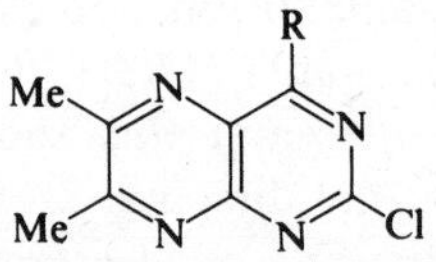

(**52**)

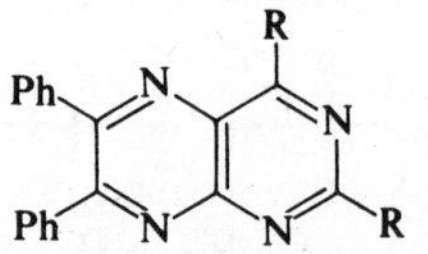

(**53**)

The conclusion respecting the relative reactivities of 6- and 7-chloro substituents is upheld in the facile conversion of 6,7-dichloro-1,3-dimethyl-2,4(1*H*,3*H*)-pteridinedione (**45**, X = Y = Cl) into 7-amino-6-chloro-1,3-dimethyl- (**45**, X = Cl, Y = NH_2), 6-chloro-7-dimethylamino-1,3-dimethyl- (**45**, X = Cl, Y = NMe_2), 6-chloro-1,3-dimethyl-7-methylamino- (**45**, X = Cl, Y = NHMe), 7-anilino-6-chloro-1,3-dimethyl- (**45**, X = Cl, Y = NHPh), 6-chloro-7-hydrazino-1,3-dimethyl- (**45**, X = Cl, Y = $NHNH_2$), 6-chloro-1,3-dimethyl-7-piperidino-, 7-azido-6-chloro-1,3-dimethyl- (**45**, X = Cl, Y = N_3), and even 6-chloro-7-(imidazol-1′-yl)-1,3-dimethyl-2,4(1*H*,3*H*)-pteridinedione. In contrast, the second stage aminolysis of these products proved all but impossible except for the transformation of (**45**, X = Cl, Y = NH_2) into 7-amino-1,3-dimethyl-6-piperidino-2,4(1*H*,3*H*)-pteridinedione by prolonged refluxing in neat piperidine.[633] Other difficult aminolyses include the conversion of 2,4-dichloro-6,7-diphenylpteridine (**53**, R = Cl), by refluxing in the appropriate neat amine for >1 h, into 2,4-bisbenzylamino- (**53**, R = $NHCH_2Ph$) (87%),[467, 1315] 2,4-dihydrazino- (**53**, R = $NHNH_2$) (91%),[467] 2,4-bis-β-hydroxyethylamino- (**53**, R = $NHCH_2CH_2OH$) (87%),[467] 2,4-dimorpholino- (91%),[467] 2,4-bis-β-dimethylaminoethylamino- (92%),[467] and similar 2,4-di(substituted-amino)-6,7-diphenylpteridines[467]; of 2,4-dichloro-6,7-dimethylpteridine into 2,4-bisdimethylamino-6,7-dimethylpteridine (ethanolic dimethylamine at 110°C for 8 h[1315]; of 4,6-dichloropteridine into 4,6-pteridinediamine (aqueous ammonia at 140°C for 4 h: ~25%)[41]; and of 2,4-dichloro-5,6,7,8-tetrahydropteridine into 5,6,7,8-tetrahydro-2,4-pteridinediamine (liquid ammonia at 200°C for 10 h: 85%; because 200°C was above liquid ammonia's critical temperature of ~133°C, the resulting pressure must have been extremely high).[486]

Examples for the aminolysis of *trichloropteridines* unfortunately contain no reports of partial or stepwise reactions. Thus appropriate trichloropteridines gave 2,4,7-pteridinetriamine (aqueous ammonia at 140°C for 4 h: 50%),[41] 4,6,7-pteridinetriamine (**54**, R = H) (similarly),[41] 2-phenyl-4,6,7-pteridinetriamine (**54**, R = Ph),[1398] 4,6,7-trismethylamino-2-phenylpteridine,[1398] 4-phenyl-2,6,7-pteridinetriamine (liquid ammonia at 140°C for 6 h: 42%; see previous comment),[522] and other such triamines.[1398]

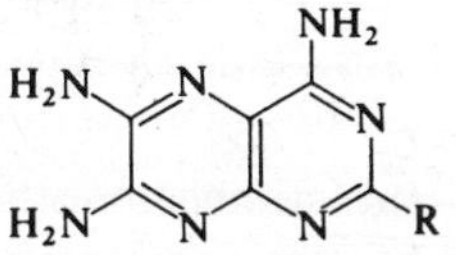

(54)

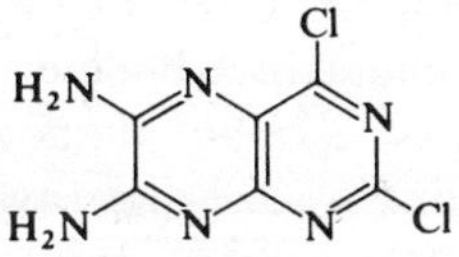

(55)

The aminolysis of 2,4,6,7-*tetrachloropteridine* with ammonia has been reported in three stages: An etherial solution with ammonia gas passed for 2 h at 25°C gave a yet unidentified trichloropteridinamine (maybe 2,4,6-trichloro-7-pteridinamine?)[441]; liquid ammonia at −70°C for 6 h gave 2,4-dichloro-6,7-pteridinediamine (**55**) (92%), thus providing some evidence that a 6-chloro is more reactive than a 4-chloro substituent[486]; and liquid ammonia at 150°C (see previous comment) for 8 h gave 2,4,6,7-pteridinetetramine (**54**, R = NH_2) (95%).[486] Stepwise alkylaminolysis of tetrachloropteridine has also been mentioned,[486] but details are not readily available.[1529]

With the usual reservations about mutual activation of chloro substituents in di- or polychloropteridines, it may be tentatively concluded that data highlighted in the previous paragraphs indicates the positional order of reactivity for chloro substituents on pteridine (towards aminolysis and perhaps other nucleophilic processes) as 7 > 6 > 4 > 2.

D. Hydrolysis and Alcoholysis

The *acidic hydrolysis* of chloropteridines has seldom been used but it is represented by the treatment of 4-chloro-8-methyl- (**56**, R = H) or 4-chloro-6,8-dimethyl-7(8*H*)-pteridinone (**56**, R = Me) in 2.5*M* hydrochloric acid at 100°C to give 8-methyl- or 6,8-dimethyl-4,7(3*H*,8*H*)-pteridinedione, respectively, in 60–70% yield.[248] It is also represented by the treatment of 6-chloro-1,3,5-trimethyl-2,4-dioxo-1,2,3,4-tetrahydro-5-pteridinium tetrafluoroborate in water at 25°C for 10 min to give 1,3,5-trimethyl-2,4,6(1*H*,3*H*,5*H*)-pteridinetrione (73%).[1680]

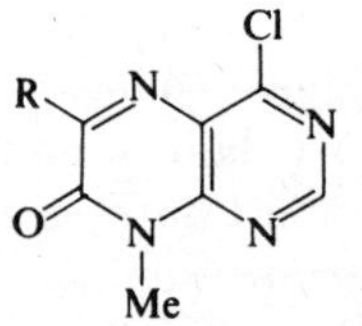

(56)

Alkaline hydrolysis is better represented, for example, in the conversion of 7-chloropteridine into 7(8*H*)-pteridinone (0.5*M* sodium carbonate at 20°C for 15 min: 86%),[35] of 6-chloropteridine into 6(5*H*)-pteridinone (1*M* sodium carbonate at 20°C for 15 min: 66%),[35] of 4-chloro-6,7-dimethylpteridine into 6,7-dimethyl-4(3*H*)-pteridinone (in boiling water maintained at pH 6 by the addition of sodium bicarbonate: 48%),[170] of 2-chloro-6,7-dimethylpteridine into 6,7-dimethyl-2(1*H*)-pteridinone (likewise: 39%),[170] of 2,4-dichloro-6,7-dimethylpteridine into 6,7-dimethyl-2,4(1*H*,3*H*)-pteridinedione (boiling aqueous sodium carbonate for 3 h: 76%),[170] of 2,4,6,7-tetrachloropteridine into 2,4-dichloro-6,7(5*H*,8*H*)-pteridinedione (boiling aqueous sodium carbonate for 3 h: 76%),[170] of 2,4,6,7-tetrachloropteridine into 2,4-dichloro-6,7(5*H*,8*H*)-pteridinedione (dissolution in 0.8*M* sodium hydroxide at 80°C),[441] of 6,7-dichloropteridine into 6-chloro-7(8*H*)-pteridinone (**57**) (1*M* sodium carbonate at 25°C for 1 h: 67% yield),[36] of 6,7-dichloro-1,3-dimethyl-2,4(1*H*,3*H*)-pteridinedione into 6-chloro-1,3-dimethyl-2,4,7(1*H*,3*H*,8*H*)-pteridinetrione (aqueous acetonic sodium hydroxide at 25°C: 51%),[1681] of 6-chloro-7-pteridinamine into 7-amino-6(5*H*)-pteridinone (**58**) (aqueous sodium carbonate at 100°C for 1 h: 56%; cf. conditions in the two previous examples for the hydrolysis of a 7-chloro substituent),[36] of 2-amino-6-chloro-4(3*H*)-pteridinone into 2-amino-

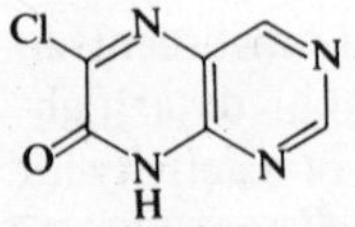

(57)

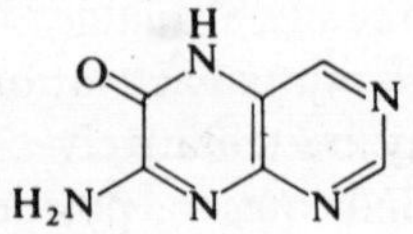

(58)

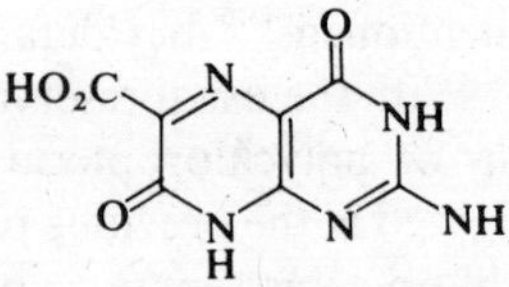

(59)

4,6(3*H*,5*H*)-pteridinedione (1*M* sodium hydroxide at 100°C for 15 min: 91%),[1307] and of 7-chloro-2,4-bisdimethylaminomethyleneamino-6-pteridinecarbonitrile (**24**) into 2-amino-4,7-dioxo-3,4,7,8-tetrahydro-6-pteridinecarboxylic acid (**59**) by a multihydrolysis process in 2.5*M* sodium hydroxide under reflux for 24 h.[469]

The *alcoholysis* of chloropteridines is usually done by treatment with the appropriate alcoholic sodium alkoxide. Such a process gave 7-methoxypteridine (reflux for 1 h: 70%),[33] 6-methoxypteridine (20°C for 30 min: 53%),[31] 4-methoxy-6,7-dimethylpteridine (20°C for 3 h: 30%),[170] 2,4-dimethoxy-5,6,7,8-tetrahydropteridine (reflux under nitrogen for 15 h: 50%),[486] 2-ethoxy-4-methyl-6,7-diphenyl-7,8-dihydropteridine (**60**) (90%),[398, 1230] 4-ethoxy-6,7-diphenyl-2-pteridinamine,[891] 7-methoxy-4-pteridinamine (reflux for 1 h: 65%),[445] 6-ethoxy-2,7-diphenyl- and 7-ethoxy-2,6-diphenyl-4-pteridinamine (reflux for 3 h: ~25%),[526] 6-methoxy-4(3*H*)-pteridinone (reflux for 4 h: 54%),[974] 6-methoxy-1,3-dimethyl-2,4(1*H*,3*H*)-pteridinedione (25°C for 8 h: 49%),[108] 2-amino-6-methoxy-4(3*H*)-pteridinone (reflux for 24 h: 63%),[1307] 2-amino-7-methoxy-4(3*H*)-pteridinone (**61**, R=H) (reflux for 4 h: ~90%),[380] its 3-methyl derivative (**61**, R=Me) (exothermic reaction was sufficient heating: 55%),[383] and 7-methoxy-6-phenyl-2,4-pteridinediamine (reflux for 2.5 h: 63%).[525] These reaction conditions appear to have little logic and some of the poor yields might perhaps be improved by rather less vigorous treatment.

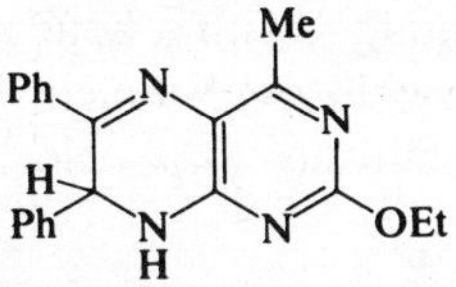

(60)

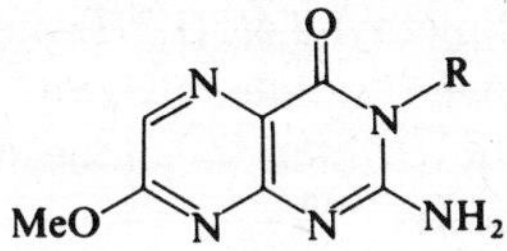

(61)

E. Thiolysis, Alkanethiolysis, and Arenethiolysis

Direct thiolysis by treatment with sodium hydrogen sulfide in ethanol or water is commonly used to convert chloropteridines into the corresponding pteridinethiones. The indirect route via a thiouronium salt does not appear to have been used. Although both alkylthio- and arylthiopteridines may be made from chloropteridines with a sodium alkanethiolate or (free) arenethiol, respectively, this route to the arylthiopteridines is the more important because, unlike alkylthiopteridines, they are not available by *S*-arylation of pteridinethiones.

Thiolysis is exemplified in the conversion of the corresponding chloropteridine into 6(5*H*)-pteridinethione (ethanolic sodium sulfide at 35°C: 84%),[36] 6,7(5*H*,8*H*)-pteridinedithione (sodium hydrogen sulfide in water at 10°C for 48 h: ~85%),[36] 6,7-dimethyl-2(1*H*)- and 6,7-dimethyl-4(3*H*)-pteridinethione (**62**) (aqueous sodium hydrogen sulfide at 95°C for 2 h: 80, 65%),[170] 4,6,7-

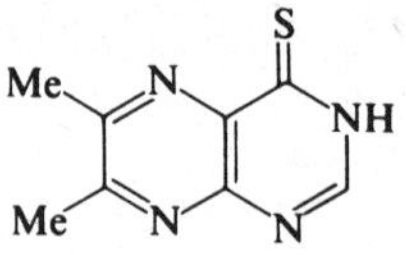

(62)

triphenyl-2(1*H*)-pteridinethione (aqueous alcoholic sodium hydrogen sulfide under reflux for 10 min),[1120] 6,7-dimethyl-2,4(1*H*,3*H*)-pteridinedithione (aqueous reagent at 95°C for 2 h: 82%),[170] 2-amino-4-pentyloxy-7(8*H*)-pteridinethione (64%),[95, 1681] 2-amino-4-oxo-7-thioxo-3,4,7,8-tetrahydro-6-pteridinecarboxylic acid (**63**) (aqueous reagent at 25°C for 3 h: 39%),[906] 1- and 3-methyl-7-thioxo-7,8-dihydro-2,4(1*H*,3*H*)-pteridinedione,[354, 1709] 1,3,6-trimethyl-7-thioxo-7,8-dihydro-2,4(1*H*,3*H*)-pteridinedione,[1709] 1,3-dimethyl-6-thioxo-5,6-dihydro- and 1,3-dimethyl-7-thioxo-7,8-dihydro-2,4(1*H*,3*H*)-pteridinedione (ethanolic sulfide),[1681] 2,4-diamino-6-phenyl-7(8*H*)pteridinethione (ethanolic reagent under reflux for 2 h: 34%),[525] and 2,4-diamino-7(8*H*)-pteridinethione.[95] In addition, 6,7-dichloropteridine has been converted (a) by initial preferential hydrolysis at the 7-position followed by thiolysis, into 6-thioxo-5,6-dihydro-7(8*H*)-pteridinone (**64**, X = S, Y = O) in ~40% yield; and (b) by initial preferential thiolysis followed by hydrolysis, into 7-thioxo-7,8-dihydro-6(5*H*)-pteridinone (**64**, X = O, Y = S) in 50% yield.[36] Similarly, 6,7-dichloro-1,3-dimethyl-2,4(1*H*,3*H*)-pteridinedione with aqueous ethanolic sodium hydrogen sulfide at 25°C gave 6-chloro-1,3-dimethyl-7-thioxo-7,8-dihydro-2,4(1*H*,3*H*)-pteridinedione preferentially in 90% yield.[1681] Other examples are known.[1680]

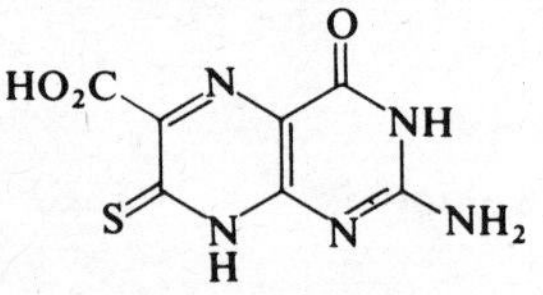

(63)

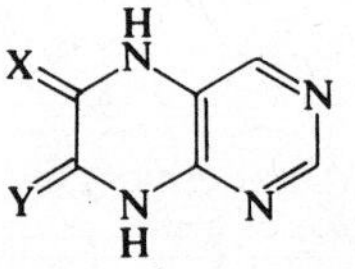

(64)

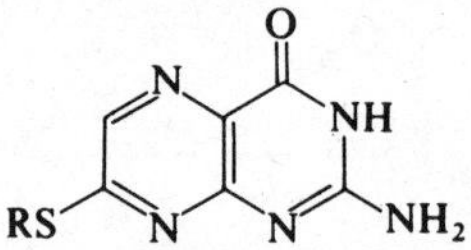

(65)

The process of *alkanethiolysis*, using sodium alkyl sulfides (sodium alkanethiolates) is represented in the formation of 6- and 7-benzylthiopteridine (benzyl mercaptan and potassium acetate in acetone under reflux for 10 min: 70–75%),[146] 6-methylthio-4(3*H*)-pteridinone (aqueous sodium methyl sulfide at 100°C for 20 min: 36%),[974] its 6-benzylthio analog (68%),[974] 7-isopropylthio-1,3,6-trimethyl-2,4(1*H*,3*H*)-pteridinedione (sodium isopropyl sulfide in boiling toluene for 30 min: 77%),[1709] 2-amino-6-methylthio-4(3*H*)-pteridinone (**65**, R = Me) (methanolic sodium methyl sulfide under reflux for 24 h: 98%),[1307] 2-amino-6-benzylthio-4(3*H*)-pteridinone (**65**, R = CH_2Ph) (methanolic sodium benzyl sulfide under reflux for 30 h: 83%),[1307] 6-methylthio-2,4-pteridinediamine (methanolic sodium methyl sulfide under reflux for 3 h: 72%),[1307] 6-ethoxycarbonylmethylthio-2,4-pteridinediamine (sodium salt of ethyl mercaptoacetate in refluxing methanol for 16 h: 51%),[1307] and 6-benzylthio-2,4-pteridinediamine (methanolic sodium benzyl sulfide under reflux for 16 h: 79%).[1307] See also ref. 1759.

Arenethiolysis is illustrated in the conversion of appropriate chloropteridines, by heating with an arenethiol in a high boiling polar solvent, into 2-amino-6-phenylthio-4(3*H*)-pteridinone (thiophenol in DMF under reflux for 3 h: 85%; thiophenol plus sodium bisulfite in DMF at 25°C for 24 h: 97%)[1307] 6-phenylthio-2,4-pteridinediamine (thiophenol in refluxing DMF for 20 min: 47%),[1307] 2-amino-6-*p*-carboxyphenylthio-4(3*H*)-pteridinone (*p*-mercaptobenzoic acid in refluxing DMF for 5 h: 80%),[1307] 6-(2′,4′,5′-trichlorophenylthio)-2,4-pteridinediamine (2,4,5-trichlorobenzenethiol in dimethyl sulfone at 200°C for 20 min: 51%),[849] 6-α- and 6-β-naphthylthio-2,4-pteridinediamine (α- or β-naphthalenethiol under similar conditions: 88, 54%),[849] and other 6-(substituted-phenyl)thio-2,4-pteridinediamines.[849, 1307] It is strange that the reaction has not been applied other than at the 6-position of pteridine; note that sodium arenethiolates have not been used.

F. Conversion into Sulfonic Acids

The displacement of a chloro substituent by sulfite ion to give the corresponding sulfonic acid as its salt has been rarely used in the pteridine series. However, 7-chloro-1,3-dimethyl-2,4(1*H*,3*H*)-pteridinedione has been so converted under gentle conditions into sodium 1,3-dimethyl-2,4-dioxo-1,2,3,4-tetrahydro-7-pteridinesulfonate (65%) and the preferential reaction of 6,7-dichloro-1,3-dimethyl-2,4(1*H*,3*H*)-pteridinedione with aqueous sodium sulfite to give only sodium 6-chloro-1,3-dimethyl-2,4-dioxo-1,2,3,4-tetrahydro-7-pteridinesulfonate (59%) has been reported.[1681] 1,3,6-Trimethyl-2,4-dioxo-1,2,3,4-tetrahydro-7-pteridinesulfonic acid was made similarly (sodium salt: 63%).[1709]

3. PREPARATION OF EXTRANUCLEAR HALOGENOPTERIDINES

Pteridines with halogenated substituents may be made by primary syntheses, by direct halogenation, from pteridines with hydroxylated substituents, or by derivatization using an halogenated reagent.

A. By Primary Syntheses

Syntheses for a number of extranuclear halogenopteridines have been included in Chapters II and III. This method is applicable only to the formation of products in which the halogeno substituent is sufficiently stable to survive the procedure(s) required. Thus few halogenoalkyl- but many halogenoarylpteridines have been so made. Typical examples *from pyrimidines* include the condensation of 6-trifluoromethyl-4,5-pyrimidinediamine (**66**) with polyglyoxal in bis-β-methoxyethyl ether at 155°C to give 4-trifluoromethylpteridine (**67**, R = H) or with biacetyl in refluxing *t*-butyl alcohol to give its 6,7-dimethyl derivative (**67**, R = Me)[156]; of 2,5,6-triamino-4(3*H*)-pyrimidinone with α,β-dibromopropionaldehyde, by shaking at 25°C in a mixture of water and benzene for 30 min followed by aerial oxidation during work-up, to give 2-amino-6-bromomethyl-4(3*H*)-pteridinone in 70% yield[731]; of 6-*p*-chloroanilino-2,4,5-pyrimidinetriamine with biacetyl to give 4-*p*-chloroanilino-6,7-dimethyl-2-pteridinamine (~75%)[1206]; of 2,4,5,6-pyrimidinetetramine with *p*-trifluoromethylbenzaldehyde and then hydrogen cyanide to give 5-(α-cyano-*p*-trifluoromethylbenzyl)amino-2,4,6-pyrimidinetriamine (**68**) and then on oxidative cyclization, 6-*p*-trifluoromethylphenyl-2,4,7-pteridinetriamine[523]; and of 2,5,6-triamino-4(3*H*)-pyrimidinone with α,β-dibromopropionaldehyde and *N*-[4-amino-2-chloro(or fluoro)benzoyl]glutamic acid to give 2′-chloro(or fluoro)pteroylglutamic acid in low yield.[775] In addition, the use of a Boon-type synthesis is exemplified in the reductive cyclization of 6-*p*-chlorophenacylamino-5-*p*-chlorophenylazo-2-dimethylamino-4(3*H*)-pyrimidinone to afford 6-*p*-chlorophenyl-2-dimethylamino-7,8-dihydro-4(3*H*)-pteridinone.[110]

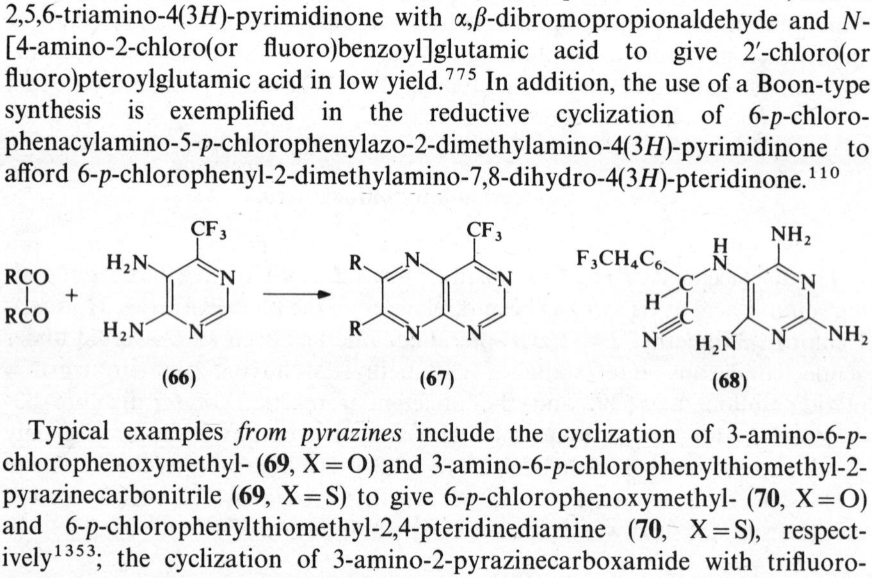

(66) **(67)** **(68)**

Typical examples *from pyrazines* include the cyclization of 3-amino-6-*p*-chlorophenoxymethyl- (**69**, X = O) and 3-amino-6-*p*-chlorophenylthiomethyl-2-pyrazinecarbonitrile (**69**, X = S) to give 6-*p*-chlorophenoxymethyl- (**70**, X = O) and 6-*p*-chlorophenylthiomethyl-2,4-pteridinediamine (**70**, X = S), respectively[1353]; the cyclization of 3-amino-2-pyrazinecarboxamide with trifluoroacetic acid–trifluoroacetic anhydride to give 2-trifluoromethyl-4(3*H*)-pteridinone[20]; and the similar use of 3-amino-2-pyrazinecarbothioamide to give 2-trifluoromethyl-4(3*H*)-pteridinethione.[20]

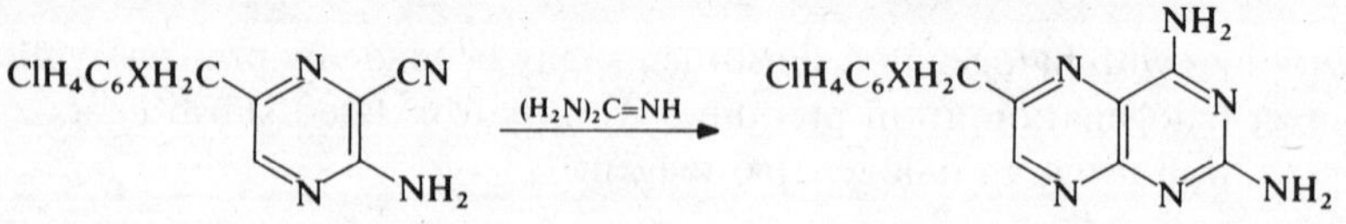

(69) **(70)**

B. By Direct Halogenation

The direct halogenation of methyl groups attached to pteridine has been covered in Chapter IV, Section 2.B(3). Typical cases are the bromination of 6- or

7-methyl-4(3*H*)-pteridinone in hot acetic acid to give 6- or 7-bromomethyl-4(3*H*)-pteridinone, respectively[121]; the conversion of 2-amino-7-methyl- into 2-amino-7-bromomethyl- or 2-amino-7-dibromomethyl-4(3*H*)-pteridinone by bromine in hydrobromic acid, according to conditions and the amount of bromine used[112, 1359, 1528]; and the formation of 2,4-diamino-6-chloromethyl-7(3*H*)-pteridinone 8-oxide, using chlorine in acetic acid under pressure at 100°C.[1296]

Halogenation of other groups attached to pteridine is exemplified in the formation of 2-amino-7-α-bromoacetonyl-4,6(3*H*,5*H*)-pteridinedione (**71**) (dioxane dibromide in pyridine at 50°C: 57%),[498] of the 7-γ-benzoyloxy-α-bromoacetonyl analog (similarly: crude product only),[498] 2-amino-6-α-bromoacetonyl-4,7(3*H*,8*H*)-pteridinedione (bromine in acetic acid at 25°C: 54%),[499] 2,4-diamino-6-α-bromo-α-carboxymethyl-7(8*H*)-pteridinone (**72**) (bromine in sulfuric acid plus acetic acid at 25°C: ~85%),[506] and 2-amino-6-α-bromo-α-carboxymethyl-4(3*H*)-pteridinone (similarly: 70%)[1308]; halogenation of the benzene ring in pteroylglutamic acid is also possible, to give products such as 3′-iodopteroylglutamic acid (**73**, R=I, Y=H, Z=NH_2) (iodine monochloride in hydrochloric acid at 30°C: ~80%)[746]; iodine monochloride in DMF at 25°C: 65%),[1104] 3′,5′-dibromopteroylglutamic acid (**73**, R=Y=Br, Z=NH_2) (bromine in hydrochloric acid at 5°C),[745] 3′,5′-dichloropteroylglutamic acid (**73**, R=Y=Cl, Z=NH_2) (likewise: >90%),[745] and analogous compounds.[745, 746, 1107] 2-Acetamido-6-methyl-4(3*H*)-pteridinone with bromine in acetic acid at 95°C has been reported to give a *perbromide*, corresponding in composition to the addition of two bromine atoms.[929]

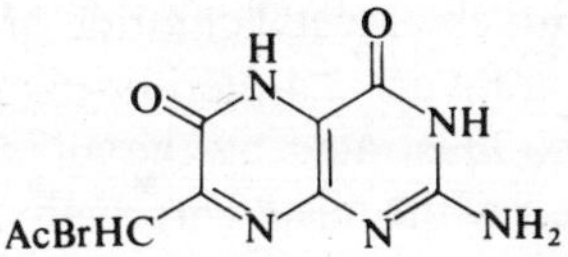

(71)

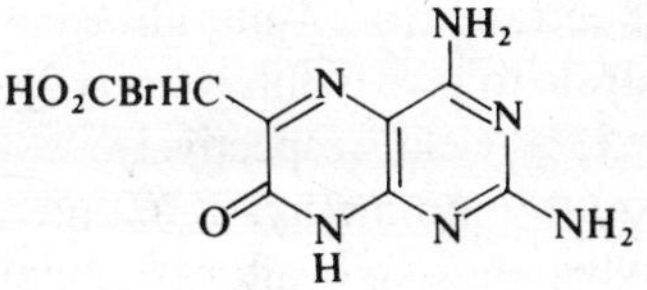

(72)

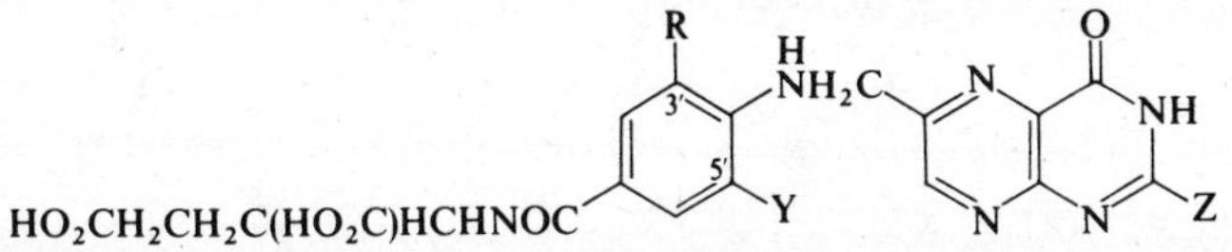

(73)

C. From Pteridinyl Alcohols

The formation of extranuclear halogenopteridines from the corresponding alcohols is poorly represented, probably because such substrates are themselves uncommon. However, 6-hydroxymethyl-2,4-pteridinediamine with thionyl chloride gave 6-chloromethyl-2,4-pteridinediamine (**74**, X=Cl)[818, 1522]; the same substrate with dibromotriphenylphosphorane (Ph_3PBr_2, from triphenylphosphine and bromine) in *N*,*N*-dimethylacetamide at 25°C for 2 h gave 6-bromomethyl-2,4-pteridinediamine (**74**, X=Br) as hydrobromide in 49% yield[390, 595, 1137, 1465]; and 2-amino-6-[L-*threo*]-α,β-dihydroxypropyl-4(3*H*)-pteridinone (**75**, R=OH, Y=H) with 2-acetoxy-2-methylpropionyl chloride in acetonitrile at 25°C for 3 days gave 6-[L-*threo*]-β-acetoxy-α-chloropropyl-2-amino-4(3*H*)-pteridinone (**75**, R=Cl, Y=Ac) (57% as hydrochloride) and then by treatment with methanolic hydrogen chloride, 2-amino-6-[L-*threo*]-α-chloro-β-hydroxypropyl-4(3*H*)-pteridinone (**75**, R=Cl, Y=H) (72%).[1439] In addition, 2-amino-6-bis(β-hydroxyethyl)aminomethyl-4(3*H*)-pteridinone with thionyl chloride in warm chloroform gave 2-amino-6-bis(β-chloroethyl)aminomethyl-4(3*H*)-pteridinone (50%).[408]

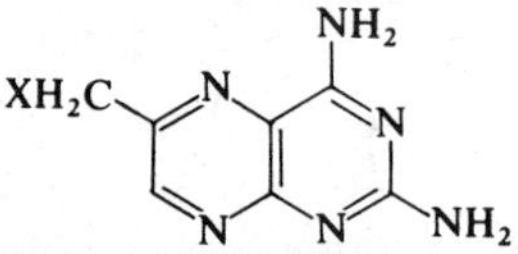

(74)

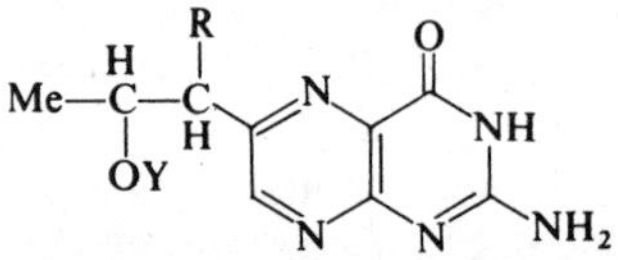

(75)

D. By Derivatization with a Halogenated Reagent

This potentially useful process is illustrated by the chloroacetylation of 5,6,7,8-tetrahydro-2,4-pteridinediamine with chloracetyl chloride or chloracetic anhydride to give 5-chloroacetyl-5,6,7,8-tetrahydro-2,4-pteridinediamine (**76**) in 48 or 33% yield, respectively.[267] It is also illustrated by the aminolysis of 2-amino-6-bromo-methyl-4(3*H*)-pteridinone with bis(β-chloroethyl)amine in triethylamine to afford 2-amino-6-bis(β-chloroethyl)aminomethyl-4(3*H*)-pteridinone (63%).[408, 895]

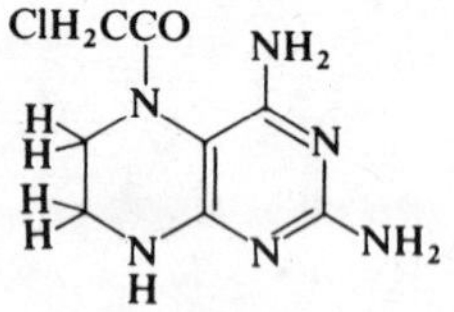

(76)

4. REACTIONS OF EXTRANUCLEAR HALOGENOPTERIDINES

Halogeno substituents attached indirectly to the pteridine nucleus are only mildly affected by the latter and hence tend, either to react simply as do alkyl halides, or to resist reacting as do most aryl halides. The range of reactions reported for extranuclear halogenopteridines is quite limited.

A. Reductive Dehalogenation

The process for reductive removal of an extranuclear halogeno substituent is represented only in the preferential partial dehalogenation of 2-amino-6,7-bisbromomethyl-4(3*H*)-pteridinone (**77**, R = Br) by warming with aqueous hydriodic acid to give (mainly) 2-amino-6-bromomethyl-7-methyl-4(3*H*)-pteridinone (**77**, R = H) (~85%),[112, 680] and in the conversion of 2-amino-7-dibromomethyl-4(3*H*)-pteridinone, by hydriodic acid in acetic acid, into 2-amino-7-methyl-4(3*H*)-pteridinone (60%).[1359]

B. Aminolysis

This process has been fairly well used, but mainly to prepare pteroylglutamic acid analogs. Thus 6-bromomethyl-4(3*H*)-pteridinone (**78**, R = Br) reacted in refluxing ethanolic solution with aniline, *p*-anisidine, *p*-aminobenzoic acid, sodium sulfanilate, or potassium *p*-aminobenzoylglutamate to give 6-anilinomethyl- (**78**, R = NHPh) (67%), 6-*p*-anisidinomethyl- (70%), 6-*p*-carboxyanilinomethyl- (28%), or 6-*p*-sulfoanilinomethyl-4(3*H*)-pteridinone (33%) and 2-deaminopteroylglutamic acid (**73**, R = Y = Z = H) (44%).[121, cf. 314] 6-Bromomethyl-4,7(3*H*,8*H*)-pteridinedione reacted with *p*-aminobenzoic acid in formic acid at 25°C for 2 days to give 6-*p*-carboxyanilinomethyl-4,7(3*H*,8*H*)-pteridinedione (25%).[279] 2-Amino-6-bromomethyl-4(3*H*)-pteridinone reacted with "dilute diethanolamine" at 120°C during 20 h to give 2-amino-6-bis(*β*-hydroxyethyl)aminomethyl-4(3*H*)-pteridinone (**79**) (54%).[408] The same substrate reacted with diethyl *N*-*p*-aminobenzoylglutamate in glycol at 100°C for 1 to 3 h to give (after cold alkaline saponification) pteroylglutamic acid (**73**, R = Y = H, Z = NH_2) (15%).[679, cf. 928] 2-Amino-6-α-bromo-α-carboxymethyl-4(3*H*)-pteridinone reacted with the same ester in glycol containing anhydrous sodium

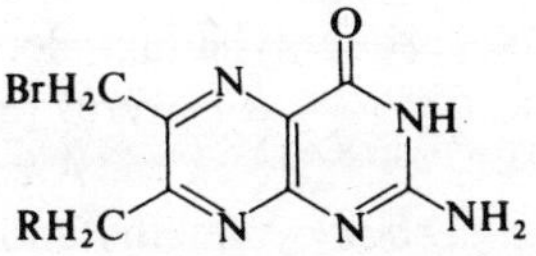

(**77**)

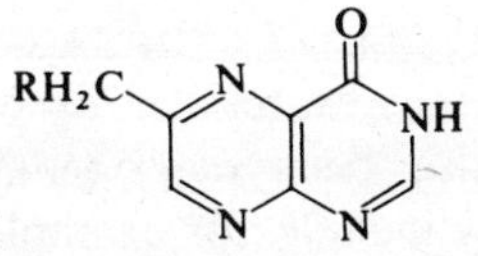

(**78**)

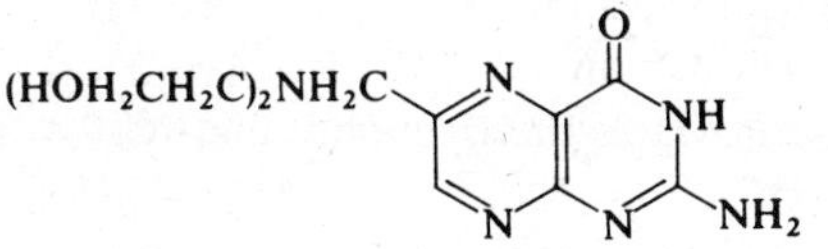

(**79**)

acetate at 25°C for 2 h, to give (after decarboxylation by boiling in water at pH 4 during 15 min) diethyl pteroylglutamate (~40%) and then by saponification, pteroylglutamic acid (65% from the ester).[1308] 2-Amino-6-bromomethyl-4,7(3*H*,8*H*)-pteridinedione reacted with diethyl *N*-*p*-aminobenzoylglutamate similarly at ~120°C to give (after saponification) 7-oxo-7,8-dihydropteroylglutamic acid (**80**) in 30% yield.[1061, 1371] The same substrate reacted with *p*-aminobenzoic acid (or its ester, if followed by saponification) to give 2-amino-6-*p*-carboxyanilinomethyl-4,7(3*H*,8*H*)-pteridinedione (8–10%).[1061] Finally, 2,4-diamino-6-α-bromo-α-carboxymethyl-7(8*H*)-pteridinone reacted with *p*-aminobenzoylglutamic acid in glycol containing sodium acetate during 17 h at 25°C to give (after decarboxylation in boiling water) 2,4-diamino-6-*p*-(*N*-α,γ-dicarboxypropylcarbamoyl)anilinomethyl-7(8*H*)-pteridinone (**81**) in ~60% yield.[506]

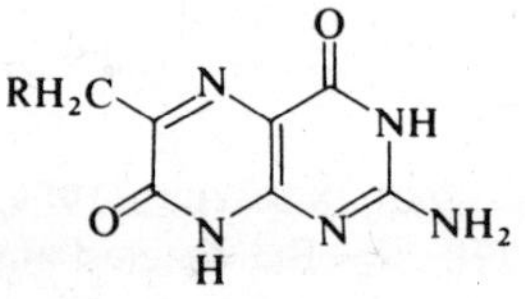

(**80**)

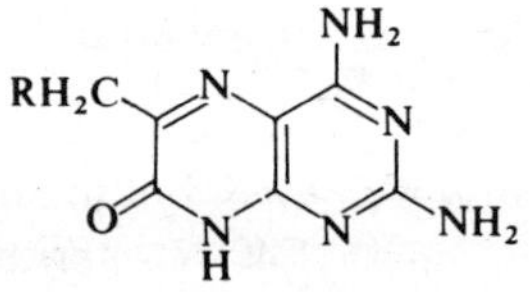

(**81**)

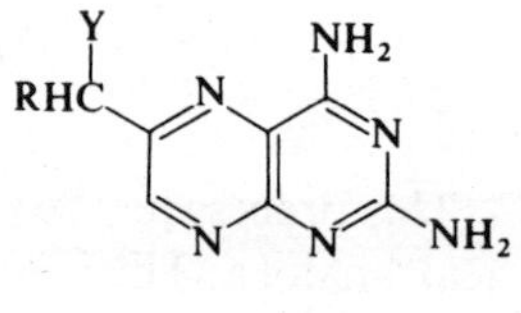

(**82**)

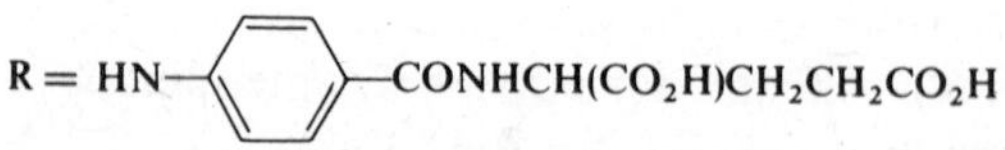

In addition, the anticancer substances, "Aminopterin" (**82**, Y=H) and its α-methyl derivative, methotrexate or "A-Methopterin" (**82**, Y=Me), as well as numerous analogs have been prepared (mainly by John Montgomery and colleagues) by aminolyses of appropriate bromomethylpteridines. Thus 6-bromomethyl-2,4-pteridinediamine and *N*-*p*-aminobenzoylglutamic acid in *N*,*N*-dimethylacetamide at 25°C for 18 h gave 6-*p*-(*N*-α,γ-dicarboxypropylcarbamoyl)anilinomethyl-2,4-pteridinediamine (**82**, Y=H) in 72% yield.[390, 1137] The same substrate with diethyl *N*-*p*-methylaminobenzoylglutamate in *N*,*N*-dimethylacetamide at 25°C for 5 days gave (after saponification of the ester) 6-1′-[*p*-(*N*-α,γ-dicarboxypropylcarbamoyl)anilino]ethyl-2,4-pteridinediamine (methotrexate) (**82**, Y=Me) in 58% yield (several procedural variations have also been described).[390, cf. 595, 818, 1137, 1522] Analogs prepared rather similarly include, for example, 6-*p*-carboxy-*N*-methylanilinomethyl-2,4-pteridinediamine (60%),[390, 1102] its 6-*p*-ethoxycarbonyl analog (89%),[1465] 6-anilinomethyl-2,4-pteridinediamine (**83**, R=Y=H) (38%),[1102] 6-*p*-acetylanilinomethyl-2,4-pteridinediamine (**83**, R=Ac, Y=H) (68%),[1102] 6-*p*-[*N*-(carboxymethyl)carbamoyl]-*N*-methylanilinomethyl-2,4-pteridinediamine (**83**, R=$CONHCH_2CO_2H$, Y=Me),[1271] 6-[*p*-(*N*-α-carboxy-γ-methoxycarbonylpropylcarbamoyl)-*N*-methylanilino]methyl-2,4-pteridinediamine [**83**, R=$CONHCH(CO_2H)CH_2CH_2CO_2Me$, Y=Me] (84%),[1139] 6-[4′-(*N*-α,γ-dicarboxypropylcarbamoyl)-2′-fluoroanilino]methyl-2,4-pteridinediamine (40%),[975] 6-*p*-[*N*-(α,γ-dicarboxypropyl)formamidomethyl]anilinomethyl-2,4-pteridinediamine (**84**),[1186] and many others.[1141, 1687]

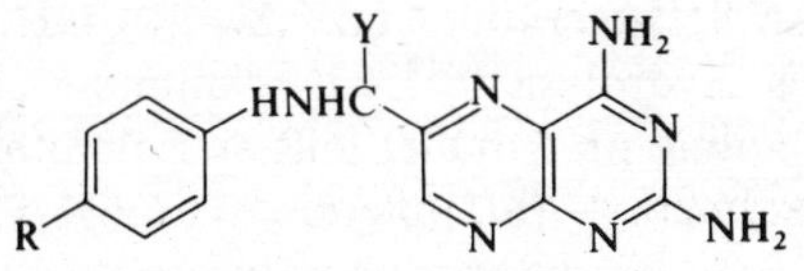

(83)

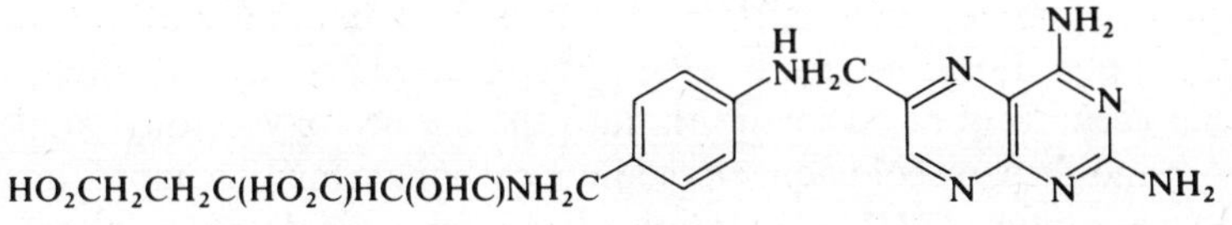

(84)

A few unrelated aminolyses include the conversion of 6-[L-*threo*]-β-acetoxy-α-chloropropyl-2-amino-4(3*H*)-pteridinone (**85**, R=Cl, Y=OAc), by stirring in liquid ammonia for 2 h, into 6-[L-*threo*]-α-acetamido-β-hydroxypropyl-2-amino-4(3*H*)-pteridinone (**85**, X=NHAc, Y=OH) (53%) by a mechanism involving a final *O*- to *N*-acetyl shift[1439]; of 2-amino-6-[L-*threo*]-α-chloro-β-hydroxypropyl-4(3*H*)-pteridinone (**85**, X=Cl, Y=OH), by stirring in aqueous ammonia for 3 days, into 2-amino-6-[L-*threo*]-α-amino-β-hydroxypropyl-4(3*H*)-pteridinone (**85**, X=NH_2, Y=OH) (33%)[1439]; of 7-bromomethyl- or 6,7-bisbromomethyl-1-methyl-2,4(1*H*,3*H*)-pteridinedione, by warming in anhydrous pyridine, into 1-methyl-7-pyridiniomethyl-2,4(1*H*,3*H*)-pteridinedione bromide (>95%) or the 6,7-bispyridiniomethyl analog (**86**) (>95%), respectively[232]; and of 5-chloroacetyl- into 5-pyridinioacetyl-5,6,7,8-tetrahydro-2,4-pteridinediamine (chloride: 82%).[267]

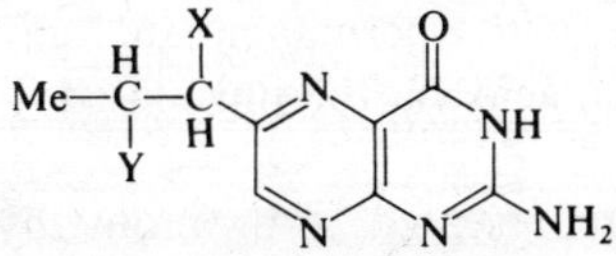

(85)

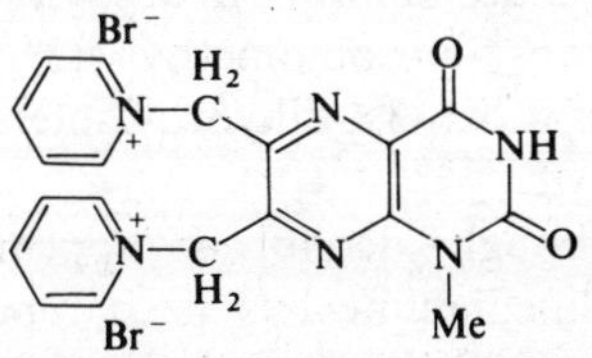

(86)

C. Hydrolysis and Alcoholysis

The hydrolysis of a halogenomethyl- or a dihalogenomethylpteridine gives a hydroxymethylpteridine or a pteridinecarbaldehyde, respectively; although no examples have been reported, hydrolysis of a trihalogenomethylpteridine should give a pteridinecarboxylic acid.

Thus 2-amino-6-bromomethyl-8-methyl-4,7(3*H*,8*H*)-pteridinedione (**87**, R = Br) in 0.5*M* sodium hydroxide at 80°C for 15 min gave 2-amino-6-hydroxymethyl-8-methyl-4,7(3*H*,8*H*)-pteridinedione (**87**, R = OH) (22% yield overall on bromination plus hydrolysis)[1214]; 2,4-diamino-6-chloromethyl-7(3*H*)-pteridinone 8-oxide in boiling pH 7 buffer followed by dithionite reduction gave 2,4-diamino-6-hydroxymethyl-7(8*H*)-pteridinone (66%)[1296]; 2-amino-7-α-bromoacetonyl-4,6(3*H*,5*H*)-pteridinedione in saturated sodium acetate solution for 1 h gave 2-amino-7-α-hydroxyacetonyl-4,6(3*H*,5*H*)-pteridinedione (**88**) (~60% crude material)[498, cf. 1313]; and 2-amino-6-α-bromoacetonyl-4,7(3*H*,8*H*)-pteridinedione (**89**, R = Br) was converted first, by potassium acetate in glacial acetic acid at 60°C for 2 h, into the 6-α-acetoxyacetonyl analog (**89**, R = OAc) and then, by cold aqueous sodium carbonate for 12 h, into 2-amino-6-α-hydroxyacetonyl-4,7(3*H*,8*H*)-pteridinedione (**89**, R = OH) (51%).[499]

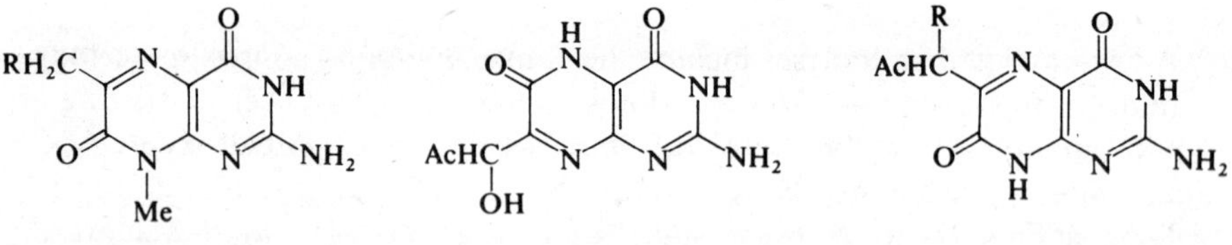

(**87**) (**88**) (**89**)

6-Dibromomethyl-4,7(3*H*,8*H*)-pteridinedione hydrobromide in boiling water for 20 min gave 4,7-dioxo-3,4,7,8-tetrahydro-6-pteridinecarbaldehyde (**90**) (85%)[29]; 2-amino-6-dibromomethyl-4(3*H*)-pteridinone likewise gave 2-amino-4-oxo-3,4-dihydro-6-pteridinecarbaldehyde (~50%)[1358] cf. an integrated procedure for dibromination of 2-acetamido-6-methyl-4(3*H*)-pteridinone, hydrolysis, and deacetylation to give the same product in 75% yield[943]); 2-acetamido-6-dibromomethyl-4(3*H*)-pteridinone with hot aqueous hydroxylamine gave 2-acetamido-7-hydroxyiminomethyl-4(3*H*)-pteridinone (80%)[1739]; and 2-amino-7-dibromomethyl-4(3*H*)-pteridinone with permanganate gave 2-amino-4-oxo-3,4-dihydro-7-pteridinecarboxylic acid via the unisolated 7-carbaldehyde.[1358]

Although alcoholysis appears to be unrepresented, 6-bromomethyl-2,4-pteridinediamine has been reported to react (in *N*,*N*-dimethylacetamide at 25°C) with phenol or *p*-hydroxybenzamide (in the presence of potassium *t*-butoxide during 6 days) and with thiophenol (in the presence of potassium

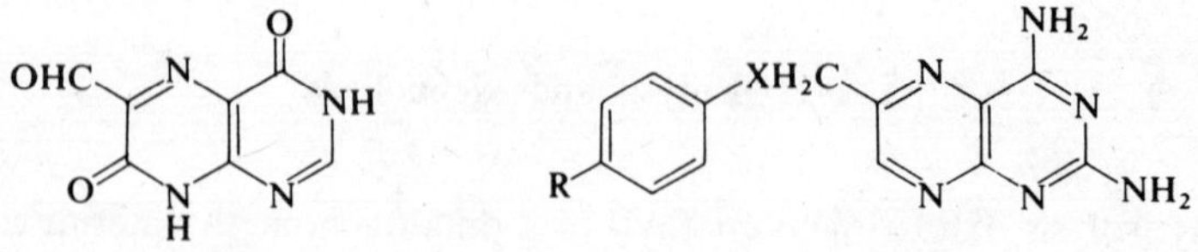

(**90**) (**91**)

carbonate during 24 h) to give 6-phenoxymethyl- (**91**, R=H, X=O) (70%), 6-*p*-carbamoylphenoxymethyl- (**91**, R=$CONH_2$, X=O) (61%), and 6-phenylthiomethyl-2,4-pteridinediamine (**91**, R=H, X=S) (87%), respectively.[1102]

D. Other Reactions

The mass spectral fragmentation patterns for 4-trifluoromethylpteridine and nine derivatives have been reported. The stage, at which a trifluoromethyl radical was lost from each, depended on the nature and position(s) of the other substituents.[780] 4-Trifluoromethylpteridine and 7-methyl-4-trifluoromethylpteridine both exist in aqueous solution as their respective 3,4-hydrates (**92**, R=H or Me), which have been isolated as stable solids; 6,7-dimethyl-4-trifluoromethylpteridine is only partly so hydrated in aqueous solution but all three cations are completely mono- or dihydrated in aqueous solution.[156] 6-Bromomethyl-2,4-pteridinediamine reacted with triphenylphosphine in *N,N*-dimethylacetamide at 60°C to give 6-triphenylphosphoranylidenemethyl-2,4-pteridinediamine (**93**) (in solution), which reacted subsequently with

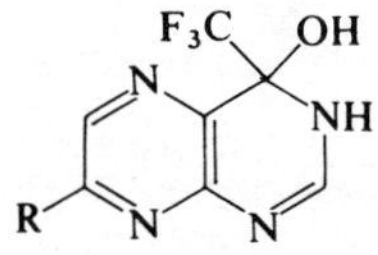

(92)

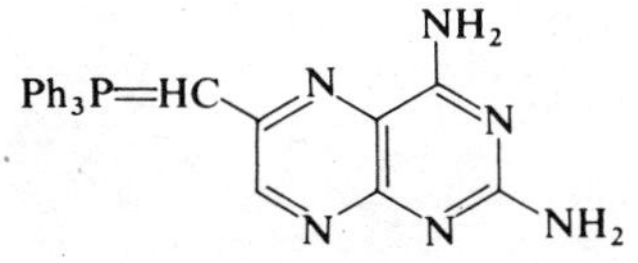

(93)

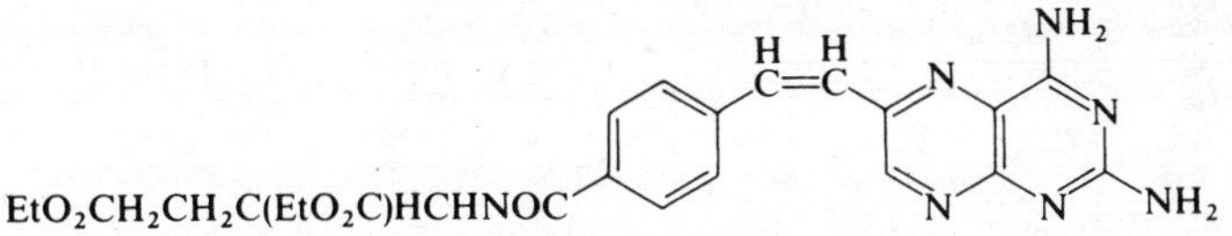

(94)

diethyl *N*-*p*-formylbenzoylglutamate in the presence of sodium methoxide to give an "Aminopterin" analog, 6-*p*-(*N*-α,γ-diethoxycarbonylpropylcarbamoyl)styryl-2,4-pteridinediamine (**94**) in 78% overall yield[1142]; other types of phosphorus derivative have been made and were used rather similarly.[569]

CHAPTER VI

Tautomeric Pteridinones and Extranuclear Hydroxypteridines

Since all naturally occurring pteridines are pteridinones and some also have extranuclear hydroxy groups, it is not surprising that this wide group of compounds have been investigated extensively from the earliest days of the pteridine era.

All the pteridinones capable of tautomerism undoubtedly exist, whether as solids or in solution, as lactams, for example (**1**), rather than as lactims, for example (**2**)[cf. 869]; this appears to be true of the ionized molecules also, whether cations or anions (see Ch. XII, Sects. 2–4).[1713] In contrast, nearly all extranuclear hydroxypteridines (mainly pteridinyl alcohols) exist as true hydroxy compounds. This distinction is obscured in the literature by the extensive use of terms such as hydroxypteridine or pteridinol, simply as convenient and less cumbersome alternatives to the correct pteridinone (with an indicated hydrogen position) or oxodihydropteridine . . . (when followed by a senior suffix such as carboxylic acid).

1. PREPARATION OF TAUTOMERIC PTERIDINONES

A. By Primary Syntheses

More pteridinones have been made by primary syntheses than by all other methods put together. Innumerable examples will be found in Chapters II and III. Such syntheses are illustrated simply in the condensation of 4,5-diamino-2(1*H*)-pyrimidinone (**3**) with glyoxal to give 2(1*H*)-pteridinone (**1**)[14, 30] or of 6-isopropoxy-2,4,5-pyrimidinetriamine with diethyl oxalate to give 2-amino-4-isopropoxy-6,7(5*H*,8*H*)-pteridinedione[371] (Gabriel and Colman syntheses); in the condensation of 5-nitroso-4,6-pyrimidinediamine with ethyl phenylacetate

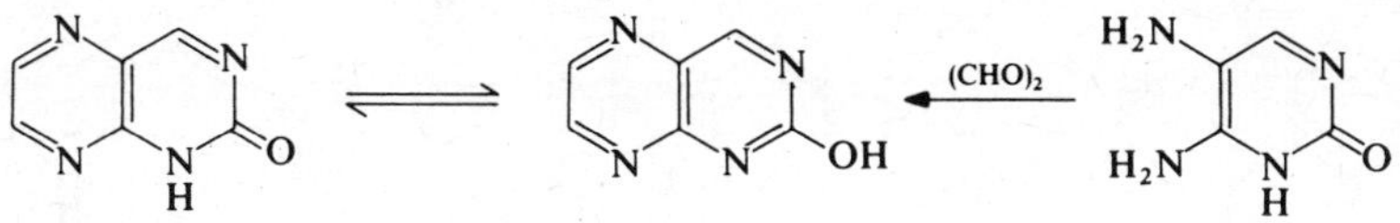

(**1**) (**2**) (**3**)

to give 4-amino-6-phenyl-7(8*H*)-pteridinone (**4**) (a Timmis synthesis)[224]; in the condensation of 2-amino-6-chloro-5-nitro-4(3*H*)-pyrimidinone with glycine ethyl ester to give the 6-(ethoxycarbonylmethyl)amino analog that underwent reductive cyclization to 2-amino-7,8-dihydro-4,6(3*H*,5*H*)-pteridinedione and subsequent oxidation to 2-amino-4,6(3*H*,5*H*)-pteridinedione (**5**) (a Boon synthesis)[451]; in the condensation of 3-amino-2-pyrazinecarboxamide with triethyl orthoacetate to give 2-methyl-4(3*H*)-pteridinone (**6**) (one of several types of synthesis from a pyrazine)[20]; in the incubation of 2(1*H*)-purinone with glyoxal to give 2(1*H*)-pteridinone (**1**) via pyrimidine intermediates (a synthesis from another heterocyclic system)[14]; and in numerous other minor types of synthesis outlined in earlier chapters.

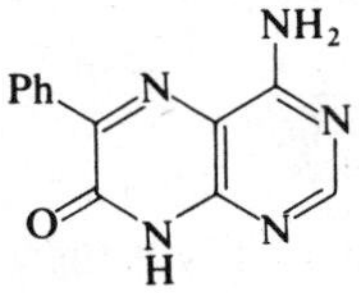

(4)

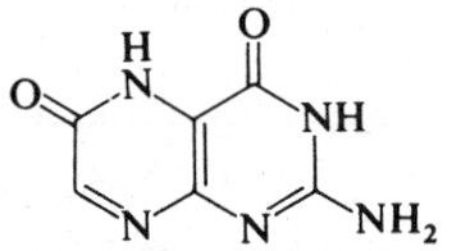

(5)

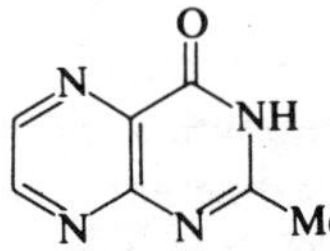

(6)

B. By Oxidative Means

The oxidation of a pteridine or a hydropteridine often results in the introduction of an oxo substituent at an apparently unoccupied 2-, 4-, 6-, or 7-site. In some cases it is clear that this simply represents the dehydrogenation of a covalent hydrate existing in equilibrium with the parent substrate within the solution; in other cases, the mechanism is far from clear. It should be borne in mind that the very formation of a covalent hydrate represents the nonoxidative introduction of an hydroxy (as distinct from an oxo) substituent into the pteridine nucleus.[147–149, 156, 157, 159]

The oxidative procedure is represented in the conversion of pteridine into 4(3*H*)-pteridinone (**7**, R = H) by treatment with hydrogen peroxide (pH 9, 20°C),[1145] with alkaline permanganate (20°C: 17% yield),[134] or with peroxyphthalic acid (etherial chloroform: 72% yield; product initially thought to be an isomeric *N*-oxide)[30]; of 7-methylpteridine into 7-methyl-4(3*H*)-pteridinone (**7**, R = Me) (peroxyphthalic acid in etherial chloroform: 70%)[33]; of 2(1*H*)-pteridinone into 2,4(1*H*,3*H*)-pteridinedione (**8**) (alkaline solution aerated at 20°C for 11 days: 67%[56]; alkaline permanganate, hot or cold: ~75%)[134]; of the

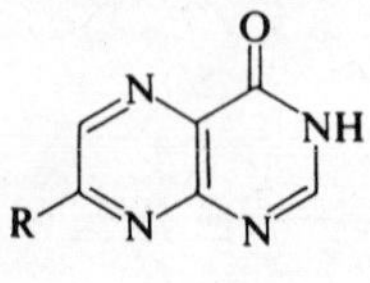

(7)

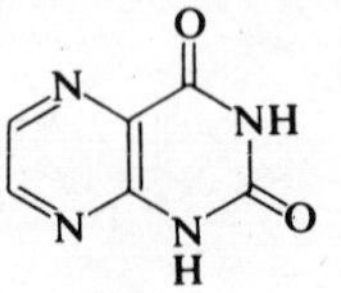

(8)

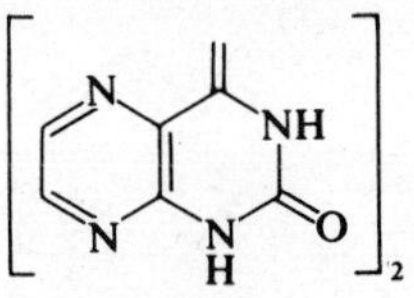

(9)

dimer (**9**) of 2(1*H*)-pteridinone into the same product (**8**) (by aeration of an alkaline solution at 98°C for 2 h: 83%)[56]; of 4(3*H*)-pteridinone into 4,6,7(3*H*,5*H*,8*H*)-pteridinetrione (**10**, R = H) (hydrogen peroxide in glacial acetic acid at 20°C for 7 days: ~35%)[155]; of 6(5*H*)-pteridinone into 6,7(5*H*,8*H*)-pteridinedione (**11**, R = Y = H) (aqueous alkali under reflux for 15 min: ~25% along with a separable second product[19]; aqueous sodium carbonate under reflux for 4 h: ~15% along with two other separable products[19]; alkaline permanganate at 25°C: 65%)[134]; of 7(8*H*)-pteridinone into the same dione (**11**, R = Y = H) (fuming nitric acid at 20°C for 12 h: 30%)[31]; of 5-methyl-6(5*H*)-pteridinone into 5-methyl-6,7(5*H*,8*H*)-pteridinedione (**11**, R = Me, Y = H) (aqueous permanganate at 70°C: ~65%?)[134]; of 2-pteridinamine into 2-amino-4(3*H*)-pteridinone (potassium dichromate in hot dilute sulfuric acid: 25%)[39]; of 8-methyl-7(8*H*)-pteridinone into 8-methyl-6,7(5*H*,8*H*)-pteridinedione (**11**, R = H, Y = Me) (fuming nitric acid in sulfuric acid at 25°C for 2 h: 70%)[34]; of 8-methyl-4(8*H*)-pteridinone into 8-methyl-4,7(3*H*,8*H*)-pteridinedione (alkaline permanganate: 61%)[248]; of 2-phenylpteridine into 2-phenyl-4(3*H*)-pteridinone (hydrogen peroxide in acetic acid for 6 h at 20°C: ~75%) or into 2-phenyl-4,6,7(3*H*,5*H*,8*H*)-pteridinetrione (**10**, R = Ph) (same reagents for 7 days)[152]; of 7-methyl- or 6,7-dimethyl-2-phenylpteridine into 7-methyl- or 6,7-dimethyl-2-phenyl-4(3*H*)-pteridinone, respectively (same reagents for 6 h: >80%)[152]; of 4-phenylpteridine into 4-phenyl-6,7(5*H*,8*H*)-pteridinedione (same reagents for 4 days: ~25%)[151]; of 2-amino-4,6(3*H*,5*H*)-pteridinedione (xanthopterin) into 2-amino-4,6,7(3*H*,5*H*,8*H*)-pteridinetrione (leucopterin) (**10**, R = NH_2) (shaking with oxygen over platinum in dilute acetic acid for 36 h: 62%[404, 533]; sodium periodate in dilute hydrochloric acid for 48 h at 20°C: ~8%)[1056]; of 2,6(1*H*,5*H*)- or 2,7(1*H*,8*H*)-pteridinedione into 2,6,7(1*H*,5*H*,8*H*)-pteridinetrione [hydrogen peroxide in dilute sulfuric acid (78%) or cold nitric acid for 24 h (28%), respectively; the authors were uneasy about the structure of this trione, even when made unambiguously][41]; of 2,4,6(1*H*,3*H*,5*H*)-pteridinetrione into 2,4,6,7(1*H*,3*H*,5*H*,8*H*)-pteridinetetrone (**12**, R = H) (bromine in dilute sulfuric acid under reflux for 10 min: 50%)[101]; of 1,3-dimethyl-2,4,7(1*H*,3*H*,8*H*)-pteridinetrione into 1,3-dimethyl-2,4,6,7(1*H*,3*H*,5*H*,8*H*)-pteridinetetrone (**12**, R = Me) (hydrogen peroxide in formic acid at <40°C for 4 days: 51%)[238]; and of 2-amino-6,8-dimethyl-7,8-dihydro-4(3*H*)-pteridinone into 2-amino-6-hydroxymethyl-8-methyl-4,7(3*H*,8*H*)-pteridinedione (oxygenation in an aqueous solution containing platinum oxide at 35°C for 12 h: 30%; note the additional extranuclear hydroxylation).[1370] Treatment of 6-methyl-2,4-

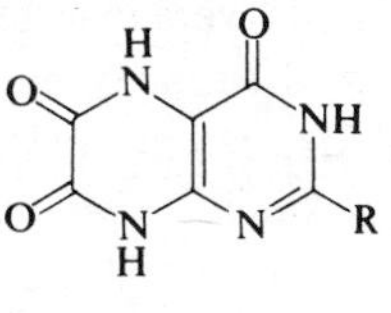

(**10**)

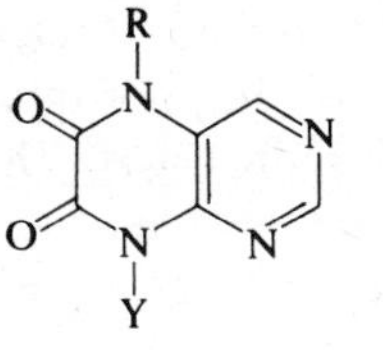

(**11**)

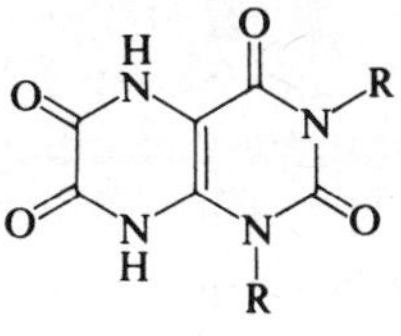

(**12**)

pteridinediamine 8-oxide with chlorine in acetic acid at 60°C gave an acetyl derivative, formulated as 8-acetoxy-2,4-diamino-6-methyl-7(8*H*)-pteridinone, which on deacylation afforded 2,4-diamino-6-methyl-7(3*H*)-pteridinone 8-oxide (86%); the 6-ethyl, propyl, isopropyl, isobutyl, and pentyl homologs were made similarly (>80%).[1298]

Pteridine derivatives with a *C*-methyl group located at the site attacked by the hydroxyl group in the process of covalent hydration, have been shown to undergo a facile demethylation on oxidation with permanganate, to afford pteridinones bearing an oxo substituent at the same position.[248] Thus, under appropriate conditions, 6,7,8-trimethyl-2,4(3*H*,8*H*)-pteridinedione gave 6,8-dimethyl-2,4,7(1*H*,3*H*,8*H*)-pteridinetrione (48%), 6,7,8-trimethyl-2(8*H*)-pteridinone gave 6,8-dimethyl-2,7(1*H*,8*H*)-pteridinedione (63%), 4-methyl-2(1*H*)-pteridinone gave 2,4(1*H*,3*H*)-pteridinedione (65%), 6,7,8-trimethyl-2-methylimino-2,8-dihydropteridine gave 6,8-dimethyl-2-methylamino-7(8*H*)-pteridinone (57%), and analogous examples have been reported.[248]

Although the acetone adduct of 6(5*H*)-pteridinone, viz. 7-acetonyl-7,8-dihydro-6(5*H*)-pteridinone (**13**), resisted oxidation by acidic ferric chloride or potassium dichromate and was destroyed by permanganate, it underwent oxidative hydrolysis in refluxing aqueous sodium hydroxide to afford 6,7(5*H*,8*H*)-pteridinedione (**11**, R=Y=H) in 50% yield.[55] It is interesting that 7(8*H*)-pteridinone in 1*M* sulfuric acid at 37°C for 5 days was transformed into 6(5*H*)-pteridinone in 93% yield,[19] presumably by ring fission to 4,5-pyrimidinediamine and glyoxylic acid followed by recombination in the preferred orientation under acidic conditions: compare Chapter II, Section 1.C(1).

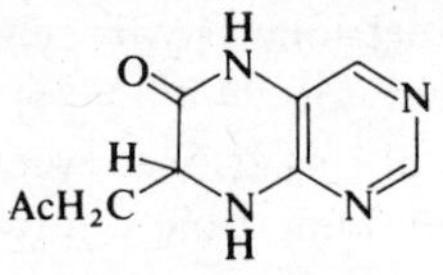

(**13**)

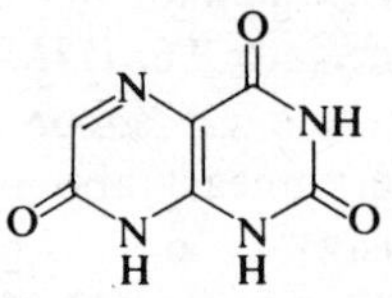

(**14**)

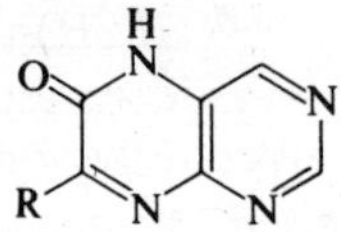

(**15**)

Following early indications that xanthine oxidase and related enzymes could convert 2-amino-4,6(3*H*,5*H*)-pteridinedione (xanthopterin) into 2-amino-4,6,7(3*H*,5*H*,8*H*)-pteridinetrione (leucopterin) and 2-amino-4(3*H*)-pteridinone (pterin) into 2-amino-4,7(3*H*,8*H*)-pteridinedione (isoxanthopterin),[1369, 1530–1532] Felix Bergmann and his colleagues undertook a kinetic study of the oxidation of pteridine, pteridinones, pteridinediones, and pteridinetriones by milk xanthine oxidase.[98, 99] With the exception of 6(5*H*)-pteridinone and 2,6(1*H*,5*H*)-pteridinedione that did not react, oxo substituents were inserted into any vacant 2-, 4-, or 7-position but not into a vacant 6-position on each substrate. Accordingly, the final product was either 2,4,7(1*H*,3*H*,8*H*)-pteridinetrione (**14**) (from those substrates lacking a 6-oxo substituent) or 2,4,6,7(1*H*,3*H*,5*H*,8*H*)-pteridinetetrone (**12**, R=H) (from substrates already carrying a 6-oxo substituent). Rates varied considerably, products were identified spectrally from chromatographically isolated material, and in most cases the partially oxidized intermediate stages were identified satisfactorily.[98, 99, 1567]

It was later shown that, unlike 6(5*H*)-pteridinone itself (**15**, R = H), the 7-methyl derivative (**15**, R = Me) *was* prone to enzymatic oxidation to give 7-methyl-4,6(3*H*,5*H*)-pteridinedione and subsequently 7-methyl-2,4,6(1*H*,3*H*,5*H*)-pteridinetrione[717]; that 5,6,7,8-tetrahydro-2,4(1*H*,3*H*)-pteridinedione in fact gave 2,4,6(1*H*,3*H*,5*H*)-pteridinetrione rather than the 2,4,7-isomer[1568, 1569]; that 2-phenyl-4(3*H*)-pteridinone gave 2-phenyl-4,7(3*H*,8*H*)-pteridinedione[101]; that 7-azapteridines (pyrimido[5,4-*e*]-*as*-triazines) (**16**) were also prone to such oxidation even at the 3-position (equivalent to the 6-position in pteridines) so that, for example, 5(6*H*)-pyrimido[5,4-*e*]-*as*-triazinone gave 3,5,7(4*H*,6*H*,8*H*)-pyrimido[5,4-*e*]-*as*-triazinetrione[100]; and that *N*-methylated pteridinones also underwent oxidation. For example, 1-methyl-4(1*H*)- and 3-methyl-4(3*H*)-pteridinone gave 1-methyl-2,4(1*H*,3*H*)- and 3-methyl-4,7(3*H*,8*H*)-pteridinedione, respectively,[600] while 8-methyl-4(8*H*)-pteridinone gave 8-methyl-2,4,7(1*H*,3*H*,8*H*)-pteridinetrione.[248]

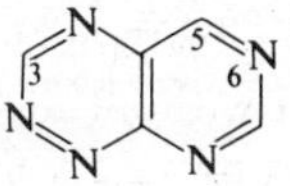

(**16**)

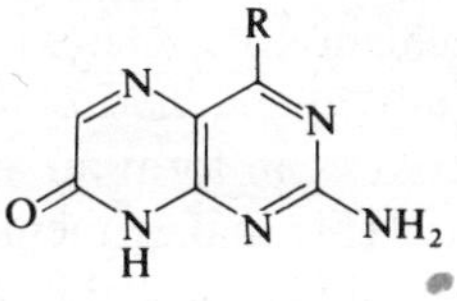

(**17**)

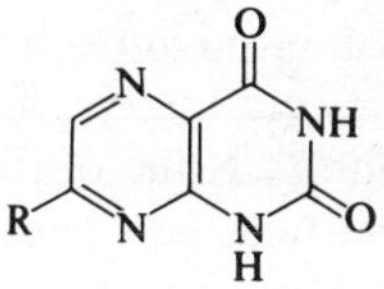

(**18**)

Pteridines bearing amino groups are also oxidized in this way. For example, 2-pteridinamine, 2-amino-4(3*H*)-pteridinone, and 2-amino-7(8*H*)-pteridinone all afford 2-amino-4,7(3*H*,8*H*)-pteridinedione (isoxanthopterin)[1350]; 4-pteridinamine, 4-amino-2(1*H*)-pteridinone, and 4-amino-7(8*H*)-pteridinone all gave 4-amino-2,7(1*H*,8*H*)-pteridinedione, albeit more slowly[1350]; and although 2,4-pteridinediamine and 4-methyl-2-pteridinamine resisted both milk and liver xanthine oxidases, they were converted by liver aldehyde oxidase into 2,4-diamino-7(8*H*)-pteridinone (**17**, R = NH_2) and 2-amino-4-methyl-7(8*H*)-pteridinone (**17**, R = Me), respectively.[979] 7-*p*-Bromo-, 7-*p*-cyano-, and 7-*p*-nitrophenyl-4(3*H*)-pteridinone have been converted into the corresponding 7-*p*-substituted-phenyl-2,4(1*H*,3*H*)-pteridinediones (**18**) with either free or immobilized (i.e., adsorbed to a suitable Sepharose) milk xanthine oxidase[1287]; appropriate 7-alkyl-4(3*H*)-pteridinones have been converted similarly into 7-ethyl-, 7-isopropyl-, 7-methyl-, and 7-propyl-2,4(1*H*,3*H*)-pteridinedione (**18**, R = Et, etc.) in good yield on a small preparative scale[1286]; and the chemical mechanisms have been discussed.[1486] The electrochemical oxidation of 6(5*H*)- and 7(8*H*)-pteridinone into 6,7(5*H*,8*H*)-pteridinedione and other products has been studied.[892]

C. From Pteridinamines

The conversion of pteridinamines into pteridinones can be done by alkaline hydrolysis, by acidic hydrolysis, or by the use of nitrous acid (the last for primary amines only). Of these methods, alkaline hydrolysis has been used most widely.

(1) *By Alkaline Hydrolysis*

The formation of 2-*pteridinones* from 2-pteridinamines under alkaline conditions is usually done in boiling aqueous alkali. Under such conditions, 2-amino-1,6-dimethyl-4(1*H*)-pteridinone gave 1,6-dimethyl-2,4(1*H*,3*H*)-pteridinedione (**19**, R=Me, Y=H) (8 min: 50%)[80]; 2-amino-1,6,7-trimethyl-4(1*H*)-pteridinone gave 1,6,7-trimethyl-2,4(1*H*,3*H*)-pteridinedione (**19**, R=Y=Me) (5 min: 70%)[166]; and 2-amino-1-methyl-6,7-diphenyl-4(1*H*)-pteridinone gave 1-methyl-6,7-diphenyl-2,4(1*H*,3*H*)-pteridinedione (**19**, R=Y=Ph) (4 h: 32%[111]; 30 min: 80%[68]; it might be concluded that the dione was destroyed slowly by prolonged treatment). Under similar conditions, 2-imino-1,3-dimethyl-6,7-diphenyl-1,2-dihydro-4(3*H*)-pteridinone (**20**, X=NH) gave, not the expected 1,3-dimethyl-6,7-diphenyl-2,4(1*H*,3*H*)-pteridinedione (**20**, X=O), but 1-methyl-6,7-diphenyl-2,4(1*H*,3*H*)-pteridinedione (**19**, R=Y=Ph) (33%), presumably by an initial Dimroth rearrangement to 1-methyl-2-methylamino-6,7-diphenyl-4(1*H*)-pteridinone (**21**) and subsequent hydrolytic loss of the methylamino group.[68] None of these substrates can form an anion in alkali without at the same time becoming an imine. This makes them unusually prone to alkaline hydrolysis and the conditions required probably give a false idea of the ease of hydrolysis for regular 2-pteridinamines, for which no data are available.

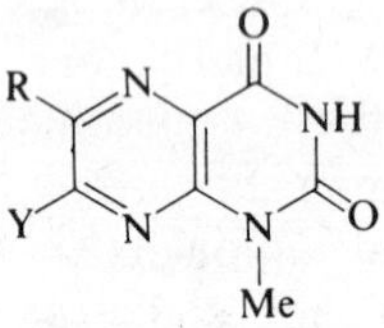

(**19**)

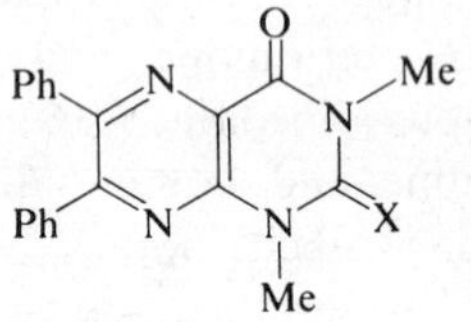

(**20**)

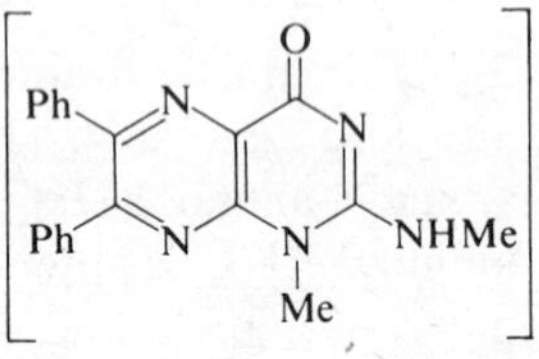

(**21**)

The formation of 4-*pteridinones* from 4-pteridinamines is a far more used process. Thus 4-pteridinamine in 1*M* sodium hydroxide at 100°C for 5 min gave 4(3*H*)-pteridinone (80%)[30]; 4-amino-6-methyl-2(1*H*)-pteridinone in refluxing 1*M* sodium hydroxide for 6 h gave 6-methyl-2,4(1*H*,3*H*)-pteridinedione (55%)[168]; 4-amino-6-hydroxymethyl-2(1*H*)-pteridinone in refluxing 0.1*M* alkali gave 6-hydroxymethyl-2,4(1*H*,3*H*)-pteridinedione[96, 1546]; and 4-amino-2,7(1*H*,8*H*)-pteridinedione in refluxing 6*M* sodium hydroxide for 5 h gave 2,4,7(1*H*,3*H*,8*H*)-pteridinetrione (75%).[41] However, one of the most useful aspects of this process is its ability to bring about a preferential hydrolysis of the 4-amino group in a 2,4-diaminopteridine to give only a 2-amino-4(3*H*)-pteridinone, that is, a pterin with a substituent configuration in the pyrimidine ring common to many naturally occurring pteridines. This is exemplified in the formation, from the corresponding 2,4-diamine, of 2-amino-6-methyl-4(3*H*)-pteridinone (**22**, R=Me) (refluxing 1*M* sodium hydroxide for 6 h: 60–80%),[1263, 1394] 2-amino-7-deutero-6-methyl (or trideuteromethyl)-4(3*H*)-pteridinone (refluxing 2*M* sodium deuteroxide for 9–18 h: ~60%),[84]

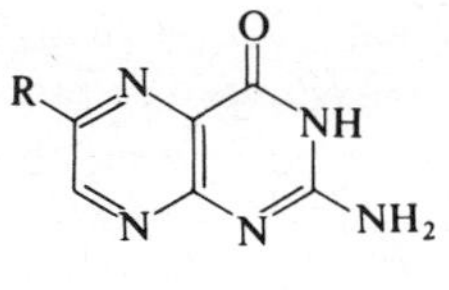

(22)

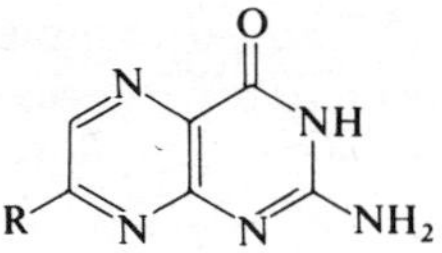

(23)

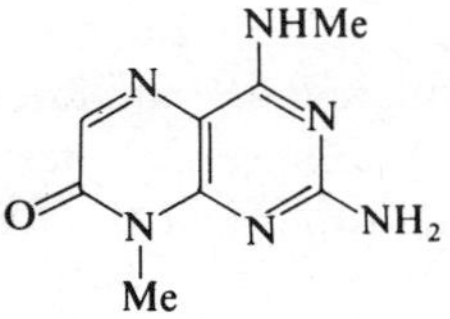

(24)

2-amino-6-phenylethynyl-4(3*H*)-pteridinone (aqueous alcoholic alkali under reflux for 8 h: 87%; also analogs),[1753] 2-amino-6-hydroxymethyl-4(3*H*)-pteridinone (**22**, R = CH_2OH) (0.5*M* sodium hydroxide under reflux for 1 h: 69%),[113] 2-amino-6-dimethoxymethyl-4(3*H*)-pteridinone [**22**, R = $CH(OMe)_2$] (1*M* sodium hydroxide under reflux for 10 min: 94%),[1306] 2-amino-4-oxo-3,4-dihydro-6-pteridinecarboxylic acid (**22**, R = CO_2H) (2*M* sodium hydroxide at 70°C for 2.5 h: 86%),[178] 6-1′-adamantyl-2-amino-4(3*H*)-pteridinone (0.1*M* sodium hydroxide under reflux for 2 h),[973] 2-amino-6-*p*-carboxyphenylthiomethyl-4(3*H*)-pteridinone (refluxing ethanolic 2*M* sodium hydroxide for 5 h: 71%),[261] 2-amino-7-phenyl-4(3*H*)-pteridinone (**23**, R = Ph) (refluxing aqueous alcoholic 0.7*M* sodium hydroxide for 5 h: 40%),[541] 2-amino-7-dimethoxymethyl-4(3*H*)-pteridinone [**23**, R = $CH(OMe)_2$] (refluxing aqueous sodium hydroxide for 5 min: 90%),[1310] 2-amino-6-butoxymethyl-4(3*H*)-pteridinone (also its 8-oxide and several 6-alkoxymethyl analogs of each) (refluxing aqueous or aqueous ethanolic 0.5*M* alkali for 1–6 h: <80%),[1728] 2-amino-7-methyl-4-oxo-3,4-dihydro-6-pteridinecarboxylic acid (refluxing 1*M* sodium hydroxide for 3 h: 59%),[307] 7-1′-adamantyl-2-amino-4(3*H*)-pteridinone (refluxing 0.1*M* sodium hydroxide for 96 h),[973] 2-amino-4(3*H*)-pteridinone 8-oxide (refluxing 1*M* alkali for 30 min: 98%),[1301] 2-amino-7-phenyl-4(3*H*)-pteridinone 5-oxide (refluxing aqueous ethanolic 0.25*M* sodium hydroxide for 50 min: 68%),[541] the isomeric 2-amino-6-phenyl-4(3*H*)-pteridinone 8-oxide (similarly: 88%),[541] and 2-amino-6,7-diphenyl-4(3*H*)-pteridinone 5,8-dioxide (similar reagent for 3.5 h: 53%).[541] Less simple examples include the conversion of 2,4-diamino-6(5*H*)-pteridinone, by aqueous alkali at 100°C for 24 h, into 2-amino-4,6(3*H*,5*H*)-pteridinedione (85%)[1293, cf. 811]; of 7-methoxy-2,4-pteridinediamine, by similar treatment, into 2-amino-4,7(3*H*,8*H*)-pteridinedione (82%)[1301]; of 2,4-diamino-6-hydroxymethyl-7(8*H*)-pteridinone, by 20% aqueous potassium hydroxide at 75°C for 2 days under nitrogen, into 2-amino-6-hydroxymethyl-4,7(3*H*,8*H*)-pteridinedione (50%) (without nitrogen, oxidation occurred to afford mainly 2-amino-4,7-dioxo-3,4,7,8-tetrahydro-6-pteridinecarboxylic acid)[1296]; of 2-amino-8-methyl-4-methylamino-7(8*H*)-pteridinone (**24**), by refluxing 10% alkali for 24 h, into 2-amino-8-methyl-4,7(3*H*,8*H*)-pteridinedione (90%)[1296]; and of 2,4,6,7-pteridinetetramine, by refluxing aqueous alkali for 12 h, into a single triaminopteridinone (93%) of undetermined structure.[486] Other examples are known.[1733]

The formation of 7-*pteridinones* by alkaline hydrolysis of the corresponding 7-pteridinamines is represented in the conversion of 2,7-diamino-4,6(3*H*,5*H*)-

pteridinedione into 2-amino-4,6,7(3*H*,5*H*,8*H*)-pteridinetrione (88%) by refluxing in 0.1*M* sodium hydroxide for 20 h.[97] It also occurs in the conversion of 7-pyrrolidino-2,4-pteridinediamine into 2-amino-4,7(3*H*,8*H*)-pteridinedione by refluxing in 1*M* alkali for 4 days (85% yield).[1301]

(2) *By Acidic Hydrolysis*

Although acidic hydrolysis of pteridinamines to pteridinones appears to be less satisfactory than alkaline hydrolysis, the method has been applied in the 2-, 4-, and 7-position.

Such hydrolysis of 2-*pteridinamines* is exemplified in the transformation of 2-amino-4-oxo-3,4-dihydro-6-pteridinecarboxylic acid (**22**, R=CO_2H) (itself made by the alkaline hydrolysis of 2,4-diamino-6-pteridinecarboxylic acid: just cited) into 2,4-dioxo-1,2,3,4-tetrahydro-6-pteridinecarboxylic acid (85%) by refluxing in 6*M* hydrochloric acid for 30 h[178]; of 2-acetamido-1,6,7-trimethyl-4(1*H*)-pteridinone into 1,6,7-trimethyl-2,4(1*H*,3*H*)-pteridinedione (**19**, R=Y=Me) (36%) by refluxing in 0.1*M* hydrochloric acid for 20 min[69]; and of 2-amino-4,7(3*H*,8*H*)-pteridinedione into 2,4,7(1*H*,3*H*,8*H*)-pteridinetrione (**25**) (95%) by refluxing in 6*M* hydrochloric acid for 28 h.[1267]

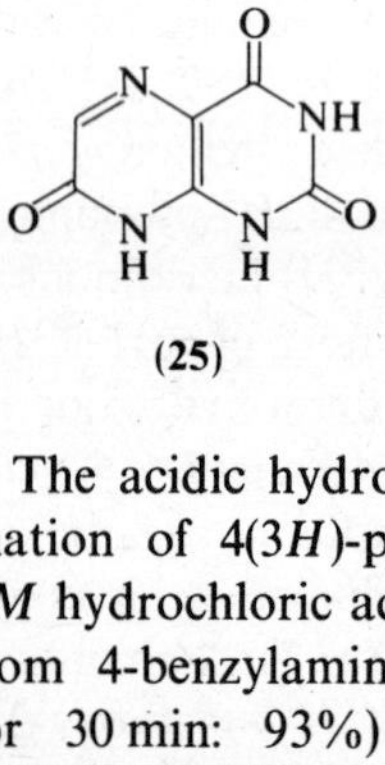

(25)

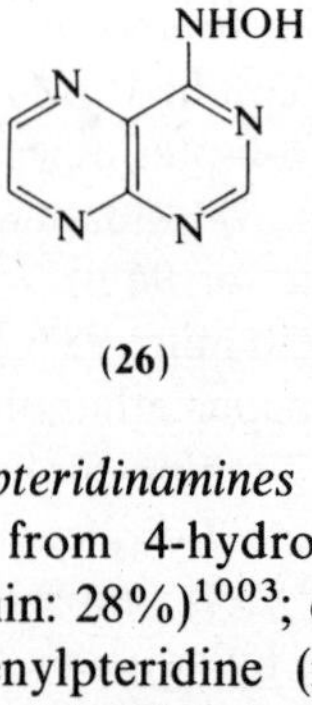

(26)

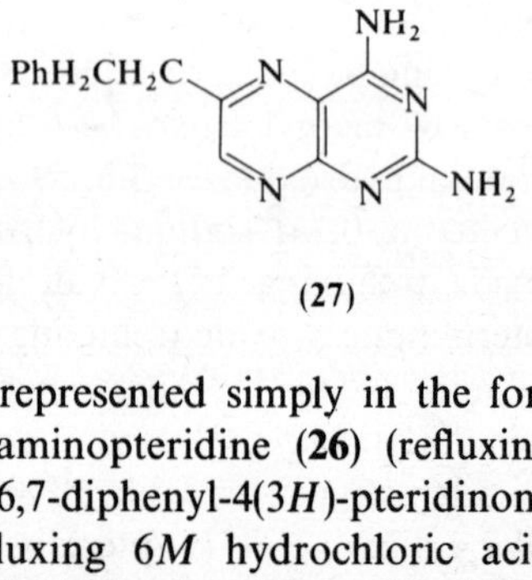

(27)

The acidic hydrolysis of 4-*pteridinamines* is represented simply in the formation of 4(3*H*)-pteridinone from 4-hydroxyaminopteridine (**26**) (refluxing 5*M* hydrochloric acid for 10 min: 28%)[1003]; of 6,7-diphenyl-4(3*H*)-pteridinone from 4-benzylamino-6,7-diphenylpteridine (refluxing 6*M* hydrochloric acid for 30 min: 93%) or from 4-butylamino-6,7-diphenylpteridine (similarly: 88%)[474, cf. 1167]; of 7-*t*-butyl-, 7-phenyl-, or 6,7-diphenyl-4(3*H*)-pteridinone from the respective 4-pteridinamines (6*M* hydrochloric acid under reflux for 15 min: 50–70%)[1740]; and of 6-hydroxymethyl-2,4(1*H*,3*H*)-pteridinedione from 4-amino-6-hydroxymethyl-2(1*H*)-pteridinone (refluxing 0.1*M* hydrochloric acid for 30 min).[96] The process has also been used for the selective 4-hydrolysis of 2,4-pteridinediamines. For example, 2,4-bismethylamino-7-phenylpteridine in refluxing 6*M* hydrochloric acid for 20 h gave 2-methylamino-7-phenyl-4(3*H*)-pteridinone (52%)[110]; 2,4-bisdimethylamino-6-phenylpteridine likewise gave 2-dimethylamino-6-phenyl-4(3*H*)-pteridinone (90%)[110]; 6-phenethyl-2,4-pteridinediamine (**27**) in refluxing 6*M* hydrochloric acid for 30 min gave 2-amino-6-phenethyl-4(3*H*)-pteridinone (**22**, R=CH_2CH_2Ph) (58%)[90]; and 2-amino-4-methylamino-7-oxo-7,8-dihydro-6-pteridinecarboxylic acid (**28**) in refluxing 6*M* hydrochloric acid for 4 h gave 2-amino-4,7-dioxo-3,4,7,8-tetrahydro-6-pteridinecarboxylic acid (**29**) (35%).[882] A general procedure for the 4-hydrolysis of 4-alkylamino-2-pteridinamines (such as 4-benzylamino-6,7-diphenyl-2-pteridinamine) has been detailed without specific examples.[1315] Acidic hydrolysis of 4-diethylamino-1-methyl-2(1*H*)-pteridinone gave 1-methyl-2,4(1*H*,3*H*)-pteridinedione (57%).[1733]

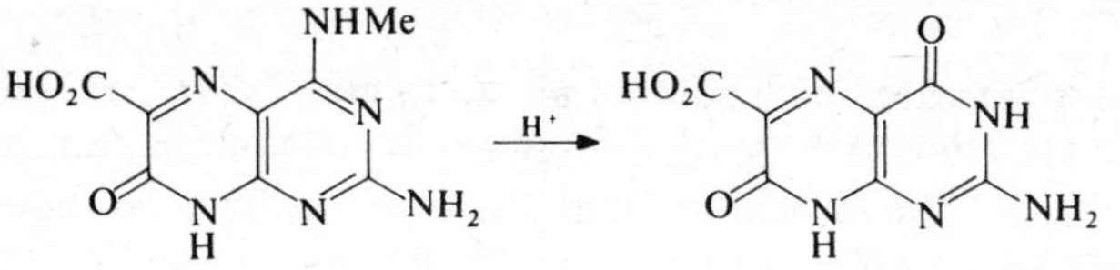

(28) **(29)**

The similar hydrolysis of 7-*pteridinamines* has been used to change 7-pteridinamine into 7(8*H*)-pteridinone (dilute acid: 95%)[131]; the 2,7-diamino-4,6(3*H*,5*H*)-pteridinedione into 2-amino-4,6,7(3*H*,5*H*,8*H*)-pteridinetrione (1*M* hydrochloric acid under reflux for 20 h: 87%)[97]; 6-phenyl-2,4,7-pteridinetriamine (**30**) into a separable mixture of 2,7-diamino-6-phenyl-4(3*H*)-pteridinone (**31**) and 2,4-diamino-6-phenyl-7(8*H*)-pteridinone (**32**) (refluxing 5*M* hydrochloric acid for 1 h: 65% and an unrecorded yield, respectively; the substrate was unaffected by refluxing in 1*M* hydrochloric acid for several hours and by treatment with nitrous acid)[1281]; and 7-amino-1,3-dimethyl-2,4(1*H*,3*H*)-pteridinedione 5-oxide into 1,3-dimethyl-2,4,6,7(1*H*,3*H*,5*H*,8*H*)-pteridinetetrone refluxing 2*M* hydrochloric acid for 90 min: 60% plus a separable by-product).[238]

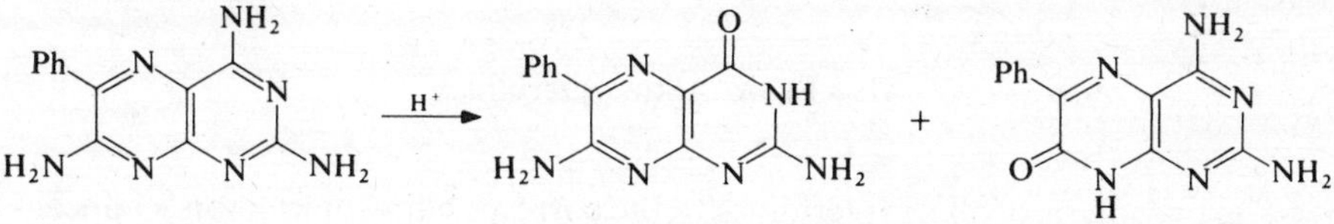

(30) **(31)** **(32)**

(3) *By Treatment with Nitrous Acid*

There is a small number of cases in which the preparation of pteridinones from (primary) pteridinamines has been done by initial diazotization of the amine by nitrous acid. Most of the successful examples involved 2-amino-substrates; failures have been recorded for some 4-amino- and 2,4-diamino-substrates. Thus 4,6,7-triphenyl-2-pteridinamine in glacial acetic acid was treated with sodium nitrite over 15 min to give 4,6,7-triphenyl-2(1*H*)-pteridinone (70%).[1120] 2-Amino-4(3*H*)-pteridinone (**33**, R = Y = H) in boiling 20%

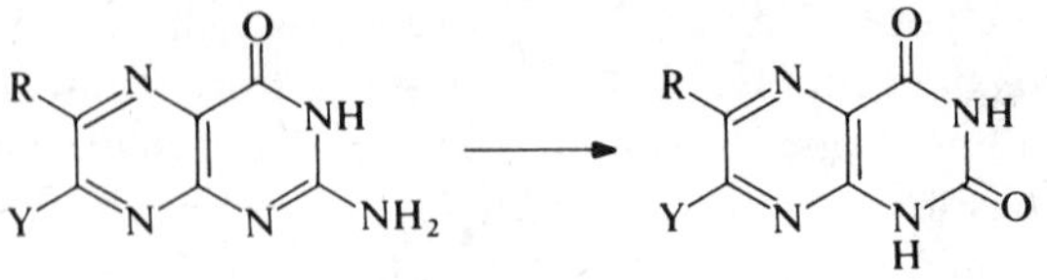

(33) (34)

sulfuric acid was treated with sodium nitrite to give 2,4(1*H*,3*H*)-pteridinedione (**34**, R=Y=H) (49%).[471] 2-Amino-6-phenyl-4(3*H*)-pteridinone (**33**, R=Ph, Y=H) in boiling 2*M* hydrochloric acid was treated with solid sodium nitrite to give 6-phenyl-2,4(1*H*,3*H*)-pteridinedione (**34**, R=Ph, Y=H) (67–77%).[67, 367] 2-Amino-7-phenyl-4(3*H*)-pteridinone (**33**, R=H, Y=Ph) similarly gave 7-phenyl-2,4(1*H*,3*H*)-pteridinedione (**34**, R=H, Y=Ph) (80%).[67] 2-Amino-6,7-diphenyl-4(3*H*)-pteridinone (**33**, R=Y=Ph) *resisted* similar treatment and even resisted diazotization in concentrated sulfuric acid at 50°C.[471] 2-Amino-6-methyl-4,7(3*H*,8*H*)-pteridinedione in 12% sulfuric acid at 80 to 90°C was treated with sodium nitrite to give 6-methyl-2,4,7(1*H*,3*H*,8*H*)-pteridinetrione (30%).[1304] 2-Amino-4,6,7(3*H*,5*H*,8*H*)-pteridinetrione in ~80% sulfuric acid was treated with nitrosylsulfuric acid (from sodium nitrite and cold concentrated sulfuric acid) at room temperature and later poured into ice to give 2,4,6,7(1*H*,3*H*,5*H*,8*H*)-pteridinetetrone (>95%).[531, cf. 403, 533] Similar treatment of 2,6-diamino-7-methyl-4(3*H*)-pteridinone or 2-amino-6-methyl-4,7(3*H*,8*H*)-pteridinedione gave 7-methyl-2,4,6(1*H*,3*H*,5*H*)- or 6-methyl-2,4,7(1*H*,3*H*,8*H*)-pteridinetrione, respectively.[279] Attempts to apply the diazotization technique to 2,4-pteridinediamine, its 6,7-dimethyl and 6,7-diphenyl derivatives, 4-amino-2(1*H*)-pteridinone, and its 6,7-diphenyl derivative all failed.[471]

D. From Alkoxypteridines

The hydrolysis of alkoxypteridines to the corresponding pteridinones appears to be little affected by the position or nature of the alkoxy group(s); although usually done in alkali, there are also examples of acidic hydrolysis. The reaction is exemplified in the transformation of 2-, 4-, or 6-methoxypteridine into 2(1*H*)-, 4(3*H*)-, or 6(5*H*)-pteridinone, respectively, in ~90% yield by shaking in 1*M* alkali at 25°C for 2 h[32]; of 2-methoxy-6,8-dimethyl-7(8*H*)-pteridinone into 6,8-dimethyl-2,7(1*H*,8*H*)-pteridinedione (**35**) in 33% yield by refluxing in 2.5*M*

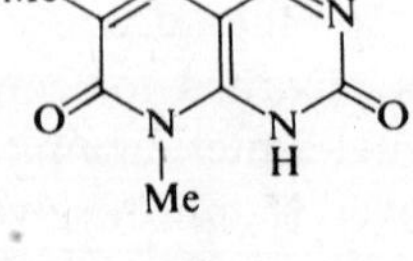

(35)

hydrochloric acid for 1 h[248]; of 7-methoxy-2,4-pteridinediamine into 2,4-diamino-7(8*H*)-pteridinone in >95% yield by refluxing in 1*M* alkali for 10 min[1301]; of the isomeric 6-methoxy-2,4-pteridinediamine into 2-amino-4,6(3*H*,5*H*)-pteridinedione in 93% yield by refluxing in 2.5*M* alkali for 21 h (note the hydrolysis of the 4-amino group under the vigorous conditions employed)[1301]; of 6-methoxy-4(3*H*)-pteridinone (**36**) into 4,6(3*H*,5*H*)-pteridinedione in 55% yield by heating in 1*M* alkali at 95°C for 18 h[794]; of 2-amino-8-β-hydroxyethyl-4-isopropoxy-6-methyl-7(8*H*)-pteridinone into 2-amino-8-β-hydroxyethyl-6-methyl-4,7(3*H*,8*H*)-pteridinedione in ~45% yield by heating in 0.2*M* alkali at 95°C for 3 h[372]; of 2-amino-7-isobutyl-4-isopropoxy-6(5*H*)-pteridinone into 2-amino-7-isobutyl-4,6(3*H*,5*H*)-pteridinedione (**37**) in 31% yield by boiling in 2*M* alkali for 20 min[1379]; of 7-isobutyl-2,4-dimethoxy-6(5*H*)-pteridinone into 7-isobutyl-2,4,6(1*H*,3*H*,5*H*)-pteridinetrione using similar conditions[1379]; of 6-hydroxymethyl-7-methoxy-2,4-pteridinediamine into 2,4-diamino-6-hydroxymethyl-7(8*H*)-pteridinone in 46% yield by stirring in 1*M* alkali at 100°C until solution was complete[1296]; of 7-methoxy-1,3-dimethyl-2,4(1*H*,3*H*)-pteridinedione 5-oxide into 1,3-dimethyl-2,4,7(1*H*,3*H*,8*H*)-pteridinetrione 5-oxide in ~60% yield by stirring in 0.1*M* alkali at 25°C for 5 h[238]; and of 2-dimethylamino-4,7-bis(2′,3′,5′-tri-*O*-benzoyl-β-D-ribofuranosyloxy)pteridine into 2-dimethylamino-4,7(3*H*,8*H*)-pteridinedione (**38**) in 61% yield by refluxing in methanolic sodium methoxide for 1 h, followed by an aqueous work-up.[246]

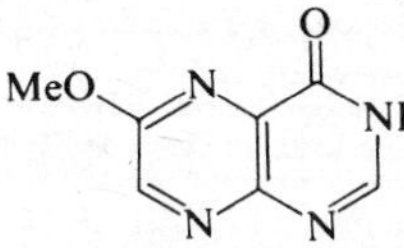

(**36**)

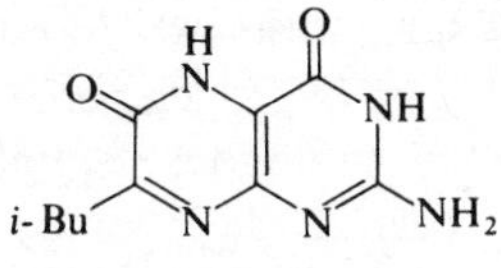

(**37**)

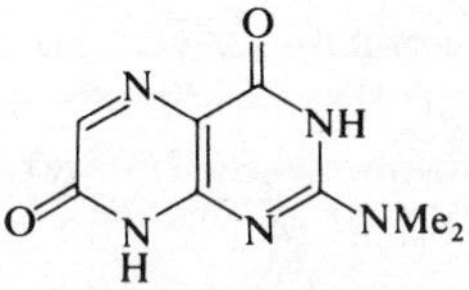

(**38**)

Hydrogenation of 6,7-dimethyl-4-trimethylsiloxy-2-trimethylsilylaminopteridine over platinum in benzene gave 2-amino-6,7-dimethyl-5,6,7,8-tetrahydro-4(3*H*)-pteridinone. The same transformation has been attained over a rhodium complex without hydrogen.[1695]

E. From Pteridinethiones or Alkylthiopteridines

The formation of pteridinones from the corresponding pteridinethiones can sometimes be done directly by acidic hydrolysis under fairly vigorous conditions, but it is usually better to oxidize the thione with hydrogen peroxide and then allow the (unisolated) sulfinic or sulfonic acid to hydrolyze under relatively mild conditions. Another practical route is to *S*-alkylate the pteridinethione initially. The resulting alkylthiopteridine undergoes smooth acidic hydrolysis to the pteridinone (preferably in a good fume hood).

The *direct hydrolysis of pteridinethiones* is illustrated in the conversion of 6-thioxo-5,6-dihydro-7(8*H*)-pteridinone (**39**, X=S, Y=O), 7-thioxo-7,8-dihydro-6(5*H*)-pteridinone (**39**, X=O, Y=S), or 6,7(5*H*,8*H*)-pteridinedithione (**39**, X=Y=S) into 6,7(5*H*,8*H*)-pteridinedione (**39**, X=Y=O) in >50% yield by refluxing in 1*M* hydrochloric acid for 1 h[36]; also in the conversion of 2-formylthio-6-phenyl-4(3*H*)-pteridinone (**40**) or substituted-phenyl analogs into 6-phenyl-2,4(1*H*,3*H*)-pteridinedione or its substituted-phenyl analogs by refluxing in aqueous ethanolic 2.5*M* hydrochloric acid for 1 h (yields were not reported).[543]

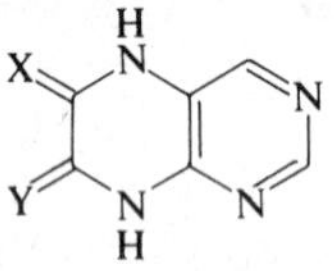

(39)

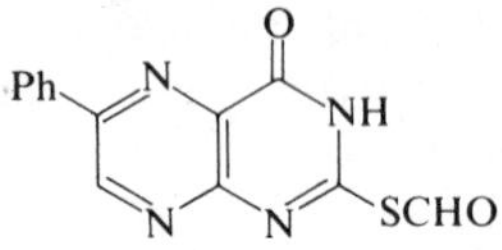

(40)

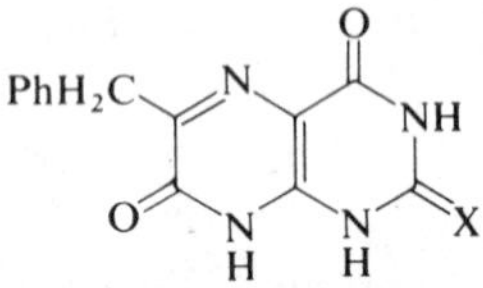

(41)

The *oxidative hydrolysis of pteridinethiones* is exemplified in the treatment of 6-benzyl-2-thioxo-1,2-dihydro-4,7(3*H*,8*H*)-pteridinedione (**41**, X=S) with alkaline hydrogen peroxide, followed after 24 h by acidification, to give 6-benzyl-2,4,7(1*H*,3*H*,8*H*)-pteridinetrione (**41**, X=O) (91%)[92]; of 2-thioxo-1,2-dihydro-4(3*H*)-pteridinone with ammoniacal hydrogen peroxide for several hours to give 2,4(1*H*,3*H*)-pteridinedione (~33%)[399, 1169]; of 6,7-diphenyl-2,4(1*H*,3*H*)-pteridinedithione with alkaline hydrogen peroxide at 25°C for 24 h to give 6,7-diphenyl-2,4(1*H*,3*H*)-pteridinedione (90%)[699]; and of 4,7-diamino-6-phenyl-2(1*H*)-pteridinethione with alkaline hydrogen peroxide for 12 h to 4,7-diamino-6-phenyl-2(1*H*)-pteridinone.[1281] Analogous examples have been described briefly.[92, 699, 900] Use has also been made of the peroxides in gewöhnlich THF to convert 6,7-diphenyl-4-thioxo-3,4-dihydro-2(1*H*)-pteridinone (**42**, X=S) into 6,7-diphenyl-2,4(1*H*,3*H*)-pteridinedione (**42**, X=O) in 80% yield by simply refluxing the former in the solvent for 30 min and then stirring in air at 25°C for 24 h.[446] In addition, irradiation of an alkaline solution of 1,3-dimethyl-7-thioxo-7,8-dihydro-2,4(1*H*,3*H*)-pteridinedione (**43**, X=S) gave 1,3-dimethyl-2,4,7(1*H*,3*H*,8*H*)-pteridinetrione (**43**, X=O) via the 7-sulfonic acid.[227]

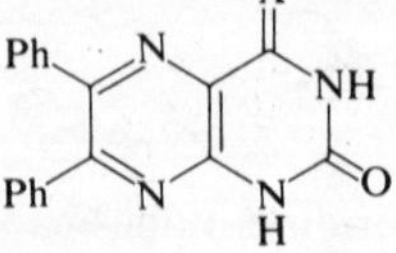

(42)

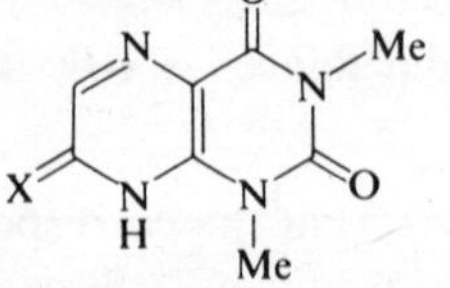

(43)

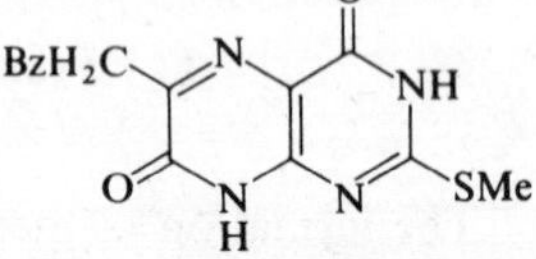

(44)

The *hydrolysis of alkylthiopteridines* to pteridinones is usually done in acid but alkali has also proven effective. Thus 4-methylthiopteridine in 4% ethanolic sodium hydroxide at 70°C for 30 min gave 4(3*H*)-pteridinone in 60% yield,[305] while 4-benzylthio-2-pteridinamine likewise gave 2-amino-4(3*H*)-pteridinone (55%).[305] The more usual acidic conditions are illustrated in the conversion of 4-methylthiopteridine into 4(3*H*)-pteridinone (1*M* acetic acid under reflux for 1 h: ~65%)[1317]; of 2-methylthio-6-phenacyl-4,7(3*H*,8*H*)-pteridinedione (**44**) into 6-phenacyl-2,4,7(1*H*,3*H*,8*H*)-pteridinetrione (aqueous ethanolic hydrochloric acid under reflux for 1 h: 50%)[1524]; of 3,6,7-trimethyl-2-methylthio-4(3*H*)-pteridinone into 3,6,7-trimethyl-2,4(1*H*,3*H*)-pteridinedione (6*M* hydrochloric acid under reflux for 1–2 h: 50–60%)[166, 362]; of 3-methyl-2-methylthio-6-phenyl-4(3*H*)-pteridinone into 3-methyl-6-phenyl-2,4(1*H*,3*H*)-pteridinedione (refluxing 6*M* hydrochloric acid for 2 h: 90%)[367]; of 2-ethylthio-6,7-diphenyl-4(3*H*)-pteridinone into 6,7-diphenyl-2,4(1*H*,3*H*)-pteridinedione (refluxing 6*M* hydrochloric acid for 3 h)[699]; and of 3,*N*-dimethyl-2-methylthio-4,6-dioxo-3,4,5,6-tetrahydro-7-pteridinecarboxamide into 3,*N*-dimethyl-2,4,6-trioxo-1,2,3,4,5,6-hexahydro-7-pteridinecarboxamide (refluxing 0.5*M* sulfuric acid for 4 h: 50%; note the survival of the amide group).[342] Initial oxidation of the alkylthio to an alkylsulfonylpteridine facilitates alkaline hydrolysis.[621]

F. By Rearrangement of Pteridine *N*-Oxides

Several 6- or 7-pteridinones have been made by rearrangement of pteridine 5- or 8-oxides in acetic anhydride followed by hydrolysis of the (frequently unisolated) acetoxy intermediate. Thus 1-methyl-2,4(1*H*,3*H*)-pteridinedione 5-oxide (**45**, R = H) was heated under reflux in acetic anhydride to give 6-acetoxy-1-methyl-2,4(1*H*,3*H*)-pteridinedione (**46**, R = H) (59%), which underwent hydrolysis on boiling in 1*M* hydrochloric acid for 1 min to afford 1-methyl-2,4,6(1*H*,3*H*,5*H*)-pteridinetrione (**47**, R = H) (74%).[367] In a rather similar manner, 1,3-dimethyl-2,4(1*H*,3*H*)-pteridinedione 5-oxide (**45**, R = Me) was stirred in a mixture of trifluoroacetic acid and acetic anhydride for 18 h and the crude product was then submitted to brief acidic hydrolysis to give 1,3-dimethyl-2,4,6(1*H*,3*H*,5*H*)-pteridinetrione (**47**, R = Me) in 85% overall yield.[1275] The first of these procedures was also used to convert 3-methyl-7-phenyl-2,4(1*H*,3*H*)-pteridinedione 5,8-dioxide into 6-acetoxy-3-methyl-7-phenyl-2,4(1*H*,3*H*)-pteridinedione 8-oxide (65%) and then into 3-methyl-7-phenyl-2,4,6(1*H*,3*H*,5*H*)-pteridinetrione 8-oxide (66%; replacement of the acidic hydrolysis by brief alkaline hydrolysis or by treatment with trifluoroacetic acid in acetonitrile did not affect the yield appreciably).[331] Broadly similar procedures were employed

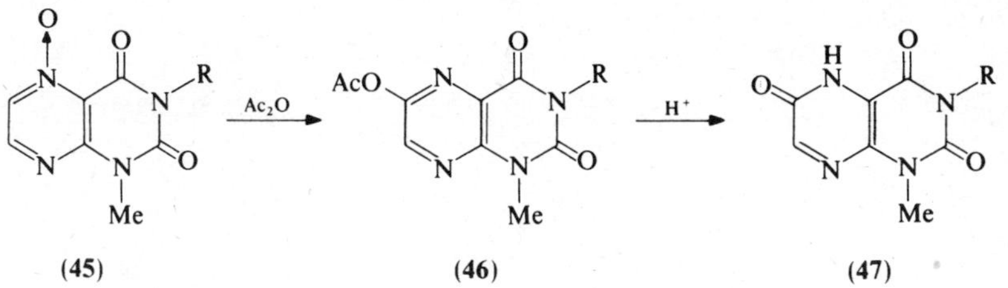

(45) **(46)** **(47)**

to convert 3-methyl-7-phenyl-2,4(1*H*,3*H*)-pteridinedione 5-oxide into 6-acetoxy-3-methyl-7-phenyl-2,4(1*H*,3*H*)-pteridinedione (60%) and then into 3-methyl-7-phenyl-2,4,6(1*H*,3*H*,5*H*)-pteridinetrione (92% or, without isolation of the intermediate, in 90% overall yield; an alternative procedure involved refluxing the 5-oxide in 1*M* potassium hydroxide for 1 h followed by acidification to give a 75% yield of the trione);[331, 367] 1,3-dimethyl-2,4,7(1*H*,3*H*,8*H*)-pteridinetrione 5-oxide into 1,3-dimethyl-2,4,6,7(1*H*,3*H*,5*H*,8*H*)-pteridinetetrone (55%) (alternatively, refluxing the *N*-oxide in 2*M* hydrochloric acid for 1 h gave the same product in 75% yield!)[238]; 7-methoxy-1,3-dimethyl-2,4(1*H*,3*H*)-pteridinedione 5-oxide into 7-methoxy-1,3-dimethyl-2,4,6(1*H*,3*H*,5*H*)-pteridinetrione (59%) (hydrolysis was carried out gently in aqueous ethanol to preserve the methoxy group)[238]; 7-amino-1,3-dimethyl-2,4(1*H*,3*H*)-pteridinedione 5-oxide into 7-acetamido-1,3-dimethyl-2,4,6(1*H*,3*H*,5*H*)-pteridinetrione (56%) and then by brief alkaline hydrolysis to the 7-amino anolog (57%)[238]; 1,3-dimethyl-7-phenyl-2,4(1*H*,3*H*)-pteridinedione 5-oxide into 1,3-dimethyl-7-phenyl-2,4,6(1*H*,3*H*,5*H*)-pteridinetrione (60%)[367]; and 2-amino-6-phenyl-4(3*H*)-pteridinone 8-oxide into 2-amino-6-phenyl-4,7(3*H*,8*H*)-pteridinedione (25%).[541]

However, not all such rearrangements are necessarily regular. In 1973, Wolfgang Pfleiderer and colleagues observed that 2-amino-4(3*H*)-pteridinone 8-oxide (**49**) in refluxing acetic anhydride containing trifluoroacetic acid, followed by hydrolysis, gave 2-amino-4,6(3*H*,5*H*)-pteridinedione (**50**) (51%) rather than the expected isomer (**48**).[541] This β-rearrangement (unprecedented in the pteridine but not in other series) was confirmed in 1975 by Ted Taylor and colleagues, who used a mixture of trifluoroacetic acid and trifluoroacetic anhydride at 50°C for 20 min to give (after hydrolysis) a >90% yield of xanthopterin (**50**) and who suggested a rational mechanism for the reaction.[1301] 2,4-Pteridinediamine 8-oxide behaved similarly to afford 2,4-diamino-6(5*H*)-pteridinone,[541, 1301] confirmed in identity by alkaline hydrolysis to 2-amino-4,6(3*H*,5*H*)-pteridinedione (65%).[1301] A different type of β-rearrangement is seen in the reaction of 3,6,7-trimethyl-2,4(1*H*,3*H*)-pteridinedione 5,8-dioxide in boiling acetic anhydride for 5 or 45 min to give 7-acetoxymethyl-3,6-dimethyl-2,4(1*H*,3*H*)-pteridinedione 5-oxide (4%) or 6,7-bis(acetoxymethyl)-3-methyl-2,4(1*H*,3*H*)-pteridinedione (5%), respectively.[367] However, the reported conversion of 6-phenyl-2,4-pteridinediamine 5,8-dioxide, by refluxing acetic anhydride, into 2,4-diacetamido-6-phenyl-7(8*H*)-pteridinone (58%),[541] has yet to be explained.

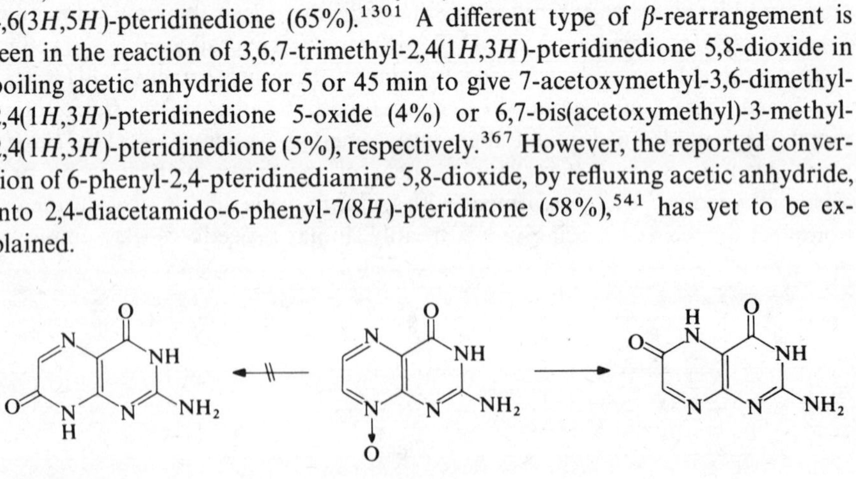

(**48**) (**49**) (**50**)

2. PROPERTIES AND REACTIONS OF TAUTOMERIC PTERIDINONES

The most noticable physical properties of the tautomeric pteridinones are their resistance to melting and their poor solubilities in water and indeed in all other solvents. Unlike pteridine, which melts below 140°C and dissolves in seven parts of cold water, the pteridinones either melt with decomposition above 250°C or simply decompose without melting at even higher temperatures, while their solubilities in water decrease markedly as the number of oxo substituents increases (see Table X). These effects clearly result from the intermolecular hydrogen bonding being exceptionally powerful within the crystal lattice, thereby discouraging melting and any tendency to form hydrogen bonds to water as a first step towards dissolution therein.[15, 675–677] The crystal structure of dilumazine trihydrate, 2,4(1*H*,3*H*)-pteridinedione sesquihydrate, confirms that it is built up of parallel layers of dimers. The molecules in each dimer are held together unsymmetrically by two powerful hydrogen bonds between an oxo substituent in each molecule and an NH group in the other. The dimers are themselves held together by hydrogen bonding involving the water-of-hydration

TABLE X. DECOMPOSITION TEMPERATURES AND SOLUBILITIES[15, 32, 41] IN WATER FOR SIMPLE PTERIDINONES

Compound	Decomposition (°C)	Parts of Water (20°C)
Pteridine	137–140[a]	7
2(1*H*)-Pteridinone	>240	600
4(3*H*)-Pteridinone	>340	200
6(5*H*)-Pteridinone	>250	3,500[b]
7(8*H*)-Pteridinone	>250	900
2,4(1*H*,3*H*)-Pteridinedione	350	800
2,6(1*H*.5*H*)-Pteridinedione	>250	110,000[b]
2,7(1*H*,8*H*)-Pteridinedione	>275	1,400
4,6(3*H*,5*H*)-Pteridinedione	>350	5,000
4,7(3*H*,8*H*)-Pteridinedione	>350	4,000
6,7(5*H*,8*H*)-Pteridinedione	>360	3,000
2,4,6(1*H*,3*H*,5*H*)-Pteridinetrione	380	7,400
2,4,7(1*H*,3*H*,8*H*)-Pteridinetrione	>350	12,000
2,6,7(1*H*,5*H*,8*H*)-Pteridinetrione	>200	⩾1,200[c]
4,6,7(3*H*,5*H*,8*H*)-Pteridinetrione	>360	27,000
2,4,6,7(1*H*,3*H*,5*H*,8*H*)-Pteridinetetrone	>360	58,000

[a] True melting point.
[b] Exists exclusively as covalent hydrate and hence more akin to next group down.
[c] Dissolves in 1200 parts of water at 100°C but structure still in some doubt.

molecules.[796] Even the highly substituted lumazine, 6,7-dimethyl-1-β-D-ribofuranosyl-2,4(1*H*,3*H*)-pteridinedione, has a rather comparable crystal structure in which the pteridine parts are stacked in endless ribbons, which are separated by regions containing the sugar parts and in which the oxo substituents and NH group of each molecule are involved in intermolecular hydrogen bonding.[426, cf. 1044]

Stability constants for some complexes between selected heavy metals (Cu, Fe, etc.) and several pteridines (including pteroylglutamic acid) have been measured. Despite its apparent structural similarity to oxine (8-hydroxyquinoline), 4(3*H*)-pteridinone has a far lower affinity for such metallic ions. However, since its pK_a ensures that it is almost entirely present as the chelating species, viz. anion, at physiological pH 7.3, it could conceivably compete successfully with the abundant common amino acids for such metallic ions, a fact not devoid of potential biological significance in respect of natural pteridinones.[13, 37, 673] A variety of metal complexes from 8-alkyl-2,4(3*H*,8*H*)-pteridinones have been studied[583]; also complexes from 2,4(1*H*,3*H*)-pteridinedione with Mn, Co, Ni, Cu, and Zn salts have been prepared and characterized.[1225] A persistent second spot on paper chromatograms of pure 4(3*H*)-pteridinone and some other pteridinones was eventually shown to be due to chelation with a heavy metal present in some filter papers (perhaps from the metallic gauze screens on which they were dried) The difficulty was obviated simply by the addition of a trace of sodium sulfide to the chromatographic fluid, thereby converting the metal(s) to unreactive sulfide(s).[32] The paper chromatography of pteridinones has been studied in detail to permit the unequivocal detection of each pteridinone in the presence of others. The criteria used were the R_f values in nine solvents, the appearance and/or fluorescence of the spots in uv light of two frequencies, and the color of the spots after complex formation following spraying with copper acetate and diphenylcarbazide. In addition, the quantitative recoveries from paper chromatography in various solvent systems were recorded.[277] 7(8*H*)-Pteridinone has been shown to undergo an interesting photoreduction when a spot on paper is irradiated with 254-nm light. The product was identified as 5,6-dihydro-7(8*H*)-pteridinone, identical in all respects with authentic material. The same reaction occurred in aqueous solution only when a hydrogen donor such as *N*-allylthiourea was added prior to irradiation.[23]

A. Degradation of Pteridinone Nuclei

The ring fission of pteridinones by alkali, acids, amines, reduction, oxidation, and so on, under vigorous conditions may give pyrimidines, pyrazines, or other miscellaneous products.

Degradation to pyrimidines is quite rare and most known examples up to 1985 have been collected elsewhere.[1534] These comprise mainly 6- or 7-pteridinones without substituents in their pyrimidine ring, as well as a few 4-pteridinones. For example, reduction of 7(8*H*)-pteridinone (**51**) with hot alkaline dithionite gave 5-

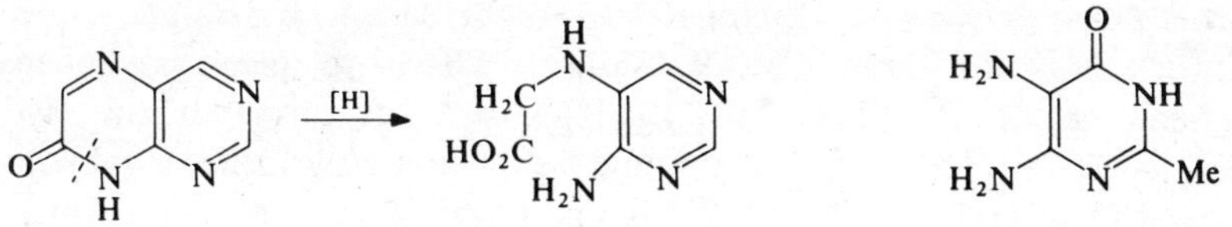

(51) (52) (53)

carboxymethylamino-4-pyrimidinamine (**52**) in 95% yield[31]; 7,8-dihydro-6(5*H*)-pteridinone in boiling 1*M* alkali for 90 min gave the isomeric 4-carboxymethylamino-5-pyrimidinamine (75%)[19]; and 2-methyl-4(3*H*)-pteridinone in hydrazine hydrate at 95°C for 1 h gave 5,6-diamino-2-methyl-4(3*H*)-pyrimidinone (**53**) (~70%).[153]

Degradation to pyrazines is the usual result of ring fission of a great many 4-pteridinones or 2,4-pteridinediones, with or without other substituents. Innumerable examples were first collected and discussed[466] in 1954 and an excellent tabulation of such reactions has since appeared in 1982.[1480b] Accordingly, it is unnecessary to repeat the information here, apart from the following typical examples. The ammonium salt of 2,4(1*H*,3*H*)-pteridinedione (**55**) with 25% sodium hydroxide in a bomb at 170°C for 2 h gave 3-amino-2-pyrazinecarboxylic acid (**54**) in 94% yield but the same substrate (**55**) in concentrated sulfuric acid at 240°C for 15 min gave 2-pyrazinamine (**56**) in 80% yield[1438, cf. 30]; 6- or 7-methyl-4(3*H*)-pteridinone in 10*M* sodium hydroxide under reflux (~140°C) for 3–4 h gave 3-amino-6-methyl- (80%) or 3-amino-5-methyl-2-pyrazinecarboxylic acid (70%), respectively[32]; 2-amino-6-phenyl-4(3*H*)-pteridinone in 4*M* sodium hydroxide at 170°C for 24 h gave 3-oxo-6-phenyl-3,4-dihydro-2-pyrazinecarboxylic acid (**57**)[176]; the sodium salt of 7-dimethylamino-6-phenyl-4(3*H*)-pteridinone 5-oxide (**58**) underwent facile degradation in boiling water during 6 h to give 3-amino-5-dimethylamino-6-phenyl-2-pyrazinecarboxamide 1-oxide (**59**) (20%)[764]; and others.[152, 974, 1084, 1437] Degradation in nitrogenous

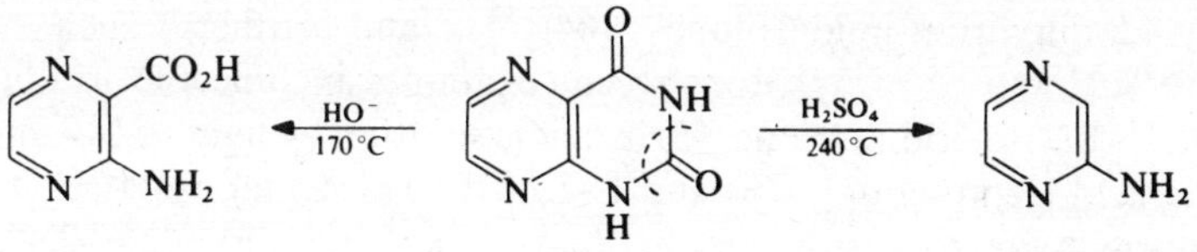

(54) (55) (56)

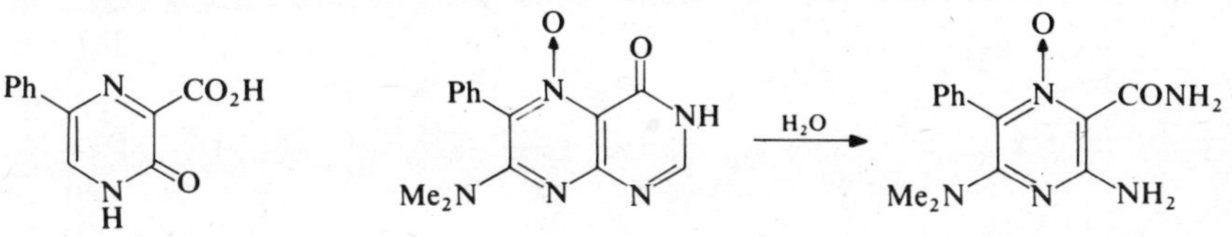

(57) (58) (59)

bases is illustrated by the reactons of 6,7-diphenyl-2,4(1*H*,3*H*)-pteridinedione (**60**) with aqueous ammonia at 185°C for 16 h to give 5,6-diphenyl-2-pyrazinamine (**61**, R=H, Y=NH_2) (94%), with aqueous hydrazine hydrate under reflux for 6 h to give 3-amino-5,6-diphenyl-2-pyrazinecarbohydrazide (**61**, R=$CONHNH_2$, Y=NH_2) (73%), and with refluxing piperidine in DMF to give *N*,*N*-pentamethylene-5,6-diphenyl-3-piperidinocarbonylamino-2-pyrazinecarboxamide [**61**, R=$CON(CH_2)_5$, Y=$NHCON(CH_2)_5$] (40%)[468]; also by the reaction of 4(3*H*)-pteridinone with benzylamine under reflux for 2 h to give mainly 3-amino-*N*-benzyl-2-pyrazinecarboxamide (**62**, R=CH_2Ph, Y=H) (>90%),[155] and of 7-methyl-4(3*H*)-pteridinone with hydrazine hydrate at 95°C for 1 h to give only 3-amino-5-methyl-2-pyrazinecarbohydrazide (**62**, R=NH_2, Y=Me).[153]

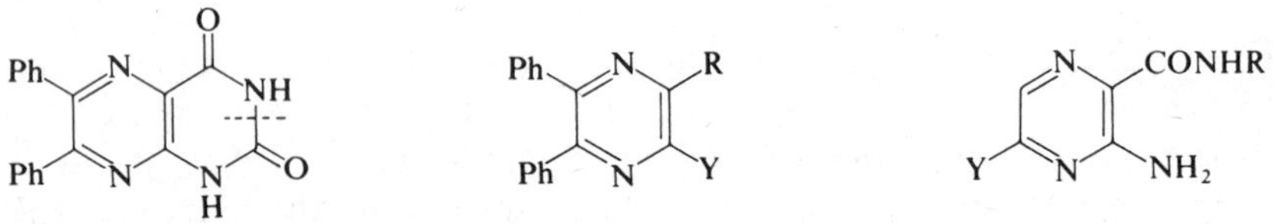

(**60**) (**61**) (**62**)

The formation of *miscellaneous products by degradation* of pteridinones is recorded mainly in the older literature, which has been covered adequately elsewhere.[466] For example, heating 2-amino-4,6(3*H*,5*H*)-pteridinedione with 5*M* hydrochloric acid at 160 to 200°C, with or without added zinc chloride, gave >50% of glycine, isolated as hippuric acid after *N*-benzoylation[438]; treatment of the same substrate with concentrated hydriodic acid and red phosphorus at 170°C produced a similar result[438]; the oxidation of 2-amino-4-oxo-3,4-dihydro-6-pteridinecarboxylic acid (**63**) with sodium chlorate in hydrochloric acid at 80°C gave guanidine, isolated as its picrate[1363]; oxidation of 2-amino-4,6(3*H*,5*H*)-pteridinedione with hydrogen peroxide gave a small yield of iminooxonic acid, probably best formulated as 2-amino-5-carboxyamino-5-hydroxy-4,5-dihydro-4-imidazolone (**64**)[533]; and other such records abound.[440, 531, 532, 534, 1182] More recent examples include the irradiation of 2,4,7(1*H*,3*H*,8*H*)-pteridinetrione (**65**) in acidic solution by light of 247 nm to give glyoxylic acid (representing C-6 and C-7 of the substrate), isolated as its 2,4-dinitrophenylhydrazone[1267]; the reaction of 3-benzenesulfonyloxy-2,4(1*H*,3*H*)-pteridinedione (**66**) with methanolic sodium methoxide to give, by a logical if complicated mechanism, methyl 3-oxo-2,3-dihydro-*s*-triazolo[4,3-*a*]pyrazine-8-

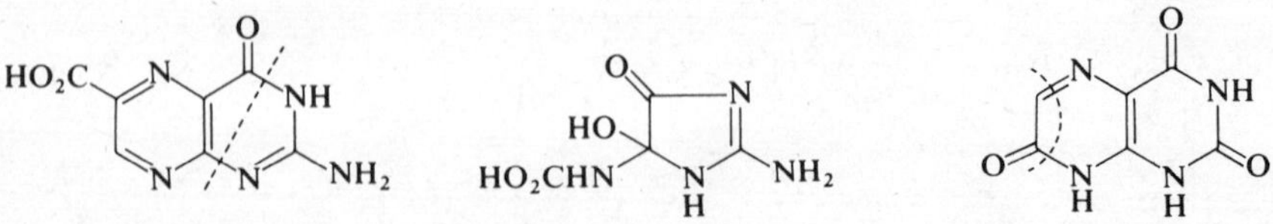

(**63**) (**64**) (**65**)

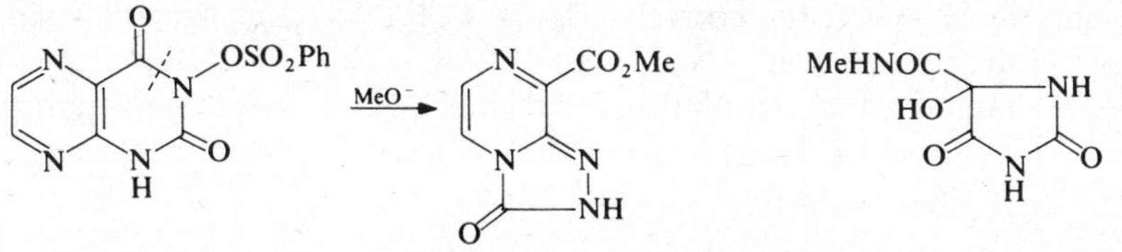

(66) (67) (68)

carboxylate (**67**) in 85% yield[1288]; and a kinetic study of the conversion of 8-methyl-6,7-diphenyl-2,4(3*H*,8*H*)-pteridinedione into the alloxanic acid derivative, 4-hydroxy-*N*-methyl-2,5-dioxo-2,3,4,5-tetrahydro-4-imidazolecarboxamide (**68**).[1284] Miscellar effects induced by the presence of surfactants during the irradiation of folic acid have been studied without specifying products formed.[1076]

The mass spectral fragmentation patterns of simple pteridinones are very diverse.[1707] Thus 2(1*H*)-pteridinone lost HCN initially, followed by CO[207, 1246]; 4(3*H*)-pteridinone lost the same fragments but in both orders[207, 779, cf. 1391]; 6(5*H*)- and 7(8*H*)-pteridinones both lost CO initially from the pyrazine ring, probably to give the purine cations (**69**) and (**70**), respectively, and thereafter their spectra closely resembled that of purine itself[207, 1246]; 2,4(1*H*,3*H*)-pteridinedione behaved rather like 2(1*H*)-pteridinone by initially losing HNCO (representing N-3 and C-4 with its attached O in this case) and then CO[207]; and 2,4,7(1*H*,3*H*,8*H*)-pteridinetrione first lost HNCO followed by CO and then probably lost a second CO and HCN.[1411] These patterns were disturbed, in some cases radically, by the presence of any *C*-methyl group(s) in the molecule.[779, 1246, 1391] A great many data have been provided for the fragmentation of *C*- and/or 8-alkylated 2,4-pteridinediones.[1553, 1554] Of the naturally occurring pteridinones, 2-amino-4(3*H*)-pteridinone lost HCN and then CO, 2-amino-4,6(3*H*,5*H*)-pteridinedione did likewise, 2-amino-4,7(4*H*,8*H*)-pteridinedione lost CO first, and 2-amino-4,6,7(3*H*,5*H*,8*H*)-pteridinetrione did likewise.[778] *N*-Acylated derivatives of this compound initially lost a ketene fragment[1535] and both biopterin and sepiapterin have been examined in some detail.[976]

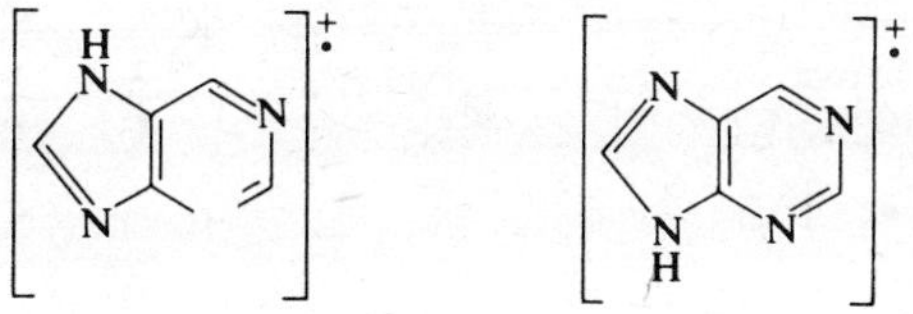

(69) (70)

Difficulties in the ultimate quantitative degradation of the pteridinones to carbon dioxide, water, and nitrogen caused great uncertainty in the empirical formulas for the early-discovered natural pteridines.[26] For example, that for xanthopterin was given first as $C_{19}H_{19}N_{15}O_7$,[1238] then as $C_{19}H_{18}N_{16}O_6$,[438] and only in 1940 was the correct formula, $C_6H_5N_5O_2$, reported.[404] Modifications to the classical methods of Dumas, Pregl, Kjeldahl, and Zimmermann to make them suitable for pteridinones and related highly nitrogenous heterocycles have been outlined elsewhere.[573, 1277] Modern analytical instruments appear to handle such analyses reasonably satisfactorily.

B. Formation of Covalent Adducts

The now famous hysteresis loop in plotted pH values, which was produced by Adrien Albert and colleagues during titration of 6(5*H*)-pteridinone (**71**), first with one equivalent of acid and then with one equivalent of alkali,[31, 1537] was eventually explained correctly by postulating rapid reversible covalent hydration of the neutral molecule (**71**) to give 7-hydroxy-7,8-dihydro-6(5*H*)-pteridinone (**72**).[17, 19] This immediately rationalized the following observed phenomena[55]: (a) the hysteresis loop; (b) the long wavelength uv absorption peak, which was 67 nm lower than that of the anion,[31] suggesting loss of a double bond; (c) experimental proof[1537] that neither ring-fission nor keto–enol tautomerism occurred; (d) the identity[134] of the uv spectrum of the neutral molecule with that[31] of 7,8-dihydro-6(5*H*)-pteridinone (**73**); and (e) the facile oxidation[134] to 6,7(5*H*,8*H*)-pteridinedione (**74**). The hypothesis also explained an apparent slow tautomerism evident in the spectrum of xanthopterin in alkaline solution,[1256] and it was shown[31] that xanthopterin also gave a loop on titration–back titration.

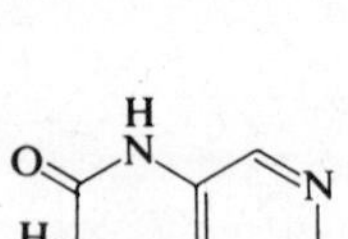

(**71**)

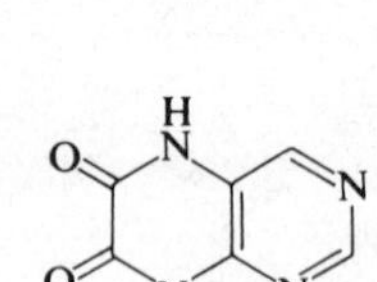

(**72**)

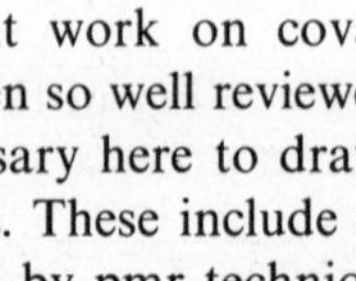

(**73**)

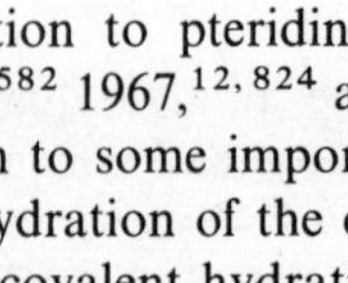

(**74**)

The subsequent work on covalent addition to pteridinones and related molecules has been so well reviewed in 1962,[582] 1967,[12, 824] and 1976,[11] that it seems only necessary here to draw attention to some important aspects with leading references. These include covalent hydration of the cations of pteridinones as revealed by pmr techniques[27]; the covalent hydration of 2(1*H*)- and

6(5*H*)-pteridinone and their *C*-methyl derivatives[954]; the dianions from 2(1*H*)- and 6(5*H*)-pteridinone covalent hydrates[140]; kinetic aspects of the reversible hydration of 2(1*H*)-pteridinones,[241, 243, 245] 6(5*H*)-pteridinones,[241, 244] 2,6(1*H*,5*H*)-pteridinediones,[40, 241] 4,6(3*H*,5*H*)-pteridinediones,[241, 244] and xanthopterin;[241, 244, 961] the covalent hydration of 2-amino-4(3*H*)-pteridinone and its 7,8-dihydro derivative[453, 584, 606]; the existence of 6,7-bistrifluoromethyl-4(3*H*)-pteridinone as its stable covalent dihydrate, 6,7-dihydroxy-6,7-bistrifluoromethyl-5,6,7,8-tetrahydro-4(3*H*)-pteridinone,[1571] and of 8-benzyl-6,7-bistrifluoromethyl-2,4(3*H*,8*H*)-pteridinedione as a monohydrate[1570]; Michael-type additions of acetone, acetylacetone, ethyl acetoacetate, diethyl malonate, ethyl cyanoacetate, sodium bisulfite, barbituric acid, 2-thiobarbituric acid, dimedone, amines, and so on to the 3,4-bond of 2(1*H*)-pteridinones,[38] to the 5,6- and 7,8-bond of 4(3*H*)-pteridinones,[47] to the 7,8-bond of 6(5*H*)-pteridinones,[55] to the 5,6-bond of 7(8*H*)-pteridinones,[44] to the 5,6- and 7,8-bond of 2,4(1*H*,3*H*)-pteridinone,[47] to the 5,6-bond of 2-amino-7,8-dihydro-4(3*H*)-pteridinone,[453] and to the 7,8-bond of 2-amino-4,6(3*H*,5*H*)-pteridinedione,[47] all of which occurred with reasonable ease to give characterized adducts; and the addition of primary alcohols to pteridinones, often by several recrystallizations of the corresponding covalent hydrate from the appropriate alcohol.[32, 38, 134] All such additions were discouraged, if not completely prevented, by the presence of a *C*-alkyl group at each potential addition site.

C. Removal of Oxo Substituents from Pteridinones

Perhaps the easiest ways to deoxygenate a tautomeric pteridinone are (a) to convert it into the corresponding chloropteridine (Ch. V, Sect. 1.B) followed by dechlorination (Ch. V, Sect. 2.B) or (b) to convert it into the corresponding pteridinethione, either directly (Ch. VI, Sect. 2.G) or via the chloropyrimidine (Ch. V, Sect. 2.E), followed by desulfurization of the thione or alkylthio derivative (Ch. VIII, Sects. 2.A and 7.A).

However, deoxygenation can sometimes be achieved more directly by chemical or electrolytic reduction followed, if necessary, by rearomatization of the nucleus. Thus reduction of 2-amino-4,6,7(3*H*,5*H*,8*H*)-pteridinetrione (**75**) by sodium amalgam in water gave exclusively 2-amino-7,8-dihydro-4,6(3*H*,5*H*)-pteridinedione (**76**) (55%), which underwent dehydrogenation by treatment with alkaline permanganate at room temperature to give 2-amino-4,6(3*H*,5*H*)-pteridinedione (**77**) in 80% yield.[60, cf. 877, 1268] The same product (**76**) was obtained on electrolytic reduction in basic media,[605] but in strongly acid media, only 2-amino-4,7(3*H*,8*H*)-pteridinedione (**78**) was isolated in 13% yield.[1369] However, this selectivity was not maintained in analogs submitted to electrolytic reduction: 6,7(5*H*,8*H*)-pteridinedione in alkali gave 7,8-dihydro-6(5*H*)-pteridinone (47%); 4,6,7(3*H*,5*H*,8*H*)-pteridinetrione in alkali or in concentrated perchloric acid gave only 7,8-dihydro-4,6(3*H*,5*H*)-pteridinedione (58 or 75%, respectively); 4-methoxy-6,7(5*H*,8*H*)-pteridinedione under mildly alkaline

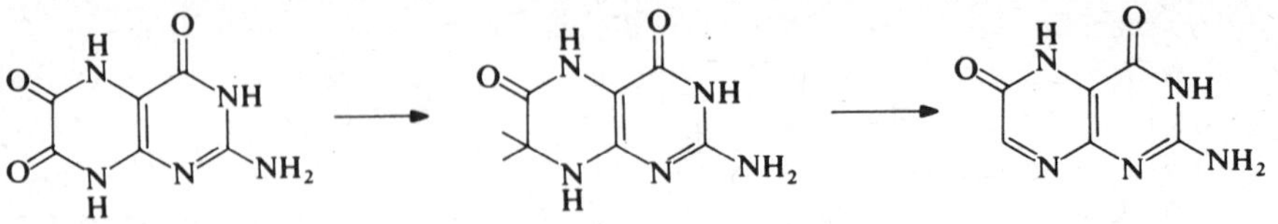

(75) (76) (77)

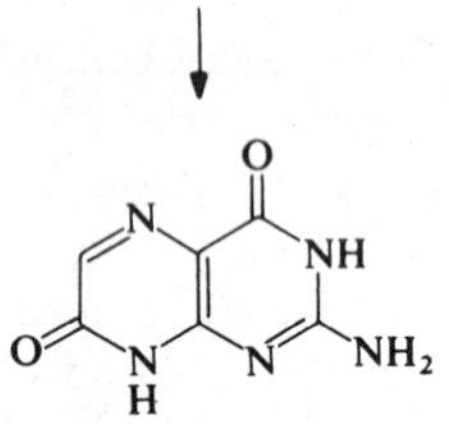

(78)

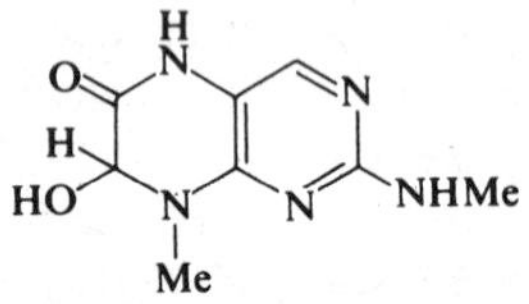

(79)

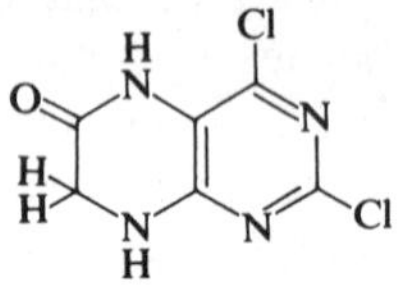

(80)

conditions gave 4-methoxy-7,8-dihydro-6(5*H*)-pteridinone (68%); 4-amino-6,7(5*H*,8*H*)-pteridinedione in alkali or in 60% perchloric acid gave only 4-amino-7,8-dihydro-6(5*H*)-pteridinone that underwent aerial oxidation on prolonged work-up to give good yields of 4-amino-6(5*H*)-pteridinone; and 2,4-diamino-6,7(5*H*,8*H*)-pteridinedione in alkali or perchloric acid gave 2,4-diamino-7,8-dihydro-6(5*H*)-pteridinone or 2,4-diamino-6(5*H*)-pteridinone in up to 80% yield, according to the work-up.[208, cf. 605] No deoxygenation occurred in similarly treated 8-alkyl analogs of these compounds. For example, 8-methyl-2-methylamino-6,7(5*H*,8*H*)-pteridinedione in alkali or acid gave only the reduced product 7-hydroxy-8-methyl-2-methylamino-7,8-dihydro-6(5*H*)-pteridinone (**79**) in 50 to 80% yield.[208] Other examples of chemical deoxygenations include the conversion of 4,6,7(3*H*,5*H*,8*H*)-pteridinetrione by sodium amalgam into 7,8-dihydro-4,6(3*H*,5*H*)-pteridinedione (90%) and then, by treatment with alkaline permanganate, to 4,6(3*H*,5*H*)-pteridinedione (70%)[29]; also the conversion of 2,4-dichloro-6,7(5*H*,8*H*)-pteridinedione by sodium amalgam into 2,4-dichloro-7,8-dihydro-6(5*H*)-pteridinone (**80**) in very small yield.[486]

D. Conversion into Halogenopteridines

The formation of chloro- and other halogenopteridines from tautomeric pteridinones has been discussed in Chapter V, Section 1.B.

E. Conversion into Pteridinamines

Pteridinones are usually converted into the corresponding pteridinamines by transformation into the chloropteridine (Ch. V, Sect. 1.B) and subsequent aminolysis (Ch. V, Sect. 2.C); alkylthiopteridines are used occasionally as intermediates. However, if the molecule is sufficiently stable, the conversion can be done directly, although the reactions involved have not been developed yet to their full potential. Thus treatment of 6,7-dimethyl-2-thioxo-1,2-dihydro-4(3*H*)-pteridinone with ethanolic methylamine in a bomb at 180°C for 18 h gave 6,7-dimethyl-2,4-bismethylaminopteridine (**81**, R = Me) in 85% yield.[1315] Similarly, 6,7-dimethyl-2-methylthio-4(3*H*)-pteridinone fused with butylamine acetate at 140°C for 2 h yielded mainly the expected 2-butylamino-6,7-dimethyl-4(3*H*)-pteridinone (~60%) but 2,4-bisbutylamino-6,7-dimethylpteridine (**81**, R = Bu) (~5%) was also obtained.[169] Substituted phosphoramides appear to be more effective reagents. Thus, a mixture of 2-amino-4(3*H*)-pteridinone, its 6,7-dimethyl derivative, or its 6,7-diphenyl derivative with phenyl phosphorodiamidate [$PhOP(:O)(NH_2)_2$] containing a trace of dimethylamine hydrochloride as catalyst, was heated at 215°C for 3 h to give 2,4-pteridinediamine (27%), its 6,7-dimethyl derivative (**81**, R = H) (34%), or the corresponding 6,7-diphenyl derivative (42%), respectively.[602] 2-Amino-4,7(3*H*,8*H*)-pteridinedione and hexamethylphosphoramide [$OP(NMe_2)_3$] containing traces of stearic anhydride and phosphoric acid as catalysts, were heated at 215°C for 80 min to give 2,4,7-trisdimethylaminopteridine as the main isolable product (note transamination at the 2-position).[864] 1,3-Dimethyl-2,4,6(1*H*,3*H*,5*H*)-pteridinetrione (**82**) was heated at 230°C for 30 min with phenyl phosphorodiamidate [$PhOP(:O)(NH_2)_2$], phenyl *N*,*N*′-dimethylphosphorodiamidate [$PhOP(:O)(NHMe)_2$], or phenyl *N*,*N*,*N*′,*N*′-tetramethylphosphorodiamidate [$PhOP(:O)(NMe_2)_2$] to give, respectively, 6-amino-1,3-dimethyl- (**83**, R = Y = H) (20%), 1,3-dimethyl-6-methylamino- (**83**, R = H, Y = Me) (44%), or 6-dimethylamino-1,3-dimethyl-2,4(1*H*,3*H*)-pteridinedione (**83**, R = Y = Me) (52%).[622, 1275] Similar procedures with 1,3-dimethyl-2,4,7(1*H*,3*H*,8*H*)-pteridinetrione as substrate afforded 7-amino-1,3-dimethyl- (**84**, R = Y = H) (65%), 1,3-dimethyl-7-methylamino- (**84**, R = H, Y = Me) (68%), or 7-dimethylamino-1,3-dimethyl-2,4(1*H*,3*H*)-pteridinedione (**84**, R = Y = Me) (75%).[622, 1275]

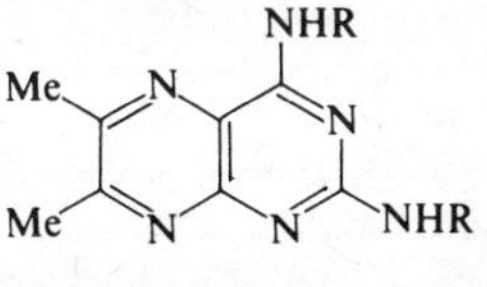

(**81**)

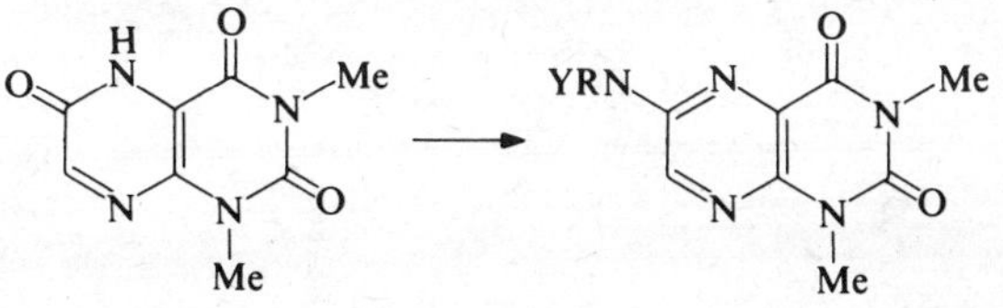

(**82**) (**83**)

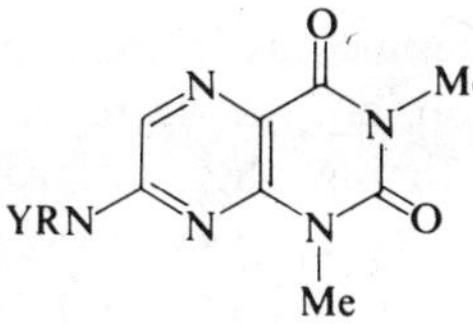

(**84**)

F. Conversion into Alkoxypteridines or Trimethylsiloxypteridines

The most reliable way to convert a pteridinone into the corresponding alkoxypteridine is to transform it first into a chloropteridine (Ch. V, Sect. 1.B) and then with alcoholic alkoxide into the alkoxypteridine (Ch. V, Sect. 2.D). However, in suitable cases, which have usually been 6(5*H*)- or 7(8*H*)-pteridinones to date, it has been found possible to convert *pteridinones into alkoxypteridines* directly by several methods, which can utilize the reactivity of a small proportion of hydroxypteridine tautomer existing in the solution of a pteridinone.

The *alcoholic hydrogen chloride* method is somewhat akin to the classical sulfuric acid method for making diethyl ether. It is illustrated in the treatment of 2,4-diamino-6(5*H*)-pteridinone with methanolic hydrogen chloride under reflux for 4 h to give 6-methoxy-2,4-pteridinediamine in 63% yield[268]; of 2-amino-4,6(3*H*,5*H*)-pteridinedione (**85**, R = H) with ethanolic or propanolic hydrogen chloride under reflux to give only 2-amino-6-ethoxy- (**86**, R = H, Y = Et) (~70%) or 2-amino-6-propoxy-4(3*H*)-pteridinone (**86**, R = H, Y = Pr) (~50%), respectively, thereby exhibiting a clear position preference in ether formation[369]; of 2-amino-3-methyl-4,6(3*H*,5*H*)-pteridinedione (**85**, R = Me) with methanolic hydrogen chloride under reflux to give 2-amino-6-methoxy-3-methyl 4(3*H*)-pteridinone (**86**, R = Y = Me) (~85%)[369]; of 1,3-dimethyl-2,4,6(1*H*,3*H*,5*H*)-pteridinetrione with methanolic hydrogen chloride under reflux for 7 h to give 6-methoxy-1,3-dimethyl-2,4(1*H*,3*H*)-pteridinedione (50%)[369]; and of 2-amino-4,6(3*H*,5*H*)-pteridinedione (**85**, R = H) with hydrogen chloride in 1,2-ethanediol (ethylene glycol) or in 1,2-propanediol at 100 to 110°C for 1 h to give (from the first reagent) 2-amino-6-β-hydroxyethoxy-4(3*H*)-pteridinone (**86**, R = H, Y = CH_2CH_2OH) (~40%) or (from the second reagent) a separable mixture of 2-amino-6-β-hydroxypropoxy- (**86**, R = H, Y = $CH_2CHOHMe$) and 2-amino-6-(β-hydroxy-α-methylethoxy)-4(3*H*)-pteridinone (**86**, R = H, Y = $CHMeCH_2OH$).[299]

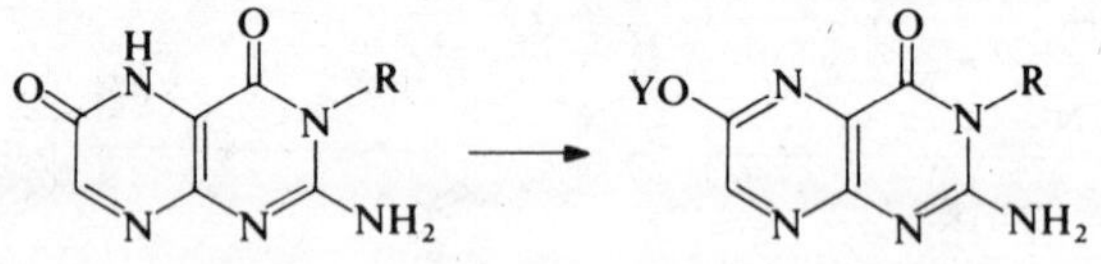

(85) **(86)**

The use of *diazomethane* sometimes favors the *O*-methylation of a pteridinone at the expense of *N*-methylation, but the degree of preference is very variable and difficult to forecast. Thus stirring etherial diazomethane with a methanolic suspension of 7(8*H*)-pteridinone gave 32% of an *N*-methyl-7-pteridinone [later identified[34] as 8-methyl-7(8*H*)-pteridinone] but only a trace of 7-methoxypteridine.[31] In contrast, similar treatment of 1,3-dimethyl-2,4,6(1*H*,3*H*,5*H*)-pteridinetrione gave nearly 40% of 6-methoxy-1,3-dimethyl-2,4(1*H*,3*H*)-pteridinedione (**87**, R=H)[340]; 1,3-dimethyl-2,4,6,7(1*H*,3*H*,5*H*,8*H*)-pteridinetetrone gave 6,7-dimethoxy-1,3-dimethyl-2,4(1*H*,3*H*)-pteridinedione (**87**, R=OMe) (~30%)[343]; 1,3,6-trimethyl-2,4,7(1*H*,3*H*,8*H*)-pteridinetrione gave 7-methoxy-1,3,6-trimethyl-2,4(1*H*,3*H*)-pteridinedione (~90%)[339]; 3,8-dimethyl-2,4,7(1*H*,3*H*,8*H*)-pteridinetrione gave 2-methoxy-3,8-dimethyl-4,7(3*H*,8*H*)-pteridinedione (**88**) (25%)[1192]; 1-methyl-2,4,7(1*H*,3*H*,8*H*)-pteridinetrione gave a separable mixture of the *O*,*O*-dimethyl derivative, 4,7-dimethoxy-1-methyl-2(1*H*)-pteridinone (**89**) (15%), and the *O*,*N*-dimethyl derivative, 7-methoxy-1,3-dimethyl-2,4(1*H*,3*H*)-pteridinedione (43%)[1192]; 2-dimethylamino-4(3*H*)-pteridinone gave 2-dimethylamino-4-methoxypteridine (33%)[511]; 1,3-dimethyl-2-thioxo-1,2-dihydro-4,7(3*H*,8*H*)-pteridinedione (**90**, X=S, Y=O) gave 7-methoxy-1,3-dimethyl-2-thioxo-1,2-dihydro-4(3*H*)-pteridinone (**91**, X=S, Y=O) (32%)[1000]; 1,3-dimethyl-4-thioxo-3,4-dihydro-2,7(1*H*,8*H*)-pteridinedione (**90**, X=O, Y=S) gave 7-methoxy-1,3-dimethyl-4-thioxo-3,4-dihydro-2(1*H*)-pteridinone (**91**, X=O, Y=S) (40%)[1000]; and 1,3-dimethyl-2,4-dithioxo-1,2,3,4-tetrahydro-7(8*H*)-pteridinone (**90**, X=Y=S) gave 7-methoxy-1,3-dimethyl-2,4(1*H*,3*H*)-pteridinedithione (**91**, X=Y=S) (29%).[1000] Other examples have been reported.[344]

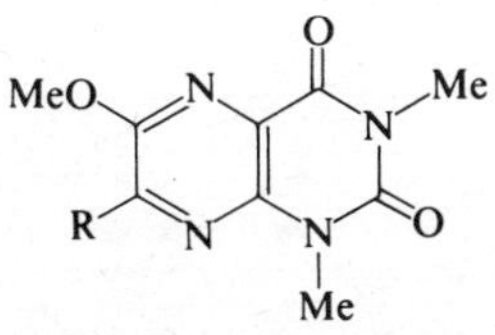

(87)

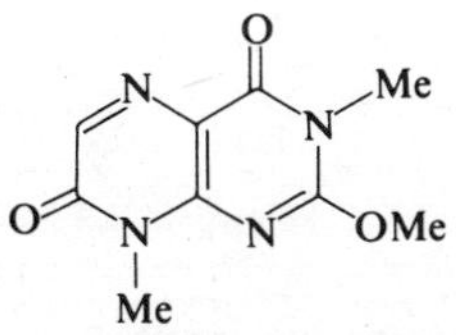

(88)

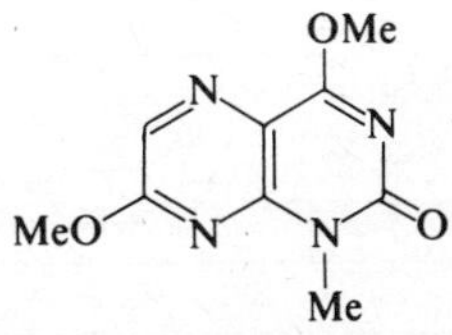

(89)

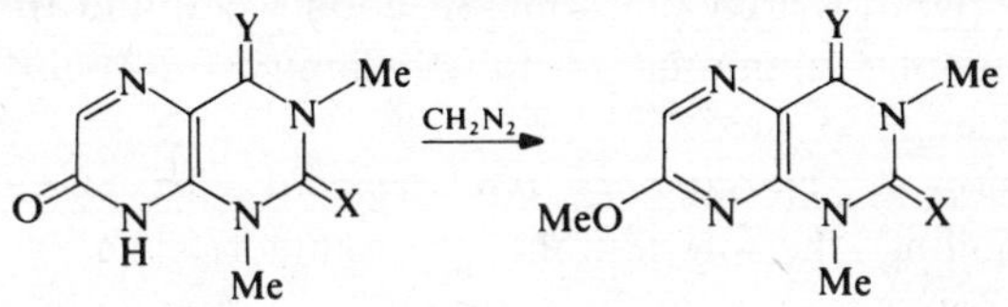

(90) **(91)**

The use of *methyl iodide or dimethyl sulfate*, in alkali or with the preisolated sodium salt of the substrate, seldom favors *O*-methylation to *N*-methylation of pteridinones, although sometimes a mixture is formed. Thus 1,3-dimethyl-2,4,6(1*H*,3*H*,5*H*)-pteridinetrione with dimethyl sulfate in an aqueous medium maintained at pH 9 and 40°C gave a separable mixture of 6-methoxy-1,3-dimethyl-2,4(1*H*,3*H*)-pteridinedione (**87**, R=H) (~20% yield) and 1,3,5-trimethyl-2,4,6(1*H*,3*H*,5*H*)-pteridinetrione (~10% yield).[340] 2-Amino-3-methyl-4,6,7(3*H*,5*H*,8*H*)-pteridinetrione on similar treatment gave a separable mixture of 2-amino-6-methoxy-3,8-dimethyl-4,7(3*H*,8*H*)-pteridinedione (~10% yield) and 2-amino-3,5,8-trimethyl-4,6,7(3*H*,5*H*,8*H*)-pteridinetrione (~65% yield).[384]

4-Dimethylamino-7(8*H*)-pteridinone likewise gave 4-dimethylamino-7-methoxypteridine (~10% yield) and 4-dimethylamino-8-methyl-7(8*H*)-pteridinone (~55% yield).[445, 530, 799] Other such reactions have been described.[344] Exceptionally satisfactory in producing methoxy products were the reactions of dimethyl sulfate on 1,3,6-trimethyl-2,4,7(1*H*,3*H*,8*H*)-pteridinetrione at pH 9 to give 7-methoxy-1,3,6-trimethyl-2,4(1*H*,3*H*)-pteridinedione in ~65% yield[399]; of methyl iodide on the sodium salt of 1,3-dimethyl-2,4,7(1*H*,3*H*,8*H*)-pteridinetrione (**90**, X=Y=O) in DMF at 50°C to give 7-methoxy-1,3-dimethyl-2,4(1*H*,3*H*)-pteridinedione (**91**, X=Y=O) in >50% yield[238]; and of dimethyl sulfate on 3,6-dimethyl-1-phenyl-2,4,7(1*H*,3*H*,8*H*)-pteridinetrione at pH 9 to give 7-methoxy-3,6-dimethyl-1-phenyl-2,4(1*H*,3*H*)-pteridinedione (50%).[238] When 2-dimethylamino-4(3*H*)-pteridinone was treated with dimethyl sulfate in a pH 7 medium, the only isolable product proved to be 2-dimethylamino-4-methoxy-8-methyl-7(8*H*)-pteridinone (**92**).[511] The suggested mechanism involved hydroxylation of an 8-quaternized intermediate followed by oxidation but it might have involved methylation of a covalent 7,8-hydrate followed by oxidation?

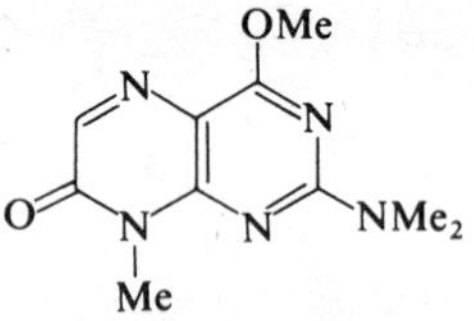

(**92**)

Treatment of the silver salt of 4-dimethylamino-7(8*H*)-pteridinone with acetobromoglucose in refluxing xylene gave 4-dimethylamino-7-(2′,3′,4′,6′-tetra-*O*-acetyl-*β*-D-glucopyranosyloxy)pteridine (in good yield) and then by treatment in warm methanolic ammonia, 4-dimethylamino-7-*β*-D-glucopyranosyloxypteridine, again in good yield.[386]

The conversion of *pteridinones into trimethylsiloxypteridines* is usually achieved by boiling the pteridinone in commercial hexamethyldisilazane [$(Me_3Si)_2NH$] containing a catalytic amount of ammonium sulfate for up to 48 h under anhydrous conditions, distilling off the excess of reagent, and then purifying the product by distillation in a vacuum or by recrystallization, as appropriate. Although the resulting trimethylsilyl groups are invariably attached through the oxygen rather than the nitrogen of the cyclic amides involved, the reagent will attack other NH, NH_2, SH (or potential SH) groups in the molecule with equal relish to form *N*- or *S*-trimethylsilyl derivatives, respectively. Unfortunately, there is a widespread tendency to use trimethylsiloxy and other trimethylsilyl derivatives without the customary purification and analysis.

The following properly characterized and analyzed trimethylsiloxypteridines have been made as just described: 2-trimethylsiloxy-4-trimethylsilylaminopteridine (**93**) (distillation: 98%),[225, 557] 4-dimethylamino-7-trimethylsiloxy-2-

trimethylsilylaminopteridine (distillation: ~75% yield),[357] and 4,7-bistrimethylsiloxy-2-trimethylsilylaminopteridine (distillation: 97%).[436]

The following trimethylsiloxypteridines have been described and characterized without given analyses: 2,4-bistrimethylsiloxypteridine (**94**, R=H) (distillation),[283, 414, 557] its 6,7-dimethyl derivative (**94**, R=Me) (distillation),[414, 557] the 6,7-diphenyl analog (**94**, R=Ph),[414, 557, 1022] 6-phenyl-2,4-bistrimethylsiloxypteridine (distillation: 85%),[412] 2-dimethylamino-4,7-bistrimethylsiloxypteridine (distillation: 65%),[246] and 2-methylthio-7-trimethylsiloxy-4-trimethylsilylaminopteridine (distillation).[1023]

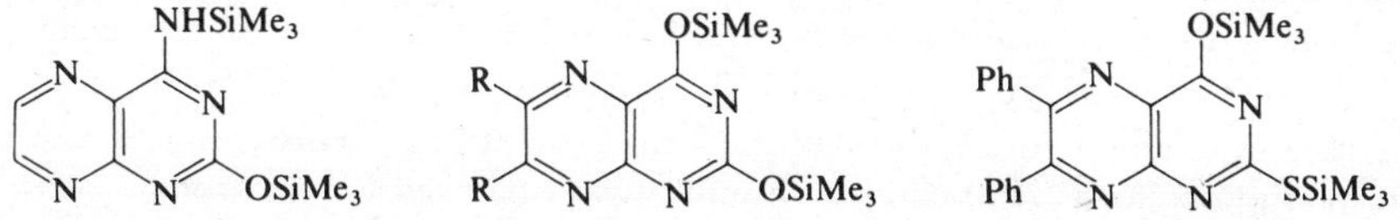

(93) **(94)** **(95)**

The following trimethylsiloxypteridines are examples of many such compounds that have been prepared and used without characterization or details: 7-phenyl-2,4-bistrimethylsiloxypteridine (99%),[412] 6,7-diphenyl-4-trimethylsiloxy-2-trimethylsilylthiopteridine (**95**),[447] 4-dimethylamino-2-trimethylsiloxypteridine,[557] 6,7-diphenyl-2-trimethylsiloxy-4-trimethylsilylaminopteridine,[557] 2-dimethylamino-4-methoxy-7-trimethylsiloxypteridine,[435] 4-isopropoxy-7-trimethylsiloxy-2-trimethylsilylaminopteridine,[435] 7-trimethylsiloxypteridine,[356] 4-dimethylamino-7-trimethylsiloxypteridine,[325] methyl 2-methylthio-7-trimethylsiloxy-4-trimethylsilylamino-6-pteridinecarboxylate,[325] and others.[325, 356, 435, 612, 1233]

Trimethylsilylation of pteridinones has also been done with *N*,*O*-bistrimethylsilylacetamide and/or chlorotrimethylsilane; tris(trideuteromethyl)silylations have been done with the fully deuterated analogs of these reagents.[987] A peculiar way to make 2,4,6,7-tetrakis(trimethylsiloxy)pteridine (**97**) has been described using the condensation of 2,4-bistrimethylsiloxy-5,6-bistrimethylsilylaminopyrimidine (**96**) with oxalyl chloride in toluene containing triethylamine at 0°C. The mechanism must have involved two *N* to *O* rearrangements and the yield was good.[701]

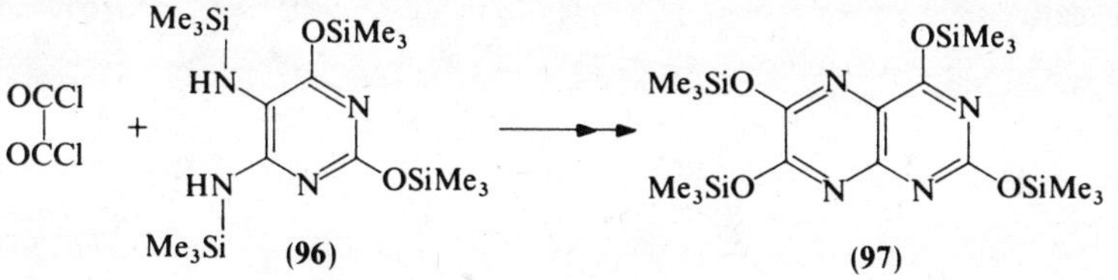

(96) **(97)**

As well as their classical use in mass spectral studies, trimethylsiloxypteridines have been employed extensively as substrates in the Silyl Hilbert–Johnson reaction to produce *N*-glycosidation of pteridinones (see Ch. VII, Sect. 2.C); the same substrates may also be used to produce *O*-glycosidation of pteridinones under different conditions.[246]

G. Conversion into Pteridinethiones

Pteridinones may be converted into chloropteridines (Ch. V, Sect. 1.B) and then into pteridinethiones (Ch. V, Sect. 2.E). However, the transformation can often be achieved directly by treatment with phosphorus pentasulfide in pyridine, dioxane, or another suitable solvent. Since both tautomeric and nontautomeric (fixed) pteridinones undergo this reaction with reasonable facility, the thiation of both types is included here. Although such thiations have been carried out at all positions, those at the 2-position appear to proceed less easily. Purified phosphorus pentasulfide should always be used for thiations.

For example, 6,7-dimethyl- (**98**, R = Me, X = O) or 6,7-diphenyl-4(3*H*)-pteridinone (**98**, R = Ph, X = O), on heating with phosphorus pentasulfide in refluxing anhydrous pyridine for 2 h, gave 6,7-dimethyl-4(3*H*)-pteridinethione (**98**, R = Me, X = S) (60%)[313] or 6,7-diphenyl-4(3*H*)-pteridinethione (**98**, R = Ph, X = S) (75%),[474] respectively; 1-methyl-, 3-methyl-, or 1,3-dimethyl-6,7-diphenyl-2,4(1*H*,3*H*)-pteridinedione with phosphorus pentasulfide in boiling dioxane gave 1-methyl-, 3-methyl-, or 1,3-dimethyl-6,7-diphenyl-4-thioxo-3,4-dihydro-2(1*H*)-pteridinone, respectively, each in ~80% yield[1679]; 6,7(5*H*,8*H*)-pteridinedione (**99**, X = Y = O) similarly (pyridine) gave a separable mixture of 7-thioxo-7,8-dihydro-6(5*H*)-pteridinone (**99**, X = O, Y = S) (9% yield) and 6-thioxo-5,6-dihydro-7(8*H*)-pteridinone (**99**, X = S, Y = O) (12% yield)[36]; 6,7-dimethyl- or 6,7-diphenyl-2-thioxo-1,2-dihydro-4(3*H*)-pteridinone gave 6,7-dimethyl- (45%) or 6,7-diphenyl-2,4(1*H*,3*H*)-pteridinedithione (72%), respectively,

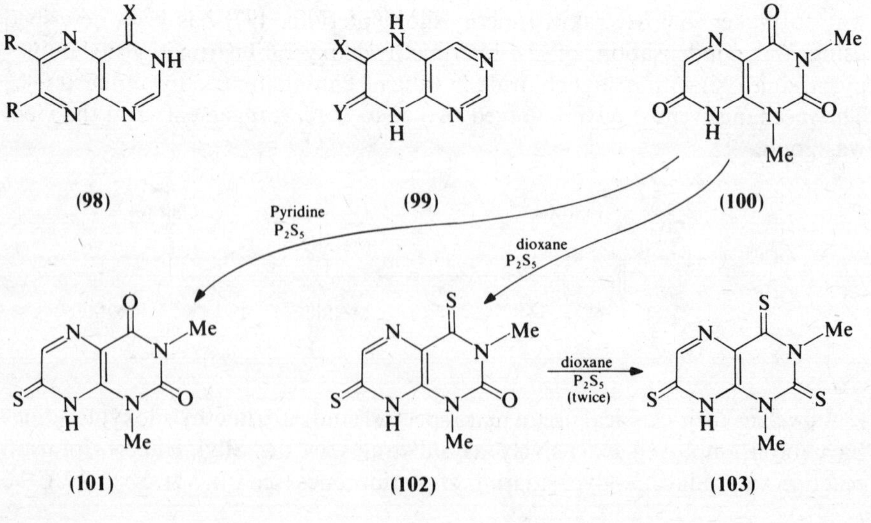

(98) (99) (100)

(101) (102) (103)

using phosphorus pentasulfide in refluxing pyridine[434] or (for better yields) in refluxing dioxane[927, cf. 699]; 1,3-dimethyl-2,4,7(1*H*,3*H*,8*H*)-pteridinetrione (**100**) gave 1,3-dimethyl-7-thioxo-7,8-dihydro-2,4(1*H*,3*H*)-pteridinedione (**101**) (phosphorus pentasulfide in pyridine: 74%) or 1,3-dimethyl-4,7-dithioxo-3,4,7,8-tetrahydro-2(1*H*)-pteridinone (**102**) (in dioxane: 59%) while the latter on repeated treatments with the same reagents gave 1,3-dimethyl-2,4,7(1*H*,3*H*,8*H*)-pteridinetrithione (**103**) (44%)[1000]; 2-methylthio-6,7-diphenyl-4(3*H*)-pteridinone or its 3-methyl derivative gave 2-methylthio-6,7-diphenyl-4(3*H*)-pteridinethione (pyridine: 86%) or its 3-methyl derivative (similarly: 44%)[437]; 3-methyl-2-thioxo-1,2-dihydro-4(3*H*)-pteridinone gave 3-methyl-2,4(1*H*,3*H*)-pteridinedithione (pyridine for 96 h: 79%)[446]; 1,3-dimethyl-2-thioxo-1,2-dihydro-4,7(3*H*,8*H*)-pteridinedione gave 1,3-dimethyl-2,7-dithioxo-1,2,7,8-tetrahydro-4(3*H*)-pteridinone (pyridine: 61%)[1000]; 4-amino-6,7-diphenyl-2(1*H*)-pteridinone gave the corresponding thione on refluxing with phosphorus pentasulfide in dichloroethane (yield unstated)[699]; 2-amino-1-methyl- and 1-methyl-2-methylamino-6,7-diphenyl-4(1*H*)-pteridinone, with phosphorus pentasulfide in pyridine, gave the respective thiones (50%, 16%)[111]; 1-methyl-6,7-diphenyl-2-thioxo-1,2-dihydro-4(3*H*)-pteridinone likewise gave the corresponding, 2,4-dithione[111]; 2-amino-4-pentyloxy-7(8*H*)-pteridinone with phosphorus pentasulfide in boiling dioxane gave the corresponding thione (57%)[1681]; 2-acetamido-6-α,β-diacetoxypropyl- or 2-benzamido-6-α,β-dibenzoyloxypropyl-4(3*H*)-pteridinone, with phosphorus pentasulfide in refluxing pyridine in the dark, gave the respective thiones (19, 42%)[1439]; and 7(8*H*)-pteridinone gave 7(8*H*)-pteridinethione (phosphorus pentasulfide–pyridine at 100°C: 48%).[33]

H. Formation of Dimers

Pteridinones form several types of dimeric or near-dimeric compounds. Of these, the best characterized are an orange dimer (substance O) from 6(5*H*)-pteridinone[19] and a purple dimer (purple substance II) from 2(1*H*)-pteridinone.[56]

The orange dimer was formed on refluxing 6(5*H*)-pteridinone in aqueous ammonium hydrogen sulfide for 10 min (~ 50% yield)[19] but a better yield (95%) was later obtained by heating the same pteridinone in formamide containing sodium carbonate at 150°C under nitrogen.[590, cf. 56] It is probably best formulated as the hydrogen-bonded form[590] of 6,6′-dioxo-5,5′,6,6′-tetrahydro-[8*H*,8′*H*]-7,7′-bipteridinylidene (**104**). The structure was based initially[19] on elemental analysis, its conversion in alkaline solution by a stream of air into 6,7(5*H*,8*H*)-pteridinedione, and the intense absorption of the anion and cation in the 450 to 490 nm region. It was later confirmed[590] by an nmr spectrum in fluorosulfonic acid, by a mass spectral molecular weight of 476 for its 5,5′-dibenzyl derivative, and by detailed studies of the ir and uv spectra of the dimer and its derivatives.

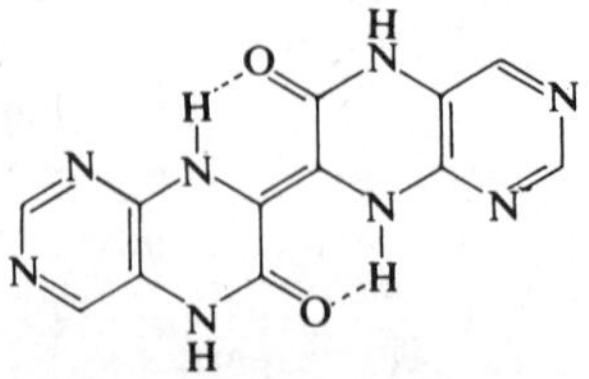

(104)

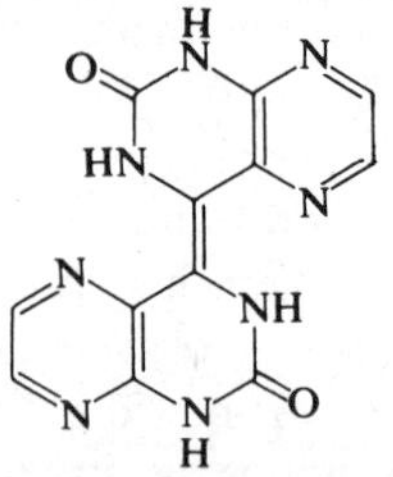

(105)

An early observation,[30] that 2(1*H*)-pteridinone in warm acid or alkali gave a deep purple color, was eventually traced to the formation of a dimer akin to that just described. *Purple substance II* was best prepared by refluxing a solution of 2(1*H*)-pteridinone in 0.5*M* sodium carbonate for 30 h to give an 80% yield of product.[56] Its structure, as 2,2′-dioxo-1,1′,2,2′-tetrahydro-[3*H*,3′*H*]-4,4′-bipteridinylidene (**105**), was confirmed by aerial oxidation in alkaline solution to 2,4(1*H*,3*H*)-pteridinedione, its massive absorption as cation in sulfuric acid at 500 to 700 nm, elemental analysis, and the fact that 6,7-dimethyl- but not 4,6,7-trimethyl-2(1*H*)-pteridinone produced an homologous purple product.[56]

The autoxidation of 2-amino-5,6,7,8-tetrahydro-4(3*H*)-pteridinone appears to lead to a near dimer, formulated as 2,2′-diamino-4,4′-dioxo-3,3′,4,4′,5,5′,6,6′,7,7′,8,8′-duodecahydro-6,6′-bipteridinyl (**106**) or its tautomer(s).[1328] The formation of dipteridinylmethane derivatives[57, 58] from appropriate pteridinones has been discussed in Chapter IV, Section 2.B(7).

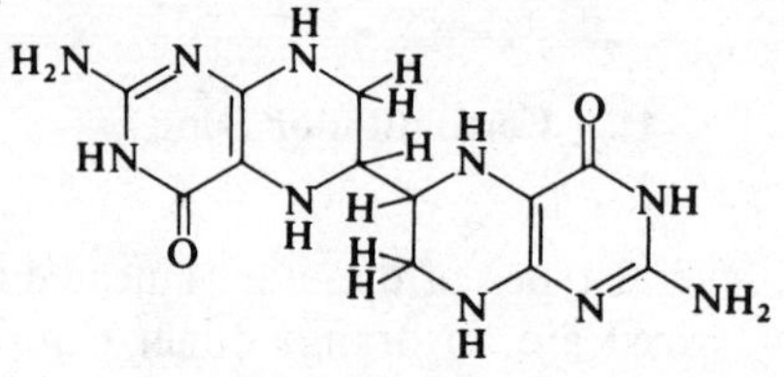

(106)

3. PREPARATION OF EXTRANUCLEAR HYDROXYPTERIDINES

A. By Primary Syntheses

Many hydroxyalkylpteridines have been made by primary syntheses from pyrimidines (Ch. II) or from pyrazines or other systems (Ch. III). Such syntheses are exemplified briefly in the condensation of 2,5,6-triamino-4(3*H*)-pyrimidinone with hydroxymethylglyoxal to give (mainly) 2-amino-7-hydroxymethyl-4(3*H*)-pteridinone (**107**)[776, 1212]; in the reaction of 5-amino-6-*β*-

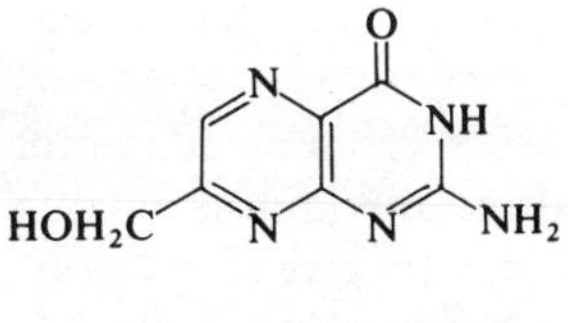

(107)

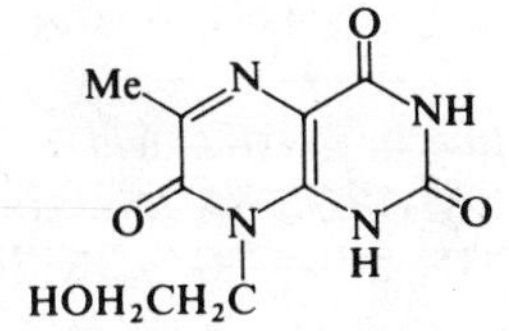

(108)

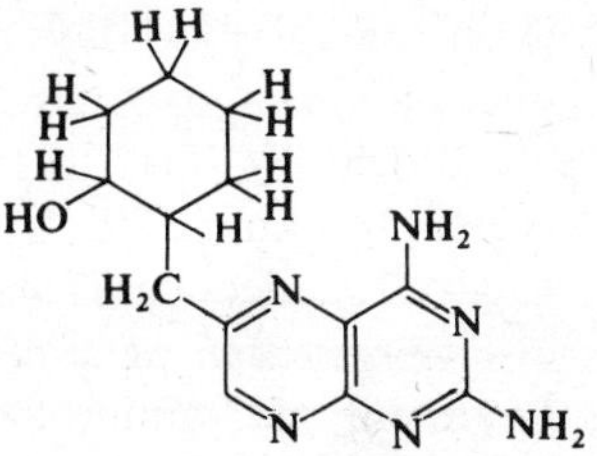

(109)

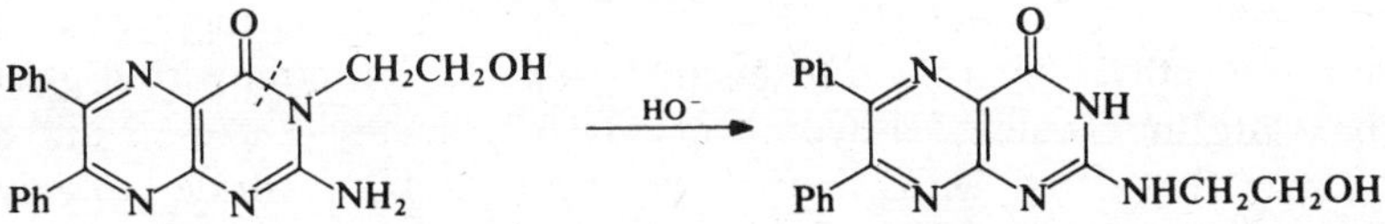

(110) (111)

hydroxyethylamino-2,4(1*H*,3*H*)-pyrimidinedione with ethyl pyruvate to give 8-β-hydroxyethyl-6-methyl-2,4,7(1*H*,3*H*,8*H*)-pteridinetrione (**108**)[1082]; and in the condensation of 3-amino-6-2′-hydroxycyclohexylmethyl-2-pyrazinecarbonitrile with guanidine to afford 6-2′-hydroxycyclohexylmethyl-2,4-pteridinediamine (**109**).[1299] An unusual route from one to another hydroxyalkylpteridine is seen in the Dimroth rearrangement of 2-amino-3-β-hydroxyethyl- (**110**) in alkali to yield 2-β-hydroxyethylamino-6,7-diphenyl-4(3*H*)-pteridinone (**111**) (64%).[1500]

B. By Reductive Processes

The reduction of a pteridinecarbaldehyde or a pteridinecarboxylic acid (or ester) will furnish a primary alcohol, whereas reduction of a pteridinyl ketone will afford a secondary alcohol.

The *reduction of a pteridinecarbaldehyde* is illustrated in the treatment of 2-amino-4-oxo-3,4-dihydro-6-pteridinecarbaldehyde (**112**) in dilute alkali with sodium borohydride to give 2-amino-6-hydroxymethyl-4(3*H*)-pteridinone (**113**) in 63 to 75% yield[1270, 1336]; in the Cannizzaro reaction of the same substrate

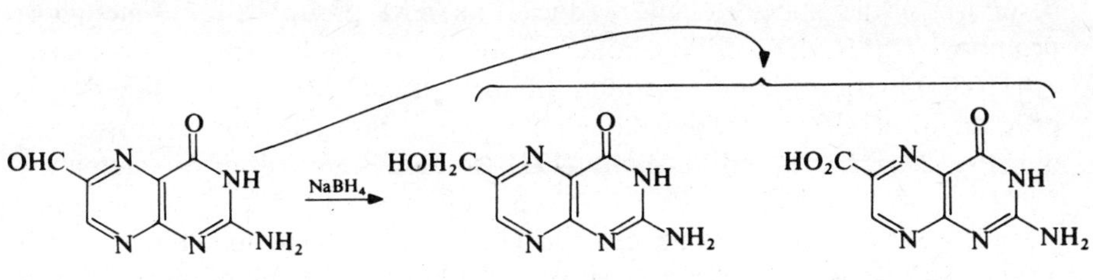

(112) (113) (114)

(**112**) in alkali for 5 days at 5° C to give a separable mixture of the alcohol (**113**) plus 2-amino-4-oxo-3,4-dihydro-6-pteridinecarboxylic acid (**114**), each in almost 50% yield[1358]; and in the peculiar reaction of 2-amino-6-hydrazonomethyl-4(3*H*)-pteridinone [the hydrazone of (**112**)] on dissolution in trifluoroacetic acid to give the alcohol (**113**).[513]

The *reduction of a pteridinecarboxylic acid or ester* is represented in the treatment of 2-amino-8-methyl-4-oxo-3,4,7,8-tetrahydro-6-pteridinecarboxylic acid (**115**, $R = CO_2H$) in THF at <5°C with diborane to give 2-amino-6-hydroxymethyl-8-methyl-5,6,7,8-tetrahydro-4(3*H*)-pteridinone (**116**) as the hydrochloride (>75%) and then by oxygenation of a solution in aqueous sodium hydrogen carbonate, the required 2-amino-6-hydroxymethyl-8-methyl-7,8-dihydro-4(3*H*)-pteridinone (**115**, $R = CH_2OH$).[684] It is also represented in the treatment of ethyl 4-hydroxy-6,7-dimethyl-2-oxo-1,2,3,4-tetrahydro-4-pteridinecarboxylate (the covalent 3,4-hydrate of ethyl 6,7-dimethyl-2-oxo-1,2-dihydro-4-pteridinecarboxylate) with methanolic sodium borohydride to give 4-hydroxymethyl-6,7-dimethyl-3,4-dihydro-2(1*H*)-pteridinone (**117**) in 65% yield.[1440]

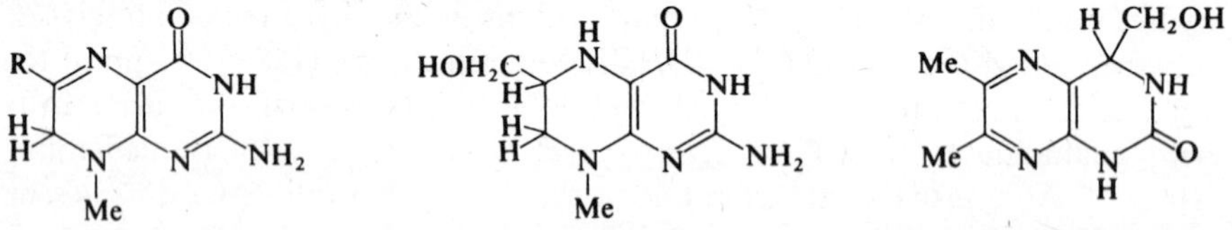

(**115**) (**116**) (**117**)

The *reduction of a pteridinyl ketone* is seen in the conversion of 6-acetyl-4,7-pteridinediamine (**118**, R = H), its 2-phenyl derivative (**118**, R = Ph), or its 2-piperidino derivative [**118**, $R = N(CH_2)_5$] into the corresponding secondary alcohols, viz. 6-α-hydroxyethyl-4,7-pteridinediamine (**119**, R = H) (75%), its 2-phenyl derivative (**119**, R = Ph) (90%), or its 2-piperidino derivative [**119**, $R = N(CH_2)_5$] (85%), respectively, by treatment with methanolic sodium borohydride at <50°C[328]; of 7-propionyl- or 7-isobutyryl-2,4-pteridinediamine into 7-α-hydroxypropyl- or 7-α-hydroxy-β-methylpropyl-2,4-pteridinediamine, respectively, also with borohydride[1678]; of 1-methyl-7-methylthio-6-propionyl-2,4(1*H*,3*H*)-pteridinedione into 7-α-hydroxypropyl-1-methyl-2,4(1*H*,3*H*)-pteridinedione by treatment with aluminum–copper alloy in alkali, which both desulfurized the substrate and reduced its acyl group[354]; of 3-methyl-6-propionyl-2,4(1*H*,3*H*)-pteridinedione into 6-α-hydroxypropyl-3-methyl-2,4(1*H*,3*H*)-pteridinedione (sodium borohydride)[354]; of 1,3-dimethyl-6-propionyl-2,4(1*H*,3*H*)-pteridinedione (**120**, R = H) into 6-α-hydroxypropyl-1,3-dimethyl-2,4(1*H*,3*H*)-pteridinedione (**121**) (sodium borohydride in methanol: 80%)[108]; of 1,3-dimethyl-7-methylthio-6-propionyl-2,4(1*H*,3*H*)-pteridinedione (**120**, R = SMe) into the same product (**121**) by treatment in refluxing ethyl acetate with Raney-nickel (65% yield)[108]; of 7-acetonyl-2-amino- into 2-amino-7-β-hydroxypropyl-4,6(3*H*,5*H*)-pteridinedione (sodium borohydride in dilute

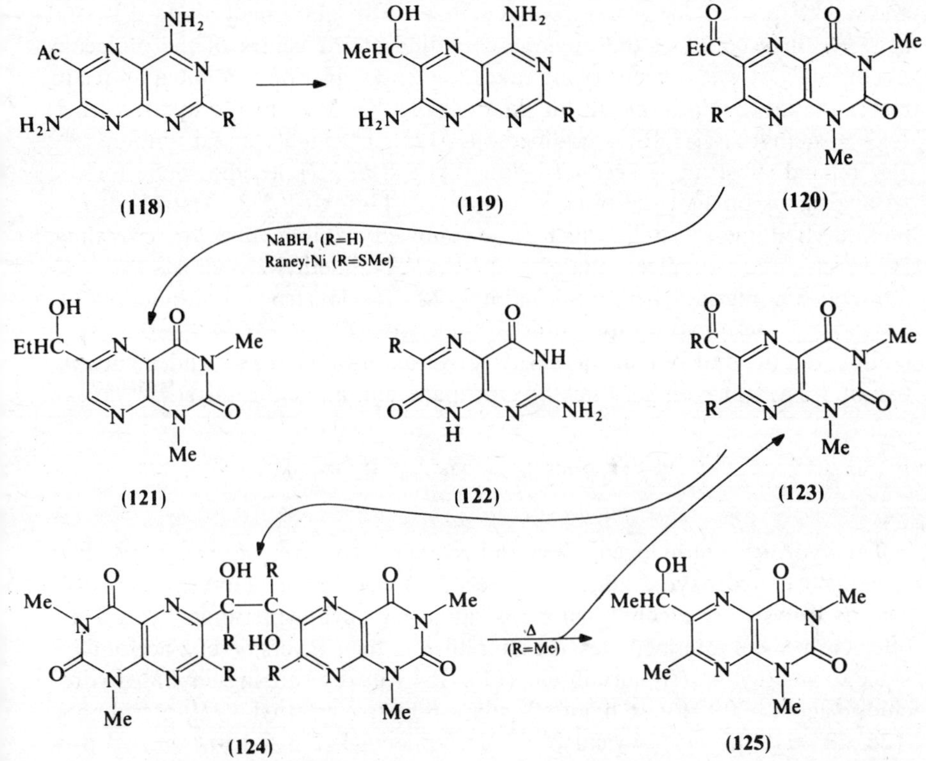

(118) (119) (120) (121) (122) (123) (124) (125)

alkali for 48 h at 20°C: 30%)[496]; of 6-acetonyl-2-amino- into 2-amino-6-β-hydroxypropyl-4,7(3*H*,8*H*)-pteridinedione (sodium borohydride as just cited: ~70%)[906, 1083]; and of 2-amino-7-α-(methylthio)acetonyl- into 2-amino-7-β-hydroxy-α-methylthiopropyl-7,8(?)-dihydro- (sodium borohydride) and then into 2-amino-7-β-hydroxy-α-methylthiopropyl-4,6(3*H*,5*H*)-pteridinedione (platinum oxide–oxygen).[1295] 2-Amino-6-α,β-dihydroxypropyl-4,7(3*H*,8*H*)-pteridinedione (**122**, R = CHOHCHOHMe) has been made by two relevant and interesting ways: (a) 2-amino-6-α-bromoacetonyl-4,7(4*H*,8H)-pteridinedione (**122**, R = CHBrAc) was heated with potassium acetate in glacial acetic acid to give the 6-α-acetoxyacetonyl analog [**122**, R = CH(OAc)Ac], which on treatment with sodium borohydride in aqueous sodium carbonate gave the required product in 47% yield; and (b) 6-acetonyl-2-amino-4,7(3*H*,8*H*)-pteridinedione (**122**, R = CH_2Ac) was oxidized by selenium dioxide to give the 6-acetylcarbonyl analog [**122**, R = C(:O)Ac] in which both carbonyl groups were reduced subsequently by sodium borohydride to afford the product in 35% yield.[499]

When 6-acetyl-1,3,7-trimethyl-2,4(1*H*,3*H*)-pteridinedione (**123**, R = Me) was submitted to cathodic reduction in propanol containing potassium chloride at pH 9, a one-electron reduction and stereospecific dimerization occurred to furnish the *threo*-pinacol, 2,3-bis(1′,3′,7′-trimethyl-2′4′-dioxo-1′,2′,3′,4′-tetrahydro-6′-pteridinyl)butane-2,3-diol (**124**, R = Me), in 80% yield[365]; an x-ray analysis[798] proved the structure and indicated an abnormally long 2,3-bond (1.59 Å) on account of the obvious crowding at the centre of the molecule. Accordingly, it was not surprising that the pinacol disproportionated above its melting point to afford the starting ketone (**123**, R = Me) and 6-α-hydroxyethyl-1,3,7-trimethyl-2,4(1*H*,3*H*)-pteridinedione (**125**) in roughly equal amounts.[365] The related substrate, 7-acetyl-1,3-dimethyl-2,4(1*H*,3*H*)-pteridinedione formed an analogous pinacol, 2,3-bis(1′,3′-dimethyl-2′,4′-dioxo-1′,2′,3′,4′-tetrahydro-7′-pteridinyl)butane-2,3-diol, which also showed lability due to crowding. However, the simpler analog, 2,3-bis(1′,3′-dimethyl-2′,4′-dioxo-1′,2′,3′,4′-tetrahydro-6′-pteridinyl)ethane-1,2-diol (**124**, R = H) from 1,3-dimethyl-2,4-dioxo-1,2,3,4-tetrahydro-6-pteridinecarbaldehyde (**123**, R = H), showed no such lability and the diol system was easily converted into its isopropylidene derivative by treatment with β,β-dimethoxypropane and an acid catalyst.[365, 622]

C. By Hydrolytic or Alcoholytic Processes

The hydrolysis of halogenoalkyl and related derivatives of pteridine to give extranuclear hydroxypteridines has been discussed in Chapter V, Section 4.C. The hydrolysis (or alcoholysis) of extranuclear acyloxypteridines, which are often convenient intermediates; is exemplified in the conversion of 2-acetamido-6-acetoxymethyl-4(3*H*)-pteridinone (**126**, R = Ac), by stirring in 3*M* hydrochloric acid at 20°C for 16 h, into 2-amino-6-hydroxymethyl-4(3*H*)-pteridinone (**126**, R = H) in 95% yield[1443]; of 6-phenyl-1-(2′,3′,5′-tri-*O*-benzoyl-β-D-ribofuranosyl)-2,4,7(1*H*,3*H*,8*H*)-pteridinetrione, by methanolic sodium methoxide at room temperature, into 6-phenyl-1-β-D-ribofuranosyl-2,4,7(1*H*,3*H*,8*H*)-pteridinetrione (62%)[413]; of 2-benzamido-6-α,β-dibenzoyloxypropyl-4(3*H*)-pteridinethione (**127**, R = Bz), by a similar procedure, into 2-amino-6-α,β-dihydroxypropyl-4(3*H*)-pteridinethione (**127**, R = H) (87%)[1439]; of 4-dimethylamino-7-(2′,3′,4′,6′-tetra-*O*-acetyl-β-D-glucopyranosyloxy)pteridine, by saturated ethanolic ammonia, into 4-dimethylamino-7-β-D-glucopyranosyloxypteridine (~45%)[386]; and of 6-acetoxymethyl- into 6-hydroxymethyl-1,3-dimethyl-2,4(1*H*,3*H*)-pteridinedione (warm dilute acid: 85%).[1704]

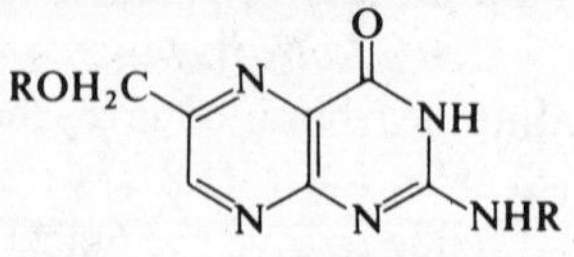

(126)

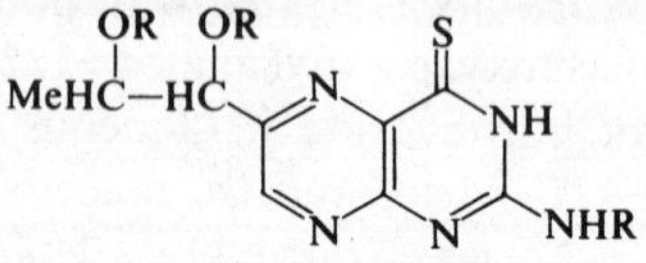

(127)

D. By Other Means

There are several other ways to make extranuclear hydroxypteridines.

The *oxidative insertion* of such a hydroxy group is illustrated in the treatment of 2-acetamido-6-methyl-4(3*H*)-pteridinone (**128**, R = H, Y = Ac) with lead tetraacetate in glacial acetic acid at 90°C for 24 h to give 2-acetamido-6-acetoxymethyl-4(3*H*)-pteridinone (**128**, R = OAc, Y = Ac) (84%) and then by acidic hydrolysis, 2-amino-6-hydroxymethyl-4(3*H*)-pteridinone (**128**, R = OH, Y = H)[1443]; also in the conversion of 2-amino-6,8-dimethyl-7,8-dihydro-4(3*H*)-pteridinone into 2-amino-6-hydroxymethyl-8-methyl-4(8*H*)-pteridinone (**129**) (55%) by treatment with alkaline permanganate,[1370] or into 2-amino-6-hydroxymethyl-8-methyl-4,7(3*H*,8*H*)-pteridinedione (30%) by aeration at pH 7 in the presence of platinum oxide.[1370]

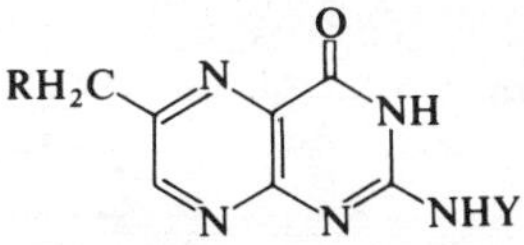

(**128**)

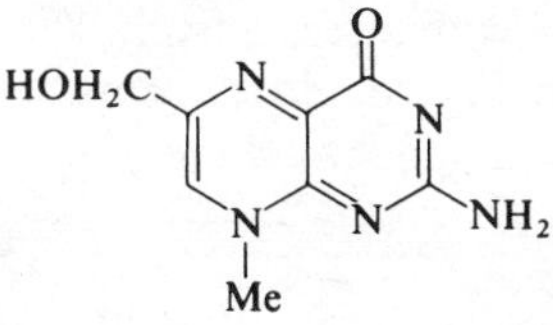

(**129**)

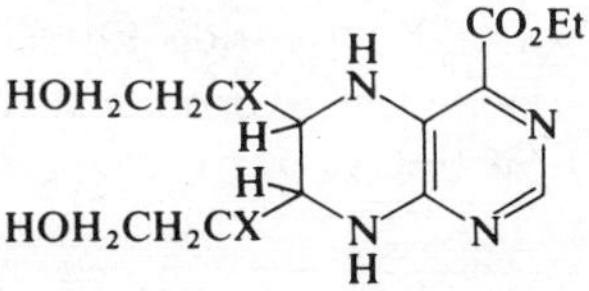

(**130**)

The *covalent addition* of ethylene glycol to ethyl 4-pteridinecarboxylate afforded ethyl 6,7-bis-β-hydroxyethoxy-5,6,7,8-tetrahydro-4-pteridinecarboxylate (**130**, X = O) (~70%)[158]; the similar use of ethanolamine gave ethyl 6,7-bis-β-hydroxyethylamino-5,6,7,8-tetrahydro-4-pteridinecarboxylate (**130**, X = NH) (~75%)[158]; but use of 1,2-ethanedithiol or β-mercaptoethanol gave only the tricyclic products, ethyl 5,5a,7,8,9a,10-hexahydro-[1,4]-dithiino[2,3-*g*]pteridine-4-carboxylate or ethyl 5,5a,7,8,9a,10-hexahydro-[1,4]-thioxino[2,3-*g*]pteridine-4-carboxylate, respectively.[158] The addition of phloroglucinol to 2(1*H*)-pteridinethione or 7(8*H*)-pteridinone gave 4-(2′,4′,6′-trihydroxyphenyl)-3,4-dihydro-2(1*H*)-pteridinethione or 6-(2′,4′,6′-trihydroxyphenyl)-5,6-dihydro-7(8*H*)-pteridinone, respectively.[46]

The *homolytic α-hydroxyalkylation* of at least one pteridine has been reported. Thus, treatment of 2-amino-8-methyl-4,7(3*H*,8*H*)-pteridinedione in an appropriate alcohol (buffered by aqueous ammonium phosphate to pH ~6) with ammonium persulfate under reflux gave the 6-hydroxymethyl (53%), 6-α-hydroxyethyl (36%), 6-α-hydroxypropyl (54%), 6-α-hydroxybutyl (41%), 6-α-hydroxy-α-methylethyl (45%), or 6-α-hydroxy-α-methylpropyl derivative (22%).[1711]

The *derivatization* of 6,7-bis-*p*-aminophenyl-2,4-pteridinediamine, by treatment of an acidic solution with formaldehyde, gave 6,7-bis-*p*-[*N*-(hydroxymethyl)amino]phenyl-2,4-pteridinediamine (**131**) (>95%).[141]

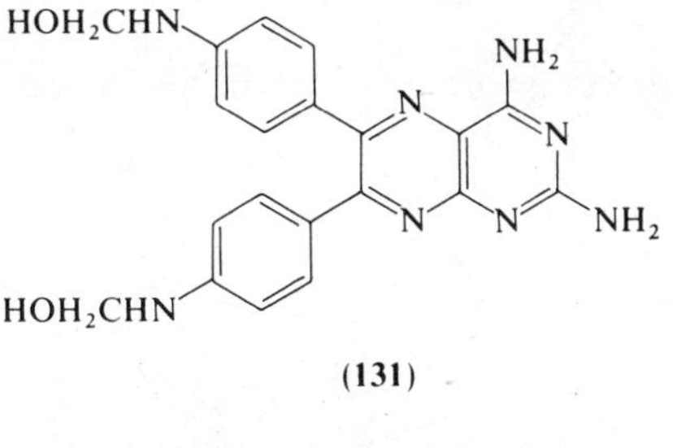

(**131**)

The *rearrangement of an N-oxide* in acetic anhydride can give an acetoxymethyl- and then an hydroxymethylpteridine. For example, 3,6,7-trimethyl-2,4(1*H*,3*H*)-pteridinedione 8-oxide (**132**) in refluxing acetic anhydride for 45 min gave 7-acetoxymethyl-3,6-dimethyl-2,4(1*H*,3*H*)-pteridinedione (**133**) (24%)[367]; 1,3,6-trimethyl-2,4(1*H*,3*H*)-pteridinedione 5-oxide in refluxing acetic anhydride for 7 h gave 6-acetoxymethyl-1,3-dimethyl-2,4(1*H*,3*H*)-pteridinedione (66%)[1704]; and 3,6,7-trimethyl-2,4(1*H*,3*H*)-pteridinedione 5,8-dioxide gave either 7-acetoxymethyl-3,6-dimethyl-2,4(1*H*,3*H*)-pteridinedione 5-oxide or 6,7-bisacetoxymethyl-3-methyl-2,4(1*H*,3*H*)-pteridinedione by refluxing in acetic anhydride for 5 or 45 min, respectively.[367] Hydrolysis to the corresponding hydroxymethyl analogs could be done as previously described in Section C.

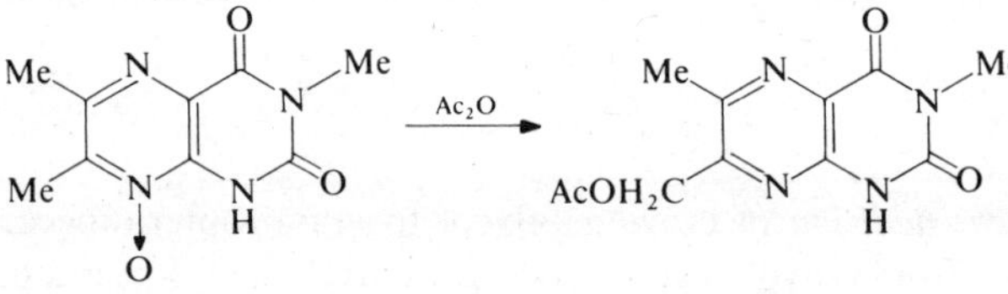

(132) (133)

4. REACTIONS OF EXTRANUCLEAR HYDROXYPTERIDINES

Extranuclear hydroxypteridines undergo most of the reactions to be expected of alcohols. See, for example, ref. 1689.

A. Conversion into Chloroalkylpteridines

This conversion, using reagents such as thionyl chloride, has been discussed in Chapter V, Section 3.C.

B. Oxidation to Aldehydes or Carboxylic Acids

The formation of *pteridinecarbaldehydes* from extranuclear hydroxypteridines is typified in the oxidation of 6-hydroxymethyl-4-methylthio-2-pteridinamine with manganese dioxide to give 2-amino-4-methylsulfonyl-6-pteridine-

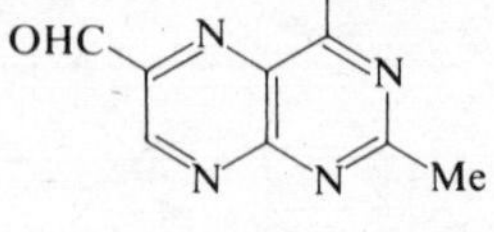

(134)

carbaldehyde (**134**)[815]; of 2-amino-6-[L-*erythro*]-α,β,γ-trihydroxypropyl-4(3*H*)-pteridinone[1344] or its 6-[D-*arabino*]-α,β,γ,δ-tetrahydroxybutyl analog[1373] with sodium periodate at 70 to 90°C for 20 to 40 min to give 2-amino-4-oxo-3,4-dihydro-6-pteridinecarbaldehyde (85%); and of 6-[D-*arabino*]-α,β,γ,δ-tetrahydroxybutyl-2,4-pteridinediamine with sodium periodate at 20°C for several hours to give 2,4-diamino-6-pteridinecarbaldehyde (~50%).[1058] In addition, anaerobic photolysis of 2-amino-6-α,β-dihydroxypropyl-4(3*H*)-pteridinone (**135**, R = H) or its 2-dimethylamino analog (**135**, R = Me) has been reported to give the unstable 5,8-dihydropteridinecarbaldehydes (**136**, R = H or Me), which were oxidized in air to afford 2-amino- (**137**, R = H) or 2-dimethylamino-4-oxo-3,4-dihydro-6-pteridinecarbaldehyde (**137**, R = Me), respectively.[1463] See also Chapter X, Section 9.B.

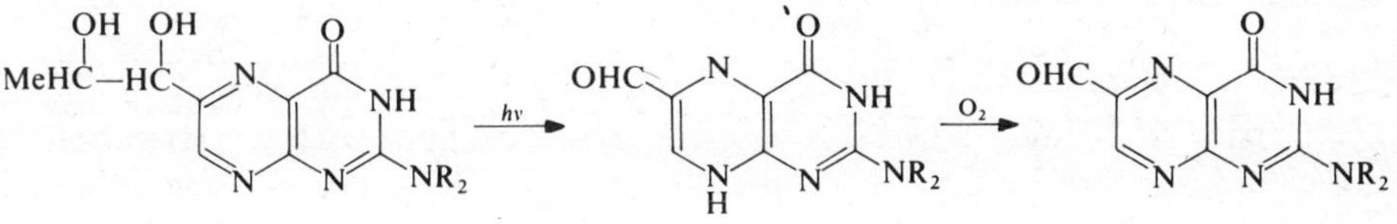

(**135**) (**136**) (**137**)

The formation of *pteridinecarboxylic acids* from pteridinyl alcohols has been used largely as an analytical tool to confirm the 6- and/or 7-orientation of such hydroxymethylpteridines. The 6- and 7-carboxylic acids differ markedly in uv spectra, whereas there is no such difference between the corresponding alcohols. Thus 7-hydroxymethyl-2,4-pteridinediamine (**138**, R = CH_2OH), of initially doubtful orientation, was oxidized by permanganate in hot water to afford 2,4-diamino-7-pteridinecarboxylic acid (**138**, R = CO_2H) (61%), which underwent alkaline hydrolysis to 2-amino-4-oxo-3,4-dihydro-7-pteridinecarboxylic acid (**139**),[123] easily identified with authentic material[1084] by spectra; 2-amino-7-hydroxymethyl-4(3*H*)-pteridinone (**140**) underwent a similar oxidation to give the same product (**139**) (~ 85%)[774]; 2-amino-6-[D-*arabino*]-α,β,γ,δ-tetrahydroxybutyl-4(3*H*)-pteridinone likewise gave 2-amino-4-oxo-3,4-dihydro-6-pteridinecarboxylic acid[1147, 1373]; 2-amino-6-hydroxymethyl-4(3*H*)-pteridinone gave the same product,[96] which was identified chromatographically[1066] with authentic material; 2-dimethylamino-6-[L-*threo*]-α,β,γ-trihydroxypropyl-4(3*H*)-pteridinone with alkaline permanganate gave 2-dimethylamino-4-oxo-3,4-dihydro-6-pteridinecarboxylic acid (50%)[1710]; and 4-amino-6-hydroxymethyl-2(1*H*)-pteridinone underwent acidic or alkaline hydrolysis to 6-hydroxymethyl-

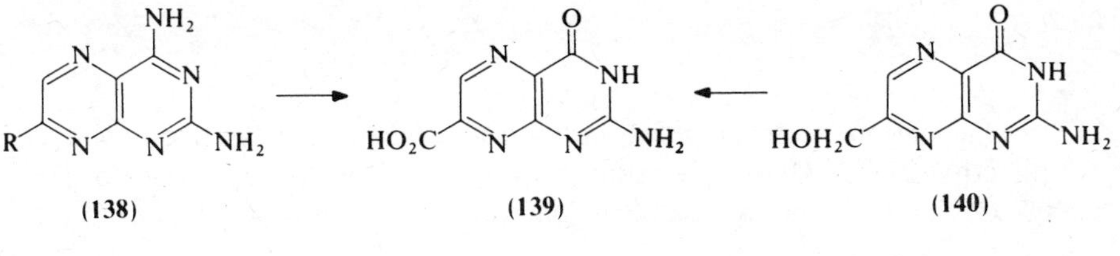

(**138**) (**139**) (**140**)

2,4(1*H*,3*H*)-pteridinedione (**141**, R = CH_2OH) followed by permanganate oxidation to 2,4-dioxo-1,2,3,4-tetrahydro-6-pteridinecarboxylic acid (**141**, R = CO_2H),[96] which could be identified with authentic material[655] by its spectra.

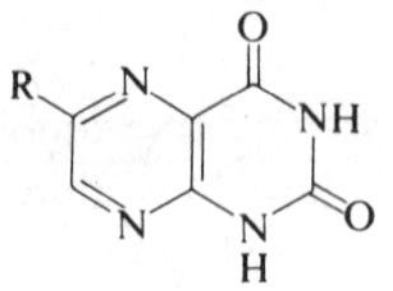

(**141**)

C. Acylation

The acetylation of ω-hydroxypteridines is fairly well represented in the conversion of 2-amino-6-hydroxymethyl-4(3*H*)-pteridinone (**142**, R = H) into 2-acetamido-6-acetoxymethyl-4(3*H*)-pteridinone (**142**, R = Ac) (refluxing acetic anhydride for 3 h: 70%)[113, 1358]; of 6-hydroxymethyl-2,4-pteridinediamine into 2,4-diacetamido-6-acetoxymethylpteridine (refluxing acetic anhydride for 6 h: 20%)[113]; of 2-amino-7-hydroxymethyl-4(3*H*)-pteridinone into 2-acetamido-7-acetoxymethyl-4(3*H*)-pteridinone (similarly for 1 h: ~70%)[776]; of 8-β-hydroxyethyl-6,7-diphenyl-2,4(3*H*,8*H*)-pteridinedione into the 8-β-acetoxyethyl analog (refluxing acetic anhydride for 30 min: 60%[409]; acetic anhydride at 100°C for 4 h: 74%[240]; although both products had the correct analysis, they differed in mp by 40°C); of 2-amino-7-β-hydroxy-α-methylthiopropyl- into 2-acetamido-7-β-acetoxy-α-methylthiopropyl-4,6(3*H*,5*H*)-pteridinedione (boiling acetic anhydride: 44%)[1295]; of 1-β-D-glucopyranosyl-5,6,7,8-tetrahydro-2,4(1*H*,3*H*)-pteridinedione into 5-acetyl-1-(2′,3′,4′,6′-tetra-*O*-acetyl-β-D-glucopyranosyl-5,6,7,8-tetrahydro-2,4(1*H*,3*H*)-pteridinedione (warm acetic anhydride: 60%)[265]; and of 2-amino-4-oxo-3,4-dihydro-6-pteridinecarbaldehyde into 2-acetamido-6-diacetoxymethyl-4(3*H*)-pteridinedinone (**143**, R = Ac), derived in theory from the hydrated aldehyde (**143**, R = H) (acetic anhydride under reflux for 3 h: ~50%).[1451] Hints for performing *N*- without *O*-deacetylations on some of these products have been given.[113, 265]

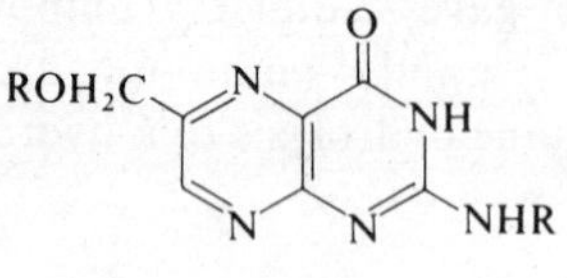

(**142**)

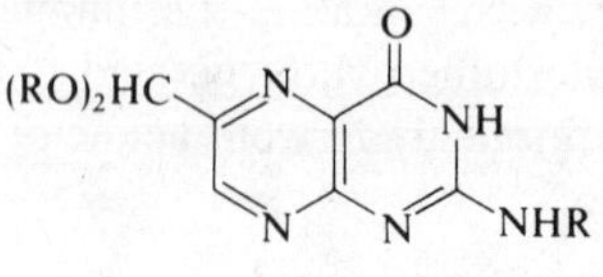

(**143**)

Other acylations are exemplified in the transformation of 1-β-hydroxyethyl-6,7-dimethyl-2,4(1*H*,3*H*)-pteridinedione or its 6,7-diphenyl analog into 1-β-mesyloxyethyl-6,7-dimethyl-2,4(1*H*,3*H*)-pteridinedione (**144**, R = Me) or its diphenyl analog (**144**, R = Ph), respectively, by the addition of mesyl chloride to a solution of the substrate in pyridine at 3°C for 1 h (90% yield in each case)[1339]; of 2-amino-6-hydroxymethyl-4(3*H*)-pteridinone (**142**, R = H) into 2-amino-4-oxo-3,4-dihydropteridine-6-ylmethyl pyrophosphate (**145**) (80%) by treatment with (molten) pyrophosphoric acid at 60°C for 2 h in the dark[978, cf. 1269, 1336]; of 6-hydroxymethyl-2,4(1*H*,3*H*)-pteridinedione into 2,4-dioxo-1,2,3,4-tetrahydro-pteridin-6-ylmethyl pyrophosphate (similarly)[1546]; and of several pteridine *N*-1 and *N*-8 nucleosides into the corresponding pteridine nucleoside-5′-monophosphates, using phosphoryl chloride in trimethyl phosphate.[936]

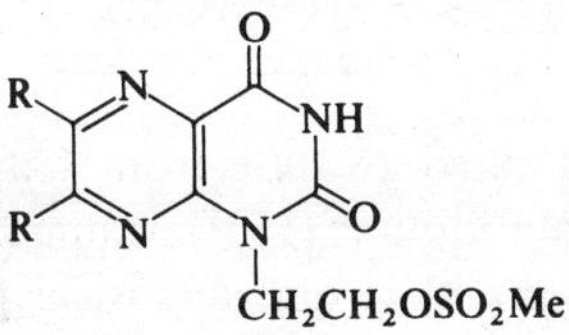

(144)

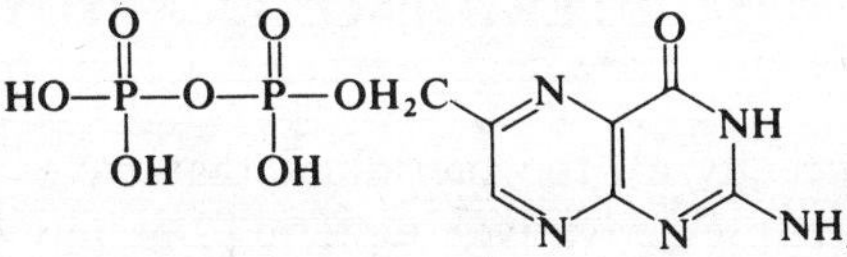

(145)

D. Other Reactions

The fragmentation patterns of 2-amino-6-hydroxymethyl-4(3*H*)-pteridinone, 6-*β*,*γ*-dihydroxypropyl-2,4(1*H*,3*H*)-pteridinedione, 6-*α*,*β*,*γ*,*δ*-tetrahydroxybutyl-2,4(1*H*,3*H*)-pteridinedione, and 2-amino-6-*α*,*β*,*γ*-trihydroxypropyl-4(3*H*)-pteridinone, all as polytrimethylsilyl derivatives, have been recorded and discussed.[1249] The removal of an extranuclear hydroxy group has been reported. Treatment of 2-amino-7-*α*-hydroxyacetonyl-4,6(3*H*,5*H*)-pteridinedione (**146**, R = OH) with sodium borohydride in very dilute alkali for 36 h gave 7-acetonyl-2-amino-4,6(3*H*,5*H*)-pteridinedione (**146**, R = H).[498] 2-Amino-6-hydroxymethyl-4(3*H*)-pteridinone was condensed with ethyl 4-amino-5-nitrobenzoate in

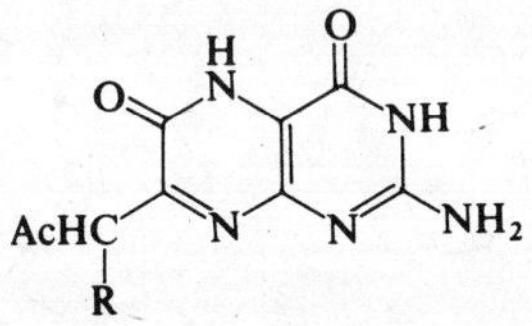

(146)

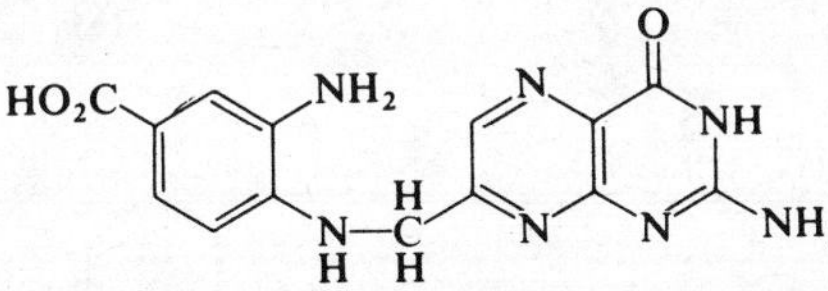

(147)

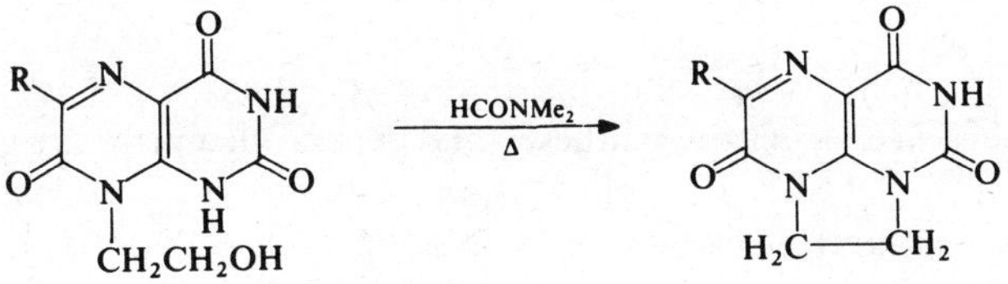

(148) **(149)**

formic acetic anhydride to give, after saponification in alkali and reduction of the nitro group, 2-amino-6-(2′-amino-4′-carboxyanilino)methyl-4(3*H*)-pteridinone (**147**).[1395] When heated in DMF for 10 h, 8-*β*-hydroxyethyl-2,4,7(1*H*,3*H*,8*H*)-pteridinetrione (**148**, R = H), its 6-methyl derivative (**148**, R = Me), and its 6-phenyl derivative (**148**, R = Ph) gave 1,8-ethano-2,4,7(1*H*,3*H*,8*H*)-pteridinetrione (**149**, R = H), its 6-methyl derivative (**149**, R = Me), and its 6-phenyl derivative (**149**, R = Ph), respectively, in 40 to 60% yield.[1339]

5. PTERIDINONES FROM NATURAL SOURCES

Virtually all the pteridines that have been isolated to date from natural sources are derivatives of pterin [2-amino-4(3*H*)-pteridinone (**153**)] or lumazine [2,4(1*H*,3*H*)-pteridinedione (**154**)]. However, for practical and historical purposes, it seems better to classify natural pteridines (albeit somewhat illogically) as derivatives of six parent structures that themselves occur naturally: xanthopterin (**150**), leucopterin (**151**), isoxanthopterin (**152**), pterin (**153**), lumazine (**154**), and violapterin (**155**). A few miscellaneous natural pteridines remain outside this classification and there are some natural products, for example, pterobillin,[535] which are not pteridines at all, despite their names. The possible abiogenic synthesis of pteridines has been investigated under conditions thought to have existed on earth during the prebiological era: Irradiation of guanine and sodium pyruvate under nitrogen did in fact give a mixture containing pteridines.[1035]

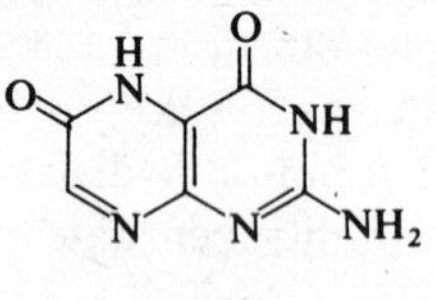

(**150**)

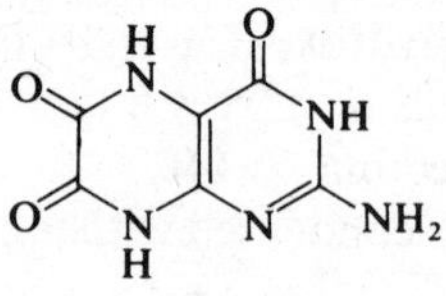

(**151**)

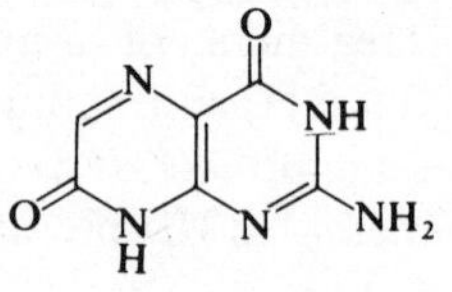

(**152**)

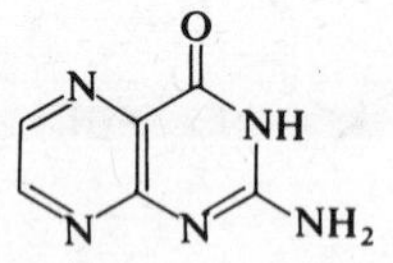

(**153**)

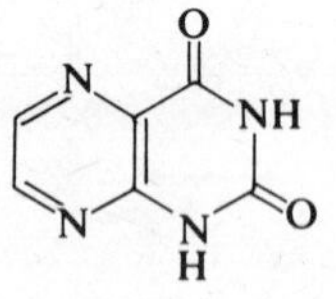

(**154**)

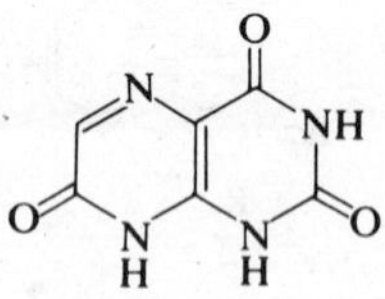

(**155**)

In treating each of the following natural products, brief information on history, occurrence, structure, syntheses, and general chemistry are given but not necessarily in that order. For information on biosynthesis, biochemistry, and bioactivity of natural pteridines, as well as on all aspects of folic acid(s) and folates in general, an appropriate volume of the excellent treatise, *Folates and Pterins*, edited by Ray Blakley and Stephen Benkovic[1572] or other relevant review-type publications should be consulted.[18, 222, 223, 334, 393, 554, 661, 698, 808, 844, 1536] For good reviews of the biological sites in which natural pteridines occur within various species of insect, fish, and so on, the publications of Irmgard Ziegler,[785, 831, 1566, 1590] Henri Descimon,[613, 739, 740, 756, 1580] and others[217, 1477] are of particular use. In addition, there are numerous reviews on general procedures for separating and isolating the natural pteridines,[397, 565, 603, 635, 711, 739, 740, 786, 845, 1072, 1164, 1202, 1276, 1477, 1697] as well as general papers on the identification of natural pteridines by spectral and other physical means.[561, 572, 1039, 1478]

A. The Xanthopterin Group

(1) *Xanthopterin*: 2-*Amino*-4,6(3*H*,5*H*)-*pteridinedione*

Although Frederick Hopkins suggested[794] that xanthopterin (**150**) had been produced on heating uric acid in water, as early as 1857 by Friedrich Wöhler[1509] and later by himself[2, 5, 6], this eventually proved incorrect[345, 352] (see Ch. III, Sect. 2 for details). In fact, xanthopterin was isolated first from the English brimstone butterfly about 1889[1–5] and later from other species of butterfly.[6] Despite intensive German work from 1924 to 1939 in the laboratories of Heinrich Wieland and Clemens Schöpf (exemplified in original papers[440, 719, 1238, 1239, 1381] and reviews[574, 594, 794, 824, 1173]), the structure of xanthopterin remained unknown because of difficulties with purification and elemental analyses.[cf. 672] Only in 1940 was the correct structure suggested[523, 1172] and subsequently proven by Robert Purrmann's synthesis,[404, cf.528] involving acylation of 2,5,6-triamino-4(3*H*)-pyrimidinone (**156**) with dichloroacetic acid at 120°C to give 2,6-diamino-5-α,α-dichloroacetamido-4(3*H*)-pyrimidinone (**157**) followed by cyclization of the silver salt to give xanthopterin (**158**, R = H).

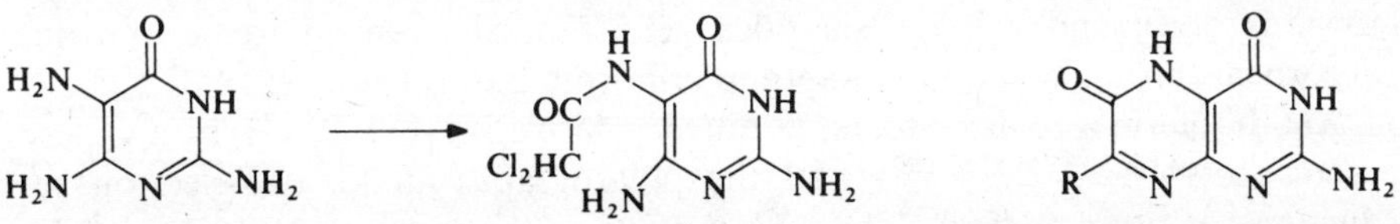

(156) **(157)** **(158)**

The isolation of xanthopterin (also once known as uropterin) from various sources (e.g., human urine,[1410, 1574] goldfish skin or scales,[834] wasp stripes,[896] crab skin,[1573, cf.551] and other sites[720, 830, 897, 931, 1431, 1574]) continued and x-ray structures for xanthopterin[1073] and its hydrochloride[106, 618] have been reported. In addition, many other syntheses (in most cases, see the appropriate section of Chs. II or III for details) have been carried out by the Gabriel and Colman synthesis[188, 269, 271, 273, 1011, 1410]; by the Boon synthesis[451, 727]; by the sodium amalgam reduction of 2-amino-4,6,7(3*H*,5*H*,8*H*)-pteridinetrione, followed by oxidation of the resulting 7,8-dihydroxanthopterin[60, 71, 800, 877, 965, 1268]; by hydriodic acid reduction of 2-amino-4,6-dioxo-3,4,5,6-tetrahydro-7-pteridinecarboxylic acid (**158**, R = CO_2H) to its 3,4,5,6,7,8-hexahydro analog (**159**, R = CO_2H), followed by decarboxylation to 7,8-dihydroxanthopterin (**159**, R = H) and platinum oxide–oxygen oxidation to xanthopterin (**158**, R = H)[405]; by the direct decarboxylation of this acid (**158**, R = CO_2H)[342]; from guanine and ethyl glyoxylate hemiacetal[14]; by an unusual β-rearrangement of 2-amino-4(3*H*)-pteridinone 8-oxide (**160**) in trifluoroacetic anhydride[477, 1079, 1301]; by alkaline hydrolysis of 6-methoxy-2,4-pteridinediamine (93%) or of 2,4-diamino-6(5*H*)-pteridinone (**161**) (85%)[811, 1293, 1301]; or by degradation of pteridines bearing a six-side chain, for example, pteroylglutamic acid.[452, 1125]

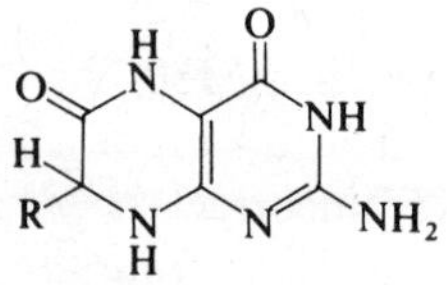

(159)

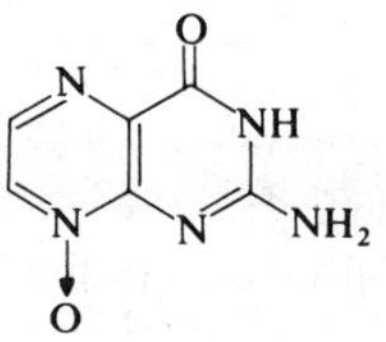

(160)

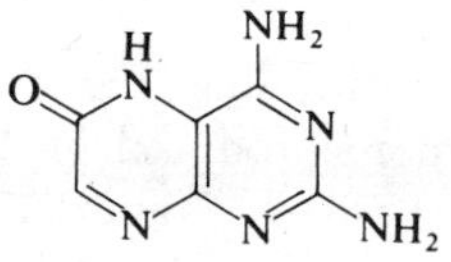

(161)

The ordinary metathetical reactions of xanthopterin are discussed in appropriate sections of this book and are not listed here. The physical properties, information on salts, and degradations of xanthopterin have been extracted from the older literature and summarized[913]; the hydrochloride has proven particularly useful for purification.[656] The mass spectral fragmentation of xanthopterin begins with concomitant losses of HCN and CO from the molecular ion[778, 1535]; its chelation with cupric ions is marked but apparently without biological significance[28]; it gives a blue coloration with the Benedict reagent[804] and the Folin–Ciocalteau reagent,[1053] whereas isoxanthopterin is negative towards both; and a magnesia–silica gel, "Florisil," proved to be a useful absorbent during isolation of xanthopterin from natural sources.[637]

Xanthopterin is reduced to its 7,8-dihydro derivative (**159**, R = H) by a variety of reagents[536, 1381, 1575, 1576] and the polarography of its two-electron reduction has been studied[639]; because it exists as the covalent 7,8-hydrate (**159**, R = OH) under normal circumstances,[241] xanthopterin may be oxidized readily to leucopterin by platinum oxide–oxygen (63% yield)[533] or by periodate (~20% yield).[1056] Treatment of xanthopterin or its barium salt with glacial acetic acid containing some aqueous 30% hydrogen peroxide has long been known[532] to give a well-characterized compound, xanthopterin peroxide, analyzing for a 1:1 mixture (**162**) of xanthopterin and hydrogen peroxide. Xanthopterin could be recovered by boiling the peroxide in water,[532] while further treatment with aqueous hydrogen peroxide gave a separable mixture of leucopterin and melanurenic acid, 6-amino-1,3,5-triazine-2,4(1*H*,3*H*)-dione.[94] In contrast with earlier suggestions as to its structure,[533] it is tempting in the light of modern work[94] to consider the peroxide as a covalent 7,8-adduct (**163**) of xanthopterin and hydrogen peroxide. This, however, remains unproven.* The existence of xanthopterin as an equilibrium mixture of two forms in aqueous solution (as shown by clear changes, with time and pH, of its uv and fluorescence spectra, as well as by only one form being oxidized to leucopterin by xanthine oxidase) was explained initially as resulting from a slow keto–enol tautomerism.[1256] Subsequent application of rapid reaction spectral techniques, however, showed clearly that an equilibrium between anhydrous (**158**, R = H) and covalently 7,8-hydrated (**159**, R = OH) molecules was the real basis of the observed phenomena.[241] A hydrogen cyanide adduct has been described but its proposed structure (**164**) appears to be less likely than the isomer (**159**, R = CN).[514, cf. 925]

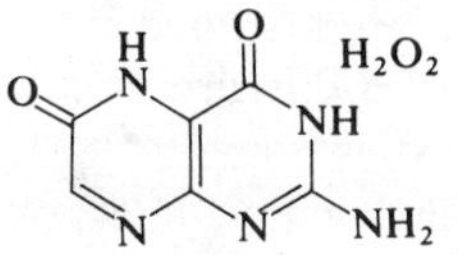

(162)

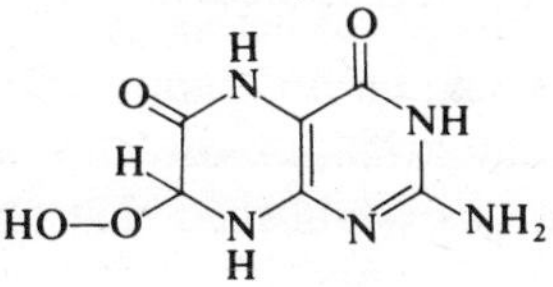

(163)

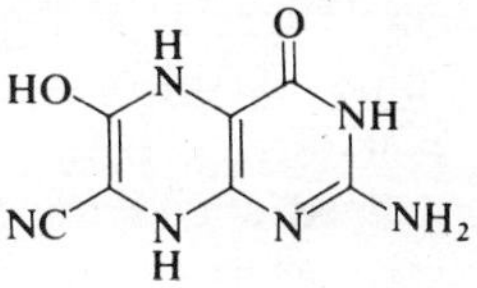

(164)

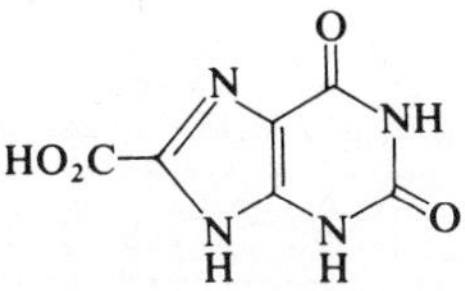

(165)

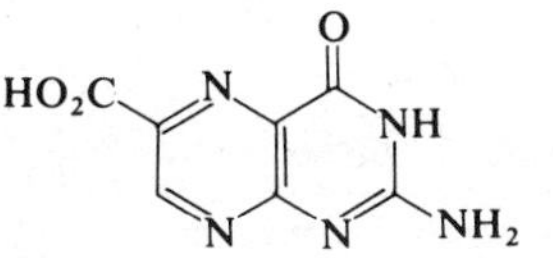

(166)

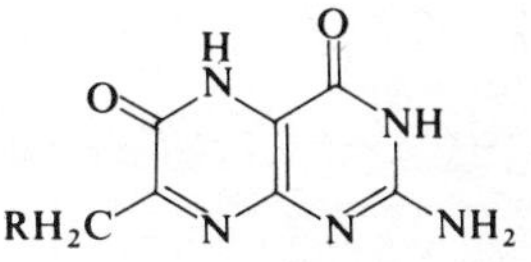

(167)

The following in vitro biotransformations of xanthopterin are included as examples. Although the normal biosynthesis of folic acids does not involve xanthopterin, it has been shown that the latter is transformed by several bacteria or a yeast into 5-formyl-5,6,7,8-tetrahydropteroic acid [272] Certain soil bacteria have been shown to utilize xanthopterin to afford xanthine-8-carboxylic acid (**165**), probably via lumazine and 2,4,6,7(1*H*,3*H*,5*H*,8*H*)-pteridinetetrone.[570] In contrast, *Streptococcus faecalis* and several other microorganisms converted xanthopterin into 2-amino-4-oxo-3,4-dihydro-6-pteridinecarboxylic acid (**166**) among other unidentified products.[275] Incubation of xanthopterin with pig liver homogenate gave leucopterin (by addition of water and oxidation by the xanthine oxidase present). When acetone was present, the main product was 7-acetonyl-2-amino-4,6(3*H*,5*H*)-pteridinedione (**167**, R = Ac) by a similar mechanism, but when ethyl acetoacetate was added the main products were 7-acetonyl- (**167**, R = Ac) and 7-ethoxycarbonylmethylxanthopterin (**167**, R = CO_2Et), rather than that expected.[1540] Finally, xanthopterin strongly

* The conversion of 1,3,8-trimethyl-6-phenyl-2,4,7(1*H*,3*H*,8*H*)-pteridinetrione into its isolable 6,8a-endoperoxide by singlet oxygen (oxygenation in light) may be relevant.[1755]

stimulated cell proliferation in bone marrow cultures while related 6-substituted analogs did so only weakly and 7-substituted analogs proved inhibitory.[1421]

(2) *Dihydroxanthopterin*: *2-amino-7,8-dihydro-4,6(3H,5H)-pteridinedione*

Although dihydroxanthopterin (**159**, R = H) was prepared as early as 1940 by decarboxylation[405] of the carboxylic acid (**159**, R = CO_2H) and subsequently by sodium amalgam reduction[60, 877, 1268] of leucopterin (**151**) or by the Boon synthesis,[450, 451, 727] it was not until 1967 that Henri Descimon extracted it (in the presence of thiodiglycol as an antioxidant) from the wing pigments of a single species of Pierid butterfly.[784] Whether it may have been formed from some of the cooccurring sepiapterin, 2-amino-6-lactoyl-7,8-dihydro-4(3*H*)-pteridinone (**168**), by an aerial oxidation already described in vitro,[1221] remained undetermined but it was established that any such reaction had not taken place as a post mortem decomposition.[784]

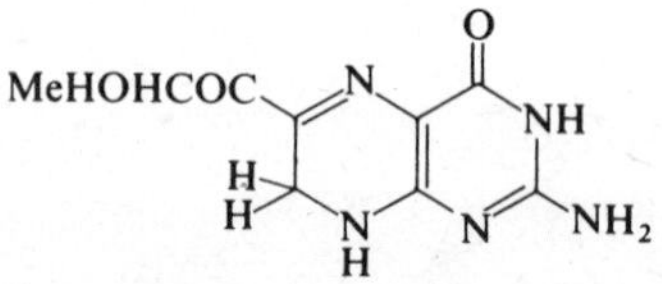

(168)

(3) *Chrysopterin*: *2-Amino-7-methyl-4,6(3H,5H)-pteridinedione*

Chrysopterin (**169**, R = H) was probably isolated in a reasonably pure state from the butterfly, *Gonepteryx rhamni*, by Clemens Schöpf and Erich Becker[719, 1237, cf. 1238] prior to 1936, but it was not until 1951 that its structure was established by Rudolf Tschesche and Friedhelm Korte.[1314] It has been isolated from other insects, the milkweed and red fire bugs,[1446, 1477] and it has been synthesized by the Gabriel and Colman procedure,[874, 878] by hydrolysis of 2-amino-7-sulfomethyl-4,6(3*H*,5*H*)-pteridinedione (**169**, R = SO_3H),[504] by peroxide oxidation of 6-acetyl-2-amino-7,8-dihydro-4,6(3*H*,5*H*)-pteridine-

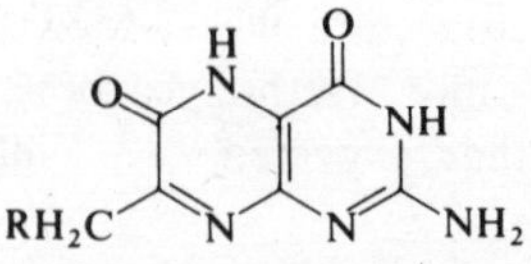

(169)

dione,[1453] and by mild alkaline hydrolysis of erythropterin (**169**, R$=COCO_2H$),[348] with which it often occurs in nature.

(4) *Ekapterin*: 2-*Amino*-7-*β*-*carboxy*-*β*-*hydroxyethyl*-4, 6(3*H*,5*H*)-*pteridinedione*

Although chromatographic traces had probably been noticed earlier, (e.g. ref. 1525) ekapterin was first isolated as fraction-G1 from the meal moth, *Ephestia kühniella*, and subsequently characterized spectrally by Max Viscontini, Alfred Kühn, and A. Engelhaaf in 1956.[1059] It has been isolated also from *Plodia interpunctella* (another meal moth).[816] Its close association with erythropterin (**169**, R$=COCO_2H$) suggested the structure (**169**, R$=CHOHCO_2H$) for ekapterin[515, 516] and this was proven subsequently by a synthesis involving catalytic hydrogenation[406, 515] or sodium borohydride reduction of erythropterin to give ekapterin in 40% yield, probably via aerial oxidation of its 7,8-dihydro derivative.[515,816,1479] The qualitatively observed optical rotation[515] of natural ekapterin was later measured[516]: $[\alpha]_D^{20}$ = ca. − 144°C.

(5) *Lepidopterin*: 2-*Amino*-7-*β*-*amino*-*β*-*carboxyvinyl*-4,6(3*H*,5*H*)-*pteridinedione*

About 1925, the term *lepidopterin* was introduced[1381] as a group name for all wing pigments derived from *Lepidoptera* butterflies.[cf. 917] However, in 1962 the word lepidopterin was redefined as a specific name for one natural product derived from the meal moth, *Ephestia kühniella*[515]; it has been isolated also from the related species, *Plodia interpunctella*.[816] Like ekapterin (just cited), it was associated with erythropterin and its reactions and spectrum suggested to Max Viscontini and H. Stierlin that it had structure (**170**),[516] an hypothesis proven subsequently by its synthesis from erythropterin (**169**, R$=COCO_2H$) in aqueous ammonia for several days (25% yield).[816, 1479] It does not appear to have been isolated from any other source.[1477]

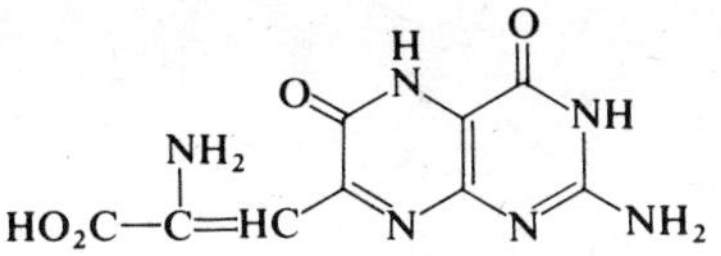

(170)

(6) *Erythropterin*: 2-*Amino*-7-*oxalomethyl*-4,6(3*H*,5*H*)-*pteridinedione*

An orange-red pteridine, erythropterin, was extracted from the wing pigments of several species of butterfly in 1936 by Clemens Schöpf and Erich Becker.[719, 1237] It was subsequently isolated from other butterflies and moths,[355, 515, 516, 613, 1059, 1245, 1477, 1479, 1525, 1577–1580] from milkweed and red fire bugs,[1446, 1477] and from other organisms.[816, 1477, 1548, cf. 998, 1539]

The mildly basic properties of erythropterin did not escape early workers[1237] and in 1940 close but incorrect analyses were reported for its sulfate and perchlorate.[406] These figures, combined with an unfortunate failure to observe the quite strongly acidic properties of erythropterin, led to much fruitless work on its structure.[406, 496, 498, 1313] However, in 1961 Wolfgang Pfleiderer repeated its alkaline degradation to chrysopterin (**167**, R = H) with loss of oxalic acid and its diazomethane methylation to 2-amino-7-methoxalylmethyl-3,5-dimethyl-4,6(3*H*,5*H*)-pteridinedione (**171**), which he then synthesized by the action of dimethyl oxalate on 2-amino-3,5,7-trimethyl-4,6(3*H*,5*H*)-pteridinedione (**172**) in methanolic methoxide. This clearly identified free erythropterin as having structure (**169**, R = $COCO_2H$).[333, 348] A few months later this was followed by two formal syntheses of erythropterin. The first synthesis was by aerial oxidation in alkali, or (better) formic acid, of the covalent 7,8-adduct (**159**, R = CH_2COCO_2H) of pyruvic acid and xanthopterin (**158**, R = H).[515, 816, 952, 1479] The second synthesis was by monodecarboxylation and aerial oxidation of the covalent 7,8-adduct [**159**, R = $CH(CO_2H)COCO_2H$] of oxaloacetic acid and xanthopterin at pH 4.2 and 80°C (35% yield).[439] In like manner, xanthopterin and methyl pyruvate in formic acid gave the methyl ester of erythropterin, 2-amino-7-methoxalylmethyl-4,6(3*H*,5*H*)-pteridinedione (**169**, R = $COCO_2Me$).[1446] The uv and nmr spectra of erythropterin suggest strongly that it exists in the intramolecularly hydrogen-bonded structure (**173**).[1323]

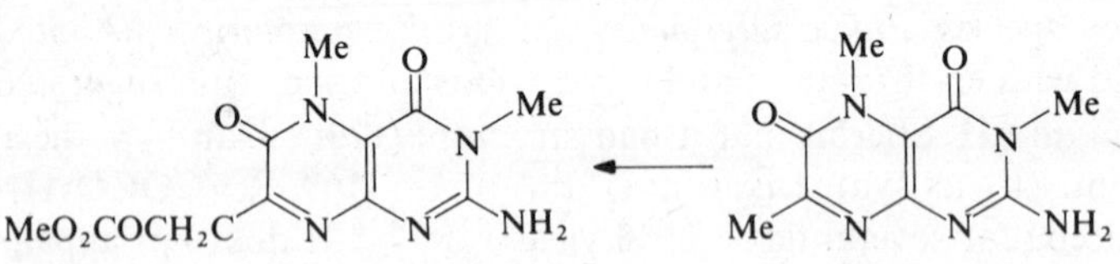

(**171**) (**172**)

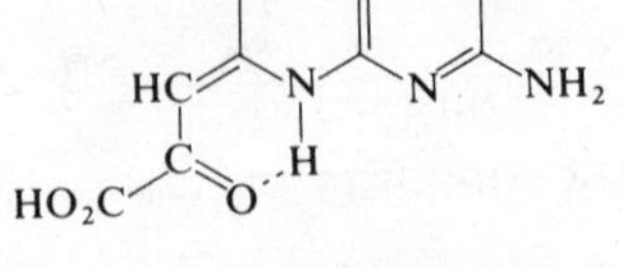

(**173**)

(7) *Pterorhodin*

Pterorhodin and related compounds have been discussed in Chapter IV, Section 2.B(7). It has also been called rhodopterin.

B. Leucopterin
[2-Amino-4,6,7(3*H*,5*H*,8*H*)-pteridinetrione]

Although Frederick Hopkins certainly isolated leucopterin from the wing pigments of white cabbage butterflies and related species about 1889, at the time he mistook it for uric acid.[1, 5, 6, 794] This error was quite understandable in view of the extreme insolubility and infusibility of both compounds and the great difficulty in obtaining good analytical figures for either compound, even to this day. Despite the subsequent isolation of leucopterin from a variety of sources,[531, 550, 613, 719, 769, 896, 1237, 1238, 1314, 1382, 1577–1580] and considerable work on its constitution in Germany from 1924 to 1939,[195, 531, 532, 719, 1238, 1368, 1382, cf. 574, 1073, 1173] it was not until 1940 that Robert Purrmann proved its structure (**175**) by synthesis from 2,5,6-triamino-4(3*H*)-pyrimidinone (**174**) and oxalic acid at 260°C.[403] The possible alternative isomeric product, 2-amino-6-oxo-1,6-dihydro-8-purinecarboxylic acid (**176**),[1172] seemed unlikely because a closely related 8-purinecarboxylic acid underwent facile decarboxylation,[1583] whereas both the corresponding leucopterin derivative and leucopterin itself did not.[404] Moreover, leucopterin was converted by nitrous acid into 2,4,6,7(1*H*,3*H*,5*H*,8*H*)-pteridinetetrone[403] and then with phosphorus pentachloride in phosphoryl chloride into a tetrachloro derivative (**177**),[441] whereas the purine (**176**) could at best provide a trichloro derivative by similar treatment.

The original synthesis of leucopterin (just cited) has been modified in several ways[71, 800, 988, 1152, 1158, 1268] culminating in two excellent procedures[60, 343]

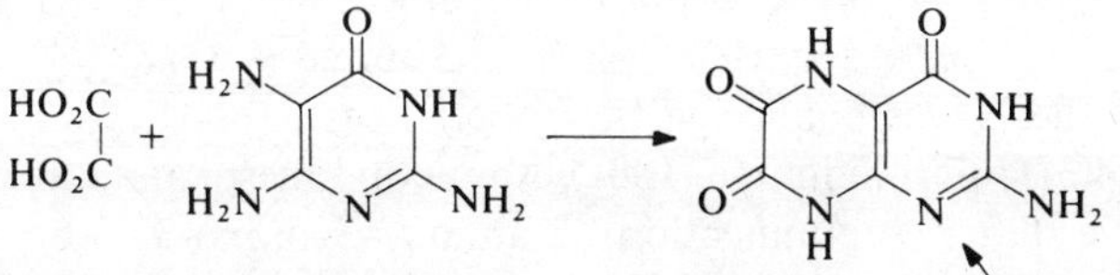

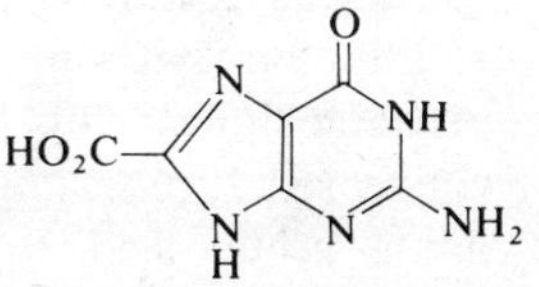

(**174**) (**175**) (**176**)

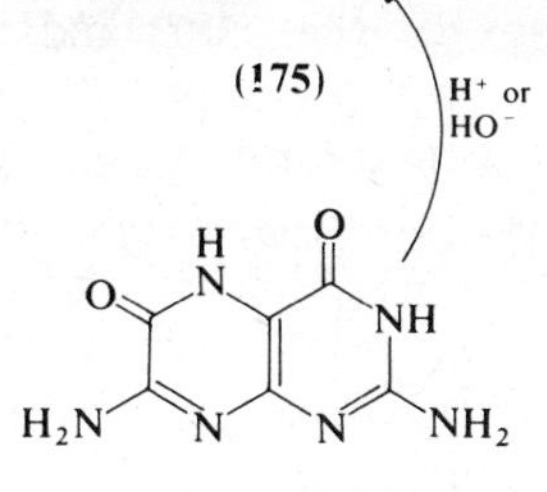

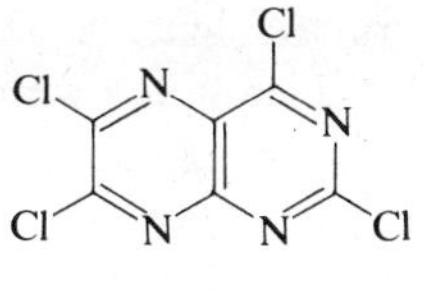

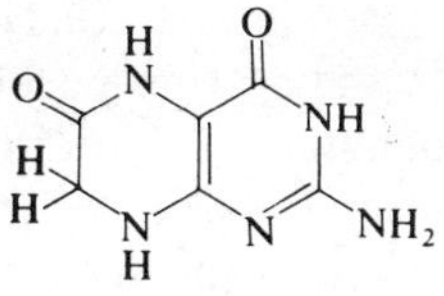

(**177**) (**178**) (**179**)

giving >80% yield. In addition, leucopterin has been made by acidic (47% yield) or alkaline (88%) hydrolyses of 2,7-diamino-4,6(3*H*,5*H*)-pteridinedione (**178**)[97]; by treatment of the covalent 7,8-hydrate of xanthopterin with platinum–oxygen (63% yield)[404, 533] or with periodate (<20%)[1056]; by treatment of xanthopterin peroxide [see Sect. 5.A(1)] with hydrogen peroxide[94]; and by catalytic oxygenation of 7,8-dihydroxanthopterin.[1125]

Leucopterin, which has been purified conveniently through its potassium salt,[441] underwent quantitative deamination on treatment with nitrous acid to give 2,4,6,7(1*H*,3*H*,5*H*,8*H*)-pteridinetetrone.[531–533] It reacted with phosphorus pentachloride–phosphoryl chloride to give 2-amino-4-chloro-6,7(5*H*,8*H*)-pteridinedione,[531] proven in structure by hydriodic acid dechlorination to 2-amino-6,7(5*H*,8*H*)-pteridinedione, which was identical with unambiguously synthesized material.[536] It has been 7-deoxygenated by sodium amalgam to give 7,8-dihydroxanthopterin (**179**) and then by permanganate oxidation to give xanthopterin.[60, 71, 800, 877, 1268] Finally, its mass spectral fragmentation (unlike that of xanthopterin or pterin) begins with the loss of CO from the molecular ion.[778]

C. The Isoxanthopterin Group

(1) *Isoxanthopterin*: 2-*Amino*-4,7(3*H*,8*H*)-*pteridinedione*

Isoxanthopterin is probably the most widespread of all the simple natural pteridines. Although Frederick Hopkins must have handled it at the end of the last century,[6] it remained unrecognized as an entity until Heinrich Wieland, Clemens Schöpf et al. isolated anhydroleucopterin in 1933 from impure leucopterin derived from white cabbage butterfly wings[531] or from those of the lemon-yellow butterfly, *Gonepteryx rhamni.*[1238] After considerable chemical work,[536, 1368] it was found that isoxanthopterin was a deoxyleucopterin.[405] Since it was neither 4-deoxyleucopterin [2-amino-6,7(5*H*,8*H*)-pteridinedione], which was already known[536] nor 7-deoxyleucopterin, which was well known as xanthopterin, it appeared to be 6-deoxyleucopterin [2-amino-4,7(3*H*,8*H*)-pteridinedione]. This hypothesis was confirmed by Robert Purrmann[405] who condensed 2,5,6-triamino-4(3*H*)-pyrimidinone (**180**) with diethyl mesoxalate in dilute acetic acid to give (after saponification) 2-amino-4,7-dioxo-3,4,7,8-tetrahydro-6-pteridinecarboxylic acid (**181**) and then by decarboxylation in nitrogen at 260°C[cf. 469] to give a product (**182**), which was isomeric with xanthopterin and identical with natural isoxanthopterin.

Isoxanthopterin (**182**) (also known as mesopterin, ranachrome-4, leucopterin-B, and cyprino-purple-A1) has also been detected in, or isolated from, the wings of numerous butterfly species,[1314, 1399, 1477, 1577, 1578, 1580] the eye pigments of several species of fruit fly (*Drosophila*),[969, 1054, 1477, 1585, cf. 967] the eggs or larvae of the silkworm (*Bombyx mori*),[833, 1034] the scales of carp,[502, 1033] the skins of snakes and salamanders,[697, 1550] the fresh water crayfish (*Astacus*

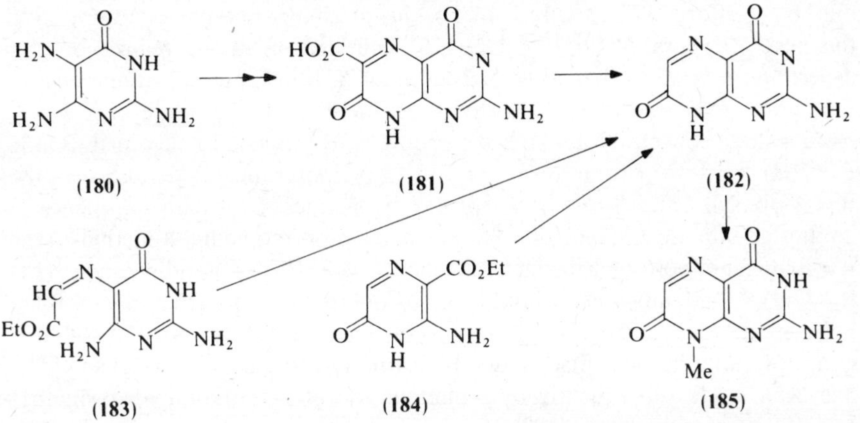

fluviatilis),[1065] the skins of several frogs,[219, 1034] the microorganism, *Serratia indica*,[593] human urine,[1584] and other sources.[791(?), 1566]

In addition to the synthesis just mentioned, isoxanthopterin has best been made by the condensation of 2,5,6-triamino-4(3*H*)-pyrimidinone with ethyl glyoxylate hemiacetal to give the intermediate Schiff base (**183**), which underwent cyclization in hot aqueous ammonia to give (**182**) in 50% overall yield.[61, 62, 1267] other syntheses involved alkaline hydrolysis of 7-methoxy-2,4-pteridinediamine (82% yield), cyclization of ethyl 3-amino-5-oxo-4,5-dihydro-2-pyrazinecarboxylate (**184**) with guanidine (14% yield) or the hydrolysis of 7-pyrrolidino-2,4-pteridinediamine in alkali (85% yield).[1301] The fine tautomeric structure of isoxanthopterin has been investigated by comparing its ionization constants and uv spectra with those of various methylated derivatives. It does in fact exist as the structure (**182**).[383] Although isoxanthopterin with dimethyl sulfate at pH 7 gave only the 8-methylated product, 2-amino-8-methyl-4,7(3*H*,8*H*)-pteridinedione (**185**), under other conditions a further three products could be isolated, viz. 2-amino-3-methyl-, 2-amino-3,8-dimethyl- and 3,8-dimethyl-2-methylamino-4,7(3*H*,8*H*)-pteridinedione.[383, 457] Like leucopterin, the mass spectral fragmentation begins by loss of CO from the molecular ion.[778] An interesting soil microorganism, *Alcaligenes faecalis*, which can use isoxanthopterin as its sole source of carbon and nitrogen, has been shown to convert the pteridine successively into 2,4,7(1*H*,3*H*,8*H*)-pteridinetrione, 2,4,6,7(1*H*,3*H*,5*H*,8*H*)-pteridinetetrone, and (the isomeric) 2,6-dioxo-1,2,3,6-tetrahydro-8-purinecarboxylic acid.[1116]

(2) *Asperopterin-B*: 2-*Amino*-6-*hydroxymethyl*-8-*methyl*-4,7(3*H*,8*H*)-*pteridinedione*
Asperopterin-A: *Its* 6-*β*-D-*Ribofuranosyloxymethyl Analog*

The presence of two or more pteridinelike compounds in the fermentation broth of *Aspergillus* fungi was strongly held in Japan prior to 1949.[1586] Mainly due to the efforts of Yasuyuki Kaneko, this gradually became a certainty over the next 20 years,[254, 255, 1070, 1587, 1588] culminating in the realization that asperopterin-B was a derivative of 2-amino-4,7(3*H*,8*H*)-pteridinedione bearing a 6-hydroxymethyl and an *N*-methyl substituent, while asperopterin-A was its D-riboside.[256] Several years later, asperopterin-B was finally identified as (**186**, R = OH) and asperopterin-A as its *β*-D-ribofuranosyl derivative (**186**, R = *O*-ribosyl). The former was promptly synthesized by the sequence 2-amino-6,8-dimethyl- (**186**, R = H)→2-amino-6-bromomethyl-8-methyl- (**186**, R = Br)→2-amino-6-hydroxymethyl-8-methyl-4,7(3*H*,8*H*)-pteridinedione (**186**, R = OH).[1214] Meanwhile, without being recognized as such, asperopterin-B had been made by catalytic oxygenation of 2-amino-6,8-dimethyl-7,8-dihydro-4(3*H*)-pteridinone (**187**) (itself made by a Boon synthesis), in 30% yield.[1370] It has been made subsequently by 8-methylation of 2-amino-6-hydroxymethyl-4,7(3*H*,8*H*)-pteridinedione (**188**), in 72% yield.[1296] It was also made by direct *C*-hydroxymethylation of 2-amino-8-methyl-4,7(3*H*,8*H*)-pteridinedione in methanol by the addition of ammonium persulfate (53%).[1711]

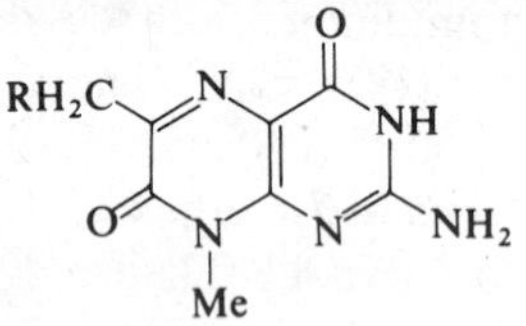

(**186**)

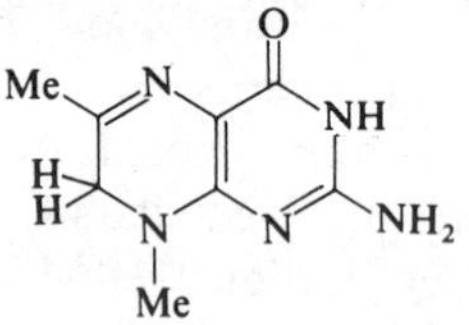

(**187**)

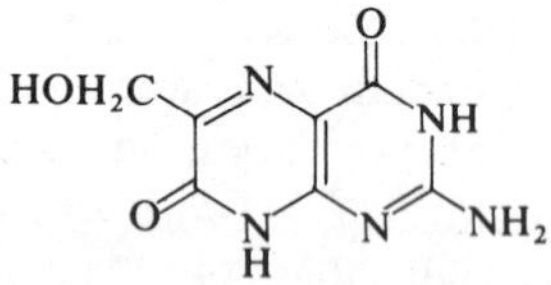

(**188**)

(3) *Ichthyopterin*: 2-*Amino*-6-*α*,*β*-*dihydroxypropyl*-4,7(3*H*,8*H*)-*pteridinedione*

Although there had been considerable interest in the fluorescent materials in fish scales and skin from about the year 1930, it was Rudolf Hüttel and G. Sprengling who reported the isolation of a pteridine derivative, ichthyopterin, from the skin and scales of a mixed bag of fresh water white fish (all *Cyprinidae*) caught in Bodensee.[235] In the same year, Michel Polonovski also described the isolation from carp scales of a pteridine, fluorocyanine,[1589] which clearly resembled Hüttel's compound apart from its fluorescence spectrum. The identity of both products and of similar ones isolated later[801] from carp scales and silkworm eggs, has never been resolved in a totally satisfactory way.[270, 502, 930, 1274, 1590]

Meanwhile, it was reported in 1951 that natural ichthyopterin was identical with synthetic 2-amino-6-carboxymethyl-4,7(3*H*,8*H*)-pteridinedione (**189**).[1304] Following a serious challenge to this fact,[1033] it emerged that the comparison had been made with an old and very colored specimen of natural ichthyopterin, probably photochemically altered [1590] by storage under poor conditions.[499, 995] Accordingly, new rigorous comparisons were made with 2-amino-6-*α*,*β*-

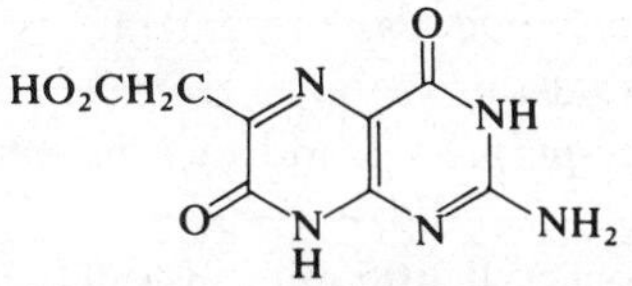

(189)

dihydroxypropyl- **(190)** and 2-amino-β,γ-dihydroxypropyl-4,7(3*H*,8*H*)-pteridinedione, indicating that ichthyopterin had indeed the structure **(190)**.[499, 995] This was confirmed subsequently by the preparation (with analysis) of a new specimen of natural material, followed by recomparisons with these synthetic materials.[1549] The side-chain stereoconfiguration of natural ichthyopterin remains unknown.[995] One synthesis of racemic material began with the condensation of 2,5,6-triamino-4(3*H*)-pyrimidinone and ethoxalylacetone to give 6-acetonyl-2-amino-4,7(3*H*,8*H*)-pteridinedione (**191**, R = H), which underwent bromination to the 6-α-bromoacetonyl analog (**191**, R = Br) and then acetoxylation to the 6-α-acetoxyacetonyl analog (**191**, R = OAc). This, on sodium borohydride reduction in aqueous sodium carbonate, gave the product **(190)**.[499] A second route from the same substrate (**191**, R = H) involved selenium dioxide oxidation to 6-acetylcarbonyl-2-amino-4,7(3*H*,8*H*)-pteridinedione followed by borohydride reduction to **(190)**.[499] Ichthyopterin has also been known as cyprino-purple-A2.[292, 555]

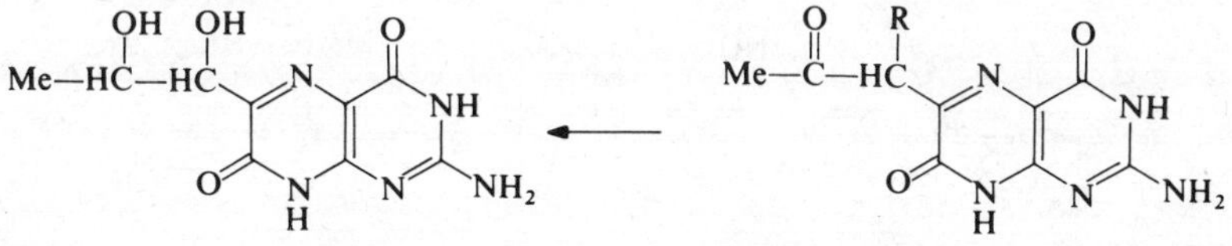

(190) **(191)**

(4) *Cyprino-purple-B*: 2-*Amino*-4,7-*dioxo*-3,4,7,8-*tetrahydro*-6-*pteridinecarboxylic Acid*
*Cyprino-purple-C*1: 6-β-*Acetoxy*-α-*hydroxypropyl*-2-*amino*-4,7(3*H*,8*H*)-*pteridinedione*
*Cyprino-purple-C*2: 6-α-*Acetoxy*-β-*hydroxypropyl Isomer*

By 1954 it was clear that the skin and scales of adult carp species contained at least one purple-fluorescing pteridine other than ichthyopterin and isoxanthopterin.[276, cf. 291, 292, 998, 1590] The following year, this compound (now known as *cyprino-purple-B*) was identified as the carboxylic acid (**192**),[1033] which was already a known compound.[405] In addition, it was realized in 1956 that at least one other pteridine was present in larvae of the same fish.[292] In 1960, it became evident that this material consisted of two distinct but closely related pteridines, *cyprino-purple-C*1 and *cyprino-purple-C*2, which both gave cyprino-purple-B on manganese dioxide oxidation.[218] In 1968 they were identified as the isomeric monoacetylichthyopterins (**193**) and (**194**), respectively.[1029]

Cyprino-purple-B (**192**) has been synthesized by two distinct Gabriel and Colman syntheses[405, 481] and by an unambiguous route involving the cyclization of a pyrazine *N*-oxide to give 6-hydroxyiminomethyl-2,4-pteridinediamine 8-oxide (**195**), treatment with phosphoryl chloride–DMF to give 7-chloro-2,4-bisdimethylaminomethyleneamino-6-pteridinecarbonitrile (**196**), and an alkaline hydrolysis to the product (**192**)[469]; its fine structure is in fact represented in formula (**192**).[385] A separable mixture of cyprino-purple-C1 and C2 was synthesized by partial hydrolysis of 2-amino-6-α,β-diacetoxypropyl-4,7(3*H*,8*H*)-pteridinedione (**197**), made by acetylation of ichthyopterin.[1029]

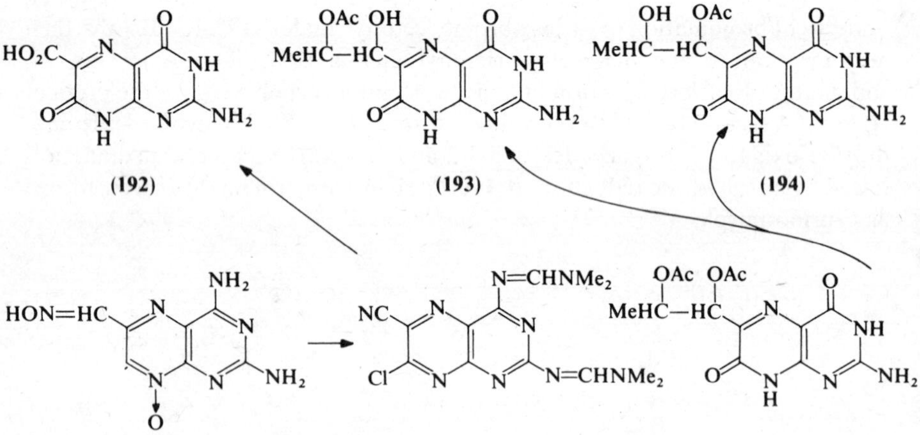

(**192**) (**193**) (**194**)

(**195**) (**196**) (**197**)

(5) 2-*Amino-6-α-hydroxypropyl-8-methyl*-4,7(3*H*,8*H*)-*pteridinedione*

In 1981 a new derivative of 8-methylisoxanthopterin was isolated from the firefly, *Photinus pyralis*, and careful examination revealed that it was probably 2-amino-6-α-hydroxypropyl-8-methyl-4,7(3*H*,8*H*)-pteridinedione, a homolog of asperopterin-B.[1712] This formulation was confirmed subsequently by synthesis through treatment of 2-amino-8-methyl-4,7(3*H*,8*H*)-pteridinedione (**185**) in propanol with ammonium persulfate to give a product identical with that from the firefly (presumably, apart from optical activity).[1711] No trivial name has been suggested.

D. The Pterin Group

(1) *Pterin: 2-Amino-4(3H)-pteridinone*

Pterin (**198**) was synthesized first in 1946 by Ted Taylor et al., using a Gabriel and Colman synthesis from 2,5,6-triamino-4(3*H*)-pyrimidinone and glyoxal,[747, cf. 1084] but almost a decade passed before it was recognized as a natural product. Thus in 1955 it was isolated from wild strains of the fruit fly, *Drosophila melanogaster*, initially by Hugh Forrest and H. Mitchell[1585] and later in the same year by Max Viscontini et al.,[1049] who also found it in the fresh water crayfish, *Astacus fluviatalis*.[1065] It has been isolated subsequently from a variety of sources such as amphibia,[212, 1597] butterflies and moths,[516, 613, 1059, 1479, 1525, 1580, 1593] the honeybee,[1197] a wasp,[896] several fruit flies,[1032, 1052, 1594] tsetse flies and related *Diptera*,[221, 1591] ants,[1539] and other insects.[1592, 1595, 1596]

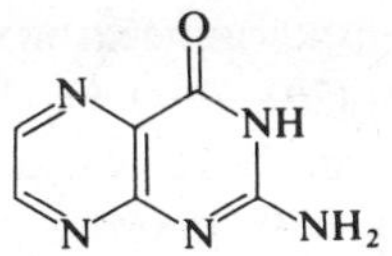

(198)

Pterin has been made by various routes as well as by the previously mentioned synthesis [see the index: 2-Amino-4(3*H*)-pteridinone] and it undergoes the usual reactions of a pteridinamine and a pteridinone. Its fragmentation pattern begins with loss of HCN or CO from the molecular ion.[778] The conformations of 5,6,7,8-tetrahydropterins have been studied.[1694] The nmr of pterins has been reported.[741, 762]

(2) *6-Pterincarboxylic Acid: 2-Amino-4-oxo-3,4-dihydro-6-pteridinecarboxylic Acid*

6-Pterincarboxylic acid (**201**, R$=CO_2H$) was first reported in 1946 by Robert Angier and co-workers,[1174] who synthesized it by preferential 7-chlorination of 2-amino-4,7-dioxo-3,4,7,8-tetrahydro-6-pteridinecarboxylic acid (**199**) and subsequent dechlorination of the resulting 2-amino-7-chloro-4-oxo-3,4-dihydro-6-pteridinecarboxylic acid (**200**, R$=$Cl).[1084] The site of chlorination–dechlorination was confirmed by decarboxylation of the product (**201**, R$=CO_2H$) to give pterin (**198**).[1174] It was 6 years later before 6-pterincarboxylic acid was recognized as a natural product in its own right when Tadao Hama isolated it from the frog, *Rana nigromaculata*,[1604] under the name ranachrome-5.[213, 219] Later it was recognized in, and/or isolated from, a lemon-

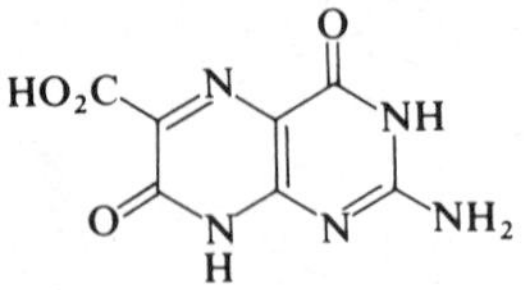

(199)

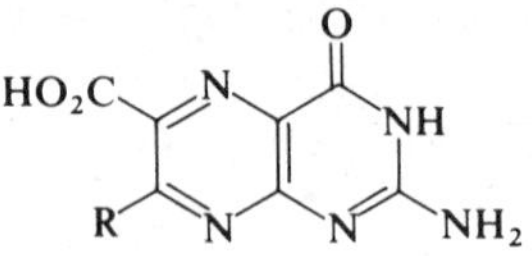

(200)

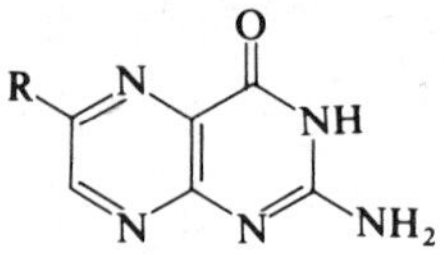

(201)

colored mutant of the silkworm,[833] a variety of butterfly species,[613, 1578–1580, 1600] fruit flies,[1049] the honeybee,[1197, cf. 1564] crayfish,[1065] snakeskin,[697] amphibia,[1597] and even the pineal glands of sheep.[1447]

As well as the synthesis just mentioned, 6-pterincarboxylic acid has been made by a direct Gabriel and Colman synthesis,[1084] by monodecarboxylation of 2-amino-4-oxo-3,4-dihydro-6,7-pteridinedicarboxylic acid (**200**, $R=CO_2H$),[307] by alkaline hydrolysis of 2,4-diamino-6-pteridinecarboxylic acid,[178] by reductive desulfonation of 2-amino-4-oxo-7-sulfo-3,4-dihydro-6-pteridinecarboxylic acid (**200**, $R=SO_3H$),[906] and by acidic hydrolysis of 2-amino-4-oxo-3,4-dihydro-6-pteridinecarboxamide (**201**, $R=CONH_2$).[925] In addition, the carboxylic acid (**201**, $R=CO_2H$) has been made frequently by permanganate oxidation of pterins that bear a side chain linked through the carbon atom to the 6-position, usually in order to confirm a 6- as distinct from a 7-linkage in the substrate. Thus, for example, it has been derived from 2-amino-6-methyl-4(3*H*)-pteridinone (**201**, R=Me),[193, 389, 1174] 2-amino-6-hexyl (or heptyl or nonyl)-4(3*H*)-pteridinone,[420] 2-amino-6-hydroxymethyl-4(3*H*)-pteridinone (**201**, $R=CO_2OH$),[1174] the 6-β-hydroxyethyl homolog (**201**, $R=CH_2CH_2OH$),[502] the 6-α,β-dihydroxypropyl analog (**201**, R=CHOH-CHOHMe),[500] the 6-α,β,γ,δ-tetrahydroxybutyl analog,[192, 1373] various Michael-type 5,6-adducts of 2-amino-7,8-dihydro-4(3*H*)-pteridinone,[453] 2-amino-4-oxo-3,4-dihydro-6-pteridinecarbaldehyde (**201**, R=CHO) (by oxidation[943] or the Cannizzaro reaction[1358]), 2-amino-6-pyridiniomethyl-4(3*H*)-pteridinone iodide,[1174] 6-methyl-2,4-pteridinediamine (additional hydrolysis),[1263] pteroylglutamic acid,[1263, 1376] and various derivatives thereof.[744, 745, 964, 966, 1104, 1188]

(3) *Ranachrome*-3: 2-*Amino*-6-*hydroxymethyl*-4(3*H*)-*pteridinone*

2-Amino-6-hydroxymethyl-4(3*H*)-pteridinone (**201**, $R=CH_2OH$) was synthesized first by Paul Karrer and a colleague in 1949, using a Gabriel and Colman synthesis.[1212, cf. 193] Ranachrome-3 was recognized as a succinct chromatographic spot in extracts of the skin of the frog, *Rana nigromaculata*, in 1952 by Tadao Hama,[1604] but despite continuing work on this substance from frog[206, 212, 213, 219, 968] or reptilian skin,[318] it was 1965 before chemical work on ranachrome-3 was reported[214] and its identity with the synthetic hydroxymethylpterin (**201**, $R=CH_2OH$) was established.[1603]

Meanwhile, 2-amino-6-hydroxymethyl-4(3*H*)-pteridinone (**201**, R$=CH_2OH$) had been implicated as an intermediate in the biochemical pathway in microorganisms from purines to (folate) pteridines[1602] and it had been isolated from cultures of a *Corynebacterium* species,[1601] from a *Pseudomonas* variety,[1423, 1617] and from spinach chloroplasts.[986, 1627, 1628] In addition, a glucoside was isolated from one species of blue-green algae.[1615] This stimulated further work on the synthesis of (**201**, R$=CH_2OH$), which has been made subsequently by borohydride reduction of 2-amino-4-oxo-3,4-dihydro-6-pteridinecarbaldehyde (**201**, R=CHO) derived from folic acid or neopterin[1270, 1336]; by an improved Gabriel and Colman synthesis[96]; by hydrolysis of 6-hydroxymethyl-2,4-pteridinediamine (**202**)[96, 113]; by an excellent method involving lead tetraacetate oxidation of 2-acetamido-6-methyl-4(3*H*)-pteridinone (**203**, R=H) to give the 6-acetoxymethyl analog (**203**, R=OAc) followed by deacetylation in acid at 20°C (95% yield)[1443]; and by other routes.[513, 1358] Its mass spectrum has been examined.[778]

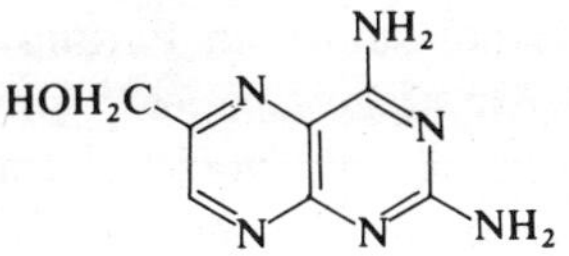

(202)

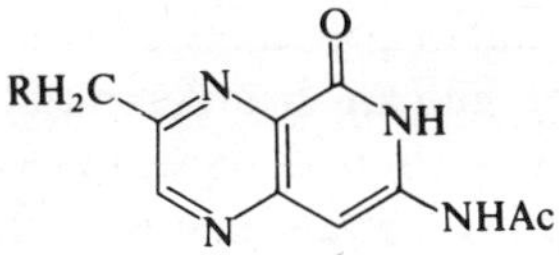

(203)

(4) *Biopterin*: 2-*Amino*-6-[L-*erythro*]-α,β-*dihydroxypropyl*-4(3*H*)-*pteridinone*
Biopterin Glucoside: *Biopterin* 1′-α-D-*Glucoside*
Ciliapterin: *the* [L-*threo*]-*Isomer of Biopterin*

Biopterin was probably a major component in 'pyrrolechrome' isolated in 1941 from two species of frog by Tokunosuke Goda[1605] and subsequently described in some detail.[1606] It may well have been identical with a fluorescyanine isolated from silkworms,[1607] and certainly with ranachrome-1, isolated from frog skin by Tadao Hama and colleagues[213, 215, 216, 219] but only identified with biopterin (**204**) in 1958.[206] Meanwhile, in 1955, Ernest Patterson et al. reported the isolation from human urine of a pteridine that promoted the growth of the protozoan, *Crithidia fasciculata*, and which they called biopterin.[1557] Its structure (**204**) was correctly deduced from the outset[1018, 1557] and its characterization as the L-*erythro* isomer was achieved by a Gabriel and Colman synthesis from 5-deoxy-L-arabinose and 2,5,6-triamino-4(3*H*)-pyrimidinone.[1043] In the same year, two other groups headed by Max Viscontini[1054] and Hugh Forrest[1585] reported the isolation of biopterin (under the temporary names HB2 and F14a, respectively) from the fruit fly, *Drosophila melanogaster*. The first group initially arrived at an incorrect structure[1049] but soon realized the error and amended it correctly.[1050] The second group gave the correct structure from the outset and confirmed it by a synthesis akin to that just

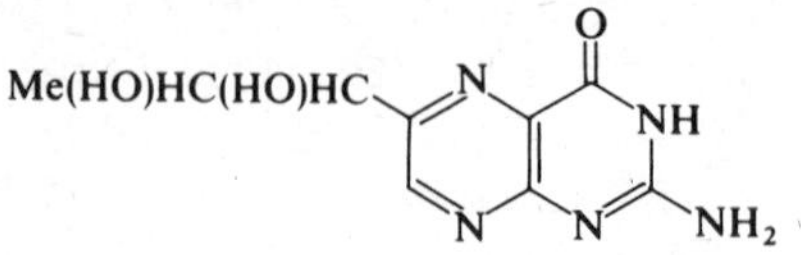

(204)

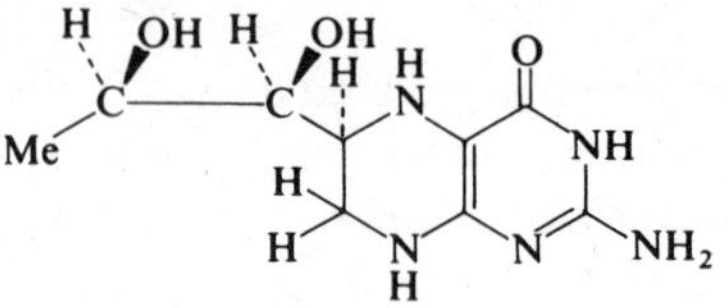

(205)

cited.[1585] There followed a spate of papers on the isolation of biopterin from diverse sources such as the skins of several amphibia,[212, 1552, 1559, 1597] reptiles,[318] and a crayfish[1065]; the royal jelly of honeybees[1197, 1564, 1663]; some ants[1539]; fruit fly mutants[1559]; several species of Diptera fly[221, 1591]; and a meal moth.[516, 1059]

The natural 6(*R*), α(*R*), β(*S*)-diastereoisomer of 5,6,7,8-tetrahydrobiopterin (**205**) is the cofactor for enzymes that hydroxylate tyrosine, phenylalanine, its 3,4-dihydro derivative (dopa), and other such molecules in vivo, and hence it is of enormous importance in neurological disfunctions, inherited metabolic diseases such as phenylketonuria, in Parkinson's disease, and in other human disorders (see Wilf Armarego's paper[82] for a summary with leading references beyond the scope of this book). Because of this medical interest, a great many syntheses of biopterin (and then its tetrahydro derivative) have been devised. The main route to biopterin, using a sugar as a proaldehydoketone with 2,5,6-triamino-4(3*H*)-pyrimidinone under a variety of conditions, has been reviewed in Chapter II, Section 1.B(3)(e).[429, 454, 455, 500, 564, 571, 645, 860, 1043, 1324, 1326, 1327, 1439, 1456, 1497, 1564] A second (indirect) Gabriel and Colman synthesis involved condensation of 5,6-diamino-2-methylthio-4(3*H*)-pteridinone with 5-deoxy-L-arabinose phenylhydrazone followed by oxygenation to give 6-[L-*erythro*]-α,β-dihydroxypropyl-2-methylthio-4(3*H*)-pteridinone and subsequent aminolysis to biopterin.[458] A third type of synthesis employed the intermediate, benzyl 3-amino-6-[L-*erythro*]-α,β-dihydroxypropyl-2-pyrazinecarboxylate 4-oxide (**206**), which underwent cyclization with guanidine to give biopterin 8-oxide and then by dithionite reduction, to give biopterin (82%).[1285, 1318, cf. 772, 1401] Finally, a fourth procedure used a Boon synthesis involving reductive cyclization of 2-amino-6-*N*-[L-*erythro*-α,β-dihydroxybutyryl)methyl]amino-5-nitro-4(3*H*)-pyrimidinone (**207**) and subsequent dehydrogenation.[653]

In 1977, Max Viscontini and colleagues reported that catalytic hydrogenation of biopterin over platinum oxide in trifluoroacetic acid gave a mixture of two

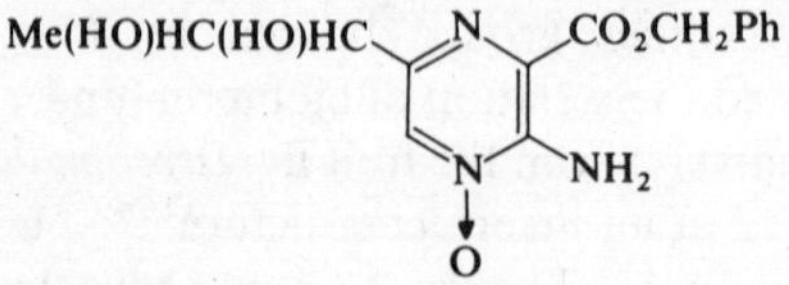

(206)

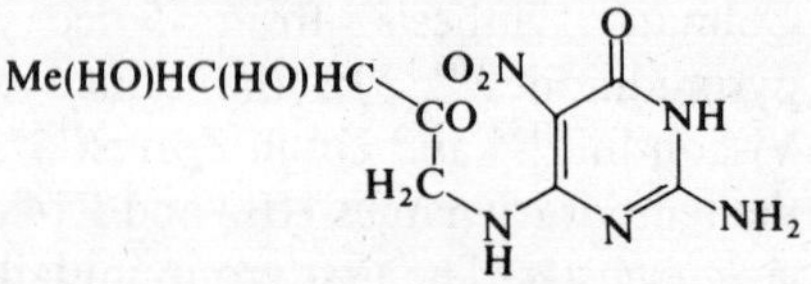

(207)

diastereoisomers, as indicated by ^1H and ^{13}C nmr measurements.[429,1608] A similar mixture of (6*R*)- (**205**) and (6*S*)-5,6,7,8-tetrahydrobiopterin was obtained on dithionite or borohydride reduction[430,1608] and these were separated by hplc (high performance liquid chromatography) on a strong cation exchanger.[1608, cf. 685] Separation was also achieved by crystallization of a derived mixture of diastereoisomeric tetraacetyl compounds (**208**, R = H), followed by deacylation of the individual isomers.[82,199] The ratio, natural (*R*)-isomer (**205**) to unnatural (*S*)-isomer, obtained on hydrogenation of biopterin, may be controlled by variation in the pH of the medium: at pH 2.3, the ratio was 2.6:1, whereas at pH 11.8, it was 7.3:1.[1474,1518,1692] The absolute configuration of natural tetrahydrobiopterin has been confirmed as (6*R*) by x-ray crystallographic analysis of the dihydrochloride[1519] and its pentaacetyl derivative (**208**, R = Ac).[401] It was also confirmed by an appropriate Cotton effect in its cd (circular dichroism) spectrum, as viewed in comparison with those of well authenticated naphthalene[1519] and quinazoline models.[298,645] The conformation(s) of the reduced pyrazine ring and of the side chain in natural and unnatural tetrahydrobiopterin, as well as of their acetyl derivatives, have been investigated by high field nmr techniques.[82]

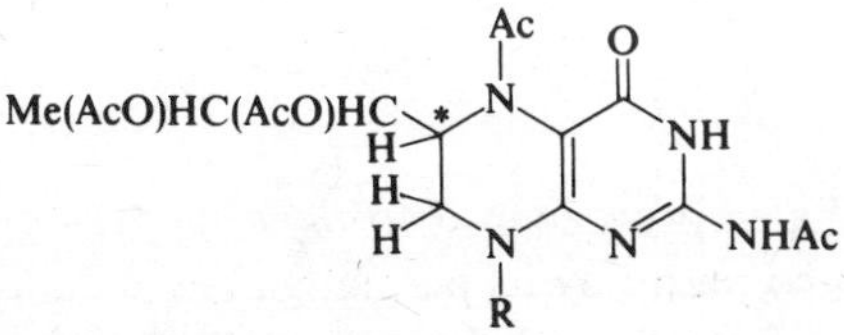

(**208**)

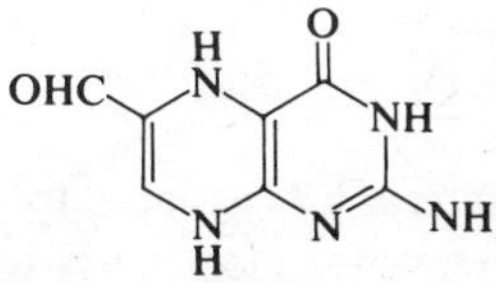

(**209**)

The detection and estimation of biopterin–tetrahydrobiopterin in minute amounts within the body fluids has become medically important[1685] and may be done in several ways, for example, by the glc/ms (gas–liquid chromatography/mass spectra) of trimethylsilylated material to 10^{-6} g[1111]; by a radioimmunoassay based on the ^{125}I-labeled conjugate of biopterin and tyramine together with specific antibodies against biopterin (prepared in rabbits) (as little as 0.5×10^{-12} mol of biopterin in 10^{-6} L of urine could be estimated in the presence of tetrahydrobiopterin, dihydrobiopterin, or related compounds)[1133]; or by a fluoroimmunoassay based on fluorescein-labeled biopterin (this technique was less sensitive but usually adequate at 3×10^{-11} mol of biopterin per assay).[1454] These should be compared with more troublesome bioassay methods which, however, approached the 0.5×10^{-13} mol level.[1609] The mass spectra of trimethylsilylated biopterin,[1111] neat biopterin,[976] and reduced products[1074] are recorded. The photodegradation of biopterin to 2-amino-4-oxo-3,4,5,8-tetrahydro-6-pteridinecarbaldehyde (**209**) with subsequent oxidation to 2-amino-4-oxo-3,4-dihydro-6-pteridinecarbaldehyde (in the presence of air) has been examined.[1094] The redox sequence, biopterin–dihydrobiopterin–tetrahydrobiopterin, is an immensely important biochemical process quite beyond the scope of this book[1572, cf. 410, 411, 462, 619, 685, 695, 1168, 1552, 1559]; for a discussion of quinonoid dihydrobiopterin, and so on, see Chapter IX, Section 7.

Biopterin glucoside was isolated (as compound-C) in 1957 from a blue-green alga, *Anacystis nidulans*.[1561] Its chemical behavior, elemental analysis, and so on, suggested that it was a glucoside of biopterin with the sugar attached at the 1′- or 2′-position on the side chain.[1560] This was broadly upheld by an ambiguous synthesis from biopterin, glucose, and *p*-toluenesulfonic acid in DMF that gave an identical product in low yield.[1562] The balance of evidence appears to favor the 1′-glucosyl structure (**210**),[1614] which has been accepted by subsequent reviewers.[1477, 1536] There is also some evidence that this,[1618] and other glycosides in several species of such algae,[1615] may actually exist in the cells as tetrahydro derivatives.[1612–1614] An unidentified biopterin glucoside has been isolated from sheep pineal glands.[1447]

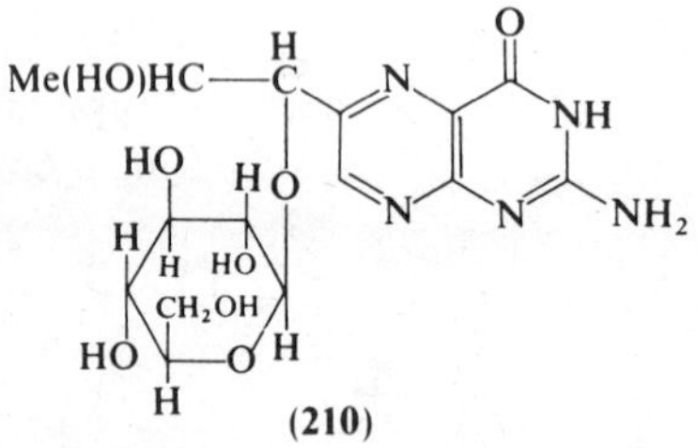

(210)

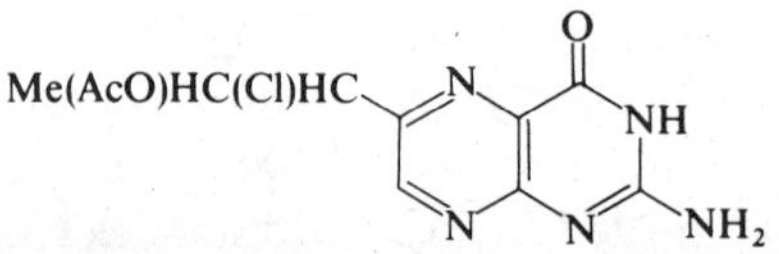

(211)

Ciliapterin was isolated from a ciliated protozoan, *Tetrahymena pyriformis*, in 1968[1611]: it was clearly a 6-substituted pterin by its permanganate oxidation to 6-pterincarboxylic acid, a dihydroxypropyl side chain was indicated by the formation of acetaldehyde on periodate oxidation, its spectra were almost identical with those of biopterin, and certain biochemical and biological evidence suggested that it was the [L-*threo*] isomer of biopterin.[1610, 1611] Labeling experiments showed that it was derived entirely from guanosine.[1611]

In fact, 2-amino-6-[L-*threo*]-α,β-dihydroxypropyl-4(3*H*)-pteridinone had been synthesized by an appropriate Gabriel and Colman synthesis[1555] prior to the isolation of ciliapterin, as well as on a subsequent occasion.[454] In addition, it has been made by treatment of biopterin (**204**) with α-acetoxy-α-methylpropionyl chloride to give 6-[L-*threo*]-β-acetoxy-α-chloropropyl-2-amino-4(3*H*)-pteridinone (**211**) followed by hydrolysis in dilute ammonia.[1439] However, a direct comparison of synthetic with natural ciliapterin does not appear to have been reported.

(5) *Neopterin: 2-Amino-6-[D-erythro]-α,β,γ-trihydroxypropyl-4(3H)-pteridinone* *Bufochrome: the [L-erythro]-Isomer of Neopterin* *Monapterin: the [L-threo]-Isomer of Neopterin*

Although 2-amino-6-[D-*erythro*]-α,β,γ-trihydroxypropyl-4(3*H*)-pteridinone (**212a**) was synthesized in 1958 by a Gabriel and Colman synthesis,[330] it was not until 1963 that Heinz Rembold and a colleague isolated *neopterin* from honey-bee pupae and deduced that it was a 6-trihydroxypropylpterin of unknown stereoconfiguration.[1197] Later in the year he resynthesized the [D-*erythro*]-isomer (**212a**) and showed that it was identical with natural neopterin.[1563] Subsequently, neopterin has been isolated from frog skin and fruit flies,[911] from cultures of microorganisms such as *Serratia indica* or *Pseudomonas ovalis*,[463] and (more importantly) from sheep pineal glands[1462] and human urine,[1422, 1620] in which it probably occurs as a di- or tetrahydro derivative.[1620, 1621] The levels of neopterin (and biopterin) in urine were found to be considerably elevated in most cancer patients,[1619, 1622] thus constituting a useful noninvasive diagnostic tool when coupled with a sufficiently sensitive glc/ms or radioimmunoassay method.[1484, 1623] The connection of urinary pteridine levels with disease has been summarized with leading biological references[1477] and the even more important role of pteridines in the immune response has been discussed.[1685] Derivatives of neopterin have also been isolated from microorganisms, for example, its 3′-*β*-D-glucuronide from *Azotobacter agilis*[1030] or its 2′:3′-cyclic phosphate (**213b**) from *Photobacterium phosphoreum*.[1424]

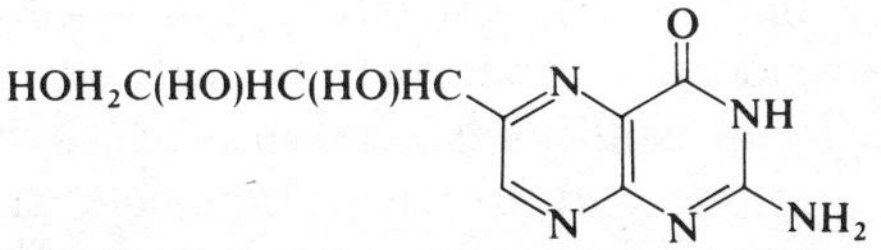

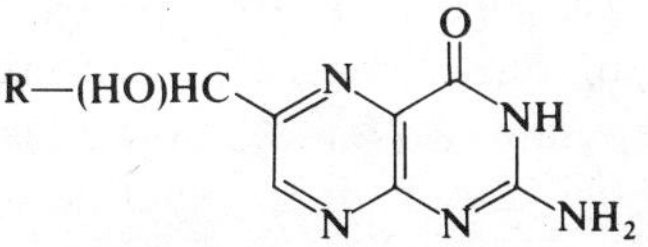

(**212a**) D-*erythro*
(**212b**) L-*erythro*
(**212c**) L-*threo*

(**213a**) R = $H_2O_3POH_2C(HO)HC–$
(**213b**) R =

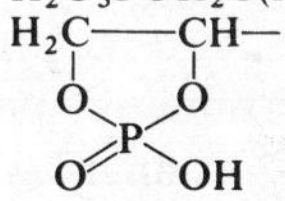

As well as the synthesis just mentioned, neopterin has been made by several variations of the Gabriel and Colman route,[509, 905, 911, 1378, 1422, 1496] by reductive cyclization of 2-amino-5-nitro-6-*N*-[D-*erythro*-(*α*,*β*,*γ*-trihydroxybutyryl)methyl]amino-4(3*H*)-pyrimidinone and subsequent oxidation,[653] from a pyrazine intermediate,[1401] and from mild aminolysis of 2-methylthio-6-[D-*erythro*]-*α*,*β*,*γ*-trihydroxypropyl-4(3*H*)-pteridinone.[458] Phosphorylation of neopterin gave initially the *γ*-phosphate ester, 2-amino-6-*γ*-dihydroxyphosphinyloxy-*α*,*β*-dihydroxypropyl-4(3*H*)-pteridinone (**213a**),[905, 1216, 1335]; which on treatment with dicyclohexylcarbodiimide gave the cyclic phosphate (**213b**)[1216, 1335]; polyphosphates have been made.[73] Reduction of neopterin gave a 7,8-dihydro derivative and several other products, some from aerial oxidation of the dihydro derivative[1215]; hydrogenation over platinum gave the 5,6,7,8-tetrahydro derivative.[428]

Early reports of fluorescent materials from toad skins suggested that there might be several unique pteridines, provisionally named bufochrome-1, -2, and so on,[1624] but by 1958 one had been identified as pterin, two were suspect as nonpteridines, and only bufochrome-1 remained of interest.[220] This later became known simply as *bufochrome* and occasionally as L-neopterin after it had been identified with synthetic 2-amino-6-[L-*erythro*]-α,β,γ-trihydroxypropyl-4(3*H*)-pteridinone (**212b**).[204, 205, 1603]

Although preparations of compound (**212b**) by the Gabriel and Colman syntheses had been reported in 1947 and several years later,[962, 1210] it emerged subsequently that the product was mainly the 7-isomer in each case.[330, cf. 862] The first such synthesis of bufochrome, in which appropriate precautions were taken to force the side chain into the 6-position, was performed in 1958[330] and a modified procedure was later adopted.[1498, cf. 1196, 1346,1556] These results were confirmed by two independent syntheses, one via cyclization of 2,5-diamino-6-*N*-[L-*erythro*-(α,β,γ-trihydroxybutyryl)methyl]amino-4(3*H*)-pyrimidinone with subsequent oxidation[278, 653]; the other by aminolysis of 2-methylthio-6-[L-*erythro*]-α,β,γ-trihydroxypropyl-4(3*H*)-pteridinone.[458] The 5,6,7,8-tetrahydro derivative of bufochrome has been prepared[428] and its (6*R*)- and (6*S*)-diastereoisomers have been separated.[198] A γ-phosphate ester, akin to that (**213b**) from neopterin, has been made.[1216]

In 1946, Max Viscontini and colleagues isolated (from *Pseudomonas roseus fluorescens*) a third isomer of 2-amino-6-α,β,γ-trihydroxypropyl-4(3*H*)-pteridinone,[1423] which they were able to identify immediately as the [L-*threo*] compound (**212c**) by comparison with a synthetic specimen made by a Gabriel and Colman synthesis the previous year by Heinz Rembold and L. Buschmann.[1563] More details of the new product were provided later and it was called *monapterin* after the organism from which it had been isolated.[1616] Shortly afterwards it was identified as a product from *Serratia indica* under the name L-*threo*-neopterin,[262, 593] from *Escherichia coli* (in growth phase),[1626] and from germinating potatoes.[1626] Subsequently, it was made by several variations of the original synthesis but yields of pure material were never very high.[262, 509, 1216, 1496] A γ-phosphorylated derivative (**213a**) has been prepared[1216] as well as a 5,6,7,8-tetrahydro derivative.[428]

(6) *Euglenapterin*: 2-*Dimethylamino*-6-[L-*threo*]-α,β,γ-*trihydroxypropyl*-4(3*H*)-*pteridinone*

In 1976, Erich Elstner and Adelheid Heupel isolated three new pteridines from the phytoflagellate, *Euglena gracilis*.[1625] Two proved to be phosphate derivatives of the parent, which broadly resembled the neopterin family in chromatographic and spectrographic properties.[1625] With the help of Wolfgang Pfleiderer and co-workers, the parent euglenapterin was shown to be 2-dimethylamino-6-[L-*threo*]-α,β,γ-trihydroxypropyl-4(3*H*)-pteridinone (**214a**) by preparing all four possible isomers by the Gabriel and Colman syntheses for comparison.[109, 1710] This aroused great interest among pteridine workers because euglenapterin was the first natural pteridine without a 2-(primary-amino) group or a 2-oxo substituent. Its physical properties have been discussed[335] and

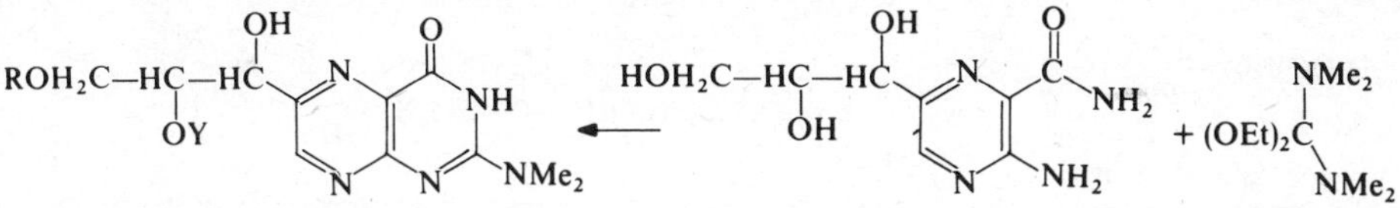

(214a) R = Y = H
(214b) R = H_2O_3P-, Y = H
(214c) R + Y = -OP(O)(OH)O-

(215)

(216)

its structure has been confirmed by an unequivocal synthesis involving cyclization of 3-amino-6-[L-*threo*]-α,β,γ-trihydroxypropyl-2-pyrazinecarboxamide (**215**) with *N,N,N′, N′*-tetramethylurea diethyl acetal (**216**).[985] The phosphorylated pteridines wree shown to be the γ-phosphate ester (**214b**) and the β,γ-cyclophosphate (**214c**).[109, 1710] An x-ray analysis of the parent compound has been reported.[1710]

(7) *Sepiapterin: 2-Amino-6(S)-lactoyl-7,8-dihydro-4(3H)-pteridinone*

The eye pigments of the fruit fly, *Drosophila melanogaster*, have fascinated geneticists for many years (e.g., refs. 1178 and 1642–1644) but their chemistry has proven difficult. As early as 1941, Edgar Lederer suggested that they might contain some pteridines,[1630, cf. 1631] but only in 1951 did a paper chromatographic procedure allow the separation of red from yellow pigments and reveal that a much larger proportion of the yellow material(s) occurred in a *sepia* mutant of the fly.[1632, cf. 894] Hugh Forrest and colleagues proceeded to isolate one of these[191, 880, cf. 1634] and to show that it was a lactoyldihydropterin,[1598] although their suggested formula[1598, 1612] eventually proved to be incorrect. In 1960, Saburo Nawa produced strong evidence for the structure of the recently named[1633] sepiapterin as 2-amino-6-lactoyl-7,8-dihydro-4(3*H*)-pteridinone (**217**),[1221] a formulation (cf. ref. 566) confirmed subsequently by chemical[1217] and nmr evidence,[875] as well as by a synthesis of the racemic material (along with a separable structural isomer) by condensation of 2-amino-7,8-dihydro-4(3*H*)-pteridinone (**219**) with 3-hydroxy-2-oxobutyric acid (**218**) in the presence of zinc chloride.[1222] A more satisfactory synthesis was achieved on autoxidation of 5,6,7,8-tetrahydrobiopterin (**220**) by stirring at pH 4 in air with the exclusion of light for 60 h. After chromatographic separation from other products, the yield of sepiapterin was ~20%.[350, 430, 615] This synthesis appeared to give the natural enantiomer because the product showed a positive Cotton effect in the long wave transition of its cd spectrum,[350] whereas the unnatural enantiomer, obtained by treating the synthetic racemate with the enzyme, sepiapterin reductase,[1222] showed a negative Cotton effect. Since there should be a configurational identity of C-2′ in both the substrate (**220**) and the product (**217**), the absolute configuration of natural sepiapterin at C-2′ appeared to be (*S*)[350, 615]; in fact, this conclusion coincided with that reached much earlier on rather insubstantial evidence.[1634, 1635] Independent confirmation was

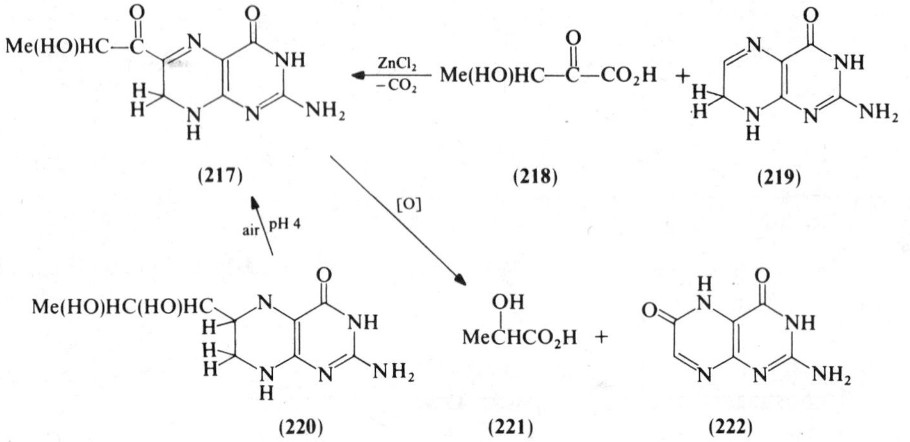

provided by Sadao Matsuura, Takashi Sugimoto, and Modoo Tsusue, who degraded natural sepiapterin (**217**) to xanthopterin (**222**) plus lactic acid (**221**) by stirring it in a borate buffer under air in the dark for 2 days and then assayed the solution for both D- and L-lactic acid. Only the latter was present.[301]

As well as from the sources just listed, sepiapterin has been isolated from the amphibian, *Rhacophorus schlegelii*, under the name rhacophorus-yellow,[1629, cf. 214] from other amphibia,[212, 1460, 1461] from silkworms (initially under the name xanthopterin-B1),[1637, cf. 833] from various fish,[212] from several species of butterfly,[784, 1577–1580, 1600] and from other winged insects.[568, 846, 1063, 1591] The physiology has been discussed.[812]

The mass spectral fragmentation pattern of sepiapterin has been reported in some detail,[976] and its very evident instability to light has been studied.[350, 617] Among its biochemical transformations (mainly beyond the scope of this treatment) may be noted its enzymatic reduction to a tetrahydro analog with subsequent reactions[1062]; its conversion into deoxysepiapterin via dihydrobiopterin[1558]; its catabolic conversion by *Bacillus subtilis* into 2-amino-6-α-carboxyethoxy-4(3*H*)-pteridinone, which retained the (*S*) configuration at Cα[299, 302]; and its deamination by sepiapterin deaminase present in the fat bodies of the silkworm.[1581]

(8) *Deoxysepiapterin*: 2-*Amino*-6-*propionyl*-7,8-*dihydro*-4(3*H*)-*pteridinone*

Of the two major yellow spots on the chromatograms of the extracts of fruit fly eyes,[1632] one proved to be sepiapterin (cited previously); the second spot was not investigated until 1959 when Max Viscontini and E. Möhlmann isolated a yellow crystalline solid and called it isosepiapterin, apparently in the belief that it was isomeric with sepiapterin.[1634, 1635] Despite continuing work on this[566] and on identical material[858, 1612] isolated from a blue-green alga, *Anacystis nidulans,*[1560, 1561] only in 1962 was the true relationship of isosepiapterin to sepiapterin clarified by Hugh Forrest and Saburo Nawa.[567, 875, 1127] In fact, isosepiapterin proved to be deoxysepiapterin, 2-amino-6-propionyl-7,8-dihydro-4(3*H*)-pteridinone (**223**, R = Et), as indicated by nmr work[875] and by its synthesis from (fresh) 2-amino-7,8-dihydro-4(3*H*)-pteridinone (**225**) and 2-oxobutyric acid (**224**, R = Et) in the presence of thiamine, followed by isolation of the product (**223**, R = Et) in 22% yield by chromatography.[1127] Deoxysepiapterin has also been made chemically as a by-product in the aerial oxidation of 5,6,7,8-tetrahydrobiopterin at pH 4 (~1% yield)[350]; by stirring 7,8-dihydrobiopterin in 25% acetic acid under nitrogen for 20 min at 60°C and then allowing the solution to stand in air to oxidize the 5,6-dihydrodeoxysepiapterin formed: yield ~35%[430, cf. 1558] and by homolytic acylation (propionaldehyde–Fe^{2+}–*t*-butylhydroperoxide) of 7-methylthio-2,4-pteridinediamine (**228**) to the 6-propionyl derivative (**227**), which underwent selective hydrolysis in 6*M* hydrochloric acid to 2-amino-7-methylthio-6-propionyl-4(3*H*)-pteridinone (**226**) and subsequent reduction and desulfurization by aluminum–copper alloy in alkaline ethanol to give deoxysepiapterin (**223**, R = Et) in 28% yield after separation from other products (alternative sequences along similar lines were also reported).[95]

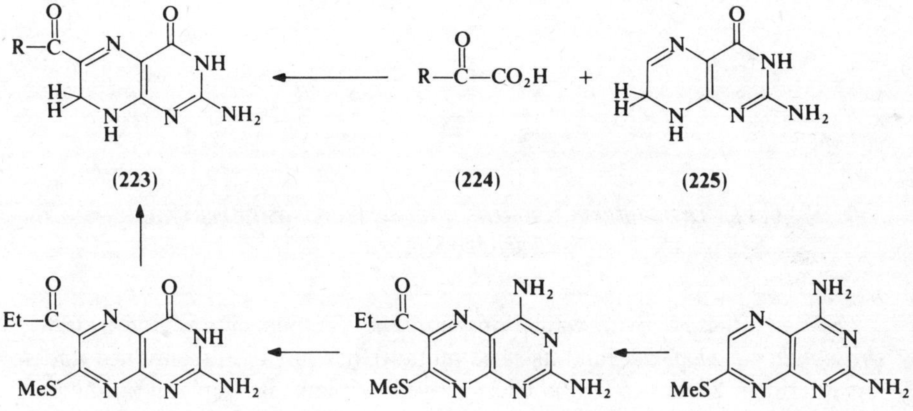

(**223**) (**224**) (**225**)

(**226**) (**227**) (**228**)

(9) *Sepiapterin-C*: 6-*Acetyl*-2-*amino*-7,8-*dihydro*-4(3*H*)-*pteridinone*

A minor pteridine, apparently visible as a third yellow spot on chromatograms of material from the *sepia* mutant of *Drosophila melanogaster*, has been isolated and characterized by uv, ms, nmr, and its chemical behavior as 6-acetyl-2-amino-7,8-dihydro-4(3*H*)-pteridinone (**223**, R = Me), under the rather poor name, sepiapterin-C.[1426] It proved identical[1426] with synthetic material made by treatment of 2-amino-7,8-dihydro-4(3*H*)-pteridinone (**225**) with pyruvic acid (**224**, R = Me) in the presence of thiamine.[1127]

(10) 6-*Aminopterin*: 2,6-*Diamino*-4(3*H*)-*pteridinone*

In 1959, C. Van Baalen and Hugh Forrest reported the isolation of 6-aminopterin from the fruit fly, *Drosophila melanogaster*, and from two blue-green algae, *Anacystis nidulans* and *Nostoc muscorum*.[507] They produced reasonably satisfactory evidence for its structure (**229**) and suggested that it might be an artifact arising by covalent addition of ammonia to 2-amino-7,8-dihydro-4(3*H*)-pteridinone during the isolation procedures, with subsequent oxidation.[226, 507] It was later suggested that the compound was really the 5,8-dihydro isomer (**230**),[1325] but this was withdrawn after further work.[1332] Moreover, independent confirmations of the structure (**229**) emerged in the form of an unambiguous Boon synthesis,[450, 584] a study of the addition of ammonia to authentic 2-amino-7,8-dihydro-4(3*H*)-pteridinone followed by oxidation to (**229**),[453, 584] and oxidation of (**229**) by xanthine oxidase or air to afford 2,6-diamino-4,7(3*H*,8*H*)-pteridinedione (**231**).[1221, 1754] Quantum chemical aspects of 6-aminopterin have been studied and discussed.[1448]

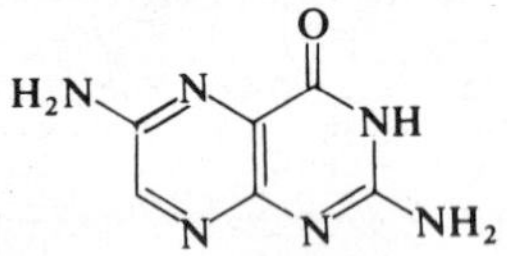

(**229**)

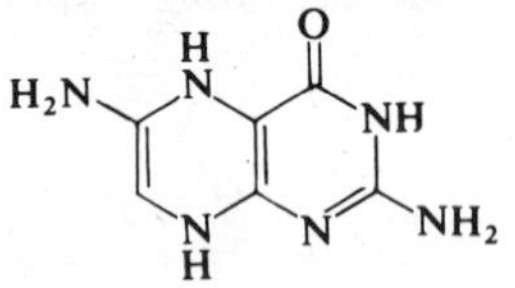

(**230**)

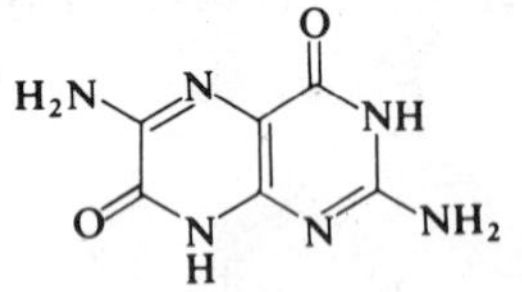

(**231**)

(11) *Drosopterin, Isodrosopterin, Neodrosopterin, and Aurodrosopterin*

The early history of investigations into the eye pigments of the fruit fly, *Drosophila melanogaster* (and its *sepia* mutant), has just been mentioned under sepiapterin in Section 5.D (7). The new yellow pigments (sepiapterin, deoxysepiapterin, and sepiapterin-C) proved reasonably amenable to study [see Sections 5.D (7)–(9)] but the red pigments did not. Thus Max Viscontini and others in Zürich wrestled with the problem for several years from 1955 onwards.[1054] It soon became evident that there were at least three such pigments, all of which gave 2-amino-4-oxo-3,4-dihydro-6-pteridinecarboxylic acid (ranachrome-3) on oxidation: these were named drosopterin, isodrosopterin, and neodrosopterin.[1638] The optical rotations of drosopterin and isodrosopterin were approximately equal but opposite in sign, while neodrosopterin showed no optical rotation[508, 1639, 1640]; Isodrosopterin was obtained quite pure as fine orange colored acicular plates[1640] and all three gave rather similar elemental analyses, akin to that of biopterin.[508] On the available evidence, they were allotted isomeric biopterinlike formulas.[1635] However, these structures did not explain their long wavelength absorption at >500 nm, some of their anomalous reactions, or the formation of all three when 2-amino-7,8-dihydro-4(3*H*)-pteridinone was allowed to stand at room temperature with 3-hydroxy-2-oxobutyric acid or with 2-hydroxy-3-oxobutyric acid.[309, 567, 1429]

By 1969, these formulations came under reasoned criticism from Wolfgang Pfleiderer,[351] who suggested a dimeric constitution for drosopterin and its isomers, along the lines of a 7,7′-bipteridinyl with two α-hydroxypropionyl side chains at the 6- and 6′-position, for example, (**232**); this idea was welcomed by other workers.[1428] There followed some quite intense studies at Konstanz, initially on drosopterin and isodrosopterin,[431–434, 1636] and later on neodrosopterin and aurodrosopterin,[558, 614, 1641] of which the last has escaped much chemical attention since its isolation from a rare *Drosophila* mutant[1642–1644]: the bipteridinyl-type structure (**232**) was discarded in favor of a 6,6′-dipteridinylmethane basis, on which drosopterin was formulated as (**233**) and the others along similar lines. This work included the synthesis of a mixture of drosopterin and isodrosopterin by the action of ethyl 2-acetoxy-3-oxobutyrate on 2-amino-6-hydroxy-5,6,7,8-tetrahydro-4(3*H*)-pteridinone in which it was envisaged that the terminal (C-4) methyl group of the ester formed the bridging carbon between the two pteridine nuclei of (**233**).[433] However, by using (^{13}C-4)-tagged ester and subsequently oxidizing the product to 2-amino-4-oxo-3,4-dihydro-6-pteridinecarboxylic acid (**234**), which proved to be devoid of the tag, it became quite evident that the terminal methyl group did *not* form the bridging carbon.[1280] This called into question the structure (**233**) and subsequent nmr and other spectral work on the parent and specifically methylated derivatives suggested a new basic structure for drosopterin (and its isomers), viz. the

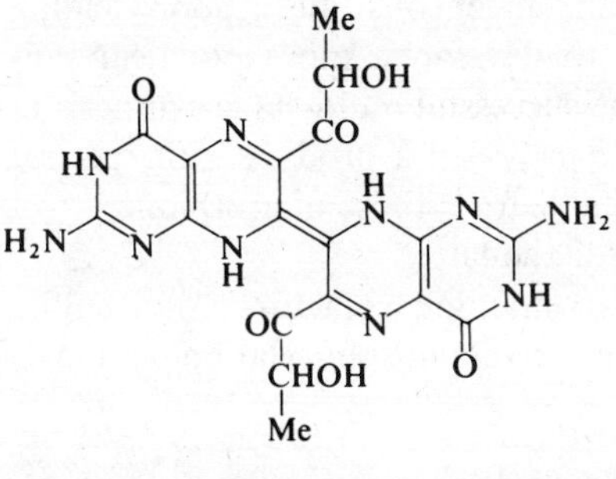

(**232**)

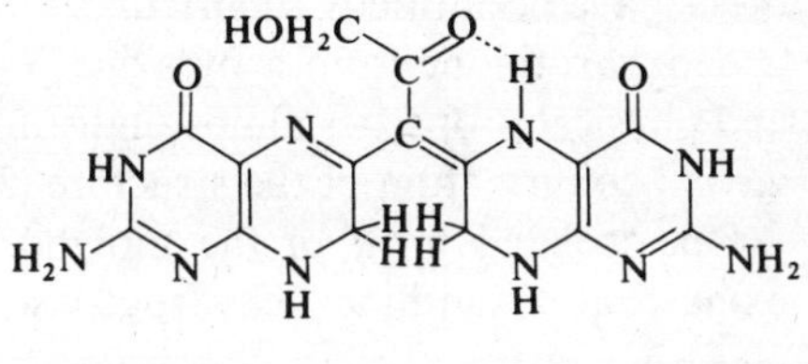

(**233**)

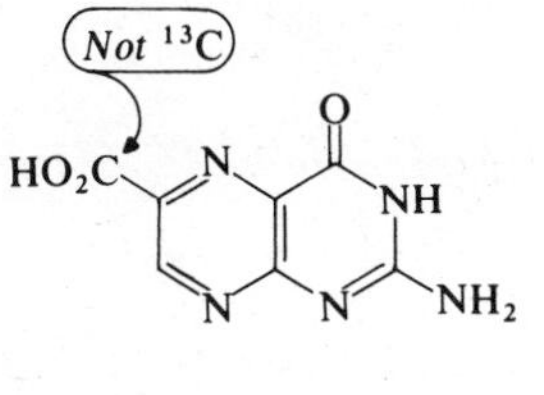

(**234**)

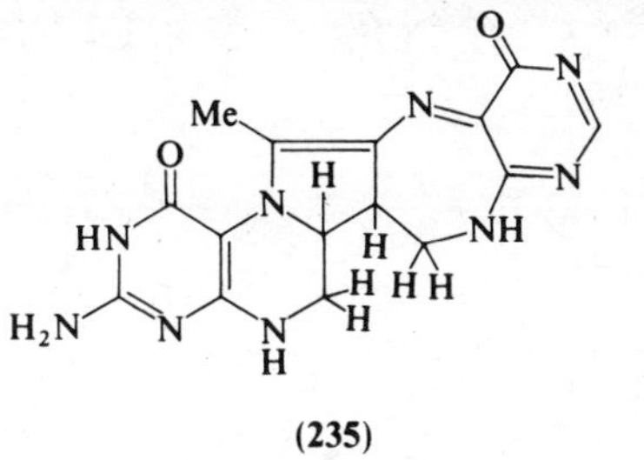

(**235**)

pentacyclic system (**235**).[491, 492] Although the current formulation appears to account broadly for the spectral properties, chemical reactions, and ionization data of the drosopterin family of compounds, only time will tell whether it is the final answer. Some preliminary synthetic work has been briefly reported.[1699]

Drosopterin, isodrosopterin, and neodrosopterin have also been isolated from the dewlaps of certain Puerto Rican lizards of the genus *Anolis*[319] and have been identified in products from salamandas, newts, frogs, and one toad.[1477]

(12) *Urothione*: 2-*Amino*-7-α,β-*dihydroxyethyl*-6-*methylthio*-4(3*H*)-*thieno*[3,2-*g*]*pteridinone*

In 1940, Walter Koschara described the isolation from human urine of an orange-red pteridine derivative that contained sulfur and was optically active.[1646] Its spectrum, chemical properties, and analysis (which, with hindsight, was correct except for two extra hydrogen atoms) suggested that urothione was a pteridine bearing an α,β-dihydroxyethyl side chain and an additional $C_3H_{3-5}S_2$ portion.[1646–1648] In 1955, Rudolf Tschesche and colleagues synthesized several simple derivatives of the 2-amino-7,8-dihydro-4(3*H*)-thieno[2,3-*g*]pteridine system (**236**), which spectrally so resembled natural urothione, that the formula (**236**, R = $CHOHCH_2OH$, Y = SMe) was suggested for the latter.[495, 503, 1295] However, this was incorrect and only in 1967–1969 did Miki Goto, Atsushi Sakurai, and their colleagues show that urothione was 2-amino-7-α,β-dihydroxyethyl-6-methylthio-4(3*H*)-thieno[3,2-*g*]pteridinone (**237**), based on the first correct analysis of the natural material, its spectra, its chemical reactions, and an ingeneous but poor-yielding synthesis.[906, 1028, 1262, 1427, 1645] Attempts to develop a more practical synthesis have been briefly reported.[624, 625] The successful multi-stage synthesis of 2′-deoxyurothione from a pyrazine *N*-oxide provided a product with spectral and physical properties almost indistinguishable from those of urothione,[1649] a virtual confirmation of the structure (**237**) for the latter.

One isolated form of the molybdenum-containing cofactor[1650] of sulfite oxidase and xanthine dehydrogenase gave, on dephosphorylation and oxi-

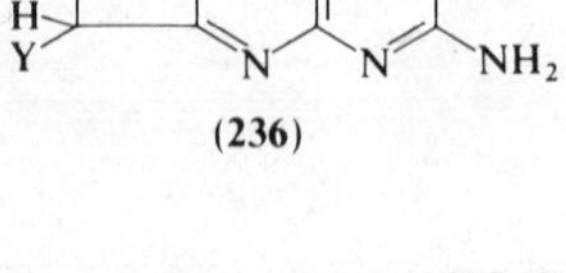

(236)

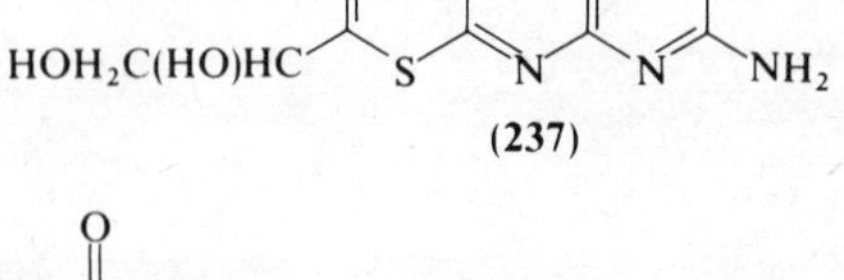

(237)

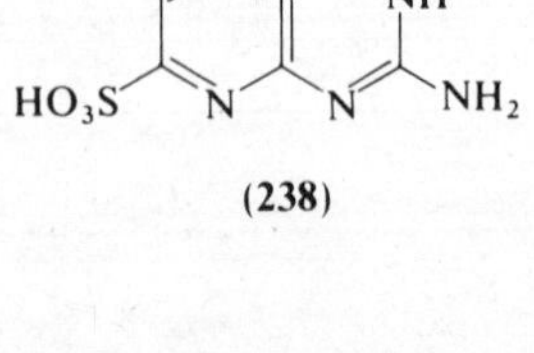

(238)

dation, 2-amino-4-oxo-7-sulfo-3,4-dihydro-6-pteridinecarboxylic acid (**238**),[1526] which was the same product as that derived from the oxidation of urothione.[906] This may provide a clue to the origin of the latter in urine.[1526]

E. The Lumazine Group

(1) *Lumazine*: 2,4(1*H*,3*H*)-*Pteridinedione*

Contrary to popular belief, lumazine (**239**, R=H) is a most uncommon natural product. However, it has been isolated from male ants of the species, *Formica polyctena*, along with several of its derivatives.[1041, cf. 1244, 1258]

Lumazine was synthesized long before it was isolated from nature. In 1894, Otto Kühling oxidized the benzo[*g*]pteridine, tolualloxazin, to 2,4-dioxo-1,2,3,4-tetrahydro-6,7-pteridinedicarboxylic acid (**239**, R=CO_2H), which gave lumazine on decarboxylation (see Ch. III, Sect. 3)[7, 8]; in 1907, 2,3-pyrazinedicarboxamide (**240**) was converted by sodium hypobromite into lumazine (see Ch. III, Sect. 1.E)[902]; in 1937, the first Gabriel and Colman synthesis of lumazine from 5,6-diamino-2,4(1*H*,3*H*)-pyrimidinedione and glyoxal was performed,[1012] followed by improved procedures[30, 367, 747, 1438] (see Ch. II, Sect. 1.A); and it has been made in several other ways, for example, by oxidation of the covalent 3,4-hydrate (**241**) of 2(1*H*)-pteridinone,[134] by treatment of 2-amino-4(3*H*)-pteridinone with nitrous acid,[471] or by oxidative hydrolysis of 2-thioxo-1,2-dihydro-4(3*H*)-pteridinone (**242**) by ammoniacal hydrogen peroxide.[399, 1169] Lumazine undergoes degradation to pyrazines and all the other reactions expected of pteridinones (Sect. 2.A–G). The name, lumazine, was first used by Richard Kuhn and Arthur Cook in 1937 to avoid the previously used confusing name alloxazin.[1012] The new name appears to have been based on the marked fluorescence of their specimen in solution but it should be noted that highly purified lumazine has only a very mild fluorescence in solution and, of course, no obvious luminescent properties.[9]

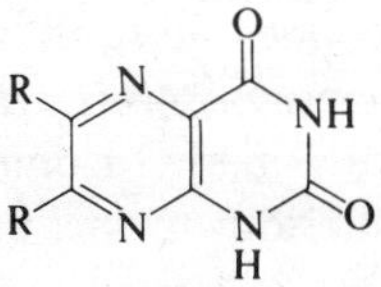

(239)

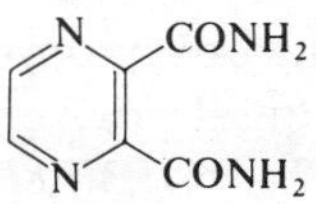

(240)

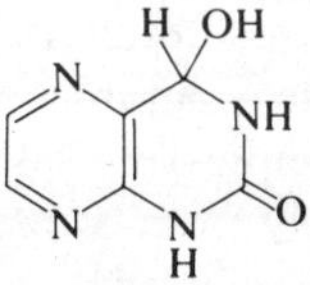

(241)

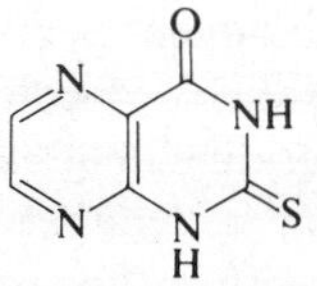

(242)

(2) 6-*Hydroxymethyllumazine*: 6-*Hydroxymethyl*-2,4(1*H*,3*H*)-*pteridinedione*

In 1966, Katsura Sugiura and Miki Goto isolated a simple pteridine from spinach (*Spinacea oleracea*) in a yield of 1 mg/25 kg of leaves.[1031] The compound underwent oxidation to 2,4-dioxo-1,2,3,4-tetrahydro-6-pteridinecarboxylic acid (**243**, R = CO_2H) and its uv spectrum closely resembled that of 6-methyl-2,4(1*H*,3*H*)-pteridinedione (**243**, R = Me). Its structure, 6-hydroxymethyl-2,4(1*H*,3*H*)-pteridinedione (**243**, R = CH_2OH), was confirmed by a Gabriel and Colman synthesis from 5,6-diamino-2,4(1*H*,3*H*)-pteridinedione and dihydroxyacetone in the presence of cystein, followed by chromatographic purification.[1031, 1546] It was also confirmed by the alkaline hydrolysis of authentic 4-amino-6-hydroxymethyl-2(1*H*)-pteridinone.[1546] Two phosphate esters have been prepared.[1546]

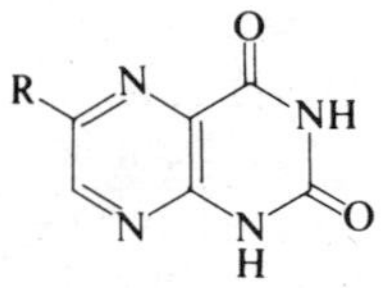

(243)

(3) *Leucettidine*: 6-(*S*)-α-*Hydroxypropyl*-1-*methyl*-2,4(1*H*,3*H*)-*pteridinedione*

In 1981, a novel optically active pteridine was isolated from the calcareous sponge, *Leucetta microraphis*, and on the basis of its analysis and its ms, uv, and nmr spectra, it was allotted a 6-α-hydroxypropyl-3-methyllumazine structure; by comparison with related optically active compounds, the configuration at Cα was probably (*S*).[751] However, an anomaly in its uv spectrum suggested to Wolfgang Pfleiderer that the methyl group may have been misplaced.[354] The matter was resolved by unambiguous syntheses of both the 1- and 3-methyl isomers: Thus 1-methyl-2,4,7(1*H*,3*H*,8*H*)-pteridinetrione was converted successively into 7-chloro-1-methyl-, 1-methyl-7-thioxo-7,8-dihydro-, and 1-methyl-7-methylthio-2,4(1*H*,3*H*)-pteridinedione that underwent 6-propionylation and subsequent reduction and desulfurization by aluminum–copper alloy in an alkaline medium, to afford 6-α-hydroxypropyl-1-methyl-2,4(1*H*,3*H*)-pteridinedione (**244**), identical with the natural product; the 3-methyl isomer, prepared rather similarly from 3-methyl-2,4,7(1*H*,3*H*,8*H*)-pteridinetrione, differed from the natural product.[354]

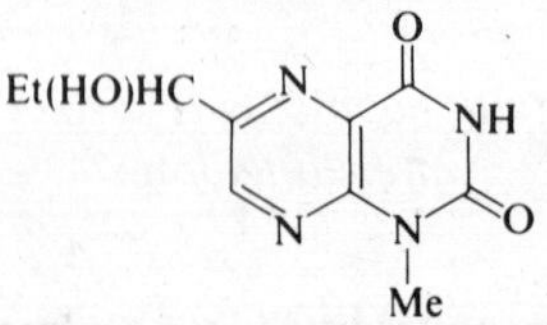

(244)

(4) 6-*Lumazinecarboxylic Acid*: 2,4-*Dioxo*-1,2,3,4-*tetrahydro*-6-*pteridinecarboxylic Acid*

Like lumazine, its 6-carboxylic acid (**243**, R$=CO_2H$) was first isolated from nature when Max Viscontini and G. Schmidt extracted it from the ant, *Formica polyctena*, in 1967.[1041] Its structure became evident on comparison with synthetic material made much earlier by the hydrolysis of 2-amino-4-oxo-3,4-dihydro-6-pteridinecarboxylic acid,[70, 178] or by permanganate oxidation of pteroylglutamic acid.[655]

(5) *Dimethylribolumazine*: 6,7-*Dimethyl*-8-D-*ribityl*-2,4(3*H*,8*H*)-*pteridinedione*

Being established as a precursor in the biosynthesis of riboflavine,[1652, 1658, 1659, 1684] dimethylribolumazine (**245**) (sometimes called ribolumazine) naturally has a large literature. It was isolated first in 1955 by Toru Masuda, who obtained it as compound-G from the wet mycelium of the mould, *Eremothecium ashbyii*,[1665] and soon recognized it (from its analysis, spectra, and chemistry) as a dimethyllumazine bearing a ribityl group.[1651, 1654] The suspected location of the sugar at N-8 was confirmed subsequently by a synthesis of the 6,7,8-trimethyl-2,4(3*H*,8*H*)-pteridinedione for uv spectral comparison.[290, 1027] There followed a spate of syntheses of 6,7-dimethyl-8-D-ribityl-2,4(3*H*,8*H*)-pteridinedione (**245**), all involving condensation of biacetyl with 5-amino-6-D-ribitylamino-2,4(1*H*,3*H*)-pyrimidinedione obtained by in situ reduction of 5-nitroso-,[172, 1437, 1543, 1652] 5-nitro-,[164, 165, 422] or 5-phenylazo-6-D-ribitylamino-2,4(1*H*,3*H*)-pyrimidinedione.[378]

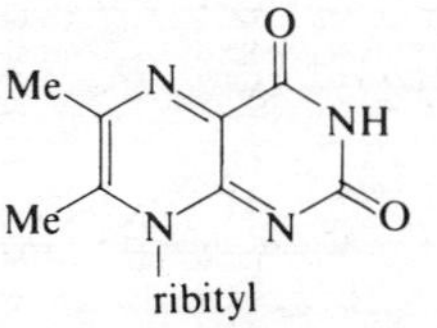

(**245**)

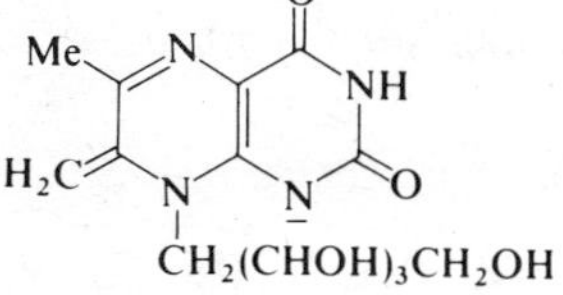

(**245a**)

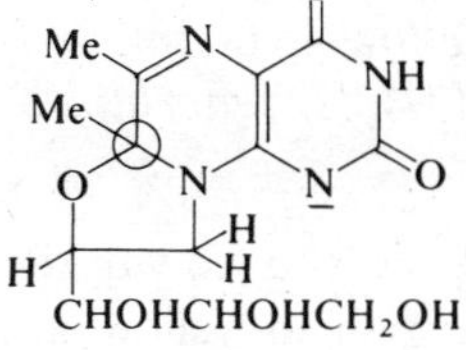

(**245b**)

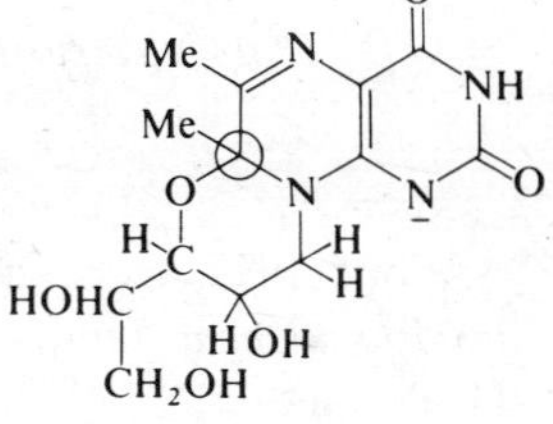

(**245c**)

Dimethyribolumazine has also been isolated from the mold, *Ashbya gossypii*,[1437, 1653] and its role in the biosynthesis of riboflavine has been investigated extensively.[422–424, 537, 1437, 1652] It is interesting that none of the many synthetic analogs of dimethylribolumazine can be converted enzymatically to riboflavine like products.[537]

A very detailed examination of the ^1H and ^{13}C nmr spectra of dimethylribolumazine (**245**) as anion (along with those of model compounds) has revealed that it exists in alkaline solution in five main forms: there is one in the methylene form (**245a**) (17%); there are two C-7-stereoisomeric internal adducts with a five-membered ring (**245b**) (together, 62%); and there are two C-7-stereoisomeric internal adducts with a six-membered ring (**245c**) (together, 21%).[1686]

(6) *Sepialumazine*: 6-*Lactoyl*-7,8-*dihydro*-2,4(1*H*,3*H*)-*pteridinedione*

The xanthopterin-B, isolated by Yoshimasa Hirata from the lemon mutant of the silkworm (*Bombyx mori*) in 1950,[1223, cf. 671] eventually proved to be a mixture of xanthopterin-B1 (sepiapterin) and xanthopterin-B2.[833, 1637, 1656] The latter was formulated initially as an *N*-methylated lactoyldihydropterin[1425] but this structure was revised the following year to 6-lactoyl-7,8-dihydro-2,4(1*H*,3*H*)-pteridinedione (**246**), the 2-oxo analog of sepiapterin, on the basis of spectra and degradation experiments.[907] The discovery of a sepiapterin deaminase in silkworms explained the occurrence of both analogs side by side.[1581] In 1980, the name sepialumazine was introduced for the L-stereoisomer of (**246**), this time isolated from the *kiuki* (i.e., faint yellow) mutant of *B. mori*,[1655] and its structure was rigorously confirmed by nmr, uv, ms, ord, and chemical means[1655, 1657]; no chemical synthesis of sepialumazine appears to have been reported.

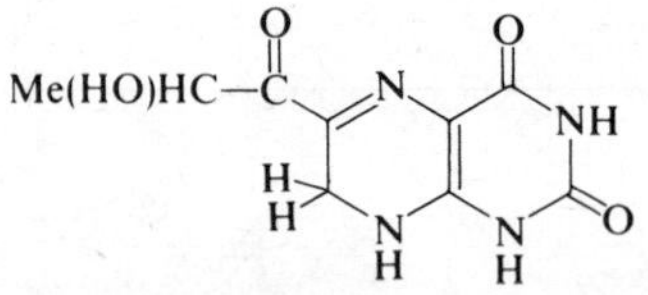

(**246**)

(7) *Russupteridine-Yellow*-1: 6-*Amino*-7-*formamido*-8-D-*ribityl*-2,4(3*H*,8*H*)-*pteridinedione*

About 1970, Conrad Eugster and co-workers realized that toadstools of the *Russulaceae* family were rich in yellow and red pteridines.[1660] Among those isolated[1400, 1599, 1660–1662] was russupteridine-yellow-1 [100 mg from *R. sardonia* (180 kg) or from *R. paludosa* (20 kg)] which, on analysis and examination by uv, ir, cd, ^1H nmr, and ^{13}C nmr, emerged as a most unusual structure, 6-amino-7-formamido-8-D-ribityl-2,4(3*H*,8*H*)-pteridinedione (**247a**), formulated by the workers as the tautomer (**247b**).[1400] No synthesis has been reported.

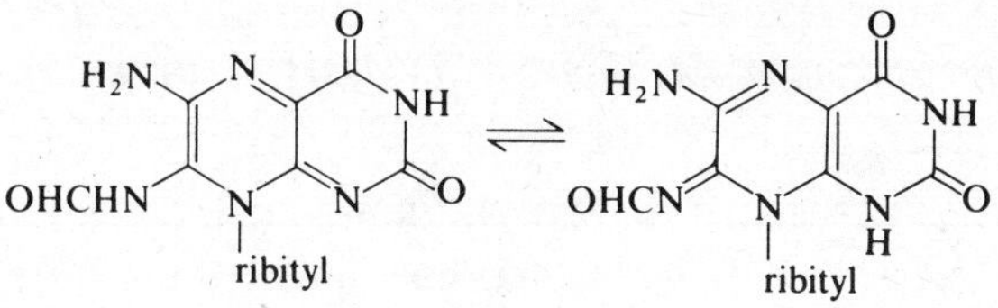

(**247a**) (**247b**)

F. The Violapterin Group

(1) *Violapterin*: 2,4,7(1*H*,3*H*,8*H*)-*Pteridinetrione*

As part of a broader study on the pteridines in honey-bees,[571, 786, 1563, 1564, 1663] Heinz Rembold isolated a pteridine in 1963, not hitherto reported to occur in nature, which he called violapterin and which subsequently turned out to be 2,4,7(1*H*,3*H*,8*H*)-pteridinetrione (**249**, R = H) by comparison with an authentic sample.[1197] Several years later, the same material was isolated from the ant, *Formica polyctena*, and an alternative name, isoxantholumazine,* was suggested.[1041]

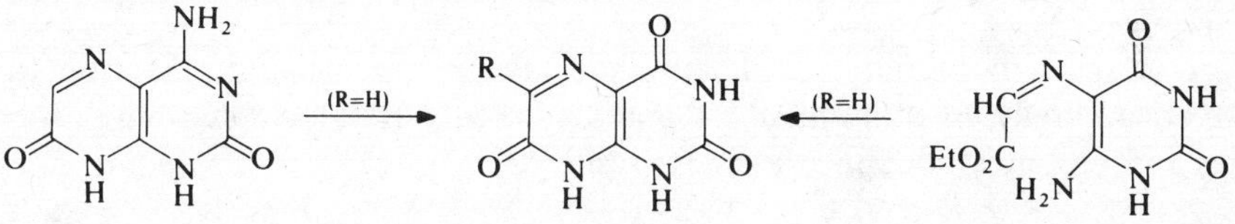

(**248**) (**249**) (**250**)

As pointed out previously (Ch. II, Sect. 1), violapterin (**249**, R = H) was made by Friedrich Wöhler from uric acid in 1857,[1509] although this was not entirely clear until a century later.[345] It has also been synthesized by the alkaline hydrolysis of 4-amino-2,7(1*H*,8*H*)-pteridinedione (**248**),[41] by the acidic hydrolysis of 2-amino-4,7(3*H*,8*H*)-pteridinedione,[1267] by the decarboxylation of 2,4,7-trioxo-1,2,3,4,7,8-hexahydro-6-pteridinecarboxylic acid (**249**, R = CO_2H),[1304]

* There is a certain logic in such a change, in that violapterin is *not* a pterin. However, the old name does suggest in a poetic way the marked violet fluorescence of the compound, whereas the new name suggests a luminescent yellow compound that is quite untrue. It seems a pity to complicate the literature further by changing established names, which in any case are of only pragmatic utility and have no real place in organic chemical nomenclature.

by cyclization of 6-amino-5-ethoxycarbonylmethyleneamino-2,4(1*H*,3*H*)-pyrimidinedione (**250**),[339] and by the in vitro oxidation of lumazine by mammalian xanthine oxidase.[98, 797] The ionization, as well as the uv,[979] nmr,[716] and ms properties[1411] have been reported.

(2) *Luciopterin*: 8-*Methyl*-2,4,7(1*H*,3*H*,8*H*)-*pteridinetrione*

In 1968, a group of scientists in Nagoya showed that extracts of the Japanese firefly, *Luciola cruciata*, contained a pteridine as well as the more usual luciferin [2-(6′-hydroxybenzothiazol-2′-yl)-4,5-dihydro-4-thiazolecarboxylic acid].[1538] Analysis and spectral investigation of the pteridine suggested an *N*-methylpteridinetrione and comparison with authentic synthetic materials revealed that luciopterin was 8-methyl-2,4,7(1*H*,3*H*,8*H*)-pteridinetrione (**251**).[1538] The synthetic material (**251**) was made by a (unambiguous) Gabriel and Colman synthesis from 5-amino-6-methylamino-2,4(1*H*,3*H*)-pyrimidinedione and ethyl glyoxylate hemiacetal[339]; it has also been made by the methylation of 2,4,7(1*H*,3*H*,8*H*)-pteridinetrione[344] and its properties are well documented.[317, 716]

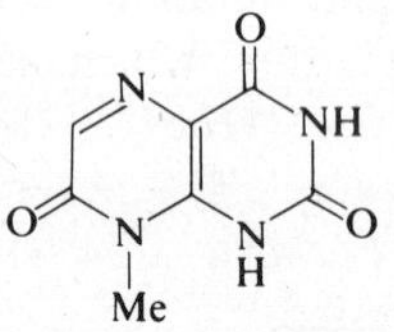

(251)

A (suspected) pteridine, isolated under the name luciferesceine from an American firefly, *Photinus pyralis*, differed markedly from luciopterin in properties and probably contained a ribityl residue.[1664]

(3) *Photolumazine-C*: 8-D-*Ribityl*-2,4,7(1*H*,3*H*,8*H*)-*pteridinetrione*

Although 8-D-ribityl-2,4,7(1*H*,3*H*,8*H*)-pteridinetrione (**252**, R = H) had been synthesized in 1960 by an unambiguous Gabriel and Colman procedure,[172] it was not until 1972 that it was isolated in Japan from cultures of *Pseudomonas*

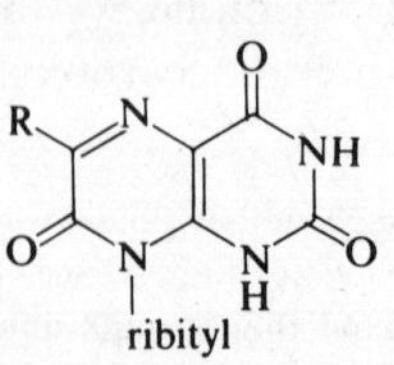

(252)

ovalis and identified by comparison with synthetic material (made as previously described but differing somewhat in melting point).[463, cf. 1673] In the following year, the same material was extracted independently from *Photobacterium phosphoreum* by one group[295] in Nagoya and by another[1430] in Tokyo: It was named photolumazine-C by one of these.[1430] In 1984, it was also isolated from several species of *Russulaceae* (toadstools).[1400]

(4) *Compound-V*: 6-*Methyl*-8-D-*ribityl*-2,4,7(1*H*,3*H*,8*H*)-*pteridinetrione*

In 1955, Toru Masuda isolated a violet-fluorescent pteridine from the mould (yeast), *Eremothecium ashybii*, which he called compound-V.[1665] Over the next few years, structural evidence was gradually collected[1666, 1668] until it was possible to confidently advance the structure (**252**, R = Me),[1109, 1110] subsequently proven by synthesis from 5-amino-6-D-ribitylamino-2,4(1*H*,3*H*)-pyrimidinedione and pyruvic acid.[1544, cf. 1667] An alternative name 6-methyl-7-hydroxyribolumazine,[1027] has not been used in subsequent literature. Meanwhile, Hugh Forrest and Walter McNutt had arrived at a similar structure (**252**, R = Me) for their compound-A from the same mould,[1082, 1669, 1670] as did Gerhard Plaut and Gladys Maley for one of their fluorescent compounds from *Ashbya gossypii*.[1441, 1667] Compound-V was subsequently isolated from the ant, *Formica polyctena*,[1041] from *Pseudomonas ovalis*,[463, 1218] from *Photobacterium phosphoreum*,[295, 1430] and from several *Russula* toadstools.[1400, 1599] Modified procedures for the synthesis of compound-V from 5-amino-6-ribitylamino-2,4(1*H*,3*H*)-pyrimidinedione with pyruvic acid (21% yield),[422] methyl pyruvate (33%),[378] or ethyl pyruvate (59%),[674] have been reported. Recorded optical rotations have varied from +2.4 to +11.45°.[378]

(5) *Photolumazine-B*: 6-*Hydroxymethyl*-8-D-*ribityl*-2,4,7(1*H*,3*H*,8*H*)-*pteridinetrione*
Photolumazine-A: 6-L-α,β-*Dihydroxyethyl*-8-D-*ribityl*-2,4,7(1*H*,3*H*,8*H*)-*pteridinetrione*

Among other products isolated from *Photobacterium phosphoreum* in 1973 were photolumazine-B[1430] and photolumazine-A (also called photopterin-A).[295, 1430]

Reduction of *photolumazine-B* gave compound-V (**252**, R = Me) and some photolumazine-C (**252**, R = H). The uv spectra of all three compounds were very similar and the structure was confirmed by the reaction of 5-amino-6-D-ribitylamino-2,4(1*H*,3*H*)-pyrimidinedione with hydroxypyruvic acid to give 6-hydroxymethyl-8-D-ribityl-2,4,7(1*H*,3*H*,8*H*)-pteridinetrione, identical with photolumazine-B (**253**).[1430]

Similar reduction of *photolumazine-A* gave only photolumazine-C (**252**, R = H)[1430]; periodate oxidation followed by borohydride reduction gave 8-β-

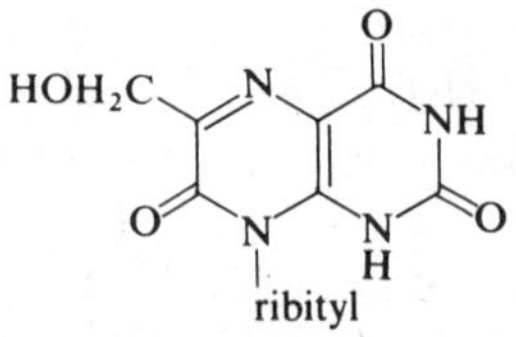

(253)

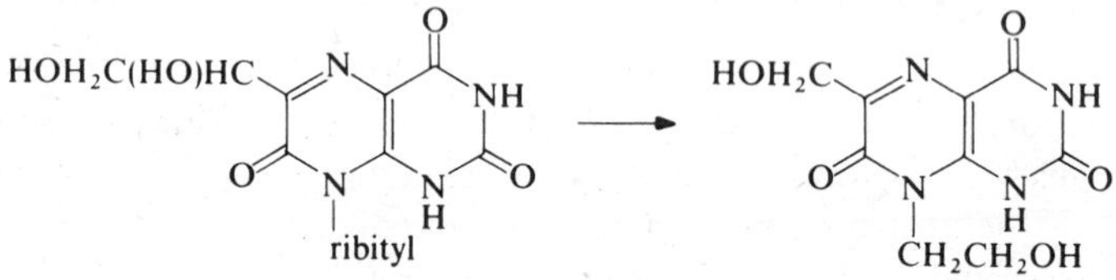

(254) (255)

hydroxyethyl-6-hydroxymethyl-2,4,7(1*H*,3*H*,8*H*)-pteridinetrione (**255**), consistent with 6-α,β-dihydroxyethyl and 8-ribityl substituents[295, 1430]; the uv and nmr spectra were also consistent with such a substituted 2,4,7-pteridinetrione[295]; and syntheses from 5-amino-6-D-ribitylamino-2,4(1*H*,3*H*)-pteridinedione with α-oxo-D-erythronate or α-oxo-L-erythronate showed that photolumazine-A was 6-L-α,β-dihydroxyethyl-8-D-ribityl-2,4,7(1*H*,3*H*,8*H*)-pteridinetrione (**254**).[1430] However, it has been suggested on ord (optical rotatory dispersion) evidence that the natural material may have a 6-DL-dihydroxyethyl side chain, that is, it is racemic.[295]

(6) *Putidolumazine*: 6-β-*Carboxyethyl*-8-D-*ribityl*-2,4,7(1*H*,3*H*,8*H*)-*pteridinetrione*

A new ribitylpteridinecarboxylic acid was isolated first from *Pseudomonas putida* in 1970 by Akira Suzuki and Miki Goto, who called it putidolumazine[1248]; in the following year, the same workers isolated it also from *P. ovalis.*[1218, cf. 463] Its spectra and chemical reactions suggested that it was 6-β-carboxyethyl-8-D-ribityl-2,4,7(1*H*,3*H*,8*H*)-pteridinetrione (**256**),[1248] a structure proven by synthesis from 5-amino-6-D-ribitylamino-2,4(1*H*,3*H*)-pyrimidinedione with 2-oxoglutaric acid.[1218]

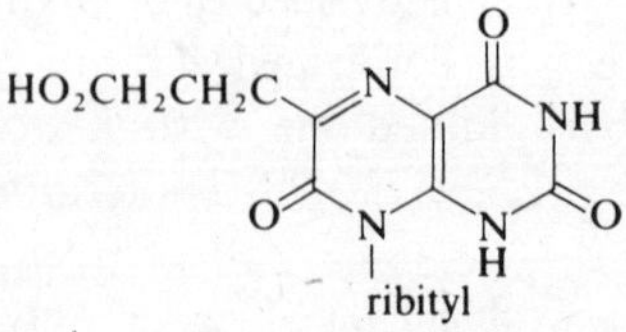

(256)

(7) *6-Indol-3′-yl-8-*D*-ribityl-2,4,7(1H,3H,8H)-pteridinetrione*

During a study of soil microorganisms,[1675] Isao Takeda noticed (in 1968) the formation of two fluorescent products during the growth of certain mutant strains of *Achromobacter petrophylum* on hexadecane in the presence of purine bases and the usual trace requirements.[1671] They were soon recognized as 6-substituted-8-ribitylpteridinetriones[1672] and one was shown subsequently by analysis, spectra, and reactions to be the 6-indol-3′-yl derivative (**257**), a diagnosis promptly confirmed by a synthesis from 5-amino-6-ribitylamino-2,4(1*H*,3*H*)-pyrimidinedione with 3-oxaloindole.[1673, 1674] The structure of the other compound was deduced to be the corresponding 6-*p*-hydroxyphenyl analog (**258**), partly by biochemical incorporation experiments.[1674] This was confirmed later by synthesis from this pyrimidine with *p*-oxalophenol.[1218]

The compounds (**257**) and (**258**) were also isolated from the culture medium of *Pseudomonas ovalis*, along with other pteridines.[1218]

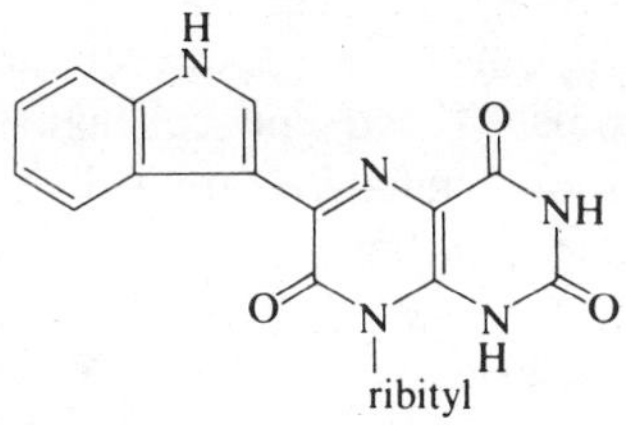

(**257**)

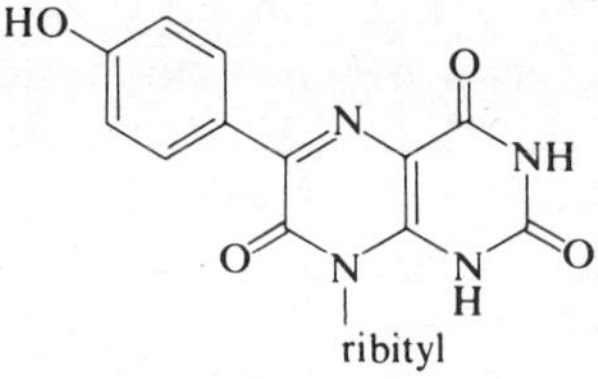

(**258**)

(8) *Russupteridine-Yellow-4*: *4-*D*-Ribityl-1H-imidazo[4,5-g]pteridine-2,6,8(4H,5H,7H)-trione*

Among the other products isolated by Conrad Eugster and others in 1979 from the toadstool, *Russula sardonia*, was russupteridine-yellow-4 (10 mg from 120 kg of material).[1662] Spectral examination and analysis suggested a linear tricyclic molecule, 4-D-ribityl-1*H*-imidazo[4,5-*g*]pteridine-2,6,8(4*H*,5*H*,7*H*)-trione (**259**), a structure proven by synthesis from 5-amino-6-D-ribitylamino-2,4(1*H*,3*H*)-pyrimidinedione with parabanic acid.[1400, 1662] Two related pteridines, russupteridine-yellow-2 and -5, have been isolated from toadstools without full structural elucidation to date.[1400]

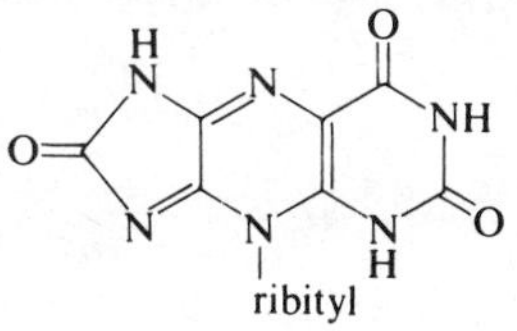

(**259**)

G. Other Natural Pteridines

(1) *Pteroylglutamic Acid and Related Molecules*

As mentioned in the introduction to Section 5 in this chapter, a treatment of folic acid(s) and *folates* is considered to be beyond the scope of this book. However, the *folic acid-type synthesis* has been discussed in some detail in Chapter II, Sect. 1.B(3)(d). For all other information, the following references should prove adequate: general biochemistry, therapeutics, and biology[1572]; chemistry[1419, 1476]; historical aspects[18, 26, 251, 556, 913]; rhizopterin[1187, 1362]; methanopterin[1002, 1459]; sarcinapterin[1459]; and brief overviews.[25, 822, 920, 1021, 1536]

(2) *Riboflavine and Related Molecules*

Like the folates, riboflavine is considered outside the scope of this book. For a general overview, the article of Hamish Wood and colleagues is unsurpassed.[1135] Details of the biosynthesis of riboflavine are still being investigated.[1684] Riboflavine and related molecules have some fascinating aspects.[230, 231, 665, 681, 809, 919, 948, 996, 1015, 1068, 1108, 1114, 1160, 1282, 1283, cf. 734]

(3) *Surugatoxin*

The investigation of a new marine toxin followed a serious outbreak of distressing symptoms (akin to those caused by atropine) in people who had eaten a quite rare carnivorous gastropod, *Babylonia japonica*, gathered in Suruga Bay during 1965. The crystalline toxin (10 mg from 1 kg of a mid-gut gland in the organism) was analyzed, examined spectrally, and finally submitted to x-ray analysis. The complex bromine-containing structure (**260**) appears to be well supported by the brief details available.[1676]

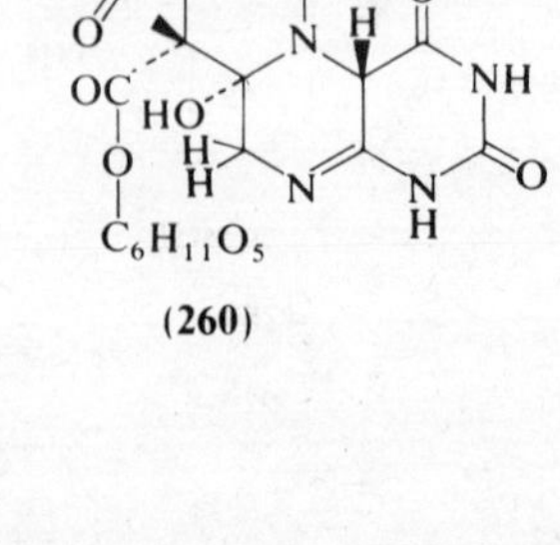

(**260**)

(4) 2,4,6(1*H*,3*H*,5*H*)-*Pteridinetrione and Its* 7,8-*Dihydro Derivative*

In 1971, Jost Dustmann reported that the eyes, hemolymph, gut, and feces of honeybees yielded 2,4,6(1*H*,3*H*,5*H*)-pteridinetrione and its 7,8-dihydro derivative, as well as leucopterin; they were identified with synthetic materials by their uv spectra, fluorescences, and chromatographic R_f values.[179]

(5) *The Acrasin of a Cellular Slime Mould*

The factor (or acrasin) inducing cell-aggregation of the slime mould, *Dictyostelium lacteum*, has been partially purified and identified on spectral and biochemical grounds as a derivative of 2-amino-4(3*H*)-pteridinone; the property is shared by several natural pteridines such as biopterin and neopterin.[1433] However, the acrasin is species-specific: The related mould, *Polyphondylium violaceum*, which has its own nonheterocyclic acrasin, is unaffected by these pteridines and vice versa.[1677]

CHAPTER VII

Alkoxypteridines, *N*-Alkylpteridinones, and Pteridine *N*-Oxides

Alkoxypteridines and *N*-alkylpteridinones may be envisaged as tautomeric pteridinones that have been fixed in one or another of their tautomeric forms by *O*- or *N*-alkylation. Thus the 4-pteridinone, usually written as 4(3*H*)-pteridinone (**2**, R = H), exists in a tautomeric equilibrium with 4-hydroxypteridine (**1**, R = H), 4(1*H*)-pteridinone (**3**, R = H), and 4(8*H*)-pteridinone (**4**, R = H). All four forms can be fixed by alkylation (or arylation) to give, for example, the methylated derivatives (**1**–**4**; R = Me). Likewise, both 2- and 7-pteridinone can each give rise to four monoalkylated derivatives but, because of the arrangement of nitrogen atoms in pteridine, 6-pteridinone can give rise to only two such derivatives, for example, (**5**, R = Me) and (**6**, R = Me). Fortunately, these alkoxypteridines and *N*-alkylpteridinones need not necessarily be made by alkylation of the parent (tautomeric) pteridinones.

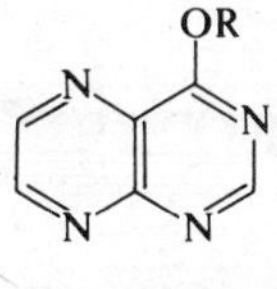

(**1**)

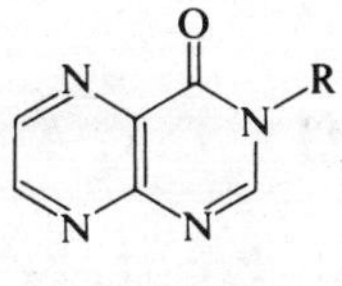

(**2**)

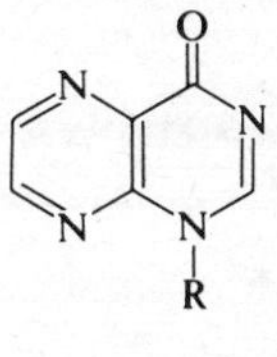

(**3**)

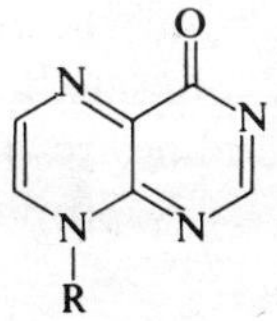

(**4**)

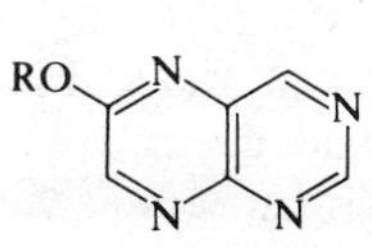

(**5**)

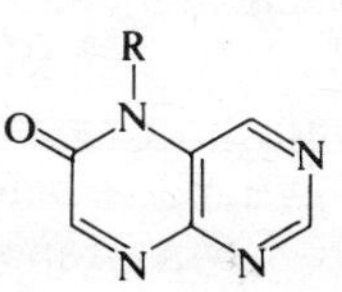

(**6**)

The pteridine *N*-oxides can exist as their *N*-hydroxy tautomers only when there is an appropriately placed tautomeric substituent capable of supplying a proton. Thus 2-pteridinone 1-oxide can exist as 2(3*H*)-pteridinone 1-oxide (**7**, R = H) or as 1-hydroxy-2(1*H*)-pteridinone (**8**, R = H), both of which could (at least in theory) be fixed, for example, by methylation, to give the 3-methylated *N*-oxide (**7**, R = Me) and the *N*-methoxy derivative (**8**, R = Me), respectively. Whether such potentially tautomeric *N*-oxides exist predominantly as such or as *N*-hydroxy compounds appears to be unsettled at present.[366,485] For pragmatic reasons, all pteridine *N*-oxides are here named as such, except in those rare cases in which a pteridinone would have to be written as a *C*-hydroxypteridine in order to accommodate the formulation of its *N*-oxide as such. For example, 'lumazine 3-oxide' is named 3-hydroxy-2,4(1*H*,3*H*)-pteridinedione (**9**) rather than either of the other possibilities (**10**) or (**11**), each involving a *C*-hydroxy group. The chemistry of pteridine *N*-oxides was reviewed expertly and in some detail by Wolfgang Pfleiderer in 1973.[1067]

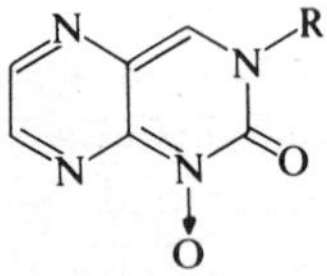

(**7**)

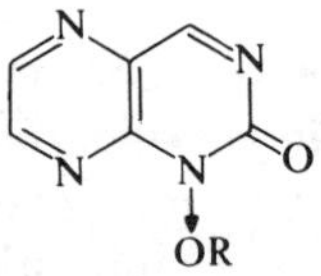

(**8**)

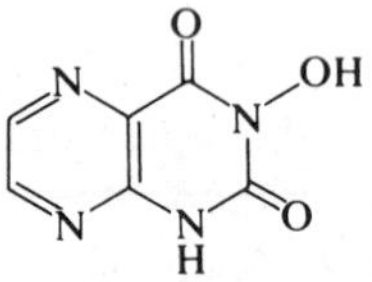

(**9**)

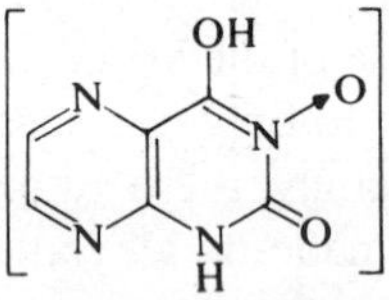

(**10**)

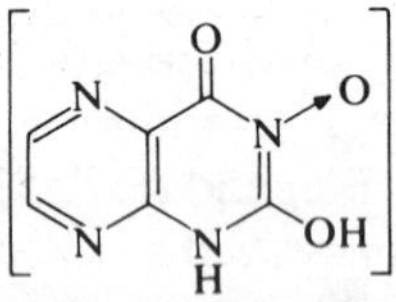

(**11**)

1. PREPARATION OF ALKOXY- AND ARYLOXYPTERIDINES

A. By Primary Syntheses

There are numerous examples in Chapters II and III of the passive formation of nuclear *C*-, nuclear *N*- and extranuclear *C*-alkoxy pteridines from synthones already bearing the alkoxy substituent. For example, 2-methoxy- or 6-methoxy-4,5-pyrimidinediamine with glyoxal gave 2- or 4-methoxypteridine, respectively[32]; 6-isopropoxy-2,4,5-pyrimidinetriamine with glyoxal gave 4-isopropoxy-

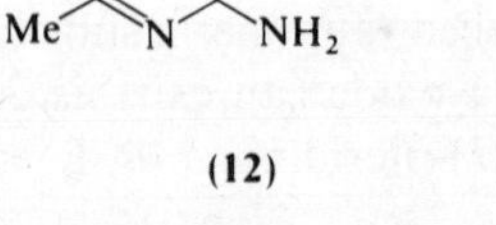

(**12**)

2-pteridinamine[370]; 6-amino-1,3-dimethyl-5-nitroso-2,4(1*H*,3*H*)-pyrimidinedione with *p*-methoxyphenylacetone gave 6-*p*-methoxyphenyl-1,3,7-trimethyl-2,4(1*H*,3*H*)-pteridinedione[1348]; methyl 3-amino-5-methyl-2-pyrazinecarbohydroxamate (**12**) with formic acid in acetic anhydride gave 3-methoxy-7-methyl-4(3*H*)pteridinone[759]; 3-amino-5-β-methoxypropyl-2-pyrazinecarbonitrile with guanidine gave 7-β-methoxypropyl-2,4-pteridinediamine[1305]; and methyl 3-amino-6-cyano-5-methoxy-2-pyrazinecarboximidate with triethyl orthoformate gave 4,7-dimethoxy-6-pteridinecarbonitrile.[1136]

B. By Covalent Addition of Alcohols

The covalent alcoholation of appropriate pteridines can afford alkoxydihydro- or even dialkoxytetrahydropteridines, which should sometimes be oxidizable to the alkoxy- or dialkoxypteridines (as are similar adducts in the closely related pyrimido[5,4-*e*]-*as*-triazine series[1702, 1703]). For example, pteridine in methanol slowly became the isolable di-adduct,[48] 6,7-dimethoxy-5,6,7,8-tetrahydropteridine (**13**) (see also Ch. IV, Sect. 1.C); 6,7-dimethylpteridine in methanol containing hydrogen chloride gave the mono-adduct,[50] 4-methoxy-6,7-dimethyl-3,4-dihydropteridine, isolable as its hydrochloride [see also Ch. IV, Sect. 2.B(1)]; 2(1*H*)- or 6(5*H*)-pteridinone in methanol gave the mono-adduct,[38] 4-methoxy-3,4-dihydro-2(1*H*)- or 7-methoxy-7,8-dihydro-6(5*H*)-pteridinone, respectively (see also Ch. VI, Sect. 2.B); ethyl 4-pteridinecarboxylate with an excess of ethylene glycol gave the di-adduct, ethyl 6,7-bis-β-hydroxyethoxy-5,6,7,8-tetrahydro-4-pteridinecarboxylate (**14**)[158]; and 2(3*H*)-pteridinone 1-oxide in boiling ethanol gave 4-ethoxy-3,4-dihydro-2(8*H*)-pteridinone 1-oxide (**15**).[1098]

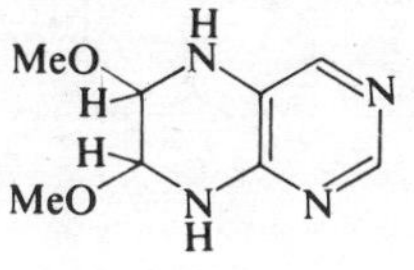

(13)

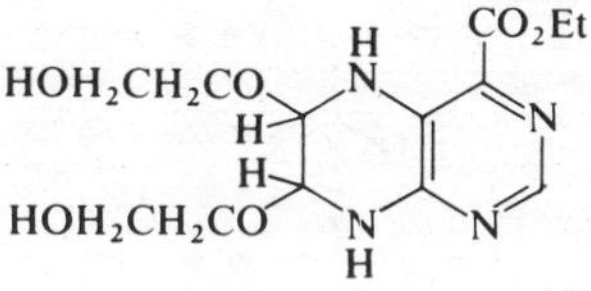

(14)

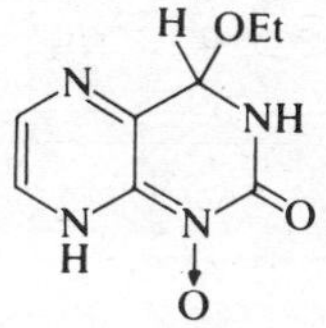

(15)

C. From Halogenopteridines

The formation of alkoxy- and aryloxypteridines from the corresponding halogenopteridines is exemplified briefly in the treatment of 7-chloropteridine with methanolic sodium methoxide to give 7-methoxypteridine (70%),[31] and in the treatment of 6-bromomethyl-2,4-pteridinediamine with phenol in *N*,*N*-dimethylacetamide containing potassium *t*-butoxide to give 6-phenoxymethyl-2,4-pteridinediamine (70%)[1102]; such processes are discussed further in Chapter V, Sections 2.D and 4.C.

D. From Pteridinones and Extranuclear Hydroxypteridines

Pteridinones may be converted into alkoxypteridines via chloropteridines or sometimes by direct alkylation with alcoholic hydrogen chloride, diazomethane, or methyl iodide–alkali. The direct approach is more often successful with 6- or 7- than with 2- or 4-pteridinones; the related conversion of pteridinones into trimethylsiloxypteridines by hexamethyldisilazane is invariably successful. These processes are fully discussed in Chapter VI, Section 2.F. Although extranuclear hydroxypteridines do not appear to have been *O*-alkylated directly, their acylation is well represented (see Chapter VI, Section 4.C).

E. Miscellaneous Methods

When 1-(2′,3′,5′-tri-*O*-benzoyl-β-D-ribofuranosyl)-2,4(1*H*,3*H*)-pteridinedione 5-oxide (**16**) was treated with acetyl chloride, 6-chloro-1-(2′,3′,5′-tri-*O*-benzoyl-β-D-ribofuranosyl)-2,4(1*H*,3*H*)-pteridinedione (**17**) resulted. In methanolic methoxide this compound underwent both deacylation and methoxylation to give 6-methoxy-1-β-D-ribofuranosyl-2,4(1*H*,3*H*)-pteridinedione (**18**). The same overall result (in better yield) was achieved by treating the related substrate, 1-(2′,3′,5′-tri-*O*-acetyl-β-D-ribofuranosyl)-2,4(1*H*,3*H*)-pteridinedione 5-oxide (**16**) with methanolic hydrogen chloride to go directly to the product (**18**).[210]

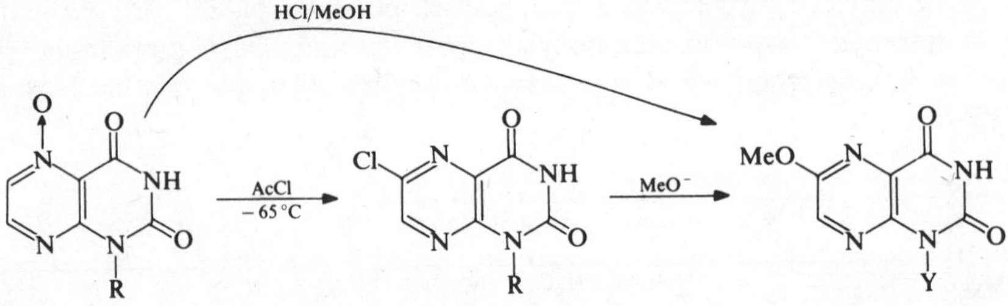

(R = tri-*O*-acylribofuranosyl) (Y = ribofuranosyl)

(**16**) (**17**) (**18**)

The nucleophilic displacement of an alkylsulfonyl group by methoxide ion has been used to convert 2-methylsulfonyl-7-phenyl-4-pteridinamine (**19**, R = SO_2Me) into 2-methoxy-7-phenyl-4-pteridinamine (**19**, R = OMe) (at 25°C for 1 h). It also converts 4-amino-8-methyl-2-methylsulfonyl- (**20**, R = SO_2Me) into 4-amino-2-methoxy-8-methyl-7(8*H*)-pteridinone (**20**, R = OMe) under similar conditions.[621]

Although seldom applicable in the pteridine series, *transetherification* has been used to prepare some alkoxypteridines from others. Thus boiling 4-methoxypteridine with a suspension of silver oxide in propanol gave 4-propoxypteridine (88%),[138] although similar treatment of 2- or 6-methoxypteridine

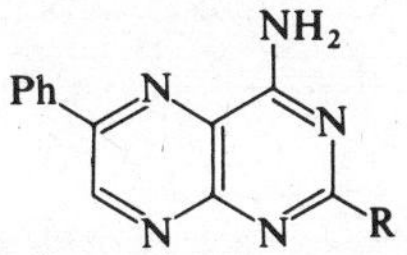

(19)

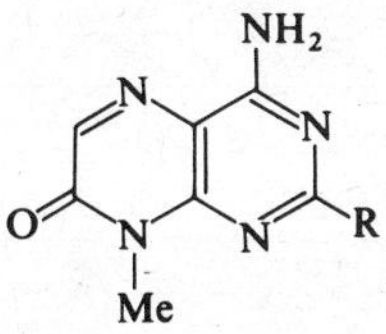

(20)

gave only unidentified products. Using a more traditional reagent, sodium benzyl oxide in refluxing benzyl alcohol, 2,4-dimethoxy-7(8*H*)-pteridinone and its 6-methyl derivative were both converted in good yield into 2,4-dibenzyloxy-7(8*H*)-pteridinone and its 6-methyl derivative, respectively.[1192]

2. REACTIONS OF ALKOXY- AND ARYLOXYPTERIDINES

The fragmentation of 3-methoxy-4(3*H*)-pteridinone and several of its *C*-methylated derivatives has been studied. All lost CH_3O initially but the ions so produced broke down differently from the isomeric molecular ions from the corresponding 4(3*H*)-pteridinones.[779]

A. Hydrolysis and Alcoholysis

The facile hydrolysis of a 2-, 4-, 6-, or 7-alkoxypteridine to the corresponding pteridinone has been discussed in Chapter VI, Section 1.D. Extranuclear alkoxypteridines would require vigorous acidic conditions for cleavage (akin to those needed for aliphatic ethers) and the reaction does not appear to have been used; however, the relatively easy cleavage of extranuclear acyloxypteridines to the corresponding hydroxypteridines has been covered in Chapter VI, Section 3.C. Likewise, *N*-alkoxypteridinones do not appear to have been converted into *N*-hydroxypteridinones (as in other series) but an *N*-acyloxypteridinone, 3-benzenesulfonyloxy-2,4(1*H*,3*H*)-pteridinedione, has been converted by alkaline hydrolysis into 3-hydroxy-2,4(1*H*,3*H*)-pteridinedione.[1445]

The rarely successful alcoholysis of an alkoxypteridine to give another alkoxypteridine (transetherification) has just been discussed in Section 1.E.

B. Aminolysis

Although alkoxy groups have been recognized in other series as excellent substrates for aminolysis, alkoxypteridines have been rarely so used. The reaction is represented, however, by the formation of 2-amino-6-hydrazino-4(3*H*)-pteridinone (**21**, R$=$NHNH$_2$) in ~70% yield by boiling either 2-amino-

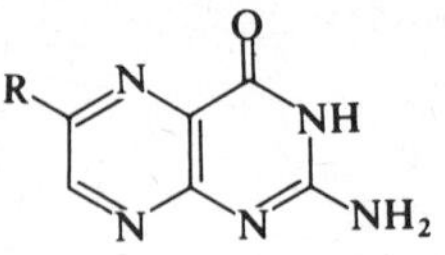

(21)

6-methoxy-4(3*H*)-pteridinone (**21**, R=OMe)[369] or its 6-ethoxy homolog (**21**, R=OEt)[1332] in hydrazine hydrate.

C. The Silyl Hilbert–Johnson Reaction

The potential (base-catalyzed) thermal rearrangement of an alkoxypteridine to an *N*-alkylpteridinone does not appear to have been achieved; nor has the regular Hilbert–Johnson reaction of an alkoxypteridine with an alkyl iodide to give an *N*-alkylpteridinone. However, the Silyl Hilbert–Johnson versions of the latter reaction have been used extensively, especially by Wolfgang Pfleiderer and colleagues, to convert trimethylsiloxypteridines into the corresponding pteridinone *N*-glycosides (often accompanied by *O*-glycosides as by-products) by treatment (a) with an acylated halogeno-sugar, for example, 2,3,5-tri-*O*-benzoyl-D-ribofuranosyl bromide (**22**), usually in boiling toluene with mercuric bromide–mercuric oxide as a catalyst, or (b) with a fully (differentially) acylated sugar, for example, 1-*O*-acetyl-2,3,5-tri-*O*-benzoyl-*β*-D-ribofuranose (**25**), usually as a melt of the two reactants at ~140°C without added solvent but with the addition of a little zinc chloride as catalyst.

The first of these procedures is illustrated simply by the reaction of 6,7-diphenyl-2,4-bistrimethylsiloxypteridine (**23**) with the sugar bromide (**22**) in refluxing benzene (containing mercuric bromide and mercuric oxide) for 4 h. Chromatographic separation of the products gave 6,7-diphenyl-1-(2′,3′,5′-tri-*O*-benzoyl-*β*-D-ribofuranosyl)-2,4(1*H*,3*H*)-pteridinedione (**24a**) (74%), its 3-tribenzoylribofuranosyl isomer (5%), and 6,7-diphenyl-1,3-bis(2′,3′,5′-tri-*O*-benzoyl-

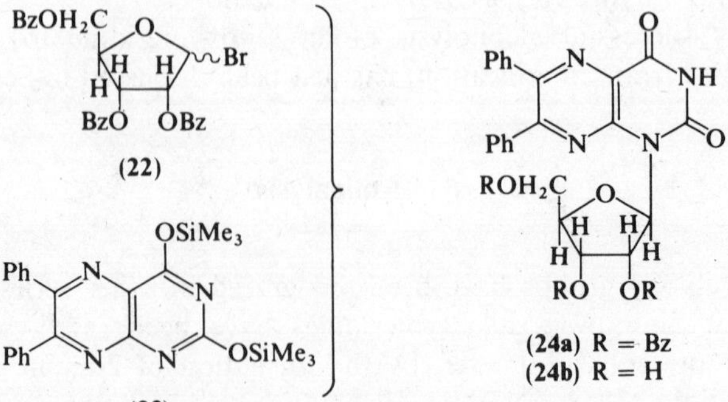

(22)

(23)

(**24a**) R = Bz
(**24b**) R = H

β-D-ribofuranosyl)-2,4(1*H*,3*H*)-pteridinedione (5%). The main product (**24a**) was then deacylated in methanolic sodium methoxide at room temperature to give 6,7-diphenyl-1-β-D-ribofuranosyl-2,4(1*H*,3*H*)-pteridinedione (**24b**) in 78% yield.[1022] The second of these procedures is exemplified in the fusion of 2-methylthio-7-trimethylsiloxy-4-trimethylsilylaminopteridine (**26**) with the 1-acetylated sugar (**25**) and zinc chloride at 140°C for 80 min under a partial vacuum to remove volatiles. Chromatographic purification gave 4-amino-2-methylthio-8-(2′,3′,5′-tri-*O*-benzoyl-β-D-ribofuranosyl)-7(8*H*)-pteridinone (**27a**) (75% yield), which in methanolic sodium methoxide at 25°C gave 4-amino-2-methylthio-8-β-D-ribofuranosyl-7(8*H*)-pteridinone (**27b**) in 86% yield.[1023]

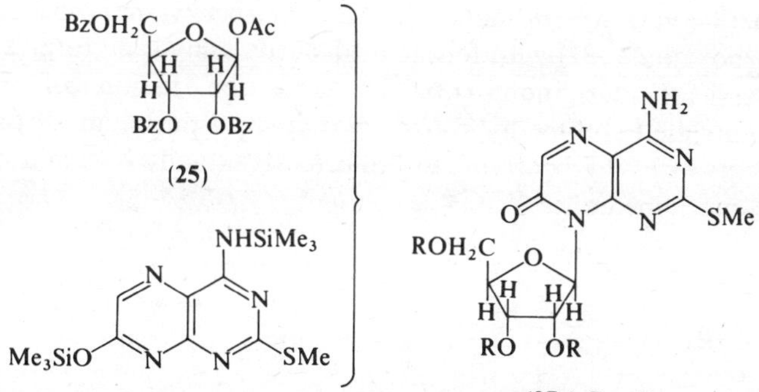

(**25**)

(**26**)

(**27a**) R = Bz
(**27b**) R = H

As might be expected, application of these methods to diverse pteridine substrates was not without problems, exceptions, and odd by-products. For example, 2-dimethylamino-4,7-bistrimethylsiloxypteridine with the sugar bromide (**22**) gave a separable mixture of 2-dimethylamino-4,7-bis(2′,3′,5′-tri-*O*-benzoyl-β-D-ribofuranosyloxy)pteridine (16% yield) and 2-dimethylamino-8-(2′,3′,5′-tri-*O*-benzoyl-β-D-ribofuranosyl)-4-(2′,3′,5′-tri-*O*-benzoyl-β-D-ribofuranosyloxy)-7(8*H*)-pteridinone (22% yield), neither of which was to be expected.[246] Fortunately, a mass of data relevant to these procedures was summarized in five successive papers in 1971,[325, 356, 435, 557, 1542] although all the background details, including some subsequent experiments, appeared only several years afterwards in a remarkable series of papers from Konstanz.[209, 224, 225, 237, 246, 266, 283, 357, 382, 412, 414, 436, 447, 1022, 1023, 1048, 1126, cf. 143, 144]

3. PREPARATION OF *N*-ALKYLPTERIDINONES

A. By Primary Syntheses

A great many (nontautomeric) *N*-alkylpteridinones have been made conveniently by primary syntheses from pyrimidines (see Ch. II) or from pyrazines or other systems (see Ch. III). Such syntheses are typified by the condensation of 4,5-diamino-1-methyl-2(1*H*)-pyrimidinone with glyoxal to give 3-methyl-2(3*H*)-pteridinone (80%)[34]; of 5-amino-6-methylamino-4-(3*H*)-pyrimidinone (**28**) with glyoxal to give 8-methyl-4(8*H*)-pteridinone (**29**)[248]; of 5,6-diamino-1,3-dimethyl-2,4(1*H*,3*H*)-pyrimidinedione with biacetyl to give 1,3,6,7-tetramethyl-2,4(1*H*,3*H*)-pteridinedione[1096]; of 5-methylamino-4-pyrimidinamine with diethyl oxalate at 180°C to give 5-methyl-6,7(5*H*,8*H*)-pteridinedione (**30**)[134]; of 6-amino-1,3-dimethyl-5-nitroso-2,4(1*H*,3*H*)-pyrimidinedione with phenylacetaldehyde to give (by a Timmis reaction) 1,3-dimethyl-6-phenyl-2,4(1*H*,3*H*)-pteridinedione[176]; of 2-amino-6-chloro-5-*p*-chlorophenylazo-4(3*H*)-pyrimidinone with *N*-methylaminoacetaldehyde to give (by a Boon sequence) 2-amino-8-methyl-4(8*H*)-pteridinone[422]; of 3-benzylamino-5,6-dimethyl-2-pyrazinecarboxamide (**31**) with formic acid–acetic anhydride to give 1-benzyl-6,7-dimethyl-4(1*H*)-pteridinone (**32**)[1008]; and the fission of 2-amino-7-ethoxycarbonylmethyl-9-methyl-6-oxo-1,6-dihydro-7-purinium bromide (**33**) and subsequent recyclization to afford eventually 2-amino-8-methyl-4,7(3*H*,8*H*)-pteridinedione.[181]

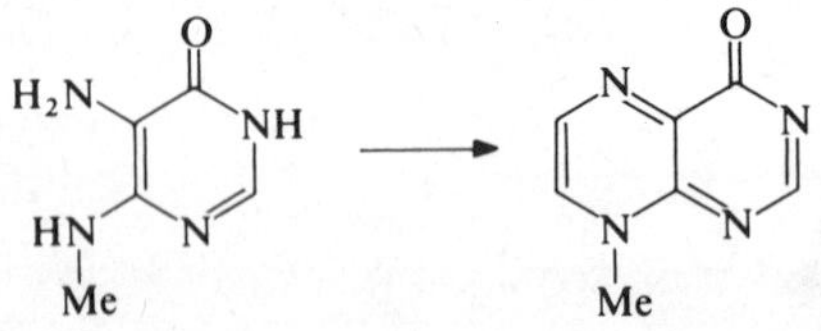

(**28**) (**29**)

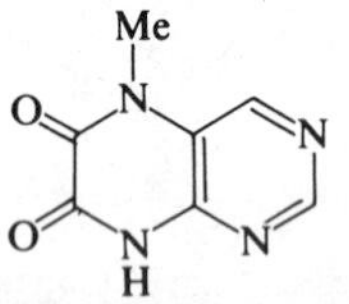

(**30**)

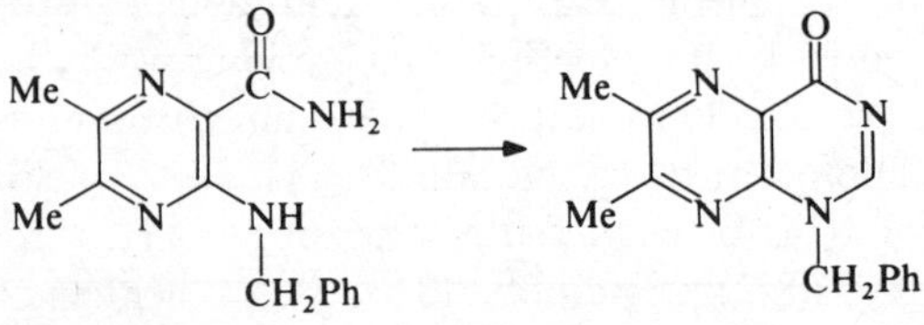

(**31**) (**32**)

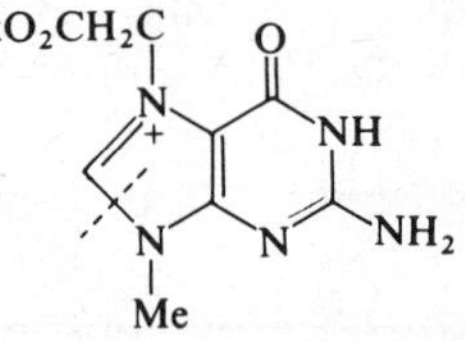

(**33**)

B. By *N*-Alkylation of Pteridinones

As mentioned in the discussion on *O*-alkylation of pteridinones (Ch. VI, Sect. 2.F), it is not easy to forecast whether alkylation of a pteridinone will afford one or more *N*-alkyl derivative(s), an *O*-alkyl derivative, or a mixture of both types. Probably the best advice that can be given after analysis of existing data, is to proceed by analogy with the nearest known example(s) and to expect a mixture on principle.

The *N*-alkylation of *simple monopteridinones* is exemplified in the conversion of 2(1*H*)-pteridinone into 1-methyl-2(1*H*)-pteridinone (**34**) (methyl iodide–

methanolic sodium methoxide under reflux: 70%, apparently without isomeric products)[34]; of 4(3*H*)-pteridinone into (a) a mixture of 1-methyl-4(1*H*)-pteridinone (**35**) (12% yield) and 3-methyl-4(3*H*)-pteridinone (**36**) (45% yield), by using dimethyl sulfate at pH 8, or (b) a mixture of 3-methyl-4(3*H*)-pteridinone (56% yield) and 4-methoxypteridine (21% yield), by using diazomethane in methanolic ether[34,538]; of 6(5*H*)-pteridinone into 5-methyl-6(5*H*)-pteridinone (**37**) (dimethyl sulfate–methanolic sodium methoxide at 20°C; chromatographic purification from some unchanged substrate was required)[34]; of 7(8*H*)-pteridinone into (a) a separable mixture of 8-methyl-7(8*H*)-pteridinone (**38**, R = H) (50% yield) and 6,8-dimethyl-7(8*H*)-pteridinone (**38**, R = Me) (<5% yield), by dimethyl sulfate at pH 8,[34,538] or (b) a mixture of 8-methyl-7(8*H*)-pteridinone (**38**, R = H) (32% yield) and 7-methoxypteridine (<1% yield), by diazomethane in methanolic ether at 0°C[31]; of 6,7-diphenyl-4(3*H*)-pteridinone into 3-benzyl-6,7-diphenyl-4(3*H*)-pteridinone (benzyl chloride–methanolic potassium hydroxide under reflux: 25% yield; also a comparable amount of its degradation product, 2-amino-*N*-benzyl-5,6-diphenyl-2-pyrazinecarboxamide)[1316]; and of 7-methyl-4(3*H*)-pteridinone into 3-β-cyanoethyl-7-methyl-4(3*H*)-pteridinone (**39**) (acrylonitrile in refluxing aqueous pyridine: 32%; a

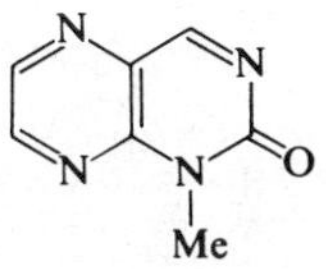

(34)

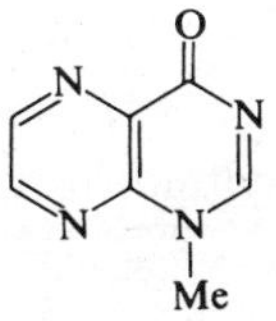

(35)

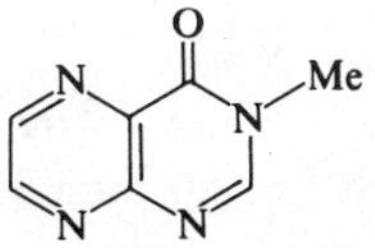

(36)

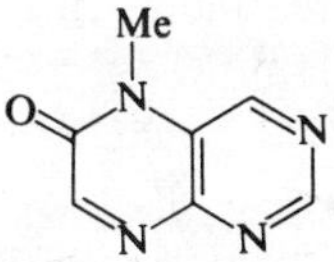

(37)

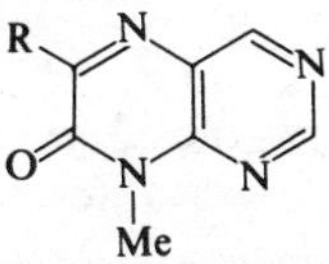

(38)

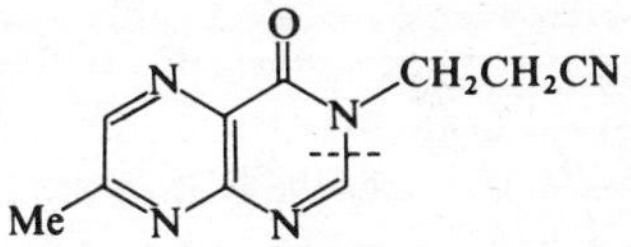

(39)

ring-fission product, *N*-β-cyanoethyl-3-formamido-5-methyl-2-pyrazinecarboxamide, was also obtained.[167] It is interesting that treatment of 4(3*H*)-pteridinone, its 1-methyl derivative (**35**), or its 3-methyl derivative (**36**) with an excess of methyl iodide in DMF gave 1,3-dimethyl-4-oxo-3,4-dihydro-1-pteridinium iodide, perhaps best formulated as (**40**, R = H)[587,1130]; 6,7-dimethyl- and 6,7-diphenyl-4(3*H*)-pteridinone (as well as their respective 1- and 3-methylated derivatives) likewise afforded either 1,3,6,7-tetramethyl-4-oxo- (**40**, R = Me) or 1,3-dimethyl-4-oxo-6,7-diphenyl-3,4-dihydro-1-pteridinium iodide (**40**, R = Ph), as appropriate.[1130]

The *N*-alkylation of *less simple monopteridinones* is illustrated by the conversion of 4-methylthio-6,7-diphenyl-2(1*H*)-pteridinone (**41**, R = H) into its 1-methyl derivative (**41**, R = Me) [methyl iodide and potassium carbonate in

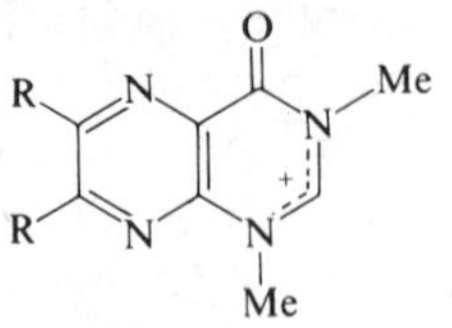

(40)

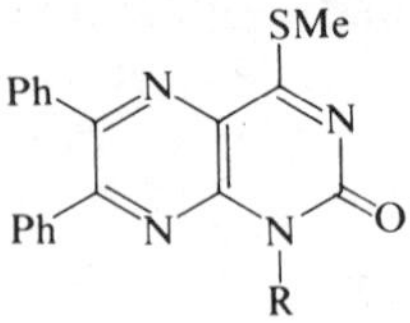

(41)

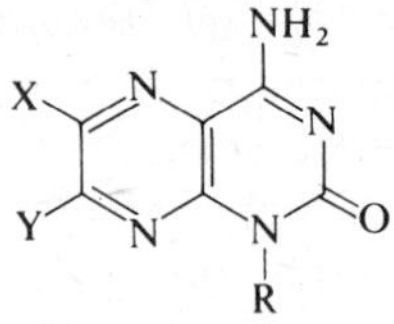

(42)

dioxane at 20°C: 83%; the same product (**41**, R = Me) was also obtained from 6,7-diphenyl-4-thioxo-3,4-dihydro-2(1*H*)-pteridinone in a similar way but with 2 equivalents of methyl iodide][1679]; of 6,7-diphenyl-2-thioxo-1,2-dihydro-4(3*H*)-pteridinone into 3-methyl-2-methylthio-4(3*H*)-pteridinone (methyl iodide–potassium carbonate–DMF–methanol at 25°C: 65%)[437]; of 6,7-dimethyl-2-methylthio-4(3*H*)-pteridinone into its 3-*β*-ethoxycarbonylethyl derivative (ethyl acrylate in boiling aqueous pyridine: 21% yield after purification)[168]; of 4-dimethylamino-6,7-dimethyl-2(1*H*)-pteridinone into its 1,6,7-trimethyl analog (dimethyl sulfate–alkali at 40°C: ~60%)[362]; of 2-dimethylamino-6,7-dimethyl-4(3*H*)-pteridinone into its 6,7,8-trimethyl analog (dimethyl sulfate–acetic acid–DMF)[563]; of 4-dimethylamino-7(8*H*)-pteridinone into a separable mixture of its 8-methyl derivative (~55% yield) and 4-dimethylamino-7-methoxypteridine (~6% yield), using dimethyl sulfate in alkali[445]; of 4-dimethylamino-2-methylthio-7(8*H*)-pteridinone into its 8-methyl derivative (dimethyl sulfate–alkali at 25°C: 21%)[1126]; and of 4-benzyloxy-2-dimethylamino-7(8*H*)-pteridinone into its 8-(2′,3′5′-tri-*O*-benzoyl-*β*-D-ribofuranosyl) derivative (2,3,5-tri-*O*-benzoyl-D-ribofuranosyl bromide and mercuric cyanide in refluxing nitromethane: 68%) and then, by treatment with methanolic sodium methoxide, into 4-benzyloxy-2-dimethylamino-8-*β*-D-ribofuranosyl-7(8*H*)-pteridinone.[1024]

A special case within the last category is the *N*-alkylation of *amino-2-, 4-, or 7-pteridinones*. The first and third of such substrates behave quite regularly and their alkylations are typified in the conversion of 4-amino-2(1*H*)-pteridinone (**42**, R = X = Y = H) into its 1-methyl derivative (**42**, R = Me, X = Y = H) (methyl iodide–methanolic sodium methoxide under reflux: 75%)[130]; of 4-amino-6-methyl-2(1*H*)-pteridinone (**42**, R = Y = H, X = Me) into its 1,6-dimethyl analog (**42**, R = X = Me, Y = H) (dimethyl sulfate–alkali: 50%) or into its 1-*β*-cyanoethyl derivative (**42**, R = CH_2CH_2CN, X = Me, Y = H) (acrylonitrile in refluxing aqueous pyridine: 50% after purification)[168]; of 4-amino-6,7-dimethyl-2(1*H*)-pteridinone (**42**, R = H, X = Y = Me) into its 1,6,7-trimethyl analog (**42**, R = X = Y = Me) (methyl iodide in refluxing methanolic sodium methoxide: 84%)[130]; of 2-amino-7(8*H*)-pteridinone (**43**, R = Y = H) into its 8-methyl derivative (**43**, R = Me, Y = H) (dimethyl sulfate–alkali: ~55%)[383]; of 2-amino-4-pentyloxy-7(8*H*)-pteridinone (**43**, R = H, Y = OPe) into its 8-methyl derivative (**43**, R = Me, Y = OPe) (dimethyl sulfate–alkali in aqueous DMF: 63%)[436]; and of 4-amino-2-phenyl-7(8*H*)-pteridinone into its 8-ethyl derivative (ethyl iodide–aqueous ethanolic potassium hydroxide).[1390]

The amino-4-pteridinones undergo less predictable alkylations.[563] Heating 2-amino-4(3*H*)-pteridinone with methanolic methyl iodide in a sealed tube at 115°C gave 2-amino-8-methyl-4(8*H*)-pteridinone (**44**, R = Y = H) as hydriodide (~50%)[130]; similar treatment of 2-amino-6-methyl- or 2-amino-6,7-dimethyl-4(3*H*)-pteridinone afforded the hydriodide of 2-amino-6,8-dimethyl- (**44**, R = Me, Y = H)[80] or 2-amino-6,7,8-trimethyl-4(8*H*)-pteridinone (**44**, R = Y = Me),[130] respectively; and treatment of 2-amino-4-oxo-3,4-dihydro-6-pteridinecarboxylic acid with dimethyl sulfate in DMF containing acetic acid gave only 2-amino-8-methyl-4-oxo-4,8-dihydro-6-pteridinecarboxylic acid (58%)[70].

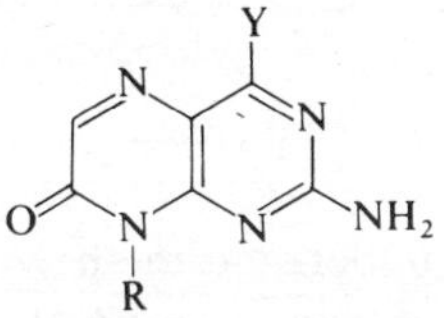
(43)

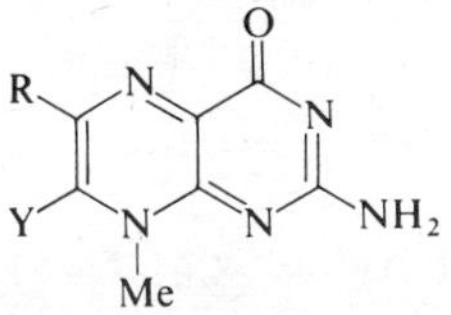
(44)

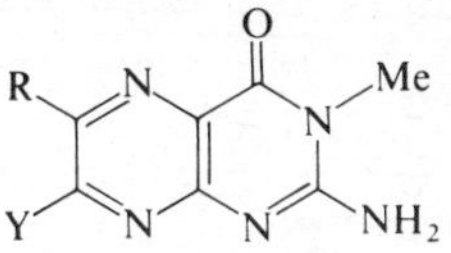
(45)

In contrast, methylation of 2-amino-6-methyl-4(3*H*)-pteridinone by methyl iodide in refluxing methanolic sodium hydroxide gave a separable mixture of 2-amino-3,6-dimethyl-4(3*H*)-pteridinone (**45**, R = Me, Y = H) (~55% yield) and 2-amino-1,6-dimethyl-4(1*H*)-pteridinone (**46**, R = Me, Y = H) (~20% yield)[80]; 2-amino-6,7-dimethyl-4(3*H*)-pteridinone with dimethyl sulfate in a hot aqueous medium maintained at pH 7, gave a difficultly separable mixture of 2-amino-3,6,7-trimethyl-4(3*H*)- (**45**, R = Y = Me) (22% yield) and 2-amino-1,6,7-trimethyl-4(1*H*)-pteridinone (**46**, R = Y = Me) (16% yield)[69]; 2-amino-6,7-diphenyl-4(3*H*)-pteridinone with dimethyl sulfate (in aqueous DMF containing alkali) gave 2-amino-3-methyl-6,7-diphenyl-4(3*H*)- (**45**, R = Y = Ph) (10%) and the isomeric 2-amino-1-methyl-6,7-diphenyl-4(1*H*)-pteridinone (**46**, R = Y = Ph) (52%)[68]; 2-amino-4-oxo-3,4-dihydro-6-pteridinecarboxylic acid with dimethyl sulfate in aqueous alkali (cf. just cited) gave a separable mixture of 2-amino-1-methyl-4-oxo-1,4-dihydro-6-pteridinecarboxylic acid, 2-amino-3-methyl-4-oxo-3,4-dihydro-6-pteridinecarboxylic acid, and 2-imino-3,8-dimethyl-4-oxo-2,3,4,8-tetrahydro-6-pteridinecarboxylic acid (20% as its hydrochloride)[70]; and 2-amino-4-oxo-3,4-dihydro-7-pteridinecarboxylic acid likewise gave a separable mixture of 2-amino-1-methyl-4-oxo-1,4-dihydro- and 2-amino-3-methyl-4-oxo-3,4-dihydro-7-pteridinecarboxylic acid (both in low yield with no indication of any 8-methylated material).[70]

Besides the ionic state of the substrate (neutral molecule or anion) in these methylations, it appears that steric hindrance can play a role in the position of methylation. Thus methylation of 2-amino-6-phenyl-4(3*H*)-pteridinone with dimethyl sulfate in DMF–acetic acid gave 2-amino-8-methyl-6-phenyl-4(8*H*)-pteridinone (**44**, R = Ph, Y = H) as the main product (44% yield), whereas similar treatment of 2-amino-7-phenyl-4(3*H*)-pteridinone (in which the phenyl group might be expected to discourage 8-alkylation on steric grounds) gave no 8-

methylated product (**44**, R = H, Y = Ph) but only a small yield of 2-imino-1,3-dimethyl-7-phenyl-1,2-dihydro-4(3*H*)-pteridinone (**47**), isolated as its hydriodide.[68] Likewise, methylation of 2-amino-6,7-diphenyl-4(3*H*)-pteridinone under such acidic conditions gave no 8-methylated derivative (**44**, R = Y = Ph) as expected, but only a small amount of 2-amino-1-methyl-6,7-diphenyl-4(1*H*)-pteridinone (**46**, R = Y = Ph).[68]

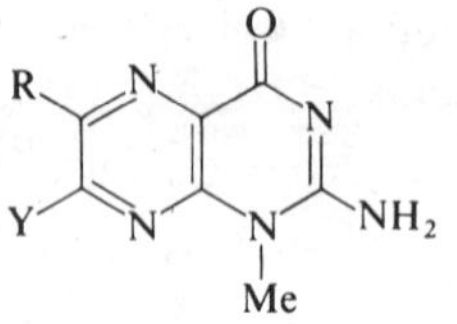

(46)

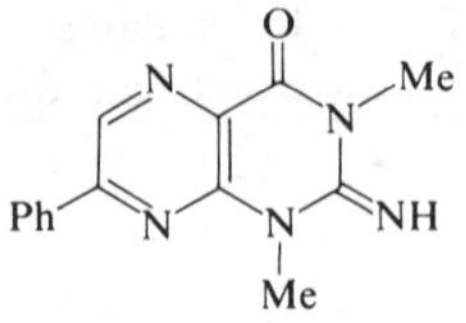

(47)

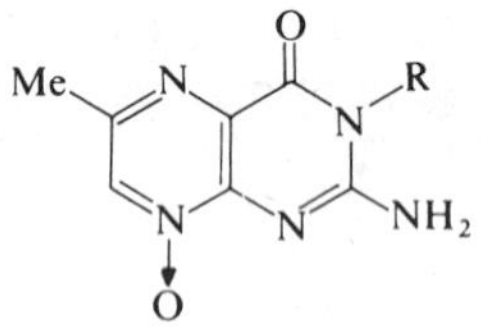

(48)

2-Amino-6-methyl-4(3*H*)-pteridinone 8-oxide (**48**, R = H) underwent methylation with dimethyl sulfate in aqueous alkali to give only its 3,6-dimethyl analog (**48**, R = Me) in 63% yield.[485] The treatment of 2-amino-6-methyl-5,6,7,8-tetrahydro-4(3*H*)-pteridinone with an excess of methyl iodide and a limited amount of sodium hydroxide in methanol gave 2-imino-1,3,6-tri- or 2-imino-1,3,5,6-tetramethyl-1,2,5,6,7,8-hexahydro-4(3*H*)-pteridinone dihydrochloride, according to the proportion of base and the conditions used.[83] In addition, acylation of the same substrate has been studied.[607]

The *N*-alkylation of *pteridinediones* is seen in the conversion of 2,4(1*H*,3*H*)-pteridinedione (**49**, R = H) into its 1,3-dimethyl derivative (**49**, R = Me) (dimethyl sulfate at pH 8: ~40%[34]; diazomethane in methanolic ether: ~50%),[338] its 1,3-diethyl derivative (**49**, R = Et) (ethyl iodide and potassium carbonate in DMF: 69%),[108] and its 1,3-dibenzyl derivative (**49**, R = CH_2Ph) (benzyl bromide and potassium carbonate in DMF: 57%)[108]; of the same substrate (**49**, R = H) into its 1-butyl derivative (butyl iodide and potassium carbonate in dimethyl sulfoxide at 90°C for 3 h: 64%, apparently without other products)[876]; of 6,7(5*H*,8*H*)-pteridinedione (**50**, R = Y = H) or its 8-methyl derivative (**50**, R = H, Y = Me) into the 5,8-dimethyl derivative (**50**, R = Y = Me) (dimethyl sulfate in methanolic sodium methoxide: 35%)[34]; of 6-methyl-2,4(1*H*,3*H*)-pteridinedione (**51**, R = H, Y = Me) into its 1,3,6-trimethyl analog (**51**, R = Y = Me) (dimethyl sulfate–alkali: ~25%)[80] and into its 1,3-bis-*β*-cyanoethyl derivative (**51**, R = CH_2CH_2CN, Y = Me) (acrylonitrile in aqueous pyridine: 50%)[168]; of 6,7-dimethyl-2,4(1*H*,3*H*)-pteridinedione into its 1,3,6,7-tetramethyl analog (dimethyl sulfate–alkali: 49%)

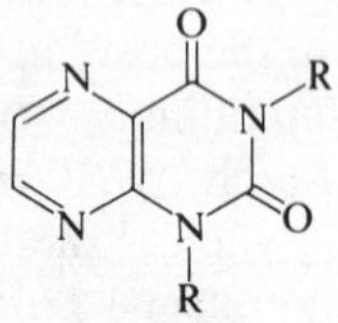

(49)

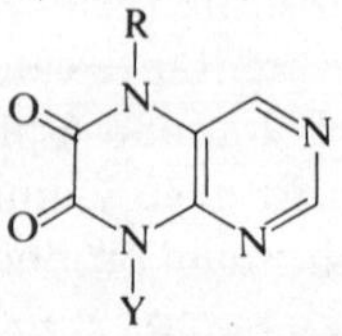

(50)

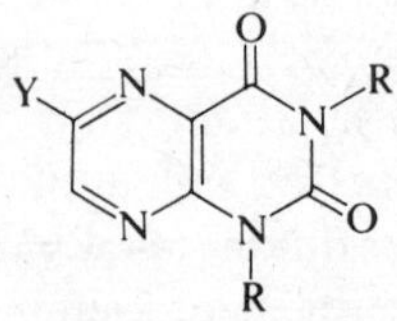

(51)

or into its 1,3-bis-β-cyanoethyl (72%) or 1,3-bis-β-carbamoylethyl derivative (38%) (acrylonitrile or acrylamide, respectively, in aqueous pyridine)[168]; of 2,4-dioxo-1,2,3,4-tetrahydro-6-pteridinecarboxylic acid (**51**, R = H, Y = CO_2H) into a separable mixture of 1,3-dimethyl-2,4-dioxo-1,2,3,4-tetrahydro- (**51**, R = Me, Y = CO_2H) (48%) and 3,8-dimethyl-2,4-dioxo-2,3,4,8-tetrahydro-6-pteridinecarboxylic acid (**52**) (9%), using dimethyl sulfate in aqueous alkali[70]; of 6-phenyl- (**51**, R = H, Y = Ph) and 7-phenyl-2,4(1*H*,3*H*)-pteridinedione into their respective 1,3-dimethyl derivatives (dimethyl sulfate–sodium hydroxide in aqueous DMF: 50 and 81%, respectively)[67]; of 7-propionyl-2,4(1*H*,3*H*)-pteridinedione into its 1,3-diethyl (30%) or 1,3-dibenzyl derivative (60%), using ethyl iodide or benzyl bromide, respectively, in aqueous DMF containing potassium carbonate[108]; of 6-phenyl-2,4(1*H*,3*H*)-pteridinedione 8-oxide (**53**, R = Y = H, X = Ph) into its 3-methyl derivative (**53**, R = Me, X = Ph, Y = H) (dimethyl sulfate in aqueous methanolic potassium hydroxide: 78%)[331]; of 6,7-dimethyl-2,4(1*H*,3*H*)-pteridinedione 8-oxide (**53**, R = H, X = Y = Me) into its 3,6,7-trimethyl analog (**53**, R = X = Y = Me) (similarly: 50%)[331]; and of 3-methyl-7-phenyl-2,4(1*H*,3*H*)-pteridinedione 5-oxide (**54**, R = H) into its 1,3-dimethyl analog (**54**, R = Me) (similarly: 80%).[331] In addition, the methylation of 2-amino-4,7(3*H*,8*H*)-pteridinedione (**55**, R = X = Y = H) (anionic pK_a values: 7.34 and 10.05) by an excess of dimethyl sulfate at three pH values has been studied: at pH 7, the substrate had partially lost the 8-proton and the only product was the 8-methyl derivative (**55**, R = X = H, Y = Me); at pH 9, the substrate had largely lost the 8-proton and partially lost the 3-proton, so that the 8-methyl (**55**, R = X = H, Y = Me), the 3-methyl (**55**, R = Y = H, X = Me), and the 3,8-dimethyl derivative (**55**, R = H, X = Y = Me) were all formed (in ratio 3:1:1); and at pH > 13, the substrate had lost both protons entirely, so that the main product was initially the 3,8-dimethyl derivative (**55**, R = H, X = Y = Me) and much of this was further methylated, presumably to the imine (**56**), which underwent

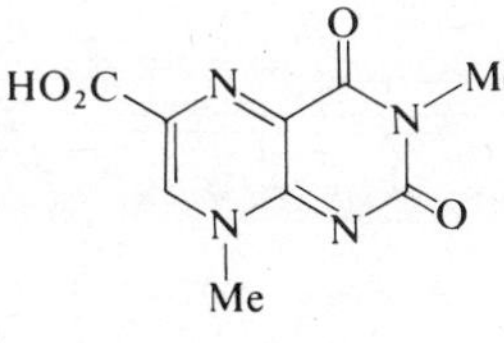

(52)

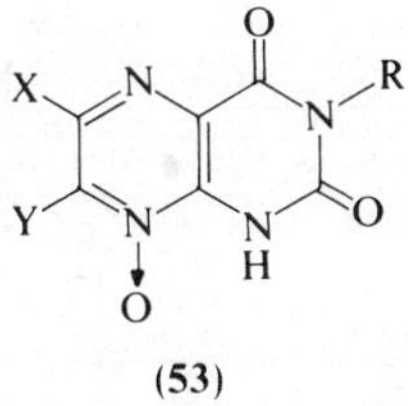

(53)

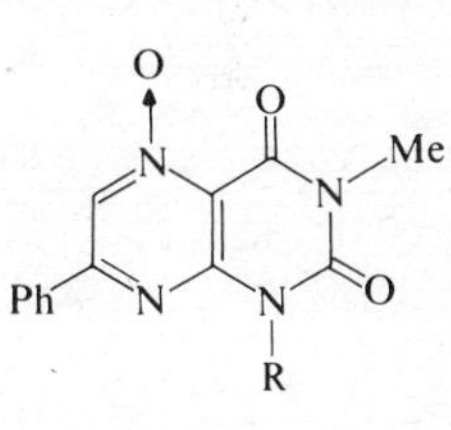

(54)

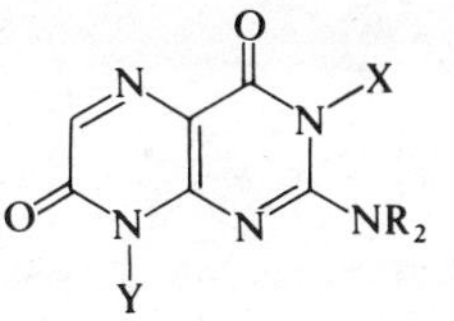

(55)

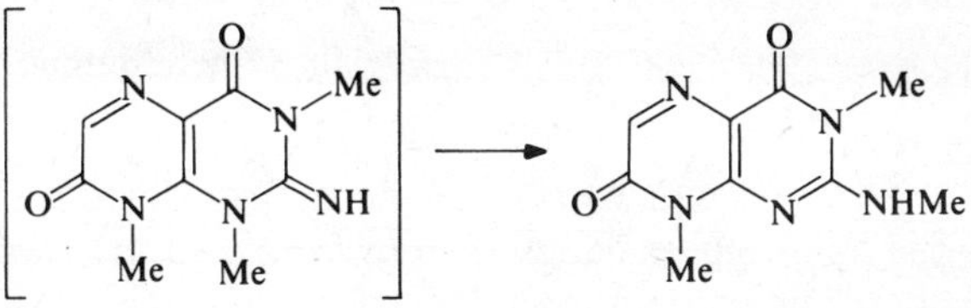

(56) **(57)**

Dimroth rearrangement to afford 3,8-dimethyl-2-methylamino-4,7(3*H*,8*H*)-pteridinedione (**57**) in appreciable amount.[457] 2-Dimethylamino-4,7(3*H*,8*H*)-pteridinedione (**55**, R = Me, X = Y = H) behaved rather similarly under appropriate conditions to give isolable amounts of the 8-methyl (**55**, R = Y = Me, X = H) and 3,8-dimethyl derivative (**55**, R = X = Y = Me).[457] 2-Amino-3,7-dimethyl-4,6(3*H*,5*H*)-pteridinedione gave its 3,5,7-trimethyl analog.[348]

An interesting reaction (catalyzed by mercuric bromide) between 6,7-diphenyl-2,4(1*H*,3*H*)-pteridinedione and its 1,3-bis-(2′,3′,5′-tri-*O*-benzoyl-β-D-ribofuranosyl) derivative gave a mixture of 6,7-diphenyl-1- and 6,7-diphenyl-3-(2′,3′,5′-tri-*O*-benzoyl-β-D-ribofuranosyl)-2,4(1*H*,3*H*)-pteridinedione, but not in equal amounts.[239]

The *N*-alkylation of *pteridinetriones* is not well represented but it is exemplified in the transformation of 2,4,7(1*H*,3*H*,8*H*)-pteridinetrione (**58**, R = X = Y = H), by treatment in aqueous methanolic alkali with dimethyl sulfate, into its 1,3-dimethyl derivative (**58**, R = H, X = Y = Me) (a minor product: ~12%) and its 8-methyl derivative (**58**, R = Me, X = Y = H) (the major product: ~60%), the second of which underwent further methylation on similar treatment at a higher temperature to afford 3,8-dimethyl-2,4,7(1*H*,3*H*,8*H*)-pteridinetrione (**58**, R = X = Me, Y = H) in ~40% yield[344]; of 3-methyl-2,4,7(1*H*,3*H*,8*H*)-pteridinetrione (**58**, R = Y = H, X = Me), by treatment with dimethyl sulfate–alkali, into a separable mixture of the just cited 3,8- and 1,3-dimethylpteridinetriones (20 and 15%, respectively) as well as 7-methoxy-1,3-dimethyl-2,4(1*H*,3*H*)-pteridinedione (~5%)[344]; of 2-amino-3-methyl-4,6,7(3*H*,5*H*,8*H*)-pteridinetrione (**59**, R = H) into its 3,5,8-trimethyl analog (**59**, R = Me) [dimethyl sulfate–alkali: ~70% yield, plus some 2-amino-6-methoxy-3,8-dimethyl-4,7(3*H*,8*H*)-pteridinedione][384]; and of 1,3-dimethyl-2,4,6(1*H*,3*H*,5*H*)-pteridinetrione (**60**, R = H) into its 1,3,5-trimethyl analog (**60**, R = Me) [dimethyl sulfate–alkali: ~10% yield, plus much more of the isomeric 6-methoxy-1,3-dimethyl-2,4(1*H*,3*H*)-pteridinedione].[340]

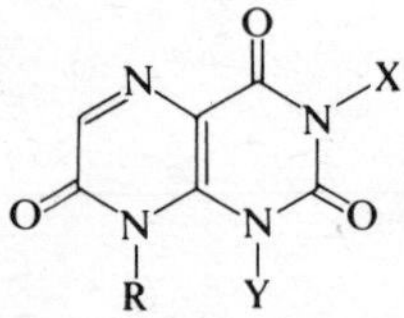

(58)

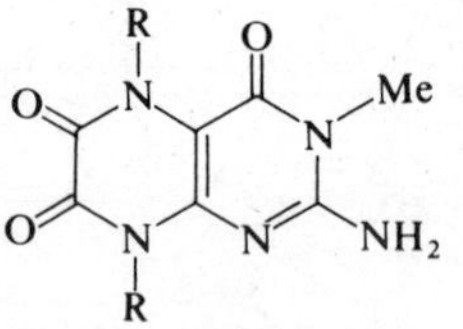

(59)

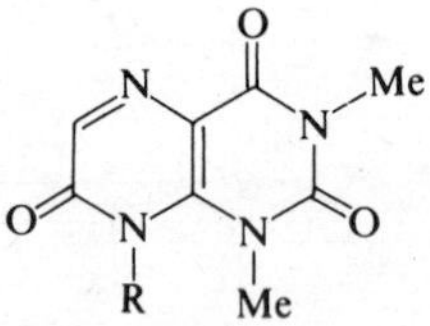

(60)

C. By Other Methods

Several minor ways to produce *N*-alkylated pteridinones are represented in the following reactions. The *hydrolysis* of *pteridinimines* can afford the corresponding pteridinones. Thus 1-methyl- (**61**, R = Y = H) and 1,6,7-trimethyl-4(1*H*)-pteridinimine (**61**, R = H, Y = Me) gave 1-methyl (**62**, Y = H) (85%) and

1,6,7-trimethyl-4(1*H*)-pteridinone (**62**, Y = Me) (79%), respectively, on heating in 2.5*M* hydrochloric acid at 100°C for 30 to 120 min[129]; likewise, 1-methyl- (**61**, R = Me, Y = H) and 1,6,7-trimethyl-4-methylimino-1,4-dihydropteridine (**61**, R = Y = Me) gave the same two respective products, (**62**, Y = H) (82%) and (**62**, Y = Me) (50%), but only in more concentrated acid under more vigorous conditions[129] (see also: Ch. IX, Sect. 6.A.)

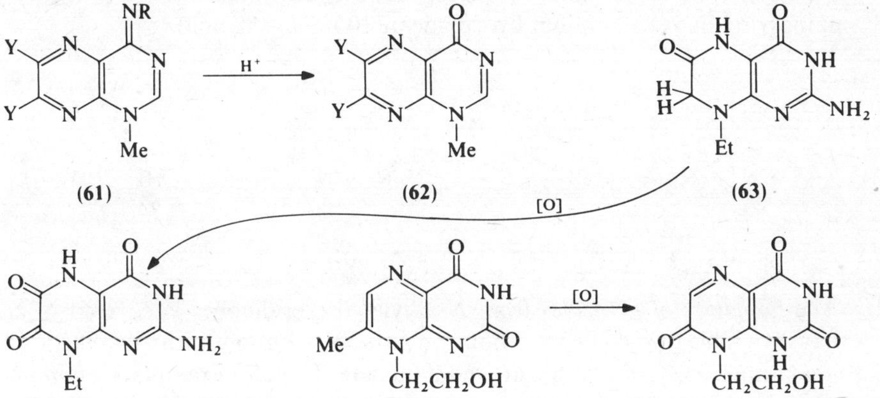

(61) (62) (63) (64) (65) (66)

The *oxidative insertion of an oxo substituent* can furnish pteridinones: for example, treatment of 2-amino-8-ethyl-7,8-dihydro-4,6(3*H*,5*H*)-pteridinedione (**63**) in acetic acid with oxygen in the presence of platinum oxide, gave 2-amino-8-ethyl-4,6,7(3*H*,5*H*,8*H*)-pteridinetrione (**64**)[183]; treatment of ethyl 2-imino-3,8-dimethyl-7-oxo-2,3,7,8-tetrahydro-6-pteridinecarboxylate with ferricyanide at pH 7 gave ethyl 2-amino-3,8-dimethyl-4,7-dioxo-3,4,7,8-tetrahydro-6-pteridinecarboxylate[376]; in a rather different way, treatment of 8-β-hydroxyethyl-7-methyl-2,4(3*H*,8*H*)-pteridinedione (**65**) with permanganate caused replacement of the methyl group by an oxo substituent to afford 8-β-hydroxyethyl-2,4,7(1*H*,3*H*,8*H*)-pteridinetrione (**66**) in 26% yield[1144]; and somewhat similarly, 6,7,8-trimethyl-2,4(3*H*,8*H*)-pteridinedione apparently gave 6,8-dimethyl-2,4,7(1*H*,3*H*,8*H*)-pteridinetrione.[1131]

The *photolysis of a pteridinethione* can give a pteridinone: for example, irradiation (> 320 nm) of 1,3,5-trimethyl-6-thioxo-5,6-dihydro- or 1,3,8-trimethyl-7-thioxo-7,8-dihydro-2,4(1*H*,3*H*)-pteridinedione gave 1,3,5-trimethyl-2,4,6(1*H*,3*H*,5*H*)- or 1,3,8-trimethyl-2,4,7(1*H*,3*H*,8*H*)-pteridinetrione.[632]

4. REACTIONS OF *N*-ALKYLPTERIDINONES

A. Degradative Reactions

Like the tautomeric pteridinones, the nontautomeric pteridinones sometimes undergo ring fission to give pyrimidines but more often to give pyrazines.

Pyrimidine formation is illustrated in the conversion of 8-methyl-7(8*H*)-pteridinone (**67**, R = H) or 6,8-dimethyl-7(8*H*)-pteridinone (**67**, R = Me) into 4-methylamino-5-pyrimidinamine (**68**) (1*M* sodium hydroxide or 0.05*M* sulfuric acid at 100°C for 2 h: ~70% yield in each case)[34]; of 8-methyl-6,7-diphenyl-2(8*H*)-pteridinone into 5-amino-4-methylamino-2(1*H*)-pyrimidinone (1*M* hydrochloric acid at 100°C for 12 h: >95%; 5*M* sodium hydroxide under reflux for 7 h: 66%)[189]; and of 5,8-dimethyl-6,7(5*H*,8*H*)-pteridinedione into 4,5-bismethylaminopyrimidine (1*M* sodium hydroxide at 100°C for 90 min).[34]

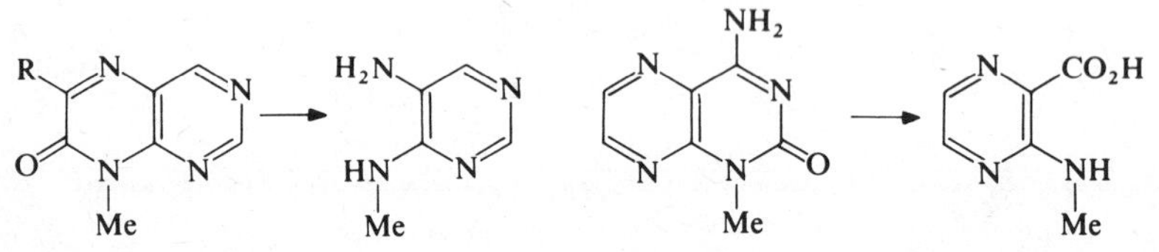

(**67**) (**68**) (**69**) (**70**)

The *formation of pyrazines* from *N*-alkylated pteridinones is covered in an exhaustive table of such degradations prepared by Gordon Barlin in 1982,[1480b] so that only a few typical cases are required here. Thus, for example, 4-amino-1-methyl-2(1*H*)-pteridinone (**69**) gave 3-methylamino-2-pyrazinecarboxylic acid (**70**) (1*M* sodium hydroxide at 100°C for 5 h: 39%)[130]; 7-amino-1-methyl-4(1*H*)-pteridinone (**71**) gave 5-amino-3-methylamino-2-pyrazinecarboxamide (1*M* sodium hydroxide at 100°C for 15 min: ~65%)[131]; 1,3-dimethyl-7-phenyl-2,4(1*H*,3*H*)-pteridinedione (**72**) gave *N*-methyl-3-methylamino-5-phenyl-2-pyrazinecarboxamide (**73**) (ethanolic alkali under reflux for 15 min: 99%),[176] 3-methylamino-5-phenyl-2-pyrazinecarboxylic acid (**76**) (ethanolic alkali at 200°C for 16 h: 80%),[176] 3-methylamino-5-phenyl-2-pyrazinecarboxamide (**74**) (ethanolic ammonia at 210°C for 20 h),[174] or 3-methylamino-5-phenyl-2-pyrazinecarbohydrazide (**75**) (85% hydrazine hydrate under reflux for 6 h:

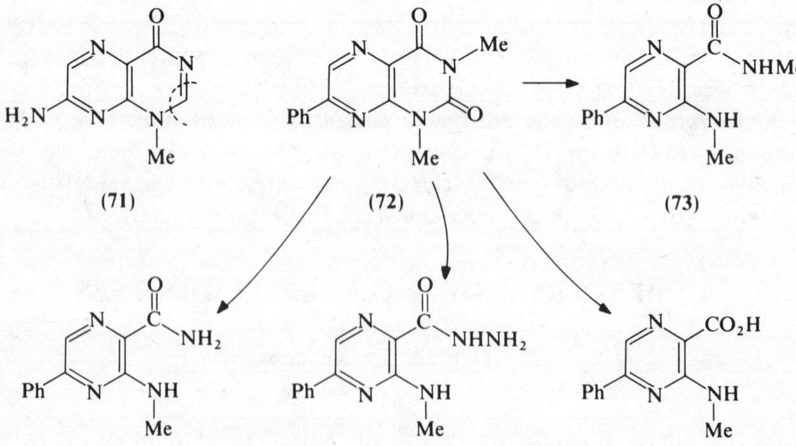

(**71**) (**72**) (**73**)

(**74**) (**75**) (**76**)

80%)[174]; 3-benzyl-2,6,7-triphenyl-4(3*H*)-pteridinone gave 3-amino-*N*-benzyl-5,6-diphenyl-2-pyrazinecarboxamide (ethanolic sodium ethoxide under reflux for 3 h: 93%)[476]; and many other such cases might be quoted.[167, 168, 209, 266, 311, 337, 364, 759, 1130, 1316]

B. Thiation

The thiation of *N*-alkylated pteridinones has been included in the coverage of the thiation of tautomeric pteridinones (Ch. VI, Sect. 2.G). For example, treatment of 1,3-dimethyl-6,7-diphenyl-2,4(1*H*,3*H*)-pteridinedione (**77**, X=O) with phosphorus pentasulfide in refluxing dioxane for 1 h gave 1,3-dimethyl-6,7-diphenyl-4-thioxo-3,4-dihydro-2(1*H*)-pteridinone (**77**, X=S) in 77% yield.[1679]

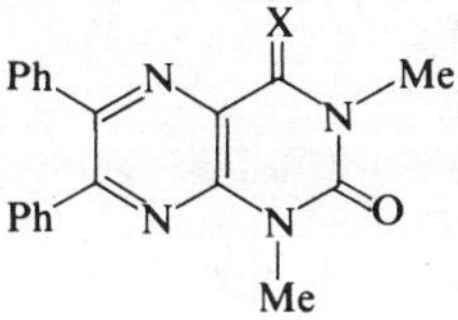

(**77**)

C. Other Reactions

4-Amino-8-benzyl-6-methyl-2,7(1*H*,8*H*)-pteridinedione (**78**) underwent an interesting ring fission and recyclization in boiling 5*M* sodium hydroxide for 7 h to give 4-benzylamino-6-methyl-2,7(1*H*,8*H*)-pteridinedione (**79**) in 50% yield.[363] The photolysis (in uv light) of 6,7-diphenyl-8-[β-(2′,3′,5′-tri-*O*-benzoyl-

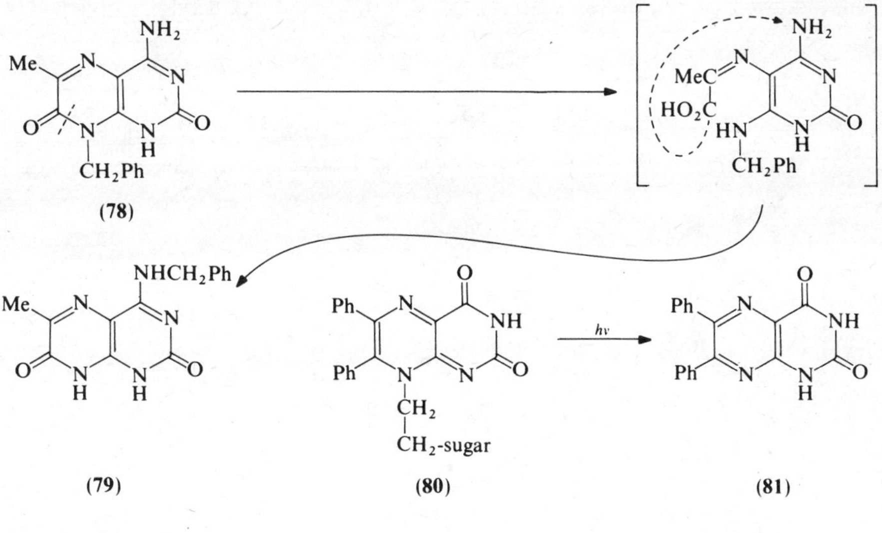

(**78**) (**79**) (**80**) (**81**)

β-D-ribofuranosyl)ethyl]-2,4(3*H*,8*H*)-pteridinedione (**80**) in methanol under nitrogen removed the 8-group completely to afford 6,7-diphenyl-2,4(1*H*,3*H*)-pteridinedione (**81**) in 66% yield[240]; 3-methyl-6,7-diphenyl-2,4(1*H*,3*H*)-pteridinedione was formed similarly in 76% yield from 8-β-hydroxyethyl-3-methyl-6,7-diphenyl-2,4(3*H*,8*H*)-pteridinedione.[409] Although the neutral molecule of 4(3*H*)-pteridinone resisted deuteration of its *C* protons, both 1-methyl-4(1*H*)- and 3-methyl-4(3*H*)-pteridinone underwent 2-deuteration with $t_{1/2}$ values of 2 and 80 h, respectively, under unstated conditions.[610]

5. PREPARATION OF PTERIDINE *N*-OXIDES

A. By Primary Syntheses

The primary synthesis of pteridine *N*-oxides *from pyrimidines* already carrying an *N*-oxide or potential *N*-oxide substituent is quite possible but has been little used. Known cases, which have been included in Chapter II, are exemplified in the condensation of 5,6-diamino-1-hydroxy-2,4(1*H*,3*H*)-pyrimidinedione (**82**, R=H) or its 1-benzyloxy analog (**82**, R=CH_2Ph) with glyoxal to give 2,4(3*H*,8*H*)-pteridinedione 1-oxide (**83**) or 1-benzyloxy-2,4(1*H*,3*H*)-pteridinedione, respectively, in 18 or 50% yield (a Gabriel and Colman synthesis)[1098]; of 6-methylthio-5-nitroso-2,4-pyrimidinediamine with *N*-(α-cyanobenzyl)pyridinium benzenesulfonate to give 4-methylthio-6-phenyl-2,7-pteridinediamine 5-oxide (**84**) (a Timmis synthesis)[328]; and of 2-dimethylamino-5-nitro-4-pyrimidinamine (**85**) with cyanoacetyl chloride to give 4-α-cyanoacetamido-2-dimethylamino-5-nitropyrimidine (**86**) and then (by cyclization in alkali at 40°C) 2-dimethylamino-7-oxo-7,8-dihydro-6-pteridinecarbonitrile 5-oxide (**87**).[490]

The primary synthesis of pteridine *N*-oxides *from pyrazine N-oxides*, and so on, has been used more extensively (see Ch. III and E. C. Taylor's review[599]).

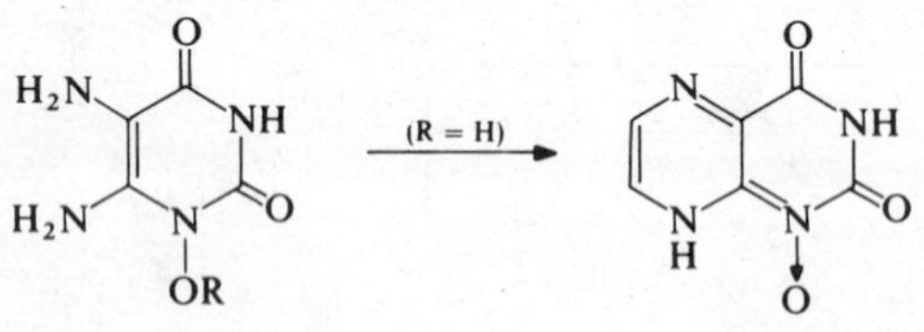

(82) **(83)**

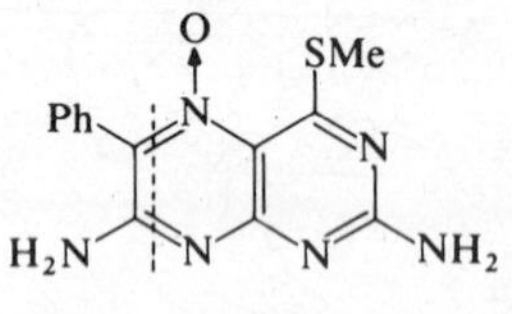

(84)

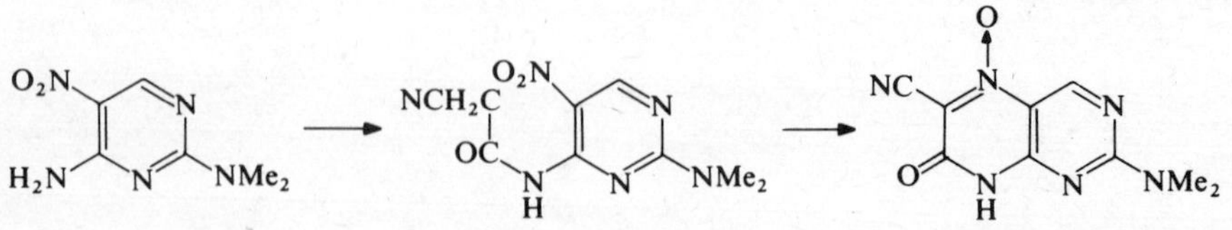

(85) **(86)** **(87)**

For example, treatment of 3-amino-2-pyrazinecarboxamide oxime (**88**) with triethyl orthoformate gave a good yield of 4-pteridinamine 3-oxide (**89**)[999, 1003]; similar treatment of 3-amino-6-methyl-2-pyrazinecarboxamide 4-oxide gave 6-methyl-4(3*H*)-pteridinone 8-oxide (**90**) in 64% yield[475, 485]; and 3-amino-2-pyrazinecarbonitrile 4-oxide with guanidine gave 2,4-pteridinediamine 8-oxide (**91**) in 90% yield.[477, 484] The reaction of 7-[1,2,5]oxadiazolo[3,4-*d*]pyrimidinamine 1-oxide (**92**) with methyl levulinate (i.e., methyl *β*-acetylpropionate) in methanolic ammonia at 40°C gave 7-methoxycarbonylmethyl-6-methyl-4-pteridinamine 5,8-dioxide (**93**) in 17% yield[723]; several other pteridines were made similarly (Ch. III, Sect. 3).

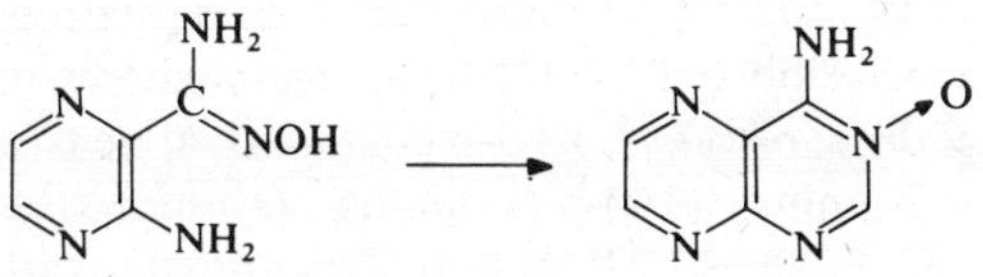

(**88**) (**89**)

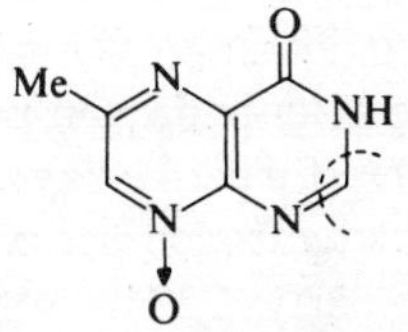

(**90**)

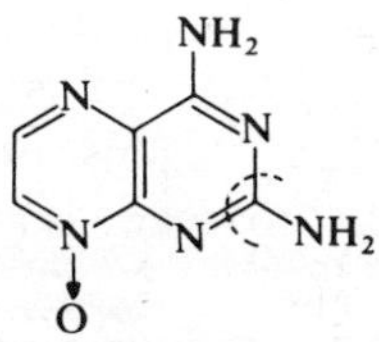

(**91**)

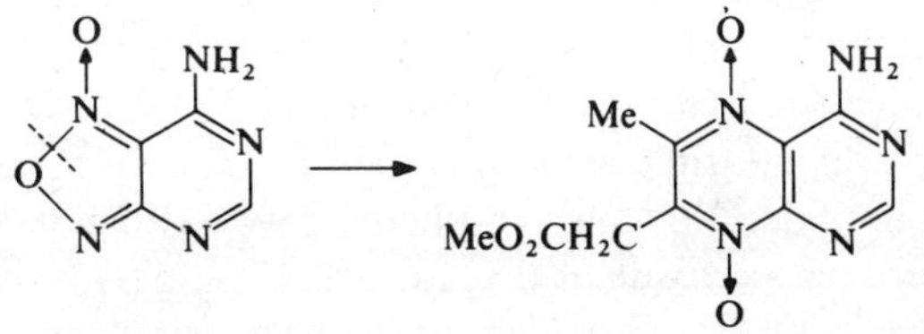

(**92**) (**93**)

B. By Direct *N*-Oxidation

Although the action of hydrogen peroxide on pteridine was first reported[30] to give a pteridine *N*-oxide, it was later realized[26, 1145] that the product was 4(3*H*)-pteridinone, formed by oxidation of the covalent 3,4-hydrate of pteridine present in the reaction mixture; this proved to be a common phenomenon in pteridine chemistry. However, many derivatives of 2,4(1*H*,3*H*)-pteridinedione, 2-amino-4(3*H*)-pteridinone, and 2,4-pteridinediamine have been converted successfully into *N*-oxides by peroxycarboxylic acids, often made in situ. The site of oxidation is normally N-8, but if it is sterically crowded by appropriately bulky groups in the 1- and/or 7-position(s), oxidation occurs at N-5. Naturally, 5-oxidation is likewise affected adversely by such 4- and/or 6-substituents.

Thus 8-*oxides* have been so made from 2,4(1*H*,3*H*)-pteridinedione [hydrogen peroxide in formic acid at 25°C for 7 days: 23%; the product (**94**) was originally described incorrectly[366] as the 5-oxide][367]; 3-methyl-2,4(1*H*,3*H*)-pteridinedione (trifluoroacetic acid–hydrogen peroxide for 4 h: 29%)[367]; 6,7-dimethyl-2,4(1*H*,3*H*)-pteridinedione (formic acid–hydrogen peroxide for 4 days: 18%; originally described incorrectly[366] as the 5-oxide);[367] 3,6,7-trimethyl-

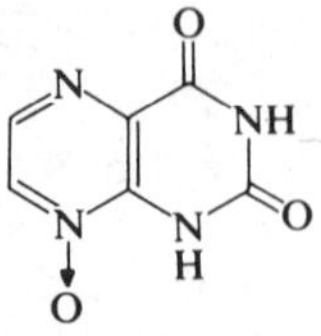

(94)

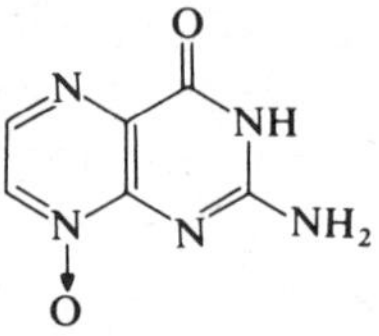

(95)

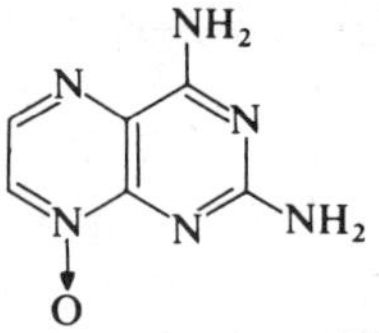

(96)

2,4(1*H*,3*H*)-pteridinedione (similarly: 28%; originally described incorrectly[366] as the 5-oxide)[367]; 6-phenyl-2,4(1*H*,3*H*)-pteridinedione (trifluoroacetic acid–hydrogen peroxide for 7 h: 75%)[367]; its 3-methyl derivative (formic acid–hydrogen peroxide for 6 days: 19%[367]; trifluoroacetic acid–hydrogen peroxide at 25°C for 8 h: 67%)[331]; 6,7-diphenyl-2,4(1*H*,3*H*)-pteridinedione (formic acid–hydrogen peroxide for 9 days: 64%)[367]; its 3-methyl derivative (similarly for 14 days: 63%)[367]; 2-amino-4(3*H*)-pteridinone [trifluoroacetic acid–hydrogen peroxide for 12 h: 33% of (**95**)][541]; its 6- or 7-methyl derivative (similarly: 33% or 43%, respectively)[541]; 2-amino-6,7-dimethyl-4(3*H*)-pteridinone (similarly for 24 h at 0°C: 42%)[541]; 2-amino-6-phenyl-4(3*H*)-pteridinone (similarly for 7 h at 0°C: 69%)[541]; 2-amino-6,7-diphenyl-4(3*H*)-pteridinone (similarly for 8 h at 25°C: 67%)[541]; 2,4-pteridinediamine [trifluoroacetic acid–hydrogen peroxide for 12 h at 0°C: 28% of (**96**)][541]; its 6,7-dimethyl derivative (formic acid–hydrogen peroxide for 3 days at 25°C: 20%, plus some 5,8-dioxide)[541]; and 6-phenyl-2,4-pteridinediamine (trifluoroacetic acid–hydrogen peroxide at 0°C for 9 h: 17%, plus some 5,8-dioxide).[541]

In contrast, 5-*oxides* have been similarly made (using trifluoroacetic acid–hydrogen peroxide at 25°C for the stated time, unless otherwise indicated) from 1-methyl-2,4(1*H*,3*H*)-pteridinedione [2 h: 49% of (**97**)][366, 367]; its 1,3-dimethyl analog (15 h: 37%)[367]; its 1,3,6-trimethyl analog [trichloroacetic acid–hydrogen peroxide: 48%; plus the 8-oxide (small amount, unisolated) and the 5,8-dioxide (7%)][1704]; its 1,6,7-trimethyl analog (3 h: 89%)[366, 367]; its 1,3,6,7-tetramethyl analog (45 min: 79%)[366, 367] 6,7-diisopropyl-1-methyl-2,4(1*H*,3*H*)-pteridinedione (2 h: 55%)[367]; 6,7-diisopropyl-1,3-dimethyl-2,4(1*H*,3*H*)-pteridinedione (1.5 h: 53%)[367]; 7-*t*-butyl-3-methyl-2,4(1*H*,3*H*)-pteridinedione (5 h: 40%)[367]; 7-phenyl-2,4(1*H*,3*H*)-pteridinedione (4 h: 24%; plus some 5,8-dioxide)[367]; its 3-methyl derivative (7 h: 44%; plus some 5,8-dioxide)[367]; its 1,3-dimethyl analog (formic acid–hydrogen peroxide at 60°C for 15 h: 82%)[367]; 1,3-dimethyl-6-phenyl-2,4(1*H*,3*H*)-pteridinedione (2 h: 26%)[367]; 1-methyl-6,7-diphenyl-2,4(1*H*,3*H*)-pteridinedione (24 h: 85%)[367]; its 1,3-dimethyl analog (2

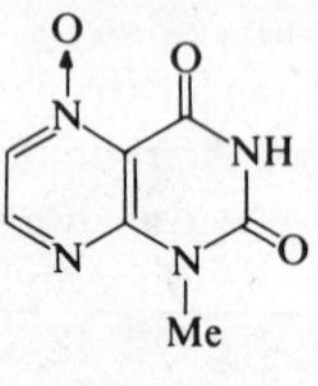

(97)

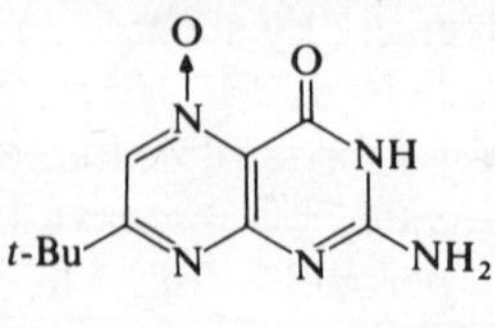

(98)

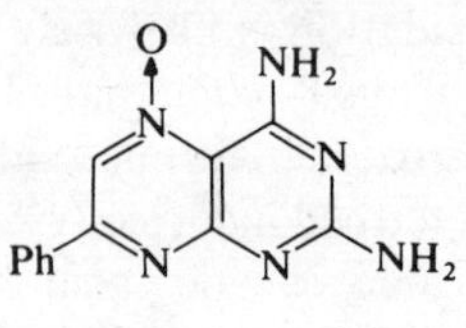

(99)

days: 66%)[367]; 2-amino-7-*t*-butyl-4(3*H*)-pteridinone [7 h: 49% of (**98**)][541]; its 7-phenyl analog (18 h: 42%)[541]; 2-amino-1,6,7-trimethyl-4(3*H*)-pteridinone (35 min: 70%)[541]; 7-phenyl-2,4-pteridinediamine [10 h at 0–5°C: 52% of (**99**)][541]; 7-methoxy-1,3-dimethyl-2,4(1*H*,3*H*)-pteridinedione (10 h at 0°C: 95%)[238]; its 1,3,6-trimethyl analog (formic acid–hydrogen peroxide for 2 days: 70%)[238]; 7-methoxy-3,6-dimethyl-1-phenyl-2,4(1*H*,3*H*)-pteridinedione (1 h at 35°C: 81%)[238]; 7-amino-1,3-dimethyl-2,4(1*H*,3*H*)-pteridinedione (3.5 h: 65%)[238]; 1,3,6-trimethyl-2,4,7(1*H*,3*H*,8*H*)-pteridinetrione 30 min at <40°C: 29%)[238]; 3,6-dimethyl-1-phenyl-2,4,7(1*H*,3*H*,8*H*)-pteridinetrione (1.5 h: 24%)[238]; and other pteridinones.[210]

In addition, 5,8-*dioxides* have been made (using trifluoroacetic acid–hydrogen peroxide at 25°C for the stated time) from 6,7-dimethyl-2,4(1*H*,3*H*)-pteridinedione [12 h: 15% of (**100**)][367]; its 3,6,7-trimethyl analog (3 h: 38%)[367]; 7-phenyl-2,4(1*H*,3*H*)-pteridinedione (see 5-oxide just cited: 32%)[367]; its 3-methyl derivative (see 5-oxide just cited: 28%[367]; or for 2 days: 59%[331]; also from the 5-oxide similarly: 66%)[331]; 2-amino-6,7-dimethyl-4(3*H*)-pteridinone (3 days at 0–5°C: 54%)[541]; its 6,7-diphenyl analog [1 day: 34% of (**101**); also from the 8-oxide: 22%][541]; 2,4-pteridinediamine [26 h at 5°C: 33% of (**102**, R = H)][541]; 6,7-dimethyl-2,4-pteridinediamine (8 h at 0°C: 23%; formic acid–hydrogen peroxide for 7 days at 25°C: 57%)[541]; 7-phenyl-2,4-pteridinediamine (16 h at 0–5°C: 11%; cf. 5-oxide cited previously)[541]; its 6-phenyl isomer [5 h: 65% of (**102**, R = Ph)][541]; the 6,7-diphenyl analog (12 h at 5°C: 68%)[541]; and 6-phenyl-2,4,7-pteridinetriamine (15 h at 5°C: 65%).[541]

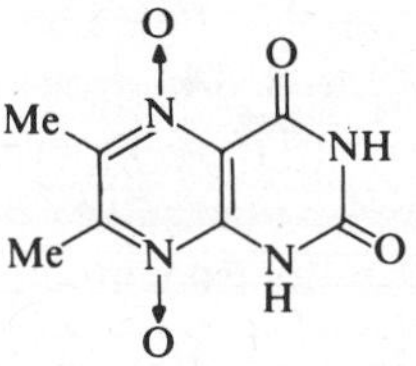

(100)

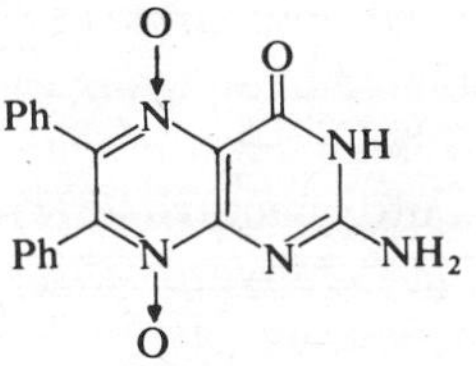

(101)

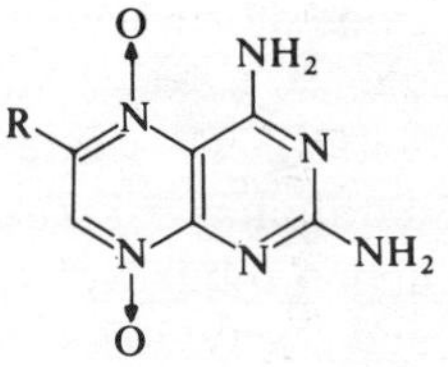

(102)

C. By Other Means

The conversion of an *O*-acylated pteridine *N*-oxide into the parent *N*-oxide is illustrated in the deacylation of benzenesulfonyloxy-2,4(1*H*,3*H*)-pteridinedione (**103**) by brief alkaline hydrolysis to give 3-hydroxy-2,4(1*H*,3*H*)-pteridinedione.[1445]

An interesting route to some pteridine *N*-oxides involves treatment of 4(3*H*)-pteridinones with hydroxylamine. Thus 7-methyl-4(3*H*)-pteridinone appeared to undergo cleavage of its 3,4-bond by hydroxylamine to give an intermediate pyrazine (**104**), which recyclized spontaneously to afford 7-methyl-4(1*H*)-pteridinone 3-oxide (**105**) in ~80% yield[154]; 6-methyl-, 2,7-dimethyl-, and 6,7-dimethyl-4(3*H*)-pteridinone likewise gave their respective 3-oxides in reasonable

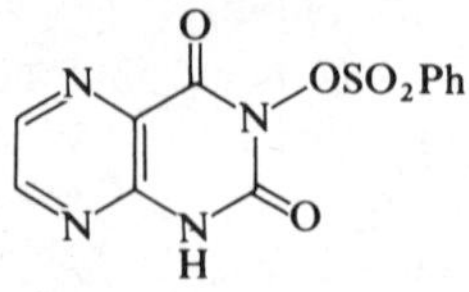

(103)

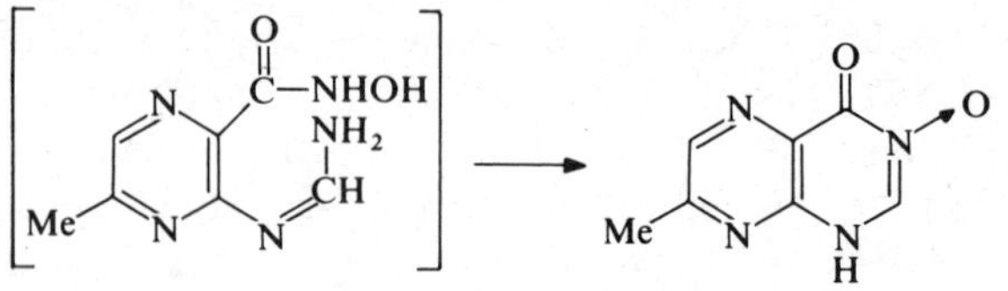

(104)

(105)

yield but, although similar products from the parent 4(3*H*)-pteridinone and its 2-methyl derivative could be detected in solution, they reacted further with hydroxylamine to afford 5,6-diamino-4(3*H*)-pyrimidinone and its 2-methyl derivative, respectively, among other degradation products.[154]

6. REACTIONS OF PTERIDINE *N*-OXIDES

A. Deoxygenation

The reductive removal of oxygen from an *N*-oxide is exemplified in the treatment of 2-amino-6-methyl-4(3*H*)-pteridinone 8-oxide (**106**) with dithionite to give 2-amino-6-methyl-7,8-dihydro-4(3*H*)-pteridinone and then, by permanganate oxidation, 2-amino-6-methyl-4(3*H*)-pteridinone.[478] Both 7-methoxy-1,3,6-trimethyl- (**107**, R = Me) and 7-methoxy-3,6-dimethyl-1-phenyl-2,4(1*H*,3*H*)-pteridinedione 5-oxide (**107**, R = Ph) underwent removal of their *N*-oxide substituent in boiling acetic anhydride to afford the parent pteridines in 52 to 53% yield.[238] When irradiated in aqueous solution at pH 7 to 8, 3-hydroxy-2,4(1*H*,3*H*)-pteridinedione (**108**) underwent deoxygenation to 2,4(1*H*,3*H*)-pteridinedione.[1099] 2-Amino-6-butoxymethyl-4(3*H*)-pteridinone 8-oxide gave 2-amino-6-butoxymethyl-5,6,7,8-tetrahydro-4(3*H*)-pteridinone on hydrogenation and several analogs were made thus.[1728]

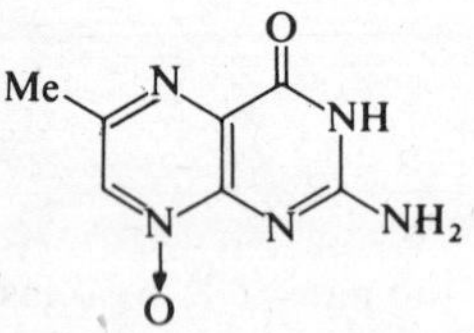

(106)

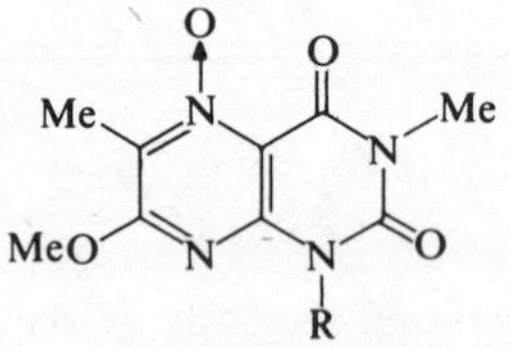

(107)

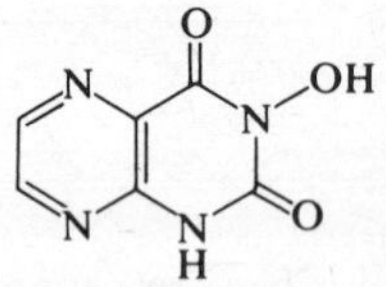

(108)

B. Covalent Adduct Formation

As might be expected, pteridine *N*-oxides are prone to covalent addition. For example, 2(3*H*)-pteridinone 1-oxide (**109**), on boiling in methanol or ethanol for 1 h, gave 4-methoxy- (**110**, R = Me) or 4-ethoxy-3,4-dihydro-2(8*H*)-pteridinone

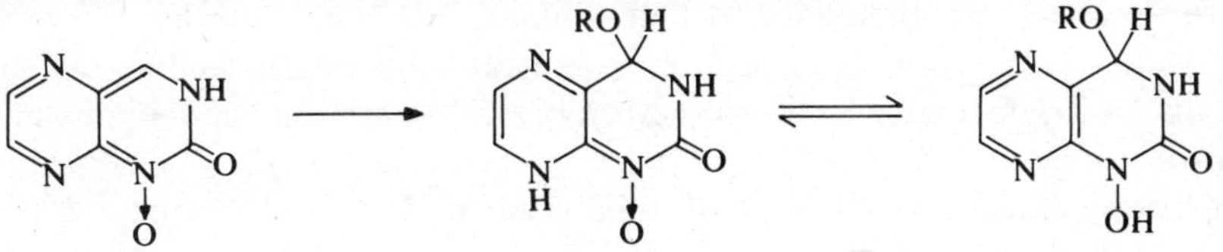

(109) **(110)** **(110a)**

1-oxide (**110**, R = Et)[1098]; the corresponding covalent hydrate (**110**, R = H) and the just cited alcoholates have been formulated as their tautomeric *N*-hydroxy isomers (**110a**), apparently on nmr evidence.[1098]

C. Degradation Reactions

Pteridine *N*-oxides undergo a variety of degradation reactions to give usually either pyrimidines or (more often) pyrazines. Such processes are illustrated in the conversion of 4(1*H*)-pteridinone 3-oxide (**111**, R = Y = H) or its 6-methyl derivative (**111**, R = H, Y = Me) into 5,6-diamino-4(1*H*)-pyrimidinone 3-oxide (**112**, R = H) by heating with aqueous hydroxylamine at 90°C for 5 and 20 h, respectively[154]; of 2-methyl-4(1*H*)-pteridinone 3-oxide (**111**, R = Me, Y = H) into 5,6-diamino-2-methyl-4(1*H*)-pyrimidinone 3-oxide (**112**, R = Me) by similar means[154]; of 4-pteridinamine 3-oxide (**113**, R = H) or its 2-methyl derivative (**113**, R = Me) into 2-formamido-3-(5′-methyl-1′,2′,4′-oxadiazol-3′-yl)pyrazine (**114**, R = Y = H) (52%) or its 2-diacetylamino analog (**114**, R = Me, Y = Ac) (37%), respectively, by stirring with acetic anhydride at 25°C for 8 h[999]; of the similar substrates, 4-dimethylaminomethyleneaminopteridine 3-oxide (**115**, R = H), its 2-methyl derivative (**115**, R = Me), or its 2-phenyl derivative (**115**, R = Ph) into 3-(1′,2′,4′-oxadiazol-3′-yl)-2-pyrazinamine (**116**, Y = NH_2) in

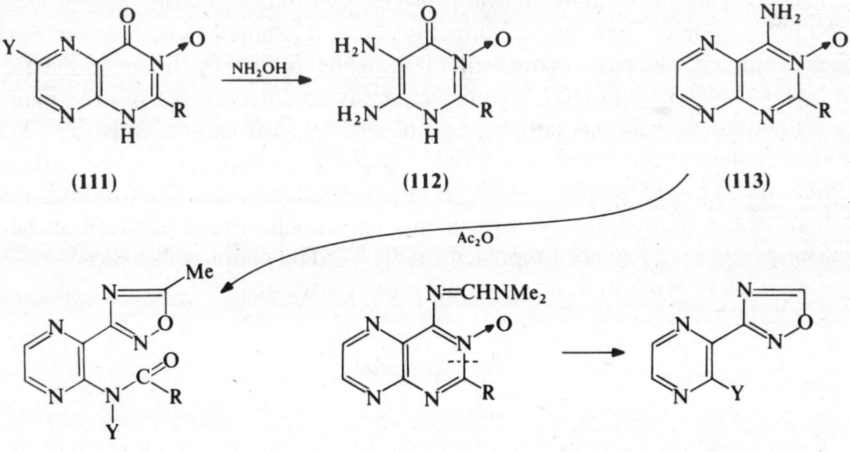

(111) **(112)** **(113)**

(114) **(115)** **(116)**

50 to 80% yield by stirring in 5*M* hydrochloric acid at 25°C for 8 h[1004]; of this substrate (**115**, R=H) into 2-dimethylaminomethyleneamino-3-(1′,2′,4′-oxadiazol-3′-yl)pyrazine (**116**, Y = Me_2NCH:N) (methanolic dimethylamine hydrochloride at 25°C for 5 h: 41%), into 2-formamido-3-(1′,2′,4′-oxadiazol-3′-yl)pyrazine (**116**, Y=NHCHO) (formic acid at 25°C for 35 min: 80%), or into 2-(hydroxyiminomethyl)amino-3-(1′,2′,4′-oxadiazol-3′-yl)pyrazine (**116**, Y=HON:CHNH) (aqueous hydroxylamine hydrochloride at 25°C for 2 h: 77%)[1004]; of the substrate (**115**, R=Ph) into 2-benzamido-3-(1′,2′,4′-oxadiazol-3′-yl)pyrazine (**116**, R=NHBz) (5*M* hydrochloric acid at 25°C for 35 min: 80%)[1004]; of 2-phenyl-4-pteridinamine 3-oxide (**113**, R=Ph) into (mainly) 3-(α-hydroxyiminobenzyl)amino-2-pyrazinecarboxylic acid (**117**) (46%) by brief heating in alkali[1004]; of 1,6,7-trimethyl- (**118**, R=H) or 1,3,6,7-tetramethyl-2,4(1*H*,3*H*)-pteridinedione 5-oxide (**118**, R=Me) into 5,6-dimethyl-3-methylamino-2-pyrazinecarboxylic acid 1-oxide (**119**, Y=OH) [refluxing 1*M* alkali for 5 h: 45%; presumably via the amide (**119**, Y=NH_2)] or 5,6,*N*-trimethyl-3-methylamino-2-pyrazinecarboxamide 1-oxide (**119**, Y=NHMe) (0.1*M* alkali under reflux for 30 min: 68%), respectively[367]; and of 6-phenyl-2,4-pteridinediamine 5,8-dioxide (**102**, R=Ph) into 6-amino-1,3,5-triazine-2,4(1*H*,3*H*)-dione (30%) on treatment with hydrogen peroxide in formic acid at 25°C for 6 days by a fascinating but complicated mechanism.[542]

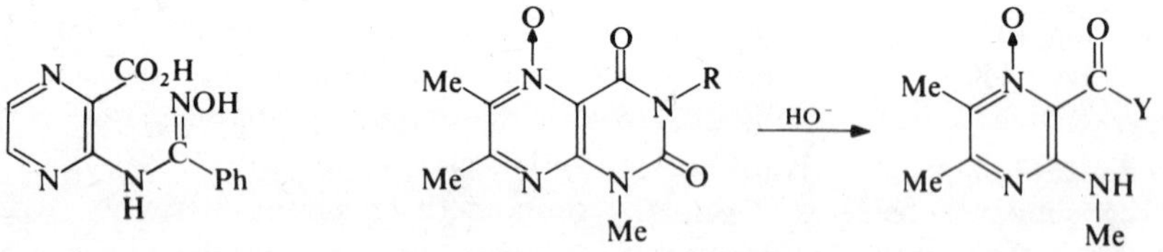

(**117**) (**118**) (**119**)

The mass spectral fragmentation of 4(1*H*)-pteridinone 3-oxide and of its 2-methyl, 6-methyl, 7-methyl, 2,7-dimethyl, and 6,7-dimethyl derivatives have been examined. The parent compound fragmented mainly by the successive loss of NO, CO, HCN, and HCN again; the methylated derivatives showed major peaks consistent with this pattern, as modified by their substituent(s).[1226] The initial fragmentations of the 5- or 8-oxides and the 5,8-dioxides of some thirty 2-amino-4(3*H*)-pteridinones, 2,4(1*H*,3*H*)-pteridinediones, and 2,4-pteridinediamines (some bearing a methyl or phenyl substituent at one or more of their vacant positions) have been reported briefly.[427] The major peaks shown by 3-hydroxy-2,4(1*H*,3*H*)-pteridinedione have been discussed.[1289]

D. Acylation

Those pteridine *N*-oxides that are capable of tautomerism to *N*-hydroxypteridines may be *O*-acylated. Thus 4(1*H*)-pteridinone 3-oxide (**120**)/(**121**, R=H), prepared from a pyrazine in the presence of an excess of acetic anhydride, was accompanied by 3-acetoxy-4(3*H*)-pteridinone (**121**, R = Ac) of unconfirmed structure[540]; more convincingly, 2,4(3*H*,8*H*)-pteridinedione 1-oxide, tautomeric with 1-hydroxy-2,4(1*H*,3*H*)-pteridinedione (**122**, R = H), gave 1-acetoxy- (**122**, R = Ac) (89%), 1-mesyloxy- (**122**, R = $MeSO_2$) (70%), or 1-tosyloxy-2,4(1*H*,3*H*)-pteridinedione (**122**, R = *p*-$MeC_6H_4SO_2$) (96%) on heating in acetic anhydride–acetic acid, mesyl chloride–pyridine, or tosyl chloride–alkali, respectively.[1098] When there is an unoccupied carbon atom suitably placed, however, rearrangement to the isomeric *C*-acyloxy derivative usually ensues. For example, 2-amino-6-phenyl-4(3*H*)-pteridinone 8-oxide, which can be written as the *N*-hydroxy tautomer (**123**), in boiling acetic anhydride gave an uncharacterized acetyl derivative, probably (**124**) on account of its alkaline hydrolysis to afford 2-amino-6-phenyl-4,7(3*H*,8*H*)-pteridinedione (**125**).[541] Likewise 2,4(1*H*,3*H*)-pteridinedione 8-oxide in boiling acetic anhydride gave an acylated intermediate that underwent gentle acidic hydrolysis to afford 2,4,6(1*H*,3*H*,5*H*)-pteridinetrione (rather than the expected 2,4,7-trione) by a rare but explainable *β*-rearrangement.[367]

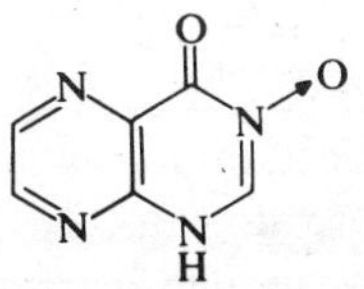

(120)

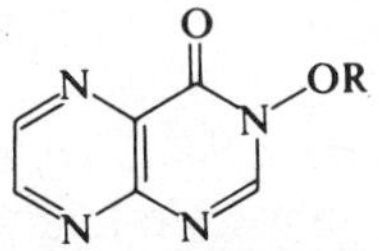

(121)

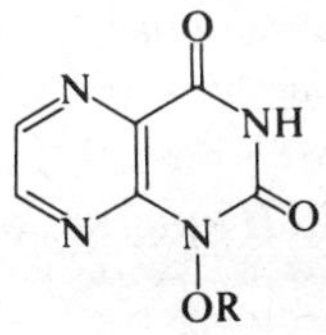

(122)

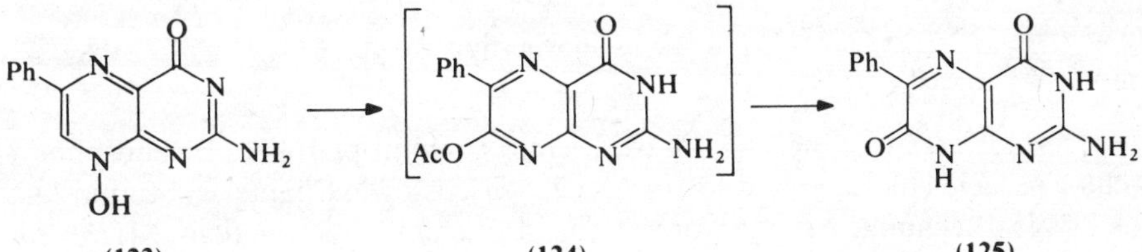

(123) **(124)** **(125)**

Pteridine *N*-oxides, which cannot exist as *N*-hydroxy tautomers, have no choice but to rearrange if they are to react with acetic anhydride, and so on. Thus 1-methyl-2,4(1*H*,3*H*)-pteridinedione 5-oxide (**126**) in boiling acetic anhydride for 2 h gave 6-acetoxy-1-methyl-2,4(1*H*,3*H*)-pteridinedione (**127**) in 59% yield[367]; 1,3,6-trimethyl-2,4(1*H*,3*H*)-pteridinedione 5-oxide similarly gave 6-acetoxymethyl-1,3-dimethyl-2,4(1*H*,3*H*)-pteridinedione (66%; note the final extranuclear location of the acetoxy group, attained by a mechanism usually known as the Pachter rearrangement[1705] or the Boekelheide rearrangement[1706])[1704]; 3-methyl-7-phenyl-2,4(1*H*,3*H*)-pteridinedione 5,8-dioxide in boiling acetic anhydride gave 6-acetoxy-3-methyl-7-phenyl-2,4(1*H*,3*H*)-pteridinedione 8-oxide (**128**) (65%; note the impossibility of further rearrangement and hence the survival of the 8-oxide substituent)[331]; in contrast to the last

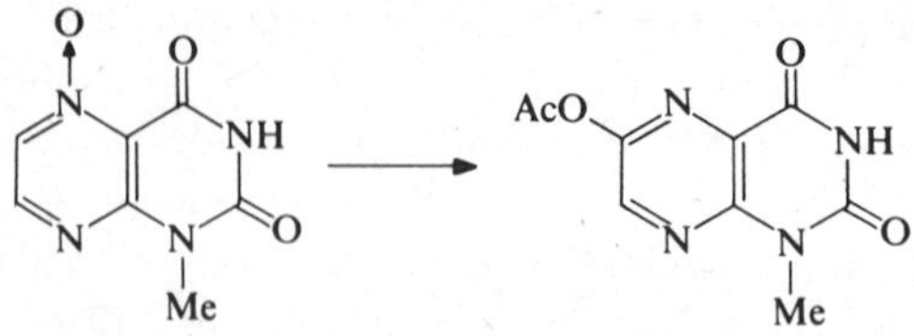

(126) (127)

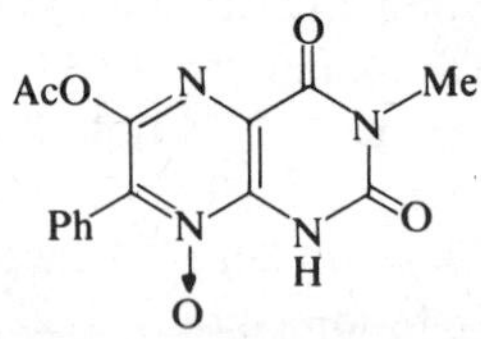

(128)

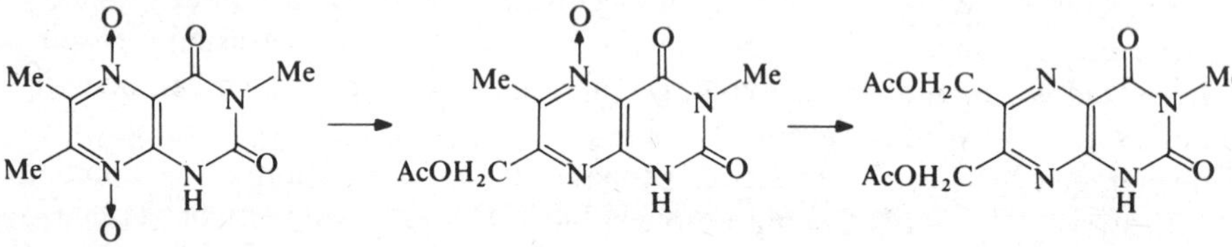

(129) (130) (131)

example, and for the obvious reason, 3,6,7-trimethyl-2,4(1*H*,3*H*)-pteridinedione 5,8-dioxide (**129**) in boiling acetic anhydride rapidly gave 7-acetoxymethyl-3,6-dimethyl-2,4(1*H*,3*H*)-pteridinedione 5-oxide (**130**) and then more slowly, 6,7-bisacetoxymethyl-3-methyl-2,4(1*H*,3*H*)-pteridinedione (**131**).[367] The conversion of 6-phenyl-2,4-pteridinediamine 5,8-dioxide into 2,4-diacetamido-6-phenyl-7(8*H*)-pteridinone by acetic anhydride followed by an aqueous work-up, is a reminder that acetic anhydride can deoxygenate as well as induce rearrangement of *N*-oxides.[541]

E. Chlorination

The conversion of pteridine *N*-oxides into *C*-chloropteridines by phosphoryl chloride, acid chlorides, and so on, has been discussed in Chapter V, Section 1.C. A typical example is the formation of 6-chloro-1,3-dimethyl-2,4(1*H*,3*H*)-pteridinedione (89%) by treatment of 1,3-dimethyl-2,4(1*H*,3*H*)-pteridinedione 5-oxide with acetyl chloride at −65°C followed by trifluoroacetic acid at −30°C.[1275]

F. Dimroth-Type Rearrangements

Pteridine *N*-oxides may undergo Dimroth rearrangements with or without the direct involvement of the *N*-oxide substituent. Thus 2-amino-3,6-dimethyl-4(3*H*)-pteridinone 8-oxide (**132**) underwent a perfectly normal Dimroth rearrangement in hot dilute alkali to give 6-methyl-2-methylamino-4(3*H*)-pteridinone 8-oxide (**133**) in 90% yield.[485] In contrast, 4-pteridinamine 3-oxide (**134**, R = H) or its 2-phenyl derivative (**134**, R = Ph) on prolonged boiling in

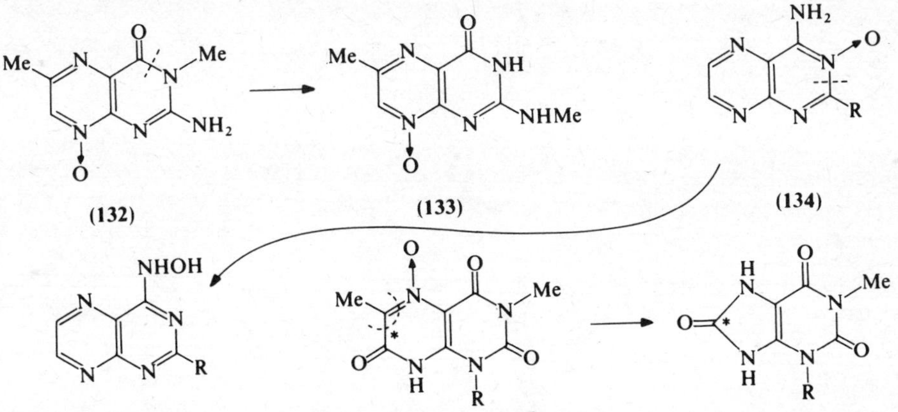

(132) **(133)** **(134)**

(135) **(136)** **(137)**

water gave 4-hydroxyaminopteridine (**135**, R = H) (73%) or its 2-phenyl derivative (**135**, R = Ph) (31%), respectively[1003, 1004]

G. Transformation into Purines

When 1,3,6-trimethyl-2,4,7(1*H*,3*H*,8*H*)-pteridinetrione 5-oxide (**136**, R = Me) or its 1-phenyl analog (**136**, R = Ph) was dissolved in refluxing acetic anhydride, the expected Pachter–Boekelheide rearrangement did not take place. Instead, ring contraction occurred by a mechanism involving fission of the 6,7-bond, subsequent loss of C-6 together with its attached methyl group, and final reclosure to afford 1,3-dimethyl- (**137**, R = Me) (84%) or 1-methyl-3-phenyl-2,6,8(1*H*,3*H*,7*H*)-purinetrione (**137**, R = Ph) (62%), respectively.[236, 238] A tagging experiment showed clearly that C-7 in the pteridine became C-8 in the purine.[238]

H. Amination

Treatment of 2,4-pteridinediamine 8-oxide in pyrrolidine at 95°C for 6 h gave 7-pyrrolidino-2,4-pteridinediamine in 62% yield. It was identified by its subsequent alkaline hydrolysis to 2-amino-4,7(3*H*,8*H*)-pteridinedione.[1301]

CHAPTER VIII

The Sulfur-Containing Pteridines

This chapter includes the tautomeric and nontautomeric pteridinethiones; the extranuclear mercaptopteridines; the dipteridinyl sulfides and disulfides; the pteridine sulfenic, sulfinic, and sulfonic acids; the alkylthio- and arylthiopteridines; and the alkylsulfinyl- and alkylsulfonylpteridines (sulfoxides and sulfones). Some of these categories are poorly represented in the pteridine literature up to the present.

1. PREPARATION OF PTERIDINETHIONES

A. By Primary Syntheses

Many of the tautomeric pteridinethiones and probably all of the few known (nontautomeric) *N*-alkylpteridinethiones have been made by primary syntheses from pyrimidines (Ch. II) or from pyrazines or other systems (Ch. III). Typical examples include the condensation of 4,5-diamino-2(1*H*)-pyrimidinethione (**1**, R=H) or its 6-methyl derivative (**1**, R=Me) with glyoxal to give 2(1*H*)-pteridinethione (**2**, R=H) in 52% yield[878] (the first monosubstituted pteridine ever made[22]) or its 4-methyl derivative (**2**, R=Me) (57%),[46] respectively; of 4,5,6-triamino-2(1*H*)-pyrimidinethione with benzil,[472] or of 3-chloro-5,6-

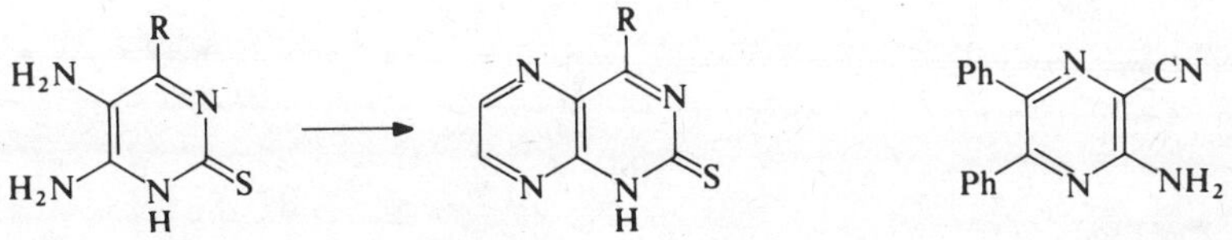

(1) (2) (3)

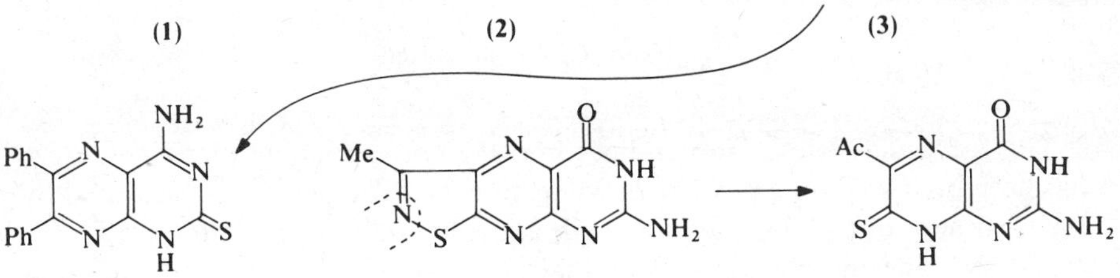

(4) (5) (6)

diphenyl-2-pyrazinecarbonitrile (**3**) with thiourea,[483] to give 4-amino-6,7-diphenyl-2(1*H*)-pteridinethione (**4**) in 84 or 51% yield, respectively; of 5,6-diamino-4(3*H*)-pyrimidinethione with glyoxal,[33] or of 3-amino-2-pyrazinecarbothioamide with triethyl orthoformate,[30] to give 4(3*H*)-pteridinethione in 83 or 84% yield, respectively; and of 5-acetamido-3-methyl-4-nitrosoisothiazole with 2,6-diamino-4(3*H*)-pyrimidinone to give 7-amino-3-methyl-5(6*H*)-isothiazolo[4,5-*g*]pteridinone (**5**), which underwent dithionite reduction to afford 6-acetyl-2-amino-7-thioxo-7,8-dihydro-4(3*H*)-pteridinone (**6**) in >90% yield.[1309] The formation of a nontautomeric product is exemplified in the condensation of 5-amino-4-methylamino-2(1*H*)-pyrimidinethione with biacetyl to give 6,7,8-trimethyl-2(8*H*)-pteridinethione (**7**) (~80%)[248]; also of 3-amino-*N*-butyl-5,6-diphenyl-2-pyrazinecarbothioamide (**8**) with formic acid (or triethyl orthoformate) in acetic anhydride to give 3-butyl-6,7-diphenyl-4(3*H*)-pteridinethione (**9**) in 60–70% yield.[474]

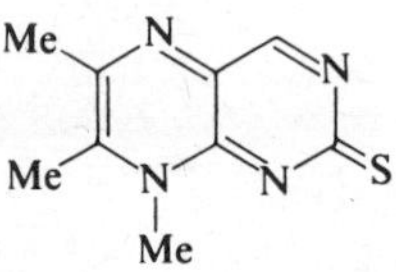

(**7**)

(**8**) (**9**)

B. By Covalent Addition

The covalent addition of hydrogen sulfide to a pteridine with subsequent aerial oxidation occurred when a stream of hydrogen sulfide was passed into a solution of 6(5*H*)-pteridinone in dilute hydrochloric acid contained in an open vessel to afford up to 70% yield (after 2 days) of 7-thioxo-7,8-dihydro-6(5*H*)-pteridinone (**10**).[36]

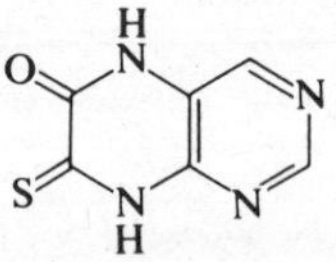

(**10**)

C. From Chloropteridines

The conversion of chloropteridines into the corresponding pteridinethiones has been discussed fully in Chapter V, Section 2.E. The process is illustrated by the treatment of 6-chloropteridine with aqueous sodium sulfide to give 6(5*H*)-pteridinethione (initially as its sodium salt) in 84% yield.[36] Likewise, 2-amino-7-chloro-4-oxo-3,4-dihydro-6-pteridinecarboxylic acid with aqueous sodium hydrogen sulfide gave 2-amino-4-oxo-7-thioxo-3,4,7,8-tetrahydro-6-pteridinecarboxylic acid.[906]

D. By Thiation of Pteridinones

The thiation of both tautomeric and nontautomeric pteridinones has been covered in Chapter VI, Section 2.G. A typical example from each type of substrate is the conversion of 6,7-dimethyl-2-thioxo-1,2-dihydro-4(3*H*)-pteridinone into 6,7-dimethyl-2,4(1*H*,3*H*)-pteridinedithione (phosphorus pentasulfide in refluxing pyridine: 45%) and 3-methyl-2-methylthio-6,7-diphenyl-4(3*H*)-pteridinone into the corresponding pteridinethione (similarly: 44%).[437]

E. By Thiolysis of Alkylthiopteridines

An occasionally useful reaction is to convert an alkylthiopteridine into its parent pteridinethione. Thus 4-benzylthio-2-pteridinamine, made by primary synthesis, was treated with ethanolic sodium hydrogen sulfide for 30 min to give 2-amino-4(3*H*)-pteridinethione in 50% yield.[305] Rather similarly, 4-pentyloxy-6-propionyl-7-propylthio-2-pteridinamine with sodium hydrogen sulfide in DMF gave 2-amino-4-pentyloxy-6-propinoyl-7(8*H*)-pteridinethione.[629] 6,7-Dimethyl-4-methylthiopteridine with ammonium hydrogen sulfide gave 6,7-dimethyl-4(3*H*)-pteridinethione.[313]

The conversion of an alkylthiopteridine into a nontautomeric pteridinethione is possible. Thus quaternization of 6,7-dimethyl-4-methylthiopteridine with methyl iodide gave 1,6,7-trimethyl-4-methylthio-1-pteridinium iodide (**11**), which underwent thiolysis in aqueous ammonium sulfide to give 1,6,7-trimethyl-4(1*H*)-pteridinethione (**12**) in 40% yield.[313]

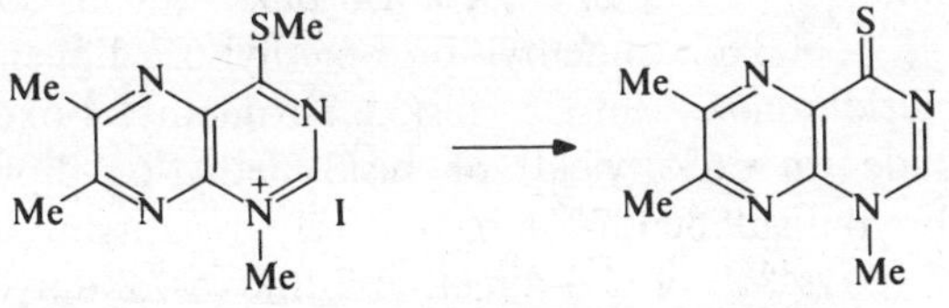

(**11**) (**12**)

F. By Reductive Processes

Reduction of 1,3,6-trimethyl-2,4-dioxo-1,2,3,4-tetrahydro-7-pteridinesulfenic acid by sodium borohydride in dilute aqueous sodium hydrogen carbonate during 12 h gave 1,3,6-trimethyl-7-thioxo-7,8-dihydro-2,4(1*H*,3*H*)-pteridinedione in 54% yield.[1709]

2. PROPERTIES AND REACTIONS OF PTERIDINETHIONES

As in other series, the potentially tautomeric pteridinethiones are assumed to exist predominantly as such, rather than as pteridinethiols. While there appears to be only incomplete experimental evidence, such as the uv spectra of Michel Polonovski,[1414] to uphold this assumption, there really seems little reason to doubt it. The cupric complex of 4(3*H*)-pteridinethione,[1103] as well as the cupric and cobaltous complexes of 6,7-diphenyl-4(3*H*)-pteridinethione,[754] have been accorded cursory examination.

A. Desulfurization

Possibly because of facile ring reduction and/or complex formation, no successful desulfurizations of pteridinethiones by Raney-nickel have been reported.[629, 1309] However, treatment of 2-amino-4-pentyloxy-6-propionyl-7(8*H*)-pteridinethione with Raney-cobalt in aqueous ethanol did give 4-pentyloxy-6-propionyl-2-pteridinamine in 44% yield.[629] Desulfurization of alkylthiopteridines has been marginally more successful (see Sect. 7.A next).

B. Oxidative Reactions

The oxidation of (tautomeric) pteridinethiones to dipteridinyl disulfides, pteridinesulfenic acids, pteridinesulfinic acids, or pteridinesulfonic acids has been used all too little, although the mechanism of the photooxidation of 1,3-dimethyl-6-thioxo-5,6-dihydro-2,4(1*H*,3*H*)-pteridinedione has been studied in some detail.[1708] The *formation of disulfides* is exemplified in the conversion of 2-amino-4-pentyloxy-7(8*H*)-pteridinethione into bis(2-amino-4-pentyloxy-7-pteridinyl) disulfide (**13**) (iodine in chloroform plus aqueous sodium hydrogen carbonate: 57%)[1681]; of 3,6,7-trimethyl- or 3-methyl-6,7-diphenyl-2-thioxo-1,2-dihydro-4(3*H*)-pteridinone into bis(3,6,7-trimethyl-4-oxo-3,4-dihydro-2-pteridinyl) disulfide (in 56% yield) or bis(3-methyl-6,7-diphenyl-4-oxo-3,4-dihydro-2-pteridinyl) disulfide (22%), respectively, by stirring in THF open to the air for >5 days[446]; of 1,3-dimethyl-7-thioxo-7,8-dihydro-2,4(1*H*,3*H*)-pteridinedione into bis(1,3-dimethyl-2,4-dioxo-1,2,3,4-tetrahydro-7-pteridinyl) disulfide (irradiation in acid solution with access of air)[632]; of the isomeric 6-

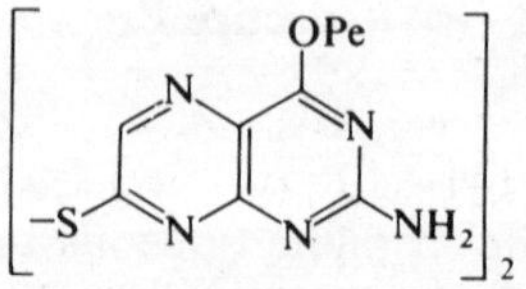

(**13**)

thioxo-5,6-dihydro substrate (**14**) into the corresponding bis(1,3-dimethyl-2,4-dioxo-1,2,3,4-tetrahydro-6-pteridinyl) disulfide (similarly)[632, 1463]; of 1- or 3-methyl-7-thioxo-7,8-dihydro-2,4(1*H*,3*H*)-pteridinedione into bis(1- or 3-methyl-2,4-dioxo-1,2,3,4-tetrahydro-7-pteridinyl) disulfide (61 and 51% yield, respectively, using iodine in aqueous potassium iodide as the oxidant)[1709]; and of 1,3,6-trimethyl-7-thioxo-7,8-dihydro-2,4(1*H*,3*H*)-pteridinedione into bis(1,3,6-trimethyl-2,4-dioxo-1,2,3,4-tetrahydro-7-pteridinyl) disulfide (iodine in dichloromethane: 75%).[1709]

The *formation of a pteridinesulfenic acid* is seen in the treatment of 1,3,6-trimethyl-7-thioxo-7,8-dihydro-2,4(1*H*,3*H*)-pteridinedione with hydrogen peroxide in aqueous acetonic sodium hydrogen carbonate solution for 2 h at room temperature to give 1,3,6-trimethyl-2,4-dioxo-1,2,3,4-tetrahydro-7-pteridinesulfenic acid, isolated in 20% yield as its silver salt.[1709]

The *formation of pteridinesulfinic acids* is illustrated in the treatment of 1,3-dimethyl-6-thioxo-5,6-dihydro-2,4(1*H*,3*H*)-pteridinedione (**14**) in aqueous sodium hydrogen carbonate with manganese dioxide, to give 1,3-dimethyl-2,4-dioxo-1,2,3,4-tetrahydro-6-pteridinesulfinic acid (**15**) in 32% yield[1681]; in the irradiation of the same substrate (**14**) at pH 9 in the presence of oxygen to give the product (**15**)[227, 1681]; and in the treatment of 2-thioxo-1,2-dihydro-4(3*H*)-pteridinone (**16**, R=H), its 6,7-dimethyl derivative (**16**, R=Me), or the 6,7-diphenyl derivative (**16**, R=Ph) with alkaline hydrogen peroxide at <0°C to give 4-oxo-3,4-dihydro-2-pteridinesulfinic acid (**17**, R=H, $x=2$) (42%), its 6,7-dimethyl derivative (**17**; R=Me, $x=2$) (54%), or the 6,7-diphenyl derivative (**17**, R=Ph, $x=2$), respectively, each isolated as its potassium salt.[629]

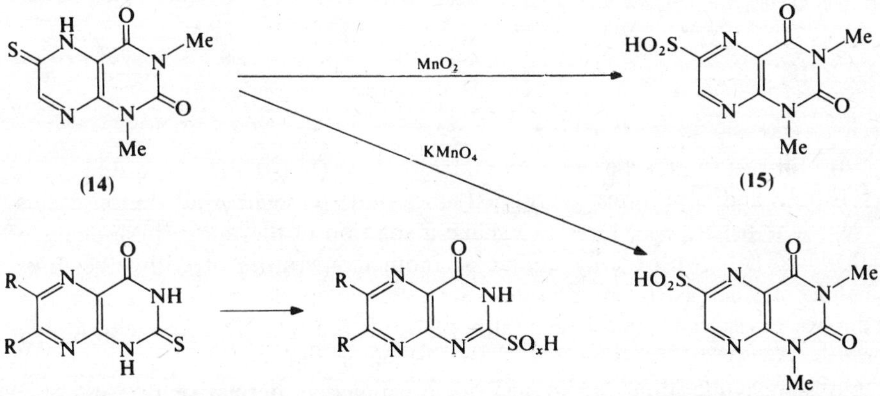

(**14**) (**15**)

(**16**) (**17**) (**18**)

The *formation of pteridinesulfonic acids* is rather better represented in the conversion of this substrate (**14**) into 1,3-dimethyl-2,4-dioxo-1,2,3,4-tetrahydro-6-pteridinesulfonic acid (**18**) (permanganate: 56% as potassium salt[1681]; or prolonged irradiation in the presence of oxygen[227, 632]); of 2-amino-4-pentyloxy-7(8*H*)-pteridinethione into 2-amino-4-pentyloxy-7-pteridinesulfonic acid (*m*-chloroperoxybenzoic acid–chloroform: 56%)[1681]; of the substrates (**16**, R = H, Me, or Ph) into 4-oxo-3,4-dihydro-2-pteridinesulfonic acid (**17**, R = H, $x = 3$), its 6,7-dimethyl derivative (**17**, R = Me, $x = 3$), or the 6,7-diphenyl derivative (**17**, R = Ph, $x = 3$), respectively (permanganate: each in ~50% yield as the potassium salt)[629]; and of 2-amino-4-oxo-7-thioxo-3,4,7,8-tetrahydro-6-pteridinecarboxylic acid into 2-amino-4-oxo-7-sulfo-3,4-dihydro-6-pteridinecarboxylic acid (permanganate: 58% as dipotassium salt).[906]

The *oxidation of nontautomeric pteridinethiones* is possible. Thus treatment of 1,3,5-trimethyl-6-thioxo-5,6-dihydro-2,4(1*H*,3*H*)-pteridinedione (**19**, X = S) with *m*-chloroperoxybenzoic acid in a well-shaken mixture of ether and water at room temperature gave the corresponding (blue!) 6-sulfinyl compound (**19**, X = S:O) in 57% yield[1680]; likewise, 3-methyl- (**20**, R = H, X = S) and 1,3-dimethyl-6,7-diphenyl-4-thioxo-3,4-dihydro-2(1*H*)-pteridinone (**20**, R = Me, X = S) with peroxycarboxylic acids gave the corresponding 4-sulfinyl products (**20**, R = H or Me, X = S:O), which were identified in solution by their spectra although they could not be isolated in the pure state.[1679] The former product (**20**, R = H, X = S:O) may of course be written as the tautomeric sulfenic acid (**21**).

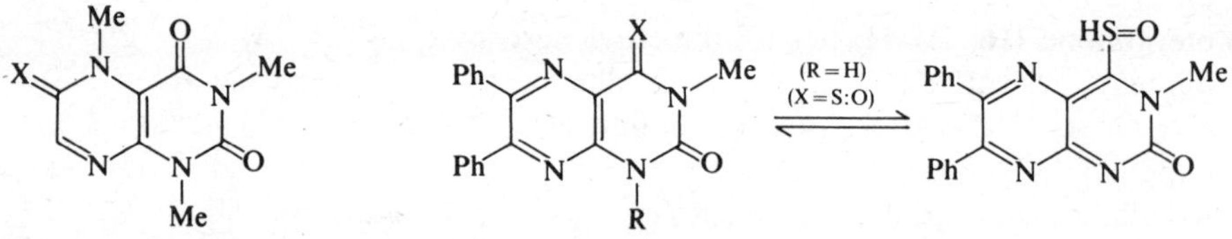

(**19**) (**20**) (**21**)

C. *S*-Alkylation

In contrast to pteridinones that undergo *N*- or *O*-alkylation according to the substrate and conditions, pteridinethiones undergo facile *S*-alkylation exclusively. The usual procedure is to shake a solution of the thione in aqueous or alcoholic base with methyl iodide at room temperature or with other alkyl halides at increased temperatures. Such treatment of appropriate pteridinethiones gave 2-, 4-, and 7-methylthiopteridine (80–90%)[33]; 6-methylthiopteridine (77%)[36]; 6,7-dimethyl-2-methylthiopteridine (40%)[313]; 4-methylthio-2-phenylpteridine (92%) and its 6,7-dimethyl derivative (97%)[1436]; 2-methylthio-4,6,7-triphenylpteridine (30%)[1120]; 6,7-dimethyl-2-methylthio-4(3*H*)-pteridinone (100°C for 15 min: 63%)[1508]; 4-methylthio-6,7-diphenyl-2(1*H*)-pteridinone (in dioxane at 25°C: 79%)[1679]; 1- and 3-methyl-7-methylthio-2,4(1*H*,3*H*)-pteridinedione (good yields)[354]; 1,3-dimethyl-6-methylthio-2,4(1*H*,3*H*)-pteridinedione (70%)[1681]; 1,3-dimethyl-7-methylthio-2,4(1*H*,3*H*)-pteridinedione (74%)[1000]; its 1,3,6-trimethyl homolog (78%)[1709]; 1,3-dimethyl-7-methylthio-2,4(1*H*,3*H*)-pteridinedithione (67%)[1000]; 1,3-dimethyl-7-

methylthio-2-thioxo-1,2-dihydro-4(3*H*)-pteridinone (43%)[1000]; 1,3-dimethyl-7-methylthio-4-thioxo-3,4-dihydro-2(1*H*)-pteridinone (60%)[1000]; 2-ethylthio-4(3*H*)-pteridinone (ethyl bromide in ethanolic ethoxide under reflux for 2 h)[399, 1169]; 2-ethylthio-6,7-diphenyl-4(3*H*)-pteridinone (similarly)[1169]; 2-ethylthio-6-phenyl-4,7-pteridinediamine (similarly)[1281]; 4-amino-6-benzyl-2-ethylthio-7(8*H*)-pteridinone (similarly: 47%)[93]; 7-isopropylthio-1,3,6-trimethyl-2,4(1*H*,3*H*)-pteridinedione (isopropyl iodide in acetonic potassium carbonate: 22%)[1709]; 1,3,6-trimethyl-7-*p*-nitrophenethylthio-2,4(1*H*,3*H*)-pteridinedione (*p*-nitrophenethyl iodide and pyridine in DMF: 85%)[1709]; 6-benzyl-2-isopropylthio-4,7(3*H*,8*H*)-pteridinedione (isopropyl bromide in butanolic butoxide under reflux: 64%)[93]; 7-benzyl-2-isopropylthio-4,6(3*H*,5*H*)-pteridinedione (similarly)[990]; and related compounds.[93] For an interesting *S*-alkylation by phenacyl bromide, see Section 2.F to follow. Treatment of 3,6,7-trimethyl- (**22**, R = Me) or 3-methyl-6,7-diphenyl-2-thioxo-1,2-dihydro-4(3*H*)-pteridinone (**22**, R = Ph) with dibromomethane (in DMF containing potassium carbonate at 60°C) gave bis(3,6,7-trimethyl-4-oxo-3,4-dihydro-2-pteridinylthio)methane (**23**, R = Me) (87%) or bis(3-methyl-4-oxo-6,7-diphenyl-3,4-dihydro-2-pteridinylthio)methane (**23**, R = Ph) (53%).[446] For reactions of 1,2-dibromoethane and 1,3-dibromopropane with 2-thioxo-1,2-dihydro-4(3*H*)-pteridinone and its 6,7-dimethyl derivative, see Chapter III, Section 3.

The partly tautomeric 6,7,8-trimethyl-2,4(3*H*,8*H*)-pteridinedithione (**24**, R = Me) has been reported briefly to undergo *S*,*S*′-dimethylation by dimethyl sulfate–alkali to give 6,8-dimethyl-7-methylene-2,4-bismethylthio-7,8-dihydropteridine (**25**).[634] In contrast, the related substrate, 8-methyl-6,7-diphenyl-2,4(3*H*,8*H*)-pteridinedithione (**24**, R = Ph), was clearly unable to give an analogous product on similar treatment but it did give 8-methyl-2,4-bismethylthio-6,7-diphenyl-7,8-dihydro-7-pteridinol (**26**, R = H), which on acid-catalysed recrystallization from methanol gave 7-methoxy-8-methyl-2,4-bismethylthio-6,7-diphenyl-7,8-dihydropteridine (**26**, R = Me), proven in structure by an x-ray analysis.[634]

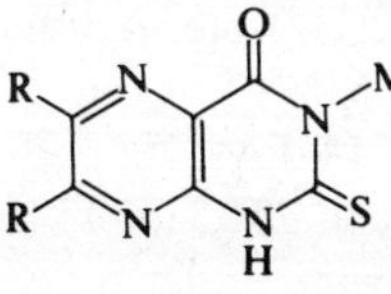

(22)

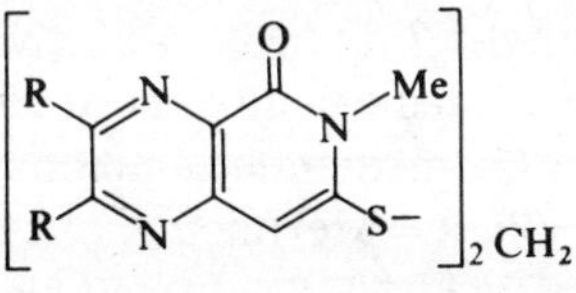

(23)

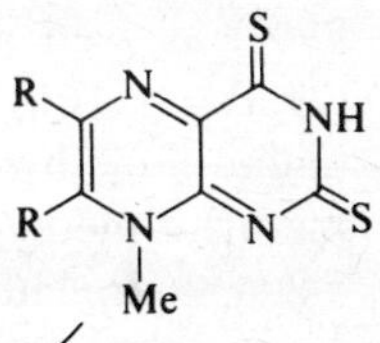

(24)

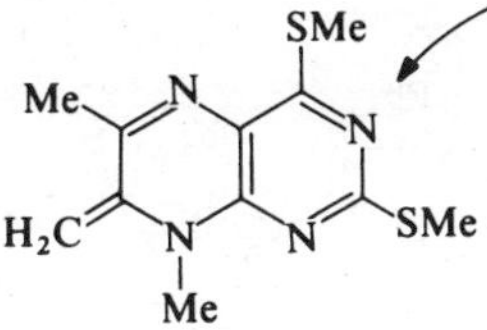

(25)

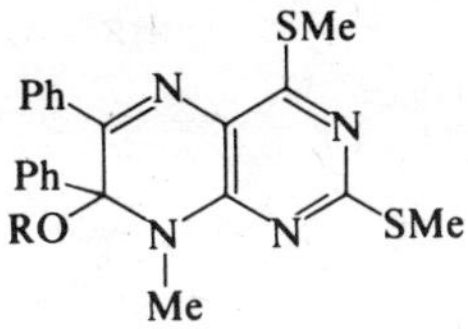

(26)

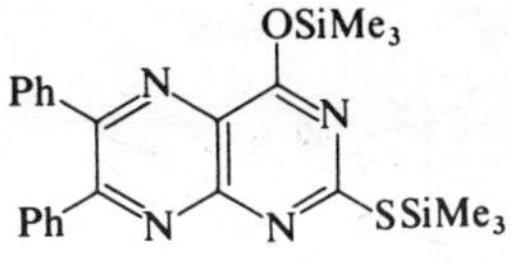

(27)

Tautomeric pteridinethiones undergo *S*-trimethylsilylation by hexamethyldisilazane (catalyzed by ammonium sulfate) to give, for example, 6,7-diphenyl-4-trimethylsiloxy-2-trimethylsilylthiopteridine (**27**), which on treatment with appropriate halogeno-sugars gave 6,7-diphenyl-2-(2′,3′,5′-tri-*O*-benzoyl-*β*-D-ribofuranosylthio)-4(3*H*)-pteridinone and related products.[447] Other examples have been reported.[1457]

D. Hydrolysis to Pteridinones

Although the direct hydrolysis of (tautomeric) pteridinethiones to pteridinones is possible sometimes,[36] it is usually better to proceed via *S*-methylation or via oxidation to a sulfinic or sulfonic acid, followed by hydrolysis. These routes have been discussed in Chapter VI, Section 1.E.

E. Aminolysis

As might be expected, aminolysis of pteridinethiones often requires quite vigorous conditions, which may bring about unwanted side reactions. *Uncomplicated aminolyses* are exemplified in the conversion of the corresponding pteridinethiones into 6,7-diphenyl-2-pteridinamine (**28**, R = H) (alcoholic ammonia at 130°C for 10 h: 84%), 4-butylamino-6,7-diphenylpteridine (**28**, R = Bu) (ethanolic butylamine containing mercuric oxide under reflux for 2 h: 74%), and 4-benzylamino-6,7-diphenylpteridine (**28**, R = CH_2Ph) (ethanolic benzylamine containing mercuric oxide under reflux for 5 h: 99%)[474]; into 4-hydrazino-6,7-dimethylpteridine (ethanolic hydrazine hydrate under reflux for 20 min: 67%), 4-hydrazino-2-methylpteridine (similarly for 40 min: 70%), and 4-hydrazino-2,6,7-trimethylpteridine (similarly: 53%)[135]; into 4-amino-1-methyl- (**29**, R = H) (methanolic ammonia at 25°C for 2 days: 78%) and 1-methyl-4-methylamino-6,7-diphenyl-2(1*H*)-pteridinone (**29**, R = Me) (ethanolic methylamine similarly: 79%) as well as 1-methyl-6,7-diphenyl-4-phenylhydrazino-2(1*H*)-pteridinone (**29**, R = NHPh) (phenylhydrazine in dioxane at 25°C for 24 h: 62%)[1679]; into 4-amino-3-methyl- (**30**, R = H) (methanolic ammonia at 25°C for 14 days: 20%), 3-methyl-4-methylamino- (**30**, R = Me) (methanolic methylamine at 25°C for 15 h: 74%), 4-ethylamino-3-methyl- (**30**, R = Et) (ethylamine in dioxane under reflux for 3 days: 32%), and 4-butylamino-3-methyl-6,7-diphenyl-2(3*H*)-pteridinone (**30**, R = Bu) (butylamine in dioxane at 25°C for 15 h: 70%) as well as 3-methyl-

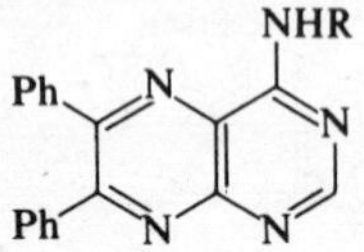

(28)

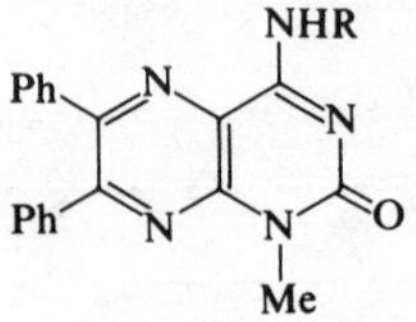

(29)

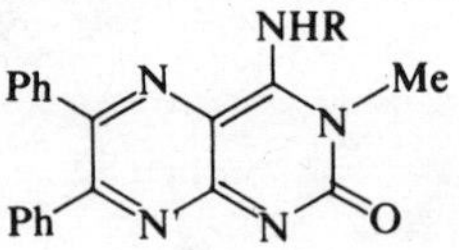

(30)

6,7-diphenyl-4-phenylhydrazino-2(3*H*)-pteridinone (**30**, R = NHPh) (phenylhydrazine in dioxane–DMF at 25°C for 4 h: 71%)[1679]; and into 2-amino-1-methyl- (butanolic ammonia containing mercuric oxide under reflux for 6 h) and 1-methyl-2-methylamino-6,7-diphenyl-4(1*H*)-pteridinone (methylamine similarly: 40%).[111] In addition, 4-amino-6,7-diphenyl-2(1*H*)-pteridinethione gave 6,7-diphenyl-2,4-pteridinediamine (ethanolic ammonia at 190°C for 10 h; 84%), 6,7-diphenyl-2-piperidino-4-pteridinamine (piperidine in DMF under reflux for 10 h: 65%), and 2-morpholino-6,7-diphenyl-4-pteridinamine (neat morpholine under reflux for 10 h: 45%)[467, 472]; 7(8*H*)-pteridinethione gave 7-hydrazinopteridine (ethanolic hydrazine under reflux for 20 min: 42%)[136]; and other examples have been reported.[177, 227, 632, 927, 1162]

Similar *aminolyses complicated by transaminations* are exemplified by the conversion of 4-amino-6,7-diphenyl-2(1*H*)-pteridinethione (by prolonged refluxing in neat amine) into 2,4-bisbenzylamino- (79%), 2,4-dihydrazino- (ethanol added to mixture: >95%), 2,4-bis(β-hydroxyethylamino)- (>95%; cf. ref. 1315), 2,4-bis(γ-dimethylaminopropylamino)- (84%), 2,4-bis(γ-diethylaminopropylamino)- (90%), and 2,4-bis(γ-diisopropylaminopropylamino)-6,7-diphenylpteridine[467]; of the same substrate with ethanolic amine at 180°C for 10 h into 2,4-bismethylamino- or 2,4-bisdimethylamino-6,7-diphenylpteridine[110, cf. 472]; and of 4-amino-6,7-dimethyl-2(1*H*)-pteridinethione into 2,4-bisdimethylamino-6,7-dimethylpteridine [ethanolic dimethylamine at 180 to 220°C for 16 h: 21% after separation from 2-dimethylamino-6,7-dimethyl-4-pteridinamine,[1315] which was the main product (93%) if 180°C was not exceeded[472]].

The *aminolysis of nontautomeric thiones* is illustrated in the conversion of 1,3-dimethyl-6,7-diphenyl-4-thioxo-3,4-dihydro-2(1*H*)-pteridinone (**31**) into 4-imino-1,3-dimethyl- (**32**, R = H) (ammonium acetate in refluxing dioxane for 8 h: 56%; methanolic ammonia plus dioxane at 25°C for 3 days: 77%), 1,3-dimethyl-4-methylimino- (**32**, R = Me) (ethanolic methylamine in dioxane at 25°C for 3 days: 81%), 4-butylimino-1,3-dimethyl- (**32**, R = Bu) (butylamine in dioxane at 25°C for 8 h: 64%), 4-ethoxyimino-1,3-dimethyl- (**32**, R = OEt) (*O*-ethylhydroxylamine in dioxane under reflux for 3 days: 70%), 1,3-dimethyl-4-*N*-methyl-*N*-phenylhydrazono- (**32**, R = NMePh) (*N*-methyl-*N*-phenylhydrazine in dioxane under reflux for 2 days: 24%), and 1,3-dimethyl-4-methylhydrazono-6,7-diphenyl-3,4-dihydro-2(1*H*)-pteridinone (**32**, R = NHMe) (methylhydrazine in dioxane at 25°C for 30 min: 79%)[1679]; of the same substrate (**31**) into 1,3-dimethyl-6,7-diphenyl-4-phenylhydrazono-3,4-dihydro-2(1*H*)-pteridinone (**32**, R = NHPh) (phenylhydrazine in refluxing dioxane for 8 h: 52%)[1679]; and of 3-butyl-6,7-diphenyl-4(3*H*)-pteridinethione (**33**, X = S), by passing ammonia through an ethanolic solution containing mercuric oxide under reflux for 8 h, into a product (70%), which was formulated originally as 3-butyl-6,7-diphenyl-4(3*H*)-pteridinimine (**33**, X = NH) but which was probably the isomeric 4-butylamino-6,7-diphenylpteridine, formed by Dimroth rearrangement of the imine under the basic conditions pertaining during preparation. Both the postulated imine (**33**, R = NH) and the authentic butylaminopterdine were reported as having the same melting point.[474]

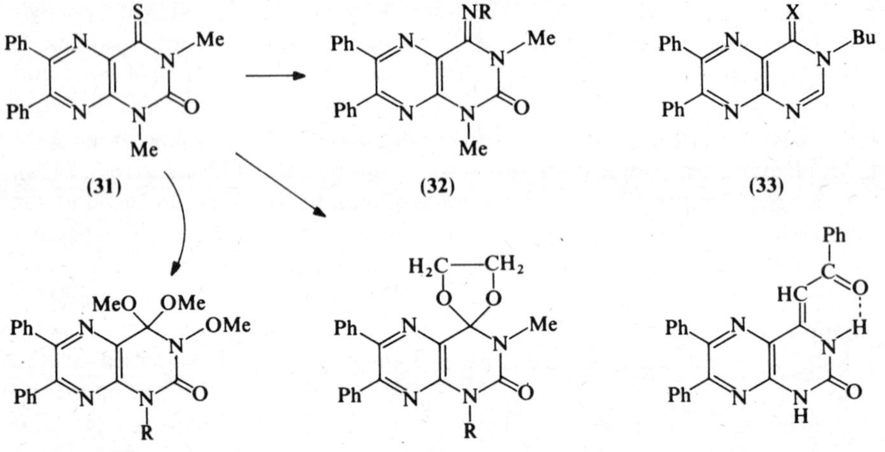

(31) (32) (33)

(34) (35) (36)

F. Other Reactions

Treatment of 1,3-dimethyl-6,7-diphenyl-4-thioxo-3,4-dihydro-2(1*H*)-pteridinone (**31**) with dioxane containing methanolic sodium methoxide and mercuric bromide at room temperature gave 4,4-dimethoxy-1,3-dimethyl-6,7-diphenyl-3,4-dihydro-2(1*H*)-pteridinone (**34**, R = Me) (74%). The same substrate (**31**) with disodium 1,2-ethanediolate in dioxane containing mercuric bromide gave 4,4-ethylenedioxy-1,3-dimethyl-6,7-diphenyl-3,4-dihydro-2(1*H*)-pteridinone (**35**, R = Me) (7%). Similarly, the related substrate, 3-methyl-6,7-diphenyl-4-thioxo-3,4-dihydro-2(1*H*)-pteridinone (**20**, R = H, X = S), afforded 4,4-dimethoxy- (**34**, R = H) (69%) and 4,4-ethylenedioxy-3-methyl-6,7-diphenyl-3,4-dihydro-2(1*H*)-pteridinone (**35**, R = H) (64%).[1679]

6,7-Diphenyl-4-thioxo-3,4-dihydro-2(1*H*)-pteridinone underwent *S*-alkylation by phenacyl bromide in dioxane containing triethylamine to give 4-phenacylthio-6,7-diphenyl-2(1*H*)-pteridinone (?) (cf. Sect. 2.C), which on prolonged refluxing in DMF underwent sulfur extrusion to afford 4-phenacyl-6,7-diphenyl-2(1*H*)-pteridinone (45%), shown to exist predominantly as the hydrogen-bonded tautomer, 4-phenacylidene-6,7-diphenyl-3,4-dihydro-2(1*H*)-pteridinone (**36**).[1679]

3. EXTRANUCLEAR MERCAPTOPTERIDINES

Very little has been reported on pteridinylalkanethiols or related compounds. A rather specialized route, exemplified in the ring fission of 7,8-dihydro-10*H*-thiazolo[2,3-*b*]pteridin-10-one (**37**) to give 3-*β*-mercaptoethyl-2,4(1*H*,3*H*)-pteridinedione (**38**),[876] has been discussed in Chapter III, Section 3. The covalent 3,4-

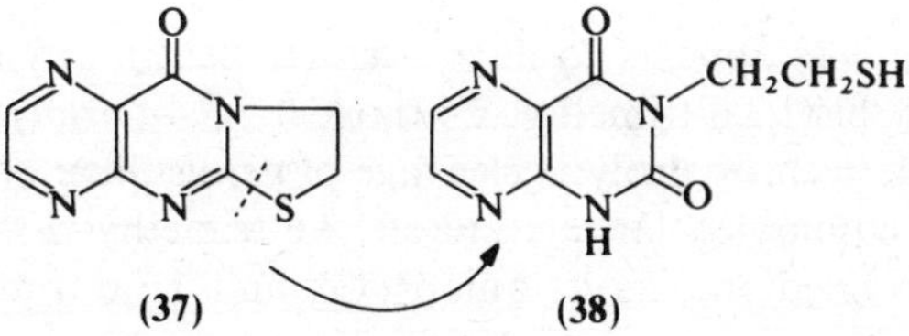

(37) (38)

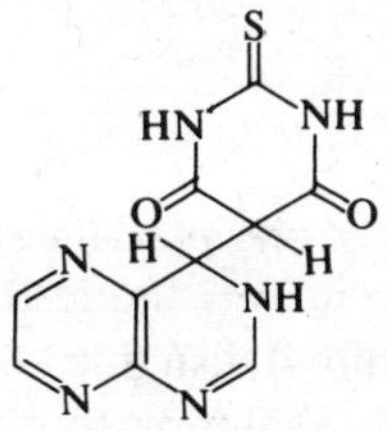

(39)

adduct of 2-thiobarbituric acid with pteridine, 4-(4′,6′-dioxo-2′-thioxohexahydro-5′-pyrimidinyl)-3,4-dihydropteridine (**39**) is an extranuclear mercaptopteridine when written as its appropriate tautomer.[49]

4. THE (SYMMETRICAL) DIPTERIDINYL SULFIDES AND DISULFIDES

The *dipteridinyl sulfides* are represented only by bis(6-oxo-5,6,7,8-tetrahydro-7-pteridinyl) sulfide (**40**), which was made in >90% yield by treatment of 6(5*H*)-pteridinone in alkaline solution with sodium sulfide.[36] Treatment of thiamine with methanolic methoxide has been reported to give a cyclic sulfide of pteridine.[1064]

The *dipteridinyl disulfides* are a little better represented. Most known examples have been made by oxidation of pteridinethiones, as discussed previously in Section 2.B. However, bis(1,3,6-trimethyl-2,4-dioxo-1,2,3,4-tetrahydro-7-pteridinyl) disulfide was made in 71% yield by simply boiling 1,3,6-trimethyl-2,4-dioxo-1,2,3,4-tetrahydro-7-pteridinesulfenic acid in aqueous buffer of pH 3, although this reaction was reversed (in 52% yield) under alkaline conditions.[1709] The reversible disproportionation of several disulfides is discussed next in Section 5.

The related but unsymmetrical compound, 7-ethyldithio-1,3,6-trimethyl-2,4(1*H*,3*H*)-pteridinedione (**41**) was made (13%) by treatment of 1,3,6-trimethyl-2,4-dioxo-1,2,3,4-tetrahydro-7-pteridinesulfenic acid with ethanethiol under acidic conditions.[228, 1709] It was also obtained by the oxidative coupling of 1,3,6-trimethyl-7-thioxo-2,4(1*H*,3*H*)-pteridinedione and ethanethiol in the presence of diethyl azodicarboxylate (33% yield).[1709]

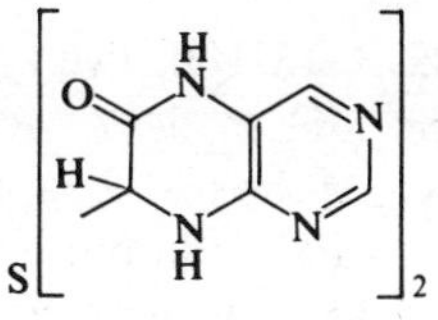

(40)

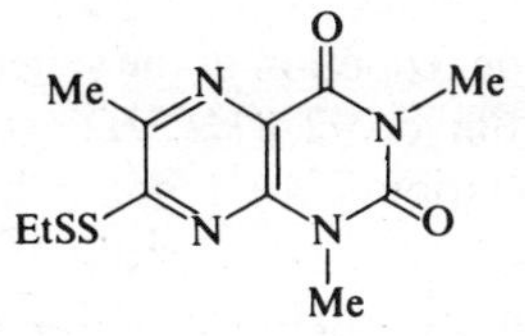

(41)

5. THE PTERIDINE SULFENIC, SULFINIC, AND SULFONIC ACIDS

Only one *pteridinesulfenic acid* appears to have been studied in any detail. Thus it was noticed that bis(1,3,6-trimethyl-2,4-dioxo-1,2,3,4-tetrahydro-7-pteridinyl) disulfide (**42**) underwent hydrolytic cleavage of its disulfide bond in alkaline solution to give an equimolecular mixture of 1,3,6-trimethyl-7-thioxo-7,8-dihydro-2,4(1*H*,3*H*)-pteridinedione, as its anion (**43**), and 1,3,6-trimethyl-2,4-dioxo-1,2,3,4-tetrahydro-7-pteridinesulfenic acid, also as its anion (**44**), a reaction reversed in acidic media. Both products could be isolated as their sparingly soluble silver salts.[228] A better route to the sulfenic acid proved to be *S*-alkylation of the 1,3,6-trimethyl-7-thioxo-7,8-dihydro-2,4(1*H*,3*H*)-pteridinedione anion (**43**) to give 7-isopropylthio-1,3,6-trimethyl-2,4(1*H*,3*H*)-pteridinedione (**45**, R = S-*i*-Pr) followed by oxidation with *m*-chloroperoxybenzoic acid to afford the 7-isopropylsulfinyl analog [**45**, R = S(O)-*i*-Pr], which underwent thermal cleavage in boiling toluene to produce 1,3,6-trimethyl-2,4-dioxo-1,2,3,4-tetrahydro-7-pteridinesulfenic acid [**45**, R = S(O)H] as the pure free acid in 57% yield.[228, 1709] The first of these methods, followed by spontaneous disproportionation was used to convert appropriate disulfides into 1-methyl-, 3-methyl-, and 1,3-dimethyl-2,4-dioxo-1,2,3,4-tetrahydro-7-pteridinesulfinic acid in solution (for spectroscopic examination?) but several other disulfides decomposed under such reaction conditions.[228]

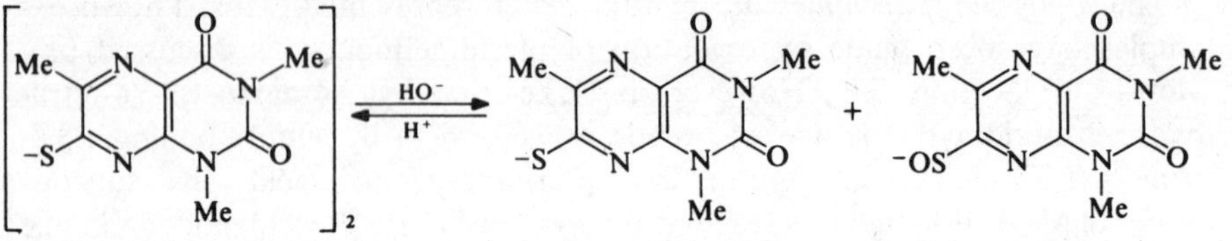

(42) **(43)** **(44)**

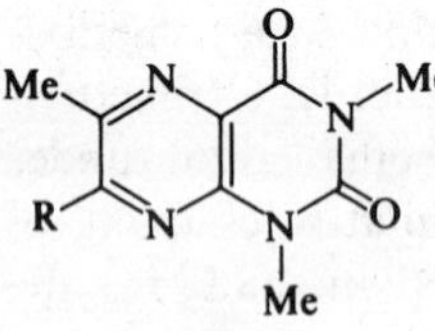

(45)

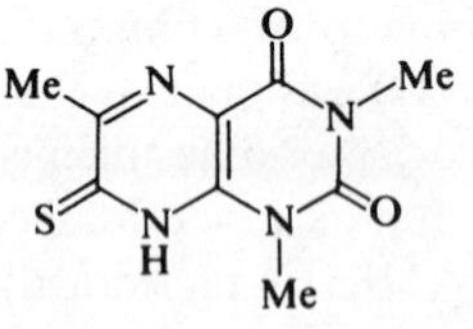

(46)

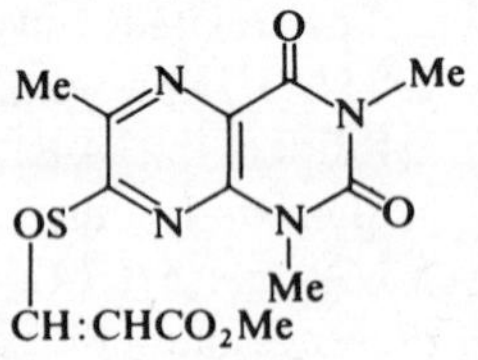

(47)

Borohydride reduction of the sulfenic acid (**45**, R = SHO) gave 1,3,6-trimethyl-7-thioxo-7,8-dihydro-2,4(1*H*,3*H*)-pteridinedione (**46**) in 54% yield; hydrogen peroxide oxidation gave crude 1,3,6-trimethyl-2,4-dioxo-1,2,3,4-tetrahydro-7-pteridinesulfinic acid (**45**, R = SO_2H); permanganate oxidation gave the corresponding sulfonic acid (**45**, R = SO_3H) in 49% yield; and treatment with ethanethiol gave 7-ethyldithio-1,3,6-trimethyl-2,4(1*H*,3*H*)-pteridinedione (**45**,

R = SSEt).[228, 1709] In addition, the same sulfenic acid with hydrogen peroxide–sodium hydrogen carbonate followed by methyl iodide gave 1,3,6-trimethyl-7-methylsulfonyl-2,4(1*H*,3*H*)-pteridinedione (**45**, R = SO_2Me) (43%), which was identified by an unambiguous synthesis from 1,3,6-trimethyl-7-methylthio-2,4(1*H*,3*H*)-pteridinedione (**45**, R = SMe) and *m*-chloroperoxybenzoic acid.[1709] Other reactions of the sulfenic acid (**45**, R = SHO) include (a) the complete desulfurization by treatment with peroxide followed by acidic conditions to give 1,3,6-trimethyl-2,4(1*H*,3*H*)-pteridinedione (**45**, R = H) in 41% yield, and (b) an unusual *S*-alkylation by treatment with methyl propiolate in boiling toluene to afford 7-β-methoxycarbonylvinylsulfinyl-1,3,6-trimethyl-2,4(1*H*,3*H*)-pteridinedione (**47**) (33%).[1709]

Several *pteridinesulfinic acids* have been prepared, mainly by controlled oxidation of the corresponding thiones (see Sect. 2.B).[227, 629, 1681] In addition, 1,3,6-trimethyl-2,4-dioxo-1,2,3,4-tetrahydro-7-pteridinesulfenic acid has been oxidized to the corresponding sulfinic acid (**45**, R = SO_2H)[228] and several disulfides have been converted into (unisolated) sulfinic acids (see the sulfenic acid part of this section).[228, 1709]

Some of the potential reactions of pteridinesulfinic acids have been employed in the desulfurization of 1,3-dimethyl-2,4-dioxo-1,2,3,4-tetrahydro-6-pteridinesulfinic acid (**48**) by boiling in glacial acetic acid to give 1,3-dimethyl-2,4(1*H*,3*H*)-pteridinedione (89%)[1681]; in the *S*-methylation of the same substrate (**48**) by using methyl iodide–potassium carbonate in acetone to afford 1,3-dimethyl-6-methylsulfonyl-2,4(1*H*,3*H*)-pteridinedione (**49**) (81%), identical with material prepared by the peroxycarboxylic acid oxidation of the corresponding methylthio compound[1681]; in the (inferred) oxidation of 4-oxo-3,4-dihydro-2-pteridinesulfinic acid and related compounds into the corresponding sulfonic acids by permanganate[629]; and in the aminolysis of potassium 4-oxo-6,7-diphenyl-3,4-dihydro-2-pteridinesulfinate in neat morpholine to give 2-morpholino-6,7-diphenyl-4(3*H*)-pteridinone (> 70%).[629]

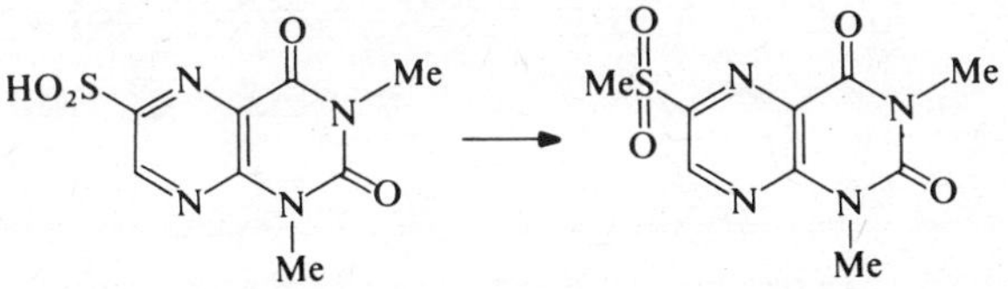

(48) **(49)**

The known *pteridinesulfonic acids* have been prepared by several routes, illustrated in the following examples. Oxidation of the thioxopteridine, 1,3-dimethyl-6-thioxo-5,6-dihydro-2,4(1*H*,3*H*)-pteridinedione, by aqueous permanganate gave 1,3-dimethyl-2,4-dioxo-1,2,3,4-tetrahydro-6-pteridinesulfonic acid that was isolated as its potassium salt in 56% yield[1681]; for analogous cases, see Section 2.B. Oxidation of 1,3,6-trimethyl-2,4-dioxo-1,2,3,4-tetrahydro-7-pteridinesulfenic acid by permanganate gave the corresponding sulfonic acid (**45**, R = SO_3H).[228, 1709] Oxidative degradation of urothione, 2-amino-7-α,β-dihydroxyethyl-6-methylthio-4(3*H*)-thieno[3,2-*g*]pteridinone (**50**), afforded a primary synthesis of 2-amino-4-oxo-7-sulfo-3,4-dihydro-6-pteridinecarboxylic acid (**51**).[1526] Treatment of the chloropteridine, 7-chloro-1,3,6-trimethyl-2,4(1*H*,3*H*)-pteridinedione (**45**, R = Cl), with boiling aqueous sodium sulfite gave sodium 1,3,6-trimethyl-2,4-dioxo-1,2,3,4-tetrahydro-7-pteridinesulfonate (**45**, R = SO_3Na) in 63% yield[1709]; for other such cases, see Chapter V, Section 2.F. Pteridinesulfonic acids may also be made by covalent addition of sodium hydrogen sulfite to appropriate pteridines. The following compounds were prepared in this manner: 4-methyl-5,6,7,8-tetrahydro-6,7-pteridinedisulfonic acid (as monosodium salt: 56%),[64] 2-amino-3,4-dihydro-4-pteridinesulfonic acid (sodium salt: 30%),[45] 4-oxo-3,4,5,6,7,8-hexahydro-6,7-pteridinedisulfonic acid (**52**) (disodium salt: 52%),[47] 2-thioxo-1,2,3,4-tetrahydro-4-pteridinesulfonic acid (sodium salt: 50%),[46] 2-methylthio-3,4-dihydro-4-pteridinesulfonic acid (as the hydrated bisulfite of the sodium salt: 55%),[46] and adducts from folic acid and dihydrofolic acid.[1319] Aerial oxidation of 2-amino-5,6,7,8-tetrahydro-4(3*H*)-pteridinone sulfite[1052] has been reported to give 2-amino-4-oxo-3,4-dihydro-6-pteridinesulfonic acid hydrate.[453, 584, 1051] This compound was shown later to be the sulfate, 2-amino-6-sulfo-oxy-4(3*H*)-pteridinone (**53**).[1431]

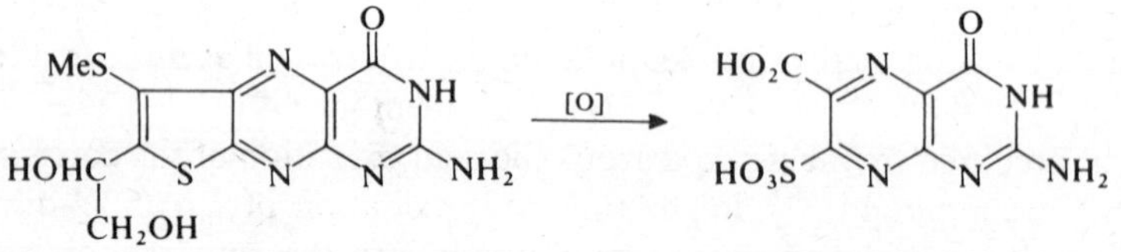

(**50**) (**51**)

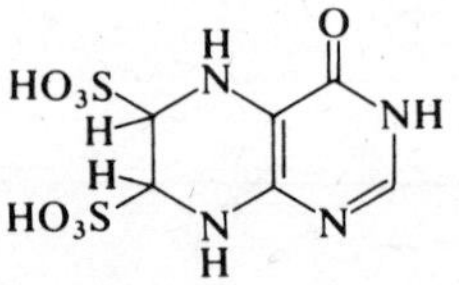

(**52**)

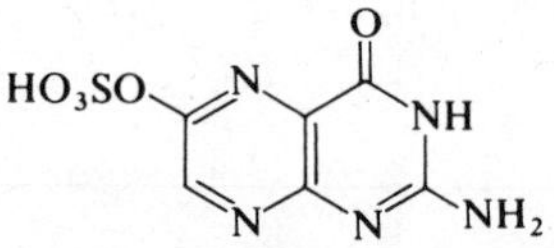

(**53**)

Only a few of the potential reactions of pteridinesulfonic acids have been used in the literature: 2-amino-4-oxo-7-sulfo-3,4-dihydro-6-pteridinecarboxylic acid (**51**) underwent hydrolysis (and incidental decarboxylation) in boiling alkali to afford 2-amino-4,7(3*H*,8*H*)-pteridinedione[1526]; potassium 4-oxo-6,7-diphenyl-3,4-dihydro-2-pteridinesulfonate underwent facile aminolysis in boiling neat morpholine during 2 min to give 2-morpholino-6,7-diphenyl-4(3*H*)-pteridinone (> 70%)[629]; and the fragmentations of the fully trimethylsilylated derivatives of 2-amino-4-oxo-3,4-dihydro- and 2,4-dioxo-1,2,3,4-tetrahydro-6-pteridinesulfonic acid have been recorded.[1242]

Although rare, *extranuclear sulfinic and sulfonic acids* have been made in the pteridine series. Thus treatment of 7-acetonyl-2-amino-4,6(3*H*,5*H*)-pteridinedione (**54**, R = Ac) with concentrated sulfuric acid at 110°C for 2 h gave 2-amino-7-sulfomethyl-4,6(3*H*,5*H*)-pteridinedione (**54**, R = SO_3H) (36%) [the same product was obtained by similar treatment of 2-amino-7-methyl- (**54**, R = H), 2-amino-7-ethoxycarbonylmethyl- (**54**, R = CO_2Et), 2-amino-7-γ-benzoyloxyacetonyl- (**54**, R = $COCH_2OBz$), and 2-amino-7-γ-phthalimido-acetonyl-4,6(3*H*,5*H*)-pteridinedione]. Subsequent boiling of the product in *dilute* sulfuric acid induced complete desulfurization to furnish 2-amino-7-methyl-4,6(3*H*,5*H*)-pteridinedione (50%).[504] The analogous substrate, 2-amino-7-ethyl-4,6(3*H*,5*H*)-pteridinedione (**54**, R = Me) underwent similar α-sulfonation to afford 2-amino-7-α-sulfoethyl-4,6(3*H*,5*H*)-pteridinedione (18%).[504] Treatment of 6,7-bis-*m*-aminophenyl-2,4-pteridinediamine (**55**, R = H) with ethanolic sodium formaldehyde sulfoxylate ($NaSO_2CH_2OH$) under reflux gave 6,7-bis-*m*-sulfinomethylaminophenyl-2,4-pteridinediamine (**55**, R = CH_2SO_2H), isolated as its trihydrated disodium salt in 85% yield.[141] Preparations of related extranuclear derivatives include 1-β-mesyloxyethyl- (**56**, R = SO_2Me) by mesylation of 1-β-hydroxyethyl-6,7-dimethyl-2,4(1*H*,3*H*)-pteridinedione (**56**, R = H)[1336]; the condensation of 2-*p*-acetamidobenzenesulfonamido-5,6-diamino-4(3*H*)-pyrimidinone with glyoxal or biacetyl to afford 2-*p*-acetamidobenzenesulfonamido-4(3*H*)-pteridinone (**57**, R = H) or its 6,7-dimethyl derivative (**57**, R = Me), respectively[186]; other primary syntheses leading to different types of extranuclear sulfonamides[926, 1213]; and the nucleophilic displacement reaction of 6-bromomethyl-4(3*H*)-pteridinone with sodium sulfanilate in refluxing ethanol to give 6-*p*-sulfoanilinomethyl-4(3*H*)-pteridinone (**58**), isolated as its dihydrated sodium salt in 33% yield.[121]

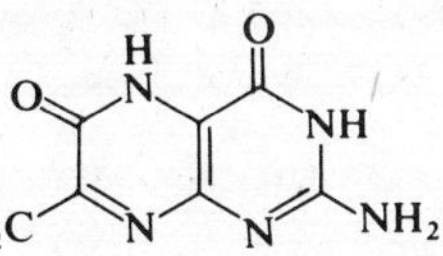

(54)

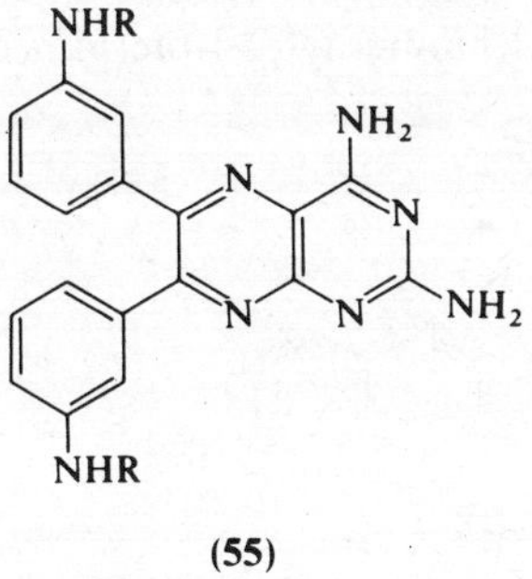

(55)

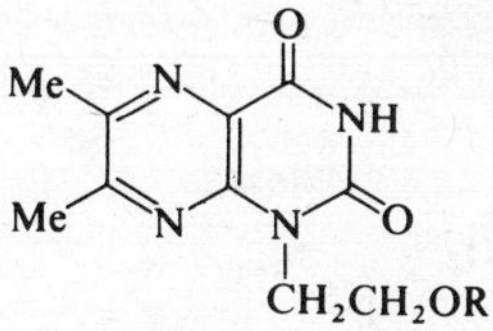

(56)

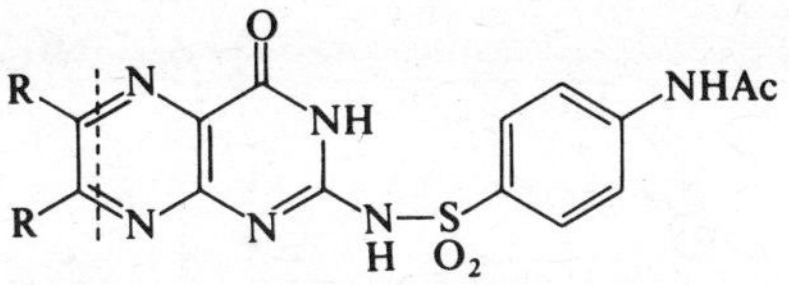

(57)

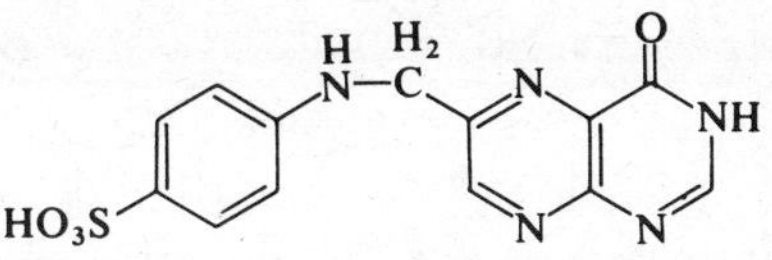

(58)

6. PREPARATION OF ALKYLTHIO- AND ARYLTHIOPTERIDINES

Both alkylthio- and arylthiopteridines may be made by several methods. However, the convenient route to alkylthiopteridines by alkylation of pteridinethiones is inapplicable to the formation of arylthiopteridines.

A. By Primary Syntheses

The primary synthetic methods for alkylthio- and arylthiopteridines have been covered in Chapters II and III. A few typical examples include the condensation of 6-methyl-2-methylthio-4,5-pyrimidinediamine (**59**) with glyoxal to give 4-methyl-2-methylthiopteridine (**60**) in >56% yield (a Gabriel and Colman synthesis)[46]; of 6-methylthio-4,5-pyrimidinediamine with biacetyl to give 6,7-dimethyl-4-methylthiopteridine (62%)[129]; of 2-ethylthio-4,5,6-pyrimidinetriamine with benzil to give 2-ethylthio-6,7-diphenyl-4-pteridinamine (70%)[955]; of 2-methylthio-5-nitroso-4,6-pyrimidinediamine (**61**) with cyanoacetic acid to give 4,7-diamino-2-methylthio-6-pteridinecarboxylic acid (**62**) (a Timmis synthesis)[324]; and of 4,6-diacetamido-5-nitroso-2-pyrimidinamine with 1-cyanomethylpyridinium chloride and thiophenol in the presence of alkali to give 6-phenylthio-2,4,7-pteridinetriamine in 71% yield.[329] In addition, cyclization of 6-amino-5-benzylideneamino-2-thioxo-1,2-dihydro-4(3*H*)-pyrimidinone with triethyl orthoformate in DMF gave the acylthio product, 2-formylthio-6-phenyl-4(3*H*)-pteridinone (**63**) (73%)[543]; and treatment of 3-amino-6-*p*-ethoxycarbonylphenylthiomethyl-2-pyrazinecarbonitrile with ethanolic guanidine completed the pyrimidine ring to give the extranuclear thioether, 6-*p*-ethoxycarbonylphenylthiomethyl-2,4-pteridinediamine (**64**) in 67% yield.[261, cf. 303, 304]

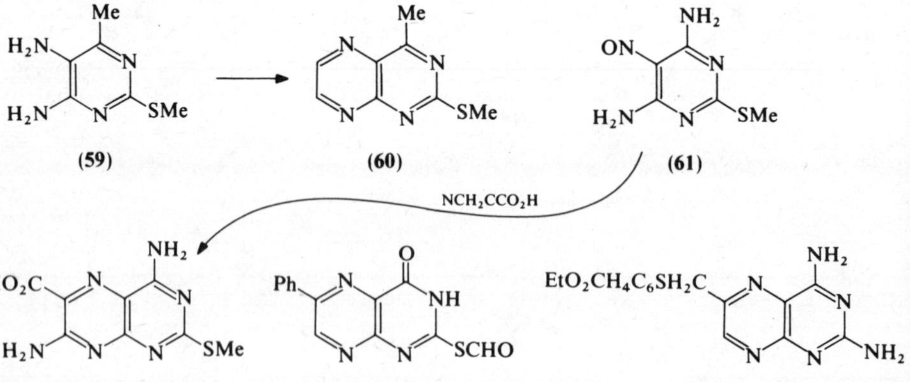

(**59**) (**60**) (**61**)

(**62**) (**63**) (**64**)

B. By *S*-Alkylation

The formation of alkylthiopteridines from pteridinethiones has been covered in Section 2.C. It is an immensely useful process.

C. By Nucleophilic Displacement

The alkane- and arenethiolysis of nuclear chloropteridines to afford alkylthio- or arylthiopteridines have been discussed in Chapter V, Section 2.E; extranuclear chloropteridines react similarly to give extranuclear thioethers (see Ch. V, Sect. 4.C). Displacement of an alkylsulfonyl group by an alkanethiol or arenethiol is also possible: 2-amino-6-methylsulfonyl-4(3*H*)-pteridinone with benzyl mercaptan (phenylmethanethiol) in refluxing DMF for 22 h gave 2-amino-6-benzylthio-4(3*H*)-pteridinone (70%) or with thiophenol in refluxing DMF for 2 h gave 2-amino-6-phenylthio-4(3*H*)-pteridinone (71%).[1307] Likewise, 6-methylsulfonyl-2,4,7-pteridinetriamine in refluxing neat thiophenol or with *p*-chlorothiophenol in refluxing DMF, gave 6-phenylthio- (85%) or 6-*p*-chlorophenylthio-2,4,7-pteridinetriamine (86%), respectively.[249]

D. By Covalent Addition

Although alkane- and arenethiols should add readily to appropriate pteridines to afford dihydroalkylthio- and dihydroarylthiopteridines, perhaps the stench of the reagents has discouraged such experiments. However, 7(8*H*)-pteridinone has been reported to react smoothly with thiophenol in aqueous ethanol during 72 h to give 6-phenylthio-5,6-dihydro-7(8*H*)-pteridinone (24%)[44]; likewise 6(5*H*)-pteridinethione and phenylmethanethiol gave 7-benzylthio-7,8-dihydro-6(5*H*)-pteridinethione (79%)[36] and less simple examples have been reported.[453]

7. REACTIONS OF ALKYLTHIOPTERIDINES

A. Desulfurization

Although desulfurization of thioethers by Raney-nickel has become a standard reaction in most heterocyclic series, it is by no means so in the pteridines, possibly on account of progressive nuclear reduction and consequent instability. However, under carefully controlled conditions, desulfurization can be successful with alkylthiopteridines. For example, 1,3-dimethyl-7-methylthio-6-propionyl-2,4(1*H*,3*H*)-pteridinedione (**65**) with Raney-nickel in acetone (i.e., deactivated by prereducing some of the large excess of solvent) gave the expected 1,3-

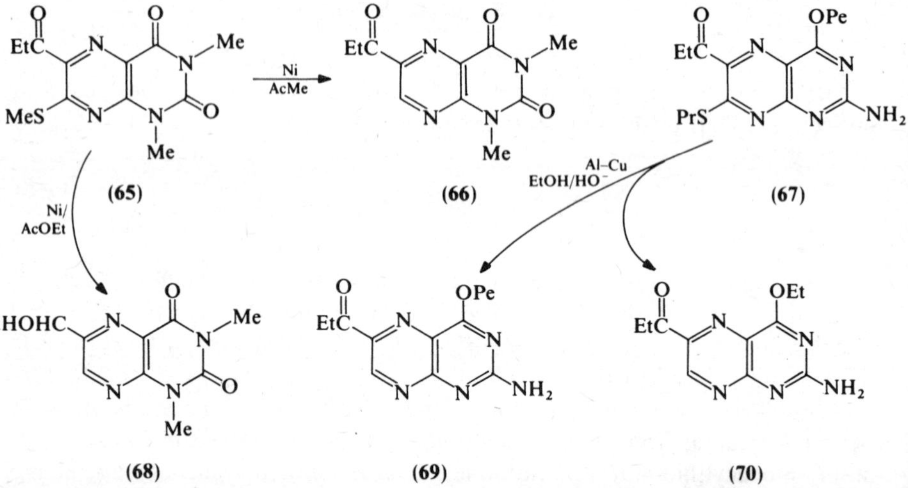

dimethyl-6-propionyl-2,4(1*H*,3*H*)-pteridinedione (**66**) in 64% yield but the similar use of (undeactivated) nickel in ethyl acetate gave 6-α-hydroxypropyl-1,3-dimethyl-2,4(1*H*,3*H*)-pteridinedione (**68**) in 65% yield.[108] The use of an aluminum–copper alloy appears to be less logical. When 1-methyl-7-methylthio-6-propionyl-2,4(1*H*,3*H*)-pteridinedione and its 3-methyl isomer were submitted separately to the action of the alloy in an alkaline medium, the first gave only 6-α-hydroxypropyl-1-methyl-2,4(1*H*,3*H*)-pteridinedione, whereas the second gave only the expected product, 3-methyl-6-propionyl-2,4(1*H*,3*H*)-pteridinedione.[354] Treatment of 4-pentyloxy-6-propionyl-7-propylthio-2-pteridinamine (**67**) with the same alloy in ethanolic alkali gave yet a different result, viz a mixture (71% yield) of 4-pentyloxy- (**69**) and 4-ethoxy-6-propionyl-2-pteridinamine (**70**) (the latter component by transetherification). On alkaline hydrolysis, the mixture gave a single product, 2-amino-6-propionyl-4(3*H*)-pteridinone.[629]

B. Hydrolysis to Pteridinones

The hydrolysis of alkylthiopteridines to the corresponding pteridinones under acidic or alkaline conditions has been covered in Chapter VI, Section 1.E. A warning is contained in the behavior of 4-methylthiopteridine: Under reflux in dilute acetic acid, only 4(3*H*)-pteridinone was formed in ~70% yield; however, even in aqueous sodium hydrogen carbonate, the pteridinone (in low yield) was accompanied by 3-amino-2-pyrazinecarbonitrile, 3-amino-2-pyrazinecarboxamide, and 3-amino-2-pyrazinecarboxylic acid, in amounts that depended on the conditions used.[1317] Other examples (not previously cited) include the acidic hydrolysis of 2-methylthio-6-phenyl-4,7-pteridinediamine to 4,7-diamino-6-phenyl-2(1*H*)-pteridinone (80%),[1281] of 7-amino-2-methylthio-6-phenyl-4(3*H*)-

pteridinone to 7-amino-6-phenyl-2,4(1*H*,3*H*)-pteridinedione,[1281] of 3-β-ethoxycarbonylethyl-6,7-dimethyl-2-methylthio-4(3*H*)-pteridinone to 3-β-carboxyethyl-6,7-dimethyl-2,4(1*H*,3*H*)-pteridinedione (60%),[168] and of 1,6,7-trimethyl-4-methylthio-1-pteridinium iodide (**71**) to (nontautomeric) 1,6,7-trimethyl-4(1*H*)-pteridinone (**72**).[313]

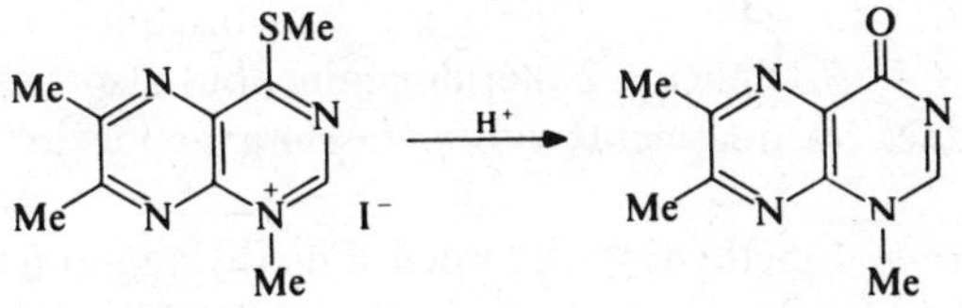

(71) **(72)**

C. Aminolysis

The aminolysis of alkylthiopteridines to furnish 2-, 4-, or 7-amino (or substituted-amino)-pteridines is quite well represented in the literature. Thus in the 2-*position*, 2-methylthio- gave 2-hydrazinopteridine (ethanolic hydrazine hydrate under reflux for 3 h: 65%)[136]; 3,6,7-trimethyl-2-methylthio- gave 2-amino-3,6,7-trimethyl- (ethanolic ammonia at 150°C for 5 h: 35%) or 3,6,7-trimethyl-2-propylamino-4(3*H*)-pteridinone (ethanolic propylamine under reflux for 11 h: 81%)[166]; 6,7-dimethyl-2-methylthio- (**73**, R = SMe) gave 2-amino-6,7-dimethyl- (**73**, R = NH_2) (neat ammonium acetate at 160°C for 1 h: 93%; cf. 50% yield from the 2-thioxo-1,2-dihydro substrate under similar conditions), 2-butylamino-6,7-dimethyl- (**73**, R = NHBu) (butylamine acetate at 140°C for 2 h: ~70%; plus 2,4-bisbutylamino-6,7-dimethylpteridine: < 10%), 2-dibutylamino-6,7-dimethyl- (**73**, R = NBu_2) (dibutylamine acetate at 165°C for 1 h: ~30%), 2-dimethylamino-6,7-dimethyl- (**73**, R = NMe_2) (dimethylamine acetate at 160°C for 1 h: ~70%), or 2-anilino-6,7-dimethyl-4(3*H*)-pteridinone (**73**, R = NHPh) (neat aniline under reflux for 8 h: ~60%; aniline acetate at 150°C for 1 h: ~85%)[169]; 6,7-dimethyl-2-methylthio-4-pteridinamine gave 6,7-dimethyl-2,4-pteridinediamine (ammonium acetate at 165°C for 3 h: ~60%) or 2-dimethylamino-6,7-dimethyl-4-pteridinamine (dimethylamine acetate at 170°C for 2.5 h: 90%)[169]; 6-acetyl- or 6-benzoyl-2-methylthio-4,7-pteridinediamine gave 6-acetyl- or 6-benzoyl-2-piperidino-4,7-pteridinediamine, respectively (refluxing piperidine for 26 h: ~80%)[328]; 6-[L-*erythro*]-α,β-dihydroxypropyl-2-

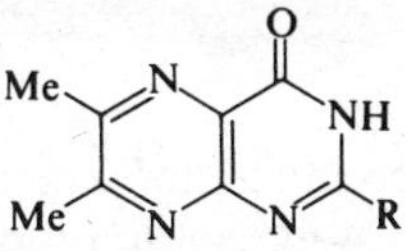

(73)

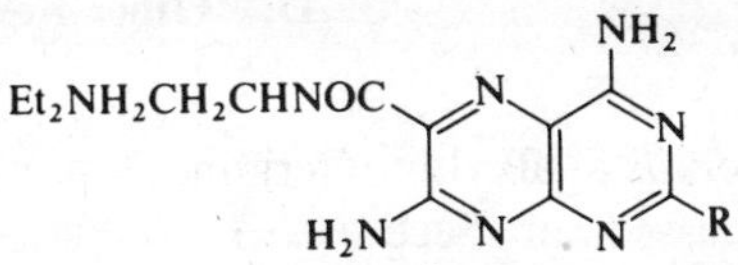

(74)

methylthio-4(3*H*)-pteridinone gave 2-*β*-aminoethylamino-6-[L-*erythro*]-α,*β*-dihydroxypropyl-4(3*H*)-pteridinone (ethylenediamine at 110°C for 4 h: ~40%)[300]; the same or related substrates gave biopterin or analogs[458, 1484]; 4,7-diamino-*N*-*β*-diethylaminoethyl-2-methylthio-6-pteridinecarboxamide (**74**, R = SMe) gave 4,7-diamino-*N*-*β*-diethylaminoethyl-2-diethylaminoethylamino-6-pteridinecarboxamide (**74**, R = $NHCH_2CH_2NEt_2$) (neat *β*-diethylaminoethylamine under reflux for 24 h: ~4%)[321]; and 2-methylthio-4,6,7-triphenylpteridine gave mainly 4,6,7-triphenyl-2-pteridinamine, but also some 2-methylthio-6,8-diphenylpurine, by treatment with potassium amide in liquid ammonia at −33°C.[1118, 1120, 1122]

At the 4-*position*, 4-methylthio- afforded 4-methylamino (ethanolic methylamine under reflux for 2 h: characterized also as its hydriodide) or 4-dimethylaminopteridine (dimethylamine similarly: 94% as base)[129]; 6,7-dimethyl-4-methylthio- gave 6,7-dimethyl-4-methylamino- (methylamine similarly: 95% as base) or 4-dimethylamino-6,7-dimethylpteridine (dimethylamine similarly: 90% as base)[129]; 4-methylthio- gave 4-hydrazinopteridine (methanolic hydrazine hydrate under reflux for 15 min: 90%)[33]; 4-benzylthio-2-pteridinamine gave 4-hydroxyamino-2-pteridinamine (ethanolic hydroxylamine hydrochloride under reflux for 10 min: 75%), 4-hydrazino-2-pteridinamine (methanolic hydrazine hydrate under reflux for 15 min: 40%), or 2,4-pteridinediamine (aqueous ethanolic ammonia at 100°C for 6 h: 50%)[305]; 6-hydroxymethyl-4-methylthio- gave 4-hydroxyamino-6-hydroxymethyl-2-pteridinamine (hydroxylamine)[815]; 4-benzylthio- gave 4-guanidino-2-pteridinamine (**75**) (ethanolic guanidine under reflux for 30 min: 52%)[1101]; and 4-methylthio-6-phenyl- gave 6-phenyl-4-piperidino-2,7-pteridinediamine 5-oxide (neat piperidine under reflux for 16 h: ~70%).[328]

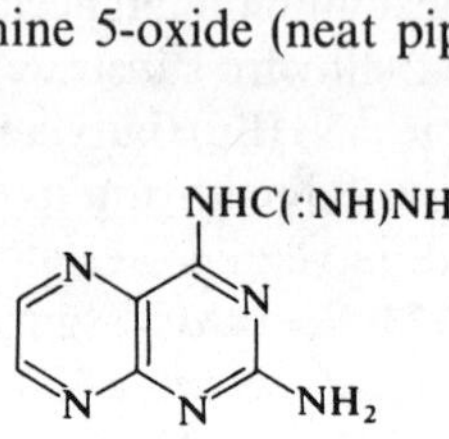

(75)

At the 7-*position*, 7-methylthiopteridine afforded 7-pteridinamine (ethanolic ammonia at 125°C for 6 h: 75%) or 7-dimethylaminopteridine (methanolic dimethylamine at 20°C for 12 h and then under reflux for 30 min: 85%).[33]

D. Other Reactions

The *thiolysis* of alkylthiopteridines to the corresponding pteridinethiones has been discussed in Section 1.E. Treatment of 4-methylthio-6-phenyl-2,7-pteridinediamine 5-oxide with refluxing methanolic sodium methoxide, not only removed the *N*-oxide function but also caused *alcoholysis* of the methylthio function to give 4-methoxy-6-phenyl-2,7-pteridinediamine (~60%).[328] An unusual *thermal rearrangement* of 6,7-diphenyl-2-tri-*O*-benzoylribofuranosylthio-4(3*H*)-pteridinone into the isomeric 6,7-diphenyl-2-thioxo-3-tri-*O*-benzoylribofuranosyl-1,2-dihydro-4(3*H*)-pteridinone (**76**, R = tribenzoylribosyl) has been reported to occur on prolonged refluxing in benzene containing mercuric bromide as a catalyst.[612] The *oxidation* of alkylthio- to alkylsulfinyl- or alkylsulfonylpteridines is discussed next in Section 8.

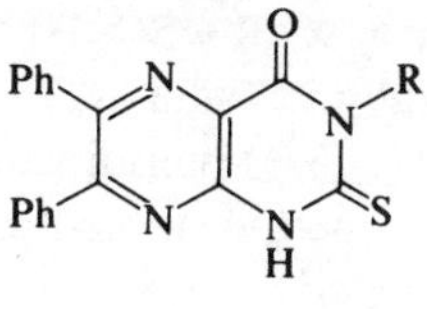

(76)

8. PREPARATION OF PTERIDINYL SULFOXIDES AND SULFONES

Only a few pteridinyl *sulfoxides* have been prepared. Thus controlled oxidation of 1,3,6-trimethyl-7-methylthio- (**77**, R = SMe), 7-isopropylthio-1,3,6-trimethyl- (**77**, R = S-*i*-Pr), or 1,3,6-trimethyl-7-*p*-nitrophenethylthio-2,4-(1*H*,3*H*)-pteridinedione (**77**, R = $SCH_2CH_2C_6H_4NO_2$-*p*) with *m*-chloroperoxybenzoic acid in dichloromethane at 25°C gave the corresponding 7-methylsulfinyl (**77**, R = SOMe) (50%), 7-isopropylsulfinyl (**77**, R = SO-*i*-Pr) (77%), or 7-*p*-nitrophenethylsulfinyl derivative (**77**, R = $SOCH_2CH_2C_6H_4NO_2$-*p*) (40%), respectively.[1709] Similar oxidation of 1,3-dimethyl-6-methylthio-2,4(1*H*,3*H*)-pteridinedione (in ethanol) gave the 6-methylsulfinyl analog (60%).[1681] A second route to a sulfoxide involved *S*-alkylation of 1,3,6-trimethyl-2,4-dioxo-1,2,3,4-tetrahydro-7-pteridinesulfenic acid (**77**, R = SHO) with methyl propiolate to afford 7-*β*-methoxycarbonylvinylsulfinyl-1,3,6-trimethyl-2,4(1*H*,3*H*)-pteridinedione (**77**, R = $SOCH{:}CHCO_2Me$) in 43% yield.[1709]

Known pteridinyl *sulfones* are slightly more numerous. Thus, oxidation of appropriate alkylthio- or arylthiopteridines by *m*-chloroperoxybenzoic acid (>2 mol) or other oxidant (when indicated) gave 2-methylsulfonyl-7-phenyl-4-pteridinamine[621]; 4-amino-8-methyl-2-methylsulfonyl-7(8*H*)-pteridinone[621]; 2-amino-6-methylsulfonyl- (**78**, R = Me) (aqueous permanganate: 74%), 2-amino-6-phenylsulfonyl- (**78**, R = Ph) (hydrogen peroxide–glacial acetic acid: 75%), 2-

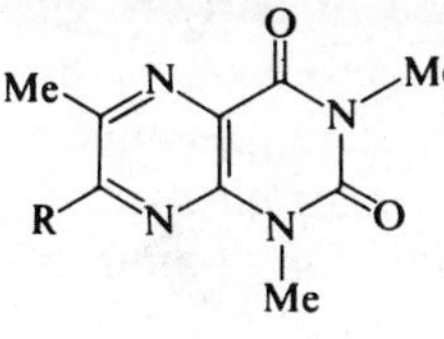

(77)

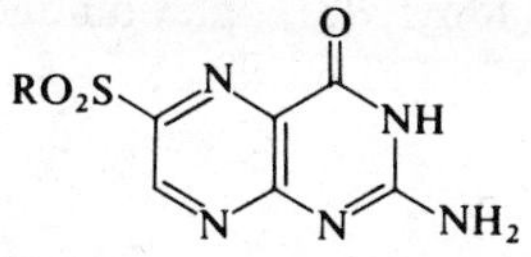

(78)

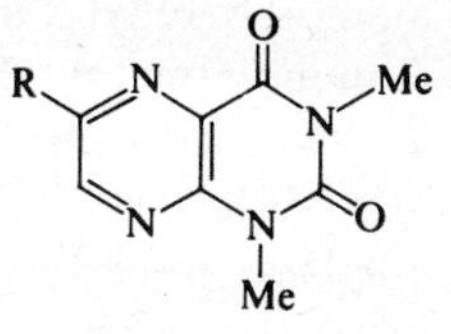

(79)

amino-6-*p*-tolylsulfonyl- (**78**, R = C_6H_4Me-*p*) (likewise: 90%), and 2-amino-6-*p*-chlorophenylsulfonyl-4(3*H*)-pteridinone (**78**, R = C_6H_4Cl-*p*) (likewise: 94%)[1307]; 2-amino-4-methylsulfonyl-6-pteridinecarbaldehyde (from 6-hydroxymethyl-4-methylthio-2-pteridinamine and manganese dioxide)[815]; 1,3-dimethyl-6-methylsulfonyl-2,4(1*H*,3*H*)-pteridinedione (**79**, R = SO_2Me) (57%)[1681]; and 1,3,6-trimethyl-7-methylsulfonyl-2,4(1*H*,3*H*)-pteridinedione (**77**, R = SO_2Me) (83%).[1709] A second route to sulfones is illustrated in the *S*-alkylation (methyl iodide/base) of 1,3-dimethyl-2,4-dioxo-1,2,3,4-tetrahydro-6-pteridinesulfinic acid (**79**, R = SO_2H) or 1,3,6-trimethyl-2,4-dioxo-1,2,3,4-tetrahydro-7-pteridinesulfinic acid (**77**, R = SO_2H) to afford 1,3-dimethyl-6-methylsulfonyl-2,4(1*H*,3*H*)-pteridinedione (**79**, R = SO_2Me) (81%)[1681] or 1,3,6-trimethyl-7-methylsulfonyl-2,4(1*H*,3*H*)-pteridinedione (**77**, R = SO_2Me) (43%),[1709] respectively.

An *extranuclear sulfoxide* and an *extranuclear sulfone* have been made by oxidation of 6-*p*-chlorophenylthiomethyl-2,4-pteridinediamine with hydrogen peroxide–glacial acetic acid to give 6-*p*-chlorophenylsulfinylmethyl (**80**, *n* = 1) (64%) or 6-*p*-chlorophenylsulfonylmethyl-2,4-pteridinediamine (**80**, *n* = 2) (61%), respectively, according to the amount of oxidant and the conditions.[1353]

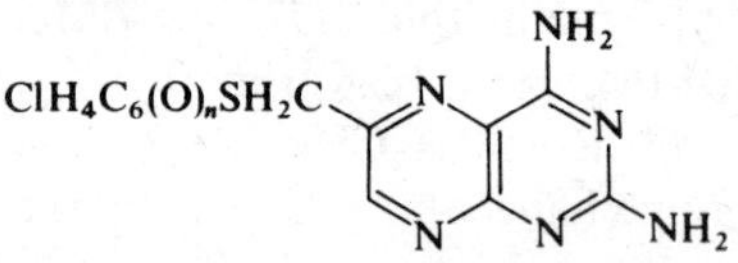

(80)

9. REACTIONS OF PTERIDINYL SULFOXIDES AND SULFONES

Few of the potential displacement or other reactions of pteridinyl sulfoxides and sulfones have been reported but enough are known to confirm the utility of such substituents as excellent leaving groups.[1069, 1161] For example, 2-methylsulfonyl-7-phenyl-4-pteridinamine (**81**, R = SO_2Me) underwent rapid hydrolysis in aqueous dioxane containing a little alkali at 25°C to give 4-amino-7-phenyl-2(1*H*)-pteridinone or equally rapid methanolysis in methanolic sodium methoxide to give 2-methoxy-7-phenyl-4-pteridinamine (**81**, R = OMe)[621]; 4-amino-8-methyl-2-methylsulfonyl-7(8*H*)-pteridinone behaved similarly to give 4-amino-8-methyl-2,7(1*H*,8*H*)-pteridinedione or 4-amino-2-methoxy-8-methyl-7(8*H*)-pteridinone, according to the reagent used[621]; and 2-amino-6-methylsulfonyl-4(3*H*)-pteridinone (**78**, R = Me) with thiophenol or phenylmethanethiol in refluxing DMF gave 2-amino-6-phenylthio- or 2-amino-6-benzylthio-4(3*H*)-pteridinone, respectively, each in >70% yield.[1307] The attempted aminolysis of 6-methylsulfonyl-2,4,7-pteridinetriamine (**82**, R = H,

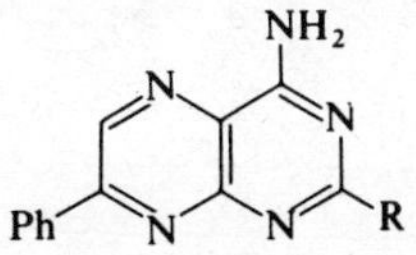

(81)

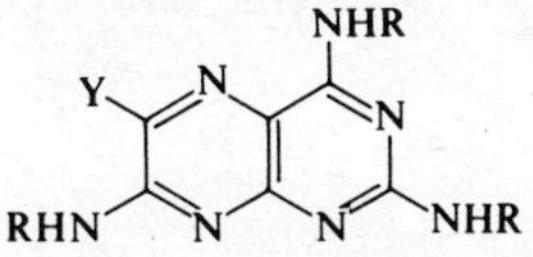

(82)

Y = SO_2Me) with benzylamine and related amines led only to transamination in each case, to give 2,4,7-trisbenzylamino- (**82**, R = CH_2Ph, Y = SO_2Me) (44%), 2,4,7-tris-*p*-methylbenzylamino- (20%), or 2,4,7-tris(2′,4′-dimethylbenzylamino)-6-methylsulfonylpteridine as well as 6-methylsulfonyl-2,4,7-trisphenethylamino-pteridine (**82**, R = CH_2CH_2Ph, Y = SO_2Me). This reversal of the expected reactivities resulted from (a) extreme deactivation of the methylsulfonyl leaving group by electron donation from three amino groups and (b) enhanced activation of the amino groups towards transamination through powerful electron-withdrawal by the methylsulfonyl group.[249] In contrast, treatment of the same substrate (**82**, R = H, Y = SO_2Me) with refluxing thiophenol gave 6-phenylthio-2,4,7-pteridinetriamine (**82**, R = H, Y = SPh) in 86% yield.[249] The Cope like elimination reaction[1709] of 7-isopropylsulfinyl-1,3,6-trimethyl-2,4(1*H*,3*H*)-pteridinedione in boiling toluene to afford 1,3,6-trimethyl-2,4-dioxo-1,2,3,4-tetrahydro-7-pteridinesulfenic acid (57%) has been noted in Section 5 of this chapter.

CHAPTER IX

Pteridine Amines and Imines

This chapter covers primary, secondary, and tertiary aminopteridines, both nuclear and extranuclear. It also covers the iminodihydropteridines, fixed as such by alkylation on a ring nitrogen. As pointed out elsewhere, the IUPAC rules of nomenclature have been modified so that, whereas a nuclear primary amino or primary imino group is designated as the suffix within an *x*-pteridinamine or an *x*(*yH*)-pteridinimine (providing there is no more senior suffix available in the molecule), all other amino or imino groups appear invariably as prefixes.

1. PREPARATION OF NUCLEAR AMINOPTERIDINES

A. By Primary Syntheses

Primary syntheses of a great many aminopteridines, either from pyrimidines or from pyrazines or other systems, have been discussed in Chapters II and III, respectively. Such processes from *pyrimidines* are illustrated simply in the *Gabriel and Colman* condensation of 2,4,5-pyrimidinetriamine (**1**, $R=NH_2$), 2-methylamino-4,5-pyrimidinediamine (**1**, R = NHMe), or 2-dimethylamino-4,5-pyrimidinediamine (**1**, $R=NMe_2$) with glyoxal to give 2-pteridinamine (**2**, $R=NH_2$), 2-methylaminopteridine (**2**, R = NHMe), or 2-dimethylaminopteridine (**2**, $R=NMe_2$), respectively[30]; of 2,4,5,6-pyrimidinetetramine with glyoxal to give 2,4-pteridinediamine (**3**, R = H),[267, 284, 475, 730] or with biacetyl to give 6,7-dimethyl-2,4-pteridinediamine (**3**, R = Me)[284, 475, 541, 749, 884]; and of 2,5,6-triamino-4(3*H*)-pyrimidinone with oxalic acid at ~250°C to give 2-amino-4,6,7-(3*H*,5*H*,8*H*)-pteridinetrione.[60, 243] They are also exemplified in the *Timmis* condensation of 5-nitroso-2-phenyl-4,6-pyrimidinediamine (**4**) with phenylacetone to give 7-methyl-2,6-diphenyl-4-pteridinamine (**5**),[326] or of the same substrate with (a) cyanoacetamide[1466] or (b) benzoylacetamide plus sodium cyanide,[326] to give 4,7-diamino-2-phenyl-6-pteridinecarboxamide (**6**). They are illustrated further in the *Boon* condensation of 2,4-dichloro-5-nitropyrimidine with aminoacetone to give 4-acetonylamino-2-chloro-5-nitropyrimidine (**7**, R = Cl) followed successively by aminolysis to 4-acetonylamino-5-nitro-2-pyrimidinamine (**7**, $R=NH_2$), reductive cyclization to 6-methyl-7,8-dihydro-2-pteridinamine (**8**), and oxidation to afford 6-methyl-2-pteridinamine (**9**).[39, 688]

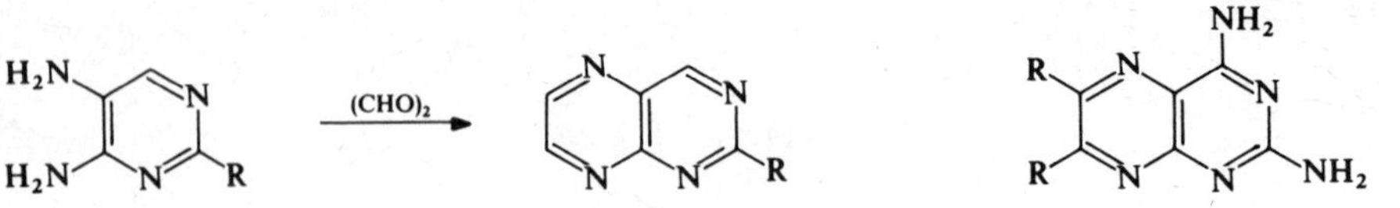

(1) (2) (3)

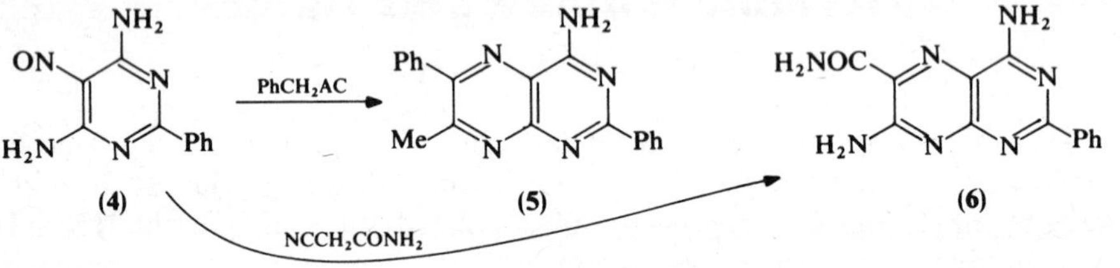

(4) (5) (6)

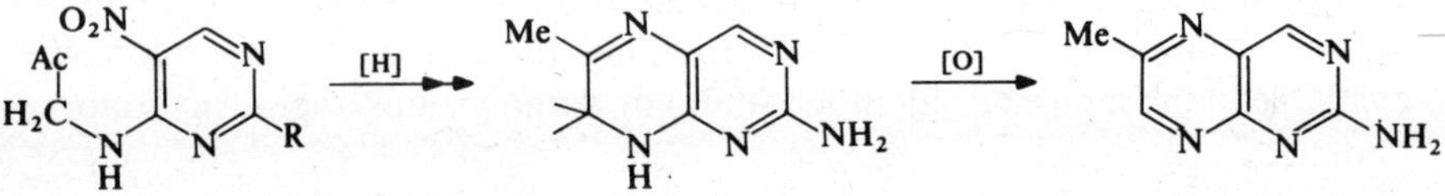

(7) (8) (9)

Finally several *minor* condensations are exemplified in that of 6-chloro-5-nitro-4-pyrimidinamine with *N*,*N*-dimethyl-α-phenylacetamidine to give 6-α-dimethylaminophenethylideneamino-5-nitro-4-pyrimidinamine (**10**), which underwent direct cyclization in ethoxide to give 7-dimethylamino-6-phenyl-4-pteridinamine 5-oxide (**11**) and subsequent loss of the *N*-oxide function on reduction to give 7-dimethylamino-6-phenyl-4-pteridinamine (**12**).[764]

Synthetic processes from *pyrazines*, and so on, are illustrated briefly in the condensation of 3-amino-2-pyrazinecarbonitrile (**14**, R = H) with formamidine acetate to give 4-pteridinamine (**13**),[53] or of 3-amino-5-methyl-2-pyrazinecarbonitrile (**14**, R = Me) with guanidine to give 7-methyl-2,4-pteridinediamine (**15**)[1305]; in the conversion of the quaternary purine, 2-amino-7-ethoxycarbonyl-

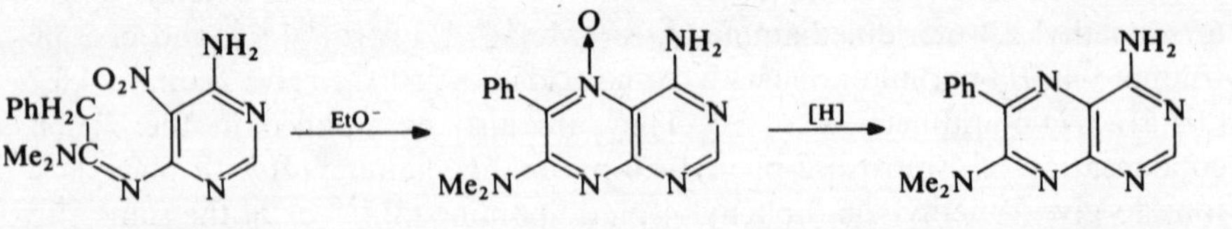

(10) (11) (12)

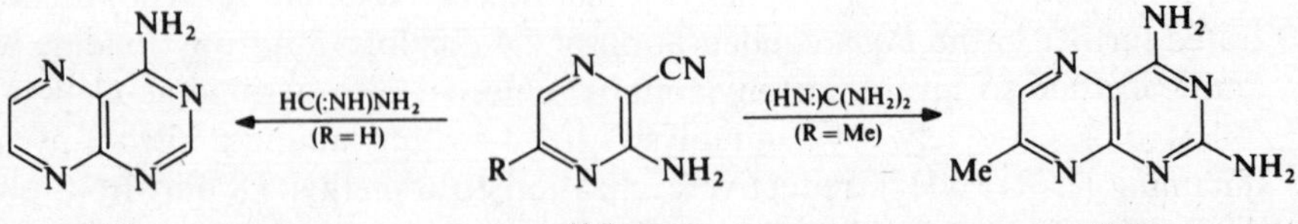

(13) (14) (15)

methyl-9-methyl-6-oxo-1,6-dihydro-7-purinium bromide, by alkali in the presence of air, into 2-amino-8-methyl-4,7(3*H*,8*H*)-pteridinedione (**16**) in 63% yield (for mechanism see Ch. III, Sect. 2)[181]; and in the conversion of 7-α-ethoxycarbonylmethylamino-5-[1,2,5]oxadiazolo[3,4-*d*]pyrimidinamine (**17**), by hydrogenation, into 2,4-diamino-7,8-dihydro-6(5*H*)-pteridinone (**18**) and then by iodine oxidation to 2,4-diamino-6(5*H*)-pteridinone.[1293]

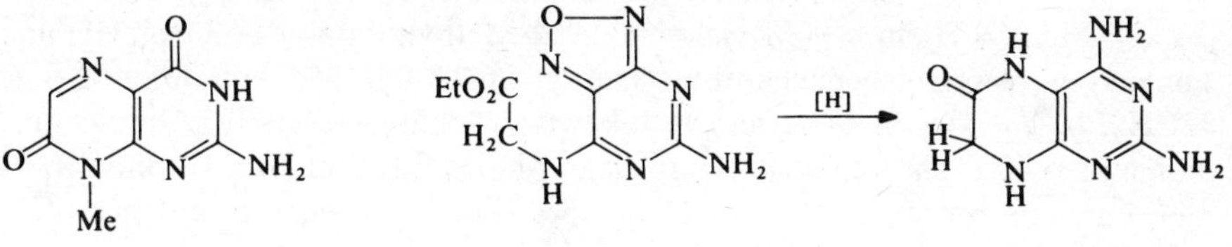

(**16**) (**17**) (**18**)

B. By Nucleophilic Displacement Reactions

Because of the plan followed in this book, most of the important routes to aminopteridines involving nucleophilic displacements by ammonia or amines, have been discussed already. Thus, for the aminolysis of *halogenopteridines*, see Chapter V, Section 2.C; of *pteridinones*, see Chapter VI, Section 2.E; of *alkoxypteridines*, see Chapter VII, Section 2.B; of *pteridinethiones* or *alkylthiopteridines*, see Chapter VIII, Sections 2.E and 7.C, respectively; of *pteridinesulfinic acids* and *pteridinesulfonic acids*, or salts thereof, see Chapter VIII, Section 5; and of a pteridine sulfone (unsuccessful), see Chapter VIII, Section 9. The interconversion of aminopteridines by *transamination* is discussed later in Section 2.F.

C. By Covalent Addition of Ammonia or Amines

When pteridine was treated with ammonia in pH 10 buffer (or simply in liquid ammonia at −40°C) there was good spectral evidence[184, 1138] for the formation of 3,4-dihydro-4-pteridinamine (**19**) and this was confirmed by dissolving pteridine in liquid ammonia containing one redox equivalent of potassium permanganate to give, within a few minutes, a 50% yield of 4-pteridinamine (**20**, R = H)[934]; on using diethylamine instead of ammonia in the latter experiment, 4-diethylaminopteridine (**20**, R = Et) resulted.[934] However, this simple

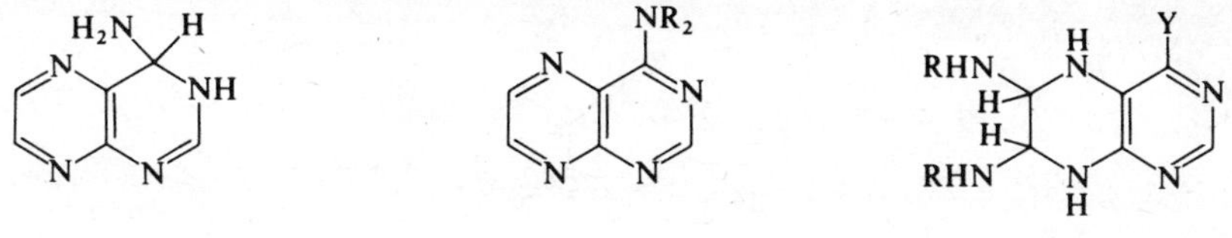

(**19**) (**20**) (**21**)

picture did not apply when ammonia was passed into a solution of pteridine in ethanolic ammonia or when a solution of pteridine in liquid ammonia was allowed to come to room temperature in a sealed tube. The spectral evidence suggested[52, 1138] the formation of 5,6,7,8-tetrahydro-6,7-pteridinediamine (**21**, R = Y = H), although it could not be isolated in a pure state[52]; confirmation was provided on replacing ammonia by methylamine to give the spectrally similar but isolable di-adduct, 6,7-bismethylamino-5,6,7,8-tetrahydropteridine (**21**, R = Me, Y = H), in >90% yield.[52, 1700] 4-Methylpteridine and benzylamine similarly gave 6,7-bisbenzylamino-4-methyl-5,6,7,8-tetrahydropteridine (**21**, R = CH_2Ph, Y = Me) in 60%yield.[50] Likewise, 2-chloropteridine in liquid ammonia at −40°C showed a spectrum clearly indicating 2-chloro-3,4-dihydro-4-pteridinamine but at 25°C (sealed) the spectrum indicated the presence of 2-chloro-5,6,7,8-tetrahydro-6,7-pteridinediamine.[1138, 1700] The addition of 2-chloropteridine to liquid ammonia or ethylamine (at −70°C), each containing permanganate, gave 2-chloro-4-pteridinamine (94%) or 2-chloro-4-ethylaminopteridine (58%), respectively.[934] Likewise, appropriate substrates in liquid amines containing permanganate gave 7-methyl- (88%), 7-*t*-butyl- (92%), 7-phenyl- (90%), 7-*p*-methoxyphenyl- (61%), 2-phenyl- (50%), 6,7-dimethyl- (47%), and 6,7-diphenyl-4-pteridinamine (70%)[1740]; also 4-ethylamino- (26%) and 4-*t*-butylamino-7-phenylpteridine (70%) as well as other such products.[1167] Both 6-chloro- and 7-chloropteridine reacted with ammonia or an appropriate amine in benzene at <5°C to give the isolable adducts, 6-chloro-3,4-dihydro-4-pteridinamine (80%), its 7-chloro isomer (73%), 6-chloro-4-cyclohexylamino-3,4-dihydropteridine (50%), and 4-benzylamino-7-chloro-3,4-dihydropteridine (46%), respectively.[146]

The addition of ammonia to 6(5*H*)-pteridinone was far less simple and resulted in a highly insoluble product of high molecular weight, possibly di(6-oxo-5,6,7,8-tetrahydro-7-pteridinyl)amine; however, when potassium ferricyanide was added to the reaction mixture at the outset, 7-amino-6(5*H*)-pteridinone (**22**) was isolated in 55% yield as its sodium salt.[19] In contrast, aqueous hydroxylamine and 6(5*H*)-pteridinone (under reflux but without any added oxidizing agent) gave an easily separable mixture of a highly insoluble material (differing from the one just cited) and 7-amino-6(5*H*)-pteridinone (**22**), the latter in 65% yield.[19] No rational mechanism for its formation has been advanced. The reaction of morpholine with 2-methyl-6(5*H*)- and 2-methyl-7(8*H*)-pteridinone, in the presence of sulfur as an oxidant, afforded 2-methyl-7-morpholino-6(5*H*)- and 2-methyl-6-morpholino-7(8*H*)-pteridinone, respectively.[591]

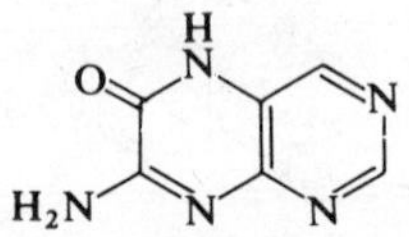

(22)

2-Amino-7,8-dihydro-4(3*H*)-pteridinone in aqueous ammonia gave an isolable 1:1 adduct, which on oxidation with permanganate or manganese dioxide gave 2,6-diamino-4(3*H*)-pteridinone (**23**, $R=NH_2$) in ~20% overall yield.[453, 507] 2-Amino-5,6,7,8-tetrahydro-4(3*H*)-pteridinone behaved similarly with ammonia or an amine in DMF open to air, to give 2,6-diamino- (**23**, $R=NH_2$), 2-amino-6-methylamino- (**23**, R = NHMe), 2-amino-6-dimethylamino- (**23**, $R=NMe_2$), 2-amino-6-morpholino-, and 2-amino-6-γ-dimethylaminopropylamino-4(3*H*)-pteridinone [**23**, $R=Me_2N(CH_2)_3NH$].[1332, cf. 1325, 1328]

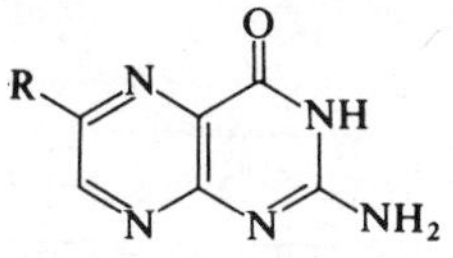

(**23**)

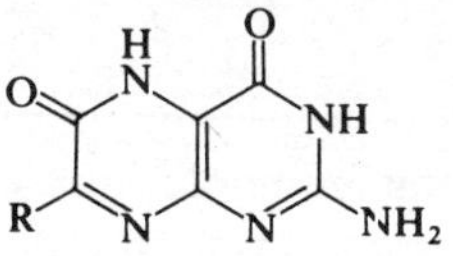

(**24**)

The conversion of 2-amino-4,6-dioxo-3,4,5,6-tetrahydro-7-pteridinecarboxylic acid (**24**, $R=CO_2H$) into 2,7-diamino-4,6(3*H*,5*H*)-pteridinedione (**24**, $R=NH_2$) (58%), by stirring with aqueous ammonia containing manganese dioxide for 5 days, appears to proceed by 7,8-addition of ammonia followed by decarboxylation and oxidation.[97] Oxidative amination of xanthopterin gave 2,7-diamino-4,6(3*H*,5*H*)-pteridinedione in 80% yield.[1754]

D. From Acylaminopteridines

It is sometimes useful to transform an acetamido- or other acylaminopteridine into the parent pteridinamine. The process is exemplified in the conversion of 7-acetamido- (**25**, R = Ac) into 7-amino-1,3-dimethyl-2,4,6(1*H*,3*H*,5*H*)-pteridinetrione (**25**, R = H) (2*M* potassium hydroxide at 25°C for 5 min: 57% via the isolated potassium salt)[238]; of 2-acetamido- into 2-amino-6-methoxy-4(3*H*)-pteridinone (1*M* sodium hydroxide under reflux for 1 h: ~70%)[369]; of 2-acetamido-4,6(3*H*,5*H*)-pteridinedione into the same product (methanolic hydrogen chloride: ~60%; note the *O*-methylation)[369]; of 2*N*-acetylpteroylglutamic acid into pteroylglutamic acid (0.1*M* alkali at 90°C for 30 min under nitrogen: ~85%)[1300]; of 2-acetamido-6-acetoxymethyl-4(3*H*)-pteridinone into 2-amino-6-hydroxymethyl-4(3*H*)-pteridinone (3*M* hydrochloric acid at 25°C for 16 h: 95%)[1443]; of 2,4-diacetamidopteridine into 2,4-

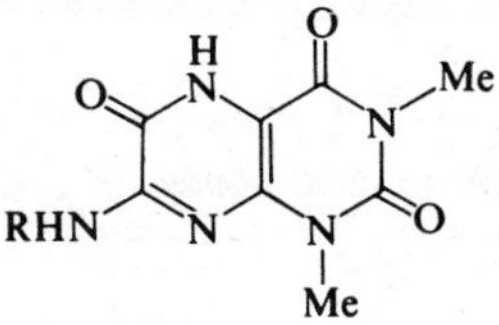

(**25**)

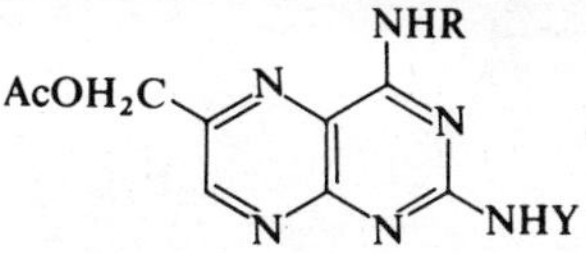

(**26**)

pteridinediamine (refluxing ethanolic sodium ethoxide for 25 min: 50%)[123]; and of 2,4-diacetamido-6-acetoxymethylpteridine (**26**, R = Y = Ac) into a separable mixture of 2-acetamido-6-acetoxymethyl-4-pteridinamine (**26**, R = H, Y = Ac) (17% yield) and 6-acetoxymethyl-2,4-pteridinediamine (**26**, R = Y = H) (20% yield) (by stirring with Merck "Kieselgel-60" in methanolic chloroform for 2 days).[113] Other examples are known.[1739, 1753]

E. From Pteridinecarboxamides

Although many carboxamides of pteridine are known, their conversion into the corresponding aminopteridines by the Hofmann or a similar reaction is rare indeed. Such a process is exemplified in the treatment of 4-amino-2,7-diphenyl-6-pteridinecarboxamide (**27**, R = Ph) in DMF with aqueous potassium hypobromite to give ~80% of 2,7-diphenyl-4,6-pteridinediamine (**28**, R = Ph).[327] Similar treatment of 2,4-diamino-7-phenyl-6-pteridinecarboxamide (**27**, R = NH_2) gave 7-phenyl-2,4,6-pteridinetriamine (**28**, R = NH_2) in ~50% yield.[327]

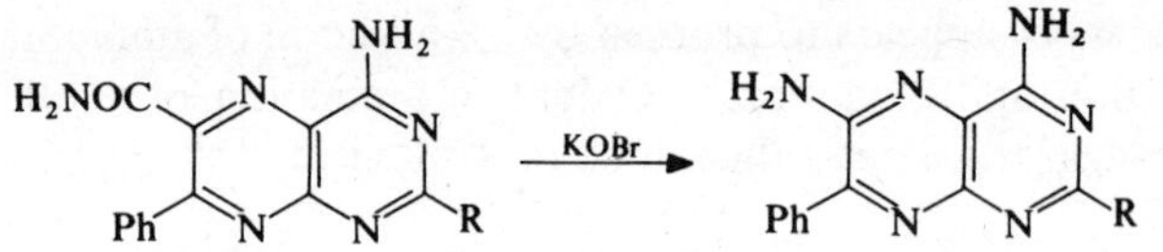

(27) (28)

F. By the Dimroth Rearrangement

The formation of alkylaminopteridines by rearrangement of suitable *N*-alkylated pteridinimines (or potential imines) can be a most useful process. It is discussed under the Dimroth rearrangement in Section 6.B to follow.

G. From *N*-Oxides

Treatment of 2,4-pteridinediamine 8-oxide with pyrrolidine at 95°C for 16 h gave a single product, 7-pyrrolidino-2,4-pteridinediamine, in 62% yield.[1301] The product was identified beyond doubt by analysis, nmr, and alkaline hydrolysis to 2-amino-4,7(3*H*,8*H*)-pteridinedione,[1301] but the mechanism appears to remain unelucidated.

H. Formation of *N*-Aminopteridinones

Some of the potential methods for making *N*-aminopteridinones or related compounds are illustrated in the following sequences. Cyclization of 3-amino-

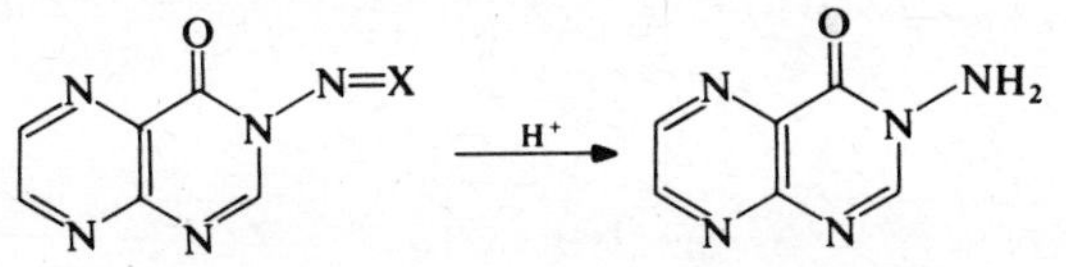

(29) (30)

N'-benzylidene-2-pyrazinecarbohydrazide or its N'-isopropylidene analog by triethyl orthoformate in acetic anhydride gave 3-benzylideneamino- (**29**, X=CHPh) or 3-isopropylideneamino-4(3*H*)-pteridinone (**29**, R=CMe_2), respectively, each in >90% yield. The former underwent rapid hydrolysis in ethanolic hydrochloric acid at room temperature to give 3-amino-4(3*H*)-pteridinone (**30**) in 15% yield, but similar treatment of the isopropylidene compound gave no less than 83% of the same product (**30**).[487] Treatment of pteroylglutamic acid in alkali at 0°C with hydroxylamine-*O*-sulfonic acid gave a modest yield of 3-aminopteroylglutamic acid, used to make the 2-hydrazino analog of pteroylglutamic acid by a Dimroth rearrangement under acidic conditions.[488] The peculiar reaction of 3-methyl-7-phenyl-2,4(1*H*,3*H*)-pteridinedione (**31**) or its 5-oxide (**33**) with boiling hydrazine hydrate to give 3-amino-7-phenyl-2,4(1*H*,3*H*)-pteridinedione (**32**) in 67 or 48% yield, respectively, has not been explained adequately.[331]

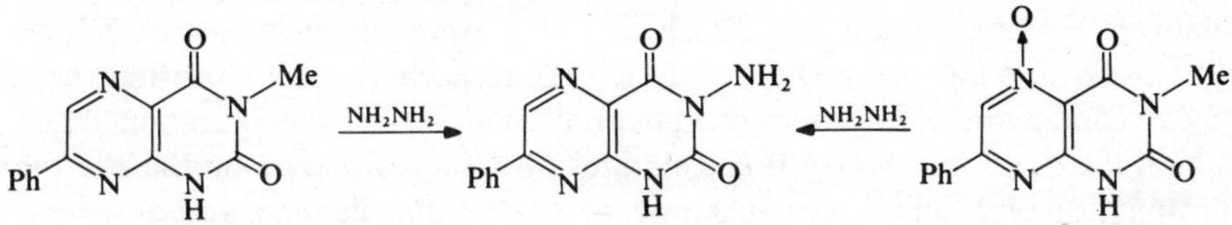

(31) (32) (33)

2. PROPERTIES AND REACTIONS OF NUCLEAR AMINOPTERIDINES

Although the tautomeric state of primary pteridinamines has not been investigated thoroughly, there seems every reason to believe that they exist predominantly as such (as do analogs in related heterocyclic series[1713]) rather than as dihydroiminopteridines. Thus, in comparing the uv spectra of 2-, 4-, and 7-pteridinamine with those of the corresponding dimethylaminopteridines as early as 1952,[26] Adrien Albert concluded that "these spectra make it unlikely that any of the three aminopteridines exist as an imino form"; a little later he reached a similar conclusion in respect to 6-pteridinamine.[31] Subsequent work did provide some of the pteridinimines fixed as such by methylation on one or another ring nitrogen, for example, 1-methyl-4(1*H*)-pteridinimine (**34**), which would be required for rigorous spectral studies of tautomerism in 4-pteridinamine,[129] 7-pteridinamine,[131] 4,7-pteridinediamine,[131] 2-amino-4(3*H*)-pteri-

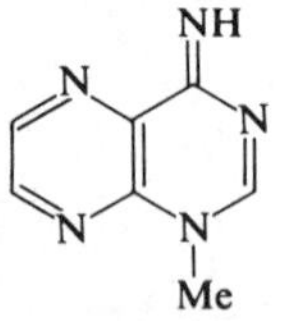

(34)

dinone,[130] and 2,4-pteridinediamine.[130, 267] While the strongly basic imines were stable as cations, they proved quite unstable in buffers of sufficiently high pH to measure spectra of the free bases. Thus, mainly for lack of appropriate rapid-reaction facilities at the time, we lost the opportunity to study properly the tautomerism of the parent amines. However, the relatively high pK_a values for such imines, compared with those for the corresponding pteridinamines and dimethylaminopteridines, did provide limited confirmation for predominance of the amino tautomers in solutions of the pteridinamines. So too did the good agreement between the variable electronegativity self consistent field molecular orbital configuration interaction (VESCF–MO–CI) calculated values for π–π^* electronic transitions for all pteridinamines (in the amino form) and the experimental values.[1013, 1016]

It has been suggested on less firm spectral grounds that 7-pteridinamine probably protonates at N-3 to give the cation (**35**),[131] whereas 4,7-pteridinediamine[131] and 4-pteridinamine itself[129, cf. 1536] probably protonate on N-1 to give the cation (**36**, R = NH_2) or (**36**, R = H), respectively; self consistent field linear combination of atomic orbitals molecular orbital configuration interaction (SCF–LCAO–MO–CI) calculations also suggested N-1 as the site for protonation of 2- and 4-pteridinamine, of 2,4-pteridinediamine, and of several derivatives.[947] Some protonation sites have been detected by ^{13}C nmr studies.[826]

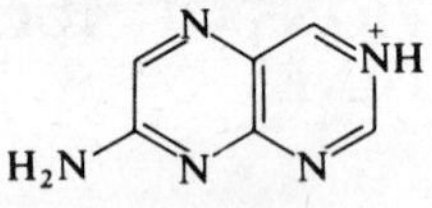

(35)

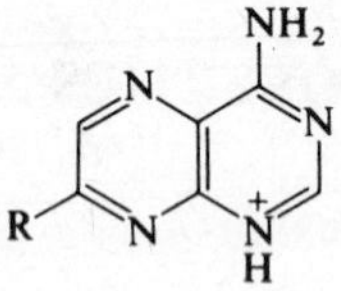

(36)

A potentially useful method to extend the pH range in which normally insoluble aminopteridinecarboxamides may be retained in solution for administration, and so on, has been reported. It involved the addition of ~10% of deoxyribonucleic acid prior to pH adjustment.[1115, cf. 1404]

A. Degradative Reactions

Treatment of 4-pteridinamine (**38**) in 15% aqueous ammonia at 170°C under pressure for 1 h gave 3-amino-2-pyrazinecarboxylic acid (**37**), initially as its ammonium salt in high yield[185]; the same substrate (**38**) in refluxing hydrazine hydrate gave 3-aminomethyleneamino-2-pyrazinecarboxamidrazone (**39**).[185] More complicated reactions occurred when 4-pteridinamine (**38**) was treated with aqueous chloroacetaldehyde at 60°C for 5 h to afford mainly 2-formamido-3-imidazol-2′-ylpyrazine (**40**), some 3-(1′-formylimidazol-2′-yl)-2-pyrazinamine (**41**), and (when heating was prolonged to 24 h) some 2-β-chloroethylideneamino-3-imidazol-2′-ylpyrazine (**42**).[459] As might be expected, 2-methyl-, 6,7-dimethyl- or 2,6,7-trimethyl-4-pteridinamine likewise gave mainly 2-acetamido-3-imidazol-2′-yl-, 2-formamido-3-imidazol-2′-yl-5,6-dimethyl-, and 2-acetamido-3-imidazol-2′-yl-5,6-dimethylpyrazine, respectively.[459]

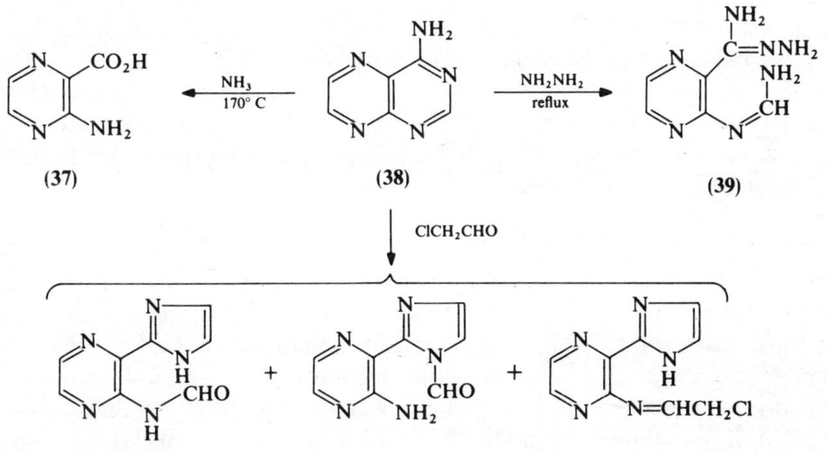

(**37**) (**38**) (**39**)

(**40**) (**41**) (**42**)

Some interesting mass spectral fragmentation patterns have been reported for 4-pteridinamine,[1391] 4-dimethylamino-6,7-dimethylpteridine,[1384] 2-amino-7-phenacyl-4,6(3*H*,8*H*)-pteridinedione and some *p*-substituted derivatives,[958] 6,7-dimethyl-2,4-pteridinediamine,[1391] 4-trifluoromethyl-2-pteridinamine and related compounds,[780] 2,4-diamino-7-phenacyl-6(5*H*)-pteridinone and some *p*-substituted derivatives,[958] 2-acetamido-4(3*H*)-pteridinone and derivatives bearing trimethylsilyl and/or additional acetyl groups,[589] and other aminopteridinones.[778, 1707]

B. Formation of Covalent Adducts

Although the neutral molecules of simple pteridinamines show no tendency to form covalent hydrates or alcoholates, at least 2-pteridinamine does form adducts with powerful nucleophiles to give the following isolable products: sodium 2-amino-3,4-dihydro-4-pteridinesulfonate (**43**) (shaking with aqueous sodium hydrogen sulfite for 10 min: 48%), 4-(4′,4′-dimethyl-2′,6′-dioxocyclohexyl-3,4-dihydro-2-pteridinamine (stirring in aqueous dimedone for 48 h:

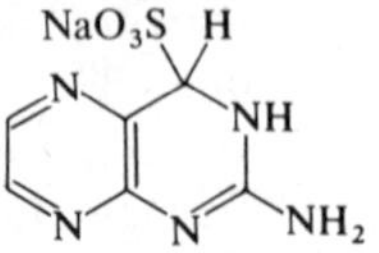

(43)

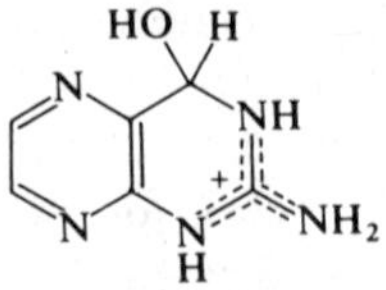

(44)

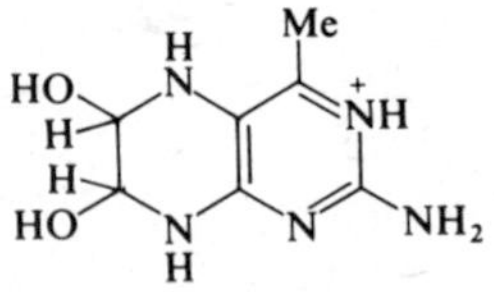

(45)

60%), 4-(4′,6′-dioxo-3′,4′,5′,6′-tetrahydro-5′-pyrimidinyl)-3,4-dihydro-2-pteridinamine (93%), similar adducts with barbituric or thiobarbituric acid, and 4-nitromethyl-3,4-dihydro-2-pteridinamine (stirring with aqueous nitromethane for 48 h: 52%).[45]

In contrast, both uv[39] and subsequent nmr[27] studies of the cations of simple 2-pteridinamines clearly showed that the parent amine existed in aqueous acid as the 2-amino-4-hydroxy-3,4-dihydropteridinium cation (**44**); its 6- and 7-methyl derivatives existed largely as similar 3,4-hydrates; 2-methylamino- and (possibly) 2-dimethylaminopteridine did likewise; 4,6-dimethyl-, 4,7-dimethyl-, 6,7-dimethyl-, and 4,6,7-trimethyl-2-pteridinamine all had virtually anhydrous cations; and 4-methyl-2-pteridinamine initially showed about equal amounts of a 3,4-hydrated and an anhydrous cation, both of which slowly gave a 5,6,7,8-dihydrate such as the 2-amino-6,7-dihydroxy-4-methyl-5,6,7,8-tetrahydro-3-pteridinium cation (**45**).[27, 39] In addition, 6-pteridinamine appeared to form a cationic 3,4-hydrate.[27] It proved possible to isolate and characterize the guanidinium-resonance stabilized cationic alcoholates, 2-amino-4-methoxy-3,4-dihydropteridinium chloride,[45] its 4-ethoxy homolog,[45] and 2-amino-4-ethoxy-3,4-dihydropteridinium bromide,[766] of which the last was conclusively proven in structure by an x-ray crystal analysis.[766]

C. Acylation Reactions

The acylation of primary or secondary aminopteridines proceeds quite readily: Acetylation has been, by far, the most commonly used process. Acylation of *monoaminopteridines* is exemplified in the formation of 2-acetamidopteridine (**46**, R = H) (refluxing acetic anhydride for 30 min: 70%),[30] 4-acetamidopteridine (likewise),[30] 7-acetamidopteridine (likewise: 70%),[33] 2-acetamido-4-methylpteridine (**46**, R = Me) (acetic anhydride at 100°C for 30 min: 75%),[282] 2-acetamido-4-methoxypteridine (**46**, R = OMe) (acetic anhydride–pyridine under reflux for 1 h: ~50%),[368] and 4-acetamido-6,*N*,*N*-trimethyl-7-pteridinecarboxamide 5,8-dioxide (acetic anhydride–pyridine at 100°C for 1 h: 67%; note that the *N*-oxide functions remained unaffected because of the substitution pattern).[723]

The acylation of *diaminopteridines* is illustrated in the formation of 2,4-diacetamidopteridine (**46**, R = NHAc) (acetic anhydride under reflux for 1 h: 60%[123]; acetic anhydride at 25°C for 1 week: 80%[267]; acetic anhydride–pyridine under reflux for 5 min: 50%[267]), its 6-methyl derivative

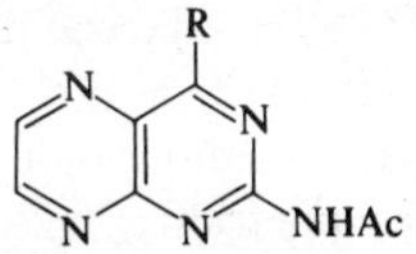

(46)

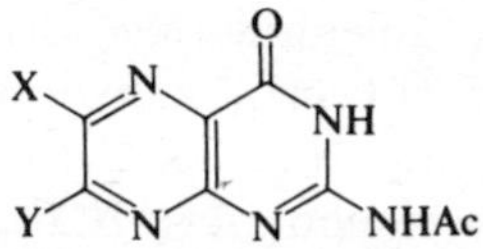

(47)

(acetic anhydride under reflux),[1263, cf. 123] the 7-methyl isomer (brief boiling in acetic anhydride: 65%),[1263] 2,4-diacetamido-6-phenylpteridine (refluxing acetic anhydride for 4 h: 63%),[541] its 6-phenethyl analog (boiling acetic anhydride),[90] 2,4-diacetamido-6,7-dimethylpteridine (refluxing acetic anhydride: 60%),[648] 2,4-diacetamido-7-propionylpteridine (refluxing acetic anhydride for 20 min: 54%),[1678] 4,7-diacetamido-2-methylthio-6-pteridinecarboxamide (hot acetic anhydride),[324] and 4,7-diacetamido-2-phenyl-6-pteridinecarboxamide (acetic anhydride–DMF–pyridine under reflux for 15 min: 31%).[1357]

Acylation of *aminopteridinones* and related compounds is well represented in the preparation of 2-acetamido-4(3*H*)-pteridinone (**47**, X = Y = H) (refluxing acetic anhydride for 3 h: ~70%),[368] its 6-methyl derivative (**47**, X = Me, Y = H) [refluxing acetic anhydride for 1 (27%),[389] 4.5,[929] or 12 h (33%)[1394]; 1:1 acetic acid–acetic anhydride under reflux for 90 min (80–90%)[1443]],[cf. 993] the 7-methyl isomer (**47**, X = H, Y = Me) (acetic anhydride under reflux for 3 h: 58%),[389, 1683, cf. 993] the 6,7-dimethyl analog (**47**, X = Y = Me) (refluxing acetic anhydride for 4.5 h: 75%),[69] 6,7-dimethyl-2-*N*-methylacetamido-4(3*H*)-pteridinone (refluxing acetic anhydride for 3 h: 36%),[69] 2-acetamido-7(8*H*)-pteridinone (acetic anhydride–pyridine under reflux for 3 h: ~90%),[381] 2-acetamido-1,6,7-trimethyl-4(1*H*)-pteridinone (refluxing acetic anhydride for 4 h: 75%),[69] the isomeric 2-acetamido-3,6,7-trimethyl-4(3*H*)-pteridinone (similarly: 45%),[654] 2-acetamido-6-methoxy-4(3*H*)-pteridinone (**47**, X = OMe, Y = H) (refluxing acetic anhydride for 2 h: ~90%),[369] 2-acetamido-4-pentyloxy-7(8*H*)-pteridinone (acetic anhydride–DMF–pyridine under reflux for 30 min: 51%),[436] 2-acetamido-4-oxo-3,4-dihydro-6-pteridinecarbaldehyde (**47**, X = CHO, Y = H) [acetic anhydride at 130°C under nitrogen until the substrate dissolved: 80 to 90%[704, 755, 1310]; under slightly different conditions, 2-acetamido-6-diacetoxymethyl-4(3*H*)-pteridinone was isolated in ~50% yield and could be converted into the aldehyde by brief heating in very dilute hydrochloric acid[1451]], 2-acetamido-6-diethoxymethyl-4(3*H*)-pteridinone [**47**, X = $CH(OEt)_2$, Y = H] (acetic anhydride at 90–100°C for 4 h: > 95%),[464, 1300, 1403] the homologous 6-diethoxymethyl-2-propionamido-4(3*H*)-pteridinone (propionic anhydride),[1403] 6-chloro-2-pivalamido-4(3*H*)-pteridinone (pivalic anhydride plus an organic base under reflux: 72%),[1753] 2-acetamido-4-oxo-3,4-dihydro-6-pteridinecarboxylic acid (**47**, X = CO_2H, Y = H) (acetic anhydride plus a little concentrated sulfuric acid under reflux for 1 h: 19%),[389] the isomeric 7-carboxylic acid (**47**, X = H, Y = CO_2H) (similarly for 30 min: 11%),[389] dimethyl 2-acetamido-4-oxo-3,4-dihydro-6,7-pteridinedicarboxylate (**47**, X = Y = CO_2Me)

(refluxing acetic anhydride for 20 min: 84%),[307] 2-benzamido-4,6(3*H*,5*H*)-pteridinedione (neat benzoic anhydride at 200°C for 90 min),[1362] 6-acetoxy-1,3-dimethyl-7-*N*-methylacetamido-2,4(1*H*,3*H*)-pteridinedione (**48**) [by heating 1,3-dimethyl-7-methylamino-2,4,6(1*H*,3*H*,5*H*)-pteridinetrione in acetic anhydride containing potassium acetate; note the additional *O*-acetylation],[1294] and 2-acetamido-7-α-cyano-α-ethoxycarbonylmethyl-4(3*H*)-pteridinone [as a 15% by-product in the reaction of 2-amino-4(3*H*)-pteridinone 8-oxide with ethyl cyanoacetate in acetic anhydride to give (mainly) 2-amino-7-α-cyano-α-ethoxycarbonylmethyl-4(3*H*)-pteridinone].[1301]

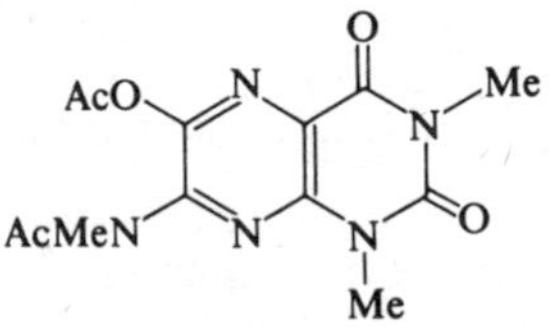

(48)

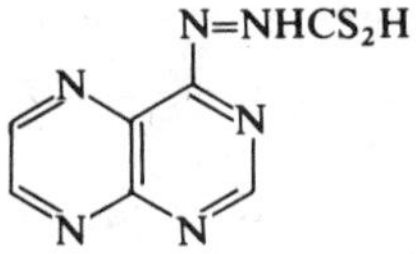

(49)

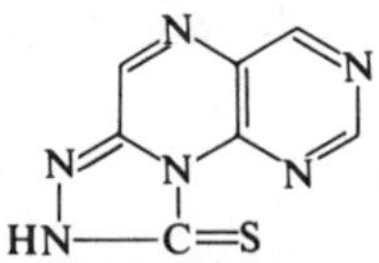

(50)

Appropriate acylation of a hydrazinopteridine may result in spontaneous cyclization. For example, treatment of 4-hydrazinopteridine with carbon disulfide (equivalent to dithiocarbonic acid, $HOCS_2H$) in pyridine gave the expected 4-*N'*-(dithiocarboxy)-hydrazinopteridine (**49**),[137] whereas 7-hydrazinopteridine gave no such isolable derivative but the cyclized product, *s*-triazolo[3,4-*h*]pteridine-9(8*H*)-thione (**50**) in 59% yield.[136, 137] The acylation of ring-NH groups in reduced pteridines is discussed in Chapter XI, Sections 2.C and 4.B.

D. Alkylation

The alkylation of pteridinamines to afford initially ring-*N*-alkylated pteridinimines and subsequently the corresponding alkylaminopteridines (by Dimroth rearrangement) is discussed later in Sections 5 and 6.B, respectively. In addition, primary and (possibly) secondary aminopteridines can undergo direct *N*-trimethylsilylation by prolonged boiling with hexamethyldisilazane containing a little ammonium sulfate to give, for example, 2-methylthio-7-trimethylsiloxy-4-trimethylsilylaminopteridine (**51**) [88% from 4-amino-2-methylthio-7(8*H*)-pteridinone],[1023] 2-trimethylsiloxy-4-trimethylsilylaminopteridine [98% from 4-amino-2(1*H*)-pteridinone],[225] its 6,7-diphenyl derivative,[225] and 4-

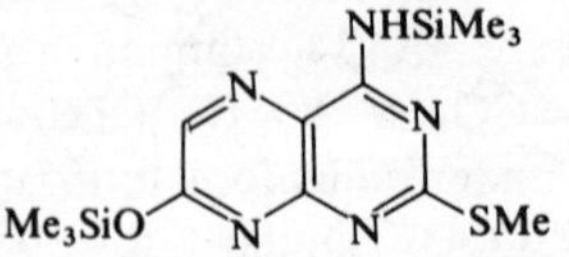

(51)

dimethylamino-7-trimethylsiloxy-2-trimethylsilylaminopteridine.[356] Unlike the trimethylsiloxy group that is useful for the Silyl Hilbert–Johnson reaction (Ch. VII, Sect. 2.C), the trimethylsilylamino group appears to have no particular use except to increase volatility for mass spectral purposes.

E. Conversion into Schiff Bases

The reaction of primary amino or N'-unsubstituted hydrazinopteridines with aldehydes, ketones, or their equivalents to form Schiff bases are illustrated rather meagrely in the pteridine series by the conversion of 2-hydrazino- into 2-isopropylidenehydrazino-1,6,7-trimethyl-4(1*H*)-pteridinone (**52**) (warm aqueous acetone for 5 min: 69%)[177]; of 4-pteridinamine 3-oxide, its 2-methyl derivative, or its 2-phenyl derivative into 4-dimethylaminomethyleneaminopteridine 3-oxide (**53**, R = H) (68%), the 2-methyl derivative (**53**, R = Me) (58%), or the 2-phenyl derivative (**53**, R = Ph) (84%), respectively (neat DMF dimethyl acetal at 105°C for 3.5 h)[1004]; of 2-methylthio-4-pteridinamine or its 6,7-dimethyl derivative into 6-methylthioimidazo[1,2-*c*]pteridine (**54**, R = H) or its 2,3-dimethyl derivative (**54**, R = Me), presumably via 4-β-chloroethylideneamino-2-methylthiopteridine or its 6,7-dimethyl derivative, respectively (chloroacetaldehyde in aqueous methanol at 65°C for 20 h: ~40% yield in each case)[460]; of 4- or 7-pteridinamine into 4- (73%)[135] or 7-dimethylaminomethyleneaminopteridine (91%),[136] respectively (DMF dimethyl acetal at 100°C for 10–20 min); of 4-hydrazinopteridine (**55**) into *s*-triazolo[4,3-*c*]pteridine (**57**), presumably via cyclization of 4-ethoxymethylenehydrazinopteridine (**56**) (refluxing triethyl orthoformate–dioxane for 1 h: 96%)[135]; and of numerous substrates akin to (**55**) into products resembling (**57**).[135–137]

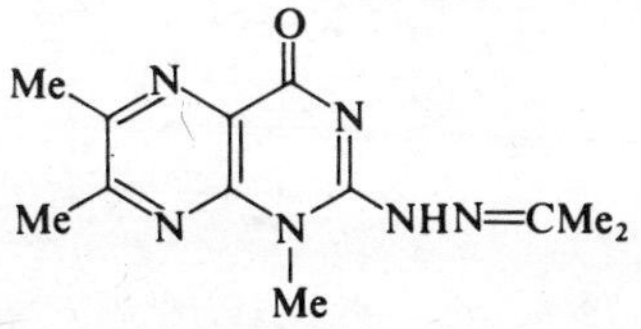

(**52**)

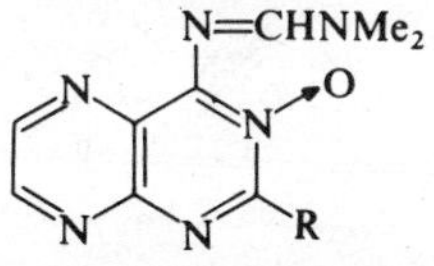

(**53**)

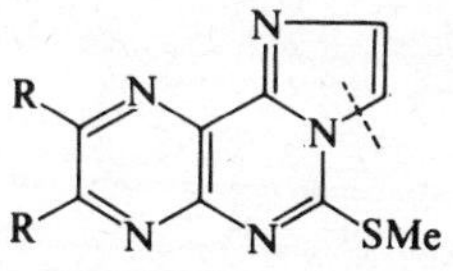

(**54**)

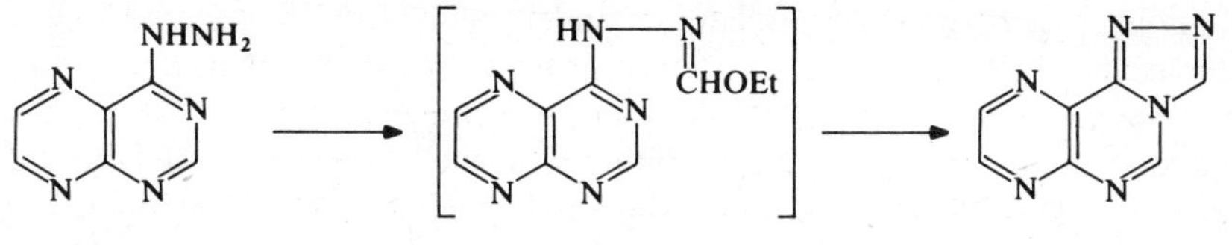

(**55**) (**56**) (**57**)

F. Transamination

The direct interchange of amino or substituted-amino groups attached to the pteridine nucleus (transamination) is usually carried out simply by heating with ammonia or an amine in a sealed vessel, by refluxing with neat amine if the latter has a sufficiently high boiling point, or by heating with amine in a high boiling solvent. Transamination appears to occur more readily at the 4- than at the 2-position and to be significantly catalyzed by the addition of a little mineral acid.[467] This process is illustrated in the conversion of 2,4-pteridinediamine into 4-hydrazino-2-pteridinamine (refluxing 85% hydrazine hydrate for 1 h: 80%)[305]; of 6,7-diphenyl-2,4-pteridinediamine (**58**, R = H) into 2,4-dihydrazino- (**58**, R = NH_2) (refluxing hydrazine hydrate plus acid: 66%),[467] 2,4-bis-benzylamino- (**58**, R = CH_2Ph) [refluxing benzylamine plus acid: 91%[467]; refluxing benzylamine without the addition of acid gave 4-benzylamino-6,7-diphenyl-2-pteridinamine (68%)[1315]], 2,4-bis-γ-dimethylaminopropylamino- [**58**, R = $(CH_2)_3NMe_2$] [γ-dimethylaminopropylamine plus acid at 180°C (sealed) for 20 to 24 h: 86 to 90%[400, 467]; under similar conditions without the addition of acid, the product was 4-γ-dimethylaminopropylamino-6,7-diphenyl-2-pteridinamine (91%)[467]], and 2,4-bis-β-diethylaminoethylamino-6,7-diphenylpteridine (**58**, R = $CH_2CH_2NEt_2$) (β-diethylaminoethylamine plus acid at 180°C for 24 h: 68%),[400] as well as analogs[400, 467, 1315]; of 6,7-diisopropyl-2,4-pteridinediamine into 2,4-bis-β-dimethylaminoethylamino-6,7-diisopropylpteridine (β-dimethylaminoethylamine plus acid at 180°C for 24 h: 84%) and analogs[400]; and of 6,7-dimethyl-2,4-bismethylaminopteridine into 4-β-hydroxyethylamino-6,7-dimethyl-2-methylaminopteridine (refluxing ethanolamine for 36 h: 64%).[1315]

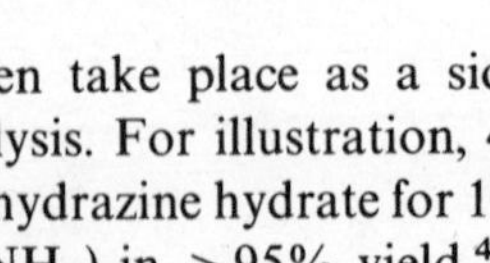

(58)

Transamination may often take place as a side reaction during, or even instead of, a desired aminolysis. For illustration, 4-amino-6,7-diphenyl-2(1*H*)-pteridinethione in refluxing hydrazine hydrate for 10 h gave 2,4-dihydrazino-6,7-diphenylpteridine (**58**, R = NH_2) in >95% yield.[467] The same substrate with ethanolic γ-diethylaminopropylamine under reflux for 7 h underwent transamination to give 4-γ-diethylaminopropylamino-6,7-diphenyl-2(1*H*)-pteridinethione as the major product,[400] and only on prolonged refluxing[400] or by using neat amine under reflux,[467] did the aminolytic reaction occur as well to give 2,4-bis-γ-diethylaminopropylamino-6,7-diphenylpteridine in good yield. 4-Chloro-2-methylamino-6,7-diphenylpteridine underwent both aminolysis and transamination on heating in methanolic ammonia at 155°C for 16 h, to give 6,7-diphenyl-2,4-pteridinediamine (**58**, R = H).[1189] Likewise, 4-amino-6,7-diphenyl-2(1*H*)-pteridinethione with ethanolic methylamine or dimethylamine at 180°C for 10 h gave 2,4-bismethylamino- (**58**, R = Me) or 2,4-bisdimethylamino-6,7-diphenylpteridine, respectively.[110, cf. 472] The same substrate with neat benzylamine, initially under reflux and then at 150°C in an open vessel, gave 2,4-bisbenzylamino-6,7-diphenylpteridine (**58**, R = CH_2Ph).[1315] 7-Methylsulfonyl-2,4,7-pteridinetriamine (**59**, R = H) with benzylamine under reflux gave 2,4,7-trisbenzylamino-6-methylsulfonylpteridine (**59**, R = CH_2Ph) without any aminolysis at the sulfone grouping[249] and other such examples have been reported.[249, 400, 467]

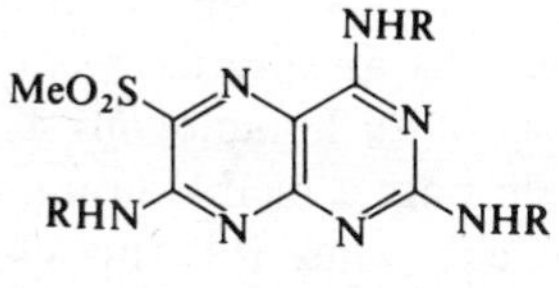

(59)

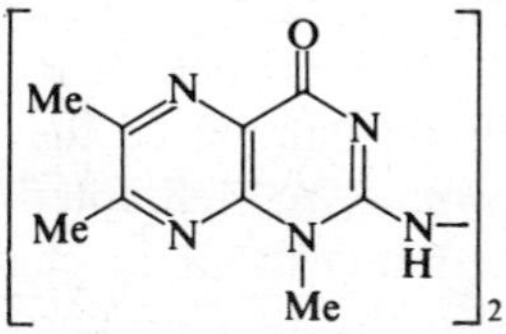

(60)

Several de facto transaminations should also be mentioned as occasionally applicable. Reduction of 4-hydroxyamino-2-pteridinamine with dithionite afforded 2,4-pteridinediamine[305]; 4-pteridinamine 3-oxide in boiling water for 7 h gave 4-hydroxyaminopteridine (73%)[1003]; 2-amino-4,7(3*H*,8*H*)-pteridinedione (isoxanthopterin), heated in a sealed tube with hexamethylphosphoramide and stearic anhydride containing a trace of phosphoric acid gave 2,4,7-trisdimethylaminopteridine among other products[864]; and 2-hydrazino-1,6,7-trimethyl-4(1*H*)-pteridinone in refluxing hydrazine hydrate for 3 h gave the red *N*,*N*′-bis(1,6,7-trimethyl-4-oxo-1,4-dihydro-2-pteridinyl)hydrazine (**60**) in 81% yield.[177]

G. Hydrolysis to Pteridinones

This important transformation has been discussed in Chapter VI, Section 1.C.

H. Pteridinamines as Drugs

Apart from some synthetic aspects treated, for example, in Chapter II, Section 1.B(3.d) or Chapter V, Section 4.B, information on the vast history,[251, also: 835, 839, 946, 971, 1025, 1181–1184] biochemistry and purification (e.g., refs. 666, 686, 814, 855, 909, 970, 1124, 1166, 1203, 1260, 1272, 1514), biosynthesis (e.g., refs. 494, 630, 650, 866, 868, 903, 1134, 1157, 1201, 1259, 1406), analogs (e.g., refs. 630, 650, 866, 868, 1134, 1259), and indeed all aspects of the aminopteridinone, folic acid (pteroylglutamic acid), or on folates or tetrahydrofolates in general should be sought elsewhere.[1572] Papers on potential pteridine drugs (e.g., refs. 171, 767, 768, 820, 1077, 1163) are not reviewed. However, it seems appropriate to mention here some relevant facets of two pteridinamines presently in use as drugs.

(1) *Methotrexate*

In 1947, Doris Seeger and colleagues reported a new pteroylglutamic acid antagonist in which the 4-oxo substituent of pteroylglutamic acid had been replaced by an amino group.[1265] The product (**61**, R = H), confusingly known as "Aminopterin", was shown subsequently to produce significant remissions in some acute leukemias of childhood.[1420] Shortly afterwards, a similar American Cyanamid group announced the synthesis of an homologous antagonist,[1263] initially known as "A-Methopterin", which bore a methyl group on *N*-10 and also produced remissions in childhood leukemias.[1714] This compound (**61**, R = Me) eventually became known as methotrexate and overtook "Aminopterin" as a clinically useful antineoplastic agent for several common acute leukemias and some less common cancerous conditions, such as gestational choriocarcinoma.[1715–1717]

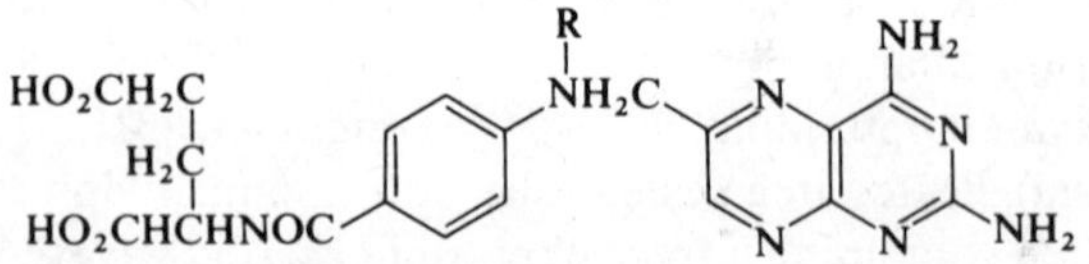

(61)

Both these pteridinediamines were made originally by *folic acid-type* syntheses [see Ch. II, Sect. 1.B(3.d)]. Thus, for example, 2,4,5,6-pyrimidinetetramine, α,β-dibromopropionaldehyde, and *p*-amino- or *p*-methylaminobenzoylglutamic acid in the presence of iodine as an oxidant gave "Aminopterin" (**61**, R = H) or methotrexate (**61**, R = Me), respectively, each in quite low yield after rigorous purification.[1263] More recently, the diethyl ester of methotrexate has been made in 34% yield by condensation of guanidine with an appropriately 6-substituted-3-amino-2-pyrazinecarbonitrile.[1533] A third synthesis of methotrexate, involving condensation of 6-bromomethyl-2,4-pteridinediamine with diethyl *N*-(*p*-methylaminobenzoyl)glutamate and subsequent hydrolysis of the ester functions, has been reported[390] and is an established commercial procedure for preparing the drug.[595] A similar route also afforded "Aminopterin",[390] while a related procedure, using *p*-aminobenzoylglutamic acid directly, has been used to make the latter (68%) and homologs.[1137] The reduction and ^{13}C nmr spectrum of methotrexate have been studied,[641, 761, 888] as was the basis of its photosensitization.[1075]

Although effective, methotrexate does not have a high therapeutic index and its necessarily prolonged administration is frequently accompanied by distressing and dangerous toxic symptoms. Moreover, leukemic cells can (and often do) develop resistance to the drug. For these reasons it is usually better to employ multiple drug therapy. For example, treatment may be started with vincristine followed by more prolonged treatment with methotrexate and 6-mercaptopurine [6(1*H*)-purinethione] as a synergistic pair. In addition, it has been found that folinic acid (**62**) (also known as leucovorin,[669, 744, 1081, 1148, 1191, 1504] citrovorum factor, or 5-formyltetrahydrofolic acid) and thymidine both have marked antitoxic effects when administered on alternate days to methotrexate. By using this so-called *rescuing* technique, very large doses of methotrexate may be given to extend its efficacy to tumors, such as osteogenic sarcoma, which do not respond to lower doses. Further information and leading references to such general aspects of methotrexate therapy may be found in three succinct reviews.[1715–1717]

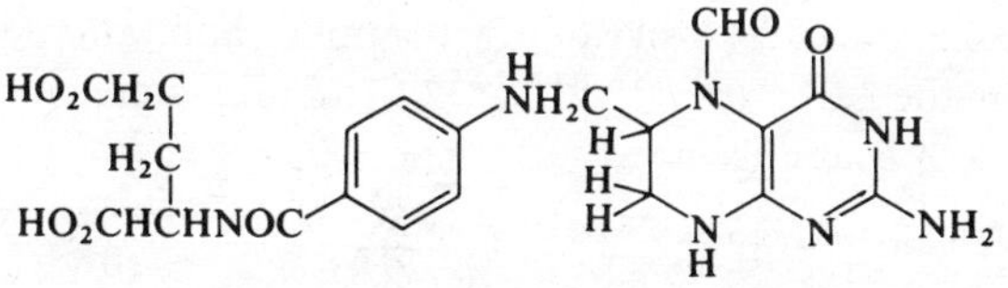

(62)

As might be expected, a great many analogs of methotrexate have been prepared over the years.[753, 850, 857, 975, 1102, 1139, 1141, 1186, 1271, 1355, 1465, 1533, 1724, 1725, 1760] Some,[306] including a few as simple as 2,4-pteridinediamine and its 6- and/or 7-alkylated derivatives,[123, 160, 749, 973] have shown promising antineoplastic activity at preliminary stages but none has survived as a useful chemotherapeutic agent.

(2) *Triamterene*

Although 6-phenyl-2,4,7-pteridinetriamine (**63**) was made in 1954 by a Timmis synthesis from 5-nitroso-2,4,6-pyrimidinetriamine and phenylacetonitrile,[1281] it was not until 1961 that its diuretic properties were reported in animals[1720] and in man[1721–1723]; it was known initially as "SKF-8542" and subsequently as triamterene.

Triamterene is characterized pharmacologically as a potassium-conserving diuretic,[819] which not only inhibits the excretion of potassium ions but also the

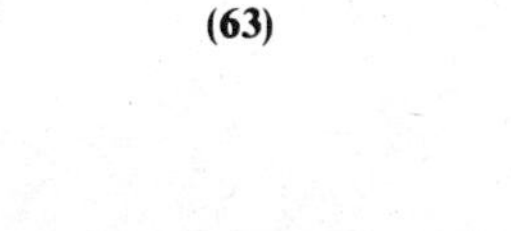

(63)

resorption of sodium ions. These desirable properties can occasionally lead to an increase in serum potassium to the point of hyperkalaemia.[1719] Although it can be used alone, triamterene may also be used to advantage with another diuretic (such as chlorothiazide) to counter the potassium-wasting effect of the latter.[1718] It may be determined in trace amounts within body fluids by polarography.[1159]

Although a differently based synthesis of triamterene via 5-α-cyanobenzylideneamino-2,4,6-pyrimidinetriamine (**64**) was reported in 1963;[1415] it appears to offer little advantage over the original method,[1281] which was used subsequently to prepare [7-^{14}C]- and [3′-^{3}H]-tagged specimens for metabolic studies.[771] Some isomers[522] and hundreds of analogs of triamterene have been made,[523–525, 577, 729, 847, 884, 1151, 1281, 1409, 1415] but none has shown much tendency to survive as a useful diuretic, save perhaps for "Furterene" (6-fur-2′-yl-2,4,7-pteridinetriamine) (**65**), which was made[524,cf. 682] via 5-α-cyanofurfurylideneamino-2,4,6-pyrimidinetriamine and subsequently caused some interest in the literature.[682, 840, 1156] Structure–activity relationships of such diuretics have been discussed.[1356]

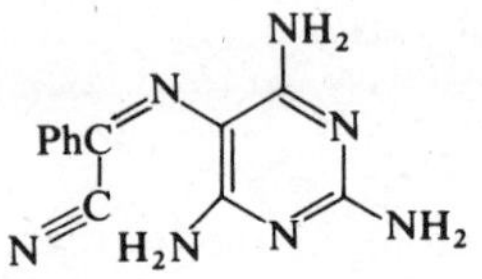

(64)

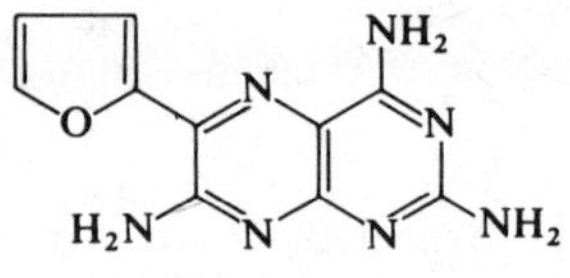

(65)

3. PREPARATION OF EXTRANUCLEAR AMINOPTERIDINES

A. By Primary Syntheses

Numerous extranuclear aminopteridines have been made by primary syntheses from pyrimidine (see Ch. II) or pyrazine intermediates (see Ch. III). For example, 6-β-dimethylaminoethylamino-2-phenyl-4,5-pyrimidinediamine with aqueous alcoholic glyoxal gave 4-β-dimethylaminoethylamino-2-phenylpteridine (**66**) in 61% yield (a Gabriel and Colman synthesis)[1436]; practically all the myriad products from the *folic acid-type* synthesis [Ch. II, Sect. 1.B(3.d)], for example, pteroylglutamic acid or methotrexate (**61**, R = Me), are such extra-

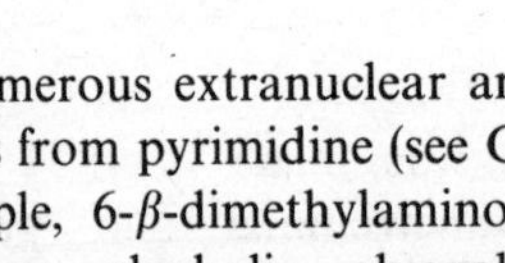

(66)

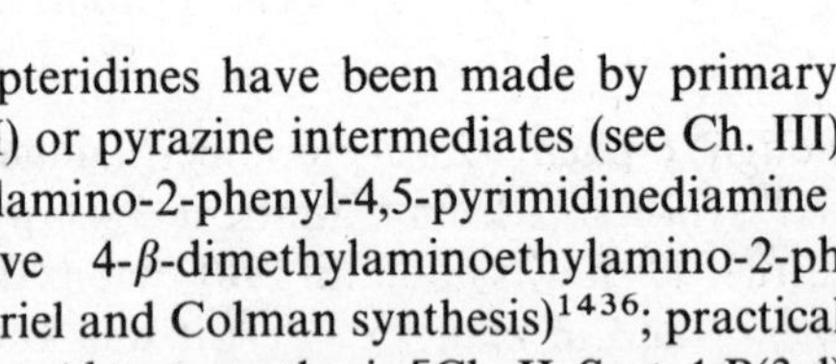

(67)

nuclear aminopteridines; 2-morpholino-5-nitroso-4,6-pyrimidinediamine with α-cyano-*N*-γ-dimethylaminopropylacetamide gave 4,7-diamino-*N*-γ-dimethylaminopropyl-2-morpholino-6-pteridinecarboxamide (**67**) in 88% yield (a Timmis synthesis)[321]; cyclization of 6-amino-5-*p*-dimethylaminobenzylideneamino-1,3-dimethyl-2,4(1*H*,3*H*)-pyrimidinedione with triethyl orthoformate gave 6-*p*-dimethylaminophenyl-1,3-dimethyl-2,4(1*H*,3*H*)-pteridinedione (**68**) in 70% yield (a minor primary synthesis)[543]; and 3-amino-6-(*p*-chloro-*N*-ethylanilino)methyl-2-pyrazinecarbonitrile with guanidine gave 6-(*p*-chloro-*N*-ethylanilino)methyl-2,4-pteridinediamine (**69**) (60%).[1355]

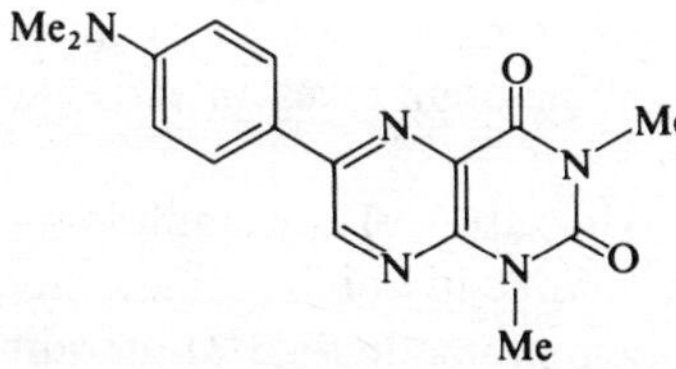

(68)

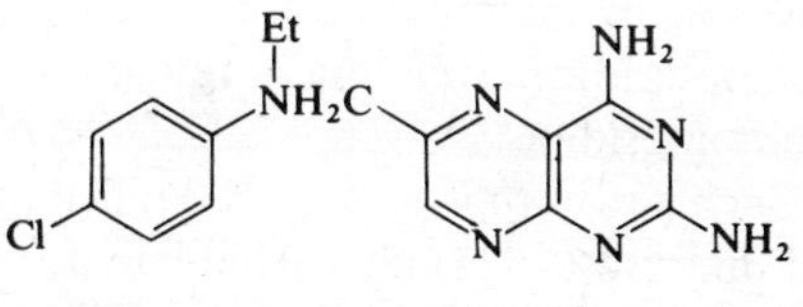

(69)

B. From Pteridinecarbonitriles

The conversion of a cyano- into an aminomethyl group requires vigorous reduction that would be expected to reduce the pteridine ring as well.* Probably for this reason, examples of such transformations are confined to tetrahydropyrimidinecarbonitriles in which the ring is immune to further reduction. Thus 2-amino-6-methyl-4-oxo-3,4,5,6,7,8-hexahydro-6-pteridinecarbonitrile (**70**, R=X=H) underwent hydrogenation over platinum (initially as its oxide) in trifluoroacetic acid to afford 2-amino-6-aminomethyl-6-methyl-5,6,7,8-tetrahydro-4(3*H*)-pteridinone (**71**, R=X=H), isolated as pure trihydrochloride

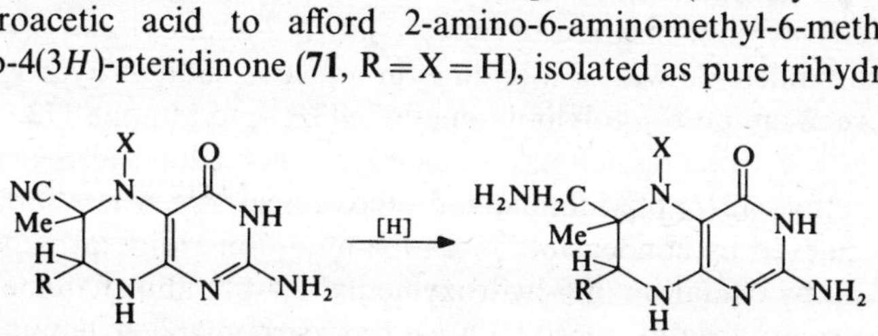

(70) **(71)**

in 30% yield[644]; reduction of the corresponding 5-acetylated nitrile (**70**, R=H, X=Ac) likewise gave the 5-acetyl intermediate (**71**, R=H, X=Ac) in 70% yield (isolated as dihydrochloride) and then, on deacetylation by methanolic hydrogen chloride, this product (**71**, R=X=H) in 53% yield as

* For example, hydrogenation of 2-acetamido-4-oxo-3,4-dihydro-6-pteridinecarbonitrile over platinum gave 2-acetamido-6-aminomethyl-5,6,7,8-tetrahydro-4(3*H*)-pteridinone in good yield.[1739]

trihydrochloride[1340, cf. 1345]; and the homologous substrate 5-acetyl-2-amino-6,7-dimethyl-4-oxo-3,4,5,6,7,8-hexahydro-6-pteridinecarbonitrile (**70**, R = Me, X = Ac) similarly gave 2-amino-6-aminomethyl-6,7-dimethyl-5,6,7,8-tetrahydro-4(3*H*)-pteridinone (**71**, R = Me, X = H), via its 5-acetyl derivative (**71**, R = Me, X = Ac).[1337, cf. 1341]

C. From Pteridinecarbaldehydes

The reaction of a pteridinecarbaldehyde with a primary amine to give a Schiff base (or hydrazone) must be considered as the formation of an extranuclear iminopteridine. Subsequent reduction of the double bond gives an extranuclear aminopteridine.

Such a *two stage process* is typified in the condensation of 2-acetamido-4-oxo-3,4-dihydro-6-pteridinecarbaldehyde with *p*-toluidine in boiling ethanol to give a mixture of *syn* and *anti* 2-acetamido-6-*p*-tolyliminomethyl-4(3*H*)-pteridinone (**72**, R = Ac) (92% yield), which underwent reduction by sodium borohydride in aqueous ethanol at 60°C to afford 2-acetamido-6-*p*-toluidinomethyl-4(3*H*)-pteridinone (**73**, R = Ac) (84%) and then, by hydrolysis in warm alkali, 2-amino-6-*p*-toluidinomethyl-4(3*H*)-pteridinone (**73**, R = H) (88%).[260, 596] Similar procedures using appropriate aromatic amines gave 2-amino-6-(2′,5′-dimethylanilino)methyl- (71% overall), 2-amino-6-(2′,4′,6′-trimethylanilino)methyl- (61%), 6-*p*-acetamidoanilinomethyl-2-amino- (63%), and 2-amino-6-*p*-carboxyanilinomethyl-4(3*H*)-pteridinone (pteroic acid; methyl *p*-aminobenzoate was used and the ester group was converted into an acid group during the final deacetylation process: 49% overall)[260]; pteroylglutamic acid,[704] pteroyldiglutamic acid,[1547] and pteroyltriglutamic acid[443] were made similarly. In addition, 2-amino-6-*p*-toluidinomethyl-4(3*H*)-pteridinone (**73**, R = H) has been made more directly by condensation of (unacetylated) 2-amino-4-oxo-3,4-dihydro-6-pteridinecarbaldehyde with *p*-toluidine (in aqueous sodium hydrogen carbonate) to give 2-amino-6-*p*-tolyliminomethyl-4(3*H*)-pteridinone (**72**, R = H) followed by borohydride reduction[1344]; analogous procedures also gave 2-amino-6-anilinomethyl-4(3*H*)-pteridinone and pteroic acid.[1344] A less satisfactory result was achieved by condensing[595] 2,4-diamino-6-pteridinecarbaldehyde (prepared in situ by oxidation of 6-hydroxymethyl-2,4-pteridinediamine) with ethyl *p*-aminobenzoate to afford (crude) 6-*p*-ethoxycarbonylphenyliminomethyl-2,4-pteridinediamine (47%) followed by borohydride reduction to 6-*p*-ethoxycarbonylanilinomethyl-7,8-dihydro-2,4-pteridinediamine (~25%),[1290] which required permanganate oxidation[857] to yield ethyl pteroate.

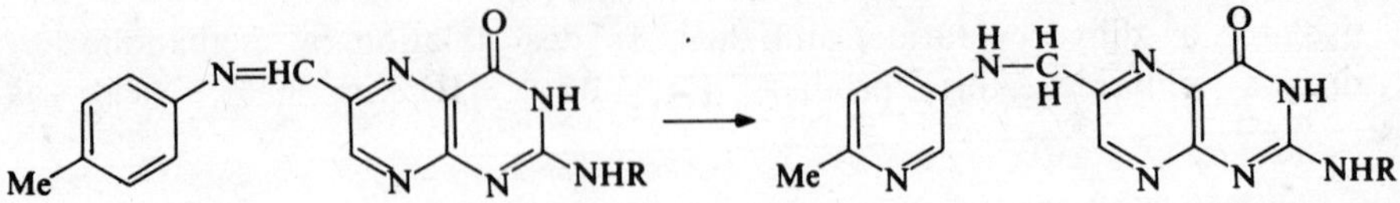

(**72**) (**73**)

The two essential stages in this process may be combined into a *one-pot procedure.* For illustration, 2-acetamido-4-oxo-3,4-dihydro-6-pteridinecarbaldehyde, *p*-aminobenzoylglutamic acid, *p*-thiocresol, and *β*-methoxyethanol were heated under reflux in a nitrogen atmosphere for 3.5 h. Formation of the Schiff base and reduction by the thiocresol took place to afford 2-acetylated pteroylglutamic acid in 54% yield.[1300] A similar condensation of the same substrate with 4-amino-3-nitrobenzoic acid, followed by alkaline deacetylation and subsequent hydrogenation of the nitro function over Raney-nickel, gave *m*-aminopteroic acid in poor yield.[1395, 1402] Although details are not readily available, the 2′,5′-diaza analog of pteroylglutamic acid (in which the benzene ring was replaced by a pyrazine ring) appears to have been made similarly.[817] A subtly different procedure involved the hydrogenation of a mixture of 2-amino-4,7-dioxo-3,4,7,8-tetrahydro-6-pteridinecarbaldehyde, *p*-aminobenzoylglutamic acid, sodium hydrogen carbonate, and sodium acetate in aqueous ethanol over Raney-nickel at 130°C under pressure to furnish 7-oxo-7,8-dihydropteroylglutamic acid in ~50% yield as crude product.[1371]

The following extranuclear iminopteridines (Schiff bases or hydrazones) have been prepared without any reported attempt to reduce them to the corresponding amines, at least in the same paper: 2-acetamido-6-phenyliminomethyl-4(3*H*)-pteridinone (condensation in ethanol containing a zinc salt as catalyst),[1482] 2-acetamido-6-*p*-ethoxycarbonylphenyliminomethyl-4(3*H*)-pteridinone (similarly; name in original paper is incorrect),[1482] 2-acetamido-6-*p*-methoxyphenyliminomethyl-4(3*H*)-pteridinone (similarly; name in original paper also misleading),[1482] 2-amino-6-(2′,4′-dinitrophenyl)hydrazonomethyl-4(3*H*)-pteridinone,[192, 940, 943] and 6-(2′,4′-dinitrophenyl)hydrazonomethyl-2,4-pteridinediamine.[1290] A peculiar reaction that may be mentioned here is that of 2-amino-5,6,7,8-tetrahydro-4(3*H*)-pteridinone with DMF containing hydrazine to give a separable mixture of *N*,*N*′-bis(2-amino-4-oxo-3,4,7,8-tetrahydro-6-pteridinylmethylene)hydrazine (**74**) and 2-amino-6-hydrazonomethyl-7,8-dihydro-4(3*H*)-pteridinone (**75**)[513]; on standing in air for several weeks, a solution of the latter gave 2-amino-6-hydrazonomethyl-4(3*H*)-pteridinone (**76**).[513]

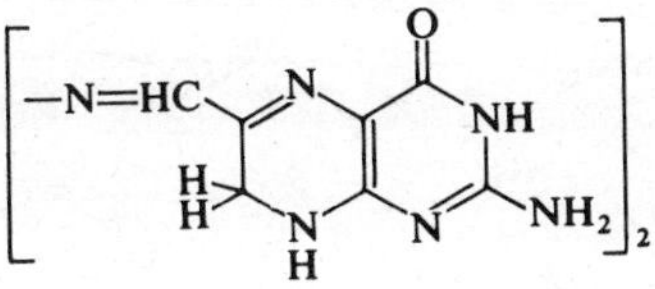

(**74**)

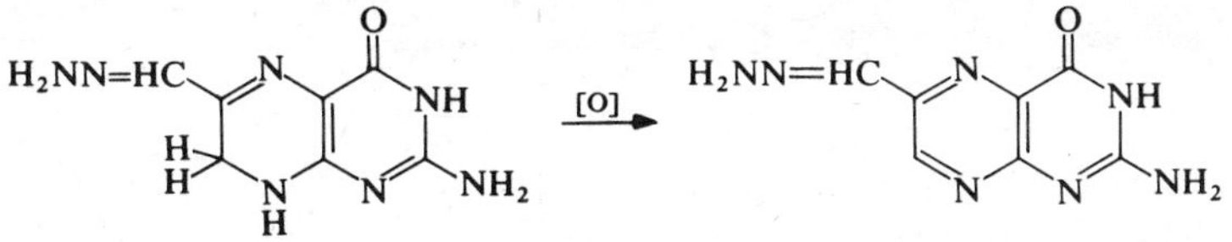

(**75**) (**76**)

D. By Aminolysis of Halogenoalkylpteridines

This useful aminolytic process has been discussed in Chapter V, Section 4.B. A typical example is the conversion of 6-bromomethyl-4(3*H*)-pteridinone (**77**, R = Br) into 6-anilinomethyl-4(3*H*)-pteridinone (**77**, R = NHPh) (60%) by boiling in ethanolic aniline for 2 h.[121]

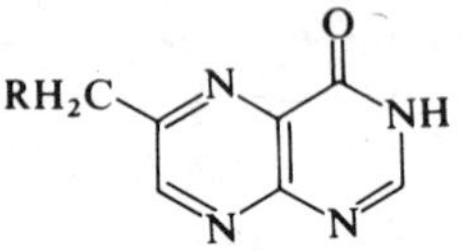

(77)

E. By Deacylation Reactions

Extranuclear acylaminopteridines usually arise from the necessity to protect a primary or secondary amino group during primary synthesis of the pteridine. The subsequent deacylation process is illustrated in the conversion of 2-amino-6-β-(*N*-phenylacetamido)ethyl- (**78**, R = Ac) into 2-amino-6-β-anilinoethyl-4(3*H*)-pteridinone (**78**, R = H) (2.5*M* sodium hydroxide under reflux for 3 h: 68%)[743]; of 2-amino-6-β-(*N*-3′,4′,5′-trimethoxyphenylbenzamido)ethyl- into 2-amino-6-β-(3′,4′,5′-trimethoxyanilino)ethyl-4(3*H*)-pteridinone (2*M* alkali at 150°C in a sealed tube for 6 h: 22%)[743]; of 2-amino-6-benzamidomethyl- into 2-amino-6-aminomethyl-4,7(3*H*,8*H*)-pteridinedione (5*M* alkali at 100°C for 1 h)[1083]; of 7-β-phthalimidoethyl- (**79**, R = H) into 7-β-aminoethyl-2,4,6(1*H*,3*H*,5*H*)-pteridinetrione (40% hydrazine hydrate under reflux for 2 h: 65% yield)[829]; of 1,3-dimethyl-7-β-phthalimido- (**79**, R = Me) into 7-β-aminoethyl-1,3-dimethyl-2,4,6(1*H*,3*H*,5*H*)-pteridinetrione (similarly for 1 h: 60%)[829]; and others.[743]

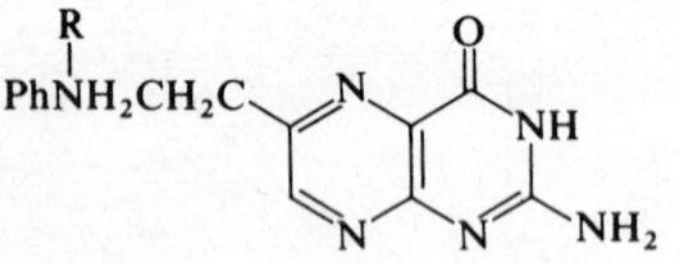

(78)

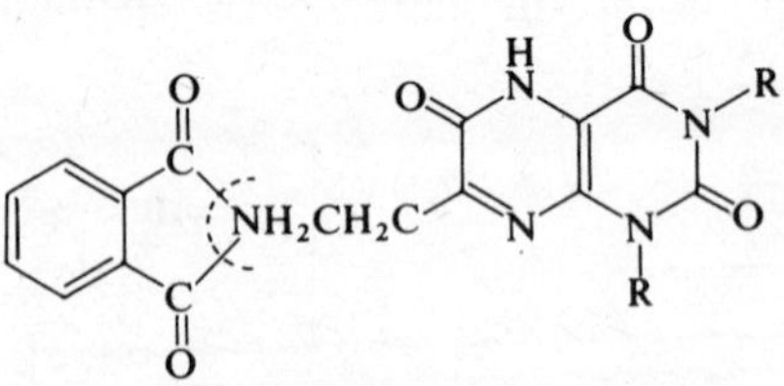

(79)

F. By Covalent Addition

Although perhaps marginally relevant, the covalent addition of 6-methyl-7(8*H*)-pteridinone across the 3,4-bond of quinazoline to give 6-(3′,4′-dihydroquinazolin-4′-yl)methyl-7(8*H*)-pteridinone (**80**) (85%),[57] the reversed

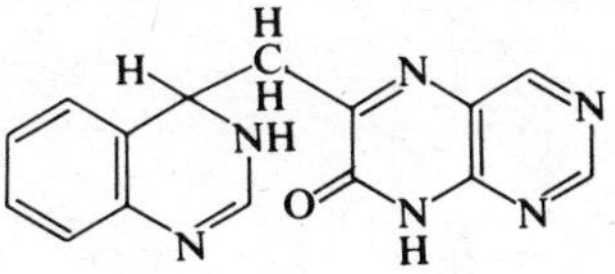

(80)

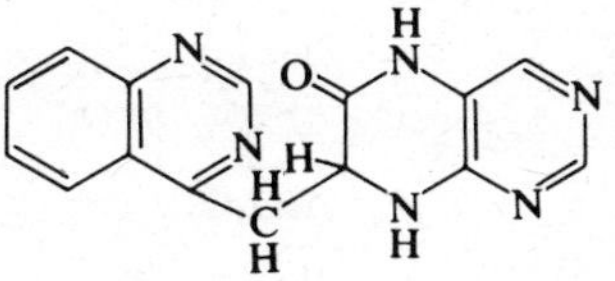

(81)

addition of 4-methylquinazoline to 6(5*H*)-pteridinone to give 7-quinazolin-4′-ylmethyl-7,8-dihydro-6(5*H*)-pteridinone (**81**),[57] and similar facile additions[44, 47, 57] do represent the formation of extranuclear aminopteridines.

4. REACTIONS OF EXTRANUCLEAR AMINOPTERIDINES

Although an extranuclear aminopteridine should surely undergo all the reactions to be expected of an aromatic or aliphatic primary, secondary, or tertiary amine as the case may be, few such reactions have been reported.

Acylation is represented in the conversion of 6,7-bis-*p*-aminophenyl-2,4-pteridinediamine (**82**, R = H) into 2,4-diacetamido-6,7-bis(*p*-acetamidophenyl)-pteridine (**82**, R = Ac) (acetic anhydride–acetic acid at 80°C for 45 min: 56%)[141]; of pteroylglutamic acid (**61**, R = H) into its 10-formyl derivative (**61**, R = CHO) (formic acid–acetic anhydride at 100°C for 1 h[1726, cf. 867, 879, 1148]; or neat 98% formic acid at 20°C for 2 days[700]); of 5,6,7,8-tetrahydropteroylglutamic acid (**83**, R = H) into its 5,10-diformyl derivative (**83**, R = CHO) (similarly)[1148]; and of 2-amino-6-*p*-carboxyanilinomethyl-4(3*H*)-pteridinone (pteroic acid) into 2-amino-6-(*N*-*p*-carboxyphenylformamido)methyl-4(3*H*)-pteridinone (rhizopterin) (98% formic acid at 100°C for 1 h).[1362]

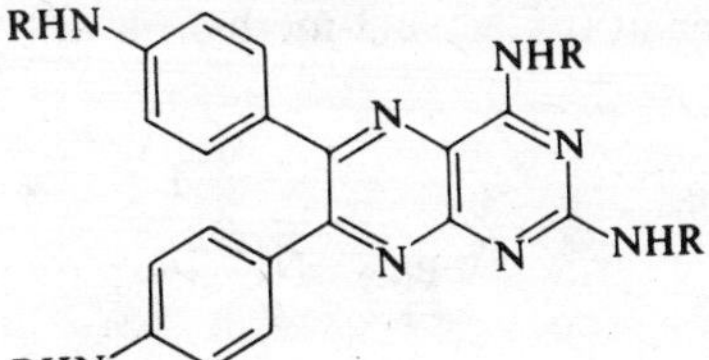

(82)

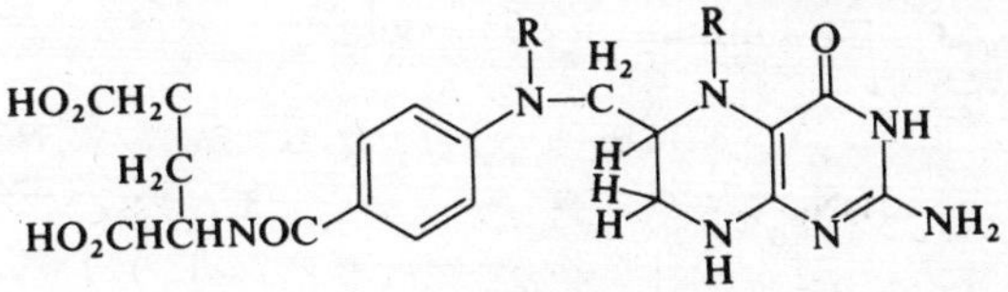

(83)

The process of *N-nitrosation* is exemplified in the treatment of a solution of pteroylglutamic acid (**61**) in 5–10*M* hydrochloric acid at 5°C with sodium nitrite to give the 10-nitroso derivative (**61**, R = NO).[732, cf. 259] It is also seen in the rather similar treatment of 6-*p*-carboxyanilinomethyl-2,4(1*H*,3*H*)-pteridinedione (**84**, R = H) to afford 6-(*p*-carboxy-*N*-nitrosoanilino)methyl-2,4(1*H*,3*H*)-pteridinedione (**84**, R = NO).[1362]

The aerial *oxidation* of a solution of 2-amino-6-aminomethyl-6-methyl-5,6,7,8-tetrahydro-4(3*H*)-pteridinone (**85**) resulted in the eventual formation of

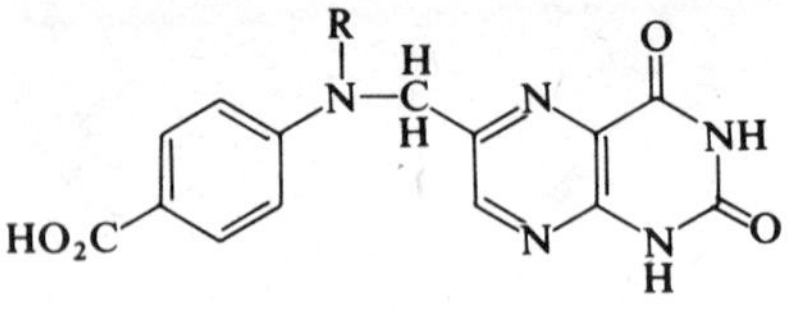

(84)

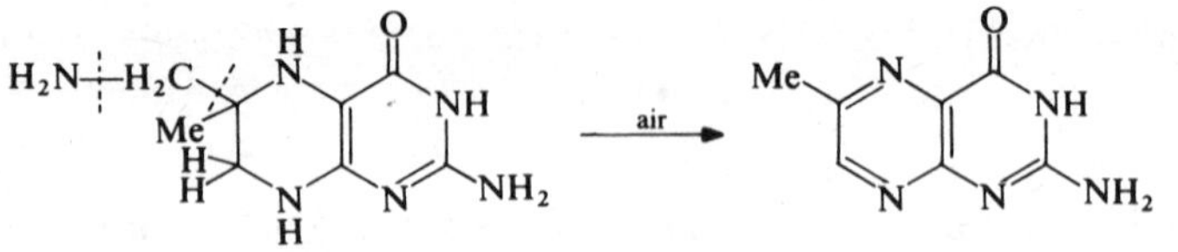

(85) **(86)**

2-amino-6-methyl-4(3*H*)-pteridinone (**86**) in 48% yield and the isolation of ammonia (35%) and formaldehyde (30%, as its dimedone derivative)[1322, 1341]; a mechanism has been proposed.[1322]

5. PREPARATION OF (FIXED) PTERIDINIMINES

A few pteridinimines have been made by *primary syntheses*. For example, condensation of 2,4-bisbenzylamino-5-pyrimidinamine (**87**) with glyoxal or biacetyl gave 8-benzyl-2-benzylimino-2,8-dihydropteridine (**88**, R = H) or its 6,7-dimethyl derivative (**88**, R = Me), respectively.[189] Likewise, 4,6-bismethylamino-5-pyrimidinamine with biacetyl in refluxing methanol gave 6,7,8-trimethyl-4-methylimino-4,8-dihydropteridine.[129] And 3-ethoxymethyleneamino-2-pyrazinecarbonitrile with ethanolic methylamine at 0°C gave 3-methyl- 4(3*H*)-pteridinimine in 65% yield.[53]

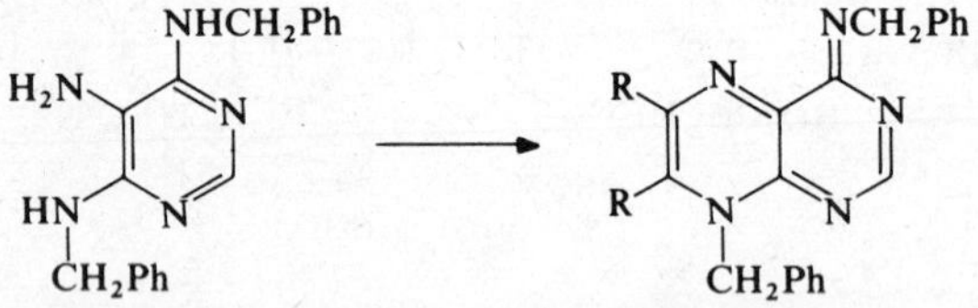

(87) **(88)**

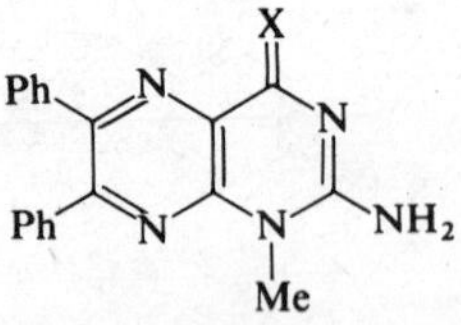

(89)

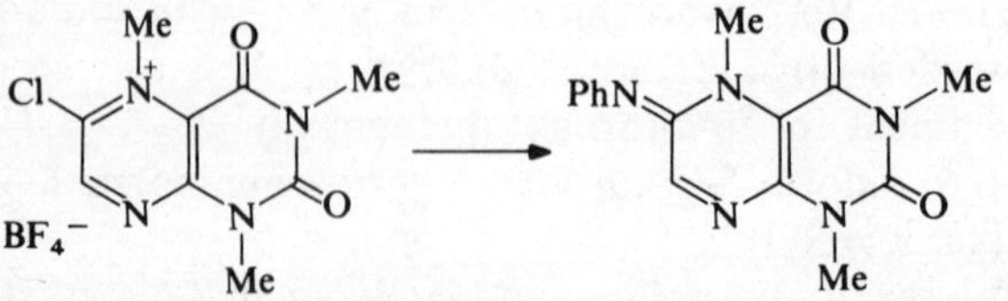

(90) **(91)**

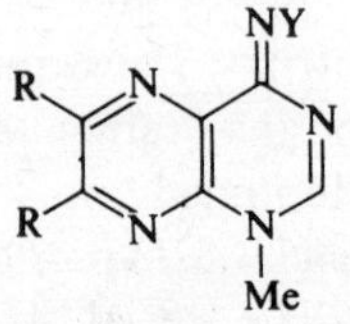

(92)

Some pteridinimines have been made by *aminolytic displacements*. Thus, heating a solution of 2-amino-1-methyl-6,7-diphenyl-4(1*H*)-pteridinethione (**89**, X = S) in ethanolic chloroform under reflux for 6 h, while a stream of ammonia or methylamine gas was passed, gave 4-imino-1-methyl-6,7-diphenyl-1,4-dihydro-2-pteridinamine (**89**, X = NH) ($\sim$50%) or its 4-methylimino homolog (**89**, X = NMe) (21%), respectively[111]; treatment of 6-chloro-1,3,5-trimethyl-2,4-dioxo-1,2,3,4-tetrahydro-5-pteridinium tetrafluoroborate (**90**) with aniline in refluxing dichloromethane gave 1,3,5-trimethyl-6-phenylimino-5,6-dihydro-2,4(1*H*,3*H*)-pteridinedione (**91**) in 38% yield[1680]; and treatment of a refluxing solution of 3-butyl-6,7-diphenyl-4(3*H*)-pteridinethione in ethanolic chloroform containing mercuric oxide with a stream of ammonia for 6 h gave either 3-butyl-6,7-diphenyl-4(3*H*)-pteridinimine or its rearranged isomer, 4-butylamino-6,7-diphenylpteridine in 70% yield.[474]

Most known pteridinimines were made by *nuclear-N-alkylation* of aminopteridines. The process is exemplified in the conversion of 4-pteridinamine or its 6,7-dimethyl derivative into 1-methyl- (**92**, R = Y = H) (78% as hydriodide) or 1,6,7-trimethyl-4(1*H*)-pteridinimine (**92**, R = Me, Y = H) (89% as hydriodide), respectively (methyl iodide at 140°C in a sealed tube for 4 h)[129]; of 4-methylamino- or 6,7-dimethyl-4-methylaminopteridine into 1-methyl- (**92**, R = H, Y = Me) (hydriodide: 75%) or 1,6,7-trimethyl-4-methylimino-1,4-dihydropteridine (**92**, R = Y = Me) (hydriodide: 86%), respectively (methyl iodide similarly)[129]; of 7-pteridinamine into a separable mixture of equal parts of 1-methyl-7(1*H*)-pteridinimine (**93**) (partly as hydriodide and partly as the hydriodide + I_2) and 3-methyl-7(3*H*)-pteridinimine (**94**) (as hydriodide) (methanolic methyl iodide in a rocked tube at 100°C for 40 min)[131]; and of 4,7-pteridinediamine into 7-imino-1-methyl-1,7-dihydro-4-pteridinamine (**95**, R = H) (or tautomer; as hydriodide) (methanolic methyl iodide at 70°C for 4 h: 93%) or, by prolonged treatment, into 1-methyl-4-methylamino-7(1*H*)-pteridinimine (**95**, R = Me) (or tautomer; as hydriodide), presumably by an additional direct methylation or by an additional 3-methylation followed by rearrangement (the latter product was also formed in $\sim$70% yield by similar treatment of 4-methylamino-7-pteridinamine).[131] When 2,4-pteridinediamine was heated with methanolic methyl iodide at 110°C, two isomeric products were formed. The minor product was 4-imino-8-methyl-4,8-dihydro-2-pteridinamine (**96**) isolated as its hydriodide, while the major product must have been 4-imino-1-methyl-1,4-dihydro-2-pteridinamine (**97**, R = H, X = NH) because, although it could not be isolated as such or as a salt in the pure state, alkaline hydrolysis gave initially 2-amino-1-

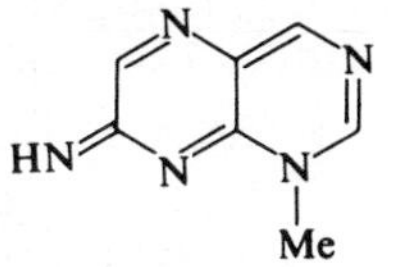

(93)

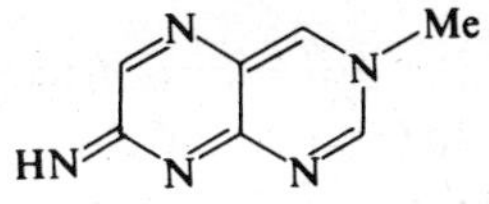

(94)

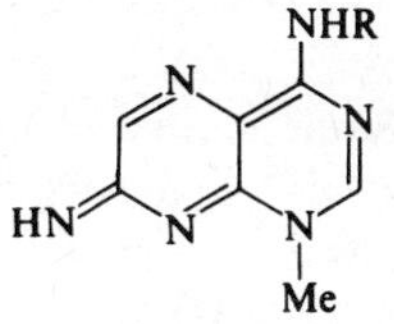

(95)

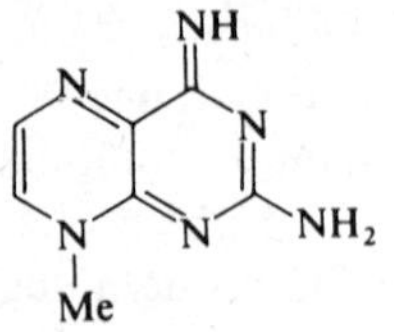

(96)

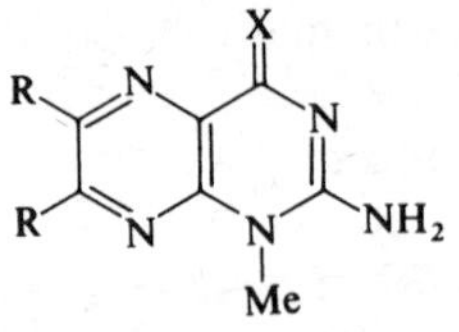

(97)

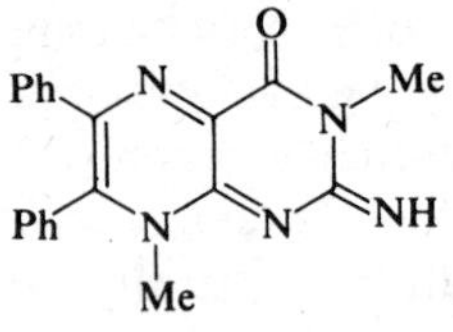

(98)

methyl-4(1*H*)-pteridinone (**97**, R=H, X=O) and later 1-methyl-2,4(1*H*,3*H*)-pteridinedione.[130] In contrast, rather similar treatment of 6,7-diphenyl- and 6,7-dimethyl-2,4-pteridinediamine gave 4-imino-1-methyl-6,7-diphenyl-1,4-dihydro-2-pteridinamine (**97**, R=Ph, X=NH)[111] and its 1,6,7-trimethyl analog (**97**, R=Me, X=NH),[130] respectively, as the only identifiable product in each case, although from the mother liquors of the latter an unidentified isomer was eventually obtained.[130] In like manner, treatment of 2-amino-3-methyl-6,7-diphenyl-4(3*H*)-pteridinone with dimethyl sulfate in acetic acid–DMF gave both a 1- and a (minor) 8-methylated product, viz. 2-imino-1,3-dimethyl-6,7-diphenyl-1,2-dihydro-4(3*H*)-pteridinone and 2-imino-3,8-dimethyl-6,7-diphenyl-2,8-dihydro-4(3*H*)-pteridinone (**98**); the former was the only isolable product from similar methylation of 2-amino-1-methyl-6,7-diphenyl-4(1*H*)-pteridinone.[68, 563] Moreover, it was clear that the 6- and/or 7-substituent affected (sterically?) the ratio of products.[563] Thus 2-amino-3-methyl-6-phenyl-4(3*H*)-pteridinone gave mainly 2-imino-3,8-dimethyl-6-phenyl-2,8-dihydro-4(3*H*)-pteridinone plus a little 2-imino-1,3-dimethyl-6-phenyl-1,2-dihydro-4(3*H*)-pteridinone, whereas 2-amino-3-methyl-7-phenyl-4(3*H*)-pteridinone gave only 2-imino-1,3-dimethyl-7-phenyl-1,2-dihydro-4(3*H*)-pteridinone.[68] Methyl 2-imino-3,8-dimethyl-7-oxo-2,3,7,8-tetrahydro-6-pteridinecarboxylate and three analogs have been characterized as *p*-toluenesulfonate salts.[376, 1311]

6. REACTIONS OF (FIXED) PTERIDINIMINES

Being bases of pK_a > 10, fixed pteridinimines are quite labile as free bases that undergo rearrangement, hydrolysis, and/or degradation to pyrazines with considerable ease. The salts of pteridinimines are normally stable, although they do undergo hydrolysis on heating in mineral acids.

A. Hydrolysis and/or Degradation

Some simple examples of the *hydrolysis* of fixed pteridinimines to pteridinones[129] have been discussed in Chapter VII, Section 3.C. The process is further illustrated in the conversion of 4-imino-8-methyl-4,8-dihydro-2-pteridinamine (**99**, X=NH) into 2-amino-8-methyl-4(8*H*)-pteridinone (**99**, X=O) (1*M* hydrochloric acid at 100°C for 30 min: 80% yield as hydrochloride)[130]; of 2-imino-1,3-

dimethyl-6,7-diphenyl-1,2-dihydro-4(3*H*)-pteridinone into 1,3-dimethyl-6,7-diphenyl-2,4(1*H*,3*H*)-pteridinedione (methoxyethanolic hydrochloric acid under reflux for 46 h: 88%)[68]; of 4-imino-1-methyl-1,4-dihydro-2-pteridinamine (**97**, R = H, X = NH) into 2-amino-1-methyl-4(1*H*)-pteridinone (**97**, R = H, X = O) (by mild alkaline hydrolysis),[133] and then into 1-methyl-2,4(1*H*,3*H*)-pteridinedione (**100**, R = H) (1*M* sodium hydroxide at 100°C for 15 min: 90%)[130]; of 4-imino-1,6,7-trimethyl-1,4-dihydro-2-pteridinamine (**97**, R = Me, X = NH) into 1,6,7-trimethyl-2,4(1*H*,3*H*)-pteridinedione (1*M* alkali at 100°C for 15 min: 92%)[130]; and of 4-imino-1,3-dimethyl-3,4-dihydro-2(1*H*)-pteridinone into 1,3-dimethyl-2,4(1*H*,3*H*)-pteridinedione (2*M* hydrochloric acid in propionitrile: 95%).[1733]

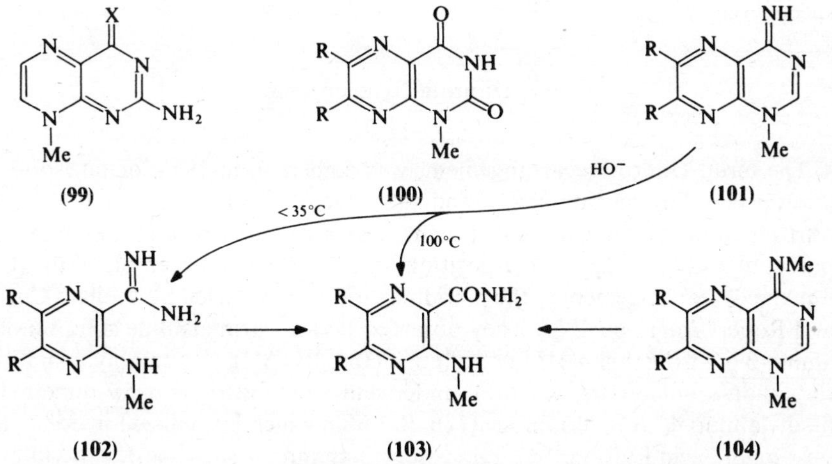

The *degradation* of pteridinimines,[cf. 1480b] with or without prior (or subsequent) hydrolysis of the imino function, is exemplified in the transformation of 1-methyl-4(1*H*)-pteridinimine (**101**, R = H) into 3-methylamino-2-pyrazinecarboxamidine (**102**, R = H) (2.5*M* alkali at 0°C for 20 min: 97%) or 3-methylamino-2-pyrazinecarboxamide (**103**, R = H) (1*M* alkali at 100°C for 10 min: 64%)[129]; of 1,6,7-trimethyl-4(1*H*)-pteridinimine (**101**, R = Me) into 5,6-dimethyl-3-methylamino-2-pyrazinecarboxamidine (**102**, R = Me) (2.5*M* alkali at 35°C for 10 min: 75%) or 5,6-dimethyl-3-methylamino-2-pyrazinecarboxamide (**103**, R = Me) (1*M* alkali at 100°C for 40 min: 70%)[129]; of 1-methyl- (**104**, R = H) or 1,6,7-trimethyl-4-methylimino-1,4-dihydropteridine (**104**, R = Me) into 3-methylamino-2-pyrazinecarboxamide (**103**, R = H) (1*M* alkali at 100°C for 10 min: 60%) or its 5,6-dimethyl derivative (**103**, R = Me) (1*M* alkali at 100°C for 40 min: 70%)[129]; of 4-imino-1-methyl-1,4-dihydro-2-pteridinamine (**97**, R = H, X = NH) into 3-methylamino-2-pyrazinecarboxylic acid (in several stages by refluxing 1*M* alkali for 3 h)[130, 133]; of 1-methyl-7(1*H*)-pteridinimine (**93**) into 3-methylamino-5-oxo-4,5-dihydro-2-pyrazinecarbaldehyde (**105**, R = Me) (1*M* alkali at 100°C for 1 h: ~45%)[131]; of 3-methyl-7(3*H*)-pteridinimine (**94**) into 3,5-diamino-2-pyrazinecarbaldehyde (**106**) (pH 4 buffer at 100°C for 15 min: 73%)

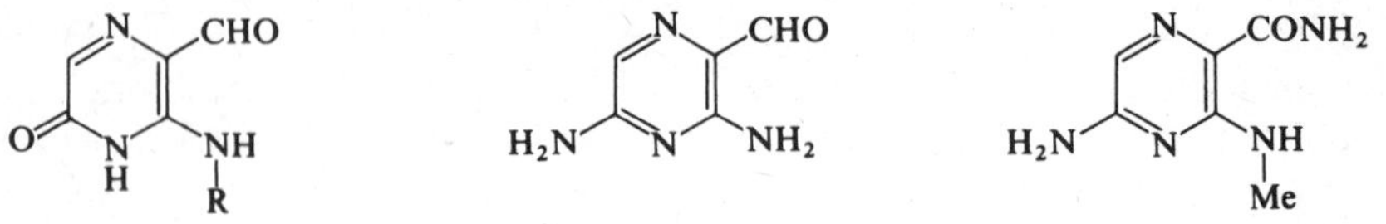

(105) (106) (107)

or 3-amino-5-oxo-4,5-dihydro-2-pyrazinecarbaldehyde (**105**, R = H) (1*M* alkali at 100°C for 1 h: ~50%)[131]; and of 7-imino-1-methyl-1,7-dihydro-4-pteridinamine (**95**, R = H) or its homolog, 1-methyl-4-methylamino-7(1*H*)-pteridinimine (**95**, R = Me), into 5-amino-3-methylamino-2-pyrazinecarboxamide (**107**) (1*M* alkali at 100°C for 15–30 min: 66% or ~45%, respectively).[131]

B. Dimroth Rearrangement

The term, Dimroth rearrangement, was coined[127] in 1963 for an isomerization proceeding by ring fission and subsequent recyclization, whereby a ring nitrogen and its attached substituent exchanged places with an imino (or potential imino) group in an α-position to it.[1727] The first example of the (then unnamed) rearrangement in the pteridine series was reported by William Curran and Robert Angier in 1958. They observed that a warm alkaline solution of 2-amino-3,6,7-trimethyl-4(3*H*)-pteridinone (**108**, R = Me), presumably containing the anionic imine (**109**, R = Me), underwent isomerization to 6,7-dimethyl-2-methylamino-4(3*H*)-pteridinone (**110**, R = Me), which crystallized in 55% yield on mild acidification.[166] Likewise, 2-amino-3-methyl-4(3*H*)-pteridinone (**108**, R = H) gave 2-methylamino-4(3*H*)-pteridinone (**110**, R = H) in ~60% yield,[133, 368] and 2-amino-3-methyl-6,7-diphenyl-4(3*H*)-pteridinone (**108**, R = Ph) gave 2-methylamino-6,7-diphenyl-4(3*H*)-pteridinone (**110**, R = Ph) in ~85% yield.[68]

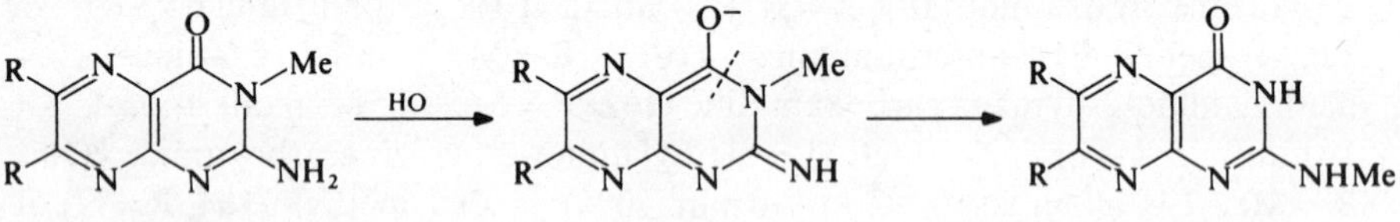

(108) (109) (110)

The simplest of these rearrangements, (**108**, R = H)→(**110**, R = H), was studied kinetically in some detail by Doug Perrin, employing rapid reaction spectrophotometric techniques.[332] On adding the substrate (**108**, R = H) to alkali at 20°C, its anion (**109**, R = H) underwent two successive first-order reactions: (a) ring fission to a pyrazine intermediate, presumably (**111**, R = H) or tautomer, with $t_{1/2}$ = 11 s at pH 14.7, and (b) subsequent recyclization to the product (**110**, R = H), as its anion, with $t_{1/2}$ = 81 s; the second reaction was mildly

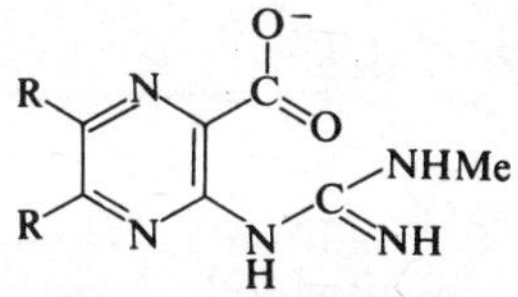

(111)

base catalyzed but the first was not so catalyzed.[128, 332, 1727] Similar studies on rearrangement of the 6,7-dimethyl substrate (**108**, R = Me) revealed that the fission step, (**109**, R = Me)→(**111**, R = Me), had $t_{1/2} = 47$ s while the reclosure, (**111**, R = Me)→ (**110**, R = Me), had $t_{1/2} = 960$ s[128, 332]: this suggested stabilization of both the substrate and intermediate through electron-release by the methyl groups to the π layer, a concept subsequently explored in depth and established satisfactorily, albeit in the pyrimidine rather than the pteridine series for reasons of synthetic facility.[1727]

Other examples of Dimroth rearrangement in the pteridine series are provided in the isomerization of 2-imino-1,3-dimethyl-6,7-diphenyl-1,2-dihydro-4(3*H*)- into 1-methyl-2-methylamino-6,7-diphenyl-4(1*H*)-pteridinone (aqueous methoxyethanolic alkali at 100°C for 2 min: 67%)[68]; 2-imino-3,8-dimethyl-6,7-diphenyl-2,8-dihydro-4(3*H*)- into 8-methyl-2-methylamino-6,7-diphenyl-4(8*H*)-pteridinone (similarly for 5 min: 65%; the structure was confirmed by primary synthesis of the product in 61% yield from 5-amino-2,6-bismethylamino-4(3*H*)-pyrimidinone and benzil)[68]; 2-amino-3-methyl- into 2-methylamino-7-phenyl-4(3*H*)-pteridinone (aqueous methoxyethanolic alkali at 100°C for 90 min: 45%)[67]; 2-imino-3,8-dimethyl-4-oxo-2,3,4,8-tetrahydro-6-pteridinecarboxylic acid (**112**) into 8-methyl-2-methylamino-4-oxo-4,8-dihydro-6-pteridinecarboxylic acid (**113**) (aqueous sodium hydrogen carbonate at 100°C for 30 min: 50%; also 1*M* alkali at 25°C for 24 h) or into 8-methyl-2-methylamino-4,7-dioxo-3,4,7,8-tetrahydro-6-pteridinecarboxylic acid (**114**) [air-free 2.5*M* alkali at

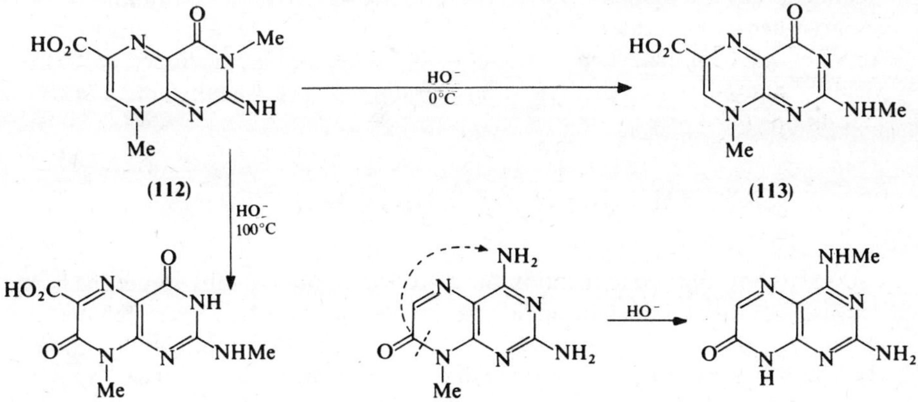

(112) **(113)** **(114)** **(115)** **(116)**

100°C for 45 min: 25%; a disproportionation rather than aerial oxidation was postulated to account for the 7-oxo substituent; the product's structure was checked by an unambiguous synthesis from the condensation of 5-amino-2,6-bismethylamino-4(3*H*)-pyrimidinone with diethyl mesoxalate, followed by saponofication][70]; 2-amino-3-methyl- into 2-methylamino-4-oxo-3,4-dihydro-6 (or 7)-pteridinecarboxylic acid (0.5*M* alkali at 100°C for 30 min: 70 or 80%, respectively)[70]; 2-amino-3,6-dimethyl- into 6-methyl-2-methylamino-4(3*H*)-pteridinone (1*M* sodium hydroxide at 100°C for 2 min: ~60% as sodium salt)[80]; 2-amino-3,8-dimethyl- into 8-methyl-2-methylamino-4,7(3*H*,8*H*)-pteridinedione (1*M* potassium hydroxide at 15°C for 1 h: ~10%)[457]; ethyl 2-imino-3,8-dimethyl-7-oxo-2,3,7,8-tetrahydro- into ethyl 8-methyl-2-methylamino-7-oxo-7,8-dihydro-6-pteridinecarboxylate (aqueous bicarbonate at room temperature for 3 h: 42%)[1311]; 2-amino-3,6-dimethyl-4(3*H*)-pteridinone 8-oxide into 6-methyl-2-methylamino-4(3*H*)-pteridinone 8-oxide (2*M* alkali at 100°C for 5 min: 90%)[485]; 2-amino-3-β-hydroxyethyl- into 2-β-hydroxyethylamino-6,7-diphenyl-4(3*H*)-pteridinone (3*M* aqueous methanolic alkali at 100°C for 2 h: 64%)[1500]; dimethyl 3-methylpteroylglutamate into the 2-methylamino analog of pteroylglutamic acid (0.1*M* alkali at 20°C under nitrogen for 18 h: ~60%; note the saponification of the ester functions)[488]; 3-aminopteroylglutamic acid into the 2-hydrazino analog of pteroylglutamic acid (warm 1*M* hydrochloric acid: 65%; note the use of acidic conditions, not without precedent[1727] in other series)[488]; and 2-amino-3-β-carboxyethyl-4-oxo-3,4-dihydro-6-pteridinecarboxylic acid into 2-(β-carboxyethyl)amino-4-oxo-3,4-dihydro-6-pteridinecarboxylic acid (0.5*M* alkali at 60°C for 1 h: >80%; for another example and the context of this rearrangement, see the end of Ch. III, Sect. 3).[654]

Examples of an interesting abnormal Dimroth rearrangement have been reported in the pteridine series. Treatment of 2,4-diamino-8-methyl-7(8*H*)-pteridinone (**115**) with 1.25*M* alkali at 100°C for 40 min gave 2-amino-4-methylamino-7(8*H*)-pteridinone (**116**) in 75% yield, while prolonged alkaline treatment gave 2-amino-4,7(3*H*,8*H*)-pteridinedione (77%).[1296] Similarly, 4-amino-8-benzyl-6-methyl-2,7(1*H*,8*H*)-pteridinedione in 5*M* alkali under reflux for 7 h gave 4-benzylamino-6-methyl-2,7(1*H*,8*H*)-pteridinedione in 50% yield,[363] and 4-amino-8-methyl-7(8*H*)-pteridinone gave 4-methylamino-7(8*H*)-pteridinone (45%).[445]

7. THE *QUINONOID* DIHYDROPTERINS

It seems appropriate to mention the so-called quinonoid dihydropterins here because they are still often formulated as imines. Tetrahydrobiopterin, that is, 2-amino-6-[L-*erythro*]-α,β-dihydroxypropyl-5,6,7,8-tetrahydro-4(3*H*)-pteridinone (**117**), is the naturally occurring coenzyme used by hydroxylases that catalyze the hydroxylation of aromatic amino acids such as phenylalanine, tyrosine, or tryptophan in the presence of oxygen–hydrogen peroxide.[1729, 1730] Fortunately, tetrahydrobiopterin may be replaced, albeit less effectively, by simpler unnatural tetrahydropterin cofactors, such as 2-amino-5,6,7,8-tetrahydro-4(3*H*)-pteridinone (**118**, X=Y=H), the 6-methyl derivative (**118**, X=Me, Y=H), or the *cis*-6,7-dimethyl derivative (**118**, X=Y=Me). Using the last of these, it was shown more than 25 years ago that the tetrahydropterin underwent a cyclic process of oxidation to a dihydropterin and subsequent reduction to the tetrahydropterin.[1731] It became clear to Seymour Kaufman by 1964 that this essential dihydropterin was not the 5,6-, 7,8-, or 5,8-dihydro isomer but the imino- (**119**) or one of the aminoquinonoid dimethyldihydropterins (**120**) or (**121**)[1009, cf. 1732] When its reduction was precluded or delayed, this entity proved labile under natural conditions and rapidly gave the 7,8-dihydro isomer among other products akin to those from nonenzymatic oxidation of tetrahydropterins.[74, 76, 353, 510, 560, 695, 842, 933, 1320, 1329] The electrochemical oxidation of 2-amino-5-methyl-,[1487] 2-amino-6-methyl-[871] and 2-amino-6,7-dimethyl-5,6,7,8-tetrahydro-4(3*H*)-pteridinone[861] as well as related compounds[872] appears to have given in each case a quinonoid dihydropterin that underwent quite rapid first-order isomerization into a 7,8-dihydropterin. Eventually, the quinonoid 6-methyl- and 6,7-dimethyldihydropterins[1485] as well as the even less stable[81] quinonoid dihydrobiopterin[1517] were isolated in 1985–1986 as solid salts by using an exceptionally mild oxidation procedure (hydrogen peroxide–potassium iodide at 0°C for 5 min) on the hydrochlorides of 6-methyl- (**118**, X=Me, Y=H), 6,7-dimethyl- (**118**, X=Y=Me), and 6-α,β-dihydroxypropyl-5,6,7,8-tetrahydropterin (**117**), respectively. The dry solids were stable in the refrigerator over a long period but began to change rapidly on dissolution; unaccountably, all were formulated in their imino-quinonoid forms, for example, (**119**).[1485, 1517, 1693]

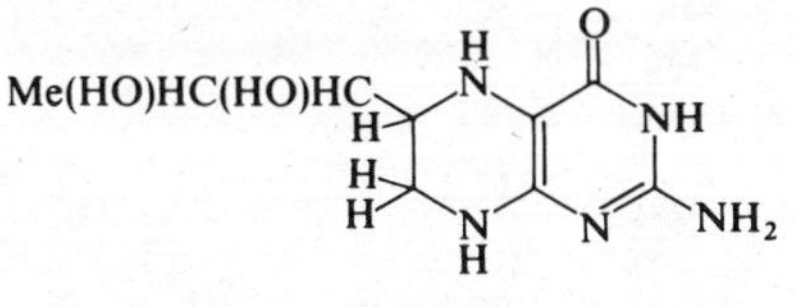

(**117**)

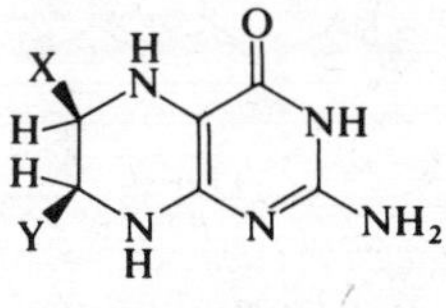

(**118**)

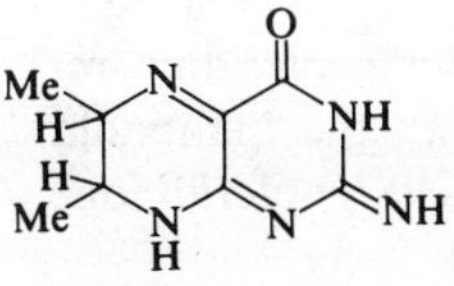

(**119**)

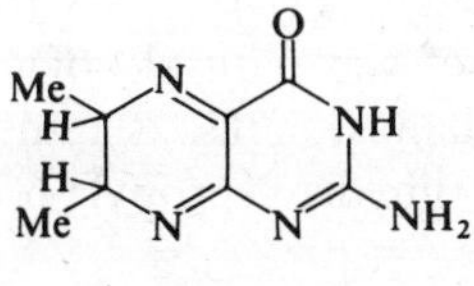

(**120**)

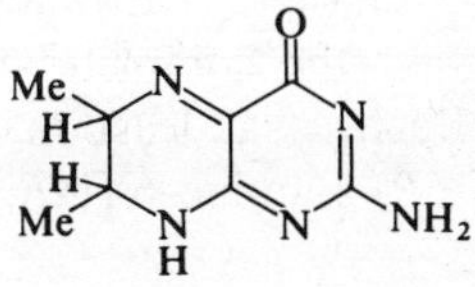

(**121**)

Meanwhile, considerable effort was being made along diverse lines to find which form of the quinonoid dihydropterins predominated in solution. (a) Extensive ab initio SCF-calculations suggested that the imino- and both amino-quinonoid forms were all equally reactive, some 26 kcal/mol higher in energy than the relatively stable 7,8-dihydropterin form.[1516] (b) Electrochemical studies strongly suggested that the model tetrahydro-4(3*H*)-pteridinones underwent reversible oxidation to an ortho-quinonoid dihydropteridinone type of structure.[1014] (c) Proton nmr spectral studies of the quinonoid-6-methyldihydropterin virtually precluded the amino form akin to (**120**) but could not distinguish between the other amino form akin to (**121**) and the imino form akin to (**119**).[1087] (d) A ^{15}N nmr spectral study of quinonoid-6,7-dimethyldihydropterin (deuterated at the 6- and 7-position to slow rearrangement) clearly indicated the predominance of the amino form (**121**) over the other amino form (**120**) or the imino form (**119**), at least in solution.[1469] (e) Additional evidence for an amino- (**120–121**) as distinct from the iminoquinonoid form (**119**) has been provided by syntheses of 6,7-dimethyl- (**122**, R = H), 2,6,7-trimethyl- (**122**, R = Me), and 6,7-dimethyl-2-methylthio-5,6,7,8-tetrahydro-4(3*H*)-pteridinone (**122**, R = SMe), as well as other such analogs of tetrahydropterin. On oxidation, all of these gave biologically active quinonoid–dihydropterin analogs, which of their nature could adopt a configuration akin to that of the amino forms (**120–121**) but not to that of the imino form (**119**)[1508]. This theme was explored further.[85, 88, 1683] Such work has been assisted by the availability of 2-amino-6,6-dimethyl-5,6,7,8-tetrahydro-4(3*H*)-pteridinone (**123**, R = H)[644] and its 6,6,8-trimethyl analog (**123**, R = Me)[1442] from which may be generated abnormally stable quinonoid dihydro species that cannot rearrange to their biologically inactive 7,8-dihydro isomers[725, cf. 696]; other highly relevant contributions have been reported.[75, 86, 560, 611, 647, 687, 690–692, 738, 889, 908, 910, 1040, 1047, 1080, 1086, 1334, 1450, 1455, 1470, 1488]

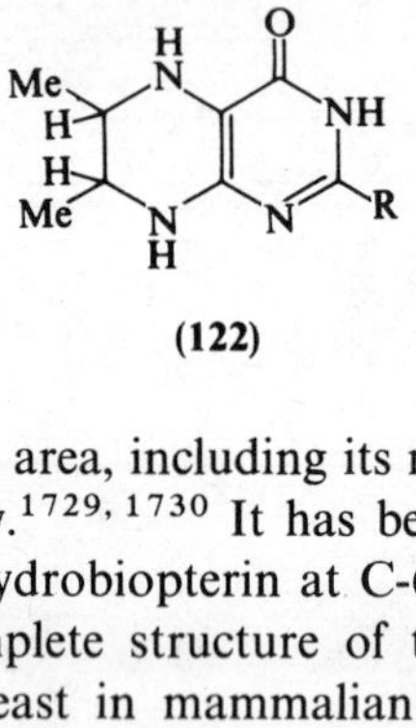

(122)

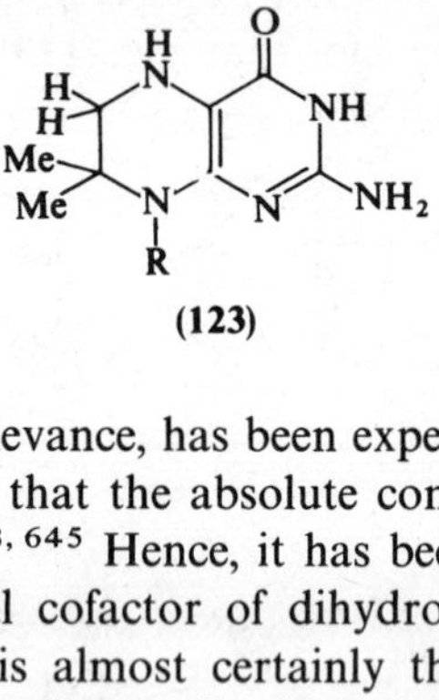

(123)

This whole area, including its medical relevance, has been expertly reviewed quite recently.[1729, 1730] It has been shown that the absolute configuration of 5,6,7,8-tetrahydrobiopterin at C-6 is (*R*).[298, 645] Hence, it has been concluded that the complete structure of the natural cofactor of dihydropteridine reductase, at least in mammalian systems, is almost certainly the quinonoid (6*R*)-2-amino-6-[(α*R*, β*S*)-α,β-dihydroxypropyl]-6,7-dihydro-4(8*H*)-pteridinone

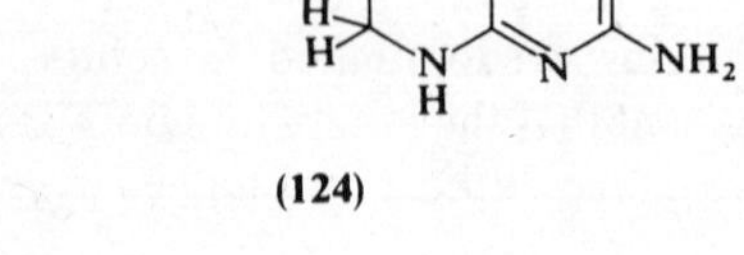

(124)

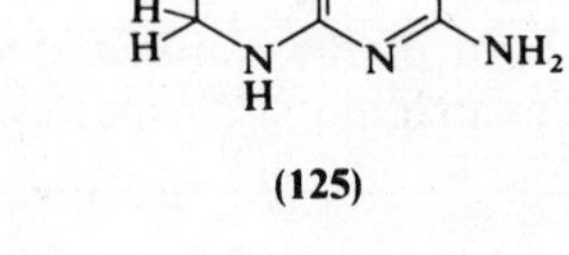

(125)

(**124**).[1730] Some members of a series of 6-alkoxymethyl-2-amino-5,6,7,8-tetrahydro-4(3*H*)-pteridinones (**125**) behaved as bio-analogs of tetrahydrobiopterin. For example, 2-amino-6-ethoxymethyl-5,6,7,8-tetrahydro-4(3*H*)-pteridinone (**125**, R = Et) was an excellent cofactor for phenylalanine hydroxylase and it showed promise as a cofactor replacement in neurological and psychiatric diseases involving biopterin deficiency.[1728] Unreduced biopterin has been discussed in Chapter VI, Section 5.D(4).

CHAPTER X

Pteridinecarboxylic Acids and Related Compounds

This chapter embraces pteridinecarboxylic acids, alkyl pteridinecarboxylates (esters), pteridinecarboxamides, pteridinecarbonitriles, pteridinecarbaldehydes, and pteridine ketones; extranuclear carboxy, alkoxycarbonyl, carbamoyl, cyano, *C*-formyl, and *C*-acyl derivatives are included, usually at the end of each appropriate nuclear category.

1. PREPARATION OF PTERIDINECARBOXYLIC ACIDS

A. By Primary Syntheses

A limited number of pteridinecarboxylic acids have been made directly by primary syntheses from pyrimidines (see Ch. II) or from other heterocyclic intermediates (see Ch. III). Such processes are illustrated briefly in the Gabriel and Colman condensation of 2,4-bisethylamino-5-pyrimidinamine (**1**) with disodium mesoxalate or alloxan (in alkali) to give 8-ethyl-2-ethylamino-7-oxo-7,8-dihydro-6-pteridinecarboxylic acid (**2**),[387] and of 2,5,6-triamino-4(3*H*)-pyrimidinone with ethyl α-(diethoxyacetyl)propionate to give 2-amino-6-α-

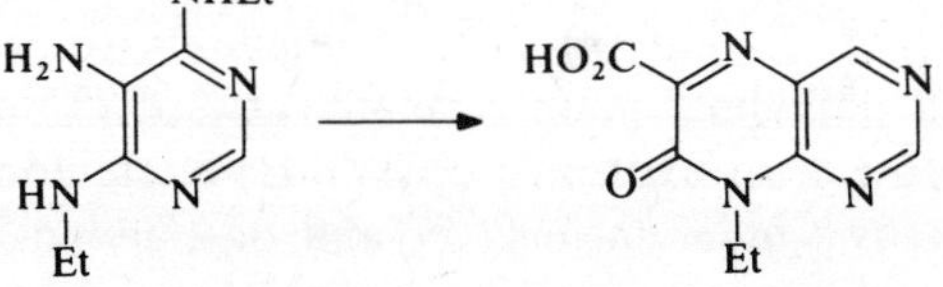

(**1**) (**2**)

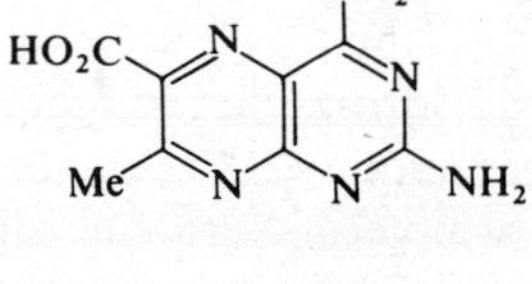

(**3**)

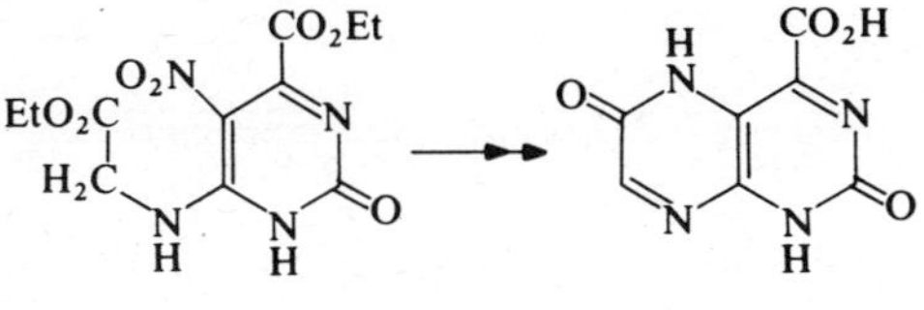

(**4**) (**5**)

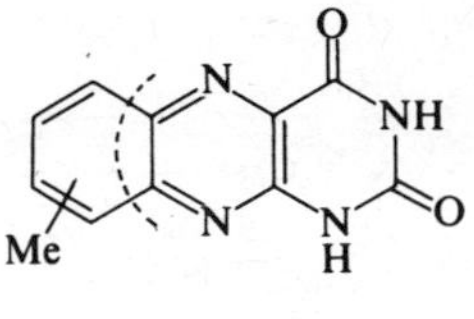

(**6**)

carboxyethyl-4(3*H*)-pteridinone[906]; in the Timmis condensation of 5-nitroso-2,4,6-pyrimidinetriamine with methyl acetoacetate (in methoxyethanolic methoxyethoxide) to give 2,4-diamino-7-methyl-6-pteridinecarboxylic acid (**3**)[307]; in the Boon reductive cyclization of ethyl 6-ethoxycarbonylmethylamino-5-nitro-2-oxo-1,2-dihydro-4-pyrimidinecarboxylate (**4**) by sodium dithionite in aqueous bicarbonate to give sodium 2,6-dioxo-1,2,5,6,7,8-hexahydro-4-pteridinecarboxylate and then on oxidation, 2,6-dioxo-1,2,5,6-tetrahydro-4-pteridinecarboxylic acid (**5**)[150]; and in the classical oxidative degradation of tolualloxazin (**6**) to give 2,4-dioxo-1,2,3,4-tetrahydro-6,7-pteridinedicarboxylic acid.[7, 8]

B. By Oxidative Means

Almost any substituent or side chain that is attached to the pteridine nucleus by a carbon-to-carbon linkage may be oxidized to furnish a pteridinecarboxylic acid. Thus the oxidation of *alkylpteridines* is exemplified in the conversion of 6-methyl-2,4-pteridinediamine into 2,4-diamino-6-pteridinecarboxylic acid (hot aqueous permanganate for 30 min: ~55%),[123] of 6-ethyl-1,3-dimethyl-2,4(1*H*,3*H*)-pteridinedione into 1,3-dimethyl-2,4-dioxo-1,2,3,4-tetrahydro-6-pteridinecarboxylic acid (similarly: 45%),[1291] and of 2-amino-6,7-dimethyl-4(3*H*)-pteridinone into 2-amino-4-oxo-3,4-dihydro-6,7-pteridinedicarboxylic acid (**7**) (alkaline permanganate at 100°C for 90 min).[748, cf. 1363] Other examples are discussed in Chapter IV, Section 2.B(6).

The oxidation of *pteridinyl alcohols* is typified in the transformation of 7-hydroxymethyl-2,4-pteridinediamine into 2,4-diamino-7-pteridinecarboxylic acid (hot aqueous permanganate: 61%)[123]; of 2-amino-6-α,β-dihydroxyethyl-7-methyl-4(3*H*)-pteridinone (**9**) into 2-amino-7-methyl-4-oxo-3,4-dihydro-6-pteridinecarboxylic acid (**8**) (alkaline permanganate at 100°C for a few minutes; or alkaline sodium periodate followed by peroxide: 30%) or, under more vigorous conditions, into 2-amino-4-oxo-3,4-dihydro-6,7-pteridinedicarboxylic acid (**7**) (alkaline permanganate at 100°C for 3 h; note that the dihydroxyethyl side chain was more easily oxidized than the methyl group)[1378]; of 2-amino-6-α,β-dihydroxypropyl-4(3*H*)-pteridinone (biopterin) into 2-amino-4-oxo-3,4-dihydro-6-pteridinecarboxylic acid (alkaline permanganate at 100°C for 30 min)[1018, cf. 274]; of 2-amino-6-α,β,γ-trihydroxypropyl-4(3*H*)-pteridinone (neopterin) into the same product by similar means[1422]; and of 2-amino-7-α,β,γ,δ-tetrahydroxybutyl-4(3*H*)-pteridinone into 2-amino-4-oxo-3,4-dihydro-7-pteridinecarboxylic acid (alkaline permanganate at 70°C for 1 h).[1147] Other

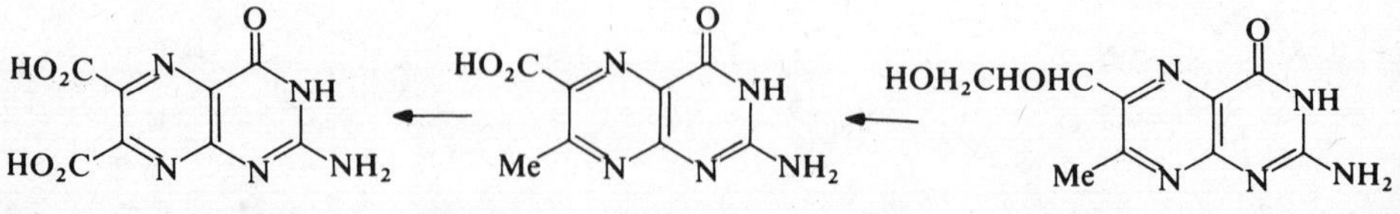

(**7**) (**8**) (**9**)

examples have been given in Chapter VI, Section 4.B. In addition, treatment of crude 7-bromomethyl-2-dimethylamino-4(3*H*)-pteridinone (equivalent to the 7-hydroxymethyl analog) with alkaline permanganate at 80°C gave 2-dimethylamino-4-oxo-3,4-dihydro-7-pteridinecarboxylic acid (25%).[1710] Oxidation of an alcohol side chain, linked to the nucleus through oxygen rather than carbon, naturally should give an extranuclear carboxy derivative. Thus 2-amino-6-β-hydroxy-α-methylethoxy-4(3*H*)-pteridinone (**10**, R = CH_2OH) with alkaline permanganate at 50°C gave 2-amino-6-α-carboxyethoxy-4(3*H*)-pteridinone (**10**, R = CO_2H) in ~25% yield.[299, 302]

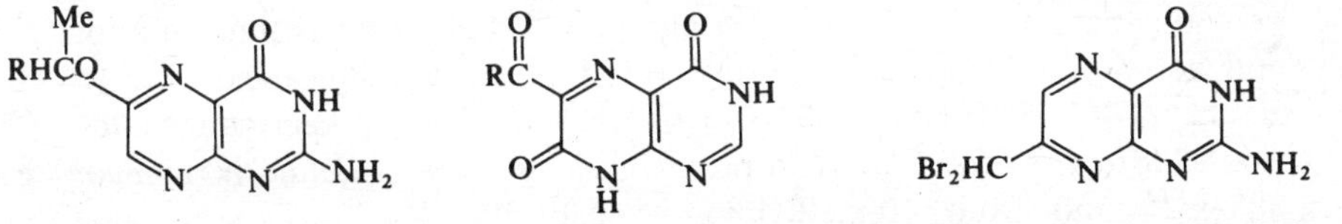

(10) **(11)** **(12)**

Oxidation of *pteridinecarbaldehydes* is represented by the conversion of 2-amino-4-oxo-3,4-dihydro-6-pteridinecarbaldehyde into the corresponding pteridinecarboxylic acid (alkaline hydrogen peroxide at 25°C for >2 h: ~75%, as sodium salt[192]; alkaline permanganate at 25°C: >95%)[943]; of 4,7-dioxo-3,4,7,8-tetrahydro-6-pteridinecarbaldehyde (**11**, R = H) into the 6-pteridinecarboxylic acid (**11**, R = OH) (alkaline permanganate at 25°C: 75%)[29]; of 1,3-dimethyl-2,4,7-trioxo-1,2,3,4,7,8-hexahydro-6-pteridinecarbaldehyde into the corresponding pteridinecarboxylic acid (likewise: ~50%)[337]; and of 2-amino-4,6-dioxo-3,4,5,6-tetrahydro-7-pteridinecarbaldehyde (made in situ) into the corresponding carboxylic acid (alkaline permanganate).[1313] The direct oxidation of the aldehyde equivalent, 2-amino-7-dibromomethyl-4(3*H*)-pteridinone (**12**), with alkaline permanganate gave 2-amino-4-oxo-3,4-dihydro-7-pteridinecarboxylic acid in 85% yield.[1359] 2-Amino-4-oxo-3,4-dihydro-6-pteridinecarbaldehyde underwent the Cannizaro reaction in 2.5*M* alkali at 5°C for 5 days to give 2-amino-4-oxo-3,4-dihydro-6-pteridinecarboxylic acid and 2-amino-6-hydroxymethyl-4(3*H*)-pteridinone, both in good yield.[1358]

The oxidation of pteridines bearing other *carbonyl-containing side chains* is illustrated in the conversion of 2-amino-6-lactoyl-7,8-dihydro-4(3*H*)-pteridinone (**13**) (sepiapterin) into 2-amino-4-oxo-3,4-dihydro-6-pteridinecarboxylic acid (permanganate)[1062]; of 6-acetyl-2-amino-7-methyl-7,8-dihydro-4(3*H*)-pteridinone (**14**) into 2-amino-7-methyl-4-oxo-3,4-dihydro-6-pteridine-

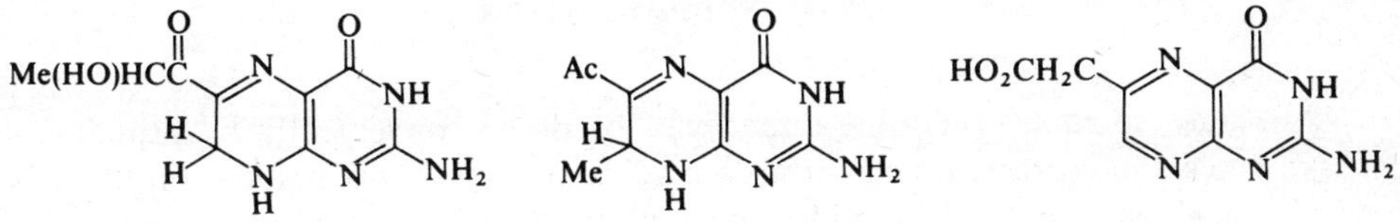

(13) **(14)** **(15)**

carboxylic acid (**8**) (iodine) or 2-amino-4-oxo-3,4-dihydro-6,7-pteridinedicarboxylic acid (**7**) (alkaline permanganate)[1453]; of 2-amino-6-carboxymethyl-4(3*H*)-pteridinone (**15**) into 2-amino-4-oxo-3,4-dihydro-6-pteridinecarboxylic acid (alkaline permanganate at 75°C for 2 h: ~60%)[1084, cf. 502, 1174]; and of 2-amino-7-α-cyano-α-ethoxycarbonylmethyl-4(3*H*)-pteridinone (**16**, R = H) or its 2-acetamido analog (**16**, R = Ac) into 2-amino-4-oxo-3,4-dihydro-7-pteridinecarboxylic acid (alkaline permanganate at 80°C for 2 h: ~20%).[1301]

The similar oxidation of *folate-type* molecules is exemplified in the formation of 2-amino-4-oxo-3,4-dihydro-6-pteridinecarboxylic acid from pteroylglutamic acid [**17**, R = X = H, Y = $NHCH(CO_2H)CH_2CH_2CO_2H$] (alkaline permanganate,[500, 745, 1376] sodium chlorate,[1363] aerated alkaline solution,[745] and aeration in light[865, 899]), from *m*-iodopteroylglutamic acid [**17**, R = H, X = I, Y = $NHCH(CO_2H)CH_2CH_2CO_2H$] (alkaline permanganate),[1104] from 5-formyl-5,6,7,8-tetrahydropteroylglutamic acid (alkaline permanganate: 68%),[744] and from 10-methylpteroic acid (**17**, R = Me, X = H, Y = OH) (alkaline permanganate but not aeration of an alkaline solution).[161] It is also evident in the formation of 2,4-dioxo-1,2,3,4-tetrahydro-6-pteridinecarboxylic acid from the 2-oxo analog of pteroylglutamic acid (alkaline permanganate at 90°C: ~40%),[655] and in the formation of 2,4-diamino-6-pteridinecarboxylic acid from 4-amino-*m*-iodo-4-deoxypteroylglutamic acid (**18**, R = X = H, Y = I)[1107] or 4-amino-*m,m'*-dichloro-10-methyl-4-deoxypteroylglutamic acid (**18**, R = Me, X = Y = Cl)[178] (alkaline permanganate: ~69% yield from the former substrate).

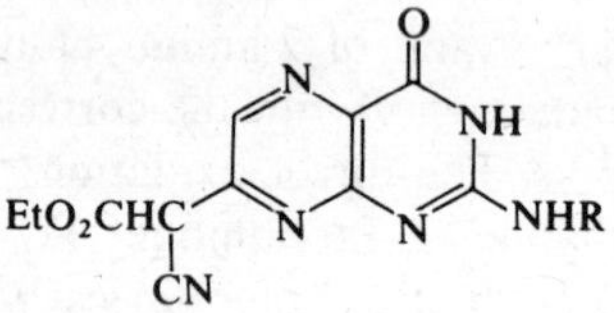

(**16**)

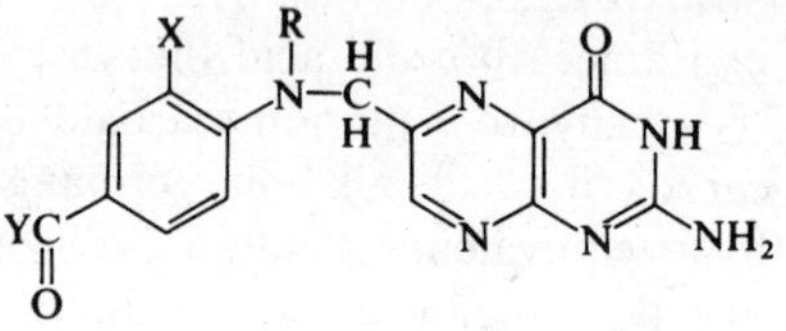

(**17**)

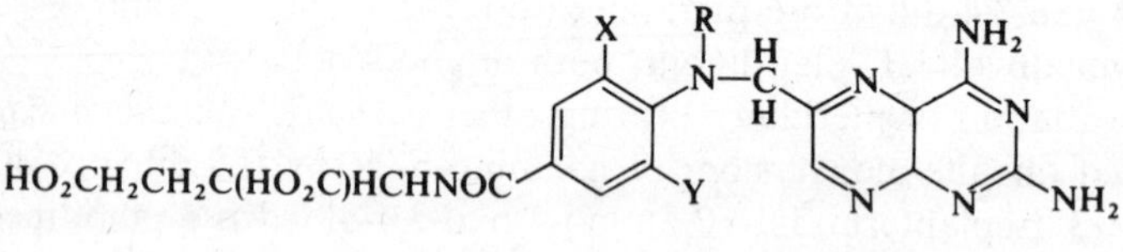

(**18**)

C. By Hydrolytic Means

Because so many pteridine esters may be made conveniently by primary syntheses, the corresponding carboxylic acids are commonly made from these by hydrolysis; a few carboxamides and a carbonitrile have also been so used to make carboxylic acids.

The hydrolysis of *regular pteridine esters* may be done under acidic or basic conditions, although the latter has been favored. Thus ethyl 4-pteridinecarboxylate, as its covalent dihydrate (**19**, R = Et), gave 4-pteridinecarboxylic acid, also as its dihydrate (**19**, R = H) (aqueous sodium hydrogen carbonate at 60°C for 15 min: ~65%)[157]; dimethyl 1,3-dimethyl-2,4-dioxo-1,2,3,4-tetrahydro-6,7-pteridinedicarboxylate (**20**, R = Me) gave, not the corresponding dicarboxylic acid (**20**, R = H), but 1,3-dimethyl-2,4-dioxo-1,2,3,4-tetrahydro-6-pteridinecarboxylic acid after spontaneous 7-decarboxylation (3*M* hydrogen chloride in aqueous acetic acid under reflux for 8 h: 34%)[1388]; ethyl 8-ethyl-2-ethylamino-7-oxo-7,8-dihydro-6-pteridinecarboxylate gave its parent carboxylic acid (1*M* sodium hydrogen carbonate under reflux for 10 min: ~70%)[387]; ethyl 2-amino-4-isopropoxy-7-oxo-7,8-dihydro-6-pteridinecarboxylate (**21**, R = Et, Y = *i*-Pr, Z = H) gave the acid (**21**, R = Z = H, Y = *i*-Pr) (aqueous bicarbonate at 100°C for 4 h: ~70%)[371]; ethyl 2-amino-4-methoxy (or isopropoxy)-8-methyl-7-oxo-7,8-dihydro-6-pteridinecarboxylate (**21**, R = Et, Y = Me or *i*-Pr, Z = Me) gave the corresponding acid (**21**, R = H, Y = Me or *i*-Pr, Z = Me) (similarly for 2 h: 90% yield of the latter product)[371]; ethyl or methyl 4,8-dimethyl-2-methylamino-7-oxo-7,8-dihydro-6-pteridinecarboxylate gave their parent acid (0.1*M* alkali at 100°C for 30 min)[1311]; ethyl 8-ethyl-4-ethylamino-7-oxo-7,8-dihydro-6-pteridinecarboxylate gave its acid (0.4*M* alkali at 25°C for 15 h: 83%)[1311]; ethyl 2-amino-3,8-dimethyl-4,7-dioxo-3,4,7,8-tetrahydro-6-pteridinecarboxylate (**22**, R = Et, X = Z = Me, Y = NH_2) gave its acid (**22**, R = H, X = Z = Me, Y = NH_2) (1*M* sodium hydrogen carbonate under reflux for 30 min: ~75% yield)[385]; ethyl 2-dimethylamino-8-methyl-4,7-dioxo-3,4,7,8-tetrahydro-6-pteridinecarboxylate (**22**, R = Et, X = H, Y = NMe_2, Z = Me) gave its acid (**22**, R = X = H, Y = NMe_2, Z = Me) (1*M* bicarbonate at 100°C for 1 h: ~90%)[360]; ethyl 2-methylamino-4,7-dioxo-3,4,7,8-tetrahydro-6-pteridinecarboxylate (**22**, R = Et, X = Z = H, Y = NHMe) and its 3-methyl, 8-methyl, and 3,8-dimethyl derivatives all gave their respective carboxylic acids (**22**, R = X = Z = H, Y = NHMe; R = Z = H, X = Me, Y = NHMe; R = X = H,

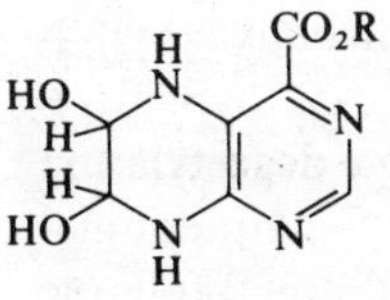

(19)

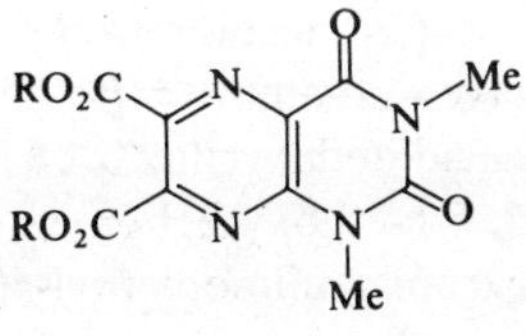

(20)

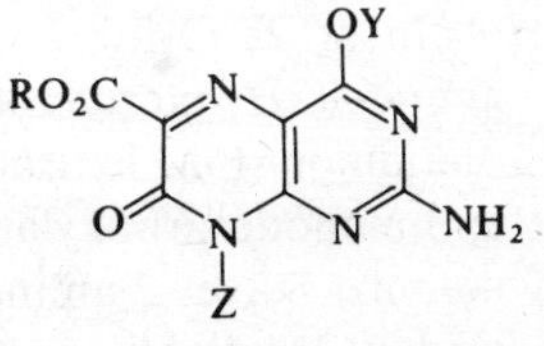

(21)

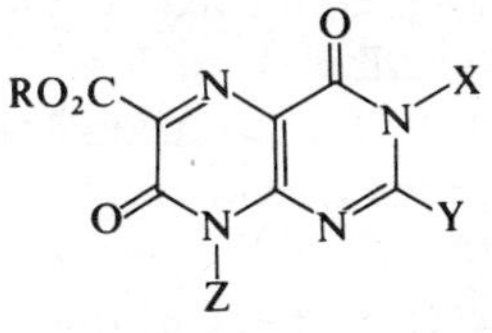

(22)

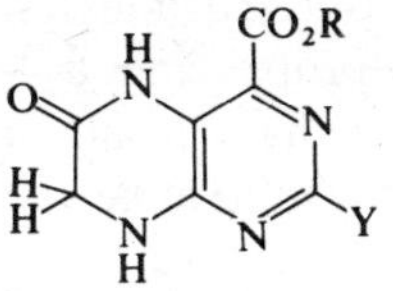

(23)

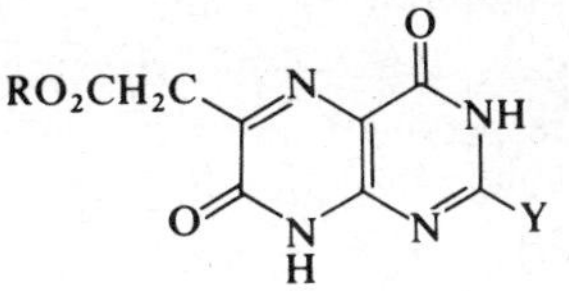

(24)

Y=NHMe, Z=Me; and R=H, X=Z=Me, Y=NHMe) (1*M* bicarbonate under reflux for 45 min: >85% each)[1565]; ethyl 4-amino-2,7-dioxo-1,2,7,8-tetrahydro-6-pteridinecarboxylate gave its parent acid (1*M* bicarbonate under reflux for 3 h: ~85%)[363]; methyl 7-methoxy-1,3-dimethyl-2,4-dioxo-1,2,3,4-tetrahydro-6-pteridinecarboxylate gave its acid (1*M* bicarbonate incubated at 40°C for 2 days: ~65%)[341]; and ethyl 1,3,8-trimethyl-2,4,7-trioxo-1,2,3,4,7,8-hexahydro-6-pteridinecarboxylate gave its acid (likewise for 12 h: ~55%).[341]

The hydrolysis of *dihydropteridine esters* is at least equally easy. For example, ethyl 3,4-dihydro-2-pteridinecarboxylate gave the corresponding acid (45%) by simply stirring in 0.1*M* hydrochloric acid for 2 days[51]; ethyl 2-chloro-6-oxo-5,6,7,8-tetrahydro-4-pteridinecarboxylate (**23**, R=Et, Y=Cl) and its 2-amino, 2-dimethylamino, and 2-ethoxy analogs (**23**; R=Et; Y=NH_2, NMe_2, or OEt, respectively) all gave the corresponding acids (**23**; R=H; Y=Cl, NH_2, NMe_2, or OEt) in unstated yields by warming in 2*M* alkali for 2 min[750]; ethyl 2,6-dioxo-1,2,5,6,7,8-hexahydro-4-pteridinecarboxylate gave its parent acid under similar conditions[750]; and ethyl 6-oxo-2-thioxo-1,2,5,6,7,8-hexahydro-4-pteridinecarboxylate did likewise.[750]

Hydrolysis of *extranuclear pteridine esters* is well represented in the literature. Thus 6-ethoxycarbonylmethyl-4,7(3*H*,8*H*)-pteridinedione (**24**, R=Et, Y=H) gave the 6-carboxymethyl analog (**24**, R=Y=H) (2.5*M* alkali under reflux for 1 h: 80%)[29]; 2-amino-6-ethoxycarbonylmethyl-4,7(3*H*,8*H*)-pteridinedione (**24**, R=Et, Y=NH_2) gave its 6-carboxymethyl analog (**24**, R=H, Y=NH_2) (0.5*M* alkali for 1 h[1195]; cf. boiling aqueous acetic acid: 4% yield[497]); 4-di(ethoxycarbonyl)methyl-3,4-dihydro-2(1*H*)-pteridinone gave the 4-dicarboxymethyl analog (1*M* alkali at 25°C for 4 h: 80%)[38]; 4-α-cyano-α-ethoxycarbonylmethyl-3,4-dihydro-2(1*H*)-pteridinone gave the 4-α-carboxy-α-cyanomethyl analog (1*M* alkali at 25°C for 2.5 h)[38]; 2-amino-7-ethoxycarbonylmethyl-4,6(3*H*,5*H*)-pteridinedione gave the 7-carboxymethyl analog (0.5*M* alkali at 20°C)[497]; 2-amino-7-(2′-methoxycarbonyl-1′-propenyl)-4,6(3*H*,5*H*)-pteridinedione gave its 7-(2′-carboxy-1′-propenyl) analog 2.5*M* alkali at 25°C for 2 days)[1055]; 6-(β-acetamido-β,β-diethoxycarbonylethyl)-2-amino-4(3*H*)-pteridinone gave 2-amino-6-β-amino-β-carboxyethyl-4(3*H*)-pteridinone (6*M* hydrochloric acid under reflux for 6 h: 63%; note deacetylation and monodecarboxylation)[408]; 6-*p*-ethoxycarbonylanilinomethyl-2,4-pteridinediamine gave 2-amino-6-*p*-carboxyanilinomethyl-4(3*H*)-pteridinone (pteroic acid) (0.2*M* alkali under reflux for 90 min: 83%; note hydrolysis of the 4-amino group during saponification of the ester)[1299]; 6-β-(*p*-ethoxycarbonylphenyl)propyl- and 6-β-(*p*-methoxycarbonylphenyl)butyl-2,4-pteridinediamine gave the 6-β-*p*-carboxyphenylpropyl and 6-β-*p*-carboxyphenylbutyl analogs, respectively (aqueous methoxyethanolic alkali at 100°C for 15 min: 80 and 84%)[1582]; and dimethyl pteroylglutamate gave pteroylglutamic acid (0.1*M* alkali at 25°C for 12 h).[1363]

Formation of carboxylic acids by hydrolysis of *amides* is best illustrated by the conversion of *N*-methyl-2,4,6-trioxo-1,2,3,4,5,6-hexahydro-7-pteridinecarbox-

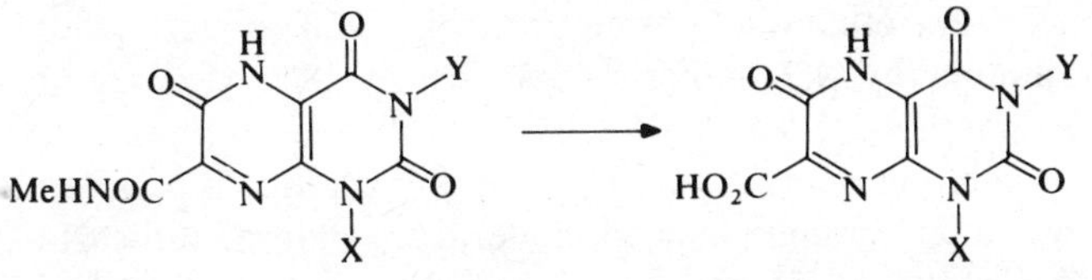

(25) (26)

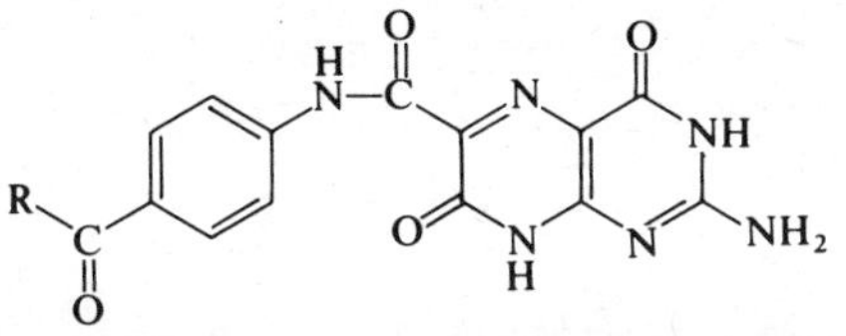

(27)

amide (**25**, X = Y = H) and its *N*,1- (**25**, X = Me, Y = H) and *N*,3-dimethyl (**25**, X = H, Y = Me) analogs into the corresponding 7-carboxylic acids (**26**, X = Y = H; X = Me, Y = H; and X = H, Y = Me) (2*M* alkali under reflux for 2 h: ~50% in each case).[342] Other examples include the hydrolysis of 2-amino-*N*-methyl-4,6-dioxo-3,4,5,6-tetrahydro-7-pteridinecarboxamide into 2-amino-4,6-dioxo-3,4,5,6-tetrahydro-7-pteridinecarboxylic acid (similarly: ~40%)[342]; the so-called "oxidation" of 2-amino-4-oxo-3,4-dihydro-6-pteridinecarboxamide by alkaline permanganate at 25°C for 1 h to give 2-amino-4-oxo-3,4-dihydro-6-pteridinecarboxylic acid, an odd conception, partly clarified by admissions that the same result was obtained by alkali alone or by heating with 2*M* hydrochloric acid for 1 h[925]; the oxidative hydrolysis of 2-amino-4,6-dioxo-3,4,5,6,7,8 (?)-hexahydro-7-pteridinecarboxamide to 2-amino-4,6-dioxo-3,4,5,6-tetrahydro-7-pteridinecarboxylic acid[514]; and the acidic hydrolysis of the *extranuclear* substituted-amide function in the pteroylglutamic acid analog [**27**, R = $NHCH(CO_2H)CH_2CH_2CO_2H$] to give 2-amino-*N*-(*p*-carboxyphenyl)-4,7-dioxo-3,4,7,8-tetrahydro-6-pteridinecarboxamide (**27**, R = OH).[539]

Complete hydrolysis of a *nitrile* to the corresponding carboxylic acid has been represented only in the conversion of 2-amino-8-methyl-4-oxo-3,4,5,6,7,8-hexahydro-6-pteridinecarbonitrile (or its 5-acetyl derivative ?) into 2-amino-8-methyl-4-oxo-3,4,5,6,7,8-hexahydro-6-pteridinecarboxylic acid by stirring with dilute alkali under nitrogen for 4 days (~90%),[684] and of 2-amino-5,6,7,7-tetramethyl-4-oxo-3,4,5,6,7,8-hexahydro-6-pteridinecarbonitrile into the corresponding acid (hydrobromide) by heating in hydrobromic acid for 17 h (59% yield).[644]

D. Miscellaneous Methods

Extranuclear carboxy groups may be carried into a pteridine molecule on the back of a reagent. For example, aminolysis of 6-bromomethyl-4(3*H*)-

pteridinone with ethanolic *p*-aminobenzoic acid under reflux gave 6-*p*-carboxyanilinomethyl-4(3*H*)-pteridinone (**28**) in 28% yield.[121] In a different way, 1,3-dimethyl-2,4,7-trioxo-1,2,3,4,7,8-hexahydro-6-pteridinecarbaldehyde reacted with malonic acid (in pyridine containing a little piperidine at 100°C for 2 h) to give, after spontaneous monodecarboxylation, 6-β-carboxyvinyl-1,3-dimethyl-2,4,7(1*H*,3*H*,8*H*)-pteridinetrione (**29**) in ~45% yield.[337]

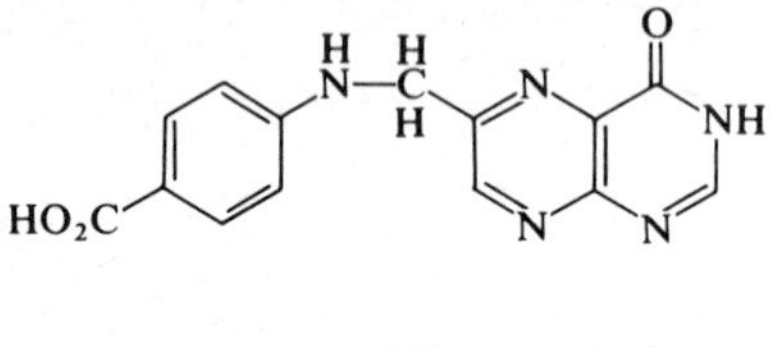

(**28**)

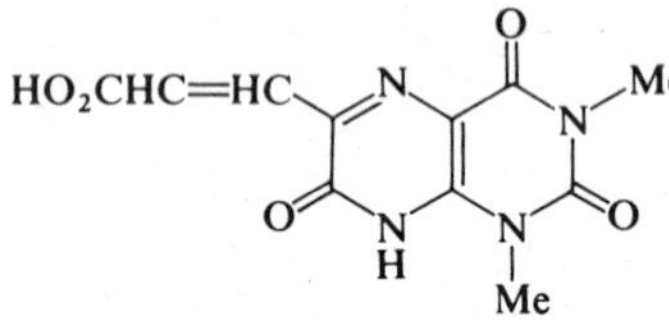

(**29**)

2. REACTIONS OF PTERIDINECARBOXYLIC ACIDS

The fluorescence spectra of several pteridinecarboxylic acids have been compared with those of 2-amino-4(3*H*)-pteridinone and its 6- and 7-formyl derivatives.[1200]

A. Decarboxylation

The conditions for decarboxylation in the pteridine series vary considerably. The usual *thermal process* is illustrated by decarboxylation of the following substrates: 4,7-diamino-6-pteridinecarboxylic acid (refluxing quinoline for 1 h),[324] its 2-phenyl derivative (refluxing quinoline for 5 h: 53%),[1357] 2-amino-4-oxo-3,4-dihydro-6,7-pteridinedicarboxylic acid hydrochloride (**30**) [250°C for 5 min gave a separable mixture of 2-amino-4-oxo-3,4-dihydro-6 (and 7)-carboxylic acid (**31**) and (**32**)[307, 601]: the former (**31**) gave 2-amino-4(3*H*)-pteridinone (**33**) (65%) on heating at 280°C for 5 h[1376]; its isomer (**32**) gave the same product (**33**) (61%) under similar conditions[529]], 4,7-dioxo-3,4,7,8-tetrahydro-6-pteridinecarboxylic acid (265°C under nitrogen for 1 h: 70%),[29] 8-ethyl-2-ethylamino-7-oxo-3,4,7,8-tetrahydro-6-pteridinecarboxylic acid (175°C in a vacuum for 3 days: 36% plus by-product),[1311] 2,4-diamino-7-oxo-7,8-dihydro-6-pteridinecarboxylic acid (sublimation at 350°C in a vacuum: ~65% yield),[1229] 2-amino-4,6-dioxo-3,4,5,6-tetrahydro-7-pteridinecarboxylic acid (300°C under nitrogen for 7 h: ~25%),[342] 2,4,6-trioxo-1,2,3,4,5,6-hexahydro-7-pteridinecarboxylic acid (270°C under nitrogen for 6 h: ~60%),[342] its 3-methyl derivative (250°C under nitrogen for 5 h: ~65%),[342] and 2-amino-8-β-hydroxyethyl-4,7-dioxo-3,4,7,8-tetrahydro-6-pteridinecarboxylic acid (250°C for 1 h: ~75%).[882] In contrast to the dicarboxylic acids (**20**, R = H) and (**30**), 2,4-dioxo-1,2,3,4-tetrahydro-6,7-pteridinedicarboxylic acid underwent selective 6-

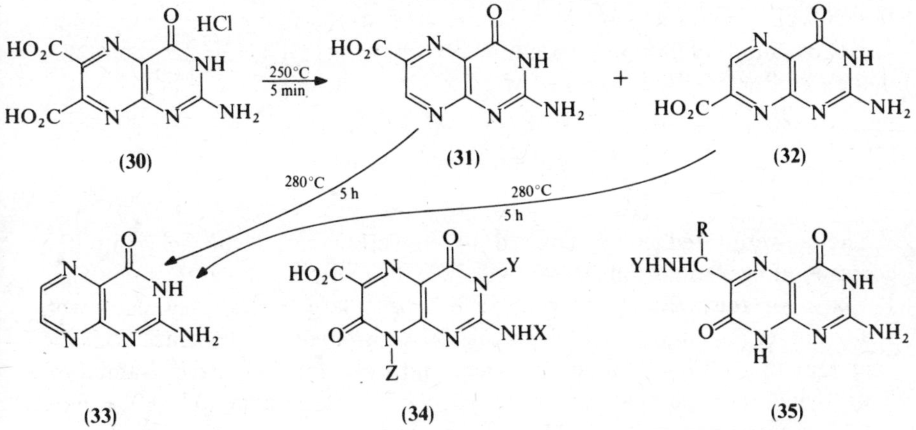

decarboxylation in refluxing quinoline to give only 2,4-dioxo-1,2,3,4-tetrahydro-7-pteridinecarboxylic acid.[748] The mass spectrum of 4-pteridinecarboxylic acid dihydrate showed a massive peak at M–CO_2 of the anhydrous molecule.[777]

It is sometimes possible to remove a carboxyl group (and indeed some other substituents) from the 6-position by reduction. Such *reductive elimination* of CO_2 is seen in the conversion of 2-amino-4,7-dioxo-3,4,7,8-tetrahydro-6-pteridinecarboxylic acid (**34**, X=Y=Z=H) into 2-amino-4,7(3*H*,8*H*)-pteridinedione (aluminum amalgam in 1.2*M* alkali at 100°C: 89%,[1034, 1128] aluminum amalgam in 5% aqueous ammonia at 80°C for 35 min: ~95%),[1565] and also in carbon dioxide elimination (using aluminum amalgam in ammonia as just described to avoid potential rearrangements in alkali) from 2-methylamino- (**34**, X=Me, Y=Z=H) (~70%), 2-amino-3-methyl- (**34**, X=Z=H, Y=Me) (~60%), 8-methyl-2-methylamino- (**34**, X=Z=Me, Y=H) (~70%), 2-amino-3,8-dimethyl- (**34**, X=H, Y=Z=Me) (~70%), and 3,8-dimethyl-2-methylamino-4,7-dioxo-3,4,7,8-tetrahydro-6-pteridinecarboxylic acid (**34**, X=Y=Z=Me) (~80%).[1565] In addition, such eliminations have been done by hydrogenation over palladium–carbon, followed by reoxidation of the resulting 5,6-dihydro derivative with ferricyanide or a stream of air. In this way 8-ethyl-2-ethylamino-7-oxo-7,8-dihydro-6-pteridinecarboxylic acid gave 8-ethyl-2-ethylamino-7(8*H*)-pteridinone (60%) while appropriate 6-carboxy substrates gave the analogous products, 4-dimethylamino-8-methyl-7(8*H*)-pteridinone (79%), 8-ethyl-4-ethylamino-7(8*H*)-pteridinone (60%), and 2-amino-8-ethyl-4,7(3*H*,8*H*)-pteridinedione (78%).[1311]

Some *extranuclear decarboxylations* yielding simple alkylpteridines have been discussed in Chapter IV, Section 2.A(3). Less simple examples include the partial decarboxylation of 4-dicarboxymethyl- to 4-carboxymethyl-3,4-dihydro-2(1*H*)-pteridinone (~30%) by refluxing in propionic acid containing a little sulfuric acid[38]; and the formation of 2-amino-6-aminomethyl-4,7(3*H*,8*H*)-pteridinedione (**35**, R=Y=H), either by hydrolysis and decarboxylation of 6-α-acetamido-α-ethoxycarbonylmethyl-2-amino-4,7(3*H*,8*H*)-pteridinedione (**35**,

R=CO_2Et, Y=Ac) in 2*M* hydrochloric acid at 100°C for 1 h, or by simple decarboxylation of 2-amino-6-α-amino-α-carboxymethyl-4,7(3*H*,8*H*)-pteridinedione (**35**, R=CO_2H, Y=H) in 1*M* hydrochloric acid at 50°C.[1083]

B. Reduction to an Alcohol

The powerful reduction required to convert a carboxylic acid into the corresponding alcohol will inevitably reduce the pteridine ring at the same time. Perhaps for this reason, the process has been largely neglected. However, treatment of 2-amino-8-methyl-4-oxo-3,4,7,8-tetrahydro-6-pteridinecarboxylic acid (**36**) in THF with diborane under nitrogen for 24 h gave 2-amino-6-hydroxymethyl-8-methyl-5,6,7,8-tetrahydro-4(3*H*)-pteridinone (**37**) (80%; note the concomitant 5,6-reduction) which, on oxygenation in sodium bicarbonate solution, gave 2-amino-6-hydroxymethyl-8-methyl-7,8-dihydro-4(3*H*)-pteridinone (**38**) (70%; note reversion to the original oxidation state of the ring that resists further dehydrogenation on account of the 8-methyl group).[684]

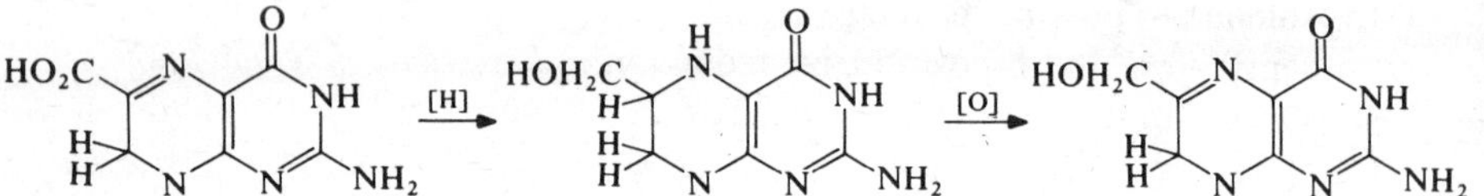

(**36**) (**37**) (**38**)

C. Esterification

The literature on esterification of *nuclear pteridinecarboxylic acids* holds few surprises. Thus esterification of appropriate acids gave methyl 2,4-diamino-6-pteridinecarboxylate (methanolic hydrogen chloride for 1.5 h: 44%),[1354] methyl 2,4-diamino-7-pteridinecarboxylate (similarly for 24 h: 77%),[748] dimethyl 2,4-diamino-6,7-pteridinedicarboxylate (likewise: 67%),[748] methyl 2,4-diamino-7-methyl-6-pteridinecarboxylate (likewise for 12 h),[884] methyl 2-amino-4-oxo-3,4-dihydro-6-pteridinecarboxylate (methanolic sulfuric acid under reflux for 6 h: 35%[389]; methanolic hydrogen chloride at 25°C for 24 h: ~75%),[1363] methyl 2-amino-4-oxo-3,4-dihydro-7-pteridinecarboxylate (methanolic hydrogen chloride at 25°C for 3 days: 28%[389]; unspecified conditions: 83%[748]), its homologous ethyl ester (ethanolic hydrogen chloride under reflux for 4 h: 82% as hydrochloride or 65% as base),[389] ethyl 8-ethyl-2-ethylamino-7-oxo-7,8-dihydro-6-pteridinecarboxylate (**39**, R=Et) (ethanolic sulfuric acid under reflux for 5 h: ~80%),[387] the homologous methyl ester (**39**, R=Me) (methanolic sulfuric acid under reflux for 5 h: ~80%; diazomethane in methanolic ether: ~35%),[387] ethyl 2-amino-8-methyl-4,7-dioxo-3,4,7,8-tetrahydro-6-pteridinecarboxylate

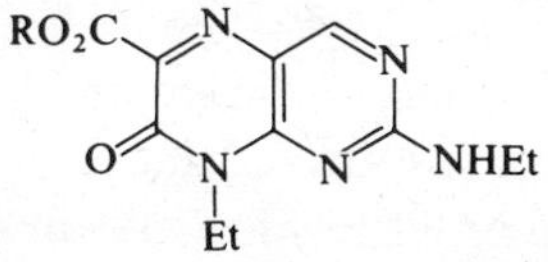

(39)

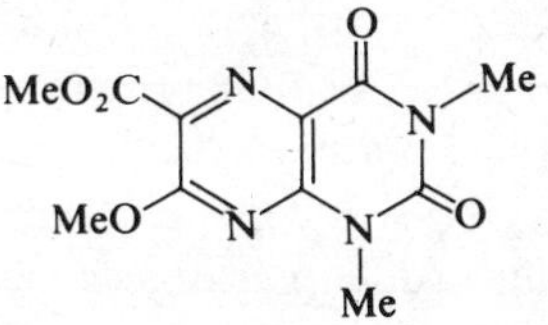

(40)

(ethanolic sulfuric acid under reflux for 4 h: ~55%),[385] methyl 2,4-dioxo-1,2,3,4-tetrahydro-7-pteridinecarboxylate (methanolic hydrogen chloride for 24 h: 48%),[748] dimethyl 2,4-dioxo-1,2,3,4-tetrahydro-6,7-pteridinedicarboxylate (likewise: 69%),[748] and ethyl 2,4,6-trioxo-1,2,3,4,5,6-hexahydro-7-pteridinecarboxylate (ethanolic sulfuric acid under reflux for 4 h: ~65%).[347] When 1,3-dimethyl-2,4,7-trioxo-1,2,3,4,7,8-hexahydro-6-pteridinecarboxylic acid was treated with diazomethane in methanolic ether, both esterification and *O*-methylation at the 7-position took place to afford methyl 7-methoxy-1,3-dimethyl-2,4-dioxo-1,2,3,4-tetrahydro-6-pteridinecarboxylate (**40**) (~50%); similar treatment of the ethyl ester of this substrate gave the same product (**40**) (~65%), thus involving transesterification as well as the expected *O*-methylation.[341]

The process of *extranuclear esterification* is exemplified in the formation, from appropriate carboxy substrates, of 8-β-hydroxyethyl-6-β-methoxycarbonylethyl-2,4,7(1*H*,3*H*,8*H*)-pteridinetrione (**41**, R = Me, Y = H) (methanolic sulfuric acid under reflux for 4 h: 65%)[1449]; 6-β-ethoxycarbonylethyl-8-β-D-ribityl-2,4,7(1*H*,3*H*,8*H*)-pteridinetrione [**41**, R = Et, Y = CH(OH)CH(OH)CH_2OH] (no details)[1248]; and methotrexate diethyl ester (ethanolic hydrogen chloride at 25°C for 3 days (~80%), diisopropyl ester (hydrogen chloride in isopropyl alcohol and benzene under reflux for 4 h with a water-removal device: ~80%), or several other dialkyl esters using one or the other procedure.[1725, 1765]

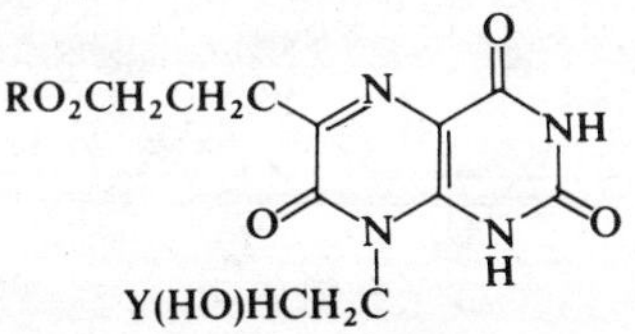

(41)

D. Conversion into Carboxamides

Although one esoteric method has been described for the direct coupling of oligoglutamyl esters with 4-amino-10-methyl-4-deoxypteroic acid [6-(*p*-carboxy-*N*-methylanilino)methyl-2,4-pteridinediamine] in the presence of *N*,*N*-dicyclohexylcarbodiimide–1-hydroxybenzotriazole dissolved in dimethyl sulfoxide at 25°C,[1691] most acid→amide conversions are still done indirectly via an acid chloride, ester, or other intermediate.

For example, 1,3-dimethyl-2,4-dioxo-1,2,3,4-tetrahydro-6-pteridinecarboxylic acid (**42**, R = OH) was converted, by thionyl chloride at 100°C for 3 h, into the corresponding (crude) pteridinecarbonyl chloride (**42**, R = Cl) which underwent aminolysis in ethanolic ammonia at 25°C to give 1,3-dimethyl-2,4-dioxo-1,2,3,4-tetrahydro-6-pteridinecarboxamide (**42**, R = NH_2) in 82% overall yield.[1388] 2-Amino-4,7-dioxo-3,4,7,8-tetrahydro-6-pteridinecarboxylic acid was converted, by trituration with phosphorus pentachloride–phosphoryl chloride, into the (crude) carbonyl chloride which, on treatment with *p*-aminobenzoylglutamic acid at 0°C in cold alkali, afforded a pteroylglutamic acid analog [**27**, R = $NHCH(CO_2H)CH_2CH_2CO_2H$].[539] 2-ε-Carboxypentylamino-6,7-dimethyl-4(3*H*)-pteridinone (**43**, R = OH) was converted, by stirring with ethyl chloroformate–triethylamine–DMF, into the crude acyl chloride (**43**, R = Cl), which underwent aminolysis by cold butylamine to give 2-ε-(*N*-butylcarbamoyl)pentylamino-6,7-dimethyl-4(3*H*)-pteridinone (**43**, R = NHBu) (~30%).[461] The same substrate (**43**, R = OH) was esterified) (methanolic hydrogen chloride) and the resulting crude methyl ester (**43**, R = OMe) was treated with aqueous ammonia for 20 h to give 2-ε-carbamoylpentylamino-6,7-dimethyl-4(3*H*)-pteridinone (**43**, R = NH_2) in ~60% yield.[461] Finally, treatment of sodium 2-amino-4-oxo-3,4-dihydro-6-pteridinecarboxylate (**44**) with trifluoroacetic anhydride gave an (unisolated) mixed anhydride (2-trifluoroacetamido-6-trifluoroacetoxycarbonyl-4(3*H*)-pteridinone?) (**45**), which reacted with *p*-aminobenzoic acid to give (after deacylation of the 2-trifluoroacetamido group) 2-amino-*N*-(*p*-carboxyphenyl)-4-oxo-3,4-dihydro-6-pteridinecarboxamide (**46**) in 45% yield.[1123]

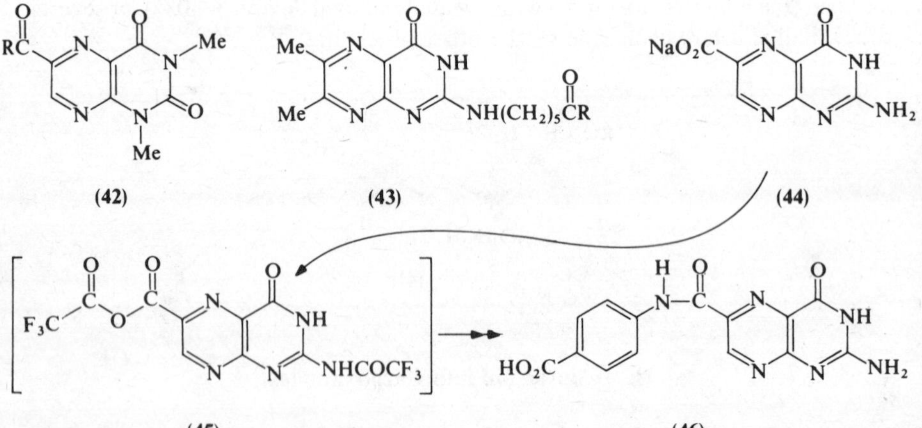

(**42**) (**43**) (**44**)

(**45**) (**46**)

A related sequence involved conversion of 2-acetamido-6-*N*-(*p*-carboxyphenyl)acetamidomethyl-4(3*H*)-pteridinone (**47**, R = CO_2H) into 2-acetamido-6-*N*-(*p*-azidocarbonylphenyl)acetamidomethyl-4(3*H*)-pteridinone (**47**,

R = CON_3) (diphenylphosphoryl azide–triethylamine–DMF at 25°C for 20 h: 83%), rearrangement with loss of nitrogen to give crude 2-acetamido-6-*N*-(*p*-isocyanatophenyl)acetamidomethyl-4(3*H*)-pteridinone (**47**, R = NCO) (boiling xylene for 25 min), condensation with dimethyl glutamate hydrochloride to give the protected urea derivative (**48**, R = Me, Y = Ac) (in DMF containing triethylamine at 25°C for 2 h: 40%), and a final alkaline hydrolysis to afford the subtle pteroylglutamic acid analog (**48**, R = Y = H) (10%).[1106]

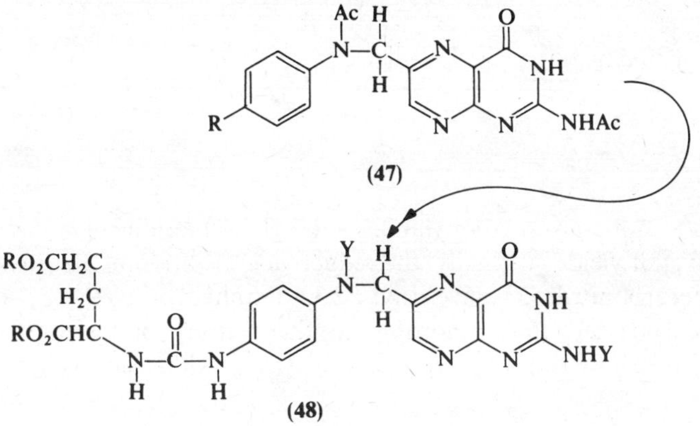

(47)

(48)

3. PREPARATION OF PTERIDINECARBOXYLIC ESTERS

A. By Primary Syntheses

Many pteridine esters have been made by primary syntheses from pyrimidines (Ch. II) or from other heterocyclic intermediates (Ch. III). For example, the Gabriel and Colman condensation of ethyl 5,6-diamino-4-pyrimidinecarboxylate with biacetyl in refluxing ethanol gave ethyl 6,7-dimethyl-4-pteridinecarboxylate (**49**) (~70%).[147] The Timmis condensation of 2-methylthio-5-nitroso-4,6-pyrimidinediamine with dimethyl malonate in methanolic methoxide gave methyl 4-amino-2-methylthio-7-oxo-7,8-dihydro-6-pteridinecarboxylate (**50**) in 68% yield.[1126] The Boon reductive cyclization of 6-α′-(*p*-ethoxycarbonylanilino)acetonylamino-5-nitro-2,4-pyrimidinediamine (**51**) afforded 6-*p*-ethoxycarbonylanilinomethyl-7,8-dihydro-2,4-pteridinediamine and then, by

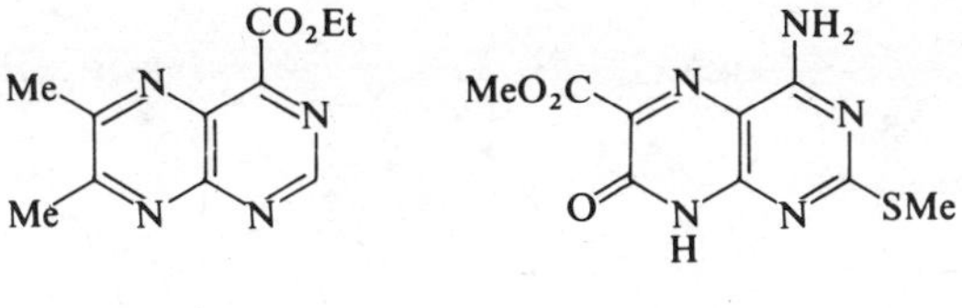

(49) **(50)**

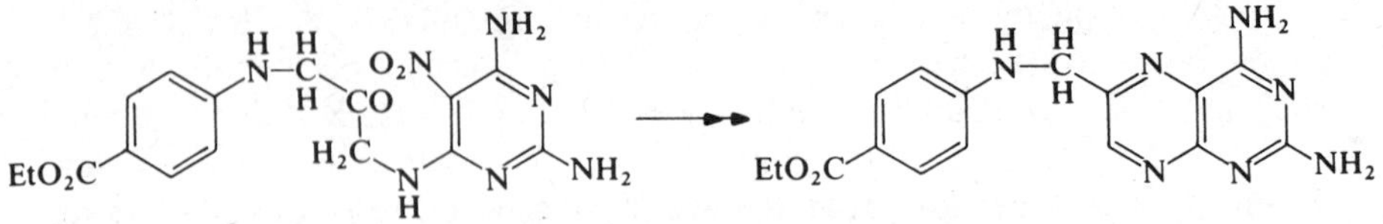

(51) (52)

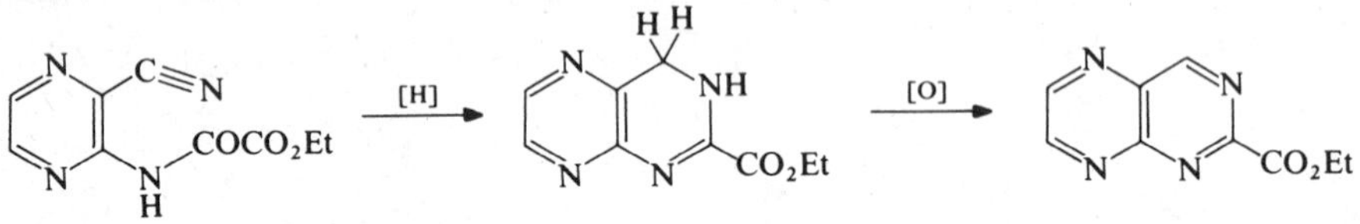

(53) (54) (55)

oxidation, 6-*p*-ethoxycarbonylanilinomethyl-2,4-pteridinediamine (**52**) in ~40% overall yield.[857] Finally, the reductive cyclization of 3-ethoxalylamino-2-pyrazinecarbonitrile (**53**) gave ethyl 1,2,3,4-tetrahydro- or ethyl 3,4-dihydro-2-pteridinecarboxylate (**54**), according to the severity of hydrogenation, and subsequent oxidation of the latter gave ethyl 2-pteridinecarboxylate (**55**).[51]

B. By Esterification or Transesterification

The formation of esters from carboxylic acids has been discussed in Section 2.C. An additional unusual example is the esterification of 4-*N*′-(dithiocarboxy)hydrazinopteridine (**56**, R = H) as its sodium salt, by reaction with methyl iodide or *β*-chloro-*N*,*N*-dimethylethylamine, to give 6-*N*′-[(methylthio)thiocarbonyl]hydrazino- (**56**, R = Me) or 6-*N*′-(*β*-dimethylaminoethylthio)thiocarbonyl]hydrazinopteridine (**56**, R = $CH_2CH_2NMe_2$), respectively.[137]

Although there may be no evident reasons, known cases of *transesterification* appear to be confined to the 6-position in pteridines and to involve only the transformation of ethyl to methyl esters. Thus boiling the corresponding ethyl esters in methanol containing a little thorium nitrate for 24 h gave methyl 2,8-dimethyl- (**57**, X = Me, Y = H) (80%), methyl 8-methyl-2-phenyl- (**57**, X = Ph, Y = H) (>95%), methyl 4,8-dimethyl-2-methylamino- (**57**, X = NHMe, Y = Me), and methyl 8-methyl-2-methylamino-7-oxo-7,8-dihydro-6-pteridine-

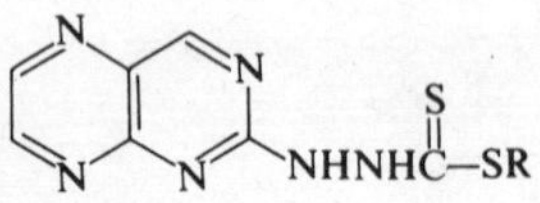

(56)

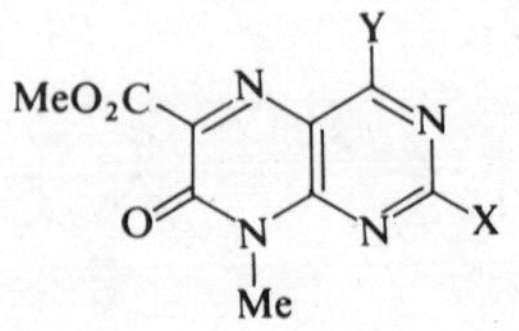

(57)

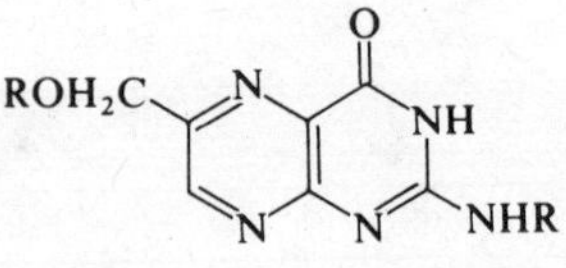

(58)

carboxylate (**57**, X=NHMe, Y=H).[1311] Another example involved trans-esterification during the *O*-methylation of ethyl 1,3-dimethyl-2,4,7-trioxo-1,2,3,4,7,8-hexahydro-6-pteridinecarboxylate with diazomethane in methanolic ether to give methyl 7-methoxy-1,3-dimethyl-2,4-dioxo-1,2,3,4-tetrahydro-6-pteridinecarboxylate (65%).[341]

It must be remembered that there are many known back-to-front extra-nuclear pteridine esters, which may be prepared by esterification of an aliphatic or aromatic acyl chloride or an anhydride with a pteridinylalkanol: For example, 2-amino-6-hydroxymethyl-4(3*H*)-pteridinone (**58**, R=H) in refluxing acetic anhydride for 3 h gave 2-acetamido-6-acetoxymethyl-4(3*H*)-pteridinone (**58**, R=Ac) in 70% yield[113]; the process is discussed in Chapter VI, Section 4.C. The nuclear versions of such esters are usually made from pteridine *N*-oxides by the well-known acylation–rearrangement reaction (Ch. VII, Sect. 6.D).

C. By Covalent Addition

Several extranuclear dihydropteridine esters have been made by covalent additions. Thus pteridine gave 4-diethoxycarbonylmethyl- (**59**, R=CO_2Et) (ethanolic diethyl malonate at 25°C for 4 days: 32%) and 4-α-ethoxy-carbonylphenacyl-3,4-dihydropteridine (**59**, R=Bz) (ethyl benzoylacetate in THF at 25°C for 5 days: 20%)[49]; 2(1*H*)-pteridinone gave 4-α-ethoxycarbonyl-acetonyl- (**60**, R=Ac, X=O) (ethyl acetoacetate in aqueous potassium carbonate at 25°C for 24 h: 80%), 4-diethoxycarbonylmethyl- (**60**, R=CO_2Et, X=O) (diethyl malonate similarly: 90%), and 4-α-cyano-α-ethoxycarbonylmethyl-3,4-dihydro-2(1*H*)-pteridinone (**60**, R=CN, X=O) (neat ethyl α-cyanoacetate at 100°C for 4 h: 90%)[38]; 2(1*H*)-pteridinethione gave 4-α-ethoxycarbonylacetonyl-3,4-dihydro-2(1*H*)-pteridinethione (**60**, R=Ac, X=S) (aqueous ethyl acetoacetate at 25°C for 7 days: 51%)[46]; 6(5*H*)-pteridinone gave 7-diethoxycarbonyl-methyl- (**61**, R=CO_2Et) (diethyl malonate in aqueous sodium carbonate at 20°C for 15 h: 60%) and 7-α-cyano-α-ethoxycarbonylmethyl-7,8-dihydro-6(5*H*)-pteridinone (**61**, R=CN) [ethyl cyanoacetate in aqueous alkali at 20°C for 15 h; boiling in water for 5 min gave 7-α-carboxy-α-cyanomethyl-7,8-dihydro-6(5*H*)-pteridinone and prolonged boiling gave 7-cyanomethyl-7,8-dihydro-6(5*H*)-pteridinone][55]; and 7(8*H*)-pteridinone gave 6-α-ethoxycarbonylacetonyl- (**62**, R=Ac) (ethyl acetoacetate in water at 25°C for 24 h: 76%), 6-α,γ-diethoxy-

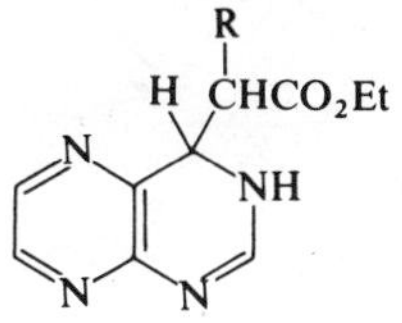

(59)

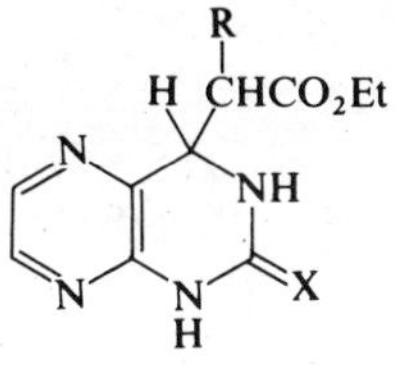

(60)

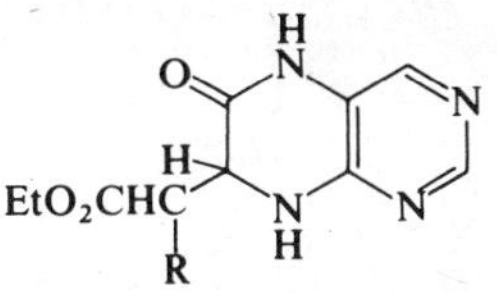

(61)

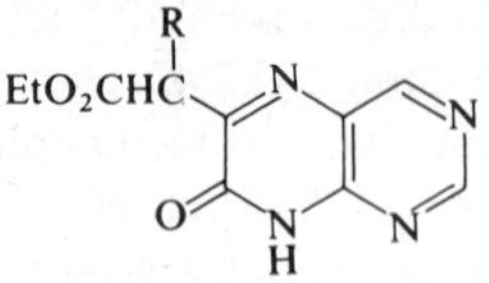

(62)

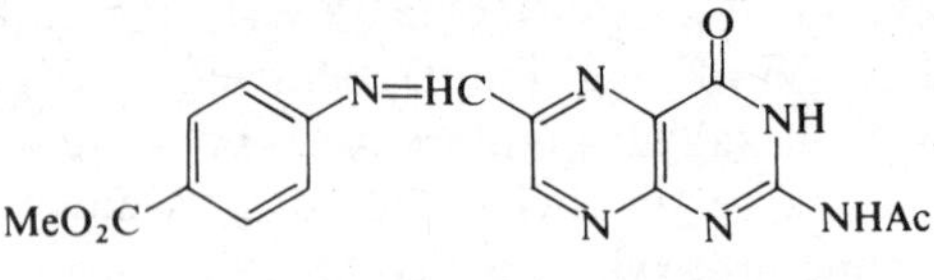

(63)

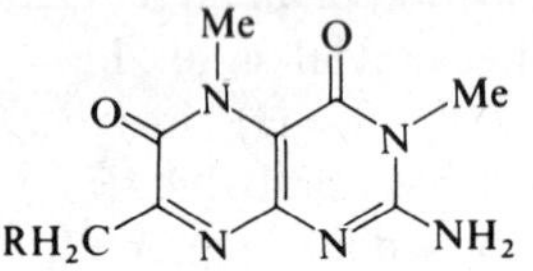

(64)

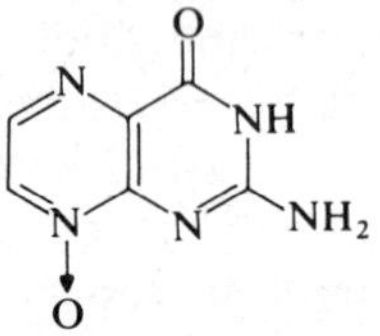

(65)

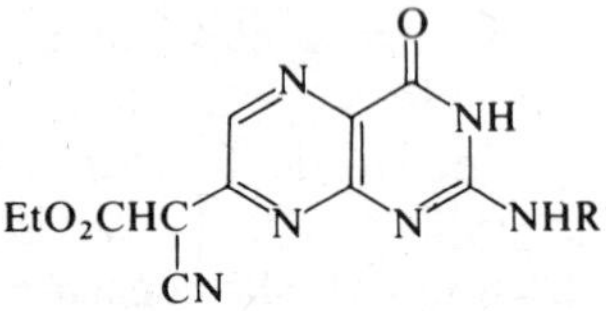

(66)

carbonylacetonyl- (**62**, R=$COCH_2CO_2Et$) (diethyl α,γ-acetonedicarboxylate in aqueous ethanol at 20°C for 7 days: 51%), and 6-diethoxycarbonylmethyl-5,6-dihydro-7(8*H*)-pteridinone (**62**, R=CO_2Et) (diethyl malonate plus tributylamine at 20°C for 8 days: 62%).[44]

D. By Other Routes

Condensation of 2-acetamido-4-oxo-3,4-dihydro-6-pteridinecarbaldehyde with methyl *p*-aminobenzoate in boiling ethanol gave the Schiff base, 2-acetamido-6-*p*-methoxycarbonylphenyliminomethyl-4(3*H*)-pteridinone (**63**), in 88% yield and then by borohydride reduction, 2-acetamido-6-*p*-methoxycarbonylanilinomethyl-4(3*H*)-pteridinone, which on alkaline hydrolysis gave pteroic acid.[260] Treatment of 2-amino-3,5,7-trimethyl-4,6(3*H*,5*H*)-pteridinedione (**64**, R=H) with dimethyl oxalate in methanolic potassium methoxide gave the Claisen product, 2-amino-7-methoxalylmethyl-3,5-dimethyl-4,6(3*H*,5*H*)-pteridinedione (**64**, R=$COCO_2Me$) in ~40% yield.[348] The reaction of 2-amino-4(3*H*)-pteridinone 8-oxide (**65**) with ethyl cyanoacetate and acetic anhydride in hexamethylphosphoramide at 90°C for 12 h gave a separable mixture of 2-amino- (**66**, R=H) and 2-acetamido-7-α-cyano-α-ethoxycarbonylmethyl-4(3*H*)-pteridinone (**66**, R=Ac). A single example of the conversion of a nitrile into an ester is given later in Section 8.

4. REACTIONS OF PTERIDINECARBOXYLIC ESTERS

A. Formation of Covalent Adducts

Because of its marked electron-withdrawing nature, an alkoxycarbonyl substituent further depletes the already meagre π-electron layer of pteridine and renders it even more prone to covalent additions. Thus the anhydrous neutral molecule, ethyl 4-pteridinecarboxylate (**67**, R = H), gradually changed into its covalent dihydrate, ethyl 6,7-dihydroxy-5,6,7,8-tetrahydro-4-pteridinecarboxylate (**68**, R = H), either by shaking with water at 25°C for 2 h (~75% yield)[147] or by standing in water for 2 days and then freeze drying (>95% yield).[148] The same dihydration was achieved more quickly by dissolution of the substrate (**67**, R = H) in dilute hydrochloric acid (to give the cation that underwent immediate dihydration) followed by neutralization in buffer to give the neutral dihydrate (**68**, R = H) in ~85% yield.[148] The anhydrous molecule (**67**, R = H) was recovered in ~60% yield from the dihydrate (**68**, R = H) by boiling in *t*-butyl alcohol under reflux for 8 h.[148] Because of steric hindrance by its methyl groups, neutral ethyl 6,7-dimethyl-4-pteridinecarboxylate (**67**, R = Me) showed little tendency to hydrate in water but the dihydrate, ethyl 6,7-dihydroxy-6,7-dimethyl-5,6,7,8-tetrahydro-4-pteridinecarboxylate (**68**, R = Me) could be obtained in ~45% yield via its cation as just described.[148] Ethyl 7-methyl-4-pteridinecarboxylate hydrated slowly in water but the dihydrate, ethyl 6,7-dihydroxy-7-methyl-5,6,7,8-tetrahydro-4-pteridinecarboxylate, was obtained in a pure state only via its cation.[148] Ethyl 4-pteridinecarboxylate (**67**, R = H) formed a diethanolate, ethyl 6,7-diethoxy-5,6,7,8-tetrahydro-4-pteridinecarboxylate (**69**) by heating in ethanol under reflux for >4 h.[147]

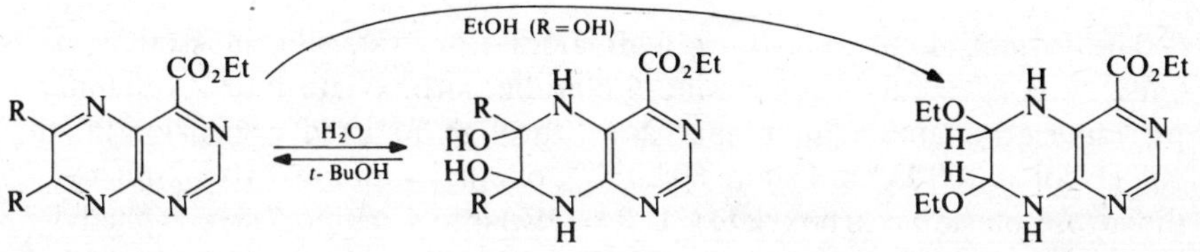

(67) **(68)** **(69)**

The mass spectral fragmentation of anhydrous ethyl 4-pteridinecarboxylate and its *C*-methyl derivatives involved an initial loss of the ester group in two stages and subsequent degradation of the pteridine ion beginning with the pyrimidine ring.[148] Under normal conditions the hydrates of these substrates lost their water during vaporization and consequently showed spectra closely similar to those of the anhydrous molecules. However, spectra at ~11 eV did show molecular ions attributable to the hydrates.[777]

B. Reduction

The reduction of a pteridine ester to an alcohol appears to be represented only by the conversion of ethyl 6,7-dimethyl-2-oxo-1,2-dihydro-4-pteridinecarboxylate (as its 3,4-hydrate) into 4-hydroxymethyl-6,7-dimethyl-3,4-dihydro-2(1*H*)-pteridinone (**70**) (sodium borohydride in methanol; note the additional nuclear reduction).[1440]

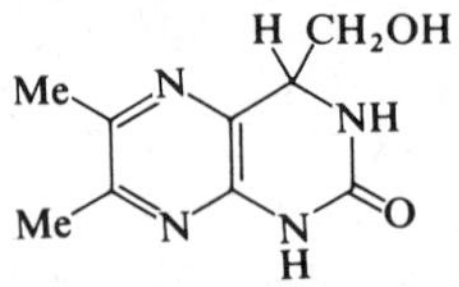

(70)

C. Hydrolysis

The hydrolysis of esters to carboxylic acids has been covered thoroughly in Section 1.C.

D. Transesterification

Existing cases of transesterification have been detailed in Section 3.B.

E. Aminolysis

The formation of amides or hydrazides from esters is illustrated in the conversion of methyl 2,4-diamino-6-pteridinecarboxylate into 2,4-diamino-6-pteridinecarboxamide (ammonia passed through a stirred suspension in ethylene glycol at 100°C for 45 min: 60%)[1354]; of ethyl 4-amino-7-oxo-2-phenyl-7,8-dihydro-6-pteridinecarboxylate (**71**, R = OEt) into 4-amino-7-oxo-2-phenyl-7,8-dihydro-6-pteridinecarboxamide (**71**, R = NH_2) (ammonia through ethanolic ester under reflux for 1 h: 20%), its *N*-ethyl derivative (**71**, R = NHEt) (ester plus ethanolic ethylamine under reflux for 2.5 h), and the *N*-*β*-dimethylaminoethyl derivative (**71**, R = $NHCH_2CH_2NMe_2$) (ester plus *β*-dimethylaminoethylamine in DMF at 25°C for 18 h: 40%)[1357]; of the same ester (**71**, R = OEt) into 2-amino-*N*′-methyl-7-oxo-2-phenyl-7,8-dihydro-6-pteridinecarbohydrazide (**71**, R = NHNHMe) [methylhydrazine in DMF at 5°C for 9 h and then at 25°C; the isomeric structure (**71**, R = $NMeNH_2$) was precluded by failure of the product to react with benzaldehyde][1357]; of ethyl 4,7-diamino-2-phenyl-6-pteridinecarboxylate into 4,7-diamino-*N*,*N*-dimethyl-2-phenyl-6-pteridinecarboxamide (dimethylamine gas into a refluxing solution of the ester in *β*-ethoxyethanol), its *N*-*β*-

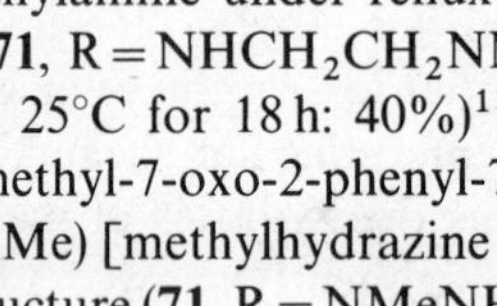

(71)

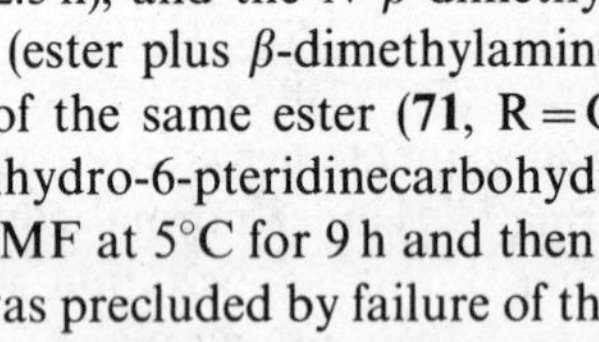

(72)

aminoethyl analog (ethylenediamine in DMF at 25°C for >18 h: 89%), and 10 other such analogs (similarly)[1357]; of ethyl 2,4,6-trioxo-1,2,3,4,5,6-hexahydro-7-pteridinecarboxylate into 2,4,6-trioxo-1,2,3,4,5,6-hexahydro-7-pteridinecarboxamide (ethanolic ammonia at 100°C in a sealed vessel for 1 h: 80%) and its *N*,*N*-dimethyl derivative (neat dimethylamine at 100°C in a sealed vessel for 2 h: ~80%)[347]; of 3-β-ethoxycarbonylethyl- into 3-β-carbamoylethyl-6,7-dimethyl-2-methylthio-4(3*H*)-pteridinone (methanolic ammonia at <14°C for 15 h: 52%)[168]; of 8-β-hydroxyethyl-6-β-methoxycarbonylethyl- (**72**, R = OMe) into 6-β-hydrazinocarbonylethyl-8-β-hydroxyethyl-2,4,7(1*H*,3*H*,8*H*)-pteridinetrione (**72**, R = $NHNH_2$) (neat hydrazine hydrate at 25°C for 30 min: 60%)[1449]; and of the 8-β-D-ribityl analog of this substrate into 6-β-hydrazinocarbonylethyl-8-β-D-ribityl-2,4,7(1*H*,3*H*,8*H*)-pteridinetrione (similarly: 60%).[1449] Other examples are known.[1739]

5. PREPARATION OF PTERIDINECARBOXAMIDES

A. By Primary Syntheses

Although it may not be the most important route to pteridinecarboxamides, primary synthesis from pyrimidine (see Ch. II) or other heterocyclic intermediates (see Ch. III) has been used quite extensively. For instance, condensation of 5,6-diamino-4-pyrimidinecarboxamide or its *N*-ethyl derivative with biacetyl in ethanol gave 6,7-dimethyl-4-pteridinecarboxamide (**73**, R = H) or its *N*-ethyl derivative (**73**, R = Et) (~60%), respectively[157]; condensation of 5-nitroso-2-phenyl-4,6-pyrimidinediamine with α-cyano-*N*-cyclohexylacetamide in ethanolic ethoxide gave 4,7-diamino-*N*-cyclohexyl-2-phenyl-6-pteridinecarboxamide (**74**) in 92% yield (some 200 analogs were made similarly!)[321]; and treatment of [1,2,5]oxadiazolo[3,4-*d*]pyrimidin-7-amine 1-oxide (**75**) with α-acetyl-*N*,*N*-dimethylacetamide in methanolic ammonia under reflux gave 4-

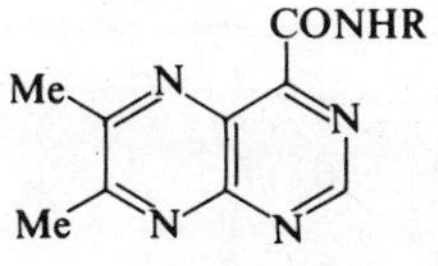

(**73**)

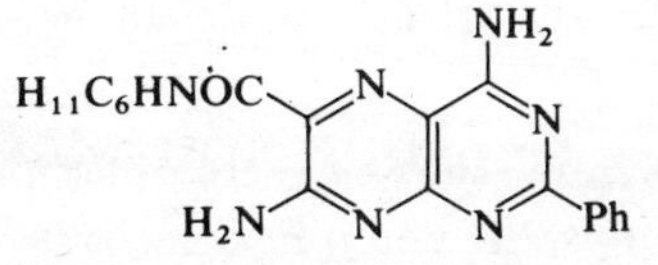

(**74**)

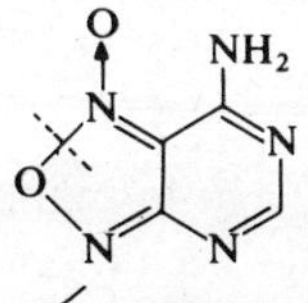

(**75**)

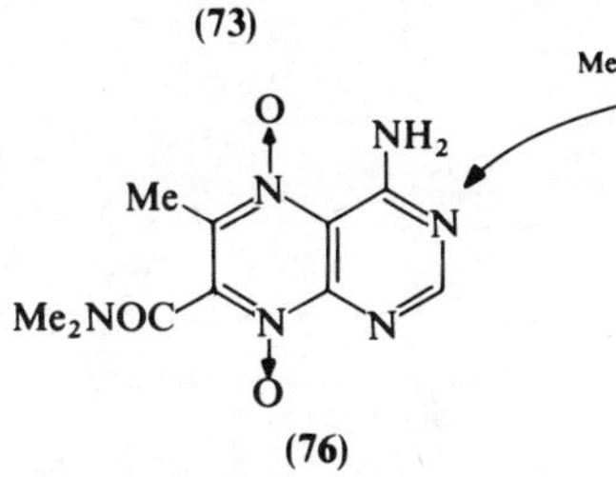

(**76**)

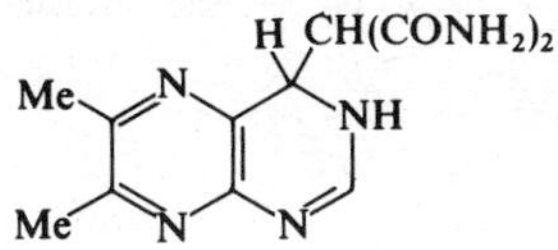

(**77**)

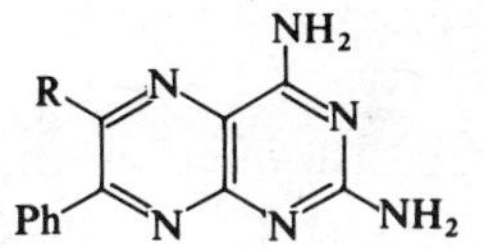

(**78**)

amino-6,*N*,*N*-trimethyl-7-pteridinecarboxamide 5,8-dioxide (**76**) in 30% yield.[723]

B. By Covalent Addition

The formation of extranuclear pteridine amides by covalent addition is represented by that of 4-dicarbamoylmethyl-6,7-dimethyl-3,4-dihydropteridine (**77**) (29%), from 6,7-dimethylpteridine and malondiamide in water for 5 days.[50]

C. By Hydration of Nitriles

Although controlled hydrolysis of pteridinecarbonitriles to pteridinecarboxamides has been little used, it is clearly quite possible. Thus heating 2,4-diamino-7-phenyl-6-pteridinecarbonitrile (**78**, R = CN) in concentrated sulfuric acid at 100°C for 15 min gave the corresponding carboxamide (**78**, $R = CONH_2$) in >90% yield[327]; 2-cyanomethyl-6-phenyl-4,7-pteridinediamine in sulfuric acid at 100°C for 1 h gave 2-carbamoylmethyl-6-phenyl-4,7-pteridinediamine (51%)[523]; and treatment of the HCN adduct of 2-amino-7,8-dihydro-4(3*H*)-pteridinone, presumably 2-amino-4-oxo-3,4,5,6,7,8-hexahydro-6-pteridinecarbonitrile, with a limited quantity of alkaline permanganate gave 2-amino-4-oxo-3,4,7,8-tetrahydro-6-pteridinecarboxamide (49%), possibly a rudimentary Radziszewski reaction.[453, cf. 925] There appears to be some parallel to the last reaction in the confusing series of changes reported to occur on treatment of 2-amino-4,6(3*H*,5*H*)-pteridinedione or its 7,8-dihydro derivative with potassium cyanide–ammonia in the presence of air to afford 2-amino-4,6-dioxo-3,4,5,6-tetrahydro-7-pteridinecarboxamide inter alia.[514, cf. 616] Treatment of 2-amino-5,6,7,7-tetramethyl-4-oxo-3,4,5,6,7,8-hexahydro-6-pteridinecarbonitrile with aqueous ethanolic hydrogen chloride at 25°C for 15 h gave the corresponding carboxamide as hydrochloride (~25%).[644]

D. By Aminolytic Reactions

The formation of amides from esters, acid chlorides, or (indirectly) from acids has been discussed in Sections 2.D and 4.E.

E. By Other Means

The insertion of a side chain already bearing a passenger carbamoyl group is illustrated by the *N*-alkylation of 6,7-dimethyl-2,4(1*H*,3*H*)-pteridinedione by acrylamide in aqueous pyridine to afford 1,3-bis-*β*-carbamoylethyl-6,7-dimethyl-2,4(1*H*,3*H*)-pteridinedione in 38% yield.[168]

6. REACTIONS OF PTERIDINECARBOXAMIDES

Pteridinecarboxamides undergo few reactions. Their *hydrolysis* to carboxylic acids has been covered in Section 1.C in this chapter. Their simple *dehydration to nitriles* is represented only by the conversion of 4-pteridinecarboxamide (**79**) into 4-pteridinecarbonitrile (**80**) in 1% yield by boiling in propionic anhydride for 5 min.[157] The *aminolysis of thioamides* has been exemplified in treatment of 4-amino-*N*-β-methoxyethyl-2-phenyl-7-thioxo-7,8-dihydro-6-pteridinecarbothioamide (**81**) with refluxing butylamine for 10 h to give 4-amino-*N*-butyl-7-butylamino-*N*′-β-methoxyethyl-2-phenyl-6-pteridinecarboxamidine (**82**) (note the additional aminolysis of the thioxo function; analogs were made similarly).[1162] The polarographic *half-wave potentials* for large series of 4- amino-2-aryl-7,*N*-disubstituted-6-pteridinecarboxamides (**83**) and 4,7-diamino-2-aryl-*N*-substituted-6-pteridinecarboxamides (**83**, X$=$$NH_2$) have been reported,[1113] possibly in connection with structure–activity studies of their diuretic properties. Conversion of pteridinecarboxamides into pteridinamines by the *Hofmann reaction* has been covered in Chapter IX, Section 1.E.

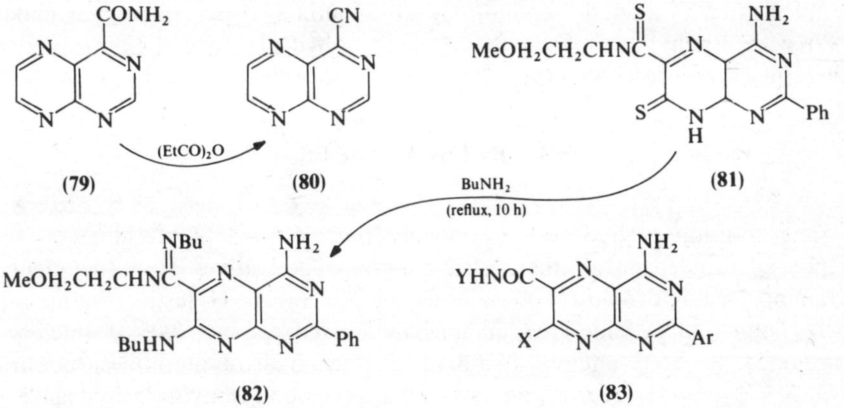

(**79**) (**80**) (**81**) (**82**) (**83**)

7. PREPARATION OF PTERIDINECARBONITRILES

A. By Primary Syntheses

Although relatively few pteridinecarbonitriles have been made by primary syntheses discussed in Chapters II and III, several of the available methods have been so used. For example, condensation of 5,6-diamino-2,4(1*H*,3*H*)-pyrimidinedione hydrochloride with *p*-cyanophenylglyoxal in ethanol gave 7-*p*-cyanophenyl-2,4(1*H*,3*H*)-pteridinedione (**84**)[1287, cf. 367]; condensation of 6-isopropoxy-5-nitroso-2,4-pyrimidinediamine with malononitrile in glycol at 100°C gave 2,7-diamino-4-isopropoxy-6-pteridinecarbonitrile (**85**) in

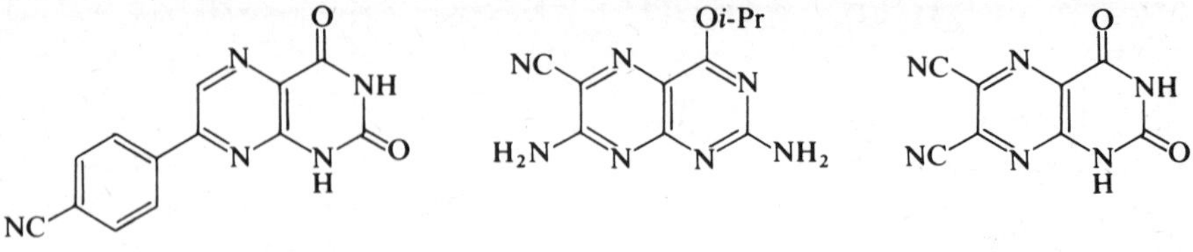

(84) (85) (86)

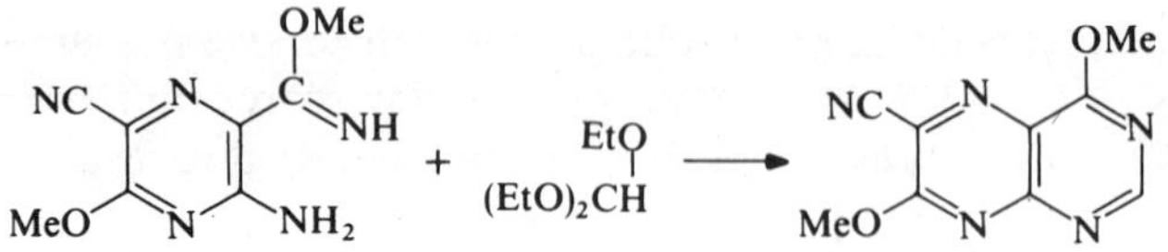

(87) (88)

~50% yield[371]; condensation of alloxan with diaminomaleonitrile [$H_2N(NC)C{:}C(CN)NH_2$] in acetic acid containing boric acid at 100°C gave 2,4-dioxo-1,2,3,4-tetrahydro-6,7-pteridinedicarbonitrile (**86**) in 40% yield[1198]; and condensation of methyl 3-amino-6-cyano-5-methoxy-2-pyrazinecarboximidate (**87**) with triethyl orthoformate in acetic anhydride gave 4,7-dimethoxy-6-pteridinecarbonitrile (**88**) (96%).[1136]

B. By Covalent Addition

The addition of hydrogen cyanide to the 5,6-bond of pteridines is well represented, at least in a rather restricted series of derivatives. Thus, treatment of 2-amino-7,8-dihydro-4(3*H*)-pteridinone (**89**, X=Y=Z=H) (as its bisulfite salt) with potassium cyanide gave an uncharacterized product (90%), which was presumed to be 2-amino-4-oxo-3,4,5,6,7,8-hexahydro-6-pteridinecarbonitrile (**90**, X=Y=Z=H) because on mild alkaline oxidative hydrolysis it gave 2-amino-4-oxo-3,4,7,8-tetrahydro-6-pteridinecarboxamide (**89**, X=$CONH_2$, Y=Z=H).[453] Strangely enough, the homologous substrate, 2-amino-6-methyl-7,8-dihydro-4(3*H*)-pteridinone (**89**, X=Me, Y=Z=H) (as hydrochloride), in which the 6-methyl group might be expected to impede addition, gave with potassium cyanide a more amenable product (74%), which was characterized as 2-amino-6-methyl-4-oxo-3,4,5,6,7,8-hexahydro-6-pteridinecarbonitrile (**90**,

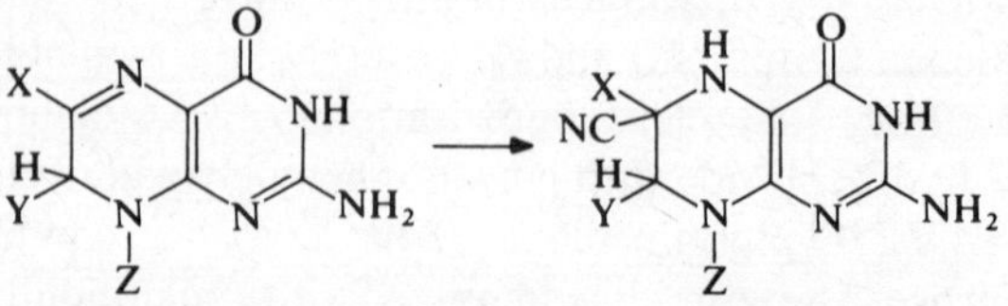

(89) (90)

X=Me, Y=Z=H).[1340] The same product was made subsequently in 94% yield by heating the dry substrate as free base in neat acetone cyanohydrin at 110°C for 1 h.[644] Likewise, 2-amino-6,7-dimethyl-7,8-dihydro-4(3*H*)-pteridinone (**89**, X=Y=Me, Z=H) with potassium cyanide gave 2-amino-6,7-dimethyl-4-oxo-3,4,5,6,7,8-hexahydro-6-pteridinecarbonitrile (**90**, X=Y=Me, Z=H) (84%)[1337]; 2-amino-8-methyl-7,8-dihydro-4(3*H*)-pteridinone (**89**, X=Y=H, Z=Me) gave 2-amino-8-methyl-4-oxo-3,4,5,6,7,8-hexahydro-6-pteridinecarbonitrile (**90**, X=Y=H, Z=Me) in ~60% yield, characterized as its hydrochloride[684]; and 2-amino-5,6,7,7-tetramethyl-4-oxo-3,4,7,8-tetrahydro-5-pteridinium *p*-toluenesulfonate with potassium cyanide in boiling acetonitrile gave 2-amino-5,6,7,7-tetramethyl-4-oxo-3,4,5,6,7,8-hexahydro-6-pteridinecarbonitrile (crude: ~60%).[644]

There are also some examples of the formation of extranuclear nitriles by the addition of a Michael-type reagent bearing a cyano group. Thus 2(1*H*)- and 6(5*H*)-pteridinone reacted with ethyl cyanoacetate to afford 4-(α-cyano-α-ethoxycarbonylmethyl)-3,4-dihydro-2(1*H*)-pteridinone (90%)[38] and 7-(α-cyano-α-ethoxycarbonylmethyl)-7,8-dihydro-6(5*H*)-pteridinone,[55] respectively. On prolonged boiling in water, the latter underwent hydrolysis and decarboxylation to give 7-cyanomethyl-7,8-dihydro-6(5*H*)-pteridinone.[55]

C. By Other Means

2-Amino-4-oxo-3,4-dihydro-6-pteridinecarboxamide in thionyl chloride–DMF gave 2-dimethylaminomethyleneamino-4-oxo-3,4-dihydro-6-pteridinecarbonitrile (47%).[1739] Somewhat similarly, 2-acetamido-6-hydroxyiminomethyl-4(3*H*)-pteridinone in acetic anhydride gave 2-acetamido-4-oxo-3,4-dihydro-6-pteridinecarbonitrile (70%).[1739]

Other routes to extranuclear nitriles are suggested in the *β*-cyanoethylation of appropriate 6- and/or 7-substituted 2,4(1*H*,3*H*)-pteridinediones or related substrates by prolonged boiling with acrylonitrile in aqueous pyridine to give, 1,3-bis-*β*-cyanoethyl-6-methyl-2,4(1*H*,3*H*)-pteridinedione (**91**, X=Me, Y=H) (50%), 1,3-bis-*β*-cyanoethyl-6,7-dimethyl-2,4(1*H*,3*H*)-pteridinedione (**91**, X=Y=Me) (72%), 1,3-bis-*β*-cyanoethyl-2,4-dioxo-1,2,3,4-tetrahydro-6-pteridinecarboxylic acid (**91**, X=CO_2H, Y=H) (52%), the 2-oxo analog of 1,3-bis-*β*-cyanoethylpteroylglutamic acid (84%), and 4-amino-1-*β*-cyanoethyl-6-methyl-2(1*H*)-pteridinone (50%).[168] A similar process gave 3-*β*-cyanoethyl-

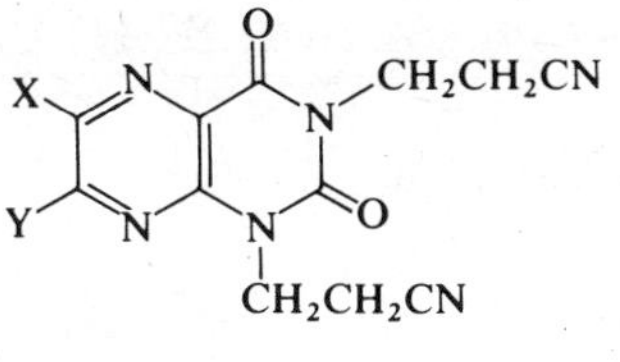

(91)

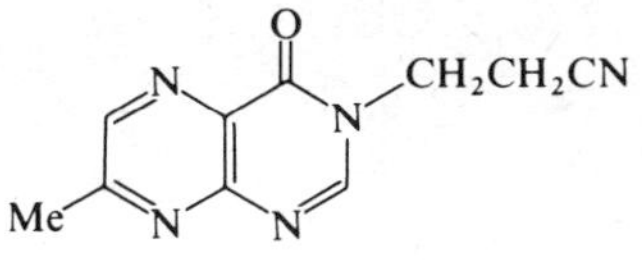

(92)

7-methyl-4(3*H*)-pteridinone (**92**) (32%), accompanied in this case by a separable degradation product, *N*-*β*-cyanoethyl-3-formamido-5-methyl-2-pyrazinecarboxamide (10%).[167]

The pseudoacetal, 4,4-dimethoxy-1,3-dimethyl-6,7-diphenyl-3,4-dihydro-2(1*H*)-pteridinone (**93**) (prepared by methanolysis of the corresponding 4-thioxopteridine as outlined in Ch. VIII, Sect. 2.F), reacted with malononitrile in refluxing dioxane for 1 h to afford 4-dicyanomethylene-1,3-dimethyl-6,7-diphenyl-3,4-dihydro-2(1*H*)-pteridinone (**94**) in 47% yield.[1679] 4-Pteridinamine has been converted indirectly into 4-cyanoaminopteridine.[135]

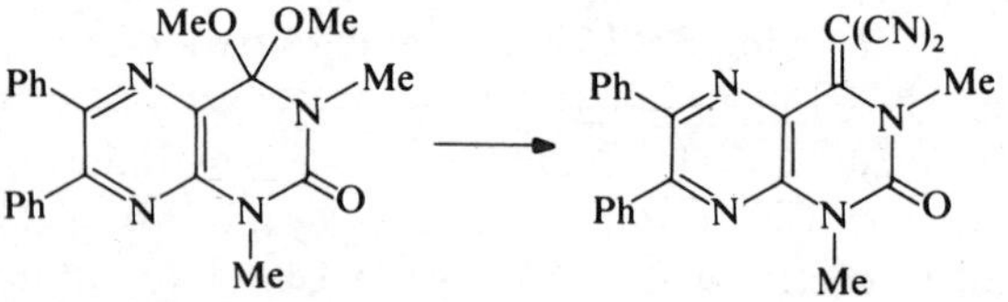

(93) **(94)**

8. REACTIONS OF PTERIDINECARBONITRILES

Almost all of the few reported reactions of nitriles in the pteridine series have been discussed already: for *reduction* (to aminomethyl derivatives), see Chapter IX, Section 3.B; for *hydrolysis* (to amides or acids), see Section 5.C in this chapter. The *alcoholysis* of a nitrile has been represented by the reaction of 2-amino-8-methyl-4-oxo-3,4,7,8-tetrahydro-6-pteridinecarbonitrile with ethanolic sodium ethoxide at 25°C to give ethyl 2-amino-8-methyl-4-oxo-3,4,7,8-tetrahydro-6-pteridinecarboximidate (**95**, X=NH) (~70%)[684]; brief treatment with warm dilute hydrochloric acid gave the corresponding 6-pteridinecarboxylate (**95**, X=O) in 70% yield.[684]

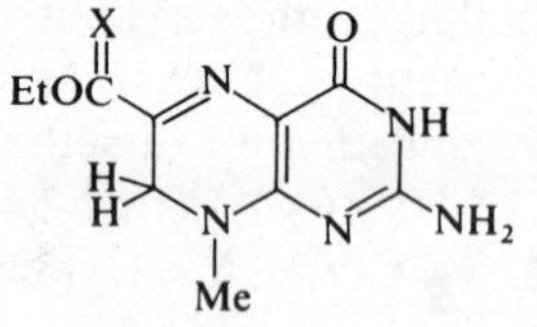

(95)

9. PREPARATION OF PTERIDINECARBALDEHYDES

A. By Primary Syntheses

For fairly evident reasons, the main types of primary synthesis (Chs. II and III) have proven inapplicable to the direct formation of pteridinecarbaldehydes.

However, the occasional derivative of an aldehyde has been so made: for example, 3-amino-6-dimethoxymethyl-2-pyrazinecarbonitrile in refluxing methanolic guanidine for 18 h gave 6-dimethoxymethyl-2,4-pteridinediamine (84%)[1306]; 3-amino-6-hydroxyiminomethyl-2-pyrazinecarbonitrile 3-oxide (**96**) in refluxing ethanolic guanidine for 16 h gave 6-hydroxyiminomethyl-2,4-pteridinediamine 8-oxide (**97**) (89%)[484]; and 6-allylamino-1,3-dimethyl-2,4(1*H*,3*H*)-pyrimidinedione (**98**) reacted with an excess of nitrous acid to afford a 33% yield of 6-hydroxyiminomethyl-1,3-dimethyl-2,4(1*H*,3*H*)-pteridinedione (**100**), probably via the diisonitroso intermediate (**99**).[196] The last reaction does not appear to have general applicability.

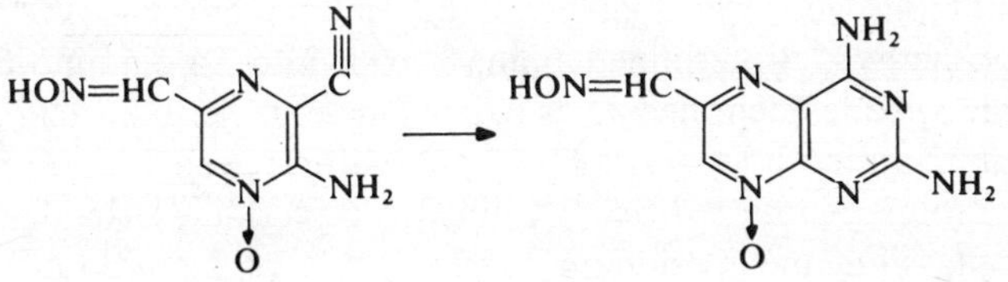

(**96**) (**97**)

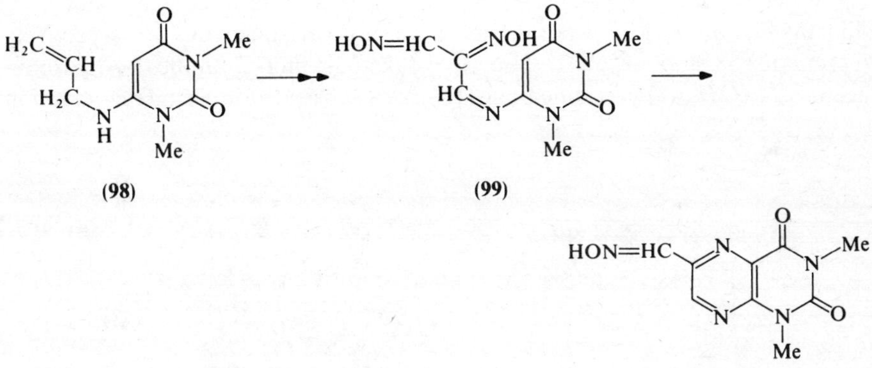

(**98**) (**99**) (**100**)

B. By Oxidative Means

The direct oxidation of *methylpteridines* by selenium dioxide to give pteridinecarbaldehydes has been covered in Chapter IV, Section 2.B(6).[940, 1061] An interesting indirect route involved the sequence 1,7-dimethyl- → 7-bromomethyl-1-methyl- → 1-methyl-7-pyridiniomethyl-2,4(1*H*,3*H*)-pteridinedione (as Br^-) (**101**), followed by reaction with *N*,*N*-dimethyl-*p*-nitrosoaniline in aqueous methanolic potassium carbonate to give an intermediate nitrone, 7-*p*-dimethylaminophenyliminomethyl-1-methyl-2,4(1*H*,3*H*)-pteridinedione *N*-oxide (**102**) (80%), which on shaking with dilute sulfuric acid gave 1-methyl-2,4-

dioxo-1,2,3,4-tetrahydro-7-pteridinecarbaldehyde (**103**, R = H) (60%).[232] The corresponding 6,7-dicarbaldehyde (**103**, R = CHO) was made similarly.[232]

The oxidation of *pteridine alcohols* to aldehydes has been discussed in Chapter VI, Section 4.B but hitherto unmentioned examples include the conversion of 2-amino-6-hydroxymethyl-7,7-dimethyl-7,8-dihydro-4(3*H*)-pteridinone (**104**, R = CH_2OH) into 2-amino-7,7-dimethyl-4-oxo-3,4,7,8-tetrahydro-6-pteridinecarbaldehyde (**104**, R = CHO) (by passing a mixture of oxygen and sulfur dioxide into an aqueous solution of the substrate: the resulting bisulfite gave the free aldehyde in 63% yield)[1501]; of 2-amino-6-hydroxymethyl-7-methyl-7-phenethyl-7,8-dihydro-4(3*H*)-pteridinone into 2-amino-7-methyl-4-oxo-7-phenethyl-3,4,7,8-tetrahydro-6-pteridinecarbaldehyde (by stirring in aqueous butanolic acetic acid open to the air for 15 h: 80%)[1501]; of 6-α,β-dihydroxyphenethyl-2,4-pteridinediamine 8-oxide into 2,4-diamino-6-pteridinecarbaldehyde 8-oxide, identified as its oxime (aqueous periodic acid at 25°C for 15 h)[484]; of 2-amino-6-[D-*erythro*]-α,β,γ-trihydroxypropyl-4(3*H*)-pteridinone (neopterin) into 2-amino-4-oxo-3,4-dihydro-6-pteridinecarbaldehyde (**105**) (2 mol of potassium metaperiodate at pH 3)[1422]; of 2-amino-6-[D-*arabino*]-α,β,γ,δ-tetrahydroxybutyl-4(3*H*)-pteridinone into the same product (**105**) (periodate: 82%)[192, 1060]; and of 1-methyl- or 1,3-dimethyl-6-[D-*arabino*]-α,β,γ,δ-tetrahydroxybutyl-2,4(1*H*,3*H*)-pteridinedione into 1-methyl- or 1,3-dimethyl-2,4-dioxo-1,2,3,4-tetrahydro-6-pteridinecarbaldehyde, respectively (periodate: 58 and 64%).[962] Use of the ketoalcohol, 2-amino-7-α-hydroxyacetonyl-4,6(3*H*,5*H*)-pteridinedione (**106**), as a substrate for periodate oxidation

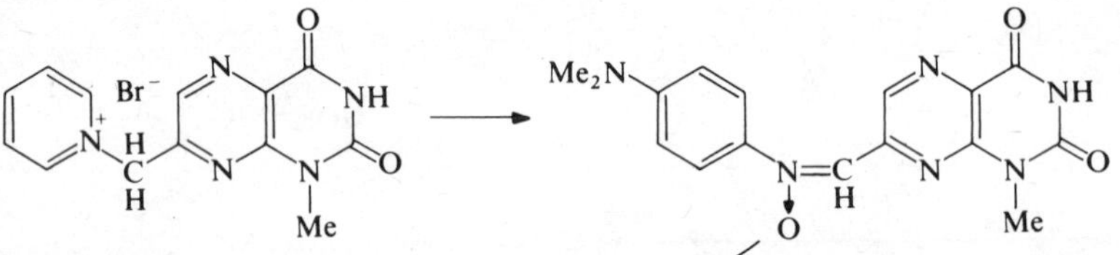

(101) (102)

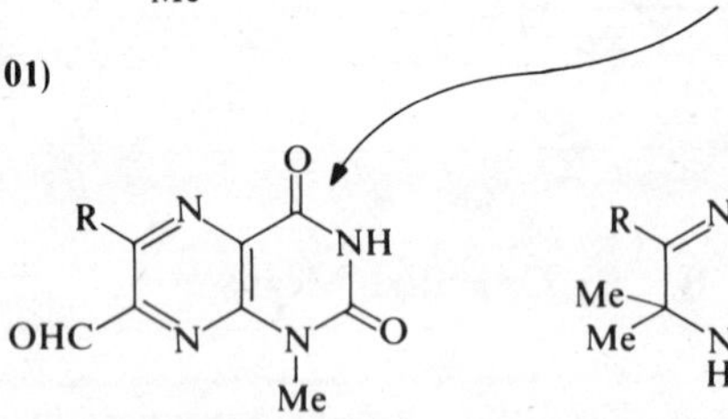

(103)

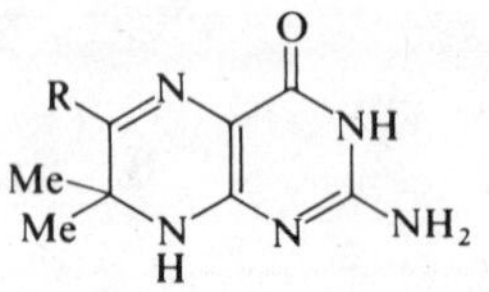

(104)

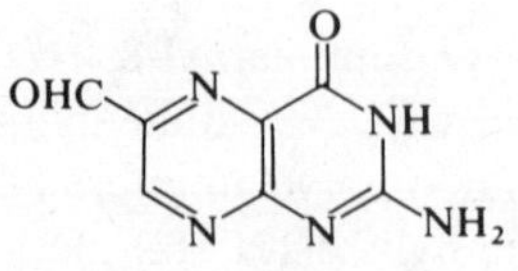

(105)

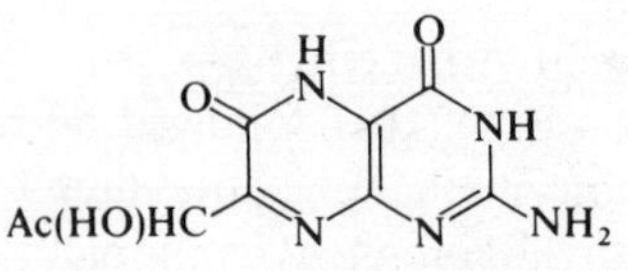

(106)

gave 2-amino-4,6-dioxo-3,4,5,6-tetrahydro-7-pteridinecarbaldehyde, which was characterized by permanganate oxidation to the corresponding carboxylic acid.[1313]

C. By Cleavage of Pteroylglutamic Acid, and so on

The cleavage of pteroylglutamic acid to yield 2-amino-4-oxo-3,4-dihydro-6-pteridinecarbaldehyde (**105**) was first done by heating in aqueous sodium hydrogen sulfite containing free acetic acid at 90°C for 13 h and resulted in a quite low yield after purification.[1358] It later proved more effective to carry out the cleavage by heating in an excess of hydrobromic acid containing free bromine to give the aldehyde in 58% yield after purification.[1270] Not surprisingly, iodination of pteroylglutamic acid by treatment with iodine monochloride in DMF gave, not only the required 3′-iodopteroylglutamic acid (65%), but also an appreciable amount of the aldehyde (**105**).[1104] The similar iodination of 4-amino-4-deoxypteroylglutamic acid likewise gave both the 3′-iodo derivative (38%) and 2,4-diamino-6-pteridinecarbaldehyde (52% yield).[1107] The aldehyde (**105**) was also formed among the products from irradiation of aqueous pteroylglutamic acid in air with visible light.[899]

D. From Dibromomethylpteridines

The conversion of dibromomethylpteridines into the corresponding pteridinecarbaldehydes has been covered in Chapter V, Section 4.C.

E. From Acetals or Other Derivatives

The recovery of free aldehydes from their acetals, oximes, and so on, is sometimes necessary or useful. Thus 2-amino-7-dimethoxymethyl-4(3*H*)-pteridinone (**107**) was briefly heated in 1*M* hydrochloric acid to give 2-amino-4-oxo-3,4-dihydro-7-pteridinecarbaldehyde (**108**) in >95% yield.[1310] The isomeric substrate, 2-amino-6-dimethoxymethyl-4(3*H*)-pteridinone, gave 2-amino-4-oxo-3,4-dihydro-6-pteridinecarbaldehyde (91%) by dissolution in 90% formic acid at 25°C for 30 min.[1306] 2-Acetamido-6-diethoxymethyl-4(3*H*)-pteridinone gave 2-acetamido-4-oxo-3,4-dihydro-6-pteridinecarbaldehyde (95%) by dissolution in 88% formic acid for 2 h.[1300, cf. 464] 2-amino-6-diethoxymethyl-7,8-dihydro-4(3*H*)-pteridinone (**109**) gave 2-amino-4-oxo-3,4,7,8-tetrahydro-6-pteridinecarbaldehyde (**111**) by hydrolysis in 0.1*M* hydrochloric acid.[1393] Finally, 2-amino-6-hydroxyiminomethyl-4(3*H*)-pteridinone 8-oxide (**110**) (prepared by primary synthesis) gave 2-amino-4-oxo-3,4-dihydro-6-pteridinecarbaldehyde (**112**) by initial reductive deoxygenation and oxime removal by sodium hydrogen sulfite to give (presumably) the dihydro aldehyde

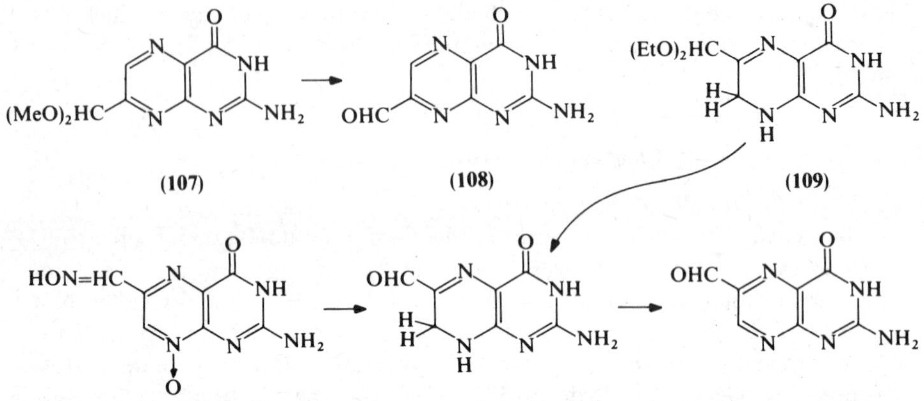

(107) (108) (109)

(110) (111) (112)

(**111**) followed by mild iodine oxidation of this intermediate. The yield of product (**112**) was 50% overall.[479]

10. REACTIONS OF PTERIDINECARBALDEHYDES

A. Oxidation and Reduction

The oxidation of pteridinecarbaldehydes to the corresponding acids has been covered in Section 1.B in this chapter; the reduction of aldehydes to hydroxymethylpteridines or to a pinacol type of dimeric molecule has been discussed in Chapter VI, Section 3.B. Background polarographic studies on the reduction of 2-amino-4-oxo-3,4-dihydro-6-pteridinecarbaldehyde to a dimeric and/or regular product have been reported.[640, 989]

B. Conversion into Styryl-Type Compounds

Although styryl-type compounds may be made from methylpteridines and appropriate aldehyde reagents [see Ch. IV, Sect. 2.B(5)], it is also possible to make them in the reverse fashion from pteridinecarbaldehydes and appropriately activated methylene reagents. Thus 1,3-dimethyl-2,4,7-trioxo-1,2,3,4,7,8-hexahydro-6-pteridinecarbaldehyde with malonic acid in boiling pyridine containing piperidine gave (after spontaneous decarboxylation) 6-β-carboxyvinyl-1,3-dimethyl-2,4,7(1*H*,3*H*,8*H*)-pteridinetrione.[337] Unfortunately, the methyl group of toluene, for example, is insufficiently active to react with pteridine aldehydes directly but this has been overcome by employing a Wittig reagent to achieve the same purpose. For example, treatment of diethyl *p*-chlorobenzylphosphonate in dimethyl sulfoxide with sodium hydride followed by the addition of 2-amino-7,7-dimethyl-4-oxo-3,4,7,8-tetrahydro-6-

pteridinecarbaldehyde (**113**) afforded 2-amino-6-*p*-chlorostyryl-7,7-dimethyl-7,8-dihydro-4(3*H*)-pteridinone (**114**, R=Cl) in 67% yield.[1503] The 6-*p*-methoxystyryl analog (**114**, R=OMe) and the 6-(3′,4′-methylenedioxystyryl) analog were made similarly using appropriate Wittig reagents.[1503]

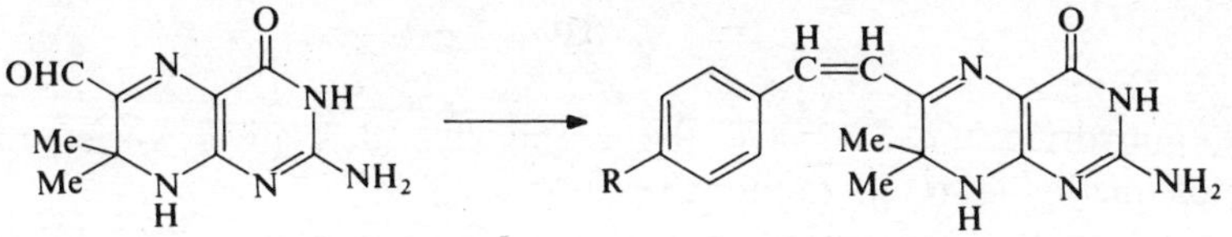

(**113**) (**114**)

C. Conversion into Schiff Bases

The formation of Schiff bases from pteridinamines and aldehyde or ketone reagents has been discussed in Chapter IX, Section 2.E; the reverse process, condensation of pteridinecarbaldehydes with primary amino reagents to give Schiff bases or hydrazones, has been largely covered in Chapter IX, Section 3.C. Another typical example of this process is the condensation of 2-amino-4-oxo-3,4-dihydro-7-pteridinecarbaldehyde with ethyl *p*-aminobenzoate in hot DMF to give 2-amino-7-*p*-ethoxycarbonylphenyliminomethyl-4(3*H*)-pteridinone (**115**) in 93% yield.[1310]

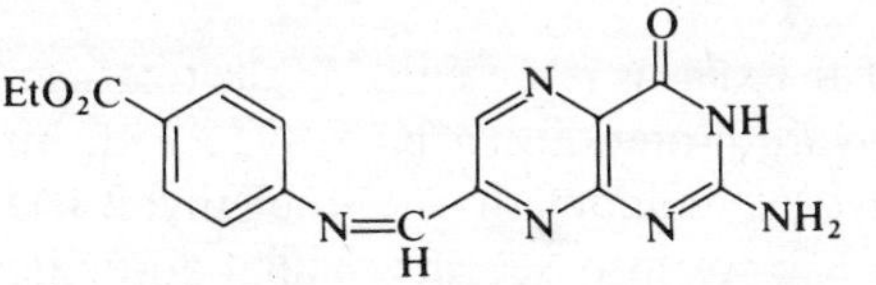

(**115**)

D. Formation of Other Functional Derivatives

Most known *acetals* in the pteridine series have in fact been intermediates for, rather than derivatives of, pteridinecarbaldehydes. However, 2-amino-4-oxo-3,4-dihydro-6-pteridinecarbaldehyde was converted into its monothiohemiacetal, 2-amino-6-α-hydroxy-α-(methylthio)methyl-4(3*H*)-pteridinone (**116**), by heating with methanethiol in concentrated hydrochloric acid for 10 h (~60% yield).[503] 2-Amino-4,7-dioxo-3,4,7,8-tetrahydro-6-pteridinecarbaldehyde was likewise converted into 2-amino-6-α-hydroxy-α-(methylthio)methyl- and 2-amino-6-α-(β′,γ′-dihydroxypropylthio)-α-hydroxymethyl-4,7(3*H*,8*H*)-pteridinedione by using methanethiol and 3-mercapto-1,2-propanediol (α-thioglycerol), respectively.[503] Heating the same substrate in

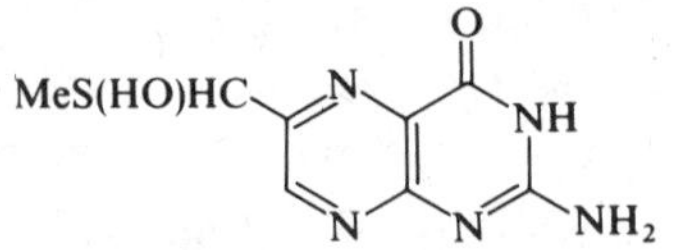

(116)

acetic anhydride for 3 h gave the acetal like derivative, 2-acetamido-6-diacetoxymethyl-4(3*H*)-pteridinone (~50% yield).[1451]

The formation of *semicarbazones* is typified in the conversion of 2,4-diamino-6-pteridinecarbaldehyde into 6-semicarbazonomethyl- (**117**, X=O) or 6-thiosemicarbazonomethyl-2,4-pteridinediamine (**117**, X=S) (semicarbazide or thiosemicarbazide in hot very dilute acid for 15 min: each was characterized as its hydrochloride)[977]; the similar formation of 2-amino-6-semicarbazonomethyl- and 2-amino-6-thiosemicarbazonomethyl-4(3*H*)-pteridinone (in almost neutral solution: characterized as free bases) also occurs.[977]

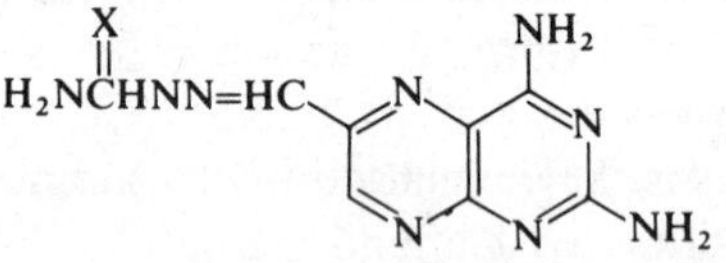

(117)

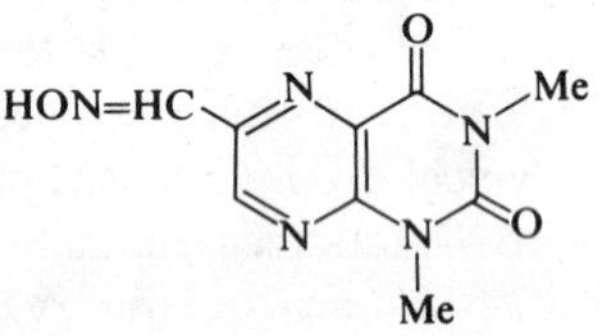

(118)

The formation of an *oxime* is represented in the treatment of 1,3-dimethyl-2,4-dioxo-1,2,3,4-tetrahydro-6-pteridinecarbaldehyde with hydroxylamine (no details) to give 6-hydroxyiminomethyl-1,3-dimethyl-2,4(1*H*,3*H*)-pteridinedione (**118**) in 71% yield[196]; we also see the similar formation of 2-acetamido-6-hydroxyiminomethyl-4(3*H*)-pteridinone (75%).[1739]

11. PREPARATION OF PTERIDINE KETONES

A. By Primary Syntheses

In contrast to the situation with aldehydes, many pteridine ketones have been made directly by primary syntheses, both from pyrimidine (Ch. II) and from other heterocyclic intermediates (Ch. III). Typical examples include the condensation of 2,5,6-triamino-4(3*H*)-pyrimidinone (**120**) with pentane-2,3,4-trione in dilute acid to give 6-acetyl-2-amino-7-methyl-4(3*H*)-pteridinone (**119**),[1217] or with ethyl acetylpyruvate at pH 1 for 30 min to give 7-acetonyl-2-amino-4,6(3*H*,5*H*)-pteridinedione (**121**) (63% yield) plus a separable purine by-product[712]; of 5-nitroso-2-phenyl-4,6-pyrimidinediamine with benzoylacetonitrile in refluxing ethanolic sodium cyanide to give 6-benzoyl-2-phenyl-4,7-

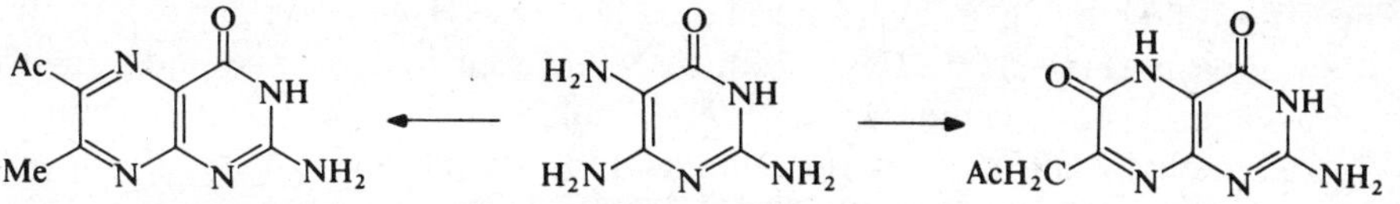

(119) (120) (121)

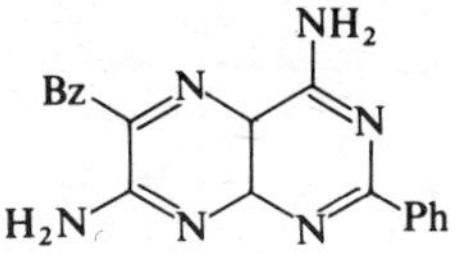

(122)

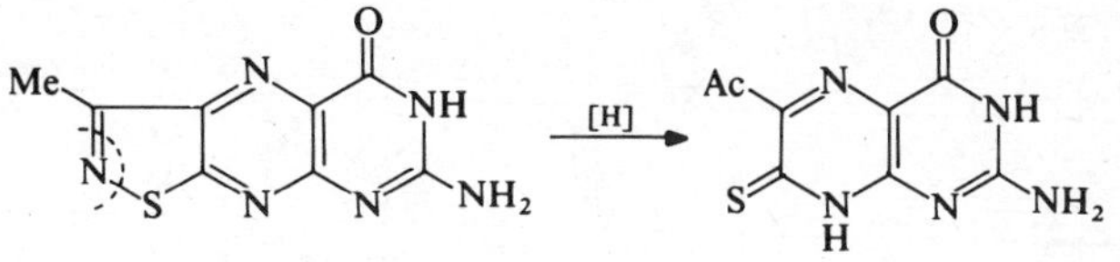

(123) (124)

pteridinediamine (**122**) (>90% yield)[328]; and of 2,6-diamino-4(3*H*)-pyrimidinone with 5-acetamido-3-methyl-4-nitrosoisothiazole to give 7-amino-3-methyl-5(6*H*)-isothiazolo[4,5-*g*]pteridinone (**123**) which, on dithionite reductive fission, gave 6-acetyl-2-amino-7-thioxo-7,8-dihydro-4(3*H*)-pteridinone (**124**) in ~80% yield.[1309]

B. By Covalent Addition

The formation of pteridine ketones by covalent additions is illustrated in the conversion of 7(8*H*)-pteridinone into 6-diacetylmethyl-5,6-dihydro-7(8*H*)-pteridinone (**125**) (aqueous acetylacetone at 20°C for 90 min: 65%)[44]; of 2(1*H*)-pteridinone into 4-diacetylmethyl-3,4-dihydro-2(1*H*)-pteridinone (acetylacetone in aqueous sodium hydrogen carbonate at 25°C for 5 h: 75%; deacetylation occurred in 1*M* alkali at 20°C for 1 day to give 4-acetonyl-3,4-dihydro-2(1*H*)-pteridinone: 95%)[38]; of 2(1*H*)-pteridinethione into 4-diacetylmethyl-3,4-dihydro-2(1*H*)-pteridinethione (aqueous acetylacetone at 25°C for 7 days: 68%)[46]; of pteridine into 4-α-ethoxycarbonylphenacyl-3,4-dihydropteridine (**126**) (ethyl benzoylacetate in THF at 20°C for 5 days: 20%)[49]; and of 2-methyl- or 6,7-dimethylpteridine into 4-(4′,4′-dimethyl-2′,6′-dioxocyclohexyl)-2-methyl- (**127**, X=Me, Y=H) or 4-(4′,4′-dimethyl-2′,6′-dioxocyclo-

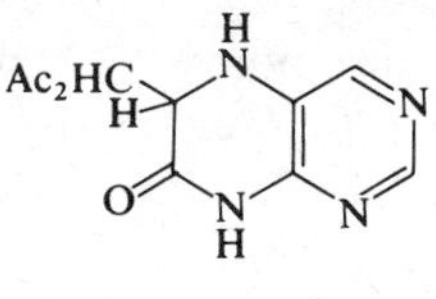

(125)

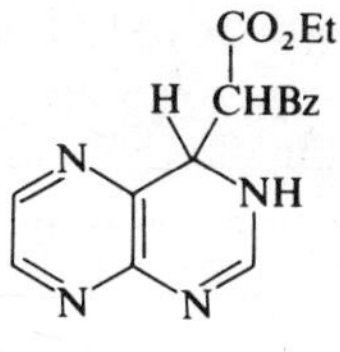

(126)

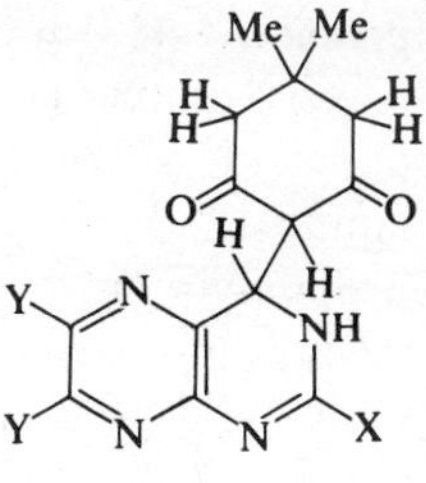

(127)

hexyl)-6,7-dimethyl-3,4-dihydropteridine (**127**, X=H, Y=Me), respectively (dimedone in THF at 25°C for 15 h: 69% in each case).[50] The addition of pyruvic acid to 2-amino-4,6(3*H*,5*H*)-pteridinedione (**128**) gave 2-amino-7-oxalomethyl-4,6(3*H*,5*H*)-pteridinedione (**130**) via the aerial oxidation of an intermediate adduct (**129**, R=H).[515, 816, 952, 1479] Addition of oxaloacetic acid to the same substrate (**128**) gave the same final product (**130**) via aerial oxidation and spontaneous decarboxylation of an intermediate adduct (**129**, R=CO_2H).[439]

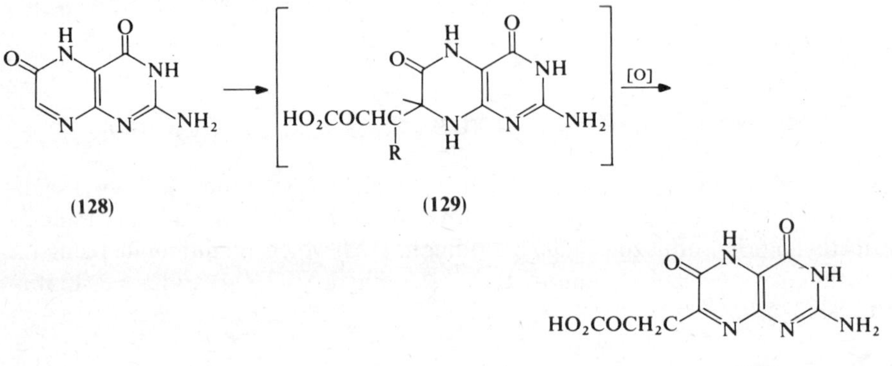

(**128**) (**129**) (**130**)

C. By Direct *C*-Acylation

The direct homolytic *C*-acylation of appropriate pteridines is possible by using a "radical nucleophilic substitution" procedure,[629] apparently often known as the Minisci reaction.[1735, 1736] For example, a stirred solution of 2,4(1*H*,3*H*)-pteridinedione and propionaldehyde in aqueous acetic acid below room temperature was treated slowly and simultaneously with aqueous ferrous sulfate and aqueous *t*-butyl hydroperoxide to give 7-propionyl-2,4(1*H*,3*H*)-pteridinedione (**131a**) in 37% yield.[108] This substrate and its 1- and/or 3-alkylated derivatives all underwent 7- rather than 6-acylation, but when there was an additional 7-substituent present, acylation occurred at the 6-position. The use of appropriate substrates and aldehydes thus afforded 1-methyl-7-propionyl- (**131b**) (40%), 3-methyl-7-propionyl- (**131c**) (24%), 1-phenyl-7-propionyl- (**131d**) (14%), 7-acetyl-1,3-dimethyl- (**132a**) [32%; plus a little of the 6-acetyl isomer (**133a**)], 1,3-dimethyl-7-propionyl- (**132b**) (30%), 7-butyryl-1,3-dimethyl- (**132c**) (53%), 1,3-dimethyl-7-phenethylcarbonyl- (**132d**) (25%), 6-amino-1,3-dimethyl-7-propionyl- (**132e**) (40%), 1,3-dimethyl-6-methylamino-7-propionyl- (**132f**) (46%), 6-dimethylamino-1,3-dimethyl-7-propionyl- (**132g**) (69%), 6-methoxy-1,3-dimethyl-7-propionyl- (**132h**) (31%), 6-chloro-1,3-dimethyl-7-propionyl- (**132i**) (crude: 62%; pure: 21%), 6-acetyl-1,3,7-trimethyl- (**133b**) (18%), 1,3,7-trimethyl-6-propionyl- (**133c**) (44%), 6-butyryl-1,3,7-trimethyl- (**133d**) (49%), 1,3,7-trimethyl-6-phenethylcarbonyl- (**133e**) (44%),

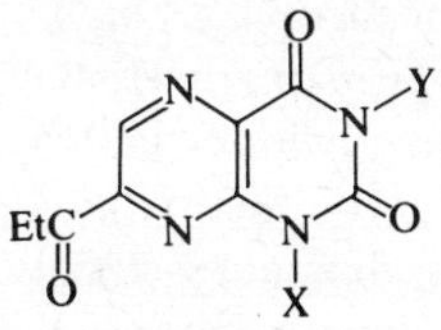

(131)

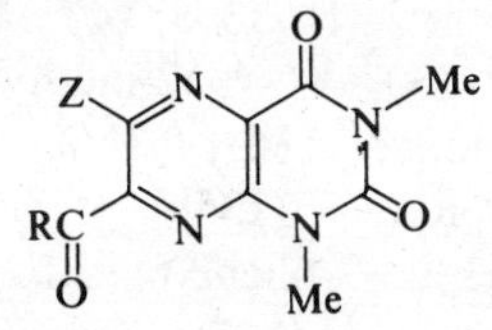

(132)

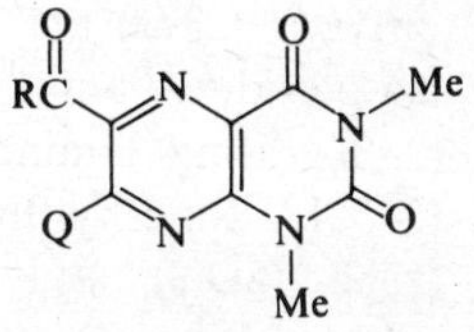

(133)

	X	Y
a	H	H
b	Me	H
c	H	Me
d	Ph	H

	R	Z
a	Me	H
b	Et	H
c	Pr	H
d	$PhCH_2CH_2$	H
e	Et	NH_2
f	Et	NHMe
g	Et	NMe_2
h	Et	OMe
i	Et	Cl

	R	Q
a	Me	H
b	Me	Me
c	Et	Me
d	Pr	Me
e	$PhCH_2CH_2$	Me
f	Et	Pr
g	Et	NH_2
h	Et	NHMe
i	Et	NMe_2
j	Et	OMe
k	Et	SMe

1,3-dimethyl-6-propionyl-7-propyl- (**133f**) (31%), 7-amino-1,3-dimethyl-6-propionyl- (**133g**) (32%), 1,3-dimethyl-7-methylamino-6-propionyl- (**133h**) (72%), 7-dimethylamino-1,3-dimethyl-6-propionyl- (**133i**) (71%), 7-methoxy-1,3-dimethyl-6-propionyl- (**133j**) (69%), and 1,3-dimethyl-7-methylthio-6-propionyl-2,4(1*H*,3*H*)-pteridinedione (**133k**) (68%).[108]

As might be expected, insertion of an electron-donating group into the 6- or 7-position of these substrates appears to have improved the yield of acylated product.

In like manner, appropriate 2,4-pteridinediamines and 2-methylthio-4-pteridinamines gave 7-acetyl- (**134a**) (30%), 7-propionyl- (**134b**) (55%), 7-

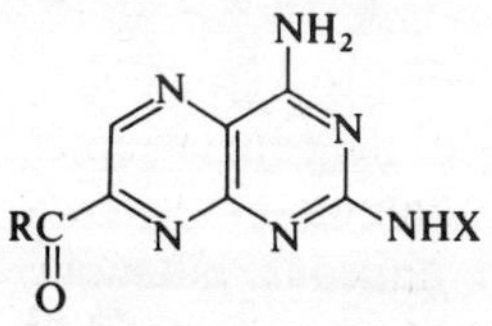

(134)

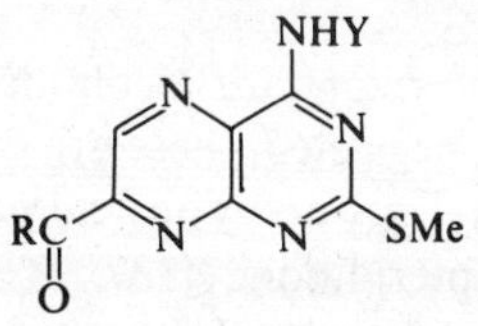

(135)

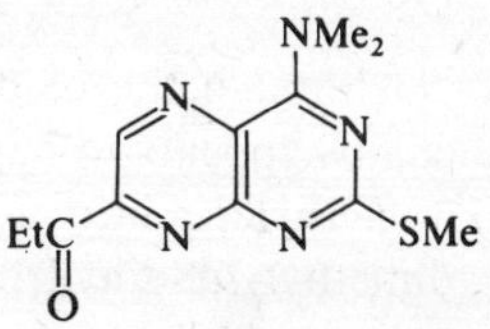

(136)

	R	X
a	Me	H
b	Et	H
c	Pr	H
d	*i*-Pr	H
e	$PhCH_2CH_2$	H
f	Et	Me
g	Et	Bu

	R	Y
a	Me	H
b	Et	H
c	Pr	H
d	$PhCH_2CH_2$	H
e	Et	Me

butyryl- (**134c**) (32%), 7-isobutyryl- (**134d**) (47%), and 7-phenethylcarbonyl-2,4-pteridinediamine (**134e**) (30%)[1678]; 2-methylamino-7-propionyl- (**134f**) (33%), 2-butylamino-7-propionyl- (**134g**) (51%), 7-acetyl-2-methylthio- (**135a**) (75%), 2-methylthio-7-propionyl- (**135b**) (56%), 7-butyryl-2-methylthio- (**135c**) (58%), and 2-methylthio-7-phenethylcarbonyl-4-pteridinamine (**135d**) (61%)[1678]; and 4-methylamino- (**135e**) (53%) and 4-dimethylamino-2-methylthio-7-propionylpteridine (**136**) (86%).[1678]

A similar method was used to produce 1,3-dimethyl-7-propionyl-2,4,6(1*H*,3*H*,5*H*)-pteridinetrione (14%)[108]; 1,3-dimethyl-6-propionyl-2,4,7(1*H*,3*H*,8*H*)-pteridinetrione (47%)[108]; 1- and 3-methyl-7-methylthio-6-propionyl-2,4(1*H*,3*H*)-pteridinedione (both 78%)[354]; 7-methoxycarbonylmethylthio-, 7-(ethoxycarbonylmethyl)amino-, and 7-*N*-(ethoxycarbonylmethyl)-*N*-methylamino-1,3-dimethyl-6-propionyl-2,4(1*H*,3*H*)-pteridinedione (no details)[629]; 6-methoxycarbonylmethylthio-1,3-dimethyl-7-propionyl-2,4(1*H*,3*H*)-pteridinedione[629]; 7-methylthio-6-propionyl-2,4-pteridinediamine (85%)[95]; and 4-pentyloxy-6-propionyl-7-propylthio-2-pteridinamine (79%).[95] Application of the same procedure to 1,3-dimethyl-2,4-dioxo- 1,2,3,4-tetrahydro-7-pteridinecarboxylic acid afforded 1,3-dimethyl-7-propionyl-2,4(1*H*,3*H*)-pteridinedione (50%), a reaction involving loss of carbon dioxide initially.[108] The 6-acylation of 2-amino-7,8-dihydro-4(3*H*)-pteridinone by α-oxobutyric acid in aqueous thiamine at 37°C occurred during 2 days in the dark to give 2-amino-6-propionyl-7,8-dihydro-4(3*H*)-pteridinone in up to 22% yield, presumably by a radical mechanism.[1127]

Acylation of a *C*-methyl group attached to a pteridine occurred in the Claisen reaction of 2-amino-3,5,7-trimethyl-4,6(3*H*,5*H*)-pteridinedione with dimethyl oxalate in methanolic potassium methoxide to give the 7-methoxalylmethyl analog.[348]

D. By Oxidative Means

Although the oxidation of secondary alcoholic side chains to ketonic side chains appears to have been neglected, there are examples of the conversion of a methylene into a carbonyl group. Thus aeration of 2-amino-6-ethyl-7,7-dimethyl-7,8-dihydro-4(3*H*)-pteridinone (**137**, R = Et) in aqueous butanolic acetic acid at 60°C during 6 h with the exclusion of light gave 6-acetyl-2-amino-7,7-dimethyl-7,8-dihydro-4(3*H*)-pteridinone (**137**, R = Ac) in 83% yield[1501]; and treatment of 6-acetonyl-2-amino-4,7(3*H*,8*H*)-pteridinedione

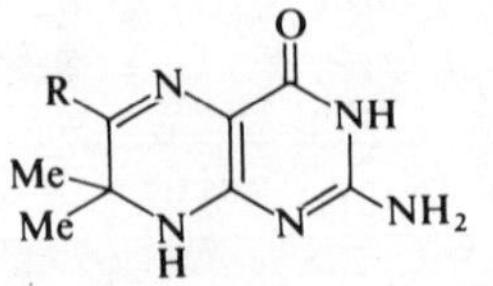

(137)

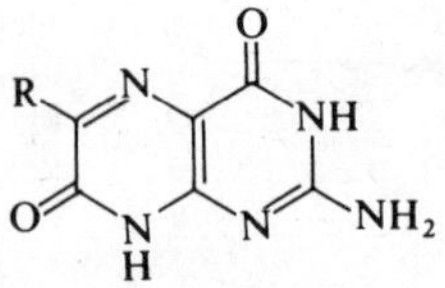

(138)

(**138**, R = CH_2Ac) with selenium dioxide in acetic acid at 90°C for 2 h gave the diketone, 6-acetylcarbonyl-2-amino-4,7(3*H*,8*H*)-pteridinedione (**138**, R = COAc) in 35% yield.[499]

E. From Acetals

The recovery of ketones from the corresponding cycloacetals (prepared by primary syntheses) is illustrated in the dissolution of 6-γ-ethylenedioxybutyl-2,4-pteridinediamine (**139**) in trifluoroacetic acid with a little sulfuric acid at 0°C to give an 81% yield of 6-acetonylmethyl-2,4-pteridinediamine (**140**),[1302] and in the similar treatment of 6-(2′-ethylenedioxycyclohexyl)methyl-2,4-pteridinediamine to give 6-(2′-oxocyclohexyl)methyl-2,4-pteridinediamine (81%).[1299]

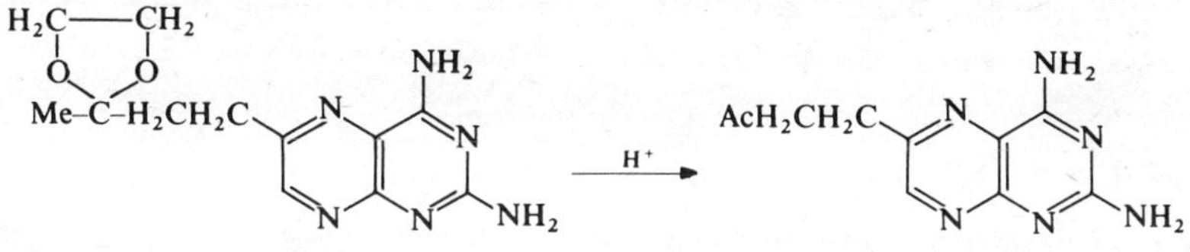

(**139**) (**140**)

12. REACTIONS OF PTERIDINE KETONES

A careful examination by Yasuo Iwanami and colleagues of the mass spectra of 7-acetonyl-2-amino-4,6(3*H*,5*H*)-pteridinedione (**141**, R = Me) and the isomeric 6-acetonyl-2-amino-4,7(3*H*,8*H*)-pteridinedione (**143**) strongly suggested that the former exists predominantly as the hydrogen-bonded tautomer (**142**, R = Me) even in the vapor state, whereas its isomer (**143**) exists as such. In brief, fragmentation of the stabilized molecule (**142**, R = Me) began with loss of C-6 as CO without rupture of the ketonic side chain but fragmentation of the isomer (**143**) began with a more normal disintegration of the side chain by loss of $H_2C{:}CO$, followed by loss of C-7 as CO.[620] In contrast, the four related ketones, 2-amino-6-phenacyl-4,7(3*H*,8*H*)-pteridinedione, 2-amino-7-phenacyl-4,6-(3*H*,5*H*)-pteridinedione (**141**, R = Ph), 2,4-diamino-6-phenacyl-7(8*H*)-pteridinone, and 2,4-diamino-7-phenacyl-6(5*H*)-pteridinone (**144**) all appeared

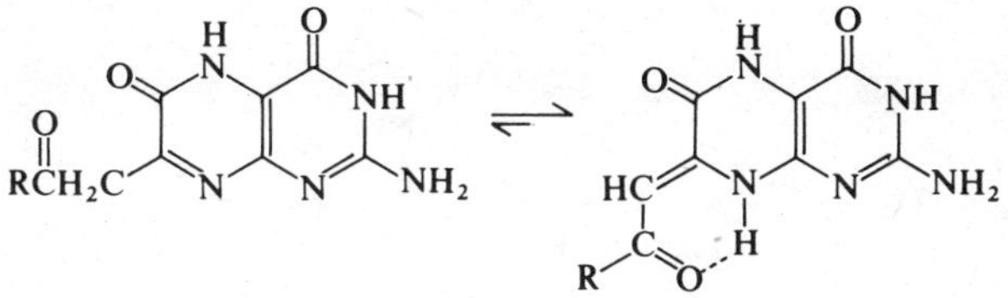

(**141**) (**142**)

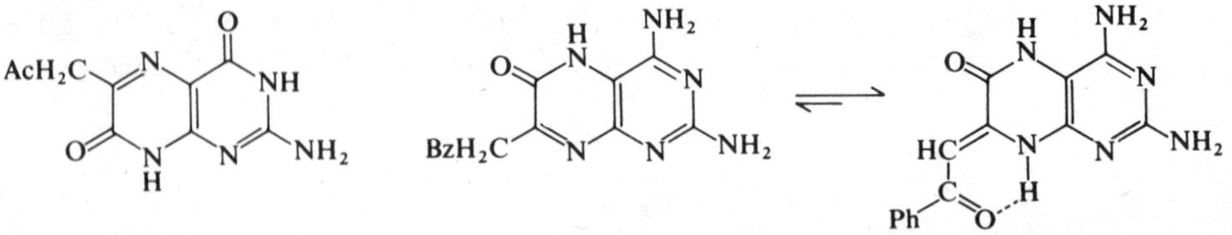

(143) (144) (145)

to exist (on nmr evidence) as their phenacylidene tautomers, for example, molecules (**142**, R = Ph) or (**145**), irrespective of whether the side chain occupied the 6- or 7-position.[982] This postulate has been confirmed in the case of 2-amino-7-phenacyl-4,6(3*H*,5*H*)-pteridinedione (**141**, R = Ph) and several 7-*p*-substituted-phenacyl derivatives, which all showed mass spectra consistent with predominance of the hydrogen-bonded phenacylidene form (**142**, R = Ph).[958]

A. Oxidation

The oxidation of pteridine ketones to carboxylic acids has been covered in Section 1.B in this chapter.

B. Reductive Reactions

Apart from nuclear reduction, pteridine ketones undergo several types of reductive reaction according to the reagent used. The *complete removal* of a ketonic substituent occurred when 6-benzoyl-1,3-dimethyl-7-phenyl-2,4(1*H*,3*H*)-pteridinedione (**146**, R = Bz) was electrochemically reduced in propanolic potassium chloride to give 1,3-dimethyl-7-phenyl-2,4(1*H*,3*H*)-pteridinedione (**146**, R = H) in 71% yield, plus the 7,8-dihydro derivative of the substrate (15%)[365]; when 6-acetonyl-2-amino-4,7(3*H*,8*H*)-pteridinedione (**147**, R = CH_2Ac) was treated in alkali with aluminum amalgam to give 2-amino-4,7(3*H*,8*H*)-pteridinedione (**147**, R = H) (80%)[1034, 1128]; and when the isomeric substrate, 7-acetonyl-2-amino-4,6(3*H*,5*H*)-pteridinedione was treated similarly to give initially (in this case) 2-amino-7,8-dihydro-4,6(3*H*,5*H*)-pteridinedione (?) and then by manganese dioxide oxidation, 2-amino-4,6(3*H*,5*H*)-pteridinedione.[1128]

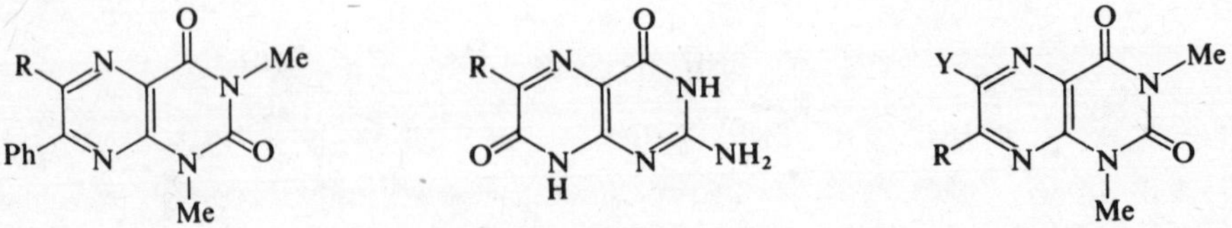

(146) (147) (148)

Reduction of the ketonic group *to an alkyl group* with the same number of carbon atoms occurred when 1,3-dimethyl-7-propionyl-2,4(1*H*,3*H*)-pteridinedione [**148**, R = C(:O)Et, Y = H] was treated with hydriodic acid, iodine, and red phosphorus in refluxing acetic acid to give the 7-propyl analog (**148**, R = Pr, Y = H) in 86% yield.[108]

Smooth reduction *to a secondary alcohol* occurred most commonly with sodium borohydride in aqueous or alcoholic media. In this way, 1,3-dimethyl-6-propionyl- (**148**, R = H, Y = C(:O)Et] gave 6-α-hydroxypropyl-1,3-dimethyl-2,4(1*H*,3*H*)-pteridinedione [**148**, R = H, Y = CH(OH)Et] in 80% yield[108]; 6-acetyl-2-amino- gave 2-amino-6-α-hydroxyethyl-7-methyl-4(3*H*)-pteridinone (~60%; zinc–alkali or hydrogenation over platinum oxide gave the same product in unstated yield)[1217]; and many other examples have been discussed in Chapter VI, Section 3.B. The use of reducing agents other than borohydride to achieve the same result or to prepare pinacol-type products has been covered in the same section.

C. Other Reactions

The conversion[504] of 7-acetonyl-2-amino- into 2-amino-7-sulfomethyl-4,6(3*H*,5*H*)-pteridinedione (36%) by sulfuric acid at 110°C and its subsequent reactions have been discussed in Chapter VIII, Section 5.

The formation of functional derivatives is exemplified in the conversion of 7-propionyl-2,4-pteridinediamine (**149**, X = O) into 7-α-(hydroxyimino)propyl- (**149**, X = NOH) (aqueous hydroxylamine hydrochloride under reflux: 69%), 7-α-(methoxyimino)propyl- (**149**, X = NOMe) (*O*-methylhydroxylamine hydrochloride similarly: 61%), and 7-α-(*t*-butylhydrazono)propyl-2,4-pteridinediamine (**149**, X = NNH*t*-Bu) (*t*-butylhydrazine hydrochloride in refluxing ethanol: 80%).[1678] Other examples are known.[1759]

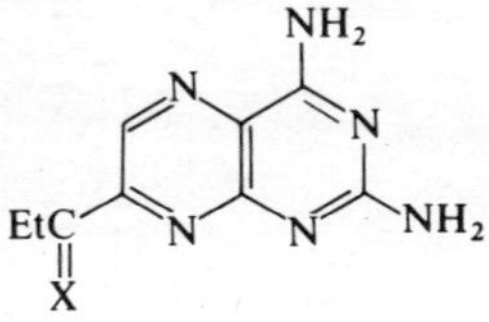

(149)

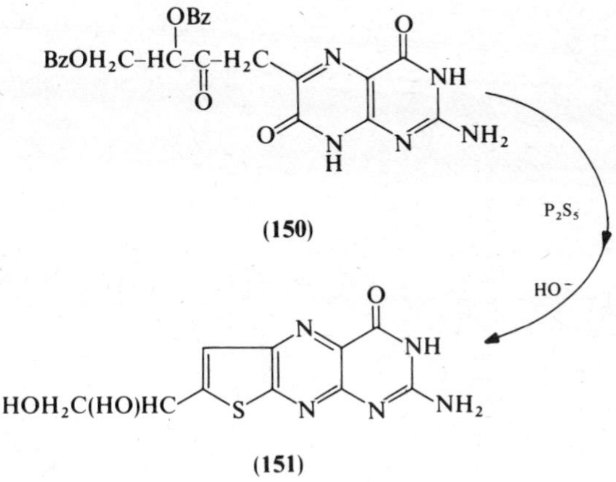

(150)

(151)

Thiation of 2-amino-6-(α,β-dibenzoyloxypropionyl)methyl-4,7(3*H*,8*H*)-pteridinedione (**150**) by phosphorus pentasulfide in thioacetamide followed by saponification of the ester functions gave 2-amino-7-α,β-dihydroxyethyl-4(3*H*)-thieno[3,2-*g*]pteridinone (**151**) in 14% yield[1262]; the site(s) of thiation and course of the reaction are unproven.

CHAPTER XI

The Hydropteridines

Although it is possible to envisage reduction of the pteridine ring to the decahydro stage, in fact only a few di- and tetrahydropteridine systems are represented in the literature. It is widely held that most of the other systems would be thermodynamically too unstable for more than transitory existence,[1419, 1478] even when stabilized by the attachment of electron-donating substituents. Owing to the exigencies of nomenclature, names of individual compounds may contain a polyhydro term that is not always indicative of the degree of nuclear reduction. For example, 2,4,6-trioxo-1,2,3,4,5,6-hexahydro-7-pteridinecarboxamide (**1**) is *not* a reduced system at all and 2,6-dioxo-1,2,5,6,7,8-hexahydro-4-pteridinecarboxylic acid (**2**) is only a dihydropteridine system. A good rule of thumb for determining the oxidation state of a pteridine is to add up the double bonds that are a part of, or directly attached to, the pteridine nucleus: five such double bonds indicate that the compound, for example (**1**), is an unreduced pteridine system; four such bonds, as in (**2**), indicate a dihydropteridine; and three such bonds, as in (**3**), indicate a tetrahydropteridine.

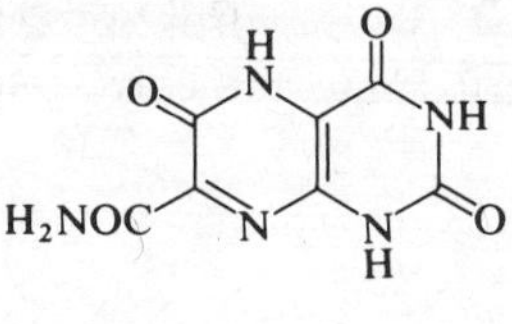

(**1**)

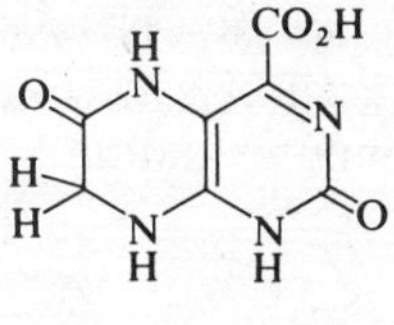

(**2**)

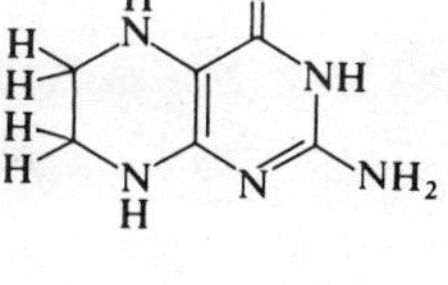

(**3**)

The preparative sections of this chapter summarize the scope of methods available for making the several known dihydro- and tetrahydropteridine systems, irrespective of attached substituents. The sections on reactivity include only those reactions peculiar to the reduced systems, for example, nuclear oxidation–reduction or acylation of a ring NH grouping, because the vast majority of substituent reactions (metatheses) have been covered in Chapters IV–X without particular advertence to the degree of aromaticity of the nucleus.

1. PREPARATION OF DIHYDROPTERIDINES

Of the 16 intraannular and 8 interannular dihydropteridine systems that are possible, only 5 of the former and none of the latter category are actually represented in the literature by characterized compounds of proven structure. Most such compounds have been made by primary synthesis or by ring reduction (with or without concomitant reductive modification or removal of a substituent), although a few have been made by covalent additions, oxidation of tetrahydropteridines, or rearrangement of isomeric dihydropteridines.

A. By Primary Syntheses

Primary syntheses have been used to prepare 3,4-, 5,6-, and 7,8-dihydropteridines. The following examples are typical of those contained in Chapters II and III.

Pyrazine intermediates were used to make 3,4-*dihydropteridines*. Thus reduction of 3-amino-2-pyrazinecarbonitrile gave 3-aminomethyl-2-pyrazinamine (**4**), which cyclized on heating in triethyl orthoformate or triethyl orthoacetate to give 3,4-dihydropteridine (**5**, R=H) or its 2-methyl derivative (**5**, R=Me), respectively, each in >70% yield.[51, 54, 585] Essentially similar routes gave 6-methyl-3,4-dihydropteridine, 3,4-dihydro-2(1*H*)-pteridinone, 3,4-dihydro-2-pteridineamine (**5**, R=NH_2), and ethyl 3,4-dihydro-2-pteridinecarboxylate (**5**, R=CO_2Et), all of which were converted into the corresponding aromatic pteridines by mild oxidation.[51]

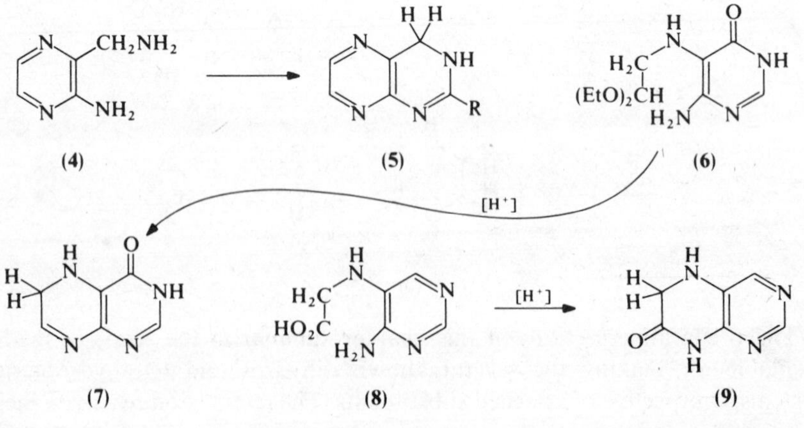

(**4**) (**5**) (**6**) (**7**) (**8**) (**9**)

Some 5,6-*dihydropteridines* have been made by an initial appropriate alkylation of the 5-amino group in a 4,5-pyrimidinediamine followed by cyclization [Ch. II, Sect. 4.A(1)]. Thus cyclization of 6-amino-5-β,β-diethoxyethylamino-4(3*H*)-pyrimidinone (**6**) in boiling dilute acid gave 5,6-dihydro-4(3*H*)-pteridinone (**7**) in 88% yield[43]; rather similarly, 5-carboxymethylamino-4-pyrimidinamine (**8**) gave 5,6-dihydro-7(8*H*)-pteridinone (**9**).[31] Modified Gabriel and Colman syntheses have also been used. Condensation of 2,6-diamino-5-methylamino-4(3*H*)-pyrimidinone with desyl chloride (α-benzoylbenzyl chloride, an equivalent of benzoin) gave 2-amino-5-methyl-6,7-diphenyl-5,6-dihydro-4(3*H*)-pteridinone (28%),[509] and condensation of 2,5,6-triamino-4(3*H*)-pyrimidinone (**11**) with benzoin in ethanol gave 2-amino-6,7-diphenyl-5,6-dihydro-4(3*H*)-pteridinone (**10**) (78% yield), whereas the same reaction in acetic acid gave the 7,8-dihydro isomer (**12**) (60% yield), which was also formed by slow rearrangement of the 5,6-dihydro isomer in hot dilute acetic acid.[1333]

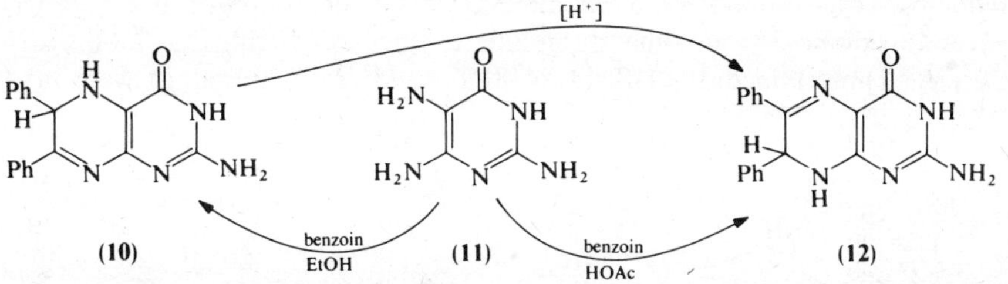

A great many 7,8-*dihydropteridines* have been made from pyrimidines, either by a modified Gabriel and Colman synthesis using a partially reduced α,β-dicarbonyl reagent as just exemplified or (much more commonly) by the first stage of a Boon synthesis. The former route is illustrated further in the condensation of 2,5-diamino-6-methylamino-4(3*H*)-pteridinone with benzoin in ethanolic acetic acid to give 2-amino-8-methyl-6,7-diphenyl-7,8-dihydro-4(3*H*)-pteridinone (**13**) in 70% yield[190]; in the condensation of 6-methyl-4,5-pyrimidinediamine with benzoin to give 4-methyl-6,7-diphenyl-7,8(?)-dihydropteridine[1227]; in the condensation of 2-ethoxy-6-methyl-4,5-pyrimidinediamine with benzoin to give successive products, the second of which was identified as 2-ethoxy-4-methyl-6,7-diphenyl-7,8-dihydropteridine by an unambiguous Boon synthesis[398]; in the similar condensation of 5,6-diamino-2-ethylthio-4(3*H*)-pyrimidinone with benzoin or symmetrically disubstituted benzoins to give 2-ethylthio-6,7-diphenyldihydro-4(3*H*)-pteridinone (or appropriately substituted-phenyl derivatives), each in an α-form or a β-form according to conditions, presumably representing 5,6- and 7,8-dihydro isomers[955]; and in the condensation of 2,5-diamino-6-methylamino-4(3*H*)-pyrimidinone (**14**) with chloroacetaldehyde (equivalent to hydroxyacetaldehyde, that is, a partially reduced glyoxal) to give 2-amino-8-methyl-7,8-dihydro-4(3*H*)-pteridinone (**15**).[511]

The Boon route to 7,8-dihydropteridines has been fully discussed in Chapter II, Section 3. Two hitherto unmentioned typical examples leading to oxidizable

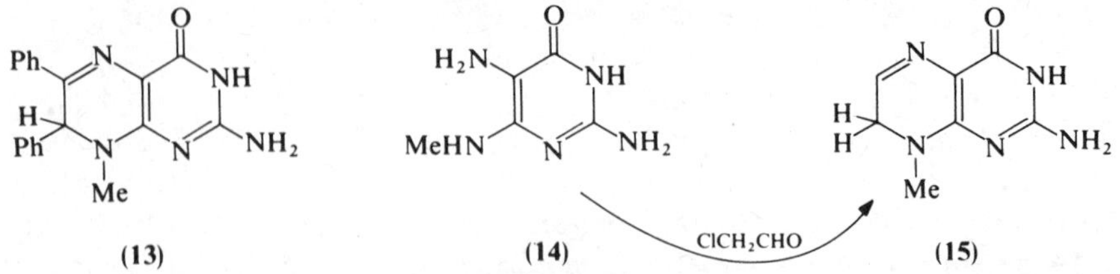

(13) (14) (15)

7,8-dihydro derivatives involved the reductive cyclization of 6-acetonylamino-2-amino-5-phenylazo-4(3*H*)-pyrimidinone (**16**) to give 2-amino-6-methyl-7,8-dihydro-4(3*H*)-pteridinone (**17**),[1394] and treatment of 2-amino-5-nitro-6-(4′-pyridinylcarbonylmethyl)amino-4(3*H*)-pyrimidinone with dithionite to give 2-amino-6-(4′-pyridinyl)-7,8-dihydro-4(3*H*)-pteridinone (53%), which underwent aerial oxidation to 2-amino-6-(4′-pyridinyl)-4(3*H*)-pteridinone (65%).[1396] Boon syntheses of blocked (nonoxidizable) analogs are typified in the reductive cyclization of 6-(α,α-dimethylacetonyl)amino-5-nitro-2,4(1*H*,3*H*)-pyrimidinedione by alkaline dithionite to give 6,7,7-trimethyl-7,8-dihydro-2,4(1*H*,3*H*)-pteridinedione (**18**) in 74–81% yield[375, 1503]; analogs were made similarly.[388, 450, 628]

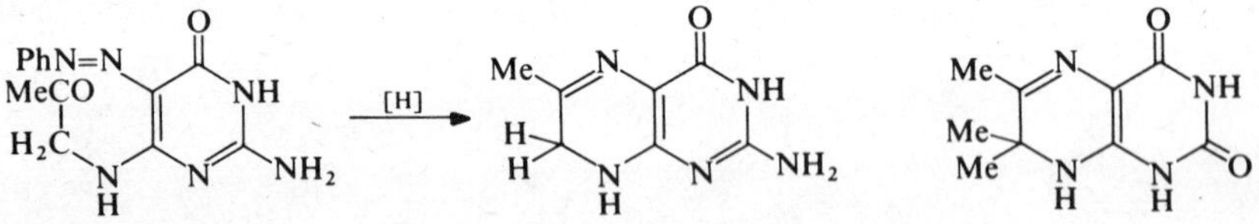

(16) (17) (18)

B. By Reductive Processes

The reduction of aromatic pteridines has been used extensively to make 3,4-, 5,6-, 5,8- (?), and 7,8-dihydropteridines. It has been observed[45] that pteridines tend to reduce at the same bond(s), which are prone to covalent hydration and other additions.

(1) *Reduction to* 3,4-*Dihydropteridines*

The reductive formation of 3,4-dihydropteridines is illustrated in the conversion of 2(1*H*)-pteridinone into 3,4-dihydro-2(1*H*)-pteridinone (hydrogenation over palladium in dilute alkali: 67%; sodium borohydride in dilute alkali: 55%; sodium dithionite in dilute alkali: 25%)[42] and of 2-pteridinamine into its 3,4-dihydro derivative (potassium borohydride in dilute alkali: 80%, as its *p*-toluenesulfonate salt).[45] Other examples appear to have been confined to treatment of 8-alkyl-7-oxo-7,8-dihydro-6-pteridinecarboxylic acid derivatives, having only hydrogen or a methyl group at the 4-position, with sodium borohydride. Such substrates were thereby converted into bright yellow dihydro derivatives that exhibited uv absorption maxima some 50 nm above those of their parent compounds. On the basis of this and other unusual properties, the products were diagnosed initially as 5,8-dihydro derivatives,[387, 562] but subsequent work in the same laboratories led to the firm conclusion that they were 3,4-dihydro derivatives,[376, 1311] a finding consistent with the spectral properties of unambiguously synthesized 3,4-dihydropteridines.[51] The reductive process involved treatment of each substrate in DMF with sodium borohydride at 25°C for 30 min to give methyl 2,8-dimethyl-7-oxo-3,4,7,8-tetrahydro-6-pteridinecarboxylate (**19a**) (75%), methyl 8-methyl-7-oxo-2-phenyl-3,4,7,8-tetrahydro-6-pteridinecarboxylate (**19b**) (30%), methyl 4,8-dimethyl-2-methylamino-7-oxo-3,4,7,8-tetrahydro-6-pteridinecarboxylate (**19c**) (90%), the corresponding carboxylic acid (**19d**) (78%), methyl 8-methyl-2-methylamino-7-oxo-3,4,7,8-tetrahydro-6-pteridinecarboxylate (**19e**) (35%), the corresponding carboxylic acid (**19f**) (60%), ethyl 2-methoxy-8-methyl-7-oxo-3,4,7,8-tetrahydro-6-pteridinecarboxylate (**19g**) (36%), ethyl 8-methyl-2,7-dioxo-1,2,3,4,7,8-hexahydro-6-pteridinecarboxylate (**20**, R = Et) (75%), the corresponding carboxylic acid (**20**, R = H) (71%), and ethyl 8-ethyl-2-ethylamino-7-oxo-3,4,7,8-tetrahydro-6-pteridinecarboxylate (exceptionally by catalytic hydrogenation over platinum: 19%; with several similar substrates, reductive methods other than borohydride gave 5,6-dihydro derivatives: see the following subsection).[1311] The analogous use of borohydride in ethanol, methanol, water, or a mixture converted the tosylate salts of the imines, methyl or ethyl 2-imino-3,8-dimethyl-7-oxo-2,3,7,8-tetrahydro-6-pteridinecarboxylate (**21**, R = Me or Et, Y = H) and methyl or ethyl 3,8-dimethyl-2-methylimino-7-oxo-2,3,7,8-tetrahydro-6-pteridinecarboxylate (**21**, R = Me or Et, Y = Me), into methyl or ethyl 2-amino-3,8-dimethyl-7-oxo-3,4,7,8-tetrahydro-6-pteridinecarboxylate (**22**, R = Me or Et, Y = H) and methyl or ethyl 3,8-dimethyl-2-methylamino-

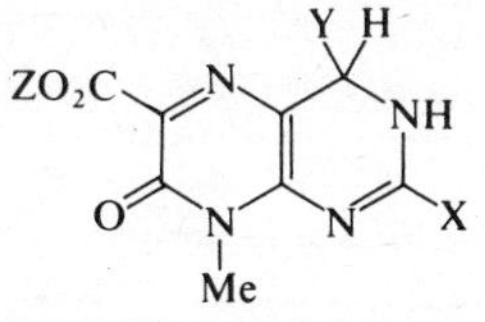

(19)

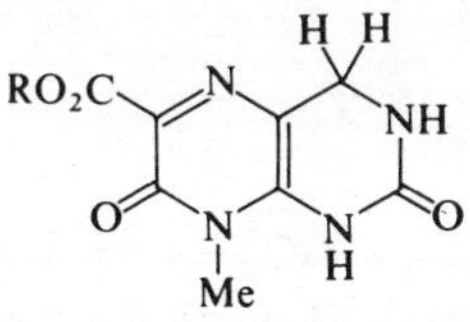

(20)

	X	Y	Z
a	Me	H	Me
b	Ph	H	Me
c	NHMe	Me	Me
d	NHMe	Me	H
e	NHMe	H	Me
f	NHMe	H	H
g	OMe	H	Et

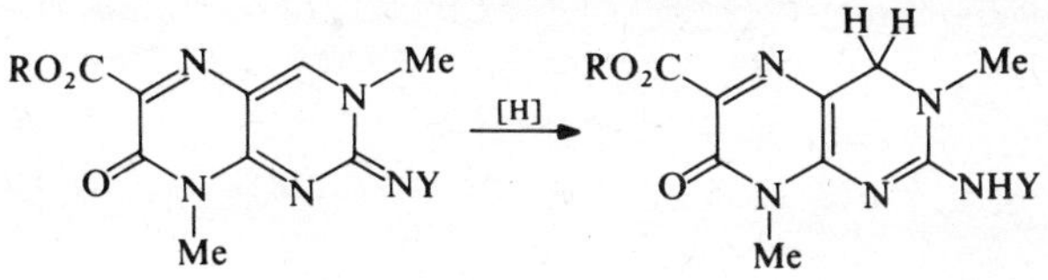

(21) **(22)**

7-oxo-3,4,7,8-tetrahydro-6-pteridinecarboxylate (**22**, R = Me or Et, Y = Me), respectively, in 66 to 95% yield.[376, 1311]

(2) *Reduction to 5,6-Dihydropteridines*

Reductive preparations of 5,6-dihydropteridines are represented in the products 5,6-dihydro-7(8*H*)-pteridinone (**23**) (potassium borohydride in dilute aqueous potassium carbonate at 25°C for 15 h: 80%; potassium amalgam in water at 5°C for 10 min: 43%;* sodium reagents were avoided because the substrate had a sparingly soluble sodium salt),[43] 5,6-dihydro-4,7(3*H*,8*H*)-pteridinedione (potassium borohydride–aqueous potassium carbonate: 85%),[43] 6-methyl- and 6-carboxymethyl-5,6-dihydro-4,7(3*H*,8*H*)-pteridinedione (sodium amalgam in water: 90 and 80%, respectively),[29] 8-ethyl-2-ethylamino-7-oxo-5,6,7,8-tetrahydro-6-pteridinecarboxylic acid (**24**, R = H) (hydrogenation over palladium in dilute alkali: 27%),[1311] the corresponding ethyl ester (**24**, R = Et) (hydrogenation over palladium in ethanol: >95% yield),[1311] ethyl 2-methoxy-8-methyl-7-oxo-5,6,7,8-tetrahydro-6-pteridinecarboxylate (zinc dust in glacial acetic acid: 80%),[1311] ethyl 4-dimethylamino-8-methyl-7-oxo-5,6,7,8-tetrahydro-6-pteridinecarboxylate (hydrogenation over palladium in ethanol: >95%),[1311] and 2-amino-6,7-diphenyl-5,6-dihydro-4(3*H*)-pteridinone (hydrogenation over platinum in ethanolic acetic acid: 64%; plus the 7,8-dihydro isomer: 14%).[72] Other examples include the conversion of 2-amino-8-methyl-4,7(3*H*,8*H*)-pteridinedione into its 5,6-dihydro derivative (**25**, R = H), which was isolated as 2-amino-5-formyl-8-methyl-5,6-dihydro-4,7(3*H*,8*H*)-pteridinedione (**25**, R = CHO) (hydrogenation over platinum in formic acid: 37%)[181]; of 7-

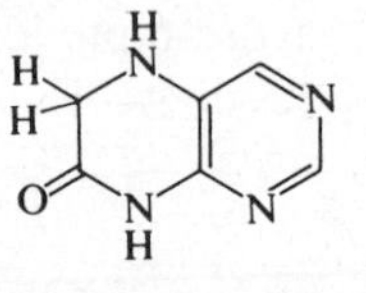

(23)

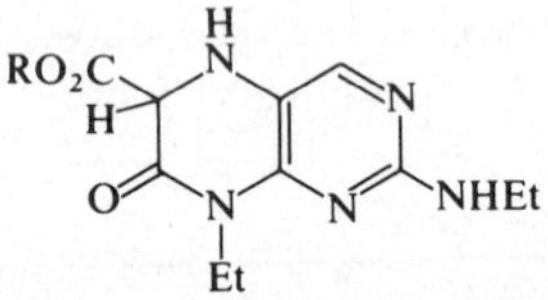

(24)

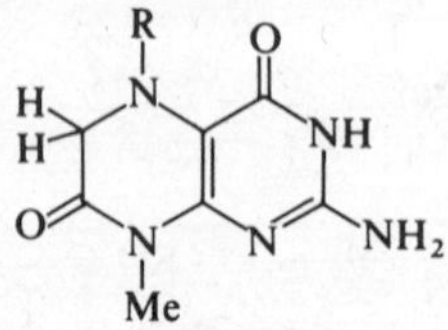

(25)

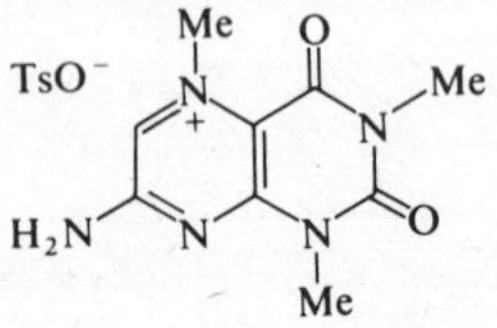

(26)

---

* An attempted dithionite reduction gave 5-carboxymethylamino-4-pyrimidinamine that underwent cyclization in refluxing dilute hydrochloric acid to give 5,6-dihydro-7(8*H*)-pteridinone in 60% overall yield.[31]

amino-1,3,5-trimethyl-2,4-dioxo-1,2,3,4-tetrahydro-5-pteridinium tosylate (**26**) into 7-amino-1,3,5-trimethyl-5,6-dihydro-2,4(1*H*,3*H*)-pteridinedione (sodium borohydride in water: 70%)[1275]; and of 7(8*H*)-pteridinone into its 5,6-dihydro derivative (**23**), either by uv irradiation on paper[23] or by electrochemical reduction,[893] although neither method is of preparative value.

(3) *Reduction to 5,8-Dihydropteridines*

The formation of 5,8-dihydropteridines as transient species during the reduction of pteridines is well documented, especially in the polarographic–electrochemical literature[114–116, 604, 638–643, 670, 859, 890, 932, 1014, 1204, 1208, 1489] and such evidence has been expertly reviewed by Glen Dryhurst.[1419] Despite the presentation of spectroscopic and other data gleaned from solutions purporting to contain 5,8-dihydropteridines,[391, 392, 925] it must be kept in mind that the only isolated and properly characterized derivatives reported to belong to this system, eventually proved to be their 3,4-dihydro isomers [see Sect. 1.B(1) in this chapter]. Thus from a preparative point of view, 5,8-dihydropteridines remain virtually unknown apart from one stabilized derivative, 5,8-diacetyl-1,3-dimethyl-5,8-dihydro-2,4(1*H*,3*H*)-pteridinedione,[365] which was trapped during electrolytic reduction of 1,3-dimethyl-2,4(1*H*,3*H*)-pteridinedione in acetic anhydride and was subsequently proven in structure by an x-ray analysis.[795] However, the recent successful isolation of quinonoid dihydropterins (Ch. IX, Sect. 7) suggests that 5,8-dihydropteridines may yet prove amenable to isolation.

(4) *Reduction to 7,8-Dihydropteridines*

The reduction of pteridines to 7,8-dihydropteridines is well represented, possibly because the 7,8-dihydro system is thermodynamically more stable than isomeric systems[1478] and therefore constitutes the end point of any prototropic rearrangement(s) associated with the reductive process. Thus reduction of appropriate parent pteridines gave 7,8-dihydro-6(5*H*)-pteridinone (**27**, R=H) [sodium amalgam in water at 45°C: 85%[31]; sodium borohydride–dilute alkali: 90%[43]; sodium dithionite–aqueous sodium carbonate: 80%[43]; hydrogenation over palladium in dilute alkali: 63%[43]; electrolytic reduction[893]; it was also obtained from 6,7(5*H*,8*H*)-pteridinedione by electrolytic reductive deoxygenation in alkali: 47%[208]]; 7,8-dihydro-4(3*H*)-pteridinone (potassium borohydride in water: 16%);[43] 7,8-dihydro-2,6(1*H*,5*H*)-pteridinedione (potassium borohydride: 80%; hydrogenation over palladium or platinum in dilute alkali: 73 or 85%, respectively);[43] 7,8-dihydro-4,6(3*H*,5*H*)-pteridinedione (potassium borohydride: 96%[43]; potassium amalgam: 72%[43]; hydrogenation over palladium: 79%[43]; electrolytic reduction in 60% perchloric acid or dilute alkali: 75% or 58%,[208] respectively); 7-hydroxy-7,8-dihydro-6(5*H*)-pteridinone (**27**, R=OH) (electrolytic reduction in 50% sulfuric acid: 79%)[208]; 2-amino-6-methyl-

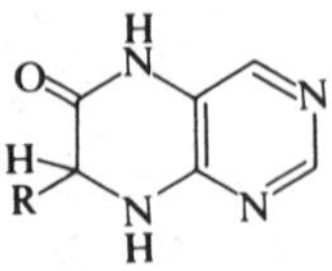

(27)

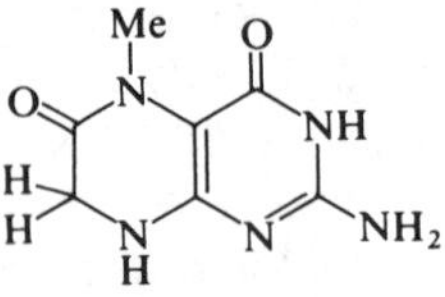

(28)

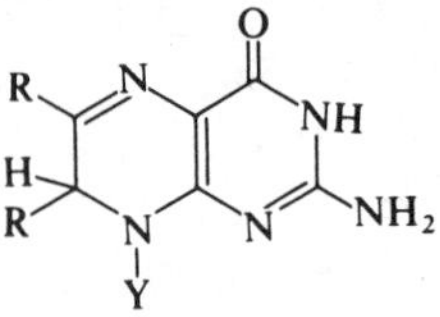

(29)

7,8-dihydro-4(3*H*)-pteridinone (zinc dust–alkali: 30 to 95%, as hydrochloride[388, 509, 1394]; sodium dithionite–dilute alkali: 30 to 87%, as hydrochloride;[202, 388, 813, 1394] hydrogenation over Raney-nickel in dilute alkali: ~85%, as hydrochloride[388]); 2-amino-6,7-dimethyl-7,8-dihydro-4(3*H*)-pteridinone (sodium dithionite–dilute alkali: ~40%, as hydrochloride[388]; it is probably better made by oxidation of the tetrahydro analog[509]); 2-amino-6,7-diphenyl-7,8-dihydro-4(3*H*)-pteridinone (zinc dust–dilute alkali: ~75%)[388]; 2-amino-6-benzyl- and 2-amino-6,7-dibenzyl-7,8-dihydro-4(3*H*)-pteridinone (zinc dust–dilute alkali: 84 and 78%, respectively, as hydrochlorides)[419]; 2,4-diamino-7,8-dihydro-6(5*H*)-pteridinone (hydrogenation over platinum in dilute alkali: 75%)[268]; 6-methoxy-7,8-dihydro-2,4-pteridinediamine (hydrogenation over platinum in water: 68%)[268]; 2-amino-5-methyl-7,8-dihydro-4,6(3*H*,5*H*)-pteridinedione (**28**) (similarly: 28%)[349]; 2-amino-6-α,β-dihydroxypropyl-7,8-dihydro-4(3*H*)-pteridinone (dihydrobiopterin) (sodium dithionite–dilute alkali: 72%)[430]; the phosphate ester, 2-amino-6-dihydroxyphosphinyloxymethyl-7,8-dihydro-4(3*H*)-pteridinone (sodium dithionite–aqueous sodium bicarbonate: ~75%)[1336]; 2-amino-6,7-dimethyl-8-phenyl-7,8-dihydro-4(3*H*)-pteridinone (**29**, R = Me, Y = Ph) (potassium borohydride in water: ~30%, as hydrochloride)[373]; 2-amino-8-methyl-6,7-diphenyl-7,8-dihydro-4(3*H*)-pteridinone (**29**, R = Ph, Y = Me) (sodium borohydride in water: 56%)[373]; 8-methyl-2-methylamino-6,7-diphenyl-7,8-dihydro-4(3*H*)-pteridinone (sodium borohydride in water: 68%)[374]; 2-dimethylamino-8-methyl-6,7-diphenyl-7,8-dihydro-4(3*H*)-pteridinone (sodium borohydride in aqueous ethanol: 46%)[373]; 2-amino-3,8-dimethyl-6,7-diphenyl-7,8-dihydro-4(3*H*)-pteridinone (sodium borohydride in water: 55%)[373]; 8-methyl- and 3,8-dimethyl-6,7-diphenyl-7,8-dihydro-2,4(1*H*,3*H*)-pteridinedione (sodium borohydride in aqueous methanol: 62 and 48%, respectively)[375]; 8-β-hydroxyethyl-6,7-dimethyl-7,8-dihydro-2,4(1*H*,3*H*)-pteridinedione (hydrogenation over platinum in water: 75%[422]; similarly, but followed by prolonged aeration to remove any tetrahydro by-product: 50%)[375]; and more complicated analogs.[685, 1215]

Other electrolytic reductions converted 4-dimethylamino-8-methyl-6,7(5*H*,8*H*)-pteridinedione (**30**) into 4-dimethylamino-7-hydroxy-8-methyl-7,8-dihydro-6(5*H*)-pteridinone (**31**) (in dilute hydrochloric acid: 90%), 4,8-dimethyl-2,6,7(1*H*,5*H*,8*H*)-pteridinetrione into 7-hydroxy-4,8-dimethyl-7,8-dihydro-2,6(1*H*,5*H*)-pteridinedione (in dilute acid: 77%; in dilute alkali: 60%), 8-methyl-2-methylamino-6,7(5*H*,8*H*)-pteridinedione into 7-hydroxy-8-methyl-2-methylamino-7,8-dihydro-6(5*H*)-pteridinone (in dilute alkali: 50%; in dilute acid:

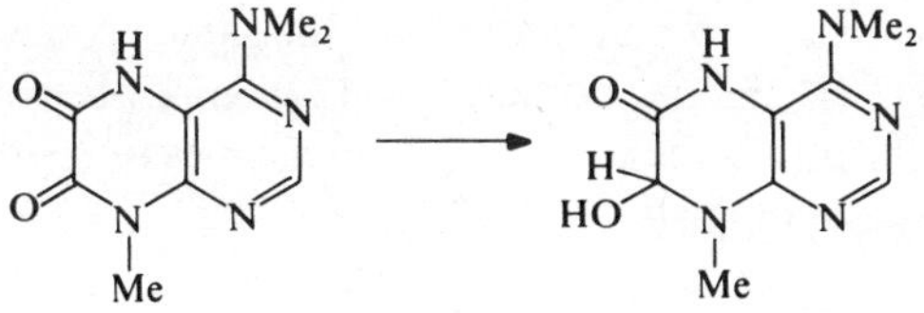

(30) (31)

82%), and 2-dimethylamino-8-ethyl-6,7(5*H*,8*H*)-pteridinedione into 2-dimethylamino-8-ethyl-7-hydroxy-7,8-dihydro-6(5*H*)-pteridinone (in dilute acid: 35%).[208] Pteroic acid, 2-amino-6-*p*-carboxyanilinomethyl-4(3*H*)-pteridinone, has been reduced to its 7,8-dihydro derivative (dithionite: 74%)[197]; and pteroylglutamic acid has been reduced to its 7,8-dihydro derivative by hydrogenation over palladium in dilute alkali,[1125] by titanous chloride,[1001] or sodium dithionite in aqueous media (77–82% yield).[259, 1504, cf. 78]

The reductive deoxygenation of 2-amino-4,6,7(3*H*,5*H*,8*H*)-pteridinetrione (leucopterin) to 2-amino-7,8-dihydro-4,6(3*H*,5*H*)-pteridinedione (dihydroxanthopterin) is best done by sodium amalgam in water at 50°C (55% yield).[60, cf. 71, 877, 1268] Similar treatment of 4,6,7(3*H*,5*H*,8*H*)-pteridinetrione gave 7,8-dihydro-4,6(3*H*,5*H*)-pteridinedione in 90% yield[29]; 2,4,6,7(1*H*,3*H*,5*H*,8*H*)-pteridinetetrone likewise gave 7,8-dihydro-2,4,6(1*H*,3*H*,5*H*)-pteridinetrione in >90% yield[41]; treatment of 2-amino-8-*β*,*γ*-dihydroxypropyl-4,7-dioxo-3,4,7,8-tetrahydro-6-pteridinecarboxylic acid with zinc amalgam in dilute hydrochloric acid gave 2-amino-8-*β*,*γ*-dihydroxypropyl-4-oxo-3,4,7,8-tetrahydro-6-pteridinecarboxylic acid (65%, as the ammonium salt)[480]; the 8-D-sorbityl and 8-D-ribityl analogs were made similarly[480]; and 2-amino-8-*β*-hydroxyethyl-4,6,7(3*H*,5*H*,8*H*)-pteridinetrione with sodium amalgam in water gave 2-amino-8-*β*-hydroxyethyl-7,8-dihydro-4,6(3*H*,5*H*)-pteridinedione.[183] In a different way, treatment of 6-methyl-2,4-pteridinediamine 8-oxide with sodium dithionite in boiling water gave 6-methyl-7,8-dihydro-2,4-pteridinediamine, initially as its bisulfite salt (65% yield)[484]; the 6-propyl analog was made similarly[484]; 2-amino-6-methyl-4(3*H*)-pteridinone 8-oxide gave 2-amino-6-methyl-7,8-dihydro-4(3*H*)-pteridinone (dithionite: 90%; hydrogenation over Raney-nickel: 42%)[485]; and the 6-phenyl analog (88%) resulted similarly from a dithionite reduction, this time in dilute alkali.[485] Reduction of 2,4,6,7-tetrachloropteridine (**32**) by hydrogenation over palladium or platinum in benzene gave 2,4-dichloro-7,8-dihydro-6(5*H*)-pteridinone (**33**) (82 and >95%, respectively)[486]; treatment of 2,4-dichloro-6,7(5*H*,8*H*)-pteridinedione

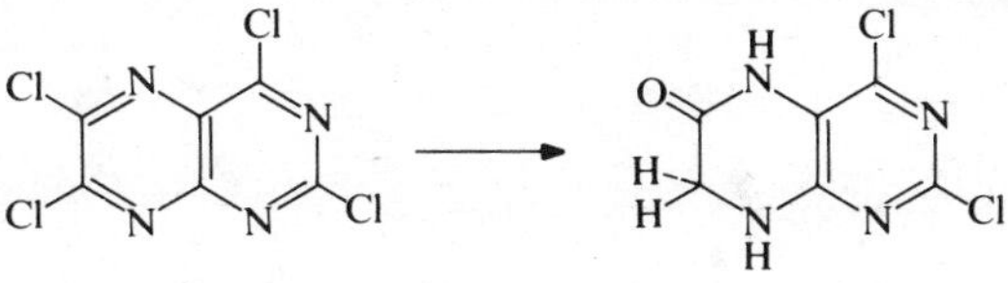

(32) (33)

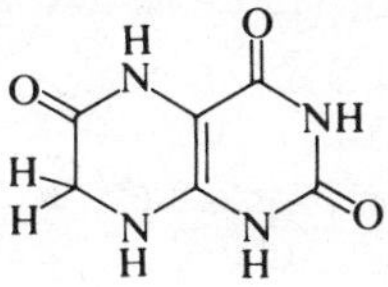

(34)

with sodium amalgam in water gave a little of the same product (**33**) but mainly 7,8-dihydro-2,4,6(1*H*,3*H*,5*H*)-pteridinetrione (**34**) in 47% yield.[486] A general procedure[1352] for the dithionite reduction of pteridines to 7,8-dihydropteridines may be of use to enzymologists.

C. By Oxidation of Tetrahydropteridines

The biochemically important oxidation of tetrahydrobiopterin, the naturally occurring coenzyme used by essential hydroxylases, and of related tetrahydropterins to the corresponding quinonoid 6,7-*dihydropterins* has been discussed in Chapter IX, Section 7. In a more chemical sense, the oxidation of tetra- to dihydropteridines is not a very important route to the latter class of compound, although appropriate 5,6,7,8-tetrahydro derivatives have been so used to make the following 7,8-*dihydropteridines*: 2-amino-7-methyl-7,8-dihydro-4(3*H*)-pteridinone (by stirring the substrate in aqueous diethylamine for 70 min: 39%),[201] 2-amino-6,7-dimethyl-7,8-dihydro-4(3*H*)-pteridinone (**35**) (oxygenation of the substrate in aqueous diethylamine for 20 min: 83%, as hydrochloride),[1038] 2-amino-8-methyl-7,8-dihydro-4(3*H*)-pteridinone (aeration in refluxing aqueous alcohol for 45 min: 76%),[373] its 8-phenyl analog (similarly: 50%),[373] 2-amino-6,7,8-trimethyl-7,8-dihydro-4(3*H*)-pteridinone (aeration for 15 h: 50%),[373] its 6,7-dimethyl-8-phenyl analog (oxygenation for 6 h: 55%),[373] 6,7,8-trimethyl-7,8-dihydro-2,4(1*H*,3*H*)-pteridinedione (**36**, R=H) (aeration in aqueous ethanol for 90 min: 61%),[375] the 3,6,7,8-tetramethyl analog (**36**, R=Me) (aeration in water for 2.5 h: 46%),[375] 1,3,6,7-tetramethyl-7,8-dihydro-2,4(1*H*,3*H*)-pteridinedione (aeration in acetonitrile for 15 h: 43%; the filtrate gave some fully oxidized product),[1097] 1,3-dimethyl-6,7-diphenyl-7,8-dihydro-2,4(1*H*,3*H*)-pteridinedione (similarly: 75%),[1097] dimethyl 2-amino-4-oxo-3,4,7,8-tetrahydro-6,7-pteridinedicarboxylate (**37**, R=H, X=Y=CO_2Me) (aeration of a refluxing aqueous solution of the 3,4,5,6,7,8-hexahydro substrate for 3.5 h: 44%),[307] its 8-methyl derivative (**37**, R=Me, X=Y=CO_2Me) (aeration of a refluxing methanolic solution for 5.5 h: 34%),[307] 2-amino-8-methyl-4-oxo-3,4,7,8-tetrahydro-6-pteridinecarboxylic acid (**37**, R=Me, X=CO_2H, Y=H) (aeration of the hexahydro substrate in aqueous bicarbonate for 12 h: 80%),[684] the corresponding 6-pteridinecarbonitrile (**37**, R=Me, X=CN, Y=H) (oxygenation of the substrate in DMF for 12 h: 75%),[684] and 5-methyl-5,6(?)-dihydropteroylglutamic acid (potassium ferricyanide in phos-

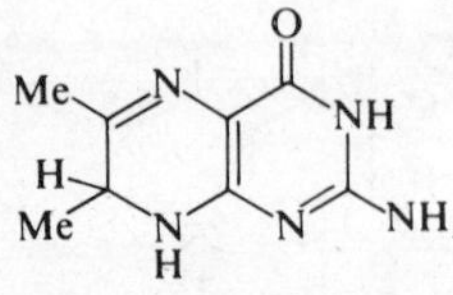

(**35**)

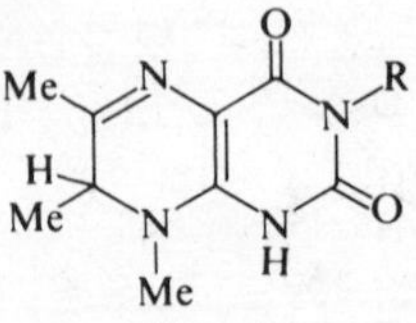

(**36**)

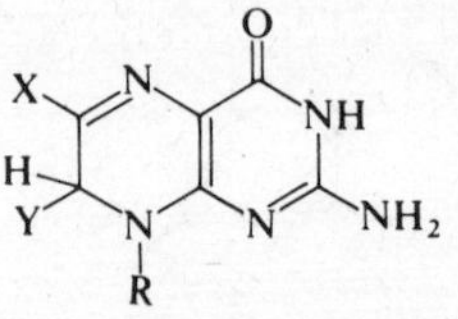

(**37**)

phate buffer).[77, 598] In addition, 2-amino-6-aminomethyl-6-methyl-5,6,7,8-tetrahydro-4(3*H*)-pteridinone (**38**) underwent aerial oxidation in phosphate buffer to give 2-amino-6-methyl-7,8-dihydro-4(3*H*)-pteridinone, formaldehyde, and ammonia[1340]; and (dark) aerial oxidation at pH 4 of 2-amino-6-α,β-dihydroxypropyl-5,6,7,8-tetrahydro-4(3*H*)-pteridinone (tetrahydrobiopterin) (**39**) gave a separable mixture of 2-amino-6-lactoyl-7,8-dihydro-4(3*H*)-pteridinone (**40**, R = OH) (sepiapterin) and 2-amino-6-propionyl-7,8-dihydro-4(3*H*)-pteridinone (**40**, R = H) (deoxysepiapterin), both in low yield, while in aqueous bicarbonate (pH 8–9), the same substrate gave 2-amino-7,8-dihydro-4,6(3*H*,5*H*)-pteridinedione in < 30% yield.[350]

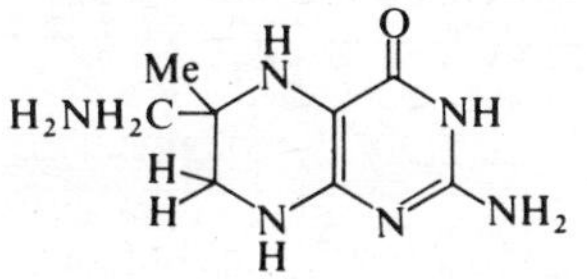

(**38**)

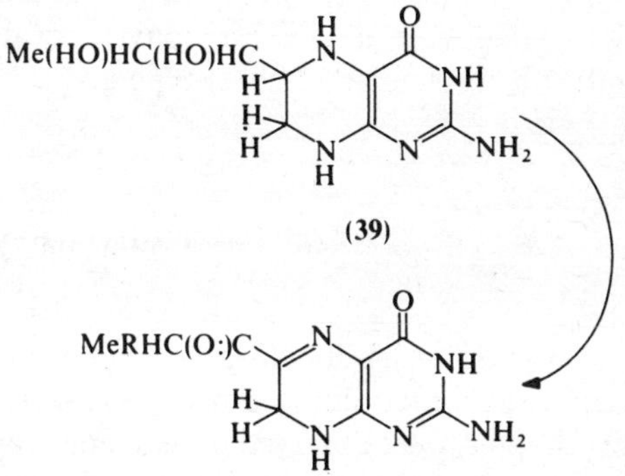

(**39**)

(**40**)

D. By Covalent Addition

The covalent addition of water, alcohols, amines, Michael reagents, and so on, to a susceptible pteridine results in the formation of a di- or tetrahydropteridine, as represented in a mono- or di-adduct. The formation of such mono-adducts from various *types of substrate* has been covered already, for example, those from pteridine, see Chapter IV, Section 1.C; those from alkylpteridines, see Chapter IV, Section 2.B(1); those from halogenopteridines, see Chapter V, Section 2.A; those from pteridinones, see Chapter VI, Section 2.B; those from pteridine *N*-oxides, see Chapter VII, Section 6.B; those from pteridinamines, see Chapter IX, Section 2.B; and those from pteridine esters, see Chapter X, Section 4.A.

The *types of dihydropteridine* produced by covalent addition are illustrated in the conversion of 4-trifluoromethylpteridine into 4-trifluoromethyl-3,4-dihydro-4-pteridinol (**41**) (water at 20°C for 12 h: 55%)[156]; of 2-pteridinamine into 4-ethoxy-3,4-dihydro-2-pteridinamine hydrochloride (**42**) (ethanolic hydrochloric acid at 20°C: 68%)[45] (for other similarly formed alkoxydihydropteridines, see Ch. VII, Sect. 1.B); of 6-chloropteridine into 4-benzylamino-6-chloro-3,4-dihydropteridine (benzylamine in dry benzene at 5°C: 71%)[146] (see also Ch. IX, Sect. 1.C); of 6(5*H*)-pteridinethione into 7-benzylthio-7,8-dihydro-6(5*H*)-pteri-

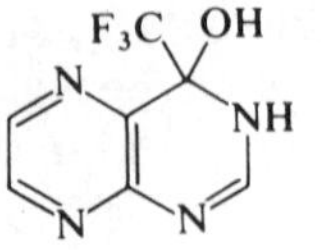

(41)

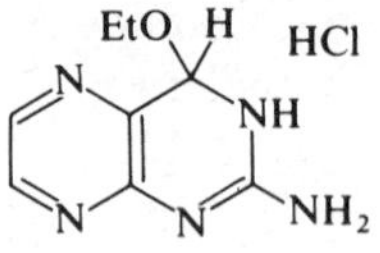

(42)

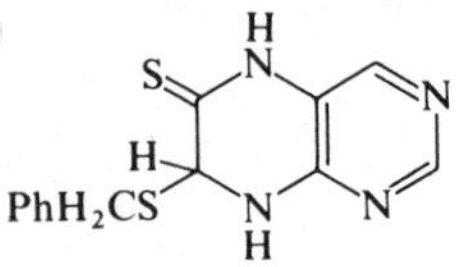

(43)

dinethione (**43**) (α-toluenethiol in ethanol: 79%)[36] (see also Ch. VIII, Sect. 6.D); of 2-pteridinamine into sodium 2-amino-3,4-dihydro-4-pteridinesulfonate (aqueous sodium hydrogen sulfite: 48%)[45] (see also Ch. VIII, Sect. 5); and of pteridine into 4-(diethoxycarbonyl)methyl-3,4-dihydropteridine (ethanolic diethyl malonate at 20°C for 4 days: 32%)[49] (see also Ch. X, Sects. 3.C, 5.B, 7.B, and 11.B for similar Michael-type additions).

E. By Prototropic Rearrangement

The various 3,4-dihydropteridines within the class of 8-alkyl-7-oxo-3,4,7,8-tetrahydro-6-pteridinecarboxylates (**19a–g**; **20**; **22**), described in Section 1.B(1) of this chapter, were observed to rearrange rapidly, quantitatively, and irreversibly into the corresponding 5,6-dihydropteridines, that is, into their 5,6,7,8-tetrahydro isomers, by dissolution in trifluoroacetic acid even at room temperature.[1311] No experimental details have been reported.

The rearrangement of 5,6- into 7,8-dihydropteridines is represented in the conversion of 2-amino-6,7-diphenyl-5,6-dihydro-4(3*H*)-pteridinone (**44**) into the 7,8-dihydro isomer (**45**) by heating in glacial acetic acid at 100°C (within an evacuated sealed tube) for 8 h. The ferric chloride oxidation of both substrate and product gave 2-amino-6,7-diphenyl-4(3*H*)-pteridinone (**46**).[1333] The mechanism of this rearrangement has been discussed.[72, 444]

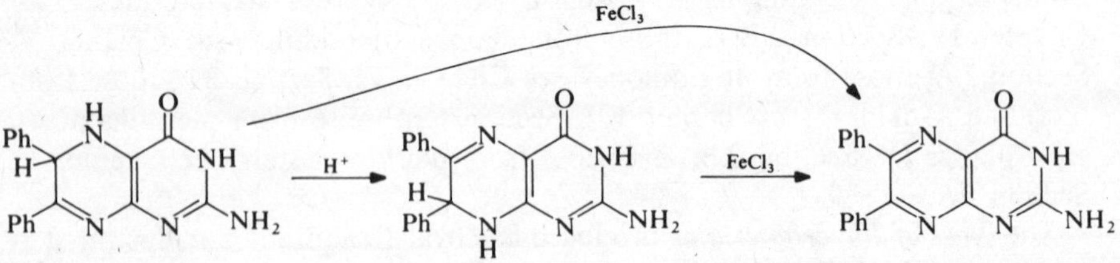

(44) (45) (46)

The rearrangement of quinonoid 6,7-dihydropteridines into 7,8-dihydropteridines has been mentioned in Chapter IX, Section 7; that of 5,8- into 7,8-dihydropteridines during electrolytic reductions has been mentioned in Section 1.B(3) of this chapter.

2. REACTIONS OF DIHYDROPTERIDINES

The crystal structure of 2-amino-6-methyl-7,8-dihydro-4(3*H*)-pteridinone hydrochloride has been determined by x-ray analysis. The molecule was protonated at N-5 and it proved virtually planar apart from the 7-protons, one of which was 0.82 Å above the plane and the other 0.71 Å below the plane.[788]

A. Oxidation (Aromatization)

The aromatization of 3,4-*dihydropteridines* is exemplified in the formation of pteridine (manganese dioxide and barium oxide in THF at 25°C for 48 h: 52%), 2-methylpteridine (similarly: 80%), 6-methylpteridine (manganese dioxide and anhydrous magnesium sulfate in THF at 5°C for 48 h: 74%), 2(1*H*)-pteridinone (alkaline ferricyanide for 5 h: 78%), 2-pteridinamine (permanganate in pyridine: 74%), and ethyl 2-pteridinecarboxylate (manganese dioxide and anhydrous magnesium sulfate in THF).[51] Aromatization also occurs in the conversion of the covalent hydrate, 4-trifluoromethyl-3,4-dihydro-4-pteridinol (**41**), into 4-trifluoromethylpteridine by dehydration in boiling *t*-butyl alcohol (41% yield) or by sublimation at 150°C in vacuo (65%), and in similar reactions.[156]

The aromatization of 5,6-*dihydropteridines* is represented in the conversion of 5,6-dihydro-7(8*H*)-pteridinone (**47**) into 7(8*H*)-pteridinone by treatment of its sodium salt in aqueous suspension with potassium permanganate at room temperature[31]; in the rapid aerial oxidation of ethyl 8-ethyl-4-ethylamino-7-oxo-5,6,7,8-tetrahydro-6-pteridinecarboxylate (**48**) in solution to give its 7,8-dihydro analog[1311]; in the peculiar loss of benzamide from the adduct, 4-amino-6-benzamido-5,6-dihydro-7(8*H*)-pteridinone (**49**), in neat trifluoroacetic acid at 20°C to give 4-amino-7(8*H*)-pteridinone (91%)[1386]; and possibly in the reported oxidative loss of methane from 2-amino-5-methyl-6,7-diphenyl-5,6-dihydro-4(3*H*)-pteridinone in acidic solution to give 2-amino-6,7-diphenyl-4(3*H*)-pteridinone.[1347]

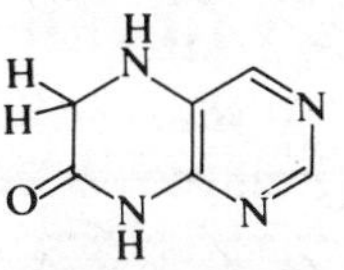

(**47**)

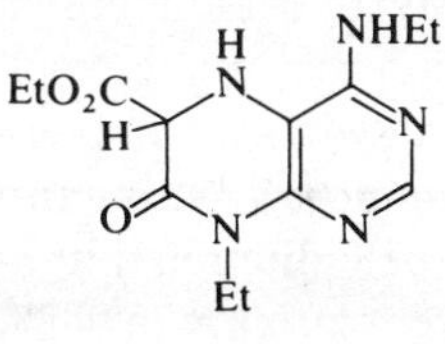

(**48**)

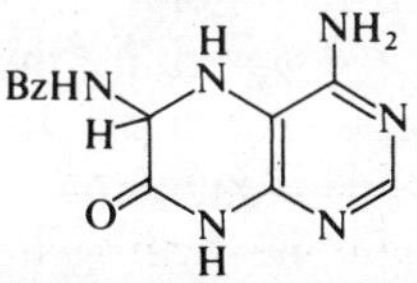

(**49**)

The more usual aromatization of 7,8-*dihydropteridines* is exemplified in the formation of 6(5*H*)-pteridinone (alkaline permanganate at 25°C: 80%),[31] 6-methyl-2(1*H*)-pteridinone (similarly: 67%),[42] 2-amino-4(3*H*)-pteridinone (similarly),[453] 2-amino-6-methyl-4(3*H*)-pteridinone (similarly: 94%),[485, 727] 4,6(3*H*,5*H*)-pteridinedione (similarly: 70%),[29] 2-amino-4,6(3*H*,5*H*)-pteridinedione [alkaline permanganate: 80%[60, 71, 727, 877]; silver nitrate: <50%[1268]; also by aeration of 2-amino-6-lactoyl-7,8-dihydro-4(3*H*)-pteridinone in borate buffer[301, cf. 619]], 2,4,6(1*H*,3*H*,5*H*)-pteridinetrione (alkaline permanganate),[41] 6-methyl-2,4-pteridinediamine (similarly: >95%),[484] 2,6-diamino-4(3*H*)-pteridinone (**50**) (similarly: 45%, as sodium salt),[450] 2-ethylthio-6,7-diphenyl-4(3*H*)-pteridinone (ferric chloride in boiling ethanol or acetic acid: 60 to 80%; similarly from the 5,6-dihydro isomer),[955] 2,6-dioxo-1,2,5,6-tetrahydro-4-pteridinecarboxylic acid [from the 1,2,5,6,7,8-hexahydro analog (**51**) with alkaline permanganate: ~50%],[150] pteroylglutamic acid (ferricyanide at pH 9),[145, cf. 1266] and 6-chloropteridine (oxidation of 7,8-dihydro-6(5*H*)-pteridinone during treatment with phosphoryl chloride containing phosphorus pentachloride: 36%).[35] Other examples will be found under the Boon synthesis in Chapter II, Section 3.A.

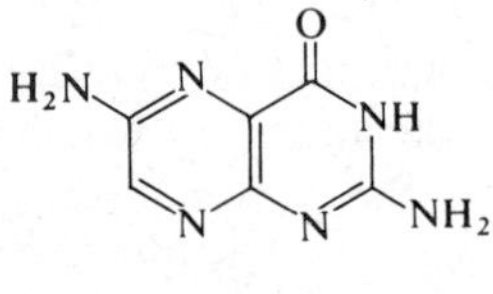

(**50**)

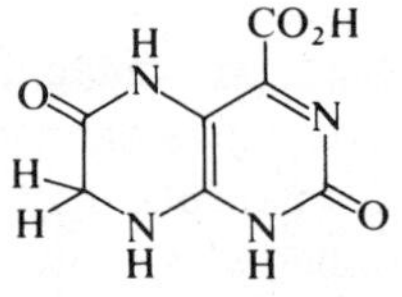

(**51**)

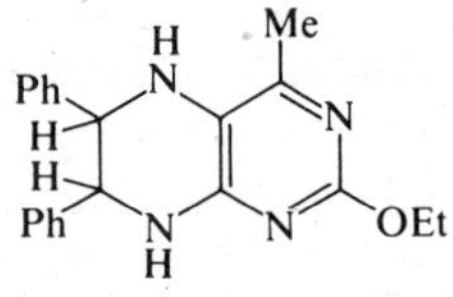

(**52**)

B. Reduction to 5,6,7,8-Tetrahydropteridines

Examples of the further reduction of dihydropteridines are largely confined to the conversion of 7,8-dihydro- into 5,6,7,8-tetrahydropteridines such as 2-ethoxy-4-methyl-6,7-diphenyl-5,6,7,8-tetrahydropteridine (**52**) (hydrogenation over platinum in acetic acid or Raney-nickel in ethanol),[398] 2-chloro-4,6-dimethyl-5,6,7,8-tetrahydropteridine (hydrogenation over platinum in acetic acid: 77%; subsequent hydrogenation over palladium in ethanol gave 4,6-dimethyl-5,6,7,8-tetrahydropteridine: 56%),[1088] 4,6-dimethyl-5,6,7,8-tetrahydro-2-pteridinamine (similarly: 60%),[1088] 4,6-dimethyl-5,6,7,8-tetrahydro-2(1*H*)-pteridinone (similarly),[1088] 2-dimethylamino-4,6-dimethyl-5,6,7,8-tetrahydropteridine (hydrogenation over Raney-nickel in ethanol: 84%),[1088] 2-amino-6-(4′-pyridinyl)-5,6,7,8-tetrahydro-4(3*H*)-pteridinone (hydrogenation over palladium: 66%),[1396] 2-amino-6,7,7-trimethyl-5,6,7,8-tetrahydro-4(3*H*)-pteridinone (hydrogenation over palladium in dilute hydrochloric acid: 78%, as hydrochloride; hydrogenation over platinum in 98% formic acid: 74%, as the 5-formyl derivative),[1503] 2-amino-8-methyl-5,6,7,8-tetrahydro-4(3*H*)-pteridinone (no details),[511] 2-amino-8-methyl-6,7-diphenyl-5,6,7,8-tetrahydro-4(3*H*)-pteridinone (sodium borohydride in aqueous methanol: 31%),[1330] 4-methyl-6,7-diphenyl-5,6,7,8-tetrahydropteridine (hydrogenation over platinum in acetic acid: 67%),[1227] 2-amino-6-*p*-carboxyanilinomethyl-5,6,7,8-tetrahydro-4(3*H*)-pteridinone (tetrahydropteroic acid) (sodium borohydride in aqueous ethanol: 50% yield),[197] 5,6,7,8-tetrahydropteroylglutamic acid [sodium borohydride in aqueous media at pH 10 (60% yield, as dihydrochloride)[194] or in water (75%)[295]; enzymic[1504]], 2-amino-6-methyl-5,6,7,8-tetrahydro-4(3*H*)-pteridinone (enzymic),[89, 298] and 2-amino-6,6-dimethyl-5,6,7,8-tetrahydro-4(3*H*)-pteridinone [from 2-amino-6-methyl-7,8-dihydro-4(3*H*)-pteridinone, first protected by trimethylsilylation and then treated with methyllithium and later with methanolic hydrogen chloride: 28%, as hydrochloride].[644] Likewise, reduction of 2-amino-5,6,7-trimethyl-4-oxo-8-phenyl-3,4,7,8-tetrahydro-5-pteridinium tosylate (**53**, R = Ph) by borohydride in water gave 2-amino-5,6,7-trimethyl-8-phenyl-5,6,7,8-tetrahydro-4(3*H*)-pteridinone (**54**) (60%),[374] 2-amino-3,5,6,7,7-pentamethyl-5,6,7,8-tetrahydro-4(3*H*)-pteridinone (**55**) (34%) was made similarly,[374] and hydrogenation of 1,3,8-trimethyl-6,7-diphenyl-7,8-dihydro-2,4(1*H*,3*H*)-pteridinedione over palladium in ethyl acetate gave its 5,6,7,8-tetrahydro analog (79%).[1096]

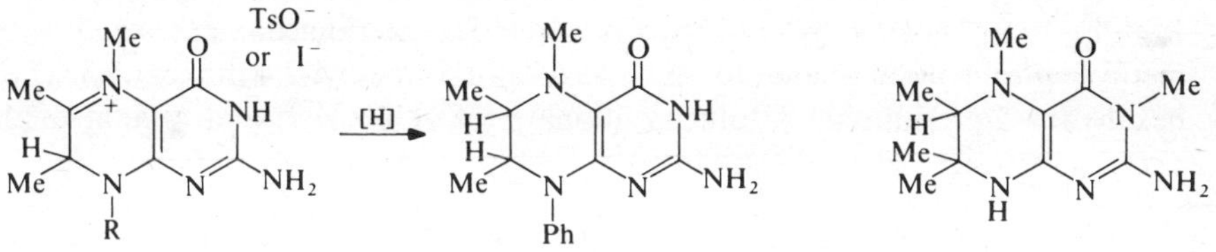

(53) (54) (55)

In addition, the 5,6-dihydropteridine, 2-amino-6,7-diphenyl-5,6-dihydro-4(3*H*)-pteridinone, underwent reduction by sodium borohydride in dilute alkali to give the corresponding 5,6,7,8-tetrahydro analog in 25% yield[1330]; likewise, 2-amino-5-methyl-6,7-diphenyl-5,6-dihydro-4(3*H*)-pteridinone gave its 5,6,7,8-tetrahydro analog (borohydride in aqueous methanol: 46–52%;[509, 981, 1330] hydrogenation over palladium in acetic acid: 51%).[981]

C. Alkylation or Acylation at a Ring Nitrogen

The ring-NH grouping produced in forming a dihydropteridine may be alkylated or acylated. For example, treatment of 6,7,7-trimethyl-7,8-dihydro-2,4-pteridinediamine (**56**, R = H, Y = Me) with butyllithium in dimethyl sulfoxide–hexane followed by benzyl chloride gave the 8-benzyl derivative (**56**, R = CH_2Ph, Y = Me) in 13% yield.[1503] An essentially similar procedure (except for the alkyl halide involved) converted 6-methyl-7,8-dihydro-2,4-pteridinediamine (**56**, R = Y = H) into 6,8-dimethyl- (**56**, R = Me, Y = H) (71%), 8-ethyl-6-methyl- (**56**, R = Et, Y = H) (60%), 8-isopropyl-6-methyl- (**56**, R = *i*-Pr, Y = H) (24%), and 8-benzyl-6-methyl-7,8-dihydro-2,4-pteridinediamine (**56**, R = CH_2Ph, Y = H) (80%).[752, 753] The hydrogenation of 2-amino-8-methyl-4,7(3*H*,8*H*)-pteridinedione over platinum in formic acid, followed by treatment with acetic anhydride at 100°C gave 5-formyl-8-methyl-5,6-dihydro-4,7(3*H*,8*H*)-pteridinedione (**57**) in 37% yield.[181] In contrast, treatment of 2-amino-6,7,8-

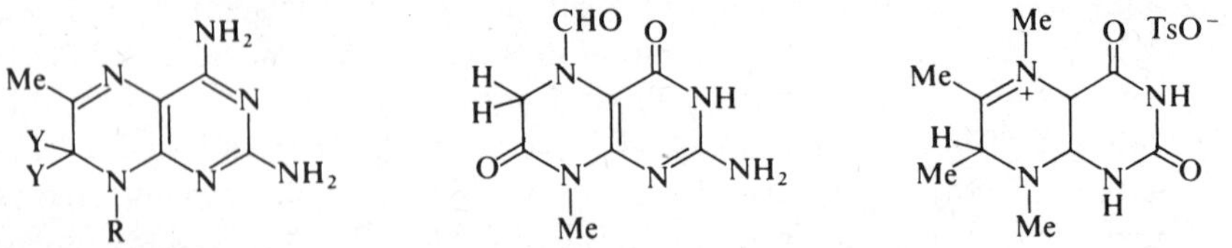

(56) (57) (58)

trimethyl-7,8-dihydro-4(3*H*)-pteridinone with methyl *p*-toluenesulfonate at 110°C resulted in quaternization at N-5 to give 2-amino-5,6,7,8-tetramethyl-4-oxo-3,4,7,8-tetrahydro-5-pteridinium *p*-toluenesulfonate (**53**, R = Me) (45%)[374]; 2-amino-6,7-dimethyl-8-phenyl-7,8-dihydro-4(3*H*)-pteridinone reacted similarly with methanolic methyl iodide at 100°C in a sealed vessel to give 2-amino-5,6,7-trimethyl-4-oxo-8-phenyl-3,4,7,8-tetrahydro-5-pteridinium iodide (**53**, R = Ph)[374]; 6,7,8-trimethyl-7,8-dihydro-2,4(1*H*,3*H*)-pteridinedione reacted with methyl *p*-toluenesulfonate to give 5,6,7,8-tetramethyl-2,4-dioxo-1,2,3,4,7,8-hexahydro-5-pteridinium *p*-toluenesulfonate (**58**) (30%)[375]; and several analogous substrates behaved similarly.[374, 375]

3. PREPARATION OF TETRAHYDROPTERIDINES

Of the large number of possible tetrahydropteridine nuclei, only the 1,2,3,4- and 5,6,7,8-tetrahydro systems are represented in the literature to date. Both the known 1,2,3,4-tetrahydropteridines were made by primary synthesis from pyrazine intermediates (Ch. III) and one of them was made also by reduction. In contrast, the ubiquitous 5,6,7,8-tetrahydropteridines have been made by several primary syntheses from pyrimidines (Ch. II), by reductive processes, and by covalent additions.

A. By Primary Syntheses

Treatment of 3-aminomethyl-2-pyrazinamine (**59**) with aqueous formaldehyde in dilute alkali under reflux gave 1,2,3,4-tetrahydropteridine (**60**, R = H) in 43% yield.[51] Although no attempt appears to have been made to extend this reaction by using a further substituted pyrazine intermediate and/or other aldehydes, ethyl 1,2,3,4-tetrahydro-2-pteridinecarboxylate (**60**, R = CO_2Et) was

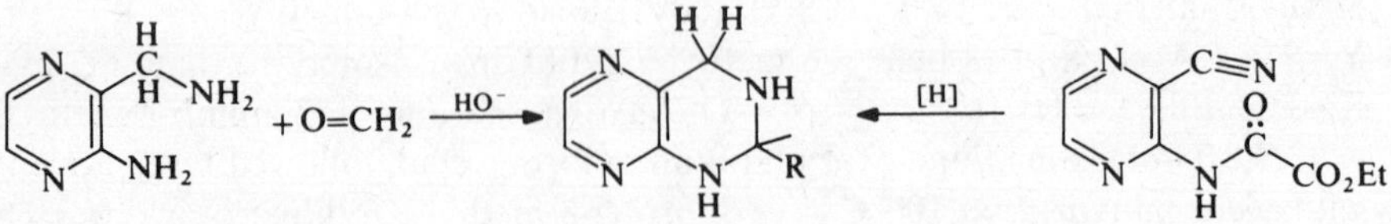

(59) (60) (61)

made in low yield by a quite different method involving the vigorous reductive cyclization of 3-ethoxalylamino-2-pyrazinecarbonitrile (**61**).[51]

The use of a Boonlike synthesis to make 5,6,7,8-tetrahydropteridines directly has been discussed in Chapter II, Section 3.C. Further examples include the cyclization of 2,5-diamino-6-(*N*-β-amino-β-methylpropyl-*N*-methylamino)-4(3*H*)-pyrimidinone (**62**) by successive treatment with bromine in methanolic trifluoroacetic acid, methanolic sodium methoxide, and β-mercaptoethanol (for reduction of a dihydro intermediate?) to give 2-amino-6,6,8-trimethyl-5,6,7,8-tetrahydro-4(3*H*)-pteridinone (**63**) in 58% yield;[1442] and the conversion of 6-acetonylamino-2-amino-5-phenylazo-4(3*H*)-pteridinone into 2-amino-6-methyl-5,6,7,8-tetrahydro-4(3*H*)-pteridinone by vigorous reductive cyclization (hydrogenation over platinum: 70%).[1458] Other useful primary syntheses from 5-halogenopyrimidines have been covered in Chapter II, Section 4.E. For example, 5-bromo-2,4(1*H*,3*H*)-pyrimidinedione with ethylenediamine gave 5,6,7,8-tetrahydro-2(1*H*)-pteridinone in 70% yield[886]; 2,6-diamino-5-bromo-4(3*H*)-pyrimidinone with α,β-propanediamine gave a separable mixture of 2-amino-6- and 2-amino-7-methyl-5,6,7,8-tetrahydro-4(3*H*)-pteridinone[296]; and 5-bromo-6-chloro-1,3-dimethyl-2,4-(1*H*,3*H*)-pyrimidinedione with β-methylaminopropylamine gave only 1,3,4,6-tetramethyl-5,6,7,8-tetrahydro-2,4(1*H*,3*H*)-pteridinedione (**64**) in ~60% yield.[456]

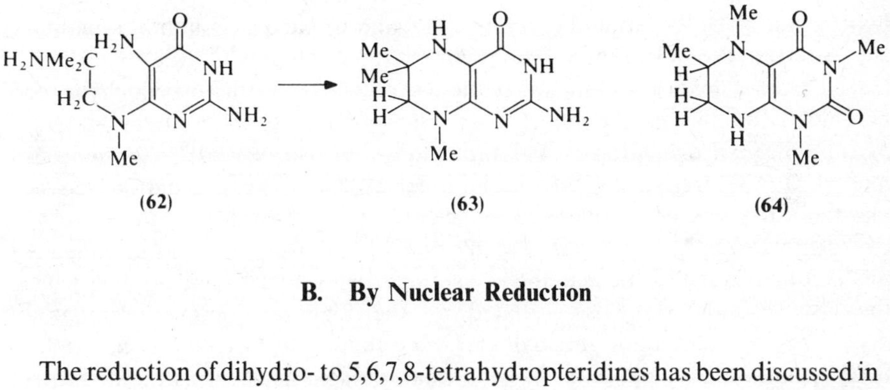

(62) **(63)** **(64)**

B. By Nuclear Reduction

The reduction of dihydro- to 5,6,7,8-tetrahydropteridines has been discussed in Section 2.B in this chapter. The apparently direct reduction of aromatic pteridines to tetrahydropteridines undoubtedly proceeds via dihydropteridines in most cases, an assumption for which there is some experimental evidence.[201, 202]

The reductive formation of *simple tetrahydropteridines* is illustrated in the conversion of pteridine into a separable mixture of 1,2,3,4- (**65**) and 5,6,7,8-tetrahydropteridine (**66**) (sodium borohydride in THF with the gradual addition of trifluoroacetic acid: 38 and 58% yield, respectively)[721]; only the latter product (**66**) was obtained on reducing pteridine in ether with lithium aluminum hydride (58% yield) or by the hydrogenolysis over palladium in methanol of 2,4-dichloro-5,6,7,8-tetrahydropteridine, itself prepared by treatment of 2,4-dichloro- or 2,4,6,7-tetrachloropteridine with lithium aluminum hydride.[473, 486] Other ex-

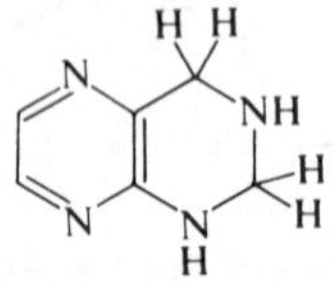

(65)

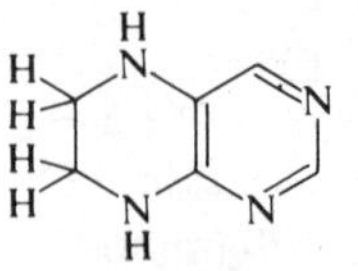

(66)

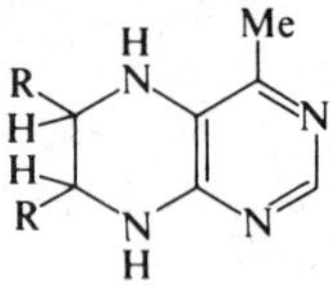

(67)

amples include the conversion of appropriate pteridines into 4-methyl-5,6,7,8-tetrahydropteridine (**67**, R = H) (lithium aluminum hydride in ether: ~60%),[683] 4-methyl-6,7-diphenyl-5,6,7,8-tetrahydropteridine (**67**, R = Ph) (hydrogenation over platinum in acetic acid),[1227] and 2-ethoxy-6,7-dimethyl-4-trifluoromethyl-5,6,7,8-tetrahydropteridine (sodium cyanoborohydride in dilute hydrochloric acid: 16%).[1440] In addition, sodium borohydride in methanol was so used to produce ethyl 2-amino-6,7-dimethyl-5,6,7,8-tetrahydro-4-pteridinecarboxylate (**68**, R = NH_2) (37%), as well as its 2-dimethylamino (**68**, R = NMe_2) (38%), 2-methylthio (**68**, R = SMe) (30%), 2-chloro (**68**, R = Cl) (26%), and 2-ethoxy analog (**68**, R = OEt) (40%).[1440]

The reduction of 2-amino-4(3*H*)-pteridinone and derivatives to 5,6,7,8-*tetrahydropterins* is exemplified in the formation of 2-amino-5,6,7,8-tetrahydro-4(3*H*)-pteridinone itself (**69a**) [hydrogenation over platinum in trifluoroacetic acid and isolated as sulfate (70%)[509, 693] or hydrochloride (> 90%),[646, 693] or by hydrogenation over platinum in dilute alkali and isolated as 2-amino-5-methyl-5,6,7,8-tetrahydro-4(3*H*)-pteridinone (**69b**) (~75%, as hydrochloride) or as the 5-benzyl analog (**69c**) (~75%) by reductive alkylation with formaldehyde or benzaldehyde,[297] respectively]; 2-amino-3-methyl-5,6,7,8-tetrahydro-4(3*H*)-pteridinone (hydrogenation over platinum in trifluoroacetic acid)[693]; 2-amino-6-methyl-5,6,7,8-tetrahydro-4(3*H*)-pteridinone (**69d**) [hydrogenation over platinum in acetic acid (low yield as acetate salt),[1394] over palladium in 6*M* hydrochloric acid (63%, as dihydrochloride),[939] or over platinum in aqueous methanolic trifluoroacetic acid (66%, as dihydrochloride)[202]; the product from the last reaction was also isolated as the 2-amino-5,6-dimethyl analog (**69e**)[1392]][1367]; 2-amino-7-methyl-5,6,7,8-tetrahydro-4(3*H*)-pteridinone (**69f**) (hydrogenation over platinum in trifluoroacetic acid: isolated as the dihydrochloride)[201]; 2-amino-6,7-dimethyl-5,6,7,8-tetrahydro-4(3*H*)-pteridinone (**69g**) [hydrogenation over platinum in 5*M* hydrochloric acid (~80%, as hydrochloride),[1148] in trifluoroacetic acid (77%, as sulfate),[1343] in 0.1*M* hydrochloric acid,[1224] or over palladium in 5*M* hydrochloric acid (as the hydrochloride)[1038]; it may also be isolated as its 2-amino-5,6,7-trimethyl (**69h**) or 2-amino-5-benzyl-6,7-dimethyl analog (**69i**) after an appropriate alkylation[297]; hydrogenation over platinum in formic acid gave 2-amino-5-formyl-6,7-dimethyl-5,6,7,8-tetrahydro-4(3*H*)-pteridinone (**69j**) (~80%)[1148]]; 2-amino-6-phenyl-5,6,7,8-tetrahydro-4(3*H*)-pteridinone (**69k**) (hydrogenation over platinum in 0.1*M* hydrochloric acid)[1224]; 2-amino-6,7-diphenyl-5,6,7,8-tetrahydro-4(3*H*)-pteridinone (**69l**) [hydrogenation over platinum in aqueous or alcoholic hydrochloric acid: Isolated as the hydrochloride[1224] or as the 5-methyl derivative (**69m**) after reductive

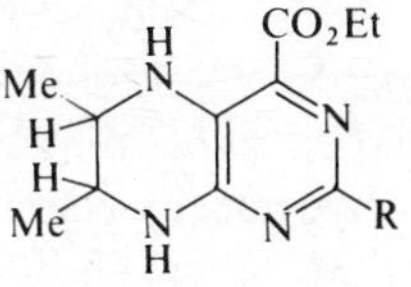

(68)

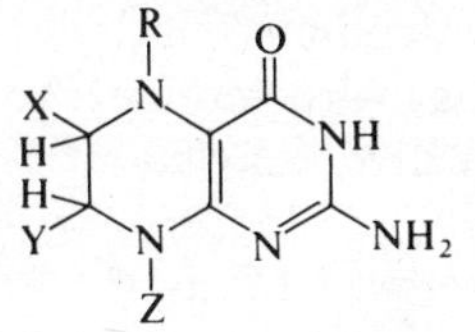

(69)

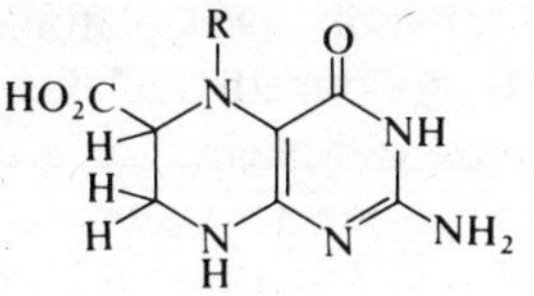

(70)

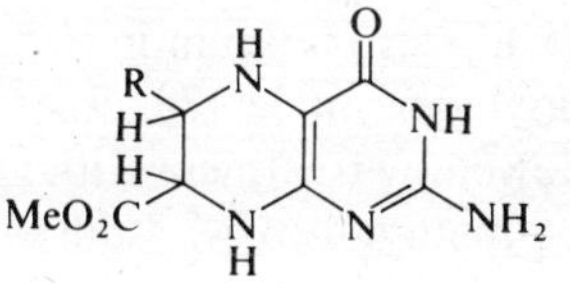

(71)

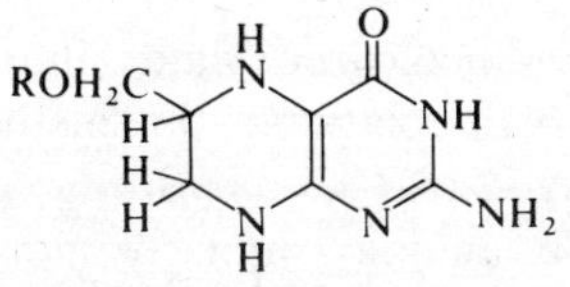

(72)

	R	X	Y	Z
a	H	H	H	H
b	Me	H	H	H
c	CH_2Ph	H	H	H
d	H	Me	H	H
e	Me	Me	H	H
f	H	H	Me	H
g	H	Me	Me	H
h	Me	Me	Me	H
i	CH_2Ph	Me	Me	H
j	CHO	Me	Me	H
k	H	Ph	H	H
l	H	Ph	Ph	H
m	Me	Ph	Ph	H
n	H	CH_2Ph	CH_2Ph	H
o	H	H	H	Me
p	Me	H	H	Me
q	CH_2Ph	H	H	Me
r	H	H	H	*i*-Pr
s	H	H	H	CH_2Ph
t	H	H	H	Ph
u	H	Me	Me	Me
v	Me	Me	Me	Me
w	H	Me	Me	CH_2Ph
x	H	Me	Me	Ph
y	H	Ph	Ph	Me
z	Me	Ph	Ph	Me

alkylation[297, 981]]; 2-amino-6,7-dibenzyl-5,6,7,8-tetrahydro-4(3*H*)-pteridinone (**69n**) (hydrogenation over platinum in trifluoroacetic acid: 64%, as hydrochloride)[419]; 2-amino-8-methyl-5,6,7,8-tetrahydro-4(3*H*)-pteridinone (**69o**) [from 2-amino-8-methyl-4(8*H*)-pteridinone by sodium borohydride in water (52%, as base),[373] by hydrogenation over platinum in trifluoroacetic acid (69%, as dihydrochloride),[373] or by hydrogenation over platinum in methanol and isolated after reductive alkylation as the 5,8-dimethyl (**69p**) or 5-benzyl-8-methyl analog (**69q**) in poor yields[297]]; 2-amino-8-isopropyl-5,6,7,8-tetrahydro-4(3*H*)-pteridinone (**69r**) (hydrogenation over platinum in trifluoroacetic acid: 40% as hydrochloride)[373]; 2-amino-8-benzyl-5,6,7,8-tetrahydro-4(3*H*)-pteridinone (**69s**) (sodium borohydride in water: 73%)[373]; 2-amino-8-phenyl-5,6,7,8-tetrahydro-4(3*H*)-pteridinone (**69t**) (similarly: 50%)[373]; 2-amino-8-*p*-tolyl- and 2-amino-8-*p*-chlorophenyl-5,6,7,8-tetrahydro-4(3*H*)-pteridinone (sodium borohydride: 52% and 78%, respectively)[373]; 2-amino-6,7,8-trimethyl-5,6,7,8-tetrahydro-4(3*H*)-

pteridinone (**69u**) [sodium borohydride in water: 35%[373]; hydrogenation over platinum: 56%, as hydrochloride[373] or ~60% as the 5,6,7,8-tetramethyl analog (**69v**) after reductive alkylation[297]]; 2-amino-8-benzyl-6,7-dimethyl-5,6,7,8-tetrahydro-4(3*H*)-pteridinone (**69w**) (sodium borohydride in water: 60%)[373]; 2-amino-6,7-dimethyl-8-phenyl-5,6,7,8-tetrahydro-4(3*H*)-pteridinone (**69x**) (similarly: 50%)[373]; 2-amino-6,7-dimethyl-8-*p*-tolyl- and 2-amino-8-*p*-chlorophenyl-6,7-dimethyl-5,6,7,8-tetrahydro-4(3*H*)-pteridinone (sodium borohydride in water: 52 and 67%, respectively)[373]; 2-amino-8-methyl-6,7-diphenyl-5,6,7,8-tetrahydro-4(3*H*)-pteridinone (**69y**) (hydrogenation over platinum in aqueous ethanolic hydrochloric acid: ~80%, as the 5,8-dimethyl analog (**69z**) after reductive methylation)[297]; 2-amino-3,6,7-trimethyl-8-phenyl (or *p*-tolyl)-5,6,7,8-tetrahydro-4(3*H*)-pteridinone (sodium borohydride in water: 94 and 58%, respectively)[373]; and 2-amino-8-*p*-chlorophenyl-3-methyl-5,6,7,8-tetrahydro-4(3*H*)-pteridinone (similarly: 34%).[373]

Similar reductions were involved in the transformation of 2-amino-4-oxo-3,4-dihydro-6-pteridinecarboxylic acid into 2-amino-4-oxo-3,4,5,6,7,8-hexahydro-6-pteridinecarboxylic acid (**70**, R = H) (hydrogenation over platinum in trifluoroacetic acid: 71%, as sulfate) or into its 5-formyl derivative (**70**, R = CHO) (hydrogenation over platinum in dilute alkali followed by treatment with formic acid–acetic anhydride: 40%)[307]; of methyl 6-methyl-4-oxo-3,4-dihydro-7-pteridinecarboxylate into its 3,4,5,6,7,8-hexahydro analog (**71**, R = Me) hydrogenation over platinum in trifluoroacetic acid: 91% as dihydrochloride)[307]; of dimethyl 2-amino-4-oxo-3,4-dihydro-6,7-pteridinedicarboxylate into its 3,4,5,6,7,8-hexahydro analog (**71**, R = CO_2Me) (similarly: 72%)[307]; of dimethyl 2-acetamido-4-oxo-3,4-dihydro-6,7-pteridinedicarboxylate into its 3,4,5,6,7,8-hexahydro analog (hydrogenation over platinum in methanol: 54%)[307]; and of dimethyl 2-amino-8-methyl-4-oxo-4,8-dihydro-6,7-pteridinedicarboxylate into its 3,4,5,6,7,8-hexahydro analog (similarly: not isolated as such but oxidized to the 3,4,7,8-tetrahydro analog before characterization).[307] In addition, a series of 6-alkoxymethyl-2-amino-4(3*H*)-pteridinone 8-oxides has been reduced by hydrogenation over platinum in trifluoroacetic acid (with concomitant deoxygenation) to give 2-amino-6-methoxymethyl-5,6,7,8-tetrahydro-4(3*H*)-pteridinone (**72**, R = Me) (75%) and its 6-ethoxy- (**72**, R = Et) (64%), 6-propoxy- (**72**, R = Pr) (68%), 6-isopropoxy- (**72**, R = *i*-Pr) (48%), 6-butoxy- (**72**, R = Bu) (77%), 6-isobutoxy- (**72**, R = *i*-Bu) (46%), 6-*t*-butoxy- (**72**, R = *t*-Bu) (50%), 6-pentyloxy- (**72**, R = Pe) (67%), 6-*β*-methoxyethoxy- (**72**, R = CH_2CH_2OMe) (77%), and 6-octyloxy-methyl analog (84%).[1728] It is not at all clear whether the same products were (or could be) obtained from the corresponding substrates lacking the *N*-oxide function.

Several 5,6,7,8-*tetrahydro*-2,4-*pteridinediamines* and related compounds have been made by reduction of their aromatic counterparts. The products include, for example, 5,6,7,8-tetrahydro-2,4-pteridinediamine itself (**73**, R = H) [hydrogenation over platinum in methanol (>80%, as base) or in water (67%, as dihydrochloride; or lower yields as the 5-formyl or other acyl derivatives)][267]; 2,4-bisdimethylamino-5,6,7,8-tetrahydropteridine (hydrogenation over platinum in methanol: 37% as its 5-benzoyl derivative after treatment with benzoyl chloride in pyridine; 23% as its 5,8-diacetyl derivative after treatment with acetic anhydride)[267]; 6-methyl-5,6,7,8-tetrahydro-2,4-pteridinediamine (**73**, R = Me) (hydrogenation over platinum in 3*M* hydrochloric acid: >95% as hydrochloride)[84]; and 2,4-diacetamido-6,7-dimethyl-5,6,7,8-tetrahydropteridine (hydrogenation over platinum in methanol: 77%).[648]

A variety of 5,6,7,8-*tetrahydrolumazines* have been prepared from 2,4(1*H*,3*H*)-pteridinediones. Such reduced lumazines are represented by 5,6,7,8-tetrahydro-2,4(1*H*,3*H*)-pteridinedione (**74**, R = Y = H) (sodium amalgam in the presence of β,γ-dimercaptopropanol as an antioxidant: 63%; sodium dithionite in dilute alkali under nitrogen: 54%),[43] its 6,7,8-trimethyl derivative (**74**, R = Y = Me) (sodium borohydride in water: 20%),[375] the 6,7-diisopropyl-8-methyl homolog (**73**, R = Me, Y = *i*-Pr) (hydrogenation over platinum in ethanol: 69%),[375] the 8-isopropyl-6,7-dimethyl analog (**74**, R = *i*-Pr, Y = Me) (hydrogenation over Raney-nickel in methanol: 50%, as its 5-formyl derivative after treatment with formic acid–acetic anhydride),[375] the 8-β-hydroxyethyl-6,7-dimethyl analog (**74**, R = CH_2CH_2OH, Y = Me) (hydrogenation over platinum in water: 48%),[375] 3,6,7,8-tetramethyl-5,6,7,8-tetrahydro-2,4(1*H*,3*H*)-pteridinedione (similarly: 17%),[375] 6,7-diisopropyl-3,8-dimethyl-5,6,7,8-tetrahydro-2,4(1*H*,3*H*)-pteridinedione (hydrogenation over platinum in methanol: 44%),[375] 8-isopropyl-3,6,7-trimethyl-5,6,7,8-tetrahydro-2,4(1*H*,3*H*)-pteridinedione (similarly: 44%, as its 5-formyl derivative after treatment with formic acid–acetic anhydride),[375] 6,7-diphenyl-5,6,7,8-tetrahydro-2,4(1*H*,3*H*)-pteridinedione (**74**, R = H, Y = Ph) [hydrogenation over platinum in methanol: isolated initially after treatment with acetic anhydride as its 5-acetyl derivative (58%), which was then deacetylated by methanolic hydrogen chloride (72% yield)],[265] 1-methyl-6,7-diphenyl-5,6,7,8-tetrahydro-2,4(1*H*,3*H*)-pteridinedione [hydrogenation over platinum in acetic acid followed by acetylation to the 5-acetyl derivative (48%) and subsequent deacetylation (79%)],[265] 1,3,5-trimethyl-6,7-diphenyl-5,6,7,8-tetrahydro-2,4(1*H*,3*H*)-pteridinedione [from 1,3-dimethyl-6,7-diphenyl-2,4(1*H*,3*H*)-pteridinedione by hydrogenation over platinum in ethanol containing formaldehyde: 68%; note the 5-methylation by concomitant reductive alkylation],[1097] 5-acetyl-1,3-dimethyl-6,7-diphenyl-5,6,7,8-tetrahydro-2,4(1*H*,3*H*)-pteridinedione (from the same substrate with zinc dust in acetic acid containing acetic anhydride: ~70%; note the additional 5-acetylation),[1096] and 1,3,6,7-tetramethyl-5,6,7,8-tetrahydro-2,4(1*H*,3*H*)-pteridinedione (hydrogenation over palladium in acetic acid: 95%, as its 5-acetyl derivative after treatment with acetic anhydride).[1096]

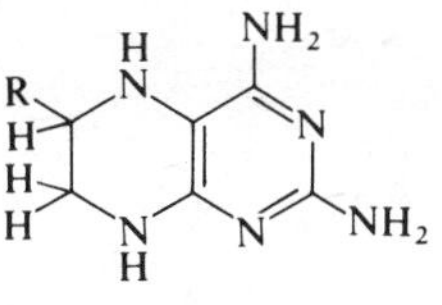

(**73**)

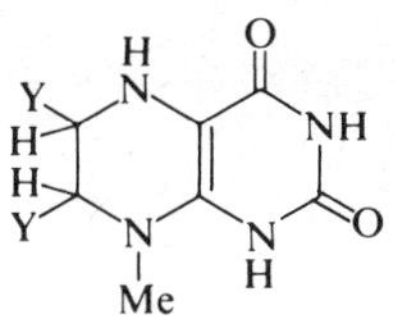

(**74**)

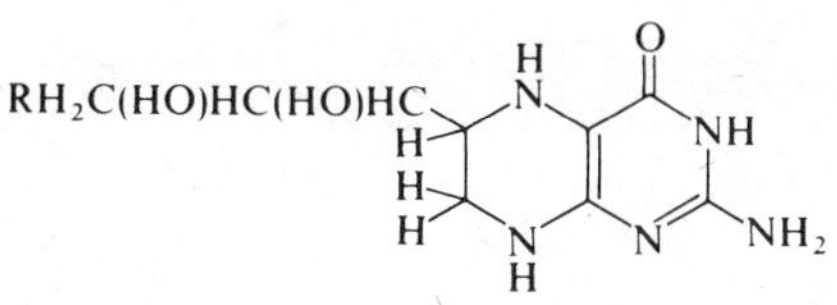

(**75**)

Some *complicated* 5,6,7,8-*tetrahydropteridines* have been made by nuclear reduction. The reduction of biopterin to the natural and unnatural stereoisomers of tetrahydrobiopterin, 2-amino-6-α,β-dihydroxypropyl-5,6,7,8-tetrahydro-4(3*H*)-pteridinone (**75**, R = H), has been discussed in Chapter VI, Section 5.D(4). Two naturally occurring stereoisomers of 2-amino-6-α,β,γ-trihydroxypropyl-4(3*H*)-pteridinone (**75**, R = OH), viz neopterin and monapterin, have been reduced to their respective 5,6,7,8-tetrahydro derivatives by hydrogenation over platinum in trifluoroacetic acid. Each product was isolated as a dihydrochloride in 96% yield.[428] As might be expected, reduction of folic acid to 5,6,7,8-tetrahydrofolic acid has long been known. Although titanous chloride,[1001] sodium dithionite,[78] and hydrogenation over platinum in acetic acid[744, 1081, 1125] or in formic acid (best isolated as 5-formyltetrahydrofolic acid, i.e., folinic acid, by subsequent treatment of the solution with acetic anhydride)[1148, 1191] have all been used, the definitive method appears to be hydrogenation over platinum in trifluoroacetic acid with isolation as the sulfate (57%).[509, 693]

C. By Covalent Addition

The addition of water, alcohols, amines, or Michael reagents to both the 5,6- and the 7,8-bonds of appropriate pteridines leads to 5,6,7,8-tetrahydropteridines. For example, ^{1}H nmr studies have indicated that pteridine, at equilibrium in aqueous solution of pH 2, exists as 5,6,7,8-tetrahydro-6,7-pteridinediol (**76**, R = H) (80%) and as 3,4-dihydro-4-pteridinol (20%).[27] Pteridine in the lower alcohols gave similar mixtures from which 6,7-dimethoxy- (**76**, R = Me) and 6,7-diethoxy-5,6,7,8-tetrahydropteridine (**76**, R = Et) were isolated, analyzed, and characterized.[48] Finally, dissolution of pteridine in ethanolic methylamine gave 6,7-bismethylamino-5,6,7,8-tetrahydropteridine (**77**, R = Y = H) (92%).[52] Likewise, 4-methylpteridine with benzylamine in THF gave 6,7-bisbenzylamino-4-methyl-5,6,7,8-tetrahydropteridine (**77**, R = Me, Y = Ph)[50]; 4(3*H*)-pteridinone with aqueous sodium hydrogen sulfite gave disodium 4-oxo-3,4,5,6,7,8-hexahydro-6,7-pteridinedisulfonate (52%) or with aqueous dimedone gave 6,7-bis(4′,4′-dimethyl-2′,6′-dioxocyclohexyl)-5,6,7,8-tetrahydro-4(3*H*)-pteridinone (48%)[47]; 2-methyl-4(3*H*)-pteridinone or 2,4(1*H*,3*H*)-pteridinedione with aqueous sodium hydrogen sulfite gave disodium 2-methyl-4-oxo-3,4,5,6,7,8-hexahydro- (70%) or disodium 2,4-dioxo-1,2,3,4,5,6,7,8-octahydro-6,7-pteridinedisulfonate (**78**) (70%), respectively[47]; ethyl 4-pteridinecarboxylate with

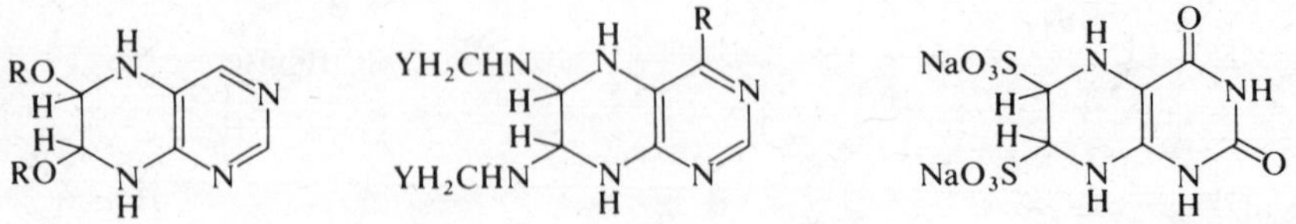

(76) (77) (78)

water or ethanol gave ethyl 6,7-dihydroxy- (> 70%) or ethyl 6,7-diethoxy-5,6,7,8-tetrahydro-4-pteridinecarboxylate[147, 148]; ethyl 7-methyl-4-pteridinecarboxylate with water gave ethyl 6,7-dihydroxy-7-methyl-5,6,7,8-tetrahydro-4-pteridinecarboxylate (~40%)[148]; and condensation of hexafluorobiacetyl with 4,5-pyrimidinediamine, followed by recrystallization from aqueous ethanol, gave 6,7-bistrifluoromethyl-5,6,7,8-tetrahydro-6,7-pteridinediol (9%).[1571]

4. REACTIONS OF TETRAHYDROPTERIDINES

Because of the importance of three-dimensional shape in essential enzyme reactions involving tetrahydrobiopterin and tetrahydrofolic acid, much nmr, x-ray, and other physical study has gone into determining the chirality and preferred conformation(s) of the natural coenzymes and of simplified model tetrahydropterins that still retain activity as such. These areas of tetrahydropterins have been reviewed recently[1434, 1476, 1478, 1536] along with their biochemical background.[1572] Some important aspects of such work (with leading original references) include: stereochemistry and conformation of 2-amino-6-methyl- and 2-amino-6,7-dimethyl-5,6,7,8-tetrahydro-4(3*H*)-pteridinone and some methyl derivatives[84, 89, 298, 518, 519, 521, 607, 648, 705, 789, 1471]; the ^{13}C nmr and crystal structure of 2-amino-5-formyl-5,6,7,8-tetrahydro-4(3*H*)-pteridinone[102, 104, 609]; absolute configuration and preferred conformation of 2-amino-6-α,β-dihydroxypropyl-5,6,7,8-tetrahydro-4(3*H*)-pteridinone (tetrahydrobiopterin)[298, 401, 429, 645, 1608]; use of polyacetylated tetrahydrobiopterins in stereochemical studies[199, 203, 401]; the similar use of pentaacetylated neopterin, 2-acetamido-5-acetyl-6-α,β,γ-triacetoxypropyl-5,6,7,8-tetrahydro-4(3*H*)-pteridinone[198]; and the configuration and conformation of 5,6,7,8-tetrahydropteroylglutamic acid and its derivatives.[87, 197, 442, 1737] The geometries and relevant properties of a number of tetrahydropteridines have been calculated and discussed.[1515] The fluorescent properties of tetrahydrofolic acid and related compounds have been recorded with a view to detecting and estimating such compounds in very low concentration.[1338] Mass spectra have been recorded and used in analyses.[1367]

A. Oxidative reactions

The oxidation of tetrahydro- to dihydropteridines has been discussed already in Section 1.C in this chapter. The more specialized biochemical oxidation of tetrahydrobiopterin to *quinonoid* dihydrobiopterin, as well as some related oxidations, have been covered in Chapter IX, Section 7. In addition, the chemical oxidation of dihydropteridines to pteridines has been summarized in Section 2.A in this chapter. Although it must therefore be possible to oxidize many tetrahydro- to dihydropteridines and then to pteridines, the one-pot aromatization of tetrahydropteridines to pteridines has rarely been reported.[cf. 854, 1342]

Indeed, attempts to prepare pteridine from its 5,6,7,8-tetrahydro derivative by no less than 10 differing methods, all failed.[473-486] However, using a more stable substrate to give a more stable product, successful aromatizations have been reported from 5,6,7,8-tetrahydro-2,4-pteridinediamine (**79**) (fresh manganese dioxide in DMF at 100°C for 11 h: 79%),[486] and from 2-amino-6,7-dimethyl-5,6,7,8-tetrahydro-4(3*H*)-pteridinone (aeration in strongly alkaline solution).[1038] Successful oxidations involving additional changes to the molecule include the conversion of 2-amino-5,6,7,8-tetrahydro-4(3*H*)-pteridinone sulfite by prolonged aerial oxidation (or better by permanganate treatment) into a product, which was formulated initially as 2-amino-4-oxo-3,4-dihydro-6-pteridinesulfonic acid hydrate, but which was later shown to be 2-amino-6-sulfo-oxy-4(3*H*)-pteridinone (**80**), that is, the sulfate ester of xanthopterin[1051, 1431]; of 6,7-dihydroxy-5,6,7,8-tetrahydro-4-pteridinecarboxamide (**81**) into 4-pteridinecarboxamide in low yield by removal of two molecules of water on refluxing in bis-β-methoxyethyl ether for 10 min[157]; and of 2-amino-6-α,β,γ-trihydroxypropyl-5,6,7,8-tetrahydro-4(3*H*)-pteridinone (**75**, R = OH) into 2-amino-6-α,β,γ-trihydroxypropyl- (neopterin: 53% yield) and 2-amino-6-α,γ-dihydroxypropyl-4(3*H*)-pteridinone (**82**) in small yield (along with two 7,8-dihydropteridines) by dark aeration during 5 days.[73, cf. 560] Such so-called *autoxidations* of tetrahydropterins clearly proceed by complicated mechanisms, probably involving hydroperoxide intermediates[953, 981, 1038, 1097] or even dimeric compounds.[1328] Further leading references have been cited in a brief review of autoxidation.[1478]

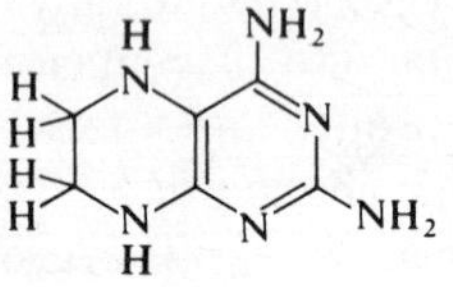

(**79**)

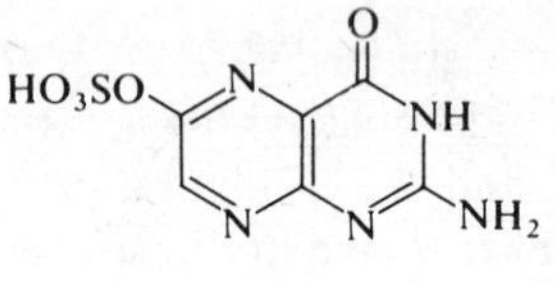

(**80**)

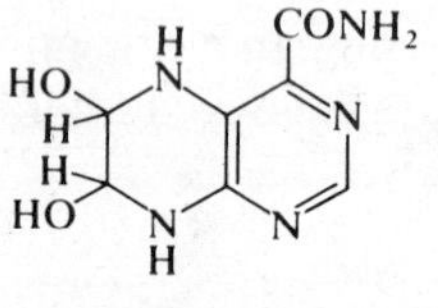

(**81**)

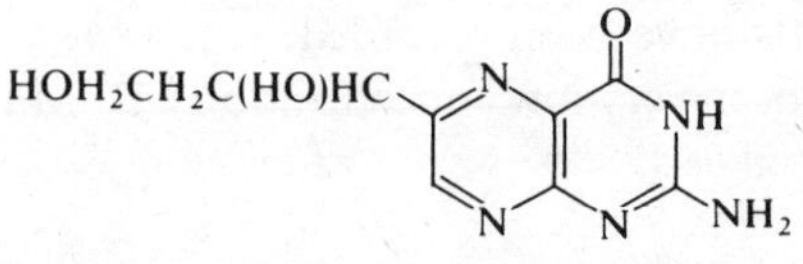

(**82**)

B. *N*-Acylation and *N*-Alkylation

Those 5,6,7,8-tetrahydropteridines with free 5-NH and 8-NH groupings undergo acylation initially at the 5- and then at the 8-position (with subsequent acylation of any amino substituent present). Reductive alkylation (hydrogenation in the presence of an aldehyde) appears to follow suit by attacking the 5-NH grouping first.

Thus appropriate 5,6,7,8-tetrahydropteridine substrates (each present as salt, as free base, or simply prepared in a suitable solvent) gave 5-formyl-4-methyl-5,6,7,8-tetrahydropteridine (**83**, X = CHO, Y = H) (formic acid–acetic anhydride at 25°C for 24 h: 70%),[83, 118] the 5,8-diformyl analog (**83**, X = Y = CHO) (acetic formic anhydride at 25°C for 15 h),[118] the 5,8-diacetyl analog (**83**, X = Y = Ac) (boiling acetic anhydride for 10 min: 75%),[118] 8-benzyl-2-chloro-5-formyl-4-methyl-5,6,7,8-tetrahydropteridine (formic acid–acetic anhydride at 25°C for 15 h),[118] the 5-acetyl homolog (brief boiling in acetic anhydride),[118] 2-amino-5-formyl-6-methyl-5,6,7,8-tetrahydro-4(3*H*)-pteridinone (formic acid–acetic anhydride at 22°C for 24 h: 33%),[200] 2-acetamido-5,8-diacetyl-6-methyl-5,6,7,8-tetrahydro-4(3*H*)-pteridinone (from the 2-amino-6-methyltetrahydropteridinone with acetic anhydride containing a catalytic amount of trifluoroacetic anhydride at 25°C for 15 h: 70%),[520] 6-methyl-2-trifluoroacetamido-3,5,8-tristrifluoroacetyl-5,6,7,8-tetrahydro-4(3*H*)-pteridinone (from the 2-amino-6-methyltetrahydropteridinone with trifluoroacetic anhydride at 25°C, twice: 75%),[520] 2-amino-5-formyl-6,7-dimethyl-5,6,7,8-tetrahydro-4(3*H*)-pteridinone (formic acid–acetic anhydride at 22°C for 16 h: 42%),[703] 2-acetamido-5-acetyl-6,7-dimethyl-5,6,7,8-tetrahydro-4(3*H*)-pteridinone (from the 2-amino-6,7-dimethyltetrahydropteridinone with acetic anhydride plus a little trifluoroacetic anhydride at 25°C for 15 h: 60%, plus a separable triacetyl derivative),[520] 2-amino-6,7-dimethyl-5-trifluoroacetyl-5,6,7,8-tetrahydro-4(3*H*)-pteridinone (trifluoroacetic anhydride at 25°C for 15 h: 66%),[520] 6,7-dimethyl-2-trifluoroacetamido-3,5-bistrifluoroacetyl-5,6,7,8-tetrahydro-4(3*H*)-pteridinone (from the 2-amino-6,7-dimethyltetrahydropteridinone with trifluoroacetic anhydride at 25°C, twice: 53%),[520] 5,6,7-trimethyl-2-trifluoroacetamido-3,8-bistrifluoroacetyl-5,6,7,8-tetrahydro-4(3*H*)-pteridinone (from the 2-amino-5,6,7-trimethyltetrahydropteridinone with trifluoroacetic anhydride at 25°C, twice),[520] 2-amino-5-formyl-8-methyl-5,6,7,8-tetrahydro-4(3*H*)-pteridinone (formic acid–acetic anhydride at 25°C for 15 min: 87%),[373] its 8-phenyl analog (similarly: 54%),[373] its 8-*p*-chlorophenyl analog (similarly: 70%),[373] 2-amino-5-formyl-6,8-dimethyl-5,6,7,8-tetrahydro-4(3*H*)-pteridinone (formic acid–acetic anhydride at 25°C for 3 h: 17%),[373] 2-amino-5-formyl-6,7,8-trimethyl-5,6,7,8-tetrahydro-4(3*H*)-pteridinone (formic acid–acetic anhydride at 25°C for 1 h: 48%),[373] its 6,7-dimethyl-8-phenyl analog (similarly: 40%),[373] the 8-benzyl-6,7-dimethyl analog (similarly: 28%),[373] 5-acetyl-2-amino-6,7-dimethyl-8-phenyl-5,6,7,8-tetrahydro-4(3*H*)-pteridinone (brief warming in acetic anhydride: ~85%),[373] the 8-*p*-chlorophenyl analog (similarly: 44%),[373] 5-acetyl-2-amino-8-

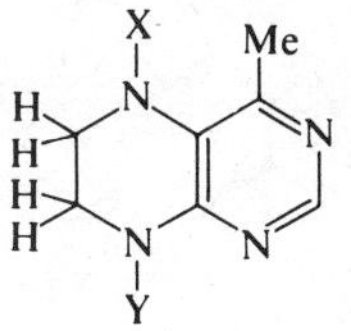

(83)

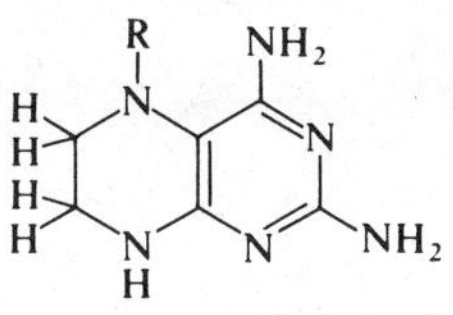

(84)

benzyl-6,7-dimethyl-5,6,7,8-tetrahydro-4(3*H*)-pteridinone (similarly: 45%),[373] dimethyl 2-amino-5-formyl-4-oxo-3,4,5,6,7,8-hexahydro-6,7-pteridinedicarboxylate (formic acid–acetic anhydride at 25°C for 18 h: ~60%),[307] 5-formyl-5,6,7,8-tetrahydropteroylglutamic acid (methyl formate in pyridine–dimethyl sulfoxide at 22°C for 24 h: 84%),[259] 5-formyl-5,6,7,8-tetrahydro-2,4-pteridinediamine (**84**, R = CHO) (formic acid–acetic anhydride at 25°C for 15 h: 37%),[267] the 5-acetyl analog (**84**, R = Ac) (acetic anhydride at 25°C for several hours: 50%),[267] the 5-chloroacetyl analog (**84**, R = $COCH_2Cl$) (chloroacetyl chloride–acetic acid–anhydrous sodium acetate for 3 min: 48%; chloroacetic anhydride in dioxane at 25°C for 5 h: 33%),[267] the 5-benzoyl analog (**84**, R = Bz) (benzoyl chloride–pyridine at 25°C for 7 h: 39%, as hydrochloride),[267] 2,4-diacetamido-5,8-diacetyl-5,6,7,8-tetrahydropteridine (from the tetrahydropteridinediamine with refluxing acetic anhydride for 15 min: 63%),[267] 5-formyl-5,6,7,8-tetrahydro-2,4(1*H*,3*H*)-pteridinedione (acetic formic anhydride at 20°C for 15 h: 28%),[43] 5-formyl-6,7,8-trimethyl-5,6,7,8-tetrahydro-2,4(1*H*,3*H*)-pteridinedione (formic acid–acetic anhydride at 20°C for 12 h: 65%),[375] and 5-formyl-1,3-dimethyl-6,7-diphenyl-5,6,7,8-tetrahydro-2,4(1*H*,3*H*)-pteridinedione (formic acid–acetic anhydride at 20°C for 30 min: 47%).[1097] Other examples, in which tetrahydropteridines have been prepared by reduction and subsequently isolated as their 5-acyl derivatives, have been given in Section 3.B in this chapter.

The 6(*R*)- and 6(*S*)-diastereoisomers of 5-formyltetrahydrofolic acid (leucovorin) have been prepared neatly by 5-acylation of 6(*RS*)-tetrahydrofolic acid with (−)menthyl chloroformate, separation of the menthyloxycarbonyl products into diastereoisomers by extraction with butanol, and conversion of each isomer separately (by hydrogen bromide in formic acid followed by mild hydrolysis of the cyclic intermediate) into the required 6(*R*)- or 6(*S*)-leucovorin, respectively.[1748]

When polyacetylation occurs (as in some previous examples), a subsequent process of partial deacylation (which appears to take place in the reverse order to acylation) may provide the required product. Thus treatment of 5,8-diacetyl-2,4-bisdimethylamino-5,6,7,8-tetrahydropteridine (**85**, R = Ac) in dilute aqueous *p*-toluenesulfonic acid under reflux for 16 h gave the corresponding 5-(mono)acetylpteridine (**85**, R = H) (24%)[267]; 2-acetamido-5,8-diacetyl-6-methyl-5,6,7,8-tetrahydro-4(3*H*)-pteridinone in boiling water gave the 5-(mono)acetyl analog (~90%)[520]; and 2-acetamido-6-methyl-3,5,8-tristrifluoroacetyl-5,6,7,8-tetrahydro-4(3*H*)-pteridinone in aqueous acetonitrile for 15 h gave 2-amino-6-methyl-5-trifluoroacetyl-5,6,7,8-tetrahydro-4(3*H*)-pteridinone (86%).[520]

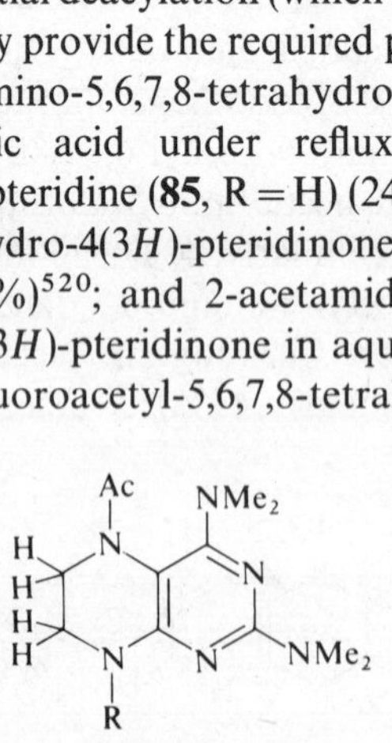

(**85**)

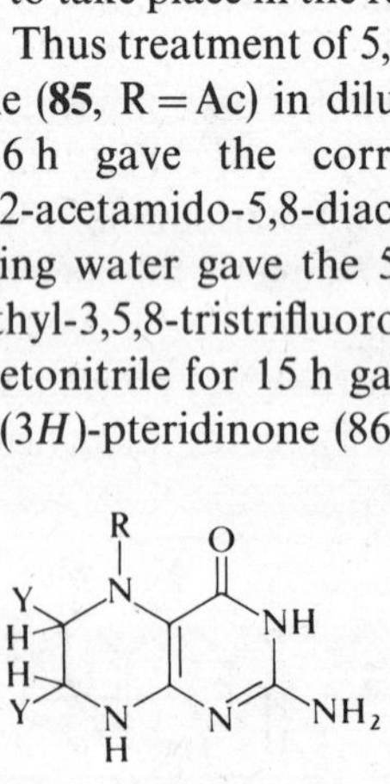

(**86**)

Most known examples of the reductive alkylation of 5,6,7,8-tetrahydropteridines at the 5-NH grouping have already been given in Section 3.B in this chapter, in which the process served as a convenient method to isolate and characterize tetrahydropteridines prepared by nuclear reduction. Thus treatment of a freshly reduced solution of 2-amino-5,6,7,8-tetrahydro-4(3*H*)-pteridinone with formaldehyde or benzaldehyde followed by renewed hydrogenation over platinum gave its 5-methyl (**86**, R = Me, Y = H) or 5-benzyl derivative (**86**, R = CH_2Ph, Y = H), respectively, each in good yield.[297, 588] Likewise, 1,3,6-trimethyl-5,6,7,8-tetrahydro-2,4(1*H*,3*H*)-pteridinedione gave its 1,3,5,6-tetramethyl analog in >80% yield.[456] A rational mechanism for the process is evident from the report that treatment of 2-amino-6,7-dimethyl-5,6,7,8-tetrahydro-4(3*H*)-pteridinone with aqueous formaldehyde, initially at pH 2 and raised gradually to pH 6 during 15 min at 25°C, gave the 5-hydroxymethyl derivative (**86**, R = CH_2OH, Y = Me) (83%), identified by analysis and an appropriate nmr spectrum. Subsequent hydrogenation over platinum in trifluoroacetic acid gave 2-amino-5,6,7-trimethyl-5,6,7,8-tetrahydro-4(3*H*)-pteridinone (**86**, R = Y = Me) in 89% yield.[1343] Other 5-hydroxymethyl intermediates were also isolated and characterized.[1343]

A subtly different and little-used method for achieving 5-methylation is exemplified in the simple hydrogenation of 2-amino-5-formyl-6,7-dimethyl-5,6,7,8-tetrahydro-4(3*H*)-pteridinone (**86**, R = CHO, Y = Me) over platinum in trifluoroacetic acid to give 2-amino-5,6,7-trimethyl-5,6,7,8-tetrahydro-4(3*H*)-pteridinone (**86**, R = Y = Me) (84%).[703] The 5,6-dimethyl homolog was made similarly.[83]

During studies on the polyacetylation of 5,6,7,8-tetrahydroneopterin,[1701] it was observed that further vigorous acetylation of 2-acetamido-6-α,β,γ-triacetoxypropyl-5,6,7,8-tetrahydro-4(3*H*)-pteridinone (**87**) in acetic anhydride–pyridine at 135°C for 90 min gave inter alia a crystalline heptaacetyl derivative that was proven by the usual criteria including x-ray analysis to be 4-acetoxy-8-acetyl-2-diacetylamino-5-ethyl-6-α,β,γ-triacetoxypropyl-5,6,7,8-tetrahydropteridine (**88**).[1701] The 5-ethyl group did not come from ethanol during the workup because no deuteration occurred on replacing the ethanol with perdeuteroethanol: Suggested mechanisms remain unproven.[1696]

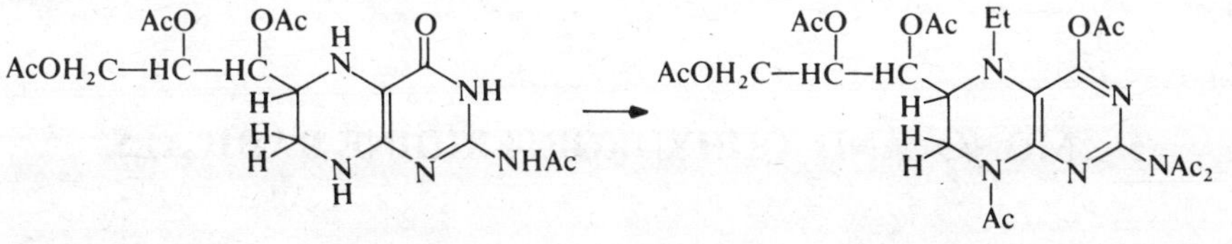

(**87**) (**88**)

The various essential biochemical reactions in which 5,6,7,8-tetrahydropteroylglutamic acid takes part, all involve its presence in one or more of six "one carbon factor forms": these comprise 5-formyl-, 10-formyl-, 5-methyl-, and 5-iminomethyltetrahydrofolic acid, as well as two 5:10-bridged forms, 5,10-methylene- (**89**) and 5,10-methenyltetrahydrofolic acid (**90**).[253, 568, 608, 694, 700, 1343] The biochemical roles of such cofactors have been briefly summarized[1412, 1738] and their chemistry and biochemistry have been reviewed in more depth.[1476, 1478, cf. 91, 728]

C. Other Reactions

The process of *N*-dealkylation (with an incidental dechlorination) is represented in the treatment of 8-benzyl-2-chloro-4-methyl-5,6,7,8-tetrahydropteridine (**91**, R = H) or its 5-acetyl derivative (**91**, R = Ac) in liquid ammonia with sodium to give 4-methyl-5,6,7,8-tetrahydropteridine (**92**, R = H) (~50%) or its 5-acetyl derivative (**92**, R = Ac) in low yield.[118] The *N*-nitrosation of tetrahydropteridines is illustrated in the formation of a crystalline 5-nitroso derivative (**91**, R = NO) by the action of nitrous acid on 8-benzyl-2-chloro-4-methyl-5,6,7,8-tetrahydropteridine (**91**, R = H).[118]

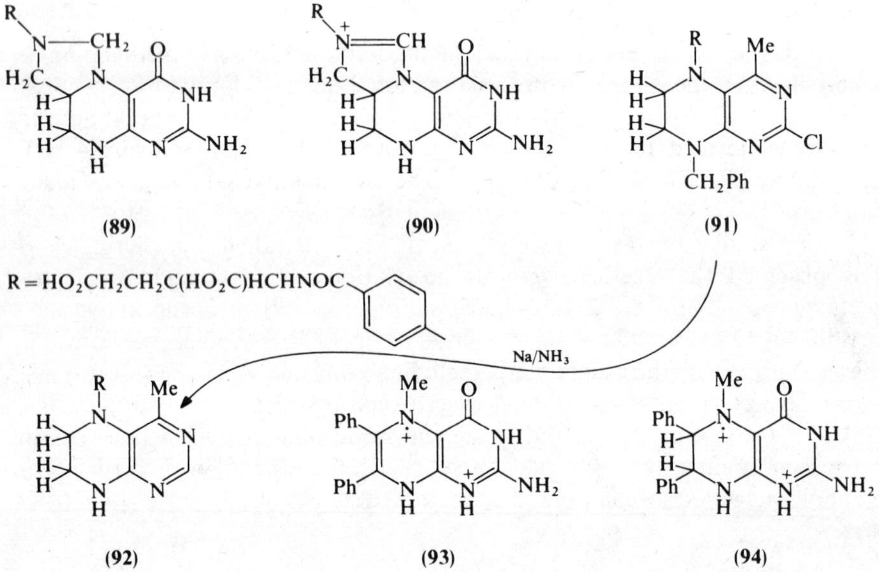

(**89**) (**90**) (**91**) (**92**) (**93**) (**94**)

5. MONO- AND TRIHYDROPTERIDINE RADICALS

Pteridines or dihydropteridines in acidic media undergo a one-electron reduction, for example, by the addition of a trace of zinc, to give strongly colored monohydropteridine or trihydropteridine radical cations, respectively. Likewise, dihydropteridines or tetrahydropteridines in acidic media undergo one-electron oxidation, for example, by the addition of a little hydrogen peroxide, to give the same radical cations. Although transient, such radicals are sufficiently stable for quantitative investigation by esr spectral means.[180, 691, 887, 1383] Typical mono- and trihydropterin structures are represented in the radical cations (**93**) and (**94**), although it has been calculated that in monohydropteridines the unpaired electron is delocalized over the 5-, 6-, 7-, and 8-positions, whereas in trihydropteridines it is largely localized at N-5.[180] Neutral mono- and trihydrolumazine radicals have also been obtained and characterized by their esr spectra.[1383] One-electron reductions of pterin, folic acid, and lumazines have been studied by pulse radiolysis and kinetic absorption spectrophotometry over the pH range 0 to 14.[308, 1046, 1119] Radical anions have apparently been detected during some electrochemical reductions of lumazine derivatives[365] but little is known of them. Possible biochemical implications of pteridine radicals have been discussed.[180, 887]

CHAPTER XII

Ionization and Spectra

The ionization and spectral properties of pteridine and its derivatives are of little fundamental interest, arising as they do from a rather esoteric, unsymmetrical, highly π-deficient,[821] polyaza heteroaromatic system. However, such properties do have considerable value in the structural elucidation and characterization of individual pteridines and classes of pteridine. Moreover, they reflect and are therefore useful in the diagnosis and quantitative measurement of important phenomena, such as covalent hydration, tautomerism, or reversible nuclear reduction, associated with susceptible pteridines and are often of great biological significance.

1. IONIZATION CONSTANTS

Because of an instantaneous covalent hydration of the pteridine cation, the basic strength of anhydrous pteridine cannot be measured even with sophisticated rapid reaction techniques.[1145] The value obtained, pK_a ~4.1,[30] evidently represents that of the 3,4-hydrate (**1**) and/or the 5,6,7,8-dihydrate (**2**), both of which might be expected to have a pK_a not too greatly lower than those of 3,4-dihydropteridine (pK_a 6.4),[51] 5,6,7,8-tetrahydropteridine (pK_a 6.6),[43] or 4,5-pyrimidinediamine (**3**) (pK_a 6.0).[31] Efforts to calculate the intrinsic pK_a of pteridine from the measured value[33] for 4-methylpteridine (2.94), in which at least 3,4-hydration was sterically discouraged, led to a value for pteridine of ~2.6.[1145] However, with later evidence based on photoelectronic spectra, the figure has been reduced drastically to ~ −2,[1513] a value much closer to that expected for a triaza derivative of quinoline (pK_a 4.9). Pteridine also shows a well-defined

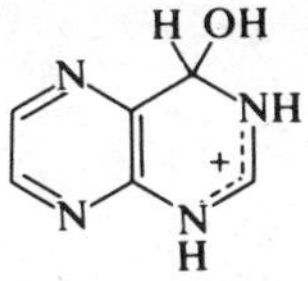

(1)

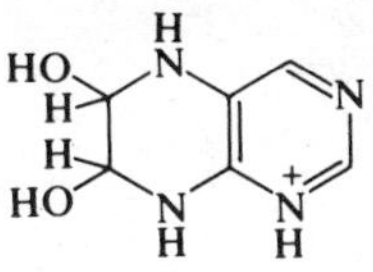

(2)

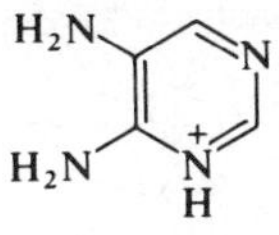

(3)

anionic pK_a at 11.2, probably representing the formation of an anion (**4**) from the 3,4-monohydrate.[24, 1145]

The pK_a values for more than 1000 pteridines have been measured (Table XI), mainly by potentiometric titration or spectrophotometric means.[1741] Many of those values proved to be reasonably predictable,[1742] simply by following some well-established principles governing the ionization of heterocyclic compounds and π-deficient nitrogenous heteroaromatic derivatives in particular.[24, 1743, 1744, cf. 18, 26] However, values for those like pteridine, which form covalent hydrates, must be interpreted in terms of the actual species present in solution at a particular pH and by taking account of the rates for their formation and/or subsequent change(s), often by no means instantaneous. Such discussion should be sought in the original references to individual compounds in the table. A typical example, involving such phenomena in the reversible formation of the 6(5*H*)-pteridinone cation, has been discussed in Chapter VI, Section 2.B. In addition, although 6(5*H*)-pteridinone 7,8-hydrate (**5**) formed a quite regular anhydrous anion (pK_a 6.7),[31] which naturally could not anionize further, rapid reaction techniques have clearly revealed the formation of a transient di-anion (**6**) (pK_a 14.2) arising from the neutral hydrate (**5**)[140]; 2(1*H*)-pteridinone also formed such a di-anion (pK_a 14.1).[140]

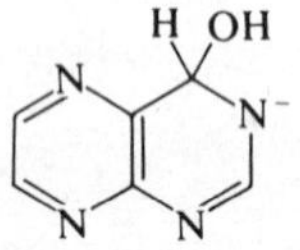

(4)

(5) **(6)**

TABLE XI. MEASURED pK_a VALUES FOR PTERIDINES IN WATER[a]

Pteridine	Acidic pK_a	Basic pK_a	Reference(s)
2-Acetamido-6-acetoxymethyl-4-pteridinamine	6.99	−1.06	113
2-Acetamido-6-α,β-diacetoxypropyl-4(3*H*)-pteridinone	6.93	−1.25	1439
2-Acetamido-4-methoxypteridine		1.95	114, 368
2-Acetamido-6-methoxy-4(3*H*)-pteridinone	7.89		369
2-Acetamido-6-methyl-4-oxo-3,4-dihydro-7-pteridinecarboxylic acid	8.0	1.5	307
2-Acetamido-6-methyl-4(3*H*)-pteridinone	7.81	−0.5	389
2-Acetamido-7-methyl-4(3*H*)-pteridinone	8.00	−0.2	389
2-Acetamido-4-oxo-3,4-dihydro-6-pteridinecarboxylic acid	2.75, 7.10	−1.65	389
2-Acetamido-4-oxo-3,4-dihydro-7-pteridinecarboxylic acid	2.20, 7.44	−1.8	389
2-Acetamidopteridine		2.67	33
4-Acetamidopteridine		1.21	33
2-Acetamido-4(3*H*)-pteridinone		7.37	114, 368
2-Acetamido-7(8*H*)-pteridinone	6.79	1.12	381

TABLE XI. (*Contd.*)

Pteridine	Acidic pK_a	Basic pK_a	Reference(s)
7-Acetonyl-2-amino-5-benzyl-4,6(3*H*,5*H*)-pteridinedione	8.59	1.04	712
6-Acetonyl-2-amino-8-methyl-4,7(3*H*,8*H*)-pteridinedione	8.36	−1.35	712
6-Acetonyl-2-amino-8-phenyl-4,7(3*H*,8*H*)-pteridinedione	8.47	−0.97	712
4-Acetonyl-3,4-dihydro-2(1*H*)-pteridinone	12.38		38, 45
7-Acetonyl-7,8-dihydro-6(5*H*)-pteridinone	10.79	4.73	55
6-Acetonyl-8-*β*-hydroxyethyl-2,4,7(1*H*,3*H*,8*H*)-pteridinetrione	3.02		712
8-*β*-Acetoxyethyl-2,4,7(1*H*,3*H*,8*H*)-pteridinetrione	3.20		317
6-Acetoxymethyl-2-amino-4(3*H*)-pteridinone	7.88	2.07	113
7-Acetoxymethyl-3,6-dimethyl-2,4(1*H*,3*H*)-pteridinedione 5-oxide	7.20		367
6-Acetoxymethyl-2,4-pteridinediamine		4.66	113
5-Acetyl-2-amino-8-benzyl-6,7-dimethyl-5,6,7,8-tetrahydro-4(3*H*)-pteridinone	10.55	1.8	373
5-Acetyl-2-amino-8-*p*-chlorophenyl-6,7-dimethyl-5,6,7,8-tetrahydro-4(3*H*)-pteridinone	10.16	1.7	373
6-Acetyl-2-amino-7,8-dihydro-4(3*H*)-pteridinone (deoxysepiapterin)	10.05	1.35	335
5-Acetyl-2-amino-6,7-dimethyl-8-phenyl-5,6,7,8-tetrahydro-4(3*H*)-pteridinone	10.06	1.71	373
5-Acetyl-8-benzyl-4-methyl-5,6,7,8-tetrahydropteridine		5.49	118
5-Acetyl-2,4-bisdimethylamino-5,6,7,8-tetrahydropteridine		5.84	267
5-Acetyl-6,7-dimethyl-1-(2′,3′,5′-tri-*O*-acetyl-*β*-D-ribofuranosyl)-5,6,7,8-tetrahydro-2,4(1*H*,3*H*)-pteridinedione	10.04		365
5-Acetyl-6,7-dimethyl-1-(2′,3′,5′-tri-*O*-benzoyl-*β*-D-ribofuranosyl)-5,6,7,8-tetrahydro-2,4(1*H*,3*H*)-pteridinedione	9.70		265
5-Acetyl-6,7-diphenyl-5,6,7,8-tetrahydro-2,4(1*H*,3*H*)-pteridinedione	6.83		265
5-Acetyl-6,7-diphenyl-1-(2′,3′,5′-tri-*O*-acetyl-*β*-D-ribofuranosyl)-5,6,7,8-tetrahydro-2,4(1*H*,3*H*)-pteridinedione	8.89		265
1-(5′-*O*-Acetyl-2′,3′-isopropylidene-*β*-D-ribofuranosyl)-6,7-diphenyl-2,4(1*H*,3*H*)-pteridinedione	8.25		263
1-(5′-*O*-Acetyl-2′,3′-isopropylidene-*β*-D-furanosyl)-6.7-diphenyl-2-thioxo-1,2-dihydro-4(3*H*)-pteridinone	7.17		263
5-Acetyl-1-methyl-6,7-diphenyl-5,6,7,8-tetrahydro-2,4(1*H*,3*H*)-pteridinedione	10.17		265
5-Acetyl-4-methyl-5,6,7,8-tetrahydropteridine		5.26	118
5-Acetyl-1-methyl-5,6,7,8-tetrahydro-2,4(1*H*,3*H*)-pteridinedione	10.71		265
7-Acetyl-2-methylthio-4-pteridinamine		2.31	1678

TABLE XI. (*Contd.*)

Pteridine	Acidic pK_a	Basic pK_a	Reference(s)
5-Acetyl-1-(2′,3′,4′,6′-tetra-*O*-acetyl-*β*-D-glucopyranosyl)-5,6,7,8-tetrahydro-2,4(1*H*,3*H*)-pteridinedione	9.48		265
5-Acetyl-5,6,7,8-tetrahydro-2,4-pteridinediamine		5.84	267
5-Acetyl-5,6,7,8-tetrahydro-2,4(1*H*,3*H*)-pteridinedione	7.52		335
5-Acetyl-1-(2′,3′,5′-tri-*O*-acetyl-*β*-D-ribofuranosyl)-5,6,7,8-tetrahydro-2,4(1*H*,3*H*)-pteridinedione	9.62		365
5-Acetyl-1-(2′,3′,5′-tri-*O*-benzyl-*β*-D-ribofuranosyl)-5,6,7,8-tetrahydro-2,4(1*H*,3*H*)-pteridinedione	9.57		265
5-Acetyl-1,6,7-trimethyl-5,6,7,8-tetrahydro-2,4(1*H*,3*H*)-pteridinedione	10.69		265
2-Allylamino-4,6,7-trimethylpteridine		3.82	124
2-Amino-6-[L-*threo*]-α-amino-*β*-hydroxypropyl-4(3*H*)-pteridinone	7.19	1.63[b]	1439
2-Amino-8-benzyl-6,7-dimethyl-4(8*H*)-pteridinone	10.90	5.94	377
4-Amino-8-benzyl-6,7-dimethyl-2(8*H*)-pteridinone	11.44	5.05	362
2-Amino-5-benzyl-6,7-dimethyl-5,6,7,8-tetrahydro-4(3*H*)-pteridinone	11.35	5.35, 0.97	297, 588
2-Amino-8-benzyl-5-formyl-6,7-dimethyl-5,6,7,8-tetrahydro-4(3*H*)-pteridinone	10.75	1.7	373
4-Amino-8-benzyl-6-methyl-2,7(1*H*,8*H*)-pteridinedione	8.34	1.29	363
2-Amino-5-benzyl-8-methyl-5,6,7,8-tetrahydro-4(3*H*)-pteridinone	11.2	5.50, 0.08	297, 588
4-Amino-8-benzyl-2,7(1*H*,8*H*)-pteridinedione	8.02	0.87	363
2-Amino-8-benzyl-4(8*H*)-pteridinone		5.42	377
2-Amino-5-benzyl-5,6,7,8-tetrahydro-4(3*H*)-pteridinone	11.34	5.31, 1.04	297, 588
2-Amino-7-*t*-butyl-4(3*H*)-pteridinone	8.54	2.54	541
2-Amino-7-*t*-butyl-4(3*H*)-pteridinone 5-oxide	7.52	1.62	541
2-Amino-6-α-carboxyethoxy-4(3*H*)-pteridinone	8.53	2.65	299, 302
2-Amino-8-*p*-chlorophenyl-6,7-dimethyl-4(8*H*)-pteridinone	10.35	5.32	377
2-Amino-8-*p*-chlorophenyl-5-formyl-3-methyl-5,6,7,8-tetrahydro-4(3*H*)-pteridinone		1.35	373
2-Amino-8-*p*-chlorophenyl-5-formyl-5,6,7,8-tetrahydro-4(3*H*)-pteridinone	10.05	1.5	373
2-Amino-8-*p*-chlorophenyl-3-methyl-7,8-dihydro-4(3*H*)-pteridinone		2.66	373
2-Amino-8-*p*-chlorophenyl-4(8*H*)-pteridinone	10.16	4.35	377
2-Amino-7-chloro-4(3*H*)-pteridinone	7.65	1.60	380
4-Amino-8-(2′-deoxy-α-D-ribofuranosyl)-2-phenyl-7(8*H*)-pteridinone		1.67	224
4-Amino-1-[2′-deoxy-α(*β*)-D-ribofuranosyl]-2(1*H*)-pteridinone		2.30	225
2-Amino-6-[L-*erythro*]-α,*β*-diacetoxypropyl-4(3*H*)-pteridinone	7.82	1.88	1439
2-Amino-7,8-dihydro-4,6(3*H*,5*H*)-pteridinedione	9.62	1.54	349
2-Amino-7,8-dihyro-4(3*H*)-pteridinone	10.85	4.16, 3.09	81, 353
2-Amino-6-(α*R*)-α,*β*-dihydroxyethyl-4(3*H*)-pteridinone	8.00	2.41	455
2-Amino-6-(α*S*)-α,*β*-dihydroxyethyl-4(3*H*)-pteridinone	8.05	2.45	455
2-Amino-6-α,*β*-dihydroxypropyl-4(3*H*)-pteridinethione	6.37	1.58	1439

TABLE XI. (*Contd.*)

Pteridine	Acidic pK_a	Basic pK_a	Reference(s)
2-Amino-6-[D-*erythro*]-α,β-dihydroxypropyl-4(3*H*)-pteridinone	7.90	2.23	454
2-Amino-6-[L-*erythro*]-α,β-dihydroxypropyl-4(3*H*)-pteridinone (biopterin)	7.98	2.25	335, 350, 1439
	7.89	2.23	454
	7.70		1474
2-Amino-6-[D-*threo*]-α,β-dihydroxypropyl-4(3*H*)-pteridinone	7.92	2.20	454
2-Amino-6-[L-*threo*]-α,β-dihydroxypropyl-4(3*H*)-pteridinone	7.87	2.24	454
2-Amino-4-dimethylamino-8-methyl-7(8*H*)-pteridinone		3.01	357
2-Amino-4-dimethylamino-6(5*H*)-pteridinone	7.50	4.90	268
2-Amino-4-dimethylamino-7(8*H*)-pteridinone	8.45	2.69	268, 1541
4-Amino-2-dimethylamino-6(5*H*)-pteridinone	7.36	4.40	268
4-Amino-2-dimethylamino-7(8*H*)-pteridinone	8.81	2.57	268
2-Amino-4-dimethylamino-8-α-D-ribofuranosyl-7(8*H*)-pteridinone		2.69	357, 1541
2-Amino-4-dimethylamino-8-β-D-ribofuranosyl-7(8*H*)-pteridinone		2.69	357
2-Amino-3,8-dimethyl-7,8-dihydro-4,6(3*H*,5*H*)-pteridinedione		1.06	349
2-Amino-7,7-dimethyl-7,8-dihydro-4,6(3*H*,5*H*)-pteridinedione	9.60	1.61	349
2-Amino-6,7-dimethyl-7,8-dihydro-4(3*H*)-pteridinone	11.09	4.16	115, 388, 1459
2-Amino-6,8-dimethyl-7,8-dihydro-4(3*H*)-pteridinone	>9.70	4.40, <1.0	1370
2-Amino-7,7-dimethyl-7,8-dihydro-4(3*H*)-pteridinone	11.1	2.89	115
2-Amino-3,8-dimethyl-4,7-dioxo-3,4,7,8-tetrahydro-6-pteridinecarboxylic acid	3.74		385, 1214
2-Amino-3,8-dimethyl-6,7-diphenyl-7,8-dihydro-4(3*H*)-pteridinone		2.17	373
2-Amino-5,8-dimethyl-6,7-diphenyl-5,6,7,8-tetrahydro-4(3*H*)-pteridinone	~10.6	4.59, −0.09	297, 588
2-Amino-6,7-dimethyl-8-phenyl-7,8-dihydro-4(3*H*)-pteridinone	11.00	3.37	373
		3.81	376
2-Amino-7,7-dimethyl-6-phenyl-7,8-dihydro-4(3*H*)-pteridinone	11.1	2.89	388
2-Amino-6,7-dimethyl-8-phenyl-4(8*H*)-pteridinone	10.69	5.60	377
4-Amino-6,7-dimethyl-8-phenyl-2(8*H*)-pteridinone		4.89	362
6-Amino-1,3-dimethyl-7-propionyl-2,4(1*H*,3*H*)-pteridinedione		−1.31	108
7-Amino-1,3-dimethyl-6-propionyl-2,4(1*H*,3*H*)-pteridinedione		−3.57	108
2-Amino-3,6-dimethyl-4,7(3*H*,8*H*)-pteridinedione	8.00		383
2-Amino-6,8-dimethyl-4,7(3*H*,8*H*)-pteridinedione	8.40		383
	8.54	−0.20	712
4-Amino-1,6-dimethyl-2,7(1*H*,8*H*)-pteridinedione	5.62	2.18	363
4-Amino-6,8-dimethyl-2,7(1*H*,8*H*)-pteridinedione	8.21	1.95	363
6-Amino-1,3-dimethyl-2,4(1*H*,3*H*)-pteridinedione		0.30	1275

TABLE XI. (*Contd.*)

Pteridine	Acidic pK_a	Basic pK_a	Reference(s)
7-Amino-1,3-dimethyl-2,4(1*H*,3*H*)-pteridinedione		−1.05	238, 1275
7-Amino-1,3-dimethyl-2,4(1*H*,3*H*)-pteridinedione 5-oxide		−1.59	238
2-Amino-3,8-dimethyl-4,6,7(3*H*,5*H*,8*H*)-pteridinetrione	8.75		384
7-Amino-1,3-dimethyl-2,4,6(1*H*,3*H*,5*H*)-pteridinetrione	8.09	−2.59	238
2-Amino-1,6-dimethyl-4(1*H*)-pteridinone		3.24	80
2-Amino-3,6-dimethyl-4(3*H*)-pteridinone		2.82	80
2-Amino-6,7-dimethyl-4(3*H*)-pteridinone		2.6	69, 114, 741, 762
4-Amino-1,6-dimethyl-7(1*H*)-pteridinone	13.82	1.97	132
4-Amino-6,7-dimethyl-2(1*H*)-pteridinone	10.69	3.49	130
	10.53	3.41	362
2-Amino-6,7-dimethyl-4(3*H*)-pteridinone 5,8-dioxide	6.64	0.22	541
2-Amino-6,7-dimethyl-4(3*H*)-pteridinone 8-oxide	7.24	0.41	541
2-Amino-5,8-dimethyl-5,6,7,8-tetrahydro-4(3*H*)-pteridinone	11.3	6.3, 0.50	297, 588
2-Amino-6,7-dimethyl-5,6,7,8-tetrahydro-4(3*H*)-pteridinone	10.40	5.60, 1.37	997
2-Amino-6,8-dimethyl-5,6,7,8-tetrahydro-4(3*H*)-pteridinone	9.70	5.85, 0.80	1370
2-Amino-6,7-dimethyl-8-*p*-tolyl-4(3*H*)-pteridinone	10.76	5.66	377
2-Amino-4,7-dioxo-8-D-sorbityl-3,4,7,8-tetrahydro-6-pteridinecarboxylic acid	3.57, 8.40		385
2-Amino-4,7-dioxo-3,4,7,8-tetrahydro-6-pteridinecarboxylic acid	8.00, 9.50		385
2-Amino-6,7-diphenyl-7,8-dihydro-4(3*H*)-pteridinone	10.5	<2	115, 388
2-Amino-6,7-diphenyl-4(3*H*)-pteridinone	8.60	1.82	541
4-Amino-2,6-diphenyl-7(8*H*)-pteridinone	7.99	1.91	224
2-Amino-6,7-diphenyl-4(3*H*)-pteridinone 5,8-dioxide	4.61	−0.99	541
2-Amino-6,7-diphenyl-4(3*H*)-pteridinone 8-oxide	6.86	0.08	541
4-Amino-6,7-diphenyl-1-*β*-D-ribofuranosyl-2(1*H*)-pteridinone		2.40	225
4-Amino-2,6-diphenyl-8-*β*-D-ribofuranosyl-7(8*H*)-pteridinone		1.84	224
2-Amino-6-ethoxy-4(3*H*)-pteridinone	8.23	2.37	299
	8.23	2.52	369
2-Amino-6-ethyl-7-methyl-4(3*H*)-pteridinone	8.70	2.80	1459
2-Amino-7-ethyl-4(3*H*)-pteridinone	8.04[c]	2.28[c]	1092
6-Amino-7-formamido-8-D-ribityl-2,4(3*H*,8*H*)-pteridinedione (russupteridine yellow-1)	12.10	(?) 4.80, 0.25	1400
2-Amino-5-formyl-6,7-dimethyl-8-phenyl-5,6,7,8-tetrahydro-4(3*H*)-pteridinone	9.85	1.62	373
2-Amino-5-formyl-6,8-dimethyl-5,6,7,8-tetrahydro-4(3*H*)-pteridinone	10.65	2.20	373
2-Amino-5-formyl-8-methyl-5,6-dihydro-4,7(3*H*,8*H*)-pteridinedione	8.31		181
2-Amino-5-formyl-8-methyl-5,6,7,8-tetrahydro-4(3*H*)-pteridinone	10.45	1.70	373
2-Amino-5-formyl-4-oxo-3,4,5,6,7,8-hexahydro-6-pteridinecarboxylic acid	10.15?	1.78, −3.09	307

TABLE XI. (*Contd.*)

Pteridine	Acidic pK_a	Basic pK_a	Reference(s)
2-Amino-5-formyl-8-phenyl-5,6,7,8-tetrahydro-4(3*H*)-pteridinone	9.70	1.50	373
2-Amino-5-formyl-6,7,8-trimethyl-5,6,7,8-tetrahydro-4(3*H*)-pteridinone	10.6	1.40	373
2-Amino-7-β-D-glucopyranosyloxy-4(3*H*)-pteridinone	8.04		380
4-Amino-1-β-D-glucopyranosyl-2(1*H*)-pteridinone		2.28	382
2-Amino-8-D-glucosyl-4,7-dioxo-3,4,7,8-tetrahydro-6-pteridinecarboxylic acid	3.21, 8.14		1090
2-Amino-8-D-glucosyl-6-methyl-4,7(3*H*,8*H*)-pteridinedione	8.35		1090
2-Amino-8-D-glucosyl-4,7(3*H*,8*H*)-pteridinedione	7.96		1090
2-Amino-6-α-hydroxyacetyl-7,8-dihydro-4(3*H*)-pteridinone (sepiapterin)	9.95	1.27	335, 350
2-Amino-6-β-hydroxyethoxy-4(3*H*)-pteridinone	8.26	2.63	299
2-Amino-8-β-hydroxyethyl-4,7-dioxo-3,4,7,8-tetrahydro-6-pteridinecarboxylic acid	3.50, 8.29		385
2-Amino-8-β-hydroxyethyl-4-isopropoxy-6-methyl-7(8*H*)-pteridinone		0.74	371
2-Amino-8-β-hydroxyethyl-6-methyl-4,7(3*H*,8*H*)-pteridinedione	7.80		372
	8.43		1090
4-Amino-8-β-hydroxyethyl-6-methyl-2,7(1*H*,8*H*)-pteridinedione	8.61	2.88	363
2-Amino-6-(*R*)-α-hydroxyethyl-4(3*H*)-pteridinone	8.14	2.57	455
2-Amino-6-(*S*)-α-hydroxyethyl-4(3*H*)-pteridinone	8.10	2.52	455
2-Amino-6-β-hydroxy-α-methylethoxy-4(3*H*)-pteridinone	8.38	2.67	299, 302
2-Amino-6-hydroxymethyl-8-methyl-4,7(3*H*,8*H*)-pteridinedione (asperopterin-B)	8.09		1214
	6.70		1370
2-Amino-6-hydroxymethyl-4(3*H*)-pteridinone	7.93	2.30	113
2-Amino-6-β-hydroxypropoxy-4(3*H*)-pteridinone	8.28	2.68	299, 302
2-Amino-6-α-hydroxypropyl-7-methylthio-4(3*H*)-pteridinone	8.39	2.64	95
2-Amino-6-α-hydroxypropyl-4(3*H*)-pteridinone	7.97	2.27	95
2-Amino-4-isopropoxy-6,8-dimethyl-7(8*H*)-pteridinone		0.40	371
2-Amino-4-isopropoxy-8-methyl-7-oxo-7,8-dihydro-6-pteridinecarboxylic acid	3.53	−0.76	371
2-Amino-4-isopropoxy-8-methyl-6,7(5*H*,8*H*)-pteridinedione	8.53	0.53	371
2-Amino-4-isopropoxy-6-methyl-7(8*H*)-pteridinone	7.80	1.14	371
2-Amino-4-isopropoxy-8-methyl-7(8*H*)-pteridinone		0.17	371
2-Amino-4-isopropoxy-7-oxo-7,8-dihydro-6-pteridinecarbonitrile	5.95	−0.17	371
2-Amino-4-isopropoxy-7-oxo-7,8-dihydro-6-pteridinecarboxylic acid	3.71, 8.32	0.35	371
2-Amino-4-isopropoxy-6,7(5*H*,8*H*)-pteridinedione	8.46, 12.2	0.82	371
2-Amino-4-isopropoxy-7(8*H*)-pteridinone	7.60	0.74	371
2-Amino-6-isopropoxy-4(3*H*)-pteridinone	8.40	2.74	299
2-Amino-4-isopropoxy-8-β-D-ribofuranosyl-7(8*H*)-pteridinone		−0.44	436

TABLE XI. (*Contd.*)

Pteridine	Acidic pK_a	Basic pK_a	Reference(s)
2-Amino-8-isopropyl-6,7-dimethyl-4(8*H*)-pteridinone	12.30	6.00	347
2-Amino-1-(2′,3′-isopropylidene-*β*-D-ribofuranosyl)-6,7-diphenyl-4(1*H*)-pteridinone	11.62	2.07	263
2-Amino-8-isopropyl-4(8*H*)-pteridinone	13.38	5.29	377
2-Amino-6-lactoyl-7,8-dihydro-4(3*H*)-pteridinone (sepiapterin)	9.95	1.27	250, 615
2-Amino-4-methoxy-6,8-dimethyl-7(8*H*)-pteridinone		0.55	371
2-Amino-4-methoxy-8-methyl-7-oxo-7,8-dihydro-6-pteridinecarboxylic acid	3.36	−0.4	371
2-Amino-4-methoxy-8-methyl-7(8*H*)-pteridinone		0.21	371
2-Amino-6-methoxy-1-methyl-4(1*H*)-pteridinone		2.50	369
2-Amino-4-methoxy-7(8*H*)-pteridinone	7.8		380
2-Amino-6-methoxy-4(3*H*)-pteridinone	8.15	2.37	299, 302, 369
2-Amino-7-methoxy-4(3*H*)-pteridinone	8.33		380
4-Amino-2-methoxy-7(8*H*)-pteridinone	7.42	1.80	363
2-Amino-4-methoxy-8-*β*-D-ribofuranosyl-7(8*H*)-pteridinone		−0.51	436
2-Amino-3-methyl-7,8-dihydro-4,6(3*H*,5*H*)-pteridinedione		1.76	349
2-Amino-5-methyl-7,8-dihydro-4,6(3*H*,5*H*)-pteridinedione	9.92	1.45	349
2-Amino-7-methyl-7,8-dihydro-4,6(3*H*,5*H*)-pteridinedione	9.66	1.58	349
2-Amino-8-methyl-7,8-dihydro-4,6(3*H*,5*H*)-pteridinedione	10.01	1.04	349
2-Amino-6-methyl-7,8-dihydro-4(3*H*)-pteridinone	10.85	4.17	115, 335, 359, 388
2-Amino-8-methyl-7,8-dihydro-4(3*H*)-pteridinone	11.40	3.08	373
2-Amino-3-methyl-4,7-dioxo-3,4,7,8-tetrahydro-6-pteridinecarboxylic acid	3.40, 8.05		385
2-Amino-8-methyl-4,7-dioxo-3,4,7,8-tetrahydro-6-pteridinecarboxylic acid	3.50, 8.40		385
	3.45, 8.18		1214
2-Amino-8-methyl-6,7-diphenyl-7,8-dihydro-4(3*H*)-pteridinone	11.05	1.98	373, 374
2-Amino-8-methyl-6,7-diphenyl-4(8*H*)-pteridinone	11.76	5.34	359
2-Amino-5-methyl-6,7-diphenyl-5,6,7,8-tetrahydro-4(3*H*)-pteridinone	11.16	4.39, 0.61	297, 588
2-Amino-8-methyl-6,7-diphenyl-5,6,7,8-tetrahydro-4(3*H*)-pteridinone		0.13	588
2-Amino-6-methyl-7-*β*-D-glucopyranosyloxy-4(3*H*)-pteridinone	8.56		372
2-Amino-6-methyl-4-oxo-3,4-dihydro-7-pteridinecarboxylic acid	3.01, 8.43	1.23	307
2-Amino-7-methyl-4-oxo-3,4-dihydro-6-pteridinecarboxylic acid	3.20, 8.42	1.39	307
2-Amino-8-methyl-4-oxo-4,8-dihydro-6-pteridinecarboxylic acid	5.05, 11.80	2.1	307

TABLE XI. (*Contd.*)

Pteridine	Acidic pK_a	Basic pK_a	Reference(s)
2-Amino-8-methyl-4-oxo-3,4,7,8-tetrahydro-6-pteridinecarboxylic acid	3.86, 10.5	0.40	307
2-Amino-3-methyl-6-phenyl-7,8-dihydro-4(3*H*)-pteridinone		2.70	373
4-Amino-6-methyl-8-phenyl-2,7(1*H*,8*H*)-pteridinedione	8.35	2.31	363
2-Amino-3-methyl-4,6(3*H*,5*H*)-pteridinedione	6.70	1.60	369
2-Amino-3-methyl-4,7(3*H*,8*H*)-pteridinedione	7.45		383
2-Amino-6-methyl-4,7(3*H*,8*H*)-pteridinedione	7.98, 10.15		383
	9.94		1418
2-Amino-7-methyl-4,6(3*H*,5*H*)-pteridinedione	7.56, 9.60	2.17	348
	10.00		1418
2-Amino-8-methyl-4,7(3*H*,8*H*)-pteridinedione	8.10		383
4-Amino-1-methyl-2,7(1*H*,8*H*)-pteridinedione	5.36	2.02	363
4-Amino-6-methyl-2,7(1*H*,8*H*)-pteridinedione	—, 12.79	2.60	363
4-Amino-8-methyl-2,7(1*H*,8*H*)-pteridinedione	7.90	1.56	363
4-Amino-7-methyl-2(1*H*)-pteridinethione	8.62	2.35	1247
2-Amino-3-methyl-4,6,7(3*H*,5*H*,8*H*)-pteridinetrione	7.60, 10.95		384
2-Amino-8-methyl-4,6,7(3*H*,5*H*,8*H*)-pteridinetrione	7.95, 10.20		384
4-Amino-1-methyl-2,6,7(1*H*,5*H*,8*H*)-pteridinetrione	5.40, 10.91	1.8	363
2-Amino-1-methyl-4(1*H*)-pteridinone		2.86	114, 368
	~11.5	2.83	130, 741
		2.84	762
2-Amino-3-methyl-4(3*H*)-pteridinone		2.18	114, 368, 741
		2.25	130
		2.27	308
	13.41		332
		2.86	416
		2.22	762
2-Amino-6-methyl-4(3*H*)-pteridinone		2.48	80
	8.10	2.31	114, 389, 741, 762
	8.40		1418
2-Amino-7-methyl-4(3*H*)-pteridinone	8.24	2.19	114, 389, 741, 762
2-Amino-8-methyl-4(8*H*)-pteridinone	12.37	5.32	116, 359
	~11.5	5.42	130, 741, 762
2-Amino-8-methyl-7(8*H*)-pteridinone		2.05	381, 383, 1126
4-Amino-1-methyl-2(1*H*)-pteridinone	~12.0	2.96	130
		2.79	225, 362
4-Amino-1-methyl-7(1*H*)-pteridinone	13.24	1.70	132
4-Amino-6-methyl-7(8*H*)-pteridinone	8.13	2.80	445
4-Amino-8-methyl-7(8*H*)-pteridinone		2.15	445, 1126
7-Amino-1-methyl-4(1*H*)-pteridinone	>14.0	2.38	131
7-Amino-8-methyl-4(8*H*)-pteridinone	>14.0	6.33	131
2-Amino-6-methyl-4(3*H*)-pteridinone 8-oxide	7.07	−0.05	541
2-Amino-7-methyl-4(3*H*)-pteridinone 8-oxide	6.94	0.03	541
2-Amino-6-methyl-8-(pyridin-2′-yl)-4,7(3*H*,8*H*)-pteridinedione	8.15		139

TABLE XI. (*Contd.*)

Pteridine	Acidic pK_a	Basic pK_a	Reference(s)
2-Amino-8-methyl-6-β-D-ribofuranosylmethyl-4,7(3*H*,8*H*)-pteridinedione (asperopterin-A)	8.05		1214
2-Amino-6-methyl-7-(2′,3′,4′,6′-tetraacetyl-β-D-glucopyranosyloxy)-4(3*H*)-pteridinone	8.38		372
2-Amino-5-methyl-5,6,7,8-tetrahydro-4(3*H*)-pteridinone	11.33	5.99, 1.05	297, 588
2-Amino-8-methyl-5,6,7,8-tetrahydro-4(3*H*)-pteridinone	10.65	5.70	373
2-Amino-3-methyl-7-α,β,γ,δ-tetrahydroxybutyl-7,8-dihydro-4,6(3*H*,5*H*)-pteridinedione	12.96	1.63	358
2-Amino-7-methylthio-6-propionyl-4(3*H*)-pteridinone	7.81	2.03	95
4-Amino-2-methylthio-7(8*H*)-pteridinone		1.56	445
	7.38	1.43	1126
4-Amino-2-methylthio-8-β-D-ribofuranosyl-7(8*H*)-pteridinone		1.08	1126
2-Amino-7-oxalomethyl-4,6(3*H*,5*H*)-pteridinedione (erythropterin)	2.45, 7.95, 10.00		348
2-Amino-4-oxo-3,4-dihydro-6-pteridinecarboxylic acid	2.88, 7.72	1.43	389
(pterincarboxylic acid)	8.00		1682
2-Amino-4-oxo-3,4-dihydro-7-pteridinecarboxylic acid	3.32, 7.67	1.54	389
2-Amino-4-oxo-3,4-dihydro-6,7-pteridinedicarboxylic acid	2.10, 3.90, 8.10	1.00	307, 601
2-Amino-3,5,6,7,7-pentamethyl-4-oxo-3,4,7,8-tetrahydro-5-pteridinium *p*-toluenesulfonate		8.57	374, 376
2-Amino-3,5,6,7,7-pentamethyl-5,6,7,8-tetrahydro-4(3*H*)-pteridinone		5.75, 1.73	358, 374
2-Amino-4-pentyloxy-7(8*H*)-pteridinethione	6.18	0.15	1681
2-Amino-4-pentyloxy-7(8*H*)-pteridinone	7.85	0.48	1541
2-Amino-6-phenyl-7,8-dihydro-4(3*H*)-pteridinone	10.98	2.55	373
2-Amino-8-phenyl-7,8-dihydro-4(3*H*)-pteridinone	10.50	2.65	373
3-Amino-7-phenyl-2,4(1*H*,3*H*)-pteridinedione	7.43		331
4-Amino-8-phenyl-2,7(1*H*,8*H*)-pteridinedione	8.00	1.91	363
4-Amino-7-phenyl-2(1*H*)-pteridinethione	11.07		1247
2-Amino-6-phenyl-4(3*H*)-pteridinone	8.22	2.17	541
2-Amino-7-phenyl-4(3*H*)-pteridinone	8.14	2.30	541
2-Amino-8-phenyl-4(8*H*)-pteridinone	9.99	4.73	377
4-Amino-2-phenyl-7(8*H*)-pteridinone	7.45	1.97	224
4-Amino-6-phenyl-7(8*H*)-pteridinone	7.69	2.27	224
2-Amino-6-phenyl-4(3*H*)-pteridinone 8-oxide	6.99	−0.54	541
2-Amino-7-phenyl-4(3*H*)-pteridinone 5-oxide	5.21	1.35	541
4-Amino-2-phenyl-8-β-D-ribofuranosyl-7(8*H*)-pteridinone		1.73	224
4-Amino-6-phenyl-8-β-D-ribofuranosyl-7(8*H*)-pteridinone		1.56	224
2-Amino-6-propionyl-7,8-dihydro-4(3*H*)-pteridinone (deoxysepiapterin or isosepiapterin)	10.05	1.35	95, 350
2-Amino-6-propionyl-4(3*H*)-pteridinone	7.12	1.44	95
2-Amino-6-propoxy-4(3*H*)-pteridinone	8.10	2.37	299, 369

TABLE XI. (*Contd.*)

Pteridine	Acidic pK_a	Basic pK_a	Reference(s)
2-Amino-4,6(3*H*,5*H*)-pteridinedione (xanthopterin)	6.3, 9.23	1.6	13, 32, 338
	6.25, 9.23		30, 1418
	6.59, 9.31 (anhyd)		241, 961
	8.65, 9.99 (hyd)		
2-Amino-4,7(3*H*,8*H*)-pteridinedione	10.2		61
(isoxanthopterin)	7.34, 10.06	−0.5	268, 335, 383
(mesopterin?)			
4-Amino-2,7(1*H*,8*H*)-pteridinedione	5.54, 12.60	1.91	363
4-Amino-6,7(5*H*,8*H*)-pteridinedione	6.62, 11.00	2.06	208
2-Amino-4,6,7(3*H*,5*H*,8*H*)-pteridinetrione (leucopterin)	7.56, 9.78, 13.6	−1.66	335, 384
4-Amino-2,6,7(1*H*,5*H*,8*H*)-pteridinetrione	5.67, 8.79, 14.00	1.95	363
2-Amino-4(3*H*)-pteridinone (pterin)		2.20[d]	757
	~8.0		30
	7.92	2.31	32, 59, 130, 241
	7.86	2.20	114, 335, 368 416, 541, 762
	7.92	2.27	308, 1046, 1418
	8.10	2.31	353, 741
2-Amino-7(8*H*)-pteridinone	7.50	1.50	381, 383, 1126
4-Amino-2(1*H*)-pteridinone (isopterin)	9.97	3.21	32
	9.97	2.99	130
	10.24	3.05	362
4-Amino-6(5*H*)-pteridinone		3.39	445
4-Amino-7(8*H*)-pteridinone	7.52	2.27	224, 445, 1126
7-Amino-4(3*H*)-pteridinone	8.87	1.23	131
7-Amino-6(5*H*)-pteridinone		~8.0	19
2-Amino-4(3*H*)-pteridinone 8-oxide	6.94	−0.63	541
"Aminopterin"	5.70		761
	5.50		1682
2-Amino-8-β-D-ribofuranosyl-4,7(3*H*,8*H*)-pteridinedione	8.10		436, 856
2-Amino-8-β-D-ribofuranosyl-7(8*H*)-pteridinone		1.88	1126
4-Amino-1-β-D-ribofuranosyl-2(1*H*)-pteridinone		2.75	225
4-Amino-8-β-D-ribofuranosyl-7(8*H*)-pteridinone		1.74	224, 1126
2-Amino-7-(2′,3′,4′,6′-tetraacetyl-β-D-glucopyranosyloxy)pteridine		2.56	381
2-Amino-7-(2′,3′,4′,6′-tetraacetyl-β-D-glucopyranosyloxy)-4(3*H*)-pteridinone	7.94		380

TABLE XI. (*Contd.*)

Pteridine	Acidic pK_a	Basic pK_a	Reference(s)
2-Amino-5,6,7,8-tetrahydro-4(3*H*)-pteridinone	10.60	5.60, 1.30	353
		5.81,[d] 1.50[d]	757
2-Amino-3,6,7,7-tetramethyl-7,8-dihydro-4(3*H*)-pteridinone		4.13	115, 376, 388
2-Amino-3,5,7,7-tetramethyl-4-oxo-6-phenyl-3,4,7,8-tetrahydro-5-pteridinium *p*-toluenesulfonate		8.06, −0.10	374
2-Amino-5,6,7,7-tetramethyl-4-oxo-3,4,7,8-tetrahydro-5-pteridinium *p*-toluenesulfonate		8.50	374, 376
2-Amino-5,6,7,8-tetramethyl-4-oxo-3,4,7,8-tetrahydro-5-pteridinium *p*-toluenesulfonate		8.40, −0.89	374, 376
2-Amino-5,6,7,7-tetramethyl-5,6,7,8-tetrahydro-4(3*H*)-pteridinone		5.75	376
2-Amino-5,6,7,8-tetramethyl-5,6,7,8-tetrahydro-4(3*H*)-pteridinone	11.32	6.34, 0.27	297, 588
2-Amino-8-*p*-tolyl-4(8*H*)-pteridinone	10.06	4.80	377
4-Amino-7-(L-*lyxo*)-α,β,γ-trihydroxybutyl-2(1*H*)-pteridinone	10.52		1247
2-Amino-3,7,7-trimethyl-7,8-dihydro-4,6(3*H*,5*H*)-pteridinedione	13.51	1.55	349
		1.34	358
2-Amino-6,7,7-trimethyl-7,8-dihydro-4(3*H*)-pteridinone	11.05	4.24	115, 376
2-Amino-6,7,7-trimethyl-7,8-dihydro-4(3*H*)-pteridinone	11.05	4.24	388
2-Amino-6,7,8-trimethyl-7,8-dihydro-4(3*H*)-		4.30	115
pteridinone	11.22	4.13	373, 376
2-Amino-3,5,8-trimethyl-6,7-diphenyl-5,6,7,8-tetrahydro-4(3*H*)-pteridinone		4.70, 0.00	374
2-Amino-3,5,8-trimethyl-4-oxo-6,7-diphenyl-3,4,7,8-tetrahydro-5-pteridinium *p*-toluenesulfonate		8.48	374
2-Amino-5,6,7-trimethyl-4-oxo-8-phenyl-3,4,7,8-tetrahydro-5-pteridinium *p*-toluenesulfonate		7.55	374, 376
2-Amino-3,7,7-trimethyl-6-phenyl-7,8-dihydro-4(3*H*)-pteridinone		3.27	373
2-Amino-5,6,7-trimethyl-8-phenyl-5,6,7,8-tetrahydro-4(3*H*)-pteridinone	11.00	5.85	374, 376
2-Amino-3,5,7-trimethyl-4,6(3*H*,5*H*)-pteridinedione		2.00	348
2-Amino-1,6,7-trimethyl-4(1*H*)-pteridinone		3.20	69, 541
2-Amino-3,6,7-trimethyl-4(3*H*)-pteridinone		2.60	69, 741, 762
	14.20		332
	12.00	(?) 5.85 or 6.10	563
2-Amino-6,7,8-trimethyl-4(8*H*)-pteridinone	11.94	6.13	116, 359
	11.97	6.10	130, 741, 762
4-Amino-1,6,7-trimethyl-2(1*H*)-pteridinone		3.66	130
		3.38	362
2-Amino-1,6,7-trimethyl-4(3*H*)-pteridinone 5-oxide		2.95	541
2-Amino-5,6,7-trimethyl-5,6,7,8-tetrahydro-4(3*H*)-pteridinone	11.43	6.23, 1.05	297, 588
2-Amino-6,7,8-trimethyl-5,6,7,8-tetrahydro-4(3*H*)-		5.20, 0.62	297, 588
pteridinone	11.00	5.90	373

TABLE XI. (*Contd.*)

Pteridine	Acidic pK_a	Basic pK_a	Reference(s)
2-Anilino-6-[L-*erythro*]-α,β-dihydroxypropyl-4(3*H*)-pteridinone	7.14	1.43	458
6-Anilinomethyl-4(3*H*)-pteridinone	7.92	<3.0	13
4-Anilino-2(1*H*)-pteridinone	11.49	1.76	362
2-Anilino-6-[D-*erythro*]-α,β,γ-trihydroxypropyl-4(3*H*)-pteridinone	7.21	1.43	458
2-Anilino-6-[L-*erythro*]-α,β,γ-trihydroxypropyl-4(3*H*)-pteridinone	7.20	1.44	458
1-α-D-Arabinofuranosyl-2,4(1*H*,3*H*)-pteridinone	8.40		237
1-β-D-Arabinofuranosyl-2,4(1*H*,3*H*)-pteridinedione	8.42		237
3-α-D-Arabinofuranosyl-2,4(1*H*,3*H*)-pteridinedione	8.00		237
8-D-Arabinityl-6,7-dimethyl-2,4(3*H*,8*H*)-pteridinedione	8.29		1686
8-D-Arabinityl-6-methyl-2,4,7-(1*H*,3*H*,8*H*)-pteridinetrione	3.60, 12.85		295
8-L-Arabinityl-6-methyl-2,4,7-(1*H*,3*H*,8*H*)-pteridinetrione	3.60, 12.92		295
8-D-Arabinityl-2,4,7(1*H*,3*H*,8*H*)-pteridinetrione	3.16, 12.54		295
8-L-Arabinityl-2,4,7(1*H*,3*H*,8*H*)-pteridinetrione	3.17, 12.42		295
2-Benzamido-6-α,β-dibenzoyloxypropyl-4(3*H*)-pteridinone	6.07	−2.43	1439
5-Benzoyl-2,4-bisdimethylamino-5,6,7,8-tetrahydropteridine		5.66	267
5-Benzoyl-5,6,7,8-tetrahydro-2,4-pteridinediamine		5.91	267
4-Benzylamino-6-methyl-2,7(1*H*,8*H*)-pteridinedione	5.20, 13.20	2.34	363
4-Benzylamino-2(1*H*)-pteridinone	12.41	2.04	362
8-Benzyl-6,7-dimethyl-2,4(3*H*,8*H*)-pteridinedione	9.04	0.25	378, 1545
8-Benzyl-3-ethyl-6,7-dimethyl-2,4(3*H*,8*H*)-pteridinedione	9.50	0.15	1545
8-Benzyl-5-formyl-4-methyl-5,6,7,8-tetrahydropteridine		5.46	118
8-Benzyl-3-methyl-6,7-diphenyl-2,4(3*H*,8*H*)-pteridinedione	10.10	−0.57	1545
8-Benzyl-3-methyl-2,4(3*H*,8*H*)-pteridinedione	9.95	−0.69	1545
8-Benzyl-6-methyl-2,4,7(1*H*,3*H*,8*H*)-pteridinetrione	3.67, 13.10		378
8-Benzyl-3-methyl-2,4,7-trioxo-1,2,3,4,7,8-hexahydro-6-pteridinecarboxylic acid	1.65, 4.58		317
4-Benzyloxy-2-dimethylamino-7(8*H*)-pteridinone	8.47	0.55	246
4-Benzyloxy-6,7-dimethyl-2-pteridinamine		3.90 3.92 (10% EtOH)	370
4-Benzyloxy-2-pteridinamine		3.59 (H_2O) 3.45 (10% EtOH)	370, 741
8-Benzyl-2,4,7(1*H*,3*H*,8*H*)-pteridinetrione	3.05, 12.98		317
8-Benzyl-7(8*H*)-pteridinone		0.79	125
8-Benzyl-3,6,7-trimethyl-2,4(3*H*,8*H*)-pteridinedione	9.37	0.10	1545
8-Benzyl-2,4,7-trioxo-1,2,3,4,7,8-hexahydro-6-pteridinecarboxylic acid	1.69, 4.77, 12.9		317
6-7-Bis-*p*-chlorophenyl-2,4(1*H*,3*H*)-pteridinedione	7.81, 12.83		412
6,7-Bis-*p*-chlorophenyl-1-α-ribofuranosyl-2,4(1*H*,3*H*)-pteridinedione	8.12		412

TABLE XI. (*Contd.*)

Pteridine	Acidic pK_a	Basic pK_a	Reference(s)
6,7-Bis-*p*-chlorophenyl-1-*β*-D-ribofuranosyl-2,4(1*H*,3*H*)-pteridinedione	8.25		412
2,4-Bisdimethylamino-7,8-dihydro-6(5*H*)-pteridinone		5.16	268
2,4-Bisdimethylaminopteridine		6.15	267, 947
2,4-Bisdimethylamino-6(5*H*)-pteridinone	8.02	5.10	268
2,4-Bisdimethylamino-7(8*H*)-pteridinone	9.37	2.78	268
4-Butylamino-3-methyl-6,7-diphenyl-2(3*H*)-pteridinone	11.36	5.44	1679
2-Butylamino-4-pteridinamine		5.06	1678
2-Butylaminopteridine		3.98	124
2-*s*-Butylaminopteridine		3.80	124
2-*t*-Butylaminopteridine		3.38	124
2-*s*-Butylamino-4,6,7-trimethylpteridine		3.30	124
2-ε-(*N*-Butylcarbamoyl)pentylamino-6,7-dimethyl-4(3*H*)-pteridinone	8.50	2.51, −2.42	461
7-*t*-Butyl-3-methyl-2,4(1*H*,3*H*)-pteridinedione	8.50		367
7-*t*-Butyl-3-methyl-2,4(1*H*,3*H*)-pteridinedione 5-oxide	7.44		367
7-*t*-Butyl-4(3*H*)-pteridinone	8.32		1286
2-ε-Carbamoylpentylamino-6,7-dimethyl-4(3*H*)-pteridinone	8.46	2.59, −2.45	461
2-α-D-Carboxyethylamino-6,7-dimethyl-4(3*H*)-pteridinone	1.57, 9.02	3.60, −3.02	461
2-α-L-Carboxyethylamino-6,7-dimethyl-4(3*H*)-pteridinone	1.46, 8.97	3.20, −2.95	461
2-*β*-Carboxyethylamino-6,7-dimethyl-4(3*H*)-pteridinone	—, 8.89	2.22, −2.55	461
2-Carboxymethylamino-6,7-dimethyl-4(3*H*)-pteridinone	1.30, 8.45	3.36, −2.91	461
2-Carboxymethylamino-4(3*H*)-pteridinone	0.88, 7.90	3.26, −3.20	461
2-Carboxymethylamino-6-[L-*erythro*]-α,*β*,γ-trihydroxypropyl-4(3*H*)-pteridinone	1.01, 7.81	3.10, −1.60	461
6-Carboxymethyl-5,6-dihydro-4,7(3*H*,8*H*)-pteridinedione	4.49, 8.59, 11.00		29
4-Carboxymethyl-3,4-dihydro-2(1*H*)-pteridinone	4.14		38
2-Carboxymethylthio-7-methyl-4-pteridinamine	2.62		1247
2-Carboxymethylthio-7-phenyl-4-pteridinamine	2.42		1247
2-Carboxymethylthio-7-[L-*lyxo*]-α,*β*,γ-trihydroxybutyl-4-pteridinamine	2.54		1247
2-ε-Carboxypentylamino-6-[L-*erythro*]-α,*β*-dihydroxypropyl-4(3*H*)-pteridinone	—, 7.95	2.13	461
2-ε-Carboxypentylamino-6,7-dimethyl-4(3*H*)-pteridinone	—, 8.59	2.60, −2.36	461
2-ε-Carboxypentylamino-6-[L-*erythro*]-α,*β*,γ-trihydroxypropyl-4(3*H*)-pteridinone	—, 7.96	2.10	461
7-(2′-Carboxyprop-1′-enyl)-6(5*H*)-pteridinone	2.93	5.69	55
2-γ-Carboxypropylamino-6,7-dimethyl-4(3*H*)-pteridinone	—, 8.92	2.40, −2.45	461

TABLE XI. (*Contd.*)

Pteridine	Acidic pK_a	Basic pK_a	Reference(s)
5-Chloroacetyl-5,6,7,8-tetrahydro-2,4-pteridinediamine		5.70	267
1-(5′-Chloro-5′-deoxy-2′,3′-isopropylidene-β-D-ribofuranosyl)-6,7-diphenyl-2,4(1*H*,3*H*)-pteridinedione	7.89		263
4-Chloro-6,8-dimethyl-7(8*H*)-pteridinone	12.50		248
6-Chloro-1,3-dimethyl-7-thioxo-7,8-dihydro-2,4(1*H*,3*H*)-pteridinedione	0.88		1681
2-Chloro-8-ethyl-4-methyl-5,6,7,8-tetrahydropteridine		3.80	683
6-Chloro-3-methyl-7-phenyl-2,4(1*H*,3*H*)-pteridinedione	7.53		331
7-Chloro-3-methyl-6-phenyl-2,4(1*H*,3*H*)-pteridinedione	7.16		331
6-Chloro-3-methyl-7-phenyl-2,4(1*H*,3*H*)-pteridinedione 8-oxide	4.86		331
4-Chloro-8-methyl-7(8*H*)-pteridinone	12.0		248
2-Chloro-6-oxo-5,6,7,8-tetrahydro-4-pteridinecarboxylic acid	−0.19, 12.31	2.94	150
8-*p*-Chlorophenyl-2-imino-3-methyl-2,3-dihydro-4(3*H*)-pteridinone		4.60	377
6-*p*-Chlorophenyl-2,4(1*H*,3*H*)-pteridinedione	7.65, 12.78		412
7-*p*-Chlorophenyl-2,4(1*H*,3*H*)-pteridinedione	7.96, 13.30		412
6-*p*-Chlorophenyl-1-β-D-ribofuranosyl-2,4(1*H*,3*H*)-pteridinedione	8.12		412
7-*p*-Chlorophenyl-1-β-D-ribofuranosyl-2,4(1*H*,3*H*)-pteridinedione	8.06		412
4-Chloro-7-pteridinamine		0.81	131
7-Chloro-4-pteridinamine		2.75	131
6-Chloropteridine		3.72	35
7-Chloropteridine		3.26	35
7-Cyanomethyl-7,8-dihydro-6(5*H*)-pteridinone	9.89	4.18	55
8-Cyclohexyl-2-dimethylamino-6-methyl 7(8*H*)-pteridinone		2.77	1150
8-Cyclohexyl-3-methyl-2,4(3*H*,8*H*)-pteridinedione	11.61	−0.28	409
8-Cyclopropyl-3-methyl-2,4(3*H*,8*H*)-pteridinedione		0.52	409
5-Deoxy-5-(2′,4′-dioxo-2′,3′,4′,8′-tetrahydropteridin-8′-yl)-D-ribofuranose	9.52	−1.12	264
5-Deoxy-5-(2′,4′-dioxo-2′,3′,4′,8′-tetrahydropteridin-8′-yl)-D-ribose diethylthioacetal	8.68	−0.95	264
1-(5′-Deoxy-5′-iodo-2′,3′-isopropylidene-β-D-ribofuranosyl)-6,7-dimethyl-2,4(1*H*,3*H*)-pteridinedione	8.52		263
1-(5′-Deoxy-5′-iodo-2′,3′-isopropylidene-β-D-ribofuranosyl)-6,7-diphenyl-2,4(1*H*,3*H*)-pteridinedione	8.31		263
1-(5′-Deoxy-5′-iodo-2′,3′-isopropylidene-β-D-ribofuranosyl)-2,4(1*H*,3*H*)-pteridinedione	8.02		263
4-(2′-Deoxy-α-D-ribofuranosylamino)-7(8*H*)-pteridinone	6.89	1.16	224

TABLE XI. (*Contd.*)

Pteridine	Acidic pK_a	Basic pK_a	Reference(s)
1-(2′-Deoxy-α/β-D-ribofuranosyl)-6,7-dimethyl-2,4-(1*H*,3*H*)-pteridinedione	8.82		414
1-(2′-Deoxy-α-D-ribofuranosyl)-6,7-dimethyl-2,4(1*H*,3*H*)-pteridinedione (3′-ammonium phosphate)	9.00		143
1-(2′-Deoxy-β-D-ribofuranosyl)-6,7-dimethyl-2,4(1*H*,3*H*)-pteridinedione (3′-ammonium phosphate)	9.24		143
1-(2′-Deoxy-α/β-D-ribofuranosyl)-6,7-diphenyl-2,4(1*H*,3*H*)-pteridinedione	8.32		414
1-(2′-Deoxy-α-D-ribofuranosyl)-6,7-diphenyl-2,4(1*H*,3*H*)-pteridinedione (3′-ammonium phosphate)	9.15		143
1-(2′-Deoxy-β-D-ribofuranosyl)-6,7-diphenyl-2,4(1*H*,3*H*)-pteridinedione (3′-ammonium phosphate)	9.33		143
1-(2′-Deoxy-α/β-D-ribofuranosyl)-2,4(1*H*,3*H*)-pteridinedione	8.14		414
1-(2′-Deoxy-α-D-ribofuranosyl)-2,4(1*H*,3*H*)-pteridinedione (3′-ammonium phosphate)	7.84		143
1-(2′-Deoxy-β-D-ribofuranosyl)-2,4(1*H*,3*H*)-pteridinedione (3′-ammonium phosphate)	7.92		143
2,4-Diacetamido-5,8-diacetyl-5,6,7,8-tetrahydropteridine		2.12	267
2,4-Diacetamido-6-methylpteridine		1.55	123
2,4-Diacetamido-7-propionylpteridine		1.51	1678
2,4-Diacetamidopteridine		2.02	267
5,8-Diacetyl-2,4-bisdimethylamino-5,6,7,8-tetrahydropteridine		3.74	267
4-Diacetylmethyl-3,4-dihydro-2(1*H*)-pteridinone	8.02		38
7-Diacetylmethyl-7,8-dihydro-6(5*H*)-pteridinone	7.57	4.45	57
5,8-Diacetyl-4-methyl-5,6,7,8-tetrahydropteridine		2.15	118
2,4-Diamino-7,8-dihydro-6(5*H*)-pteridinone		5.34	268
2,7-Diamino-4-isopropoxy-6-pteridinecarbonitrile		4.10	371
2,4-Diamino-7-methyl-6-pteridinecarboxylic acid	2.09?	5.52?	307
2,4-Diamino-6-pteridinecarboxylic acid	1.94	5.17	123
	1.40	5.00	178
2,4-Diamino-7-pteridinecarboxylic acid	2.00	5.68	123
2,4-Diamino-6,7(5*H*,8*H*)-pteridinedione	7.44, 11.34	2.91, −4.19	208
2,4-Diamino-6(5*H*)-pteridinone	7.1	4.49	268
2,4-Diamino-7(8*H*)-pteridinone	7.89	2.74	268
2-Dibenzylamino-4(3*H*)-pteridinone	7.04	0.86	182
2,4-Dibenzyloxy-6-methyl-7(8*H*)-pteridinone	7.20		1192
2,4-Dibenzyloxy-7(8*H*)-pteridinone	6.11		1192
6,7-Dichloropteridine		3.27	35
4-Di(ethoxycarbonyl)methyl-3,4-dihydro-2(1*H*)-pteridinone	11.43		38

TABLE XI. (*Contd.*)

Pteridine	Acidic pK_a	Basic pK_a	Reference(s)
7-Di(ethoxycarbonyl)methyl-7,8-dihydro-6(5*H*)-pteridinone	10.17	4.12	55
2-Diethylaminopteridine		3.18	124
3,4-Dihydro-2-pteridinamine		7.73	45
3,4-Dihydropteridine		6.36	49, 50, 51
5,6-Dihydro-4,7(3*H*,8*H*)-pteridinedione	8.45		43
7,8-Dihydro-2,6(1*H*,5*H*)-pteridinedione	10.22	2.80	40, 43
7,8-Dihydro-4,6(3*H*,5*H*)-pteridinedione	9.14		43
3,4-Dihydro-2(1*H*)-pteridinethione	10.95		46
7,8-Dihydro-2,4,6(1*H*,3*H*,5*H*)-pteridinetrione	7.07		43
3,4-Dihydro-2(1*H*)-pteridinone	12.6	0.0	38, 42, 43, 45
5,6-Dihydro-4(3*H*)-pteridinone	10.29	2.94	43, 64
5,6-Dihydro-7(8*H*)-pteridinone	9.94	3.36	31, 43, 44, 46, 134
7,8-Dihydro-2(1*H*)-pteridinone	~12	3.50	42, 43
7,8-Dihydro-4(3*H*)-pteridinone	12.13	0.32	43
7,8-Dihydro-6(5*H*)-pteridinone	10.54	4.78	31, 43, 55, 134
	10.56		40
Dihydropteroylglutamic acid	9.54		1682
6-(3′,4′-Dihydroquinazolin-4′-ylmethyl)-7(8*H*)-pteridinone	9.26	6.42, 0.83	57
6-D-α,β-Dihydroxyethyl-8-D-ribityl-2,4,7(1*H*,3*H*,8*H*)-pteridinetrione [Protopterin-A = 6-DL-· · ·]	3.37, 12.40		295
6-L-α,β-Dihydroxyethyl-8-D-ribityl-2,4,7(1*H*,3*H*,8*H*)-pteridinetrione	3.36, 12.41		295
6,7-Dihydroxy-4-(6′-hydroxy-4′-methyl-5′,6′,7′,8′-tetrahydropteridin-7′-ylmethyl)-5,6,7,8-tetrahydropteridine		6.28, 4.19	64
4-(4′,6′-Dihydroxy-2′-mercaptopyrimidin-5′-yl)-3,4-dihydro-2-pteridinamine		8.36	45
8-(DL)-β,γ-Dihydroxypropyl-6,7-dimethyl-2,4(3*H*,8*H*)-pteridinedione	8.66		1686
6-[L-*erythro*]-α,β-Dihydroxypropyl-2-β-(*p*-hydroxyphenyl)ethylamino-4(3*H*)-pteridinone	8.02, 10.2	1.86	458
6-[L-*erythro*]-α,β-Dihydroxypropyl-2-*p*-methoxyanilino-4(3*H*)-pteridinone	7.40	1.70	458
6-[L-*erythro*]-α,β-Dihydroxypropyl-2-*p*-methoxybenzylamino-4(3*H*)-pteridinone	7.95	1.66	458
6-[L-*erythro*]-α,β-Dihydroxypropyl-2-methylthio-4(3*H*)-pteridinone	6.55	−1.01	458
6-[L-*erythro*]-α,β-Dihydroxypropyl-2,4(1*H*,3*H*)-pteridinedione (biolumazine)	7.84, 12.68		1439
4-(4′,6′-Dihydroxypyrimidin-5′-yl)-3,4-dihydro-2-pteridinamine		8.54	45
6,7-Diisopropyl-3,8-dimethyl-5,6,7,8-tetrahydro-2,4(1*H*.3*H*)-pteridinedione	8.65	4.17	375
6,7-Diisopropyl-8-methyl-2-methylimino-2,8-dihydropteridine		6.68	248

TABLE XI. (*Contd.*)

Pteridine	Acidic pK_a	Basic pK_a	Reference(s)
6,7-Diisopropyl-1-methyl-2,4(1*H*,3*H*)-pteridinedione	9.45		367
6,7-Diisopropyl-8-methyl-2,4(3*H*,8*H*)-pteridinedione	8.98	0.72	1545[e]
6,7-Diisopropyl-1-methyl-2,4(1*H*,3*H*)-pteridinedione 5-oxide	8.94		367
6,7-Diisopropyl-8-methyl-2(8*H*)-pteridinone	11.27	2.39	248
6,7-Diisopropyl-8-methyl-5,6,7,8-tetrahydro-2,4(1*H*,3*H*)-pteridinedione	8.06	4.00	375
4,4-Dimethoxy-3-methyl-6,7-diphenyl-3,4-dihydro-2(1*H*)-pteridinone	10.77		1679
2,4-Dimethoxy-6-methyl-7(8*H*)-pteridinone	7.38		1192
2,4-Dimethoxy-7(8*H*)-pteridinone	6.65		1192
Dimethyl 2-acetamido-4-oxo-3,4-dihydro-6,7-pteridinedicarboxylate	6.08		307, 601
2-Dimethylamine-7,8-dihydro-4,6(3*H*,5*H*)-pteridinedione	9.97	1.36	349
2-Dimethylamino-6-[L-*erythro*]-α,β-dihydroxypropyl-4(3*H*)-pteridinone	7.79	1.76	1439
2-Dimethylamino-7,7-dimethyl-7,8-dihydro-4,6(3*H*,5*H*)-pteridinedione	9.79	1.11	349
6-Dimethylamino-1,3-dimethyl-7-propionyl-2,4(1*H*,3*H*)-pteridinedione		0.63	108
7-Dimethylamino-1,3-dimethyl-6-propionyl-2,4(1*H*,3*H*)-pteridinedione		−2.53	108
4-Dimethylamino-6,7-dimethylpteridine		4.84	129
6-Dimethylamino-1,3-dimethyl-2,4(1*H*,3*H*)-pteridinedione		−0.47	1275
7-Dimethylamino-1,3-dimethyl-2,4(1*H*,3*H*)-pteridinedione		−0.47	1275
2-Dimethylamino-6,8-dimethyl-7(8*H*)-pteridinone		2.62	1150
4-Dimethylamino-6,7-dimethyl-2(1*H*)-pteridinone	10.58	3.26	362
4-Dimethylamino-6,8-dimethyl-7(8*H*)-pteridinone		2.42	445
Dimethyl 2-amino-3,8-dimethyl-3,4,7,8-tetrahydro-6,7-pteridinedicarboxylate		0.13	307
2-Dimethylamino-4,7-dioxo-3,4,7,8-tetrahydro-6-pteridinecarboxylic acid	4.05, 7.90, 10.80		360
2-Dimethylamino-8-ethyl-7-oxo-7,8-dihydro-6-pteridinecarboxylic acid	3.31		387
2-Dimethylamino-8-ethyl-7-oxo-3,4,7,8-tetrahydro-6-pteridinecarboxylic acid	7.25		387, cf. 1311
2-Dimethylamino-8-ethyl-6,7(5*H*,8*H*)-pteridinedione	8.31	2.92	208
2-Dimethylamino-8-ethyl-7(8*H*)-pteridinone		2.14	381
		2.30	387
Dimethyl 2-amino-5-formyl-4-oxo-3,4,5,6,7,8-hexahydro-6-pteridinecarboxylate	9.4	1.02, −1.7?	307
2-Dimethylamino-8-β-D-galactopyranosyl-6-methyl-7(8*H*)-pteridinone		2.14	1150
2-Dimethylamino-8-β-D-glucopyranosyl-6-methyl-7(8*H*)-pteridinone		1.99	1150
2-Dimethylamino-8-β-hydroxyethyl-6,7-dimethyl-4(8*H*)-pteridinone	12.59	5.80	116, 359

TABLE XI. (*Contd.*)

Pteridine	Acidic pK_a	Basic pK_a	Reference(s)
2-Dimethylamino-8-β-hydroxyethyl-6,7-diphenyl-4(8*H*)-pteridinone	11.66	4.91	359
2-Dimethylamino-8-β-hydroxyethyl-4(8*H*)-pteridinone	12.89	5.17	116, 359
2-Dimethylamino-4-methoxy-6,7-dimethylpteridine		4.05	416
2-Dimethylamino-4-methoxy-6-methyl-7(8*H*)-pteridinone	8.92		417
2-Dimethylamino-4-methoxypteridine		3.39	416
4-Dimethylamino-7-methoxypteridine		2.00	445, 1126
2-Dimethylamino-4-methoxy-7(8*H*)-pteridinone	8.48		417
2-Dimethylamino-8-methyl-7,8-dihydro-4,6(3*H*,5*H*)-pteridinedione	10.21	0.45	349
2-Dimethylamino-8-methyl-4,7-dioxo-3,4,7,8-tetrahydro-6-pteridinecarboxylic acid	3.89, 8.85		360
2-Dimethylamino-8-methyl-6,7-diphenyl-7,8-dihydro-4(3*H*)-pteridinone	11.35	2.20	373
2-Dimethylamino-8-methyl-6,7-diphenyl-4(8*H*)-pteridinone	13.01	5.20	359
4-Dimethylamino-8-methyl-2-methylthio-7(8*H*)-pteridinone		0.78	1126
Dimethyl 2-amino-8-methyl-4-oxo-4,8-dihydro-6,7-pteridinedicarboxylate	8.90	1.22	307, 601
Dimethyl 2-amino-8-methyl-4-oxo-3,4,7,8-tetrahydro-6,7-pteridinedicarboxylate	9.30	0.00	307
2-Dimethylamino-6-methyl-4,7(3*H*,8*H*)-pteridinedione	7.64, 10.06		360
2-Dimethylamino-8-methyl-4,7(3*H*,8*H*)-pteridinedione	7.92		246, 360
4-Dimethylamino-6-methyl-2,7(1*H*,8*H*)-pteridinedione	5.73, 13.24	2.52	363
4-Dimethylamino-8-methyl-6,7(5*H*,8*H*)-pteridinedione	6.82	1.36	208
2-Dimethylamino-8-methyl-4,6,7(3*H*,5*H*,8*H*)-pteridinetrione	7.50, 11.15		360
2-Dimethylamino-3-methyl-4(3*H*)-pteridinone		1.20	416
2-Dimethylamino-8-methyl-4(8*H*)-pteridinone	13.32	5.09	116, 359, 416, 741, 762
4-Dimethylamino-8-methyl-7(8*H*)-pteridinone		2.10	445, 1126
2-Dimethylamino-6-methyl-8-(pyridin-2′-yl)-7(8*H*)-pteridinone		2.04	139
2-Dimethylamino-6-methyl-8-(2′,3′,4′,5′-tetra-*O*-acetyl-β-D-galactopyranosyl)-7(8*H*)-pteridinone		1.83	1150
2-Dimethylamino-6-methyl-8-(2′,3′,4′,6′-tetra-*O*-acetyl-β-D-glucopyranosyl)-7(8*H*)-pteridinone		1.68	1150
4-Dimethylamino-2-methylthio-7(8*H*)-pteridinone	7.78	2.24	445, 1126
4-Dimethylamino-2-methylthio-8-β-D-ribofuranosyl-7(8*H*)-pteridinone		0.35	1126
2-Dimethylamino-4-oxo-3,4-dihydro-6-pteridinecarboxylic acid	2.86, 7.97	0.79	1710
2-Dimethylamino-4-oxo-3,4-dihydro-7-pteridinecarboxylic acid	2.74, 8.16	1.08	1710
Dimethyl 2-amino-4-oxo-3,4-dihydro-6,7-pteridinedicarboxylate	6.60	0.78	307, 601

TABLE XI. (*Contd.*)

Pteridine	Acidic pK_a	Basic pK_a	Reference(s)
2-Dimethylamino-6-oxo-5,6,7,8-tetrahydro-4-pteridinecarboxylic acid	0.57, >13	7.31	150
Dimethyl 2-amino-4-oxo-3,4,7,8-tetrahydro-6,7-pteridinedicarboxylate	9.50	0.5	307
2-Dimethylamino-6-phenyl-7,8-dihydro-4(3*H*)-pteridinone	11.05	2.71	373
2-Dimethylamino-4-pteridinamine		5.34	267, 947
4-Dimethylamino-2-pteridinamine		6.06	267, 947
2-Dimethylaminopteridine		3.03	30, 39, 285, 286, 947
4-Dimethylaminopteridine		4.33	32, 129, 285, 286
6-Dimethylaminopteridine		4.31	31, 285, 286
7-Dimethylaminopteridine		2.53	33, 285, 286
2-Dimethylamino-4,7(3*H*,8*H*)-pteridinedione	7.57, 9.95		246, 360
4-Dimethylamino-2,7(1*H*,8*H*)-pteridinedione	5.30, 12.93	2.15	363
2-Dimethylamino-4,6,7(3*H*,5*H*,8*H*)-pteridinetrione	7.93, 10.33, 13.55		360
2-Dimethylamino-4(3*H*)-pteridinone	7.81	2.26	114, 368, 416, 741, 762
	8.15	2.05	1710
2-Dimethylamino-7(8*H*)-pteridinone	7.97	2.02	381, 417
4-Dimethylamino-2(1*H*)-pteridinone	10.43	2.72	362
4-Dimethylamino-6(5*H*)-pteridinone		3.85	445
4-Dimethylamino-7(8*H*)-pteridinone	7.50	2.18	445
4-Dimethylamino-2-α-D-ribofuranosylamino-7(8*H*)-pteridinone	8.87	2.24	1541
4-Dimethylamino-2-β-D-ribofuranosylamino-7(8*H*)-pteridinone	8.75	2.26	1541
2-Dimethylamino-8-β-D-ribofuranosyl-4,7(3*H*,8*H*)-pteridinedione	8.22		246
4-Dimethylamino-1-β-D-ribofuranosyl-2(1*H*)-pteridinone		1.98	225
4-Dimethylamino-8-β-D-ribofuranosyl-7(8*H*)-pteridinone		1.90	1126
2-Dimethylamino-7-(2′,3′,4′,6′-tetraacetyl-β-D-glucopyranosyloxy)pteridine		1.81	381
2-Dimethylamino-6-[L-*threo*]-α,β,γ-trihydroxypropyl-4(3*H*)-pteridinone (euglenapterin; also three isomers)	7.90	1.80	335, 1710
2-Dimethylamino-6-[L-*threo*]-α,β,γ-trihydroxypropyl-4(3*H*)-pteridinone γ-phosphate	8.17	1.97	1710
2-Dimethylamino-3,6,7-trimethyl-4(3*H*)-pteridinone		1.69	416
4-Dimethylamino-1,6,7-trimethyl-2(1*H*)-pteridinone		3.28	362
2-Dimethylamino-6,7,8-trimethyl-4(8*H*)-pteridinone	12.30(?)	5.40	563
4,6-Dimethyl-7,8-dihydropteridine		6.00	43, 64
1,3-Dimethyl-7,8-dihydro-2,4,6(1*H*,3*H*,5*H*)-pteridinetrione	11.40		340
4,6-Dimethyl-7,8-dihydro-2(1*H*)-pteridinone	12.50	3.99	42

TABLE XI. (*Contd.*)

Pteridine	Acidic pK_a	Basic pK_a	Reference(s)
7-(6′,7′-Dimethyl-3′,4′-dihydropteridin-4′-ylmethyl)-4-hydroxy-6-methyl-3′,4′-dihydropteridine		4.79, 3.22	50
4-(4′,4′-Dimethyl-2′,6′-dioxocyclohexyl)-3,4-dihydro-2-pteridinamine		8.88	45
4-(4′,4′-Dimethyl-2′,6′-dioxocyclohexyl)-3,4-dihydropteridine	7.69	3.55	49, 50
4-(4′,4′-Dimethyl-2′,6′-dioxocyclohexyl)-3,4-dihydro-2(1*H*)-pteridinethione	12.20		46
6-(4′,4′-Dimethyl-2′,6′-dioxocyclohexyl)-5,6-dihydro-7(8*H*)-pteridinone		3.35	44
4-(4′,4′-Dimethyl-2′,6′-dioxocyclohexyl)-6,7-dimethyl-3,4-dihydropteridine	8.32	3.62	50
4-(4′,4′-Dimethyl-2′,6′-dioxocyclohexyl)-2-methyl-3,4-dihydropteridine	8.51	3.59	50
3,8-Dimethyl-6,7-diphenyl-7,8-dihydro-2,4(1*H*,3*H*)-pteridinedione	6.70	1.15	375
1,3-Dimethyl-6,7-diphenyl-2,4(1*H*,3*H*)-pteridinedione		−3.25	1339
3,8-Dimethyl-6,7-diphenyl-2,4(3*H*,8*H*)-pteridinedione	10.30	0.20	1545[e]
1,3-Dimethyl-6,7-diphenyl-2-thioxo-1,2-dihydro-4(3*H*)-pteridinone		−3.46	437
1,3-Dimethyl-2,4-dithioxo-1,2,3,4-tetrahydro-7(8*H*)-pteridinone	2.58	−2.07	1000
1,3-Dimethyl-2,7-dithioxo-1,2,7,8-tetrahydro-4(3*H*)-pteridinone	1.38	−3.85	1000
1,3-Dimethyl-4,7-dithioxo-3,4,7,8-tetrahydro-2(1*H*)-pteridinone	1.53	−2.86	1000
Dimethyl 2-imino-3,8-dimethyl-4-oxo-2,3,4,8-tetrahydro-6,7-pteridinedicarboxylate		1.57	307, 601
3,8-Dimethyl-2-methylamino-4,7-dioxo-3,4,7,8-tetrahydro-6-pteridinecarboxylic acid	3.69		1214
1,3-Dimethyl-6-methylamino-7-propionyl-2,4(1*H*,3*H*)-pteridinedione		−1.71	108
6,7-Dimethyl-4-methylaminopteridine		4.17	129
1,3-Dimethyl-6-methylamino-2,4(1*H*,3*H*)-pteridinedione		−0.24	1275
1,3-Dimethyl-7-methylamino-2,4(1*H*,3*H*)-pteridinedione		−0.62	1275
5,8-Dimethyl-4-methylamino-6,7(5*H*,8*H*)-pteridinedione		1.53	132
6,7-Dimethyl-2-methylamino-4(3*H*)-pteridinone		2.40	69
6,8-Dimethyl-2-methylamino-7(8*H*)-pteridinone)		2.71	248
1,3-Dimethyl-4-methylhydrazono-6,7-diphenyl-3,4-dihydro-2(1*H*)-pteridinone		3.62	1679
1,3-Dimethyl-4-methylimino-6,7-diphenyl-3,4-dihydro-2(1*H*)-pteridinone		6.14	1679
1,3-Dimethyl-4-*N*-methyl-*N*-phenylhydrazono-6,7-diphenyl-3,4-dihydro-2(1*H*)-pteridinone		2.72	1679
6,7-Dimethyl-4-methylthiopteridine		0.90	313
1,3-Dimethyl-7-methylthio-2,4(1*H*,3*H*)-pteridinedione		−2.49	1000
1,3-Dimethyl-7-methylthio-2,4(1*H*,3*H*)-pteridinedithione		−3.12	1000
6,7-Dimethyl-2-methylthio-4(3*H*)-pteridinone	7.12	0.24	437

TABLE XI. (*Contd.*)

Pteridine	Acidic pK_a	Basic pK_a	Reference(s)
1,3-Dimethyl-7-methylthio-2-thioxo-1,2-dihydro-4(3*H*)-pteridinone		−2.87	1000
1,3-Dimethyl-7-methylthio-4-thioxo-3,4-dihydro-2(1*H*)-pteridinone		−1.67	1000
3,8-Dimethyl-6-*N*-methylureido-2,4,7(1*H*,3*H*,8*H*)-pteridinetrione	4.98		1282
6,7-Dimethyl-4-oxo-3,4-dihydro-2-pteridinesulfinic acid	—, 7.68		629
6,7-Dimethyl-4-oxo-3,4-dihydro-2-pteridinesulfonic acid	—, 6.02		629
4,7-Dimethyl-2-phenylpteridine		1.40	152
6,7-Dimethyl-2-phenylpteridine		1.92	152
1,3-Dimethyl-6-phenyl-2,4,7(1*H*,3*H*,8*H*)-pteridinetrione	3.50		413
3,6-Dimethyl-1-phenyl-2,4,7(1*H*,3*H*,8*H*)-pteridinetrione	3.86		238
3,6-Dimethyl-1-phenyl-2,4,7(1*H*,3*H*,8*H*)-pteridinetrione 5-oxide	2.84		238
6,7-Dimethyl-2-phenyl-4(3*H*)-pteridinone	8.15		152
1,3-Dimethyl-6-propionyl-2,4,7(1*H*,3*H*,8*H*)-pteridinetrione	3.55		108
1,3-Dimethyl-7-propionyl-2,4,6(1*H*,3*H*,5*H*)-pteridinetrione	5.20		108
4,6-Dimethyl-2-pteridinamine		2.70	39
4,7-Dimethyl-2-pteridinamine		2.63	39
6,7-Dimethyl-2-pteridinamine		3.41	39
6,7-Dimethyl-4-pteridinamine		3.80	129
6,7-Dimethylpteridine		2.93	33, 285, 286
6,7-Dimethyl-2,4-pteridinediamine		5.44	541
6,7-Dimethyl-2,4-pteridinediamine 5,8-dioxide		1.36	541
6,7-Dimethyl-2,4-pteridinediamine 8-oxide		3.04	541
1,3-Dimethyl-2,4(1*H*,3*H*)-pteridinedione		−3.04	1339
3,8-Dimethyl-2,4(3*H*,8*H*)-pteridinedione	10.41	0.06	116, 1545[e]
6,7-Dimethyl-2,4(1*H*,3*H*)-pteridinedione	8.40, 13.39	−1.32	367, 437
		−1.65	1339
6,8-Dimethyl-2,7(1*H*,8*H*)-pteridinedione	7.41		248
6,8-Dimethyl-4,7(3*H*,8*H*)-pteridinedione	7.81		248
7,8-Dimethyl-2,4(3*H*,8*H*)-pteridinedione	10.10	0.15	1278
6,7-Dimethyl-2,4(1*H*,3*H*)-pteridinedione 5,8-dioxide	4.98		367
6,7-Dimethyl-2,4(1*H*,3*H*)-pteridinedione 8-oxide	6.02		367 (cf. 366)
6,7-Dimethyl-2,4(1*H*,3*H*)-pteridinedithione	5.83, 11.45		437
1,3-Dimethyl-2,4,6,7(1*H*,3*H*,5*H*,8*H*)-pteridinetetrone	3.63, 10.73		238, 343
6,7-Dimethyl-4(3*H*)-pteridinethione	7.10	0.80	313
1,3-Dimethyl-2,4,6(1*H*,3*H*,5*H*)-pteridinetrione	5.83		209, 340
1,3-Dimethyl-2,4,7(1*H*,3*H*,8*H*)-pteridinetrione	3.47	−2.20	238, 339, 716, 1000
1,6-Dimethyl-2,4,7(1*H*,3*H*,8*H*)-pteridinetrione	3.65, 10.63		339, 413
1,7-Dimethyl-2,4,6(1*H*,3*H*,5*H*)-pteridinetrione	6.52, 10.25		340
1,8-Dimethyl-2,4,7(1*H*,3*H*,8*H*)-pteridinetrione	8.00		716
3,6-Dimethyl-2,4,7(1*H*,3*H*,8*H*)-pteridinetrione	4.17, 10.42		339, 413
3,7-Dimethyl-2,4,6(1*H*,3*H*,5*H*)-pteridinetrione	6.85, 9.90		340

TABLE XI. (*Contd.*)

Pteridine	Acidic pK_a	Basic pK_a	Reference(s)
3,8-Dimethyl-2,4,7(1*H*,3*H*,8*H*)-pteridinetrione	3.83		344, 716
4,8-Dimethyl-2,6,7(1*H*,5*H*,8*H*)-pteridinetrione	6.72, 10.78	0.60	208
6,8-Dimethyl-2,4,7(1*H*,3*H*,8*H*)-pteridinetrione	4.26, 13.20		339, 378
	4.26, 13.00		1090, 1192
1,3-Dimethyl-2,4,7(1*H*,3*H*,8*H*)-pteridinetrione 5-oxide	2.30		238
1,3-Dimethyl-2,4,7(1*H*,3*H*,8*H*)-pteridinetrithione	1.41	−3.20	1000
2,6-Dimethyl-4(3*H*)-pteridinone	8.97		153
2,7-Dimethyl-4(3*H*)-pteridinone	8.80		153
6,7-Dimethyl-2(1*H*)-pteridinone	11.15 (hyd), 7.95 (anhyd)		38, 241, 954
6,7-Dimethyl-4(3*H*)-pteridinone	8.39		30, 285
	8.50		935
2,7-Dimethyl-4(1*H*)-pteridinone 3-oxide	5.31	−0.10	154
6,7-Dimethyl-4(1*H*)-pteridinone 3-oxide	5.08	−0.87	154
6,7-Dimethyl-8-D-ribityl-2,4(3*H*,8*H*)-pteridinedione	9.86	0.56	335
	8.29	0.56	378, 1545[e]
	8.31		1686
6,7-Dimethyl-1-β-D-ribofuranosyl-2,4(1*H*,3*H*)-pteridinedione	8.60		414
6,7-Dimethyl-3-β-ribofuranosyl-2,4(1*H*,3*H*)-pteridinedione	8.30		283
6,7-Dimethyl-1-β-D-ribofuranosyl-2,4-(1*H*,3*H*)-pteridinedione 5-oxide	7.74		210
1,3-Dimethyl-2-thioxo-1,2-dihydro-4,7-(3*H*,8*H*)-pteridinedione	3.14	−2.97	1000
1,3-Dimethyl-4-thioxo-3,4-dihydro-2,7-(1*H*,8*H*)-pteridinedione	2.87	−1.52	1000
1,3-Dimethyl-6-thioxo-5,6-dihydro-2,4(1*H*,3*H*)-pteridinedione	3.58		1681
1,3-Dimethyl-7-thioxo-7,8-dihydro-2,4-(1*H*,3*H*)-pteridinedione	1.67	−3.97	1000, 1681
1,3-Dimethyl-2-thioxo-1,2-dihydro-4(3*H*)-pteridinone		−2.56	437
6,7-Dimethyl-2-thioxo-1,2-dihydro-4(3*H*)-pteridinone	6.70, 12.80		437
6,8-Dimethyl-2-thioxo-1,2-dihydro-7(8*H*)-pteridinone	6.15		248
1,*N*-Dimethyl-2,4,6-trioxo-1,2,3,4,5,6-hexahydro-7-pteridinecarboxamide	4.90, 9.67		347
3,*N*-Dimethyl-2,4,6-trioxo-1,2,3,4,5,6-hexahydro-7-pteridinecarboxamide	5.22, 10.38		347
N,*N*-Dimethyl-2,4,6-trioxo-1,2,3,4,5,6-hexahydro-7-pteridinecarboxamide	4.76, 9.23		347
1,3-Dimethyl-2,4,7-trioxo-1,2,3,4,7,8-hexahydro-6-pteridinecarboxylic acid	2.10, 6.30		341
1,8-Dimethyl-2,4,7-trioxo-1,2,3,4,7,8-hexahydro-6-pteridinecarboxylic acid	2.32, 8.54		317
6,7-Dimethyl-8-D-xylityl-2,4(3*H*,8*H*)-pteridinedione	8.33		1686
2,6-Dioxo-1,2,5,6,7,8-hexahydro-4-pteridinecarboxylic acid	12.04		150
4,7-Dioxo-3,4,7,8-tetrahydro-6-pteridinecarbaldehyde	5.93, 9.31		29, 1418

TABLE XI. (*Contd.*)

Pteridine	Acidic pK_a	Basic pK_a	Reference(s)
4,6-Dioxo-3,4,5,6-tetrahydro-7-pteridinecarboxylic acid	2.30, 6.60, 9.85		346
4,7-Dioxo-3,4,7,8-tetrahydro-6-pteridinecarboxylic acid	3.00, 6.69, 10.05		29, 1418
6,7-Diphenyl-2,4-pteridinediamine		5.12	541
6,7-Diphenyl-2,4-pteridinediamine 5,8-dioxide		1.08	541
6,7-Diphenyl-2,4(1*H*,3*H*)-pteridinedione	8.09		367, 437
		−3.89	1339
6,7-Diphenyl-2,4(1*H*,3*H*)-pteridinedione 8-oxide	5.69, 12.12		367
6,7-Diphenyl-2,4(1*H*,3*H*)-pteridinedithione	5.12, 11.24 (aq MeOH)		437
6,7-Diphenyl-1-β-D-ribofuranosyl-2,4(1*H*,3*H*)-pteridinedione	8.48		414
6,7-Diphenyl-3-β-D-ribofuranosyl-2,4(1*H*,3*H*)-pteridinedione	7.86		283, 414
6,7-Diphenyl-1-β-D-ribofuranosyl-2,4(1*H*,3*H*)-pteridinedione 5-oxide	7.75		210
6,7-Diphenyl-2-β-D-ribofuranosylthio-1,2-dihydro-4(3*H*)-pteridinone	5.90		447
6,7-Diphenyl-3-β-D-ribofuranosyl-2-thioxo-1,2-dihydro-4(3*H*)-pteridinone	6.00		447
6,7-Diphenyl-5,6,7,8-tetrahydro-2,4(1*H*,3*H*)-pteridinedione		2.86	265
6,7-Diphenyl-2-thioxo-1,2-dihydro-4(3*H*)-pteridinone	6.45, 12.22		437
6,7-Diphenyl-1-(2′,3′,5′-tri-*O*-acetyl-β-D-ribofuranosyl)-2,4(1*H*,3*H*)-pteridinedione	8.00		265
6-Ethoxycarbonylmethyl-4,7(3*H*,8*H*)-pteridinedione	6.23, 9.62		29
6-(1′-Ethoxycarbonyl-2′-oxopropyl)-5,6-dihydropteridine		3.16	44
2-Ethoxy-6-oxo-5,6,7,8-tetrahydro-4-pteridinecarboxylic acid	0.81, 12.83	5.31	150
Ethyl 2-acetamido-4-oxo-3,4-dihydro-7-pteridinecarboxylate	7.06	−1.60	389
Ethyl 2-amino-3,8-dimethyl-7-oxo-3,4,7,8-tetrahydro-6-pteridinecarboxylate		1.85	376, 1311
2-Ethylamino-7-hydroxy-5,6-dihydro-6-pteridinecarboxylic acid	7.16		387
Ethyl 2-amino-8-β-hydroxyethyl-4,7-dioxo-3,4,7,8-tetrahydro-6-pteridinecarboxylate	7.57		385
Ethyl 2-amino-4-isopropoxy-8-methyl-7-oxo-7,8-dihydro-6-pteridinecarboxylate		−0.46	371
Ethyl 2-amino-4-isopropoxy-7-oxo-7,8-dihydro-6-pteridinecarboxylate	7.80	0.35	371
Ethyl 2-amino-4-methoxy-8-methyl-7-oxo-7,8-dihydro-6-pteridinecarboxylate		−0.60	371
Ethyl 2-amino-3-methyl-4,7-dioxo-3,4,7,8-tetrahydro-6-pteridinecarboxylate	6.95		385
Ethyl 2-amino-8-methyl-4,7-dioxo-3,4,7,8-tetrahydro-6-pteridinecarboxylate	7.55		385

TABLE XI. (*Contd.*)

Pteridine	Acidic pK_a	Basic pK_a	Reference(s)
Ethyl 4-amino-2-methylthio-7-oxo-7,8-dihydro-6-pteridinecarboxylate	6.94	0.93	1126
2-Ethylamino-7-oxo-7,8-dihydro-6-pteridinecarboxylic acid	3.00, 5.89		387
2-Ethylamino-7(8*H*)-pteridinone	7.50		387
Ethyl 8-benzyl-3-methyl-2,4,7-trioxo-1,2,3,4,7,8-hexahydro-6-pteridinecarboxylate	2.16		317
Ethyl 6,7-bis-β-hydroxyethoxy-5,6,7,8-tetrahydro-4-pteridinecarboxylate		3.73	158
Ethyl 2-chloro-6-oxo-5,6,7,8-tetrahydro-4-pteridinecarboxylate		−0.54(?)	150
Ethyl 6,7-diethoxy-5,6,7,8-tetrahydro-4-pteridinecarboxylate		3.60	147
Ethyl 3,4-dihydro-2-pteridinecarboxylate		3.74	51
Ethyl 6,7-dihydroxy-6-methyl-5,6,7,8-tetrahydro-4-pteridinecarboxylate		~3.90	148
Ethyl 6,7-dihydroxy-7-methyl-5,6,7,8-tetrahydro-4-pteridinecarboxylate		3.88	148
Ethyl 6,7-dihydroxy-5,6,7,8-tetrahydro-4-pteridinecarboxylate		3.64 3.73	148 158
Ethyl 2-dimethylamino-4,7-dioxo-3,4,7,8-tetrahydro-6-pteridinecarboxylate	7.18, 10.23		360
Ethyl 2-dimethylamino-8-methyl-4,7-dioxo-3,4,7,8-tetrahydro-6-pteridinecarboxylate	7.19		360
Ethyl 2-dimethylamino-6-oxo-5,6,7,8-tetrahydro-4-pteridinecarboxylate		4.71	150
Ethyl 3,8-dimethyl-2-methylamino-7-oxo-3,4,7,8-tetrahydro-6-pteridinecarboxylate		0.65	376, 1311
Ethyl 3,8-dimethyl-2-methylimino-7-oxo-2,3,7,8-tetrahydro-6-pteridinecarboxylate		6.06	376, 1311
8-Ethyl-6,7-dimethyl-2,4(3*H*,8*H*)-pteridinedione	10.12	0.90	378, 1545
Ethyl 1,8-dimethyl-2,4,7-trioxo-1,2,3,4,7,8-hexahydro-6-pteridinecarboxylate	7.74		317
Ethyl 2,6-dioxo-1,2,5,6,7,8-hexahydro-4-pteridinecarboxylate		1.75	150
Ethyl 2-ethoxy-6-oxo-5,6,7,8-tetrahydro-4-pteridinecarboxylate		3.14	150
8-Ethyl-2-ethylamino-5,6-dihydro-7(8*H*)-pteridinone		5.13	387
8-Ethyl-2-ethylamino-4-methyl-7-oxo-7,8-dihydro-6-pteridinecarboxylic acid	3.82		387
8-Ethyl-2-ethylamino-4-methyl-7-oxo-3,4,7,8-tetrahydro-6-pteridinecarboxylic acid	6.96		387, cf. 1311
8-Ethyl-2-ethylamino-4-methyl-7(8*H*)-pteridinone	2.97		387
Ethyl 2-ethylamino-7-oxo-7,8-dihydro-6-pteridine-carboxylate	7.00		387
8-Ethyl-2-ethylamino-7-oxo-7,8-dihydro-6-pteridinecarboxylic acid	3.35		387
8-Ethyl-2-ethylamino-7-oxo-3,4,7,8-tetrahydro-6-pteridinecarboxylic acid	7.14		387, cf. 1311
8-Ethyl-2-ethylamino-7(8*H*)-pteridinone		2.50	387

TABLE XI. (*Contd.*)

Pteridine	Acidic pK_a	Basic pK_a	Reference(s)
Ethyl 8-ethyl-2,4,7-trioxo-1,2,3,4,7,8-hexahydro-6-pteridinecarboxylate	2.93		317
Ethyl 2-imino-3,8-dimethyl-7-oxo-2,3,7,8-tetrahydro-6-pteridinecarboxylate		5.13	376, 1311
Ethyl 4-methyl-7-oxo-7,8-dihydro-6-pteridinecarboxylate	5.74		57
Ethyl 6-methyl-4-pteridinecarboxylate		3.69	147
Ethyl 7-methyl-4-pteridinecarboxylate		3.69	147
8-Ethyl-6-methyl-2,4,7(1*H*,3*H*,8*H*)-pteridinetrione	4.39, 13.42		378
Ethyl 7-oxo-7,8-dihydro-6-pteridinecarboxylate	5.53		57
8-Ethyl-6-phenyl-2,4,7(1*H*,3*H*,8*H*)-pteridinetrione	3.86, 13.10		413
Ethyl 2-pteridinecarboxylate		2.73	51
Ethyl 4-pteridinecarboxylate		3.62	147, 149
8-Ethyl-2,4(3*H*,8*H*)-pteridinedione	10.15	−0.10	1545
8-Ethyl-2,4,7(1*H*,3*H*,8*H*)-pteridinetrione	3.87, 13.02		317
7-Ethyl-4(3*H*)-pteridinone	8.16		1286
Ethyl 1,2,3,4-tetrahydro-2-pteridinecarboxylate		3.31	51
Ethyl 2,4,6-trioxo-1,2,3,4,5,6-hexahydro-7-pteridinecarboxylate	4.65, 8.94		347
8-Ethyl-2,4,7-trioxo-1,2,3,4,7,8-hexahydro-6-pteridinecarboxylic acid	2.28, 4.85, 13.10		317
5-Formyl-8-isopropyl-3,6,7-trimethyl-5,6,7,8-tetrahydro-2,4(1*H*,3*H*)-pteridinedione	7.22		375
5-Formyl-4-methyl-5,6,7,8-tetrahydropteridine		5.45	118
5-Formyl-5,6,7,8-tetrahydropteridine		5.00	683
5-Formyl-5,6,7,8-tetrahydro-2,4-pteridinediamine		5.76	267
5-Formyl-6,7,8-trimethyl-5,6,7,8-tetrahydro-2,4(1*H*,3*H*)-pteridinedione	6.83		375
1-*β*-D-Glucopyranosyl-4-*β*-D-glucopyranosylamino-2(1*H*)-pteridinone		−0.31	382
6-(*β*-D-Glucopyranosyloxy)methyl-2,4-pteridinediamine		4.87	113
1-*β*-D-Glucopyranosyl-2,4(1*H*,3*H*)-pteridinedione	8.00		382
1-*β*-D-Glucopyranosyl-5,6,7,8-tetrahydro-2,4(1*H*,3*H*)-pteridinedione	10.68	4.46	265
8-D-Glucosyl-6-methyl-2,4,7(1*H*,3*H*,8*H*)-pteridinetrione	4.40		1090
4-Hydrazinopteridine		4.00	33
8-*δ*-Hydroxybutyl-6,7-dimethyl-2,4(3*H*,8*H*)-pteridinedione	9.84		1686
4-Hydroxy-3,4-dihydro-2(1*H*)-pteridinone	11.05		40
7-Hydroxy-7,8-dihydro-6(5*H*)-pteridinone	9.90		40
4-*β*-Hydroxyethylamino-6,7-dimethyl-2(1*H*)-pteridinone	11.19	2.71	362
8-*β*-Hydroxyethyl-6,7-dimethyl-7,8-dihydro-2,4(1*H*,3*H*)-pteridinedione	7.25	2.66	375, 1545
1-*β*-Hydroxyethyl-6,7-dimethyl-2,4-(1*H*,3*H*)-pteridinedione	9.14		1339
8-*β*-Hydroxyethyl-6,7-dimethyl-2,4-(3*H*,8*H*)-pteridinedione	9.35	0.50	378, 409, 1545[e]
	9.32		1686

TABLE XI. (*Contd.*)

Pteridine	Acidic pK_a	Basic pK_a	Reference(s)
1-β-Hydroxyethyl-6,7-diphenyl-2,4-(1*H*,3*H*)-pteridinedione	8.86		1339
8-β-Hydroxyethyl-6,7-diphenyl-2,4-(3*H*,8*H*)-pteridinedione	7.89	−0.13	409, 1545
8-β-Hydroxyethyl-3-methyl-6,7-diphenyl-2,4(3*H*,8*H*)-pteridinedione	8.50	−0.20	409, 1545[e]
8-β-Hydroxyethyl-3-methyl-2,4(3*H*,8*H*)-pteridinedione	10.35	−0.40	409, 1545
8-β-Hydroxyethyl-6-methyl-2,4,7(1*H*,3*H*,8*H*)-pteridinetrione	4.00, 13.15 4.02, 13.35	 −1.94	378, 1339 712
8-β-Hydroxyethyl-6-phenyl-2,4,7(1*H*,3*H*,8*H*)-pteridinetrione	3.31, 12.94		1339
1-β-Hydroxyethyl-2,4(1*H*,3*H*)-pteridinedione	8.54		1339
8-β-Hydroxyethyl-2,4(3*H*,8*H*)-pteridinedione	9.70	−0.50	116, 409, 1545
8-β-Hydroxyethyl-2,4,7(1*H*,3*H*,8*H*)-pteridinetrione	3.51, 12.79		317, 1339
8-β-Hydroxyethyl-3,6,7-trimethyl-2,4(3*H*,8*H*)-pteridinedione	9.93	0.36	409, 1545
8-β-Hydroxyethyl-2,4,7-trioxo-1,2,3,4,7,8-hexahydro-6-pteridinecarboxylic acid	1.94, 4.78, 12.60		317
8-6′-Hydroxyhexyl-6,7-dimethyl-2,4(3*H*,8*H*)-pteridinedione	9.97		1686
6-Hydroxy-5-methyl-5,6-dihydro-7(8*H*)-pteridinone	9.33	2.91	134
6-Hydroxymethyl-2,4-pteridinediamine		5.02	113
7-Hydroxymethyl-2,4-pteridinediamine		5.44	123
6-Hydroxy-8-methyl-4(8*H*)-pteridinone	8.67		248
8-ε-Hydroxypentyl-6,7-dimethyl-2,4(3*H*,8*H*)-pteridinedione	10.02		1686
2-β-(*p*-Hydroxyphenyl) ethylamino-6-[D-*erythro*]-α,β,γ-trihydroxypropyl-4(3*H*)-pteridinone	7.94, 9.50	1.86	458
2-β-(*p*-Hydroxyphenyl) ethylamino-6-[L-*erythro*]-α,β,γ-trihydroxypropyl-4(3*H*)-pteridinone	7.84, 9.60	1.94	458
8-β-Hydroxypropyl-6,7-dimethyl-2,4(3*H*,8*H*)-pteridinedione	8.90		1686
8-γ-Hydroxypropyl-6,7-dimethyl-2,4(3*H*,8*H*)-pteridinedione	9.72		1686
8-γ-Hydroxypropyl-3,6,7-trimethyl-2,4(3*H*,8*H*)-pteridinedione	10.12	0.41	1545
3-Hydroxy-2,4(1*H*,3*H*)-pteridinedione	5.61, 9.00		1099
2-Imino-3,8-dimethyl-2,8-dihydro-4(3*H*)-pteridinone	12.17	7.92, (8.71, <5)	116, 359
2-Imino-3,8-dimethyl-6,7-diphenyl-2,3-dihydro-4(8*H*)-pteridinone		8.64	116, 359
4-Imino-1,3-dimethyl-6,7-diphenyl-3,4-dihydro-2(1*H*)-pteridinone		5.64	1679
2-Imino-8-isopropyl-3-methyl-2,3-dihydro-4(8*H*)-pteridinone		8.70	377
4-Imino-8-methyl-4,8-dihydro-2-pteridinamine		8.88	130
7-Imino-1-methyl-1,7-dihydro-4-pteridinamine		12.12	131

TABLE XI. (*Contd.*)

Pteridine	Acidic pK_a	Basic pK_a	Reference(s)
2-Imino-3,6,7,8-tetramethyl-2,8-dihydro-4(3*H*)-pteridinone	13.72	8.25, (10.04, 4.60)	116, 359
		7.80	563
4-Imino-1,6,7-trimethyl-1,4-dihydro-2-pteridinamine		11.90	130, 267
2-Imino-3,6,7-trimethyl-8-phenyl-2,3-dihydro-4(8*H*)-pteridinone		6.30	377
2-Imino-3,6,7-trimethyl-8-*p*-tolyl-2,3-dihydro-4(8*H*)-pteridinone		6.65	377
6-Indol-3′-yl-8-D-ribityl-2,4,7(1*H*,3*H*,8*H*)-pteridinetrione	4.75		1673, 1674
2-Isobutylamino-4,6,7-trimethylpteridine		3.13	124
4-Isopropoxy-6,7-dimethyl-2-pteridinamine		4.34	370, 741
4-Isopropoxy-2-pteridinamine		3.67	370, 741, 762
4-Isopropoxy-6,7,8-trimethyl-2(8*H*)-pteridinimine		7.81	359
8-Isopropyl-6,7-dimethyl-2,4-(3*H*,8*H*)-pteridinedione	10.13	0.00	375
1-(2′,3′-Isopropylidene-5′-*O*-mesyl-*β*-D-ribofuranosyl)-6,7-dimethyl-2,4(1*H*,3*H*)-pteridinedione	8.48		263
1-(2′,3′-Isopropylidene-5′-*O*-mesyl-*β*-D-ribofuranosyl)-6,7-diphenyl-2,4(1*H*,3*H*)-pteridinedione	8.26		263
1-(2′,3′-Isopropylidene-5′-*O*-mesyl-*β*-D-ribofuranosyl-2,4(1*H*,3*H*)-pteridinedione	8.00		263
1-(2′,3′-Isopropylidene-*β*-D-ribofuranosyl)-6,7-dimethyl-2-methylamino-4(1*H*)-pteridinone	12.90	1.94	263
1-(2′,3′-Isopropylidene-*β*-D-ribofuranosyl)-6,7-dimethyl-2,4-(1*H*,3*H*)-pteridinedione	8.76		263
1-(2′,3′-Isopropylidene-*β*-D-ribofuranosyl)-6,7-diphenyl-2,4-(1*H*,3*H*)-pteridinedione	8.40		263
1-(2′,3′-Isopropylidene-*β*-D-ribofuranosyl)-2,4(1*H*,3*H*)-pteridinedione	8.17		263
1-(2′,3′-Isopropylidene-*β*-D-ribofuranosyl)-6,7-diphenyl-2-thioxo-1,2-dihydro-4(3*H*)-pteridinone	7.60		263
1-(2′,3′-Isopropylidene-*β*-D-ribofuranosyl)-2-methylamino-6,7-diphenyl-4(1*H*)-pteridinone	12.58	1.37	263
1-(2′,3′-Isopropylidene-*β*-D-ribofuranosyl)-2-methylamino-4(1*H*)-pteridinone	12.39	1.04	263
7-Isopropyl-4(3*H*)-pteridinone	8.24		1286
8-Isopropyl-3,6,7-trimethyl-2,4(3*H*,8*H*)-pteridinedione	10.80	0.15	375
8-D-Lyxityl-6-methyl-2,4,7(1*H*,3*H*,8*H*)-pteridinetrione	3.76, 13.01		295
Methanopterin	4.00, 8.80	1.80	1459
Methotrexate	5.73		761
	5.71		1682
2-*p*-Methoxyanilino-6-[D-*erythro*]-*α*,*β*,*γ*-trihydroxypropyl-4(3*H*)-pteridinone	7.35	1.66	458
2-*p*-Methoxyanilino-6-[L-*erythro*]-*α*,*β*,*γ*-trihydroxypropyl-4(3*H*)-pteridinone	7.31	1.63	458
2-*p*-Methoxybenzylamino-6-[D-*erythro*]-*α*,*β*,*γ*-trihydroxypropyl-4(3*H*)-pteridinone	7.92	1.60	458
2-*p*-Methoxybenzylamino-6-[L-*erythro*]-*α*,*β*,*γ*-trihydroxypropyl-4(3*H*)-pteridinone	7.91	1.63	458

TABLE XI. (*Contd.*)

Pteridine	Acidic pK_a	Basic pK_a	Reference(s)
6-Methoxy-7,8-dihydro-2,4-pteridinediamine	6.14		268
7-Methoxy-1,3-dimethyl-2,4-dioxo-1,2,3,4-tetrahydro-6-pteridinecarboxylic acid	2.50		341
2-Methoxy-6,7-dimethyl-4-pteridinamine		3.56	362
4-Methoxy-6,7-dimethyl-2-pteridinamine		3.99	370, 741
7-Methoxy-1,3-dimethyl-2,4(1*H*,3*H*)-pteridinedione		−2.81	1000
7-Methoxy-1,3-dimethyl-2,4(1*H*,3*H*)-pteridinedithione		−2.37	1000
2-Methoxy-6,8-dimethyl-7(8*H*)-pteridinone	13.00		248
7-Methoxy-1,3-dimethyl-2-thioxo-1,2-dihydro-4(3*H*)-pteridinone		−3.01	1000
7-Methoxy-1,3-dimethyl-4-thioxo-3,4-dihydro-2(1*H*)-pteridinone		−1.97	1000
8-*β*-Methoxyethyl-6,7-diphenyl-2,4-(3*H*,8*H*)-pteridinedione	9.59	−0.29	409
4-Methoxy-8-methyl-6,7-diphenyl-7,8-dihydro-2-pteridinamine		3.40, −0.88	374
4-Methoxy-8-methyl-2-methylamino-6,7-diphenyl-7,8-dihydropteridine		3.65, −0.90	374
4-Methoxy-7-phenyl-2-pteridinamine		4.10	1092
2-Methoxy-4-pteridinamine		3.25	225, 362
4-Methoxy-2-pteridinamine		3.50	114, 368, 370, 416, 741, 762
		3,46	130
7-Methoxy-4-pteridinamine		2.00	445
2-Methoxypteridine		2.13	32, 285, 286
4-Methoxypteridine		<1.50	32, 285
		1.04	33, 286
6-Methoxypteridine		3.60	31, 32, 285, 286
7-Methoxypteridine		1.64	33, 285, 286
6-Methoxy-2,4-pteridinediamine		5.51	268
4-Methoxy-6,7(5*H*,8*H*)-pteridinedione	7.67, 10.78		208
4-Methoxy-6,7,8-trimethyl-2(8*H*)-pteridinimine		7.33, (9.97, 3.83)	359
7-Methoxy-1,3,6-trimethyl-2-thioxo-1,2-dihydro-4(3*H*)-pteridinone		−1.79	1000
Methyl 2-acetamido-4-oxo-3,4-dihydro-6-pteridinecarboxylate	6.75	−1.60	389
Methyl 2-amino-3,8-dimethyl-7-oxo-3,4,7,8-tetrahydro-6-pteridinecarboxylate		1.66	376, 1311
Methyl 2-amino-6-methyl-4-oxo-3,4-dihydro-7-pteridinecarboxylate	8.05	2.10	307
Methyl 2-amino-8-methyl-4-oxo-4,8-dihydro-6-pteridinecarboxylate		4.36	307
Methyl 4-amino-2-methylthio-7-oxo-8-*β*-D-ribofuranosyl-7,8-dihydro-6-pteridinecarboxylate		0.57	1126
Methyl 2-amino-4-oxo-3,4-dihydro-6-pteridinecarboxylate	7.38	1.50	389

TABLE XI. (*Contd.*)

Pteridine	Acidic pK_a	Basic pK_a	Reference(s)
Methyl 2-amino-4-oxo-3,4-dihydro-7-pteridinecarboxylate	7.12	1.60	389
2-Methylamino-4-pteridinamine		4.97	1678
4-Methylamino-7-pteridinamine		5.11	131
2-Methylaminopteridine		3.62	32, 39
4-Methylaminopteridine		3.70	129
7-Methylaminopteridine		2.56	131
2-Methylamino-4(3*H*)-pteridinone	7.95	1.95	114, 368, 741, 762
	8.16	1.98	130
2-Methylamino-7(8*H*)-pteridinone	7.59	2.08	132
4-Methylamino-2(1*H*)-pteridinone	11.79	2.62	362
4-Methylamino-6(5*H*)-pteridinone		3.46	445
4-Methylamino-7(8*H*)-pteridinone	7.54	2.22	445
4-Methylamino-1-β-D-ribofuranosyl-2(1*H*)-pteridinone		1.05	225
Methyl 8-benzyl-2,4,7-trioxo-1,2,3,4,7,8-hexahydro-6-pteridinecarboxylate	2.06		317
Methyl 5-deoxy-5-(2′,4′-dioxo-2′,3′,4′,8′-pteridin-8′-yl)-2,3-isopropylidene-β-D-ribofuranoside	9.24	−1.28	2.64
2-Methyl-3,4-dihydropteridine		7.26	51
6-Methyl-3,4-dihydropteridine		6.66	51
6-Methyl-5,6-dihydro-4,7(3*H*,8*H*)-pteridinedione	8.43, 11.40		29
6-Methyl-3,4-dihydro-2(1*H*)-pteridinone	13.05	0.20	42
6-Methyl-7,8-dihydro-2(1*H*)-pteridinone	11.85	3.42	42
7-Methyl-7,8-dihydro-6(5*H*)-pteridinone	10.89	4.80	55
Methyl 3,8-dimethyl-2-methylamino-7-oxo-3,4,7,8-tetrahydro-6-pteridinecarboxylate		0.55	376, 1311
Methyl 3,8-dimethyl-2-methylimino-7-oxo-2,3,7,8-tetrahydro-6-pteridinecarboxylate		5.99	376, 1311
3-Methyl-4,6-dioxo-3,4,5,6-tetrahydro-7-pteridine-carboxylic acid	2.40, 6.75		346
8-Methyl-6,7-diphenyl-7,8-dihydro-2,4(1*H*,3*H*)-pteridinedione	6.13	0.21	375
1-Methyl-6,7-diphenyl-4-phenylhydrazino-2(1*H*)-pteridinone	10.53	2.02	1679
3-Methyl-6,7-diphenyl-4-phenylhydrazino-2(3*H*)-pteridinone	11.95		1675
1-Methyl-6,7-diphenyl-2,4(1*H*,3*H*)-pteridinedione	8.64		367, 414
		−3.64	1339
3-Methyl-6,7-diphenyl-2,4(1*H*,3*H*)-pteridinedione	8.01		367, 414
		−3.28	1339
8-Methyl-6,7-diphenyl-2,4(3*H*,8*H*)-pteridinedione	9.50	0.20	1278
	9.69	0.36	1545[e]
1-Methyl-6,7-diphenyl-2,4(1*H*,3*H*)-pteridinedione 5-oxide	8.30		210, 367
3-Methyl-6,7-diphenyl-2,4(1*H*,3*H*)-pteridinedione 8-oxide	5.74		367
3-Methyl-6,7-diphenyl-2,4(1*H*,3*H*)-pteridinedithione	5.41		446

TABLE XI. (*Contd.*)

Pteridine	Acidic pK_a	Basic pK_a	Reference(s)
3-Methyl-6,7-diphenyl-8-β-pyrrolidinoethyl-2,4(3*H*,8*H*)-pteridinedione	7.99	−1.99	409
1-Methyl-6,7-diphenyl-5,6,7,8-tetrahydro-2,4(1*H*,3*H*)-pteridinedione		2.74	265
8-Methyl-6,7-diphenyl-5,6,7,8-tetrahydro-4(3*H*)-pteridinone		—, 0.13	297
1-Methyl-6,7-diphenyl-2-thioxo-1,2-dihydro-4(3*H*)-pteridinone	8.31		437, 447
1-Methyl-6,7-diphenyl-4-thioxo-3,4-dihydro-2(1*H*)-pteridinone	7.77		1679
3-Methyl-6,7-diphenyl-2-thioxo-1,2-dihydro-4(3*H*)-pteridinone	6.64		446, 447
3-Methyl-6,7-diphenyl-4-thioxo-3,4-dihydro-2(1*H*)-pteridinone	7.49		1679
3-Methyl-6,7-dipyridin-2′-yl-2-thioxo-1,2-dihydro-4(3*H*)-pteridinone	6.08		446
Methyl 8-β-hydroxyethyl-2,4,7-trioxo-1,2,3,4,7,8-hexahydro-6-pteridinecarboxylate	2.65		317
Methyl 2-imino-3,8-dimethyl-7-oxo-2,3,7,8-tetrahydro-6-pteridinecarboxylate		5.21	376, 1311
8-Methyl-2-methylamino-6,7-diphenyl-7,8-dihydro-4(3*H*)-pteridinone	11.10	2.16	374
1-Methyl-4-methylamino-6,7-diphenyl-2(1*H*)-pteridinone		2.69	1679
3-Methyl-4-methylamino-6,7-diphenyl-2(3*H*)-pteridinone	10.84	5.55	1679
8-Methyl-2-methylamino-6,7-diphenyl-4(8*H*)-pteridinone	12.50	5.55	377
8-Methyl-2-methylamino-6-phenyl-7(8*H*)-pteridinone	12.00	2.42	248
8-Methyl-2-methylamino-6,7(5*H*,8*H*)-pteridinedione	8.39	3.08	208
1-Methyl-4-methylimino-1,4-dihydro-7-pteridinamine		>12.00	131
1-Methyl-4-methylimino-1,4-dihydropteridine		10.34	129
3-Methyl-2-methylthio-6,7-diphenyl-4(3*H*)-pteridinethione		−1.67	437
1-Methyl-2-methylthio-6,7-diphenyl-4(1*H*)-pteridinone		−0.37	437, 446
3-Methyl-2-methylthio-6,7-diphenyl-4(3*H*)-pteridinone		−0.88	437, 446
7-Methyl-4-methylthiopteridine		<2.00	33
3-Methyl-2-methylthio-4,7(3*H*,8*H*)-pteridinedione	6.47		344
1-Methyl-2-methylthio-4(1*H*)-pteridinone		−0.81	437, 446
3-Methyl-2-methylthio-4(3*H*)-pteridinone		−0.81	437, 446
1-Methyl-4-morpholino-6,7-diphenyl-2(1*H*)-pteridinone		1.46	1679
3-Methyl-8-β-morpholinoethyl-6,7-diphenyl-2,4(3*H*,8*H*)-pteridinedione	9.50	−2.76	409
3-Methyl-8-γ-morpholinopropyl-6,7-diphenyl-2,4(3*H*,8*H*)-pteridinedione	10.27	−1.05	409
N-Methyl-7-oxo-7,8-dihydro-6-pteridinecarboxamide	5.70		57
6-(7′-Methyl-6′-oxo-5′,6′,7′,8′-tetrahydropteridin-7′-ylmethyl)-7(8*H*)-pteridinone	6.79, 11.07	4.47, 0.48	57

TABLE XI. (*Contd.*)

Pteridine	Acidic pK_a	Basic pK_a	Reference(s)
2-Methyl-4-phenylpteridine		4.08	151
4-Methyl-2-phenylpteridine		3.30	152
7-Methyl-2-phenylpteridine		2.52	152
7-Methyl-4-phenylpteridine		2.06	151
3-Methyl-6-phenyl-2,4(1*H*,3*H*)-pteridinedione	8.40		367
3-Methyl-7-phenyl-2,4(1*H*,3*H*)-pteridinedione	7.99		331, 367
8-Methyl-6-phenyl-2,7(1*H*,8*H*)-pteridinedione	7.23	2.61	248
3-Methyl-7-phenyl-2,4(1*H*,3*H*)-pteridinedione 5,8-dioxide	4.38		367
3-Methyl-6-phenyl-2,4(1*H*,3*H*)-pteridinedione 8-oxide	5.12		367
3-Methyl-7-phenyl-2,4(1*H*,3*H*)-pteridinedione 5-oxide	7.08		367
1-Methyl-6-phenyl-2,4,7(1*H*,3*H*,8*H*)-pteridinetrione	3.44, 10.81		413
3-Methyl-1-phenyl-2,4,7(1*H*,3*H*,8*H*)-pteridinetrione	3.49		344
3-Methyl-6-phenyl-2,4,7(1*H*,3*H*,8*H*)-pteridinetrione	3.65, 10.71		331
	3.84, 10.60		413
3-Methyl-7-phenyl-2,4,6(1*H*,3*H*,5*H*)-pteridinetrione	7.03, 10.21		331
3-Methyl-7-phenyl-2,4,6(1*H*,3*H*,5*H*)-pteridinetrione 8-oxide	4.86, 8.22		331
7-Methyl-2-phenyl-4(3*H*)-pteridinone	7.72		152
1-Methyl-7-propionyl-2,4(1*H*,3*H*)-pteridinedione	8.39		108
3-Methyl-6-propionyl-2,4(1*H*,3*H*)-pteridinedione	7.84		108
2-Methyl-4-pteridinamine		4.30	39
4-Methyl-2-pteridinamine		2.82	39, 267, 947
6-Methyl-2-pteridinamine		4.05	39
7-Methyl-2-pteridinamine		3.76	39
2-Methylpteridine		4.87	33, 285, 286
	11.96	5.45	1145
4-Methylpteridine	5.51	2.94	33, 64, 1145
		2.90	285, 286
6-Methylpteridine		3.89	51
7-Methylpteridine		3.49	33, 285, 286
	11.76	4.91	1145
6-Methyl-2,4-pteridinediamine		5.33	113
1-Methyl-2,4(1*H*,3*H*)-pteridinedione	8.57		130
	8.45		335, 338, 367, 414
		−3.26	1339
1-Methyl-4,7(1*H*,8*H*)-pteridinedione	3.54		132
3-Methyl-2,4(1*H*,3*H*)-pteridinedione	8.00		130, 335, 338, 367, 414
		−3.09	1339
3-Methyl-4,6(3*H*,5*H*)-pteridinedione	6.26		346
3-Methyl-4,7(3*H*,8*H*)-pteridinedione	6.19		346, 383
4-Methyl-2,7(1*H*,8*H*)-pteridinedione	6.31		248
4-Methyl-6,7(5*H*,8*H*)-pteridinedione	7.12, 10.17		64
5-Methyl-6,7(5*H*,8*H*)-pteridinedione	7.02		134
6-Methyl-4,7(3*H*,8*H*)-pteridinedione	6.82, 10.02		29
8-Methyl-2,4(3*H*,8*H*)-pteridinedione	9.89	−0.10	116, 264, 1545[e]

TABLE XI. (*Contd.*)

Pteridine	Acidic pK_a	Basic pK_a	Reference(s)
8-Methyl-4,7(3*H*,8*H*)-pteridinedione	7.58		445
1-Methyl-2,4(1*H*,3*H*)-pteridinedione	8.03		210, 367
5-oxide	7.98		366
3-Methyl-2,4(1*H*,3*H*)-pteridinedione 8-oxide	5.18		367
1-Methyl-2,4,6,7(1*H*,3*H*,5*H*,8*H*)-pteridinetetrone	3.49, 9.60		343
3-Methyl-2,4,6,7(1*H*,3*H*,5*H*,8*H*)-pteridinetetrone	4.02, 9.57		343
8-Methyl-2,4,6,7(1*H*,3*H*,5*H*,8*H*)-pteridinetetrone	4.36, 9.65		343
7-Methyl-4(3*H*)-pteridinethione	7.02		33
1-Methyl-2,4,6(1*H*,3*H*,5*H*)-pteridinetrione	5.62, 9.85		340
1-Methyl-2,4,7(1*H*,3*H*,8*H*)-pteridinetrione	3.31, 10.51		339, 413, 716
3-Methyl-2,4,6(1*H*,3*H*,5*H*)-pteridinetrione	5.96, 9.72		340
3-Methyl-2,4,7(1*H*,3*H*,8*H*)-pteridinetrione	3.60, 10.26		339, 716
	3.60, 10.51		413
6-Methyl-2,4,7(1*H*,3*H*,8*H*)-pteridinetrione	4.13, 10.09		339
7-Methyl-2,4,6(1*H*,3*H*,5*H*)-pteridinetrione	6.57, 9.45		340
8-Methyl-2,4,7(1*H*,3*H*,8*H*)-pteridinetrione	3.80, 12.85,		317
	3.80, 11.70		339, 716, 1192
1-Methyl-4(1*H*)-pteridinimine		9.51	129
1-Methyl-7(1*H*)-pteridinimine		8.33	131
3-Methyl-4(3*H*)-pteridinimine		9.50	132
3-Methyl-7(3*H*)-pteridinimine		6.82	131
1-Methyl-2(1*H*)-pteridinone	11.43	<1.00	134
1-Methyl-4(1*H*)-pteridinone		1.25	134, 288
		<2.00	285
2-Methyl-4(3*H*)-pteridinone	8.54, 8.57		38, 153
2-Methyl-6(5*H*)-pteridinone	9.50 (hyd), 6.31 (anhyd)	4.67 (hyd)	55
	10.05 (hyd), 6.53 (anhyd)		241, 954, 961
2-Methyl-7(8*H*)-pteridinone	6.68	1.71	55
3-Methyl-2(3*H*)-pteridinone	11.01	<1.00	34, 134
3-Methyl-4(3*H*)-pteridinone		<1.30	32, 285
		−0.47	134, 285
4-Methyl-2(1*H*)-pteridinone	10.85 (hyd), 8.20 (anhyd)		38, 241
4-Methyl-6(5*H*)-pteridinone	9.50 (hyd), 6.40 (anhyd)	4.09 (hyd)	55
	10.00 (hyd), 6.30 (anhyd)		241, 954, 961
4-Methyl-7(8*H*)-pteridinone	6.79	<2.00	55
5-Methyl-6(5*H*)-pteridinone	10.60	3.73	134
6-Methyl-2(1*H*)-pteridinone	7.95 (anhyd), 11.00 (hyd)	−0.20	38, 42, 241
6-Methyl-4(3*H*)-pteridinone	8.19		32, 38, 153
6-Methyl-7(8*H*)-pteridinone	6.97		55, 57
7-Methyl-2(1*H*)-pteridinone	8.07 (anhyd), 10.85 (hyd)		38, 241
7-Methyl-4(3*H*)-pteridinone	8.09, 8.10		32, 38, 153, 1418, 1286

TABLE XI. (*Contd.*)

Pteridine	Acidic pK_a	Basic pK_a	Reference(s)
7-Methyl-6(5*H*)-pteridinone	7.17 (anhyd)	3.72 (hyd)	55
	7.09 (anhyd), 10.02 (hyd)		241, 954, 961
8-Methyl-4(8*H*)-pteridinone		3.26	248
8-Methyl-7(8*H*)-pteridinone		1.10	31, 134, 285
2-Methyl-4(1*H*)-pteridinone 3-oxide	5.16	−0.44	154
6-Methyl-4(1*H*)-pteridinone 3-oxide	4.89	−1.26	154
7-Methyl-4(1*H*)-pteridinone 3-oxide	4.89	−1.19	154
7-Methyl-7-(quinazolin-4′-ylmethyl)-7,8-dihydro-6(5*H*)-pteridinone	11.00	4.26	57
6-Methyl-8-D-ribityl-2,4,7(1*H*,3*H*,8*H*)-pteridinetrione	3.71, 12.94		295
	4.04, 12.87		378
6-Methyl-1-*β*-D-ribofuranosyl-2,4,7(1*H*,3*H*,8*H*)-pteridinetrione	3.34, 10.67		413
6-Methyl-3-*β*-D-ribofuranosyl-2,4,7(1*H*,3*H*,8*H*)-pteridinetrione	3.71, 10.87		413
6-Methyl-8-*β*-D-ribofuranosyl-2,4,7(1*H*,3*H*,8*H*)-pteridinetrione	3.42, 13.11		1192
4-Methyl-5,6,7,8-tetrahydropteridine		6.74	64, 118
1-Methyl-5,6,7,8-tetrahydro-2,4(1*H*,3*H*)-pteridinedione	10.74	4.85	265
6-Methyl-5,6,7,8-tetrahydro-4(3*H*)-pteridinone	9.97	3.84	43
2-Methylthio-3,4-dihydropteridine		5.01	46
2-Methylthio-6,7-diphenyl-4(3*H*)-pteridinethione	5.80	−1.29	437
2-Methylthio-6,7-diphenyl-4(3*H*)-pteridinone	6.79	−1.00	437, 447
2-Methylthio-4-pteridinamine		2.52	1678
2-Methylthiopteridine		2.20	33, 285, 286
4-Methylthiopteridine		2.59	33, 285, 286
6-Methylthiopteridine		3.27	36
7-Methylthiopteridine		2.49	33
		2.50	285, 286
2-Methylthio-4,7(3*H*,8*H*)-pteridinedione	5.78		1000
2-Methylthio-4(3*H*)-pteridinone	6.52	−0.99	437
	6.51	−1.01	458
2-Methylthio-6-[D-*erythro*]-*α*,*β*,*γ*-trihydroxypropyl-4(3*H*)-pteridinone	6.52	−1.10	458
2-Methylthio-6-[L-*erythro*]-*α*,*β*,*γ*-trihydroxypropyl-4(3*H*)-pteridinone	6.51	−1.04	458
1-Methyl-2-thioxo-1,2-dihydro-4,7(3*H*,8*H*)-pteridinedione	2.89, 9.63		1000
1-Methyl-7-thioxo-7,8-dihydro-2,4(1*H*,3*H*)-pteridinedione	1.51, 10.16		1709
3-Methyl-2-thioxo-1,2-dihydro-4,7(3*H*,8*H*)-pteridinedione	2.96		1000
3-Methyl-7-thioxo-7,8-dihydro-2,4(1*H*,3*H*)-pteridinedione	2.08, 9.58		1709
1-Methyl-2-thioxo-1,2-dihydro-4(3*H*)-pteridinone	8.25		437
3-Methyl-2-thioxo-1,2-dihydro-4(3*H*)-pteridinone	6.65		446
N-Methyl-2,4,6-trioxo-1,2,3,4,5,6-hexahydro-7-pteridinecarboxamide	4.96, 9.59		347

TABLE XI. (*Contd.*)

Pteridine	Acidic pK_a	Basic pK_a	Reference(s)
1-Methyl-2,4,6-trioxo-1,2,3,4,5,6-hexahydro-7-pteridinecarboxylic acid	1.80, 7.11, 10.32		342
1-Methyl-2,4,7-trioxo-1,2,3,4,7,8-hexahydro-6-pteridinecarboxylic acid	2.00, 6.15, 10.75		341
3-Methyl-2,4,6-trioxo-1,2,3,4,5,6-hexahydro-7-pteridinecarboxylic acid	1.90, 7.52, 10.20		342
3-Methyl-2,4,7-trioxo-1,2,3,4,7,8-hexahydro-6-pteridinecarboxylic acid	1.80, 6.00, 10.60		341
8-Methyl-2,4,7-trioxo-1,2,3,4,7,8-hexahydro-6-pteridinecarboxylic acid	2.15, 4.72, 13.06		317
	2.20, 4.72, 11.10		341
6-Methyl-8-D-xylityl-2,4,7(1*H*,3*H*,8*H*)-pteridinetrione	3.67, 12.75		295
4-Nitromethyl-3,4-dihydro-2-pteridinamine		7.25	45
7-Oxo-7,8-dihydro-6-pteridinecarboxylic acid	1.50, 6.73		57
4-Oxo-3,4-dihydro-2-pteridinesulfinic acid	—, 7.44		629
4-Oxo-3,4-dihydro-2-pteridinesulfonic acid	—, 6.73		629
7-(7′-Oxo-7′-8′-dihydropteridin-6′-ylmethylene)-3,7-dihydro-6(5*H*)-pteridinone	7.01, 9.23		57
4-Oxo-6,7-diphenyl-3,4-dihydro-2-pteridinesulfinic acid	—, 7.36		629
4-Oxo-6,7-diphenyl-3,4-dihydro-2-pteridinesulfonic acid	—, 5.94		629
6-(6′-Oxo-5′,6′,7′,8′-tetrahydropteridin-7′-ylmethyl)-7(8*H*)-pteridinone	6.85, 11.03	4.50, 0.73	57
3,5,6,7,8-Pentamethyl-2,4-dioxo-1,2,3,4,7,8-hexahydro-5-pteridinium *p*-toluenesulfonate	9.70	4.45	375, 376
4-Pentyloxy-2-D-ribofuranosylamino-7(8*H*)-pteridinone	7.71	0.08	1541
4-Pentyloxy-2-(2′,3′,5′-tri-*O*-acetyl-D-ribofuranosyl)amino-7(8*H*)-pteridinone	7.50	−0.76	1541
4-Phenacyl-6,7-diphenyl-2(1*H*)-pteridinone	9.21		1679
6-Phenyl-1,3-bis-*β*-D-ribofuranosyl-2,4,7(1*H*,3*H*,8*H*)-pteridinetrione	3.07		413
1-Phenyl-7-propionyl-2,4(1*H*,3*H*)-pteridinedione	7.70		108
2-Phenylpteridine		3.99	152
4-Phenylpteridine		3.99	151
6-Phenyl-2,4-pteridinediamine		5.07	541
7-Phenyl-2,4-pteridinediamine		5.36	541
6-Phenyl-2,4-pteridinediamine 5,8-dioxide		~1.50	541
6-Phenyl-2,4-pteridinediamine 8-oxide		2.50	541
7-Phenyl-2,4-pteridinediamine 5-oxide		1.91	541
6-Phenyl-2,4(1*H*,3*H*)-pteridinedione	7.93		367
7-Phenyl-2,4(1*H*,3*H*)-pteridinedione	8.05		367
7-Phenyl-2,4(1*H*,3*H*)-pteridinedione 5,8-dioxide	4.50		367
6-Phenyl-2,4(1*H*,3*H*)-pteridinedione 8-oxide	5.14		367
7-Phenyl-2,4(1*H*,3*H*)-pteridinedione 5-oxide	7.19		367
6-Phenyl-2,4,7-pteridinetriamine (triamterene)		6.26, −1.18	541
6-Phenyl-2,4,7-pteridinetriamine 5,8-dioxide		~1.00	541

TABLE XI. (*Contd.*)

Pteridine	Acidic pK_a	Basic pK_a	Reference(s)
1-Phenyl-2,4,7(1*H*,3*H*,8*H*)-pteridinetrione	2.95, 9.46		344
2-Phenyl-4(3*H*)-pteridinone	7.48		152
7-Phenyl-4(3*H*)-pteridinone	8.17		1287
6-Phenyl-1-α-D-ribofuranosyl-2,4(1*H*,3*H*)-pteridinedione	8.18		412
6-Phenyl-1-β-D-ribofuranosyl-2,4(1*H*,3*H*)-pteridinedione	8.23		412
7-Phenyl-1-α-ribofuranosyl-2,4(1*H*,3*H*)-pteridinedione	8.35		412
7-Phenyl-1-β-D-ribofuranosyl-2,4(1*H*,3*H*)-pteridinedione	8.38		412
6-Phenyl-1-β-D-ribofuranosyl-2,4,7(1*H*,3*H*,8*H*)-pteridinetrione	3.31, 10.49		413
6-Phenyl-3-β-D-ribofuranosyl-2,4,7(1*H*,3*H*,8*H*)-pteridinetrione	3.39, 10.39		413
7-Propionyl-2,4(1*H*,3*H*)-pteridinedione	7.49, 12.50		108
2-Propylaminopteridine		3.78	124
7-Propyl-4(3*H*)-pteridinone	8.16		1286
	8.11		1287
2-Pteridinamine		4.39	733
		4.29	30, 39, 285, 368, 947, 1250
		6.51 (hyd)	45
4-Pteridinamine		3.51	733
	14.00 (?)	3.50	28
	> 14.00	3.56	30, 39, 129, 267, 285, 947, 1250
6-Pteridinamine		4.15	31, 285, 1250
7-Pteridinamine		~2.96	33, 285, 1250
		2.24	131
Pteridine		4.15	773, 1165
		4.12	30, 33, 52, 66, 285, 286, 308
	11.21	2.60 (anhyd), 4.79 (hyd)	66, 1145
	12.20 or 12.50	5.17 (dihyd)	277
		4.10 or 3.50	277
		4.21	1250
		−2.00 (anhyd)	1513
2,4-Pteridinediamine		5.32, <0.50	32, 59, 130, 267, 947, 1678
4,6-Pteridinediamine		4.37	41
4,7-Pteridinediamine		4.97	41, 131
6,7-Pteridinediamine		3.54, −0.35	36
2,4-Pteridinediamine 5,8-dioxide		1.20	541
2,4-Pteridinediamine 5-oxide		2.17	541

TABLE XI. (*Contd.*)

Pteridine	Acidic pK_a	Basic pK_a	Reference(s)
2,4(1*H*,3*H*)-Pteridinedione (lumazine)		−3.34	1339
		−3.00	1046
	7.92, >12.00	<1.30	13
	7.91	<1.00	30, 241, 1418
	7.20, 12.00	~−2.00	100, 277
	7.95		335, 338, 367, 437
		<1.00[d]	757
2,6(1*H*,5*H*)-Pteridinedione	5.58, 8.64 (anhyd), 8.56, 9.69 (hyd)		40, 961
	6.70 (anhyd), 11.60 (hyd)		41
	5.99, 9.66 (anhyd), 8.84, 0.29 (hyd)		241
	9.50	2.00	277
2,7(1*H*,8*H*)-Pteridinedione	5.83, 10.07		41
	5.85, 9.94		241
	5.30, 11.50 (?)		277
4,6(3*H*,5*H*)-Pteridinedione	6.08, 9.73	<2.00	13, 29, 346
	6.05, 9.51 (anhyd), 8.34, 10.08 (hyd)		241, 961
	6.60, 9.70	0.70	277
4,7(3*H*,8*H*)-Pteridinedione	6.08, 9.62	<2.00	13, 29, 241, 346
	5.50, >9.00	−2.00	277
6,7(5*H*,8*H*)-Pteridinedione	6.90, 10.00	<2.00	19
	6.87, 10.00	<2.70	31, 64, 134, 208, 241
		0.90	36, 277
2,4(3*H*,8*H*)-Pteridinedione 1-oxide	6.50, 9.35		1098
2,4(1*H*,3*H*)-Pteridinedione 8-oxide	5.34		367 (cf. 366)
2,4(1*H*,3*H*)-Pteridinedithione	5.68, 11.09		437
6,7(5*H*,8*H*)-Pteridinedithione	5.16, 8.01	0.74	36
2,4,6,7-Pteridinetetramine		6.86	41
2,4,6,7(1*H*,3*H*,5*H*,8*H*)-Pteridinetetrone	~3.50, 9.50		277
2(1*H*)-Pteridinethione	9.98		30, 285
	9.72 (hyd), 6.52 (anhyd)		46, 241, 961
4(3*H*)-Pteridinethione	6.81		30, 285, 1103
6(5*H*)-Pteridinethione	5.78	3.06	36
7(8*H*)-Pteridinethione	5.50		33, 285
2,4,7-Pteridinetriamine		6.30	41
4,6,7-Pteridinetriamine		5.57	41
2,4,6(1*H*,3*H*,5*H*)-Pteridinetrione	5.73, 9.41		41
	5.20, 9.60	−2.00	100, 277
	5.85, 9.43		340

TABLE XI. (*Contd.*)

Pteridine	Acidic pK_a	Basic pK_a	Reference(s)
2,4,7(1*H*,3*H*,8)-Pteridinetrione (violapterin)	3.61		41
	3.00 (?), 9.80		277
	3.43, 9.80		339, 716
2,6,7(1*H*,5*H*,8*H*)-Pteridinetrione	3.50, 6.70, 9.10 (?)	0.50	277
4,6,7(3*H*,5*H*,8*H*)-Pteridinetrione	6.70, 9.50	−2.00	277
2(1*H*)-Pteridinone	11.13	<2.00	30, 134, 285
	11.05 (hyd)		38, 40, 241, 954,
	7.70 (anhyd)		961, 1250
	13.03 (dianion)		45, 140
	14.11 (dianion)		140
	>12.00	−2.00	277
4(3*H*)-Pteridinone	7.89	<1.30	13, 241, 368, 1418
	7.90	1.10	28, 100, 277
	7.89	<1.50	30, 37, 38, 285
	7.89	−0.17	134, 288
	8.10		935
6(5*H*)-Pteridinone	6.70	3.70	19, 277
	6.70	3.67	31, 134, 285
	6.45 (anhyd)	3.67 (hyd)	40, 55, 1250
	9.70 (trans)		134
	14.24 (dianion)		140
	6.45 (anhyd)		241, 954, 961
	9.90 (hyd)		
7(8*H*)-Pteridinone	6.41	1.20	19, 44, 55, 285, 1250
	6.41	1.20, −2.00	31, 134, 241, 277
4(1*H*)-Pteridinone 3-oxide	4.75	−1.23	154
Pteroylglutamic acid	—, 8.26		13
	8.26	~2.30 or 1.60	308
	8.25	2.40	761
	8.38		1682
7-(Quinazolin-4′-ylmethyl)-7,8-dihydro-6(5*H*)-pteridinone	10.60	4.29	57
8-D-Ribityl-2,4,7(1*H*,3*H*,8*H*)-pteridinetrione	3.28, 12.40		295
6-(β-D-Ribofuranosyloxy)methyl-2,4-pteridinediamine		4.98	113
1-β-D-Ribofuranosyl-2,4(1*H*,3*H*)-pteridinedione	8.48		237
	8.31		414
3-β-D-Ribofuranosyl-2,4(1*H*,3*H*)-pteridinedione	7.86		237, 283
1-β-D-Ribofuranosyl-2,4(1*H*,3*H*)-pteridinedione 5-oxide	7.02		210
1-β-D-Ribofuranosyl-2,4,7(1*H*,3*H*,8*H*)-pteridinetrione	3.13, 10.26		413
3-β-D-Ribofuranosyl-2,4,7(1*H*,3*H*,8*H*)-pteridinetrione	3.23, 10.24		413

TABLE XI. (*Contd.*)

Pteridine	Acidic pK_a	Basic pK_a	Reference(s)
8-β-D-Ribofuranosyl-2,4,7(1*H*,3*H*,8*H*)-pteridinetrione	3.28, 12.73		1192
3-β-D-Ribofuranosyl-4(3*H*)-pteridinone		−0.82	182
6-*p*-Sulfoanilinomethyl-4(3*H*)-pteridinone	—, 7.91	<2.50	13
6-(2′,3′,4′,6′-Tetra-*O*-acetyl-β-D-glucopyranosyloxy)methyl-2,4-pteridinediamine		4.73	113
1-(2′,3′,4′,6′-Tetra-*O*-acetyl-β-D-glucopyranosyl)-2,4(1*H*,3*H*)-pteridinedione	7.85		382
5,6,7,8-Tetrahydrofolic acid	3.50, 4.80, 10.50	4.82, 1.24, −1.25	992
1,2,3,4-Tetrahydropteridine		5.62, 0.10	51
5,6,7,8-Tetrahydropteridine		6.63	43, 49, 52, 683
5,6,7,8-Tetrahydro-2,4-pteridinediamine		6.71	267
5,6,7,8-Tetrahydro-2,4(1*H*,3*H*)-pteridinedione		~5.00	43
	7.93	4.80	335
5,6,7,8-Tetrahydro-2(1*H*)-pteridinone	12.50	4.35	42, 43
5,6,7,8-Tetrahydro-4(3*H*)-pteridinone	10.13	3.86	43
3,6,7,8-Tetramethyl-7,8-dihydro-2,4(1*H*,3*H*)-pteridinedione	7.30	2.91	375, 376
5,6,7,7-Tetramethyl-2,4-dioxo-1,2,3,4,7,8-hexahydro-5-pteridinium *p*-toluenesulfonate	9.25	5.00	375, 376
5,6,7,8-Tetramethyl-2,4-dioxo-1,2,3,4,7,8-hexahydro-5-pteridinium *p*-toluenesulfonate		4.37	375, 376
2,4,6,7-Tetramethylpteridine		2.65, 6.71 (hyd)	66
1,3,6,7-Tetramethyl-2,4(1*H*,3*H*)-pteridinedione		−1.35	1339
3,6,7,8-Tetramethyl-2,4(3*H*,8*H*)-pteridinedione	10.34	0.82	116, 1545
1,3,6,7-Tetramethyl-2,4(1*H*,3*H*)-pteridinedione 5-oxide		−2.67	366, 367
1,3,5,6-Tetramethyl-5,6,7,8-tetrahydro-2,4(1*H*,3*H*)-pteridinedione		5.39	456
1,3,6,8-Tetramethyl-2-thioxo-1,2-dihydro-4,7(3*H*,8*H*)-pteridinedione		−2.51	1000
1,3,6,7-Tetramethyl-2-thioxo-1,2-dihydro-4(3*H*)-pteridinone		−1.47	437
2-Thioxo-1,2-dihydro-4,7(3*H*,8*H*)-pteridinedione	2.85, 8.44, 13.26		1000
2-Thioxo-1,2-dihydro-4(3*H*)-pteridinone	6.48, 11.90		437
6-Thioxo-5,6-dihydro-7(8*H*)-pteridinone	5.59, 9.28	0.76	36
7-Thioxo-7,8-dihydro-6(5*H*)-pteridinone	5.95, 9.62	0.98	36
2-Thioxo-7-[L-*lyxo*]-α,β,γ-trihydroxybutyl-1,2-dihydro-4(3*H*)-pteridinone	7.19		1247
1,2,3-Tri-*O*-acetyl-5-deoxy-5-(2′,4′-dioxo-2′,3′,4′,8′-tetrahydropteridin-8′-yl)-D-ribofuranose	8.83	−2.04	264
6-(2′,3′,5′-Tri-*O*-benzoyl-β-D-ribofuranosyloxy)-methyl-2,4-pteridinediamine		4.35	113
3-(2′,3′,5′-Tri-*O*-benzyl-α-D-arabinofuranosyl-2,4(1*H*,3*H*)-pteridinedione	8.47		237
7-[L-*lyxo*]-α,β,γ-Trihydroxybutyl-2,4(1*H*,3*H*)-pteridinedione	8.38		1247

TABLE XI. (*Contd.*)

Pteridine	Acidic pK_a	Basic pK_a	Reference(s)
4-(2′,4′,6′-Trihydroxypyrimidin-5′-yl)-3,4-dihydro-2-pteridinamine		8.44	45
4-(2′,4′,6′-Trihydroxypyrimidin-5′-yl)-3,4-dihydropteridine		7.24	49
5-(3′,4′,5′-Trimethoxybenzoyl)-5,6,7,8-tetrahydro-2,4-pteridinediamine		5.77	267
6,7,7-Trimethyl-7,8-dihydro-2,4(1*H*,3*H*)-pteridinedione	8.13	2.97	375, 376
6,7,8-Trimethyl-7,8-dihydro-2,4(1*H*,3*H*)-pteridinedione	7.24	2.80	375, 376
1,3,6-Trimethyl-2,4-dioxo-1,2,3,4-tetrahydro-7-pteridinesulfenic acid	4.84		228, 1709
6,7,8-Trimethyl-2-methylamino-4(8*H*)-pteridinone	11.90	5.60	563
1,6,7-Trimethyl-4-methylimino-1,4-dihydropteridine		11.43	129
6,7,8-Trimethyl-2-methylimino-2,8-dihydropteridine		6.07	248
6,7,8-Trimethyl-4-methylimino-4,8-dihydropteridine		6.64	129
1,6,7-Trimethyl-2-methylthio-4(1*H*)-pteridinone		−0.04	437, 446
3,6,7-Trimethyl-2-methylthio-4(3*H*)-pteridinethione		0.29	437
		0.37	446
3,6,7-Trimethyl-2-methylthio-4(3*H*)-pteridinone		0.37	437
1,3,5-Trimethyl-6-phenylimino-5,6-dihydro-2,4(1*H*,3*H*)-pteridinedione		5.36	1680
4,6,7-Trimethyl-2-pteridinamine		3.03	39
2,6,7-Trimethylpteridine		3.76	33, 66, 285, 286
4,6,7-Trimethylpteridine		1.74	66
1,6,7-Trimethyl-2,4(1*H*,3*H*)-pteridinedione	9.06		130
	9.21		367, 414
		−1.84	1339
3,6,7-Trimethyl-2,4(1*H*,3*H*)-pteridinedione	8.52		367, 414
		−1.56	1339
6,7,8-Trimethyl-2,4(3*H*,8*H*)-pteridinedione	9.90	0.85	116
	9.83		130
	9.86	0.85	335, 378
	9.72	0.93	1278
	9.93		1686
3,6,7-Trimethyl-2,4(1*H*,3*H*)-pteridinedione 5,8-dioxide	5.11		367
1,6,7-Trimethyl-2,4(1*H*,3*H*)-pteridinedione 5-oxide	8.51		210, 366, 367
3,6,7-Trimethyl-2,4(1*H*,3*H*)-pteridinedione 8-oxide	6.15		367 (cf. 366)
1,3,8-Trimethyl-2,4,6,7(1*H*,3*H*,5*H*,8*H*)-pteridinetetrone	7.64		343
1,6,7-Trimethyl-4(1*H*)-pteridinethione		1.30, −1.00	313
6,7,8-Trimethyl-2-(8*H*)-pteridinethione	8.80		248
1,3,6-Trimethyl-2,4,7(1*H*,3*H*,8*H*)-pteridinetrione	3.80		238, 339, 716
1,3,7-Trimethyl-2,4,6(1*H*,3*H*,5*H*)-pteridinetrione	6.70		340
1,3,8-Trimethyl-2,4,7(1*H*,3*H*,8*H*)-pteridinetrione		−3.67	1000
1,6,8-Trimethyl-2,4,7(1*H*,3*H*,8*H*)-pteridinetrione	9.10		716
3,6,8-Trimethyl-2,4,7(1*H*,3*H*,8*H*)-pteridinetrione	4.22		344, 716
1,3,6-Trimethyl-2,4,7(1*H*,3*H*,8*H*)-pteridinetrione 5-oxide	2.57		238
1,6,7-Trimethyl-4(1*H*)-pteridinimine		10.47	129

TABLE XI. (*Contd.*)

Pteridine	Acidic pK_a	Basic pK_a	Reference(s)
3,6,7-Trimethyl-4(3*H*)-pteridinimine		10.50	132
1,6,7-Trimethyl-2(1*H*)-pteridinone	11.74		120
1,6,7-Trimethyl-4(1*H*)-pteridinone		1.73	129
2,6,7-Trimethyl-4(3*H*)-pteridinone	9.24		153
3,6,7-Trimethyl-2(3*H*)-pteridinone	11.36	<2.00	120, 134
3,6,7-Trimethyl-4(3*H*)-pteridinone		−0.05	126
4,6,7-Trimethyl-2(1*H*)-pteridinone	8.50 (anhyd), 11.50 (hyd)		38, 241, 954
6,7,8-Trimethyl-2(8*H*)-pteridinone	10.26	<2.00	120, 134
6,7,8-Trimethyl-4(8*H*)-pteridinone	9.46	4.70	134
1,6,7-Trimethyl-5,6,7,8-tertrahydro-2,4(1*H*,3*H*)-pteridinedione	10.40	4.90	265
3,8,*N*-Trimethyl-2,4,7-trioxo-1,2,3,4,7,8-hexahydro-6-pteridinecarboxamide	2.96 (?)		1282
2,4,7-Trioxo-1,2,3,4,7,8-hexahydro-6-pteridine-carboxylic acid	2.00, 5.98, 9.90		341
6,7,8-Trimethyl-5,6,7,8-tetrahydro-2,4(1*H*,3*H*)-pteridinedione	8.30	4.55	375
1,3,6-Trimethyl-2-thioxo-1,2-dihydro-4,7(3*H*,8*H*)-pteridinedione	3.31	−1.62	1000
1,3,6-Trimethyl-7-thioxo-7,8-dihydro-2,4(1*H*,3*H*)-pteridinedione	2.29		1709, cf. 228
1,3,8-Trimethyl-2-thioxo-1,2-dihydro-4,7(3*H*,8*H*)-pteridinedione		−3.92	1000
1,3,8-Trimethyl-7-thioxo-7,8-dihydro-2,4(1*H*,3*H*)-pteridinedione		−3.07	1000
1,6,7-Trimethyl-2-thioxo-1,2-dihydro-4(3*H*)-pteridinone	8.51		437
3,6,7-Trimethyl-2-thioxo-1,2-dihydro-4(3*H*)-pteridinone	7.00		446
1,3,*N*-Trimethyl-2,4,6-trioxo-1,2,3,4,5,6-hexahydro-7-pteridinecarboxamide	5.16		347
1,3,8-Trimethyl-2,4,7-trioxo-1,2,3,4,7,8-hexahydro-6-pteridinecarboxylic acid	2.82		317, 341
2,4,6-Trioxo-1,2,3,4,5,6-hexahydro-7-pteridinecarboxamide	4.43, 9.10		347
2,4,6-Trioxo-1,2,3,4,5,6-hexahydro-7-pteridine-carboxylic acid	1.70, 7.20, 9.63		342, 347
8-D-Xylityl-2,4,7(1*H*,3*H*,8*H*)-pteridinetrione	3.18, 12.45		295

[a] This compilation of reported values is complete within the bounds of human fallibility.

[b] The α-amino group will have a much higher basic pK_a but it cannot be measured spectrophotometrically because its ionic state does not affect the uv spectra of the molecule appreciably.

[c] In a mixture with 6-Et isomer.

[d] By ion-exchange hplc.

[e] Values for hydrated and anhydrous species given also.

2. ULTRAVIOLET SPECTRA

An excellent exposition of fundamental aspects of the uv absorption spectroscopy of heterocycles was presented by Stephen Mason in 1963.[1745] In introducing a more pragmatic 1971 update to that classical work, Wilf Armarego remarked[553] that the use of uv spectroscopy in experimental organic chemistry had decreased substantially after 1963, probably on account of the increased accessibility of nmr spectroscopy. In no area was this more evident than in the pteridine series where more than 80% of known uv spectra were reported in the two decades preceding 1965. Nevertheless, the uv spectral technique remains the method of choice for measurement of ionization constants and kinetics. In addition, it still constitutes a valuable secondary source for structural diagnosis and identity.

A sufficient number of full spectral curves for typical simple pteridines have been presented elsewhere,[31, 134, 286, 385, 1052] while numerical data (λ and log ε values for maxima, shoulders, and inflexions) for a less restricted range of such pteridines are collected in Table XII. The spectra of folic acid and its derivatives have been reviewed.[1476]

Without seeking to repeat discussion of a range of wide generalities in respect of the practical use of uv spectra in heterocyclic chemistry,[553] it seems justifiable to give next a similar range of headings with brief remarks, examples, and/or references specifically relevant in the pteridine series.

Spectral band origins. A band due to an $n \rightarrow \pi$ transition (387 nm) was identified in the uv spectrum of pteridine and the effect of substituents on it suggested that the electron involved was located on N-8; other bands in the spectra of pteridine and a range of simple derivatives were assigned specifically to four distinct $\pi \rightarrow \pi$ transitions.[285, 286] It is interesting that this $n \rightarrow \pi$ band became an important feature at ~500 nm in the spectrum of pyrimido[5,4-*e*]-*as*-triazine, that is, 7-azapteridine.[1703]

Ionic species. The importance of reporting spectral data for single ionic species of a solute has been recognized widely in the pteridine series since about 1950, when pK_a values for pteridines began to appear from Adrien Albert's laboratory.[30] However, the iniquitous practice of reporting spectra in 0.1M acid and alkali has persisted in isolated pockets, probably because facilities for pK_a measurement were lacking and many pteridines have proven insufficiently soluble in water or a suitable nonaqueous solvent for spectral measurement as the neutral molecules. When pK_a values cannot be measured, the estimation of such data for most fairly simple pteridines can frequently be done with sufficient accuracy[1742] to permit spectral measurement of single ionic species with reasonable confidence using appropriate buffers.

Solvent effects. As in other nitrogenous heterocyclic systems, a pteridine that will dissolve in a nonpolar solvent such as cyclohexane will exhibit much more spectral fine structure therein than it will in a highly polar solvent such as water or

TABLE XII. THE ULTRAVIOLET SPECTRA[a] OF REPRESENTATIVE SIMPLE PTERIDINES

Compound	Solvent or Species[b]	λ_{max}(nm)/log ε[c]	References
Alkylpteridines			
Pteridine (unsubstituted)	C_6H_{12}	210/4.04, 233/3.45, 292/3.80, 296/3.84, 301/3.87, 308/3.75, 374/1.91, 380/1.92, 387/1.92, 395/1.89	285, 724
	0	(CVH) 228/3.66, 269/3.69, 318/3.89	66, 281, 1145
	0	(CVDH) 258/3.68, 302/3.87	66
	0	<220/>3.83, 230/3.65, 298/3.87, 309/3.83	30, 33, cf. 52, 1146
	+	(CVH) <220/>3.84, 286/3.88, 300/3.92	30, 33
	+	(CVDH) 285/3.86, 303/3.95	66
2-Methylpteridine	C_6H_{12}	212/4.03, 236/3.48, 305/3.87, 309/2.87, 380/2.09	285, 286
	0	230/3.68, 305/3.92, 317/3.90	33
	+	248/3.59, 285/3.88, 301/3.90	33
4-Methylpteridine	C_6H_{12}	216/4.15, 243/3.43, 248/3.43, 270/3.32, 294/3.50, 300/3.85, 304/3.55, 311/3.80, 379/1.98, 387/1.99	285, 286
	0	240/3.45, 300/3.92, 312/3.86	33
	0	(CVDH) 258/3.61, 298/3.86	64
	+	(CVDH) 306/3.99	33, 64
6-Methylpteridine	0	303/3.94, 315/3.90	51
	+	306/3.96	51
7-Methylpteridine	C_6H_{12}	212/3.98, 239/3.44, 298/3.89, 302/3.89, 310/3.78, 375/2.07	285, 286
	0	235/3.50, 298/3.98, 310/3.95	33
	+	304/3.99	33
6,7-Dimethylpteridine	C_6H_{12}	<215/>4.14, 240/3.54, 301/3.99, 314/3.90, 371/2.04	286
	0	235/3.54, 301/4.02, 314/3.98	33
	+	260/3.53, 307/4.05	33
2,6,7-Trimethylpteridine	C_6H_{12}	<220/>4.14, 240/3.60, 301/3.91, 307/4.01, 312/3.95, 321/3.96, 370/2.06	286
	0	235/3.74, 307/4.05, 320/4.03	33
	+	(CVH) 255/3.65, 307/4.07	33, 66

TABLE XII. (*Contd.*)

Compound	Solvent or Species[b]	λ_{max}(nm)/log ε^c	References
4,6,7-Trimethylpteridine	0	235/3.60, 292/3.87, 303/4.03, 315/4.00, 340/2.56	66
	+	(CVH) 248/3.50, 260/3.51, 284/3.78, 305/4.04	66
2,4,6,7-Tetramethylpteridine	0	218/4.29, 240/3.68, 298/3.90, 309/4.05, 321/4.02, 350/2.73	66
	0	(CVH) 223/3.84, 270/3.64, 321/4.01	66
	+	(CVH) 250/3.62, 288/3.81, 308/4.06	66
Halogenopteridines			
2-Chloropteridine	C_6H_{12}	<215/>4.11, 237/3.72, 243/3.59, 262/3.00, 302/3.86, 309/3.98, 314/3.92, 322/3.97, 377/2.08	285, cf. 32
4-Chloropteridine	C_6H_{12}	220/4.17, 244/3.50, 252/3.52, 260/3.54, 270/3.44, 296/3.83, 303/3.90, 307/3.88, 315/3.80, 372/2.22	285, cf. 32
6-Chloropteridine	$CHCl_3$	304/3.85, 311/3.96, 317/3.87, 325/3.93	146
	C_6H_{12}	<220/>4.20, 240/3.66, 303/3.84, 309/3.94, 315/3.88, 323/3.90, 370/2.10	285, cf. 31, 146
	0	213/4.23, 236/3.71, 298/3.77, 309/3.94, 322/3.91	35
	0	(CVH) 236/3.74, 277/3.88, 330/3.93	35
	+	(CVH) 261/3.77, 311/4.01	35
7-Chloropteridine	$CHCl_3$	246/4.05, 268/3.86, 350/4.16	146
	C_6H_{12}	221/4.02, 235/3.54, 279/3.52, 285/3.69, 291/3.82, 297/3.91, 303/3.99, 309/3.96, 316/3.92, 354/2.37	146, 285
	0	281/3.60, 285/3.70, 292/3.86, 297/3.93, 304/4.04, 309/4.00, 317/4.05	35, cf. 33
	0	(CVH) 238/3.62, 276/3.64, 327/4.02	35
	+	(CVH) 255/3.56, 310/4.06	35
2,4-Dichloropteridine	C_6H_{12}	250/3.58, 258/3.45, 269/3.29, 289/3.55, 295/3.65, 301/3.85, 306/3.84, 313/4.03, 318/3.84, 327/4.01	486
6,7-Dichloropteridine	0	225/4.09, 302/3.80, 315/4.02, 328/4.01	35
	0	(CVH) 243/3.69, 281/3.82, 337/4.02	35
	+	(CVH) 264/3.70, 319/4.10	35

2,4,6,7-Tetrachloropteridine	C_6H_{12}	238/4.30, 280/3.33, 327/4.07, 337/3.96, 343/4.06, 370/2.61	286
	THF	236/4.21, 327/4.04, 338/3.99	486
4-Trifluoromethylpteridine	C_6H_{14}	231/3.38, 285/3.64, 292/3.80, 297/3.85, 303/3.89, 309/3.79	156
	0	231/3.45, 295/3.84, 302/3.92, 313/3.86	156
Pteridinones (*Tautomeric*)			
2(1*H*)-Pteridinone	0	260/3.17, 353/3.68	38, 241
	0	(CVH) 230/3.88, 281/3.57, 307/3.83	30, 38, 241
	−	224/4.31, 260/3.85, 375/3.78	30, 38, 241
	−	(CVH) 230/3.78, 266/3.68, 311/3.65	38, 140, 241
	2−	(CVH) 228/3.93, 266/3.94, 339/3.81	45, 140, 241
4(3*H*)-Pteridinone	MeOH	233/3.93, 275/3.51, 314/3.75	1130
	0	230/3.98, 265/3.54, 310/3.82	30, 1103
	+	257/3.43, 303/3.96	134
	−	242/4.23, 333/3.79	30, 1103
6(5*H*)-Pteridinone	0	(CVH) <215/>4.10, 266/3.85, 289/4.00	31, 38, 241
	+	<215/>4.27, 287/4.09	31, 241
	−	<215/>4.25, 224/4.29, 258/3.90, 289/3.69, 356/3.60	31, 55, 241
	−	(CVH) 236/3.39, 294/4.03	241
	2−	303/3.94	140
7(8*H*)-Pteridinone	0	227/3.79, 248/3.44, 256/3.45, 303/4.00	31
	+/2+	See original paper	31
	−	226/4.27, 260/3.76, 326/4.04	31
2,4(1*H*,3*H*)-Pteridinedione	0	230/4.00, 324/3.84	30
	−	215/3.97, 235/4.02, 270/3.95, 347/3.69	30, 338
	2−	252/4.23, 365/3.78	30, 747, 1042
2,6(1*H*,5*H*)-Pteridinedione	0	(CVH) 235/4.22, 299/3.79	41
	−	225/4.41, 246/4.23, 412/3.87	41
	2−	246/4.24, 282/3.62, 415/3.87	40
2,7(1*H*,8*H*)-Pteridinedione	0	258/3.99, 328/4.11	41
	−	282/3.54, 343/4.42, 359/4.43	41
	2−	224/4.50, 271/3.78, 343/4.25, 353/4.20	41

TABLE XII. (*Contd.*)

Compound	Solvent or Species[b]	λ_{max}(nm)/log ε[c]	References
4,6(3*H*,5*H*)-Pteridinedione	0	<220/>4.14, 250/4.01, 270/4.02, 356/3.38	29, 241
	0	(CVH) 270/4.02, 298/3.82	241
	−	<220/>3.90, 241/3.88, 280/4.03, 359/3.83	29, 346
	2−	<220/>3.90, 254/4.18, 367/3.84	29, 241
4,7(3*H*,8*H*)-Pteridinedione	0	<220/>4.60, 285/3.82, 328/3.94	29
	−	227/4.38, 289/3.79, 326/3.95	29
	2−	231/4.41, 329/4.00	29
6,7(5*H*,8*H*)-Pteridinedione	0	<220/>3.98, 249/3.71, 301/4.18	31
	+	222/4.22, 281/3.88, 306/4.16	36
	−	227/4.03, 268/3.71, 319/4.29	31
	2−	<220/>4.47, 240/4.13, 277/3.68, 324/4.30, 338/4.25	31, 134
2,4,6(1*H*,3*H*,5*H*)-Pteridinetrione	0	223/4.04, 248/3.99, 300/3.16, 364/3.73	41, 340
	−	222/4.15, 265/4.05, 379/3.83	41, 340, cf. 709
	2−(?)	234/4.15, 274/3.07, 394/3.80	41, 340, 1569
2,4,7(1*H*,3*H*,8*H*)-Pteridinetrione	2−(?)	225/4.42, 248/3.92, 275/3.86, 335/4.13	41, 709
2,6,7(1*H*,5*H*,8*H*)-Pteridinetrione	−(?)	354/—	277
4,6,7(3*H*,5*H*,8*H*)-Pteridinetrione	−(?)	225/4.45, 319/4.10, 330/4.11	29, 277
2,4,6,7(1*H*,3*H*,5*H*,8*H*)-Pteridinetetrone	2−(?)	233/4.27, 292/3.88, 346/4.11	41, cf. 32
2-Amino-4(3*H*)-pteridinone	0	233/4.04, 270/4.05, 340/3.76	32, 368
	+	<220/>4.10, 242/3.94, 315/3.88	32, 368
	−	252/4.30, 359/3.83	30, 32, 747
2-Amino-6(5*H*)-pteridinone	0, +, −	See original paper	1052
2-Amino-7(8*H*)-pteridinone	0	233/3.88, 290/3.69, 341/4.22	380, 383
	+	222/4.46, 258/4.11, 322/4.01	383, 1126
	−	225/4.54, 269/3.74, 341/4.22	380, 383

4-Amino-2(1*H*)-pteridinone	0	240/4.08, 286/3.63, 337/3.94, 350/3.87	59
	+	205/4.22, 237/4.08, 335/3.90	130, 362, 471
	−	221/4.04, 255/4.31, 375/3.85	130, 362, 471
4-Amino-6(5*H*)-pteridinone	+	225/4.52, 280/4.31, 345/3.20, 360/3.08	445
	−	255/4.25, 277/3.89, 367/3.86, 384/3.76	445
4-Amino-7(8*H*)-pteridinone	0	218/4.32, 249/4.00, 295/3.71, 333/3.95	224, 1126
	+	245/3.94, 290/4.07, 320/3.96	224
	−	235/4.27, 320/3.96, 330/3.98, 345/3.87	445
6-Amino-7(8*H*)-pteridinone	+	212/4.41, 230/4.02, 281/4.00, 301/4.18, 310/4.20, 325/3.86	36
7-Amino-4(3*H*)-pteridinone	0	230/4.30, 235/4.30, 280/3.74, 288/3.72, 334/3.95	131
	+	220/4.23, 227/4.25, 239/4.16, 283/3.85, 290/3.86, 340/4.00	131
	−	228/4.35, 235/4.35, 250/4.08, 337/4.01	131
7-Amino-6(5*H*)-pteridinone	0	233/4.01, 310/4.24, 323/4.35, 339/4.16	19
	−	241/4.20, 329/4.29, 343/4.20	19
2-Amino-4,6(3*H*,5*H*)-pteridinedione	0	275/4.12, 388/3.42	241, cf. 32
	0	(CVH) 277/4.17, 300/4.09	241
	+	245/4.07, 355/3.82	32
	2−	255/4.27, 275/4.07, 392/3.85	30, 32, 241, 1175, 1256
	2−	(CVH) 276/4.19	241
2-Amino-4,7(3*H*,8*H*)-pteridinedione	0	210/4.48, 286/4.00, 340/4.14	383
	−	229/4.46, 254/3.87, 280/3.81, 332/4.11	383
	2−	221/4.58, 253/4.05, 339/4.14	383
2-Amino-6,7(5*H*,8*H*)-pteridinedione	0	345/—	1350
	−	245/—, 351/—	1350
2-Amino-4,6,7(3*H*,5*H*,8*H*)-pteridinetrione	0	225/4.15, 296/4.10, 336/3.89	384, 1056
	−	237/4.19, 288/3.94, 333/4.02, 345/4.03	384
	2−	239/4.20, 280/3.86, 341/4.02	30, 32, 384
	3−	221/4.28, 240/4.27, 287/3.88, 343/4.10	384

TABLE XII. (*Contd.*)

Compound	Solvent or Species[b]	λ_{max}(nm)/log ε[c]	References
Alkoxypteridines			
2-Methoxypteridine	C_6H_{12}	213/4.29, 238/3.83, 311/3.79, 318/3.85, 324/4.04, 331/3.89, 339/4.03, 380/2.34	285, 286
	0	<220/>4.02, 325/3.92	32
4-Methoxypteridine	C_6H_{12}	226/4.34, 245/4.34, 255/3.44, 263/3.02, 273/3.47, 291/3.75, 295/3.84, 301/3.87, 307/3.92, 313/3.78, 319/3.79, 367/2.45	285, cf. 32
	MeOH	224/4.26, 259/3.47, 302/3.88	182
	0	225/4.25, 258/3.45, 304/3.89	32
6-Methoxypteridine	C_6H_{12}	216/4.32, 240/3.80, 291/3.53, 297/3.63, 303/3.80, 309/3.81, 316/4.01, 323/3.83, 331/3.98, 360/2.47	285, cf. 31
	0	<220/>4.14, 315/3.82, 327/3.79	32
7-Methoxypteridine	C_6H_{12}	214/4.26, 230/3.82, 240/3.59, 254/3.16, 279/3.65, 284/3.75, 290/3.89, 295/3.93, 302/4.07, 308/3.94, 315/4.06, 340/2.60	285
	0	(?) 313/3.99, 320/3.95, 335/3.32	33
	+	220/4.33, 276/3.80, 299/3.81, 330/3.64	33
2,4-Dimethoxypteridine	EtOH	247/3.64, 324/3.78	486, 1042
	0	228/4.32, 323/3.88	338, 1042
Pteridinones (Nontautomeric)			
1-Methyl-2(1*H*)-pteridinone	0	240/3.91, 285/3.53, 311/3.85	134
	–	(CVH?) 236/3.89, 312/3.81, 375/2,37	134
3-Methyl-2(3*H*)-pteridinone	0	230/3.97, 285/3.59, 309/3.89	134
	–	(CVH?) 236/3.84, 271/3.66, 313/3.80, 350/3.53	134

6,7,8-Trimethyl-2(8*H*)-pteridinone	0	239/4.34, 280/3.70, 327/4.25	120, cf. 248
	−	(CVH?) 240/4.23, 344/4.19	134, cf. 248
1-Methyl-4(1*H*)-pteridinone	MeOH	230/4.09, 260/3.48, 324/3.93	182, 1130
	0	232/4.14, 260/3.34, 300/3.41, 324/3.96	134
	+	225/4.11, 263/3.49, 305/4.03	134
3-Methyl-4(3*H*)-pteridinone	MeOH	235/4.07, 276/3.62, 311/3.78	182, 1130
	0	233/4.12, 276/3.57, 312/3.81	32
	+	221/4.04, 265/3.46, 304/4.00	134
8-Methyl-4(8*H*)-pteridinone	0	227/4.26, 232/4.20, 274/3.69, 280/3.72, 308/3.73	248
	+	243/3.88, 278/3.41, 286/3.37, 355/3.86	248
3-Amino-4(3*H*)-pteridinone	EtOH	240/3.94, 311/3.73	487
	+(?)	239/3.98, 311/3.75	487
5-Methyl-6(5*H*)-pteridinone	0	265/3.85, 290/4.06	134
	+	288/4.14	134
	−	(CVHP?) 270/3.78, 296/3.99	134
8-Methyl-7(8*H*)-pteridinone	0	<220/>4.13, 250/3.55, 257/3.54, 306/3.97	31
	+	<220/>4.19, 295/3.97	31
Pteridine N-oxides			
2(3*H*)-Pteridinone 1-oxide	0	227/3.83, 241/3.88, 289/3.63, 311/3.87	1098
	−	258/4.18, 275/4.24, 415/3.52	1098
4(1*H*)-Pteridinone 3-oxide	0	241/4.09, 311/3.76	154
	+	255/3.89, 310/3.70, 325/3.70, 340/3.57	154
	−	219/4.12, 273/4.44, 325/3.38	154
Pteridinethiones (Tautomeric)			
2(1*H*)-Pteridinethione	0	(CVH) <215/>3.74, 270/4.24, 315/4.22 (4.08?)	30, 46
	−	290/4.10, 344/4.01	46
4(3*H*)-Pteridinethione	0	256/4.13, 390/4.00	30, 1103
	−	265/4.22, 408/3.93	30, 1103
6(5*H*)-Pteridinethione	0	242/3.87, 331/4.24	36
	+	230/3.91, 242/3.83, 332/4.28	36
	−	237/4.03, 299/4.17, 411/3.83	36

TABLE XII. (*Contd.*)

Compound	Solvent or Species[b]	λ_{max}(nm)/log ε^c	References
7(8*H*)-Pteridinethione	0	230/3.83, 295/3.86, 386/4.13	33
	−	259/3.89, 302/3.84, 390/4.10	33
2,4(1*H*,3*H*)-Pteridinedithione	0	290/4.29, 363/4.06, 373/4.04	437
	−	242/4.09, 266/4.00, 316/4.32, 359/3.79, 418/3.82	437
	2−	253/4.24, 303/4.18, 377/3.96, 400/3.92	437
Alkylthiopteridines			
2-Methylthiopteridine	C_6H_{12}	242/4.16, 274/4.17, 277/4.18, 358/3.86, 367/3.89	285
	0	242/4.18, 270/4.11, 360/3.84	33
4-Methylthiopteridine	C_6H_{12}	231/4.03, 254/4.03, 259/4.03, 285/3.46, 295/3.41, 334/3.81, 348/3.97, 357/3.90, 364/3.88	285
	0	238/3.95, 253/3.94, 283/3.44, 354/3.93	33
	+	233/3.97, 335/4.01	33
6-Methylthiopteridine	0	235/4.05, 274/4.08, 365/3.86	36
7-Methylthiopteridine	C_6H_{12}	243/4.06, 263/3.79, 331/4.01, 341/4.16, 356/4.09	285
	0	240/3.99, 266/3.75, 350/4.15	33
	+	239/4.22, 278/3.65, 343/4.04	33
Pteridinamines			
2-Pteridinamine	0	225/4.38, 259/3.81, 370/3.82	30, 39
	+	<210/>4.04, 232/3.92, 302/3.87	30, 39
4-Pteridinamine	0	244/4.20, 335/3.82	30, 39
	+	229/4.10, 324/3.99	30, 39
	−	See original paper	30
6-Pteridinamine	0	223/4.30, 258/4.01, 362/3.75	31
7-Pteridinamine	0	228/4.26, 262/3.80, 334/4.03	33
	+	219/4.28, 230/4.15, 300/3.97, 328/4.20	131, cf. 33
2,4-Pteridinediamine	0	224/4.07, 255/4.32, 364/3.86	130, 267
	+	240/4.10, 284/3.73, 318/3.91, 332/3.98, 345/3.90	59, 267, 284, 486, 1678

4,6-Pteridinediamine	0	263/4.18, 375/3.81	41
	+	254/4.09, 284/3.87, 376/3.90	41
4,7-Pteridinediamine	0	241/4.38, 339/4.05	41, 131
	+	233/4.24, 257/4.22, 284/3.69, 343/4.14	41, 131, 324
6,7-Pteridinediamine	0	221/4.05, 242/4.08, 268/3.59, 278/3.56, 327/4.18, 337/4.27, 352/4.12	36
	+	228/4.38, 258/4.10, 342/4.25, 352/4.25	36
	2+	219/4.41, 244/4.00, 324/4.17, 337/4.35, 353/4.28	36
2,4,7-Pteridinetriamine	0	227/4.54, 257/4.13, 350/4.17	41
	+	255/4.16, 275/3.84, 342/4.28	41
4,6,7-Pteridinetriamine	0	227/4.33, 256/4.17, 284/3.63, 345/4.11	41
	+	224/4.20, 245/4.21, 353/4.26	41
2,4,6,7-Pteridinetetramine	+	235/3.19, 305/3.88, 360/4.05	486
2-Dimethylaminopteridine	C_6H_{12}	239/4.30, 279/4.13, 376/3.65, 390/3.82, 411/3.86, 434/3.70	285, 286
	0	236/2.37, 281/4.02, 410/3.82	30, 39
	+	237/4.16, 305/3.90, 370/2.59	30, 39
4-Dimethylaminopteridine	C_6H_{12}	235/4.15, 252/4.11, 256/4.11, 260/4.12, 292/3.22, 362/3.91, 378/3.77	285, 286
	0	241/4.15, 362/3.93	32
	+	239/4.19, 344/4.09, 347/4.10, 356/4.04	32
6-Dimethylaminopteridine	C_6H_{12}	232/4.24, 269/4.19, 369/3.78, 384/3.84, 403/3.66	285, 286
	0	231/4.21, 279/4.17, 399/3.74	31
7-Dimethylaminopteridine	C_6H_{12}	240/4.25, 269/3.97, 351/3.93, 362/3.97, 379/3.76	285, 286
	0	240/4.19, 279/3.98, 364/4.05	33
	+	245/3.95, 312/3.96, 361/4.16	33
Pteridinimines (Nontautomeric)			
6,7,8-Trimethyl-2(8*H*)-pteridinimine	+	235/4.31, 328/4.00	190
1-Methyl-4(1*H*)-pteridinimine	+	233/4.14, 333/4.00, 350/3.87	129
3-Methyl-4(3*H*)-pteridinimine	+	236/4.03, 311/3.78, 320/3.79, 330/3.62	132

TABLE XII. (*Contd.*)

Compound	Solvent or Species[b]	λ_{max}(nm)/log ε^c	References
1-Methyl-7(1*H*)-pteridinimine	+	218/4.24, 222/4.21, 233/4.01, 260/3.60, 349/4.12	131
3-Methyl-7(3*H*)-pteridinimine	+	215/4.22, 233/4.01, 296/3.93, 335/4.17	131
Pteridinecarboxylic Acids, and so on			
4-Pteridinecarboxylic acid	+	(CVDH) 284/3.60, 333/4.00	157
	Z	(CVDH) 257/3.61, 330/3.90	157
Ethyl 2-pteridinecarboxylate	0	300/3.97, 311/3.90	51
	0	(CVH ?) 318/4.06	51
	+	314/4.01	51
Ethyl 4-pteridinecarboxylate	C_6H_{12}	240/3.43, 295/3.82, 301/3.88, 304/3.89, 310/3.79	147
	0	245/3.49, 304/3.88, 311/3.90	147
	0	(CVDH) 233/3.85, 258/3.62, 342/3.88	147, 148
	+	(CVDH) 239/3.78, 284/3.65, 345/4.06	147, 148
4-Pteridinecarboxamide	0	304/4.09, 313/4.08	157
	0	(CVDH) 227/3.85, 258/3.54, 342/3.87	157
	+	(CVDH) 252/3.54, 341/3.89	157
4-Pteridinecarbonitrile	0	245/3.39, 301/3.52, 311/3.57, 322/3.49	157
	0	(CVDH) 232/3.53, 259/3.27, 339/3.52	157
	+	(CVDH) 231/3.44, 255/3.27, 341/3.57	157
Dihydropteridines			
3,4-Dihydropteridine	0	335/3.96	49
	+	311/3.90	49
3,4-Dihydro-2(1*H*)-pteridinone	0	248/3.72, 317/3.89	38
	+	254/3.78, 337/3.85	42
	−	281/3.98, 343/3.84	38

3,4-Dihydro-2-pteridinamine	0	282/3.89, 339/3.89	45
	+	245/3.65, 308/3.89	45
5,6-Dihydro-4(3*H*)-pteridinone	0	286/3.78	43
	+	258/3.74	43
	−	279/3.82	43
5,6-Dihydro-7(8*H*)-pteridinone	0	<215/>4.42, 271/3.58, 319/3.70	31
	+	233/4.47, 284/3.74, 352/3.71	31, 46
	−	224/4.34, 265/3.56, 325/3.93	43, 134
4,6-Dimethyl-7,8-dihydropteridine	0	218/4.27, 293/3.73	43
	+	218/4.13, 293/3.91	43
7,8-Dihydro-2(1*H*)-pteridinone	0	223/4.35, 290/3.88	42
	+	225/3.68, 282/3.75, 290/3.79, 310/3.75	42
	−	308/3.95	42
7,8-Dihydro-4(3*H*)-pteridinone	0	248/3.86, 367/3.68	43
	+	257/3.87, 374/3.80	43
	−	253/3.97, 364/3.70	43
7,8-Dihydro-6(5*H*)-pteridinone	0	<215/>4.09, 275/3.80, 293/3.94	31, 134
	+	<220/>4.20, 292/4.01	31, 689
	−	275/3.75, 305/4.07	40, 134, 689
Tetrahydropteridines			
1,2,3,4-Tetrahydropteridine	0	244/3.90, 287/3.20, 334/3.78	51
	+	237/3.91, 289/3.29, 325/3.79	51
	2+	234/3.95, 339/3.86	51
Ethyl 1,2,3,4-tetrahydro-2-pteridinecarboxylate	0	241/3.85, 288/3.26, 328/3.80	51
	+	228/3.93, 322/3.81	51
5,6,7,8-Tetrahydropteridine	0	206/4.04, 268/3.68, 304/3.81	43, 486, 683
	+	208/4.17, 304/3.89	43
4-Methyl-5,6,7,8-tetrahydropteridine	0	211/4.18, 300/3.82	118
	+	213/4.12, 305/3.93	118

TABLE XII. (*Contd.*)

Compound	Solvent or Species[b]	λ_{max}(nm)/log ε^c	References
5,6,7,8-Tetrahydro-2(1*H*)-pteridinone	0	232/4.09, 306/3.70	42
	+	229/4.00, 327/3.69	42
	−	315/3.79	42
5,6,7,8-Tetrahydro-4(3*H*)-pteridinone	0	220/4.22, 289/3.93	43
	+	219/4.33, 259/3.83	43
	−	218/4.33, 284/3.91	43

[a] CVH = covalent 3,4-hydrate
CVHP = covalent 7,8-hydrate
CVDH = covalent 5,6,7,8-dihydrate
CVDM = covalent 5,6,7,8-dimethanolate
Z = zwitterion
THF = tetrahydrofuran
Other abbreviations are self-explanatory.

[b] Indicated species are all in aqueous buffer of appropriate pH.

[c] Because of widespread anomalies in the literature, shoulders and inflexions are here included without any distinction from maxima.

methanol: See, for example, the respective data for 4-methylpteridine in Table XII.[32, 182, 285] Dissolution in a semipolar solvent such as chloroform usually will give rise to less fine structure than in cyclohexane but more than in water. See data for 6- and 7-chloropteridine which are, however, complicated by apparently incomplete spectral records in chloroform.[35, 146] It seems superfluous to point out that the formation of a cation or anion is an appropriate buffer will affect the spectrum of a pteridine profoundly. See, for example, 4(3*H*)-pteridinone in Table XII. Likewise, the formation of a covalent hydrate or alcoholate will drastically affect the spectrum of a susceptible pteridine. See data for 2(1*H*)-pteridinone.

SUBSTITUENT EFFECTS. Substituent effects in the pteridine series differ little from those in related systems. See data for pteridine and its various methylated derivatives in Table XII. The spectral effects of many added groups are complicated in this series by the degree of covalent hydration occurring in the resulting derivatives. Other complicating factors are the location of the mobile hydrogen in tautomeric pteridinones or pteridinethiones, which affects their neutral molecular spectra, and the potential variation in the protonation site for variously substituted pteridines, which affects their cationic spectra.

PROTONATION. The intrinsic effect of protonation on the spectrum of a pteridine is usually small and may result in a bathochromic or hypsochromic shift for each peak as shown by data for 7-methylthiopteridine. However, if covalent hydration is introduced in the cation, the shifts may be quite dramatic as in the data for 2-pteridinamine, which has an unhydrated neutral molecule but a 3,4-hydrated cation (Ch. IX, Sect. 2.B).

NUCLEAR REDUCTION. The deletion of one or two double bonds by nuclear reduction of a pteridine might be expected to lead to progressive hypsochromic shifts in its spectrum, but this is by no means always so. For example, compare in Table XII the data for pteridine, 3,4-dihydropteridine, and 1,2,3,4-tetrahydropteridine; also those for 7(8*H*)-pteridinone and its 5,6-dihydro derivative. The explanation of such an apparent anomaly could lie in the existence of a long-wavelength peak of very low intensity (representing a forbidden transition) in the spectrum of the parent aromatic structure. This could easily be overlooked and remain unrecorded for subsequent comparison.

COVALENT HYDRATION. Deletion of one or two double bonds by covalent hydration or alcoholation also has a marked but variable effect on the spectrum. As might be anticipated, the spectrum of 6(5*H*)-pteridinone 7,8-hydrate (**7**) and 7,8-dihydro-6(5*H*)-pteridinone (**8**) are quite similar; so too are those of 2(1*H*)-

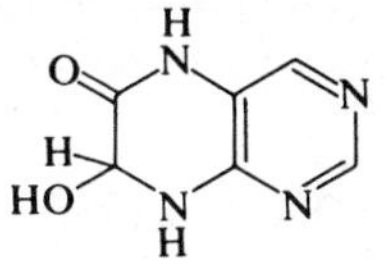

(**7**)

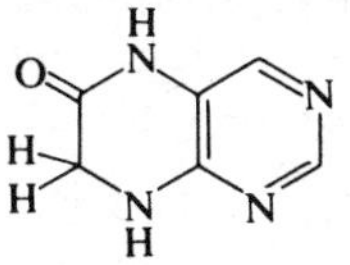

(**8**)

pteridinone 3,4-hydrate and 3,4-dihydro-2(1*H*)-pteridinone (Table XII). Although the phenomenon of covalent hydration was discovered originally by abnormalities in the potentiometric measurement of the pK_a value for cationization of 6(5*H*)-pteridinone,[17, 19, 31] the subject was subsequently developed by reference to the peculiar uv spectral changes suffered by susceptible pteridines.[12, 582, 824] The resulting body of knowledge was later confirmed and expanded further by the use of ^{1}H- and ^{13}C nmr spectra.[11, 594, 843, 959]

TAUTOMERISM. Like most *hydroxy* derivatives of other nitrogenous heteroaromatics,[287, 288] the four *hydroxypteridines* eventually proved to be pteridinones.[134] In an effort to decide between a number of possible tautomeric structures, the uv spectrum of each *hydroxypteridine* was compared with those of the corresponding methoxypteridine and as many of the *N*-methylpteridinones as could be prepared. It emerged, for example, that the spectrum of 4-hydroxypteridine closely resembled that of the 3-methyl-4(3*H*)-pteridinone (**9**, R = Me) but differed appreciably from those of the isomeric 4-methoxypteridine (**10**), 1-methyl-4(1*H*)-pteridinone (**11**), and 8-methyl-4(8*H*)-pteridinone (**12**, R = H) [as represented by its available homolog (**12**, R = Me)]. It was concluded that 4-hydroxypteridine existed, at least in aqueous solution, as 4(3*H*)-pteridinone (**9**, R = H). The other hydroxypteridines were found likewise to exist as pteridinones, although some of the comparisons were complicated by covalent hydration.[134] These conclusions were consistent with pK_a and ir spectral evidence.[287, 288]

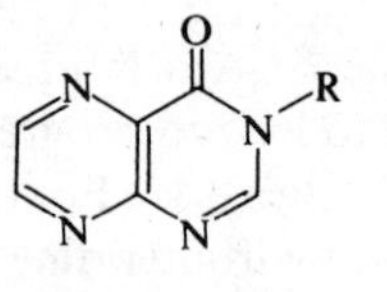

(9)

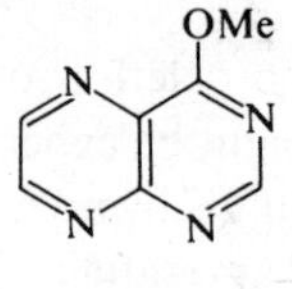

(10)

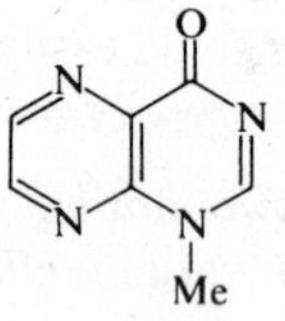

(11)

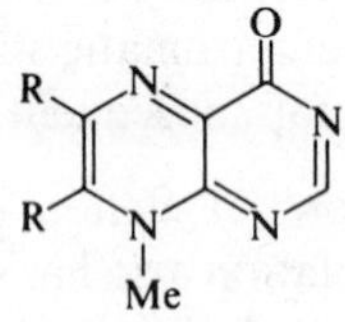

(12)

The contribution of uv spectra to the study of tautomerism in other pteridinones, aminopteridinones, pteridinamines, and pteridinethiones have been summarized briefly at appropriate points in more extensive reviews of tautomerism in heterocyclic systems.[1713, 1746]

THE JONES AND THE ALBERT–TAGUCHI RULES. Two useful spectral correlations have been found to apply in the pteridine series. The Norman Jones rule, which originally drew attention to the close spectral similarity of the neutral molecule of an aromatic amine and the anion of the corresponding phenol,[1747] emerged as equally applicable to 2-, 4-, and 7-pteridinamine with spectra almost indistinguishable from those of the anions of 2(1*H*)-, 4(3*H*)-, and 7(8*H*)-pteridinone, respectively.[26] These and other comparisons may also be made with the data in Table XII. For example, the spectra of 2,4-pteridinediamine (neutral molecule), 2-amino-4(3*H*)-pteridinone (anion), 4-amino-2(1*H*)-pteridinone (anion), and 2,4(1*H*,3*H*)-pteridinedione (dianion) are all remarkably similar.

The rule of Adrien Albert and Hiro Taguchi drew attention to the close spectral similarity of the cation of a 2-aminopyrimidine or fused pyrimidine and the neutral molecule of the corresponding 2-pyrimidinone or fused pyrimidinone[59]: compare, for example, data for the 3,4-hydrated cation of 2-pteridinamine with those for the similarly hydrated neutral molecule of 2(1*H*)-pteridinone in Table XII; also the spectrum of 2,4-pteridinediamine (monocation) with that of 4-amino-2(1*H*)-pteridinone (neutral molecule). The likely basis for this rule and the reason for it being confined to 2-substituted compounds have been discussed.[59]

3. PROTON NUCLEAR MAGNETIC RESONANCE SPECTRA

The use of ^{1}H nmr spectra to elucidate or confirm the structures of pteridines has been mentioned at innumerable points in the preceding chapters. In many cases the neutral molecules have proven insufficiently soluble in deuterochloroform or perdeuterodimethyl sulfoxide, and so on, for measurement but such substances have often been found amenable to measurement as cations in trifluoroacetic acid or a similar medium. The data for a range of simple pteridines (Table XIII) hold few surprises. The assignment of chemical shifts to appropriate protons is usually not difficult,[1750] except perhaps when the molecule contains both a 6- and a 7-proton. Although considerable assistance appears to be available in the latter case from a careful study[293] of spectra for pteridine and its methyl derivatives in deuterochloroform, including estimates for the substituent effects of each methyl group on the remaining ring protons, it is evident that some subsequent authors have chosen to ignore these guidelines. Accordingly, it is wise to treat all such assignments in the literature with some reserve. The 2- and/or 4-proton assignments are easily confirmed by their extremely large induced shift of signal on addition of the lanthanide complex, tris(dipivaloylmethanato)europium.[79] The ^{1}H nmr spectra of folic acids and tetrahydrofolic acids have been reviewed.[1476] The ^{3}H nmr spectra of folic acid has been examined after tritiation.[921]

When the aromaticity of a pteridine is decreased by nuclear reduction or covalent hydration, and so on, the signals from protons at the site of reaction undergo a profound upfield shift, while other nuclear protons undergo a smaller shift in the same direction. Since about 1965, this has constituted the preferred method of diagnosis for such reactions. For example, uv spectra and other existing evidence suggested that acidification of aqueous pteridine to pH 2 rapidly gave the 3,4-hydrated cation, which subsequently underwent ring fission to the cation of 3-aminomethyleneamino-2-pyrazinecarbaldehyde.[1145] An nmr study confirmed the first step but showed beyond a doubt that the second product was the pteridine 5,6,7,8-dihydrate cation.[27] Likewise the borohydride reduction of pteridine in THF containing trifluoroacetic acid clearly gave two isomeric tetrahydro products which, after separation, were easily identified as 1,2,3,4- and 5,6,7,8-tetrahydropteridine by their ^{1}H nmr spectra (see Table XIII).[721] But nmr spectroscopy can do much more than reveal the site(s) of reduction or addition. Its application at sufficiently high field strength can indicate stereoconfigurations and even preferred conformations within the molecule. Without going into the technical detail, it was thus shown that 2-amino-6,7-dimethyl-4(3*H*)-pteridinone and its 6,7-bistrideuteromethyl analog both underwent stereospecific catalytic hydrogenation to give the respective *cis*-5,6,7,8-tetrahydro derivatives,[84, 519] whereas the sodium–ethanol reduction of the first substrate gave a 1:1 mixture of cis- and trans-products.[84] In contrast, biopterin underwent catalytic hydrogenation in trifluoroacetic acid to give (after acetylation) a separable mixture of (6*S*)- and (6*R*)-tetraacetyl-5,6,7,8-tetrahydrobiopterin (13, R = Ac), which on hydrolysis gave (6*S*)- and (6*R*)-5,6,7,8-tetrahydrobiopterin (**13**, R = H) as hydrochlorides. Their nmr spectra not only identified each isomer but provided information about the conformation of the tetrahydropyrazine ring and the dihydroxypropyl side chain in each.[82] Other such examples have been reported.[84, 518–520]

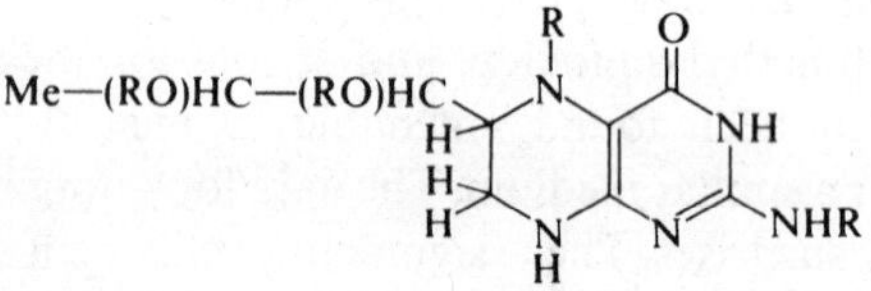

(13)

4. INFRARED SPECTRA

Fundamental aspects of the ir spectra of simple pteridines have been discussed expertly by Stephen Mason.[286] The minor roles of ir spectroscopy in elucidating the tautomeric forms of *hydroxypteridines* in the solid state[1091] and in pinpointing the site(s) of covalent hydration in tautomeric and nontautomeric pteridinones,[134] have been outlined.

Although some workers have conscientiously reported salient points from the ir spectra of their new pteridines, such data serves little purpose apart from confirming the presence of a few characteristic groups, for example, carbonyl, in the molecule and providing a means of 'fingerprinting' a compound to establish identity with other specimens. References to the ir spectra of some simple pteridines are collected in Table XIV.

TABLE XIII. THE PROTON NUCLEAR MAGNETIC RESONANCE SPECTRA OF SOME SIMPLE PTERIDINES

Compound	Medium	Chemical Shifts (δ) and Assignments[a]	References
Pteridine (unsubstituted)	$CDCl_3$	9.80 (H-4), 9.66 (H-2), 9.33 (H-7), 9.15 (H-6)	51, 293, 294, 1138
	$(CD_3)_2SO$	(CVDM) 8.11 (H-8), 7.83 (H-4), 7.31 (H-5), 4.60 (H-6 + H-7), 3.30 + 3.25 (Me_2)	48
	pD_1	(CVH) 8.77 (H-2), 8.74 (H-6), 8.67 (H-7), 6.60 (H-4)	27
	pD 1	(CVDH) 8.47 (H-2), 7.88 (H-4), 5.35 (H-6), 5.20 (H-7)	27
2-Methylpteridine	$CDCl_3$	9.66 (H-4), 9.23 (H-7), 9.03 (H-6), 3.03 (Me)	293
	pD 1	(CVH) 8.78 (H-6), 8.72 (H-7), 6.52 (H-4), 2.67 (Me)	27
	pD 1	(CVDH) 7.77 (H-4), 5.30 (H-6), 5.16 (H-7), 2.60 (Me)	27
4-Methylpteridine	$CDCl_3$	9.43 (H-2), 9.27 (H-7), 9.08 (H-6), 3.13 (Me)	293
	D_2O	9.52 (H-7), 9.42 (H-2), 9.34 (H-6), 3.12 (Me)	64
	pD 1	9.35 (H-2 + H-6), 9.20 (H-7), 3.06 (Me)	27
	pD 1	(CVDH) 8.46 (H-2), 5.27 (H-6), 5.20 (H-7), 2.52 (Me)	27, cf. 64
6-Methylpteridine	D_2O	9.69 (H-4), 9.54 (H-2), 9.44 (H-7), 2.96 (Me)	51
	pD 1	(CVH) 8.77 (H-2), 8.63 (H-7), 6.58 (H-4), 2.69 (Me)	51
	pD 1	(CVDH) 8.54 (H-2), 7.90 (H-4), 5.26 (H-7), 1.78 (Me)	51
7-Methylpteridine	$CDCl_3$	9.68 (H-4), 9.57 (H-2), 8.98 (H-6), 2.96 (Me)	51, 293, 294, 1740
	pD 1	(CVH) 8.75 (H-2), 8.68 (H-6), 6.57 (H-4), 2.68 (Me)	27
	pD 1	(CVDH) 8.45 (H-2), 7.84 (H-4), 5.00 (H-6), 1.74 (Me)	27
6,7-Dimethylpteridine	pD 1	(CVH) 8.80 (H-2), 6.58 (H-4), 2.70 (Me_2)	27
2,6,7-Trimethylpteridine	pD 1	(CVH) 6.48 (H-4), 2.70 (Me_3)	27
4,6,7-Trimethylpteridine	$CDCl_3$	9.51 (H-2), 3.12(4-Me), 2.90(7-Me), 2.88(6-Me)	66
	D_2O	9.15 (H-2), 2.88(4,7-Me_2), 2.84(6-Me)	66
	pD 1	(CVH) 8.50 (H-2), 2.59(6,7-Me_2), 1.98(4-Me)	66
2,4,6,7-Tetramethylpteridine	$CDCl_3$	3.07(4-Me), 2.96(2-Me), 2.86(7-Me), 2.83(6-Me)	66
	pD 1	(CVH) 2.58(6,7-Me_2), 2.53(2-Me), 1.96(4-Me)	66
2-Chloropteridine	$CDCl_3$	9.68 (H-4), 9.28 (H-7), 9.09 (H-6)	1138
	pD 1	9.89 (H-4), 9.45 (H-6), 9.30 (H-7)	27
	pD 1	(CVDH) 7.77 (H-4), 5.21 (H-6), 5.13 (H-7)	27
6-Chloropteridine	pD 1	(CVH) 8.82 (H-7), 8.80 (H-2), 6.58 (H-4)	27

TABLE XIII. (*Contd.*)

Compound	Medium	Chemical Shifts (δ) and Assignments[a]	References
7-Chloropteridine	pD 1	(CVH) 8.88 (H-6), 8.80 (H-2), 6.62 (H-4)	27
4-Trifluoromethylpteridine	$CDCl_3$	9.79 (H-2), 9.42 (H-7), 9.25 (H-6)	156
	pD 1	(CVH) 8.89 (H-2), 8.81 (H-6 + H-7)	156
	pD 1	(CVDH) 8.39 (H-2), 5.31 (H-7), 5.09 (H-6)	156
2(1*H*)-Pteridinone	pD 1	(CVH) 8.60 (H-6), 8.50 (H-7), 6.32 (H-4)	27
4(3*H*)-Pteridinone	$(CD_3)_2SO$	8.97 (H-6), 8.82 (H-7), 8.33 (H-2)	20, 1003
	pD 1	9.10 (H-2), 9.05 (H-6), 8.97 (H-7)	1508, cf. 20, 610
	pD 14	8.85 (H-6), 8.74 (H-7), 8.46 (H-2)	1508
6(5*H*)-Pteridinone	pD 1	(CVHP) 8.73 (H-2), 8.15 (H-4), 5.78 (H-7)	27
2-Amino-4(3*H*)-pteridinone	FSO_3H	(+ +) 10.01 + 9.22 (H-6 + H-7), 7.80 (NH_2)	762
	CF_3CO_2H	(+) 9.12 (H-6 + H-7), 8.83 (NH_2)	561, 741, 1331
	pD 14	8.55 − 8.35 (H-6 + H-7)	762
2-Amino-4,6(3*H*,5*H*)-pteridinedione	CF_3CO_2H	8.71 (H-7), 8.25 (NH_2)	561
2-Methoxypteridine	pD 1	(CVDH) 7.62 (H-4), 5.32 (H-6), 5.18 (H-7), 4.90 (Me)	27
4-Methoxypteridine	$CDCl_3$	9.27 (H-7), 9.16 (H-2), 9.05 (H-6), 4.33 (Me)	138
1-Methyl-4(1*H*)-pteridinone	D_2O (?)	9.15 (H-7), 9.07 (H-6), 8.81 (H-2)	610
	pD 1 (?)	9.51 (H-2), 9.27 (H-7), 9.07 (H-6)	610
3-Methyl-4(3*H*)-pteridinone	D_2O (?)	9.15 (H-7), 9.02 (H-6), 8.71 (H-2)	610
	pD 1 (?)	9.60 (H-2), 9.25 (H-6 + H-7)	610
3-Methoxy-4(3*H*)-pteridinone	$CDCl_3$	9.07 − 8.93 (H-6 + H-7), 8.63 (H-2), 4.28 (Me)	759
4-Pteridinamine 3-oxide	D_2O	8.62 (H-7), 8.52 (H-6), 8.49 (H-2)	999, 1003
2(3*H*)-Pteridinone 1-oxide	$(CD_3)_2SO$	See original paper	1098
	$(CD_3)_2SO$ + DCl	(CVH) 8.38 − 8.22 (H-6 + H-7), 5.59 (H-4)	1098
2(1*H*)-Pteridinethione	$(CD_3)_2SO$	(CVH) 8.30 (H-6 + H-7), 5.60 (H-4)	46

2-Methylthiopteridine	$CDCl_3$	9.44 (H-4), 9.09 (H-7), 8.87 (H-6)	1700, cf. 46, 1138
	pD 1	(CVDH) 7.70 (H-4), 5.32 (H-6), 5.13 (H-7), 2.69 (Me)	27, 46
4-Methylthiopteridine	pD 1	(CVDH) 8.43 (H-2), 5.25 (H-6), 5.22 (H-7), 2.64 (Me)	27
2-Pteridinamine	pD 1	(CVH) 8.59 (H-6 + H-7), 6.30 (H-4)	27, 46
4-Pteridinamine	$(CD_3)_2SO$	8.63 (H-7), 8.36 (H-6), 8.14 (H-2), 7.50 (NH_2)	1003
	pD 1	9.25 (H-6), 9.14 (H-7), 8.80 (H-2)	27
	CF_3CO_2H	9.24 (H-6), 9.18 (H-7), 8.93 (H-2)	934
6-Pteridinamine	pD 1	(CVH) 8.90 (H-2), 8.58 (H-7), 6.34 (H-4)	27
4-Dimethylaminopteridine	pD 1	9.20 (H-6), 9.10 (H-7), 8.80 (H-2), 4.13 + 3.72 (Me_2)	27
3-Methyl-4(3*H*)-pteridinimine	$(CD_3)_2SO$	9.05 (NH), 9.02 – 8.85 (H-6 + H-7), 8.52 (H-2), 3.52 (Me)	53
4-Pteridinecarboxylic acid	D_2O	(CVDH/Z) 8.30 (H-2), 5.28 – 5.15 (H-6 + H-7)	157
	pD 1	(CVDH) 8.35 (H-2), 5.33 – 5.22 (H-6 + H-7)	157
Ethyl 2-pteridinecarboxylate	D_2O	10.05 (H-4), 9.61 (H-6), 9.46 (H-7), 4.70 + 1.56 (Et)	51
	D_2O	(CVDH) 7.87 (H-4), 5.48 (H-6), 5.17 (H-7), 4.44 + 1.48 (Et)	51
Ethyl 4-pteridinecarboxylate	$CDCl_3$	9.66 (H-2), 9.32 (H-7), 9.13 (H-6), 4.65 + 1.50 (Et)	147
	D_2O	(CVDH) 7.94 (H-2), 5.15 (H-7), 5.08 (H-6), 4.33 + 1.30 (Et)	148
4-Pteridinecarboxamide	$(CD_3)_2SO$	9.53 (H-2), 9.40–9.23 (H-6 + H-7), 9.33 + 8.13 (NH_2)	157
	$(CD_3)_2SO$	(CVDH) 8.33 + 7.48 (NH_2), 7.90 (H-2), 4.78 (H-6 + H-7), 5.63 (H-5 + H-8)	157
	pD 1	(CVDH) 8.23 (H-2), 5.06 – 5.03 (H-6 + H-7)	157
2-Methyl-3,4-dihydropteridine	$(CD_3)_2SO$ + DCl	8.32 (H-6 + H-7), 4.88 (H-4), 2.34 (Me)	585
3,4-Dihydro-2(1*H*)-pteridinone	pD 14	7.90 (H-6), 7.80 (H-7), 4.60 (H-4)	46
3,4-Dihydro-4-pteridinamine	NH_3	8.43 (H-7), 8.33 (H-6), 7.64 (H-2), 5.52 (H-4)	1138
1,2,3,4-Tetrahydropteridine	$CDCl_3$	7.55 (H-6 + H-7), 5.95 (H-1 + H-3), 3.65 (H-2 + H-4)	721
Ethyl 1,2,3,4-tetrahydro-2-pteridinecarboxylate	$CDCl_3$	7.91 (H-6 + H-7), 4.99 (H-2), 4.19 (H-4)	585
5,6,7,8-Tetrahydropteridine	$CDCl_3$	7.80 (H-4), 7.30 (H-2), 5.80 (H-5 + H-8), 3.75 (H-6 + H-7)	721

[a] For abbreviations see the footnote to Table XII; assignments to H-6 and H-7 are often doubtful and may be reversed.

TABLE XIV. REFERENCES TO THE INFRARED, MASS, FLUORESCENCE, AND ^{13}C NUCLEAR MAGNETIC RESONANCE SPECTRA AND TO POLAROGRAPHIC DATA ON SOME SIMPLE PTERIDINES[a]

Compound	Known Data	References
Pteridine (unsubstituted)	ir	286
	ms	207, 1707
	^{13}C nmr	606, 843, 851, 852, 959, 1700
	pol	572, 604, 792, 1360, 1417, 1418
2-Methylpteridine	ir	286
	ms	207, 1707
4-Methylpteridine	ir	286
	ms	207, 1707
6-Methylpteridine	pol	792
7-Methylpteridine	ir	286
	ms	207, 1707
	^{13}C nmr	959
6,7-Dimethylpteridine	ms	207
	^{13}C nmr	959
2-Chloropteridine	ir	286
	^{13}C nmr	959, 1700
4-Chloropteridine	ir	286
2,4,6,7-Tetrachloropteridine	ir	286, 1251
4-Trifluoromethylpteridine	ms	777, 780, 1707
2(1*H*)-Pteridinone	ir	134
	ms	207, 1246, 1707
4(3*H*)-Pteridinone	ir	134, 1508
	ms	207, 779, 1246, 1391
	^{13}C nmr	851, 852, 853
	pol	572, 792, 1417, 1418
6(5*H*)-Pteridinone	ir	134
	ms	207, 1707
7(8*H*)-Pteridinone	ir	134
	ms	207, 1707
2,4(1*H*,3*H*)-Pteridinedione	ms	207, 1457, 1554, 1707
	fl	1042, 1414
	^{13}C nmr	851, 852, 853, 1095
	pol	642
6,7(5*H*,8*H*)-Pteridinedione	^{13}C nmr	851, 852
2,4,6(1*H*,3*H*,5*H*)-Pteridinetrione	ms	1457, 1554
2,4,7(1*H*,3*H*,8*H*)-Pteridinetrione	ms	1411, 1457, 1707
2,4,6,7(1*H*,3*H*,5*H*,8*H*)-Pteridinetetrone	ir	1251
	ms	1457
	^{13}C nmr	852
2-Amino-4(3*H*)-pteridinone	ir	547
	ms	778, 987, 1111, 1391, 1457, 1535, 1707
	fl	1413
	^{13}C nmr	606, 609, 851, 852, 853, 1095
	pol	114

TABLE XIV. (*Contd.*)

Compound	Known Data	References
2-Amino-6(5*H*)-pteridinone	ir	1052
2-Amino-4,6(3*H*,5*H*)-pteridinedione	ms	778, 987, 1111, 1391, 1457, 1535
	fl	984, 1007
	^{13}C nmr	851, 852, 1095
	pol	639, 792
2-Amino-4,7(3*H*,8*H*)-pteridinedione	ms	778, 987, 1111, 1391, 1457, 1535
	fl	984
	^{13}C nmr	851, 852, 1095
2-Amino-6,7(5*H*,8*H*)-pteridinedione	fl	984
2-Amino-4,6,7(3*H*,5*H*,8*H*)-pteridinetrione	ms	778
	fl	984
	^{13}C nmr	852, 1095
	pol	572, 642, 792, 1204, 1418
3-Hydroxy-2,4(1*H*,3*H*)-pteridinedione	ms	1289
2-Methoxypteridine	ir	134, 286
4-Methoxypteridine	ir	134, 286
6-Methoxypteridine	ir	134, 286
7-Methoxypteridine	ir	134, 286
2,4-Dimethoxypteridine	fl	1042
1-Methyl-2(1*H*)-pteridinone	ir	134
3-Methyl-2(3*H*)-pteridinone	ir	134
6,7,8-Trimethyl-2(8*H*)-pteridinone	ir	134
1-Methyl-4(1*H*)-pteridinone	ir	134
3-Methyl-4(3*H*)-pteridinone	ir	134
5-Methyl-6(5*H*)-pteridinone	ir	134
8-Methyl-7(8*H*)-pteridinone	ir	134
3-Methoxy-4(3*H*)-pteridinone	ms	779
4(1*H*)-Pteridinone 3-oxide	ms	1226, 1707
2-Methylthiopteridine	ir	286
	^{13}C nmr	959, 1700
4-Methylthiopteridine	ir	286
7-Methylthiopteridine	ir	286
2-Pteridinamine	^{13}C nmr	606
	pol	572, 1417, 1418
4-Pteridinamine	ir	185
	ms	1391, 1707
2,4-Pteridinediamine	^{13}C nmr	851, 852
2-Dimethylaminopteridine	ir	286
4-Dimethylaminopteridine	ir	286
6-Dimethylaminopteridine	ir	286
7-Dimethylaminopteridine	ir	286
3-Methyl-4(3*H*)-pteridinimine	ir	51
4-Pteridinecarboxylic acid	ms	157, 777
Ethyl 4-pteridinecarboxylate	ms	777
4-Pteridinecarbonitrile	ms	157
2-Amino-5,6-dihydro-4,7(3*H*,8*H*)-pteridinedione	ms	1391

TABLE XIV. (*Contd.*)

Compound	Known Data	References
2-Amino-7,8-dihydro-4,6(3*H*,5*H*)-pteridinedione	ms	1391
1,2,3,4-Tetrahydropteridine	^{13}C nmr	721
5,6,7,8-Tetrahydropteridine	^{13}C nmr	721
5,6,7,8-Tetrahydro-4(3*H*)-pteridinone	ms	1391
5,6,7,8-Tetrahydro-4-pteridinamine	ms	1391

[a] The availability of such data for other pteridines is indicated under individual entries in the Appendixed Table of Simple Pteridines.

5. MASS SPECTRA

The relatively small number of mass spectra reported for pteridines have been discussed briefly under properties or degradation reactions of pteridine (Ch. IV, Sect. 1.B), halogenopteridines (Ch. V, Sect. 4.D), tautomeric pteridinones and hydroxypteridines (Ch. VI, Sects. 2.A and 4.D), alkoxypteridines and pteridine *N*-oxides (Ch. VII, Sects. 2 and 6.C), pteridinamines (Ch. IX, Sect. 2.A), and pteridine esters and ketones (Ch. X, Sects. 4.A and 12); references to the ms of some simple pteridines are contained in Table XIV. Because of the very low volatility of pteridines bearing tautomeric groups, it is often wise to acetylate or trimethylsilylate such substances prior to mass spectral measurements.[1366, 1411, 1457] A brief review of the ms of pteridines appeared in 1985 within Quentin Porter's book.[1707]

6. FLUORESCENCE SPECTRA

The fluorescent colors associated with solutions of individual pteridines have been recorded in innumerable papers. Indeed such colors have assumed an apparent importance for diagnostic purposes that is quite unwarrented. The phenomenon has even been enshrined in such names as lumazine,[1012] which in fact exhibits no noticeable fluorescence in daylight when pure.[30] Only a few actual fluorescence spectra have ever been reported, possibly because most laboratories lack facilities for such measurements. References to some such spectra for simple pteridines appear in Table XIV.[cf. 1007]

7. ^{13}C NUCLEAR MAGNETIC RESONANCE SPECTRA

Although ^{13}C nmr spectroscopy has not been applied widely yet in the pteridine series, such spectra for most of the common natural pteridines,[851–853, 1095] related drugs,[851] a number of representative simple pteridines[847, 851–853, 959, 1095, 1700] and some reduced pteridines[102, 194, 609, 703, 721] have been reported and discussed (see also Table XIV). Little information has emerged so far which could not be obtained from ^{1}H nmr spectra or by other means. However, clear confirmation has been forthcoming in respect of protonation sites,[606, 1095] the rather complicated sequences involved in the covalent hydration of neutral and cationic pteridine,[843, 959] the covalent amination of pteridines,[959] and the structure of 1,2,3,4- and 5,6,7,8-tetrahydropteridine.[721]

8. POLAROGRAPHIC DATA

References to polarographic data on many simple pteridines have been included in Table XIV. Such data are potentially useful in devising reductive or oxidative methods for passing between pteridines, dihydropteridines, and tetrahydropteridines; in devising qualitative and quantitative analytical methods for specific pteridines in biological admixtures; and for a variety of other purposes. However, apart from some aspects already mentioned in Chapter XI and elsewhere, the mass of painstaking work reported on the polarography of pteridines (and reviewed elsewhere[1419]) appears to have been put to little practical use.

APPENDIX

Table of Simple Pteridines

This table aims to be a complete alphabetical list of simple pteridines described up to the end of 1986; it also includes some pteridines described during 1987. For each compound are recorded (a) melting or boiling point; (b) an indication of known spectra and other physical properties, apart from ionization constants that are collected in Table XI; (c) any reported salts or simple derivatives, especially when the parent compound is ill characterized; and (d) reference(s) to the original literature. For an outline of the *nomenclature* employed, see Chapter I, Section 2.

To keep the table within manageable proportions, the following categories of pteridine have been *excluded* on the grounds that they are not simple:

- Those reduced in the nucleus, that is, hydropteridines.
- Those with an heterocyclic substituent, except piperidino or morpholino.
- Those fused with another ring system.
- Those bearing any substituent with more than six carbon atoms, except for benzoyl, benzyl, phenethyl, or styryl.
- Those bearing a substituted-phenyl substituent.
- Those with a sugar substituent.
- Those with a di- or polyfunctional substituent, except for carboxymethylthio or when the functionalities are of the same type as in ribityl.

The following conventions and abbreviations have been used in the table.

The term, *melting point*, covers not only a regular melting point or melting range, but also such variations as 'decomposing at' or 'melting with decomposition at' and so on. The use of > before a melting point indicates that the substance melted or decomposed above that temperature or, alternatively, that it did not melt or decompose below that temperature. When two differing melting points are given in the literature, they are recorded in the table as, for example, '274 or 281'; when there are more than two melting points, they are recorded in a form like '210 to 241' to distinguish it from a single melting point range such as '210–213'.

The few *boiling points* in this series are distinguished from melting points by the presence of a pressure in millimeters of mercury (mmHg) after the temperature: 112/0.01.

The *abbreviations* in the table are broadly self-explanatory as exemplified in the following list:

Physical Data

anal	Analytical data (usually assumed)
crude	Compound not purified
dip	Dipole moment
fl-sp	Fluorescence spectrum
ir	Infrared spectrum
ms	Mass spectrum
nmr	Nuclear magnetic resonance spectrum (nuclei other than ^1H are given)
pol	Polarographic data
uv	Ultraviolet/visible spectrum
xl-st	Crystal structure

Salts and so on

AcOH	Acetate salt
$(CO_2H)_2$	Oxalate salt
EtOH	Ethanolate
HBr, HCl, HI	Appropriate hydrohalide salt
H_2O	Hydrate
H_2SO_4	Sulfate salt
$MeC_6H_4SO_3H$	*p*-Toluenesulfonate salt
pic	Picrate salt
$H\overset{+}{N}(CH)_5$	Pyridinium salt
K, Na	Appropriate alkali metal salt
NH_4	Ammonium salt
Hg, Co, Cu, and so on	Appropriate metal salt

Derivatives

Ac	Acetyl derivative
DNP	2,4-Dinitrophenylhydrazone
$H_2NCONHN{=}$	Semicarbazone
$HON{=}$	Oxime
$NO_2C_6H_4CH{=}$	*p*-Nitrobenzylidene derivative
$NO_2C_6H_4NHN{=}$	*p*-Nitrophenylhydrazone
$PhNHN{=}$	Phenylhydrazone

The use of cf before a *reference* number indicates that there is some suspect or incorrect information therein that has been corrected in subsequent references; a query mark (?) indicates an anomaly, apparent anomaly, or reasonable doubt associated with a reference or datum; and a starred reference number (*) indicates that the paper contains information on the pteridine bearing an isotopic label.

APPENDIX TABLE SIMPLE PTERIDINES

Compound	(°C), and so on	Reference(s)
2-Acetamido-7-acetonyl-4-acetoxy-6(5*H*)-pteridinone	uv	1295
2-Acetamido-6-acetoxymethyl-4-pteridinamine	227–229 or 230–240, nmr, uv	113, 1358
2-Acetamido-6-acetoxymethyl-4(3*H*)-pteridinone	227 to 242, ms, nmr, uv	113, 589, 776(?), 1443
2-Acetamido-7-acetoxymethyl-4(3*H*)-pteridinone	235, nmr	776(?), 1443
4-Acetamido-2-amino-6-phenylthio-7(8*H*)-pteridinone	—	329
4-Acetamido-7-*t*-butyl-6-phenylthio-2-pteridinamine	—	329
2-Acetamido-6-α,β-diacetoxyethyl-4(3*H*)-pteridinone	ms	589
2-Acetamido-6-diacetoxymethyl-4(3*H*)-pteridinone	191, nmr	1451
2-Acetamido-6-α,β-diacetoxypropyl-4,7(3*H*,8*H*)-pteridinedione	143–153	1029
2-Acetamido-6-α,β-diacetoxypropyl-4(3*H*)-pteridinethione	crude, nmr, uv	1439
2-Acetamido-6-α,β-diacetoxypropyl-4(3*H*)-pteridinone	171–172, ms	82, 589
2-Acetamido-6-β,γ-diacetoxypropyl-4(3*H*)-pteridinone	ms	589
2-Acetamido-6-dibromomethyl-4(3*H*)-pteridinone	AcOH: anal; nmr	1358, 1739
2-Acetamido-7-dibromomethyl-4(3*H*)-pteridinone	anal	1359
2-Acetamido-6-diethoxymethyl-4(3*H*)-pteridinone	191–193 or 197–200, uv	464, 1300, 1403
7-Acetamido-1,3-dimethyl-2,4,6(1*H*,3*H*,5*H*)-pteridinetrione	300, uv	238
2-Acetamido-6,7-dimethyl-4(3*H*)-pteridinone	298–299, ms, pol, uv	69, 1014, 1535
4-Acetamido-6,7-diphenyl-2-pteridinamine	140–150, uv	141
8-β-Acetamidoethyl-2,4,6,7(1*H*,3*H*,5*H*,8*H*)-pteridinetetrone	>340	674
4-Acetamido-6-hexylthio-2,7-pteridinediamine	—	329
2-Acetamido-4-methoxypteridine	215, nmr, uv	114, 368, 741
2-Acetamido-6-methoxy-4(3*H*)-pteridinone	274–275, uv	369
2-Acetamido-6-methyl-4-oxo-3,4-dihydro-7-pteridinecarboxylic acid	218, uv	307
4-Acetamido-7-methyl-6-phenylthio-2-pteridinamine	—	329
2-Acetamido-4-methylpteridine	181–183	282
2-Acetamido-6-methyl-4,7(3*H*,8*H*)-pteridinedione	ms	1535
2-Acetamido-7-methyl-4,6(3*H*,5*H*)-pteridinedione	ms	1535
2-Acetamido-6-methyl-4(3*H*)-pteridinone	220 (?), 315 to 321, ms, pol, uv; AcOH 320, nmr	389, 589, 929, 1014, 1358, 1394, 1443, 1535, 1683
2-Acetamido-7-methyl-4(3*H*)-pteridinone	292–297, nmr, uv	389, 993, 1683

APPENDIX TABLE (*Contd.*)

Compound	(°C), and so on	Reference(s)
2-Acetamido-4-oxo-3,4-dihydro-6-pteridinecarbaldehyde	$2H_2O$ anal, ms, nmr, uv; HCO_2H anal; NOH nmr	464, 704, 1300, 1403, 1739
2-Acetamido-4-oxo-3,4-dihydro-7-pteridinecarbaldehyde	crude, ir, nmr	1310, 1451
2-Acetamido-4-oxo-3,4-dihydro-6-pteridinecarbonitrile	230–233, ms, nmr; AcOH 253–255	1739
2-Acetamido-4-oxo-3,4-dihydro-6-pteridinecarboxylic acid	H_2O 210, uv	389
2-Acetamido-4-oxo-3,4-dihydro-7-pteridinecarboxylic acid	AcOH 238, uv	389
2-Acetamido-4-pentyloxy-7(8*H*)-pteridinone	275–277	436
2-Acetamido-4-pentyloxy-7-trimethylsiloxypteridine	crude	435
2-Acetamido-6-phenyliminomethyl-4(3*H*)-pteridinone	crude nmr	1482
4-Acetamido-7-phenyl-6-phenylthio-2-pteridinamine	—	329
2-Acetamido-6-phenyl-4-pteridinamine	>255	541
4-Acetamido-6-phenylthio-2,7-pteridinediamine	—	329
2-γ-Acetamidopropylamino-6-[L-*erythro*]-α,β-dihydroxypropyl-4(3*H*)-pteridinone	232–234, uv	1454
2-γ-Acetamidopropylamino-6-[D-*erythro*]-α,β,γ-trihydroxypropyl-4(3*H*)-pteridinone	205–207, uv	1454
2-Acetamidopteridine	229–231, uv	30, 33, 119
4-Acetamidopteridine	191–192, uv	30, 33, 119
7-Acetamidopteridine	>250, uv	33
2-Acetamido-4,6(3*H*,5*H*)-pteridinedione	>350, ms	369, 1535
2-Acetamido-4,7(3*H*,8*H*)-pteridinedione	ms	1535
2-Acetamido-4(3*H*)-pteridinone	310 or >350, ms, nmr, uv	114, 368, 741, 1052, 1535
2-Acetamido-7(8*H*)-pteridinone	310, uv	381
2-Acetamido-6-α,β,γ,δ-tetra-acetoxybutyl-4(3*H*)-pteridinone	ms	589
2-Acetamido-6-β,γ,δ-triacetoxybutyl-4(3*H*)-pteridinone	ms	589
2-Acetamido-6-α,β,γ-triacetoxypropyl-4(3*H*)-pteridinone	ms	589
4-Acetamido-6,*N*,*N*-trimethyl-7-pteridinecarboxamide 5,8-dioxide	228	723
2-Acetamido-1,6,7-trimethyl-4(1*H*)-pteridinone	171–172, uv	69
2-Acetamido-3,6,7-trimethyl-4(3*H*)-pteridinone	196–199, pol, uv	654, 1014
7-Acetonyl-2-amino-5-benzyl-4,6(3*H*,5*H*)-pteridinedione	>300, uv	712
6-Acetonyl-2-amino-8-methyl-4,7(3*H*,8*H*)-pteridinedione	>300, uv	712
6-Acetonyl-2-amino-8-phenyl-4,7(3*H*,8*H*)-pteridinedione	>330	712
6-Acetonyl-2-amino-4,7(3*H*,8*H*)-pteridinedione	anal, ms	620, 1083
7-Acetonyl-2-amino-4,6(3*H*,5*H*)-pteridinedione	anal, ms, nmr, uv	498, 561, 580, 620, 712, 1083, 1295, 1313, 1323, 1540
7-Acetonyl-2,4-diamino-6(5*H*)-pteridinone	uv	580
6-Acetonyl-8-β-hydroxyethyl-2,4,7(1*H*,3*H*,8*H*)-pteridinetrione	231–235, uv	712
6-Acetonylmethyl-2,4-pteridinediamine	286–287	1302

APPENDIX TABLE (*Contd.*)

Compound	(°C), and so on	Reference(s)
6-Acetonyl-8-methyl-4,7(3*H*,8*H*)-pteridinedione	304–306	712
7-Acetonyl-1-methyl-2,4,6(1*H*,3*H*,5*H*)-pteridinetrione	crude	340
6-Acetonyl-2,4,7(1*H*,3*H*,8*H*)-pteridinetrione	crude	1083
7-Acetonyl-2,4,6(1*H*,3*H*,5*H*)-pteridinetrione	anal	1083, 1304
6-Acetonyl-7(8*H*)-pteridinone	>280	55
8-Acetoxy-2,4-diamino-6-methyl-7(8*H*)-pteridinone	$MeC_6H_4SO_3H$ 218, nmr	1298
6-Acetoxy-1,3-dimethyl-7-(*N*-methylacetamido)-2,4(1*H*,3*H*)-pteridinedione	169, ir	1294
3-β-Acetoxyethyl-2-amino-6,7-diphenyl-4(3*H*)-pteridinone	260, ir, nmr, uv	1500
8-β-Acetoxyethyl-6,7-diphenyl-2,4(3*H*,8*H*)-pteridinedione	186–188 or 228, uv	240, 409
8-β-Acetoxyethyl-3-methyl-6,7-diphenyl-2,4(3*H*,8*H*)-pteridinedione	196, uv	409
8-β-Acetoxyethyl-2,4,7(1*H*,3*H*,8*H*)-pteridinetrione	273, uv	317
6-α-Acetoxy-β-hydroxypropyl-2-amino-4,7(3*H*,8*H*)-pteridinedione (cyprino-purple-C2)	R_f	1029
6-β-Acetoxy-α-hydroxypropyl-2-amino-4,7(3*H*,8*H*)-pteridinedione (cyprino-purple-C1)	R_f	1029
6-Acetoxymethyl-2-amino-4(3*H*)-pteridinone	300, nmr, uv	113
6-Acetoxymethyl-1,3-dimethyl-2,4(1*H*,3*H*)-pteridinedione	125–126, uv	1704
7-Acetoxymethyl-3,6-dimethyl-2,4(1*H*,3*H*)-pteridinedione	265, uv	367
7-Acetoxymethyl-3,6-dimethyl-2,4(1*H*,3*H*)-pteridinedione 5-oxide	235, uv	367
6-Acetoxy-3-methyl-7-phenyl-2,4(1*H*,3*H*)-pteridinedione	>330, ir, nmr	331
6-Acetoxy-3-methyl-7-phenyl-2,4(1*H*,3*H*)-pteridinedione 8-oxide	>330, ir, ms, nmr, uv	331
6-Acetoxymethyl-2,4-pteridinediamine	275–285, nmr, uv	113
6-Acetoxy-1-methyl-2,4(1*H*,3*H*)-pteridinedione	247, uv	367
1-Acetoxy-2,4(1*H*,3*H*)-pteridinedione	255, uv	1098
3-Acetoxy-4(3*H*)-pteridinone	172–173	540
6-Acetyl-2-amino-7-methyl-4(3*H*)-pteridinone	>250, nmr, uv	1217
6-Acetyl-2-amino-7-thioxo-7,8-dihydro-4(3*H*)-pteridinone	>300	1309
6-Acetylcarbonyl-2-amino-4,7(3*H*,8*H*)-pteridinedione	anal	499
6-Acetyl-1,3-dimethyl-2,4(1*H*,3*H*)-pteridinedione	173, nmr, uv	108
6(7)-Acetyl-1,7(1,6)-dimethyl-2,4(1*H*,3*H*)-pteridinedione	267	1380
7-Acetyl-1,3-dimethyl-2,4(1*H*,3*H*)-pteridinedione	177, nmr, uv	108
2-Acetylimino-6,7,8-trimethyl-2,8-dihydropteridine	165–170, uv	190
7-Acetyl-2-methylthio-4-pteridinamine	218–220, nmr, uv	1678
6-Acetyl-2-methylthio-4,7-pteridinediamine	>280, uv	328
6-Acetyl-2-phenyl-4,7-pteridinediamine	306–310, uv; PhNHN= 308	328
6-Acetyl-2-piperidino-4,7-pteridinediamine	293–298, uv	328
6-Acetyl-4,7-pteridinediamine	>340, uv	328
7-Acetyl-2,4-pteridinediamine	>170, nmr, uv	1678
6-Acetyl-1,3,7-trimethyl-2,4(1*H*,3*H*)-pteridinedione	164–165 or 167–168, nmr, uv	108, 1380
7-Allylamino-6-chloro-2,4-pteridinediamine	245–247	729
2-Allylamino-4,6,7-trimethylpteridine	184	124
"A-Methopterin" [see: methotrexate]		

APPENDIX TABLE (*Contd.*)

Compound	(°C), and so on	Reference(s)
2-Amino-6-aminomethyl-4,7(3*H*,8*H*)-pteridinedione	anal	1083
2-Amino-6-β-anilinoethyl-4(3*H*)-pteridinone	anal, uv	743
2-Amino-6-anilinomethyl-4(3*H*)-pteridinone	anal, ms, nmr	1344, 1375
2-Amino-6-benzamidomethyl-4,7(3*H*,8*H*)-pteridinedione	anal	1083
2-Amino-8-benzyl-6,7-dimethyl-4(8*H*)-pteridinone	HCl 276, uv	377
4-Amino-8-benzyl-6,7-dimethyl-2(8*H*)-pteridinone	209–211, uv	362
4-Amino-6-benzyl-2-ethylthio-7(8*H*)-pteridinone	255–257, uv	93
4-Amino-8-benzyl-6-methyl-2,7(1*H*,8*H*)-pteridinedione	297–298, uv	363
4-Amino-6-benzyl-2-methylthio-7(8*H*)-pteridinone	256–257	92
2-Amino-4-benzyloxy-6,8-dimethyl-7(8*H*)-pteridinone	216–218, uv	372
2-Amino-4-benzyloxy-6-methyl-7(8*H*)-pteridinone	310	372
2-Amino-4-benzyloxy-7(8*H*)-pteridinone	>360	380
4-Amino-6-benzyl-2,7(1*H*,8*H*)-pteridinedione	443	92
2-Amino-6-benzyl-4,7(3*H*,8*H*)-pteridinedione	441, uv	707(?), 718
2-Amino-7-benzyl-4,6(3*H*,5*H*)-pteridinedione	320, uv	710, 718
4-Amino-8-benzyl-2,7(1*H*,8*H*)-pteridinedione	314–316, uv	363
2-Amino-6-benzyl-4(3*H*)-pteridinone	>360, nmr	419
2-Amino-7-benzyl-4(3*H*)-pteridinone	>360, nmr	419
2-Amino-8-benzyl-4(8*H*)-pteridinone	HCl 230, uv	377
2-Amino-6-benzylthio-4(3*H*)-pteridinone	>330	1307
4-Amino-6-benzyl-2-thioxo-1,2-dihydro-7(8*H*)-pteridinone	322, uv	92
2-Amino-6,7-bisbromomethyl-4(3*H*)-pteridinone	anal	112, 680
2-Amino-6-bromomethyl-7-methyl-4(3*H*)-pteridinone(?)	—	112
2-Amino-7-bromomethyl-6-methyl-4(3*H*)-pteridinone(?)	—	112
2-Amino-6-bromomethyl-4,7(3*H*,8*H*)-pteridinedione	anal	1061, 1371
2-Amino-6-bromomethyl-4(3*H*)-pteridinone	anal; HBr anal	112, 408, 595, 731, 1155
2-Amino-6-bromomethyl-4(3*H*)-pteridinone	HBr anal, uv	112, 1359, 1528
2-Amino-6-butoxymethyl-4(3*H*)-pteridinone	anal, ir, nmr, uv	1728
2-Amino-6-*t*-butoxymethyl-4(3*H*)-pteridinone	anal, ir, nmr, uv	1728
2-Amino-6-butoxymethyl-4(3*H*)-pteridinone 8-oxide	crude, nmr, uv	1728
4-Amino-*N*-butyl-7-butylamino-2-phenyl-6-pteridinecarboxamide	255, pol	320, 1113
4-Amino-*N*-butyl-7-oxo-2-phenyl-7,8-dihydro-6-pteridinecarboxamide	296	320
2-Amino-7-butyl-4(3*H*)-pteridinone	—	1378
2-Amino-7-*t*-butyl-4(3*H*)-pteridinone	345–348, uv	541
2-Amino-7-*t*-butyl-4(3*H*)-pteridinone 5-oxide	H_2O >350	541
2-Amino-3-β-carboxyethyl-5,6-dimethyl-4(3*H*)-pteridinone	crude, uv	654
2-Amino-3-β-carboxyethyl-4-oxo-3,4-dihydro-6-pteridinecarboxylic acid	274–276, uv	654
2-Amino-6-α-carboxyethyl-4,7(3*H*,8*H*)-pteridinedione	anal, uv	930, 1083
2-Amino-6-β-carboxyethyl-4,7(3*H*,8*H*)-pteridinedione	280	930, 1083, 1304
2-Amino-7-β-carboxyethyl-4,6(3*H*,5*H*)-pteridinedione	anal	1379
2-Amino-6-α-carboxyethyl-4(3*H*)-pteridinone	anal, uv	906
2-Amino-6-β-carboxyethyl-4(3*H*)-pteridinone	anal, uv	906

APPENDIX TABLE (*Contd.*)

Compound	(°C), and so on	Reference(s)
4-Amino-7-carboxymethyl-2-methylthio-6(5*H*)-pteridinone	>260	699
2-Amino-6-carboxymethyl-4,7(3*H*,8*H*)-pteridinedione	>330, uv	497, 930, 1061, 1083, 1195
2-Amino-7-carboxymethyl-4,6(3*H*,5*H*)-pteridinedione	—	497
4-Amino-6-carboxymethyl-2,7(1*H*,8*H*)-pteridinedione	—	699
2-Amino-6-carboxymethyl-4(3*H*)-pteridinone	anal; Na anal	453, 649, 1084
4-Amino-6-carboxymethyl-2-thioxo-1,2-dihydro-7(8*H*)-pteridinone	>300	699
2-Amino-7-(2′-carboxyprop-1′-enyl)-4,6(3*H*,5*H*)-pteridinedione	H_2O anal, nmr, uv; H_2SO_4 anal	1055
2-Amino-6-α-carboxypropyl-4,7(3*H*,8*H*)-pteridinedione	anal	930, 1083
2-Amino-7-chloro-3-methyl-4(3*H*)-pteridinone	297	383
2-Amino-7-chloro-4-oxo-3,4-dihydro-6-pteridinecarboxylic acid	anal	1084
4-Amino-2-chloro-6,7(5*H*,8*H*)-pteridinedione	anal	1369
2-Amino-6-chloro-4(3*H*)-pteridinone	>330; HCl anal	1307
2-Amino-7-chloro-4(3*H*)-pteridinone	>330, uv	380
4-Amino-1-β-cyanoethyl-6-methyl-2(1*H*)-pteridinone	>250, uv	168
4-Amino-6-cyclohexyl-7(8*H*)-pteridinone	330–332, uv	924
4-Amino-7-cyclohexyl-6(5*H*)-pteridinone	298, uv	924
4-Amino-7-cyclopentyl-6(5*H*)-pteridinone	260, uv	924
2-Amino-6-α,β-diacetoxypropyl-4,7(3*H*,8*H*)-pteridinedione	188–196	1029
2-Amino-6-[L-*erythro*]-α,β-diacetoxypropyl-4(3*H*)-pteridinone	200, cd, nmr, uv	1439
2-Amino-6,7-dibenzyl-4(3*H*)-pteridinone	338–340	419
2-Amino-6-dibromomethyl-4(3*H*)-pteridinone	crude; HBr anal	112, 1358
2-Amino-7-dibromomethyl-4(3*H*)-pteridinone	anal; HBr anal	112, 1359
2-Amino-6-α,β-dicarboxyethyl-4,7(3*H*,8*H*)-pteridinedione	anal	930, 1083
2-Amino-6-α,β-diethoxycarbonylethyl-4,7(3*H*,8*H*)-pteridinedione	299, uv	1195
2-Amino-6-diethoxymethyl-4(3*H*)-pteridinone	anal, uv	464, 1300, 1403
4-Amino-*N*-(β-diethylaminoethyl)-7-oxo-2-phenyl-7,8-dihydro-6-pteridinecarboxamide	239–241	1357
4-Amino-2-diethylamino-8-ethyl-7(8*H*)-pteridinone	180–182	1390
2-Amino-6,7-diethyl-4(3*H*)-pteridinone	>360	32
4-Amino-6,7-diethyl-2(1*H*)-pteridinone	290	32
2-Amino-7-(3′,4′-dihydroxybut-1′-enyl)-4(3*H*)-pteridinone	—	1378
2-Amino-7-γ,δ-dihydroxybutyl-4(3*H*)-pteridinone	—	1378
2-Amino-6-α,β-dihydroxyethyl-7-methyl-4(3*H*)-pteridinone	—	1378
2-Amino-7-α,β-dihydroxyethyl-6-methyl-4(3*H*)-pteridinone	—	1378
2-Amino-6-(α*R*)-α,β-dihydroxyethyl-4(3*H*)-pteridinone	>300, uv	455

APPENDIX TABLE (*Contd.*)

Compound	(°C), and so on	Reference(s)
2-Amino-6-(α*S*)-α,β-dihydroxyethyl-4(3*H*)-pteridinone	>300, uv	455
2-Amino-8-β,γ-dihydroxypropyl-4,7-dioxo-3,4,7,8-tetrahydro-6-pteridinecarboxylic acid	306–310, uv; Ac_2 146, uv	480
2-Amino-6-α,β-dihydroxypropyl-4,7(3*H*,8*H*)-pteridinedione (ichthyopterin)	uv	499, 995
2-Amino-6-β,γ-dihydroxypropyl-4,7(3*H*,8*H*)-pteridinedione	anal, uv	499, 995
2-Amino-7-α,β-dihydroxypropyl-4,6(3*H*,5*H*)-pteridinedione	—	1295
2-Amino-6-α,β-dihydroxypropyl-4(3*H*)-pteridinethione	crude, cd	1439
2-Amino-6-[D-*erythro*]-α,β-dihydroxypropyl-4(3*H*)-pteridinone	>300, ms, uv	454, 653, 1464, 1555
2-Amino-6-[D,L-*erythro*]-α,β-dihydroxypropyl-4(3*H*)-pteridinone	anal	1326, 1327
2-Amino-6-[L-*erythro*]-α,β-dihydroxypropyl-4(3*H*)-pteridinone (biopterin)	300, cd, ir, ms, nmr, ^{13}C nmr, uv	335, 429, 454, 455, 458, 500, 564, 571*, 645, 653, 695, 772*, 976, 1018, 1043, 1050(?), 1111, 1197, 1285, 1318, 1324, 1439, 1456, 1464, 1552, 1555, 1556*, 1557, 1562, 1564
2-Amino-6-[D-*threo*]-α,β-dihydroxypropyl-4(3*H*)-pteridinone (ciliapterin)	>300, uv	454, 1324, 1555
2-Amino-6-[L-*threo*]-α,β-dihydroxypropyl-4(3*H*)-pteridinone	>300, cd, uv	454, 1439, 1555
2-Amino-6-α,γ-dihydroxypropyl-4(3*H*)-pteridinone (desoxyneopterin)	anal, uv	73, 1215
2-Amino-6-β,γ-dihydroxypropyl-4(3*H*)-pteridinone	—	1378
2-Amino-7-α,β-dihydroxypropyl-4(3*H*)-pteridinone (isobiopterin)	anal, ir, uv	500, 564, 571*, 1556*
2-Amino-7-[D-*erythro*]-α,β-dihydroxypropyl-4(3*H*)-pteridinone	—	1555
2-Amino-7-[D-*threo*]-α,β-dihydroxypropyl-4(3*H*)-pteridinone	—	1555
2-Amino-7-[L-*threo*]-α,β-dihydroxypropyl-4(3*H*)-pteridinone	—	1555
2-Amino-7-β,γ-dihydroxypropyl-4(3*H*)-pteridinone	—	1378
2-Amino-6-α,β-dihydroxypropyl-4(3*H*)-pteridinone 8-oxide	>300, ^{13}C nmr, uv	772*, 1285
2-Amino-6-[L-*erythro*]-α,β-dihydroxypropyl-4(3*H*)-pteridinone 8-oxide	anal, uv	1318

APPENDIX TABLE (*Contd.*)

Compound	(°C), and so on	Reference(s)
4-Amino-6,7-di-isopropyl-2(1*H*)-pteridinethione	276–277	400
2-Amino-6-dimethoxymethyl-4(3*H*)-pteridinone	>330, nmr, uv	1306
2-Amino-7-dimethoxymethyl-4(3*H*)-pteridinone	>330, ir, nmr	1310
4-Amino-2-dimethylamino-8-ethyl-7(8*H*)-pteridinone	241–243	1390
2-Amino-4-dimethylamino-8-methyl-7(8*H*)-pteridinone	260, nmr	357, 1541
2-Amino-6-γ-dimethylaminopropylamino-4(3*H*)-pteridinone	anal, uv; HCO_2H anal	1332
2-Amino-4-dimethylamino-6(5*H*)-pteridinone	279–281, uv	268
2-Amino-4-dimethylamino-7(8*H*)-pteridinone	>360, uv	268, 357, 1541
2-Amino-6-dimethylamino-4(3*H*)-pteridinone	H_2O anal, nmr, uv	1332
4-Amino-2-dimethylamino-6(5*H*)-pteridinone	230, uv	268
4-Amino-2-dimethylamino-7(8*H*)-pteridinone	335–338, uv	268
2-Amino-6-3′,3′-dimethylbut-1′-enyl-4(3*H*)-pteridinone	>300, ir, nmr	1753
2-Amino-3,8-dimethyl-4,7-dioxo-3,4,7,8-tetrahydro-6-pteridinecarboxylic acid	>350, uv	385, 1214
2-Amino-6,7-dimethyl-8-phenyl-4(8*H*)-pteridinone	HCl 310, uv	377
4-Amino-6,7-dimethyl-8-phenyl-2(8*H*)-pteridinone	306–308, uv	362
6-Amino-1,3-dimethyl-7-propionyl-2,4(1*H*,3*H*)-pteridinedione	240, nmr, uv	108
7-Amino-1,3-dimethyl-6-propionyl-2,4(1*H*,3*H*)-pteridinedione	210–211, nmr, uv	108
2-Amino-6,7-dimethyl-4-pteridinecarboxamide	>330	1354
2-Amino-3,6-dimethyl-4,7(3*H*,8*H*)-pteridinedione	>350, uv	383
2-Amino-3,7-dimethyl-4,6(3*H*,5*H*)-pteridinedione	>350	383
2-Amino-3,8-dimethyl-4,7(3*H*,8*H*)-pteridinedione	>350, uv	383, 457, 1565
2-Amino-6,8-dimethyl-4,7(3*H*,8*H*)-pteridinedione	>350, uv	383, 712
4-Amino-1,6-dimethyl-2,7(1*H*,8*H*)-pteridinedione	H_2O >360, uv	363
4-Amino-6,8-dimethyl-2,7(1*H*,8*H*)-pteridinedione	H_2O >360, uv	363
6-Amino-1,3-dimethyl-2,4(1*H*,3*H*)-pteridinedione	333, nmr, uv	1275
7-Amino-1,3-dimethyl-2,4(1*H*,3*H*)-pteridinedione	>360, ms, nmr, uv	227, 238, 1275, 1385, 1435
7-Amino-1,3-dimethyl-2,4(1*H*,3*H*)-pteridinedione 5-oxide	338–340, uv	238
2-Amino-6,7-dimethyl-4(3*H*)-pteridinethione	>290, nmr	1508
4-Amino-6,7-dimethyl-2(1*H*)-pteridinethione	300, uv	472, 916
2-Amino-3,8-dimethyl-4,6,7(3*H*,5*H*,8*H*)-pteridinetrione	>350, uv	384
6-Amino-1,3-dimethyl-2,4,7(1*H*,3*H*,8*H*)-pteridinetrione	—	1093
7-Amino-1,3-dimethyl-2,4,6(1*H*,3*H*,5*H*)-pteridinetrione	>360, uv	238
2-Amino-1,6-dimethyl-4(1*H*)-pteridinone	285, ir, nmr, uv	80
2-Amino-3,6-dimethyl-4(3*H*)-pteridinone	230, ir, nmr, uv, xl-st	80, 1232, 1752
2-Amino-4,6-dimethyl-7(8*H*)-pteridinone	uv	279
2-Amino-6,7-dimethyl-4(3*H*)-pteridinone	>360, ir, ms, nmr, ^{15}N nmr, pol, uv; Na >360	84*, 169, 284, 442, 471, 541, 544, 741, 747, 762, 861,

APPENDIX TABLE (*Contd.*)

Compound	(°C), and so on	Reference(s)
		933, 942, 979, 987, 1014, 1038, 1092, 1111, 1193*, 1373, 1380, 1391, 1469*
2-Amino-6,8-dimethyl-4(8*H*)-pteridinone	HI 278, ir, nmr	80
4-Amino-1,6-dimethyl-2(1*H*)-pteridinone	285–290, uv	168
4-Amino-1,6-dimethyl-7(1*H*)-pteridinone	330, uv	132
4-Amino-6,7-dimethyl-2(1*H*)-pteridinone	340 or >360, uv	130, 170, 362
2-Amino-6,7-dimethyl-4(3*H*)-pteridinone 5,8-dioxide	>360, ms, uv	427, 541
2-Amino-3,6-dimethyl-4(3*H*)-pteridinone 8-oxide	>300, nmr, uv	485
2-Amino-6,7-dimethyl-4(3*H*)-pteridinone 8-oxide	H_2O >350, ms, uv	427, 541
2-Amino-6,7-dimethyl-8-ribityl-4(8*H*)-pteridinone	190	172
2-Amino-4,7-dioxo-8-D-ribityl-3,4,7,8-tetrahydro-6-pteridinecarboxylic acid	345, uv	480
2-Amino-4,7-dioxo-8-D-sorbityl-3,4,7,8-tetrahydro-6-pteridinecarboxylic acid	343, uv	385, 480
2-Amino-4,7-dioxo-3,4,7,8-tetrahydro-6-pteridinecarbaldehyde	anal; PhNHN= >300	502, 503, 1061
2-Amino-4,6-dioxo-3,4,5,6-tetrahydro-7-pteridinecarboxylic acid	H_2O >340, uv	97, 271*, 342, 405, 406, 514, 874, 1313
2-Amino-4,7-dioxo-3,4,7,8-tetrahydro-6-pteridinecarboxylic acid (cyprino-purple-B)	>360, uv	385, 405, 469, 480, 481, 874, 882, 906, 930
4-Amino-2,7-dioxo-1,2,7,8-tetrahydro-6-pteridinecarboxylic acid	H_2O >360	363
2-Amino-6,7-diphenethyl-4(3*H*)-pteridinone	269–284	419
4-Amino-2,7-diphenyl-6-pteridinecarbonitrile	306–307, uv	327, 328
4-Amino-2,6-diphenyl-7-pteridinecarboxamide	308–309, uv	327
4-Amino-2,7-diphenyl-6-pteridinecarboxamide	322–325, uv	326, 327
3-Amino-6,7-diphenyl-2,4(1*H*,3*H*)-pteridinedione	259–260	468
4-Amino-6,7-diphenyl-2(1*H*)-pteridinethione	283 or 335	472, 483, 699
2-Amino-6,7-diphenyl-4(3*H*)-pteridinone	>360, pol, uv	175, 359, 471, 541, 747, 825, 979, 1014, 1333
4-Amino-2,6-diphenyl-7(8*H*)-pteridinone	>300, uv	224, 526
4-Amino-2,7-diphenyl-6(5*H*)-pteridinone	>310, uv	526
4-Amino-6,7-diphenyl-2(1*H*)-pteridinone	320–325 or 336–338, uv	362, 471, 483

APPENDIX TABLE (*Contd.*)

Compound	(°C), and so on	Reference(s)
7-Amino-2,6-diphenyl-4(3*H*)-pteridinone	>400	1281
2-Amino-6,7-diphenyl-4(3*H*)-pteridinone 5,8-dioxide	322–323, ms, uv	427, 541
2-Amino-6,7-diphenyl-4(3*H*)-pteridinone 8-oxide	H_2O >350, uv	541
2-Amino-7-β-ethoxycarbonylethyl-4-isopropoxy-6(5*H*)-pteridinone	anal	1379
2-Amino-6-ethoxycarbonylmethyl-4,7(3*H*,8*H*)-pteridinedione	>330, nmr, uv	247, 497, 1195
2-Amino-7-ethoxycarbonylmethyl-4,6(3*H*,5*H*)-pteridinedione	H_2O anal, nmr, uv	497, 1055
2-Amino-6-ethoxymethyl-4(3*H*)-pteridinone	anal, ir, nmr	1728
2-Amino-6-ethoxy-4(3*H*)-pteridinone	>350, uv; HCl >350	299, 369
2-β-Aminoethylamino-6-α,β-dihydroxypropyl-4(3*H*)-pteridinone	>245, uv	300
7-β-Aminoethyl-1,3-dimethyl-2,4,6(1*H*,3*H*,5*H*)-pteridinetrione	272–273	829
2-Amino-8-ethyl-6,7-dimethyl-4(8*H*)-pteridinone	HCl: >300, nmr	660
2-Amino-8-ethyl-4,7-dioxo-3,4,7,8-tetrahydro-6-pteridinecarboxylic acid	anal; NH_4 anal, uv	480, 481
4-Amino-*N*-ethyl-7-ethylamino-2-phenyl-6-pteridine-carboxamide	264, pol	320, 1113
2-Amino-6-ethyl-7-methyl-4(3*H*)-pteridinone	uv	1459
2-Amino-8-ethyl-6-methyl-7(8*H*)-pteridinone	H_2O >300, nmr	660
4-Amino-8-ethyl-2-methylthio-7(8*H*)-pteridinone	240–242	1390
4-Amino-*N*-ethyl-7-oxo-2-phenyl-7,8-dihydro-6-pteridinecarboxamide	295–299 or >360	320, 1357
4-Amino-8-ethyl-2-phenyl-7(8*H*)-pteridinone	280–283	1390
2-Amino-6-ethyl-4,7(3*H*,8*H*)-pteridinedione	anal	930, 1083
2-Amino-7-ethyl-4,6(3*H*,5*H*)-pteridinedione	anal	1083
2-Amino-8-ethyl-4,7(3*H*,8*H*)-pteridinedione	>300, uv	562, 1311
2-Amino-8-ethyl-4,6,7(3*H*,5*H*,8*H*)-pteridinetrione	anal	885
7-β-Aminoethyl-2,4,6(1*H*,3*H*,5*H*)-pteridinetrione	>340, uv	829
2-Amino-6-ethyl-4(3*H*)-pteridinone	nmr, anal	1092, 1683
2-Amino-7-ethyl-4(3*H*)-pteridinone	>360, ir, nmr, uv	942, 1092, 1475, 1491
6-Amino-7-formamido-8-D-ribityl-2,4(3*H*,8*H*)-pteridinedione (russupteridine-yellow-1)	>300, ir, nmr, ^{13}C nmr, uv	1400
2-Amino-7-formylmethyl-4(3*H*)-pteridinone	anal	1373
2-Amino-7-formyl-4-oxo-3,4-dihydro-6-pteridinecarboxylic acid	crude; PhNHN= crude	1313
4-Amino-*N*-hexyl-7-hexylamino-2-phenyl-6-pteridinecarboxamide	197	320
4-Amino-*N*-hexyl-7-oxo-2-phenyl-7,8-dihydro-6-pteridinecarboxamide	258–260	320
2-Amino-6-hexyl-4(3*H*)-pteridinone	>360, nmr, uv	420, 1683
2-Amino-7-hexyl-4(3*H*)-pteridinone	>360, nmr, uv	420, 1491, 1683

APPENDIX TABLE (*Contd.*)

Compound	(°C), and so on	Reference(s)
2-Amino-6-hex-l′-ynyl-4(3*H*)-pteridinone	>300, ir, nmr	1753
2-Amino-6-hydrazino-4(3*H*)-pteridinone	>350, nmr, uv	369, 1332
2-Amino-6-hydrazonomethyl-4(3*H*)-pteridinone	anal, nmr	513
2-Amino-6-α-hydroxybutyl-8-methyl-4,7(3*H*,8*H*)-pteridinedione	anal, nmr	1711
2-Amino-7-δ-hydroxybutyl-4(3*H*)-pteridinone	—	1378
2-Amino-6-β-hydroxyethoxy-4(3*H*)-pteridinone	>300, uv	299
2-Amino-8-β-hydroxyethyl-6,7-dimethyl-4(8*H*)-pteridinone	HCl 285–288, uv	422
2-Amino-8-β-hydroxyethyl-4,7-dioxo-3,4,7,8-tetrahydro-6-pteridinecarboxylic acid	>350, uv	385, 882, 1090
2-Amino-3-β-hydroxyethyl-6,7-diphenyl-4(3*H*)-pteridinone	260–262, ir, nmr, uv	627, 1500
2-Amino-8-β-hydroxyethyl-4-isopropoxy-6-methyl-7(8*H*)-pteridinone	229–231, uv	371, 372
2-Amino-6-α-hydroxyethyl-8-methyl-4,7(3*H*,8*H*)-pteridinedione	anal, nmr	1711
2-Amino-8-β-hydroxyethyl-6-methyl-4,7(3*H*,8*H*)-pteridinedione	328–340, uv	372, 422, 1090
4-Amino-8-β-hydroxyethyl-6-methyl-2,7(1*H*,8*H*)-pteridinedione	315–317, uv	363
2-Amino-6-α-hydroxyethyl-7-methyl-4(3*H*)-pteridinone	>250, nmr, uv	1217
2-Amino-6-α-hydroxyethyl-4,7(3*H*,8*H*)-pteridinedione	crude	1083
2-Amino-8-β-hydroxyethyl-4,7(3*H*,8*H*)-pteridinedione	>300, uv	183, 882
2-Amino-8-hydroxyethyl-4,6,7(3*H*,5*H*,8*H*)-pteridinetrione	uv	183
2-Amino-6-(*R*)-α-hydroxyethyl-4(3*H*)-pteridinone	>300, uv	455
2-Amino-6-(*S*)-α-hydroxyethyl-4(3*H*)-pteridinone	>300, uv	455
2-Amino-7-β-hydroxyethyl-4(3*H*)-pteridinone	—	1378
2-Amino-8-β-hydroxyethyl-4(8*H*)-pteridinone	HCl >300, uv	422
2-Amino-6-hydroxyiminomethyl-4(3*H*)-pteridinone 8-oxide	>360, nmr, uv	479, 1318
2-Amino-6-β-hydroxy-α-methylethoxy-4(3*H*)-pteridinone	>300, uv	299, 302
2-Amino-6-α-hydroxy-α-methylethyl-8-methyl-4,7(3*H*,8*H*)-pteridinedione	anal, nmr	1711
2-Amino-6-hydroxymethyl-8-methyl-4,7(3*H*,8*H*)-pteridinedione (asperopterin-B)	>300, nmr, uv	1214, 1296, 1370, 1711
2-Amino-6-hydroxymethyl-8-methyl-4(8*H*)-pteridinone	anal, uv	1370
2-Amino-6-α-hydroxy-α-methylpropyl-8-methyl-4,7(3*H*,8*H*)-pteridinedione	anal, nmr	1711
2-Amino-6-hydroxymethyl-4,7(3*H*,8*H*)-pteridinedione	>300, uv	499, 1296
2-Amino-6-hydroxymethyl-4(3*H*)-pteridinone	>320, ms, nmr, pol, uv	96, 113, 193, 513, 774, 776, 778, 911*, 989, 1032*, 1111, 1210, 1212, 1220, 1270, 1336, 1358, 1375, 1418, 1443, 1448

APPENDIX TABLE (*Contd.*)

Compound	(°C), and so on	Reference(s)
2-Amino-7-hydroxymethyl-4(3*H*)-pteridinone	anal	774, 776, 1210, 1212, 1375
4-Amino-6-hydroxymethyl-2(1*H*)-pteridinone	uv	96
2-Amino-6-β-hydroxypropoxy-4(3*H*)-pteridinone	>300, uv	299, 302
2-Amino-6-α-hydroxypropyl-8-methyl-4,7(3*H*,8*H*)-pteridinedione	α_D^{20}, ir, ms, nmr, uv	1711, 1712
2-Amino-6-α-hydroxypropyl-7-methylthio-4(3*H*)-pteridinone	uv	95
2-Amino-6-β-hydroxypropyl-4,7(3*H*,8*H*)-pteridinedione	anal	906, 1083
2-Amino-6-γ-hydroxypropyl-4,7(3*H*,8*H*)-pteridinedione	anal	499
2-Amino-7-β-hydroxypropyl-4,6(3*H*,5*H*)-pteridinedione	—	496, 1295
2-Amino-6-α-hydroxypropyl-4(3*H*)-pteridinone	uv	95
2-Amino-6-isobutoxymethyl-4(3*H*)-pteridinone 8-oxide	crude, ir, nmr	1728
2-Amino-7-isobutyl-4-isopropoxy-6(5*H*)-pteridinone	anal, uv	1379
2-Amino-7-isobutyl-4,6(3*H*,5*H*)-pteridinedione	anal; HCl anal	1379
2-Amino-7-isopropenyl-6-methyl-4(3*H*)-pteridinone 8-oxide	>320, nmr, uv	485
2-Amino-4-isopropoxy-6,8-dimethyl-7(8*H*)-pteridinone	243–245, uv	371, 372
2-Amino-4-isopropoxy-8-methyl-7-oxo-7,8-dihydro-6-pteridinecarboxylic acid	246–247, uv	371
2-Amino-4-isopropoxy-8-methyl-6,7(5*H*,8*H*)-pteridinedione	244–247, uv	371
2-Amino-4-isopropoxy-6-methyl-7(8*H*)-pteridinone	>360, uv	371
2-Amino-4-isopropoxy-8-methyl-7(8*H*)-pteridinone	232–234, uv	371
2-Amino-6-isopropoxymethyl-4(3*H*)-pteridinone	anal, ir, nmr, uv	1728
2-Amino-4-isopropoxy-7-oxo-7,8-dihydro-6-pteridinecarbonitrile	250, uv	371
2-Amino-4-isopropoxy-7-oxo-7,8-dihydro-6-pteridinecarboxylic acid	>360, uv	371
2-Amino-4-isopropoxy-6,7(5*H*,8*H*)-pteridinedione	>360, uv	371
2-Amino-4-isopropoxy-7(8*H*)-pteridinone	>360, uv	371
2-Amino-6-isopropoxy-4(3*H*)-pteridinone	uv	299
2-Amino-8-isopropyl-6,7-dimethyl-4(8*H*)-pteridinone	HCl 260, uv	377
2-Amino-8-isopropyl-4(8*H*)-pteridinone	HCl 260, uv	377
2-Amino-7-methoxalylmethyl-3,5-dimethyl-4,6(3*H*,5*H*)-pteridinedione	H_2O 248, uv	248
2-Amino-7-methoxalylmethyl-4,6(3*H*,5*H*)-pteridinedione	crude, nmr	1323, 1446
2-Amino-6-methoxycarbonylmethyl-4,7(3*H*,8*H*)-pteridinedione	>300	247, 930, 1304
2-Amino-7-(2′-methoxycarbonylprop-1′-enyl)-4,6(3*H*,5*H*)-pteridinedione	anal, nmr	1055
2-Amino-6-methoxy-3,8-dimethyl-4,7(3*H*,8*H*)-pteridinedione	>350, uv	384
2-Amino-4-methoxy-6,8-dimethyl-7(8*H*)-pteridinone	258–262, uv	371
2-Amino-6-(β-methoxyethoxy)methyl-4(3*H*)-pteridinone 8-oxide	anal, nmr	1728
2-Amino-3-β-methoxyethyl-6,7-diphenyl-4(3*H*)-pteridinone	261–262, ir, nmr, uv	1500
2-Amino-4-methoxy-8-methyl-7-oxo-7,8-dihydro-6-pteridinecarboxylic acid	252–254, uv	371
2-Amino-6-methoxymethyl-4,7(3*H*,8*H*)-pteridinedione	anal	1083

APPENDIX TABLE (*Contd.*)

Compound	(°C), and so on	Reference(s)
4-Amino-2-methoxy-8-methyl-2,7(1*H*,8*H*)-pteridinedione	—	621
2-Amino-4-methoxy-8-methyl-7(8*H*)-pteridinone	252–256	371, 436
2-Amino-6-methoxy-1-methyl-4(1*H*)-pteridinone	H_2O >350, uv	369
2-Amino-6-methoxymethyl-4(3*H*)-pteridinone	anal, uv	1472, 1728
2-Amino-7-methoxy-3-methyl-4(3*H*)-pteridinone	>350, uv	383
2-Amino-6-methoxymethyl-4(3*H*)-pteridinone 8-oxide	anal, ir, nmr, uv	1728
2-Amino-6-3′-methoxyprop-1′-enyl-4(3*H*)-pteridinone	>300, ir, nmr	1753
2-Amino-4-methoxy-7(8*H*)-pteridinone	>350, uv	380
2-Amino-6-methoxy-4(3*H*)-pteridinone	>350, uv; HCl >350	299, 302, 369, 1307
2-Amino-7-methoxy-4(3*H*)-pteridinone	>350, uv	380, 581
4-Amino-2-methoxy-7(8*H*)-pteridinone	>360, uv	363
4-Amino-2-methoxy-6,*N*,*N*-trimethyl-7-pteridinecarboxamide 5,8-dioxide	256, ir, nmr, uv	723
2-Amino-4-methylamino-7-oxo-7,8-dihydro-6-pteridinecarboxylic acid	anal, uv	882
2-Amino-4-methylamino-7(8*H*)-pteridinone	>300, ms, nmr	1296
2-Amino-6-methylamino-4(3*H*)-pteridinone	anal, nmr, uv	1332
2-Amino-*N*-methyl-4,6-dioxo-3,4,5,6-tetrahydro-7-pteridinecarboxamide	H_2O >340	342
2-Amino-3-methyl-4,7-dioxo-3,4,7,8-tetrahydro-6-pteridinecarboxylic acid	>350, uv	385
2-Amino-8-methyl-4,7-dioxo-3,4,7,8-tetrahydro-6-pteridinecarboxylic acid	>350, uv	385, 1214
2-Amino-1-methyl-6,7-diphenyl-4(1*H*)-pteridinethione	295	111
2-Amino-1-methyl-6,7-diphenyl-4(1*H*)-pteridinone	327–329 or 333, uv	68, 111
2-Amino-3-methyl-6,7-diphenyl-4(3*H*)-pteridinone	348–351, uv	68
2-Amino-8-methyl-6,7-diphenyl-4(8*H*)-pteridinone	H_2O >350, uv; HCl >360	190, 359, 737
4-Amino-1-methyl-6,7-diphenyl-2(1*H*)-pteridinone	340–342, uv	225, 362, 1679
4-Amino-3-methyl-6,7-diphenyl-2(3*H*)-pteridinone	267, nmr, uv	1679
2-Amino-8-methyl-4-methylamino-7-oxo-7,8-dihydro-6-pteridinecarboxylic acid	uv	882
4-Amino-*N*-methyl-7-methylamino-2-phenyl-6-pteridinecarboxamide	258, pol	320, 1113
2-Amino-8-methyl-4-methylamino-7(8*H*)-pteridinone	260–280, uv	1296
4-Amino-8-methyl-2-methylsulfonyl-7(8*H*)-pteridinone	—	621
2-Amino-1-methyl-4-oxo-1,4-dihydro-6-pteridinecarboxylic acid	295–299, uv; HCl anal	70, 1363
2-Amino-1-methyl-4-oxo-1,4-dihydro-7-pteridinecarboxylic acid	anal, uv	70
2-Amino-3-methyl-4-oxo-3,4-dihydro-6-pteridinecarboxylic acid	300–301, uv	70
2-Amino-3-methyl-4-oxo-3,4-dihydro-7-pteridinecarboxylic acid	anal, uv	70
2-Amino-6-methyl-4-oxo-3,4-dihydro-7-pteridinecarboxylic acid	>310, uv; Na crude	307, 991

APPENDIX TABLE (*Contd.*)

Compound	(°C), and so on	Reference(s)
2-Amino-7-methyl-4-oxo-3,4-dihydro-6-pteridinecarboxylic acid	300, uv	307, 1453
2-Amino-8-methyl-4-oxo-4,8-dihydro-6-pteridinecarboxylic acid	300, uv; HCl anal	70, 307
2-Amino-8-methyl-7-oxo-7,8-dihydro-6-pteridinecarboxylic acid	258, ir, uv	562
4-Amino-*N'*-methyl-7-oxo-2-phenyl-7,8-dihydro-6-pteridinecarbohydrazide	282, uv	1357
4-Amino-*N*-methyl-7-oxo-2-phenyl-7,8-dihydro-6-pteridinecarboxamide	320	320
2-Amino-8-methyl-4-pentyloxy-7(8*H*)-pteridinone	179–180	436
2-Amino-6-methyl-8-phenyl-4,7(3*H*,8*H*)-pteridinedione	>350	1090
4-Amino-6-methyl-8-phenyl-2,7(1*H*,8*H*)-pteridinedione	H_2O >360, uv	363
2-Amino-3-methyl-6-phenyl-4(3*H*)-pteridinone	355–358	67
2-Amino-3-methyl-7-phenyl-4(3*H*)-pteridinone	352–355, uv	67
2-Amino-8-methyl-6-phenyl-4(8*H*)-pteridinone	HCl anal, uv	68
4-Amino-6-methyl-2-phenyl-7(8*H*)-pteridinone	282–284 or 289–291	1312, 1354
7-Amino-2-methyl-6-phenyl-4(3*H*)-pteridinone	>350, uv; HCl anal	1281
2-Amino-6-(2'-methylprop-1'-enyl)-4(3*H*)-pteridinone 8-oxide	anal	478
2-Amino-3-methyl-4,6(3*H*,5*H*)-pteridinedione	>350, uv	369
2-Amino-3-methyl-4,7(3*H*,8*H*)-pteridinedione	>350, uv	383, 457, 1565
2-Amino-6-methyl-4,7(3*H*,8*H*)-pteridinedione	>350, pol, uv	339, 383, 506, 792, 874, 878, 930, 982, 991, 994, 1061, 1083, 1195, 1204, 1371, 1391, 1418
2-Amino-7-methyl-4,6(3*H*,5*H*)-pteridinedione (chrysopterin)	H_2O anal, ms, nmr, pol, uv	348, 504, 561, 792, 874, 878, 1204, 1391, 1418, 1453
2-Amino-8-methyl-4,7(3*H*,8*H*)-pteridinedione	>350, uv	181, 383, 422, 436, 457, 581, 1296
4-Amino-1-methyl-2,7(1*H*,8*H*)-pteridinedione	H_2O >360, uv	363
4-Amino-2-methyl-6,7(5*H*,8*H*)-pteridinedione	>300, uv	918
4-Amino-6-methyl-2,7(1*H*,8*H*)-pteridinedione	>360, uv	279, 363
4-Amino-8-methyl-2,7(1*H*,8*H*)-pteridinedione	>360, uv	363, 621
4-Amino-7-methyl-2(1*H*)-pteridinethione	320, uv	1247
2-Amino-3-methyl-4,6,7(3*H*,5*H*,8*H*)-pteridinetrione	>350, uv	384
2-Amino-8-methyl-4,6,7(3*H*,5*H*,8*H*)-pteridinetrione	>350, uv	384

APPENDIX TABLE (*Contd.*)

Compound	(°C), and so on	Reference(s)
4-Amino-1-methyl-2,6,7(1*H*,5*H*,8*H*)-pteridinetrione	H_2O >360, uv	363
2-Amino-1-methyl-4(1*H*)-pteridinone	335–337, nmr, uv	130, 133, 741, 762
2-Amino-3-methyl-4(3*H*)-pteridinone	319–320 or 322, nmr, uv	130, 133, 332, 368, 416, 511, 687, 693, 741, 762
2-Amino-6-methyl-4(3*H*)-pteridinone	>360, fl-sp, ir, ms, nmr, ^{13}C nmr, pol, uv; Na >360	80, 84*, 193, 202, 285, 449, 478, 485, 509, 544, 548, 638, 679, 741, 744, 762, 806, 939*, 940, 942, 944, 966, 973, 1014, 1084, 1095, 1111, 1147, 1154, 1208, 1212, 1253, 1254, 1277, 1338, 1340, 1373, 1394, 1418, 1443, 1448, 1455, 1475, 1482,
2-Amino-6-methyl-7(8*H*)-pteridinone	>360, ms, uv	279, 987
2-Amino-7-methyl-4(3*H*)-pteridinone	>360, ir, ms, nmr, ^{13}C nmr, pol, uv	112, 175, 192, 193, 201, 389, 449, 544, 545, 548, 561, 642, 649, 741, 762, 799, 941, 942, 944, 973, 987, 1014, 1084, 1092, 1095, 1154, 1212, 1231, 1235, 1253, 1254, 1373, 1375, 1377, 1378,

APPENDIX TABLE (*Contd.*)

Compound	(°C), and so on	Reference(s)
		1443, 1475, 1528
2-Amino-8-methyl-4(8*H*)-pteridinone	>350, nmr, uv; HCl 285 or >340; HI 265	130, 133, 259, 422, 741, 762
2-Amino-8-methyl-7(8*H*)-pteridinone	297, uv	381, 383, 581
3-Amino-2-methyl-4(3*H*)-pteridinone	228, ir	763
4-Amino-1-methyl-2(1*H*)-pteridinone	324–325 or 328–330, uv	130, 225, 362
4-Amino-1-methyl-7(1*H*)-pteridinone	345–350, uv	132
4-Amino-6-methyl-2(1*H*)-pteridinone	uv	168
4-Amino-6-methyl-7(8*H*)-pteridinone	>360, uv	279, 445, 1312
4-Amino-7-methyl-6(5*H*)-pteridinone	H_2O >320	445
4-Amino-8-methyl-7(8*H*)-pteridinone	325	445
7-Amino-1-methyl-4(1*H*)-pteridinone	>340, uv	131
7-Amino-8-methyl-4(8*H*)-pteridinone	290, uv	131
2-Amino-6-methyl-4(3*H*)-pteridinone 8-oxide	>360, ms, nmr, uv	427, 478, 485, 541
2-Amino-7-methyl-4(3*H*)-pteridinone 8-oxide	>350, ms	427, 541
2-Amino-6-methyl-8-ribityl-4,7(3*H*,8*H*)-pteridinedione	283–284	172
2-Amino-6-methyl-8-sorbityl-4,7(3*H*,8*H*)-pteridinedione	229–231	172
2-Amino-4-methylsulfonyl-6-pteridine-carbaldehyde	anal; $H_2NCSNHN=$ anal	815
2-Amino-6-methylsulfonyl-4(3*H*)-pteridinone	>330	1307
7-Amino-2-methylthio-6-phenyl-2(1*H*)-pteridinone	uv	1281
2-Amino-7-methylthio-6-propionyl-4(3*H*)-pteridinone	uv	95
4-Amino-2-methylthio-6,7(5*H*,8*H*)-pteridinedione	anal	885
2-Amino-6-methylthio-4(3*H*)-pteridinone	>330	1307
4-Amino-2-methylthio-7(8*H*)-pteridinone	305–307 or 308–310, uv	445, 1126, 1390
4-Amino-6-methyl-2-thioxo-1,2-dihydro-7(8*H*)-pteridinone	225, uv	279, 916
4-Amino-7-methyl-2-thioxo-1,2-dihydro-6(5*H*)-pteridinone	217, uv	916
2-Amino-7-methyl-6-α,β,γ-trihydroxypropyl-4(3*H*)-pteridinone	—	1378
2-Amino-6-morpholino-4(3*H*)-pteridinone	anal, uv	1332
2-Amino-6-neopentyl-4(3*H*)-pteridinone	anal, nmr	1683
2-Amino-7-neopentyl-4(3*H*)-pteridinone	anal, nmr	1683
2-Amino-7-oxalomethyl-4,6(3*H*,5*H*)-pteridinedione (erythropterin)	H_2O anal, uv	333, 348, 406, 439, 816, 952, 1323, 1379, 1446
2-Amino-4-oxo-3,4-dihydro-6-pteridinecarbaldehyde	>360, fl-sp, ir, ms, nmr, pol, uv; HON= >360; DNP anal; $H_2NCSNHN=$	192, 205, 479, 503, 640, 686, 899, 940, 943, 977, 989,

APPENDIX TABLE (*Contd.*)

Compound	(°C), and so on	Reference(s)
	300	1060*, 1104, 1220(?), 1270, 1306, 1338, 1344, 1358, 1373, 1418, 1448, 1451, 1463
2-Amino-4-oxo-3,4-dihydro-7-pteridine-carbaldehyde	H_2O >330, ir, nmr	1310, 1463
2-Amino-4-oxo-3,4-dihydro-6-pteridinecarbonitrile	>300, ir, nmr	1739
2-Amino-4-oxo-3,4-dihydro-6-pteridinecarboxamide	>300, nmr	925, 1739
2-Amino-4-oxo-3,4-dihydro-6-pteridinecarboxylic acid (pterincarboxylic acid)	>360, fl-sp, ir, ms, pol, uv; Na crude	178, 192, 193, 205, 279, 307, 309, 389, 420, 453, 464, 500, 502, 529*, 642, 697, 744, 745, 899, 906, 921*, 925, 943, 964, 966, 989, 1049, 1084, 1104, 1111, 1123, 1174, 1188, 1211, 1263, 1338, 1358, 1373, 1376*, 1418, 1463
2-Amino-4-oxo-3,4-dihydro-7-petridinecarboxylic acid	H_2O >360, ir, ^{13}C nmr, uv; Na anal	112, 123, 192, 193, 307, 389, 420, 500, 642, 748, 853, 994, 1084, 1263, 1301, 1313, 1373
2-Amino-4-oxo-3,4-dihydro-6,7-pteridinedicarboxylic acid	>340, uv; Na anal	307, 601, 748, 991, 1313, 1453
2-Amino-4-oxo-3,4-dihydro-6-pteridinesulfonic acid [see also: 2-amino-6-sulfo-oxy-4(3*H*)-pteridinone]	H_2O anal, uv; Na anal	453, 1051; cf. 1431
4-Amino-7-oxo-2-phenyl-7,8-dihydro-6-pteridinecarbox-amide	340, uv	320, 1357

APPENDIX TABLE (*Contd.*)

Compound	(°C), and so on	Reference(s)
4-Amino-7-oxo-2-phenyl-7,8-dihydro-6-pteridinecarboxylic acid	H_2O 276	320
4-Amino-7-oxo-2-phenyl-*N*-propyl-7,8-dihydro-6-pteridinecarboxamide	301	320
2-Amino-4-oxo-7-sulfo-3,4-dihydro-6-pteridinecarboxylic acid	K anal, uv	906
2-Amino-4-oxo-7-thioxo-3,4,7,8-tetrahydro-6-pteridinecarboxylic acid	crude	906
2-Amino-4-oxo-6(7)-[L-*threo*]-α,β,γ-trihydroxypropyl-7(6)-pteridinecarboxylic acid	anal	1057
2-Amino-4-oxo-7-α,β,γ-trihydroxypropyl-6-pteridinecarboxylic acid	~210 ($-CO_2$)	1313
2-Amino-6-pentyloxymethyl-4(3*H*)-pteridinone 8-oxide	anal, ir, nmr	1728
2-Amino-4-pentyloxy-7-pteridinesulfonic acid	>300, nmr, uv	1681
2-Amino-4-pentyloxy-7(8*H*)-pteridinethione	214–216, nmr, uv	1681
2-Amino-4-pentyloxy-7(8*H*)-pteridinone	>300, uv	436, 1541
2-Amino-7-pentyl-4(3*H*)-pteridinone	>360, nmr, uv	1491
2-Amino-6-phenacyl-4,7(3*H*,8*H*)-pteridinedione	H_2SO_4 >300	982
2-Amino-7-phenacyl-4,6(3*H*,5*H*)-pteridinedione	H_2O >300, ms	958, 982
2-Amino-6-phenethyl-4(3*H*)-pteridinone	>360, ir, nmr, uv	90, 420, 1194, 1683
2-Amino-7-phenethyl-4(3*H*)-pteridinone	>360, nmr, uv	90, 420, 1683
2-Amino-6-phenylethynyl-4(3*H*)-pteridinone	>300, ir, nmr	1753
2-Amino-6-phenyliminomethyl-4(3*H*)-pteridinone	anal, ms, nmr	1344, 1358
4-Amino-2-phenyl-*N*-propyl-7-propylamino-6-pteridinecarboxamide	293, pol	320, 1113
4-Amino-2-phenyl-6-pteridinecarboxylic acid	284–286, uv	1354
3-Amino-7-phenyl-2,4(1*H*,3*H*)-pteridinedione	ms, nmr, uv	331
2-Amino-6-phenyl-4,7(3*H*,8*H*)-pteridinedione	>360, uv	541, 1011, 1195
4-Amino-2-phenyl-6,7(5*H*,8*H*)-pteridinedione	>300, uv	522
4-Amino-8-phenyl-2,7(1*H*,8*H*)-pteridinedione	H_2O >360, uv	363
7-Amino-6-phenyl-2,4(1*H*,3*H*)-pteridinedione	>330	1281
4-Amino-7-phenyl-2(1*H*)-pteridinethione	295, uv	1247
2-Amino-6-phenyl-4(3*H*)-pteridinone	>360, uv	67, 176, 449, 478, 541, 1006(?), 1092, 1257
2-Amino-7-phenyl-4(3*H*)-pteridinone	>360, pol, uv	67, 176, 449, 541, 1006(?), 1014
2-Amino-8-phenyl-4(8*H*)-pteridinone	HCl 300, uv	377
3-Amino-2-phenyl-4(3*H*)-pteridinone	280, ir	763
4-Amino-2-phenyl-7(8*H*)-pteridinone	330–332 or >340, uv	224, 1312, 1390
4-Amino-6-phenyl-7(8*H*)-pteridinone	>320, uv	224
4-Amino-7-phenyl-2(1*H*)-pteridinone	—	621

APPENDIX TABLE (*Contd.*)

Compound	(°C), and so on	Reference(s)
7-Amino-6-phenyl-4(3*H*)-pteridinone	>350, uv	1281
2-Amino-6-phenyl-4(3*H*)-pteridinone 8-oxide	>360, nmr, uv	478, 485, 541
2-Amino-7-phenyl-4(3*H*)-pteridinone 5-oxide	345–350, uv	541
2-Amino-6-phenylsulfonyl-4(3*H*)-pteridinone	>330	1307
2-Amino-6-phenylthio-4(3*H*)-pteridinone	>330	1307
2-Amino-6-propionylmethyl-4,7(3*H*,8*H*)-pteridinedione	anal	1083
2-Amino-6-propionyl-4(3*H*)-pteridinone	anal, uv	95, 1463
2-Amino-6-propoxymethyl-4(3*H*)-pteridinone	anal, ir, nmr, uv	1728
2-Amino-6-propoxy-4(3*H*)-pteridinone	HCl >350, uv	299, 369
2-γ-Aminopropylamino-6-[L-*erythro*]-α,β-dihydroxypropyl-4(3*H*)-pteridinone	crude; Ac 234	1454
2-γ-Aminopropylamino-6-[D-*erythro*]-α,β,γ-trihydroxypropyl-4(3*H*)-pteridinone	crude; Ac 207	1454
2-Amino-6-propyl-4,7(3*H*,8*H*)-pteridinedione	anal	930, 1083
3-Amino-2-propyl-4(3*H*)-pteridinone	157–158, ir	763
2-Amino-6-propyl-4(3*H*)-pteridinone 8-oxide	>320, nmr, uv	485
2-Amino-4,6(3*H*,5*H*)-pteridinedione (xanthopterin)	anal; H_2O anal, fl-sp, ms, nmr, ^{13}C nmr, pol, uv; H_2O_2 anal; HCl >320, xl-str; H_2SO_4 >200	14, 16, 30, 32, 60, 71*, 94*, 103, 145, 188, 241, 244, 269, 271*, 273, 335, 342, 369, 404, 405, 440, 450–452, 477, 528, 533, 541, 561, 618, 639, 656, 687, 719, 727, 778, 792, 800, 811, 851, 852, 877, 878, 984, 1011, 1095, 1111, 1172, 1238, 1256, 1268, 1293, 1301, 1307, 1350, 1381, 1391, 1410
2-Amino-4,7(3*H*,8*H*)-pteridinedione (isoxanthopterin; leucopterin-B)	>300, ms, ^{13}C nmr, uv	61, 62, 145, 204, 262, 268, 335, 339, 383,

APPENDIX TABLE (*Contd.*)

Compound	(°C), and so on	Reference(s)
		405, 469, 502, 687, 726*, 778, 851, 852, 856, 878, 930, 979, 1095, 1111, 1128, 1267*, 1301, 1350, 1391, 1399, 1550, 1565
2-Amino-6,7(5*H*,8*H*)-pteridinedione	anal, fl, uv	41, 536, 984, 1350
4-Amino-2,7(1*H*,8*H*)-pteridinedione	H_2O >360, uv	41, 363, 979, 1350
4-Amino-6,7(5*H*,8*H*)-pteridinedione	>360, ir, uv	208, 1350, 1369(?)
2-Amino-4(3*H*)-pteridinethione	anal, uv	305
4-Amino-2(1*H*)-pteridinethione	300, uv	916
2-Amino-4,6,7(3*H*,5*H*,8*H*)-pteridinetrione (leucopterin)	>350, fl-sp, ms, ^{13}C nmr, pol, uv; Ac_3 anal; K anal; Na anal	30, 32, 60, 71*, 94*, 97, 145, 195, 335, 343, 384, 403, 404, 441, 530, 531, 532, 533, 642, 719, 778, 792, 800, 852, 984, 988, 1056, 1095, 1125, 1152, 1158, 1172, 1204, 1238, 1268, 1350, 1382, 1399, 1418
4-Amino-2,6,7(1*H*,5*H*,8*H*)-pteridinetrione	>360, uv	363, 1350, 1369
6-Amino-2,4,7(1*H*,3*H*,8*H*)-pteridinetrione	anal	1093
2-Amino-4(3*H*)-pteridinone (pterin)	>360, fl-sp, ir, ms, nmr, ^{13}C nmr, pol, uv	30, 32, 39, 59, 102, 130, 145, 175, 204, 284, 305, 335, 368, 402,

APPENDIX TABLE (*Contd.*)

Compound	(°C), and so on	Reference(s)
		416, 453, 471, 509, 529, 546, 547, 561, 609, 646*, 686, 687, 741, 747, 762, 778, 787, 851–853, 883, 911*, 941, 979, 1014, 1032*, 1037, 1049, 1084, 1095, 1111, 1307, 1331, 1338, 1350, 1376*, 1391, 1413, 1432, 1448, 1462, 1483
2-Amino-6(5*H*)-pteridinone	anal, ir, uv	686, 1052, 1539
2-Amino-7(8*H*)-pteridinone	>350, uv	381, 383, 581, 979, 1350
3-Amino-4(3*H*)-pteridinone	242–245, uv	487
4-Amino-2(1*H*)-pteridinone (isopterin)	>360, uv	32, 59, 362, 471, 979, 1350
4-Amino-6(5*H*)-pteridinone	H_2O >300 or >350, uv	208, 445
4-Amino-7(8*H*)-pteridinone	>350, uv	224, 445, 979, 1350, 1386
6-Amino-7(8*H*)-pteridinone	>250, uv	36, 1093
7-Amino-4(3*H*)-pteridinone	>360, uv	131
7-Amino-6(5*H*)-pteridinone	315	19, 36
2-Amino-4(3*H*)-pteridinone 8-oxide	>360, ms, uv	427, 477, 1301
"Aminopterin" (not simple: see text)		
2-Amino-8-ribityl-4,7(3*H*,8*H*)-pteridinedione	342–344	172
2-Amino-8-ribityl-4,6,7(3*H*,5*H*,8*H*)-pteridinetrione	330–335	172
2-Amino-8-sorbityl-4,7(3*H*,8*H*)-pteridinedione	320–330	172
2-Amino-8-sorbityl-4,6,7(3*H*,5*H*,8*H*)-pteridinetrione	331	172
2-Amino-4-styryl-6,7(5*H*,8*H*)-pteridinedione	380	421
2-Amino-7-styryl-4(3*H*)-pteridinone	crude, nmr, uv	252
2-Amino-6-styryl-4(3*H*)-pteridinone 8-oxide	>320, nmr, uv	478, 485
2-Amino-7-α-sulfoethyl-4,6(3*H*,5*H*)-pteridinedione	anal	504

APPENDIX TABLE (*Contd.*)

Compound	(°C), and so on	Reference(s)
2-Amino-7-sulfomethyl-4,6(3*H*,5*H*)-pteridinedione	anal	504
2-Amino-6-sulfo-oxy-4(3*H*)-pteridinone (previously: 2-Amino-4-oxo-3,4-dihydro-6-pteridinesulfonic acid)	H_2O anal, uv; Na anal	1431, cf. 453, 1051
2-Amino-6-α,β,γ,δ-tetrahydroxybutyl-4(3*H*)-pteridinone	anal	192, 193, 642, 937, 962, 1060*, 1143, 1219, 1378
2-Amino-6-[D-*arabino*]-α,β,γ,δ-tetrahydroxybutyl-4(3*H*)-pteridinone.	>220, uv	1060*, 1147, 1176, 1210(?), 1211, 1318, 1373
2-Amino-6-[D-*lyxo*]-α,β,γ,δ-tetrahydroxybutyl-4(3*H*)-pteridinone	>350	962, 1210(?)
2-Amino-6-[L-*xylo*]-α,β,γ,δ-tetrahydroxybutyl-4(3*H*)-pteridinone	anal, uv	1147
2-Amino-7-α,β,γ,δ-tetrahydroxybutyl-4(3*H*)-pteridinone	anal	193, 937, 1143, 1210(?), 1219, 1378
2-Amino-6-[D-*arabino*]-α,β,γ,δ-tetrahydroxybutyl-4(3*H*)-pteridinone 8-oxide	>300, uv	1318
4-Amino-2-thioxo-1,2-dihydro-6,7(5*H*,8*H*)-pteridinedione	anal, uv	916, 1152, 1369
4-Amino-2-thioxo-1,2-dihydro-7(8*H*)-pteridinone	H_2O >350	445
2-Amino-6-α,β,γ-trihydroxybutyl-4(3*H*)-pteridinone	anal	330
2-Amino-6-β,γ,δ-trihydroxybutyl-4(3*H*)-pteridinone	anal	193, 1378
2-Amino-7-β,γ,δ-trihydroxybutyl-4(3*H*)-pteridinone	anal	193, 1378
2-Amino-7-[D-*erythro*]-β,γ,δ-trihydroxybutyl-4(3*H*)-pteridinone	anal	1373
4-Amino-7-[L-*lyxo*]-α,β,γ-trihydroxybutyl-2(1*H*)-pteridinone	265, uv	1247
2-Amino-6-α,β,γ-trihydroxypropyl-4(3*H*)-pteridinone [see also specific stereoisomers to follow]	ir, nmr, uv	73, 204, 205, 262, 278, 330, 428, 429, 458, 463, 509, 653, 904, 1197, 1210, 1378
2-Amino-6-[D-*erythro*]-α,β,γ-trihydroxypropyl-4(3*H*)-pteridinone (neopterin)	anal, cd, ir, uv	330, 458, 463, 509, 653, 905*, 911*, 1111, 1210(?), 1215, 1216, 1422, 1439, 1496, 1563

APPENDIX TABLE (*Contd.*)

Compound	(°C), and so on	Reference(s)
2-Amino-6-[L-*erythro*]-α,β,γ-trihydroxypropyl-4(3*H*)-pteridinone (bufochrome)	>350, cd, uv	458, 653, 962, 1210(?), 1216, 1439, 1498, 1563
2-Amino-6-[D-*threo*]-α,β,γ-trihydroxypropyl-4(3*H*)-pteridinone	>230, cd, uv	1210(?), 1215, 1216, 1318, 1439, 1563
2-Amino-6-[L-*threo*]-α,β,γ-trihydroxypropyl-4(3*H*)-pteridinone (monapterin)	anal, cd, fl-sp, ir, uv, $\alpha_D^{30°}$	262, 509, 593, 1210(?), 1216, 1439, 1563
2-Amino-7-α,β,γ-trihydroxypropyl-4(3*H*)-pteridinone	—	1378
2-Amino-6-[D-*threo*]-α,β,γ-trihydroxypropyl-4(3*H*)-pteridinone 8-oxide	220–300, uv	1318
4-Amino-6,*N*,*N*-trimethyl-7-pteridinecarboxamide 5,8-dioxide	286, ir, nmr, uv	723
2-Amino-3,5,7-trimethyl-4,6(3*H*,5*H*)-pteridinedione	280, uv	348
2-Amino-3,6,8-trimethyl-4,7(3*H*,8*H*)-pteridinedione	>350, uv	383
2-Amino-3,5,8-trimethyl-4,6,7(3*H*,5*H*,8*H*)-pteridinetrione	>350, uv	384
2-Amino-1,6,7-trimethyl-4(1*H*)-pteridinone	anal, ir, uv	69, 166, 541
2-Amino-3,6,7-trimethyl-4(3*H*)-pteridinone	>370, nmr, pol, uv	166, 741, 762, 1014, 1189
2-Amino-6,7,8-trimethyl-4(8*H*)-pteridinone	>330, pol, nmr, uv; Na nmr; HCl 260; HI 270 or >360	130, 190, 359, 422, 660, 741, 762, 1014
4-Amino-1,6,7-trimethyl-2(1*H*)-pteridinone	308–310 or 314–316, uv	130, 362
2-Amino-3,6,7-trimethyl-4(3*H*)-pteridinone	ir	69
2-Amino-1,6,7-trimethyl-4(3*H*)-pteridinone 5-oxide	335, uv	541
2-Anilino-6-[L-*erythro*]-α,β-dihydroxypropyl-4(3*H*)-pteridinone	270–271, uv	458
2-Anilino-6,7-dimethyl-4-pteridinamine	258	1354
2-Anilino-6,7-dimethyl-4(3*H*)-pteridinone	317–320, uv	169
2-Anilino-6,7-diphenyl-4(3*H*)-pteridinethione	261–262	476
4-Anilino-6,7-diphenyl-2(1*H*)-pteridinethione	266–267	927
6-Anilinomethyl-2,4-pteridinediamine	anal	1102
6-Anilinomethyl-4(3*H*)-pteridinone	>250	121
2-Anilino-6-phenyl-4,7-pteridinediamine	320–322, uv	523
4-Anilino-6-phenyl-2,7-pteridinediamine	208–209, uv	523
4-Anilino-2(1*H*)-pteridinone	296–298, uv	362
2-Anilino-6-[D-*erythro*]-α,β,γ-trihydroxypropyl-4(3*H*)-pteridinone	>260, uv	458
2-Anilino-6-[L-*erythro*]-α,β,γ-trihydroxypropyl-4(3*H*)-pteridinone	>260, uv	458
2-Anilino-3,6,7-triphenyl-4(3*H*)-pteridinone	323–324	476
8-D-Arabinityl-6,7-dimethyl-2,4(3*H*,8*H*)-pteridinedione	H_2O 134–150, uv	537
8-L-Arabinityl-6,7-dimethyl-2,4(3*H*,8*H*)-pteridinedione	H_2O 220–224, uv	537
8-D-Arabinityl-6-methyl-2,4,7(1*H*,3*H*,8*H*)-pteridinetrione	—	295

APPENDIX TABLE (*Contd.*)

Compound	(°C), and so on	Reference(s)
8-L-Arabinityl-6-methyl-2,4,7(1*H*,3*H*,8*H*)-pteridinetrione	—	295
8-D-Arabinityl-2,4,7(1*H*,3*H*,8*H*)-pteridinetrione	—	295
8-L-Arabinityl-2,4,7(1*H*,3*H*,8*H*)-pteridinetrione	—	295
Asperopterin-A (not simple: see text)		
Asperopterin-B [see: 2-Amino-6-hydroxymethyl-8-methyl-4,7(3*H*,8*H*)-pteridinedione]		1214, 1296
Aurodrosopterin (not simple: see text)		
Azin-purin (see: pteridine)		1380
2-Benzamido-6(7)-hydroxymethyl-4(3*H*)-pteridinone	anal	776
2-Benzamido-4,6(3*H*,5*H*)-pteridinedione	270–272	1362
3-Benzenesulfonyloxy-2,4(1*H*,3*H*)-pteridinedione	249–251, ms, nmr	1289, 1445
6-Benzoyl-2-methyl-4,7-pteridinediamine	307, uv	328
6-Benzoyl-2-methylthio-4,7-pteridinediamine	335	328
6-Benzoyl-2-phenyl-4,7-pteridinediamine	327–328	328
6-Benzoyl-2-piperidino-4,7-pteridinediamine	300–301, uv	328
6-Benzoyl-4,7-pteridinediamine	291–293, uv	328
6-Benzoyl-2,4,7-pteridinetriamine	338–339, uv	328
4-Benzylamino-6,7-diphenyl-2-pteridinamine	237–238	1315
4-Benzylamino-6,7-diphenylpteridine	178–179	474
4-Benzylamino-6,7-diphenyl-2(1*H*)-pteridinone	298–300	927
4-Benzylamino-6-methyl-2,7(1*H*,8*H*)-pteridinedione	348–351, uv	363
6-Benzylamino-3-methyl-4(3*H*)-pteridinone	212–214	974
4-Benzylamino-7-phenylpteridine	231	1100
2-Benzylamino-6-phenyl-4,7-pteridinediamine	235–236, uv	523
7-Benzylamino-6-phenyl-2,4-pteridinediamine	HCl >320, uv	525
4-Benzylaminopteridine	160–161	280
6-Benzylaminopteridine	199	146
7-Benzylaminopteridine	202–203	146
4-Benzylamino-2(1*H*)-pteridinone	258–260, uv	362
2-Benzylamino-4-trifluoromethylpteridine	190, ms, nmr	159, 780
8-Benzyl-2-benzylimino-2,8-dihydropteridine	H_2O anal, uv	189
8-Benzyl-2-benzylimino-6,7-dimethyl-2,8-dihydropteridine	181–185, uv	189
8-Benzyl-4-benzyloxy-2-dimethylamino-7(8*H*)-pteridinone	176–178, nmr, uv	246
1-Benzyl-6,7-bistrifluoromethyl-2,4(1*H*,3*H*)-pteridinedione	246–247, ir, ms, nmr, uv	1570
8-Benzyl-6,7-bistrifluoromethyl-2,4(3*H*,8*H*)-pteridinedione	210–211, ir, ms, nmr, uv	1570
8-Benzyl-2-dimethylamino-4-methoxy-7(8*H*)-pteridinone	148–152, nmr, uv	246
1-Benzyl-6,7-dimethyl-2,4(1*H*,3*H*)-pteridinedione	235–236, ir, nmr, uv	1008
8-Benzyl-6,7-dimethyl-2,4(3*H*,8*H*)-pteridinedione	264, uv	378
6-Benzyl-1,3-dimethyl-2,4,7(1*H*,3*H*,8*H*)-pteridinetrione (?)	205, uv	707
7-Benzyl-1,3-dimethyl-2,4,6(1*H*,3*H*,5*H*)-pteridinetrione	205, uv	710
1-Benzyl-6,7-dimethyl-4(1*H*)-pteridinone	205–207, ir, nmr, uv	1008
8-Benzyl-6,7-dimethyl-2(8*H*)-pteridinone	240, uv	189
1-Benzyl-6,7-diphenyl-2,4(1*H*,3*H*)-pteridinedione	197–200, ir, nmr, uv	1008
3-Benzyl-6,7-diphenyl-2,4(1*H*,3*H*)-pteridinedione	194–195	474

APPENDIX TABLE (*Contd.*)

Compound	(°C), and so on	Reference(s)
1-Benzyl-6,7-diphenyl-4(1*H*)-pteridinone	269–270, ir, nmr, uv	1008
3-Benzyl-6,7-diphenyl-4(3*H*)-pteridinone	248	468, 474, 1316
8-Benzyl-3-ethyl-6,7-dimethyl-2,4(3*H*,8*H*)-pteridinedione	229, uv	1545
7-Benzyl-1-ethyl-3-methyl-2,4,6(1*H*,3*H*,5*H*)-pteridinetrione	209, uv	710
7-Benzyl-1-ethyl-2,4,6(1*H*,3*H*,5*H*)-pteridinetrione	268, uv	710
6-Benzyl-8-β-hydroxyethyl-2,4,7(1*H*,3*H*,8*H*)-pteridinetrione	236–260, nmr, uv	674
3-Benzylideneamino-4(3*H*)-pteridinone	203–204	487
3-Benzyl-2-isopropylamino-6,7-diphenyl-4(3*H*)-pteridinone	305–307	476
6-Benzyl-2-isopropylthio-4,7(3*H*,8*H*)-pteridinedione	>385, uv	93
7-Benzyl-2-isopropylthio-4,6(3*H*,5*H*)-pteridinedione	>385	990
6-Benzyl-7-methoxy-3-methyl-2-methylthio-4(3*H*)-pteridinone	193	93
6-Benzyl-7-methoxy-2-methylthio-4-pteridinamine	225–226	93
7-Benzyl-4-methoxy-2-methylthio-6(5*H*)-pteridinone	234–235, uv	92
8-Benzyl-3-methyl-6,7-diphenyl-2,4(3*H*,8*H*)-pteridinedione	250–252, uv	1545
6-Benzyl-3-methyl-1-phenyl-2,4,7(1*H*,3*H*,8*H*)-pteridinetrione	168–170, uv	708
7-Benzyl-3-methyl-1-phenyl-2,4,6(1*H*,3*H*,5*H*)-pteridinetrione	>300, uv	710
6-Benzyl-3-methyl-1-phenyl-2-thioxo-1,2-dihydro-4,7(3*H*,8*H*)-pteridinedione	crude 323	92
7-Benzyl-3-methyl-1-phenyl-2-thioxo-1,2-dihydro-4,6(3*H*,5*H*)-pteridinedione	302	990
8-Benzyl-3-methyl-2,4(3*H*,8*H*)-pteridinedione	217, uv	1545
6-Benzyl-1-methyl-2,4,7(1*H*,3*H*,8*H*)-pteridinetrione (?)	>300, uv	707
7-Benzyl-1-methyl-2,4,6(1*H*,3*H*,5*H*)-pteridinetrione	>300, uv	710, 990
8-Benzyl-6-methyl-2,4,7(1*H*,3*H*,8*H*)-pteridinetrione	301–302, uv	378
3-Benzyl-2-methyl-4(3*H*)-pteridinone	—	980
6-Benzyl-7-methyl-8-D-ribityl-2,4(3*H*,8*H*)-pteridinedione	236–238, nmr, uv	674
7-Benzyl-2-methylthio-4,6(3*H*,5*H*)-pteridinedione	322–324, uv	92
6-Benzyl-1-methyl-2-thioxo-1,2-dihydro-4,7(3*H*,8*H*)-pteridinedione	crude 286	92
7-Benzyl-1-methyl-2-thioxo-1,2-dihydro-4,6(3*H*,5*H*)-pteridinedione	286	990
8-Benzyl-3-methyl-2,4,7-trioxo-1,2,3,4,7,8-hexahydro-6-pteridinecarboxylic acid	188–190, uv	317
4-Benzyloxy-6-diethoxymethyl-2-pteridinamine	anal, uv	1403
4-Benzyloxy-2-dimethylamino-7(8*H*)-pteridinone	250–252, uv	246
4-Benzyloxy-6,7-dimethyl-2-pteridinamine	235–236, uv	370
4-Benzyloxy-2-pteridinamine	220–223, nmr, uv	370, 741
1-Benzyloxy-2,4(1*H*,3*H*)-pteridinedione	212, uv	1098
4-Benzyloxy-7-trimethylsiloxy-2-trimethylsilylamino-pteridine	crude	435
2-Benzyl-6-phenyl-4,7-pteridinediamine	296–299, nmr, uv	523
6-Benzyl-2-phenyl-4,7-pteridinediamine	280–281, uv	326, 1415
7-Benzyl-1-phenyl-2,4,6(1*H*,3*H*,5*H*)-pteridinetrione	> 300	710, 990
6-Benzyl-1-phenyl-2-thioxo-1,2-dihydro-4,7(3*H*,8*H*)-pteridinedione	crude 317	92

APPENDIX TABLE (*Contd.*)

Compound	(°C), and so on	Reference(s)
7-Benzyl-1-phenyl-2-thioxo-1,2-dihydro-4,6(3*H*,5*H*)-pteridinedione	> 300	810, 990
2-Benzyl-4-pteridinamine	188–189, uv	185
6-Benzyl-2,4-pteridinediamine	287–299	742
8-Benzyl-2,7(1*H*,8*H*)-pteridinedione	238–240	189
1-Benzyl-2,4(1*H*,3*H*)-pteridinedione	235–237, ir, nmr, uv	1008
6-Benzyl-2,4,7-pteridinetriamine	332, uv	1415
6-Benzyl-2,4,7(1*H*,3*H*,8*H*)-pteridinetrione (?)	400, uv	707
7-Benzyl-2,4,6(1*H*,3*H*,5*H*)-pteridinetrione	> 300, uv	710, 990
8-Benzyl-2,4,7(1*H*,3*H*,8*H*)-pteridinetrione	288, uv	317
1-Benzyl-4(3*H*)-pteridinone	214–215, ir, nmr, uv	1008
8-Benzyl-2(8*H*)-pteridinone	H_2O 240, uv	189
8-Benzyl-7(8*H*)-pteridinone	111–112, uv	125
6-Benzyl-8-D-ribityl-2,4,7(1*H*,3*H*,8*H*)-pteridinetrione	258–259	674
4-Benzylthio-2-pteridinamine	179 or 182–183, ms, nmr, uv	305, 1101
6-Benzylthiopteridine	98–100, uv	146
7-Benzylthiopteridine	112, uv	146
6-Benzylthio-2,4-pteridinediamine	> 300	1307
6-Benzylthio-4(3*H*)-pteridinone	231–233	974
6-Benzyl-2-thioxo-1,2-dihydro-4,7(3*H*,8*H*)-pteridinedione	287	92
7-Benzyl-2-thioxo-1,2-dihydro-4,6(3*H*,5*H*)-pteridinedione	300	990
8-Benzyl-3,6,7-trimethyl-2,4(3*H*,8*H*)-pteridinedione	237, uv	1545
8-Benzyl-2,4,7-trioxo-1,2,3,4,7,8-hexahydro-6-pteridinecarboxylic acid	268, uv	317
3-Benzyl-2,6,7-triphenyl-4(3*H*)-pteridinone	227–228	476
Biolumazine [see: 6-[L-*erythro*]-α,β-Dihydroxypropyl-2,4(1*H*,3*H*)-pteridinedione]		1439
Biopterin [see: 2-Amino-6-[L-*erythro*]-α,β-dihydroxypropyl-4(3*H*)-pteridinone]		1557
Biopterin (1′-α-D) glucoside (not simple)		1562
6,7-Bisacetoxymethyl-3-methyl-2,4(1*H*,3*H*)-pteridinedione	240, uv	367
2,4-Bisbenzylamino-6,7-diphenylpteridine	220–221 or 226	467, 1315
6,7-Bisbromomethyl-1-methyl-2,4(1*H*,3*H*)-pteridinedione	197–198	232
6,7-Bisbromomethyl-2,4-pteridinediamine	HBr anal, uv	850
2,4-Bisbutylamino-6,7-dimethylpteridine	169–170, uv	169
1,3-Bis-β-carbamoylethyl-6,7-dimethyl-2,4(1*H*,3*H*)-pteridinedione	265–267, uv	168
1,3-Bis-β-cyanoethyl-6,7-dimethyl-2,4(1*H*,3*H*)-pteridinedione	134–137, ir, uv	168
1,3-Bis-β-cyanoethyl-2,4-dioxo-1,2,3,4-tetrahydro-6-pteridinecarboxylic acid	anal, uv	168
1,3-Bis-β-cyanoethyl-6-methyl-2,4(1*H*,3*H*)-pteridinedione	143–147, ir, uv	168
6,7-Bisdibromomethyl-1-methyl-2,4(1*H*,3*H*)-pteridinedione	254–256	232
2,4-Bis-β-diethylaminoethylamino-6,7-di-isopropylpteridine	95–97	400
2,4-Bis-β-diethylaminoethylamino-6,7-diphenylpteridine	125–126	400
2,4-Bis-β-dimethylaminoethylamino-6,7-di-isopropylpteridine	106–107	400

APPENDIX TABLE (*Contd.*)

Compound	(°C), and so on	Reference(s)
2,4-Bis-β-dimethylaminoethylamino-6,7-diphenylpteridine	140–141	400
2,4-Bisdimethylamino-6,7-di-isopropylpteridine	150	110
2,4-Bisdimethylamino-6,7-dimethylpteridine	165–166	1315
2,4-Bisdimethylamino-6,7-diphenylpteridine	211	110; cf. 472
6,7-Bisdimethylamino-2-phenyl-4-pteridinamine	anal	1398
2,4-Bisdimethylamino-6-phenylpteridine	188	110
2,4-Bisdimethylamino-7-phenylpteridine	191	110
2,4-Bis-γ-dimethylaminopropylamino-6,7-diisopropylpteridine	152–153	400
2,4-Bis-γ-dimethylaminopropylamino-6,7-diphenylpteridine	137–138 or 143–144	400, 467
2,4-Bisdimethylaminopteridine	169–171, uv	267
2,4-Bisdimethylamino-6(5*H*)-pteridinone	238–240, uv	268
2,4-Bisdimethylamino-7(8*H*)-pteridinone	286–287, uv	268
2,4-Bisethylaminopteridine	123–124, nmr	934
6,7-Bis-α-ethylpropyl-2,4-pteridinediamine	181	749
2,4-Bisethylthio-6,7-diphenylpteridine	157	699
2,4-Bis-γ-isopropylaminopropylamino-6,7-diphenylpteridine	141–142	467
2,4-Bismethylamino-6,7-diphenylpteridine	262 or 264–265	110; cf. 472
2,4-Bismethylamino-6-phenylpteridine	264	110
2,4-Bismethylamino-7-phenylpteridine	256, uv	110
2,4-Bismethylaminopteridine	214	110
2,4-Bismethylthio-6,7-diphenylpteridine	216–218	699
6,7-Bistrifluoromethylpteridine	$2H_2O$ > 250, ir, ms, nmr, uv	1571
6,7-Bistrifluoromethyl-2,4(1*H*,3*H*)-pteridinedione	> 300, ir, ms, nmr, uv	1571
6,7-Bistrifluoromethyl-4(3*H*)-pteridinone	> 275; $2H_2O$ > 270, ir, ms, nmr, uv	1571
4,6-Bistrimethylsiloxy-2,7-bistrimethylsilylaminopteridine	crude, ms	1457, 1754
6-β,γ-Bistrimethylsiloxypropyl-2,4(1*H*,3*H*)-pteridinedione	crude, ms	589
2,4-Bistrimethylsiloxypteridine	95, ms	283, 414, 1457
2,4-Bistrimethylsiloxy-6-trimethylsiloxycarbonylpteridine	crude, ms	1457
4,6-Bistrimethylsiloxy-7-trimethylsiloxycarbonyl-2-trimethylsilylaminopteridine	crude, ms	1457
2,4-Bistrimethylsiloxy-6-trimethylsiloxysulfonylpteridine	crude, ms	1242
4,6-Bistrimethylsiloxy-2-trimethylsilylaminopteridine	ms	589, 1249, 1457
4,7-Bistrimethylsiloxy-2-trimethylsilylaminopteridine	155–170/0.001, anal, ms	435, 436, 1457
6-Bromo-1,3-dimethyl-2,4(1*H*,3*H*)-pteridinedione	107–108	1275
7-Bromo-1,3-dimethyl-2,4(1*H*,3*H*)-pteridinedione	209–210	1275
7-Bromomethyl-1-methyl-2,4(1*H*,3*H*)-pteridinedione	236–237	232
6-Bromomethyl-2,4-pteridinediamine	HBr anal, nmr, uv	390, 595, 975, 1137
6-Bromomethyl-4(3*H*)-pteridinone	HBr > 200	121

APPENDIX TABLE (*Contd.*)

Compound	(°C), and so on	Reference(s)
6-Bromo-2-methyl-3-phenyl-4(3*H*)-pteridinone	195	187
6-Bromo-2-methyl-4(3*H*)-pteridinone	dec, nmr	20
7-Bromomethyl-4(3*H*)-pteridinone	> 170	121
6-Bromo-2-phenyl-4(3*H*)-pteridinone	277, nmr	20
6-Bromo-2,4-pteridinediamine	> 300	729
Bufochrome [see: 2-amino-6-[L-*erythro*]-α,β,γ-trihydroxypropyl-4(3*H*)-pteridinone]		
6-Butoxymethyl-2,4-pteridinediamine	242, ir, nmr	1728
6-*t*-Butoxymethyl-2,4-pteridinediamine	286–287, ir, nmr	1728
6-Butoxymethyl-2,4-pteridinediamine 8-oxide	> 250, ir, nmr, uv	1728
4-Butoxy-6-phenyl-2,7-pteridinediamine	221–222, nmr, uv	523
6-Butoxy-2-trifluoromethyl-4(3*H*)-pteridinone	229, ir	20
2-Butylamino-6,7-dimethyl-4(3*H*)-pteridinone	238–241, uv	169
4-Butylamino-6,7-diphenylpteridine	150–151	474
4-*t*-Butylamino-6,7-diphenylpteridine	204–206, ir, ms, nmr	1167
4-Butylamino-3-methyl-6,7-diphenyl-2(3*H*)-pteridinone	251, nmr, uv	1675
4-*t*-Butylamino-7-phenylpteridine	135–137, ir, ms, nmr	1167
2-Butylamino-7-propionyl-4-pteridinamine	152, nmr, uv	1678
2-Butylamino-4-pteridinamine	190, nmr, uv	1678
2-Butylaminopteridine	123, uv	124
2-*s*-Butylaminopteridine	136, uv	124
2-*t*-Butylaminopteridine	170, uv	124
2-*s*-Butylamino-4,6,7-trimethylpteridine	141, uv	124
7-*t*-Butyl-4-t-butylaminopteridine	105–106, ir, ms, nmr	1167
4-*t*-Butyl-2-chloro-6-phenylpteridine	crude, ^{13}C nmr	959
4-*t*-Butyl-2-chloro-7-phenylpteridine	174–176, ^{13}C nmr	959
6-Butyl-1,3-dimethyl-7-phenyl-2,4(1*H*,3*H*)-pteridinedione	159, nmr, uv	1348
7-*t*-Butyl-1,3-dimethyl-2,4(1*H*,3*H*)-pteridinedione	151–152	108
8-*s*-Butyl-6,7-dimethyl-2,4(3*H*,8*H*)-pteridinedione	242, uv	409
3-Butyl-6,7-diphenyl-2,4(1*H*,3*H*)-pteridinedione	246–247	474
3-Butyl-6,7-diphenyl-4(3*H*)-pteridinethione	193–195	474
3-Butyl-6,7-diphenyl-4(3*H*)-pteridinimine (?)	149–151	474
3-Butyl-6,7-diphenyl-4(3*H*)-pteridinone	194–195	474
3-Butyl-6,7-diphenyl-4-thioxo-3,4-dihydro-2(1*H*)-pteridinone	205–209	474
4-Butylimino-1,3-dimethyl-6,7-diphenyl-3,4-dihydro-2(1*H*)-pteridinone	101–102, nmr, uv	1679
4-*t*-Butyl-2-methoxy-6-phenylpteridine	crude, ^{13}C nmr	959
4-*t*-Butyl-2-methoxy-7-phenylpteridine	142–144, ^{13}C nmr	959
7-*t*-Butyl-3-methyl-2,4(1*H*,3*H*)-pteridinedione	242, uv	367
8-*s*-Butyl-3-methyl-2,4(3*H*,8*H*)-pteridinedione	215, uv	409
7-*t*-Butyl-3-methyl-2,4(1*H*,3*H*)-pteridinedione 5-oxide	270, uv	367
3-Butyl-6-phenyl-4(3*H*)-pteridinone	103–104, nmr	1758
3-Butyl-6-phenyl-4(3*H*)-pteridinone 8-oxide	230–232, nmr	1758
3-*s*-Butyl-6-phenyl-4(3*H*)-pteridinone 8-oxide	231–232, nmr	1758
2-Butyl-4-pteridinamine	136–137, uv	185

APPENDIX TABLE (*Contd.*)

Compound	(°C), and so on	Reference(s)
7-*t*-Butyl-4-pteridinamine	265–266, nmr	1740
6-Butyl-2,4-pteridinediamine	277–285	742
1-Butyl-2,4(1*H*,3*H*)-pteridinedione	174–176, ms, uv	876
3-Butyl-2,4(1*H*,3*H*)-pteridinedione	214–215, ms, uv	876
7-*t*-Butyl-4(3*H*)-pteridinone	> 300, ir, ms, nmr, uv	1167, 1286, 1740
7-Butyryl-1,3-dimethyl-2,4(1*H*,3*H*)-pteridinedione	111–112, nmr, uv	108
7-Butyryl-2-methylthio-4-pteridinamine	225–227, nmr, uv	1678
7-Butyryl-2,4-pteridinediamine	223–224, nmr, uv	1678
6-Butyryl-1,3,7-trimethyl-2,4(1*H*,3*H*)-pteridinedione	98–99, nmr, uv	108
8-*s*-Butyl-3,6,7-trimethyl-2,4(3*H*,8*H*)-pteridinedione	203–204, uv	409
3-*β*-Carbamoylethyl-6,7-dimethyl-2-methylthio-4(3*H*)-pteridinone	261–264, uv	168
2-Carbamoylmethyl-6-phenyl-4,7-pteridinediamine	292–293, nmr, uv	523
6-*β*-Carboxyethyl-1,3-dimethyl-2,4,7(1*H*,3*H*,8*H*)-pteridinetrione	247–248	674
6-*β*-Carboxyethyl-3,8-dimethyl-2,4,7(1*H*,3*H*,8*H*)-pteridinetrione	299–304	674
6-*β*-Carboxyethyl-8-*β*-hydroxyethyl-1,3-dimethyl-2,4,7(1*H*,3*H*,8*H*)-pteridinetrione	195–196	674
6-*β*-Carboxyethyl-8-*β*-hydroxyethyl-2,4,7(1*H*,3*H*,8*H*)-pteridinetrione	262–265 or 305–310, ms, nmr, uv	1218, 1248, 1449
6-*α*-Carboxyethyl-2,4,7(1*H*,3*H*,8*H*)-pteridinetrione	anal	930, 1083
6-*β*-Carboxyethyl-2,4,7(1*H*,3*H*,8*H*)-pteridinetrione	259–263, uv	1218, 1248
6-*β*-Carboxyethyl-8-D-ribityl-2,4,7(1*H*,3*H*,8*H*)-pteridinetrione (putidolumazine)	155–158 or > 267, uv	463, 674, 1218, 1248
6-*β*-Carboxyethyl-1,3,8-trimethyl-2,4,7(1*H*,3*H*,8*H*)-pteridinetrione	249–253, nmr, uv	674
6-Carboxymethyl-1,3-dimethyl-2,4,7(1*H*,3*H*,8*H*)-pteridinetrione	305–308	337
6-Carboxymethyl-4,7(3*H*,8*H*)-pteridinedione	< 100	29
8-Carboxymethyl-2,4,6,7(1*H*,3*H*,5*H*,8*H*)-pteridinetetrone	> 300	674
6-Carboxymethyl-2,4,7(1*H*,3*H*,8*H*)-pteridinetrione	anal	1304
2-Carboxymethylthio-7-methyl-4-pteridinamine	262, uv	1247
2-Carboxymethylthio-7-phenyl-4-pteridinamine	254, uv	1247
2-Carboxymethylthio-7-[L-*lyxo*]-*α*,*β*,*γ*-trihydroxybutyl-4-pteridinamine	208, uv	1247
8-*ε*-Carboxypentyl-2,4,6,7(1*H*,3*H*,5*H*,8*H*)-pteridinetetrone	> 330	674
7-(2′-Carboxyprop-l′-enyl)-1,3-dimethyl-2,4,6(1*H*,3*H*,5*H*)-pteridinetrione	260, uv	337
7-(2′-Carboxyprop-1′-enyl)-6(5*H*)-pteridinone	240, uv	55
6-*α*-Carboxypropyl-2,4,7(1*H*,3*H*,8*H*)-pteridinetrione	crude	930, 1083
7-*β*-Carboxyvinyl-1,3-dimethyl-2,4,7(1*H*,3*H*,8*H*)-pteridinetrione	260–261, uv	337
7-Chloro-2,4-bisdimethylaminomethyleneamino-6-pteridinecarbonitrile	—	469
7-Chloro-4-dichloromethylpteridine	84–85, uv	64

APPENDIX TABLE (*Contd.*)

Compound	(°C), and so on	Reference(s)
6-Chloro-7-diethylamino-2,4-pteridinediamine	268–271	729
7-Chloro-2-dimethylaminopteridine	170–172, uv	381
6-Chloro-7-dimethylamino-2,4-pteridinediamine	263	729
6-Chloro-1,3-dimethyl-2,4-dioxo-1,2,3,4-tetrahydro-7-pteridinesulfonic acid	Na: >300, uv	1681
2-Chloro-6,7-dimethyl-4-phenylpteridine	185–186, nmr	1436
6-Chloro-1,3-dimethyl-7-phenyl-2,4(1*H*,3*H*)-pteridinedione	203–205	1279
6-Chloro-1,3-dimethyl-7-propionyl-2,4(1*H*,3*H*)-pteridinedione	157, nmr, uv; HON=173	108, 1759
2-Chloro-6,7-dimethyl-4-pteridinamine	>300, uv	170
2-Chloro-4,6-dimethylpteridine	uv	281
2-Chloro-6,7-dimethylpteridine	135–137, uv	170
4-Chloro-6,7-dimethylpteridine	149–151	170
6-Chloro-1,3-dimethyl-2,4(1*H*,3*H*)-pteridinedione	154	1275
7-Chloro-1,3-dimethyl-2,4(1*H*,3*H*)-pteridinedione	184	1275
6-Chloro-1,3-dimethyl-2,4,7(1*H*,3*H*,8*H*)-pteridinetrione	>200, nmr, uv	1681
4-Chloro-6,8-dimethyl-7(8*H*)-pteridinone	155–156, uv	248
4-Chloro-6,7-dimethyl-2-styrylpteridine	208, nmr, uv	105
6-Chloro-1,3-dimethyl-7-thioxo-7,8-dihydro-2,4(1*H*,3*H*)-pteridinedione	+ pyridine: >300, nmr, uv	1681
4-Chloro-6,7-diphenyl-2-pteridinamine	HCl: crude	141
6-Chloro-2,7-diphenyl-4-pteridinamine	257–258, uv	526
7-Chloro-2,6-diphenyl-4-pteridinamine	274–276	526
2-Chloro-4,7-diphenylpteridine	198–199, ^{13}C nmr	959
2-Chloro-4-ethylaminopteridine	141–142, nmr	934
6-Chloro-7-ethylamino-2,4-pteridinediamine	256–259	729
2-Chloro-4-ethyl-6,7-diphenylpteridine	179–181	415
6-Chloro-7-isopropoxy-2,4-pteridinediamine	238–240	729
6-Chloro-7-isopropylamino-2,4-pteridinediamine	233–235	729
4-Chloro-2-methylamino-6,7-diphenylpteridine	crude	1189
6-Chloro-4-methylaminopteridine	221, ir, nmr, uv	53
6-Chloro-3-methyl-7-phenyl-2,4(1*H*,3*H*)-pteridinedione	258–260, ms, nmr, uv	331
7-Chloro-3-methyl-6-phenyl-2,4(1*H*,3*H*)-pteridinedione	243, ms, nmr, uv	331
6-Chloro-3-methyl-7-phenyl-2,4(1*H*,3*H*)-pteridinedione 8-oxide	321–324, ms, nmr, uv	331
2-Chloro-4-methylpteridine	155–157, ^{13}C nmr, uv	281, 959
7-Chloro-6-methylpteridine	103	55
6-Chloromethyl-2,4-pteridinediamine	—	1522, 1757
7-Chloro-1-methyl-2,4(1*H*,3*H*)-pteridinedione	284–286, nmr, uv	354, 1709
7-Chloro-3-methyl-2,4(1*H*,3*H*)-pteridinedione	308–310, nmr, uv	354, 1709
7-Chloro-4-pentyloxy-2-pteridinamine	168, nmr, uv	1681
4-Chloro-8-methyl-7(8*H*)-pteridinone	177, uv	248

APPENDIX TABLE (*Contd.*)

Compound	(°C), and so on	Reference(s)
6-Chloro-3-methyl-4(3*H*)-pteridinone	217–219	974
2-Chloro-4-phenylpteridine	164–165, nmr, ^{13}C nmr	959, 1138
7-Chloro-6-phenyl-2,4-pteridinediamine	300, uv; HCl >320	525
2-Chloro-4-phenyl-6,7(5*H*,8*H*)-pteridinedione	>300	522
6-Chloro-2-pivalamido-4(3*H*)-pteridinone	272–273, ir, nmr	1753
2-Chloro-4-pteridinamine	245–249, nmr	934
4-Chloro-2-pteridinamine	crude	668
4-Chloro-7-pteridinamine	195, uv	131
6-Chloro-4-pteridinamine	186	53
6-Chloro-7-pteridinamine	>250	36
7-Chloro-4-pteridinamine	>60 or 204, uv	131, 445
2-Chloropteridine	106–107, ir, nmr, ^{13}C nmr, uv	27, 30, 32, 119, 124, 285, 286, 959, 1138
4-Chloropteridine	140, ir, uv	32, 119, 285, 286
6-Chloropteridine	146–147 or 149–150, ir, nmr, uv	27, 31, 35, 119, 146, 285, 286
7-Chloropteridine	95 or 100–101, ir, nmr, uv	27, 33, 35, 119, 146, 285, 286
6-Chloro-2,4-pteridinediamine	295 or >330	729, 1307
6-Chloro-2,4,7-pteridinetriamine	>310	729
6-Chloro-4(3*H*)-pteridinone	268–270	974
6-Chloro-7(8*H*)-pteridinone	232	36
2-Chloro-4-trifluoromethylpteridine	128–129, ms, nmr, uv; H_2O 145	159, 780
2-Chloro-4,6,7-trimethylpteridine	129	124
7-Chloro-1,3,6-trimethyl-2,4(1*H*,3*H*)-pteridinedione	161, nmr, uv	1709
2-Chloro-4,6,7-triphenylpteridine	209–210	1118, 1120*
Chrysopterin [see: 2-amino-7-methyl-4,6(3*H*,5*H*)-pteridinedione]		1314
Ciliapterin [see: 2-amino-6-[L-*threo*]-α,β-dihydroxypropyl-4(3*H*)-pteridinone]		1477
Compound-C (see: biopterin glucoside)		1560–1562
Compound-V [see: 6-methyl-8-D-ribityl-2,4,7(1*H*,3*H*,8*H*)-pteridinetrione]		
4-Cyanoaminopteridine	>210	135
3-β-Cyanoethyl-7-methyl-4(3*H*)-pteridinone	172–173, uv	167
2-Cyanomethyl-6-phenyl-4,7-pteridinediamine	314–315, nmr, uv	523
6-Cyclohex-1′-enyl-2,4,7-pteridinetriamine	>350	682

APPENDIX TABLE (*Contd.*)

Compound	(°C), and so on	Reference(s)
6-Cyclohex-3′-enyl-2,4,7-pteridinetriamine	>350, uv	1415
4-Cyclohexylamino-7-dimethylamino-6-phenylpteridine	175–176, nmr	764
4-Cyclohexylamino-7-dimethylamino-6-phenylpteridine 5-oxide	222–223, nmr	764
6-Cyclohexylaminomethyl-2,4-pteridinediamine	anal	1102
4-Cyclohexylaminopteridine	105–106	280
6-Cyclohexylaminopteridine	170–171	146
7-Cyclohexylaminopteridine	252–254	146
8-Cyclohexyl-2-dimethylamino-6-methyl-7(8*H*)-pteridinone	162–164, uv	1150
1-Cyclohexyl-6,7-diphenyl-4(1*H*)-pteridinone	286–287, ir, nmr, uv	1008
3-Cyclohexyl-2-ethyl-4(3*H*)-pteridinone	148–150	187
8-Cyclohexyl-3-methyl-2,4(3*H*,8*H*)-pteridinedione	275, uv	409
3-Cyclohexyl-2-methyl-4(3*H*)-pteridinone	203–204	187
6-Cyclohexyl-2-phenyl-4,7-pteridinediamine	338–340, uv	326
8-Cyclohexyl-2,4(3*H*,8*H*)-pteridinedione	292, uv	409
1-Cyclohexyl-4(1*H*)-pteridinone	270–271, ir, nmr, uv	1008
6-Cyclopent-1′-enyl-2,4,7-pteridinetriamine	265–270	682
7-Cyclopentylamino-6-phenyl-2,4-pteridinediamine	267–269, uv	525
3-Cyclopentyl-2-methyl-4(3*H*)-pteridinone	165–167	187
8-Cyclopropyl-3-methyl-2,4(3*H*,8*H*)-pteridinedione	285, uv	409
6-Cyclopropyl-2,4-pteridinediamine	>300	729
Cyprino-purple-B (see: 2-amino-4,7-dioxo-3,4,7,8-tetrahydro-6-pteridinecarboxylic acid)		
Cyprino-purple-C1 [see: 6-β-acetoxy-α-hydroxypropyl-2-amino-4,7(3*H*,8*H*)-pteridinedione]		
Cyprino-purple-C2 [see: 6-α-acetoxy-β-hydroxypropyl-2-amino-4,7(3*H*,8*H*)-pteridinedione]		
Deoxysepiapterin (= isosepiapterin; not simple)		95, 350
8-(2′-Deoxy-D-ribityl)-6,7-dimethyl-2,4(3*H*,8*H*)-pteridinedione	uv	713
8-(5′-Deoxy-D-ribityl)-6,7-dimethyl-2,4(3*H*,8*H*)-pteridinedione	267–268, uv	1444
8-(3′-Deoxy-D-ribityl)-6,7-dimethyl-2,4(3*H*,8*H*)-pteridinedione	uv	713
2,4-Diacetamido-6-acetoxymethylpteridine	205, nmr, uv	113
2,4-Diacetamido-6-dibromomethylpteridine	>300	123
2,4-Diacetamido-6,7-dimethylpteridine	anal, ms, uv	648, 1535
2,4-Diacetamido-6,7-diphenylpteridine	>190, uv	141
4,7-Diacetamido-2-ethylthio-6-phenylpteridine	209–210	1281
2,4-Diacetamido-6-methylpteridine	234–237 (?) or 257, uv	123; cf. 1263
2,4-Diacetamido-7-methylpteridine	236–237	1263
4,7-Diacetamido-2-methylthio-6-phenylpteridine	230	1281
4,7-Diacetamido-2-methylthio-6-pteridinecarboxamide	anal	324
2,4-Diacetamido-6-phenethylpteridine	179–181, ir, uv	90
2,4-Diacetamido-6-phenylpteridine	283–286	541

APPENDIX TABLE (*Contd.*)

Compound	(°C), and so on	Reference(s)
4,7-Diacetamido-2-phenyl-6-pteridinecarboxamide	>300	1357
2,4-Diacetamido-6-phenyl-7(8*H*)-pteridinone	330–340	541
2,4-Diacetamido-7-propionylpteridine	225–228, nmr, uv	1678
2,4-Diacetamidopteridine	235 or 237–239, ms, uv	123, 267, 1535
4,7-Diamino-*N*-(β-aminoethyl)-2-phenyl-6-pteridinecarboxamide	265–270, uv	1357
4,7-Diamino-*N*-(β-amino-β-methylpropyl)-2-phenyl-6-pteridinecarboxamide	>300, uv	1357
4,7-Diamino-*N*-(β-aminopropyl)-2-phenyl-6-pteridinecarboxamide	280–283, uv	1357
2,7-Diamino-4-anilino-6-pteridinecarboxamide	>300, uv	1357
4,7-Diamino-2-anilino-6-pteridinecarboxamide	>300, uv	1357
4,7-Diamino-*N*-benzyl-2-phenyl-6-pteridinecarboxamide	314–318, uv	1357
4,7-Diamino-2-benzyl-6-pteridinecarboxamide	334–335	1357
2,4-Diamino-6-carboxymethyl-7(8*H*)-pteridinone	anal	506, 850
2,4-Diamino-6-chloromethyl-7(3*H*)-pteridinone 8-oxide	HCl >300, nmr, uv	1296
4,7-Diamino-*N*-cyclohexyl-2-phenyl-6-pteridinecarboxamide	350	321
4,7-Diamino-*N*-cyclopentyl-2-phenyl-6-pteridinecarboxamide	344	321
4,7-Diamino-*N*-cyclopropyl-2-phenyl-6-pteridinecarboxamide	342, pol	321, 1113
2,4-Diamino-6-α,β-diethoxycarbonylethyl-7(8*H*)-pteridinone	>330, uv	1195
4,7-Diamino-*N*-(β-diethylaminoethyl)-2-phenyl-6-pteridinecarboxamide	262–265, uv	1357
4,7-Diamino-*N*-(β-dimethylaminoethyl)-2-phenyl-6-pteridinecarboxamide	264–266, uv	1357
4,7-Diamino-*N*-(γ-dimethylaminopropyl)-2-phenyl-6-pteridinecarboxamide	275–282, uv	1357
2,7-Diamino-4-dimethylamino-6-pteridinecarboxamide	>300, uv	1357
4,7-Diamino-2-dimethylamino-6-pteridinecarboxamide	>300, uv	1357
4,7-Diamino-*N*,*N*-dimethyl-2-phenyl-6-pteridinecarboxamide	320, uv	321, 1357, 1466
4,7-Diamino-2,*N*-diphenyl-6-pteridinecarboxamide	>300, uv	1357
2,4-Diamino-6-ethoxycarbonylmethyl-7(8*H*)-pteridinone	>330, uv	247, 850, 1195
4,7-Diamino-*N*-ethyl-2-phenyl-6-pteridinecarboxamide	336–338, 346, pol, uv	321, 1113, 1357, 1466
2,4-Diamino-6-ethyl-7(3*H*)-pteridinone 8-oxide	>300, nmr, uv	1298
2,4-Diamino-6-hydroxymethyl-8-methyl-7(8*H*)-pteridinone	>300, nmr, uv	1296
2,4-Diamino-6-hydroxymethyl-7(8*H*)-pteridinone	anal, uv	1296
2,4-Diamino-6-isobutyl-7(3*H*)-pteridinone 8-oxide	>300, nmr, uv	1298
2,7-Diamino-4-isopropoxy-6-pteridinecarbonitrile	220, uv	371
4,7-Diamino-*N*-isopropyl-2-phenyl-6-pteridinecarboxamide	347–349, uv	1357, 1466
2,4-Diamino-6-isopropyl-7(3*H*)-pteridinone 8-oxide	291, nmr, uv	1298
2,4-Diamino-6-methoxycarbonylmethyl-7(8*H*)-pteridinone	>300	247
2,7-Diamino-4-methylamino-6-pteridinecarboxamide	>300, uv	1357

APPENDIX TABLE (*Contd.*)

Compound	(°C), and so on	Reference(s)
2,4-Diamino-8-methyl-7-oxo-7,8-dihydro-6-pteridine-carboxylic acid	anal, uv	882
4,7-Diamino-*N*-methyl-2-phenyl-6-pteridinecarboxamide	352, pol, uv	321, 1113, 1357, 1466
4,7-Diamino-2-methyl-6-pteridinecarboxamide	>300, uv	1357
2,4-Diamino-6-methyl-7-pteridinecarboxylic acid	anal, uv	884
2,4-Diamino-7-methyl-6-pteridinecarboxylic acid	250, uv	307, 847, 884
2,4-Diamino-6-methyl-7(8*H*)-pteridinone	>330, uv	506, 979, 1195, 1281
2,4-Diamino-7-methyl-6(5*H*)-pteridinone	anal, uv	874, 1281
2,4-Diamino-8-methyl-7(8*H*)-pteridinone	280–300, nmr, uv	1296
2,6-Diamino-7-methyl-4(3*H*)-pteridinone	>360, uv	279
2,4-Diamino-6-methyl-7(3*H*)-pteridinone 8-oxide	H_2O >300, nmr; $MeC_6H_4SO_3H$ 288	1298
4,7-Diamino-2-methylthio-6-pteridinecarboxamide	>300	324
4,7-Diamino-2-methylthio-6-pteridinecarboxylic acid	>300	324
4,7-Diamino-*N*-(*β*-morpholinoethyl)-2-phenyl-6-pteridinecarboxamide	277–279, uv	1357
4,7-Diamino-2-morpholino-6-pteridinecarboxamide	>360	321
2,4-Diamino-7-oxo-7,8-dihydro-6-pteridinecarboxylic acid	>360, uv	874, 1229, 1234
4,7-Diamino-*N*,*N*-pentamethylene-2-phenyl-6-pteridinecarboxamide	273	321
2,4-Diamino-6-pentyl-7(3*H*)-pteridinone 8-oxide	275, nmr, uv	1298
2,4-Diamino-7-phenacyl-6(5*H*)-pteridinone	H_2SO_4 >300	982
2,4-Diamino-7-phenyl-6-pteridinecarbonitrile	328–330, uv	327
2,4-Diamino-7-phenyl-6-pteridinecarboxamide	318, uv	327
4,7-Diamino-2-phenyl-6-pteridinecarboxamide	>360, uv	320, 321, 326, 328, 1357, 1466
4,7-Diamino-2-phenyl-6-pteridinecarboxylic acid	256, uv; Na >320	1357
2,4-Diamino-6-phenyl-7(8*H*)-pteridinethione	315, uv	525
4,7-Diamino-6-phenyl-2(1*H*)-pteridinethione	>310, uv	1281
2,4-Diamino-6-phenyl-7(8*H*)-pteridinone	406–408, uv	541, 1195, 1281
2,4-Diamino-7-phenyl-6(5*H*)-pteridinone	>310, uv	1281
2,7-Diamino-6-phenyl-4(3*H*)-pteridinone	H_2O >350, uv; HCl anal	1281
4,7-Diamino-6-phenyl-2(1*H*)-pteridinone	anal, uv	1281
4,7-Diamino-2-propyl-6-pteridinecarboxamide	>360	321
2,4-Diamino-6-propyl-7(3*H*)-pteridinone 8-oxide	295, nmr, uv	1298
2,4-Diamino-6-pteridinecarbaldehyde	anal, uv, nmr; DNP >260; $H_2NCONHN=$ +HCl anal; $H_2NCSNHN=$ +HCl anal	595, 686, 977, 1058, 1107, 1290

APPENDIX TABLE (*Contd.*)

Compound	(°C), and so on	Reference(s)
2,4-Diamino-6-pteridinecarboxamide	>300, uv	1354
4,7-Diamino-6-pteridinecarboxamide	>360, uv	321, 324, 1357
2,4-Diamino-6-pteridinecarboxylic acid	>300, uv; H_2O >340	123, 178
2,4-Diamino-7-pteridinecarboxylic acid	H_2O >300, uv	123, 730, 748
4,7-Diamino-6-pteridinecarboxylic acid	292, uv	324
2,4-Diamino-6,7-pteridinedicarboxylic acid	>300, uv	748
2,4-Diamino-6,7(5*H*,8*H*)-pteridinedione	>350, uv	208, 730
2,6-Diamino-4,7(3*H*,8*H*)-pteridinedione	ms, uv	1754
2,7-Diamino-4,6(3*H*,5*H*)-pteridinedione	^{13}C nmr, ms, uv	97, 616, 1754
2,4-Diamino-6(5*H*)-pteridinone	H_2O >360, ir, uv; H_2SO_4 >360	208, 268, 541, 686, 730, 1293
2,4-Diamino-7(8*H*)-pteridinone	>400, nmr, uv	268, 883, 979, 1229, 1301
2,6-Diamino-4(3*H*)-pteridinone	crude; Na anal, nmr, uv	450, 453, 507, 584, 1221, 1332, 1448
4,7-Diamino-2-trifluoromethyl-6-pteridinecarboxamide	>360	321
2,4-Dibenzyloxy-6-methyl-7(8*H*)-pteridinone	217–222, uv	1192
2,4-Dibenzyloxy-7(8*H*)-pteridinone	200–205, uv	1192
1,3-Dibenzyl-7-propionyl-2,4(1*H*,3*H*)-pteridinedione	137, nmr, uv	108
6,7-Dibenzyl-2,4-pteridinediamine	258	749
1,3-Dibenzyl-2,4(1*H*,3*H*)-pteridinedione	140	108
7-Dibromomethyl-1-methyl-2,4(1*H*,3*H*)-pteridinedione	205–206	232
6-Dibromomethyl-4,7(3*H*,8*H*)-pteridinedione	HBr >250	29
6,7-Di-*t*-butyl-l-methyl-2,4(1*H*,3*H*)-pteridinedione	185, uv	367
6,7-Dibutyl-2,4-pteridinediamine	180	749
6,7-Di-*s*-butyl-2,4-pteridinediamine	210	142, 749
2,4-Dichloro-6,7-dimethylpteridine	146–148	170, 1315
6,7-Dichloro-1,3-dimethyl-2,4(1*H*,3*H*)-pteridinedione	152–153, nmr	1681
2,4-Dichloro-6,7-diphenylpteridine	crude 189–192	1315
2,4-Dichloropteridine	~150, uv	486
4,6-Dichloropteridine	crude	41
4,7-Dichloropteridine	127–128	131
6,7-Dichloropteridine	141–142, uv	35
2,4-Dichloro-6,7-pteridinediamine	>360, uv	486
2,4-Dichloro-6,7(5*H*,8*H*)-pteridinedione	266–270	441
4-Dicyanomethylene-1,3-dimethyl-6,7-diphenyl-3,4-dihydro-2(1*H*)-pteridinone	281, nmr, uv	1679
6,7-Dicyclohexyl-2,4-pteridinediamine	270	749
6-Diethoxymethyl-2-propionamido-4(3*H*)-pteridinone	anal, uv	1403
4-Diethylamino-7-dimethylamino-6-phenylpteridine	oil, anal, nmr	764
4-Diethylamino-7-dimethylamino-6-phenylpteridine 5-oxide	169–170, nmr	764
2-Diethylamino-4,6-dimethylpteridine	87–88, uv	281
4-Diethylamino-6,7-dimethylpteridine	85, uv	688

APPENDIX TABLE (*Contd.*)

Compound	(°C), and so on	Reference(s)
2-Diethylamino-6,7-diphenylpteridine	147, uv	688
4-Diethylamino-6,7-diphenylpteridine	157–158, uv	688
4-*β*-Diethylaminoethylamino-6,7-diisopropyl-2-pteridinamine	136–137	400
4-*β*-Diethylaminoethylamino-6,7-diphenyl-2-pteridinamine	194–196	400
4-*β*-Diethylaminoethylamino-6-phenyl-2,7-pteridinediamine	160, uv	523
2-Diethylamino-4-methylpteridine	122, uv	281
6-Diethylaminomethyl-2,4-pteridinediamine	271–272	1299
4-Diethylamino-1-methyl-2(1*H*)-pteridinone	102–103, ir, ms, nmr	1733
2-Diethylaminopteridine	68, uv	124
4-Diethylaminopteridine	112, uv; pic 169	688
6,7-Diethyl-8-*β*-hydroxyethyl-2,4(3*H*,8*H*)-pteridinedione	254–260, uv	537
6,7-Diethyl-8-methyl-2,4(3*H*,8*H*)-pteridinedione	273–282, uv	537
1,3-Diethyl-7-propionyl-2,4(1*H*,3*H*)-pteridinedione	109–110, nmr, uv	108
6,7-Diethyl-2-pteridinamine	230–234	32
6,7-Diethyl-4-pteridinamine	240	32
6,7-Diethylpteridine	50–52	32
6,7-Diethyl-2,4-pteridinediamine	268 or 280	142, 749
1,3-Diethyl-2,4(1*H*,3*H*)-pteridinedione	87–88	108
6,7-Diethyl-2,4(1*H*,3*H*)-pteridinedione	217–219, ir	32, 134, 1091
6,7-Diethyl-2(1*H*)-pteridinone	H_2O 230, ir, uv; EtOH 160	32, 134, 243
6,7-Diethyl-4(3*H*)-pteridinone	245, ir	32, 134, 1091
6,7-Diethyl-8-D-ribityl-2,4(3*H*,8*H*)-pteridinedione	129–145 or 229–231, uv	172, 537
6,7-Dihexyl-2,4-pteridinediamine	140	749
2,4-Dihydrazino-6,7-diphenylpteridine	>230	467
Dihydroxanthopterin (not simple)		784
6-D-*α*,*β*-Dihydroxyethyl-8-D-ribityl-2,4,7(1*H*,3*H*,8*H*)-pteridinetrione	H_2O >265, nmr, uv	295, 1430
6-L-*α*,*β*-Dihydroxyethyl-8-D-ribityl-2,4,7(1*H*,3*H*,8*H*)-pteridinetrione (Photolumazine-A)	anal, nmr, uv	295, 1430
4-*β*,*γ*-Dihydroxypropoxy-6,7-diphenyl-2-pteridinamine	300, nmr, uv	1500
6-[L-*erythro*]-*α*,*β*-Dihydroxypropyl-2-methylthio-4(3*H*)-pteridinone	>210, uv	458
6-[L-*erythro*]-*α*,*β*-Dihydroxypropyl-2,4-pteridinediamine	crude	1058
6-[L-*erythro*]-*α*,*β*-Dihydroxypropyl-2,4(1*H*,3*H*)-pteridinedione (biolumazine)	216–218, cd, nmr, uv	1439
6,7-Diisobutyl-2,4-pteridinediamine	218	749
6,7-Diisopropyl-1,3-dimethyl-2,4(1*H*,3*H*)-pteridinedione	159, uv	367
6,7-Diisopropyl-1,3-dimethyl-2,4(1*H*,3*H*)-pteridinedione 5-oxide	157, uv	367
6,7-Diisopropyl-3,8-dimethyl-2,4(3*H*,8*H*)-pteridinedione	253	375

APPENDIX TABLE (*Contd.*)

Compound	(°C), and so on	Reference(s)
6,7-Diisopropyl-8-methyl-2-methylimino-2,8-dihydropteridine	H_2O 94–96, uv	248
6,7-Diisopropyl-3-methyl-8-phenyl-2,4(3*H*,8*H*)-pteridinedione	260, uv	409
6,7-Diisopropyl-1-methyl-2,4(1*H*,3*H*)-pteridinedione	198, uv	367
6,7-Diisopropyl-8-methyl-2,4(3*H*,8*H*)-pteridinedione	267 or 272–275, uv	248, 537, 1545
6,7-Diisopropyl-1-methyl-2,4(1*H*,3*H*)-pteridinedione 5-oxide	234–235, ms, uv	367, 427
6,7-Diisopropyl-8-methyl-2(8*H*)-pteridinone	212–213, uv	248
6,7-Diisopropyl-2-pteridinamine	189–190	400
6,7-Diisopropyl-4-pteridinamine	175–176	400
6,7-Diisopropyl-2,4-pteridinediamine	246	142, 749
2,4-Dimethoxy-6,7-dimethylpteridine	184–186, uv	170
6,7-Dimethoxy-1,3-dimethyl-2,4(1*H*,3*H*)-pteridinedione	226, uv	343
6-Dimethoxymethyl-2,4-pteridinediamine	254–255, nmr, uv	1306
7-Dimethoxymethyl-2,4-pteridinediamine	208–209, ir, nmr	1310
2,4-Dimethoxy-6-methyl-7(8*H*)-pteridinone	253–254, uv	1192
4,7-Dimethoxy-1-methyl-2(1*H*)-pteridinone	204–205	1192
2,4-Dimethoxypteridine	197–198 or 200, fl–sp, uv	338, 486, 1042
4,7-Dimethoxy-6-pteridinecarbonitrile	165–167, nmr	1136
2,4-Dimethoxy-7(8*H*)-pteridinone	240, uv	1192
2,4-Dimethoxy-8-D-ribityl-7(8*H*)-pteridinone	212, nmr, uv	1192
Dimethyl 2-acetamido-4-oxo-3,4-dihydro-6,7-pteridinedicarboxylate	214, nmr, uv	307, 601
2-Dimethylamino-4,7-bistrimethylsiloxypteridine	crude 133–137	246
4-Dimethylamino-2,7-bistrimethylsiloxypteridine	crude	356
2-Dimethylamino-6-[L-*erythro*]-α,β-dihydroxypropyl-4(3*H*)-pteridinone	>190, cd, nmr, uv	1439
6-Dimethylamino-1,3-dimethyl-7-propionyl-2,4(1*H*,3*H*)-pteridinedione	142, nmr, uv	108
7-Dimethylamino-1,3-dimethyl-6-propionyl-2,4(1*H*,3*H*)-pteridinedione	153, nmr, uv	108
2-Dimethylamino-6,7-dimethyl-4-pteridinamine	>260, uv; HCl anal	169, 472, 1189
4-Dimethylamino-6,7-dimethyl-2-pteridinamine	284–285, uv	1188
4-Dimethylamino-6,7-dimethylpteridine	138–140, ms, uv; 1-MeI 153–154, uv	129, 1384
2-Dimethylamino-3,8-dimethyl-4,7(3*H*,8*H*)-pteridinedione	184–185, uv	457
6-Dimethylamino-1,3-dimethyl-2,4(1*H*,3*H*)-pteridinedione	203, nmr, uv	1275
7-Dimethylamino-1,3-dimethyl-2,4(1*H*,3*H*)-pteridinedione	231, nmr, uv	1275
2-Dimethylamino-6,7-dimethyl-4(3*H*)-pteridinone	283–288, uv	169, 1189
2-Dimethylamino-6,8-dimethyl-7(8*H*)-pteridinone	154–156, uv	1150
4-Dimethylamino-6,7-dimethyl-2(1*H*)-pteridinone	>360, uv	362
4-Dimethylamino-6,8-dimethyl-7(8*H*)-pteridinone	164–165, uv	386, 445
2-Dimethylamino-6,7-dimethyl-4-trifluoromethylpteridine	126–127, nmr	1440

APPENDIX TABLE (*Contd.*)

Compound	(°C), and so on	Reference(s)
2-Dimethylamino-4,7-dioxo-3,4,7,8-tetrahydro-6-pteridinecarboxylic acid	>350, uv	360
2-Dimethylamino-6,7-diphenyl-4-piperidinopteridine	207	110
2-Dimethylamino-6,7-diphenyl-4-pteridinamine	239, uv	110; cf. 472
4-Dimethylamino-6,7-diphenyl-2-pteridinamine	322–325, uv	110, 1188
2-Dimethylamino-6,7-diphenyl-4(3*H*)-pteridinone	>330 or 361, ir	110, 359, 841 (?)
4-Dimethylamino-6,7-diphenyl-2(1*H*)-pteridinone	>300	927
2-Dimethylamino-4-ethoxy-6-phenylpteridine	200	110
7-Dimethylamino-4-ethoxy-6-phenylpteridine	105–106, nmr	764
7-Dimethylamino-4-ethoxy-6-phenylpteridine 5-oxide	195–196, nmr	764
4-β-Dimethylaminoethylamino-6,7-diisopropyl-2-pteridinamine	177–178	400
4-β-Dimethylaminoethylamino-6,7-dimethyl-2-phenylpteridine	157, nmr	1436
4-β-Dimethylaminoethylamino-6,7-diphenyl-2-pteridinamine	233–234	400
4-β-Dimethylaminoethylamino-2-phenylpteridine	173, nmr	1436
2-β-Dimethylaminoethylamino-6-phenyl-4,7-pteridinediamine	242–245, uv	523
2-Dimethylamino-8-ethyl-7-oxo-7,8-dihydro-6-pteridinecarboxylic acid	192, ir, uv	387, 562
2-Dimethylamino-8-ethyl-6,7(5*H*,8*H*)-pteridinedione	297–305, uv	208
2-Dimethylamino-8-ethyl-7(8*H*)-pteridinone	118, uv	381, 387, 417
2-Dimethylamino-8-β-hydroxyethyl-6,7-dimethyl-4(8*H*)-pteridinone	258, uv	359
2-Dimethylamino-8-β-hydroxyethyl-6,7-diphenyl-4(8*H*)-pteridinone	262–263, uv	359
2-Dimethylamino-8-β-hydroxyethyl-4(8*H*)-pteridinone	230, nmr, uv	359
2-Dimethylamino-5-hydroxy-6,7(5*H*,8*H*)-pteridinedione	>330	490
2-Dimethylamino-6-hydroxy-7(8*H*)-pteridinone 5-oxide (see preceding compound)		
2-Dimethylamino-4-methoxy-6,7-dimethylpteridine	147–148, nmr, uv	416, 762
2-Dimethylamino-4-methoxy-6,8-dimethyl-7(8*H*)-pteridinone	204, uv	417
2-Dimethylamino-4-methoxy-6-methyl-7(8*H*)-pteridinone	256–258, uv	417
2-Dimethylamino-4-methoxy-8-methyl-7(8*H*)-pteridinone	181–182, nmr, uv	417, 436, 511
2-Dimethylamino-4-methoxypteridine	163–164, nmr, uv	182, 416, 511, 762
4-Dimethylamino-7-methoxypteridine	284, uv	386, 445, 581
2-Dimethylamino-4-methoxy-7(8*H*)-pteridinone	239, uv	417
2-Dimethylamino-4-methoxy-7-trimethylsiloxypteridine	crude	435
2-Dimethylamino-4-methylamino-6,7-diphenylpteridine	205	110
4-Dimethylamino-2-methylamino-6,7-diphenylpteridine	306	110
2-Dimethylamino-8-methyl-4,7-dioxo-3,4,7,8-tetrahydro-6-pteridinecarboxylic acid	>350, uv	360
2-Dimethylamino-8-methyl-6,7-diphenyl-4(8*H*)-pteridinone	287, uv	359
4-Dimethylaminomethyleneamino-2-methylpteridine 3-oxide	187–189, nmr	1004
2-Dimethylaminomethyleneamino-4-oxo-3,4-dihydro-6-pteridinecarbonitrile	290–295, ir, nmr	1739

APPENDIX TABLE (*Contd.*)

Compound	(°C), and so on	Reference(s)
4-Dimethylaminomethyleneamino-2-phenylpteridine 3-oxide	197–200, ms, nmr	1004
4-Dimethylaminomethyleneaminopteridine	176–179	135
7-Dimethylaminomethyleneaminopteridine	192–194	136
4-Dimethylaminomethyleneaminopteridine 3-oxide	210–213, nmr	1004
4-Dimethylamino-8-methyl-2-methylthio-7(8*H*)-pteridinone	160–162, uv	1126
2-Dimethylamino-4-methyl-7-oxo-7,8-dihydro-6-pteridinecarbonitrile 5-oxide	283	490
4-Dimethylamino-8-methyl-7-oxo-7,8-dihydro-6-pteridinecarboxylic acid	210–211, ir, uv	562, 1311
Dimethyl 2-amino-8-methyl-4-oxo-4,8-dihydro-6,7-pteridinedicarboxylate	HCl ↑80, uv	307, 601
2-Dimethylamino-4-methyl-6,7(5*H*,8*H*)-pteridinedione	>320	490
2-Dimethylamino-6-methyl-4,7(3*H*,8*H*)-pteridinedione	>350, uv	360
2-Dimethylamino-8-methyl-4,7(3*H*,8*H*)-pteridinedione	>350, uv	246, 360, 457
4-Dimethylamino-6-methyl-2,7(1*H*,8*H*)-pteridinedione	352–353, uv	363
4-Dimethylamino-8-methyl-4,7(3*H*,8*H*)-pteridinedione	anal, uv	246
4-Dimethylamino-8-methyl-6,7(5*H*,8*H*)-pteridinedione	245, uv	208
2-Dimethylamino-8-methyl-4,6,7(3*H*,5*H*,8*H*)-pteridinetrione	>350	360
2-Dimethylamino-3-methyl-4(3*H*)-pteridinone	157–159, nmr, uv	416, 762
2-Dimethylamino-6-methyl-7(8*H*)-pteridinone	258	357
2-Dimethylamino-7-methyl-4(3*H*)-pteridinone	260, uv	1710
2-Dimethylamino-8-methyl-4(8*H*)-pteridinone	220, nmr, uv	182, 359, 741, 762
2-Dimethylamino-8-methyl-7(8*H*)-pteridinone	anal, uv	581
4-Dimethylamino-8-methyl-7(8*H*)-pteridinone	159–161 or 161–163, uv	386, 445, 581, 1311
6-Dimethylamino-3-methyl-4(3*H*)-pteridinone	256–258	974
4-Dimethylamino-2-methylthio-7-propionylpteridine	181–183, nmr, uv	1678
4-Dimethylamino-2-methylthiopteridine	178–180, nmr, uv	1678
4-Dimethylamino-2-methylthio-7(8*H*)-pteridinone	273–275 or 277–279, uv	445, 1126
4-Dimethylamino-2-methylthio-7-trimethylsiloxypteridine	crude	325
2-Dimethylamino-6-methyl-7-trimethylsiloxypteridine	crude	356
7-Dimethylamino-4-morpholino-6-phenylpteridine	173–174, nmr	764
7-Dimethylamino-4-morpholino-6-phenylpteridine 5-oxide	237–238, nmr	764
2-Dimethylamino-4-oxo-3,4-dihydro-6-pteridine-carbaldehyde	—	1463
2-Dimethylamino-7-oxo-7,8-dihydro-6-pteridinecarbonitrile	312	490
2-Dimethylamino-7-oxo-7,8-dihydro-6-pteridinecarbonitrile 5-oxide	268	490
2-Dimethylamino-4-oxo-3,4-dihydro-6-pteridinecarboxylic acid	268–269, uv	1710
2-Dimethylamino-4-oxo-3,4-dihydro-6-pteridinecarboxylic acid	H_2O 236, uv	1710
Dimethyl 2-amino-4-oxo-3,4-dihydro-6,7-pteridine-dicarboxylate	288, uv	307, 601, 748, 1363
7-Dimethylamino-6-phenyl-4-piperidinopteridine	157–158, nmr	764
7-Dimethylamino-6-phenyl-4-piperidinopteridine 5-oxide	228–229, nmr	764

APPENDIX TABLE (*Contd.*)

Compound	(°C), and so on	Reference(s)
7-Dimethylamino-6-phenyl-4-pteridinamine	220–221, nmr	764
7-Dimethylamino-6-phenyl-4-pteridinamine 5-oxide	221–222, nmr	764
2-Dimethylamino-6-phenyl-4,7-pteridinediamine	270–271, nmr, uv	523
4-Dimethylamino-6-phenyl-2,7-pteridinediamine	273–274, nmr, uv	523
7-Dimethylamino-6-phenyl-2,4-pteridinediamine	HCl >320, nmr, uv	525
2-Dimethylamino-6-phenyl-4(3*H*)-pteridinone	322 or 336, uv	110
2-Dimethylamino-7-phenyl-4(3*H*)-pteridinone	325	110
7-Dimethylamino-6-phenyl-4(3*H*)-pteridinone	275, nmr	764
7-Dimethylamino-6-phenyl-4(3*H*)-pteridinone 5-oxide	270–275, nmr	764
4-γ-Dimethylaminopropylamino-6,7-diisopropyl-2-pteridinamine	152–155	400
4-γ-Dimethylaminopropylamino-6,7-diphenyl-2-pteridinamine	217–219	400, 467
2-γ-Dimethylaminopropylamino-6-phenyl-4,7-pteridinediamine	239–243, uv	523
2-Dimethylamino-4-pteridinamine	213–214, uv	267
4-Dimethylamino-2-pteridinamine	223–225, uv	267
2-Dimethylaminopteridine	125–126, ir, uv	30, 39, 119, 285, 286
4-Dimethylaminopteridine	164 to 169, ir, nmr, uv; HI 244–245; 1-MeI 227–229, uv	27, 32, 119, 129, 280, 285, 286
6-Dimethylaminopteridine	212, ir, uv	31, 119, 285, 286
7-Dimethylaminopteridine	204 or 205, ir, uv	33, 119, 146, 285, 286
2-Dimethylamino-4,6(3*H*,5*H*)-pteridinedione	>300, uv	1710
2-Dimethylamino-4,7(3*H*,8*H*)-pteridinedione	>350, uv	246, 360
2-Dimethylamino-6,7(5*H*,8*H*)-pteridinedione	>300	490
4-Dimethylamino-2,7(1*H*,8*H*)-pteridinedione	350, uv	363
2-Dimethylamino-4,6,7(3*H*,5*H*,8*H*)-pteridinetrione	>350	360
2-Dimethylamino-4(3*H*)-pteridinone	282, nmr, uv	114, 368, 416, 511, 741, 762, 1710
2-Dimethylamino-7(8*H*)-pteridinone	262(?) or 313–318, uv; Ag 293	381, 417, 490(?)
4-Dimethylamino-2(1*H*)-pteridinone	304–305, uv	225, 362
4-Dimethylamino-6(5*H*)-pteridinone	287; MeOH 287, uv	445
4-Dimethylamino-7(8*H*)-pteridinone	287, uv	445, 581
6-Dimethylamino-4(3*H*)-pteridinone	294–296	974
2-Dimethylamino-4-trifluoromethylpteridine	163, ms, nmr, uv	159, 780
2-Dimethylamino-6-[D-*erythro*]-α,β,γ-trihydroxypropyl-4(3*H*)-pteridinone	238–240, cd, nmr, uv	1710
2-Dimethylamino-6-[L-*erythro*]-α,β,γ-trihydroxypropyl-4(3*H*)-pteridinone	242, cd, nmr, uv	1710

APPENDIX TABLE (*Contd.*)

Compound	(°C), and so on	Reference(s)
2-Dimethylamino-6-[D-*threo*]-α,β,γ-trihydroxypropyl-4(3*H*)-pteridinone	241, cd, nmr, uv	1710
2-Dimethylamino-6-[L-*threo*]-α,β,γ-trihydroxypropyl-4(3*H*)-pteridinone (euglenapterin)	243, cd, nmr, uv, xl-st	109, 335, 985, 1710
2-Dimethylamino-3,6,7-trimethyl-4(3*H*)-pteridinone	157–160, uv	416
4-Dimethylamino-1,6,7-trimethyl-2(1*H*)-pteridinone	209–211, uv	362
2-Dimethylamino-6,7,8-trimethyl-4(8*H*)-pteridinone	—	563
2-Dimethylamino-7-trimethylsiloxypteridine	—	325, 356
4-Dimethylamino-7-trimethylsiloxypteridine	—	325
4-Dimethylamino-7-trimethylsiloxy-2-trimethylsilylaminopteridine	110–115	357
6,7-Dimethyl-2,4-bismethylaminopteridine	255–256 or 266	110, 1315
1,3-Dimethyl-6,7-bistrifluoromethyl-2,4(1*H*,3*H*)-pteridinedione	172–173, ir, ms, nmr, uv	1571
6,7-Dimethyl-2,4-bistrimethylsiloxypteridine	Crude 120; 136/0.0002	414
6-3′,3′-Dimethylbut-1′-ynyl-2-pivalamido-4(3*H*)-pteridinone	327–328, ir, nmr	1753
6-3′,3′-Dimethylbut-1′-ynyl-2,4-pteridinediamine	>300, ir, nmr	1753
Dimethyl 2,4-diamino-6,7-pteridinedicarboxylate	>300, uv	748
Dimethyl 1,3-dimethyl-2,4-dioxo-1,2,3,4-tetrahydro-6,7-pteridinedicarboxylate	177, nmr	1388; cf. 315
1,3-Dimethyl-2,4-dioxo-1,2,3,4-tetrahydro-6-pteridinecarbaldehyde	171; HON= 283	196
1,3-Dimethyl-2,4-dioxo-1,2,3,4-tetrahydro-6-pteridinecarboxamide	>300, ms, nmr	1388
1,3-Dimethyl-2,4-dioxo-1,2,3,4-tetrahydro-6-pteridinecarboxylic acid	230–231 or 250, nmr, uv	70, 962, 1291, 1388
1,3-Dimethyl-2,4-dioxo-1,2,3,4-tetrahydro-7-pteridinecarboxylic acid	215–216	108
3,8-Dimethyl-2,4-dioxo-2,3,4,8-tetrahydro-6-pteridinecarboxylic acid	253–255, uv	70
Dimethyl 2,4-dioxo-1,2,3,4-tetrahydro-2,4-pteridinedicarboxylate	246–247, uv	748
1,3-Dimethyl-2,4-dioxo-1,2,3,4-tetrahydro-6-pteridinesulfinic acid	Na —, nmr, uv	227, 1681
1,3-Dimethyl-2,4-dioxo-1,2,3,4-tetrahydro-6-pteridinesulfonic acid	K >310, nmr, uv	227, 1681
1,3-Dimethyl-2,4-dioxo-1,2,3,4-tetrahydro-7-pteridinesulfonic acid	Na >310, nmr, uv	227, 1681
1,3-Dimethyl-6,7-diphenyl-4-phenylhydrazono-3,4-dihydro-2(1*H*)-pteridinone	292–293 or >300, nmr, uv	927, 1679
1,3-Dimethyl-6,7-diphenyl-4-phenylimino-2(1*H*)-pteridinone	186–187	927
2,4-Dimethyl-6,7-diphenylpteridine	207	657
1,3-Dimethyl-6,7-diphenyl-2,4(1*H*,3*H*)-pteridinedione	220 to 232, ir, nmr, uv	68, 233, 367, 414, 962, 1096, 1149, 1339, 1435 1506

APPENDIX TABLE (*Contd.*)

Compound	(°C), and so on	Reference(s)
3,8-Dimethyl-6,7-diphenyl-2,4(3*H*,8*H*)-pteridinedione	277–279, uv	1545
1,3-Dimethyl-6,7-diphenyl-2,4(1*H*,3*H*)-pteridinedione 5-oxide	249–250, uv	210, 367
1,3-Dimethyl-6,7-diphenyl-2-thioxo-1,2-dihydro-4(3*H*)-pteridinone	260, uv	437
1,3-Dimethyl-6,7-diphenyl-4-thioxo-3,4-dihydro-2(1*H*)-pteridinone	221 or 229–231, nmr, uv	927, 1679
1,3-Dimethyl-2,4-dithioxo-1,2,3,4-tetrahydro-7(8*H*)-pteridinone	283–285, uv	1000
1,3-Dimethyl-2,7-dithioxo-1,2,7,8-tetrahydro-4(3*H*)-pteridinone	275–278, uv	1000
1,3-Dimethyl-4,7-dithioxo-3,4,7,8-tetrahydro-2(1*H*)-pteridinone	315–318	1000
Dimethyl 2-imino-3,8-dimethyl-4-oxo-2,3,4,8-tetrahydro-6,7-pteridinedicarboxylate	HCl 190, uv	307, 601
6,7-Dimethyl-2-(*N*-methylacetamido-4(3*H*)-pteridinone	222–225, uv	69
3,8-Dimethyl-2-methylamino-4,7-dioxo-3,4,7,8-tetrahydro-6-pteridinecarboxylic acid	>300, uv	1214, 1565
4,8-Dimethyl-2-methylamino-7-oxo-7,8-dihydro-6-pteridinecarboxylic acid	250–260, ir, nmr, uv	1311
1,3-Dimethyl-6-methylamino-7-propionyl-2,4(1*H*,3*H*)-pteridinedione	192, nmr, uv	108
1,3-Dimethyl-7-methylamino-6-propionyl-2,4(1*H*,3*H*)-pteridinedione	204–205, nmr, uv	108
6,7-Dimethyl-2-methylamino-4-pteridinamine	281	110
6,7-Dimethyl-4-methylaminopteridine	223, uv; HI 218–220	129
1,3-Dimethyl-6-methylamino-2,4(1*H*,3*H*)-pteridinedione	265, nmr, uv	1275
1,3-Dimethyl-7-methylamino-2,4(1*H*,3*H*)-pteridinedione	348, nmr, uv	1275
3,8-Dimethyl-2-methylamino-4,7(3*H*,8*H*)-pteridinedione	crude, >300	457, 1565
5,8-Dimethyl-4-methylamino-6,7(5*H*,8*H*)-pteridinedione	uv	132
1,3-Dimethyl-7-methylamino-2,4,6(1*H*,3*H*,5*H*)-pteridinetrione	370, ir	1294
6,7-Dimethyl-2-methylamino-4(3*H*)-pteridinone	H_2O 277–281, uv	166, 1189
6,8-Dimethyl-2-methylamino-7(8*H*)-pteridinone	257–258, uv	248
6,8-Dimethyl-4-methylamino-7(8*H*)-pteridinone	171–172, uv	248
1,3-Dimethyl-4-methylhydrazono-6,7-diphenyl-3,4-dihydro-2(1*H*)-pteridinone	214, nmr, uv	1679
1,3-Dimethyl-4-methylimino-6,7-diphenyl-3,4-dihydro-2(1*H*)-pteridinone	166, nmr, uv	1679
1,3-Dimethyl-6-methylsulfinyl-2,4(1*H*,3*H*)-pteridinedione	204, nmr, uv	1681
1,3-Dimethyl-6-methylsulfonyl-2,4(1*H*,3*H*)-pteridinedione	270, nmr, uv	1681
3,*N*-Dimethyl-2-methylthio-4,6-dioxo-2,3,5,6-tetrahydro-7-pteridinecarboxamide	290–292	342
6,7-Dimethyl-2-methylthio-4-phenylpteridine	169, nmr	1436
6,7-Dimethyl-4-methylthio-2-phenylpteridine	163, nmr	1436
1,3-Dimethyl-7-methylthio-6-propionyl-2,4(1*H*,3*H*)-pteridinedione	200–201, nmr, uv; HON=113	108, 1759

APPENDIX TABLE (*Contd.*)

Compound	(°C), and so on	Reference(s)
6,7-Dimethyl-2-methylthio-4-pteridinamine	274–275	472
6,7-Dimethyl-4-methylthio-2-pteridinamine	272–274	1468
6,7-Dimethyl-2-methylthiopteridine	120–121	129
6,7-Dimethyl-4-methylthiopteridine	120, uv	313
1,3-Dimethyl-6-methylthio-2,4(1*H*,3*H*)-pteridinedione	179–180, nmr, uv	1681
1,3-Dimethyl-7-methylthio-2,4(1*H*,3*H*)-pteridinedione	219–221, uv	1000
3,8-Dimethyl-2-methylthio-4,7(3*H*,8*H*)-pteridinedione	239, uv	344
1,3-Dimethyl-7-methylthio-2,4(1*H*,3*H*)-pteridinedithione	227–229, uv	1000
6,7-Dimethyl-2-methylthio-4(3*H*)-pteridinone	280–281 or 283, nmr, uv	437, 654, 1508
1,3-Dimethyl-7-methylthio-2-thioxo-1,2-dihydro-4(3*H*)-pteridinone	232–234, uv	1000
1,3-Dimethyl-7-methylthio-4-thioxo-3,4-dihydro-2(1*H*)-pteridinone	226–228, uv	1000
3,8-Dimethyl-6-*N*-methylureido-2,4,7(1*H*,3*H*,8*H*)-pteridinetrione	anal, ir, ms, nmr	1282
6,7-Dimethyl-4-oxo-3,4-dihydro-2-pteridinesulfinic acid	K anal	629
6,7-Dimethyl-4-oxo-3,4-dihydro-2-pteridinesulfonic acid	K anal	629
1,3-Dimethyl-4-oxo-3,4-dihydro-1-pteridinium iodide	213–215, ir, nmr, uv	587, 1130
1,3-Dimethyl-4-oxo-6,7-diphenyl-3,4-dihydro-1-pteridinium iodide	278–280, ir, nmr, uv	311, 587, 1130
1,3-Dimethyl-6-phenacyl-2,4,7(1*H*,3*H*,8*H*)-pteridinetrione	275, ms, uv	1524
1,3-Dimethyl-7-phenacyl-2,4,6(1*H*,3*H*,5*H*)-pteridinetrione	290, ms, uv	1524
6,7-Dimethyl-4-phenylhydrazino-2(1*H*)-pteridinethione	274	927
6,7-Dimethyl-4-phenylhydrazino-2(1*H*)-pteridinone	247	927
6,7-Dimethyl-8-phenyl-2-phenylimino-2,8-dihydropteridine	241–242, uv	190
1,3-Dimethyl-6-phenyl-7-propyl-2,4(1*H*,3*H*)-pteridinedione	112, nmr, uv	1112, 1348
1,3-Dimethyl-7-phenyl-6-propyl-2,4(1*H*,3*H*)-pteridinedione	165, nmr, uv	1148, 1348
6,7-Dimethyl-2-phenyl-4-pteridinamine	308–310	1354
2,7-Dimethyl-4-phenylpteridine	134–135, nmr, uv	151
4,7-Dimethyl-2-phenylpteridine	168–169, nmr, uv	152
6,7-Dimethyl-2-phenylpteridine	169–170, nmr, uv	152
6,7-Dimethyl-4-phenylpteridine	116–117, nmr, uv	151
1,3-Dimethyl-6-phenyl-2,4(1*H*,3*H*)-pteridinedione	251 to 275, ms, nmr, uv	67, 176, 367, 427, 543, 552, 1292, 1387; cf. 1473
1,3-Dimethyl-7-phenyl-2,4(1*H*,3*H*)-pteridinedione	298–300 or 308–309, nmr, uv	67, 176, 367, 1279, 1389; cf. 1473
6,7-Dimethyl-1-phenyl-2,4(1*H*,3*H*)-pteridinedione	>280	797
6,7-Dimethyl-8-phenyl-2,4(3*H*,8*H*)-pteridinedione	317, uv	409
1,3-Dimethyl-6-phenyl-2,4(1*H*,3*H*)-pteridinedione 5-oxide	216, nmr, uv	367
1,3-Dimethyl-7-phenyl-2,4(1*H*,3*H*)-pteridinedione 5-oxide	299–300, ms, nmr, uv	331, 367, 427
6,7-Dimethyl-2-phenylpteridine-4(3*H*)-thione	>210, nmr	1436

APPENDIX TABLE (*Contd.*)

Compound	(°C), and so on	Reference(s)
1,3-Dimethyl-6-phenyl-2,4,7(1*H*,3*H*,8*H*)-pteridinetrione	330, uv	367, 413
1,3-Dimethyl-7-phenyl-2,4,6(1*H*,3*H*,5*H*)-pteridinetrione	327	367, 1279
3,6-Dimethyl-1-phenyl-2,4,7(1*H*,3*H*,8*H*)-pteridinetrione	354–355, uv	238
3,6-Dimethyl-1-phenyl-2,4,7(1*H*,3*H*,8*H*)-pteridinetrione 5-oxide	342, uv	236, 238
6,7-Dimethyl-2-phenyl-4(3*H*)-pteridinone	280, nmr, uv	152, 663
6,7-Dimethyl-1-phenyl-2-thioxo-1,2-dihydro-4(3*H*)-pteridinone	280–281	797
6,7-Dimethyl-4-piperidino-2-pteridinamine	214–216, uv	1188
1,3-Dimethyl-6-propionyl-7-propyl-2,4(1*H*,3*H*)-pteridinedione	85, nmr, uv	108
1,3-Dimethyl-6-propionyl-2,4(1*H*,3*H*)-pteridinedione	141–142, nmr, uv	108
1,3-Dimethyl-7-propionyl-2,4(1*H*,3*H*)-pteridinedione	136, nmr, uv	108
1,3-Dimethyl-6-propionyl-2,4,7(1*H*,3*H*,8*H*)-pteridinetrione	160–163, nmr, uv	108, 1759
1,3-Dimethyl-7-propionyl-2,4,6(1*H*,3*H*,5*H*)-pteridinetrione	204–206, nmr, uv	108, 1759
1,3-Dimethyl-7-propyl-2,4(1*H*,3*H*)-pteridinedione	107	108
6,7-Dimethyl-8-propyl-2,4(3*H*,8*H*)-pteridinedione	274–290, ms	1553
4,6-Dimethyl-2-pteridinamine	312, uv	39, 281
4,7-Dimethyl-2-pteridinamine	284, ir, uv	39, 582
6,7-Dimethyl-2-pteridinamine	308, uv	39, 979
6,7-Dimethyl-4-pteridinamine	295, nmr, uv	129, 170, 1740
6,7-Dimethyl-4-pteridinamine 5,8-dioxide	245, ir, nmr, uv	723
6,7-Dimethylpteridine	148–149, ms, nmr, ^{13}C nmr, uv	27, 33, 66, 207, 285, 286, 959
6,7-Dimethyl-4-pteridinecarboxamide	209, nmr, uv	157
6,7-Dimethyl-2,4-pteridinediamine	260(?) or 360, ir, ms, pol, uv	130, 169, 284, 475, 541, 602, 847, 884, 1014, 1193*, 1231, 1391
6,7-Dimethyl-2,4-pteridinediamine 5,8-dioxide	305–307, ms, uv	427, 541
6,7-Dimethyl-2,4-pteridinediamine 8-oxide	327–330, ms, uv	427, 541
1,3-Dimethyl-2,4(1*H*,3*H*)-pteridinedione	197 to 200, ir, ms, nmr, uv	34, 338, 367, 414, 538, 552, 1017, 1042, 1096, 1121, 1264, 1339, 1435, 1554, 1681, 1733
1,6-Dimethyl-2,4(1*H*,3*H*)-pteridinedione	306–307, ir, nmr, uv	80, 168
1,7-Dimethyl-2,4(1*H*,3*H*)-pteridinedione	279–281	962
2,6-Dimethyl-4,7(3*H*,8*H*)-pteridinedione	—, uv	279

APPENDIX TABLE (*Contd.*)

Compound	(°C), and so on	Reference(s)
3,8-Dimethyl-2,4(3*H*,8*H*)-pteridinedione	pol	116
3,8-Dimethyl-2,7(3*H*,8*H*)-pteridinedione	263	134
5,8-Dimethyl-6,7(5*H*,8*H*)-pteridinedione	203–204, ir, uv	34, 134
6,7-Dimethyl-2,4(1*H*,3*H*)-pteridinedione	340 to >360, ir, ms, nmr, uv	106, 170, 284, 367, 437, 471, 747, 960, 979, 1092, 1096, 1144, 1339, 1391, 1438, 1554
6,8-Dimethyl-2,7(1*H*,8*H*)-pteridinedione	283–284, uv	248
6,8-Dimethyl-4,7(3*H*,8*H*)-pteridinedione	330, uv	248
7,8-Dimethyl-2,4(3*H*,8*H*)-pteridinedione	anal, nmr, pol	1278
6,7-Dimethyl-2,4(1*H*,3*H*)-pteridinedione 5,8-dioxide	278, uv	367
1,3-Dimethyl-2,4(1*H*,3*H*)-pteridinedione 5-oxide	249–250, ms, uv	367, 427
6,7-Dimethyl-2,4(1*H*,3*H*)-pteridinedione 8-oxide	300, ms, uv	367 (cf. 366), 427
6,7-Dimethyl-2,4(1*H*,3*H*)-pteridinedithione	230 or 250–254, uv	170, 437, 927
1,3-Dimethyl-2,4,6,7(1*H*,3*H*,5*H*,8*H*)-pteridinetetrone	>360, uv	238, 343, 1435
6,7-Dimethyl-2(1*H*)-pteridinethione	220, uv	170
6,7-Dimethyl-4(3*H*)-pteridinethione	>300, uv	170, 313
1,3-Dimethyl-2,4,6(1*H*,3*H*,5*H*)-pteridinetrione	H_2O 282 or 300, uv	101, 209, 340, 364, 1275
1,3-Dimethyl-2,4,7(1*H*,3*H*,8*H*)-pteridinetrione	253 to 282, nmr, uv	227, 238, 339, 344, 364, 716, 1000, 1275, 1380(?)
1,6-Dimethyl-2,4,7(1*H*,3*H*,8*H*)-pteridinetrione	323–324 or 330, uv	339, 413, 1380(?)
1,7-Dimethyl-2,4,6(1*H*,3*H*,5*H*)-pteridinetrione	315, uv	340
1,8-Dimethyl-2,4,7(1*H*,3*H*,8*H*)-pteridinetrione	310, nmr	716
3,6-Dimethyl-2,4,7(1*H*,3*H*,8*H*)-pteridinetrione	>340, uv	339, 413
3,7-Dimethyl-2,4,6(1*H*,3*H*,5*H*)-pteridinetrione	>340, uv	340
3,8-Dimethyl-2,4,7(1*H*,3*H*,8*H*)-pteridinetrione	>350, nmr, uv	344, 716
4,8-Dimethyl-2,6,7(1*H*,5*H*,8*H*)-pteridinetrione	>350, uv	208
6,8-Dimethyl-2,4,7(1*H*,3*H*,8*H*)-pteridinetrione	358, uv	248, 339, 344, 1131, 1441
1,3-Dimethyl-2,4,7(1*H*,3*H*,8*H*)-pteridinetrione 5-oxide	227–228, uv	238
1,3-Dimethyl-2,4,7(1*H*,3*H*,8*H*)-pteridinetrithione	290–294, uv	1000
2,6-Dimethyl-4(3*H*)-pteridinone	>250, ms, uv	153, 779
2,7-Dimethyl-4(3*H*)-pteridinone	>250, ms	153, 779
3,7-Dimethyl-4(3*H*)-pteridinone	>300, uv	167
4,6-Dimethyl-2(1*H*)-pteridinone	anal, uv	281
6,7-Dimethyl-2(1*H*)-pteridinone	H_2O >200, uv	38, 134, 170, 241, 243

APPENDIX TABLE (*Contd.*)

Compound	(°C), and so on	Reference(s)
6,7-Dimethyl-4(3*H*)-pteridinone	355–360, ms, pol, uv	30, 170, 285, 587, 779, 979, 1014, 1130, 1391
6,8-Dimethyl-7(8*H*)-pteridinone	145–147	31, 34, 55, 538
2,6-Dimethyl-4(3*H*)-pteridinone 5,8-dioxide	>300	484
2,6-Dimethyl-4(3*H*)-pteridinone 8-oxide	>300	484
2,7-Dimethyl-4(1*H*)-pteridinone 3-oxide	245, ms, uv	154, 1226
6,7-Dimethyl-4(1*H*)-pteridinone 3-oxide	250, ms, uv	154, 1226
6,7-Dimethyl-8-D-ribityl-2,4(3*H*,8*H*)-pteridinedione (dimethylribolumazine)	270 to 275, ir, nmr, uv	164, 165, 172, 335, 378, 422, 423, 537, 642, 1027, 1144, 1437, 1543, 1545, 1651, 1652*, 1734*
Dimethylribolumazine (see preceding entry)		
6,7-Dimethyl-8-sorbityl-2,4(3*H*,8*H*)-pteridinedione	229–230	172, 1444
6,7-Dimethyl-2-styrylpteridine	183, nmr, uv	105
1,3-Dimethyl-6(7?)-[D-*arabino*]-α,β,γ,δ-tetrahydroxybutyl-2,4(1*H*,3*H*)-pteridinedione	212–214	962; cf. 862
1,3-Dimethyl-6(7?)-[D-*lyxo*]-α,β,γ,δ-tetrahydroxybutyl-2,4(1*H*,3*H*)-pteridinedione	246	962; cf. 862
1,3-Dimethyl-7-[D-*arabino*]-α,β,γ,δ-tetrahydroxybutyl-2,4(1*H*,3*H*)-pteridinedione	212, ir, nmr	862
1,3-Dimethyl-2-thioxo-1,2-dihydro-4,7(3*H*,8*H*)-pteridinedione	276, nmr, uv	1000
1,3-Dimethyl-4-thioxo-3,4-dihydro-2,7(1*H*,8*H*)-pteridinedione	299–300, nmr, uv	1000
1,3-Dimethyl-6-thioxo-5,6-dihydro-2,4(1*H*,3*H*)-pteridinedione	180, nmr, uv	1681
1,3-Dimethyl-7-thioxo-7,8-dihydro-2,4(1*H*,3*H*)-pteridinedione	285, uv; $H\overset{+}{N}(CH)_5$ 291	1000, 1681
1,3-Dimethyl-2-thioxo-1,2-dihydro-4(3*H*)-pteridinone	225, uv	437
6,7-Dimethyl-2-thioxo-1,2-dihydro-4(3*H*)-pteridinone	280–285, uv	437, 916
6,7-Dimethyl-4-thioxo-3,4-dihydro-2(1*H*)-pteridinone	>275	927
6,8-Dimethyl-2-thioxo-1,2-dihydro-7(8*H*)-pteridinone	242–244, uv	248
6,7-Dimethyl-8-D-threityl-2,4(3*H*,8*H*)-pteridinedione	uv	713
6,7-Dimethyl-8-L-threityl-2,4(3*H*,8*H*)-pteridinedione	uv	713
6,7-Dimethyl-4-trifluoromethyl-2-pteridinamine	235–236, nmr	1440
6,7-Dimethyl-4-trifluoromethylpteridine	101, ms, nmr, uv	156, 780
6,7-Dimethyl-4-trimethylsiloxy-2-trimethylsilylamino-pteridine	crude, ms	1233, 1457
1,3-Dimethyl-2,4,7-trioxo-1,2,3,4,7,8-hexahydro-6-pteridinecarbaldehyde	264; PhNHN= 288	337

APPENDIX TABLE (*Contd.*)

Compound	(°C), and so on	Reference(s)
1,*N*-Dimethyl-2,4,6-trioxo-1,2,3,4,5,6-hexahydro-7-pteridinecarboxamide	>340, uv	342, 347
3,*N*-Dimethyl-2,4,6-trioxo-1,2,3,4,5,6-hexahydro-7-pteridinecarboxamide	318, uv	342, 347
N,*N*-Dimethyl-2,4,6-trioxo-1,2,3,4,5,6-hexahydro-7-pteridinecarboxamide	>350, uv	347
1,3-Dimethyl-2,4,7-trioxo-1,2,3,4,7,8-hexahydro-6-pteridinecarboxylic acid	240 to 246, uv	337, 341, 1380
1,8-Dimethyl-2,4,7-trioxo-1,2,3,4,7,8-hexahydro-6-pteridinecarboxylic acid	220, uv	317
6,7-Dimethyl-8-D-xylityl-2,4(3*H*,8*H*)-pteridinedione	$2H_2O$ 135–140, uv	537
6,7-Dimethyl-8-L-xylityl-2,4(3*H*,8*H*)-pteridinedione	130–149, uv	1444
2,4-Dimorpholino-6,7-diphenylpteridine	200–201	467
4,7-Dioxo-3,4,7,8-tetrahydro-6-pteridinecarbaldehyde	H_2O >320, uv; HON= 320	29
2,4-Dioxo-1,2,3,4-tetrahydro-6-pteridinecarboxylic acid	345, uv	70, 178, 1041, 1110
2,4-Dioxo-1,2,3,4-tetrahydro-7-pteridinecarboxylic acid	275 or 340, uv	748, 1110
2,6-Dioxo-1,2,5,6-tetrahydro-4-pteridinecarboxylic acid	>250	150
4,6-Dioxo-3,4,5,6-tetrahydro-7-pteridinecarboxylic acid	>350; H_2O —, uv	346
4,7-Dioxo-3,4,7,8-tetrahydro-6-pteridinecarboxylic acid (6-lumazinecarboxylic acid)	>300, uv	29, 655
2,4-Dioxo-1,2,3,4-tetrahydro-6,7-pteridinedicarbonitrile	>230, ir	1198
2,4-Dioxo-1,2,3,4-tetrahydro-6,7-pteridinedicarboxylic acid	>300, uv	748
4,7-Dioxo-2-thioxo-1,2,3,4,7,8-hexahydro-6-pteridinecarboxylic acid (?)	anal	399, 1169
6,7-Dipentyl-2,4-pteridinediamine	160	749
6,7-Di-*s*-pentyl-2,4-pteridinediamine	172	749
6,7-Diphenethyl-2,4-pteridinediamine	216–218	419
6,7-Diphenyl-2,4-bistrimethylsiloxypteridine	crude 160	414, 1022
6,7-Diphenyl-2,4-dipiperidinopteridine	180–181	467
6,7-Diphenyl-4-phenylhydrazino-2(1*H*)-pteridinethione	279–280	927
6,7-Diphenyl-4-phenylhydrazino-2(1*H*)-pteridinone	245	927
6,7-Diphenyl-2-piperidino-4-pteridinamine	209, uv	467
2,6-Diphenyl-4-pteridinamine	288–289, uv	326
2,7-Diphenyl-4-pteridinamine	252–254, uv	328, 1312
6,7-Diphenyl-2-pteridinamine	240–241 or 244, uv	400, 688, 979
6,7-Diphenyl-4-pteridinamine	175 to 210, nmr	400, 474, 688, 1740
2,7-Diphenyl-4-pteridinamine 5-oxide	258–260, uv	328
4,7-Diphenylpteridine	154–155, ^{13}C nmr	959
6,7-Diphenylpteridine	170–171, nmr	983, 1167
2,6-Diphenyl-4,7-pteridinediamine	368, uv	326, 327, 523, 1312, 1415
2,7-Diphenyl-4,6-pteridinediamine	280–281, uv	327
6,7-Diphenyl-2,4-pteridinediamine	282 to 296, uv	284, 472, 475,

APPENDIX TABLE (*Contd.*)

Compound	(°C), and so on	Reference(s)
		541, 602, 1179, 1189, 1277
6,7-Diphenyl-2,4-pteridinediamine 5,8-dioxide	346–347, uv	541
2,6-Diphenyl-4,7-pteridinediamine 5-oxide	>350, uv	328
1,7-Diphenyl-2,4(1*H*,3*H*)-pteridinedione	>290	797
6,7-Diphenyl-2,4(1*H*,3*H*)-pteridinedione	310 to 325, fl-sp, uv	240, 367, 399, 437, 446, 471, 474, 483, 699, 747, 1169, 1339, 1414, 1438, 1473, 1506
6,7-Diphenyl-2,4(1*H*,3*H*)-pteridinedione 8-oxide	293–295, uv	367
6,7-Diphenyl-2,4(1*H*,3*H*)-pteridinedithione	245 or 268, uv	437, 699, 927
6,7-Diphenyl-2(1*H*)-pteridinethione	189–190	400
6,7-Diphenyl-4(3*H*)-pteridinethione	270–280; Co —; Cu —; Hg 271	474, 754
6,7-Diphenyl-4(3*H*)-pteridinone	295 to 310, ir, ms, nmr, uv	32, 474, 587, 1130, 1167, 1316
6,7-Diphenyl-8-D-ribityl-2,4(3*H*,8*H*)-pteridinedione	212–218 or 220–223, uv	537, 674
6,7-Diphenyl-4-thiosemicarbazido-2(1*H*)-pteridinone	209	927
1,7-Diphenyl-2-thioxo-1,2-dihydro-4(3*H*)-pteridinone	>280	797
6,7-Diphenyl-2-thioxo-1,2-dihydro-4(3*H*)-pteridinone	196 or 203, fl-sp, uv; Ag crude; HgBr crude	437, 447, 1169, 1414
6,7-Diphenyl-4-thioxo-3,4-dihydro-2(1*H*)-pteridinone	270, uv	699, 927, 1679
6,7-Diphenyl-4-trimethylsiloxy-2-trimethylsilylthiopteridine	crude	447
2-Dipropylaminopteridine	55	124
6,7-Dipropyl-2,4-pteridinediamine	200 or 202	142, 749
6,7-Dipropyl-8-D-ribityl-2,4(3*H*,8*H*)-pteridinedione	130–137, uv	537
Drosopterin (not simple: see text)		1280, 1428, 1429
Ekapterin (not simple: see text)		1479
8-D-Erythrityl-6,7-dimethyl-2,4(3*H*,8*H*)-pteridinedione	anal, uv	713
8-L-Erythrityl-6,7-dimethyl-2,4(3*H*,8*H*)-pteridinedione	anal, uv	713
Erythropterin [see: 2-amino-7-oxalomethyl-4,6(3*H*,5*H*)-pteridinedione]		*Attempts*: 496, 498, 561, 816, 1313, 1548
3-β-Ethoxycarbonlyethyl-6,7-dimethyl-2-methylthio-4(3*H*)-pteridinone	156–159, uv	168
3-β-Ethoxycarbonylethyl-6,7-dimethyl-2,4(1H,3*H*)-pteridinedione	287–293, uv	168

APPENDIX TABLE (*Contd.*)

Compound	(°C), and so on	Reference(s)
6-β-Ethoxycarbonylethyl-2,4-pteridinediamine	266–267	1302
6-β-Ethoxycarbonylethyl-8-D-ribityl-2,4,7(1*H*,3*H*,8*H*)-pteridinetrione	anal	1248
6-Ethoxycarbonylmethyl-1,3-dimethyl-2,4,7(1*H*,3*H*,8*H*)-pteridinetrione	183	337
6-Ethoxycarbonylmethyl-4,7(3*H*,8*H*)-pteridinedione	250, uv	29
2-Ethoxy-6,7-dimethyl-2-trifluoromethylpteridine	132–134, nmr	1440
4-Ethoxy-6,7-diphenyl-2-pteridinamine	300, nmr, uv	1500
6-Ethoxy-2,7-diphenyl-4-pteridinamine	248–249	526
7-Ethoxy-2,6-diphenyl-4-pteridinamine	241–242, uv	526
4-Ethoxyimino-1,3-dimethyl-6,7-diphenyl-3,4-dihydro-2(1*H*)-pteridinone	157, nmr, uv	1679
2-Ethoxy-4-methyl-6,7-diphenylpteridine	185–187	398, 1230
6-Ethoxymethyl-2,4-pteridinediamine	247–248, ir, nmr	1728
2-Ethoxy-6-phenyl-4,7-pteridinediamine	290, uv	1281
2-Ethoxy-4-trifluoromethylpteridine	82, ms, nmr	159, 780
6-Ethoxy-2-trifluoromethyl-4(3*H*)-pteridinone	256	20
4-Ethoxy-6,7,8-trimethyl-2(8*H*)-pteridinimine	178–180, uv	190
Ethyl 2-acetamido-4-oxo-3,4-dihydro-7-pteridinecarboxylate	293–295, uv	389
Ethyl 4-amino-2-benzylthio-7-oxo-7,8-dihydro-6-pteridinecarboxylate	243–246, uv	1126
Ethyl 2-amino-3,8-dimethyl-4,7-dioxo-3,4,7,8-tetrahydro-6-pteridinecarboxylate	305 or 308, uv	385, 1311
Ethyl 2-amino-6,7-dimethyl-4-pteridinecarboxylate	150–153, nmr	1440
Ethyl 4-amino-2,7-dioxo-1,2,7,8-tetrahydro-6-pteridinecarboxylate	H_2O 308	363
4-Ethylamino-6,7-diphenylpteridine	160–162, ir, ms, nmr	1167
4-Ethylamino-6,7-diphenyl-2(1*H*)-pteridinone	>300	927
Ethyl 2-amino-8-β-hydroxyethyl-4,7-dioxo-3,4,7,8-tetrahydro-6-pteridinecarboxylate	284 or 313–315, uv	385, 882, 1090
Ethyl 2-amino-4-isopropoxy-8-methyl-7-oxo-7,8-dihydro-6-pteridinecarboxylate	192–194, uv	371
Ethyl 2-amino-4-isopropoxy-7-oxo-7,8-dihydro-6-pteridinecarboxylate	>320, uv	371
Ethyl 2-amino-4-methoxy-8-methyl-7-oxo-7,8-dihydro-6-pteridinecarboxylate	254–257, uv	371
2-Ethylamino-4-methylamino-6,7-diphenylpteridine	249	110
Ethyl 2-amino-3-methyl-4,7-dioxo-3,4,7,8-tetrahydro-6-pteridinecarboxylate	>350, uv	385
Ethyl 2-amino-8-methyl-4,7-dioxo-3,4,7,8-tetrahydro-6-pteridinecarboxylate	295–298, uv	385
4-Ethylamino-3-methyl-6,7-diphenyl-2(3*H*)-pteridinone	228, uv	1679
Ethyl 2-amino-8-methyl-7-oxo-7,8-dihydro-6-pteridinecarboxylate	201–203, ir, uv	562, 1311
Ethyl 2-amino-4-methylthio-7-oxo-7,8-dihydro-6-pteridinecarboxylate	306–307	1468
Ethyl 4-amino-2-methylthio-7-oxo-7,8-dihydro-6-pteridinecarboxylate	267–269, uv	1126

APPENDIX TABLE (*Contd.*)

Compound	(°C), and so on	Reference(s)
Ethyl 2-amino-4-oxo-3,4-dihydro-7-pteridinecarboxylate	355; HCl 280	389
2-Ethylamino-7-oxo-7,8-dihydro-6-pteridinecarboxylic acid	348, ir, uv	387, 562
Ethyl 4-amino-7-oxo-2-phenyl-7,8-dihydro-6-pteridinecarboxylate	234 or 255, uv	320, 1357
4-Ethylamino-7-phenylpteridine	234–235, ir, ms, nmr	1167
2-Ethylaminopteridine	137–138, nmr	934
4-Ethylaminopteridine	141–142, nmr	934
Ethyl 2-amino-4-pteridinecarboxylate	189, nmr; $2H_2O$ 162, nmr	758
2-Ethylamino-7(8*H*)-pteridinone	anal, uv	387
Ethyl 8-benzyl-3-methyl-2,4,7-trioxo-1,2,3,4,7,8-hexahydro-6-pteridinecarboxylate	177, uv	317
Ethyl 2-benzylthio-7-trimethylsiloxy-4-trimethylsilylamino-6-pteridinecarboxylate	crude	325
Ethyl 2-chloro-6,7-dimethyl-4-pteridinecarboxylate	77–78, nmr	1440
Ethyl 2-chloro-4-pteridinecarboxylate	79, ms, nmr; $2H_2O$ 119, nmr	758, 777
Ethyl 2,4-diamino-8-methyl-7-oxo-7,8-dihydro-6-pteridinecarboxylate	anal, uv	882
Ethyl 4,7-diamino-2-phenyl-6-pteridinecarboxylate	288–292, uv	1357
Ethyl 2,4-diamino-6-pteridinecarboxylate	298–300, uv	1354
Ethyl 2-dimethylamino-6,7-dimethyl-4-pteridinecarboxylate	140–141, nmr	1440
Ethyl 2-dimethylamino-4,7-dioxo-3,4,7,8-tetrahydro-6-pteridinecarboxylate	>350, uv	360
Ethyl 2-dimethylamino-8-ethyl-7-oxo-7,8-dihydro-6-pteridinecarboxylate	110, ir, uv	387, 562
Ethyl 2-dimethylamino-8-ethyl-7-oxo-4-phenyl-7,8-dihydro-6-pteridinecarboxylate	132–134, ir, uv	1311
Ethyl 2-dimethylamino-8-methyl-4,7-dioxo-3,4,7,8-tetrahydro-6-pteridinecarboxylate	296–298, uv	360
Ethyl 4-dimethylamino-8-methyl-7-oxo-7,8-dihydro-6-pteridinecarboxylate	124–125, ir, uv	562, 1311
Ethyl 2-dimethylamino-4-pteridinecarboxylate	112, nmr; $2H_2O$ 67, nmr	758
Ethyl 3,8-dimethyl-2-methylamino-4,7-dioxo-3,4,7,8-tetrahydro-6-pteridinecarboxylate	>300	1565
Ethyl 4,8-dimethyl-2-methylamino-7-oxo-7,8-dihydro-6-pteridinecarboxylate	220–222, ir, uv	1311
Ethyl 3,8-dimethyl-2-methylimino-7-oxo-2,3,7,8-tetrahydro-6-pteridinecarboxylate	nmr, uv; $MeC_6H_4SO_3H$ 249	376, 1311
Ethyl 6,7-dimethyl-2-methylthio-4-pteridinecarboxylate	111–112, nmr	1440
Ethyl 2,8-dimethyl-7-oxo-7,8-dihydro-6-pteridinecarboxylate	94–96	1311
6-Ethyl-1,3-dimethyl-7-phenyl-2,4(1*H*,3*H*)-pteridinedione	131–132, nmr, uv	1112, 1348
7-Ethyl-1,3-dimethyl-6-phenyl-2,4(1*H*,3*H*)-pteridinedione	170, nmr, uv	1112, 1348
N-Ethyl-6,7-dimethyl-4-pteridinecarboxamide	169, nmr, uv	157
Ethyl 6,7-dimethyl-4-pteridinecarboxylate	75, ms, nmr, uv	147, 777
6-Ethyl-1,3-dimethyl-2,4(1*H*,3*H*)-pteridinedione	139–141, ir, nmr	1291

APPENDIX TABLE (*Contd.*)

Compound	(°C), and so on	Reference(s)
8-Ethyl-6,7-dimethyl-2,4(3*H*,8*H*)-pteridinedione	294–295, ms, uv	378, 1545, 1553
Ethyl 1,8-dimethyl-2,4,7-trioxo-1,2,3,4,7,8-hexahydro-6-pteridinecarboxylate	275–277, uv	317
1-Ethyl-6,7-diphenyl-2,4(1*H*,3*H*)-pteridinedione	239–240	699
7-Ethyldithio-1,3,6-trimethyl-2,4(1*H*,3*H*)-pteridinedione	107, nmr, uv	228, 1709
Ethyl 2-ethoxy-6,7-dimethyl-4-pteridinecarboxylate	105, nmr	1440
Ethyl 2-ethoxy-4-pteridinecarboxylate	91, nmr; $2H_2O$ 125, nmr	758
8-Ethyl-2-ethylamino-4-methyl-7-oxo-7,8-dihydro-6-pteridinecarboxylic acid	219, ir, uv	387
8-Ethyl-2-ethylamino-4-methyl-7(8*H*)-pteridinone	155, uv	387
Ethyl 2-ethylamino-7-oxo-7,8-dihydro-6-pteridine-carboxylate	290, uv	387
8-Ethyl-2-ethylamino-7-oxo-7,8-dihydro-6-pteridine-carboxylic acid	217, ir, uv	387, 481, 562
8-Ethyl-4-ethylamino-7-oxo-7,8-dihydro-6-pteridine-carboxylic acid	191–193, ir, uv	562, 1311
8-Ethyl-2-ethylamino-7(8*H*)-pteridinone	155 or 164–165, ir, uv	387, 562, 1311
8-Ethyl-4-ethylamino-7(8*H*)-pteridinone	126–127, ir, uv	562, 1311
Ethyl 8-ethyl-2-ethylamino-4-methyl-7-oxo-7,8-dihydro-6-pteridinecarboxylate	127, ir, uv	387, 562
Ethyl 8-ethyl-2-ethylamino-7-oxo-7,8-dihydro-6-pteridinecarboxylate	144, ir, uv	387, 562
Ethyl 8-ethyl-4-ethylamino-7-oxo-7,8-dihydro-6-pteridinecarboxylate	88–93, ir uv	562, 1311
Ethyl 8-ethyl-2,4,7-trioxo-1,2,3,4,7,8-hexahydro-6-pteridinecarboxylate	275, uv	317
6-Ethyl-8-*β*-hydroxyethyl-2,4,7(1*H*,3*H*,8*H*)-pteridinetrione	329–331, uv	422
Ethyl 2-imino-3,8-dimethyl-4-oxo-2,3,4,8-tetrahydro-6-pteridinecarboxylate	HCl 243	70
Ethyl 2-imino-3,8-dimethyl-7-oxo-2,3,7,8-tetrahydro-6-pteridinecarboxylate	nmr, uv; $MeC_6H_4SO_3H$ 245	376, 1311
Ethyl 2-methoxy-8-methyl-7-oxo-7,8-dihydro-6-pteridinecarboxylate	100–102, ir, uv	562, 1311
Ethyl 2-methylamino-4,7-dioxo-3,4,7,8-tetrahydro-6-pteridinecarboxylate	>300	1565
Ethyl 8-methyl-2,7-dioxo-1,2,7,8-tetrahydro-6-pteridinecarboxylate	258–260, uv	562, 1311
Ethyl 3-methyl-2-methylamino-4,7-dioxo-3,4,7,8-tetrahydro-6-pteridinecarboxylate	>300	1565
Ethyl 8-methyl-2-methylamino-4,7-dioxo-3,4,7,8-tetrahydro-6-pteridinecarboxylate	>300	1565
Ethyl 8-methyl-2-methylamino-7-oxo-7,8-dihydro-6-pteridinecarboxylate	197–198, uv	1311
Ethyl 4-methyl-7-oxo-7,8-dihydro-6-pteridinecarboxylate	211, uv	57
Ethyl 8-methyl-7-oxo-7,8-dihydro-6-pteridinecarboxylate	115–116, ir, uv	562, 1311

APPENDIX TABLE (*Contd.*)

Compound	(°C), and so on	Reference(s)
Ethyl 8-methyl-7-oxo-2-phenyl-7,8-dihydro-6-pteridinecarboxylate	anal, ir, nmr, uv	1311
Ethyl 6-methyl-4-pteridinecarboxylate	82–83, nmr, uv	147
Ethyl 7-methyl-4-pteridinecarboxylate	142–144, ms, nmr, uv	147, 148, 777
6(7)-Ethyl-7(6)-methyl-2,4-pteridinediamine	HCl >300	1354
8-Ethyl-6-methyl-2,4,7(1*H*,3*H*,8*H*)-pteridinetrione	>330	378
Ethyl 2-methylthio-4-pteridinecarboxylate	89, nmr; $2H_2O$ 131, nmr	758
Ethyl 2-methylthio-7-trimethylsiloxy-4-trimethylsilylamino-6-pteridinecarboxylate	crude	325
Ethyl 2-oxo-1,2-dihydro-4-pteridinecarboxylate	H_2O 161, nmr	758
Ethyl 7-oxo-7,8-dihydro-6-pteridinecarboxylate	140, uv	57
6-Ethyl-2-phenyl-4,7-pteridinediamine	276–280, uv	326
8-Ethyl-6-phenyl-2,4,7(1*H*,3*H*,8*H*)-pteridinetrione	>350, uv	413
3-Ethyl-6-phenyl-4(3*H*)-pteridinone	215–216, nmr	1758
3-Ethyl-6-phenyl-4(3*H*)-pteridinone 8-oxide	288–290, nmr	1758
2-Ethyl-4-pteridinamine	170, uv	185
4-Ethyl-2-pteridinamine	186–188, uv	185
Ethyl 2-pteridinecarboxylate	128–129, nmr, uv	51, 54
Ethyl 4-pteridinecarboxylate	74–76, ms, nmr, uv	147, 148, 777
6-Ethyl-2,4-pteridinediamine 8-oxide	301, nmr, uv	484
7-Ethyl-2,4(1*H*,3*H*)-pteridinedione	anal, ms, nmr, uv	1286
8-Ethyl-2,4(3*H*,8*H*)-pteridinedione	>260, ms, uv	1545, 1553
8-Ethyl-2,4,6,7(1*H*,3*H*,5*H*,8*H*)-pteridinetetrone	>335, uv	674
6-Ethyl-2,4,7(1*H*,3*H*,8*H*)-pteridinetrione	ms	1411
8-Ethyl-2,4,7(1*H*,3*H*,8*H*)-pteridinetrione	>340, uv	317
2-Ethyl-4(3*H*)-pteridinone	300, ir, ms	980
7-Ethyl-4(3*H*)-pteridinone	ms, nmr, uv	1286
6-Ethyl-8-D-ribityl-2,4,7(1*H*,3*H*,8*H*)-pteridinetrione	158–160	674
2-Ethylthio-6,7-dimethyl-4(3*H*)-pteridinone	259–261	1234
2-Ethylthio-6,7-diphenyl-4-pteridinamine	224–225	955
2-Ethylthio-6,7-diphenyl-4(3*H*)-pteridinone	261–262 or 272–274	399, 955, 1169, 1234
4-Ethylthio-6,7-diphenyl-2(1*H*)-pteridinone	261	699
2-Ethylthio-3-methyl-6,7-diphenyl-4(3*H*)-pteridinone	300	955
2-Ethylthio-6-phenyl-4,7-pteridinediamine	272–274, uv	1281
2-Ethylthio-4(3*H*)-pteridinone	233–234 or 245–246	399, 1169, 1234
Ethyl 2-thioxo-1,2-dihydro-4-pteridinecarboxylate	H_2O 180, nmr	758
Ethyl 1,3,8-Trimethyl-2,4,7-trioxo-1,2,3,4,7,8-hexahydro-6-pteridinecarboxylate	239, uv	317, 341
Ethyl 2,4,6-trioxo-1,2,3,4,5,6-hexahydro-7-pteridinecarboxylate	317, uv	347
8-Ethyl-2,4,7-trioxo-1,2,3,4,7,8-hexahydro-6-pteridinecarboxylic acid	>330, uv	317
Ethyl 2,4,7-trioxo-8-D-ribityl-1,2,3,4,7,8-hexahydro-6-pteridinecarboxylate	128–129	1218
6-Ethyl-2,4,7-tristrimethylsiloxypteridine	crude, ms	1411

APPENDIX TABLE (*Contd.*)

Compound	(°C), and so on	Reference(s)
Euglenapterin [see: 2-dimethylamino-6-α,β,γ-trihydroxypropyl-4(3*H*)-pteridinone]		985
Fluorescyanine (a mixture)		270, 1274
Folic acid(s) (not simple: see text)		
Formicapterin (structure doubtful)		512, 1539
8-D-Galactityl-6,7-dimethyl-2,4(3*H*,8*H*)-pteridinedione	244–247, uv	537
4-Guanidino-2-pteridinamine	222, ms, nmr	1101
2-Hexylamino-6-phenyl-4,7-pteridinediamine	195–198, nmr, uv	523
7-Hexylamino-6-phenyl-2,4-pteridinediamine	198–200, uv	525
6-Hex-1′-ynyl-2-pivalamido-4(3*H*)-pteridinone	242–243, ir, nmr	1753
6-Hex-1′-ynyl-2,4-pteridinediamine	263–265, ir, nmr	1753
6-β-Hydrazinocarbonylethyl-8-β-hydroxyethyl-2,4,7(1*H*,3*H*,8*H*)-pteridinetrione	240–241, nmr, uv	1449
6-β-Hydrazinocarbonylethyl-8-D-ribityl-2,4,7(1*H*,3*H*,8*H*)-pteridinetrione	194–195, nmr, uv	1449
4-Hydrazino-6,7-dimethylpteridine	>199	135
6-Hydrazino-1,3-dimethyl-2,4(1*H*,3*H*)-pteridinedione	200	1275
4-Hydrazino-2-methylpteridine	222–224	135
7-Hydrazino-6-phenyl-2,4-pteridinediamine	356, nmr, uv	525
4-Hydrazino-2-pteridinamine	anal, uv; $NO_2C_6H_4CH=$ 311	305
2-Hydrazinopteridine	159	136
4-Hydrazinopteridine	215, uv	33, 119
7-Hydrazinopteridine	280	136
4-Hydrazino-2,6,7-trimethylpteridine	190–191	135
2-Hydrazino-1,6,7-trimethyl-4(1*H*)-pteridinone	~360, nmr	177
4-Hydroxyamino-6-hydroxymethyl-2-pteridinamine	—	815
4-Hydroxyamino-2-pteridinamine	$3H_2O$ anal, uv	305
4-Hydroxyaminopteridine	>310, ms, nmr	1003
6-α-Hydroxybenzyl-2,4,7-pteridinetriamine	275, uv	328
8-δ-Hydroxybutyl-6,7-dimethyl-2,4(3*H*,8*H*)-pteridinedione	237–239, nmr, uv	1686
6-γ-Hydroxybutyl-2,4-pteridinediamine	257–258	1302
4-β-Hydroxyethoxy-6,7-diphenyl-2-pteridinamine	239–241, nmr, uv	1500
8-β-Hydroxyethyl-6,7-bistrifluoromethyl-2,4(3*H*,8*H*)-pteridinedione	285–286, ir, ms, nmr, uv	1570
1-β-Hydroxyethyl-6,7-dimethyl-2,4(1*H*,3*H*)-pteridinedione	284–286, uv	1339
8-β-Hydroxyethyl-6,7-dimethyl-2,4(3*H*,8*H*)-pteridinedione	270–273 or 280, ms, nmr, uv	165, 378, 409, 422, 424, 1144, 1553
8-β-Hydroxyethyl-3,6-dimethyl-2,4,7(1*H*,3*H*,8*H*)-pteridinetrione	251–252	674
1-β-Hydroxyethyl-6,7-diphenyl-2,4(1*H*,3*H*)-pteridinedione	239, uv	1339
8-β-Hydroxyethyl-6,7-diphenyl-2,4(3*H*,8*H*)-pteridinedione	268, uv	409, 1545
8-β-Hydroxyethyl-6-hydroxymethyl-2,4,7(1*H*,3*H*,8*H*)-pteridinetrione	—	295
8-β-Hydroxyethyl-6-β-methoxycarbonylethyl-2,4,7(1*H*,3*H*,8*H*)-pteridinetrione	255–256, nmr, uv	1449

APPENDIX TABLE (*Contd.*)

Compound	(°C), and so on	Reference(s)
8-β-Hydroxyethyl-3-methyl-6,7-diphenyl-2,4(3*H*,8*H*)-pteridinedione	233, uv	409, 1545
8-β-Hydroxyethyl-3-methyl-2,4(3*H*,8*H*)-pteridinedione	250, ir, uv	409, 841, 1545
8-β-Hydroxyethyl-7-methyl-2,4(3*H*,8*H*)-pteridinedione	287–289, nmr, uv	1144
8-β-Hydroxyethyl-3-methyl-2,4,6,7(1*H*,3*H*,5*H*,8*H*)-pteridinetetrone	>320	674
8-β-Hydroxyethyl-6-methyl-2,4,7(1*H*,3*H*,8*H*)-pteridinetrione	>350, uv	295, 378, 422, 674, 712, 1082, 1339
6-α-Hydroxyethyl-2-phenyl-4,7-pteridinediamine	274–276, uv	328
8-β-Hydroxyethyl-6-phenyl-2,4,7(1*H*,3*H*,8*H*)-pteridinetrione	>320, uv	1339
6-α-Hydroxyethyl-2-piperidino-4,7-pteridinediamine	226–227, uv	328
6-α-Hydroxyethyl-4,7-pteridinediamine	248–250, uv	328
1-β-Hydroxyethyl-2,4(1*H*,3*H*)-pteridinedione	263, uv	1339
8-β-Hydroxyethyl-2,4(3*H*,8*H*)-pteridinedione	286–288, ms, uv	116, 409, 422, 1545, 1553
8-β-Hydroxyethyl-2,4,6,7(1*H*,3*H*,5*H*,8*H*)-pteridinetetrone	>320, uv	674
8-β-Hydroxyethyl-2,4,7(1*H*,3*H*,8*H*)-pteridinetrione	325–326 or >330, uv	317, 422, 463, 1144, 1339
8-β-Hydroxyethyl-3,6,7-trimethyl-2,4(3*H*,8*H*)-pteridinedione	224, uv	409, 1545
8-β-Hydroxyethyl-2,4,7-trioxo-1,2,3,4,7,8-hexahydro-6-pteridinecarboxylic acid	222, uv	317
8–6′-Hydroxyhexyl-6,7-dimethyl-2,4(3*H*,8*H*)-pteridinedione	236–238, nmr, uv	1686
6-Hydroxyiminomethyl-2,4-pteridinediamine 8-oxide	anal, uv	469, 484
6-Hydroxymethyl-1,3-dimethyl-2,4(1*H*,3*H*)-pteridinedione	209–210, uv	1704
6-Hydroxymethyllumazine [see: 6-Hydroxymethyl-2,4(1*H*,3*H*)-pteridinedione]		
6-Hydroxymethyl-7-methoxy-2,4-pteridinediamine	280, uv	1296
6-Hydroxymethyl-7-methyl-2,4-pteridinediamine	anal, ms, uv	847
6-Hydroxymethyl-4-methylthio-2-pteridinamine	—	815
7-α-Hydroxy-β-methylpropyl-2,4-pteridinediamine	248, nmr, uv	1678
6-Hydroxymethyl-2,4-pteridinediamine	333–334, nmr, uv; HBr anal; HCl >300	96, 113, 390, 595, 979, 1299, 1465, 1522, 1757
7-Hydroxymethyl-2,4-pteridinediamine	>300, uv	123
6-Hydroxymethyl-2,4(1*H*,3*H*)-pteridinedione (6-hydroxymethyllumazine)	260–262, uv	1031, 1546
6-Hydroxy-8-methyl-4(8*H*)-pteridinone	H_2O 287–288, uv	248
6-Hydroxymethyl-8-D-ribityl-2,4,7(1*H*,3*H*,8*H*)-pteridinetrione (photolumazine-B)	>265, uv	1430
8-ε-Hydroxypentyl-6,7-dimethyl-2,4(3*H*,8*H*)-pteridinedione	254–256, nmr, uv	1686
8-ε-Hydroxypentyl-6-methyl-2,4,7(1*H*,3*H*,8*H*)-pteridinetrione	250–280	674
4-ε-Hydroxypentyloxy-6,7-diphenyl-2-pteridinamine	198–199, nmr, uv	1500
8-ε-Hydroxypentyl-2,4,6,7(1*H*,3*H*,5*H*,8*H*)-pteridinetetrone	320	674
6-α-Hydroxypropyl-1,3-dimethyl-2,4(1*H*,3*H*)-pteridinedione	166–167, nmr, uv	108
8-β-Hydroxypropyl-6,7-dimethyl-2,4(3*H*,8*H*)-pteridinedione	276–278, nmr, uv	1686

APPENDIX TABLE (*Contd.*)

Compound	(°C), and so on	Reference(s)
8-γ-Hydroxypropyl-6,7-dimethyl-2,4(3*H*,8*H*)-pteridinedione	anal, uv	713
8-γ-Hydroxypropyl-6,7-diphenyl-2,4(3*H*,8*H*)-pteridinedione	282, uv	409
6-α-Hydroxypropyl-1-methyl-2,4(1*H*,3*H*)-pteridinedione (leucettidine)	anal, ir, ms, nmr, uv	354, 751
6-α-Hydroxypropyl-3-methyl-2,4(1*H*,3*H*)-pteridinedione	anal, nmr, uv	354
7-α-Hydroxypropyl-2,4-pteridinediamine	221–223, nmr, uv	1678
8-β-Hydroxypropyl-2,4,6,7(1*H*,3*H*,5*H*,8*H*)-pteridinetetrone	>300	674
8-γ-Hydroxypropyl-2,4,6,7(1*H*,3*H*,5*H*,8*H*)-pteridinetetrone	>300	674
8-γ-Hydroxypropyl-3,6,7-trimethyl-2,4(3*H*,8*H*)-pteridinedione	221–224, uv	1545
3-Hydroxy-2,4(1*H*,3*H*)-pteridinedione	325–326, ms, nmr	1099, 1289, 1445
3-Hydroxy-4(3*H*)-pteridinone [see: 4(1*H*)-pteridinone 3-oxide]		
Ichthyopterin [see also: 2-amino-6-α,β-dihydroxypropyl-4,7(3*H*,8*H*)-pteridinedione]		218, 235, 270, 995, 1033, 1304, 1549
2-Imino-3,8-dimethyl-2,8-dihydro-4(3*H*)-pteridinone	HCl >350, nmr, uv	359, 741, 762
4-Imino-1,3-dimethyl-3,4-dihydro-2(1*H*)-pteridinone	crude 148, ir, ms, nmr	1733
2-Imino-1,3-dimethyl-6,7-diphenyl-1,2-dihydro-4(3*H*)-pteridinone	241–242, uv	68
4-Imino-1,3-dimethyl-6,7-diphenyl-3,4-dihydro-2(1*H*)-pteridinone	206–207, nmr, uv	1679
2-Imino-3,8-dimethyl-6,7-diphenyl-2,3-dihydro-4(8*H*)-pteridinone	H_2O 156; HCl 241–243, uv	68, 359, 377
2-Imino-3,8-dimethyl-4-oxo-2,3,4,8-tetrahydro-6-pteridinecarboxylic acid	anal, uv; HCl —	70
2-Imino-1,3-dimethyl-6-phenyl-1,2-dihydro-4(3*H*)-pteridinone	251–253, uv	68
2-Imino-1,3-dimethyl-7-phenyl-1,2-dihydro-4(3*H*)-pteridinone	290–292, uv	68
2-Imino-3,8-dimethyl-6-phenyl-2,8-dihydro-4(3*H*)-pteridinone	HCl anal, uv	68
2-Imino-8-isopropyl-3-methyl-2,3-dihydro-4(8*H*)-pteridinone	HCl 250–253, uv	377
4-Imino-8-methyl-4,8-dihydro-2-pteridinamine	2HI 237, uv	130
7-Imino-1-methyl-1,7-dihydro-4-pteridinamine	273, uv; HI 270–272	131
4-Imino-1-methyl-6,7-diphenyl-1,4-dihydro-2-pteridinamine	256; HI 315	111
7-Imino-8-methyl-4-methylamino-6-phenyl-7,8-dihydro-2-pteridinamine	250–251, nmr, uv	523
2-Imino-3,6,7,8-tetramethyl-2,8-dihydro-4(3*H*)-pteridinone	HCl 220, uv	359
4-Imino-1,6,7-trimethyl-1,4-dihydro-2-pteridinamine	HI 280–285	130, 267
2-Imino-1,3,8-trimethyl-1,2-dihydro-4,7(3*H*,8*H*)-pteridinedione	237, uv	718
2-Imino-3,6,7-trimethyl-8-phenyl-2,3-dihydro-4(8*H*)-pteridinone	220, uv; HCl 238	377

APPENDIX TABLE (*Contd.*)

Compound	(°C), and so on	Reference(s)
6-Isobutoxymethyl-2,4-pteridinediamine 8-oxide	>250, ir, nmr, uv	1728
2-Isobutylamino-4,6,7-trimethylpteridine	164, uv	124
7-Isobutyl-2,4-dimethoxy-6(5*H*)-pteridinone	anal	1379
6-Isobutyl-1,3-dimethyl-7-phenyl-2,4(1*H*,3*H*)-pteridinedione	160, nmr, uv	1348
7-Isobutyl-1,3-dimethyl-6-phenyl-2,4(1*H*,3*H*)-pteridinedione	147, nmr, uv	1112, 1348
2-Isobutyl-4-pteridinamine	156, uv	185
6-Isobutyl-2,4-pteridinediamine	296–299	742
6-Isobutyl-2,4-pteridinediamine 8-oxide	279, nmr, uv	484
7-Isobutyl-2,4,6(1*H*,3*H*,5*H*)-pteridinetrione	crude, uv	1379
7-Isobutyryl-2,4-pteridinediamine	240–241, nmr, uv	1678
Isodrosopterin (not simple: see text)		
Isoleucopterin [see: 4-amino-2,6,7(1*H*,5*H*,8*H*)-pteridinetrione]		1369
6-Isopentyl-2,4-pteridinediamine	297–304	742
4-Isopropoxy-6,7-dimethyl-2-pteridinamine	231–232, nmr, uv	370, 741
6-Isopropoxymethyl-2,4-pteridinediamine	285–286, ir, nmr	1728
4-Isopropoxy-2-pteridinamine	167–169, nmr, uv	370, 741, 762
4-Isopropoxy-6,7,8-trimethyl-2(8*H*)-pteridinimine	192–195	359
4-Isopropoxy-7-trimethylsiloxy-2-trimethylsilylamino-pteridine	crude	435
2-Isopropylamino-6,7-diphenyl-4(3*H*)-pteridinone	324–325	476
6-Isopropylamino-3-methyl-4(3*H*)-pteridinone	crude	974
2-Isopropylamino-6-phenyl-4,7-pteridinediamine	236–237, uv	523
4-Isopropylamino-2-phenyl-6,7-pteridinediamine	—	1398
4-Isopropylamino-6-phenyl-2,7-pteridinediamine	>300, uv	523
2-Isopropylaminopteridine	136	124
Isopropyl 4,7-diamino-2-phenyl-6-pteridinecarboxylate	285–286	1357
Isopropyl 2,4-diamino-6-pteridinecarboxylate	293–294, uv	1354
7-Isopropyl-1,3-dimethyl-6-phenyl-2,4(1*H*,3*H*)-pteridinedione	212, nmr, uv	1112, 1348
7-Isopropyl-1,3-dimethyl-2,4(1*H*,3*H*)-pteridinedione	123	108
8-Isopropyl-6,7-dimethyl-2,4(3*H*,8*H*)-pteridinedione	220 or 250, ms, uv	375, 409, 1553
3-Isopropyl-6,7-diphenyl-2-thioxo-1,2-dihydro-4(3*H*)-pteridinone	270	476
3-Isopropylideneamino-4(3*H*)-pteridinone	210–211	487
2-Isopropylidenehydrazino-1,6,7-trimethyl-4(1*H*)-pteridinone	258–260, nmr	177
7-Isopropyl-3-methyl-2,4(1*H*,3*H*)-pteridinedione 5-oxide	ms	427
3-Isopropyl-2-methyl-4(3*H*)-pteridinone	202–205	187
3-Isopropyl-6-phenyl-4(3*H*)-pteridinone	160–161, nmr	1758
3-Isopropyl-6-phenyl-4(3*H*)-pteridinone 8-oxide	286–288, nmr	1758
4-Isopropyl-2-pteridinamine	162–164, uv	185
6-Isopropyl-2,4-pteridinediamine 8-oxide	285, nmr, uv	484
7-Isopropyl-2,4(1*H*,3*H*)-pteridinedione	ms, nmr, uv	1286
8-Isopropyl-2,4(3*H*,8*H*)-pteridinedione	273–279, ms	1553
6-Isopropyl-2,4,7-pteridinetriamine	343, uv	1415
3-Isopropyl-4(3*H*)-pteridinone	204–207	187
7-Isopropyl-4(3*H*)-pteridinone	ms, nmr, uv	1286
7-Isopropylsulfinyl-1,3,6-trimethyl-2,4(1*H*,3*H*)-pteridinedione	74–76, nmr, uv	228, 1709
7-Isopropylthio-1,3,6-trimethyl-2,4(1*H*,3*H*)-pteridinedione	162–163, nmr, uv	228, 1709

APPENDIX TABLE (*Contd.*)

Compound	(°C), and so on	Reference(s)
8-Isopropyl-3,6,7-trimethyl-2,4(3*H*,8*H*)-pteridinedione	236–237, uv	375, 409
Isopterin [see: 4-amino-2(1*H*)-pteridinone]		
Isosepiapterin (=deoxysepiapterin: not simple)		1127, 1558
Isoxanthopterin [see 2-amino-4,7(3*H*,8*H*)-pteridinedione]		—
Lepidopterin (not simple)		1479
Leucettidine [see: 6-α-hydroxypropyl-1-methyl-2,4(1*H*,3*H*)-pteridinedione]		354, 751
Leucopterin [see: 2-amino-4,6,7(3*H*,5*H*,8*H*)-pteridinetrione]		
Leucopterin-B [see: 2-amino-4,7(3*H*,8*H*)-pteridinedione]		1034
Luciopterin [see: 8-methyl-2,4,7(1*H*,3*H*,8*H*)-pteridinetrione]		1538
Lumazine [see: 2,4(1*H*,3*H*)-pteridinedione]		
6-Lumazinecarboxylic acid [see: 2,4-dioxo-1,2,3,4-tetrahydro-6-pteridinecarboxylic acid]		
8-D-Lyxityl-6,7-dimethyl-2,4(1*H*,8*H*)-pteridinedione	$2H_2O$ 132–140, uv	537
8-L-Lyxityl-6,7-dimethyl-2,4(3*H*,8*H*)-pteridinedione	uv	1444
8-D-Lyxityl-6-methyl-2,4,7(1*H*,3*H*,8*H*)-pteridinetrione	—	295
8-D-Mannityl-6,7-dimethyl-2,4(3*H*,8*H*)-pteridinedione	234–235, uv	1444
1-β-Mercaptoethyl-6,7-dimethyl-2,4(1*H*,3*H*)-pteridinedione	214–216, ms, uv	876
3-β-Mercaptoethyl-6,7-dimethyl-2,4(1*H*,3*H*)-pteridinedione	258–262, ms, uv	876
1-β-Mercaptoethyl-2,4(1*H*,3*H*)-pteridinedione	213–215, ms, uv	876
3-β-Mercaptoethyl-2,4(1*H*,3*H*)-pteridinedione	212–213, ms, uv	876
1-γ-Mercaptopropyl-6,7-dimethyl-2,4(1*H*,3*H*)-pteridinedione	193–194, ms, uv	876
3-γ-Mercaptopropyl-6,7-dimethyl-2,4(1*H*,3*H*)-pteridinedione	221–223, ms, uv	876
3-γ-Mercaptopropyl-2,4(1*H*,3*H*)-pteridinedione	192–194, ms, uv	876
Mesopterin [see: 2-amino-4,7(3*H*,8*H*)-pteridinedione (?)]		1314
1-β-Mesyloxyethyl-6,7-dimethyl-2,4(1*H*,3*H*)-pteridinedione	>150	1339
1-β-Mesyloxyethyl-6,7-diphenyl-2,4(1*H*,3*H*)-pteridinedione	181–184	1339
1-Mesyloxy-2,4(1*H*,3*H*)-pteridinedione	130, uv	1098
Methanopterin (not simple)		1002, 1459
Methotrexate (not simple: see text)		
6-β-Methoxycarbonylethyl-8-D-ribityl-2,4,7(1*H*,3*H*,8*H*)-pteridinetrione	148–149, nmr, uv	1449
7-Methoxycarbonylmethyl-6-methyl-2-pteridinamine 5,8-dioxide	202, ir, nmr, uv	723
7-Methoxy-1,3-dimethyl-2,4-dioxo-1,2,3,4-tetrahydro-6-pteridinecarboxylic acid	210, uv	341
7-Methoxy-3,6-dimethyl-1-phenyl-2,4(1*H*,3*H*)-pteridinedione	281, uv	238
7-Methoxy-3,6-dimethyl-1-phenyl-2,4(1*H*,3*H*)-pteridinedione 5-oxide	271, uv	238
6-Methoxy-1,3-dimethyl-7-propionyl-2,4(1*H*,3*H*)-pteridinedione	135, nmr, uv	108
7-Methoxy-1,3-dimethyl-6-propionyl-2,4(1*H*,3*H*)-pteridinedione	163, nmr, uv	108
2-Methoxy-6,7-dimethyl-4-pteridinamine	260–261, uv	362
4-Methoxy-6,7-dimethyl-2-pteridinamine	254–256, uv	370, 741
4-Methoxy-6,7-dimethylpteridine	128–129, uv	170

APPENDIX TABLE (*Contd.*)

Compound	(°C), and so on	Reference(s)
2-Methoxy-3,8-dimethyl-4,7(3*H*,8*H*)-pteridinedione	193–194	1192
6-Methoxy-1,3-dimethyl-2,4(1*H*,3*H*)-pteridinedione	223–224, uv	108, 209, 340, 369
7-Methoxy-1,3-dimethyl-2,4(1*H*,3*H*)-pteridinedione	195–196, nmr, uv	238, 339, 344, 716, 1000
7-Methoxy-1,3-dimethyl-2,4(1*H*,3*H*)-pteridinedione 5-oxide	254, uv	238
7-Methoxy-1,3-dimethyl-2,4(1*H*,3*H*)-pteridinedithione	211–213, uv	1000
7-Methoxy-1,3-dimethyl-2,4,6(1*H*,3*H*,5*H*)-pteridinetrione	288–291	238
2-Methoxy-6,8-dimethyl-7(8*H*)-pteridinone	182–183, uv	248
3-Methoxy-6,7-dimethyl-4(3*H*)-pteridinone	145, ms, nmr	759, 779
7-Methoxy-1,3-dimethyl-2-thioxo-1,2-dihydro-4(3*H*)-pteridinone	218–220, nmr, uv	1000
7-Methoxy-1,3-dimethyl-4-thioxo-3,4-dihydro-2(1*H*)-pteridinone	194–196, nmr, uv	1000
2-Methoxy-6,7-diphenyl-4-pteridinamine	228–229, uv	225, 362
4-Methoxy-6,7-diphenyl-2-pteridinamine	300, nmr, uv	1500
2-Methoxy-6,7-diphenyl-4(3*H*)-pteridinone	220	283
8-β-Methoxyethyl-6,7-diphenyl-2,4(3*H*,8*H*)-pteridinedione	278, uv	409
6-(β-Methoxyethoxy)methyl-2,4-pteridinediamine 8-oxide	241–242, ir, nmr, uv	1728
8-β-Methoxyethyl-3-methyl-6,7-diphenyl-2,4(3*H*,8*H*)-pteridinedione	203, uv	409
2-Methoxy-3-methyl-4-oxo-3,4-dihydro-7-pteridinecarbaldehyde	205–206, ir, nmr, uv	863
7-Methoxy-3-methyl-1-phenyl-2,4(1*H*,3*H*)-pteridinedione	254, uv	344
6-Methoxymethyl-2,4-pteridinediamine	248–250 or 255–256, nmr	1297, 1472, 1728
6-Methoxymethyl-2,4-pteridinediamine 8-oxide	254–256 or 265, nmr, uv	1297, 1728
2-Methoxy-3-methyl-4(3*H*)-pteridinone	190, uv	338, 1042
3-Methoxy-2-methyl-4(3*H*)-pteridinone	183, ms, nmr	759, 779
3-Methoxy-6-methyl-4(3*H*)-pteridinone	215–217, ms, nmr	759, 779
3-Methoxy-7-methyl-4(3*H*)-pteridinone	214–215, ms, nmr	759, 779
6-Methoxy-3-methyl-4(3*H*)-pteridinone	232–234	974
2-Methoxy-3-methyl-7-[D-*arabino*]-α,β,γ,δ-tetrahydroxybutyl-4(3*H*)-pteridinone	182–183, ir, nmr, uv	863
2-Methoxy-7-phenyl-4-pteridinamine	—	621
4-Methoxy-7-phenyl-2-pteridinamine	235, uv	1092
4-Methoxy-6-phenyl-2,7-pteridinediamine	258 or 275, nmr, uv	328, 523
7-Methoxy-6-phenyl-2,4-pteridinediamine	275, nmr, uv	525
7-β-Methoxypropyl-2,4-pteridinediamine	223–224	1305
6-3′-Methoxyprop-1′-ynyl-2-pivalamido-4(3*H*)-pteridinone	225–226, ir, nmr	1753
6-3′-Methoxyprop-1′-ynyl-2,4-pteridinediamine	>300, ir, nmr	1753
2-Methoxy-4-pteridinamine	224–225, uv	185, 225, 362
4-Methoxy-2-pteridinamine	204–205 or 207–209, nmr, uv	130, 133, 368, 370, 416, 741, 762
7-Methoxy-4-pteridinamine	310, uv	445

APPENDIX TABLE (*Contd.*)

Compound	(°C), and so on	Reference(s)
2-Methoxypteridine	149–151, ir, nmr, uv	27, 32, 119, 134, 285, 286
4-Methoxypteridine	193 or 195, ir, nmr, uv	32, 33, 34, 119, 134, 138, 182, 285, 286, 538
6-Methoxypteridine	124–125, ir, uv	31, 32, 119, 285, 286
7-Methoxypteridine	130, ir, uv	33, 119, 134, 285, 286
6-Methoxy-2,4-pteridinediamine	280 or >300, nmr, uv; HCl >300	268, 1301
7-Methoxy-2,4-pteridinediamine	240–250, nmr, uv	1301
4-Methoxy-6,7(5*H*,8*H*)-pteridinedione	>350, uv	208
2-Methoxy-4(3*H*)-pteridinone	uv	1042
3-Methoxy-4(3*H*)-pteridinone	197, ms, nmr	759, 779
4-Methoxy-2(1*H*)-pteridinone	185, ir, ms, nmr	1733
6-Methoxy-4(3*H*)-pteridinone	283–286	974
2-Methoxy-4-trifluoromethylpteridine	110, ms, nmr, uv	159, 780
2-Methoxy-3,6,8-trimethyl-4,7(3*H*,8*H*)-pteridinedione	243, nmr, uv	344, 716
7-Methoxy-1,3,6-trimethyl-2,4(1*H*,3*H*)-pteridinedione	237 or 241, nmr, uv	238, 339, 716
7-Methoxy-1,3,6-trimethyl-2,4(1*H*,3*H*)-pteridinedione 5-oxide	248, uv	238
4-Methoxy-6,7,8-trimethyl-2(8*H*)-pteridinimine	194–196, uv	359
3-Methoxy-2,6,7-trimethyl-4(3*H*)-pteridinone	166–167, uv	759
4-Methoxy-7-trimethylsiloxy-2-trimethylsilylaminopteridine	crude	435
7-Methoxy-1,3,6-trimethyl-2-thioxo-1,2-dihydro-4(3*H*)-pteridinone	229–231, nmr, uv	1000
Methyl 2-acetamido-4-oxo-3,4-dihydro-6-pteridinecarboxylate	283	389
2-Methylamino-4,7-dioxo-3,4,7,8-tetrahydro-6-pteridinecarboxylic acid	>300	1214, 1565
2-Methylamino-6,7-diphenyl-4-pteridinamine	307	110, cf. 472
4-Methylamino-6,7-diphenyl-2-pteridinamine	238(?) or 272	110, 141
2-Methylamino-6,7-diphenyl-4(3*H*)-pteridinone	346 to 365, uv	68, 110, 1189
Methyl 2-amino-6-methyl-4-oxo-3,4-dihydro-7-pteridinecarboxylate	307, uv	307
Methyl 2-amino-7-methyl-4-oxo-3,4-dihydro-6-pteridinecarboxylate	anal	1363
Methyl 2-amino-8-methyl-4-oxo-4,8-dihydro-6-pteridinecarboxylate	$MeC_6H_4SO_3H$ 212–228, uv	307
Methyl 2-amino-8-methyl-7-oxo-7,8-dihydro-6-pteridinecarboxylate	237, uv	1311
Methyl 4-amino-2-methylthio-7-oxo-7,8-dihydro-6-pteridinecarboxylate	260	1126

APPENDIX TABLE (*Contd.*)

Compound	(°C), and so on	Reference(s)
4-Methylamino-2-methylthio-7-propionylpteridine	155–157, nmr, uv	1678
4-Methylamino-2-methylthiopteridine	140–142, nmr, uv	1678
Methyl 2-amino-4-oxo-3,4-dihydro-6-pteridinecarboxylate	285, uv	389
Methyl 2-amino-4-oxo-3,4-dihydro-7-pteridinecarboxylate	>300, uv; HCl 250	389, 748
2-Methylamino-4-oxo-3,4-dihydro-6-pteridinecarboxylic acid	uv	70
2-Methylamino-4-oxo-3,4-dihydro-7-pteridinecarboxylic acid	uv	70
2-Methylamino-6-phenyl-4,7-pteridinediamine	235–237, nmr, uv	523
4-Methylamino-6-phenyl-2,7-pteridinediamine	330, nmr, uv	523
7-Methylamino-6-phenyl-2,4-pteridinediamine	HCl >340, nmr, uv	525
2-Methylamino-6-phenyl-4(3*H*)-pteridinone	356, uv	110
2-Methylamino-7-phenyl-4(3*H*)-pteridinone	387, uv	67, 110
2-Methylamino-7-propionyl-4-pteridinamine	210–212, nmr, uv	1678
2-Methylamino-4-pteridinamine	240 or 242, nmr, uv	110, 1678
4-Methylamino-2-pteridinamine	248	110
4-Methylamino-7-pteridinamine	243–244, uv	131
2-Methylaminopteridine	219–220, nmr, uv	27, 32, 39, 119
4-Methylaminopteridine	250 or 251–252, ir, nmr, uv; HI 234	53, 129
7-Methylaminopteridine	302–304, uv	131
6-Methylamino-2,4-pteridinediamine	192–194	1188
2-Methylamino-4,7(3*H*,8*H*)-pteridinedione	>300	1565
2-Methylamino-4(3*H*)-pteridinone	378, nmr, uv	130, 133, 332, 368, 687, 741, 762
2-Methylamino-7(8*H*)-pteridinone	>340, uv	132
4-Methylamino-2(1*H*)-pteridinone	317–320, uv	362
4-Methylamino-6(5*H*)-pteridinone	300, uv	445
4-Methylamino-7(8*H*)-pteridinone	310–330, uv	445
Methyl 8-benzyl-2,4,7-trioxo-1,2,3,4,7,8-hexahydro-6-pteridinecarboxylate	261–262, uv	317
6-Methyl-4,7-bistrimethylsiloxy-2-trimethylsilylamino-pteridine	crude, ms	1457
Methyl 2,4-diamino-7-methyl-6-pteridinecarboxylate	—	884
Methyl 4,7-diamino-2-phenyl-6-pteridinecarboxylate	286	1357
Methyl 2,4-diamino-6-pteridinecarboxylate	>300, uv	1354
Methyl 2,4-diamino-7-pteridinecarboxylate	uv	748
Methyl 4,8-dimethyl-2-methylamino-7-oxo-7,8-dihydro-6-pteridinecarboxylate	254–255, nmr, uv	1311
Methyl 3,8-dimethyl-2-methylimino-7-oxo-2,3,7,8-tetrahydro-6-pteridinecarboxylate	$MeC_6H_4SO_3H$ 246, nmr, uv	376, 1311
Methyl 2,8-dimethyl-7-oxo-7,8-dihydro-6-pteridine-carboxylate	153–154, ir, nmr, uv	1311
1-Methyl-2,4-dioxo-1,2,3,4-tetrahydro-6-pteridine-carbaldehyde	232–233	962

APPENDIX TABLE (*Contd.*)

Compound	(°C), and so on	Reference(s)
1-Methyl-2,4-dioxo-1,2,3,4-tetrahydro-7-pteridine-carbaldehyde	252–254; PhNHN= 312; $H_2NCSNHN=$ 285	232
Methyl 2,4-dioxo-1,2,3,4-tetrahydro-7-pteridinecarboxylate	>300, uv	748
1-Methyl-2,4-dioxo-1,2,3,4-tetrahydro-6-pteridinecarboxylic acid	276–280, uv	70, 962
3-Methyl-4,6-dioxo-3,4,5,6-tetrahydro-7-pteridinecarboxylic acid	>340; $2H_2O$ anal, uv	346
8-Methyl-2,7-dioxo-1,2,7,8-tetrahydro-6-pteridinecarboxylic acid	>320, uv	562, 1311
1-Methyl-2,4-dioxo-1,2,3,4-tetrahydro-6,7-pteridinedicarbaldehyde	230–234; bis-PhNHN= 317	232
1-Methyl-6,7-diphenyl-4-phenylhydrazino-2(1*H*)-pteridinone	269, nmr, uv	1679
3-Methyl-6,7-diphenyl-4-phenylhydrazino-2(1*H*)-pteridinone	>270, nmr, uv	1679
4-Methyl-6,7-diphenyl-2-pteridinamine	283–285, uv	744
7-Methyl-2,6-diphenyl-4-pteridinamine	288–290, uv	326
4-Methyl-6,7-diphenylpteridine	184 or 187	9, 1227
1-Methyl-6,7-diphenyl-2,4(1*H*,3*H*)-pteridinedione	263 to 280, uv	68, 111, 362, 367, 414, 962, 1019, 1339, 1506
3-Methyl-6,7-diphenyl-2,4(1*H*,3*H*)-pteridinedione	307 or 309–310, uv	111, 362, 367, 409, 414, 1339, 1506
8-Methyl-6,7-diphenyl-2,4(3*H*,8*H*)-pteridinedione	276 to 303, pol, uv	537, 1278, 1284, 1545
1-Methyl-6,7-diphenyl-2,4(1*H*,3*H*)-pteridinedione 5-oxide	285, uv	210, 367
3-Methyl-6,7-diphenyl-2,4(1*H*,3*H*)-pteridinedione 8-oxide	319–320, ms, uv	367, 427
1-Methyl-6,7-diphenyl-2,4(1*H*,3*H*)-pteridinedithione	375	111
3-Methyl-6,7-diphenyl-2,4(1*H*,3*H*)-pteridinedithione	306–309, uv	446
1-Methyl-6,7-diphenyl-4(3*H*)-pteridinone	>295, nmr, uv	587, 1130
2-Methyl-6,7-diphenyl-4(3*H*)-pteridinone	356–358	1228
3-Methyl-6,7-diphenyl-4(3*H*)-pteridinone	278–279, nmr, uv; 1-MeI 278–280, ir	587, 1130
8-Methyl-6,7-diphenyl-2(8*H*)-pteridinone	H_2O 220 or >300, uv	189, 248
1-Methyl-6,7-diphenyl-2-thioxo-1,2-dihydro-4(3*H*)-pteridinone	289, uv	111, 437, 447
1-Methyl-6,7-diphenyl-4-thioxo-3,4-dihydro-2(1*H*)-pteridinone	>270, uv	1679
3-Methyl-6,7-diphenyl-2-thioxo-1,2-dihydro-4(3*H*)-pteridinone	237–239, uv	446, 447, 612
3-Methyl-6,7-diphenyl-4-thioxo-3,4-dihydro-2(1*H*)-pteridinone	>270, uv	1679
Methyl 8-ethyl-2-ethylamino-7-oxo-7,8-dihydro-6-pteridinecarboxylate	170, ir, uv	387, 562

APPENDIX TABLE (*Contd.*)

Compound	(°C), and so on	Reference(s)
Methyl 8-β-hydroxyethyl-2,4,7-trioxo-1,2,3,4,7,8-hexahydro-6-pteridinecarboxylate	287, uv	317
Methyl 2-imino-3,8-dimethyl-7-oxo-2,3,7,8-tetrahydro-6-pteridinecarboxylate	nmr, uv; $MeC_6H_4SO_3H$ 233	376, 1311
Methyl 7-methoxy-1,3-dimethyl-2,4-dioxo-1,2,3,4-tetrahydro-6-pteridinecarboxylate	245–246, uv	341
3-Methyl-2-methylamino-4,7-dioxo-3,4,7,8-tetrahydro-6-pteridinecarboxylic acid	>300	1565
8-Methyl-2-methylamino-4,7-dioxo-3,4,7,8-tetrahydro-6-pteridinecarboxylic acid	>300	70, 1565
1-Methyl-2-methylamino-6,7-diphenyl-4(1*H*)-pteridinethione	300	111
1-Methyl-2-methylamino-6,7-diphenyl-4(1*H*)-pteridinone	304–306 or 307, uv	68, 111
1-Methyl-4-methylamino-6,7-diphenyl-2(1*H*)-pteridinone	>270, uv	1679
3-Methyl-4-methylamino-6,7-diphenyl-2(3*H*)-pteridinone	>300, nmr, uv	1679
8-Methyl-2-methylamino-6,7-diphenyl-4(8*H*)-pteridinone	HCl anal, uv	68, 377
8-Methyl-2-methylamino-4-oxo-4,8-dihydro-6-pteridinecarboxylic acid	anal, uv	70
8-Methyl-2-methylamino-7-oxo-7,8-dihydro-6-pteridinecarboxylic acid	247–248, ir, nmr, uv	1311
8-Methyl-2-methylamino-6-phenyl-7(8*H*)-pteridinone	223–224, uv	248
8-Methyl-2-methylamino-4,7(3*H*,8*H*)-pteridinedione	>300	1565
8-Methyl-2-methylamino-6,7(5*H*,8*H*)-pteridinedione	>360, uv	208
6-Methyl-2-methylamino-4(3*H*)-pteridinone	>250, ir, nmr	80
6-Methyl-2-methylamino-4(3*H*)-pteridinone 8-oxide	>300, nmr, uv	485
1-Methyl-4-methylimino-1,4-dihydro-7-pteridinamine	HI 295–296, uv	131
1-Methyl-4-methylimino-1,4-dihydropteridine	HCl anal, uv; HI 274	129
1-Methyl-4-methylimino-6,7-diphenyl-2-pteridinamine	256	111
Methyl 8-methyl-2-methylamino-7-oxo-7,8-dihydro-6-pteridinecarboxylate	245–246, ir, nmr, uv	1311
Methyl 8-methyl-7-oxo-2-phenyl-7,8-dihydro-6-pteridinecarboxylate	202–203, ir, nmr, uv	1311
3-Methyl-2-methylthio-6,7-diphenyl-4(3*H*)-pteridinethione	285, uv	437
1-Methyl-2-methylthio-6,7-diphenyl-4(1*H*)-pteridinone	254, uv	437, 446
1-Methyl-4-methylthio-6,7-diphenyl-2(1*H*)-pteridinone	272, uv	1679
3-Methyl-2-methylthio-6,7-diphenyl-4(3*H*)-pteridinone	293, uv	367, 437, 446
6-Methyl-2-methylthio-7-phenyl-4-pteridinamine	268–269, uv	326
3-Methyl-2-methylthio-6-phenyl-4(3*H*)-pteridinone	233	367
1-Methyl-7-methylthio-6-propionyl-2,4(1*H*,3*H*)-pteridinedione	—	354
3-Methyl-7-methylthio-6-propionyl-2,4(1*H*,3*H*)-pteridinedione	—	354
4-Methyl-2-methylthiopteridine	140–142, uv	46
7-Methyl-4-methylthiopteridine	187–188, uv	33
6-Methyl-2-methylthio-4,7-pteridinediamine	308–310, uv	326
1-Methyl-7-methylthio-2,4(1*H*,3*H*)-pteridinedione	—	354
3-Methyl-2-methylthio-4,7(3*H*,8*H*)-pteridinedione	293–294, uv	344

APPENDIX TABLE (*Contd.*)

Compound	(°C), and so on	Reference(s)
3-Methyl-7-methylthio-2,4(1*H*,3*H*)-pteridinedione	—	354
1-Methyl-2-methylthio-4(1*H*)-pteridinone	256, uv	437, 446
3-Methyl-2-methylthio-4(3*H*)-pteridinone	213, uv	437, 446
3-Methyl-2-methylthio-7-[D-*arabino*]-α,β,γ,δ-tetrahydroxybutyl-4(3*H*)-pteridinone	210–212, ir, nmr, uv	863
Methyl 2-methylthio-7-trimethylsiloxy-4-trimethylsilylamino-6-pteridinecarboxylate	—	325
3-Methyl-8-β-morpholinoethyl-6,7-diphenyl-2,4(3*H*,8*H*)-pteridinedione	224, uv	409
2-Methyl-6-morpholino-7(8*H*)-pteridinone	—	591
2-Methyl-7-morpholino-6(5*H*)-pteridinone	—	575, 591
N-Methyl-7-oxo-7,8-dihydro-6-pteridinecarboxamide	>250, uv	57
6-Methyl-2-phenyl-4-pteridinamine	240–241 or 242–244, nmr, uv	482, 1354
7-Methyl-2-phenyl-4-pteridinamine	269–271, uv	328
7-Methyl-2-phenyl-4-pteridinamine 5-oxide	287, uv	328
2-Methyl-4-phenylpteridine	120–121, nmr, uv	151
4-Methyl-2-phenylpteridine	150, nmr, uv	152
7-Methyl-2-phenylpteridine	181–182, nmr, uv	152
7-Methyl-4-phenylpteridine	179–180, nmr, uv	151
2-Methyl-6-phenyl-4,7-pteridinediamine	316–317, uv	523
6-Methyl-2-phenyl-4,7-pteridinediamine	308–309, uv	326, 1415
7-Methyl-6-phenyl-2,4-pteridinediamine	330 or 335, nmr, uv	525, 1179
1-Methyl-6-phenyl-2,4(1*H*,3*H*)-pteridinedione	327	543
1-Methyl-7-phenyl-2,4(1*H*,3*H*)-pteridinedione	349–350	174
3-Methyl-6-phenyl-2,4(1*H*,3*H*)-pteridinedione	327–328, nmr, uv	367
3-Methyl-7-phenyl-2,4(1*H*,3*H*)-pteridinedione	345, nmr, uv	331, 367
3-Methyl-8-phenyl-2,4(3*H*,8*H*)-pteridinedione	308–310, uv	409
8-Methyl-6-phenyl-2,7(1*H*,8*H*)-pteridinedione	315–317, uv	248
3-Methyl-7-phenyl-2,4(1*H*,3*H*)-pteridinedione 5,8-dioxide	275, nmr, uv	331, 367
3-Methyl-6-phenyl-2,4(1*H*,3*H*)-pteridinedione 8-oxide	340, ms, nmr, uv	331, 367, 427
3-Methyl-7-phenyl-2,4(1*H*,3*H*)-pteridinedione 5-oxide	303–305, ms, nmr, uv	367, 427
1-Methyl-6-phenyl-2,4,7(1*H*,3*H*,8*H*)-pteridinetrione	335–340, uv	413
3-Methyl-1-phenyl-2,4,7(1*H*,3*H*,8*H*)-pteridinetrione	362, uv	344
3-Methyl-6-phenyl-2,4,7(1*H*,3*H*,8*H*)-pteridinetrione	>350, uv	331, 367, 413
3-Methyl-7-phenyl-2,4,6(1*H*,3*H*,5*H*)-pteridinetrione	>360, uv	331, 367
3-Methyl-7-phenyl-2,4,6(1*H*,3*H*,5*H*)-pteridinetrione 8-oxide	>350, ms, nmr, uv	331
2-Methyl-3-phenyl-4(3*H*)-pteridinone	237	980
3-Methyl-6-phenyl-4(3*H*)-pteridinone	235–236, nmr	1758
7-Methyl-2-phenyl-4(3*H*)-pteridinone	>300, nmr, uv	152
3-Methyl-6-phenyl-4(3*H*)-pteridinone 8-oxide	304–305, nmr	1758
3-Methyl-6-piperidino-4(3*H*)-pteridinone	207–209	974
1-Methyl-7-propionyl-2,4(1*H*,3*H*)-pteridinedione	215–217, nmr, uv	108
3-Methyl-6-propionyl-2,4(1*H*,3*H*)-pteridinedione	—	354, 1463
3-Methyl-7-propionyl-2,4(1*H*,3*H*)-pteridinedione	237–239, nmr, uv	108
7-Methyl-6-propyl-8-D-ribityl-2,4(3*H*,8*H*)-pteridinedione	267–268 (?) or 283, nmr, uv	537 (?), 674

APPENDIX TABLE (*Contd.*)

Compound	(°C), and so on	Reference(s)
2-Methyl-4-pteridinamine	234–235, uv	39, 185
4-Methyl-2-pteridinamine	289 or >290, nmr, uv	27, 39, 64, 66, 185, 267, 281, 582, 979
6-Methyl-2-pteridinamine	>250, nmr, uv	27, 39, 688
7-Methyl-2-pteridinamine	ir, nmr, uv	27, 39
7-Methyl-4-pteridinamine	246–248, nmr	1740
2-Methyl-4-pteridinamine 3-oxide	278–280, ms, nmr	999, 1003
7-Methyl-4-pteridinamine 5-oxide	250–251, uv	328
2-Methylpteridine	141–142, ir, ms, nmr, uv	27, 33, 51, 52, 119, 207, 242, 285, 286, 293
4-Methylpteridine	152–153, ir, ms, nmr, uv	27, 33, 64, 66, 119, 207, 242, 281, 285, 286, 293, 1513
6-Methylpteridine	95(?) or 130, nmr, pol, uv	51, 54, 792, 828
7-Methylpteridine	128 or 134, ir, ms, nmr, ^{13}C nmr, uv	27, 33, 51, 119, 207, 242, 285, 286, 293, 294*, 959, 1037
6-Methyl-2,4-pteridinediamine	>360, ir, nmr, uv	84*, 113, 123, 284(?), 484, 595, 945, 973, 1254, 1263
6-Methyl-4,7-pteridinediamine	314–320 or 330, uv	326, 729, 730, 979
7-Methyl-2,4-pteridinediamine	>330, uv	284(?), 730, 944, 973, 1263, 1305, 1475
6-Methyl-2,4-pteridinediamine 8-oxide	>320, nmr, uv	478, 484
7-Methyl-2,4-pteridinediamine 8-oxide	295–296	1305
1-Methyl-2,4(1*H*,3*H*)-pteridinedione	271 to 291, ir, ms, nmr, uv	130, 338, 367, 414, 1042, 1339, 1733
1-Methyl-4,7(1*H*,8*H*)-pteridinedione	275–280, uv	132
3-Methyl-2,4(1*H*,3*H*)-pteridinedione	332, uv	130, 338, 367, 414, 1042, 1339, 1463
3-Methyl-4,6(3*H*,5*H*)-pteridinedione	>340, uv	101, 346
3-Methyl-4,7(3*H*,8*H*)-pteridinedione	>340, uv	346, 383

APPENDIX TABLE (*Contd.*)

Compound	(°C), and so on	Reference(s)
4-Methyl-2,6(1*H*,5*H*)-pteridinedione	>300, uv	40
4-Methyl-2,7(1*H*,8*H*)-pteridinedione	>300, uv	248
4-Methyl-6,7(5*H*,8*H*)-pteridinedione	>285, nmr, uv	64
5-Methyl-6,7(5*H*,8*H*)-pteridinedione	anal, ir, uv; H_2O 247–249	134
6-Methyl-2,4(1*H*,3*H*)-pteridinedione	330–332, ms, uv, H_2O 274, then 320	170, 1129, 1232, 1554; mixture: 1012, 1438
6-Methyl-4,7(3*H*,8*H*)-pteridinedione	H_2O 345 or >350, uv	29, 279
7-Methyl-2,4(1*H*,3*H*)-pteridinedione	>330, ms, nmr, uv	655, 1092, 1286, 1554; mixture: 1012, 1438
7-Methyl-4,6(3*H*,5*H*)-pteridinedione	uv	717
8-Methyl-2,4(3*H*,8*H*)-pteridinedione	>260, ms, uv	116, 264, 1545, 1553
8-Methyl-4,7(3*H*,8*H*)-pteridinedione	326 or 360, uv	248, 445
8-Methyl-6,7(5*H*,8*H*)-pteridinedione	309–311	34
1-Methyl-2,4(1*H*,3*H*)-pteridinedione 5-oxide	268–269, ms, uv	210, 366, 367, 427
3-Methyl-2,4(1*H*,3*H*)-pteridinedione 8-oxide	299, ms, uv	367, 427
6-Methyl-2,4(1*H*,3*H*)-pteridinedione 8-oxide	>320, nmr, uv	478, 485
1-Methyl-2,4,6,7(1*H*,3*H*,5*H*,8*H*)-pteridinetetrone	>340, uv	343, 404, 1152
3-Methyl-2,4,6,7(1*H*,3*H*,5*H*,8*H*)-pteridinetetrone	>340, uv	343
8-Methyl-2,4,6,7(1*H*,3*H*,5*H*,8*H*)-pteridinetetrone	>340, uv	343
2-Methyl-4(3*H*)-pteridinethione	>200	135
4-Methyl-2(1*H*)-pteridinethione	H_2O >180, uv	46
7-Methyl-4(3*H*)-pteridinethione	>300, uv	33
6-Methyl-2,4,7-pteridinetriamine	>340, uv	1415
1-Methyl-2,4,6(1*H*,3*H*,5*H*)-pteridinetrione	330 to >340, uv	101, 340, 367
1-Methyl-2,4,7(1*H*,3*H*,8*H*)-pteridinetrione	>340, nmr, uv	339, 413, 716
2-Methyl-4,6,7(3*H*,5*H*,8*H*)-pteridinetrione	>300, uv	918
3-Methyl-2,4,6(1*H*,3*H*,5*H*)-pteridinetrione	>350, uv	340, 342
3-Methyl-2,4,7(1*H*,3*H*,8*H*)-pteridinetrione	>340, nmr, uv	339, 344, 413, 716
6-Methyl-2,4,7(1*H*,3*H*,8*H*)-pteridinetrione	>360, ms, uv	279, 339, 930, 1109, 1110, 1304, 1411, 1524
7-Methyl-2,4,6(1*H*,3*H*,5*H*)-pteridinetrione	>360, ir, ms, uv	279, 340, 352, 717, 1092, 1110, 1524, 1554
8-Methyl-2,4,7(1*H*,3*H*,8*H*)-pteridinetrione (luciopterin)	>350, nmr, uv	317, 339, 344, 716, 1538

APPENDIX TABLE (*Contd.*)

Compound	(°C), and so on	Reference(s)
1-Methyl-4(1*H*)-pteridinimine	HCl >300, uv; HI 255	129
1-Methyl-7(1*H*)-pteridinimine	HI 195, uv; HI_3 204	131
3-Methyl-4(3*H*)-pteridinimine	215; HI 210; nmr, uv	53, 132
3-Methyl-7(3*H*)-pteridinimine	HI 250, uv	131
1-Methyl-2(1*H*)-pteridinone	H_2O 150 or 153, ir, uv; EtOH 155	34, 120, 134
1-Methyl-4(1*H*)-pteridinone	220–222 or 223–224, ir, nmr, uv	34, 129, 131, 134, 182, 285, 288, 538, 587, 610, 1130
2-Methyl-4(3*H*)-pteridinone	334, ir, ms, nmr, uv	20, 38, 153, 187, 779, 980, 1003, 1004
2-Methyl-6(5*H*)-pteridinone	>200, ms, nmr, uv	27, 55, 241, 244, 1246
2-Methyl-7(8*H*)-pteridinone	265, ms, uv	55, 1246
3-Methyl-2(3*H*)-pteridinone	H_2O 280, ir, uv	34, 134
3-Methyl-4(3*H*)-pteridinone	286 or 294–296, ir, nmr, uv; 1-MeI 213–215, ir, nmr, uv	32, 34, 126, 134, 182, 187, 285, 288, 538, 587, 610, 1130
4-Methyl-2(1*H*)-pteridinone	>100 dec, uv	38
4-Methyl-6(5*H*)-pteridinone	>230, ms, uv	55, 241, 244, 1246
4-Methyl-7(8*H*)-pteridinone	ms, uv	55, 1246
5-Methyl-6(5*H*)-pteridinone	H_2O 190, ir, uv	34, 134
6-Methyl-2(1*H*)-pteridinone	245, ms, uv	38, 42, 243, 1246
6-Methyl-4(3*H*)-pteridinone	345, ms, uv	32, 38, 153, 779, 1246
6-Methyl-7(8*H*)-pteridinone	ms, nmr, uv	27, 55, 57, 1246
7-Methyl-2(1*H*)-pteridinone	H_2O 230, nmr, uv	27, 38, 241, 243
7-Methyl-4(3*H*)-pteridinone	>340, ms, nmr, uv	32, 38, 153, 167, 779, 1286
7-Methyl-6(5*H*)-pteridinone	ms, nmr, uv; Na anal	27, 55, 241, 244, 717, 1246

APPENDIX TABLE (*Contd.*)

Compound	(°C), and so on	Reference(s)
8-Methyl-4(8*H*)-pteridinone	>150, uv	248
8-Methyl-7(8*H*)-pteridinone	125, ir, uv	31, 34, 134, 285, 538
2-Methyl-4(1*H*)-pteridinone 3-oxide	257, ms, uv	154, 980, 1226
6-Methyl-4(1*H*)-pteridinone 3-oxide	>250, ms, uv	154, 1226
6-Methyl-4(3*H*)-pteridinone 8-oxide	>320, nmr, uv	478, 485
7-Methyl-4(1*H*)-pteridinone 3-oxide	>250, uv	154, 1226
3-Methyl-8-D-ribityl-2,4,6,7(1*H*,3*H*,5*H*,8*H*)-pteridinetetrone	259–260	674
6-Methyl-8-D-ribityl-2,4,7(1*H*,3*H*,8*H*)-pteridinetrione (compound-V)	250 to 262, ir, nmr, uv	295, 378, 422, 463, 674, 1027, 1041, 1082, 1400, 1441, 1544, 1667
2-Methylsulfonyl-7-phenyl-2-pteridinamine	—	621
1-Methyl-6(?)-[D-*arabino*]-α,β,γ,δ-tetrahydroxybutyl-2,4(1*H*,3*H*)-pteridinedione	238	962: cf. 862
1-Methyl-6(?)-[D-*lyxo*]-α,β,γ,δ-tetrahydroxybutyl-2,4(1*H*,3*H*)-pteridinedione	271–273	962; cf. 862
1-Methyl-6(&7)-α,β,γ,δ-tetrahydroxybutyl-2,4(1*H*,3*H*)-pteridinedione (a mixture)	anal, 237–238	937
1-Methyl-7-[D-*arabino*]-α,β,γ,δ-tetrahydroxybutyl-2,4(1*H*,3*H*)-pteridinedione	266–267, ir, nmr	862
3-Methyl-7-[D-*arabino*]-α,β,γ,δ-tetrahydroxybutyl-2,4(1*H*,3*H*)-pteridinedione	260–261, ir, nmr	862, 863
6-Methyl-8-β,γ,δ,ε-tetrahydroxypentyl-2,4,7(1*H*,3*H*,8*H*)-pteridinetrione [Several unidentified separated isomers]	—	295
2-Methylthio-6,7-diphenyl-4-pteridinamine	247 or 252–253	472, 699
4-Methylthio-6,7-diphenyl-2-pteridinamine	235–236	1468
2-Methylthio-4,6-diphenylpteridine	218–220, ^{13}C nmr	959, 1118
2-Methylthio-4,7-diphenylpteridine	165–166, ^{13}C nmr	959, 1118
2-Methylthio-6,7-diphenylpteridine	208–209, nmr	1138
2-Methylthio-6,7-diphenyl-4(3*H*)-pteridinethione	257, uv	437
2-Methylthio-6,7-diphenyl-4(3*H*)-pteridinone	325–328, fl-sp, uv	437, 447, 1414
4-Methylthio-6,7-diphenyl-2(1*H*)-pteridinone	209–210 or 212, uv	699, 1679
2-Methylthio-6-phenacyl-4,7(3*H*,8*H*)-pteridinedione	>300	1524
2-Methylthio-4-phenylpteridine	176 or 177–179, nmr, ^{13}C nmr	959, 1118, 1436
4-Methylthio-2-phenylpteridine	189, nmr	1436
2-Methylthio-6-phenyl-4,7-pteridinediamine	306, uv	1281
4-Methylthio-6-phenyl-2,7-pteridinediamine	252–254, uv	523
2-Methylthio-6-phenyl-4,7-pteridinediamine 5-oxide	351, uv	328
4-Methylthio-6-phenyl-2,7-pteridinediamine 5-oxide	306–308	328
2-Methylthio-7-propionyl-4-pteridinamine	236, nmr, uv	1678
7-Methylthio-6-propionyl-2,4-pteridinediamine	uv	95
2-Methylthio-4-pteridinamine	208 to 211, nmr, uv	460, 1467, 1678

APPENDIX TABLE (*Contd.*)

Compound	(°C), and so on	Reference(s)
4-Methylthio-2-pteridinamine	215–218 or 221–223, uv	305, 1468
2-Methylthiopteridine	133–136, ir, nmr, ^{13}C nmr, uv	27, 33, 46, 119, 285, 286, 959, 1138
4-Methylthiopteridine	191, ir, uv, nmr	27, 33, 119, 285, 286
6-Methylthiopteridine	163, uv	36
7-Methylthiopteridine	142–143, ir, uv	33, 119, 285, 286
6-Methylthio-2,4-pteridinediamine	>300	1307
7-Methylthio-2,4-pteridinediamine	uv	95
2-Methylthio-4,7(3*H*,8*H*)-pteridinedione	>350, uv	1000
2-Methylthio-4(3*H*)-pteridinone	267–269, fl-sp, uv	437, 458, 1414
6-Methylthio-4(3*H*)-pteridinone	289–291	974
2-Methylthio-4-trifluoromethylpteridine	130, ms, nmr, uv	159, 780
2-Methylthio-7-α,β,γ-trihydroxypropyl-4-pteridinamine	290, uv	1467
2-Methylthio-6-[D-*erythro*]-α,β,γ-trihydroxypropyl-4(3*H*)-pteridinone	H_2O 156–158	458
2-Methylthio-6-[L-*erythro*]-α,β,γ-trihydroxypropyl-4(3*H*)-pteridinone	H_2O 156–158, uv	458
2-Methylthio-7-trimethylsiloxy-4-trimethylsilylamino-pteridine	145–147/0.01	325
2-Methylthio-4,6,7-triphenylpteridine	232–234	1118, 1120*
1-Methyl-2-thioxo-1,2-dihydro-4,6(3*H*,5*H*)-pteridinedione	335, uv	101
1-Methyl-2-thioxo-1,2-dihydro-4,7(3*H*,8*H*)-pteridinedione	H_2O 328, uv	1000
1-Methyl-7-thioxo-7,8-dihydro-2,4(1*H*,3*H*)-pteridinedione	$(CH)_5\overset{+}{N}H$ >300, nmr, uv	354, 1709
3-Methyl-2-thioxo-1,2-dihydro-4,7(3*H*,8*H*)-pteridinedione	224, uv	1000
3-Methyl-7-thioxo-7,8-dihydro-2,4(1*H*,3*H*)-pteridinedione	270, nmr, uv	354, 1709
6-Methyl-2-thioxo-1,2-dihydro-4,7(3*H*,8*H*)-pteridinedione	>300, uv	916
7-Methyl-2-thioxo-1,2-dihydro-4,6(3*H*,5*H*)-pteridinedione	>300, uv	916
1-Methyl-2-thioxo-1,2-dihydro-4(3*H*)-pteridinone	295, uv	437
3-Methyl-2-thioxo-1,2-dihydro-4(3*H*)-pteridinone	>315, uv	446
7-Methyl-4-trifluoromethylpteridine	126, ms, nmr, uv	156, 780
6-Methyl-4-trimethylsiloxy-2-trimethylsilylaminopteridine	ms	589, 1233, 1249, 1457
7-Methyl-4-trimethylsiloxy-2-trimethylsilylaminopteridine	crude, ms	1233
N-Methyl-2,4,6-trioxo-1,2,3,4,5,6-hexahydro-7-pteridinecarboxamide	>330, uv	342, 347
1-Methyl-2,4,6-trioxo-1,2,3,4,5,6-hexahydro-7-pteridinecarboxylic acid	H_2O 330, uv; Na_2 >340	342
1-Methyl-2,4,7-trioxo-1,2,3,4,7,8-hexahydro-6-pteridinecarboxylic acid	>340, uv	341, 1083
3-Methyl-2,4,6-trioxo-1,2,3,4,5,6-hexahydro-7-pteridinecarboxylic acid	$2H_2O$ >350, uv	342

APPENDIX TABLE (*Contd.*)

Compound	(°C), and so on	Reference(s)
3-Methyl-2,4,7-trioxo-1,2,3,4,7,8-hexahydro-6-pteridinecarboxylic acid	>340, uv	341
8-Methyl-2,4,7-trioxo-1,2,3,4,7,8-hexahydro-6-pteridinecarboxylic acid	>360, uv	317, 341
3-Methyl-6,7,8-triphenyl-2,4(3*H*,8*H*)-pteridinedione	335, uv	409
6-Methyl-2,4,7-tristrimethylsiloxypteridine	crude, ms	1411, 1457
6-Methyl-8-D-xylityl-2,4,7(1*H*,3*H*,8*H*)-pteridinetrione	—	295
Monapterin [see: 2-amino-6-[L-*threo*]-α,β,γ-trihydroxypropyl-4(3*H*)-pteridinone]		
2-Morpholino-6,7-diphenyl-4-pteridinamine	231–232, uv	467, 472
8-β-Morpholinoethyl-6,7-diphenyl-2,4(3*H*,8*H*)-pteridinedione	205–208, uv	409
2-Morpholino-6-phenyl-4,7-pteridinediamine	274–276, uv	523
Neopterin [see: 2-amino-6-[D-*erythro*]-α,β,γ-trihydroxypropyl-4(3*H*)-pteridinone]		
Neodrosopterin (not simple: see text)		
7-Oxo-7,8-dihydro-6-pteridinecarboxylic acid	>100, uv	57
4-Oxo-3,4-dihydro-2-pteridinesulfinic acid	K anal	629
4-Oxo-3,4-dihydro-2-pteridinesulfonic acid	K anal	629
4-Oxo-6,7-diphenyl-3,4-dihydro-2-pteridinesulfinic acid	K anal	629
4-Oxo-6,7-diphenyl-3,4-dihydro-2-pteridinesulfonic acid	K anal	629
4-Oxo-2-phenyl-3,4-dihydro-7-pteridinecarboxylic acid	anal, uv	1354
7-Pentylamino-6-phenyl-2,4-pteridinediamine	HCl 308–311, uv	525
4-Pentyloxy-6-propionyl-7-propylthio-2-pteridinamine	uv	95
4-Pentyloxy-6-propionyl-2-pteridinamine	uv	95
4-Pentyloxy-7-propylthio-2-pteridinamine	uv	95
6-Pentyloxy-2,4-pteridinediamine 8-oxide	255, ir, nmr, uv	1728
4-Pentyloxy-7-trimethylsiloxy-2-trimethylsilylaminopteridine	crude	435
6-Pentyl-2,4-pteridinediamine	280–284	742
6-Pentyl-2,4-pteridinediamine 8-oxide	258, nmr, uv	484
4-Phenacyl-6,7-diphenyl-2(1*H*)-pteridinone	>250, uv	1679
6-Phenacyl-2,4,7(1*H*,3*H*,8*H*)-pteridinetrione	>300, ms, uv	1524
7-Phenacyl-2,4,6(1*H*,3*H*,5*H*)-pteridinetrione	>300, ms, uv	1524
4-Phenethylaminopteridine	159–160	280
6-Phenethyl-2,4-pteridinediamine	236–238, ir, uv	90
6-Phenethyl-2,4,7-pteridinetriamine	296–298, uv	1415
6-Phenoxymethyl-2,4-pteridinediamine	anal	1102
6-Phenyl-2,4-bistrimethylsiloxypteridine	120	412
7-Phenyl-2,4-bistrimethylsiloxypteridine	crude	412
6-Phenylethynyl-2-pivalamido-4(3*H*)-pteridinone	284–285, ir, nmr	1753
6-Phenylethynyl-2,4-pteridinediamine	>300, ir, nmr	1753
6-Phenyl-2-piperidino-4,7-pteridinediamine	246–247, uv	523
6-Phenyl-4-piperidino-2,7-pteridinediamine	278–279, uv	523
6-Phenyl-4-piperidino-2,7-pteridinediamine 5-oxide	265, uv	328
1-Phenyl-7-propionyl-2,4(1*H*,3*H*)-pteridinedione	243–246, nmr, uv	108
6-Phenyl-3-propyl-4(3*H*)-pteridinone	133–134, nmr	1758
6-Phenyl-3-propyl-4(3*H*)-pteridinone 8-oxide	247–249, nmr	1758
2-Phenyl-4-pteridinamine	239 to 251, nmr, uv	185, 482, 1354, 1740

APPENDIX TABLE (*Contd.*)

Compound	(°C), and so on	Reference(s)
7-Phenyl-4-pteridinamine	284–286, ir, nmr	1740
2-Phenyl-4-pteridinamine 3-oxide	223–226, nmr	1004
2-Phenylpteridine	132–133, nmr, ^{13}C nmr, uv	152, 959
4-Phenylpteridine	164–165, nmr, ^{13}C nmr, uv	151, 959
7-Phenylpteridine	158–160, ^{1}H and ^{13}C nmr	959, 1287, 1167
2-Phenyl-4,7-pteridinediamine	295–298 or 303, uv	1312, 1357
6-Phenyl-2,4-pteridinediamine	286(?) or 340 to 352, nmr, uv	327, 484, 541, 1006(?), 1257
6-Phenyl-4,7-pteridinediamine	340 or 347–348, nmr, uv	523, 1281
7-Phenyl-2,4-pteridinediamine	290 to 340, uv	327, 541, 730, 1006
6-Phenyl-2,4-pteridinediamine 5,8-dioxide	318, ms, uv	427, 541
7-Phenyl-2,4-pteridinediamine 5,8-dioxide	303, ms, uv	427, 541
6-Phenyl-2,4-pteridinediamine 8-oxide	310–312, uv	484, 541
7-Phenyl-2,4-pteridinediamine 5-oxide	300–301, uv	541
1-Phenyl-2,4(1*H*,3*H*)-pteridinedione	307–308	108
2-Phenyl-4,6(3*H*,5*H*)-pteridinedione	>330, uv	101
2-Phenyl-4,7(3*H*,8*H*)-pteridinedione	crude, uv	101
2-Phenyl-6,7(5*H*,8*H*)-pteridinedione	250, nmr	152
4-Phenyl-6,7(5*H*,8*H*)-pteridinedione	>300, nmr	151
6-Phenyl-2,4(1*H*,3*H*)-pteridinedione	380–382, nmr, uv	67, 367, 543, 1438(?); cf. 1473
7-Phenyl-2,4(1*H*,3*H*)-pteridinedione	374–378, nmr, uv	67, 176, 367, 1438(?); cf. 1473
7-Phenyl-2,4(1*H*,3*H*)-pteridinedione 5,8-dioxide	277, nmr, uv	367
6-Phenyl-2,4(1*H*,3*H*)-pteridinedione 8-oxide	338, ms, nmr, uv	367, 427, 478, 485
7-Phenyl-2,4(1*H*,3*H*)-pteridinedione 5-oxide	330, ms, nmr, uv	367, 427
2-Phenyl-4(3*H*)-pteridinethione	>225, nmr	1436
2-Phenyl-4,6,7-pteridinetriamine	285–287, nmr, uv	522, 1398
4-Phenyl-2,6,7-pteridinetriamine	298–304, nmr, uv	522
6-Phenyl-2,4,7-pteridinetriamine (triamterene)	316 to 327, nmr, pol, uv	522, 523, 541, 771*, 819, 1113, 1151, 1159, 1281, 1415
7-Phenyl-2,4,6-pteridinetriamine	320, nmr, uv	327, 522
6-Phenyl-2,4,7-pteridinetriamine 5,8-dioxide	H_2O 340–341	541
6-Phenyl-2,4,7-pteridinetriamine 5-oxide	340, uv	328
1-Phenyl-2,4,7(1*H*,3*H*,8*H*)-pteridinetrione	>360, uv	344
2-Phenyl-4,6,7(3*H*,5*H*,8*H*)-pteridinetrione	320 to 370, nmr, uv	101, 152, 522, 663
6-Phenyl-2,4,7(1*H*,3*H*,8*H*)-pteridinetrione	>320	413
2-Phenyl-4(3*H*)-pteridinone	264 or >310, nmr, uv	20, 101, 152, 663
7-Phenyl-4(3*H*)-pteridinone	295 or >300, ms, nmr	1167, 1287, 1740
2-Phenyl-4(1*H*)-pteridinone 3-oxide	260–263, nmr	1004
6-Phenyl-4(3*H*)-pteridinone 8-oxide	>320, nmr, uv	478, 485

APPENDIX TABLE (*Contd.*)

Compound	(°C), and so on	Reference(s)
6-Phenyl-8-D-ribityl-2,4,7(1*H*,3*H*,8*H*)-pteridinetrione	265–267	674
6-Phenylthiomethyl-2,4-pteridinediamine	anal	1102
6-Phenylthio-2,4-pteridinediamine	>330	1307
6-Phenylthio-2,4,7-pteridinetriamine	—	249, 329
6(7)-Phenyl-2-thioxo-1,2-dihydro-4(3*H*)-pteridinone	anal	1190
Photolumazine-A [see: 6-L-α,β-dihydroxyethyl-8-D-ribityl-2,4,7(1*H*,3*H*,8*H*)-pteridinetrione]		
Photolumazine-B [see: 6-hydroxymethyl-8-D-ribityl-2,4,7(1*H*,3*H*,8*H*)-pteridinetrione]		
Photolumazine-C [see: 8-D-ribityl-2,4,7(1*H*,3*H*,8*H*)-pteridinetrione]		
6-Piperidin-4′-yl-2,4,7-pteridinetriamine	>300, nmr, uv	1396
6-Prop-1′-enyl-2,4-pteridinediamine	312, nmr	1297
7-Propionyl-2,4-pteridinediamine	280, nmr, uv; oxime >253	1678
7-Propionyl-2,4(1*H*,3*H*)-pteridinedione	260–261, nmr, uv	108
6-Propoxymethyl-2,4-pteridinediamine	239–243, ir, nmr	1728
4-Propoxypteridine	66–68, nmr, uv	138
2-Propylaminopteridine	163, uv	124
2-Propyl-4-pteridinamine	145–146, uv	185
4-Propyl-2-pteridinamine	183–184, uv	185
6-Propyl-2,4-pteridinediamine	282, nmr, uv	484
6-Propyl-2,4-pteridinediamine 8-oxide	290–292, nmr, uv	484
7-Propyl-2,4(1*H*,3*H*)-pteridinedione	ms, nmr, uv	1286
8-Propyl-2,4(3*H*,8*H*)-pteridinedione	260–263, ms	1553
6-Propyl-2,4,7(1*H*,3*H*,8*H*)-pteridinetrione	ms	1411
7-Propyl-4(3*H*)-pteridinone	ms, nmr, uv	1286
6-Propyl-2,4,7-tristrimethylsiloxypteridine	crude, ms	1411
2-Pteridinamine	>275, nmr, pol, uv; H_2SO_4 anal; HBr/EtOH: xl-str	27, 30, 39, 45, 51, 119, 285, 368, 582, 585, 686, 766, 979, 1013, 1016, 1250, 1418
4-Pteridinamine	305 or 309–311, ir, ms, nmr, pol, uv; $(CO_2H)_2$ anal	27, 28, 30, 39, 53, 119, 129, 185, 267, 285, 934, 979, 1003, 1013, 1016, 1250, 1391, 1417
6-Pteridinamine	>300, nmr, uv	27, 31, 119, 146, 285, 1013, 1016, 1250
7-Pteridinamine	>320, uv	33, 119, 131, 146, 285, 1016, 1250
4-Pteridinamine 3-oxide	275–278, ms, nmr	999, 1003
Pteridine	137 to 140, dip, ir, ms, nmr, ^{13}C nmr, pol, uv, xl-st; pic 118;	27, 30, 33, 34, 51, 65, 66, 79, 119, 148, 207, 242, 250, 281, 285, 286, 293, 294*, 310,

APPENDIX TABLE (*Contd.*)

Compound	(°C), and so on	Reference(s)
	$(CO_2H)_2$ >128	312, 606, 702, 724, 765, 770, 781, 787, 792, 828, 843, 851, 852, 915, 938, 950, 959, 963, 972, 979, 1013, 1016, 1037, 1045, 1138, 1153, 1250, 1303, 1360, 1417, 1418, 1432, 1483, 1513
4-Pteridinecarbonitrile	78, nmr	157
4-Pteridinecarboxamide	>200; $2H_2O$ 165, nmr, uv	157
4-Pteridinecarboxylic acid	$2H_2O$ 110, ms, nmr, uv	157
2,4-Pteridinediamine	306 to 320, ^{13}C nmr, uv	32, 59, 123, 130, 267, 284, 305, 402, 475, 486, 602, 686, 730, 749, 787, 851, 852, 979, 1016, 1432, 1483, 1678
2,6-Pteridinediamine	—	686, 1016
2,7-Pteridinediamine	—	1016
4,6-Pteridinediamine	uv	41, 1016
4,7-Pteridinediamine	>300, uv	41, 131, 324, 1016
6,7-Pteridinediamine	>250, uv	36, 1016
2,4-Pteridinediamine 5,8-dioxide	>360, ms, uv	427, 541
2,4-Pteridinediamine 8-oxide	328–330, ms, nmr, uv	427, 477, 484, 541
2,4(1*H*,3*H*)-Pteridinedione (lumazine)	310 to 350, fl-sp, ms, ^{13}C nmr, uv, xl-st, chelates; NH_4 anal	8, 28, 30, 40, 56, 59, 130, 134, 207, 284, 338, 367, 399, 437, 471, 642, 747, 787, 796, 851–853, 902, 948, 979, 1012, 1041, 1042, 1095, 1225, 1339, 1414, 1438, 1554
2,6(1*H*,5*H*)-Pteridinedione	H_2O >250, uv	14, 16, 41, 717
2,7(1*H*,8*H*)-Pteridinedione	H_2O >275, uv; Na_2 anal	41
4,6(3*H*,5*H*)-Pteridinedione	H_2O >350, uv; Na —	14, 16, 29, 101, 241, 244, 346, 717, 974
4,7(3*H*,8*H*)-Pteridinedione	>350, uv	29, 346
6,7(5*H*,8*H*)-Pteridinedione	>360; H_2O anal, ^{13}C nmr, uv	19, 31, 36, 55, 64, 134, 208, 851, 852
2,4(3*H*,8*H*)-Pteridinedione 1-oxide	321, uv	1098
2,4(1*H*,3*H*)-Pteridinedione 8-oxide	H_2O 319, ms, uv	367 (cf. 366), 427

APPENDIX TABLE (*Contd.*)

Compound	(°C), and so on	Reference(s)
2,4(1*H*,3*H*)-Pteridinedithione	286, uv	437
6,7(5*H*,8*H*)-Pteridinedithione	>250, uv	36
2,4,6,7-Pteridinetetramine	HCl >360, uv	41, 486, 1016
2,4,6,7(1*H*,3*H*,5*H*,8*H*)-Pteridinetetrone	^{13}C nmr, ir, ms, uv	32, 41, 101, 179, 403, 404, 441, 486, 530–533, 706, 852, 1116*, 1152, 1158, 1251, 1457
2(1*H*)-Pteridinethione	>205; H_2O 200–205, uv	30, 46, 119, 243, 285, 878
4(3*H*)-Pteridinethione	290, uv	30, 33, 53, 119, 285, 1103
6(5*H*)-Pteridinethione	> 160, uv	36
7(8*H*)-Pteridinethione	260, uv	33, 119, 285
2,4,6-Pteridinetriamine	anal	1016, 1369(?)
2,4,7-Pteridinetriamine	> 300; H_2O > 250, uv	41, 324, 1016
2,6,7-Pteridinetriamine	— (calculations)	1016
4,6,7-Pteridinetriamine	uv	41, 1016
2,4,6(1*H*,3*H*,5*H*)-Pteridinetrione	360–380, ms, uv	14, 16, 41, 101, 179, 340, 342, 345, 352, 367, 717, 1116, 1129, 1554
2,4,7(1*H*,3*H*,8*H*)-Pteridinetrione (violapterin)	> 350, ms, nmr, uv	41, 98, 339, 345, 352, 716, 979, 1041, 1197, 1267*, 1304, 1411
2,6,7(1*H*,5*H*,8*H*)-Pteridinetrione	uv	99, 1152, cf. 41
4,6,7(3*H*,5*H*,8*H*)-Pteridinetrione	> 360, uv	29, 101, 155, 208
2(1*H*)-Pteridinone	H_2O 240, ir, ms, nmr, uv	14, 16, 27, 30, 38, 42, 45, 51, 52, 54, 119, 134, 140, 207, 241, 245, 285, 979, 1246, 1250, 1551
4(3*H*)-Pteridinone	340 or > 350, ir, ms, nmr, ^{13}C nmr, pol, uv	20, 27, 28, 30, 32, 34, 37, 38, 119, 134, 207, 285, 288, 305, 368, 587, 610, 673, 779, 792, 827, 851–853, 979, 1003, 1004, 1130, 1145, 1246, 1317, 1391, 1417, 1418, 1508*
6(5*H*)-Pteridinone	H_2O 248, ir, ms, nmr, uv; Na —	19, 27, 31, 38, 55, 119, 134, 140, 207, 241, 244, 285, 717, 1250
7(8*H*)-Pteridinone	256, ir, ms, uv	31, 41, 44, 119, 131, 134, 207, 285, 979, 1250
2(3*H*)-Pteridinone 1-oxide	267, nmr, uv	1098
4(1*H*)-Pteridinone 3-oxide	290, ms, uv	154, 540, 1226

APPENDIX TABLE (*Contd.*)

Compound	(°C), and so on	Reference(s)
Pterin [see: 2-amino-4(3*H*)-pteridinone]		
Pterincarboxylic acid (see: 2-amino-4-oxo-3,4-dihydro-6-pteridinecarboxylic acid)		
Pteroic acid (not simple)		390, 1235
Pterorhodin (not simple)		407, 425, 501
Pteroylglutamic acid (not simple: see text)		
Putidolumazine [see: 6-β-carboxyethyl-8-D-ribityl-2,4,7(1*H*,3*H*,8*H*)-pteridinetrione]		1218, 1248
Pyrimidazine (see: pteridine)		1190
Ranachrome-3 [see: 2-amino-6-hydroxymethyl-4(3*H*)-pteridinone]		214, 1604
Ranachromes (−1, −2, −4, −5, etc. now unused: see text and index)		
Rhizopterin (not simple)		1362
Rhodopterin (see: pterorhodin)		
8-D-Ribityl-2,4,6,7(1*H*,3*H*,5*H*,8*H*)-pteridinetetrone	284–285 or 287–288, uv	674, 1444
8-D-Ribityl-2,4,7(1*H*,3*H*,8*H*)-pteridinetrione (photolumazine-C)	228–229 or 262–263, nmr, uv	172, 295, 463, 1400, 1430
Riboflavine (not simple: see text)		
Russupteridine-yellow-1 [see: 6-amino-7-formamido-8-D-ribityl-2,4(3*H*,8*H*)-pteridinedione]		1400
Russupteridine yellow-4 (not simple: see text)		1400
Sarcinapterin (not simple: see text)		1459
Sepialumazine (not simple: see text)		
Sepiapterin (not simple: see text)		301, 335, 350, 430, 615, 875, 1062, 1063, 1168, 1217, 1221, 1222, 1464, 1558
Sepiapterin-C (not simple: see text)		
2-Styrylpteridine	136, nmr, uv	105
6-Styryl-2,4-pteridinediamine	342, nmr, uv	484
7-Styryl-2,4-pteridinediamine	303, nmr, uv	252, 1305
6-Styryl-2,4-pteridinediamine 8-oxide	275, nmr, uv	484
7-Styryl-2,4-pteridinediamine 8-oxide	273–274	1305
7-Styryl-2,4(1*H*,3*H*)-pteridinedione	> 300, nmr, uv	252, 1010
6-Styryl-2,4,7-pteridinetriamine	345–347, uv	1415
Surugatoxin (not simple: see text)		
2,4,6,7-Tetrachloropteridine	161–162, ir, uv	286, 441, 486, 1251, 1405
6-[D-*arabino*]-α,β,γ,δ-Tetrahydroxybutyl-2,4-pteridinediamine	crude	1058, 1257
7-[D-*arabino*]-α,β,γ,δ-Tetrahydroxybutyl-2,4(1*H*,3*H*)-pteridinedione	anal; Ba anal	706
7-[D-*lyxo*]-α,β,γ,δ-Tetrahydroxybutyl-2,4(1*H*,3*H*)-pteridinedione	anal; Ba anal	706
8-β,γ,δ,ε-Tetrahydroxypentyl-2,4,7(1*H*,3*H*,8*H*)-pteridinetrione (mixed stereoisomers)	—	295

APPENDIX TABLE (*Contd.*)

Compound	(°C), and so on	Reference(s)
6-α,β,γ,δ-Tetrakistrimethylsiloxybutyl-2,4(1*H*,3*H*)-pteridinedione	crude, ms	589
2,4,6,7-Tetrakistrimethylsiloxypteridine	anal, 214	701, 1457
1,3,6,7-Tetramethyl-4-oxo-3,4-dihydro-1-pteridinium iodide	243–245, ir, nmr, uv	587, 1130
2,4,6,7-Tetramethylpteridine	124–126, nmr, uv	66, 1513
1,3,6,7-Tetramethyl-2,4(1*H*,3*H*)-pteridinedione	158 to 166, ir, nmr, uv	168, 367, 378, 414, 1096, 1339, 1380, 1435
3,6,7,8-Tetramethy-2,4(3*H*,8*H*)-pteridinedione	242; Na nmr	116, 660, 1545
1,3,6,7-Tetramethyl-2,4(1*H*,3*H*)-pteridinedione 5-oxide	196, ms, uv	210, 366, 367, 427
1,3,5,8-Tetramethyl-2,4,6,7(1*H*,3*H*,5*H*,8*H*)-pteridinetetrone	268–270, uv	343
1,3,6,8-Tetramethyl-2,4,7(1*H*,3*H*,8*H*)-pteridinetrione	253, uv	339
1,3,6,8-Tetramethyl-2-thioxo-1,2-dihydro-4,7(3*H*,8*H*)-pteridinedione	212, nmr, uv	1000
1,3,6,7-Tetramethyl-2-thioxo-1,2-dihydro-4(3*H*)-pteridinone	186, uv	437
1,3,5,*N*-Tetramethyl-2,4,6-trioxo-1,2,3,4,5,6-hexahydro-7-pteridinecarboxamide	188–190	347
2-Thioxo-6,7-bistrifluoromethyl-1,2-dihydro-4(3*H*)-pteridinone	230–231, ir, ms, nmr, uv	1571
2-Thioxo-1,2-dihydro-4,6(3*H*,5*H*)-pteridinedione	crude uv	101
2-Thioxo-1,2-dihydro-4,7(3*H*,8*H*)-pteridinedione	H_2O > 350, uv	1000
2-Thioxo-1,2-dihydro-4,6,7(3*H*,5*H*,8*H*)-pteridinetrione	anal, uv	878, 916
2-Thioxo-1,2-dihydro-4(3*H*)-pteridinone	> 360, fl-sp, ^{13}C nmr, uv	399, 437, 851, 916, 1169, 1414
6-Thioxo-5,6-dihydro-7(8*H*)-pteridinone	> 250, uv	36
7-Thioxo-7,8-dihydro-6(5*H*)-pteridinone	H_2O anal	36
2-Thioxo-6,7-diphenyl-1,2-dihydro-4(3*H*)-pteridinone	195–196	399
2-Thioxo-7-[L-*lyxo*]-α,β,γ-trihydroxybutyl-1,2-dihydro-4(3*H*)-pteridinone	217, uv	1247
2,4,7-Triacetamido-6-phenylpteridine	282–284	1281
2,4,7-Triacetamido-6-pteridinecarboxamide	anal	324
2,4,7-Triamino-*N*,*N*-pentamethylene-6-pteridinecarboxamide	305	321
2,4,7-Triamino-6-pteridinecarboxamide	> 360, uv	321, 1357
2,4,7-Triamino-6-pteridinecarboxylic acid	> 300	324
2,4,7-Trichloro-6-phenylpteridine	163–164	1251
2,6,7-Trichloro-4-phenylpteridine	154–156	522
4,6,7-Trichloro-2-phenylpteridine	193–195	522, 1398
2,4,7-Trichloropteridine	133	41
4,6,7-Trichloropteridine	179–181	41
4-Trifluoromethyl-2-pteridinamine	234, ms, nmr, uv	159, 780

APPENDIX TABLE (*Contd.*)

Compound	(°C), and so on	Reference(s)
4-Trifluoromethylpteridine	95–96, ms, nmr, uv	156, 780
2-Trifluoromethyl-4(3*H*)-pteridinethione	anal, nmr	20
4-Trifluoromethyl-2(1*H*)-pteridinethione	H_2O 218, nmr	159
2-Trifluoromethyl-4(3*H*)-pteridinone	214, ir	20
7-[L-*lyxo*]-α,β,γ-Trihydroxybutyl-2,4(1*H*,3*H*)-pteridinedione	252, uv	1247
7-[D-*threo*]-α,β,γ-Trihydroxypropyl-2,4(1*H*,3*H*)-pteridinedione	anal; Ba anal	706
2-Trimethylammoniopteridine chloride	148–150	124
1,3,6-Trimethyl-2,4-dioxo-1,2,3,4-tetrahydro-7-pteridinesulfenic acid	180, ir, nmr, uv	228, 1709
1,3,6-Trimethyl-2,4-dioxo-1,2,3,4-tetrahydro-7-pteridinesulfininc acid	crude	1709
1,3,6-Trimethyl-2,4-dioxo-1,2,3,4-tetrahydro-7-pteridinesulfonic acid	nmr, uv; Na >300	228, 1709
6,7,8-Trimethyl-2-methylamino-4(8*H*)-pteridinone	—	563
6,7,8-Trimethyl-4-methylimino-4,8-dihydro-2-pteridinamine	HCl 228–230, nmr, uv	1354
1,6,7-Trimethyl-4-methylimino-1,4-dihydropteridine	HCl 250; HI 247, uv	129
6,7,8-Trimethyl-2-methylimino-2,8-dihydropteridine	197–198, uv	189, 190, 248
6,7,8-Trimethyl-4-methylimino-4,8-dihydropteridine	113–115, uv	129
1,3,6-Trimethyl-7-methylsulfinyl-2,4(1*H*,3*H*)-pteridinedione	209–210, nmr, uv	1709
1,3,6-Trimethyl-7-methylsulfonyl-2,4(1*H*,3*H*)-pteridinedione	238–239, nmr, uv	1709
1,3,6-Trimethyl-7-methylthio-2,4(1*H*,3*H*)-pteridinedione	196–197, nmr, uv	1709
1,6,7-Trimethyl-4-methylthio-1-pteridinium iodide	crude, uv	313
1,6,7-Trimethyl-2-methylthio-4(1*H*)-pteridinone	231–232, uv	437, 446
3,6,7-Trimethyl-2-methylthio-4(3*H*)-pteridinethione	193, uv	437, 446
3,6,7-Trimethyl-2-methylthio-4(3*H*)-pteridinone	184–187 or 205, uv	166, 362, 437
1,3,5-Trimethyl-6-phenylimino-5,6-dihydro-2,4(1*H*,3*H*)-pteridinedione	176–177, nmr, uv	1680
2,6,7-Trimethyl-4-phenylpteridine	150–151, nmr, uv	151
4,6,7-Trimethyl-2-phenylpteridine	192–193, nmr, uv	152
1,3,6-Trimethyl-7-phenyl-2,4(1*H*,3*H*)-pteridinedione	200–202, nmr, uv	1348; cf. 1112
1,3,7-Trimethyl-6-phenyl-2,4(1*H*,3*H*)-pteridinedione	185–187, nmr, uv	1348, cf. 1112
1,3,7-Trimethyl-6-propionyl-2,4(1*H*,3*H*)-pteridinedione	125–127, nmr, uv	108
3,6,7-Trimethyl-2-propylamino-4(3*H*)-pteridinone	296–298, uv	166
2,6,7-Trimethyl-4-pteridinamine	249–250	1354
4,6,7-Trimethyl-2-pteridinamine	312–313, uv	39, 418, 1354
2,6,7-Trimethylpteridine	134–135, nmr, uv	27, 33, 66, 285, 286
4,6,7-Trimethylpteridine	139–140, nmr, uv	66
1,3,6-Trimethyl-2,4(1*H*,3*H*)-pteridinedione	204, ir, nmr, uv	552 (cf, 1232), 1704, 1709
1,3,7-Trimethyl-2,4(1*H*,3*H*)-pteridinedione	150 or 163	108, 962, 1121
1,6,7-Trimethyl-2,4(1*H*,3*H*)-pteridinedione	328 to 342, uv	69, 130, 166,

APPENDIX TABLE (*Contd.*)

Compound	(°C), and so on	Reference(s)
		362, 367, 414, 1092, 1339, 1380(?)
3,6,7-Trimethyl-2,4(1*H*,3*H*)-pteridinedione	268–272, pol, uv	166, 362, 367, 414, 446, 1014, 1339
6,7,8-Trimethyl-2,4(3*H*,8*H*)-pteridinedione	299 to 322, ir, ms, nmr, uv; Na nmr	107, 130, 290, 335, 378, 448, 659, 841, 1144, 1273, 1283, 1553
1,3,6-Trimethyl-2,4(1*H*,3*H*)-pteridinedione 5,8-dioxide	241, uv	1704
3,6,7-Trimethyl-2,4(1*H*,3*H*)-pteridinedione 5,8-dioxide	253, uv	367
1,3,6-Trimethyl-2,4(1*H*,3*H*)-pteridinedione 5-oxide	150, ms, uv	427, 1704
1,6,7-Trimethyl-2,4(1*H*,3*H*)-pteridinedione 5-oxide	267, ms, uv	210, 366, 367, 427
3,6,7-Trimethyl-2,4(1*H*,3*H*)-pteridinedione 8-oxide	290, ms, uv	331, 367, (cf. 366), 427
1,3,8-Trimethyl-2,4,6,7(1*H*,3*H*,5*H*,8*H*)-pteridinetetrone	288–290, uv	343
1,6,7-Trimethyl-4(1*H*)-pteridinethione	200–205, uv	313
2,6,7-Trimethyl-4(3*H*)-pteridinethione	> 239	135
6,7,8-Trimethyl-2(8*H*)-pteridinethione	200, uv	248
1,3,5-Trimethyl-2,4,6(1*H*,3*H*,5*H*)-pteridinetrione	180–181, uv	209, 340, 1680
1,3,6-Trimethyl-2,4,7(1*H*,3*H*,8*H*)-pteridinetrione	308 or 313, nmr, uv	238*, 337, 339, 344, 716, 1524, 1704
1,3,7-Trimethyl-2,4,6(1*H*,3*H*,5*H*)-pteridinetrione	266–268 or 283, uv	337, 340, 1524
1,3,8-Trimethyl-2,4,7(1*H*,3*H*,8*H*)-pteridinetrione	220–222, nmr, uv	339, 716, 1000
1,6,8-Trimethyl-2,4,7(1*H*,3*H*,8*H*)-pteridinetrione	317, nmr	716
3,6,8-Trimethyl-2,4,7(1*H*,3*H*,8*H*)-pteridinetrione	331–334, nmr, uv	248, 344, 716
1,3,6-Trimethyl-2,4,7(1*H*,3*H*,8*H*)-pteridinetrione 5-oxide	230, uv	236, 238*
1,6,7-Trimethyl-4(1*H*)-pteridinimine	HCl 290; HI 264, uv	129
3,6,7-Trimethyl-4(3*H*)-pteridinimine	HI 224–225, uv	132
6,7,8-Trimethyl-2(8*H*)-pteridinimine	235–240, uv	190
1,6,7-Trimethyl-2(1*H*)-pteridinone	EtOH 150, uv	120
1,6,7-Trimethyl-4(1*H*)-pteridinone	216–217, nmr, uv	129, 313, 587, 949, 1130
2,6,7-Trimethyl-4(3*H*)-pteridinone	261–262, ms, nmr, uv	153, 170, 779
3,6,7-Trimethyl-2(3*H*)-pteridinone	H_2O 250–260, ir, uv; EtOH 163	120, 132, 134

APPENDIX TABLE (*Contd.*)

Compound	(°C), and so on	Reference(s)
3,6,7-Trimethyl-4(3*H*)-pteridinone	241, nmr, uv	126, 446, 587, 949, 1130
4,6,7-Trimethyl-2(1*H*)-pteridinone	uv	38, 56
6,7,8-Trimethyl-2(8*H*)-pteridinone	255–260, ir, uv	14, 16, 120, 134, 189, 248
6,7,8-Trimethyl-4(8*H*)-pteridinone	235–242, ir, uv	134
7-Trimethylsiloxypteridine	—	356
4-Trimethylsiloxy-6-trimethylsiloxycarbonyl-2-trimethylsilyaminopteridine	crude, ms	1457
4-Trimethylsiloxy-7-trimethylsiloxycarbonyl-2-trimethylsilylaminopteridine	crude, ms	1457
4-Trimethylsiloxy-6-trimethylsiloxymethyl-2-trimethylsilylaminopteridine	ms	589, 986, 1249, 1457
4-Trimethylsiloxy-6-trimethylsiloxysulfonyl-2-trimethylsilylaminopteridine	crude, ms	1242
2-Trimethylsiloxy-4-trimethylsilylaminopteridine	135–140/0.001	225
4-Trimethylsiloxy-2-trimethylsilylaminopteridine	ms	589, 1249, 1457
7-Trimethylsiloxy-2-trimethylsilylaminopteridine	—	325
7-Trimethylsiloxy-4-trimethylsilylaminopteridine	—	325
4-Trimethylsiloxy-2-trimethylsilylamino-6-pteridinecarbaldehyde	crude, ms	1457
4-Trimethylsiloxy-2-trimethylsilylamino-6-α,β,γ-tristrimethylsiloxylpropylpteridine	ms	589
2-Trimethylsilyaminopteridine	ms	589, 1249
1,3,5-Trimethyl-6-sulfinyl-5,6-dihydro-2,4(1*H*,3*H*)-pteridinedione	148–149, nmr, uv	1680
1,3,5-Trimethyl-6-thioxo-5,6-dihydro-2,4(1*H*,3*H*)-pteridinedione	189–190, nmr, uv	1680
1,3,6-Trimethyl-7-thioxo-7,8-dihydro-2,4(1*H*,3*H*)-pteridinedione	175–178, nmr, uv	1709
3,8,*N*-Trimethyl-2,4,7-trioxo-1,2,3,4,7,8-hexahydro-6-pteridinecarboxamide	anal, ir, nmr, uv	1282
2,4,7-Trioxo-1,2,3,4,7,8-hexahydro-6-pteridinecarboxylic acid	> 360, uv; Na 197	341, 399, 480, 481, 674
1,3,6-Trimethyl-2-thioxo-1,2-dihydro-4,7(3*H*,8*H*)-pteridinedione	315, nmr, uv	1000
1,3,6-Trimethyl-7-thioxo-7,8-dihydro-2,4(1*H*,3*H*)-pteridinedione	—	228
1,3,8-Trimethyl-2-thioxo-1,2-dihydro-4,7(3*H*,8*H*)-pteridinedione	194–196, nmr, uv	1000
1,3,8-Trimethyl-7-thioxo-7,8-dihydro-2,4(1*H*,3*H*)-pteridinedione	247–249, uv	1000
1,6,7-Trimethyl-2-thioxo-1,2-dihydro-4(3*H*)-pteridinone	290 or 320, nmr, uv	177, 437
3,6,7-Trimethyl-2-thioxo-1,2-dihydro-4(3*H*)-pteridinone	277–280, uv	446
1,3,*N*-Trimethyl-2,4,6-trioxo-1,2,3,4,5,6-hexahydro-7-pteridinecarboxamide	298–305, uv	117, 336, 347

APPENDIX TABLE (*Contd.*)

Compound	(°C), and so on	Reference(s)
1,3,*N*-Trimethyl-2,4,7-trioxo-1,2,3,4,7,8-hexahydro-6-pteridinecarboxamide	288	336
1,3,8-Trimethyl-2,4,7-trioxo-1,2,3,4,7,8-hexahydro-6-pteridinecarboxylic acid	215, uv; H_2O 162	317, 341
2,4,6-Trioxo-1,2,3,4,5,6-hexahydro-7-pteridinecarboxamide	> 350	347
2,4,6-Trioxo-1,2,3,4,5,6-hexahydro-7-pteridinecarboxylic acid	> 340, uv; Na > 340	342, 347, 1304
2,4,7-Trioxo-1,2,3,4,7,8-hexahydro-6-pteridinecarboxylic acid	> 320	930, 1234, 1304
6,7,8-Triphenyl-2-phenylimino-2,8-dihydropteridine	225–227	190
2,6,7-Triphenyl-4-pteridinamine	250–251 or 255	475, 1354
4,6,7-Triphenyl-2-pteridinamine	271–272	1118, 1122
4,6,7-Triphenylpteridine	174–175, ^{13}C nmr	959, 1122
1,6,7-Triphenyl-2,4(1*H*,3*H*)-pteridinedione	> 280	797
3,6,7-Triphenyl-2,4(1*H*,3*H*)-pteridinedione	327–328	476
6,7,8-Triphenyl-2,4(3*H*,8*H*)-pteridinedione	340, uv	409
2,6,7-Triphenyl-4(3*H*)-pteridinethione	323–324	476
2,6,7-Triphenyl-4(3*H*)-pteridinone	290 or 306	476, 663
4,6,7-Triphenyl-2(1*H*)-pteridinone	299–300	1120
1,6,7-Triphenyl-2-thioxo-1,2-dihydro-4(3*H*)-pteridinone	274–275	797
3,6,7-Triphenyl-2-thioxo-1,2-dihydro-4(3*H*)-pteridinone	301–302	476
2,4,7-Trisbenzylamino-6-methylsulfonylpteridine	280–281, ir, uv	249
4,6,7-Trisdiethylamino-2-phenylpteridine	—	1398
4,6,7-Trisdimethylamino-2-phenylpteridine	—	1398
2,4,7-Trisdimethylaminopteridine	ms	864
4,6,7-Trisethylamino-2-phenylpteridine	—	1398
4,6,7-Trisisopropylamino-2-phenylpteridine	—	1398
4,6,7-Trismethylamino-2-phenylpteridine	—	1398
2,4,7-Trisphenethylamino-6-methylsulfonylpteridine	—	249
2,4,6-Tristrimethylsiloxypteridine	ms	1457
2,4,7-Tristrimethylsiloxypteridine	crude, ms	1411, 1457
2,4,7-Tristrimethylsiloxy-6-trimethylsiloxycarbonylpteridine	crude, ms	1457
4,6,7-Tristrimethylsiloxy-2-trimethylsilylaminopteridine	crude, ms	1457
6-Vinyl-2,4-pteridinediamine	> 300, nmr	1297
Violapterin [see: 2,4,7(1*H*,3*H*,8*H*)-pteridinetrione]		1197
Uropterin [see: 2-amino-4,6(3*H*,5*H*)-pteridinedione]		1410
Urothione (not simple: See text)		503, 624, 906, 1028, 1262, 1427
Xanthopterin [see: 2-amino-4,6(3*H*,5*H*)-pteridinedione]		
Xanthopterin peroxide (structure doubtful)		94*, 532
8-D-Xylityl-2,4,7(1*H*,3*H*,8*H*)-pteridinetrione	—	295

References

In each case, information was gleaned from the original publication except where an additional reference to *Chemical Abstracts, Chemisches Zentralblatt*, or another abstract journal is included below. Each citation of a Russian journal or *Angewandte Chemie* refers to the original Russian/German version, not to the respective English translation. The abbreviations for journal titles are those recommended in the *Bibliographic Guide for Editors and Authors*, published by the American Chemical Society (Washington: 1974).

1. F.G. Hopkins, *Proc. Chem. Soc.*, 1889, **5**, 117; *Ber. Dtsch. Chem. Ges., Referate*, 1891, **24**, 724.
2. F.G. Hopkins, *Nature (London)*, 1889, **40**, 335.
3. F.G. Hopkins, *Nature (London)*, 1891, **45**, 197.
4. F.G. Hopkins, *Nature (London)*, 1892, **45**, 581.
5. F.G. Hopkins, *Proc. R. Soc. London*, 1894, **57**, 5.
6. F.G. Hopkins, *Philos. Trans. R. Soc. London, Ser. B*, 1895, **186**, 661.
7. O. Kühling, *Ber. Dtsch. Chem. Ges.*, 1894, **27**, 2116.
8. O. Kühling, *Ber. Dtsch. Chem. Ges.*, 1895, **28**, 1968.
9. S. Gabriel and J. Colman, *Ber. Dtsch. Chem. Ges.*, 1901, **34**, 1234.
10. A. Albert, *Acta Unio Int. Contra Cancrum*, 1959, **15**, 70.
11. A. Albert, *Adv. Heterocycl. Chem.*, 1976, **20**, 117.
12. A. Albert, *Angew. Chem.*, 1967, **79**, 913.
13. A. Albert, *Biochem. J.*, 1953, **54**, 646.
14. A. Albert, *Biochem. J.*, 1957, **65**, 124.
15. A. Albert, *Chem. Soc. Spec. Publ.*, 1955, **3**, 124.
16. A. Albert, *Ciba Foundation Symp., Chem. Biol. Pteridines*, 1957, 97.
17. A. Albert, *Current Trends Heterocycl. Chem., Proc. Symp., Canberra*, 1958, 20.
18. A. Albert, *Fortschr. Chem. Org. Naturst.*, 1954, **11**, 350.
19. A. Albert, *J. Chem. Soc.*, 1955, 2690.
20. A. Albert, *J. Chem. Soc. Perkin Trans. 1*, 1979, 1574.
21. A. Albert, *J. Proc. Roy. Soc. NSW.*, 1965, **98**, 11.
22. A. Albert, *Nature (London)*, 1954, **173**, 1046.
23. A. Albert, *Nature (London)*, 1956, **178**, 1072.
24. A. Albert, *Phys. Methods Heterocycl. Chem.*, 1971, **3**, 1.
25. A. Albert, *Rev. Pure Appl. Chem.*, 1951, **1**, 51.
26. A. Albert, *Q. Rev., Chem. Soc.*, 1952, **6**, 197.

27. A. Albert, T.J. Batterham, and J.J. McCormack, *J. Chem. Soc. (B)*, 1966, 1105.
28. A. Albert and D.J. Brown, *Int. Congr. Biochem., 1st, Cambridge, Abstr. Commun.*, 1949, 240; *Chem. Abstr.*, 1953, **47**, 11209.
29. A. Albert and D.J. Brown, *J. Chem. Soc.*, 1953, 74.
30. A. Albert, D.J. Brown, and G.W.H. Cheeseman, *J. Chem. Soc.*, 1951, 474.
31. A. Albert, D.J. Brown, and G.W.H. Cheeseman, *J. Chem. Soc.*, 1952, 1620.
32. A. Albert, D.J. Brown, and G.W.H. Cheeseman, *J. Chem. Soc.*, 1952, 4219.
33. A. Albert, D.J. Brown, and H.C.S. Wood, *J. Chem. Soc.*, 1954, 3832.
34. A. Albert, D.J. Brown, and H.C.S. Wood, *J. Chem. Soc.*, 1956, 2066.
35. A. Albert and J. Clark, *J. Chem. Soc.*, 1964, 1666.
36. A. Albert and J. Clark, *J. Chem. Soc.*, 1965, 27.
37. A. Albert and A. Hampton, *J. Chem. Soc.*, 1954, 505.
38. A. Albert and C.F. Howell, *J. Chem. Soc.*, 1962, 1591.
39. A. Albert, C.F. Howell, and E. Spinner, *J. Chem. Soc.*, 1962, 2595.
40. A. Albert, Y. Inoue, and D.D. Perrin, *J. Chem. Soc.*, 1963, 5151.
41. A. Albert, J.H. Lister, and C. Pedersen, *J. Chem. Soc.*, 1956, 4621.
42. A. Albert and S. Matsuura, *J. Chem. Soc.*, 1961, 5131.
43. A. Albert and S. Matsuura, *J. Chem. Soc.*, 1962, 2162.
44. A. Albert and J.J. McCormack, *J. Chem. Soc.*, 1965, 6930.
45. A. Albert and J.J. McCormack, *J. Chem. Soc. (C)*, 1966, 1117.
46. A. Albert and J.J. McCormack, *J. Chem. Soc. (C)*, 1968, 63.
47. A. Albert and J.J. McCormack, *J. Chem. Soc. Perkin Trans. 1*, 1973, 2630.
48. A. Albert and H. Mizuno, *J. Chem. Soc. (B)*, 1971, 2423.
49. A. Albert and H. Mizuno, *J. Chem. Soc. Perkin Trans. 1*, 1973, 1615.
50. A. Albert and H. Mizuno, *J. Chem. Soc. Perkin Trans. 1*, 1973, 1974.
51. A. Albert and K. Ohta, *J. Chem. Soc. (C)*, 1970, 1540.
52. A. Albert and K. Ohta, *J. Chem. Soc. (C)*, 1971, 2357.
53. A. Albert and K. Ohta, *J. Chem. Soc. (C)*, 1971, 3727.
54. A. Albert and K. Ohta, *J. Chem. Soc. Chem. Commun.*, 1969, 1168.
55. A. Albert and F. Reich, *J. Chem. Soc.*, 1961, 127.
56. A. Albert and F. Reich, *J. Chem. Soc.*, 1960, 1370.
57. A. Albert and E.P. Serjeant, *J. Chem. Soc.*, 1964, 3357.
58. A. Albert and E.P. Serjeant, *Nature (London)*, 1963, **199**, 1098.
59. A. Albert and H. Taguchi, *J. Chem. Soc. Perkin Trans. 2*, 1973, 1101.
60. A. Albert and H.C.S. Wood, *J. Appl. Chem.*, 1952, **2**, 591.
61. A. Albert and H.C.S. Wood, *J. Appl. Chem.*, 1953, **3**, 521.
62. A. Albert and H.C.S. Wood, *Nature (London)*, 1953, **172**, 118.
63. A. Albert and H. Yamamoto, *Adv. Heterocycl. Chem.*, 1973, **15**, 1.
64. A. Albert and H. Yamamoto, *J. Chem. Soc. (C)*, 1968, 1181.
65. A. Albert and H. Yamamoto, *J. Chem. Soc. (C)*, 1968, 2289.
66. A. Albert and H. Yamamoto, *J. Chem. Soc. (C)*, 1958, 2292.
67. R.B. Angier, *J. Org. Chem.*, 1963, **28**, 1398.
68. R.B. Angier, *J. Org. Chem.*, 1963, **28**, 1509.
69. R.B. Angier and W.V. Curran, *J. Org. Chem.*, 1961, **26**, 2129.
70. R.B. Angier and W.V. Curran, *J. Org. Chem.*, 1962, **27**, 892.

71. R.M. Anker and J.W. Boehne, *J. Am. Chem. Soc.*, 1952, **74**, 2431.
72. S. Antoulas, J.H. Bieri, and M. Viscontini, *Helv. Chim. Acta*, 1978, **61**, 2246.
73. S. Antoulas and M. Viscontini, *Helv. Chim. Acta*, 1981, **64**, 1134.
74. M.C. Archer and K.G. Scrimgeour, *Can. J. Biochem.*, 1970, **48**, 278.
75. M.C. Archer and K.G. Scrimgeour, *Can. J. Biochem.*, 1970, **48**, 526.
76. M.C. Archer, D.J. Vonderschmitt, and K.G. Scrimgeour, *Can. J. Biochem.*, 1972, **50**, 1174.
77. G.R. Gapski, J.M. Whiteley, and F.M. Huennekens, *Biochemistry*, 1971, **10**, 2930.
78. M. Kawai and K.G. Scrimgeour, *Can. J. Biochem.*, 1972, **50**, 1191.
79. W.L.F. Armarego, T.J. Batterham, and J.R. Kershaw, *Org. Magn. Reson.*, 1971, **3**, 575.
80. W.L.F. Armarego, and B.A. Milloy, *Aust. J. Chem.*, 1977, **30**, 2023.
81. W.L.F. Armarego, D. Randles, and H. Taguchi, *Eur. J. Biochem.*, 1983, **135**, 393.
82. W.L.F. Armarego, D. Randles, H. Taguchi, and M.J. Whittaker, *Aust. J. Chem.*, 1984, **37**, 355.
83. W.L.F. Armarego and H. Schou, *Aust. J. Chem.*, 1978, **31**, 1081.
84. W.L.F. Armarego and H. Schou, *J. Chem. Soc. Perkin Trans. 1,* 1977, 2529.
85. W.L.F. Armarego and P. Waring, *Chem. Biol. Pteridines, Proc. Int. Symp., 7th, St. Andrews*, 1982, 57.
86. W.L.F. Armarego and P. Waring, *Chem. Biol. Pteridines, Proc. Int. Symp., 7th, St. Andrews*, 1982, 429.
87. W.L.F. Armarego and P. Waring, *J. Chem. Res.*, 1980, S 318 and M 3911.
88. W.L.F. Armarego and P. Waring, *J. Chem. Soc. Perkin Trans. 2*, 1982, 1227.
89. W.L.F. Armarego, P. Waring, and J.W. Williams, *J. Chem. Soc. Chem. Commun.*, 1980, 334.
90. B.R. Baker and B.-T. Ho, *J. Pharm. Sci.*, 1965, **54**, 1261.
91. B.R. Baker and D.V. Santi, *J. Pharm. Sci.*, 1965, **54**, 1252.
92. S.N. Baranov and T.E. Gorizdra, *Zh. Obshch. Khim.*, 1962, **32**, 1220.
93. S.N. Baranov and T.E. Gorizdra, *Zh. Obshch. Khim.*, 1962, **32**, 1226.
94. G.B. Barlin and W. Pfleiderer, *Chem. Ber.*, 1971, **104**, 3069.
95. R. Baur, T. Sugimoto, and W. Pfleiderer, *Chem. Lett.*, 1984, 1025.
96. C.M. Baugh and E. Shaw, *J. Org. Chem.*, 1964, **29**, 3610.
97. K. Baumgartner and J.H. Bieri, *Helv. Chim. Acta*, 1980, **63**, 1805.
98. F. Bergmann and H. Kwietny, *Biochim. Biophys. Acta*, 1958, **28**, 613.
99. F. Bergmann and H. Kwietny, *Biochim. Biophys. Acta*, 1959, **33**, 29.
100. F. Bergmann, L. Levene, Z. Neiman, and D.J. Brown, *Biochim. Biophys. Acta*, 1970, **222**, 191.
101. F. Bergmann, M. Tamari, and H. Ungar-Waron, *J. Chem. Soc.*, 1964, 565.
102. J.H. Bieri and R.E. Geiger, *Helv. Chim. Acta*, 1975, **58**, 1201.
103. J.H. Bieri, W.-P. Hummel, and M. Viscontini, *Helv. Chim. Acta*, 1976, **59**, 2374.
104. J.H. Bieri and M. Viscontini, *Helv. Chim. Acta*, 1977, **60**, 447.
105. M.E.C. Biffin, D.J. Brown, and T.-C. Lee, *Aust. J. Chem.*, 1967, **20**, 1041.
106. A.J. Birch and C.J. Moye, *J. Chem. Soc.*, 1957, 412.
107. A.J. Birch and C.J. Moye, *J. Chem. Soc.*, 1958, 2622.
108. R. Baur, E. Kleiner, and W. Pfleiderer, *Justus Liebigs Ann. Chem.*, 1984, 1798.
109. M. Böhme, W. Pfleiderer, E.F. Elstner, and W.J. Richter, *Angew. Chem.*, 1980, **92**, 474.
110. W.R. Boon, *J. Chem. Soc.*, 1957, 2146.
111. W.R. Boon and G. Bratt, *J. Chem. Soc.*, 1956, 2159.
112. J.H. Boothe, U.S. Pat. 2,584,538 (1952).

113. P.H. Boyle and Pfleiderer, *Chem. Ber.*, 1980, **113**, 1514.
114. H. Braun and W. Pfleiderer, *Justus Liebigs Ann. Chem.*, 1973, 1082.
115. H. Braun and W. Pfleiderer, *Justus Liebigs Ann. Chem.*, 1973, 1091.
116. H. Braun and W. Pfleiderer, *Justus Liebigs Ann. Chem.*, 1973, 1099.
117. H. Bredereck and W. Pfleiderer, *Chem. Ber.*, 1954, **87**, 1268.
118. P.R. Brook and G.R. Ramage, *J. Chem. Soc.*, 1955, 896.
119. D.J. Brown, *Ciba Foundation Symp., Chem. Biol. Pteridines*, 1954, 62.
120. D.J. Brown, *J. Appl. Chem.*, 1959, **9**, 203.
121. D.J. Brown, *J. Chem. Soc.*, 1953, 1644.
122. D.J. Brown, W.B. Cowden, S.-B. Lan, and K. Mori, *Aust. J. Chem.*, 1984, **37**, 155.
123. D.J. Brown and B.T. England, *J. Chem. Soc.*, 1965, 1530.
124. D.J. Brown, B.T. England, and J.M. Lyall. *J. Chem. Soc. (C)*, 1966, 226.
125. D.J. Brown, P.W. Ford, and K.H. Tratt, *J. Chem. Soc. (C)*, 1967, 1445.
126. D.J. Brown and J.S. Harper, *J. Chem. Soc.*, 1961, 1298.
127. D.J. Brown and J.S. Harper, *J. Chem. Soc.*, 1963, 1276.
128. D.J. Brown and J.S. Harper, *Pteridine Chem., Proc. Int. Symp., 3rd, Stuttgart*. 1962, 219.
129. D.J. Brown and N.W. Jacobsen, *J. Chem. Soc.*, 1960, 1978.
130. D.J. Brown and N.W. Jacobsen, *J. Chem. Soc.*, 1961, 4413.
131. D.J. Brown and N.W. Jacobsen, *J. Chem. Soc.*, 1965, 1175.
132. D.J. Brown and N.W. Jacobsen, *J. Chem. Soc.*, 1965, 3770.
133. D.J. Brown and N.W. Jacobsen, *Tetrahedron Lett.*, 1960, No 25, 17.
134. D.J. Brown and S.F. Mason, *J. Chem. Soc.*, 1956, 3443.
135. D.J. Brown and K. Shinozuka, *Aust. J. Chem.*, 1981, **34**, 189.
136. D.J. Brown and K. Shinozuka, *Aust. J. Chem.*, 1981, **34**, 2635.
137. D.J. Brown and K. Shinozuka, *Chem. Biol. Pteridines, Proc. Int. Symp., 7th, St. Andrews*, 1982, 63.
138. D.J. Brown and T. Sugimoto, *J. Chem. Soc. (C)*, 1970, 2661.
139. E. Bühler and W. Pfleiderer, *Chem. Ber.*, 1966, **99**, 2997.
140. J.W. Bunting and D.D. Perrin, *Aust. J. Chem.*, 1966, **19**, 337.
141. C.K. Cain, E.C. Taylor, and L.J. Daniel, *J. Am. Chem. Soc.*, 1949, **71**, 892.
142. N.R. Campbell, M.E.H. Fitzgerald, and H.O.J. Collier, Brit. Pat. 656,769 (1951).
143. R. Charubala and W. Pfleiderer, *Helv. Chim. Acta*, 1979, **62**, 1171.
144. R. Charubala and W. Pfleiderer, *Helv. Chim. Acta*, 1979, **62**, 1179.
145. D. Chippel and K.G. Scrimgeour, *Can. J. Biochem.*, 1970, **48**, 999.
146. J. Clark, *J. Chem. Soc.*, 1964, 4920.
147. J. Clark, *J. Chem. Soc. (C)*, 1967, 1543.
148. J. Clark, *J. Chem. Soc. (C)*, 1968, 313.
149. J. Clark, *Tetrahedron Lett.*, 1967, 1099.
150. J. Clark, W. Kernick, and A.J. Layton, *J. Chem. Soc.*, 1964, 3215.
151. J. Clark and P.N.T. Murdoch, *J. Chem. Soc. (C)*, 1969, 1883.
152. J. Clark, P.N.T. Murdoch, and D.L. Roberts, *J. Chem. Soc. (C)*, 1969, 1408.
153. J. Clark and G. Neath, *J. Chem. Soc. (C)*, 1966, 1112.
154. J. Clark and G. Neath, *J. Chem. Soc. (C)*, 1968, 919.
155. J. Clark, G. Neath, and C. Smith, *J. Chem. Soc. (C)*, 1969, 1297.
156. J. Clark and W. Pendergast, *J. Chem. Soc. (C)*, 1969, 1751.

157. J. Clark, W. Pendergast, F.S. Yates, and A.E. Cunliffe, *J. Chem. Soc. (C)*, 1971, 375.
158. J. Clark and F.S. Yates, *J. Chem. Soc. (C)*, 1971, 371.
159. J. Clark and F.S. Yates, *J. Chem. Soc. (C)*, 1971, 2278.
160. H.O.J. Collier and M. Phillips, *Nature (London)*, 1954, **174**, 180.
161. D.B. Cosulich and J.M. Smith, *J. Am. Chem. Soc.*, 1948, **70**, 1922.
162. R.M. Cresswell, A.C. Hill, and H.C.S. Wood, *J. Chem. Soc.*, 1959, 698.
163. R.M. Cresswell, T. Neilson, and H.C.S. Wood, *J. Chem. Soc.*, 1961, 476.
164. R.M. Cresswell and H.C.S. Wood, *J. Chem. Soc.*, 1959, 387.
165. R.M. Cresswell and H.C.S. Wood, *J. Chem. Soc.*, 1960, 4768.
166. W.V. Curran and R.B. Angier, *J. Am. Chem. Soc.*, 1958, **80**, 6095.
167. W.V. Curran and R.B. Angier, *J. Org. Chem.*, 1961, **26**, 2364.
168. W.V. Curran and R.B. Angier, *J. Org. Chem.*, 1962, **27**, 1366.
169. W.V. Curran and R.B. Angier, *J. Org. Chem.*, 1963, **28**, 2672.
170. J.W. Daly and B.E. Christensen, *J. Am. Chem. Soc.*, 1956, **78**, 225.
171. L.J. Daniel, L.C. Norris, M.L. Scott, and G.F. Heuser, *J. Biol. Chem.*, 1947, **169**, 689.
172. J. Davoll and D.D. Evans, *J. Chem. Soc.*, 1960, 5041.
173. J.I. DeGraw, V.H. Brown, R.S. Kisliuk, and Y. Gaumont, *J. Med. Chem.*, 1971, **14**, 866.
174. G.P.G. Dick, D. Livingston, and H.C.S. Wood, *J. Chem. Soc.*, 1958, 3730.
175. G.P.G. Dick and H.C.S. Wood, *J. Chem. Soc.*, 1955, 1379.
176. G.P.G. Dick, H.C.S. Wood, and W.R. Logan, *J. Chem. Soc.*, 1956, 2131.
177. R.G. Dickinson and N.W. Jacobsen, *Aust. J. Chem.*, 1976, **29**, 459.
178. R.L. Dion and T.L. Loo, *J. Org. Chem.*, 1961, **26**, 1857.
179. J.H. Dustmann, *Hoppe-Seyler's Z. Physiol. Chem.*, 1971, **352**, 1599.
180. A. Ehrenberg, P. Hemmerich, F. Müller, and W. Pfleiderer, *Eur. J. Biochem.*, 1970, **16**, 584.
181. K. Eistetter and W. Pfleiderer, *Chem. Ber.*, 1973, **106**, 1389.
182. K. Eistetter and W. Pfleiderer, *Chem. Ber.*, 1976, **109**, 3208.
183. G.B. Elion, *Ciba Foundation Symp., Chem. Biol. Pteridines*, 1954, 49.
184. B.E. Evans, *J. Chem. Soc. Perkin Trans. 1*, 1974, 357.
185. R.M. Evans, P.G. Jones, P.J. Palmer, and F.F. Stephens, *J. Chem. Soc.*, 1956, 4106.
186. M.H. Fahrenbach, K.H. Collins, M.E. Hultquist, and J.M. Smith, *Ciba Foundation Symp., Chem. Biol. Pteridines*, 1954, 173.
187. E. Felder, D. Pitrè, and S. Boveri, *J. Med. Chem.*, 1972, **15**, 210.
188. W.E. Fidler and H.C.S. Wood, *J. Chem. Soc.*, 1956, 3311.
189. W.E. Fidler and H.C.S. Wood, *J. Chem. Soc.*, 1957, 3980.
190. W.E. Fidler and H.C.S. Wood, *J. Chem. Soc.*, 1957, 4157.
191. H.S. Forrest and H.K. Mitchell, *Ciba Foundation Symp., Chem. Biol. Pteridines*, 1954, 143.
192. H.S. Forrest and J. Walker, *J. Chem. Soc.*, 1949, 79.
193. H.S. Forrest and J. Walker, *J. Chem. Soc.*, 1949, 2077.
194. W. Frick, R. Weber, and M. Viscontini, *Helv. Chim. Acta*, 1974, **57**, 2658.
195. H. Fromherz and A. Kotzschmar, *Justus Liebigs Ann. Chem.*, 1938, **534**, 283.
196. H. Fuchs and W. Pfleiderer, *Heterocycles*, 1978, **11**, 247.
197. H.-J. Furrer, J.H. Bieri, and M. Viscontini, *Helv. Chim. Acta*, 1978, **61**, 2744.
198. H.-J. Furrer, J.H. Bieri, and M. Viscontini, *Helv. Chim. Acta*, 1979, **62**, 2558.
199. H.-J. Furrer, J.H. Bieri, and M. Viscontini, *Helv. Chim. Acta*, 1979, **62**, 2577.
200. A.N. Ganguly, J.H. Bieri, and M. Viscontini, *Helv. Chim. Acta*, 1981, **64**, 367.

201. A.N. Ganguly, P.K. Sengupta, J.H. Bieri, and M. Viscontini, *Helv. Chim. Acta*, 1980, **63**, 395.

202. A.N. Ganguly, P.K. Sengupta, J.H. Bieri, and M. Viscontini, *Helv. Chim. Acta*, 1980, **63**, 1754.

203. S.N. Ganguly and M. Viscontini, *Helv. Chim. Acta*, 1982, **65**, 1090.

204. T. Goto, *Jpn. J. Zool.*, 1963, **14**, 83.

205. T. Goto, *Jpn. J. Zool.*, 1963, **14**, 91.

206. T. Goto and T. Hama, *Proc. Jpn. Acad.*, 1958, **34**, 724.

207. T. Goto, A. Tatematsu, and S. Matsuura, *J. Org. Chem.*, 1965, **30**, 1844.

208. R. Gottlieb and W. Pfleiderer, *Chem. Ber.*, 1978, **111**, 1763.

209. P. Goya and W. Pfleiderer, *Chem. Ber.*, 1981, **114**, 699.

210. P. Goya and W. Pfleiderer, *Chem. Ber.*, 1981, **114**, 707.

211. S.J. Gumbley and L. Main, *Aust. J. Chem.*, 1976, **29**, 2753.

212. T. Hama, *Ann. NY Acad. Sci.*, 1963, **100**, (II), 977.

213. T. Hama, *Experientia*, 1953, **9**, 299.

214. T. Hama and S. Fukuda, *Pteridine Chem., Proc. Int. Symp., 3rd, Stuttgart*, 1962, 495.

215. T. Hama and T. Goto, *C. R. Soc. Biol.*, 1954, **148**, 1313.

216. T. Hama, T. Goto, and K. Kushibiki, *C. R. Soc. Biol.*, 1954, **148**, 754.

217. T. Hama, T. Goto, Y. Tohnoki, and Y. Hiyama, *Proc. Jpn. Acad.*, 1965, **41**, 305.

218. T. Hama, J. Matsumoto, and Y. Mori, *Proc. Jpn. Acad.*, 1960, **36**, 346.

219. T. Hama and M. Obika, *Anat. Rec.*, 1959, **134**, 25.

220. T. Hama and M. Obika, *Experientia*, 1958, **14**, 182.

221. R. Harmsen, *Acta Trop.*, 1970, **27**, 2.

222. R. Harmsen, *Chem. Biol. Pteridines, Proc. Int. Symp., 4th, Toba*, 1969, 405.

223. R. Harmsen, *J. Insect Physiol.*, 1969, **15**, 2239.

224. R. Harris and W. Pfleiderer, *Justus Liebigs Ann. Chem.*, 1981, 1457.

225. K. Harzer and W. Pfleiderer, *Helv. Chim. Acta*, 1973, **56**, 1225.

226. D.L. Hatfield, *Diss. Abstr.*, 1962, **23**, 54.

227. A. Heckel and W. Pfleiderer, *Tetrahedron Lett.*, 1981, **22**, 2161.

228. A. Heckel and W. Pfleiderer, *Tetrahedron Lett.*, 1983, **24**, 5047.

229. B. Heinz, W. Ried, and K. Dose, *Angew. Chem.*, 1979, **91**, 510.

230. C. Heizmann, P. Hemmerich, R. Mengel, and W. Pfleiderer, *Chem. Biol. Pteridines, Proc. Int. Symp., 4th, Toba*, 1969, 105.

231. C. Heizmann, P. Hemmerich, R. Mengel, and W. Pfleiderer, *Helv. Chim. Acta*, 1973, **56**, 1908.

232. G. Henseke and J. Müller, *Chem. Ber.*, 1960, **93**, 2668.

233. M. Higuchi, T. Nagamura, and F. Yoneda, *Heterocycles*, 1976, **4**, 977.

234. D.T. Hurst, *Chemistry and Biochemistry of Pyrimidines, Purines, Pteridines*, Wiley, Chichester, 1980, p. 266.

235. R. Hüttel and G. Sprengling, *Justus Liebigs Ann. Chem.*, 1943, **554**, 69.

236. W. Hutzenlaub, G.B. Barlin, and W. Pfleiderer, *Angew. Chem.*, 1969, **81**, 624.

237. W. Hutzenlaub, K. Kobayashi, and W. Pfleiderer, *Chem. Ber.*, 1976, **109**, 3217.

238. W. Hutzenlaub, H. Yamamoto, G.B. Barlin, and W. Pfleiderer, *Chem. Ber.*, 1973, **106**, 3203.

239. K. Ienaga and W. Pfleiderer, *Chem. Ber.*, 1977, **110**, 3456.

240. K. Ienaga and W. Pfleiderer, *Chem. Ber.*, 1978, **111**, 2586.

241. Y. Inoue and D.D. Perrin, *J. Chem. Soc.*, 1962, 2600.

242. Y. Inoue and D.D. Perrin, *J. Chem. Soc.*, 1963, 2648.

243. Y. Inoue and D.D. Perrin, *J. Chem. Soc.*, 1963, 3936.

244. Y. Inoue and D.D. Perrin, *J. Chem. Soc.*, 1963, 4803.
245. Y. Inoue and D.D. Perrin, *J. Phys. Chem.*, 1962, **66**, 1689.
246. T. Itoh and W. Pfleiderer, *Chem. Ber.*, 1976, **109**, 3228.
247. Y. Iwanami, *Bull. Chem. Soc. Jpn.*, 1971, **44**, 1314.
248. N.W. Jacobsen, *J. Chem. Soc. (C)*, 1966, 1065.
249. W.D. Johnston, H.S. Broadbent, and W.W. Parish, *J. Heterocycl. Chem.*, 1973, **10**, 133.
250. W.G.M. Jones, *Nature (London)*, 1948, **162**, 524.
251. T.H. Jukes and E.L.R. Stokstad, *Physiol. Rev.*, 1948, **28**, 51.
252. E.M. Kaiser and S.L. Hartzell, *J. Org. Chem.*, 1977, **42**, 2951.
253. R. Kalbermatten, W. Städeli, J.H. Bieri, and M. Viscontini, *Helv. Chim. Acta*, 1981, **64**, 2627.
254. Y. Kaneko, *Agric. Biol. Chem.*, 1965, **29**, 965.
255. Y. Kaneko, *C.R. Soc. Biol.*, 1959, **153**, 887.
256. Y. Kaneko and M. Sanada, *J. Ferment, Technol. (Japan)*, 1969, **47**, 8.
257. P. Karrer, R. Schwyzer, and B.J.R. Nicolaus, *Helv. Chim. Acta*, 1950, **33**, 557.
258. E. Khalifa, H.-J. Furrer, J.H. Bieri, and M. Viscontini, *Helv. Chim. Acta*, 1978, **61**, 2739.
259. E. Khalifa, A.N. Ganguly, J.H. Bieri, and M. Viscontini, *Helv. Chim. Acta*, 1980, **63**, 2554.
260. E. Khalifa, P.K. Sengupta, J.H. Bieri, and M. Viscontini, *Helv. Chim. Acta*, 1976, **59**, 242.
261. Y.H. Kim, Y. Gaumont, R.L. Kisliuk, and H.G. Mautner, *J. Med. Chem.*, 1975, **18**, 776.
262. M. Kobayashi and K. Iwai, *Agric. Biol. Chem.*, 1971, **35**, 47.
263. K. Kobayashi and W. Pfleiderer, *Chem. Ber.*, 1976, **109**, 3159.
264. K. Kobayashi and W. Pfleiderer, *Chem. Ber.*, 1976, **109**, 3175.
265. K. Kobayashi and W. Pfleiderer, *Chem. Ber.*, 1976, **109**, 3184.
266. K. Kobayashi and W. Pfleiderer, *Chem. Ber.*, 1976, **109**, 3194.
267. G. Konrad and W. Pfleiderer, *Chem. Ber.*, 1970, **103**, 722.
268. G. Konrad and W. Pfleiderer, *Chem. Ber.*, 1970, **103**, 735.
269. F. Korte, *Chem. Ber.*, 1954, **87**, 1062.
270. F. Korte, *Ciba Foundation Symp., Chem. Biol. Pteridines*, 1954, 159.
271. F. Korte and H. Barkemeyer, *Chem. Ber.*, 1956, **89**, 2400
272. F. Korte, H. Barkemeyer, and G. Synnatschke, *Hoppe-Seyler's Z. Physiol Chem.*, 1959, **314**, 106.
273. F. Korte and E.G. Fuchs, *Chem. Ber.*, 1953, **86**, 114.
274. F. Korte and M . Goto, *Tetrahedron Lett.*, 1961, 55.
275. F. Korte, H. Weitkamp, and H.-G. Schicke, *Chem. Ber.*, 1957, **90**, 1100.
276. K. Kushibiki, T. Hama, and T. Goto, *C. R. Soc. Biol.*, 1954, **148**, 759.
277. H. Kwietny and F. Bergmann, *J. Chromatogr.*, 1959, **2**, 162.
278. J.M. Lagowski, H.S. Forrest, and H.C.S. Wood, *Proc. Chem. Soc.*, 1963, 343.
279. P.D. Landor and H.N. Rydon, *J. Chem. Soc.*, 1955, 1113.
280. C.L. Leese and G.M. Timmis, *J. Chem. Soc.*, 1958, 4104.
281. J.H. Lister, G.R. Ramage, and E. Coates, *J. Chem. Soc.*, 1954, 4109.
282. J.H. Lister and G.M. Timmis, *J. Chem. Soc.*, 1960, 1113.
283. H. Lutz and W. Pfleiderer, *Carbohydr. Res.*, 1984, **130**, 179.
284. M.F. Mallette, E.C. Taylor, and C.K. Cain, *J. Am. Chem. Soc.*, 1947, **69**, 1814.
285. S.F. Mason, *Ciba Foundation Symp., Chem. Biol. Pteridines*, 1954, 74.
286. S.F. Mason, *J. Chem. Soc.*, 1955, 2336.
287. S.F. Mason, *J. Chem. Soc.*, 1957, 5010.

288. S.F. Mason, *J. Chem. Soc.*, 1958, 674.
289. S.F. Mason, *Nature (London)*, 1954, **173**, 1175.
290. T. Masuda, *Pharm. Bull. (Tokyo)*, 1957, **5**, 28.
291. J. Matusumoto, *Jpn. J. Zool.* 1965, **14**, No 3, 45.
292. J. Matsumoto, T. Kajishima, and T. Hama, *Genetics*, 1960, **45**, 1177.
293. S. Matsuura and T. Goto, *J . Chem. Soc.*, 1963, 1773.
294. S. Matsuura and T. Goto, *J. Chem. Soc.*, 1965, 623.
295. S. Matsuura, M. Odaka, T. Sugimoto, and T. Goto, *Chem. Lett.*, 1973, 343.
296. S. Matsuura and T. Sugimoto, *Bull. Chem. Soc. Jpn.*, 1981, **54**, 2543.
297. S. Matsuura and T. Sugimoto, *Res. Bull. Dept. Gen. Educ., Nagoya Univ.*, 1969, **13**, 9.
298. S. Matsuura, T. Sugimoto, H. Hasegawa, S. Imaizumi, and A. Ichiyama, *J. Biochem. (Tokyo)*, 1980, **87**, 951.
299. S. Matsuura, T. Sugimoto, C. Kitayama, and M. Tsusue, *J. Biochem. (Tokyo)*, 1978, **83**, 19.
300. S. Matsuura, T. Sugimoto, and T. Nagatsu, *Bull. Chem. Soc. Jpn.*, 1981, **54**, 2231.
301. S. Matsuura, T. Sugimoto, and M. Tsusue, *Bull. Chem. Soc. Jpn.*, 1977, **50**, 2163.
302. S. Matsuura, T. Sugimoto, C.K. Yokokawa, and M. Tsusue, *Chem. Biol. Pteridines, Proc. Int. Symp., 6th, La Jolla*, 1978, 135.
303. H.G. Mautner and Y -H. Kim, *J. Org. Chem.*, 1975, **40**, 3447.
304. H.G. Mautner, Y.-H,. Kim, Y. Gaumont, and R. L. Kisliuk, *Chem. Biol. Pteridines, Proc. Int. Symp., 5th, Konstanz*, 1975, 515.
305. J.J. McCormack and H.G. Mautner, *J. Org. Chem.*, 1964, **29**, 3370.
306. J.A.R. Mead, H.B. Wood, and A. Goldin, *Cancer Chemother. Rep., Part 2*, **1**, No. 2, 273 (1968).
307. R. Mengel and W. Pfleiderer, *Chem. Ber.*, 1978, **111**, 3790.
308. P.N. Moorthy and E. Hayon, *J. Org. Chem.*, 1976, **41**, 1607.
309. S. Nawa and H.S. Forrest, *Natl. Inst. Genet. Annu. Rep.*, 1962, **13**, 23.
310. Z. Neiman, *Experientia*, 1975, **31**, 996.
311. Z. Neiman, *J. Chem. Soc. (C)*, 1970, 91.
312. Z. Neiman, *J. Heterocycl. Chem.*, 1974, **11**, 7.
313. Z. Neiman, F. Bergmann, and H. Weiler-Feilchenfeld, *J. Chem. Soc. (C)*, 1969, 114.
314. R.H. Nimmo-Smith and D.J. Brown, *J. Gen. Microbiol.*, 1953, **9**, 536.
315. S. Nishigaki, K. Fukami, M. Ichiba, H. Kanazawa, K. Matsuyama, S. Ogusu, K. Senga, F. Yoneda, R. Koga, and T. Ueno, *Heterocycles*, 1981, **15**, 757.
316. S. Nishigaki, S. Fukazawa, K. Ogiwara, and F. Yoneda, *Chem. Pharm. Bull.*, 1971, **19**, 206.
317. G. Nübel and W. Pfleiderer, *Chem. Ber.*, 1962, **95**, 1605.
318. S. Odate, Y. Tatebe, M. Obika, and T. Hama, *Proc. Jpn. Acad.*, 1959, **35**, 567.
319. E. Ortiz, E. Bächli, D. Price, and H.G. Williams-Ashman, *Physiol. Zool.*, 1963, **36**, 97.
320. T.S. Osdene, A.A. Santilli, L.E. McCardle, and M.E. Rosenthale, *J. Med. Chem.*, 1966, **9**, 697.
321. T.S. Osdene, A.A. Santilli, L.E. McCardle, and M.E. Rosenthale, *J. Med. Chem.*, 1967, **10**, 165.
322. T.S. Osdene and E.C. Taylor, *J. Am. Chem. Soc.*, 1956, **78**, 5451.
323. T.S. Osdene and G.M. Timmis, *J. Chem. Soc.*, 1955, 2027.
324. T.S. Osdene and G.M. Timmis, *J. Chem. Soc.*, 1955, 2036.
325. M. Ott and W. Pfleiderer, *Angew. Chem.*, 1971, **83**, 974.
326. I.J. Pachter and P.E. Nemeth, *J. Org. Chem.*, 1963, **28**, 1187.
327. I.J. Pachter and P.E. Nemeth, *J. Org. Chem.*, 1963, **28**, 1203.

328. I.J. Pachter, P.E. Nemeth, and A.J. Villani, *J. Org. Chem.*, 1963, **28**, 1197.
329. W.W. Parish and H.S. Broadbent, *J. Heterocycl. Chem.*, 1971, **8**, 527.
330. E.L. Patterson, R. Milstrey, and E.L.R. Stockstad, *J. Am. Chem. Soc.*, 1958, **80**, 2018.
331. A. Perez-Rubalcaba and W. Pfleiderer, *Justus Liebigs Ann. Chem.*, 1983, 852.
332. D.D. Perrin, *J. Chem. Soc.*, 1963, 1284.
333. W. Pfleiderer, *Angew. Chem.*, 1961, **73**, 581.
334. W. Pfleiderer, *Angew. Chem.*, 1963, **75**, 993.
335. W. Pfleiderer, *Biochem. Clin. Aspects Pteridines*, 1982, **1**, 3.
336. W. Pfleiderer, *Chem. Ber.*, 1955, **88**, 1625.
337. W. Pfleiderer, *Chem. Ber.*, 1956, **89**, 641.
338. W. Pfleiderer, *Chem. Ber.*, 1957, **90**, 2582.
339. W. Pfleiderer, *Chem. Ber.*, 1957, **90**, 2588.
340. W. Pfleiderer, *Chem. Ber.*, 1957, **90**, 2604.
341. W. Pfleiderer, *Chem. Ber.*, 1957, **90**, 2617.
342. W. Pfleiderer, *Chem. Ber.*, 1957, **90**, 2624.
343. W. Pfleiderer, *Chem. Ber.*, 1957, **90**, 2631.
344. W. Pfleiderer, *Chem. Ber.*, 1958, **91**, 1671.
345. W. Pfleiderer, *Chem. Ber.*, 1959, **92**, 2468.
346. W. Pfleiderer, *Chem. Ber.*, 1959, **92**, 3190.
347. W. Pfleiderer, *Chem. Ber.*, 1962, **95**, 749.
348. W. Pfleiderer, *Chem. Ber.*, 1962, **95**, 2195.
349. W. Pfleiderer, *Chem. Ber.*, 1974, **107**, 785.
350. W. Pfleiderer, *Chem. Ber.*, 1979, **112**, 2750.
351. W. Pfleiderer, *Chem. Biol. Pteridines, Proc. Int. Symp., 4th, Toba*, 1969, 7.
352. W. Pfleiderer, *Ciba Foundation Symp., Chem. Biol. Pteridines*, 1957, 77.
353. W. Pfleiderer, *J. Inher. Metab. Dis.*, 1978, **1**, 54.
354. W. Pfleiderer, *Tetrahedron Lett.*, 1984, **25**, 1031.
355. W. Pfleiderer, *Z. Naturforsch., Teil B*, 1963, **18**, 420.
356. W. Pfleiderer, D. Autenrieth, and M. Schranner, *Angew. Chem.*, 1971, **83**, 971.
357. W. Pfleiderer, D. Autenrieth, and M. Schranner, *Chem. Ber.*, 1973, **106**, 317.
358. W. Pfleiderer, E. Bühler, and D. Schmidt, *Chem. Ber.*, 1968, **101**, 3794.
359. W. Pfleiderer, J.W. Bunting, D.D. Perrin, and G. Nübel, *Chem. Ber.*, 1968, **101**, 1072.
360. W. Pfleiderer and K. Deckert, *Chem. Ber.*, 1962, **95**, 1597.
361. W. Pfleiderer, K. Eistetter, and M. Shanshal, *Jerusalem Symp. Quantum Chem. Biochem.*, 1972, **4**, 469.
362. W. Pfleiderer and H. Fink, *Chem. Ber.*, 1963, **96**, 2950.
363. W. Pfleiderer and H. Fink, *Chem. Ber.*, 1963, **96**, 2964.
364. W. Pfleiderer and I. Geissler, *Chem. Ber.*, 1954, **87**, 1274.
365. W. Pfleiderer and R. Gottlieb, *Heterocycles*, 1980, **14**, 1603.
366. W. Pfleiderer and W. Hutzenlaub, *Angew. Chem.*, 1965, **77**, 1136.
367. W. Pfleiderer and W. Hutzenlaub, *Chem. Ber.*, 1973, **106**, 3149.
368. W. Pfleiderer, E. Liedek, R. Lohrmann, and M. Rukwied, *Chem. Ber.*, 1960, **93**, 2015.
369. W. Pfleiderer, E. Liedek, and M. Rukwied, *Chem. Ber.*, 1962, **95**, 755.
370. W. Pfleiderer and R. Lohrmann, *Chem. Ber.*, 1961, **94**, 12.
371. W. Pfleiderer and R. Lohrmann, *Chem. Ber.*, 1961, **94**, 2708.

372. W. Pfleiderer and R. Lohrmann, *Chem. Ber.*, 1962, **95**, 738.
373. W. Pfleiderer and R. Mengel, *Chem. Ber.*, 1971, **104**, 2293.
374. W. Pfleiderer and R. Mengel, *Chem. Ber.*, 1971, **104**, 2313.
375. W. Pfleiderer and R. Mengel, *Chem. Ber.*, 1971, **104**, 3842.
376. W. Pfleiderer and R. Mengel, *Chem. Biol. Pteridines, Proc. Int. Symp., 4th Toba*, 1969, 43.
377. W. Pfleiderer, R. Mengel, and P. Hemmerich, *Chem. Ber.*, 1971, **104**, 2273.
378. W. Pfleiderer and G. Nübel, *Chem. Ber.*, 1960, **93**, 1406.
379. W. Pfleiderer and G. Nübel, *Chem. Ber.*, 1962, **95**, 1615.
380. W. Pfleiderer and F. Reisser, *Chem. Ber.*, 1962, **95**, 1621.
381. W. Pfleiderer and F. Reisser, *Chem. Ber.*, 1966, **99**, 536.
382. W. Pfleiderer, G. Ritzmann, K. Harzer, and J.C. Jochims, *Chem. Ber.*, 1973, **106**, 2982.
383. W. Pfleiderer and M. Rukwied, *Chem. Ber.*, 1961, **94**, 1.
384. W. Pfleiderer and M. Rukwied, *Chem. Ber.*, 1961, **94**, 118.
385. W. Pfleiderer and M. Rukwied, *Chem. Ber.*, 1962, **95**, 1591.
386. W. Pfleiderer and D. Söll, *J. Heterocycl. Chem.*, 1964, **1**, 23.
387. W. Pfleiderer and E.C. Taylor, *J. Am. Chem. Soc.*, 1960, **82**, 3765.
388. W. Pfleiderer and H. Zondler, *Chem. Ber.*, 1966, **99**, 3008.
389. W. Pfleiderer, H. Zondler, and R. Mengel, *Justus Liebigs Ann. Chem.*, 1970, **741**, 64.
390. J.R. Piper and J.A. Montgomery, *J. Org. Chem.*, 1977, **42**, 208.
391. M. Piraux, *Ind. Chem. Belge*, 1962, **27**, 1188.
392. M. Piraux, *Ind. Chim. Belge*, 1963, **28**, 1.
393. M. Polonovski and R.-G. Busnel, *C. R. Acad. Sci.*, 1950, **230**, 585.
394. M. Polonovski, R.-G. Busnel, and A. Baril, *Toulouse Med.*, 1951, **52**, 766.
395. M. Polonovski, R.-G. Busnel, H. Jérôme, and M. Martinet, *Ciba Foundation Symp., Chem. Biol. Pteridines*, 1954, 165.
396. M. Polonovski and H. Jérôme, *C. R. Acad. Sci.*, 1950, **230**, 392.
397. M. Polonovski, H. Jérôme, and P. Gonnard, *Ciba Foundation Symp., Chem. Biol. Pteridines*, 1954, 124.
398. M. Polonovski, M. Pesson, and A. Puister, *C. R. Acad. Sci.*, 1950, **230**, 2205.
399. M. Polonovski, R. Vieillefosse, and M. Pesson, *Bull. Soc. Chim. Fr.*, 1945, **12**, 78.
400. M.D. Potter and T. Henshall, *J. Chem. Soc.*, 1956, 2000.
401. R. Prewo, J.H. Bieri, S.N. Ganguly, and M. Viscontini, *Helv. Chim. Acta*, 1982, **65**, 1094.
402. B. Pullman, *C. R. Acad. Sci.*, 1958, **246**, 3290.
403. R. Purrmann, *Justus Liebigs Ann. Chem.*, 1940, **544**, 182.
404. R. Purrmann, *Justus Liebigs Ann. Chem.*, 1940, **546**, 98.
405. R. Purrmann, *Justus Liebigs Ann. Chem.*, 1941, **548**, 284.
406. R. Purrmann and F. Eulitz, *Justus Liebigs Ann. Chem.*, 1948, **559**, 169.
407. R. Purrmann and M. Maas, *Justus Liebigs Ann. Chem.*, 1944, **556**, 186.
408. Z.V. Pushkareva and L.V. Alekseeva, *Zh. Obshch. Khim.*, 1962, **32**, 1058.
409. V.J. Ram, W.R. Knappe, and W. Pfleiderer, *Justus Liebigs Ann. Chem.*, 1982, 762.
410. H. Rembold, *J. Inher. Metab. Dis.*, 1978, **1**, 61.
411. H. Rembold and W.L. Gyure, *Angew. Chem.*, 1972, **84**, 1088.
412. G. Ritzmann, K. Ienaga, and W. Pfleiderer, *Justus Liebigs Ann. Chem.*, 1977, 1217.
413. G. Ritzmann, L. Kiriasis, and W. Pfleiderer, *Chem. Ber.*, 1980, **113**, 1524.
414. G. Ritzmann and W. Pfleiderer, *Chem. Ber.*, 1973, **106**, 1401.

415. R. Robinson and M.L. Tomlinson, *J. Chem. Soc.*, 1935, 1283.
416. H. Rokos and W. Pfleiderer, *Chem. Ber.*, 1971, **104**, 739.
417. H. Rokos and W. Pfleiderer, *Chem. Ber.*, 1971, **104**, 770.
418. F.L. Rose, *J. Chem. Soc.*, 1952, 3448.
419. A. Rosowsky, M. Chaykovsky, M. Lin, and E.J. Modest, *J. Med. Chem.*, 1973, **16**, 869.
420. A. Rosowsky and K.K.N. Chen, *J. Org. Chem.*, 1973, **38**, 2073.
421. W.C.J. Ross, *J. Chem. Soc.*, 1948, 1128.
422. T. Rowan and H.C.S. Wood, *J. Chem. Soc. (C)*, 1968, 452.
423. T. Rowan and H.C.S. Wood, *Proc. Chem. Soc.*, 1963, 21.
424. T. Rowan, H.C.S. Wood, and P. Hemmerick, *Proc. Chem. Soc.*, 1961, 260.
425. P.B. Russell, R. Purrmann, W. Schmitt, and G.H. Hitchings, *J. Am. Chem. Soc.*, 1949, **71**, 3412.
426. W. Saenger, G. Ritzmann, and W. Pfleiderer, *Eur. J. Biochem.*, 1972, **29**, 440.
427. S.K. Saha and W. Pfleiderer, *Tetrahedron Lett.*, 1973, 1441.
428. B. Schircks, J.H. Bieri, and M. Viscontini, *Helv. Chim. Acta*, 1976, **59**, 248.
429. B. Schircks, J.H. Bieri, and M. Viscontini, *Helv. Chim. Acta*, 1977, **60**, 211.
430. B. Schircks, J.H. Bieri, and M. Viscontini, *Helv. Chim. Acta*, 1978, **61**, 2731.
431. H. Schlobach and W. Pfleiderer, *Helv. Chim. Acta*, 1972, **55**, 2518.
432. H. Schlobach and W. Pfleiderer, *Helv. Chim. Acta*, 1972, **55**, 2525.
433. H. Schlobach and W. Pfleiderer, *Helv. Chim. Acta*, 1972, **55**, 2533.
434. H. Schlobach and W. Pfleiderer, *Helv. Chim. Acta*, 1972, **55**, 2541.
435. H. Schmid, M. Schranner, and W. Pfleiderer, *Angew. Chem.*, 1971, **83**, 972.
436. H. Schmid, M. Schranner, and W. Pfleiderer, *Chem. Ber.*, 1973, **106**, 1952.
437. H.J. Schneider and W. Pfleiderer, *Chem. Ber.*, 1974, **107**, 3377.
438. C. Schöpf, E. Becker, and R. Reichert, *Justus Liebigs Ann. Chem.*, 1939, **539**, 156.
439. C. Schöpf and K.H. Gänshirt, *Angew. Chem.*, 1962, **74**, 153.
440. C. Schöpf and A. Kottler, *Justus Liebigs Ann. Chem.*, 1939, **539**, 128.
441. C. Schöpf, R. Reichert, and K. Riefstahl, *Justus Liebigs Ann. Chem.*, 1941, 548, 82.
442. W. Schwotzer, J.H. Bieri, M. Viscontini, and W. von Philipsborn, *Helv. Chim. Acta*, 1978, **61**, 2108.
443. P.K. Sengupta, J.H. Bieri, and M. Viscontini, *Helv. Chim. Acta*, 1975, **58**, 1374.
444. P.K. Sengupta, H.A. Breitschmid, J.H. Bieri, and M. Viscontini, *Helv. Chim. Acta*, 1977, **60**, 922.
445. D. Söll and W. Pfleiderer, *Chem. Ber.*, 1963, **96**, 2977.
446. I.W. Southon and W. Pfleiderer, *Chem. Ber.*, 1978, **111**, 971.
447. I.W. Southon and W. Pfleiderer, *Chem. Ber.*, 1978, **111**, 2571.
448. R. Stewart and J.M. McAndless, *J. Chem. Soc. Perkin Trans. 2*, 1972, 376.
449. C.B. Storm, R. Shiman, and S. Kaufman, *J. Org. Chem.*, 1971, **36**, 3925.
450. A. Stuart, D.W. West, and H.C.S. Wood, *J. Chem. Soc.*, 1964, 4769.
451. A. Stuart and H.C.S. Wood, *J. Chem. Soc.*, 1963, 4186.
452. A. Stuart and H.C.S. Wood, *Proc. Chem. Soc.*, 1962, 151.
453. A. Stuart, H.C.S. Wood, and D. Duncan, *J. Chem. Soc. (C)*, 1966, 285.
454. T. Sugimoto and S. Matsuura, *Bull. Chem. Soc. Jpn.*, 1975, **48**, 3767.
455. T. Sugimoto and S. Matsuura, *Bull. Chem. Soc. Jpn.*, 1979, **52**, 181.
456. T. Sugimoto and S. Matsuura, *Bull. Chem. Soc. Jpn.*, 1980, **53**, 3385.

457. T. Sugimoto and S. Matsuura, *Res. Bull. Dept. Gen. Educ., Nagoya Univ.*, 1967, **11**, 94.

458. T. Sugimoto, S. Matsuura, and T. Nagatsu, *Bull. Chem. Soc. Jpn.*, 1980, **53**, 2344.

459. T. Sugimoto, K. Shibata, and S. Matsuura, *Bull. Chem. Soc. Jpn.*, 1977, **50**, 2744.

460. T. Sugimoto, K. Shibata, and S. Matsuura, *Bull. Chem. Soc. Jpn.*, 1979, **52**, 867.

461. T. Sugimoto, K. Shibata, S. Matsuura, and T. Nagatsu, *Bull. Chem. Soc. Jpn.*, 1979, **52**, 2933.

462. K. Sugiura and M. Goto, *Experientia*, 1973, **29**, 1481.

463. A. Suzuki, T. Miyagawa, and M. Goto, *Bull. Chem. Soc. Jpn.*, 1972, **45**, 2198.

464. D. Szłompek-Nesteruk, L. Znojek, and P. Kazmierczak, *Przemysl. Chem.*, 1963, **42**, 226.

465. E.C. Taylor, *Chem. Biol. Pteridines, Proc. Int. Symp., 4th, Toba*, 1969, 79.

466. E.C. Taylor, *Ciba Foundation Symp., Chem. Biol. Pteridines*, 1954, 2.

467. E.C. Taylor, *J. Am. Chem. Soc.*, 1952, **74**, 1648.

468. E.C. Taylor, *J. Am. Chem. Soc.*, 1952, **74**, 1651.

469. E.C. Taylor and R.F. Abdulla, *Tetrahedron Lett.*, 1973, 2093.

470. E.C. Taylor, J.W. Barton, and T.S. Osdene, *J. Am. Chem. Soc.*, 1958, **80**, 421.

471. E.C. Taylor and C.K. Cain, *J. Am. Chem. Soc.*, 1949, **71**, 2538.

472. E.C. Taylor and C.K. Cain, *J. Am. Chem. Soc.*, 1952, **74**, 1644.

473. E.C. Taylor, J.A. Carbon, R.B. Garland, D.R. Hoff, C.F. Howell, and W.R. Sherman, *Ciba Foundation Symp., Chem. Biol. Pteridines*, 1954, 104.

474. E.C. Taylor, J.A. Carbon, and D.R. Hoff, *J. Am. Chem. Soc.*, 1953, **75**, 1904.

475. E.C. Taylor and C.C. Cheng, *J. Org. Chem.*, 1959, **24**, 997.

476. E.C. Taylor, R.B. Garland, and C.F. Howell, *J. Am. Chem. Soc.*, 1956, **78**, 210.

477. E.C. Taylor and P.A. Jacobi, *J. Am. Chem. Soc.*, 1973, **95**, 4455.

478. E.C. Taylor and K. Lenard, *J. Am. Chem. Soc.*, 1968, **90**, 2424.

479. E.C. Taylor and K. Lenard, *Justus Liebigs Ann. Chem.*, 1969, **726**, 100.

480. E.C. Taylor and H.M. Loux, *J. Am. Chem. Soc.*, 1959, **81**, 2474.

481. E.C. Taylor and H.M. Loux, *Chem. Ind. (London)*, 1954, 1585.

482. E.C. Taylor, S.F. Martin, Y. Maki, and G.P. Beardsley, *J. Org. Chem.*, 1973, **38**, 2238.

483. E.C. Taylor and W.W. Paudler, *Chem. Ind. (London)*, 1955, 1061.

484. E.C. Taylor, K.L. Perlman, Y.-H. Kim, I.P. Sword, and P.A. Jacobi, *J. Am. Chem. Soc.*, 1973, **95**, 6413.

485. E.C. Taylor, K.L. Perlman, I.P. Sword, M. Séquin-Frey, and P.A. Jacobi, *J. Am. Chem. Soc.*, 1973, **95**, 6407.

486. E.C. Taylor and W.R. Sherman, *J. Am. Chem. Soc.*, 1959, **81**, 2464.

487. E.C. Taylor, O. Vogl, and P.K. Loeffler, *J. Am. Chem. Soc.*, 1959, **81**, 2479.

488. C. Temple, C.L. Kussner, and J.A. Montgomery, *J. Heterocycl. Chem.*, 1977, **14**, 885.

489. C. Temple, A.G. Laseter, J.D. Rose, and J.A. Montgomery, *J. Heterocycl. Chem.*, 1970, **7**, 1195.

490. G. Tennant and C.W.Yacomeni, *J. Chem. Soc. Chem. Commun.*, 1975, 819.

491. N. Theobald and W. Pfleiderer, *Chem. Ber.*, 1978, **111**, 3385.

492. N. Theobald and W. Pfleiderer, *Tetrahedron Lett.*, 1977, 841.

493. G.M. Timmis, D.G.I. Felton, and T.S. Osdene, *Ciba Foundation Symp., Chem. Biol. Pteridines*, 1954, 93.

494. R. Truhaut and M. de Clercq, *Rev. Fr. Etud. Clin. Biol.*, 1962, **7**, 66.

495. R. Tschesche, *Ciba Foundation Symp., Chem. Biol. Pteridines*, 1954, 135.

496. R. Tschesche and H. Barkemeyer, *Chem. Ber.*, 1955, **88**, 976.

497. R. Tschesche, H. Barkemeyer, and G. Heuschkel, *Chem. Ber.*, **88**, 1258.

498. R. Tschesche and H. Ende, *Chem. Ber.*, 1958, **91**, 2074.
499. R. Tschesche and A. Glaser, *Chem. Ber.*, 1958, **91**, 2081.
500. R. Tschesche, B. Hess, I. Ziegler, and H. Machleidt, *Justus Liebigs Ann. Chem.*, 1962, **658**, 193.
501. R. Tschesche and F. Korte, *Chem. Ber.*, 1952, **85**, 139.
502. R. Tschesche and F. Korte, *Chem. Ber.*, 1954, **87**, 1713.
503. R. Tschesche, F. Korte, and G. Heuschkel, *Chem. Ber.*, 1955, **88**, 1251.
504. R. Tschesche and H. Schäfer, *Chem. Ber.*, 1955, **88**, 81.
505. R. Tschesche and G. Sturm, *Chem. Ber.*, 1965, **98**, 851.
506. R. Tschesche, Z. Zakrzewski, and F. Korte, *Chem. Ber.*, 1953, **86**, 450.
507. C. Van Baalen and H.S. Forrest, *J. Am. Chem. Soc.*, 1959, **81**, 1770.
508. M. Viscontini, *Helv. Chim. Acta*, 1958, **41**, 1299.
509. M. Viscontini, *Methods Enzymol.*, 1971, **18B**, 678.
510. M. Viscontini and A. Bobst, *Helv. Chim. Acta*, 1965, **48**, 816.
511. M. Viscontini and A. Bobst, *Helv. Chim. Acta*, 1966, **49**, 1815.
512. M. Viscontini and M. Konishi, *Chem. Biol. Pteridines, Proc. Int. Symp., 4th, Toba*, 1969, 179.
513. M. Viscontini and G. Nasini, *Helv. Chim. Acta*, 1965, **48**, 452.
514. M. Viscontini and M. Piraux, *Helv. Chim. Acta*, 1962, **45**, 1000.
515. M. Viscontini and H. Stierlin, *Helv. Chim. Acta*, 1961, **44**, 1783.
516. M. Viscontini and H. Stierlin, *Helv. Chim. Acta*, 1962, **45**, 2479.
517. O. Vogl and E.C. Taylor, *J. Am. Chem. Soc.*, 1959, **81**, 2472.
518. R. Weber and M. Viscontini, *Helv. Chim. Acta*, 1975, **58**, 1772.
519. R. Weber and M. Viscontini, *Helv. Chim. Acta*, 1976, **59**, 2379.
520. R. Weber and M. Viscontini, *Helv. Chim. Acta*, 1977, **60**, 152.
521. R. Weber and M. Viscontini, *Helv. Chim. Acta*, 1977, **60**, 161.
522. J. Weinstock and R.Y. Dunoff, *J. Med. Chem.*, 1968, **11**, 565.
523. J. Weinstock, R.Y. Dunoff, B. Sutton, B. Trost, J. Kirkpatrick, F. Farina, and A.S. Straub, *J. Med. Chem.*, 1968, **11**, 549.
524. J. Weinstock, H. Graboyes, G. Jaffe, I.J. Pachter, K. Snader, C.B. Karash, and R.Y. Dunoff, *J. Med. Chem.*, 1968, **11**, 560.
525. J. Weinstock, I.J. Pachter, P.E. Nemeth, and G. Jaffe, *J. Med. Chem.*, 1968, **11**, 557.
526. J. Weinstock, J.P. Rosenbloom, and P.S. Hines, *J. Org. Chem.*, 1968, **33**, 3339.
527. D.I. Weisblat, B.J. Magerlein, A.R. Hanze, D.R. Myers, and S.T. Rolfson, *J. Am. Chem. Soc.*, 1953, **75**, 3625.
528. Wellcome Foundation, Brit. Pat. 579,138 (1946).
529. F. Weygand and G. Schaefer, *Chem. Ber.*, 1952, **85**, 307.
530. H. Wieland and P. Decker, *Justus Liebigs Ann. Chem.*, 1941, **547**, 180.
531. H. Wieland, H. Metzger, C. Shöpf, and M. Bülow, *Justus Liebigs Ann. Chem.*, 1933, **507**, 226.
532. H. Wieland and R. Purrmann, *Justus Liebigs Ann. Chem.*, 1939, **539**, 179.
533. H. Wieland and R. Purrmann, *Justus Liebigs Ann. Chem.*, 1940, **544**, 163.
534. H. Wieland and A. Tartter, *Justus Liebigs Ann. Chem.*, 1940, **543**, 287.
535. H. Wieland and A. Tartter, *Justus Liebigs Ann. Chem.*, 1940, **545**, 197.
536. H. Wieland, A. Tartter, and R. Purrmann, *Justus Liebigs Ann. Chem.*, 1940, **545**, 209.
537. C.L. Winestock and G.W.E. Plaut, *J. Org. Chem.*, 1961, **26**, 4456.
538. H.C.S. Wood, *Ciba Foundation Symp., Chem. Biol. Pteridines*, 1954, 35.
539. D.W. Woolley and A. Pringle, *J. Biol. Chem.*, 1948, **174**, 327.

540. W.B. Wright and J.M. Smith, *J. Am. Chem. Soc.*, 1955, **77**, 3927.
541. H. Yamamoto, W. Hutzenlaub, and W. Pfleiderer, *Chem. Ber.*, 1973, **106**, 3175.
542. H. Yamamoto and W. Pfleiderer, *Chem. Ber.*, 1973, **106**, 3194.
543. F. Yoneda and M. Higuchi, *J. Chem. Soc. Perkin Trans.* 1, 1977, 1336.
544. S.I. Zav'yalov, T.K. Budkova, and N.I. Aronova, *Izv. Akad. Nauk SSSR, Ser. Khim.*, 1973, 2136.
545. S.I. Zav'yalov, T.K. Budkova, and G.I. Ezhova, *Izv. Akad. Nauk SSSR, Ser. Khim.*, 1977, 2811.
546. S.I. Zav'yalov and L.F. Ovechkina, *Izv. Akad. Nauk SSSR, Ser. Khim.*, 1976, 1195.
547. S.I. Zav'yalov and L.F. Ovechkina, *Izv. Akad. Nauk SSSR, Ser. Khim.*, 1976, 2819.
548. S.I. Zav'yalov and G.V. Pokhvisneva, *Izv. Akad. Nauk SSSR, Ser. Khim.*, 1973, 2363.
549. W. Pfleiderer, K.-H. Schündehütte, and H. Ferch, *Justus Liebigs Ann. Chem.*, 1958, **615**, 57.
550. R.-G. Busnel and A. Drilhon, *Bull. Soc. Zool. Fr.*, 1949, **74**, 21.
551. R.-G. Busnel and A. Drilhon, *Bull. Soc. Zool. Fr.*, 1948, **73**, 143.
552. R.D. Youssefyeh and A. Kalmus, *J. Chem. Soc. Chem. Commun.*, 1969, 1426.
553. W.L.F. Armarego, *Phys. Methods Heterocycl. Chem.*, 1971, **3**, 67.
554. M. Viscontini and G.H. Schmidt, *Z. Naturforsch. Teil B*, 1965, **20**, 327.
555. Y. Mori, J. Matsumoto, and T. Hama, *Z. Vgl. Physiol.*, 1960, **43**, 531.
556. R. Tschesche, *Arch. Pharm.* (*Weinheim, Ger.*), 1950, **283**, 137.
557. G. Ritzmann, K. Harzer, and W. Pfleiderer, *Angew. Chem.*, 1971, **83**, 975.
558. K. Rokos and W. Pfleiderer, *Chem. Ber.*, 1975, **108**, 2728.
559. W. Pfleiderer, *Biochem. Clin. Aspects Pteridines*, 1983, **2**, 3.
560. W.L.F. Armarego and D. Randles, *Chem. Biol. Pteridines, Proc. Int. Symp., 7th, St. Andrews*, 1982, 423.
561. W. von Philipsborn, H. Stierlin, and W. Traber, *Pteridine Chem., Proc. Int. Symp., 3rd, Stuttgart*, 1962, 169.
562. E.C. Taylor, M.J. Thompson, and W. Pfleiderer, *Pteridine Chem., Proc. Int. Symp., 3rd, Stuttgart*, 1962, 181.
563. R.B. Angier, *Pteridine Chem., Proc. Int. Symp., 3rd, Stuttgart*, 1962, 211.
564. R. Tschesche, B. Hess, I. Ziegler, and H. Machleidt, *Pteridine Chem., Proc. Int. Symp., 3rd, Stuttgart*, 1962, 233.
565. H. Rembold and E. Buschmann, *Pteridine Chem., Proc. Int. Symp., 3rd, Stuttgart*, 1962, 243.
566. M. Viscontini, *Pteridine Chem., Proc. Int. Symp., 3rd, Stuttgart*, 1962, 267.
567. H.S. Forrest and S. Nawa, *Pteridine Chem., Proc. Int. Symp., 3rd, Stuttgart*, 1962, 281.
568. I. Ziegler, *Pteridine Chem., Proc, Int. Symp., 3rd, Stuttgart*, 1962, 295.
569. L. Jaenicke, *Pteridine Chem., Proc. Int. Symp., 3rd, Stuttgart*, 1962, 377.
570. W.S. McNutt, *Pteridine Chem., Proc. Int. Symp., 3rd, Stuttgart*, 1962, 427.
571. H. Rembold, *Pteridine Chem., Proc. Int. Symp., 3rd, Stuttgart*, 1962, 465.
572. J. Komenda, *Pteridine Chem., Proc. Int. Symp., 3rd, Stuttgart*, 1962, 511.
573. J.E. Fildes, *Pteridine Chem., Proc. Int. Symp., 3rd, Stuttgart*, 1962, 507.
574. C. Schöpf, *Pteridine Chem., Proc. Int. Symp., 3rd, Stuttgart*, 1962, 3.
575. P. Schmidt, K. Eichenberger, and M. Wilhelm, *Pteridine Chem., Proc. Int. Symp., 3rd, Stuttgart*, 1962, 29.
576. F. Weygand, H. Simon, K.D. Keil, H. Millaurer, and B. Spiess, *Pteridine Chem., Proc. Int. Symp., 3rd, Stuttgart*, 1962, 15.
577. J. Weinstock and V.D. Wiebelhaus, *Pteridine Chem., Proc. Int. Symp., 3rd, Stuttgart*, 1962, 37.

578. I.J. Pachter, *Pteridine Chem., Proc. Int. Symp., 3rd, Stuttgart*, 1962, 47.
579. T.S. Osdene, *Pteridine Chem., Proc. Int. Symp., 3rd, Stuttgart*, 1962, 65.
580. F. Korte and R. Wallace, *Pteridine Chem., Proc. Int. Symp., 3rd, Stuttgart*, 1962, 75.
581. W. Pfleiderer, R. Lohrmann, F. Reisser, and D. Söll, *Pteridine Chem., Proc. Int. Symp., 3rd, Stuttgart*, 1962, 87.
582. A. Albert, *Pteridine Chem., Proc. Int. Symp., 3rd, Stuttgart*, 1962, 111.
583. P. Hemmerich, *Pteridine Chem., Proc. Int. Symp., 3rd, Stuttgart*, 1962, 143.
584. H.C.S. Wood, T. Rowan, and A. Stuart, *Pteridine Chem., Proc. Int. Symp., 3rd, Stuttgart*, 1962, 129.
585. A. Albert and K. Ohta, *Chem. Biol. Pteridines, Proc. Int. Symp., 4th, Toba*, 1969, 1.
586. W. Ehrenstein, H. Wamhoff, A. Attar, and F. Korte, *Chem. Biol. Pteridines, Proc. Int. Symp., 4th, Toba*, 1969, 21.
587. Z. Neiman, F. Bergmann, and A.Y. Meyer, *Chem. Biol. Pteridines, Proc. Int. Symp., 4th, Toba*, 1969, 29.
588. S. Matsuura and T. Sugimoto, *Chem. Biol. Pteridines, Proc. Int. Symp., 4th, Toba*, 1969, 35.
589. K. Kobayashi and M. Goto, *Chem. Biol. Pteridines, Proc. Int. Symp., 4th, Toba*, 1969, 57.
590. A. Albert and H. Rokos, *Chem. Biol. Pteridines, Proc. Int. Symp., 4th, Toba*, 1969, 95.
591. K. Eichenberger and P. Schmidt, *Chem. Biol. Pteridines, Proc. Int. Symp., 4th, Toba*, 1969, 99.
592. W. Pfleiderer and H. Rokos, *Chem. Biol. Pteridines, Proc. Int. Symp., 4th, Toba*, 1969, 113.
593. K. Iwai, M. Kobashi, and H. Fujisawa, *Chem. Biol. Pteridines, Proc. Int. Symp., 4th, Toba*, 1969, 199.
594. A. Albert, *Chem. Biol. Pteridines, Proc. Int. Symp., 5th, Konstanz*, 1975, 1.
595. J.A. Montgomery, J.D. Rose, C. Temple, and J.R. Piper, *Chem. Biol. Pteridines, Proc. Int. Symp., 5th, Konstanz*, 1975, 485.
596. P.K. Sengupta, E. Khalifa, J.H. Bieri, and M. Viscontini, *Chem., Biol. Pteridines, Proc. Int. Symp., 5th, Konstanz*, 1975, 495.
597. M.G. Nair and Ch. M. Baugh, *Chem. Biol. Pteridines, Proc. Int. Symp., 5th, Konstanz*, 1975, 503.
598. T.L. Diets, A. Russel, K. Fujii, and J.M. Whiteley, *Chem. Biol. Pteridines, Proc. Int. Symp., 5th, Konstanz*, 1975, 525.
599. E. C. Taylor, *Chem. Biol. Pteridines, Proc. Int. Symp., 5th, Konstanz*, 1975, 543.
600. F. Bergmann, L. Levene, and I. Tamir, *Chem. Biol. Pteridines, Proc. Int. Symp., 5th, Konstanz*, 1975, 603.
601. R. Mengel and W. Pfleiderer, *Chem. Biol. Pteridines, Proc. Int. Symp., 5th, Konstanz*, 1975, 617.
602. G.R. Gapski and J.M. Whiteley, *Chem. Biol. Pteridines, Proc. Int. Symp., 5th, Konstanz*, 1975, 627.
603. S.W. Bailey and J.E. Ayling, *Chem. Biol. Pteridines, Proc. Int. Symp., 5th, Konstanz*, 1975, 633.
604. H. Lund, *Chem. Biol. Pteridines, Proc. Int. Symp., 5th, Konstanz*, 1975, 645.
605. R. Gottlieb and W. Pfleiderer, *Chem. Biol. Pteridines, Proc. Int. Symp., 5th, Konstanz*, 1975, 681.
606. U. Ewers, A. Gronenborn, H. Günther, and L. Jaenicke, *Chem. Biol. Pteridines, Proc. Int. Symp., 5th, Konstanz*, 1975, 687.
607. R. Weber, *Chem. Biol. Pteridines, Proc. Int. Symp., 5th, Konstanz*, 1975, 705.
608. J.H. Bieri, *Chem. Biol. Pteridines, Proc. Int. Symp., 5th, Konstanz*, 1975, 711.
609. J.H. Bieri, *Chem. Biol. Pteridines, Proc. Int. Symp., 5th, Konstanz*, 1975, 721.
610. F. Bergmann, I. Tamir, L. Levene, and M. Rahat, *Chem. Biol. Pteridines, Proc. Int. Symp., 5th, Konstanz*, 1975, 725.

611. H.I.X. Mager, *Chem. Biol. Pteridines, Proc. Int. Symp., 5th, Konstanz*, 1975, 753.

612. I. Southon and W. Pfleiderer, *Chem. Biol. Pteridines, Proc. Int. Symp., 5th, Konstanz*, 1975, 783.

613. H. Descimon, *Chem. Biol. Pteridines, Proc. Int. Symp., 5th, Konstanz*, 1975, 805.

614. K. Rokos and W. Pfleiderer, *Chem. Biol Pteridines, Proc. Int. Symp., 5th, Konstanz*, 1975, 931.

615. W. Pfleiderer, *Chem. Biol. Pteridines, Proc. Int. Symp., 5th, Konstanz*, 1975, 941.

616. K. Baumgartner and J. H. Bieri, *Chem. Biol. Pteridines, Proc. Int. Symp., 6th, La Jolla*, 1978, 7.

617. R. Baur, M. Kappel, R. Mengel, and W. Pfleiderer, *Chem. Biol. Pteridines, Proc. Int. Symp., 6th, La Jolla*, 1978, 13.

618. J.H. Bieri, *Chem. Biol. Pteridines, Proc. Int. Symp., 6th, La Jolla*, 1978, 19.

619. T. Fukushima and J.C. Nixon, *Chem. Biol. Pteridines, Proc. Int. Symp., 6th, La Jolla*, 1978, 31.

620. Y. Iwanami, *Chem. Biol. Pteridines, Proc. Int. Symp., 6th, La Jolla*, 1978, 43.

621. L. Kiriasis and W. Pfleiderer, *Chem. Biol. Pteridines, Proc. Int. Symp., 6th, La Jolla*, 1978, 49.

622. W. Pfleiderer, *Chem. Biol. Pteridines, Proc. Int. Symp., 6th, La Jolla*, 1978, 63.

623. E.C. Taylor, R.N. Henrie, and D.J. Dumas, *Chem. Biol. Pteridines, Proc. Int. Symp., 6th, La Jolla*, 1978, 71.

624. E.C. Taylor and L.A. Reiter, *Chem. Biol. Pteridines, Proc. Int. Symp., 6th, La Jolla*, 1978, 77.

625. E.C. Taylor, *Chem. Biol. Pteridines, Proc. Int. Symp., 7th, St. Andrews*, 1982, 23.

626. S.W. Bailey and J.E. Ayling, *Chem. Biol. Pteridines, Proc. Int. Symp., St. Andrews*, 1982, 51.

627. P.H. Boyle and R.J. Lockhart, *Chem. Biol. Pteridines, Proc. Int. Symp., 7th, St. Andrews*, 1982, 73.

628. R.C. Cameron, W.J.S. Lyall, S.H. Nicholson, D.R. Robinson, C.J. Suckling, and H.C.S. Wood, *Chem. Biol. Pteridines, Proc. Int. Symp., 7th, St. Andrews*, 1982, 79.

629. W. Pfleiderer, R. Baur, M. Bartke, and H. Lutz, *Chem. Biol. Pteridines, Proc. Int. Symp., 7th, St. Andrews*, 1982, 93.

630. M.G. Nair, M.K. Rozmyslovicz, R.L. Kisliuk, F.M. Sirotnak, and Y. Gaumont, *Chem. Biol. Pteridines, Proc. Int. Symp., 7th, St. Andrews*, 1982, 121.

631. M. Kočevar, B. Stanovnik, and M. Tišler, *Chem. Biol. Pteridines, Proc. Int. Symp., 7th, St. Andrews*, 1982, 481.

632. A. Heckel and W. Pfleiderer, *Chem. Biol. Pteridines, Proc. Int. Symp., 7th, St. Andrews*, 1982, 487.

633. K. Abou-Hadeed and W. Pfleiderer, *Chem. Biol. Pteridines, Proc. Int. Symp., 7th, St. Andrews*, 1982, 493.

634. W. Hübsch and W. Pfleiderer, *Chem. Biol. Pteridines, Proc. Int. Symp., 7th, St. Andrews*, 1982, 499.

635. M.C. Archer and L.S. Reed, *Methods Enzymol.*, 1980, **66**, 452.

636. D. Augustyn, G. Hermon, and H. Marsh, *Pap., London Int. Conf. Carbon Graphite, 4th*, 1974, 61; *Chem. Abstr.*, 1978, **89**, 11366.

637. A. Asatoor and C.E. Dalgliesh, *J. Chem. Soc.*, 1958, 1717.

638. Y. Asai, *Yakugaku Zasshi*, 1959, **79**, 1554.

639. Y. Asai, *Yakugaku Zasshi*, 1959, **79**, 1559.

640. Y. Asai, *Yakugaku Zasshi*, 1959, **79**, 1565.

641. Y. Asai, *Yakugaku Zasshi*, 1959, **79**, 1570.

642. Y. Asai, *Yakugaku Zasshi*, 1959, **79**, 1574.

643. Y. Asai, *Yakugaku Zasshi*, 1959, **79**, 1548.

644. W.L.F. Armarego and P. Waring, *Aust. J. Chem.*, 1981, **34**, 1921.

645. W.L.F. Armarego, P. Waring, and B. Paal, *Aust. J. Chem.*, 1982, **35**, 785.

646. W.L.F. Armarego, D. Randles, and H. Taguchi, *Biochem. J.*, 1983, **211**, 357.

647. M. Argentini and M. Viscontini, *Helv. Chim. Acta*, 1973, **56**, 2920.

648. R.A. Archer and H.S. Mosher, *J. Org. Chem.*, 1967, **32**, 1378.

649. R.B. Angier, C.W. Waller, J.H. Boothe, J.H. Mowat, J. Semb, B.L. Hutchings, E.L.R. Stockstad, and Y. Subba Row, *J. Am. Chem. Soc.*, 1948, **70**, 3029.

650. R.B. Angier, C.W. Waller, B.L. Hutchings, J.H. Boothe, J.H. Mowat, J. Semb, and Y. Subba Row, *J. Am. Chem. Soc.*, 1950, **72**, 74.

651. R.B. Angier, E.L.R. Stockstad, J.H. Mowat, B.L. Hutchings, J.H. Boothe, C.W. Waller, J. Semb, Y. Subba Row, D.B. Cosulich, M.J. Fahrenbach, M.E. Hultquist, E. Kuh, E.H. Northey, D.R. Seeger, J.P. Sickels, and J.M. Smith, *J. Am. Chem. Soc.*, 1948, **70**, 25.

652. R.B. Angier, A.L. Gazzola, J. Semb, S.M. Gadekar, and J.H. Williams, *J. Am. Chem. Soc.*, 1954, **76**, 902.

653. K.J.M. Andrews, W.E. Barber, and B.P. Tong, *J. Chem. Soc. (C)*, 1969, 928.

654. R.B. Angier and W.V. Curran, *J. Am. Chem. Soc.*, 1959, **81**, 5650.

655. R.B. Angier, J.H. Boothe, J.H. Mowat, C.W. Waller, and J. Semb, *J. Am. Chem. Soc.*, 1952, **74**, 408.

656. A.G. Anderson and J.A. Nelson, *J. Am. Chem. Soc.*, 1949, **71**, 3837.

657. H. Andersag and K. Westphal, *Ber. Dtsch. Chem. Ges.*, 1937, **70B**, 2035.

658. A. Amer, M. Ventura, and H. Zimmer, *J. Heterocycl. Chem.*, 1983, **20**, 359.

659. R. Addink and W. Berends, *Tetrahedron*, 1973, **29**, 879.

660. R. Addink and W. Berends, *Tetrahedron*, 1981, **37**, 833.

661. K.B. Augustinsson, *Svensk. Kem. Tidskr.*, 1956, **68**, 271; *Chem. Abstr.*, 1956, **50**, 16888.

662. L. Almirante, *Ann. Chim. (Roma)*, 1959, **49**, 333; *Chem. Abstr.*, 1959, **53**, 20078.

663. R. Andrisano and L. Maioli, *Gazz. Chim. Ital.*, 1953, **83**, 264.

664. R. Andrisano and L. Maioli, *Gazz. Chim. Ital.*, 1953, **83**, 269.

665. R. Addink, *Biolumin. Chemilumin., Int. Symp. Anal. Appl. Biolumin. Chemilumin., 2nd*, 1980, 507; *Chem. Abstr.*, 1981, **95**, 79651.

666. P.G. Abdul-Ahad and G.A. Webb, *Theochem*, 1982, **6**, 25; *Chem. Abstr.*, 1982, **97**, 211302.

667. R. Andrisano and G. Modena, *Boll. Sedute Accad. Gioenia Sci. Nat. Catania*, 1952, [4], **2**, 145; *Chem. Abstr.*, 1955, **49**, 6961.

668. L.V. Alekseeva and Z.V. Pushkareva, *Zh. Obshch. Khim.*, 1963, **33**, 1693.

669. W. Allen, R.L. Pasternak, and W. Seaman, *J. Am. Chem. Soc.*, 1952, **74**, 3264.

670. Y. Assai, *Abhandl. Dtsch. Akad. Wiss. Berlin, Kl. Chem., Geol. Biol.*, 1964, 74; *Chem. Abstr.*, 1965, **62**, 14181.

671. H. Aruga, N. Yoshitake, and S. Ishikawa, *J. Sericult. Sci. Jpn.*, 1951, **20**, 399; *Chem. Abstr.*, 1954, **48**, 2270.

672. W.C. Alford, *Anal. Chem.*, 1952, **24**, 881.

673. A. Albert and A. Hampton, *Int. Congr. Biochim., 2nd, Paris, Résumés Commun.*, 1952, 444; *Chem. Abstr.*, 1955, **49**, 4648.

674. S.S. Al-Hassan, R.J. Kulick, D.B. Livingstone, C.J. Suckling, H.C.S. Wood, R. Wrigglesworth, and R. Ferone, *J. Chem. Soc. Perkin Trans. 1*, 1980, 2643.

675. D.J. Brown, *Angew. Chem.*, 1954, **66**, 476.

676. A. Albert and D.J. Brown, *Int. Congr. Biochim., 2nd, Paris, Résumés Commun.*, 1952, 196.

677. A. Albert, D.J. Brown, and G.W.H. Cheeseman, *Chem. Ind. (London)*, 1951, 187.

678. J.H. Boothe, J. Semb, C.W. Waller, R.B. Angier, J.H. Mowat, B.L. Hutchings, E.L.R. Stockstad, and Y. Subba Row, *J. Am. Chem. Soc*, 1949, **71**, 2304.

679. J.H. Boothe, C.W. Waller, E.L.R. Stockstad, J.H. Mowat, R.B. Angier, J. Semb, Y. Subba Row, D.B. Cosulich, M.J. Fahrenbach, M.E. Hultquist, E. Kuh, E.H. Northey, D.R. Seeger, J.P. Sickels, and J.M. Smith, *J. Am. Chem. Soc.*, 1948, **70**, 27.

680. J.H. Boothe, J.H. Mowat, C.W. Waller, R.B. Angier, J. Semb, and A.L. Gazzola, *J. Am. Chem. Soc.*, 1952, **74**, 5407.

681. J.V. Berrier, *Diss. Abstr. Int. (B)*, 1973, **34**, 586; *Chem. Abstr.*, 1973, **79**, 146489.

682. G. Bernini, M. Combe, and J. David, *Therapie*, 1966, 21, 1313; *Chem. Abstr.*, 1967, **66**, 45358.

683. P.R. Brook and G. R. Ramage, *J. Chem. Soc.*, 1957, 1.

684. R. Brown, M. Joseph, T. Leigh, and M.L. Swain, *J. Chem. Soc. Perkin Trans. 1*, 1977, 1003.

685. M. Bräutigam, R. Dreesen, and H. Herken, *Hoppe-Seyler's Z. Physiol. Chem.*, 1982, **363**, 341.

686. W.B. Neeley, *Mol. Pharmacol.*, 1967, **3**, 108.

687. A. Bobst, *Helv. Chim. Acta*, 1967, **50**, 1480.

688. W.R. Boon and W.G.M. Jones, *J. Chem. Soc.*, 1951, 591.

689. W.R. Boon, W.G.M. Jones, and G.R. Ramage, *J. Chem. Soc.*, 1951, 96.

690. A. Bobst, *Helv. Chim. Acta*, 1967, **50**, 2222.

691. A. Bobst, *Helv. Chim. Acta*, 1968, **51**, 607.

692. A. Bobst and M. Viscontini, *Helv. Chim. Acta*, 1966, **49**, 884.

693. A. Bobst and M. Viscontini, *Helv. Chim. Acta*, 1966, **49**, 875.

694. R.L. Blakley, *Biochim. Biophys. Acta*, 1957, **23**, 654.

695. J.A. Blair and A.J. Pearson, *Tetrahedron Lett.*, 1973, 203.

696. J.A. Blair and A.J. Pearson, *Tetrahedron Lett.*, 1973, 1681.

697. J.A. Blair and J. Graham, *Chem. Ind. (London)*, 1955, 1158.

698. C.M. Baugh, *Diss. Abstr.*, 1963, **23**, 230; *Chem. Abstr.*, 1963, **58**, 10429.

699. S.N. Baranov, T.E. Gorizdra, and O.N. Gerasimenko, *Fiziol. Aktiv. Veshchestva, Akad. Nauk Ukr. SSR, Respub. Mezhvedom. Sb.*, 1966, 24; *Chem. Abstr.*, 1967, **67**, 21891.

700. R.L. Blakley, *Biochem. J.*, 1959, **72**, 707.

701. L. Birkofer, A. Ritter, and H.P. Kühlthau, *Chem. Ber.*, 1964, **97**, 934.

702. P.J. Black, R.D. Brown, and M.L. Heffernan, *Aust. J. Chem.*, 1967, **20**, 1305.

703. J.H. Bieri and M. Viscontini, *Helv. Chim. Acta*, 1974, **57**, 1651.

704. J.H. Bieri and M. Viscontini, *Helv. Chim. Acta*, 1973, **56**, 2905.

705. J.H. Bieri and M. Viscontini, *Helv. Chim. Acta*, 1977, **60**, 1926.

706. A. Bertho and M. Bentler, *Justus Liebigs Ann. Chem.*, 1950, **570**, 127.

707. S.N. Baranov and N.E. Tarnavskaya, *Ukr. Khim. Zh.*, 1957, **23**, 646; *Chem. Abstr.*, 1958, **52**, 10103.

708. S.N. Baranov and N.E. Tarnavskaya, *Ukr. Khim. Zh.*, 1958, **24**, 472; *Chem. Abstr.*, 1959, **53**, 11394.

709. S.N. Baranov and N.E. Tarnavskaya, *Ukr. Khim. Zh.*, 1960, **26**, 626; *Chem. Abstr.*, 1961, **55**, 12416.

710. S.N. Baranov and T.E. Gorizdra, *Zh. Obshch. Khim.*, 1957, **27**, 1703.

711. M. Barial, *J. Chromatogr.*, 1967, **28**, 492.

712. G.B. Barlin and W. Pfleiderer, *Chem. Ber.*, 1969, **102**, 4032.

713. R.L. Beach and G.W.E. Plaut, *J. Org. Chem.*, 1971, **36**, 3937.

714. J. Borkovec and J. Jonas, *Spisy Přírodovédecké Fak. Univ. Brne*, 1963, 397; *Chem. Abstr.*, 1964, **60**, 15870.

715. V.M. Berezovskii, A.M. Yurkevich, and I.K. Krivosheina, *Zh. Obshch. Khim.*, 1961, **31**, 2782.

716. F. Bergmann, I. Tamir, A. Frank, and W. Pfleiderer, *J. Chem. Soc. Perkin Trans. 2*, 1979, 35.

717. F. Bergmann, H. Burger-Rachamimov, and J. Galanter, *Biochem. Biophys. Res. Commun.*, 1964, **17**, 461.

718. S.N. Baranov and T.E. Gorizdra, *Zh. Obshch. Khim.*, 1959, **29**, 3322.

719. E. Becker and C. Schöpf, *Justus Liebigs Ann. Chem.*, 1936, **524**, 124.

720. R.-G. Busnel, G. Levy, and M. Polonovski, *C. R. Soc. Biol.* 1950, **144**, 334 and 735.

721. R.C. Bugle and R.A. Osteryoung, *J. Org. Chem.*, 1979, **44**, 1719.

722. G.R. Brunk, K.A. Martin, and A.M. Nishimura, *Biophys. J.*, 1976, **16**, 1373.

723. D. Binder, C.R. Noe, B.C. Prager, and F. Turnowsky, *Arzneim.-Forsch.*, 1983, **33**, 803.

724. G.M. Badger and I.S. Walker, *J. Chem. Soc.*, 1956, 122.

725. S.W. Bailey and J.E. Ayling, *Biochemistry*, 1983, **22**, 1790.

726. O. Brenner-Holzach and P. Leuthardt, *Helv. Chim. Acta*, 1961, **44**, 1480.

727. W.R. Boon and T. Leigh, *J. Chem. Soc.*, 1951, 1497.

728. B.R. Baker, B.-T, Ho, and D. V. Santi, *J. Pharm. Sci.*, 1965, **54**, 1415.

729. J.H. Jones and E.J. Cragoe, *J. Med. Chem.*, 1968, **11**, 322.

730. C.K. Cain, U.S. Pat. 2,667,486 (1954); *Chem. Abstr.*, 1955, **49**, 4030.

731. G. Carrara and V. D'Amato, U. S. Pat. 2,710,866 (1955); *Chem. Abstr.*, 1956, **50**, 5779.

732. D.B. Cosulich and J.M. Smith, *J. Am. Chem. Soc.*, 1949, **71**, 3574.

733. V.M. Cherkasov, N.A. Kapron, G.S. Tret'yakova, and V.A. Latenko, *Khim. Geterotsikl. Soedin.*, 1969, 124; *Chem. Abstr.*, 1969, **70**, 115104.

734. C.A. Chin and P.L. Song, *Grad. Stud., Tex. Tech. Univ.*, 1981, **24**, 175; *Chem. Abstr.*, 1982, **96**, 141890.

735. J.I. DeGraw, J.P. Marsh, E.M. Action, O.P. Crews, C.W. Mosher, A.N. Fujiwara, and L. Goodman, *J. Org. Chem.*, 1965, **30**, 3404.

736. R. Denayer, *Bull. Soc. Chim. Fr.*, 1962, 1358.

737. G.P.G. Dick, W.E. Fidler, and H.C.S. Wood, *Chem. Ind. (London)*, 1956, 1424.

738. R. Dietrich and S.J. Benkovic, *J. Am. Chem. Soc.*, 1979, **101**, 6144.

739. H. Descimon and M. Barial, *Bull. Soc. Chim. Fr.*, 1973, 87.

740. H. Descimon, *Bull. Soc. Chim. Fr.*, 1973, 145.

741. A. Dieffenbacher, R. Mondelli, and W. von Philipsborn, *Helv. Chim. Acta*, 1966, **49**, 1355.

742. J.I. DeGraw, V.H. Brown, W.T. Colwell, and N.E. Morrison, *J. Med. Chem.*, 1974, **17**, 144.

743. J.I. DeGraw, V.H. Brown, M. Cory, P. Tsakotellis, R.L. Kisliuk, and Y. Gaumont, *J. Med. Chem.*, 1971, **14**, 206.

744. D.B. Cosulich, B. Roth, J.M. Smith, M.E. Hultquist, and R.P. Parker, *J. Am. Chem. Soc.*, 1952, **74**, 3252.

745. D.B. Cosulich, D.R. Seeger, M.J. Fahrenbach, B. Roth, J.H. Mowat, J.M. Smith, and M.E. Hultquist, *J. Am. Chem. Soc.*, 1951, **73**, 2554.

746. D.B. Cosulich, D.R. Seeger, M.J. Fahrenbach, K.H. Collins, B. Roth, M.E. Hultquist, and J.M. Smith, *J. Am. Chem. Soc.*, 1953, **75**, 4675.

747. C.K. Cain, M.F. Mallette, and E.C. Taylor, *J. Am. Chem. Soc.*, 1946, **68**, 1996.

748. C.K. Cain, M.F. Mallette, and E.C. Taylor, *J. Am. Chem. Soc.*, 1948, **70**, 3026.

749. N.R. Campbell, J.H. Dunsmuir, and M.E.H. Fitzgerald, *J. Chem. Soc.*, 1950, 2743.

750. J. Clark and A.J. Layton, *J. Chem. Soc.*, 1959, 3411.

751. J.H. Cardellina and J. Meinwald, *J. Org. Chem.*, 1981, **46**, 4782.

752. M. Chaykovsky, *J. Org. Chem.*, 1975, **40**, 145.

753. M. Chaykovsky, M. Hirst, H. Lazarus, J.E. Martinelli, R.L. Kisliuk, and Y. Gaumont, *J. Med. Chem.* 1977, **20**, 1323.

754. S. Chatterjee, A.K. Gosh, and I. Abrahamsohn, *J. Indian Chem. Soc.*, 1967, **44**, 183.

755. D.J. Dumas, *Diss. Abstr. Int. (B)*, 1980, **40**, 3738; *Chem. Abstr.*, 1980, **93**, 95243.

756. H. Descimon, *Ann. Soc. Entomol. Fr.*, 1967, **3**, 827; Chem. Abstr., 1968, **68**, 1137.

757. R.J.W. De Wit, *Anal. Biochem.*, 1982, **123**, 285.

758. J. Clark and W. Pendergast, *J. Chem. Soc. (C)*, 1968, 1124.

759. J. Clark and C. Smith, *J. Chem. Soc. (C)*, 1969, 2777.

760. A.M. Patterson, L.T. Capell, and D.F. Walker, *Ring Index*, American Chemical Society, Washington, 2nd Ed., 1960.

761. L. Cocco, J.P. Groff, C. Temple, J.A. Montgomery, R.E. London, N.A. Matwiyoff, and R.L. Blakley, *Biochemistry*, 1981, **20**, 3972.

762. A. Dieffenbacher and W. von Philipsborn, *Helv. Chim. Acta*, 1969, **52**, 743.

763. F. Dallacker and G. Steiner, *Justus Liebigs Ann. Chem.*, 1962, **660**, 98.

764. B. DeCroix, M. J. Strauss, A. DeFusco, and D. C. Palmer, *J. Org. Chem.*, 1979, **44**, 1700.

765. K.R. Davis and R. Wolfenden, *J. Org. Chem.*, 1983, **48**, 2280.

766. T.J. Batterham and J.A. Wunderlich, *J. Chem. Soc. (B)*, 1969, 489.

767. R.-G. Busnel, P. Chauchard, H. Mazoué, and M. Polonovski, *C. R. Soc. Biol.*, 1946, **140**, 50.

768. R.-G. Busnel, P. Chauchard, H. Mazoué, M. Pesson, R. Vieillefosse, and M. Polonovski, *C. R. Soc. Biol.*, 1944, **138**, 171.

769. P. Decker, *Hoppe-Seyler's Z. Physiol. Chem.*, 1942, **274**, 223.

770. R.D. Brown and B.A.W. Coller, *Theor. Chim. Acta*, 1967, **7**, 259.

771. D.Blackburn and G. Burghard, *J. Labelled Compd.*, 1966, **2**, 62.

772. J.I. DeGraw, V.H. Brown, and I. Uemura, *J. Labelled Compd. Radiopharm.*, 1979, **16**, 559.

773. D.F. DeTar, C.K. Cain, and B.S. Meeks, *J. Am. Chem. Soc.*, 1953, **75**, 5118.

774. H.J. Backer and A.C. Houtman, *Rec. Trav. Chim. Pays-Bas*, 1951, **70**, 725.

775. H.J. Backer and A.C. Houtman, *Rec. Trav. Chim. Pays-Bas*, 1951, **70**, 738.

776. H.J. Backer and A.C. Houtman, *Rec. Trav. Chim. Pays-Bas*, 1948, **67**, 260.

777. J. Clark and A.E. Cunliffe, *Org. Mass Spectrom.*, 1973, **7**, 737.

778. J.A. Blair and C.D. Foxall, *Org. Mass Spectrom.*, 1969, **2**, 923.

779. J. Clark, R. Maynard, and C. Smith, *Org. Mass Spectrom.*, 1971, **5**, 993.

780. J. Clark and F.S. Yates, *Org. Mass Spectrom.*, 1971, **5**, 1419.

781. F.W. Birss and N.K. Das Gupta, *Indian J. Chem. (B)*, 1979, **17**, 610.

782. V.M. Berezovskii, *Uspekhi Khim.*, 1953, **22**, 191; *Chem. Abstr.*, 1954, **48**, 10027.

783. S.N. Baranov, *Uspekhi Khim.*, 1958, **27**, 1337; *Chem. Abstr.*, 1959, **53**, 4289.

784. H. Descimon, *Bull. Soc. Chim. Biol.*, 1967, **49**, 1164.

785. I. Ziegler and R. Harmsen, *Adv. Insect Physiol.*, 1969, **6**, 139.

786. H. Rembold and L. Buschmann, *Hoppe-Seyler's Z. Physiol. Chem.*, 1962, **330**, 132.

787. B. Pullman and A. Pullman, *Proc. Natl. Acad. Sci. US*, 1958, **44**, 1197.

788. J.H. Bieri, *Helv. Chim. Acta*, 1977, **60**, 2303.

789. R.A. Archer, *Diss. Abstr.*, 1964, **24**, 2684; *Chem. Abstr.*, 1964, **60**, 14365.

790. D.W. Bjorkquist, *Diss. Abstr. Int. (B)*, 1976, **37**, 765; *Chem. Abstr.*, 1976, **85**, 118733.

791. R.M. Blazer, *Diss. Abstr. Int. (B)*, 1977, **37**, 5079; *Chem. Abstr.*, 1977, **87**, 152130.

792. J. Komenda and L. Kisova, *Chem. Zvesti*, 1962, **16**, 368; *Chem. Zentr.*, 1964, **135**, 11-0744.

793. E.L. Smith, *Proc. R. Soc. London, Ser. B*, 1962, **156**, 312.

794. F.G. Hopkins, *Proc. R. Soc. London, Ser. B*, 1942, **130**, 359.

795. Y. Kobayashi, Y. Iitaka, R. Gottlieb, and W. Pfleiderer, *Acta Crystallogr., Sect. B*, 1977, **33**, 2911.

796. R. Norrestam, B. Stensland, and E. Söderberg, *Acta Crystallogr., Sect. B*, 1972, **28**, 659.

797. K. Ganapathi and B.N. Palande, *Proc. Indian Acad. Sci., Sect. A*, 1953, **37**, 652.

798. Y. Kobayashi, Y. Iitaka, R. Gottlieb, and W. Pfleiderer, *Acta Crystallogr., Sect. B*, 1979, **35**, 247.

799. S. Yoshida and K. Tanabe, *Annu. Rept. Takamine Lab.*, 1951, **3**, 1; *Chem. Abstr.*, 1955, **49**, 339.

800. Y. Hirata, S. Nawa, and S. Matsuura, *Res. Rept. Nagoya Ind. Sci. Res. Inst.*, 1951, **3**, 66; *Chem. Abstr.*, 1955, **49**, 346.

801. Y. Hirata and S. Nawa, *Res. Rept. Nagoya Ind. Sci. Res. Inst.*, 1951, **4**, 80; *Chem. Abstr.*, 1955, **49**, 346.

802. J.M. Smith, D.B. Cosulich, M.E. Hultquist, and D.R. Seeger, *Trans. N. Y. Acad. Sci.*, 1948, **10**, 82; *Chem. Abstr.*, 1948, **42**, 8200.

803. M. Polonovski, *Exposes Annuels Biochim. Méd.*, 1950, **11**, 229; *Chem. Abstr.*, 1952, **46**, 7629.

804. I. Teshima and Y. Inukai, *Res. Rept. Nagoya Ind. Sci. Res. Inst.*, 1956, **9**, 76; *Chem. Abstr.*, 1957, **51**, 9948.

805. S. Ghosh and S.C. Roy, *Ann. Biochem. Exptl. Med. (India)*, 1955, **15**, 93; *Chem. Abstr.*, 1956, **50**, 6691.

806. Y. Fukuda, *Osaka Daigaku Igaku Zasshi*, 1956, **8**, 151; *Chem. Abstr.*, 1956, **50**, 1111.

807. T. Korzybski, *Acta Biochim. Polon.*, 1957, **4**, 49; *Chem. Abstr.*, 1959, **53**, 18129.

808. M. Goto, *Seikagaku*, 1964, **36**, 849; *Chem. Abstr.*, 1965, **62**, 12039.

809. M.M. Jezewska, *Postepy Biochem.*, 1963, **9**, 497; *Chem. Abstr.*, 1964, **60**, 4385.

810. N.E. Tarnavskaya, *Trudy L'vov. Med. Inst.*, 1957, **12**, 99; *Chem. Abstr.*, 1960, **54**, 21110.

811. B.A. Sobchuk, *Trudy L'vov. Med. Inst.*, 1957, **12**, 87; *Chem. Abstr.*, 1960, **54**, 2348.

812. H. Langer, *Umsch. Wiss. Tech.*, 1967, **67**, 112; *Chem. Abstr.*, 1967, **67**, 98005.

813. H. Mitsuda, Y. Suzuki, and F. Kawai, *J. Vitaminol.*, 1966, **12**, 205; *Chem. Abstr.*, 1966, **65**, 16971.

814. K. Slavík and V. Slavíková, *Acta Univ. Carolinae, Med. Monogr.*, 1969, **37**, 1; *Chem. Abstr.*, 1969, **71**, 67261.

815. A. Giner-Sorolla, *Cienc. Ind. Farm.*, 1972, **4**, 154; *Chem. Abstr.*, 1975, **82**, 140079.

816. N. Kokolis, *Chem. Chron., Epistem. Ekdosis*, 1970, **35**, 17; *Chem. Abstr.*, 1970, **73**, 120590,

817. Y. Nakahara, I. Sekikawa, and S. Kakimoto, *Hokkaido Daigaku Meneki Kagaku Kenkyusho Kiyo*, 1976, **36**, 8; *Chem. Abstr.*, 1976, **85**, 78085.

818. D.C. Suster, G. Ciustea, A. Dumitrescu, L.V. Feyns, E. Tarnauceanu, G. Botez, S. Angelescu, V. Dobre, and I. Niculescu-Duvaz, *Rev. Roum. Chim.*, 1977, **22**, 1195; *Chem. Abstr.*, 1978, **88**, 89999.

819. G. Rorive and P. Bovy, *Coeur Med. Interne*, 1978, **17**, 207; *Chem. Abstr.*, 1978, **89**, 122752.

820. Y. Li, L.-Q. Li, Y.-X. Chen, D.-S. Wang, Y.-Z. Gai, P.-L. Yu, and Y.-P. Zheng, *Yao Hsueh Hsueh Pao*, 1979, **14**, 108; *Chem. Abstr.*, 1980, **82**, 76445.

821. E. Campaigne, *J. Chem. Educ.*, 1986, **63**, 860.

822. C. Krumdieck and T. Shiota, in *Methodicum Chimicum*, F. Korte and M. Goto Eds., Academic, New York, 1977, Vol. **11**, (2), p. 87.

823. H. Feichtinger, in *Methodicum Chimicum*, F. Korte and J. Falbe Eds., Georg Thieme, Stuttgart, 1980, Vol. **4**, p. 268.

824. R.C. Elderfield and A.C. Mehta, in *Heterocyclic Compounds*, R.C. Elderfield, Ed., Wiley, New York, 1967, **9**, 1.

825. M.A. Kaldrikyan, G.G. Danagulyan, and A.V. Khekoyan, *Sint. Geterotsikl. Soedin.*, 1979, **11**, 43; *Chem. Abstr.*, 1981, **94**, 15680.

826. H. Günther, A. Gronenborn, U. Ewers, and H. Seel, *Jerusalem Symp. Quantum Chem. Biochem.*, 1978, **11**, 193; *Chem. Abstr.*, 1979, **90**, 68617.

827. J. Maillard, M. Bernard, M. Vincent, Vo-Van-Tri, R. Jolly, R. Morin, Benharkate, and C. Menillet, *Chim. Ther.*, 1967, **2**, 231; *Chem. Zentr.*, 1969, **140**(25), 1401.

828. J.I. Fernández Alonso and R. Domingo Sebastián, *An. R. Soc. Esp. Fis. Quim., Ser. B*, 1960, **56**, 757; *Chem. Zentr.*, 1963, **134**, 15075.

829. J. Jonas, *Spisy Přirodovědecké Fak. Univ. Brné*, 1965, 189; *Chem. Zentr.*, 1966, **137**(44), 0960.

830. T. Goto, *Jpn. J. Zool.*, 1963, **14**, 69; *Chem. Abstr.*, 1964, **61**, 2236.

831. I. Ziegler–Günder, *Biol. Rev. Cambridge Philos. Soc.*, 1956, **31**, 313.

832. P.R. Pal, *Sci. Culture* (*India*), 1953, **19**, 160; *Chem. Abstr.*, 1954, **48**, 13692.

833. S. Nawa and T. Taira, *Proc. Jpn. Acad.*, 1954, **30**, 632; *Chem. Abstr.*, 1955, **49**, 7758.

834. T. Hama, J. Matsumoto, and M. Obika, *Proc. Jpn. Acad.*, 1960, **36**, 217; *Chem. Abstr.*, 1961, **55**, 825.

835. F.A. Robinson, *Chem. Prod.*, 1947, **10**, 24; *Chem. Abstr.*, 1947, **41**, 4207.

836. V.A. Kirsanova and A.V. Trufanov, *Biokhimiya*, 1949, **14**, 413; *Chem. Abstr.*, 1950, **44**, 1116.

837. V.A. Kirsanova and A.V. Trufanov, *Biokhimiya*, 1950, **15**, 243; *Chem. Abstr.*, 1950, **44**, 10078.

838. V.A. Kirsanova and A.V. Trufanov, *Biokhimiya*, 1951, **16**, 367; *Chem. Abstr.*, 1951, **45**, 10333.

839. H. Ninomiya, *Vitamins* (*Japan*), 1951, **4**, 86 and 231; *Chem. Abstr.*, 1951, **45**, 10324.

840. G. Nouvel and J. David, *Therapie*, 1966, **21**, 1317; *Chem. Abstr.*, 1967, **66**, 45359.

841. K. Ram, B.R. Pandey, and V.J. Ram, *Spectros, Lett.*, 1977, **10**, 149; *Chem. Abstr.*, 1977, **87**, 52324.

842. J.P. Hénichart, J.L. Bernier, J.L. Dhondt, M. Dautrevaux, and G. Biserte, *Biomed. Express*, 1979, **31**, 6; *Chem. Abstr.*, 1979, **91**, 70727.

843. U. Ewers, H. Günther, and L. Jaenicke, *Angew. Chem.*, 1975, **87**, 356.

844. E. Fischer, *Dtsch. Chem.-Ztg.*, 1950, **2**, 148; *Chem. Abstr.*, 1953, **47**, 8855.

845. T. Fukushima and J.C. Nixon, *Methods Enzymol.*, 1980, **66**, 429.

846. T. Fukushima, *Methods Enzymol.*, 1980, **66**, 508.

847. J. Eder and H. Rembold, *Arzneim.-Forsch.*, 1971, **21**, 562.

848. E.F. Elslager, J.L. Johnson, and L.M. Werbel, *J. Med. Chem.*, 1981, **24**, 140.

849. E.F. Elslager, J.L. Johnson, and L.M. Werbel, *J. Med. Chem.*, 1981, **24**, 1001.

850. D. Farquhar, T.L. Loo, and S. Vadlamudi, *J. Med. Chem.*, 1972, **15**, 567.

851. U. Ewers, H. Günther, and L. Jaenicke, *Chem. Ber.*, 1973, **106**, 3951.

852. U. Ewers, H. Günthar, and L. Jaenicke, *Chem. Ber.*, 1974, **107**, 876.

853. U. Ewers, H. Günther, and L. Jaenicke, *Chem. Ber.*, 1974, **107**, 3275.

854. D. Ege-Serpkenci, *Diss. Abstr. Int. (B)*, 1983, **43**, 2892; *Chem. Abstr.*, 1983, **99**, 21723.

855. E. Eigen and G.D. Shockman, in *Analytical Microbiology*, F. Kavanagh, Ed., Academic, New York, 1963, p. 431; *Chem. Abstr.*, 1963, **59**, 1857.

856. K. Eistetter and W. Pfleiderer, *Chem. Ber.*, 1974, **107**, 575.

857. R.D. Elliott, C. Temple, and J.A. Montgomery, *J. Org. Chem.*, 1970, **35**, 1676,

858. H.S. Forrest, C. van Baalen, and J. Myers, *Arch. Biochem. Biophys.*, 1959, **83**, 508.

859. S. Kwee and H. Lund, *Bioelectrochem. Bioenerg.*, 1979, **6**, 441; *Chem. Abstr.*, 1980, **93**, 47126.

860. H. Hirano, *Vitamins*, 1977, **51**, 544; *Chem. Abstr.*, 1978, **88**, 89541.

861. D. Ege-Serpkenci and G. Dryhurst, *Bioelectrochem. Bioenerg.*, 1982, **9**, 175; *Chem. Abstr.*, 1982, **97**, 87692.

862. M. Melgarejo Sampedro, C. Rodriguez Melgarejo, and A. Sanchez Rodrigo, *An. Quim., Ser. C*, 1981, **77**, 126; *Chem. Abstr.*, 1982, **97**, 39285.

863. M. Melgarejo Sampedro, C. Rodriguez Melgarejo, M. Nogueras Montiel, and A. Sanchez Rodrigo, *An. Quim., Ser. C*, 1982, **78**, 399; *Chem. Abstr.*, 1983, **98**, 161071.

864. G.R. Pettit, R.M. Blazer, and J.J. Einck, *J. Carbohydr., Nucleosides, Nucleotides*, 1980, **7**, 315; *Chem. Abstr.*, 1981, **94**, 121853.

865. M.J. Saxby, P.R. Smith, C.J. Blake, and L.V. Coveney, *Food Chem.*, 1983, **12**, 115; *Chem. Abstr.*, 1983, **99**, 174351.

866. P.R. Pal and S.C. Roy, *Ann. Biochem. Exptl. Med. (India)*, 1955, **15**, 59; *Chem. Abstr.*, 1956, **50**, 6578.

867. K. Slavík and V. Matoulková, *Chem. Listy*, 1954, **48**, 765; *Chem. Abstr.*, 1954, **48**, 13748.

868. K. Slavík and V. Slavíkova-Matoulková, *Chem. Listy*, 1956, **50**, 1141; *Chem. Zentr.*, 1957, **128**, 3575.

869. E.A. Mironov, G.G. Dvoryantseva, and V.S. Nabokov, *Khim.-Farm. Zh.*, 1976, **10**, 136; *Chem. Abstr.*, 1977, **86**, 55677.

870. E. Paur and E.F. Elstner, *Naturwiss. Rundsch.*, 1982, **35**, 279.

871. L.G . Karber and G. Dryhurst, *J. Electroanal. Chem. Interfacial Electrochem.*, 1982, **136**, 271.

872. J. Pradac, J. Pradacova, D. Homolka, J. Koryta, J. Weber, K. Slavik, and R. Cihar, *J. Electroanal. Chem. Interfacial Electrochem.*, 1976, **74**, 205; *Chem. Abstr.*, 1977, **86**, 23530.

873. G.R. Ramage, Brit. Pat. 619,915 (1949); *Chem. Abstr.*, 1949, 43, 9087.

874. G.B. Elion, G.H. Hitchings, and P.B. Russell, *J. Am. Chem. Soc.*, 1950, **72**, 78.

875. H.S. Forrest and S. Nawa, *Nature (London)*, 1962, **196**, 372.

876. E. Falch, *Acta Chem. Scand., Ser. B*, 1977, **31**, 167.

877. G.B. Elion, A.E. Light, and G.H. Hitchings, *J. Am. Chem. Soc.*, 1949, **71**, 741.

878. G.B. Elion and G.H. Hitchings, *J. Am. Chem. Soc.*, 1947, **69**, 2553.

879. E.H. Flynn, T.J. Bond, T.J. Bardos, and W. Shive, *J. Am. Chem. Soc.*, 1951, **73**, 1979.

880. H.S. Forrest and H.K. Mitchell, *J. Am. Chem. Soc.*, 1954, **76**, 5656.

881. E.I. Fairburn, B.J. Magerlein, L. Stubberfield, E. Stepert, and D.I. Weisblat, *J. Am. Chem. Soc.*, 1954, **76**, 676.

882. G.B. Elion and G.H. Hitchings, *J. Am. Chem. Soc.*, 1953, **75**, 4311.

883. G.B. Elion and G.H. Hitchings, *J. Am. Chem. Soc.*, 1952, **74**, 3877.

884. J. Eder and H. Rembold, *Fresenius' Z. Anal. Chem.*, 1968, **237**, 50.

885. H.S. Forrest, R. Hull, H.J. Rodda, and A.R. Todd, *J. Chem. Soc.*, 1951, 3.

886. W. Ehrenstein, H. Wamhoff, and F. Korte, *Tetrahedron*, 1970, **26**, 3993.

887. A. Ehrenberg, P. Hemmerich, F. Müller, T. Okada, and M. Viscontini, *Helv. Chim. Acta*, 1967, **50**, 411.

888. R.C. Gurira and L.D. Bowers, *J. Electroanal. Chem. Interfacial Electrochem.*, 1983, **146**, 109; *Chem. Abstr.*, 1983, **98**, 224048.

889. R. Raghaven and G. Dryhurst, *J. Electroanal. Chem. Interfacial Electrochem.*, 1981, **129**, 189.

890. D.L. McAllister and G. Dryhurst, *J. Electroanal. Chem. Interfacial Electrochem.*, 1975, **59**, 75; *Chem. Abstr.*, 1975, **83**, 17655.

891. M.A. Kaldrikyan, G.G. Danagulyan, A.V. Khekoyan, F.G. Arsenyan, and A.A. Aroyan, *Arm. Khim. Zh.*, 1976, **29**, 337; *Chem. Abstr.*, 1976, **85**, 160035.

892. D.L. McAllister and G. Dryhurst, *J. Electroanal. Chem. Interfacial Electrochem.*, 1974, **55**, 69.

893. D.L. McAllister and G. Dryhurst, *J. Electroanal. Chem. Interfacial Electrochem.*, 1973, **47**, 479.

894. C.V. Tondo and F. Lewgoy, *Rev. Brasil. Biol.*, 1965, **25**, 49; *Chem. Abstr.*, 1966, **64**, 1001.

895. Z.V. Pushkareva, L.V. Alekseeva, V.N. Kohyukhov, and E.P. Darienko, *Tr. Soveshch. Po Fiz. Metodam Issled. Organ. Soedin. i Khim. Protsessov, Akad. Nauk Kirg. SSR, Instit. Organ. Khim., Frunze*, 1962, 26; *Chem. Abstr.*, 1965, **62**, 3913.

896. G.B. Saul, *Rev. Suisse Zool.*, 1960, **67**, 270; *Chem. Abstr.*, 1961, **55**, 10728.

897. S. Okay, *Rev. Faculté* Sci., *Univ. Istanbul*, 1947, **12B**, 89; *Chem. Abstr.*, 1948, **42**, 2684.

898. S.W. Schneller and W.J. Christ, *Lect. Heterocycl. Chem.*, 1982, **6**, 139.

899. A. Watanabe and Y. Asahi, *Annu. Rept. Takeda Res. Lab.*, 1955, **14**, 11; *Chem. Abstr.*, 1956, **50**, 5987.

900. F. Weygand, *Österr. Chem.-Ztg.*, 1941, **44**, 254; *Chem. Abstr.*, 1943, **37**, 3094.

901. N.S. Vul'fson, *Uspekhi Khim.*, 1948, **17**, 249; *Chem. Abstr.*, 1948, **42**, 5458.

902. S. Gabriel and A. Sonn, *Ber. Dtsch. Chem. Ges.*, 1907, **40**, 4850.

903. M.J. Griffin and G.M. Brown, *J. Biol. Chem.*, 1964, **239**, 310.

904. M. Goto, M. Ohno, H.S. Forrest, and J.M. Lagowski, *Arch. Biochem. Biophys.*, 1965, **110**, 444.

905. M. Goto, T. Okada, and H.S. Forrest, *Arch. Biochem. Biophys.*, 1965, **110**, 409.

906. M. Goto, A. Sakurai, K. Ohta, and H. Yamakami, *J. Biochem.* (*Tokyo*), 1969, **65**, 611.

907. M. Goto, M. Konishi, K. Sugiura, and M. Tsusue, *Bull. Chem. Soc. Jpn.*, 1966, **39**, 929.

908. M. Goldstein and J.H. Joh, *Mol. Pharmacol.*, 1967, **3**, 396.

909. M.R. Gasco, *Farmaco, Ed. Sci.*, 1976, **31**, 817.

910. G. Dryhurst, R. Raghaven, D. Ege-Serpenci, and L.G. Karber, *Adv. Chem. Ser.*, 1982, 457.

911. M. Goto and K. Sugiura, *Methods Enzymol.*, 1971, **18(B)**, 746.

912. M. Goto, K. Kobayashi, H. Sato, and F. Korte, *Justus Liebigs Ann. Chem.*, 1965, **689**, 221.

913. M. Gates, *Chem. Rev.*, 1947, 41, 63.

914. A.H. Gowenlock, G.T. Newbold, and F.S. Spring, *J. Chem. Soc.*, 1948, 517.

915. T.H. Goodwin and A.L. Porter, *J. Chem. Soc.*, 1956, 3596.

916. E.M. Gal, *J. Am. Chem. Soc.*, 1950, **72**, 3532.

917. U. Grossbach, *Z. Naturforsch., Teil B*, 1957, **12**, 462.

918. E.M. Gal, *J. Am. Chem. Soc.*, 1950, **72**, 5315.

919. E.M. Gal, *Experientia*, 1951, **7**, 261.

920. J.A. Elvidge, *Annu. Rep. Prog. Chem.*, 1948, **45**, 226.

921. E.A. Evans, J.P. Kitcher, D.C. Warrell, J.A. Elvidge, J.R. Jones and R. Lenk, *J. Labelled Compd. Radiopharm.*, 1979, **16**, 697.

922. G. Favini, I. Vandoni, and M. Simonetta, *Theoret. Chim. Acta*, 1965, **3**, 418.

923. R.L. Flurry, E.W. Stout, and J.J. Bell, *Theoret. Chim. Acta*, 1967, **8**, 203.

924. J.D. Fissekis, C.G. Skinner, and W. Shive, *J. Org. Chem.*, 1959, **24**, 1722.

925. H.S. Forrest, C. van Baalen, M. Viscontini, and M. Piraux, *Helv. Chim. Acta*, 1960, **43**, 1005.

926. H.S. Forrest and J. Walker, *J. Chem. Soc.*, 1949, 2002.

927. T.E. Gorizdra, *Khim. Geterotsikl. Soedin.*, 1969, 908.

928. A.A. Goldman and C.W. Waller, U.S. Pat. 2,615,890 (1952); *Chem. Abstr.*, 1953, **47**, 10014.

929. A.A. Goldman and C.W. Waller, U.S. Pat. 2,615,891 (1952); *Chem. Abstr.*, 1953, **47**, 10014.

930. Y. Hirata, S. Nawa, S. Matsuura, and H. Kakizawa, *Experientia*, 1952, **8**, 339.

931. P.M. Good and A.W. Johnson, *Nature* (*London*), 1949, **163**, 31.

932. H. Huck, *Fresenius' Z. Anal. Chem.*, 1983, **315**, 227.

933. J.P. Hénichart, J.L. Bernier, J.L. Dhondt, M. Dautrevaux, and G. Biserte, *J. Heterocycl. Chem.*, 1979, **16**, 1489.

934. H. Hara and H.C. van der Plas, *J. Heterocycl. Chem.*, 1982, **19**, 1527.

935. P. Hemmerich and S. Fallab, *Helv. Chim. Acta*, 1958, **41**, 498.

936. M. Hattori and W. Pfleiderer, *Justus Liebigs Ann. Chem.*, 1978, 1780.

937. G. Henseke and M. Winter, *Chem. Ber.*, 1956, **89**, 956.

938. B. Tinland, *Theor. Chim. Acta*, 1967, **8**, 361.

939. R.A. Lazarus, M.A. Sulewski, and S.J. Benkovic, *J. Labelled Compound. Radiopharm.*, 1982, **19**, 1189.

940. S.I. Zav'yalov and L.F . Ovechkina, *Izv. Akad. Nauk SSSR, Ser. Khim.*, 1978, 1935.

941. S.I. Zav'yalov and L.F. Ovechkina, *Izv. Akad. Nauk SSSR, Ser. Khim.*, 1974, 2157.

942. S.I. Zav'yalov and T.K. Budkova, *Izv. Akad. Nauk SSSR, Ser. Khim.*, 1979, 1323.

943. S.I. Zav'yalov and A.G. Zavozin, *Izv. Akad. Nauk SSSR, Ser. Khim.*, 1981, 1669.

944. S.I. Zav'yalov, A.. Zavozin, and G.I. Ezhova, *Izv. Akad. Nauk SSSR, Ser. Khim.*, 1981, 2626.

945. S.I. Zav'yalov and A.G. Zavozin, *Izv. Akad. Nauk SSSR, Ser. Khim.*, 1982, 1910.

946. J.C.E. Simpson, *Annu. Rep. Prog. Chem.*, 1946, **43**, 250.

947. K.L. Kapoor, *Int. J. Quantum Chem.*, 1973, **7**, 27.

948. M.H. Palmer, I. Simpson, and R.J. Platenkamp, *J. Mol. Struct.*, 1980, **66**, 243.

949. I. Tamir, *J. Magn. Reson.*, 1976, **23**, 293.

950. J.E. Ridley and M.C. Zerner, *J. Mol. Spectrosc.*, 1974, **50**, 457.

951. S.C. Wait and J.W. Wesley, *J. Mol. Spectrosc.*, 1966, **19**, 25.

952. M. Viscontini and H. Stierlin, *Gazz. Chim. Ital.*, 1963, **93**, 391.

953. H.I.X. Mager and W. Berends, *Rec. Trav. Chim. Pays-Bas*, 1965, **84**, 1329.

954. D.D. Perrin and Y. Inoue, *Proc. Chem. Soc.*, 1960, 342.

955. M. Pesson, *Bull Soc. Chim. Fr.*, 1948, 963.

956. M. Polonovski, R. Vieillefosse, S. Guinand, and H. Jérome, *Bull. Soc. Chim. Fr.*, 1946, 80.

957. C. Párkányi and W. C. Herndon, *Phosphorus Sulfur*, 1978, **4**, 1.

958. Y. Iwanami, T. Inagaki, and H. Sakata, *Org. Mass Spectrom.*, 1977, **12**, 302.

959. J.P. Geerts, A. Nagel, and H.C. van der Plas, *Org. Magn. Reson.*, 1976, **8**, 607.

960. J.W. Triplett, S.W. Mack, S.L. Smith, and G.A. Digenis, *J. Labelled Compd. Radiopharm.*, 1978, **14**, 35.

961. Y. Inoue, *Tetrahedron*, 1964, **20**, 243.

962. G. Henseke and H.G. Patzwaldt, *Chem. Ber.*, 1956, **89**, 2904.

963. T.A. Hamor and J.M. Robertson, *J. Chem. Soc.*, 1956, 3586.

964. M.E. Hultquist, J.M. Smith, D.R. Seeger, D.B. Cosulich, and E. Kuh, *J. Am. Chem. Soc.*, 1949, **71**, 619.

965. G.H. Hitchings and G.B. Elion, *J. Am. Chem. Soc.*, 1949, **71**, 467.

966. B.L. Hutchings, E.L.R. Stockstad, J.H. Mowat, J.H. Boothe, C.W. Waller, R.B. Angier, J. Semb, and Y. Subba Row, *Ann. NY Acad. Sci.*, 1946, **48**, 273; *J. Am. Chem. Soc.*, 1948, **70**, 10.

967. E. Hadorn and I. Schwinck, *Nature (London)*, 1956, **177**, 940.

968. G. Habermehl, *Z. Naturforsch., Teil B*, 1965, **20**, 1130.

969. E. Hadorn, *Experientia*, 1954, **10**, 483.

970. M.R. Heinrich, V.C. Dewey, and G.W. Kidder, *J. Chromatogr.*, 1959, **2**, 296.

971. B.L. Hutchings, E.L.R. Stockstad, N. Bohonos, N.H. Sloane, and Y. Subba Row, *Ann. NY Acad. Sci.*, 1946, **48**, 265; *J. Am. Chem. Soc.*, 1948, **70**, 1.

972. A. Hinchliffe, M.A. Ali, and E. Farmer, *Spectrochim. Acta, Part A*, 1967, **23**, 501.

973. J.P. Jonak, S.F. Zakrzewski, L.H. Mead, and L.D. Allshouse, *J. Med. Chem.*, 1972, **15**, 1331.

974. J.A. Jones, J.B. Bicking, and E.J. Cragoe, *J . Med. Chem.*, 1967, **10**, 899.

975. J. Henkin and W.L. Washtien, *J. Med. Chem.*, 1983, **26**, 1193.

976. Y. Iwanami and M. Akino, *Tetrahedron Lett.*, 1972, 3219.

977. H.I. Hochman, K.C. Agrawal, C.W. Shansky, and A.C. Sartorelli, *J. Pharm. Sci.*, 1973, **62**, 150.

978. R.I. Ho, L. Corman, and W.O. Foye, *J. Pharm. Sci.*, 1974, **63**, 1474.

979. C.N. Hodnett, J.J. McCormack, and J.A. Sabean, *J. Pharm. Sci.*, 1976, **65**, 1150.

980. W.J. Irwin and D.G. Wibberly, *Tetrahedron Lett.*, 1972, 3359.

981. J.A. Jongejan, H.I.X. Mager, and W. Berends, *Tetrahedron*, 1975, **31**, 533.
982. Y. Iwanami and T. Seki, *Bull. Chem. Soc. Jpn.*, 1972, **45**, 2829.
983. O. Isay, *Ber. Dtsch. Chem. Ges.*, 1906, **39**, 250.
984. W. Jacobson and D.M. Simpson, *Biochem. J.*, 1946, **40**, 3.
985. P.A. Jacobi, M. Martinelli, and E.C. Taylor, *J. Org. Chem.*, 1981, **46**, 5416.
986. K. Iwai, M. Kobashi, and T. Suzuki, *Methods Enzymol.*, 1980, **66**, 512.
987. P. Haug, *Anal. Biochem.*, 1970, **37**, 285.
988. H. Heinrich and W.F. Buth, U.S. Pat. 2,673,204 (1954); *Chem. Abstr.*, 1955, **49**, 1825.
989. O. Hrdý, *Chem. Listy*, 1958, **52**, 1058; *Collect. Czech. Chem. Commun.*, 1959, **24**, 1180.
990. T.E. Gorizdra, *Trudy L'vov. Med. Inst.*, 1957, **12**, 95; *Chem. Abstr.*, 1960, **54**, 19689.
991. P. Karrer and H. Feigal, *Helv. Chim. Acta*, 1951, **34**, 2155.
992. P. Karrer and B.J.R. Nicolaus, *Helv. Chem. Acta*, 1951, **34**, 1029.
993. P. Karrer and R. Schwyzer, *Helv. Chim. Acta*, 1950, **33**, 39.
994. P. Karrer, B.J.R. Nicolaus, and R. Schwyzer, *Helv. Chim. Acta*, 1950, **33**, 1233.
995. T. Kauffmann, *Justus Liebigs Ann. Chem.*, 1959, **625**, 133.
996. F. Korte, H.U. Aldag, G. Ludwig, W. Paulus, and K. Störiko, *Justus Liebigs Ann. Chem.*, 1958, **619**, 70.
997. R.G. Kallen and W.P. Jencks, *J. Biol. Chem.*, 1966, **241**, 5845.
998. T. Kauffmann, *Z. Naturforsch., Teil B*, 1959, **14**, 358.
999. M. Kočevar, B. Stanovnik, and M. Tišler, *J. Heterocycl. Chem.*, 1982, **19**, 1397.
1000. Z. Kazimierczuk and W. Pfleiderer, *Chem. Ber.*, 1979, **112**, 1499.
1001. M. Kawai and K.G. Scrimgeour, *Can. J. Biochem.*, 1972, **50**, 1183.
1002. J.T. Keltjens, M.J. Huberts, W.H. Laarhoven, and G.D. Vogels, *Eur. J. Biochem.*, 1983, **130**, 537.
1003. M. Kočevar, B. Stanovnik, and M. Tišler, *Heterocycles*, 1981, **15**, 293.
1004. M. Kočevar, B. Stanovnik, and M. Tišler, *Tetrahedron*, 1983, **39**, 823.
1005. F.E. King and P.C. Spensley, *Nature (London)*, 1949, **164**, 574.
1006. F.E. King and P.C. Spensley, *J. Chem. Soc.*, 1952, 2144.
1007. F. Korte and H. Weitkamp, *Angew. Chem.*, 1958, **70**, 434.
1008. W.F. Keir, A.H. MacLennan, and H.C.S. Wood, *J. Chem. Soc. Perkin Trans. 1*, 1977, 1321.
1009. S. Kaufman, *J. Biol. Chem.*, 1964, **239**, 332.
1010. E.M. Kaiser, *Tetrahedron*, 1983, **39**, 2055.
1011. F. Korte, *Chem. Ber.*, 1952, **85**, 1012.
1012. R. Kuhn and A.H. Cook, *Ber. Dtsch. Chem. Ges.*, 1937, **70**, 761.
1013. M. Kamiya, *Bull. Chem. Soc. Jpn.*, 1971, **44**, 285.
1014. S. Kwee and H. Lund, *Biochim. Biophys. Acta*, 1973, **297**, 285.
1015. S. Kuwada, T. Masuda, T. Kishi, and M. Asai, *Chem. Pharm. Bull.*, 1958, **6**, 618.
1016. M. Kamiya and Y. Akahori, *Chem. Pharm. Bull.*, 1972, **20**, 918.
1017. Y. Ishiguro, M. Sawada, H. Hoshida, and K. Kawabe, *Yakugaku Zasshi*, 1982, **102**, 484.
1018. E.L. Patterson, M.H. von Saltza, and E.L.R. Stockstad, *J. Am. Chem. Soc.*, 1956, **78**, 5871.
1019. H. von Euler, K.M. Brandt, and G. Neumüller, *Biochem. Z.*, 1935, **281**, 206.
1020. N.K. Dasgupta, A. Dasgupta, and F.W. Birss, *Indian J. Chem., Sect. B*, 1982, **21**, 334; *Chem. Abstr.*, 1982, **97**, 161892.
1021. J.C.E. Simpson, in *Thorpe's Dictionery of Applied Chemistry*, *2nd Edition*, Longmans, London, 1950, Vol. 10, p. 264.

1022. G. Ritzmann and W. Pfleiderer, in *Nucleic Acid Chemistry*, L.B. Townsend and R.S. Tipson, Eds., Wiley, New York, 1978, Vol. 2, p. 729.

1023. M. Ott and W. Pfleiderer, in *Nucleic Acid Chemistry*, L.B. Townsend and R.S. Tipson, Eds., Wiley, New York, 1978, Vol. 2, p. 735.

1024. T. Itoh and W. Pfleiderer, in *Nucleic Acid Chemistry*, L.B. Townsend and R.S. Tipson, Eds., Wiley, New York, 1978, Vol. 2, p. 741.

1025. H.M. Rauen and H. Waldmann, *Experientia*, 1950, **6**, 387.

1026. F. Weygand, A. Wacker, and V. Schmied-Kowarzik, *Experientia*, 1948, **4**, 427.

1027. S. Kuwada, T. Masuda, T. Kishi, and M. Asai, *Chem. Pharm. Bull.*, 1958, **6**, 447.

1028. A. Sakurai and M. Goto, *J. Biochem. (Tokyo)*, 1969, **65**, 755.

1029. K. Ohta and M. Goto, *J. Biochem. (Tokyo)*, 1968, **63**, 127.

1030. A. Suzuki and M. Goto, *J. Biochem. (Tokyo)*, 1968, **63**, 798.

1031. K. Sugiura and M. Goto, *J. Biochem. (Tokyo)*, 1966, **60**, 335.

1032. T. Okada and M. Goto, *J. Biochem. (Tokyo)*, 1965, **58**, 458.

1033. S. Matsuura, S. Nawa, M. Goto, and Y. Hirata, *J. Biochem. (Tokyo)*, 1955, **42**, 419.

1034. S. Nawa, M. Goto, S. Matsuura, H. Kakizawa, and Y. Hirata, *J. Biochem. (Tokyo)*, 1954, **41**, 657.

1035. K.L. Gladilin, N.A. Andreeva, A. Yu. Tsygankov, A.K. Gadzhieva, A.F. Orlovskii, and D.B. Kirpotin, *Dokl. Akad. Nauk SSSR*, 1978, **240**, 1461.

1036. R. Purrmann, *Fortschr. Chem. Org. Naturst.*, 1945, **4**, 64.

1037. A. Veillard, *J. Chim. Phys.*, 1962, **59**, 1056.

1038. H.I.X. Mager, R. Addink, and W. Berends, *Recl. Trav. Chim. Pays-Bas*, 1967, **86**, 833.

1039. D.M. Simpson, *Analyst*, 1947, **72**, 382.

1040. A.J. Pearson, *Chem. Ind. (London)*, 1974, 233.

1041. G.H. Schmidt and M. Viscontini, *Helv. Chim. Acta*, 1967, **50**, 34.

1042. E.Lippert and H. Prigge, *Z. Electrochem.*, 1960, 64, 662.

1043. E.L. Patterson, R. Milstrey, and E. L. R. Stockstad, *J. Am. Chem. Soc.*, 1956, **78**, 5868.

1044. C. Mitra, A. Saran, and G. Govil, *Indian J. Chem., Sect. B*, 1978, **16**, 132.

1045. M.K. Mahanti, *Indian J. Chem., Sect. B*, 1977, **15**, 168.

1046. P.N. Moorthy and E. Hayon, *Indian J. Chem., Sect. B*, 1976, **14**, 206.

1047. M. Kawai, M.C. Archer, D. Chippel, and K.C. Scrimgeour, *Chem. Biol. Pteridines, Proc. Int. Symp., 4th, Toba*, 1969, 121.

1048. H. Rokos and G. Harzer, *Chem. Biol. Pteridines, Proc. Int. Symp., 5th, Konstanz*, 1975, 795.

1049. M. Viscontini, E. Loeser, P. Karrer, and E. Hadorn, *Helv. Chim. Acta*, 1955, **38**, 1222.

1050. M. Viscontini, E. Loeser, P. Karrer, and E. Hadorn, *Helv. Chim. Acta*, 1955, **38**, 2034.

1051. M. Viscontini and H.R. Weilenmann, *Helv. Chim. Acta*, 1959, **42**, 1854.

1052. M. Viscontini and H.R. Weilenmann, *Helv. Chim. Acta*, 1958, **41**, 2170.

1053. W.S. McNutt, *Anal. Chem.*, 1964, **36**, 912.

1054. M. Viscontini, M. Schoeller, E. Loeser, P. Karrer, and E. Hadorn, *Helv. Chim. Acta*, 1955, **38**, 397.

1055. L. Merlini, W. von Philipsborn, and M. Viscontini, *Helv. Chim. Acta*, 1963, **46**, 2597.

1056. M. Viscontini, *Helv. Chim. Acta*, 1957, **40**, 586.

1057. A. Wacker and E.-R. Lochmann, *Z. Naturforsch., Teil B*, 1959, **14**, 222.

1058. A. Wacker and L. Träger, *Z. Naturforsch., Teil B*, 1962, **17**, 369.

1059. M. Viscontini, A. Kühn, and A. Engelhaaf, *Z. Naturforsch., Teil B*, 1956, **11**, 501.

1060. F. Weygand, A. Wacker, A. Trebst, and O.P. Swoboda, *Z. Naturforsch., Teil B*, 1956, **11**, 689.

1061. R. Tschesche, C.-H. Köhncke, and F. Korte, *Z. Naturforsch., Teil B*, 1950, **5**, 132.

1062. T. Taira, *Nature (London)*, 1961, **189**, 231.

1063. A. Momzikoff, *Experientia*, 1973, **29**, 1575.

1064. G.E. Risinger, P.N. Parker, and H.H. Hsieh, *Experientia*, 1965, **21**, 434.

1065. M. Viscontini, H. Schmid, and E. Hadorn, *Experientia*, 1955, **11**, 390.

1066. F. Weygand, A. Wacker, and V. Schmied-Kowarzik, *Experientia*, 1950, **6**, 184.

1067. W. Pfleiderer, *Khim. Geterotsikl. Soedin.*, 1974, 1299.

1068. P. Kierkegaard, R. Norrestam, and P.E. Werner, *Wenner-Gren Cent. Int. Symp. Ser.*, 1970, **18**, 117; *Chem. Abstr.*, 1976, **84**, 74228.

1069. W.D. Johnston, *Diss. Abstr. Int. (B)*, 1974, **35**, 2657; *Chem. Abstr.*, 1975, **82**, 125359.

1070. Y. Kaneko, *Nippon Nogei Kagaku Kaishi*, 1966, **40**, 227; *Chem. Abstr.*, 1966, **65**, 7525.

1071. T. Hama, *Kagaku*, 1959, **29**, 458; *Chem. Abstr.*, 1960, **54**, 24929.

1072. A.G. Renfrew and P.C. Piatt, *J. Am. Pharm. Assoc.*, 1950, **39**, 657; *Chem. Abstr.*, 1951, **45**, 1840.

1073. F.P. Mazza and G. Tappi, *Arch. Sci. Biol.*, 1939, **25**, 438; *Chem. Abstr.*, 1940, **34**, 5847.

1074. M. Karobath, F. Roethler, and K. Heckl, *Recent Dev. Mass Spectrom. Biochem. Med., Proc. Int. Symp., 4th*, 1977, 493; *Chem. Abstr.*, 1978, **89**, 125533.

1075. C. Chahidi, P. Morliere, M. Aubailly, L. Dubertret, and R. Santus, *Photochem. Photobiol.*, 1983, **38**, 317; *Chem. Abstr.*, 1983, **99**, 190715.

1076. C.A. Infante and J.A. Myers, *Solution Behav. Surfactants: Theor. Appl. Aspects, Proc. Int. Symp.*, 1980, **2**, 1083; *Chem. Abstr.*, 1983, **98**, 13738.

1077. R.N. Henrie, *Diss. Abstr. Int. (B)*, 1979, **39**, 4893; *Chem. Abstr.*, 1979, **91**, 39426.

1078. E.M. Hawes and D.K.J. Gorecki, *Can. J. Pharm. Sci.*, 1977, **12**, 27; *Chem. Abstr.*, 1977, **87**, 84949.

1079. P.A. Jacobi, *Diss. Abstr. Int. (B)*, 1974, **34**, 4878; *Chem. Abstr.*, 1974, **81**, 25634.

1080. D.L. McAllister, *Diss. Abstr. Int. (B)*, 1974, **34**, 5359; *Chem. Abstr.*, 1974, **81**, 37013.

1081. M. May, T.J. Bardos, F.L. Berger, M. Lansford, J.M. Ravel, G.L. Sutherland, and W. Shive, *J. Am. Chem. Soc.*, 1951, **73**, 3067.

1082. W.S. McNutt, *J. Am. Chem. Soc.*, 1960, **82**, 217.

1083. S. Matsuura, S. Nawa, H. Kakizawa, and Y. Hirata, *J. Am. Chem. Soc.*, 1953, **75**, 4446.

1084. J.H. Mowat, J.H. Boothe, B.L. Hutchings, E.L.R. Stockstad, C.W. Waller, R.B. Angier, J. Semb, D.B. Cosulich, and Y. Subba Row, *J. Am. Chem. Soc.*, 1948, **70**, 14.

1085. J.H. Mowat, A.L. Gazzola, B.L. Hutchings, J.H. Boothe, C.W. Waller, R.B. Angier, J. Semb, and Y. Subba Row, *J. Am. Chem. Soc.*, 1949, **71**, 2308.

1086. R.A. Lazarus, C.W. DeBrosse, and S.J. Benkovic, *J. Am. Chem. Soc.*, 1982, **104**, 6869.

1087. R.A. Lazarus, C.W. DeBrosse, and S.J. Benkovic, *J. Am. Chem. Soc.*, 1982, **104**, 6871.

1088. J.H. Lister and G.R. Ramage, *J. Chem. Soc.*, 1953, 2234.

1089. S.F. Mason, *J. Chem. Soc.*, 1962, 493.

1090. R. Lohrmann and H.S. Forrest, *J. Chem. Soc.*, 1964, 460.

1091. S.F. Mason, *J. Chem. Soc.*, 1957, 4874.

1092. J. Mirza, W. Pfleiderer, A.D. Brewer, A. Stuart, and H.C.S. Wood, *J. Chem. Soc. (C)*, 1970, 437.

1093. A. McKillop, A. Henderson, P.S. Ray, C. Avendano, and E.G. Molinera, *Tetrahedron Lett.*, 1982, **23**, 3357.

1094. R. Mengel, W. Pfleiderer, and W.R. Knappe, *Tetrahedron Lett.*, 1977, 2817.

1095. G. Müller and W. von Philipsborn, *Helv. Chim. Acta*, 1973, **56**, 2680.

1096. H.I.X. Mager and W. Berends, *Recl. Trav. Chim. Pays-Bas*, 1972, **91**, 1137.

1097. H.I.X. Mager and W. Berends, *Tetrahedron*, 1976, **32**, 2303.

1098. T.-C. Lee, *J. Org. Chem.*, 1973, **38**, 703.

1099. F.L. Lam and T.-C. Lee, *J. Org. Chem.*, 1978, **43**, 167.

1100. Y. Maki, M. Suzuki, K. Kameyama, M. Kawai, M. Suzuki, and M. Sako, *Heterocycles*, 1981, **15**, 895.

1101. D. L. Ladd, *J. Heterocycl. Chem.*, 1982, **19**, 917.

1102. J.A. Montgomery, J.R. Piper, R.D. Elliot, E.C. Roberts, C. Temple, and Y.F. Shealy, *J. Heterocycl. Chem.*, 1979, **16**, 537.

1103. H.G. Mautner, *J. Org. Chem.*, 1958, **23**, 1450.

1104. T.L. Loo and R.L. Dion, *J. Org. Chem.*, 1965, **30**, 2837.

1105. C.W. Mosher, E.M. Acton, O.P. Crews, and L. Goodman, *J. Org. Chem.*, 1967, **32**, 1452.

1106. J.E. Martinelli, M. Chaykovsky, R.L. Kisliuk, and Y. Gaumont, *J. Med. Chem.*, 1979, **22**, 874.

1107. T.L. Loo, R.L. Dion, R.H. Adamson, M.A. Chirigos, and R.L. Kisliuk, *J. Med. Chem.*, 1965, **8**, 713.

1108. T. Masuda, T. Kishi, M. Asai, and S. Kuwada, *Chem. Pharm. Bull.*, 1958, **6**, 523.

1109. T. Masuda, T. Kishi, and M. Asai, *Chem. Pharm. Bull.*, 1958, **6**, 113.

1110. T. Masuda, T. Kishi, and M. Asai, *Chem. Pharm. Bull.*, 1958, **6**, 291.

1111. T. Lloyd, S. Markey, and N. Weiner, *Anal. Biochem.*, 1971, **42**, 108.

1112. N. Vinot, *Bull. Soc. Chim. Fr.*, 1971, 2708.

1113. M. Lapidus and M.E. Rosenthale, *J. Pharm. Sci.*, 1966, **55**, 555.

1114. L.M. Lhoste, A. Haug, and P. Hemmerich, *Biochemistry*, 1966, **5**, 3290.

1115. M. Lapidus, *J. Pharm. Sci.*, 1968, **57**, 878.

1116. W.S. McNutt, *J. Biol. Chem.*, 1963, **238**, 1116.

1117. E. Ohmura and M. Kawashima, *Yakugaku Zasshi*, 1953, **73**, 1145.

1118. A. Nagel and H.C. van der Plas, *Chem. Pharm. Bull.*, 1975, **23**, 2678.

1119. P.N. Moorthy and E. Hayon, *J. Phys. Chem.*, 1975, **79**, 1059.

1120. J. Nagel and H.C. van der Plas, *Heterocycles*, 1977, **7**, 205.

1121. H. Ogura, M. Sakaguchi, T. Okamoto, K. Gonda, and S. Koga, *Heterocycles*, 1979, **12**, 359.

1122. A. Nagel and H. C. van der Plas, *Tetrahedron Lett.*, 1978, 2021.

1123. M. G. Nair and C. M. Baugh, *J. Org. Chem.*, 1973, **38**, 2185.

1124. V. T. Oliverio, *Anal. Chem.*, 1961, **33**, 263.

1125. B. L. O'Dell, J. M. Vandenbelt, E. S. Bloom, and J. J. Pfiffner, *J. Am. Chem. Soc.*, 1947, **69**, 250.

1126. M. Ott and W. Pfleiderer, *Chem. Ber.*, 1974, **107**, 339.

1127. S. Nawa and H.S. Forrest, *Nature (London)*, 1962, **196**, 169.

1128. S. Nawa, S. Matsuura, and Y. Hirata, *J. Am. Chem. Soc.*, 1953, **75**, 4450.

1129. T. Neilson and H.C.S. Wood, *J. Chem. Soc.*, 1962, 44.

1130. Z. Neiman, F. Bergmann, and A.K. Meyer, *J. Chem. Soc. (C)*, 1969, 2415.

1131. K. Oyama and R. Stewart, *Can. J. Chem.*, 1974, **52**, 3879.

1132. M.G. Nair, L.P. Mercer, and C.M. Baugh, *J. Med. Chem.*, 1974, **17**, 1268.

1133. T. Nagatsu, T. Yamaguchi, T. Kato, T. Sugimoto, S. Matsuura, M. Akino. S. Tsushima, N. Nakazawa, and H. Ogawa, *Anal. Biochem.*, 1981, **110**, 182.

1134. E. Ohmura, J. Katsuragi, K. Morita, and N. Toki, *Yakugaku Zasshi*, 1953, **73**, 1150.

1135. K. Ohta, R. Wrigglesworth, and H.C.S. Wood, in *Rodd's Chemistry of Carbon Compounds*, 2nd ed., Elsevier, Amsterdam, 1980, vol. 4, Part L, p. 237.

1136. J. Perchais and J.P. Fleury, *Tetrahedron*, 1974, **30**, 999.
1137. J.R. Piper and J.A. Montgomery, *J. Heterocycl. Chem.*, 1974, **11**, 279.
1138. A. Nagel, H.C. van der Plas, and A. van Veldhuizen, *Recl. Trav. Chim. Pays-Bas*, 1975, **94**, 45.
1139. J.R. Piper, J.A. Montgomery, F.M. Sirotnak, and P.L. Chello, *J. Med. Chem.*, 1982, **25**, 182.
1140. H.A. Parish, R.D. Gilliom, W.P. Purcell, R.K. Browne, R.F. Spirk, and H.D. White, *J. Med. Chem.*, 1982, **25**, 98.
1141. J.R. Piper, G.S. McCaleb, and J.A. Montgomery, *J. Med. Chem.*, 1983, **26**, 291.
1142. J.R. Piper and J.A. Montgomery, *J. Med. Chem.*, 1980, **23**, 320.
1143. H.G. Petering and D.I. Weisblat, *J. Am. Chem. Soc.*, 1947, **69**, 2566.
1144. T. Paterson and H.C.S. Wood, *J. Chem. Soc. Perkin Trans. 1*, 1972, 1051.
1145. D.D. Perrin, *J. Chem. Soc.*, 1962, 645.
1146. Y. Pocker, D. Bjorkquist, W. Schaffer, and C. Henderson, *J. Am. Chem. Soc.*, 1975, **97**, 5540.
1147. H.G. Petering and J.A. Schmidt, *J. Am. Chem. Soc.*, 1949, **71**, 3977.
1148. A. Pohland, E.H. Flynn, R.G. Jones, and W. Shive, *J. Am. Chem. Soc.*, 1951, **73**, 3247.
1149. V.D. Orlov, I.Z. Papiashvili, and P.A. Grigorov, *Khim. Geterotsikl. Soedin.*, 1983, 671.
1150. W. Pfleiderer and E. Bühler, *Chem. Ber.*, 1966, **99**, 3022.
1151. T.S. Osdene, P.B. Russell, and L. Rane, *J. Med. Chem.*, 1967, **10**, 431.
1152. R. Purrmann, Ger. Pat. 721,930 (1942); *Chem. Zentr.*, 1942, **113** (II), 2293.
1153. G. Leroy, F. van Remoortere, and C. Aussems, *Bull. Soc. Chim. Belg.*, 1968, **77**, 191.
1154. S.I. Zav'yalov and T.K. Budkova, *Khim. Geterotsikl. Soedin.*, 1978, 270.
1155. W. Loop and R. Tschesche, Ger. Pat. 828,545 (1952); *Chem. Abstr.*, 1956, **50**, 3504.
1156. S. Naito, Y. Nagasaka, and K. Yamamoto, *Yakuzaigaku*, 1968, **28**, 93; *Chem. Abstr.*, 1968, **69**, 89694.
1157. M. Luckner and C. Wasternack, *Pharmazie*, 1967, 22, 181; *Chem. Abstr.*, 1968, **68**, 18582.
1158. R. Purrmann, U.S. Pat. 2,345,215 (1944); *Chem. Abstr.*, 1944, **38**, 4385.
1159. F. Pellerin, J.F. Letavernier, and N. Chanon, *Analusis*, 1977, **5**, 19; *Chem. Abstr.*, 1977, **86**, 164982.
1160. M.H. Palmer and R.J. Platenkamp, *Jerusalem Symp. Quantum Chem. Biochem.*, 1979, **12**, 147; *Chem. Abstr.*, 1980, **92**, 197766.
1161. C.W. McCausland, *Diss. Abstr. Int. (B)*, 1976, **37**, 772; *Chem. Abstr.*, 1976, **85**, 192670.
1162. T.S. Osdene and A.A. Santilli, U.S. Pat. 3, 516, 998 (1970); *Chem. Abstr.*, 1970, **73**, 66625.
1163. W.W. Parish, *Diss. Abstr. Int. (B)*, 1970, **31**, 2575; *Chem. Abstr.*, 1971, **75**, 76740.
1164. M. Polonovski, C. Alcantara, and R.-G. Busnel, *C.R. Acad. Sci.*, 1952, **235**, 1703.
1165. T. Nakajima and A. Pullman, *C.R. Acad. Sci.*, 1958, **246**, 1047.
1166. C. Thiéry-Cailly, *C.R. Acad. Sci., Ser. C*, 1968, **266**, 250.
1167. H. Sladowska, A. van Veldhuizen, and H.C. van der Plas, *J. Heterocycl. Chem.*, 1986, **23**, 843.
1168. C.L. Fan and G.M. Brown, *Biochem. Genet.*, 1979, **17**, 351.
1169. M. Polonovski, M. Pesson, and R. Vieillefosse, *C.R. Acad. Sci.*, 1944, **218**, 796.
1170. O. Chalvet and C. Sandorfy, *C.R. Acad. Sci.*, 1949, **228**, 566.
1171. F. Weygand and G. Schaefer, *Naturwissenschaften*, 1951, **38**, 432.
1172. C. Schöpf, *Naturwissenschaften*, 1940, **28**, 478.
1173. C. Schöpf, *Naturwissenschaften*, 1942, **30**, 359.
1174. R.B. Angier, J.H. Boothe, B.L. Hutchings, J.H. Mowat, J. Semb, E.L.R. Stockstad, Y. Subba Row, C.W. Walle , D.B. Cosulich, M.J. Fahrenbach, M.E. Hultquist, E. Kuh, E.H. Northey, D.R. Seeger, J.P. Sickels, and J.M. Smith, *Science*, 1946, **103**, 667.

1175. E.S. Bloom, J.M. Vandenbelt, S.B. Binkley, B.L. O'Dell, and J.J. Pfiffner, *Science*, 1944, **100**, 295.

1176. H.S. Forrest and J. Walker, *Nature (London)*, 1948, **161**, 308.

1177. H.S. Forrest and J. Walker, *Nature (London)*, 1948, **161**, 721.

1178. T.G. Gregg and L.A. Smucker, *Genetics*, 1965, **52**, 1023.

1179. G.M. Timmis, *Nature (London)*, 1949, **164**, 139.

1180. M.E. Hultquist, E. Kuh, D.R. Seeger, J.P. Sickels, J.M. Smith, R.B. Angier, J.H. Boothe, B.L. Hutchings, J.H. Mowat, J. Semb, E.L.R. Stockstad, Y. Subba Row, and C.W. Waller, *Ann. NY Acad. Sci.*, 1947, **48**, Suppl. article No. 5.

1181. E.L.R. Stockstad, B.L. Hutchings, and Y. Subba Row, *Ann. NY Acad. Sci.*, 1946, **48**, 261; *J. Am. Chem. Soc.*, 1948, **70**, 3.

1182. E.L.R. Stockstad, B.L. Hutchings, J.H. Mowat, J.H. Boothe, C.W. Waller, R.B. Angier, J. Semb, and Y. Subba Row, *Ann. NY Acad. Sci.*, 1946, **48**, 269; *J. Am. Chem. Soc.*, 1948, **70**, 5.

1183. W.H. Peterson, *Ann. NY Acad. Sci.*, 1946, **48**, 257.

1184. J.H. Mowat, J.H. Boothe, B.L. Hutchings, E.L.R. Stockstad, C.W. Waller, R.B. Angier, J. Semb, D.B. Cosulich, and Y. Subba Row, *Ann. NY Acad. Sci.*, 1946, **48**, 279.

1185. C.W. Waller, B.L. Hutchings, J.H. Mowat, E.L.R. Stockstad, J.H. Boothe, R.B. Angier, J. Semb, Y. Subba Row, D.B. Cosulich, M.J. Fahrenbach, M.E. Hultquist, E. Kuh, E.H. Northey, D.R. Seeger, J.P. Sickels, and J.M. Smith, *Ann. NY Acad. Sci.*, 1946, **48**, 283.

1186. A. Rosowsky and R. Forsch, *J. Med. Chem.*, 1982, **25**, 1454.

1187. E.L. Rickes, N.R. Trenner, J.B. Conn, and J.C. Keresztesy, *J. Am. Chem. Soc.*, 1947, **69**, 2751.

1188. B. Roth, J.M. Smith, and M.E. Hultquist, *J. Am. Chem. Soc.*, 1950, **72**, 1914.

1189. B. Roth, J.M. Smith, and M.E. Hultquist, *J. Am. Chem. Soc.*, 1951, **72**, 2864.

1190. R. Robinson and M.L. Tomlinson, *J. Chem. Soc.*, 1935, 467.

1191. B. Roth, M.E. Hultquist, M.J. Fahrenbach, D.B. Cosulich, H.P, Broquist, J.A. Brockman, J.M. Smith, R.P. Parker, E.L.R. Stockstad, and T.H. Jukes, *J. Am. Chem. Soc.*, 1952, **74**, 3247.

1192. G. Ritzmann, K. Ienaga, L. Kiriasis, and W. Pfleiderer, *Chem. Ber.*, 1980, **113**, 1535.

1193. H. Rembold and G. Hennings, *Heterocycles*, 1978, **10**, 247.

1194. A. Rosowsky and K.K.N. Chen, *J. Org. Chem.*, 1974, **39**, 1248.

1195. A.G. Renfrew, P.C. Piatt, and L.H. Cretcher, *J. Org. Chm.*, 1952, **17**, 467.

1196. H. Rembold, *Helv. Chim. Acta*, 1969, **52**, 333.

1197. H. Rembold and L. Buschmann, *Justus Liebigs Ann. Chem.*, 1963, **662**, 72.

1198. H.W. Rothkopf, D. Wöhrle, R. Müller, and G. Kossmehl, *Chem. Ber.*, 1975, **108**, 875.

1199. V.J. Ram, W.R. Knappe, and W. Pfleiderer, *Tetrahedron Lett.*, 1977, 3795.

1200. H.M. Rauen and W. Stamm, *Hoppe-Seyler's Z. Physiol. Chem.*, 1952, **289**, 201.

1201. J.J. Reynolds and G.M. Brown, *J. Biol. Chem.*, 1964, **239**, 317.

1202. H. Rembold, *Methods Enzymol.*, 1971, **18** (Part **B**), 652.

1203. H.M. Rauen and H. Waldmann, *Hoppe-Seyler's Z. Physiol. Chem.*, 1951, **286**, 180.

1204. J. Komenda, L. Kišová, and J. Koudelka, *Collect. Czech. Chem. Commun.*, 1960, **25**, 1020.

1205. P.R. Pal, *J. Indian Chem. Soc.*, 1955, **32**, 89.

1206. D. Roy, S. Ghosh, and B.C. Gupta, *J. Indian Chem. Soc.*, 1959, **36**, 651.

1207. P.R. Pal, *J. Indian Chem. Soc.*, 1954, **31**, 673.

1208. J. Komenda, *Chem. Listy*, 1958, 82, 1065; *Collect. Czech. Chem. Commun.*, 1959, **24**, 903.

1209. P. Karrer and R. Schwyzer, *Helv. Chim. Acta*, 1949, **32**, 1689.

1210. P. Karrer, R. Schwyzer, B. Erden, and A. Siegwart, *Helv. Chim. Acta*, 1947, **30**, 1031.

1211. P. Karrer and R. Schwyzer, *Helv. Chim. Acta*, 1949, **32**, 1041.

1212. P. Karrer and R. Schwyzer, *Helv. Chim. Acta*, 1949, **32**, 423.
1213. M. Viscontini and J. Meier, *Helv. Chim. Acta*, 1949, **32**, 877.
1214. S. Matsuura, M. Yamamoto, and Y. Kaneko, *Bull. Chem. Soc. Jpn.*, 1972, **45**, 492.
1215. K. Sugiura and M. Goto, *Bull. Chem. Soc. Jpn.*, 1973, **46**, 939.
1216. K. Sugiura, H. Yamashita, and M. Goto, *Bull. Chem. Soc. Jpn.*, 1972, **45**, 3564.
1217. K. Sugiura and M. Goto, *Bull. Chem. Soc. Jpn.*, 1969, **42**, 2662.
1218. A. Suzuki and M. Goto, *Bull. Chem. Soc. Jpn.*, 1971, **44**, 1869.
1219. P. Karrer and R. Schwyzer, *Helv. Chim. Acta*, 1948, **31**, 782.
1220. P. Karrer and R. Schwyzer, *Helv. Chim. Acta*, 1948, **31**, 777.
1221. S. Nawa, *Bull. Chem. Soc. Jpn.*, 1969, **33**, 1555.
1222. K. Sugiura and M. Goto, *Nippon Kagaku Kaishi*, 1972, 206.
1223. Y. Hirata, K. Nakanishi, and H. Kikkawa, *Bull. Chem. Soc. Jpn.*, 1950, **23**, 76.
1224. C.B. Storm and H.S. Chung, *Org. Magn. Reson.*, 1976, **8**, 361.
1225. M. Goodgame and M.A. Schmidt, *Inorg. Chim. Acta*, 1979, **36**, 151.
1226. J. Clark and C. Smith, *Org. Mass Spectrom.*, 1971, **5**, 447.
1227. M. Pesson, *Bull. Soc. Chim. Fr.*, 1951, 428.
1228. M. Pesson, *Bull. Soc. Chim. Fr.*, 1951, 423.
1229. T.S. Osdene and G.M. Timmis, *J. Chem. Soc.*, 1955, 2038.
1230. M. Polonovski, M. Pesson, and A. Puister, *Bull. Soc. Chim. Fr.*, 1951, 521.
1231. S.I. Zav'yalov and A.G. Zavozin, *Izv. Akad. Nauk SSSR, Ser. Khim.*, 1981, 1666.
1232. H. Zondler, H.S. Forrest, and J.M. Lagowski, *J. Heterocycl. Chem.*, 1967, **4**, 12.
1233. W.J.A. Vanden Heuvel, J.L. Smith, P. Haug, and J. L. Beck, *J. Heterocycl. Chem.*, 1972, **9**, 451.
1234. W. Steinbuch, *Helv. Chim. Acta*, 1948, **31**, 2051.
1235. C.W. Waller, B.L. Hutchings, J.H. Mowat, E.L.R. Stockstad, J.H. Boothe, R.B. Angier, J. Semb, Y. Subba Row, D.B. Cosulich, M.J. Fahrenbach, M.E. Hultquist, E. Kuh, E.H. Northey, D.R. Seeger, J.P. Sickels, and J.M. Smith, *J. Am. Chem. Soc.*, 1948, **70**, 19.
1236. M.E. Hultquist, E. Kuh, D.B. Cosulich, M.J. Fahrenbach, E.H. Northey, D.R. Seeger, J.P. Sickels, J.M. Smith, R.B. Angier, J.H. Boothe, B.L. Hutchings, J.H. Mowat, J. Semb, E.L.R. Stockstad, Y. Subba Row, and C.W. Waller, *J. Am. Chem. Soc.*, 1948, **70**, 23.
1237. C. Schöpf and E. Becker, *Justus Liebigs Ann. Chem.*, 1936, **524**, 49.
1238. C. Schöpf and E. Becker, *Justus Liebigs Ann. Chem.*, 1933, **507**, 266.
1239. C. Schöpf, A. Kottler, and R. Reichert, *Justus Liebigs Ann. Chem.*, 1939, **539**, 168.
1240. S. Senda, K. Hirota, T. Asao, and K. Maruhashi, *J. Am. Chem. Soc.*, 1977, **99**, 7358.
1241. S. Senda, K. Hirota, T. Asao, and K. Maruhashi, *J. Am. Chem. Soc.*, 1978, **100**, 7661.
1242. J.W. Serum, P. Haug, T. Urushibara, and H.S. Forrest, *Fresenius' Z. Anal. Chem.*, 1972, **262**, 110.
1243. K. Gonda, D. Koga, M. Sakaguchi, Y. Miyata, H. Ogura, and T. Okamoto, *Yakugaku Zasshi*, 1978, **98**, 708.
1244. G. Schmidt, *Zool. Anz.*, 1958, **161**, 304; *Chem. Abstr.*, 1959, **53**, 15385.
1245. J. G. Baust, *Zoologica*, 1967, **52**, 15; *Chem. Abstr.*, 1968, **69**, 42004.
1246. A. Tatematsu, T. Goto, and S. Matsuura, *Nippon Kagaku Zasshi*, 1966, **87**, 1226.
1247. M. Tsuchiya, *Nippon Kagaku Zasshi*, 1971, **92**, 1177.
1248. A. Suzuki and M. Goto, *Nippon Kagaku Zasshi*, 1970, **91**, 404.
1249. K. Kobayashi, T. Kinoshita, and M. Goto, *Nippon Kagaku Zasshi*, 1970, **91**, 1096.
1250. M. Kamiya, *Nippon Kagaku Zasshi*, 1969, **90**, 769.

1251. H. Iida, H. Ishikawa, and Y. Bansho, *Kogyo Kagaku Zasshi*, 1970, **73**, 749.

1252. H. Sato, *J. Chem. Soc. Jpn.*, 1951, **72**, 815.

1253. H. Sato, M. Nakajima, and H. Tanaka, *J. Chem. Soc. Jpn.*, 1951, **72**, 953.

1254. H. Sato, M. Nakajima, and H. Tanaka, *J. Chem. Soc. Jpn.*, 1951, **72**, 868.

1255. H. Sato, M. Nakajima, and H. Tanaka, *J. Chem. Soc. Jpn.*, 1951, **72**, 866.

1256. M.A. Schou, *Arch. Biochem.*, 1950, **28**, 10.

1257. Y. Sakurai and K. Yoshino, *Yakugaku Zasshi*, 1952, **72**, 1294.

1258. G.H. Schmidt, *Z. Naturforsch., Teil B*, 1969, **24**, 1153.

1259. T. Shiota and M.N. Disraely, *Biochim. Biophys. Acta*, 1961, **52**, 467.

1260. C. Thiéry, *Eur. J. Biochem.*, 1973, **37**, 100.

1261. J. Semb, J.H. Boothe, R.B. Angier, C.W. Waller, J.H. Mowat, B.L. Hutchings, and Y. Subba Row, *J. Am. Chem. Soc.*, 1949, **71**, 2310.

1262. A. Sakurai and M. Goto, *Tetrahedron Lett.*, 1968, 2941.

1263. D.R. Seeger, D.B. Cosulich, J.M. Smith, and M.E. Hultquist, *J. Am. Chem. Soc.*, 1949, **71**, 1753.

1264. M. Sakaguchi, Y. Miyata, H. Ogura, K. Gonda, S. Koga, and T. Okamoto, *Chem. Pharm. Bull.*, 1979, **27**, 1094.

1265. D.R. Seeger, J.M. Smith, and M.E. Hultquist, *J. Am. Chem. Soc.*, 1947, **69**, 2567.

1266. M.R. Summers, *Biochemistry*, 1972, **11**, 3088.

1267. E. Shaw, C.M. Baugh, and C.L. Krumdieck, *J. Biol. Chem.*, 1966, **241**, 379.

1268. J.R. Totter, *J. Biol. Chem.*, 1944, **154**, 105.

1269. T. Shiota, M.N. Disraely, and M.P. McCann, *Biochem. Biophys. Res. Commun.*, 1962, **7**, 194.

1270. H.H.W. Thijssen, *Anal. Biochem.*, 1973, **54**, 609.

1271. D. C. Suster, E. Tărnăuceanu, D. Ionescu, V. Dobre, and I. Niculescu-Duvăz, *J. Med. Chem.*, 1978, **21**, 1162.

1272. W.H. Tafolla, A.C. Sarapu, and G.R. Dukes, *J. Pharm. Sci.*, 1981, **70**, 1273.

1273. R. Stewart and K. Oyama, *Can. J. Chem.*, 1974, **52**, 3884.

1274. R. Tschesche and F. Korte, *Angew. Chem.*, 1953, **65**, 600.

1275. H. Steppan, J. Hammer, R. Baur, R. Gottlieb, and W. Pfleiderer, *Justus Liebigs Ann. Chem.*, 1982, 2135.

1276. B. Stea, R.M. Halpern, and R.A. Smith, *J. Chromatogr.*, 1979, **168**, 285.

1277. P.D. Sternglanz, R.C. Thompson, and W.L. Savell, *Anal. Chem.*, 1951, **23**, 1027.

1278. R. Stewart, R. Srinivasan, and S.J. Gumbley, *Can. J. Chem.*, 1981, **59**, 2755.

1279. K. Senga, K. Shimizu, and S. Nishigaki, *Heterocycles*, 1977, **6**, 1907.

1280. K. Sugiura and M. Goto, *Tetrahedron Lett.*, 1973, 1187.

1281. R.G.W. Spickett and G.M. Timmis, *J. Chem. Soc.*, 1954, 2887.

1282. E.B. Skibo and T.C. Bruice, *J. Am. Chem. Soc.*, 1983, **105**, 3304.

1283. M. Sun, T.A. Moore, and P.-S. Song, *J. Am. Chem. Soc.*, 1972, **94**, 1730.

1284. R. Stewart and K. Oyama, *J. Chem. Soc.*, 1975, **97**, 6510.

1285. E.C. Taylor and P.A. Jacobi, *J. Am. Chem. Soc.*, 1974, **96**, 6781.

1286. J. Tramper, W.E. Hennink, and H.C. van der Plas, *J. Appl. Biochem.*, 1982, **4**, 263.

1287. J. Tramper, A. Nagel, H.C. van der Plas, and F. Müller, *Recl. Trav. Chim. Pays-Bas*, 1979, **98**, 224.

1288. K.-Y. Tserng and L. Bauer, *J. Heterocycl. Chem.*, 1974, **11**, 163.

1289. K.-Y. Tserng, C.L. Bell, and L. Bauer, *J. Heterocycl. Chem.*, 1975, **12**, 79.

1290. C. Temple, J.D. Rose, and J.A. Montgomery, *J. Heterocycl. Chem.*, 1976, **13**, 567.

1291. E.C. Taylor and M. Inbasekaran, *Heterocycles*, 1978, **10**, 37.

1292. K. Senga, Y. Kanamori, and S. Nishigaki, *Heterocycles*, 1977, **6**, 693.

1293. E.C. Taylor and A.J. Cocuzza, *J. Org. Chem.*, 1979, **44**, 302.

1294. E.C. Taylor, Y. Maki, and A. McKillop, *J. Org. Chem.*, 1972, **37**, 1601.

1295. R. Tschesche and G. Heuschkel, *Chem. Ber.*, 1956, **89**, 1054.

1296. E.C. Taylor, R.F. Abdulla, and P.A. Jacobi, *J. Org. Chem.*, 1975, **40**, 2336.

1297. E.C. Taylor and T. Kobayashi, *J. Org. Chem.*, 1973, **38**, 2817.

1298. E.C. Taylor and P.A. Jacobi, *J. Org. Chem.*, 1975, **40**, 2332.

1299. E.C. Taylor, R.C. Portnoy, D.C. Hochstetler, and T. Kobayashi, *J. Org. Chem.*, 1975, **40**, 2347.

1300. M. Sletzinger, R. Reinhold, J. Grier, M. Beachem, and M. Tishler, *J. Am. Chem. Soc.*, 1955, **77**, 6365.

1301. E.C. Taylor, R.F. Abdulla, K. Tanaka, and P.A. Jacobi, *J. Org. Chem.*, 1975, **40**, 2341.

1302. E.C. Taylor and J.L. LaMattina, *J. Org. Chem.*, 1977, **42**, 1523.

1303. C.D. Shirrell and D.E. Williams, *J. Chem. Soc. Perkin Trans. 2*, 1975, 40.

1304. R. Tschesche and F. Korte, *Chem. Ber.*, 1951, **84**, 801.

1305. E.C. Taylor and T. Kobayashi, *J. Org. Chem.*, 1976, **41**, 1299.

1306. E.C. Taylor, R.N. Henrie, and R.C. Portnoy, *J. Org. Chem.*, 1978, **43**, 736.

1307. E.C. Taylor and R. Kobylecki, *J. Org. Chem.*, 1978, **43**, 680.

1308. R. Tschesche, F. Korte, and R. Petersen, *Chem. Ber.*, 1951, **84**, 579.

1309. E.C. Taylor and E. Wachsen, *J. Org. Chem.*, 1978, **43**, 4154.

1310. E.C. Taylor and D.J. Dumas, *J. Org. Chem.*, 1981, **46**, 1394.

1311. E.C. Taylor, M.J. Thompson, K. Perlman, R. Mengel, and W. Pfleiderer, *J. Org. Chem.*, 1971, **36**, 4012.

1312. E.C. Taylor and B.E. Evans, *J. Chem. Soc. Chem. Commun.*, 1971, 189.

1313. R. Tschesche and F. Korte, *Chem. Ber.*, 1951, **84**, 77.

1314. R. Tschesche and F. Korte, *Chem. Ber.*, 1951, **84**, 641.

1315. E.C. Taylor and C.K. Cain, *J. Am. Chem. Soc.*, 1951, **73**, 4384.

1316. E.C. Taylor, *J. Am. Chem. Soc.*, 1952, **74**, 2380.

1317. E.C. Taylor, R.J. Knopf, J.A. Cogliano, J.W. Barton, and W. Pfleiderer, *J. Am. Chem. Soc.*, 1960, **82**, 6058.

1318. E.C. Taylor and P.A. Jacobi, *J. Am. Chem. Soc.*, 1976, **98**, 2301.

1319. D.J. Vonderschmitt, K. Smith-Vitols, F.M. Huennekens, and K.G. Scrimgeour, *Arch. Biochem. Biophys.*, 1967, **122**, 488.

1320. D. J. Vonderschmitt and K. G. Scrimgeour, *Biochem. Biophys. Res. Commun.*, 1967, **28**, 302.

1321. S. Uyeo and S. Mizukami, *Yakugaku Zasshi*, 1952, **72**, 843.

1322. M. Viscontini and M. Argentini, *Justus Liebigs Ann. Chem.*, 1971, **745**, 109.

1323. W. von Philipsborn, H. Stierlin, and W. Traber, *Helv. Chim. Acta*, 1963, **46**, 2592.

1324. M. Viscontini and H. Raschig, *Helv. Chim. Acta*, 1958, **41**, 108.

1325. M. Viscontini and M. Piraux, *Helv. Chim. Acta*, 1962, **45**, 615.

1326. M. Viscontini and W.F. Frei, *Helv. Chim. Acta*, 1972, **55**, 574.

1327. M. Viscontini, R. Provenzale, and W.F. Frei, *Helv. Chim. Acta*, 1972, **55**, 570.

1328. M. Viscontini and M. Piraux, *Helv. Chim. Acta*, 1963, **46**, 1537.

1329. M. Viscontini and T. Okada, *Helv. Chim. Acta*, 1967, **50**, 1845.

1330. M. Viscontini and T. Okada, *Helv. Chim. Acta*, 1967, **50**, 1492.

1331. M. Viscontini, L. Merlini, and W. von Philipsborn, *Helv. Chim. Acta*, 1963, **46**, 1181.

1332. M. Viscontini, L. Merlini, G. Nasini, W. von Philipsborn, and M. Piraux, *Helv. Chim. Acta*, 1964, **47**, 2195.

1333. M. Viscontini and S. Huwyler, *Helv. Chim. Acta*, 1965, **48**, 764.

1334. M. Viscontini, H. Leidner, G. Mattern, and T. Okada, *Helv. Chim. Acta*, 1966, **49**, 1911.

1335. M. Viscontini and Y. Furuta, *Helv. Chim. Acta*, 1973, **56**, 1819.

1336. M. Viscontini and Y. Furuta, *Helv. Chim. Acta*, 1973, **56**, 1710.

1337. M. Viscontini and M. Cogoli-Greuter, *Helv. Chim. Acta*, 1971, **54**, 1125.

1338. K. Uyeda and J.C. Rabinowitz, *Anal. Biochem.*, 1963, **6**, 100.

1339. E. Uhlmann and W. Pfleiderer, *Heterocycles*, 1981, **15**, 437.

1340. M. Viscontini, M. Frater-Schroeder, and M. Argentini, *Helv. Chim. Acta*, 1971, **54**, 811.

1341. M. Viscontini, M. Frater-Schroeder, M. Cogoli-Greuter,and M. Argentini, *Helv. Chim. Acta*, 1970, **53**, 1434.

1342. M. Viscontini and A. Bobst, *Helv. Chim. Acta*, 1964, **47**, 2087.

1343. M. Viscontini and J. Bieri, *Helv. Chim. Acta*, 1972, **55**, 21.

1344. M. Viscontini and J. Bieri, *Helv. Chim. Acta*, 1971, **54**, 2291.

1345. M. Viscontini and M. Argentini, *Helv. Chim. Acta*, 1971, **54**, 2287.

1346. M. Viscontini, *Helv. Chim. Acta*, 1969, **52**, 335.

1347. M. Viscontini and T. Okada, *Helv. Chim. Acta*, 1969, **52**, 306.

1348. N. Vinot, *Bull. Soc. Chim. Fr.*, 1971, 3695.

1349. N. Vinot, *Bull. Soc. Chim. Fr.*, 1973, 2752.

1350. D.M. Valerino and J.J. McCormack, *Biochim. Biophys. Acta*, 1964, **184**, 154.

1351. S. Uyeo, S. Mizukami, T. Kubota, and S. Takagi, *J. Am. Chem. Soc.*, 1950, **72**, 5339.

1352. M. Zimmerman, R.L. Tolman, H. Morman, D.W. Graham, and E.F. Rogers, *J. Med. Chem.*, 1977, **20**, 1213.

1353. L.M. Werbel, J. Johnson, E.F. Elslager, and D.F. Worth, *J. Med. Chem.*, 1978, **21**, 337.

1354. J. Weinstock, R.Y. Dunoff, J.E. Carevic, J.G. Williams, and A.J. Villani, *J. Med. Chem.*, 1968, **11**, 618.

1355. D.F. Worth, J. Johnson, E.F. Elslager, and L.M. Werbel, *J. Med. Chem.*, 1978, **21**, 331.

1356. J. Weinstock, J.W. Wilson, V.D. Wiebelhaus, A.R. Maass, F.T. Brennan, and G. Sosnowski, *J. Med. Chem.*, 1968, **11**, 573.

1357. J. Weinstock, R.Y. Dunoff, and J.G. Williams, *J. Med. Chem.*, 1968, **11**, 542.

1358. C.W. Waller, A.A. Goldman, R.B. Angier, J.H. Boothe, B.L. Hutchings, J.H. Mowat, and J. Semb, *J. Am. Chem. Soc.*, 1950, **72**, 4630.

1359. C.W. Waller, M.J. Fahrenbach, J.H. Boothe, R.B. Angier, B.L. Hutchings, J.H. Mowat, J.F. Poletto, and J. Semb., *J. Am. Chem. Soc.*, 1952, **74**, 5405.

1360. K.B. Wiberg and T.P. Lewis, *J. Am. Chem. Soc.*, 1970, **92**, 7154.

1361. D.I. Weisblat, B.J. Magerlein, D.R. Myers, A.R. Hanze, E.I. Fairburn, and S.T. Rolfson, *J. Am. Chem. Soc.*, 1953, **75**, 5893.

1362. D.E. Wolf, R.C. Anderson, E.A. Kaczka, S.A. Harris, G.E. Arth, P.L. Southwick, R. Mozingo, and K. Folkers, *J. Am. Chem. Soc.*, 1947, **69**, 2753.

1363. E.L. Wittle, B.L. O'Dell, J.M. Vandenbelt, and J.J. Pfiffner, *J. Am. Chem. Soc.*, 1947, **69**, 1786.

1364. W.B. Wright, D.B. Cosulich, M.J. Fahrenbach, C.W. Waller, J.M. Smith, and M.E. Hultquist, *J. Am. Chem. Soc.*, 1949, **71**, 3014.

1365. S. Wowzonek, *J. Org. Chem.*, 1976, **41**, 310.

1366. R. Weber, W. Frick, and M. Viscontini, *Helv. Chim. Acta*, 1973, **56**, 2919.

1367. R. Weber, W. Frick, and M. Viscontini, *Helv. Chim. Acta*, 1974, **57**, 1485.

1368. H. Wieland and A. Kotzschmar, *Justus Liebigs Ann. Chem.*, 1937, **530**, 152.

1369. H. Wieland and R. Liebig, *Justus Liebigs Ann. Chem.*, 1944, **555**, 146.

1370. A-.W. Wahlefeld, L. Jaenicke, and G. Hein, *Justus Liebigs Ann. Chem.*, 1968, **715**, 52.

1371. R. Tschesche, K.-H. Köhncke, and F. Korte, *Chem. Ber.*, 1951, **84**, 485.

1372. F. Weygand, H.-J. Mann, and H. Simon, *Chem. Ber.*, 1952, **85**, 463.

1373. F. Weygand, A. Wacker, and V. Schmied-Kowarzik, *Chem. Ber.*, 1949, **82**, 25.

1374. F. Weygand and V. Schmied-Kowarzik, *Chem. Ber.*, 1949, **82**, 333.

1375. F. Weygand, V. Schmied-Kowarzik, A. Wacker, and W. Rupp, *Chem. Ber.*, 1950, **83**, 460.

1376. F. Weygand and O.P. Swoboda, *Chem. Ber.*, 1956, **89**, 18.

1377. F. Weygand and H.J. Bestmann, *Chem. Ber.*, 1955, **88**, 1992.

1378. F. Weygand, H. Simon, K.D. Keil, and H. Millauer, *Chem. Ber.*, 1964, **97**, 1002.

1379. F. Weygand and B. Spiess, *Chem. Ber.*, 1964, **97**, 3456.

1380. F. Sachs and G. Meyerheim, *Ber. Dtsch. Chem. Ges.*, 1908, **41**, 3957.

1381. H. Wieland and C. Schöpf, *Ber. Dtsch. Chem. Ges.*, 1925, **58**, 2178.

1382. C. Schöpf and H. Wieland, *Ber. Dtsch. Chem. Ges.*, 1926, **59**, 2067.

1383. J. Westerling, H.I.X. Mager, and W. Berends, *Tetrahedron*, 1977, **33**, 2587.

1384. C.P. Whittle, *Tetrahedron Lett.*, 1968, 3689.

1385. D.R. Sutherland and J. Pickard, *J. Heterocycl. Chem.*, 1974, **11**, 457.

1386. M. Sato and C.H. Stammer, *J. Heterocycl. Chem.*, 1977, **14**, 149.

1387. F. Yoneda, M. Higuchi, and M. Kawamura, *Heterocycles*, 1976, **4**, 1659.

1388. F. Yoneda, R. Koga, S. Nishigaki, and S. Fukazawa, *J. Heterocycl. Chem.*, 1982, **19**, 949.

1389. K. Senga, H. Kanazawa, and S. Nishigaki, *J. Chem. Soc. Chem. Commun.*, 1976, 588.

1390. R.D. Youssefyeh and A. Kalmus, *J. Chem. Soc. Chem. Commun.*, 1970, 1371.

1391. V.P. Williams and J.E. Ayling, *J. Heterocycl. Chem.*, 1973, **10**, 827.

1392. J.M. Whiteley, J.H. Drais, and F.M. Huennekens, *Arch. Biochem. Biophys.*, 1969, **133**, 436.

1393. J.M. Whiteley, J. Drais, J. Kirchener, and F.M. Huennekens, *Arch. Biochem. Biophys.*, 1968, **126**, 955.

1394. J.M. Whiteley and F.M. Huennekens, *Biochemistry*, 1967, **6**, 2620.

1395. J.N. Wells and M.S. Strahl, *J. Pharm. Sci.*, 1971, **60**, 533.

1396. R.J.A. Walsh and K.R.H. Wooldridge, *J. Chem. Res.*, 1980, S 38 & M 0549.

1397. R.A. Weisman and G.M. Brown, *J. Biol. Chem.*, 1964, **239**, 326.

1398. J. Weinstock, Belg. Pat. 648,743 (1964); *Chem. Abstr.*, 1965, **63**, 16368.

1399. H. Watanabe, *Nippon Sanshigaku Zasshi*, 1956, **25**, 300; *Chem. Abstr.*, 1957, **51**, 18351.

1400. P.X. Iten, H. Märki-Danzig, H. Koch, and C.H. Eugster, *Helv. Chim. Acta*, 1984, **67**, 550.

1401. J.H. Sun, *Diss. Abstr. Int. (B)*, 1983, **43**, 3990; *Chem. Abstr.*, 1983, **99**, 139618.

1402. M.S. Strahl, *Diss. Abstr. Int. (B)*, 1971, **31**, 4605; *Chem. Abstr.*, 1971, **75**, 63744.

1403. M. Sletzinger and M. Tishler, U.S. Pat. 2,740,784 (1956); *Chem. Abstr.*, 1956, **50**, 15601.

1404. P.F. Skultety, *Diss. Abstr. Int. (B)*, 1983, **43**, 3994; *Chem. Abstr.* 1983, **99**, 59619.

1405. W.R. Sherman, *Diss. Abstr.*, 1956, **16,** 457; *Chem. Abstr.*, 1956, **50**, 10726.

1406. D.W. Shivvers, *Diss. Abstr. Int. (B)*, 1972, **32**, 5598; *Chem. Abstr.*, 1972, **77**, 45300.

1407. C.D. Shirrell, *Diss. Abstr. Int. (B)*, 1974, **35**, 2127; *Chem. Abstr.*, 1975, **82**, 79123.

1408. E.C. Taylor, J.V. Berrier, A.J. Cocuzza, R. Kobylecki, and J.J. McCormack, *J. Med. Chem.*, 1977, **20**, 1215.

1409. H. Graboyes, G.E. Jaffe, I.J. Pachter, J.P. Rosenbloom, A.J. Villani, J.W. Wilson, and J. Weinstock, *J. Med. Chem.*, 1968, **11**, 568.

1410. W. Koschara, *Hoppe-Seyler's Z. Physiol. Chem.*, 1943, **277**, 159.

1411. P. Haug and T. Urushibara, *Org. Mass Spectrom.*, 1970, **3**, 1365.

1412. D.W. Young, *Chem. Ind. (London)*, 1981, 556.

1413. H. Voorhof and H. Pollak, *J. Chem. Phys.*, 1966, **45**, 3542.

1414. M. Polonovski, S. Guinand, M. Pesson, and R. Vieillefosse, *Bull. Soc. Chim. Fr.*, 1945, **12**, 924.

1415. I.J. Pachter, *J. Org. Chem.*, 1963, **28**, 1191.

1416. P.B. Russell, G.B. Elion, and G.H. Hitchings, *J. Org. Chem.*, 1949, **71**, 474.

1417. J. Komenda and D. Laskafeld, *Collect. Czech. Chem. Commun.*, 1962, **27**, 199.

1418. J. Komenda, *Collect. Czech. Chem. Commun.*, 1962, **27**, 212.

1419. G. Dryhurst, *Electrochemistry of Biological Molecules*, Academic Press, New York, 1977, p. 320.

1420. S. Farber, L.K. Diamond, R.D. Mercer, R.F. Sylvester, and J.A. Wolff, *New Engl. J. Med.*, 1948, **238**, 787.

1421. E.R. Norris and J.J. Majnarich, *Am. J. Physiol.*, 1948, **152**, 652.

1422. A. Sakurai and M. Goto, *J. Biochem. (Tokyo)*, 1967, **61**, 142.

1423. M. Viscontini, M. Pouteau-Thouvenot, R. Bühler-Moor, and M Schroeder, *Helv. Chim. Acta*, 1964, **47**, 1948.

1424. A. Suzuki and M. Goto, *Biochim. Biophys. Acta*, 1973, **304**, 222.

1425. M. Goto, M. Konishi, and Tsusue, *Bull. Chem. Soc. Jpn.*, 1965, **38**, 503.

1426. K. Sugiura, S. Takikawa, M. Tsusue, and M. Goto, *Bull. Chem. Soc. Jpn.*, 1973, **46**, 3312.

1427. M. Goto, A. Sakurai, K. Ohta, and H. Yamakami, *Tetrahedron Lett.*, 1967, 4507.

1428. K. Sugiura and M. Goto, *Tetrahedron Lett.*, 1970, 4059.

1429. K. Sugiura, M. Goto, and S. Nawa, *Tetrahedron Lett.*, 1969, 2963.

1430. A. Suzuki and M. Goto, *Biochim. Biophys. Acta*, 1973, **313**, 229.

1431. M. Goto, H.S. Forrest, L.H Dickerman, and T. Urushibara, *Arch. Biochem. Biophys.*, 1965, **111**, 8.

1432. J.E. Gready, *J. Mol. Struct. (Theochem)*, 1984, **109**, 231.

1433. P.J.M. Van Haastert, R.J.W. De Wit, Y. Grijpma, and T.M. Konijn, *Proc. Natl. Acad. Sci. USA*, 1982, **79**, 6270.

1434. M. Viscontini, *Biochem. Clin. Aspects Pteridines*, 1983, **2**, 21.

1435. F.F. Blicke and H.C. Godt, *J. Am. Chem. Soc.*, 1954, **76**, 2798.

1436. D.J. Brown and K. Mori, *Aust. J. Chem.*, 1985, **38**, 467.

1437. G.F. Maley and G.W.E. Plaut, *J. Biol. Chem.*, 1959, **234**, 641.

1438. J. Weijlard, M. Tishler, and A.E. Erickson, *J. Am. Chem. Soc.*, 1945, **67**, 802.

1439. M. Kappel, R. Mengel, and W. Pfleiderer, *Justus Liebigs Ann. Chem.*, 1984, 1815.

1440. W. Pendergast and W.R. Hall, *J. Org. Chem.*, 1985, **50**, 388.

1441. G.W.E. Plaut and G.F. Maley, *J. Biol. Chem.*, 1959, **234**, 3010.

1442. D. Randles and W.L.F. Armarego, *Eur. J. Biochem.*, 1985, **146**, 467.

1443. P. Waring and W.L.F. Armarego, *Aust. J. Chem.*, 1985, **38**, 629.

1444. C.H. Winestock, T. Aogaichi, and G.W.E. Plaut, *J. Biol. Chem.*, 1963, **238**, 2866.

1445. L. Bauer, C.N.V. Nambury, and F.M. Hershenson, *J. Heterocycl. Chem.*, 1966, **3**, 224.

1446. L. Merlini and R. Mondelli, *Gazz. Chim. Ital.*, 1962, **92**, 1251.

1447. I. Ebels, H.P.J.M. Noteborn, A. de Morée, M.G.M. Balemans, and J. van Benthem, *Biochem. Clin. Aspects Pteridines*, 1983, **2**, 71.

1448. G. Reibnegger, D. Fuchs, A. Hausen, O. Knosp, and H. Watcher, *Biochem. Clin. Aspects Pteridines*, 1983, **2**, 35.

1449. C.D. Ginger, R. Wrigglesworth, W.D. Inglis, R.J. Kulick, and H.C.S. Wood, *J. Chem. Soc, Perkin Trans. 1*, 1984, 953.

1450. C.E. Lunte and P.T. Kissinger, *Anal. Chim. Acta*, 1984, **158**, 33.

1451. P.H. Boyle and M.J. O'Mahony, *J. Heterocycl. Chem.*, 1984, **21**, 909.

1452. I.I. Naumenko and G.V. Shishkin, *Khim. Geterotsikl. Soedin.*, 1983, 1406.

1453. J.T. Keltjens, H.J. Rozie, and G.D. Vogels, *Arch. Biochem. Biophys.*, 1984, **229**, 532.

1454. M. Swada, T. Yamaguchi, T. Sugimoto, S. Matsuura, and T. Nagatsu, *Clin. Chim. Acta*, 1984, **138**, 275.

1455. C.E. Lunte and P.T. Kissinger, *Anal. Chem.*, 1984, **56**, 658.

1456. M. Viscontini, *Biochem. Clin. Aspects Pteridines*, 1984, **3**, 19.

1457. T. Kuster and A. Niederwieser, *J. Chromatogr.*, 1983, **278**, 245.

1458. J.M. Whiteley and S. Webber, *Anal. Biochem.*, 1984, **137**, 394.

1459. P. van Beelen, J.F.A. Labro, J.T. Keltjens, W.J. Geerts, G.D. Vogels, W.H. Laarhoven, W. Guiljt, and C.A.G. Haasnoot, *Eur. J. Biochem.*, 1984, **139**, 359.

1460. S.K. Frost, L.G. Epp, and S.J. Robinson, *J. Embryol. Exp. Morphol.*, 1984, **81**, 127.

1461. S.K. Frost, L.G. Epp, and S.J. Robinson, *J. Embryol. Exp. Morphol.*, 1984, **81**, 105.

1462. S.N. Ganguly and M. Viscontini, *Helv. Chim. Acta*, 1984, **67**, 166.

1463. W. Pfleiderer, M. Kappel, and R. Bauer, *Biochem. Clin. Aspects Pteridines*, 1984, **3**, 3.

1464. I. Ebels, H.P.J.M. Noteborn, A. de Morée, and M.G.M. Balemans, *Biochem. Clin. Aspects Pteridines*, 1984, **3**, 127.

1465. A. Rosowsky, J.H. Freisheim, H. Bader, R.A. Forsch, S.S. Susten, C.A. Cucchi, and E. Frei, *J. Med. Chem.*, 1985, **28**, 660.

1466. T.S. Osdene and E.C. Taylor, U.S. Pat. 2,975,280 (1961); *Chem. Abstr.*, 1961, **55**, 17665.

1467. G.M. Blackburn and A.W. Johnson, *J. Chem. Soc.*, 1960, 4358.

1468. M. Israel, H.K. Protopapa, S. Chatterjee, and E.J. Modest, *J. Pharm. Sci.*, 1965, **54**, 1626.

1469. S.J. Benkovic, D. Sammons, W.L.F. Armarego, P. Waring, and R. Inners, *J. Am. Chem. Soc.*, 1985, **107**, 3706.

1470. K.G. Karber, *Diss. Abstr. Int. B*, 1983, **44**, 488; *Chem. Abstr.*, 1984, **100**, 14360.

1471. T.C. Williams, *Diss. Abstr. Int. B*, 1984, **44**, 3070; *Chem. Abstr.*, 1984, **101**, 72153.

1472. Wellcome Foundation, Jpn. Kokai Tokkyo Koho, JP 59 76086 [84 76086] (1984); *Chem. Abstr.*, 1984, **101**, 171289.

1473. K. Ganapathy, *J. Indian. Chem. Soc.*, 1937, **14**, 627.

1474. S. Matsuura, S. Murata, and T. Sugimoto, *Chem. Lett.*, 1984, 735.

1475. S.I. Zav'yalov, G.I. Ezhova, and T.K. Budkova, *Izv. Akad. Nauk. SSSR, Ser. Khim.*, 1984, 1590.

1476. C. Temple and J.A. Montgomery, *Folates Pterins*, 1984, **1**, 61.

1477. J.C. Nixon, *Folates Pterines*, 1985, **2**, 1.

1478. W. Pfleiderer, *Folates Pterins*, 1985, **2**, 43.

1479. M. Viscontini and H. Stierlin, *Helv. Chim. Acta*, 1963, **46**, 51.

1480. G.B. Barlin, *Pyrazines*, Wiley, New York, 1982: (a) p. 220, (b) p. 40.

1481. J. Spanget-Larsen, *J. Chem. Soc. Perkin Trans. 2*, 1985, 417.

1482. S. Kwee, *Bioelectrochem. Bioenerg.*, 1983, **11**, 467.

1483. J.E. Gready, *J. Comput. Chem.*, 1984, **5**, 411.

1484. T. Nagatsu, M. Sawada, T. Yamaguchi, T. Sugimoto, S. Matsuura, M. Akino, N. Nakazawa, and H. Ogawa, *Anal. Biochem.*, 1984, **141**, 472.

1485. S. Matsuura, T. Sugimoto, and S. Murata, *Tetrahedron Lett.*, 1985, **26**, 4003.

1486. H.C. van der Plas, J. Tramper, S.A.G.F. Angelino, T.W.G. de Meester, H.S.D. Naeff, F. Müller, and W.F. Middelhoven, *Prog. Ind. Microbiol.*, 1984, **20**, 93.

1487. L.G. Karber and G. Dryhurst, *J. Electroanal. Chem. Interfacial Electrochem.*, 1984, **160**, 141.

1488. D. Ege-Serpkenci, R. Raghavan, and G. Dryhurst, *Bioelectrochem. Bioenerg.*, 1983, **10**, 357; *Chem. Abstr.*, 1984, **100**, 33933.

1489. D. Ege-Serpkenci and G. Dryhurst, *Bioelectrochem. Bioenerg.*, 1983, **11**, 51; *Chem. Abstr.*, 1984, **100**, 102558.

1490. F. Savelli, G. Pirisino, M.C. Alamanni, P. Manca, and M. Satta, *Farmaco, Ed. Sci.*, 1983, **38**, 869.

1491. S.I. Zav'yalov, A.G. Zavozin, N.E. Kravchenko, G.I. Ezhova, and I.V. Sitkareva, *Izv. Akad. Nauk. SSSR, Ser. Khim.*, 1984, 2756.

1492. J. Semb, U.S. Pat, 2,477,426 (1949); *Chem. Abstr.*, 1950, **44**, 1146.

1493. M.E. Hultquist and P.F. Dreisbach, U.S. Pat. 2,443,837 (1948); *Chem. Abstr.*, 1948, **42**, 7944.

1494. P. Griess and G. Harrow, *Ber. Dtsch. Chem. Ges.*, 1887, **20**, 281 and 2205.

1495. H. Ohle and M. Hielscher, *Ber. Dtsch. Chem. Ges.*, 1941, **74**, 13.

1496. M. Viscontini, R. Provenzale, S. Ohlgart, and J. Mallevialle, *Helv. Chim. Acta*, 1970, **53**, 1202.

1497. M. Viscontini and R. Provenzale, *Helv. Chim. Acta*, 1969, **52**, 1225.

1498. M. Viscontini and R. Provenzale, *Helv. Chim. Acta*, 1968, **51**, 1495.

1499. W. Rigby, *J. Chem. Soc.*, 1950, 1907.

1500. P.H. Boyle and R.J. Lockhart, *J. Org. Chem.*, 1985, **50**, 5127.

1501. S.S. Al-Hassan, R.J. Cameron, A.W.C. Curran, W.J.S. Lyall, S.H. Nicholson, D.R. Robinson, A. Stuart, C.J. Suckling, I. Stirling, and H.C.S. Wood, *J. Chem. Soc. Perkin Trans. 1*, 1985, 1645.

1502. R. Cameron, S.H. Nicholson, D.H. Robinson, C.J. Suckling, and H.C.S. Wood, *J. Chem. Soc. Perkin Trans. 1*, 1985, 2133.

1503. S.S. Al-Hassan, R. Cameron, S.H. Nicholson, D.H. Robinson, C.J. Suckling, and H.C.S. Wood, *J. Chem. Soc. Perkin Trans. 1*, 1985, 2145.

1504. L. Rees, E. Valente, C.J. Suckling, and H.C.S. Wood, *Tetrahedron*, 1986, **42**, 117.

1505. G.R. Ramage and G. Trappe, *J. Chem. Soc.*, 1952, 4410.

1506. W. Pfleiderer and H.-U. Blank, *Angew. Chem.*, 1968, **80**, 534.

1507. D.J. Brown, *The Pyrimidines*, Wiley, New York, 1962, p. 210; *Supplement I*, 1970, p. 156; *Supplement II*, 1985 p. 215.

1508. W.L.F. Armarego, A. Ohnishi, and H. Taguchi, *Aust. J. Chem.*, 1986, **39**, 31.

1509. F. Wöhler, *Justus Liebigs Ann. Chem.*, 1857, **103**, 117.

1510. H. Hlasiwetz, *Justus Liebigs Ann. Chem.*, 1857, **103**, 200.

1511. A. Albert and D.J. Brown, *J. Chem. Soc.*, 1954, 2060.

1512. J.E. Gready, *J. Mol. Struct.* (*Theochem.*), 1985, **124**, 1.

1513. J. Spanget-Larsen, R. Gleiter, W. Pfleiderer, and D.J. Brown, *Chem. Ber.*, 1986, **119**, 1275.

1514. J.E. Gready, *Biochemistry*, 1985, **24**, 4761.

1515. J.E. Gready, *J. Comput. Chem.*, 1985, **6**, 377.

1516. J.E. Gready, *J. Am. Chem. Soc.*, 1985, **107**, 6689.

1517. S. Matsuura, S. Murata, and T. Sugimoto, *Tetrahedron Lett.*, 1986, **27**, 585.

1518. S. Matsuura, S. Murata, and T. Sugimoto, *Heterocycles*, 1985, **23**, 3115.

1519. S. Matsuura, T. Sugimoto, S. Murata, Y. Sugawara, and H. Iwasaki, *J. Biochem.* (*Tokyo*), 1985, **98**, 1341.

1520. R.B. Angier, J.H. Boothe, B.L. Hutchings, J.H. Mowat, J. Semb, E.L.R. Stockstad, Y. Subba Row, C.W. Waller, D.B. Cosulich, M.J. Fahrenbach, M.E. Hultquist, E. Kuh, E.H. Northey, D.R. Seeger, J.P. Sickles, and J.M. Smith, *Science*, 1945, **102**, 227.

1521. G. Favini, I. Vandoni, and M. Simonetta, *Theor. Chim. Acta*, 1965, **3**, 45.

1522. I. Niculescu-Duvaz, L.V. Feyns, D. Suster, and G. Cuistea, Brit. Pat. 1,414,752 (1975); *Chem. Abstr.*, 1976, **84**, 90582.

1523. R.D. Brown and B.A.W. Coller, *Theor. Chim. Acta*, 1967, **7**, 259.

1524. S.A.L. Abdel-Hady, M.A. Badawy, M.A.N. Mosselhi, and Y.A. Ibrahim, *J. Heterocycl. Chem.*, 1985, **22**, 801.

1525. A. Kühn and A. Egelhaaf, *Z. Naturforsch., Teil B*, 1959, **14**, 654.

1526. J.L. Johnson and K.V. Rajagopalan, *Proc. Natl. Acad. Sci. USA.*, 1982, **79**, 6856.

1527. S.J. Childress and R.L. McKee, *J. Am. Chem. Soc.*, 1950, **72**, 4271.

1528. S. Avakian and G.J. Martin, U.S. Pat. 2,620,340 (1952); *Chem. Abstr.*, 1953, **47**, 10014.

1529. C.K. Cain and C. Schenker, *Abstracts of Papers, 117th American Chemical Society Meeting*, 1950, p. 41 L; from C. Schenker, Ph.D. Thesis, Cornell University, 1949.

1530. H.M. Kalckar and H. Klenow, *J. Biol. Chem.*, 1948, **172**, 349.

1531. O.H. Lowry, O.A. Bessey, and E.J. Crawford, *J. Biol. Chem.*, 1949, **180**, 389.

1532. O.H. Lowry, O.A. Bessey, and E.J. Crawford, *J. Biol. Chem.*, 1949, **180**, 399.

1533. M. Chaykovsky, A. Rosowsky, N. Papathanasopoulos, K.N. Chen, E.J. Modest, R.S. Kisliuk, and Y. Gaumont, *J. Med. Chem.*, 1974, **17**, 1212.

1534. D.J. Brown, *The Pyrimidines*, Wiley, New York, 1962, p. 106; *Supplement I*, 1970, p. 71; *Supplement II*, 1985, p. 90.

1535. H. Yamakani, A. Sakurai, and M. Goto. *Nippon Kagaku Zasshi*, 1967, **88**, 1320.

1536. W. Pfleiderer, in *Comprehensive Heterocyclic Chemistry*, A.J. Boulton and A. McKillop, Eds. Pergamon, Oxford, 1984 vol. **3**, p. 263.

1537. A. Albert, *Ciba Foundation Symp., Chem. Biol. Pteridines*, 1954, 62.

1538. Y. Kishi, S. Matsuura, S. Inoue, O. Shimomura, and T. Goto, *Tetrahedron Lett.*, 1968, 2847.

1539. G.H. Schmidt and M. Viscontini, *Helv. Chim. Acta*, 1962, **45**, 1571.

1540. F. Korte and H. Bannuscher, *Justus Liebigs Ann. Chem.*, 1959, **622**, 126.

1541. D. Autenrieth, H. Schmid, M. Ott, and W. Pfleiderer, *Justus Liebigs Ann. Chem.*, 1977, 1194.

1542. D. Autenrieth, H. Schmid, K. Harzer, M. Ott, and W. Pfleiderer, *Angew. Chem.*, 1971, **83**, 970.

1543. T. Masuda, T. Kishi, M. Asai, and S. Kuwada, *Chem. Pharm. Bull.*, 1959, **7**, 361.

1544. T. Masuda, T. Kishi, M. Asai, and S. Kuwada, *Chem. Pharm. Bull.*, 1959, **7**, 366.

1545. W. Pfleiderer, J.W. Bunting, D.D. Perrin, and G. Nübel, *Chem. Ber.*, 1966, **99**, 3503.

1546. K. Sugiura and M. Goto, *Nippon Kagaku Zasshi*, 1966, **87**, 623.

1547. E. Khalifa, J.H. Bieri, and M. Viscontini, *Helv. Chim. Acta*, 1973, **56**, 2911.

1548. R. Tschesche and F. Vester, *Chem. Ber.*, 1953, **86**, 454.

1549. R. Hüttel and D. Schreck, *Chem. Ber.*, 1960, **93**, 2439.

1550. T. Kauffmann and K. Vogt, *Chem. Ber.*, 1959, **92**, 2855.

1551. A. Albert, *Biochem. J.*, 1954, **57**, x.

1552. T. Fukushima and M. Akino, *Arch. Biochem. Biophys.*, 1968, **128**, 1.

1553. L. Uemura and M. Goto, *Nippon Kagaku Zasshi*, 1968, **89**, 520.

1554. T. Urushibara, H. Sato, and M. Goto, *Nippon Kagaku Zasshi*, 1966, **87**, 972.

1555. B. Green and H. Rembold, *Chem. Ber.*, 1966, **99**, 2162.

1556. H. Rembold and H. Metzger, *Chem. Ber.*, 1963, **96**, 1395.

1557. E.L. Patterson, H.P. Broquist, A.M. Albrecht, M.H. von Saltza, and E.L.R. Stokstad, *J. Am. Chem. Soc.*, 1955, **77**, 3167.

1558. S. Katoh and M. Akino, *Experientia*, 1966, **22**, 793.

1559. I. Ziegler, *Z. Naturforsch., Teil B*, 1960, **15**, 460.

1560. C. van Baalen, H.S. Forrest, and J. Myers, *Proc. Natl. Acad. Sci. USA.*, 1957, **43**, 701.
1561. H.S. Forrest, C. van Baalen, and J. Myers, *Science*, 1957, **125**, 699.
1562. H.S. Forrest, C. van Baalen, and J. Myers, *Arch. Biochem. Biophys.*, 1958, **78**, 95.
1563. H. Rembold and L. Buschmann, *Chem. Ber.*, 1963, **96**, 1406.
1564. A. Butenandt and H. Rembold, *Hoppe-Seyler's Z. Physiol. Chem.*, 1958, **311**, 79.
1565. T. Sugimoto and S. Matsuura, *Res. Bull. Dept. Gen. Educ., Nagoya Univ.*, 1967, **11**, 87.
1566. I. Ziegler, *Ergeb. Physiol., Biol. Chem. Exp. Pharmakol.*, 1965, **56**, 1.
1567. H.S. Forrest, E.W. Hanly, and J.M. Lagowski, *Biochim. Biophys. Acta*, 1961, **50**, 596.
1568. H. Rembold and W. Gutensohn, *Biochem. Biophys. Res. Commun.*, 1968, **31**, 837.
1569. H. Rambold, H. Metzger, and W. Gutensohn, *Biochim. Biophys. Acta*, 1971, **230**, 117.
1570. M. Cushman, W.C. Wong, and A. Bacher, *J. Chem. Soc Perkin Trans. 1*, 1986, 1051.
1571. M. Cushman, W.C. Wong, and A. Bacher, *J. Chem. Soc. Perkin Trans. 1*, 1986, 1043.
1572. R.L. Blakley and S.J. Benkovic Eds. *Folates and Pterins*, Wiley, New York, 1984.
1573. M. Polonovski and E. Fournier, *C. R. Soc. Biol.*, 1944, **138**, 357; *Chem. Abstr.*, 1945, **39**, 4696.
1574. P. Karrer, C. Manunta, and R. Schwyzer, *Helv. Chim. Acta*, 1948, **31**, 1214.
1575. W. Koschara, *Hoppe-Seyler's Z. Physiol. Chem.*, 1936, **240**, 127; *Chem. Abstr.*, 1936, **30**, 4878.
1576. W. Koschara, *Hoppe-Seyler's Z. Physiol. Chem.*, 1937, **250**, 161; *Chem. Abstr.*, 1938, **32**, 2154.
1577. W.B. Watt, *Nature (London)*, 1964, **201**, 1326.
1578. W.B. Watt and S.R. Bowden, *Nature (London)*, 1966, **210**, 304.
1579. W.B. Watt, *J. Biol. Chem.*, 1967, **242**, 565.
1580. H. Descimon, *Biochimie*, 1971, **53**, 407.
1581. M. Tsusue and T. Mazda, *Experientia*, 1977, **33**, 854.
1582. J.I. DeGraw, P.H. Christie, H. Tagawa, R.L. Kisliuk, Y. Gaumont, F.A. Schmid, and F.M. Sirotnak, *J. Med. Chem.*, 1986, **29**, 1056.
1583. W. Traube, F. Schottländer, C. Goslich, R. Peter, F.A. Meyer, H. Schülter, W. Steinbach, and K. Bredow, *Justus Liebigs Ann. Chem.*, 1923, **432**, 266.
1584. J.A. Blair, *Biochem. J.*, 1958, **68**, 385.
1585. H.S. Forrest and H.K. Mitchell, *J. Am. Chem. Soc.*, 1955, **77**, 4865.
1586. Y. Kaneko, K. Sakaguchi, and Y. Kihara, *J. Ferment. Technol.*, 1949, **27**, 156; quoted in ref. 254.
1587. Y. Kaneko, *Nippon Nogei Kagaku Kaishi*, 1957, **31**, 118; *Chem. Abstr.*, 1958, **52**, 12993.
1588. Y. Kaneko, *Nippon Nogei Kagaku Kaishi*, 1957, **31**, 122; *Chem. Abstr.*, 1958, **52**, 12993.
1589. M. Polonovski, R.-G. Busnel, and M. Pesson, *C.R. Acad. Sci.*, 1943, **217**, 163; *Chem. Abstr.*, 1944, **38**, 4322.
1590. I. Ziegler-Günder, *Z. Vgl. Physiol.*, 1956, **39**, 163.
1591. E. Cerioni, C. Contini, and U. Laudani, *Chem. Biol. Pteridines, Proc. Int. Symp., 5th, Konstanz*, 1975, 851.
1592. F. Fischer, W. Kupitza, M. Gersch, and H. Unger, *Z. Naturforsch., Teil B*, 1962, **17**, 834.
1593. F.F. de Almeida, *Z. Naturforsch., Teil B*, 1958, **13**, 687.
1594. I. Ziegler and M. Feron, *Z. Naturforsch., Teil B*, 1965, **20**, 318.
1595. A.H. Bartel, B.W. Hudson, and R. Craig, *J. Insect Physiol.*, 1958, **2**, 348; *Chem. Abstr.*, 1959, **53**, 11672.
1596. B.W. Hudson, A.H. Bartel, and R. Craig, *J. Insect Physiol.*, 1959, **3**, 63; *Chem. Abstr.*, 1959, **53**, 11672.
1597. T. Hama and M. Obika, *Nature (London)*, 1960, **187**, 326.
1598. H.S. Forrest and H.K. Mitchell, *J. Am. Chem. Soc.*, 1954, **76**, 5658.

1599. C.H. Eugster and P.X. Iten, *Chem. Biol. Pteridines, Proc. Int. Symp., 5th, Konstanz*, 1975, 881.

1600. W. Pfleiderer, *Angew. Chem.*, 1964, **76**, 757.

1601. C.L. Krumdieck, C.M. Baugh, and E.N. Shaw, *Biochim. Biophys. Acta*, 1964, **90**, 573.

1602. C.M. Baugh and E.N. Shaw, *Biochem. Biophys. Res. Commun.*, 1963, **10**, 28.

1603. J.T. Bagnara and M. Obika, *Comp. Biochem. Physiol.*, 1965, **15**, 33.

1604. T. Hama, *Zool. Mag.*, 1952, **61**, 89; quoted in ref. 213.

1605. T. Goda, *J. Fac. Sci. Imp. Univ. Tokyo, Sect. IV*, 1941, **5**, 305; quoted in refs. 213, 216.

1606. T. Goda, *Seiri Seitai*, 1947, **1**, 1; *Chem. Abstr.*, 1951, **45**, 8058.

1607. M. Polonovski, P. Gonnard, and A. Baril, *Enzymologia*, 1951, **14**, 311; *Chem. Abstr.*, 1952, **46**, 1172.

1608. S.W. Bailey and J.E. Ayling, *J. Biol. Chem.*, 1978, **253**, 1598.

1609. V.C. Dewey and G.W. Kidder, *Methods Enzymol.*, 1971, **18** (Part **B**), 618.

1610. G.W. Kidder and V.C. Dewey, *Methods Enzymol.*, 1971, **18**, (Part **B**), 739.

1611. G.W. Kidder and V.C. Dewey, *J. Biol. Chem.*, 1968, **243**, 826.

1612. H.S. Forrest, D. Hatfield, and C. van Baalan, *Nature (London)*, 1959, **183**, 1269.

1613. I. Ziegler, *Biochem. Z.*, 1963, **337**, 62.

1614. F.I. MacLean, H.S. Forrest, and J. Myers, *Arch. Biochem. Biophys.*, 1966, **114**, 404.

1615. D.L. Hatfield, C. van Baalan, and H.S. Forrest, *Plant Physiol.*, 1961, **36**, 240.

1616. M. Viscontini and R. Bühler-Moor, *Helv. Chim. Acta*, 1968, **51**, 1548.

1617. M. Viscontini and M. Frater-Schroeder, *Helv. Chim. Acta*, 1968, **51**, 1554.

1618. A. Momzikoff, *Chem. Biol. Pteridines, Proc. Int. Symp., 5th, Konstanz*, 1975, 871.

1619. H. Rokos, K. Rokos, H. Frisius, and H.J. Kirstraedter, *Clin. Chim. Acta*, 1980, **105**, 275; *Chem. Abstr.*, 1980, **93**, 130216.

1620. T. Fukushima and T. Shiota, *J. Biol. Chem.*, 1972, **247**, 4549.

1621. T. Fukushima and J.C. Nixon, *Anal. Biochem.*, 1980, **102**, 176.

1622. K. Rokos, H. Rokos, H. Frisius, and M. Hüfner, *Chem. Biol. Pteridines, Proc. Int. Symp., 7th, St. Andrews*, 1982, 153.

1623. H. Rokos and K. Rokos, *Chem. Biol. Pteridines, Proc. Int. Symp., 7th, St. Andrews*, 1982, 815.

1624. T. Hama *et al.*, *Kagaku*, 1952, **22**, 542; *Zool. Mag.*, 1957, **66**, 92; both quoted in ref. 220.

1625. E. Elstner and A. Heupel, *Arch. Biochem. Biophys.*, 1976, **173**, 614.

1626. H. Wachter, A. Hansen, E. Reider, and M. Schweiger, *Naturwissenschaften*, 1980, **67**, 610.

1627. K. Iwai, M. Kobashi, T. Suzuki, and M. Bunno, *Chem. Biol. Pteridines, Proc. Int. Symp., 6th, La Jolla*, 1978, 111.

1628. K. Iwai, M. Bunno, M. Kobashi, and T. Suzuki, *Biochim. Biophys. Acta*, 1976, **444**, 618.

1629. T. Hama and T. Goto, *C.R. Soc. Biol.*, 1955, **149**, 859.

1630. E. Lederer, *Biol. Rev. Cambridge Phil. Soc.*, 1940, **15**, 273; *Chem. Abstr.*, 1941, **35**, 206.

1631. W.K. Maas, *Genetics*, 1948, **33**, 177; *Chem. Abstr.*, 1950, **44**, 8547.

1632. E. Hadorn and H.K. Mitchell, *Proc. Natl. Acad. Sci. USA*, 1951, **37**, 650.

1633. I. Ziegler-Günder and E. Hadorn, *Z. Induct. Abstamm. Vererbungsl.*, 1958, **89**, 235; *Chem. Abstr.*, 1959, **53**, 11675.

1634. M. Viscontini and E. Möhlmann, *Helv. Chim. Acta*, 1959, **42**, 836.

1635. M. Viscontini and E. Möhlmann, *Helv. Chim. Acta*, 1959, **42**, 1679.

1636. H. Schlobach and W. Pfleiderer, *Angew. Chem.*, 1971, **83**, 440.

1637. M. Tsusue and M. Akino, *Dobutsugaku Zasshi*, 1965, **74**, 94; *Chem. Abstr.*, 1966, **64**, 20068.

1638. M. Viscontini, E. Hadorn, and P. Karrer, *Helv. Chim. Acta*, 1957, **50**, 579.

1639. M. Viscontini and P. Karrer, *Helv. Chim. Acta*, 1957, **40**, 968.

1640. M. Viscontini, *Helv. Chim. Acta*, 1958, **41**, 922.
1641. W. Pfleiderer, K. Rokos, and H. Schlobach, *Chimia*, 1973, **27**, 656.
1642. C. Baglioni, *Experientia*, 1959, **15**, 465.
1643. I. Schwinck and M. Mancini, *Arch. Genet.*, 1973, **46**, 41; *Chem. Abstr.*, 1976, **85**, 17593.
1644. I. Schwinck, *Chem. Biol. Pteridines, Proc. Int. Symp., 5th, Konstanz*, 1975, 919.
1645. M. Goto, A. Sakurai, and H. Yamakami, *Nippon Kagaku Zasshi*, 1967, **88**, 897.
1646. W. Koschara, *Hoppe-Seyler's Z. Physiol. Chem.*, 1940, **263**, 78.
1647. W. Koschara, *Hoppe-Seyler's Z. Physiol. Chem.*, 1943, **277**, 284.
1648. W. Koschara, *Hoppe-Seyler's Z. Physiol. Chem.*, 1943, **279**, 44.
1649. E.C. Taylor and L.A. Reiter, *J. Org. Chem.*, 1982, **47**, 528.
1650. J.L. Johnson, B.E. Hainline, and K.V. Rajagopalan, *J. Biol. Chem.*, 1980, **255**, 1783.
1651. T. Masuda, *Pharm. Bull. (Japan)*, 1956, **4**, 375.
1652. G.W.E. Plaut, *J. Biol. Chem.*, 1963, **238**, 2225.
1653. G.F. Maley and G.W.E. Plaut, *Fed. Proc., Fed. Am. Soc. Exp. Biol.*, 1958, **17**, 268.
1654. T. Masuda, *Pharm. Bull. (Japan)*, 1956, **4**, 71.
1655. T. Mazda, M. Tsusue, and S. Sakate, *Insect. Biochem.*, 1980, **10**, 357.
1656. H. Aruga, S. Kawase, and M. Akino, *Experientia*, 1954, **10**, 336.
1657. M. Tsusue, T. Mazda, and S. Sakate, *Chem. Biol. Pteridines, Proc. Int. Symp., 7th, St. Andrews*, 1982, 675.
1658. P. Bacher, Q. Le Van, M. Bühler, P. Keller, and H.G. Floss, *Chem. Biol. Pteridines, Proc. Int. Symp., 7th, St. Andrews*, 1982, 699.
1659. P. Nielsen and A. Bacher, *Chem. Biol. Pteridines, Proc. Int. Symp., 7th St. Andrews*, 1982, 705.
1660. C.H. Eugster, E.F. Frauenfelder, and H. Koch, *Helv. Chim. Acta*, 1970, **53**, 131.
1661. P.X. Iten, S. Arihara, and C.H. Eugster, *Helv. Chim. Acta*, 1973, **56**, 302.
1662. P.X. Iten, H. Märki-Danzig, and C.H. Eugster, *Chem. Biol. Pteridines, Proc. Int. Symp., 6th, La Jolla*, 1978, 105.
1663. G. Hanser and H. Rembold, *Hoppe-Seyler's Z. Physiol. Chem.*, 1960, **319**, 200.
1664. B.L. Strehler, *Arch. Biochem. Biophys.*, 1951, **32**, 397.
1665. T. Masuda, *Pharm. Bull. (Japan)*, 1955, **3**, 434.
1666. T. Masuda, *Pharm. Bull. (Japan)*, 1956, **4**, 72.
1667. G.W.E. Plaut and G.F. Maley, *Arch. Biochem. Biophys.*, 1959, **80**, 219.
1668. T. Masuda, T. Kishi, and M. Asai, *Pharm. Bull. (Japan)*, 1957, **5**, 598.
1669. H.S. Forrest and W.S. McNutt, *J. Am. Chem. Soc.*, 1958, **80**, 739.
1670. W.S. McNutt and H.S. Forrest, *J. Am. Chem. Soc.*, 1958, **80**, 951.
1671. I. Takeda and S. Hayakawa, *Agr. Biol. Chem.*, 1968, **32**, 873; *Chem. Abstr.*, 1968, **69**, 74691.
1672. I. Takeda, *Agr. Biol. Chem.*, 1969, **33**, 122; *Chem. Abstr.*, 1969, **70**, 75222.
1673. W.S. McNutt and I. Takeda, *Biochemistry*, 1969, **8**, 1370.
1674. I. Takeda, *Chem. Biol. Pteridines, Proc. Int. Symp., 4th, Toba*, 1969, 183.
1675. I. Takeda, *Hakko Kyokaishi*, 1969, **27**, 305; *Chem. Abstr.*, 1970, **73**, 11478.
1676. T. Kosuge, H. Zenda, A. Ochiai, N. Masaki, M. Noguchi, S. Kimura, and H. Narita, *Tetrahedron Lett.*, 1972, 2545.
1677. O. Shimomura, H.L.B. Suthers, and J.T. Bonner, *Proc. Natl. Acad. Sci. USA*, 1982, **79**, 7376.
1678. R.C. Boruah, R. Baur, and W. Pfleiderer, *Croat. Chem. Acta*, 1986, **59**, 183.
1679. H. Lutz and W. Pfleiderer, *Croat. Chem. Acta*, 1986, **59**, 199.
1680. A. Heckel and W. Pfleiderer, *Helv. Chim. Acta*, 1986, **69**, 704.
1681. A. Heckel and W. Pfleiderer, *Helv. Chem. Acta*, 1986, **69**, 708.

1682. M. Poe, *J. Biol. Chem.*, 1973, **248**, 7025.

1683. P. Waring and W.L.F. Armarego, *Eur. J. Med. Chem.*, 1987, **22**, 83.

1684. B. Bews and D.H.G. Crout, *J. Chem. Soc. Perkin Trans. 1*, 1986, 1459.

1685. I. Ziegler and H. Rokos, *Rivista Eos*, 1986, **6**, 169.

1686. D.H. Brown, P.J. Keller, H.G. Floss, H. Sedlmaier, and A. Bacher, *J. Org. Chem.*, 1986, **51**, 2461.

1687. T. Tsushima, K. Kawada, O. Shiratori, and N. Uchida, *Heterocycles*, 1985, **23**, 45.

1688. A. Albert, *Chem. Biol. Pteridines, Proc. Int. Symp., 8th*, Montreal, 1986, 1.

1689. W. Pfleiderer, Y. Kang, R. Soyka, W. Hutzenlaub, M. Wiesenfeldt, and W. Leskopf, *Chem. Biol. Pteridines, Proc. Int. Symp., 8th Montreal*, 1986, 31.

1690. M.G. Nair, T.R. Toghiyani, B. Ramamurthy, R.L. Kisliuk, and Y. Gaumont, *Chem. Biol. Pteridines, Proc. Int. Symp., 8th, Montreal*, 1986, 45.

1691. M. Przybylski, R. Renkel, and P. Fonrobert, *Chem. Biol. Pteridines, Proc. Int. Symp., 8th Montreal*, 1986, 65.

1692. S. Matsuura, T. Sugimoto, and S. Murata, *Chem. Biol. Pteridines, Proc. Int. Symp., 8th Montreal*, 1986, 77.

1693. S. Matsuura, S. Murata, and T. Sugimoto, *Chem. Biol. Pteridines, Proc. Int. Symp., 8th Montreal*, 1986, 81.

1694. J.E. Gready, *Chem. Biol. Pteridines, Proc. Int. Symp., 8th, Montreal*, 1986, 85.

1695. P.H. Boyle and M.J. Kelly, *Chem. Biol. Pteridines, Proc. Int. Symp., 8th Montreal*, 1986, 91.

1696. S. Antoulas and M. Viscontini, *Chem. Biol. Pteridines, Proc. Int. Symp., 8th, Montreal*, 1986, 95.

1697. K.B. Jacobson and J. Ferré, *Chem. Biol. Pteridines, Proc. Int. Symp., 8th, Montreal*, 1986, 107.

1698. J. Ferré, J.L. Yim, W. Pfleiderer, and K.B. Jacobson, *Chem. Biol. Pteridines, Proc. Int. Symp., 8th, Montreal*, 1986, 115.

1699. D.R. Paton and G.M. Brown, *Chem. Biol. Pteridines, Proc. Int. Symp., 8th Montreal*, 1986, 295.

1700. H.C. van der Plas, *Lect. Heterocycl. Chem.*, 1982, **6**, 1.

1701. S. Antoulas, R. Prewo, J.H. Bieri, and M. Viscontini, *Helv. Chim. Acta*, 1986, **69**, 210.

1702. M.E.C. Biffin and D.J. Brown, *Tetrahedron Lett.*, 1968, 2503.

1703. M.E.C. Biffin, D.J. Brown, and T. Sugimoto, *J. Chem. Soc. (C)*, 1970, 139.

1704. H. Zondler, H.S. Forrest, and J.M. Lagowski, *J. Heterocycl. Chem.*, 1967, **4**, 124.

1705. I.J. Pachter, *J. Am. Chem. Soc.*, 1953, **75**, 3026.

1706. V. Boekelheide and W.J. Linn, *J. Am. Chem. Soc.*, 1954, **76**, 1286.

1707. Q.N. Porter, *Mass Spectrometry of Heterocyclic Compounds*, Wiley, New York 2nd ed., 1985, p. 755.

1708. A. Heckel and W. Pfleiderer, *Helv. Chim. Acta*, 1986, **69**, 1088.

1709. A. Heckel and W. Pfleiderer, *Helv. Chim. Acta*, 1986, **69**, 1095.

1710. M. Böhme, W. Hutzenlaub, W.J. Richter, E.F. Elstner, G. Huttner, J. von Seyerl, and W. Pfleiderer, *Justus Liebigs Ann. Chem.*, 1986, 1705.

1711. T. Sugimoto, S. Murata, S. Matsuura, and W. Pfleiderer, *Tetrahedron Lett.*, 1986, **27**, 4179.

1712. M.A. Goetz, J. Meinwald, and T. Eisner, *Experientia*, 1981, **37**, 679.

1713. J. Elguero, C. Marzin, A.R. Katritzky, and P. Linda, *Adv. Heterocycl. Chem., Suppl.* 1, 1976, 1.

1714. S. Farber, *Blood*, 1952, **7**, 107.

1715. P. Calabresi and R.E. Parks, in *Pharmacological Basis of Therapeutics*, A. G. Gilman, L.S. Goodman, and A. Gilman, Eds., Macmillan, New York, 6th ed. 1980, p. 1249.

1716. A. Albert, *Selective Toxicity*, Chapman and Hall, London, 7th ed. 1985, p. 347.

1717. J. Thompson, ed., *Prescription Proprietaries Guide*, Australian Pharmaceutical Publishing Co, Melbourne, 15th Issue, 1986, p. 935.

1718. G.H. Mudge, in *Pharmacological Basis of Therapeutics*. A.G. Gilman, L.S. Goodman, and A. Gilman, Eds. Macmillan, New York, 6th ed. 1980, p. 892.

1719. J. Thompson, ed., *Prescription Proprietaries Guide*, Australian Pharmaceutical Publishing Co., Melbourne 15th Issue, 1986, p. 609.

1720. V.D. Wiebelhaus, J. Weinstock, F.T. Brennan, G. Sonsowski, and T.L. Larsen, *Fed. Proc., Fed. Am. Soc. Exp. Biol.*, 1961, **20**, 409.

1721. A.P. Crosley, L. Ronquillo, and F. Alexander, *Fed. Proc. Fed. Am. Soc. Exp. Biol.*, 1961, **20**, 410.

1722. J.H. Laragh, E.B. Reilly, J.B. Stites, and M. Angers, *Fed. Proc. Fed. Am. Soc. Exp. Biol.*, 1961, **20**, 410.

1723. R.J. Donnelly, P. Turner, and G.S.C. Sowry, *Lancet*, 1962, (I), 245.

1724. M. Chaykovsky, A. Rosowsky, and E.J. Modest, *J. Heterocycl. Chem.*, 1973, **10**, 425.

1725. A. Rosowsky, *J. Med. Chem.*, 1973, **16**, 1190.

1726. M. Gordon, J.M. Ravel, R.E. Eakin, and W. Shive, *J. Am. Chem. Soc.*, 1948, **70**, 878.

1727. D.J. Brown, in *Mechanisms of Molecular Migrations*, B.S. Thyagarajan, ed., Interscience, New York, 1968, Vol. 1, p. 209.

1728. E.C. Bigham, G.K. Smith, J.F. Reinhard, W.R. Mallory, C.A. Nichol, and R.W. Morrison, *J. Med. Chem.*, 1987, **30**, 40.

1729. W.L.F. Armarego, *Lect. Heterocycl. Chem.*, 1984, **7**, 121.

1730. W.L.F. Armarego, D. Randles, and P. Waring, *Med. Res. Rev.*, 1984, **4**, 267.

1731. S. Kaufman, *Pteridine Chem., Proc. Int. Symp., 3rd Stuttgart*, 1962, 307.

1732. P. Hemmerich, *Pteridine Chem., Proc. Int. Symp., 3rd, Stuttgart*, 1962, 323.

1733. K. Tsuzuki and M. Tada, *J. Heterocycl. Chem.*, 1986, **23**, 1299.

1734. G.W.E. Plaut and R.A. Harvey, *Methods Enzymol.*, 1971, **18 (B)**, 515.

1735. F. Minisci, *Synthesis*, 1973, 1.

1736. F. Minisci and O. Porta, *Adv. Heterocycl. Chem.*, 1974, **16**, 123.

1737. J.C. Fontecilla-Camps, C.E. Bugg, C. Temple, J.D. Rose, J.A. Montgomery, and R. Kisliuk, *J. Am. Chem. Soc.*, 1979, **109**, 6114.

1738. A. Albert, *Selective Toxicity*, Chapman and Hall, London, 7th ed. 1985, p. 345.

1739. P. Waring, *Aust. J. Chem.*, 1988, **41**, in press.

1740. H. Sladowska, J.W.G. De Meester, and H.C. van der Plas, *J. Heterocycl. Chem.*, 1986, **23**, 477.

1741. A. Albert and E.P. Serjeant, *Determination of Ionization Constants*, Chapman and Hall, London 3rd ed. 1984, pp. 218.

1742. D.D. Perrin, B. Dempsey, and E.P. Serjeant. *pK_a Prediction for Organic Acids and Bases*, Chapman and Hall, London, 1981, pp. 168.

1743. A. Albert, *Phys. Methods Heterocycl. Chem.*, 1963, **1**, 1.

1744. A. Albert, *Heterocycl Chemistry*, Athlone Press, London, 2nd ed. 1968, p. 431.

1745. S.F. Mason, *Phys. Methods Heterocyl. Chem.*, 1963, **2**, 1.

1746. A.R. Katritzky and J.M. Lagowski, *Adv. Heterocycl. Chem.*, 1963, **1**, 339.

1747. R.N. Jones, *J. Am. Chem. Soc.*, 1945, **67**, 2127.

1748. L. Rees, C.J. Suckling, and H.C.S. Wood, *J. Chem. Soc., Chem. Commun.*, 1987, 470.

1749. J. Haddow, C.J. Suckling, and H.C.S. Wood, *J. Chem. Soc. Chem. Commun.*, 1987, 478.

1750. T.J. Batterham, *NMR Spectra of Simple Heterocycles*, Wiley, New York, 1973, p. 347.

1751. J.E. Gready, *Int. J. Quantum Chem.*, 1987, **31**, 369.

1752. T.W. Hambley, and J.E. Gready, *J. Mol. Struct.* (*Theochem*), 1988, **165**, in press.

1753. E.C. Taylor and P.S. Ray, *J. Org. Chem.*, 1987, **52**, 3997.

1754. T. Netscher, and C. Strauss, *Justus Liebigs Ann Chem.*, 1987, 259.

1755. T. Nishio, T. Nishiyama, and Y. Omote, *Tetrahedron Lett.*, 1986, **27**, 5637.

1756. W.L.F. Armarego, H. Taguchi, R.G.H. Cotton, S. Battiston, and L. Leong, *Eur. J. Med. Chem.*, 1987, **22**, 283.

1757. C.F. Shey, C.T. Chen, J.M. Horng, and C.H. Wang, *Shih Ta Hsueh Pao (Taipei)*, 1984, **29**, 631; *Chem. Abstr.*, 1984, **101**, 230476.

1758. J.W.G. De Meester, W. Kraus, H.C. van der Plas, H.J. Brons, and W.J. Middelhoven, *J. Heterocycl. Chem.*, 1987, **24**, 1109.

1759. R.H. Baur and W. Pfleiderer, *Isr. J. Chem.*, 1986, **27**, 81.

1760. M.I. Dawson, D. O'Krongley, P.D. Hobbs, J.R. Barrueco, and F.M. Sirotnak, *J. Pharm. Sci.*, 1987, **76**, 635.

1761. Y. Kang, R. Soyka, and W. Pfleiderer, *J. Heterocycl. Chem.*, 1987, **24**, 597.

Index

This index covers the text and all included tables except Table XI (pK_a values) and the Appendix table (known simple pteridines), both of which are excessively long and alphabetically self-indexing.

The page number(s) immediately following each primary entry refer to synthesis or general information. Although each number indicates that the subject is treated on that and possibly subsequent pages, the actual name of the primary entry may appear there only in part, as a synonym, or even simply by inference.

The relatively few authors who are mentioned by name in the text (usually in connection with some historical or outstanding contribution) are included in this index.